HYDRAULIC FILL STRUCTURES

A speciality conference sponsored by
the Geotechnical Engineering Division of the
American Society of Civil Engineers

Co-sponsored by the
Society of Mining Engineers of AIME

Colorado State University
Fort Collins, Colorado
August 15-18, 1988

Edited by D.J.A. Van Zyl and Steven G. Vick

Geotechnical Special Publication No. 21

Published by the
American Society of Civil Engineers
345 East 47th Street
New York, New York 10017

ABSTRACT

This volume contains 53 papers covering many aspects of hydraulic fill structures with special emphasis on geotechnical issues. The papers present an excellent review of historical development, as well as art and practice of hydraulic fill structures. Papers are divided into the following main topics: keynote; excavation, transport, and deposition; material properties; static and dynamic stability; soil improvement/geosynthetics; new developments, case histories, and postnote. Papers are included on historical reviews, dredging technology and dredge spoil disposal, depositional behavior, mine backfill, tailings disposal, slurry pipelines, hydraulic fill properties and behavior and static and dynamic loading conditions, densification and improvement of hydraulic fills, artificial islands, land reclamation, and case histories of successes and failures.

Library of Congress Cataloging-in-Publication Data

Hydraulic fill structures: a specialty conference sponsored by the Geotechnical Division of the American Society of Civil Engineers, co-sponsored by the Society of the Mining Engineers of the AIME/edited by D.J.A. Van Zyl and Steven G. Vick.
p. cm.
Held at Colorado State University, Fort Collins, Colo., August 15-18, 1988.
Includes Index.
ISBN 0-87262-672-5
1. Hydraulic filling—Congresses. I. Van Zyl, Dirk J.A. II. Vick, Steven G., 1950– . III. American Society of Civil Engineers. Geotechnical Engineering Division. IV. Society of Mining Engineers of AIME.
TA750.H93 1988 88-22202
627—dc19 CIP

Library of Congress Catalog Card No.: 88-22202
ISBN 0-87262-672-5
Manufactured in the United States of America.

PREFACE

Hydraulic filling is used throughout the world for the construction of artificial islands, water retention dams, mine tailings impoundments, backfill in underground mines, disposal of dredged spoils, and the reclamation of land. Hydraulic fill structures are constructed by depositing soils transported as a slurry. The two main advantages of the use of hydraulic fills are lower construction cost and the ability to move large volumes of material over long distances.

Within the past decade, significant advances have been made to allow better prediction of depositional behavior, drainage, consolidation, and stability of low-density hydraulic fills. Advances have also been made with regard to the slurry transport, such as the transport of higher density slurries.

The purpose of this specialty conference is to review the art and practice of hydraulic fill design and construction methods. Special emphasis is placed on geotechnical aspects. This volume contains an excellent collection of papers covering a broad range of topics related to hydraulic fills. It is expected that it will serve the profession as a useful resource document and reference guide for future hydraulic fill development.

It is the practice of the Geotechnical Engineering Division that papers published in a Specialty Conference Proceedings undergo a peer review. The standards for peer review are essentially the same as those for papers being reviewed for possible publication in the ASCE *Journal of Geotechnical Engineering*. Papers must receive two positive reviews and must be revised to conform to the mandatory revisions of the reviewers. Because there is a tight schedule from receipt of papers through review and on to publication, there is not time for more than one cycle of editing and revising for specialty conference papers. All papers published in this volume are eligible for discussion in the ASCE *Journal of Geotechnical Engineering* and for ASCE awards.

Conference Organizing Committee

John D. Nelson, Chairman
Richard L. Volpe, Co-Chairman
W. David Carrier, III
Rudy J. Dietrich
William F. Marcuson, III
Janet Lee Montera
Norbert R. Morgenstern
Dirk J. A. van Zyl, Co-Editor, Proceedings
Steven G. Vick, Co-Editor, Proceedings
Richard E. Wardwell

CONTENTS

Keynote

Session 1—Excavation, Transport, and Deposition

Session 2—Material Properties

Session 3—Static and Dynamic Stability

Session 6—Soil Improvement/Geosynthetics

Session 7—New Developments/Case Histories

Postnote

HYDRAULIC FILL STRUCTURES - A PERSPECTIVE

N.R. Morgenstern* and A.A.G. Kupper**

INTRODUCTION

It is revealing to compare the treatment of hydraulic fill structures in Sherard et.al., (1963) with that given by Justin, Hinds and Creager (1945). The former reflects the state of the art in design and construction of engineered fill dams in the 1950's and early 1960's while the latter draws on the period 1930's to early 1940's. The emphasis in both is on North American practice and the difference in treatment of hydraulic fill structures is striking.

There is no systematic discussion of hydraulic fill structures in Sherard et.al. (op.cit.). Some failures are included in a table listing unsatisfactory performance of earth dams but the number of hydraulic fill entries is modest. The only reference to modern utilization of hydraulic fill techniques is to the Turtle Creek Dam which was intended to be a compacted earth fill structure with material to be excavated by draglines from a borrow pit below the water table, then loaded on trucks and subsequently compacted by either tractors or vibratory rollers. This proved impractical because the trucks were bogging down and the sand fill was too wet for conventional compaction by either tractor or vibratory rollers. The project was successfully completed by switching to hydraulic fill techniques.

Dredged sand was brought to the surface of the dam by pipeline and discharged in such a manner that the silty fines were kept in suspension and ultimately wasted by flow into a discharge pipe through the embankment. This separated the silt from the sand. The water-deposited sand was found to be adequately dense without any further compaction. The travel of a tractor only loosened it so that subsequent tractor work was minimized. More than 6 x $10^6 m^3$ of material was dredged at about $13{,}500 m^3$/day. The cost of the dredged sand in place was about one half of the cost with the previous unsatisfactory construction procedure.

The treatment of hydraulic fill structures in Justin, Hinds and Creager (op.cit.) is much more extensive. The stability of hydraulic fill dams is discussed and Gilboy's (1934) classical study on the

* Professor of Civil Engineering, University of Alberta, Edmonton, Alberta T6G 2G7 CANADA

** Graduate Student, Dept. of Civil Engineering, University of Alberta, Edmonton, Alberta T6G 2G7 CANADA mechanics of hydraulic fill dams is presented with examples. In addition to this, more than twenty pages are devoted to matters of depositional details and a discussion of some case histories. Clearly in the period between the publication of the two books, hydraulic fill methods fell out of favour and ultimately could no longer be regarded as a common method of construction in civil engineering for dams in North America.

At least one reason for this is anticipated by Justin, Hinds, and Creager (p. 782, op.cit.) who state:

> "Because of the magnitude of some hydraulic fill dams, great publicity has been given to the several slides and construction accidents which have occurred in connection with several of them. As a result some engineers exhibit a prejudice against that type of construction. As a matter of fact, since 1930 there have been several more or less serious construction accidents in connection with rolled fill earth dams but only one in connection with a hydraulic fill dam. There is no sound reason for such prejudice. All that is necessary is for the engineer to appreciate that the hydraulic fill dam is an engineering structure and that it should be given the same competent attention in investigation, design, and construction that one would give any other engineering structure".

Prejudice or not, and notwithstanding the sage comments repeated above, few engineered hydraulic fill structures were built in North America, and Western Europe in subsequent years, at least within civil engineering practice. Engineering for hydraulic fill structures was not treated in academic curricula and the concept developed an aura of lack of safety. This was accentuated in the post 1960's by the increased concern over and research into earthquake induced liquefaction. There was also more competitiveness from rolled fill dams. Following World War II, there was a dramatic increase in the scale and cost-effectiveness of motorized earth-moving equipment which contributed enormously to the feasibility and economy of large earth-moving operations for dams and other engineered fill structures.

While an exhaustive study of the recent literature on engineered fill structures in the Western world has not been undertaken, it is our experience that the prejudice identified by Justin et.al. remains intact. The contributions to the 14th Congress on Large Dams in 1982 are characteristic. Question 55 was devoted to materials and construction methods for embankment dams and cofferdams, excluding tailings dams, but otherwise singling out hydraulic fill dams for attention. In a comprehensive General Report on this Question Penman (1982) is able to devote only a few lines to hydraulic fill in Western practice, noting that it is used to build flood protection dykes in Holland and elsewhere, that it continued as a method of dam construction until the failure of Fort Peck in 1938 and that the technique was used as part of the fill for the High Aswan Dam where the fill as placed had a dry density of 15.2 kN/m^3 which was subsequently increased by vibration to 16.5 kN/m^3. Reference is also made to the more extensive use of hydraulic fill methods in China (Chang-Tse, 1976) which had obviously evolved under different technological imperatives.

Of the sixty papers published in response to this Question the only paper from a Western source discussing hydraulic fill dams was by d'Angremond et.al. (1982) who advise that hydraulic fill has been used in Holland for dams to a height of about 25m. Sand that is placed by hydraulic means has a relative density of 50-65% when placed above water and 35-40% when placed under water, which imply 100% and 85-90% of standard compaction respectively. Additional compaction can be undertaken. Construction details are provided.

The paper by Borovoi et.al. (1982), which represents the views of the USSR National Committee on Large Dams, states that the hydraulic fill method is in common use in the USSR and that it is finding increasing applications in hydro engineering. Typical material specification limits and slope inclinations for preliminary design are given. These independent developments will be discussed in more detail below.

As noted by both Justin et.al. (op.cit.) and Penman (1982) the collapse of the Fort Peck Dam constitutes a watershed in the evolution of hydraulic fill methods for dam construction in North America and marks its decline. The failure of Fort Peck Dam was subjected to extensive investigations, most of which are reviewed and referenced by Middlebrooks (1942). The consulting board appointed to investigate the causes of the slide concluded that the slide:

> "was due to the fact that the shearing resistance of the weathered shale and bentonite seams in the foundation was insufficient to withstand the shearing force to which the foundation was subjected. The extent to which the slide progressed upstream may have been due, in some degree, to a partial liquefaction of the material in the slide".

Middlebrooks (1942) emphasizes the following:

> "It can be concluded definitely that the hydraulic fill was not at fault. It was the general opinion of the investigating engineers and other engineers having an intimate knowledge of the project that the fill material performed excellently even under the most difficult circumstances existing during the slide".

While hydraulic fill methods are common in dredge disposal and reclamation, they almost disappeared from the construction of water storage dams in Western practice. It is of value to question whether this has been a matter of prejudice or prudence.

This should not imply that the design and construction of hydraulic fill received no attention elsewhere. As noted earlier, Russian practice makes much use of the technique with a variety of materials ranging from silts to gravel with a maximum particle size of 100-150mm. Moreover hydraulic fill applications to engineered structures is finding increasing application in the USSR. In the mining industry, the requirements to dispose of mill tailings in an economic manner resulted in the evolution of the tailings dam which, as mining operations became bigger and bigger, also became very substantial. To-day the largest dams in the world in terms of volume are tailings dams and it has become increasingly common for their safety requirements to be the same as for water storage dams.

Another area of activity that has encouraged the development of hydraulic fill technology is the exploration for hydrocarbons in the Beaufort Sea. As explorers for hydrocarbons in the Arctic ventured offshore in the early 1970's, they encountered ice as a dominant consideration affecting the safety and cost of the exploration system. Ice affects ship navigation and safety, limiting the usefulness year round of drillships. For fixed bottom structures, such as jack-up rigs, ice can exert very substantial horizontal loads that are not tolerated by most conventional rigs. Hydraulic fill islands were perceived as an economic, environmentally sensitive method of providing year round access for offshore drilling with structures capable of resisting the design ice forces.

The intent of this conference is to encourage a re-evaluation of hydraulic fill technology. As summarized above, it has been abandoned in some areas of civil engineering practice, but has evolved in others. To facilitate this re-evaluation we summarize in separate sections to follow:

1) the development of hydraulic fill technology in the USSR,
2) the applications of hydraulic fill to tailings dams,
and, 3) the construction of offshore islands by hydraulic fill methods.

A classification of hydraulic fill structures is presented based on the environment of deposition, and the end product from a geotechnical perspective. Certain specific design and construction issues are central to most hydraulic fill problems. The most important among them are singled out for special discussion in a later section.

HYDRAULIC FILL FOR WATER STORAGE STRUCTURES IN THE USSR

Between 1947 and 1973, over 100 hydraulic fill dams with a corresponding total volume of 800 x $10^6 m^3$ were built in the USSR. (Hydroprojeckt, 1974) Only few failures occurred. They happened during construction and were generally not very significant.

More than 10% of the Soviet dams are hydraulic fills. This is a significant number if one considers that the hydraulic fill technique is mainly adopted where alluvial foundations and wide valleys prevail. These conditions are encountered on lowland rivers and such rivers provide only 20% of Soviet hydropower potential (Energomachexport, 1974).

The hydraulic fill technique has been used in the USSR not only for water storage dams, but also for tailings disposal, artificial islands for offshore oil exploration, reclamation, and fill construction under winter conditions. In addition it has been used as an efficient and economic technique for both diversion of rivers and direct dam construction without diversion. The success of this application is related to the capability of delivering an enormous amount of material in a short period of time which is readily achieved by hydraulic filling. Finally, hydraulic fill is also used for construction on slump-prone foundations, such as collapsible soils, loess, etc. because wetting takes place during the placement of the structure, while the load is increasing. Foundation settlements occur mainly during the sluicing period when the fill material is most deformable (Perevezentseva, 1977).

An evolution of both design and construction methods can be traced in Soviet hydraulic fill technology. Cross-sections adopted in current practice are characterized by only modest permeability contrasts or no contrast at all, no positive cutoffs, and flat slopes. Typical sections are discussed in more detail below.

Techniques to attain high rates of construction are central to Soviet practice. According to Melent'ev (1980), with the appropriate work plan and equipment, hydraulic fill output has reached 300,000 m^3/day on some projects in the USSR. Such high rates of construction require specific equipment such as special dredges and pumps as well as the utilization of free draining granular material. High outputs and high rate of construction results not only in a reduction of cost but also appears to produce denser fill.

Construction of hydraulic fills are also undertaken under severe winter conditions. Development of reliable methods of maintaining holes through the ice for the working reservoir of the dredge, and heating of the faces being excavated, the slurry and the pipeline, provide for stable and economic operations at subfreezing temperatures. Shkundin (1976) points out that the difficulties related to excavation, transport and deposition under winter conditions diminish sensibly with increasing equipment capacity. Equipment with an output of more than 3000 m^3/h can be operated under winter conditions in Siberia.

Typical Cross-sections

The zoned or heterogeneous profile contains a low permeability core and granular shoulders achieved by hydraulic segregation, see Figure 1(a). The core is not consolidated and, as recognized in early Western designs, exerts high pressure on the shoulders. The thickness of the core is controlled by the dimensions of the pool and the grain size distribution of the borrow material, especially the percentage of fines. Cores are formed when the borrow material has a Uniformity Coefficient greater than 3. Zoned dams of this type with impervious cores are not very common in USSR.

Central zone profiles are more common. This type of profile resembles the zoned profile in the sense that it has a core that is less pervious than the shoulders. The difference is that in the central zone profile, the core is not formed by plastic material, the core is not thick and the permeability ratio between the core and the shoulders is only 10 to 100. In the zoned profile this ratio is 100 to 10,000. Central zone profiles are constructed when the borrow material has a Uniformity Coefficient between 2 and 3 (Yufin, 1965). Due to restraints of both time and fill volume associated with core consolidation, Soviet hydraulic fill technology moved towards more homogeneous profiles.

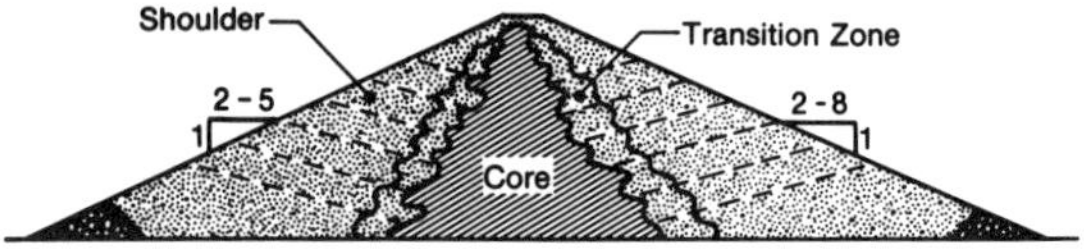

(a) Zoned Profile

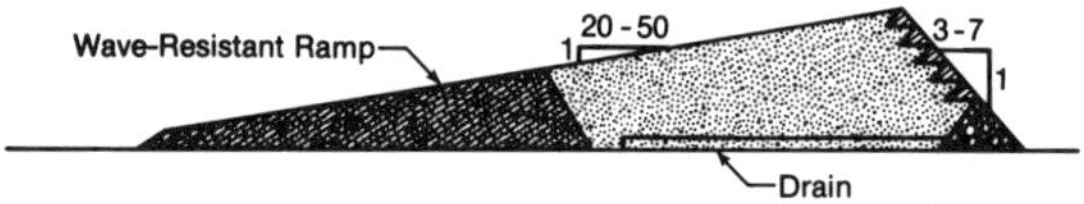

(b) Homogeneous Flattened Profile

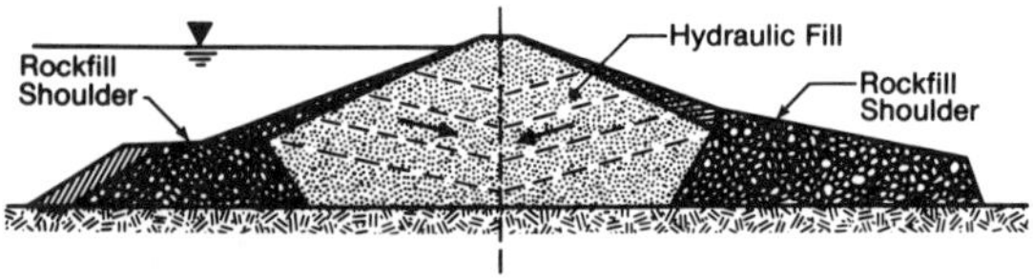

(c) Mixed Profile

Figure 1. Typical Soviet hydraulic fill sections

The homogeneous profile is characterized by the same grain size distribution throughout the whole section, see Figure 1(b). This is attainable when the soil from the borrow area has a Uniformity Coefficient less than 2. This type of dam can be built with sand, sandy silt or loess. There is no pool formation during the construction of the homogeneous profile. The slurry is allowed to flow long distances forming flat slopes, say 20-50H:1V. This profile has proved to be effective, especially for dams under 30m. For higher dams, the volumes required become huge due to the flat slopes that develop.

The mixed profile is composed of part mechanically placed

material (dumped or compacted) and part hydraulically filled material, see Figure 1(c). Usually, the construction starts with mechanical placement of massive shoulders, which provide stability for the dam during construction and operation. Then, the space between the shoulders is filled hydraulically. The mixed profile limits the width of dam and provides extra resistance to failure during earthquakes.

There are many variations of the mixed cross-section. The Playavinyas Dam, for example, has the downstream shoulder placed by trucks while the upstream and central zones were placed hydraulically.

Borrow Areas

For most, if not all, Soviet hydraulic fill projects, the borrow area is an alluvial deposit in the river channel or a flood-plain deposit which is exploited by dredging, as opposed to abutment borrow areas excavated by hydraulic monitors which was the common procedure in early Western practice.

The factors that determine the adequacy of a potential borrow area are the type of material, the water availability, its position with relation to the dam and the volume of material available. Almost every size of material can be utilized from gravel to clay. However, the utilization of very coarse or very fine materials is more difficult, more time consuming, and more expensive than the use of sandy soils. The difficulty in using coarse material is related to extraction problems, loss of hydraulic head and high wear of pipelines and equipment in general. According to Melent'ev (1980), the maximum acceptable diameter of particles is about 10 to 15 cm, but such values must be considered in economic terms to justify their utilization. The use of clayey soils leads to the need to invoke additional procedures to allow the material to dry and to adopt low rates of filling, which increase construction time and costs, unless the clay forms balls. Hydraulically deposited clay balls have also been used. (Lutovinov et al., 1975) Most hydraulic fill dams in the USSR are built with sandy materials, but there are some dams composed of sandy gravel and clayey soils.

Based on experience the Soviets have defined limiting grain size curves which are recommended for a preliminary determination of material adequacy at the design stage. The curves are presented by Melent'ev (1980) and Borovoi et.al. (1982) among others. Comparing these curves with grain size distribution curves from North American hydraulic fill dams one can see that the latter define a wider range characterized by less uniform materials.

Hydraulic Segregation Analysis

Hydraulic segregation refers to the process of deposition of particles of different sizes at different distances from the deposition point. Larger particles tend to deposit soon after being discharged, while smaller particles can be carried by the flow and deposited further downslope.

The Soviet standard specification SNiP-11-53-73 recommends consideration of segregation of fill if the borrow material has $D_{60}/D_{10} > 2.5$ and/or $D_{90}/D_{10} \geqslant 5$; although, actually, hydraulic segregation starts to be observed for $D_{60}/D_{10} \simeq 1.3$ to 1.6 if the beach is long enough.

The hydraulic segregation is usually forecast by Melent'ev's method. This method (Melent'ev et al., 1973) is a semi-empirical method based on a large set of experimental data, hydraulic flow laws and some statistical analysis which determines the average grain size distribution at any distance from the deposition point.

Construction Technique

The Soviets have several different techniques of hydraulic fill placement. However, since the 1950's they have been using almost exclusively trestleless pipelines with extendable single point discharge, see Fig. 2. In this case the slurry is discharged through the end point of a pipeline that lies directly on the downstream slope surface. The slurry flows in the upstream direction forming a very flat wave resistant slope. The end of the distributing pipeline has to be changed from time to time to form an even slope. To achieve this, the pipeline is set up with a quick joint system so a crane can install and remove segments of pipe without stopping the slurry discharge. Little foundation preparation, except for removal of top soil, is involved. In addition clean layers of gravel are treated or removed.

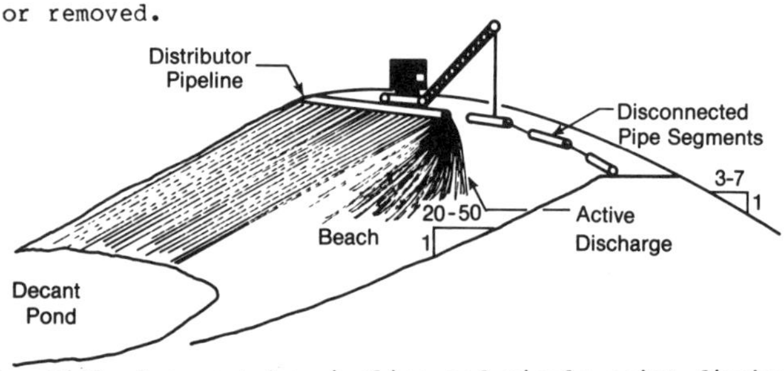

Figure 2. Fill placement by pipeline and single point discharge.

The rate of filling is kept within allowable values in order to assure stability of the structure. Excessively high rates of filling can cause instability of the structure due to a high phreatic level in the shoulders and/or piping at the point where the phreatic surface emerges on the external slope. This is especially critical when the foundation has a low permeability. An example, of such instability occurred during the construction of the Bratsk dam (Tarasenko, 1970).

In most Soviet dams within the river channel, a toe drain is built on the downstream side. For dams on flood-plains, the most common drainage system is a pipe drain. In some cases, inverted filters are installed in the lower part of the downstream slope. The drainage system also usually includes downstream relief wells.

With this technology, the Soviets have been successfully applying hydraulic fill techniques to a variety of foundation conditions, site characteristics, using a number of different materials.

Excellent performance of hydraulic fills has been reported even under relatively severe seismic conditions (Volnin and Ivanovskaya, 1977).

HYDRAULIC FILL AND TAILINGS DAMS

While there are invariably many options other than hydraulic fill for the construction of a water storage structure, in the case of the disposal of mine and mill waste there are few. Many mine and mill operations produce a wet waste stream and for all but the shortest distances, it is usually substantially more economical to transport solids hydraulically than by truck, conveyor, or other dry handling methods. Therefore hydraulic fill methods are naturally attractive to build retention structures of and for tailings.

Some of the largest earth structures in the world are tailings dams. For example, the tailings dam at Phalaborwa in South Africa is expected to have a volume of 162 x $10^6 m^3$ at completion which is to be compared with the Tarbela Dam, the largest (volume) earthfill in the world, having a volume of 142 x $10^6 m^3$. At the Syncrude oil sands project in Canada, tailings are stored in an out-of-pit tailings pond until a future date when they will be introduced into the mined out pit. Approximately 267 x $10^6 m^3$ of sand and 318 x $10^6 m^3$ of thick sludge will require storage in the Syncrude tailings pond. To accommodate the almost 600 x $10^6 m^3$ of solid waste, approximately 18.5 km of hydraulic fill dykes, ranging from 27 to 82 m in height, will have to be constructed. At completion, the tailings pond will have a surface area of 17 square kilometres. These and other mining projects that utilize hydraulic fill, involve materials handling on a grand scale in both an economic and industrial environment that differs from that of water storage dam construction.

Tailings are of no direct economic value and therefore there is a pervasive pressure to find the most economical, acceptable way to dispose of the tailings. The construction of a tailings dam is usually carried out by the mine operators with the height of the dam being increased as required to provide the needed waste storage. This has advantages when compared with water storage dams which are usually completed in a short period before filling and which are usually constructed within a less flexible contractual arrangement than tailings dams. However construction control is usually poorer and failures arise due to misapplication of sound construction practice. Economic considerations lead to the use in construction in most tailings dams of the coarser fraction of the tailings instead of importing more costly borrow material. Another important distinction is that the bulk of the material stored is not water but loose, sometimes impervious, and potentially liquefiable, tailings.

The upstream method is the oldest, simplest and most economical method of tailings dam construction, see Figure 3. It appears to have evolved empirically over many decades without much benefit of geotechnical input. The downstream shell is raised in increments as needed. In North America, for many years tailings were most commonly discharged by spigotting but single point discharge and cycloning were not unknown. A beach was formed and good construction practice minimized the pond level and kept it well off the beach. The downstream face of the dam was often at or near the angle of repose. On reflection, it appears that the structural element of the dam was really the beach which, because of its layered structure and connection to the clean sand shell, could be assured to be fully drained. Provided no water entered into the sand shell, and the starter dam did not block drainage, the low factor of safety of the face was not too critical. Deeper instability through the beach would be critical.

While many upstream tailings dams were constructed in a satisfactory manner, failures were also common due to a variety of geotechnical and construction reasons. Following the catastrophic failures at Aberfan, El Cobre, and Buffalo Creek it became clear that geotechnical engineering had much to contribute to ensure the safety of mine waste disposal systems. It was increasingly recognized that the original upstream method did not result in structures of adequate safety and departures from this traditional structure were advocated. This resulted in increased use of centreline or downstream construction procedures on the one hand, see Figure 3, and major modifications in upstream construction on the other.

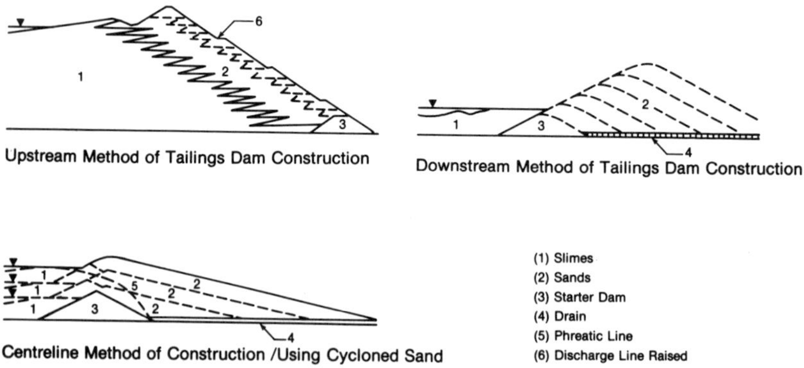

Figure 3. Tailings dams construction methods

Downstream methods have the advantage that as the dam is raised, it is not underlain by previously deposited tailings. With the provision of adequate compaction and seepage control, safe dams to substantial heights can be achieved. However, the volume of coarse tailings required for embankment construction is about twice that required in the centreline method and many tailings streams are inadequate in this regard. The configuration of many valleys or other storage areas may also be unsuited to provide abutment ties for downstream construction.

The centreline method is a variant of the downstream method, but instead of the crest moving downstream as the dam is built, it rises vertically. Placement and compaction control can be exercised as required. Provided that the tailings stream is suitable, construction can proceed quickly using coarse-grained cyclone underflow. Underdrainage systems can be installed as the dam is built, thereby controlling the line of saturation. One disadvantage is the large volume of sand needed for construction which is often a problem in the early stages of mine production. Therefore a higher starter dam of imported fill is needed to initiate centreline construction.

From the 1970's on, the design and construction of tailings dams has attracted a substantial literature. Some major references are Blight in (Morgenstern, et al, 1977), Klohn (1981), and Vick (1983). Others are noted by Morgenstern (1985). It is not intended to address details of tailings dam design and construction here.

Notwithstanding the contributions of geotechnical engineering to the rational design of tailings dams, failures still occur. The better known examples are the Mochikoshi Dams in Japan (1973), the Tyrone Dam in the USA (1980), two tailings dams in Chile (1985) and the Stava tailings dams in Italy (1985). It is of interest to note that all of these structures were constructed by the upstream method. The authors are not aware of any collapse of tailings dams constructed by the centreline or downstream method.

This should not be construed as a blanket condemnation of upstream construction. The number of tailings dams constructed by the upstream method is much greater than by the downstream and centreline methods combined and it is therefore reasonable that the statistics of failure reflect this. Safe structures can be built by the upstream method provided that their design and construction respect geotechnical principles. Modern upstream construction employs flatter downstream slopes than their traditional predecessors, beaches are made wider and compaction is sometimes employed.

Experience with the use of hydraulic fill in the construction of tailings dams has revealed certain characteristics that bear on both design and performance and that are of interest for any application of hydraulic fill techniques. The most significant are:

1) Particle size separation
2) Drainage measures

3) Compaction
4) Earthquake resistance

Particle Size Separation

The common methods of tailings disposal and dyke construction include beach deposition by either single-point discharge or multi-point spigotting, on-dam cycloning and stationary cycloning with subsequent re-handling. The grain size distribution as deposited depends on the method of placement and therefore this aspect of the composition of the resulting fill is, to some degree, controllable.

Abadjiev (1985a) sumarizes methods for determining the grain size separation along beaches composed of spigotted materials. Blight and Bentel (1983) have also developed a theoretical relation for the particle size sorting that occurs on a hydraulic beach and demonstrated at least a consistent fit with observations of beaches in gold and platinum tailings. With regard to cycloning, multiple pass cycloning can be used to control the fines content of the underflow used for fill construction. This has a significant bearing on both permeability and earthquake resistant properties and, as noted by Troncoso and Verdugo (1985), it may be worthwhile to invest in the necessary cycloning efforts required to lower the fines content of a fill material.

It will be stressed in a subsequent section that our ability to forecast and control grain size separation is still limited and a greater understanding of this process is needed. Nevertheless, it is possible to increasingly regard hydraulic fill in some applications as an engineered fill with controllable properties.

Drainage Measures

Modern tailings dam design and construction relies on internal drainage measures to assure integrity of the structure. Drains are needed to control both construction water and long-term seepage. The significant investment associated with various drainage measures was noted by Morgenstern (1985).

Drains are readily incorporated into downstream construction. The Perez Caldera No. 2 dam in Chile (Griffin et al, 1983) is an example of a comprehensive internal drainage system in a dam that has been designed with special attention to seismic stability.

Internal drains are equally desirable, if not more so, in upstream construction. Examples of internal drains used in the large tailings structures associated with the oil-sand industry in Canada are summarized by Morgenstern (op.cit.). Abadjiev (1985b) shows that the provision of an extensive drainage tongue upstream of the starter dam can improve the stability of upstream construction. The tongue projection, shown in Figure 4, is built of free draining material like the starter dam. Its length is equal to the distance from the starter dam to the decant pond at the initial stage and it has an inclination to the pond no smaller than that of the beach, thus ensuring that the coarser fractions are deposited on the tongue. This enhances both seepage control and stability in an economical manner.

Inventive use of drainage measures will extend the applications and economy of hydraulic fill construction.

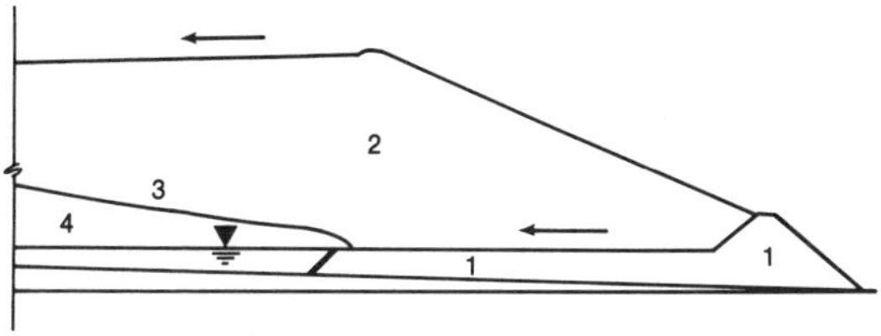

1 - Starter Dam and Tongue
2 - Beach Deposited Material
3 - Phreatic Line
4 - Decant Pond at the Initial Stage

Figure 4. Drainage for upstream construction (Abadjev, 1985b)

Compaction

The designer of a hydraulic fill structure, at least above water, is not restricted to accept the density of the fill as deposited. Clearly both downstream and centreline construction lend themselves to compaction. It is not so well known that upstream construction is also readily compacted.

To achieve compaction, the oil sands industry has adopted the hydraulic cell method of deposition, which after compaction, achieves a relative density of about 75% quite economically (Mittal and Hardy, 1977). Additional details on the hydraulic cell method and subsequent compaction are given by Handford et al. (1982).

Earthquake Resistance

The catastrophic failures of several upstream-constructed tailings embankments in Chile in 1965 highlighted the limited earthquake resistance of tailings dams constructed by upstream methods (Dobry and Alvarez, 1967). This is further emphasized by liquefaction of the Machikoshi tailings dam (Ishihara, 1984) and failures again in Chile following the March, 1985 earthquake (Troncoso, 1988).

That some of these structures were intrinsically unsafe is clear. Nevertheless many upstream hydraulic fill dams have successfully survived major earthquakes. A particularly interesting example is the Dashihe tailings dam which was subjected to the Tangshang earthquake in 1976.

A cross-section of the Dashihe dam is shown in Figure 5 (Central Research Institute of Building and Construction, 1987). The area in which the tailings dam was located experienced a magnitude 7.8 shock on July 28, 1975 and another of magnitude 7.1 some fifteen hours later. The dam was located 40km and 15km from the epicenter of these two shocks respectively. In addition there were numerous aftershocks with magnitude greater than 5.

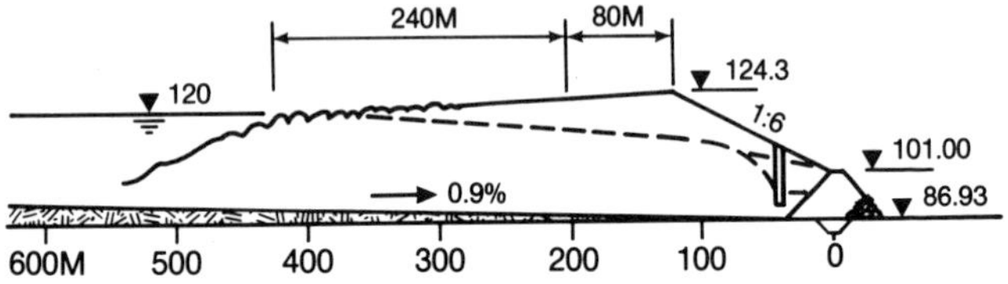

Figure 5. Dashihe Dam

As a result of the earthquake, some cracks developed in the downstream face and on the beach, near the pond, there were sand boils, waterspouts and a fissure zone. Nevertheless, the dam served its intended function and continues to do so to-day.

The circumstances controlling acceptable behaviour of an upstream tailings dam during an earthquake remain enigmatic. Experience indicates that the construction method is not intrinsically unsafe but further studies are needed to isolate those details of construction procedure, fill material, and drainage control that result in safe structures.

HYDRAULIC FILL AND ARTIFICIAL ISLANDS

Hydraulic fill placed under water behaves differently than fill placed above water. It tends to assume flatter slopes, be more sensitive to the presence of fines and end up in a looser state on average than hydraulic fill placed sub-aerially. Equilibrium slopes and density achieved in-situ by placement alone are very dependent on underwater fill placement methods.

Hydraulic fill underwater has been used for a number of applications such as earth dam and embankment construction, expansion of industrial and harbour facilities and the development of man-made islands. For example, underwater fill was placed and compacted as part of the High Aswan Dam and underwater fills were constructed as part of the Yonkers, N.Y., sewage treatment plant. Johnson et al. (1972) remain a valuable reference in this regard for experience through to the early 1970's and Whitman (1970) provides a comprehensive review of experience with hydraulic fills used to support structural loads. Not all of these latter cases deal with underwater placement. Ishihara et al. (1981) provide information on the characteristics of the fill used in the construction of Owi Island No. 1 in Tokyo Bay.

The greatest expansion in recent times in the underwater placement of hydraulic fill for engineered structures arose as a result of accelerated hydrocarbon exploration efforts in the Canadian Beaufort Sea. Offshore exploratory drilling in the Beaufort Sea began in the early 1970's with an extension of onshore techniques into very shallow water and the use of drillships in deeper water. Due to severe ice conditions, drillships were limited to a restricted drilling season and there were incentives to maintain drilling year round. The initial artificial island structures pioneered by ESSO Resources Canada Ltd. were extended to deeper and deeper water. However these structures were waterline penetrating sacrificial beach islands with gentle side slopes of 10:1 to 15:1. This concept was impractical in water depths greater than about 20m due to both economic and construction schedule considerations.

Dome Petroleum Ltd., who had leases in deeper water, took the lead in developing steep-sided submarine berms that would support a vertical sided, waterline penetrating caisson designed to resist the environmental forces and still provide workspace for the exploration drilling. while originally conceived as a temporary structure, this concept is adaptable to a permanent structure and is under consideration for use when hydrocarbon production starts.

The first topside unit comprised a multiple concrete caisson system which was subsequently replaced on later islands by various forms of steel drilling caissons that were more readily deployable. This constituted a major advance over sacrificial beach islands in terms of reduced fill volumes, minimizing the effects of wave erosion, enhanced mobility and pre-assembly of at least parts of the drilling system requirements. Sacrificial beach and waterline penetrating caisson systems are compared in Figure 6.

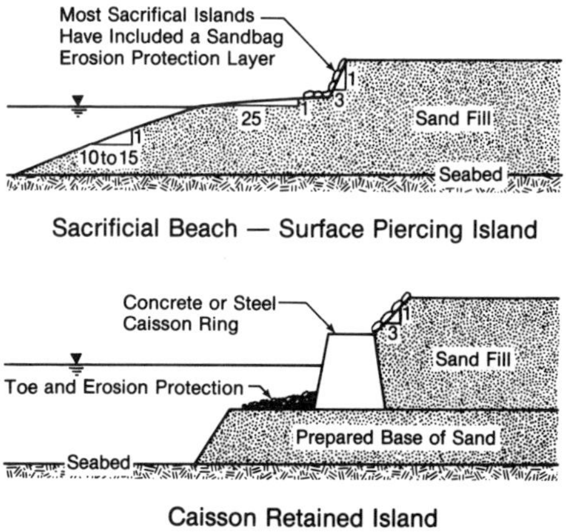

Figure 6. Concepts for large artificial islands

Artificial island construction began in 1972 and 1973 in only a few metres of water. Fill quantities were small, approximately 50,000-200,000m^3 per island, and construction methods varied from trucks hauling gravel from onshore onto the winter ice to a clamshell barge placing local material or placement with a cutter section dredge. Gravel is in short supply in this region and its availability for larger structures would ultimately be proscribed.

From these beginnings, the sacrificial beach island concept evolved, which reached its climax in 1978-79 with the construction of the Issungnak island in 19m of water. This island was 100m in diameter, and required 4.1 x $10^6 m^3$ of fill which was obtained mainly by using stationary suction dredges in an adjacent sea bottom borrow pit. The fill exhibited an average D_{50} of 200 microns and a fines content of 15%. All fill material was placed at the site by pumping the sand slurry from the dredges to discharge barges at the island through floating pipelines (Boone, 1980). Final construction slopes stabilized at 10:1 but they may have been re-worked later as a result of storm erosion.

Mass movement from sacrificial beach islands are primarily due to storm-induced erosion. However it has been suggested by Crooks et al. (1985) that there was also storm-induced liquefaction at another island (Alerk) which was also constructed of hydraulic fill containing a substantial proportion of fines.

The Tarsiut N-44 island constructed in 1980-81 marked a major departure from the sacrificial beach island concept. It was the first island to be built outside of the landfast ice zone and hence large horizontal ice forces had to be contemplated in design and it was the first caisson retained exploration island constructed in the Canadian Beaufort Sea. The island was in 22m of water and had a soft clay foundation which was sub-cut to a depth of about 2.5m-3m and backfilled with sand.

Sand berm construction which was to a height of 17.5m above the sea floor was achieved in about two months. Much of the material was dredged from a distant borrow and transported to the island in split-hull dump barges with individual capacities of 1500 m^3. This material, when dumped, was controlled by concentric bunds with sideslopes of between 5:1 and 6.5:1 which were constructed by pumping hydraulic fill through a discharge pipe mounted on an anchored barge filled with borrow (Fitzpatrick and Stenning, 1983). The specifications for the fill were D_{50} greater than 250 microns and less than 10% fines. Except for construction error when some clay balls were introduced into the fill for the caissons, the borrow readily met these specifications. About 1.8 x $10^6 m^3$ of hydraulic fill was used at Tarsiut, including fill for the caisson. Cone testing indicated relative densities of about 60% which was in excess of design requirements. While there were some construction problems, Tarsiut performed as intended. It demonstrated that steep containment slopes could be constructed and that medium dense fills were attainable by bottom dump placement methods. The construction technique is illustrated in Figure 7.

The Uviluk site constitutes the next major advance in artificial island experience. The site was characterized by a 31m water depth and a competent seabed foundation. At Uviluk, the multiple concrete caisson system adopted at Tarsiut was abandoned in favour of a single steel drilling caisson system (SSDC) which had many advantages in terms of deployment and overall economics. A comparison of the two systems is shown in Figure 8. (Mitchell, 1984).

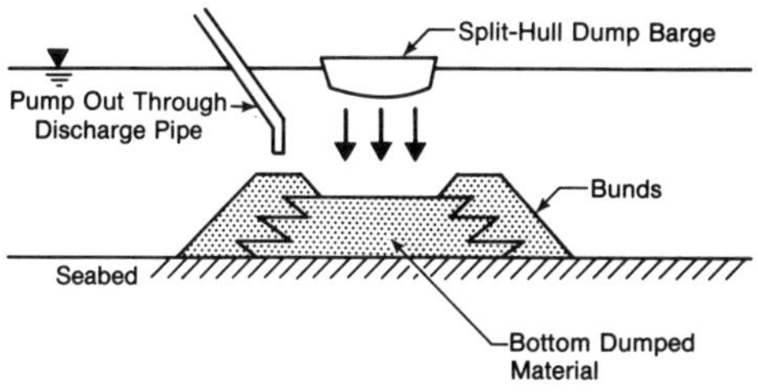

Figure 7. Construction method for Tarsiut Island

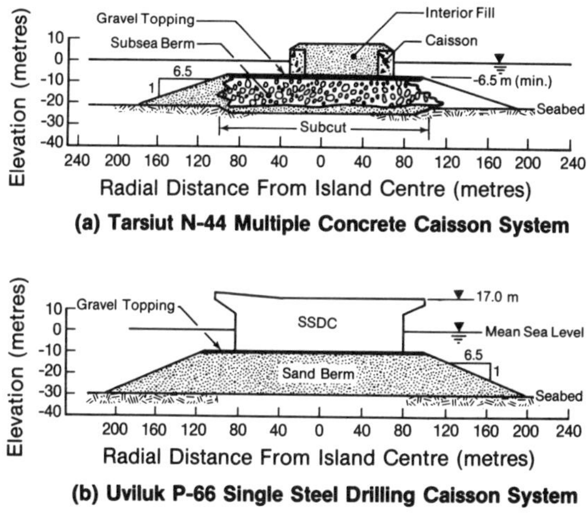

Figure 8. Cross section of hybrid exploration structures (Mitchell, 1984)

The submarine berm at Uviluk comprised $2.1 \times 10^6 m^3$ and the sand had a D_{50} of 320 microns and less than 5% fines. Fill placement was similar to Tarsiut but with improved quality control. The finished berm had nominal side slopes of 7:1. Details of caisson set-down and subsequent deformation of the hydraulic fill are beyond the scope of this presentation. Suffice it to say that the settlement of the hydraulic fill and the lateral deformations under load were generally as forecast.

Cone penetration testing of the berm revealed a substantial sensitivity to what might be regarded as relatively minor variations in material type. This was subsequently studied in more detail by (Berzins and Hewitt, 1984) who discovered a very substantial variation of cone resistance at a given depth with relatively modest variations in the fines content in otherwise comparable fills. Their correlation is shown in Figure 9.

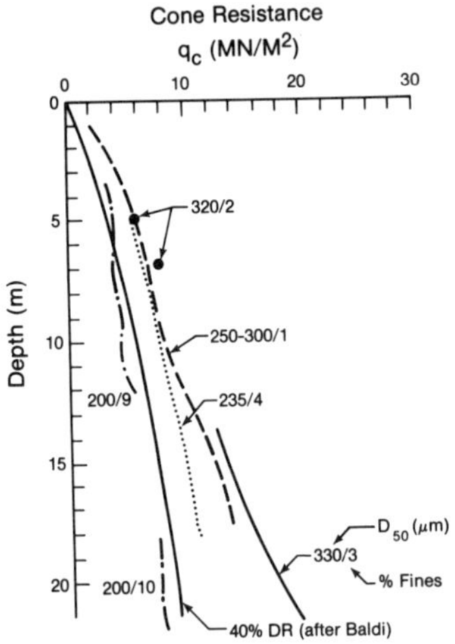

Figure 9. Cone resistance at a depth as a function of D_{50} and percent of fines (Berzins and Hewitt, 1984)

Following the success of Uviluk, and the increased understanding of the factors controlling placement density, as well as improved quality control procedures, the construction of substantial submarine berms appeared to be coming routine. This proved not to be the case.

The Nerlerk sand berm was initiated in 1982 in 45m of water. From the point of view of hydraulic fill operation, the construction at Nerlerk was made more difficult by:

1) the increased water depth and hence the added pressure to build slopes steeper and faster than before to meet the set-down schedule,
2) the need for the almost exclusive use of a deep suction exploration and pipeline placement system to meet the schedule,
3) the need to exploit a local marginal borrow source to support the above construction operation
and, 4) the uncertainty as to the in-situ densities which one could expect using these materials and placement techniques.

Approximately 4 x $10^6 m^3$ of fill were placed, almost completing the berm, when five major slope failures occurred. The site was ultimately abandoned. Details are given by (Mitchell, 1984).

Most of the fill at Nerlerk was deposited by a nozzle from a relatively stationary placement barge which was connected by floating pipeline to a deep suction dredge. At the time of construction it was not recognized that construction methods could have such an overwhelming influence on depositional density. Figure 10 from (Mitchell, op.cit.) compares nozzle deposited with hopper dredge deposited material of comparable composition to illustrate this point. A hopper dredge utilizes a bottom valve release system. The Nerlerk failures emphasize our limited understanding of the interaction between in-situ densities and placement technique.

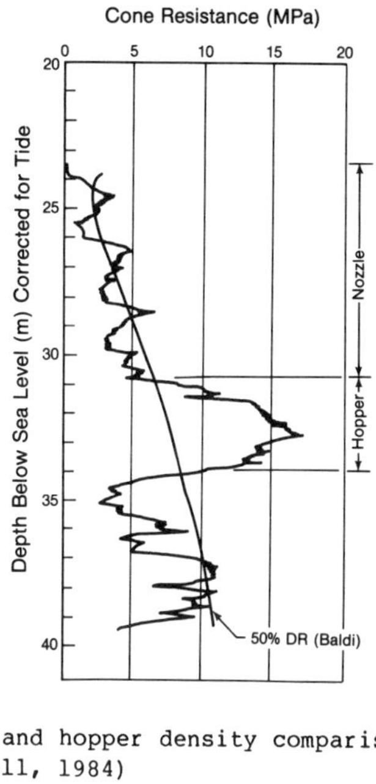

Figure 10. Nozzle and hopper density comparison at Nerlerk (Mitchell, 1984)

Following Nerlerk additional islands were built and operated with only minor problems, but never in water depths exceeding about 30m. One sand-filled caisson system was subjected to a sustained high ice loading in 1986. This ice loading is of a cyclic nature and pore pressures and movements in the core fill continued to increase until the ice movement ceased. The behaviour approached the limits of serviceability and underlines our limited ability to calculate deformation of saturated sand fill under horizontal cyclic loading. Details of subsequent analyses of this event have not yet been published.

Approximately 40 x $10^6 m^3$ of hydraulic fill have been utilized in artificial island construction in the Canadian Beaufort Sea. The program has not been without mishap but the geotechnical achievements have been substantial. However there is much more to be understood before the design and construction of large sub-aqueous hydraulic fills become routine.

CLASSIFICATION OF HYDRAULIC FILL STRUCTURES

Hydraulic fill has several applications in different areas that involve professionals with diverse backgrounds. To facilitate discussion and to emphasize some of the factors controlling the behaviour of hydraulic fill structures a classification of such structures based on the physical phenomena involved rather than the type of utilization is proposed. This classification is shown in Figure 11.

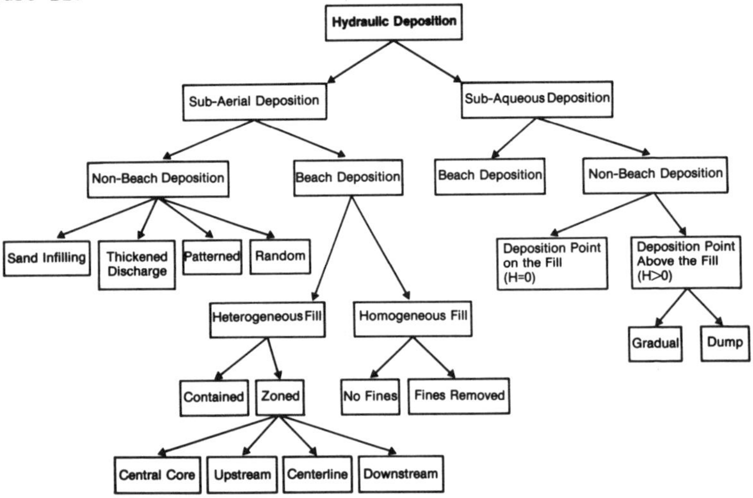

Figure 11. Classification of hydraulic deposition

The geotechnical behaviour of the fill is much affected by the environment of deposition, the mechanics of deposition and the processing of the fill. Different end products result, not only in terms of geotechnical material but also in terms of geometry. Considering these factors, the proposed classification is based on structures with similar geotechnical considerations.

The first distinction in the classification is made between sub-aerial (above water) and sub-aqueous (under water) deposition. As noted previously, sub-aqueous hydraulic fills, prior to any compaction, tend to be looser and more variable than comparable fills deposited sub-aerially.

The next distinction that is made is whether or not a beach is created. Beach deposition involves a regular segregation pattern with the coarser fraction being deposited adjacent to the discharge point and the finer fraction being deposited farther away.

For sub-aerial non-beach deposition, the final component of the classification reflects the end product in terms of degree of homogeneity. Some examples are given, but this is not intended to be an all-inclusive list.

Rock-infilling refers to the sluicing of sand into the voids of rockfill as was done at the High Aswan Dam. Patterned refers to the filling of cells to construct a structure. This method has been adopted in the USSR to build homogeneous dams using heterogeneous borrow. Random refers to the type of filling commonly used in land reclamation and the hydraulic backfilling sometimes adopted underground in mines (Mitchell, et.al., 1975). Thickened discharge refers to the procedures developed by Robinsky (1979) to inhibit segregation by adjusting the placement water content so that a stable pile can be achieved, at least for purposes of waste disposal.

With regard to sub-aerial beach deposition, the distinction is made between whether the resulting structure is composed of homogeneous fill or not, as indicated by a Uniformity Coefficient, say, less than 3 for homogeneous fill. The final component of this portion of the classification refers to the geometry of the end product and design and construction aspects such as the need to ensure separation of fines, pool design and type of drainage system required.

A contained heterogeneous fill refers simply to the beaching of the hydraulic fill between containment dykes that might be of rockfill. The classical sections of tailings dams, shown in Figure 3, are examples of zoned structures.

Homogeneous fill structures arise either from the deposition initially of fill material without fines or heterogeneous material with fines being removed during construction. An example of the latter case is current USSR practice in construction homogeneous hydraulic fills.

With regard to sub-aqueous deposition, the distinction is also made whether a beach is created or not. With regard to non-beach deposition the free drop of the slurry (H) exercises an important influence on the end product and a further distinction is made whether the hydraulic fill is placed gradually or dumped. Some zonation is possible by mixed construction modes.

ISSUES IN DESIGN AND CONSTRUCTION OF HYDRAULIC FILLS

Based on the perspective presented in the previous pages several issues that are central to improved design and construction of hydraulic fill are highlighted in the following. The issues focus on concerns that affect the reliability of utilizing hydraulic fill for an engineered structure. Other important subjects such as construction methods, consolidation of hydraulically deposited fines, and the use of hydraulic backfill in underground mining are not considered here.

Role of Laboratory and Field Experiments

It is of interest to review the elements of current practice when dealing with rolled fills. After the potential borrow area is delineated by site investigation, disturbed samples are taken and compacted in the laboratory. The compaction energy used is approximately compatible with the energy to be used in the field to produce samples with similar fabric. Laboratory tests with these samples are used as a basis for material characterization and the establishment of construction specifications. These parameters enter directly into embankment design and constitute a basis for field quality control. This is a straight forward procedure well supported by years of experience.

Hydraulic fill design poses more difficult questions right at the outset. Even knowing the borrow material, segregation and loss of fines make the fill material different than the borrow. Furthermore, hydraulic deposition creates characteristic fabric, micro and macro layering, which cannot be reproduced in a conventional soil mechanics laboratory. Flume tests probably can create similar depositional conditions, but it is still not very clear how the fill density can be evaluated from these tests.

While the potential contributions of laboratory and field experiments are yet to be worked out, any generalizations will ultimately be limited by scale effects. The observations of Valore and Giglio (1985) are useful in this regard. They note that the hydraulic deposition of dredged materials within containment basins give rise to marked spatial variations in the physical and mechanical properties of the resulting soils. These heterogeneities, in this case, govern the progress of consolidation of the hydraulic fill.

The design of hydraulic fills usually focus on geometry, lay-out and specifications for construction control based on experience. Stability and seepage analysis are done using assumed values that may depart significantly from the actual values that will exist within the fill. However, as the Nerlerk failures indicate, this can be a hazardous process. The resolution of these problems resides at the interface between sedimentary mechanics and soil mechanics and some combination of theory, laboratory experiments, and field-scale studies should prove fruitful.

The most ambitious developments in this regard are the relations developed from experience in the USSR with the construction of hydraulic fill dams and which can be applied to tailings dam design. This work is not readily accessed in the Western literature but Abadjiev (1985a) provides a convenient starting point. Empirical correlations describing particle separation in a beaching operation are described. Correlations also exist between grain size distributions, subsequent compressibility and shear strength. It is clearly easier to make observations and confirm relations for sub-aerial deposition but, as Ogawa (1969) has shown, even the problem of dispersion of dumped sand from a hopper-barge is amenable to analysis.

So far, laboratory experiments appear to have made only limited contributions to hydraulic fill studies. The most significant studies of this type have been undertaken by Blight et al. (1985) and Blight (1987) who have confirmed by both laboratory flume studies and field investigations the conclusions of Melent'ev et al. (1973) that the profile of a dam beach can be represented by a single dimensionless master profile. As a result, the profile and composition of a beach can be forecast.

We have reviewed a number of flume based experiments, mainly unpublished, whose objectives were to forecast beach angles and fill properties as a function of certain depositional parameter such as composition of fill and rate of deposition. The results from various studies have generally been disappointing in that few general trends consistent with field behaviour have been observed. As a result such experimental techniques are not used much. However we are of the view that better-controlled experiments are needed to evaluate the potential of flume studies in hydraulic fill.

Figure 12 illustrates a facility recently constructed at the University of Alberta. The intention is to study the effect of deposition on the fill charcteristics such as slope, grain size distribution, density and shear strength. The tests are carried out under uniform one-dimensional deposition at the discharge point. In order to achieve these conditions sand is fed at controlled flow rates to a water stream. The slurry goes through a vertical flexible pipe that feeds a specially designed spreader. Initial results are just being processed.

Figure 12. Test facility at University of Alberta

Characterization of Hydraulic Fills

Some hydraulic fills stored as mineral waste or used in reclamation are fine-grained. However most materials used in engineered fill structures are cohesionless soils and are much simpler to characterize than clays. But cohesionless soils traditionally are regarded as hard to sample and inspect. As a result the characterization of cohesionless soils in routine practice is actually more primitive than that of cohesive soils.

The geotechnical behaviour of a cohesionless soil is dominated by its composition, its in-situ density, its structure and the stresses acting on it. These are listed in more or less generally increasing difficulty of accessibility.

Drawing on the tradition of routine foundation engineering, the difficulties associated with realistic geotechnical characterization of cohesionless soils are circumvented by reliance on various in-situ tests, particularly the Standard Penetration Test (SPT) and the Cone Penetration Test (CPT). Interpretations of soil behaviour range from purely empirical such as for the SPT to interpretation supported by a theoretical base (e.g., Been et al., 1986). It should be widely understood that no such test is perfect and that they are all influenced to various degrees by progressive failure, rate effects, lateral stresses and operational details. This is not the place to enter into a discussion of the arcane world of penetration testing, its corrections and interpretations except to comment that improved engineering of hydraulic fill will require better undisturbed sampling and direct measurement of the in-situ density.

Ishihara (1985) has outlined recent developments in undisturbed soil sampling and it is to be hoped that these techniques will be adopted more extensively, at least on major projects. There have also been major developments in the use of down hole nuclear techniques for the direct measurement of density. The paper to this conference by Plewes et al. (1988) is particularly noteworthy in this regard. Continuous downhole density logging was undertaken with a petroleum industry gamma ray density tool to evaluate the in-situ density of a 90m high tailings dyke. Measurements were possible beyond the depth where SPT data loses meaning. Careful comparisons were made with undisturbed samples and the accuracy reported is impressive. The capacity for continuous direct density logging is a significant contribution to the evaluation of liquefaction potential.

Stability and Deformation of Hydraulic Fill

Where hydraulic fills are deformed under fully drained conditions their strength and deformation properties are usually unexceptional and amenable to determination by current laboratory testing techniques. Experience indicates that drained stability can be evaluated in a straight forward manner and that both vertical and lateral deformations can be assessed reasonably using methods of non-linear analysis that are common in practice to-day. The behaviour of hydraulic fill during undrained loading is more problematical.

Undrained loading can arise due to rapid construction, foundation deformation, earthquake and wave effects. The on-going intense studies into liquefaction and cyclic mobility are central to the evaluation of undrained stability and deformation of hydraulic fill.

The articulation of Steady State Line concepts (e.g., Poulos et al., 1985a) provides a fundamental starting point. Only specimens representative of composition are necessary to define the steady state line. Provided the in-situ void ratio is known, the undrained shear strength of the soil in the steady state is known and the stability at the steady state can be evaluated with ease. As a design concept, this takes the view that if the soil can liquefy it will, and the results constitute a bound on the stability assessment. Poulos et al. (op.cit.) advocate piston sampling for the in-situ void ratio determination but, as noted previously, nuclear methods have considerable promise. The role of initial fabric on affecting whether stress paths to steady state are much influenced deserves attention.

Design based on the steady state strength of a sand is like designing with the residual strength of a clay; there are circumstances where it is the appropriate strength, but if a higher peak resistance appears available, one is obliged to inquire whether the loading mechanism will exceed the peak strength or not. Given the complexity of the loading mechanisms under consideration and the strain-weakening characteristics of some hydraulic fills, this is a difficult question to answer and is the subject of much active research.

For monotonic undrained failure Sladen et al. (1985a) have proposed the existence of a collapse surface defined by the peak undrained shear strength of a sand at the same void ratio but at different initial confining stresses; see Figure 13. This collapse surface is essentially independent of stress path, at least for the limited data available at the time of publication. For liquefaction to occur the soil state has to reach the collapse surface and the shear stress must exceed the steady state shear strength. (Sladen et al., 1985b) applied the collapse surface concept to the analysis of the Nerlerk berm slides with encouraging results. Additional experience is needed with this fertile concept.

For the evaluation of stability under cyclic loading Poulos et al. (1985b) propose a methodology that invokes an evaluation of the strain associated with the cyclic loading of the design earthquake and an assessment of the strain required to trigger liquefaction and thereby assess whether peak strength is exceeded. The logic of this analysis has wide application to all types of hydraulic fill structures and it merits detailed study.

Sladen et al. (1985a) speculated that the collapse surface also constitutes a bound to the influence of cyclic loading and that liquefaction could occur if cyclic loading took the soil state to the collapse surface. It is intriguing to speculate whether the collapse surface is the same as the "line of phase transformation" proposed by

Ishihara et al. (1975) to explain characteristics of the undrained cyclic loading of sands.

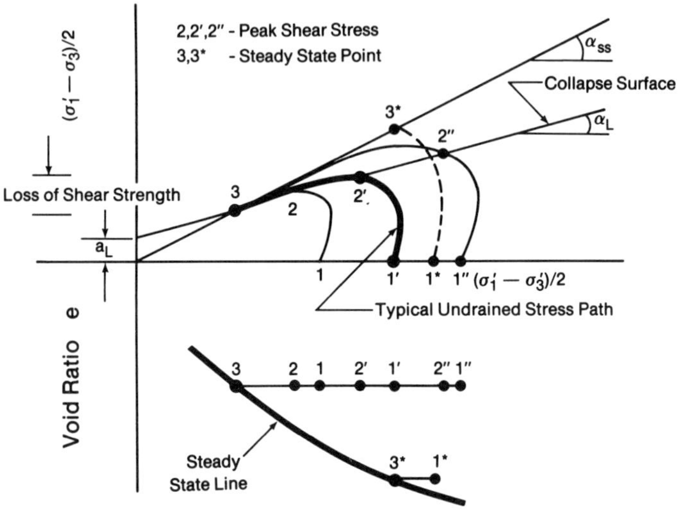

Figure 13. The collapse surface Sladen et al. (1985)

Even if collapse is avoided, large deformations may develop during undrained loading that threaten the serviceability of the earth fill structure. Seed (1987) has cautioned that large deformation may precede the mobilization of steady state strength and could control the design. The previously mentioned response of an offshore structure to cyclic ice loads is an apt example. The ability to forecast large undrained deformation of hydraulic fill is still limited but fortunately the subject has become the focus of much research.

Improvement of Hydraulic Fills

Although the collapse of the Ft. Peck Dam contributed to the decline in usage of hydraulic fill in North America, it should be noted that hydraulic fill was ultimately used to complete the dam. However this fill was compacted by crawler tracter with relative densities increasing from about 60% to 80-90%. This and other examples of hydraulic fill compaction by machine are summarized in Johnson et al. (1972). Mittal and Hardy (1977) provide a comparable example of machine compaction used in tailings dam construction.

Densification can also be achieved by blasting and pneumatic methods. Compaction is not limited to sub-aerial access but can also be undertaken sub-aqueously as well. The sea-bed compaction associated with the Oosterschelde project is noteworthy in this regard (Pladet, 1978).

The improvement of hydraulic fills is not limited to densification alone. The incorporation of geosynthetics for purposes of drainage, material separation, or reinforcement provide options that are a challenge to the imagination of designers and constructors alike.

CONCLUSIONS

Although hydraulic fill declined in use for water retaining structures in North America about 40-50 years ago, it continued to evolve as a technique for engineered structures elsewhere and in other fields of application. Extensive use has been made of hydraulic fill for dams in the USSR, for tailings dams in many countries, for sea dyking in Holland and for exploration structures in the Canadian Beaufort Sea, to single out some of the more substantial areas of application. The evolution of hydraulic fill technology has been fragmentary and it is timely to initiate a new synthesis of both theory and practice in order to evaluate afresh the potential for engineered hydraulic fill structures.

The role of experiment into the depositional processes as they affect geotechnical behaviour is not yet clear. Some of the more promising findings have been identified. Other key issues addressed that bear on design and construction of hydraulic fill are in-situ geotechnical characterization, the analysis of stability and deformation of hydraulic fills, and the improvement of hydraulic fill properties. Enhanced understanding and experience in each of these areas will lead to more reliable hydraulic fill design and construction.

However, as previous examples indicate, the behaviour of hydraulic fill is much influenced by the placement process. Hydraulic fill technology is an inter-disciplinary activity. The geotechnical engineer will have to collaborate with the hydraulic engineer and the marine construction industry in order to take full advantage of the economies offered by hydraulic fill structures.

REFERENCES

Abadjiev, C.B. (1985a). "Estimation of the physical characteristics of deposited tailings in the tailings dam of nonferrous metallurgy", Proc. 11th Int. Conf. Soil Mechs. Found. Eng., Vol. 3, 1231-1234, San Francisco.

Abadjiev, C.B. (1985b). "Improved upstream construction of tailings dam", Proc. 11th Int. Conf. Soil Mechs. Found. Eng., Vol. 3, 1235-1236, San Francisco.

d'Angremond, K., et al. (1982). "Design and construction methods of dams in the Netherlands" Transactions, 14th Int. Cong. Large Dams, 4, 643-666, Rio de Janeiro.

Been, K., et al. (1986). "The cone penetration test in sands: Part I, state parameter interpretation". Geotechnique, 36 (2), 239-249.

Berzins, W.E. and Hewitt, K.J. (1984). "The use of hydraulically placed sand in island construction". Proc. 16th Offshore Technology Conf., Vol. 1, pp. 213-222, Houston.

Blight, G.E. (1987). "The concept of the master profile for tailings dam beaches". in "Prediction and Performance in Geotechnical Engineering" ed by R.C. Joshi and F.J. Griffiths, University of Calgary, 361-368.

Blight, G. and Bentel, G.M. (1983). "The behaviour of mine tailings during hydraulic deposition". J.S. Afr. Inst. Min. Metal., 83 (4), 73-86.

Blight, G.E., Thomson, R.R. and Vorster, K. (1985). "Profiles of hydraulic-fill tailings beaches and seepage through hydraulically sorted tailings". J.S. Afr. Inst. Min. Metal., 85 (5), 157-161.

Boone, D.J. (1980). "The construction of an artificial drilling island in intermediate water depths in the Beaufort Sea". Proc. 12th Offshore Technology Conf., 1, 187-195, Houston.

Borovoi, A.A., et al. (1982). "Soils for and methods of embankment dam construction". Transactions 14th Int. Cong. Large Dams, 4, 503-516, Rio de Janeiro.

Central Research Institute of Building and Construction (1987). "Documents for Sino-Canadian Joint Investigation on Earthquake Safety of Chinese Tailings Dams". Ministry of Metallurgical Industry, People's Republic of China.

Chang-Tse, C. (1976). "Earth dams built by the sluicing-siltation method". Transactions 12th Int. Cong. Large Dams, 5, 99-103, Mexico City.

Crocks, J.H., Shinde, S.B. and Been, K. (1985). "In-situ state of underwater hydraulic sand fills". Civil Engineering in the Arctic Offshore, 85, ASCE, 84-92, San Francisco.

Dobry, R. and Alvarez, L. (1967). "Seismic failures of Chilean tailings dams". J. of Soil Mechs. and Foundations Div., ASCE, 93 (SM6), 237-260.

Energomachexport (1974). "The dams of the USSR". V/O Energomachexport, USSR Trade Representation, Moscow.

Fitzpatrick, J. and Stenning, D.G. (1983). "Design and construction of Tarsuit Island in the Canadian Beaufort Sea". Proc 15th Offshore Technology Conf., 2, 51-60, Houston.

Gilboy, G. (1934). "Mechanics of hydraulic fill dams". J. Boston Soc. Civil Engrs., Vol. 20, No. 3, July, 1934, p. 185.

Griffin, P.M., et al. (1983). "Perez Caldera Tailings Dams - a case history". Proc. 7th Pan American Conf. Soil Mechs. Found. Eng., Vol. 2, 751-761, Vancouver.

Handford, G. et al. (1982). "The use of observational approach in the design and construction of the Syncrude tailings dyke." Proc. Seminar on "Geotechnical aspects of mine design and tailings containment" University of Calgary, Calgary, Alberta, 31p.

Hydroprojeckt (1974). "Hydraulic Fills" Symposium Technique - Paris-Moscow 1974.

Ishihara, K. (1984). "Post-earthquake failure of a tailings dam due to liquefaction of the pond deposit". Proc. Int. Conf. on Case Histories in Geotech. Eng., 3, 1129-1143, St. Louis.

Ishihara, K. (1985). "Stability of natural deposits during earthquakes". Proc. 11th Int. Conf. Soil Mechs. Found. Eng., Vol. 1, 321-376, San Francisco.

Ishihara, K., Shimizu, K. and Yamada, Y. (1981). "Pore water pressures measured in sand deposit during an earthquake". Soils and Foundations, 21 (4), 85-100.

Ishihara, K., Tatsuoka, F. and Yasuda, S. (1975). "Undrained deformation and liquefaction of sand under cyclic stresses". Soils and Foundations, 15 (1), 29-44.

Johnson, S.J., Compton, J.R. and Ling, S.C. (1972). "Control for underwater construction" in "Underwater Soil Sampling, Testing, and Construction Control". ASTM, Special. Tech. Publ. 501, 122-180, Philadelphia.

Justin, J.D. Hinds, J. and Creager, W.P. (1945). "Engineering for Dams, Vol. 3". John Wiley and Sons, New York.

Klohn, E.J. (1981). "The development of current tailings dam design and construction methods" in "Design and Construction of Tailings Dams" ed. by D. Wilson, Colorado School of Mines Press, Golden, Colorado, 1-52.

Lutovinov, A.G., V.G. Terent'ev and I.I. Shipko (1975). "Hydraulic filling of earth structures using a mixture of lumpy clays and sand". Hydrotechnical Construction, No. 6, June 1975, pp. 531-535.

Melent'ev, V.A. (1980). "Development of hydraulic-fill dam construction in the Soviet Union". Hydrotechnical Construction, No. 12, December 1980, pp. 1268-1271.

Melent'ev, V.A., Kolpashnikov, N.P. and Volnin, B.A. (1973). "Hydraulic Fill Structures" Energy, Moscow (English translation).

Middlebrooks, T.A. (1942). "Fort Peck Slide". Transaction ASCE, 68 (8), 723-742.

Mitchell, D.E. (1984). "Liquefaction slides in hydraulically-placed sands". Proc. 37th Canadian Geotech. Conf., 141-146, Toronto.

Mitchell, R.J., Smith, J.D. and Libby, D.J. (1975). "Bulkhead pressures due to cemented hydraulic mine backfills". Canadian Geotechnical Journal, 12 (3), 362-371.

Mittal, H. and Hardy, R.M. (1977). "Geotechnical aspects of a tar sand tailings dyke". Proc. Conf. on Geotechnical Practice for Disposal of Solid Waste Materials, ASCE, University of Michigan, 327-347.

Morgenstern, N.R. (1985). "Geotechnical aspects of environmental control". Proc. 11th Int. Conf. Soil Mechs. Found. Eng., Vol. 1, 155-186, San Francisco.

Morgenstern, N.R., et al. (1977). "Slopes and excavations" Proc. 9th Int. Conf. Soil Mechs. Found. Eng., Vol. 3, 547-602, Tokyo.

Ogawa, H. (1969). "Dispersion of dumped sand from a hopper-barge". Trans. Japan Soc. Civ. Eng., 1 (1), 1-15.

Penman, A.D. (1982). "Materials and construction methods for embankment dams and cofferdams - general report". Transaction 14th Int. Cong. Large Dams, 4, 1105-1228, Rio de Janeiro.

Perevezentsera, I.I. (1977). "Hydraulic filling of dams using silt loam on a slump-prone foundation". Hydrotechnical Construction, No. 2, February 1977, pp. 136-141.

Pladet, A.A. (1978). "Densification in field practice - methods and requirements". Proc. Int. Symp. on Soil Mechanics Research and Foundation Design for the Oosterschelkde Storm Surge Barrier, Delft, 2, Paper V-4.

Plewes, H.D., McRoberts, E.C. and Chan, W.K. (1988). "In-situ density measurements in sand tailings using geophysical nuclear logging techniques". Conference on Hydraulic Fill Structures, ASCE, Colorado State University, in press.

Poulos, S.J., Castro, G. and France, J.W. (1985a). "Liquefaction evaluation procedure". J. Geotech. Eng., ASCE, 111 (6), 772-795.

Poulos, S.J., Robinsky, E.I. and Keller, T.O. (1985b). "Liquefaction resistance of thickened tailings". J. Geotech. Eng., ASCE, 111 (12), 1380-1394.

Robinsky, E. (1979). "Tailings disposal by the thickened discharge method for improved economy and environmental control". Proc. 2nd Int. Tailings Symp., 75-95, Miller Freeman, San Francisco.

Seed, H.B. (1987). "Design problems in soil liquefaction". J. Geotech. Eng., ASCE, 113 (8), 827-845.

Sherard, J.L., et al. (1963). "Earth and Earth-Rock Dams". J. Wiley and Sons, New York.

Shkundin, B.M. (1976). "Hydromechanization for modern hydraulic construction". Hydrotechnical Construction, No. 14, November 1976, pp. 1069-1074.

Sladen, J.A., D'Hollander, R.D. and Krahn, J. (1985a). "The liquefaction of sands, a collapse surface approach". Canadian Geotechnical Journal, 22 (4), 564-578.

Sladen, J.A., D'Hollander, R.D., Krahn, J. and Mitchell, D.E. (1985b). "Back analysis of the Nerlerk berm liquefaction slides". Canadian Geotechnical Journal, 22 (4), 579-588.

SNiP-II-53-73 (1974). "Embankment Dams, Moscow, Stroiizdat, 28p. (in Russian).

Tarasenko, S.M. (1970). "Prevention of Slope Failure during the construction of hydraulic-fill, fine sand structures on impermeable foundations" Hydrotechnical Construction, No. 1, January 1970, pp. 28-30.

Troncoso, J.H. (1988). "Behaviour of hydraulic fill structure in the March 1985 earthquakes in Chile" Conference on Hydraulic Fill Structures ASCE, Colorado State University, in press.

Troncoso, J.H. and Verdugo, R. (1985). "Silt content and dynamic behaviour of tailing sands". Proc. 11th Int. Conf. Soil Mechs. Found. Eng., Vol. 3, 1311-1314, San Francisco.

Valore, C. and Giglio, I.A. (1985). "Field consolidation of thick layers of dredged material". Proc. 11th Int Conf. Soil Mechs. Found. Eng., Vol. 3, 1315-1320, San Francisco.

Vick, S.G. (1983). "Planning, Design, and Analysis of Tailings Dams". Wiley-Interscience, New York.

Volnin, B.A. and Ivanovskaya, N.V. (1977). "Providing dynamic stability for hydraulic fill dams". Hydrotechnical construction, No. 3, March 1977, pp. 256-263.

Whitman, R.V. (1970). "Hydraulic fills to support structural loads". J. of Soil Mechs. and Foundation Div., ASCE, 86 (SM1), 23-48.

Yufin, A.P. (1965). "Hydromechanization" Moscow 496 pp. (in Russian).

SLOPES AND DENSITIES OF HYDRAULICALLY PLACED SANDS

M.B. de Groot, *
F.T. Heezen, **
D.R. Mastbergen, ***
H. Stefess, ****

ABSTRACT

Data have been collected about the slopes and density of dams of fine sand constructed in estuaries by hydraulic fill methods. Field measurements have been made during construction and laboratory tests carried out to investigate the sedimentation process.
A relationship has been found between the slope above the water level and the grain size, as well as the specific flow rate of the sand-water mixture . A relatively steep slope can be achieved by spreading the mixture over the full width of the dumping site.
The underwater slope appears to be the result of the rapid sedimentation of the sand from the sand-water mixture density current and intermittant flow slides. This slope is strongly influenced by the grain diameter. The specific sand transport rate and the water depth also influence the slope.
Relatively gentle slopes have been observed in the tidal zone.

In all cases, the relative density of the sand deposited below the water level was low and the sand might have been susceptible to liquefaction.

INTRODUCTION

Frequently sand bodies are constructed in seas, estuaries, lakes and rivers by hydraulic fill to serve as breakwaters, harbour extensions, artifical islands, closure dams, etc. The low price of the construction material is often an attractive reason for using this type of structure. Less attractive factors are however:

- the gentle slopes caused by the construction method which necessitate the use of large quantities of sand, sometimes over relatively large areas
- the low relative density of the soil which introduces the risk of flow slides.

----

* Delft Geotechnics, P.O. Box 69, 2600 AB DELFT, the Netherlands
** DHV Consultants, P.O. Box 85, 3800 AB AMERSFOORT, the Netherlands
*** Delft Hydraulics, P.O. Box 177, 2600 MH DELFT, the Netherlands
**** Rijkswaterstaat, P.O. Box 5044, 2600 GA DELFT, the Netherlands

During the last 15 years more than ten large sand dams have been constructed in the estuaries in the south-western part of the Netherlands within the framework of the "Delta-works". Most of these dams are closure dams, which had to be constructed quickly in order to minimize the loss of sand. The final closure dam, the Krammer Dam, was constructed by four large dredgers, with a combined maximum production of 18000 m^3/hr.. Two dredgers worked on each side of the dam. The water-depth was 4 to 20 m and the length of the dam about 1100 m. All dams were constructed with local sand with an average grain size varying between 150 µm to 280 µm and steep grain size curves.
In order to reduce the size of this type of dam and to increase the stability it is necessary to improve the construction method. This requires knowledge of the processes occurring after the sand-water mixture has left the pipe. In order to obtain the information required, the Rijkswaterstaat (Public Works Department of the Ministry of Transport and Public Works, the Netherlands) and the Delft University of Technology started an extensive research programme in 1984, in collaboration with Delft Hydraulics and Delft Geotechnics. Field measurements, laboratory experiments at different scales and theoretical studies have been carried out and data from many sand bodies constructed in recent years collected and analysed. The results of the research programme which relate to construction methods in which sand is released above the water level, see Figure 1, are discussed in this paper.

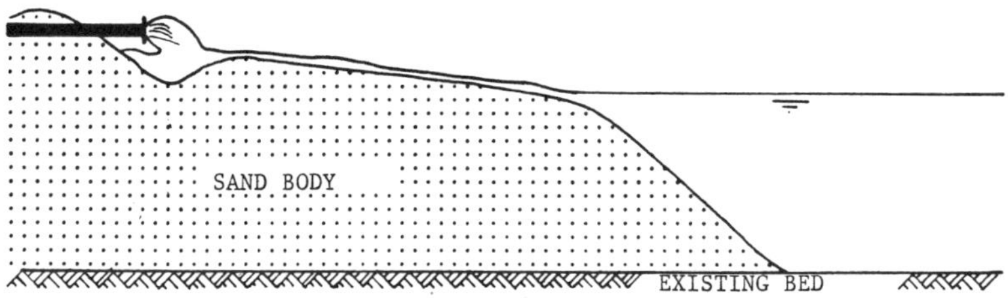

FIG 1 CONSTRUCTION METHOD, LONGITUDINAL PROFILE

A short description is given of the construction procedure and the phenomena observed and the following three subjects are discussed:
- the sand-water mixture flow above the water level and the resulting slopes above the water level and in the tidal zone
- the formation of the underwater slope
- the resulting density of the sand body.

CONSTRUCTION PROCEDURE AND OBSERVED PHENOMENA

The sand is dredged from pits in the vicinity of the works. It is then pumped to the dumping site as a sand-water mixture at concentrations up to 30% by volume. At the site the mixture is released, forming a crater underneath the mouth of the pipe and subsequently flowing over the dump into the water, see Figure 1. The width of the dump is restricted by guide bunds of sand 20 to 100 m apart, which are maintained by bulldozers. The bunds are needed to maintain the flow direction; otherwise the direction of the sand-water mixture flow would vary over 180° and a half cone-shaped delta would be formed, producing a very wide dam, see Figure 2.

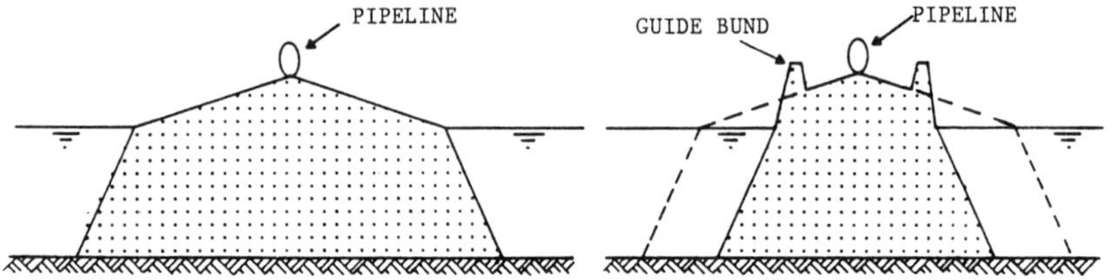

FIG 2 CROSS SECTION WITHOUT AND WITH GUIDE BUNDS

Bunds cannot be maintained below the water line and a cone-shaped delta develops. The under-water part of the dam is formed by the progress of the delta in one direction. The side slopes are equal to the slope of the cone.
The dam is extended by periodically lengthening the pipe, moving the mouth forwards.

In tidal waters, the variation in water level influences the flow properties. During periods of rising water, the mixture often spreads out over a series of terraces and cascades, Figure 3. The flow depth is generally less than 0.1 m; specific mixture flow rates are in the range 0.03 to 0.3 m^2/s. The specific flow rate is the volume discharge of sand-water mixture per unit width. Terraces and cascades (Figure 3) move slowly upstream at a velocity of 0.01 to 0.03 m/s. The sediment concentration decreases considerably to about 10% by volume during the flow from pipe to water line. Apparently sedimentation occurs before the flow meets the water level. During periods of falling water the flow often concentrates in meandering channels, a few metres wide and up to one metre deep, Figure 4. The velocity of the mixture is 1 to 2 m/s in the channels. The concentration is and remains about the same as in the pipeline. Most of the sand is transported to the water through the channels, resulting in the underwater slope building out quickly.

Slopes differ above and below the water level, slopes under water generally being steeper than above water. The most gentle slopes are observed in the tidal zone, if this is present.

The density of the sand body under water is nearly always low and the sand is, therefore, sensitive to liquefaction. The density of the sand body above water level is, in general, significantly higher and not sensitive to liquefaction.

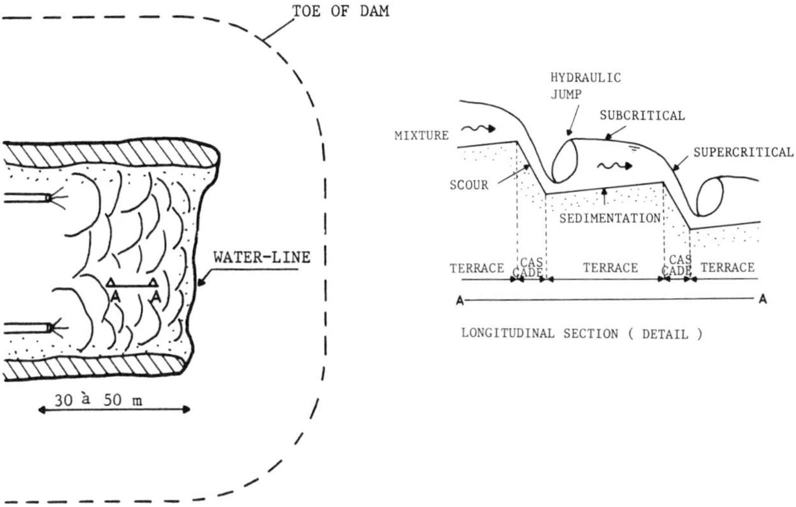

FIG 3 MIXTURE SPREADS OUT OVER A SERIES OF TERRACE-CASCADES

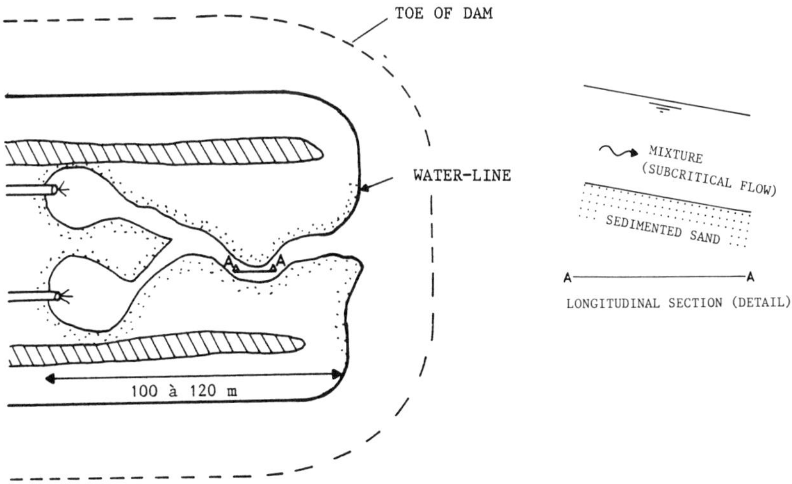

FIG 4 MIXTURE CONCENTRATES IN CHANNELS

SAND-WATER MIXTURE ABOVE THE WATER LEVEL

Description of field measurements and laboratory tests

Field measurements were made during the construction of hydraulic fill closure dams in the Oosterschelde Estuary.

The following sand-water mixture parameters were measured at the dumping site: flow velocity, flow depth, channel width, bed slope, sand concentration and grain size. From these data specific mixture flow rates and sand transport rates were calculated resulting in average values of these parameters for each dumping site at a specific time, (Delver and Verwoert 1986 and Winterwerp a.o. 1988).

A full scale experimental set up, which included special measuring instruments, has been developed to study sand-water mixture flow features in detail, (Mastbergen and Winterwerp 1987; Winterwerp a.o. 1988). Small-scale orientation tests were also carried out, (Lantsheer and Neerings 1984). The large scale facility is illustrated in Figure 5.

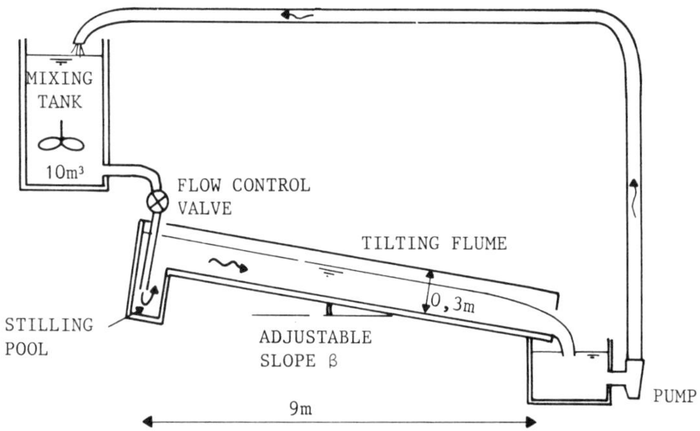

FIG 5 LARGE SCALE EXPERIMENTAL SET-UP FOR SAND-WATER MIXTURE FLOW STUDIES

Theoretical considerations

An important result of these studies is a formula for the equilibrium slope, defined as the slope at which sedimentation and erosion are in balance. There is often no equilibrium at the dumping sites studied here. These equilibrium cases still are none the less very relevant since the net sedimentation or erosion is small.

Two main flow regimes can be distinguished: laminar flow and turbulent flow. For mixture flows with 30% by volume of sand grains, laminar effects become important if the specific mixture flow rate $q \leq 0.01$ m^2/s.

Since viscosity is strongly influenced by the presence of grains this effect has to be taken into account. At a concentration of 30%, the apparent viscosity is about 4 times the viscosity of clear water.
In small scale tests, the flow is laminar and the sand grains are supported by mutual collisions. At the dumping site, however, the flow is mainly turbulent and the grains are kept in suspension by turbulent diffusion.

The turbulent flow regime can be subdivided into two morphological sub-regimes. At low specific mixture flow rates, $0.01 < q < 0.1$ m²/s, terraces and cascades are present. Subcritical flow occurs over the terraces and supercritical flow over the cascades, Figure 3. Although sedimentation takes place on the terraces and scour at the cascades, the average slope may be an equilibrium slope.
At high specific mixture flow rates, which occur when the mixture is concentrated in channels, Figure 4, the flow is constantly subcritical and more or less uniform.
The characteristics of turbulent equilibrium flow can, in principal, be described by classical transport formulas. But, for the high sand concentrations considered here, these formulas have to be adapted because of the "hindered settling", that is the reduction of the effective fall velocity of the sand grains due to the close proximity of other grains. This reduction can be expressed as a function of the concentration, (Richardson and Zaki 1954).

For a sand volume concentration (that is volume of the grains divided by the volume of the mixture) $C = 0.2$, the fall velocity is reduced to about a third of the fall velocity of an individual grain. The sedimentation rate, i.e. the product of concentration and effective fall velocity, hardly varies with concentration around $C = 0.2$. Introducing this effect in a classical sand transport formula, for instance the well-known Engelund-Hansen formula (Engelund and Hansen 1967), yields the following form for the equilibrium slope β_1 of the sandbed above water level:

$$\tan \beta_1 \sim D^{0.6} \times C^{0.6} \times (1-C)^{1.2} \times q^{-0.4} \qquad (1)$$

which is mainly a function of grain size D and specific mixture flow rate q and only a weak function of the concentration C, (Mastbergen a.o. 1986; Winterwerp a.o. 1988). The same effect appears in pipeline slurry flows, where the critical or equilibrium velocity is also only a weak function of concentration C for $C > 0.15$, (Vanoni 1977).

The fall velocity of sand particles can also be reduced by the presence of silt or clay particles.

<u>Empirical equilibrium slope</u>

The theoretical relation for the equilibrium slope according to formula (1) is very close to an empirical formula derived from the large scale laboratory tests and the field measurements:

$$\tan \beta_1 = 0.006 \left(\frac{D}{D_o} - 1\right)\left(\frac{q}{q_o}\right)^{-0.45} \qquad (2)$$

in which D_o and q_o are constants with values of 65 μm and 1 m²/s respectively. Formula (2) is valid for $D > D_o$ and $q > 0.01\ q_o$, see Figure 6. The effect of grain size in this formula is a little greater than that found from sand transport formulas, and may, in fact, be exaggerated. This may be due to the fact that very clear water was used in the laboratory tests with the coarse sand, whereas some silt was present in the water used in the tests with fine sand, which reduced the apparent fall velocity.

Formula (2) can be applied in the channel flow sub-regime. The large scale laboratory tests and a detailed mathematical description of the flow and the sedimentation and scour process (Mastbergen a.o. 1986) also indicate that it can be used for the overall bed slope of terrace-cascade flow.

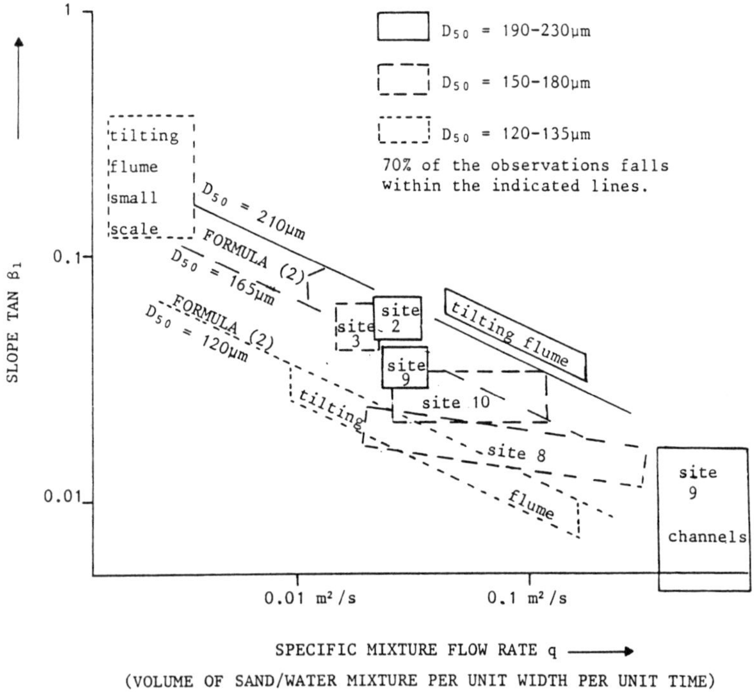

FIG 6 SLOPES OBSERVED ABOVE (HIGH) WATER

TABLE 1 SLOPES OBSERVED ABOVE THE WATER-LEVEL AND IN THE TIDAL ZONE

Location or site	D50 (μm)	Production (m^3/hr structure)	Flow rate (m^2/s mixture) q	Slope above the water level $\tan\beta_1$	Slope in the tidal zone $\tan\beta_2$
Large scale	120		0.01-0.15	0.01-0.04	
tilting flume	225		0.05-0.15	0.04-0.06	
2 Springersdiep	200	2000	0.03-0.5 ?	0.05	0.02
3 Geul	150-180	3500	0.02-0.3 ?	0.05	0.025
8 Speelmansplaten	150	2200	0.02-0.3	0.013-0.02	
9 Marollegat	210	3600	0.03-0.5	0.035	0.025
10 Slaak	180	3500	0.025	0.03	

TABLE 2 CORRESPONDENCE BETWEEN PRODUCTION AND FLOW RATE

A PRODUCTION OF 1000 m^3/hr STRUCTURE IS EQUIVALENT TO :		
Conditions	Production	Sand transport rate and specific mixture flow rate
porosity n = 0.4	$\frac{1000}{3600}$ = 0.28 m^3/s structure 0.28x(1- 0.4) = 0.17 m^3/s grains 0.17x2650 = 440 kg/s grains	
if spread out over 50 m C = 0.25		$s=\frac{440}{50}$ = 8.8 kg/ms grains $\frac{0.17}{50}$ = 0.0033 m^2/s grains $q=\frac{0.0033}{0.25}$ = 0.013 m^2/s mixture
if flowing in a 5 m wide channel C = 0.25		$s=\frac{440}{5}$ = 88 kg/ms grains $\frac{0.17}{5}$ =0.033 m^2/s grains $q=\frac{0.033}{0.25}$ = 0.13 m^2/s mixture

Slopes observed above the water level

The results of field measurements are summarized in Table 1 and areshown in Figure 6. Slope definitions are given in Figure 7 and appendix 2. An indication of how production, mixture flow rate and sand transport rate depend on each other for a given production,is given in Table 2.

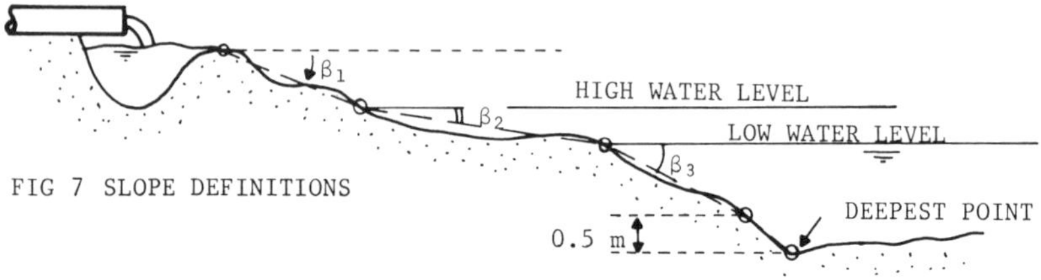

FIG 7 SLOPE DEFINITIONS

Measurements were made at Marollegat, Location (site) 9 (Delver and Verwoert 1986). During high tide, the sand-water mixture was spread out over a wide area and was flowing over terrace-cascades (Figure 3) to the water line. The mean flow rate was 0.03 m²/s. Much sand deposited in this area at an average slope tan β_1 = 0.04.

The slopes measured above water at Location 2 and 3 (Nelissen a.o. 1985) were steeper than at Location 9. Much more gentle slopes were observed at Location 8 (Koster 1985). This may have been partly due to the smaller grain diameter.

Slopes observed in the tidal zone

The formation of slopes in the tidal zone (β_2) is more complicated and a specific example is discussed below to illustrate the processes involved in slope formation. At Location 9, slopes were formed in the following way (Delver and Verwoert 1986): at the end of the period of high water, production was interrupted for some time to enable the pipeline to be lengthened. At this time the channel slope was tanβ_1 = 1:30, see Figures 6 and 8. As soon as the production was re-started a channel scoured in the newly formed sand body, because the water level had fallen much lower in the period of no production, see Figure 8. In this figure, the sand body formed during high water is indicated by AA and AC; the channel-bed is indicated with BB.

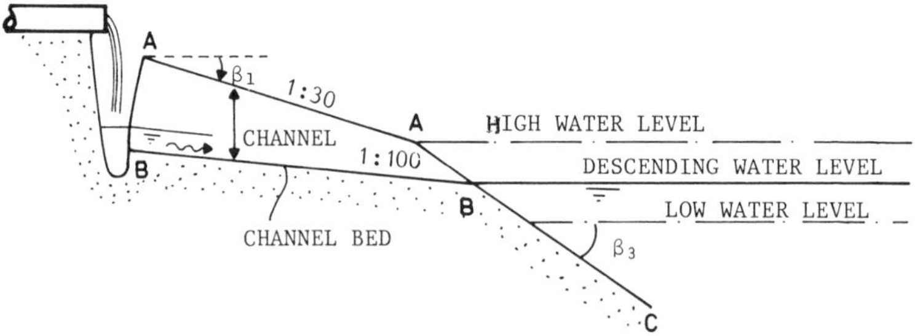

FIG. 8 CHANNEL DEVELOPMENT DURING FALLING WATER-LEVEL

The specific flow rate in the channel in the period of falling water levels after production had restarted was about ten times as much as on the terraces at high tide. The slope became much more gentle (mean slope 1 : 100).
During low water, a gently sloping plane BD, an extension of the channel bed BB, formed at the end of the period of falling water level, see Figure 9.

The slope in the tidal zone was mainly formed during the rising tide by sedimentation just below the water level on top of this gently sloping plane, see Figure 9. Sedimentation also took place simultaneously in the channel. Although sedimentation below the water level creates relatively steep slopes (see next paragraph), the quickly rising water level produced so little sedimentation that the final slope in the tidal zone (ED) was between that of the channel (BD) and that of the underwater slope (DD).

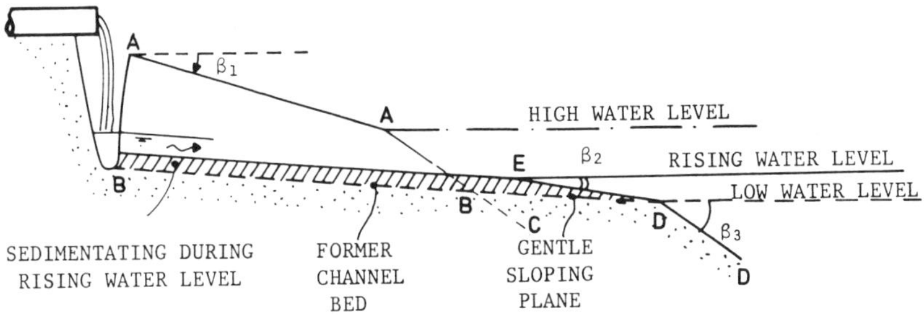

FIG. 9 FORMATION OF SLOPE IN THE TIDAL ZONE

About the same construction method was applied at Locations 2 and 3 and the average slope in the tidel zone was found to be nearly the same as that formed at Location 9 (Nelissen a.o. 1985). At Location 10, however, the production was not interrupted at high tide and the contractor managed to prevent channel formation with bulldozers. The resulting slope in the tidal zone was clearly steeper, see Table 1.

SAND-WATER MIXTURE FLOW BELOW THE WATER LEVEL

Description of field measurements

Current velocities and sediment concentrations were measured simultaneously at several locations immediately at the underwater slope downstream of the spot where the sand-water mixture reached the water line. Extensive measurements were taken at Location 9 (Koster 1986). It became clear that:

- the sand-water mixture continued to flow below the water level as a density current over a relatively short distance
- this flow took place over a limited width in a varying direction over the half cone in accordance with the process of delta-formation described above
- nearly all the sand settled within a relatively short distance (10 to 50 m) from the water line, so at the upper part of the underwater slope.

Further down slope movement can only be explained by assuming that flow slides occurred. This hypothesis was confirmed in an extensive measurement campaign at Location 10 during the closure of the Slaak (de Groot a.o. 1987). At a fixed position the sedimentation rate, concentration, movement of the seabed, density and pore-water pressure were measured continuously during the hydraulic fill operations.

Flow slides were observed (see below), but only during the short periods that the density current flowed towards the measurement position, causing deposition of sand at the higher part of the slope.

Description of laboratory tests

Density current, sedimentation and flow slide processes, have been studied in a two-dimensional situation in the laboratory at prototype scale (Mastbergen a.o. 1988; Bezuijen a.o. 1988; Fig. 12). A 2.5 m deep flume with glass windows was used, with a sand-water mixture inlet at one end and an overflow at the other which maintained the water level constant.

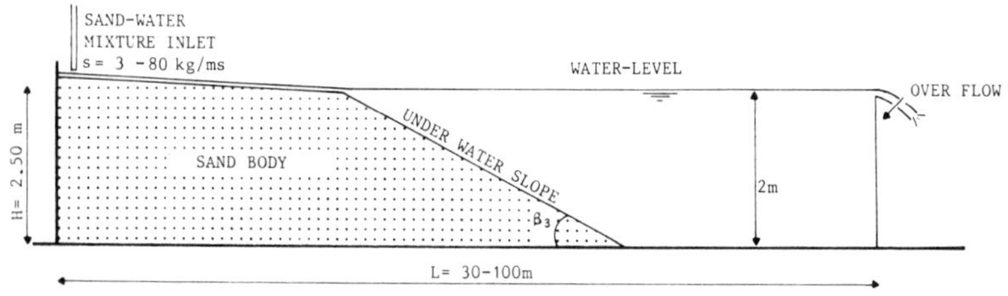

FIG. 12 LONGITUDINAL SECTION OF FLUME USED FOR LARGE SCALE LABORATORY TESTS.

Various density current parameters (flow velocity, flow depth, concentration) and parameters of the newly formed sand body (density, pore water pressure) were measured. Similar tests were also carried out at a much smaller scale (Delver and Verwoert 1986).

Theoretical considerations

Before discussing the laboratory tests in detail a general remark should be made about the occurrence of density currents. Turbulent density flows in oceans of natural origin, are well known to sedimentologists and are usually called turbidity currents. Quantitative information about the hydraulics and sediment transport phenomena is however very limited. More information has been gained from the present field measurements and laboratory tests, and from comparisons with the flow phenomena above the water level.

Similar phenomena as observed above the water level, are important for sand-water mixtures flowing below the water level. In this case we can also distinguish between laminar and turbulent flow. Laminar flow was observed in the small scale experiments whereby sheets of grains (with water in between) were flowing down steep slopes. The turbulent suspension regime can also be divided into sub-regimes with and without terraces and cascades, depending on the internal Froude number.

An essential difference between the flow above and below the water level is that the driving gravity force under water is much smaller than above water. Therefore, the underwater equilibrium slope must be much steeper, to be able to reach the same flow velocity and thus the equilibrium between sedimentation and erosion. The following relationship can be derived for the equilibrium slope under water β_3:

$$\tan \beta_3 \sim D^{0.6} \times (1-C)^{1.2} (1+\Delta\ C) \times s^{-0.4} \quad (3)$$

in which Δ = specific density of the grains relative to water and s = specific sand transport rate. In this formula sand transport rate instead of mixture flow rate appears, since density effects are important at the under water slope. The slope in this case is mainly a function of grain size and sand transport rate.
At the dumping site the length under water over which the sand settles out from the sand-water mixture flow can influence the final slope. In practice this sedimentation length is usually small compared to the length of the underwater slope. Sand accumulates on the upper part of the slope, which becomes steeper and steeper. Finally a flow slide occurs, if the soil mechanical conditions, such as porosity, allow it, which is usually the case, Figure 13A.

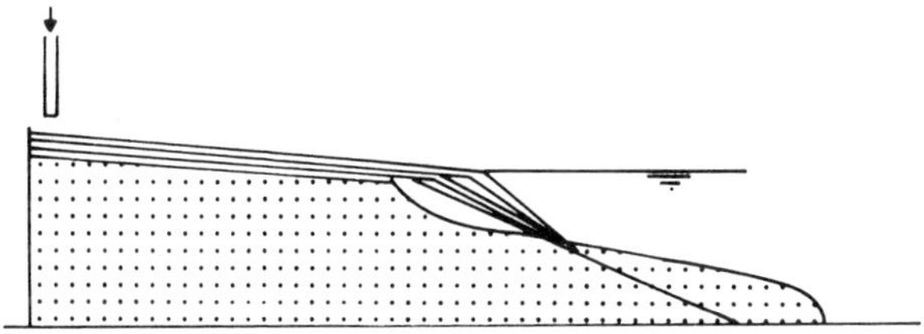

FIG. 13A SHORT SEDIMENTATION LENGTH RESULTING IN STEEPENING SLOPES AND A FLOW SLIDE

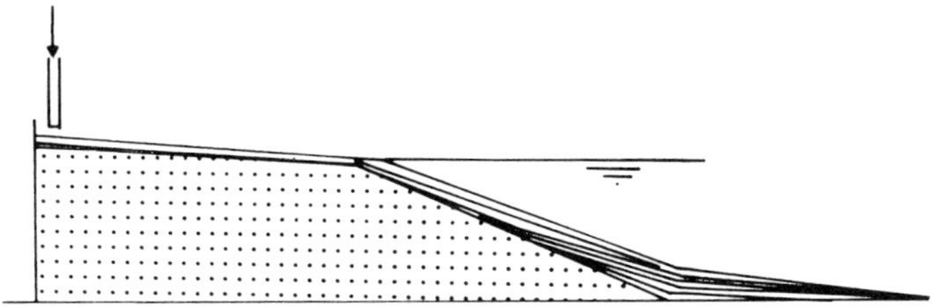

FIG. 13B LONG SEDIMENTATION LENGTH RESULTING IN FLATTENING SLOPES. NO FLOW SLIDES OBSERVED

With very high sand transport rates, s > 50 kg/ms, and small water depths, however, the sand will not settle out before reaching the toe of the sand dam, resulting in very gentle slopes and no flow slides (Figure 13B). These two different mechanisms were clearly observed during the large scale tests (Mastbergen a.o. 1988).

Liquefaction and flow slides

Some results of the field measurements at Location 10 are presented in (de Groot a.o. 1987). Very similar effects were observed in large scale laboratory tests; several times during each test excess water pressures were measured from which it could be concluded that liquefaction occurred over a depth varying between 0.1 m and 1.0 m. Moving bed material was observed through the glass windows simultaneously with the liquefaction, during the tests with low sand transport rates. More detailed results of these tests will be presented elsewhere (Bezuijen a.o. 1988).

Slopes observed underwater

A considerable amount of data has been collected from small and large scale laboratory tests (Mastbergen 1988), and also from field measurements. The most extensive field measurements were made at the Locations 2 and 3 (Nelissen a.o. 1985) and at the Locations 10, 11 and 12 (Stefess 1987). The observed slopes underwater are given in Table 3. The slope (β_3) is defined in Figure 7 and appendix II. Remark: some figures in Table 3 differ from those in Table 1, because they refer to a different construction stage.

TABLE 3 OBSERVED UNDERWATER SLOPES AND PORE-VOLUMES

Location or site	Water depth below low water (m)	D50 (μm)	Production (m^3/hr structure)	Sand transport rate s (kg/ms)	Mean slope under water $\tan\beta_3$
Laboratory, large scale	2	120-140		2-80	0.07-0.25
1. Brouwersdam	16	200-250	200		0.06
2. Springersdiep	5	200	2000	30-300 ?	0.07
3. Geul	10	150-180	3500	20-200 ?	0.05
4. Europoort	20	200	1200 ?		0.06
5. Zuiderdam	8	200	2000 ?		0.08
6. Gioia Taura	50	400-450	2000 ?		0.27
7. Markiezaat	4	155	1200 ?	.	0.05
9. Marollegat	8	210	3600	30-450	0.04
10. Slaak	2-8	190	2000-2700	10-15	0.055
11. Tholense gat	5	140-170	3000-7000	15-30	0.040
12. Krammer South	8	150-280	5000-8000	20-35	0.06
North			7000-11000	30-50	0.045

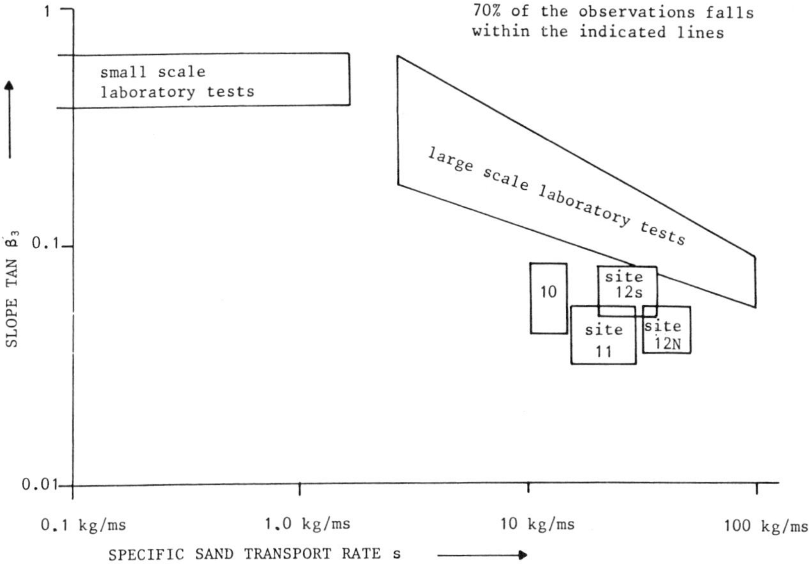

FIG. 10 SLOPES OBSERVED BELOW THE LOW WATER-LEVEL

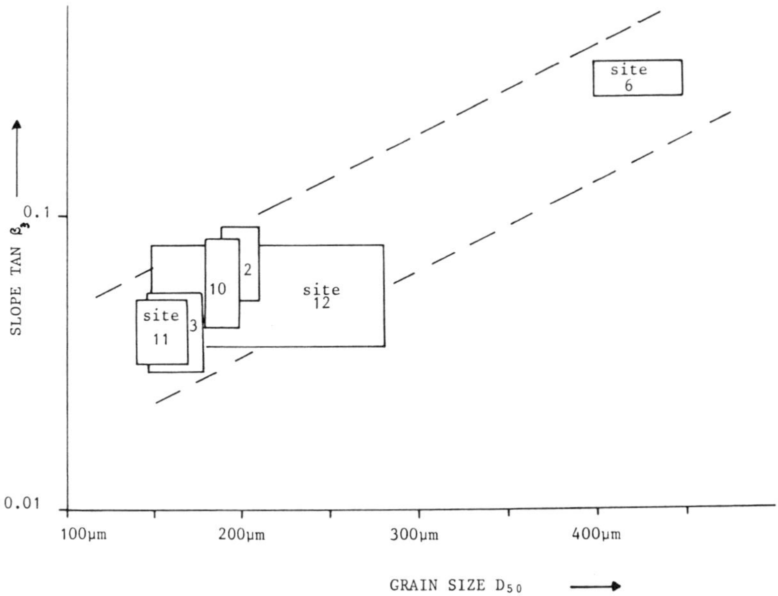

FIG. 11 SLOPES BELOW THE WATER-LEVEL VERSUS D_{50}

The main results of the laboratory and field measurements for fine sand are presented in Figure 10 as a function of the sand transport rate. The main field slopes observed are also presented in Figure 11 as a function of grain size.
In the cases where the slope in front of the pipe during construction differed from the side slopes, only the side slopes are given. At Location 3, the front slopes appeared to be about 20% more gentle; at Locations 10, 11 and 12, however, the front slopes were 20% to 40% steeper than the side slopes.

The general tendency given by the theoretical relationship (3) is confirmed by Figures 10 and 11: steeper slopes with smaller sand transport rates and coarser grain sizes. The influence of grain size, however, is stronger than (3) indicates and the influence of the sand transport rate is less clear: a wide scatter can be observed. This can be explained by the occurrence of flow slides in the case of fine sand, as mentioned above.

The grain size, therefore, is of primary importance. However, if only fine sand is available, relatively steep slopes can still be created, if the hydraulic fill is well controlled, as was the case at Locations 10, 11 and 12: careful maintenance of the guide bunds, no breaks in production, avoidance of channel formation at the dumping site in order to obtain low specific sand transport rates at the water line.

In addition to the grain size and the sand transport rate, the water depth can also influence the slope. A slight tendency for decreasing slope with increasing water depth can be observed. An exception is the situation of a very low water depth, combined with a very high sand transport rate, which produces a relatively large sedimentation length. Very gentle slopes were observed in this situation in the large scale laboratory tests and also during the last phase of constructing the Krammer Dam (Location 12)

In other fine sand cases the influence of flow slides probably predominates slope formation. Nevertheless, the slopes formed in this way during the construction of these sand dams were significantly more gentle than the slopes observed after flow slides along estuarine channels and banks in the same region.

During the measurements, no signicant influence of currents and waves on the slope formation could be found.

DENSITIES

Measurements of the density of a sand body are essential to analyse its stability. In order to understand the processes which influence the density better, small scale experiments have been carried out (Heezen and van der Stap 1985). Densities were also measured in the large scale laboratory tests mentioned above. In addition porosities and densities were determined at one or two sites at almost every location. The results are given in Table 4.
Minimum, maximum and critical porosities were measured for 2 or 3 samples from every location. The minimum and maximum porosities were

determined using the procedure described in the report "Various Properties of Natural Sands etc." (1979). The critical porosity (void ratio or density) was determined according to the Casagrande and Geuze method, described by Lindenberg and Koning, 1981 ("dry critical density"). The in-situ porosity was measured continuously over the full depth of the dam and was derived from the electrical resistance (Vermeiden and Lubking 1978).

Both laboratory and field measurements indicate that the density of the underwater part of a sand body, constructed by hydraulic fill is almost independent of dumping parameters such as concentration, or sand transport rate; only the sand characteristics seem to influence the resulting density.

The large scale laboratory measurements show a considerable increase in density during the flow slides. Nevertheless, the final porosity remains above the critical value and the sand below the water table may remain susceptible to further liquefaction. The porosity in the field is also higher than the critical porosity. So flow slides may occur in the field, also after construction is completed.

The density of the sand above the water table has been estimated using cone penetration tests. Remarkably high cone penetration values were encountered, indicating rather high densities. According to available data the sand above the water table has a 4 to 5% lower porosity than the underwater sand and is not susceptible to flow slides ($n_{in\text{-}situ} < n_{crit}$).

TABLE 4 POROSITIES OBSERVED BELOW THE WATERLEVEL

Location	D_{50} μm	n_{min}	n_{max}	n_{cr}	$n_{in\ situ}$	$D_r = \frac{e_{max} - e}{e_{max} - e_{min}}$
Laboratory small scale	150	0.37	0.50	0.45	0.48	15%
Laboratory large scale	120	0.35	0.49	0.42	0.46-0.49	0-25%
1. Brouwersdam	200-250	0.33	0.46	0.40	0.40-0.43	25-50%
3. Geul	150-180	0.30	0.46	0.41	0.41-0.43	25-35%
4. Europoort	200	0.32	0.48	0.39	0.41-0.43	35-50%
6. Gioia Taura	400-450	0.30	0.45	?	0.39-0.40	40-45%
10. Slaak	180-200	0.34	0.47	0.42	0.43-0.44	25-35%
11. Tholense Gat	170-190	0.34	0.46	0.41	0.41-0.44	20-45%
12. Krammer	170-250	0.32	0.45	0.41	0.41-0.44	10-40%

CONCLUSIONS

The following conclusions can be drawn about the construction of sand bodies in seas or rivers by hydraulic fill, when the sand is released above the water:

- The slope resulting from the sedimentation above the water line is nearly equal to the equilibrium slope, i.e. the slope at which a sand-water mixture flows if the average sedimentation is zero. This slope is mainly determined by grain size and specific mixture flow rate, i.e. the flow per unit width. From laboratory tests an empirical formula has been derived, confirmed by theoretical considerations and field measurements, to quantify these effects.
  Coarse sands and low specific flow rates, i.e. a well spread flow over the full width of the dumping site, result in relatively steep slopes.
- The equilibrium slope below the waterline is much steeper. But the slopes observed during field measurements were only slightly steeper than above the water level. This can be explained by the occurence of flow slides as observed during the field measurements and the large scale laboratory tests. The grain size is of primary importance for the underwater slope.
  From the laboratory tests a correlation could be found between observed underwater slopes and specific sand transport rate, i.e. the sand production rate per unit width. So the underwater slope will probably become steeper if the flow is well spread over the full width of the dumping site. The slope is also influenced by water depth and by other operation features, but these effects could not be quantified yet.
- The most gentle slopes often occur in the tidal zone, if this is present; particularly if the water depth is small, the tidal difference is large and if the mixture flow is concentrated in channels during low water.
- The sand which has settled below the water level generally has a low relative density and may often be susceptible to liquefaction. The construction method was not observed to influence the density.
  Above the water level, the density of the sand is relatively high and the sand is seldom susceptible to liquefaction.

ACKNOWLEDGEMENTS

The authors are grateful to the Rijkswaterstaat (Public Works Department, Ministry of Transport and Public Works) and the CSB (Dredging Research Association) in which participate Amsterdam Ballast Dredging, HAM, Royal Boskalis Westminster, Volker Stevin Dredging and Rijkswaterstaat, for the opportunity to perform the research summarised in this paper. The authors are also grateful to their colleagues of Rijkswaterstaat, Delft University of Technology, Delft Hydraulics and Delft Geotechnics for their contribution to the research.

APPENDIX I - REFERENCES

Bezuijen, A., Mastbergen, D.R. a.o. (1988). "On the construction of sand fill dams; part II: soil mechanical aspects". Int. Symp. on Mod. SOil-WAter-Structure Interactions, Delft, the Netherlands.

Delver, G. and Verwoert, H. (1986). "Research on sand-water mixture flow during hydraulic fill sand closures". Delft Un. of Techn., Fac. Civil Eng., Delft; the Netherlands (in Dutch).

Engelund, F. and Hansen, E. (1967). "A monograph on Sediment Transport in Alluvial Streams". Teknisk Forlag, Copenhagen, Denmark.

Groot, M.B. de, Silvis, F., Rossum H. van and Koster, M.J. (1987). "Liquefied sand flowing over a gentle slope". Ninth Eur. Conf. Soil Mech. Found. Eng., Dublin, Ireland.

Heezen, F.T. and Stap, A.C.M. van der (1985). "Underwater dumped sand bodies I, II". Delft Un. Techn., Fac. Civ. Eng., Delft, the Netherlands (in Dutch).

Koster, M.J. (1985). "Evaluation of measurements on dumping site Speelmansplaten 1". Rijkswaterstaat, Deltadienst, DDWT-85.002, Den Haag, the Netherlands (in Dutch).

Koster, M.J. (1986). "Measurements sand closure Marollegat" ("Meetbundel"). Rijkswaterstaat, DWW, Delft, the Netherlands (in Dutch).

Lantsheer, C. and Neerings H. (1984). "Experimental research on Sand Slurry Flow", Delft Un. of Techn., Fac. of Civ. Eng., Delft, the Netherlands (in Dutch).

Lindenberg, J. and Koning, H.L. (1981). "Critical density of sand", Geotechnique 31, No. 2, 231-245.

Mastbergen, D.R., Leeuwestein, W. and Bakker, W.T. (1986). "Behaviour of Sand Water Mixtures at Sand Closures", Delft Un. Techn., Fac. Civ., Eng., Delft, the Netherlands (in Dutch).

Mastbergen, D.R. and Winterwerp, J.C. (1987). "Experimental study on sand water mixture flow above the water table", Delft Hydraulics, Z46-02, Delft, the Netherlands (in Dutch).

Mastbergen, D.R., Bezuijen, A., Winterwerp, J.C. a.o. (1988). "On the construction of sand fill dams; part I: Hydraulic aspects" Int. Symp. on Mod. SOil-WAter-Structure Interactions, Delft, the Netherlands.

Nelissen, H.A.M., Jonker, A. and Rossum, H. van (1985). "Evaluation of Sand Closures", Delft Geotechnics, CO-416619/8, Delft, the Netherlands (in Dutch).

Richardson, J.F. and Zaki, W.N. (1954). "Sedimentation and fluidization: Part I". Trans. Inst. Chem. Eng., Vol. 32.

Stefess, H. (1987). "Evaluation dump slopes of Slaak, Tholense Gat and Krammer", Rijkswaterstaat, DWW, Bes 18, Bes 28 and Bes 62, Delft, the Netherlands (in Dutch).

Vanoni, V. (1977). "Sedimentation Engineering" ASCE, Manuals and Reports on Eng. Practice, no. 54.

"Various Properties of Natural Sands for Netherlands Highway Engineering" (1979). Study Centre for Road Construction, Arnhem, the Netherlands.

Vermeiden, J. and Lubking P. (1978). "Soil investigation", Proc. Symp. Foundation Aspects of Coastal Structures, Delft, the Netherlands.

Winterwerp, J.C., Groot M.B. de, Mastbergen, D.R. and Verwoert, H. (1988). "Hyperconcentrated sand-water mixture flow over a flat bed", in preparation, sent for publication to J. Hydr. Eng., ASCE.

APPENDIX II - NOTATION

C	-	concentration, i.e. volume of sand grains over volume of sand-water mixture	(-)
D	-	grain diameter	(L)
D_o	-	constant coefficient = 65 μm	(L)
D_{50}	-	median of grain diameter	(L)
D_r	-	relative density = $\frac{e_{max} - e}{e_{max} - e_{min}}$	(-)
e	-	void ratio	(-)
e_{cr}	-	critical void ratio	(-)
e_{max}	-	maximum void ratio	(-)
e_{min}	-	minimum void ratio	(-)
n	-	porosity	(-)
n_{cr}	-	critical porosity = $\frac{e_{cr}}{1 + e_{cr}}$	(-)
$n_{in\ situ}$	-	porosity measured in situ	(-)
n_{max}	-	maximum porosity	
n_{min}	-	minimum porosity	
q	-	specific mixture flow rate, i.e. volume discharge of sand-water mixture per unit width	$(L^2 T^{-1})$
q_o	-	constant coefficient = 1 m^2/s	$(L^2 T^{-1})$
s	-	specific sand transport rate, i.e. mass discharge of solid particles per unit width	$(ML^{-1} T^{-1})$
β_1	-	equilibrium slope of sand-water flow above the water or slope at which sand sedimentates above the water level (Figure 7), supposed to be equal	(-)
β_2	-	slope at which sand sedimentates in tidal zone (Figure 7)	(-)
β_3	-	slope at which sand sedimentates below the water level (Figure 7)	(-)
ρ_s	-	density of solid particles	(ML^{-3})
ρ_w	-	density of water	(ML^{-3})
Δ	-	relative density of solid particles = $\frac{\rho_s - \rho_w}{\rho_w}$	(-)

A REVIEW OF TAILINGS DISPOSAL PRACTICES IN NORTH AMERICA AND AUSTRALIA

Andrew Minns*
B.E. M.I.E. AUST.

ABSTRACT

This paper discusses a range of tailings disposal methods currently practiced in the mining industries in North America and Australia. These methods include various inpit disposal techniques; upstream, downstream and centreline dam construction; and zoned earth fill embankment construction. The thickened tailings disposal method and the disposal of fine clays are also discussed. The importance of improving the depositional properties of tailings to enhance storage capacity, water reclaim, dam wall stability and final reclamation is stressed.

INTRODUCTION

There are many factors influencing the way in which a mining company disposes of tailings. Some of the more important factors include physical and chemical properties of tailings, cost, availability of dam wall construction materials, type of mining operation, topography and environmental constraints.

As a result, the mining industry has seen a diverse and innovative range of disposal methods developed to satisfy the many conditions imposed at tailings disposal sites.

This paper describes tailings disposal methods selected predominantly from the North American and Australian mining industries. Examples from the gold, coal, copper, phosphate, bauxite, tar sand and diamond mining industries are presented. The paper is intended as a review paper, and as such, presents a range of examples and references, with an emphasis on recent developments.

For the purposes of this discussion, tailings disposal methods have been categorised into the following types:-

- in pit
- conventional zoned earth embankment
- upstream construction techniques
- centreline construction techniques
- downstream construction techniques
- thickened tailings disposal methods
- disposal of fine clays
- dry disposal techniques.

* Senior Planning Engineer, Comalco Aluminium Ltd, P.O. Box Weipa, Queensland 4874, Australia.

IN PIT DISPOSAL

The return of tailings for disposal in mined out areas is often practiced by large opencut or strip mining operations such as in the phosphate, coal, tar sand or bauxite mining industries. These mining operations normally effect large areas of land which must be rehabilitated.

An important objective in the design of inpit disposal systems is therefore to enhance the depositional properties of tailings to aid rehabilitation of the land. A number of inpit tailings disposal concepts are illustrated by the following examples from North American mining operations.

In pit Disposal of Coal Refuse

The combined coarse and fine coal refuse disposal facility developed for the Lamira Preparation Plant in Ohio, U.S.A. is a good example of how mine planning can be integrated with the development of a disposal facility. During development of a previous mine pit, overburden was used to construct a starter dam around the perimeter of the mine pit as mining progressed. Coarse reject was then used to raise the dam wall upstream to provide additional capacity for fine refuse storage (D'Appolonia, 1986). The currently active mine pit is being developed in the same way. Overburden is strategically placed to form the foundation and starter wall for the next combined coarse and fine refuse disposal facility.

Long Term Planning for In Pit Disposal

Long term tailings disposal planning is required at Syncrude's Tar Sand Mining operation near Ft. McMurray in northern Alberta, Canada (Handford et al 1982). In the past all tailings produced by the bitumen extraction plant have been stored in a dam sited close to the plant. Once the walls of this dam have reached design crest level an alternative disposal facility will be required for the tailings sands. In pit disposal options are being analysed in detail. Since mining operations must continue around the highwall of the pit, development of a tailings disposal facility within the same pit requires special placement techniques designed to avoid interference with the mining operations. Placement techniques which are being considered include upstream, downstream and central discharge methods.

The phosphate mining industry in Florida

The phosphate mining industry in Florida has developed methods for the disposal of phosphatic clay and sand tailings on mined land. Because of the poor consolidation and strength properties of phosphatic clay wastes, the disposal methods are being developed to enhance the depositional properties of the tailings. By encouraging drainage, desiccation and associated consolidation and strength gain of tailings surfaces, the mined land may be restored to a state suitable for a wider range of land uses. Techniques for impoving depositional properties of fine grained tailings are discussed later in the paper.

CONVENTIONAL ZONED EARTH EMBANKMENT

Where the tailings contain contaminants which could pose a threat to the surrounding environment, a common approach is to construct a dam which will minimise seepage from the embankment and the base of the dam. This style of dam design is common particularly in the uranium and gold mining industries as well as many base metal operations.

The basic features of such a structure usually include a clay core, sand and gravel filter zones, an upstream and downstream shell, erosion protection and a cut off trench and /or grout curtain.

These structures are usually costly, especially if lining of the entire impoundment is necessary and may significantly affect the economics of a proposed mining operation.

The Cannon Mine in Washington, McLaughlin Mine in California and Golden Giant Mine in Ontario, Canada are all gold operations which have constructed tailings dams in this way. Similarly, water retaining structures have been constructed at the Quirke Uranium Mine near Elliot Lake, Ontario, Canada to minimise seepage from the impoundment. Grouting beneath the Cannon, Quirke and Golden Giant embankments was also undertaken to reduce seepage through the foundation (Welch et al 1986, Caldwell et al 1986).

To defer capital expenditure, the construction of these types of dams may be staged over the life of the operation.

The embankment constructed for the Page-Williams Gold Mine near Hemlo, Ontario, Canada (Fig. 1) has been designed to accommodate further lifts by extending the central fill core and upstream and downstream shells. The construction materials required for this embankment are excavated from nearby borrow pits.

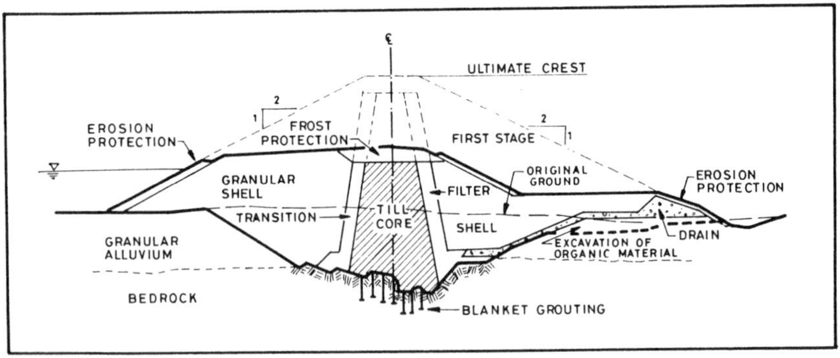

Figure 1. Typical Section of Tailings Dam at Page-Williams Gold Mine (after Welch et al, 1986).

The tailings dam constructed for the Afton copper mine in British Columbia, Canada (Fig. 2) is progressively raised by the placement of random rockfill from the nearby mine in the downstream shell, compacted rockfill in the upstream shell, glacial till in an upstream core and crushed rock and fine waste rock placed in a filter zone behind the till core (Lighthall 1987).

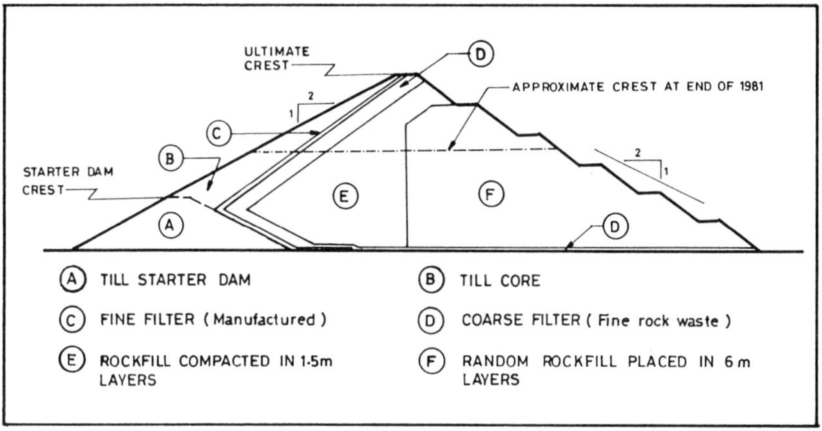

Figure 2. Typical Section of Afton Tailings Dam (after Lighthall, 1987)

UPSTREAM CONSTRUCTION TECHNIQUES

Upstream construction is the technique used to raise the height of a dam wall by progressively building new lifts upstream over previously deposited tailings. The new lifts may be constructed from tailings sands, coarse reject material or excavated borrow material. A starter wall is usually constructed to provide initial storage capacity.

This method of construction is particularly cost effective since the volume of wall construction is reduced significantly. In effect the tailings deposited against the dam wall becomes the foundation for future lifts. The strength and permeability of the deposited tailings become critical factors governing the overall stability of the wall. It is therefore necessary to pay close attention to predicted material properties at the design stage and stability monitoring during the filling operation.

There are numerous variations to the way in which tailings dam walls can be raised by upstream methods. Some of these construction techniques will be discussed briefly by example:-

Upstream Construction Using Rockfill

The tailings dam at the Westmin Resources Myra Falls Mine at Vancouver Island, Canada (Fig. 3) is designed to be raised upstream by placement of waste rockfill over deposited tailings (Knight & Haile, 1983; Lighthall 1987). This facility has been in operation since 1984 and has achieved all design objectives. The first upstream lift was constructed in 1987. Strength of the deposited tailings is maximised by encouraging drainage and evaporation from the tailings surface. A decant system is used to keep pond water levels to a minimum. By moving the discharge points around the dam wall on a regular basis, drying of sections of the tailings surface is promoted. This type of deposition is termed the subaerial method.

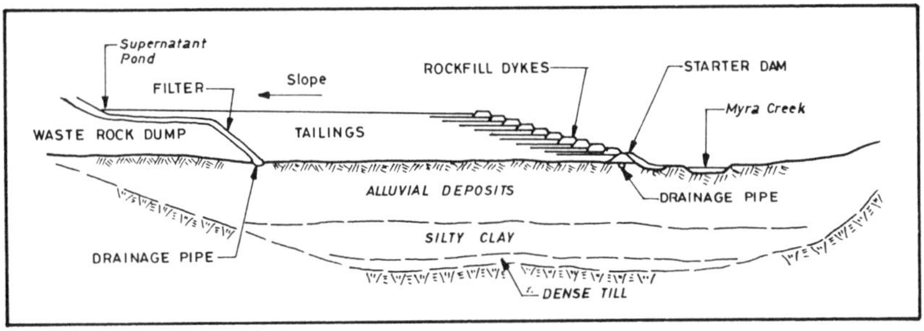

Figure 3. Typical Section Through the Westmin Myra Falls Tailings Dam Showing Upstream Construction by Constructing Rockfill Dykes (after Knight & Haile, 1983).

Upstream Construction Using Spigotted Tailings

The tailings dam developed at the Troy Mine in Montana, U.S.A. consists of a number of segments divided by internal dykes. Tailings are discharged into selected segments of the dam by spigotting through nozzles from the main dam wall. When a segment is filled, it is drained and left to dry and consolidate whilst discharge is directed to another segment. The filled segment is then raised by excavating competent tailings sand deposited close to the wall and placing it upstream to construct the new lift. The downstream face of the wall is battered and capped with topsoil to promote the growth of vegetation.

The tailings disposal facility developed at the Inco operation near Sudbury, Ontario, Canada (Fig. 4) also uses spigotted tailings to raise the dam wall upstream by pushing deposited coarse sand tailings up the beach with dozers. The upstream lifts are constructed initially from rockfill starter dams. Finger drains extend into the tailings deposit from the rockfill starter dam to promote drainage and ensure wall stability (Van Cruyningen & Puro, 1987). Seepage from the permeable starter wall is collected in drains and either pumped back into the dam or directed to a water treatment plant.

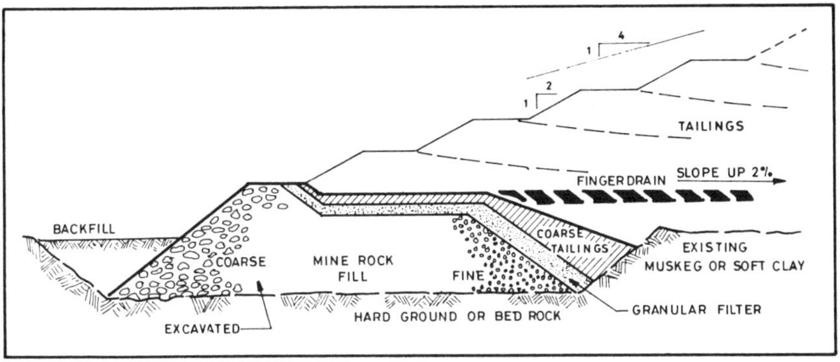

Figure 4. Typical Starter Dam Design at the Inco Operation (after Van Cruyningen & Puro, 1987).

Similar construction techniques have been developed to raise dam walls at the Weipa Bauxite Mine in Queensland, Australia (Fig. 5). Tailings (less than 1.7 mm) are discharged at 20% solids content via spigots around the 6 km dam wall perimeter to form a wide sand beach. The coarse tailings sands are pushed close to the wall with D7 dozers, spread across the wall with graders and compacted with a tamping foot compactor to form new lifts (see Fig. 5).

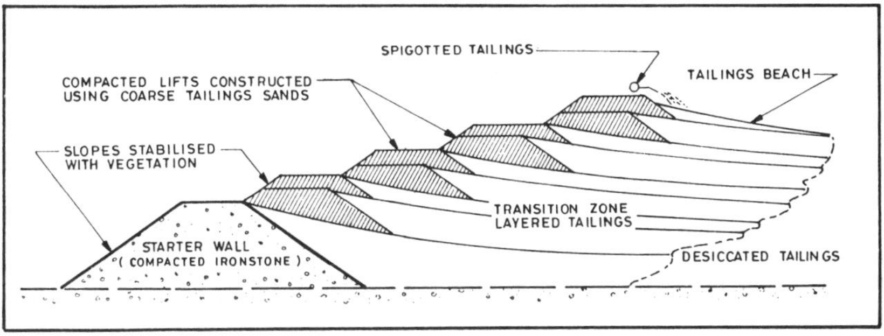

Figure 5. Section of Tailings Dam Wall at Weipa Showing Upstream Construction.

Upstream Construction Using Coarse Coal Refuse

Most coal preparation plants produce a coarse and fine coal refuse requiring disposal. At the Emerald Coal Mine in Pennsylvania, U.S.A., all coarse reject from the preparation plant is used to construct a valley containment structure for fine refuse slurry. The first three stages of the coarse reject embankment have been constructed upstream and over deposited fine tailings. The fines are discharged along the upstream face of the embankment to form a competent delta over which the coarse refuse lifts may be constructed (Fig 7). The coarse refuse is conveyed to the dam site and placed and compacted in layers, less than 2' thick (D'Appolonia, 1983).

Similarly coarse refuse from the Lamira Preparation Plant, Ohio U.S.A. is used to raise dam walls upstream over deposited fines (Fig. 6). A scraper is used to transport the coarse refuse from the plant to the dam site. The coarse refuse material is spread in thin, wide layers across the wall and compacted to ensure stability and minimise seepage through the embankment (D'Appolonia, 1986).

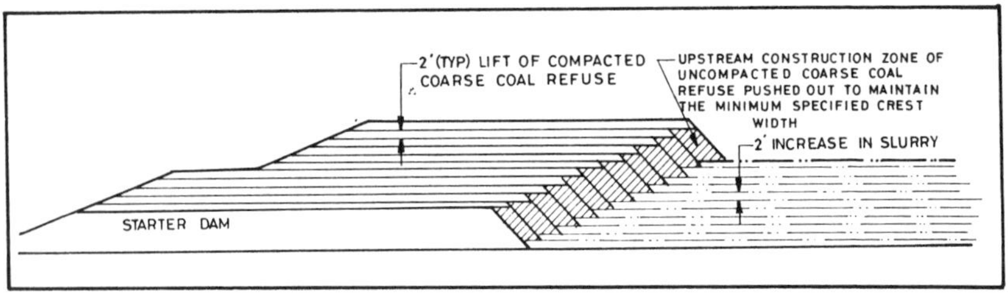

Figure 6. Typical Section of Dam used to store Coarse & Fine Coal Refuse from Lamira Preparation Plant (after D'Appolonia,1986)

CENTRELINE CONSTRUCTION TECHNIQUES

Centreline construction techniques are used to progressively raise embankments by placement of tailings upstream and downstream of the centreline of the dam wall. Cycloned tailings sand is usually placed on the downstream shell. Either spigotting of cyclone overflow or direct spigotting is used to deposit tailings on the upstream beach.

Placement of cycloned sands on the downstream shell may be carried out by direct cyclone underflow discharge or if compaction of sand fill is required various hydraulic cell placement techniques have been developed.

Centreline construction techniques are commonly used for the valley disposal of tailings from high tonnage base metal operations. The milled tailings discharged from flotation plants normally contain a sufficient sand fraction suitable for centreline construction.

Good examples of centreline construction methods are practiced by the following mining operations.

- The Cypress Bagdad Copper Mine in Arizona, U.S.A. raise their dam wall by direct discharge of cyclone underflow at the dam wall crest. Cyclone overflow is spigotted upstream into the dam.

- The Highland Valley Copper operation near Kamloops, British Columbia, Canada uses mobile secondary cycloning stations on the dam wall crest to discharge sand in hydraulic cells. Wide track dozers are used to compact the sand hydraulically placed in cells (Scott & Lo, 1984).

- The Syncrude Tar Sand Mine in Alberta, Canada uses a similar hydraulic cell construction technique to raise tailings embankments, although tailings are not cycloned. Instead the sludge is drained from the cells and discharged into the dam via spill boxes (Handford, et al 1982).

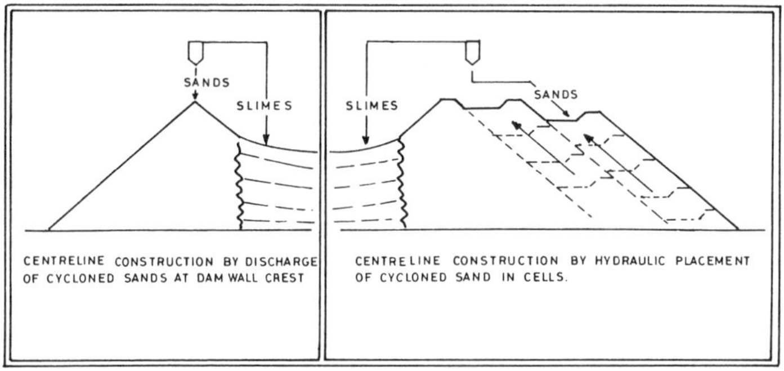

Figure 7. Schematic Showing Two Centreline Construction Techniques

DOWNSTREAM CONSTRUCTION TECHNIQUES

Downstream construction involves the placement of fill on the downstream face of the dam to raise its level. The fill material may be waste rock from the mine or cycloned sand as is commonly used for centre line type structures.

Since deposited tailings within the dam do not form part of the dam wall structure as is the case for upstream structures, the same stability problems relating to tailings strength are not encountered.

If the tailings dam embankment is located close to the mine, the use of waste rock or overburden for embankment construction may be particularly economic.

For example the tailings dam at the Jamestown Gold Mine in California, U.S.A. is designed to be constructed in stages by placement of mine rockfill on the downstream face of the dam (see Figure 10). A clay seal, drainage and filter zone will be raised on the upstream face (Haile & East, 1986).

Downstream construction techniques may be used for the construction of coarse coal refuse embankments to impound fine refuse if the coarse to fine refuse ratio is high enough. The Emerald Coal Mine in Pennsylvania, U.S.A. uses coarse refuse from the preparation plant to construct a valley containment structure in 4 stages (D'Appolonia, 1983). The first 3 stages involve upstream construction over deposited fine refuse. The final stage involves construction of a downstream lift over the 3 initial upstream lifts to ensure overall stability and provide a balance between required fine refuse storage capacity and coarse refuse placement (Figure 8).

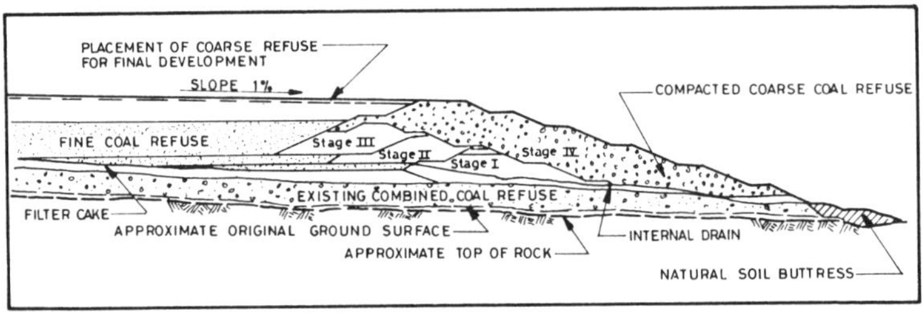

Figure 8. Typical Section of Combined Fine & Coarse Coal Refuse Disposal facility, Emerald Coal Mine (after D Appolonia, 1983)

Similar coal refuse disposal facility designs have been proposed by de Ambrosis and Seddon (1986). A conceptual design is presented in Figure 9. The downstream lifts of coarse reject are constructed at critical height increments to ensure adequate stability. The upstream lifts are constructed over deposited fine tailings. The strength of the deposited fine tailings is enhanced by alternate placement and desiccation of tailings in isolated cells within the dam.

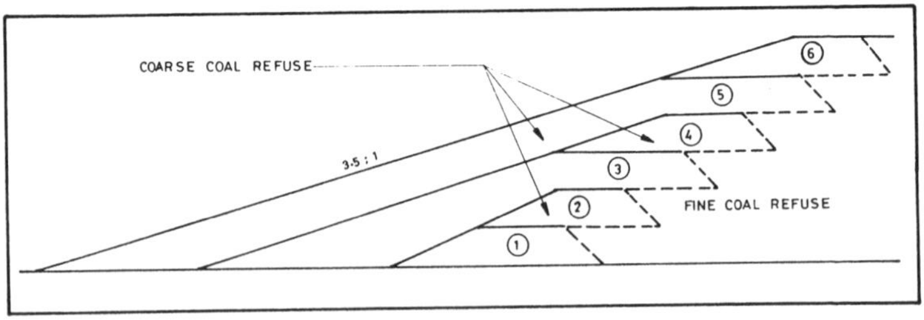

Figure 9. Conceptual Embankment Design for Combined Coarse and Fine Coal Refuse Disposal Facility. (after de Ambrosis & Seddon, 1986)

THICKENED TAILINGS DISPOSAL METHODS

This method of disposal requires the tailings slurry to be thickened to a controlled density and discharged to form a conical mound or stack. The method described by Robinsky (1978), relies on thickener underflow density control to ensure the design beach slope is achieved. The main advantage of this method of disposal is that dam wall construction is minimised. Normally only a small perimeter dam is required at the toe of the tailings stack to collect runoff water. The main disadvantage is the cost of thickening and the difficulty in tightly controlling discharge density.

A number of these thickened discharge disposal systems have been developed for mine sites in North America and described by Wood (1984), Yeomans (1985) and Lighthall (1987).

Thickened tailings disposal methods may have particular application to disposal sites where poor foundation conditions would otherwise require the construction of very expensive dam walls or where borrow material for conventional wall construction is not available in sufficient quantities. The method may also be used to provide tailings storage capacity on the surface of existing conventional tailings impoundments.

The Argyle Diamond Mine in Western Australia uses a thickened tailings disposal method, but has found it necessary to modify the original disposal concept. The Argyle Plant produces a coarse tailings (1 mm to 8 mm), a tailings sand (1 mm to 0.1 mm) and a fine tailings slurry (less than 0.1 mm). The coarse and sand tailings streams are combined and conveyed to a coarse tailings stack. The fine tailings stream is thickened to about 45% solids and pumped to the fine tailings dam. Problems associated with the disposal of fine tailings include poor beaching and settling characteristics of tailings in the dam, resulting in substantially lower deposited densities, reduced dam capacity and water reclaim. To alleviate some of these problems, Argyle now mix as much of the fine tailings component as possible with the coarse tailings for disposal at the coarse dump. The excess fine tailings are disposed of in the tailings dam. In the long term Argyle intend to pump the thickened fines from this dam and deposit them on top of a dried out area of fine/coarse tailings mixture.

DISPOSAL OF FINE CLAYS

The beneficiation of certain mineral ores produces large quantities of very fine clay tailings. The type of clay minerals present in fine tailings will effect the settling and consolidation characteristics of the material.

The phosphate industry in Florida produces large quantities of phosphatic clay waste with particularly poor settling properties. The bauxite and alumina industries produce a fine clay and red mud residue which settle to form a highly plastic material. Similarly, coal beneficiation and tar sand processing produce large quantities of very fine material requiring disposal.

Most of the disposal methods for fine clays and sludges aim at accelerating the dewatering and consolidation process to improve strength and allow eventual reclamation of the land.

Numerous techniques have been devised to improve the effectiveness of fine tailings disposal. Bromwell (1984), Somogyi et al (1984), Garlanger et al (1987), Carrier et al (1987) and Scott and Cymerman (1984) have proposed specialised techniques for fine clay disposal. The Florida Institute of Phosphate Research has funded research into numerous dewatering methods and disposal techniques for the phosphate industry. Some of these techniques are discussed:-

. Dredge Disposal Method

  This method is being developed for the Trombetas Bauxite Mine in Brazil. The method involves the discharge of fine bauxite tailings at a solids content of 7 - 9% to a large thickening pond. The fines are settled and dredged at a solids content of 25 - 30%. This thickened tailings material is then pumped to mine areas for disposal and eventual reclamation.

. Dredged Clay - Sand Mix Disposal Method

  CF Industries in Florida, U.S.A. are presently trialing a sand - clay mix disposal technique which involves the discharge of a clay slurry into a large settling dam. The settled fines are dredged and pumped to a mixing station where thickened clays are combined with cycloned sand to produce a sand/clay slurry at about 30 - 35% solids. This material is then pumped to mined out areas for final disposal. Better consolidation characteristics of the mixed tailings facilitates reclamation (Garlanger et al 1987).

. Sand - Clay Sandwich Disposal Method

  This method has been developed by the phosphate mining industry in Florida and is described by Pittman and Sweeney (1983). Alternate layers of sand and clay are placed in mine backfill areas. The sand tailings layers provide a drainage path for water during consolidation of clay layers.

DRY DISPOSAL TECHNIQUES

This form of disposal is practiced by coal mining companies for disposal of coarse reject or in some cases disposal of combined coarse and fine reject. Dry dump disposal is also commonly practiced by power generating authorities for disposal of coal combustion wastes in the United States.

The advantages of dry disposal includes ease of reclamation of the dump surface and reduced seepage from the dump (if suitably compacted). The application of this method to the disposal of mine tailings is usually unattractive however due to the high cost of mechanical dewatering of 'wet' tailings. The cost of transport and placement of dry waste material must also be considered against the costs of pumping a slurry. The problem of dust generation from dry dump surfaces must also be addressed.

METHODS TO IMPROVE DEPOSITIONAL PROPERTIES

Most tailings contain a proportion of fine clay particles which do not settle and consolidate well. For example fine tailings produced by many coal preparation plants are deposited in tailings dams in relatively low density states. By developing methods to improve depositional density of fine tailings, dam storage capacity may be increased, and a resulting higher strength tailings deposit aids final reclamation of the dam surface.

Williams and Morris (1987) are studying methods to improve the deposited properties of coal tailings for this purpose. A broader study, being undertaken by the Australian Mineral Industries Research Association, aims at investigating means of improving the deposited properties of mine tailings for bauxite, iron ore and coal mines in Australia (AMIRA, 1987). The main objective of these studies involves the development of a model to simulate and optimise the placement and desiccation of tailings in layers. This tailings placement method optimises the depositional properties of tailings at mine sites where high evaporation rates and low precipitation rates are used to advantage. These conditions prevail at many Australian mine sites.

Closely controlled subaerial deposition techniques may also be used to improve storage density and the strength of deposited tailings. The Jamestown and McLaughlin Gold Mines in California, U.S.A. and the Westmin Myra Falls Mine on Vancouver Island, Canada all use subaerial deposition techniques for this purpose. Tailings discharge points are regularly moved around the perimeter of the tailings impoundment to promote drainage and drying of deposited tailings layers.

The Jamestown dam has been designed with a permeable drainage zone at the base of the impoundment and on the upstream face of the dam wall to promote drainage from the tailings deposit. Such an underdrainage system aids consolidation of fine tailings thereby creating an impermeable barrier which will act as a seal at the base of dam in the longer term (Haile and East, 1986).

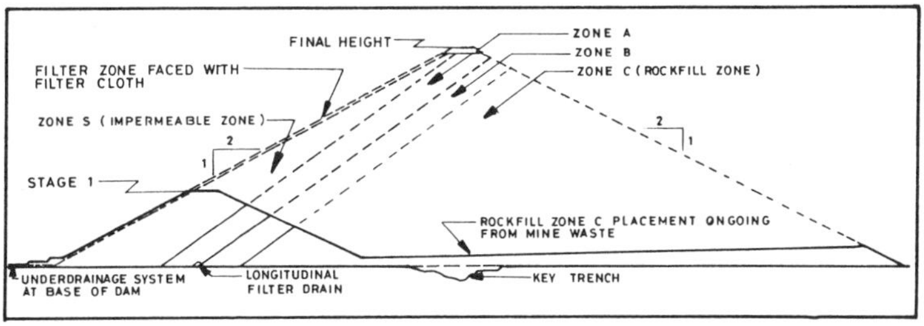

Figure 10. Typical Section of tailings Dam at Jamestown Gold Mine (after Knight & Piesold Ltd.)

The Alumina refineries operated by Alcoa, near Perth in Western Australia have also developed specialised techniques for improving depositional properties by thickening and underdrainage (Glenister & Press, 1986). The shear mass and volume of bauxite residue to be stored, its alkalinity and the location of the refineries (Pinjarra, Kwinana, Wagerup) in the most densely populated and most arable area of Western Australia means that residue disposal is a sensitive environmental issue which requires careful engineering and management.

Up until the early 1980's, conventional tailings disposal practice was followed whereby residue was pumped as a low density combined mud (fine fraction) and sand (coarse fraction) slurry to disposal areas up to 160 hectares in area. A number of improved disposal techniques have since been developed to reduce seepage from disposal areas and to improve the despositional properties of red mud residue.

Current residue disposal practice at Alcoa's refineries includes a combination of low density deposition into gravity underdrained impoundments and "dry disposal". Dry disposal involves thickening of the desanded low density mud slurry, pumping of the dense slurry and distribution into drying areas where the material is solar dried in layers.

A 90 metre diameter, high torque thickener has been installed at Alcoa's Pinjarra Refinery to thicken mud slurry to around 50% solids. Diaphragm pumps are used to pump the dense slurry to a 70 hectare drying area where the material is hydraulically placed in layers up to 1 m depth and solar dried. Drying areas are being expanded to around 200 hectares over the next 2 years to accommodate full scale dry disposal.

At the Kwinana refinery, a dredging technique has been under development to recover dense slurry from an existing underdrained disposal area to avoid the need for mechanical dewatering equipment. Some difficulties have been experienced with control of the dredging and pumping operation to maintain the slurry densities required and due to the success of the Pinjarra thickener, it is likely that a similar machine will be installed at Kwinana.

At the Wagerup refinery a new gravity underdrained disposal area is under construction, however here also it is planned to convert to a dry disposal operation within the next 4 - 5 years.

Tailings disposal methods at Comalco's bauxite mine in Queensland, Australia are also currently under review. One of the dams at this site has recently been converted to a 2 dam system by construction of an internal dividing wall within the dam. This allows tailings to be discharged and solar dried in cycles, promoting higher storage densities and tailings strength. The internal dividing wall is raised at the end of each drying cycle using coarse tailings sands (Fig. 11). Scrapers are used to transfer sand from the tailings beach to the internal wall. The main dam wall is raised upstream at the end of each drying cycle when tailings have gained sufficient strength to enable safe operation of earthmoving equipment on the tailings beach.

Pond water levels in the dam are controlled by decant systems installed in each of the 2 segments.

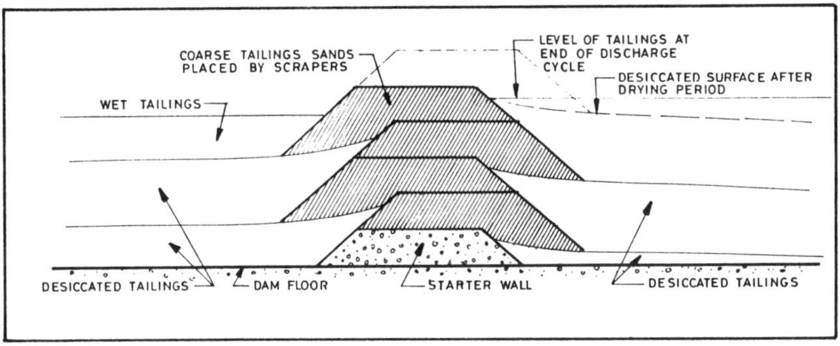

Figure 11. Method used to Raise Internal Dividing Wall, West Weipa 2 dam, Weipa.

METHODS TO IMPROVE WATER RECLAIM & CONTROL DAM WATER LEVELS

Many mining and mineral processing operations are sited in remote locations requiring the development of new water supply systems. The operational security of these operations can be enhanced and the cost of water supply reduced if water reclaim from tailings is maximised.

Methods used to enhance water recovery include:-

- controlled decanting of water from tailings ponds to minimise pond size and reduce potential for seepage through base of dam.
- improve consolidation characteristics of deposited tailings to reduce permeability and seepage losses through tailings.
- use of synthetic membrane liners or clay liners to seal reclaim storage dams or settling ponds.
- minimise the discharge of water to the dam by thickening tailings at the plant.

Application of these methods is important to many mining operations where the supply of water is costly.

At some disposal sites, a build up of excess water in the dam may occur, especially for sites where annual precipitation is high and evaporation rates are low. This water imbalance occurs for many tailings disposal facilities in North America, necessitating the installation of water treatment facilities and allowing treated water to be released to natural streams. Some tailings disposal facilities in Australia however, rely on high evaporation rates and low precipitation to minimise dam water levels. This situation exists at Mt. Isa Mines,

Queensland, Australia, where water inputs to the tailings dam are balanced by a high evaporation rate of about 3.1 metres/year.

CONCLUSIONS

A wide variety of tailings disposal practices are continuing to be developed within the mining industry to satisfy increasing environmental and cost constraints. It is important that mining companies are made aware of the full range of disposal methods and dam construction techniques available today. Such an awareness will help improve the planning, design and construction of safer and more efficient tailings disposal facilities in the future.

ACKNOWLEDGEMENTS

This paper is based on the results of a study tour undertaken by the author and made possible by the 1987 Julius Kruttschnitt Travel Grant awarded by the Queensland Chamber of Mines. Support by Comalco Aluminium Ltd and information supplied by many mining companies consultants and government authorities is acknowledged. The author is also grateful to Don Glenister (Alcoa) and Bill Hutton (Argyle) for providing information on their tailings systems.

REFERENCES

Australian Mineral Industries Research Association (AMIRA), 'Investigation of Means of Improving the Deposited Properties of Mine Tailings', 86/P233, The University of NSW, School of Civil Engineering, Progress Report, 1987.

Bromwell L.G., 'Consolidation of Mining Wastes' Proceedings of Symposium: Sedimentation Consolidation Models ASCE, October, 1984.

Caldwell J, Moore D, Hutchinson I, Sluder T, Design and Construction Outlined of the Cannon Mine Tailings Impoundment, Mining Engineering, P 813 - 815, August, 1986.

Carrier W.D. III, Scott J.D., Shaw W.H., Dusseault M.B., Reclamation of Athabasca Oil Sand Sludge, Proc. Geotechnical Practice for Waste Disposal, June 1987.

D'Appolonia, Revised Operational Plan, Coal Refuse Disposal Facility, Juzwiak Site, Lamira Coal Preparation Plant, R & F Coal Company, Belmont County, Ohio, June 1986.

D'Appolonia, Slurry Impoundment Coal Refuse Disposal Facility, Emerald Mine No. 1, Emway Resources Inc., Waynesburg, Pennsylvania, October 1983.

De Ambrosis L., Seddon K., 'Disposal of Coal Washery Waste', Mine Tailings Disposal Workshop, University of Queensland, Department of Civil Engineering, August, 1986.

Garlanger J.E., Babcock J.W., Keen P.W., Evaluation of Dredge Sand-Clay Mix Filling and Reclamation, Symposium on Consolidation and Disposal of Phosphatic and other Waste Clays, Florida, May, 1987.

Glenister D., & Press M., Engineering Characteristics of Bauxite Residue, Mine Tailings Disposal Workshop, University of Queensland, 1986.

Haile J.P. & East D.R., Recent Developments in the Design of Drained Tailings Impoundments, Geotechnical & Geohydrological Aspects of Waste Management, Fort Collins, 1986.

Handford G., Buckles J. Ellis H., Khan F.Z., 'The Use of the Observational Approach in the Design and Construction of the Syncrude Tailings Dyke', Seminar on Geotechnical Aspects of Mine Design and Tailings Containment, April, 1982.

Knight R.B., Haile J.P., 'Subaerial Deposition with Underdrainage', Proc. 7th Pan-American Soil Mechanics Conference, Vol. II Vancouver: Canadian Geotechnical Society, 1983.

Lighthall P.C., Innovative tailings disposal methods in Canada, International Journal of Surface Mining 1 (1987): 7 - 12.

Lighthall P.C., Trojan and Afton Tailings Dams, Cancold Annual Meeting, Kamloops, British Columbia, September, 1987.

Pittman W.E. Jr., Sweeney J.W., A Review of Phosphatic Clay Dewatering Research, Vol. 1, State of the Art of Phosphatic Clay Dewatering Technology and Disposal Techniques, Bureau of Mines, U.S. Department of Interior, Tuscaloosa Research Enter, Final Report, September, 1983.

Robinsky E.I., Tailings Disposal by the Thickened Discharge Method for Improved Economy & Environmental Control, 2nd International Tailings Symposium, Denver, Colorado, May, 1978.

Scott J.D., Cymerman G.J., 'Prediction of Viable Tailings Disposal Methods, Symposium on Sedimentation Consolidation Models: Predictions and validation, October,1984.

Scott M.D., Lo R.C., Evolution of Design and Construction of Lornex L - L Tailings Dam, International Conference on Case Histories in Geotechnical Engineering, University of Missouri - Rolla, May 6 - 11, 1984.

Somogyi F., Carrier W.D. III, Lawver J.E., Beckman J.F., 'Waste Phosphatic Clay Disposal in Mine Cuts', Proceedings of Symposium: Sedimentation Consolidation Models ASCE, October, 1984.

Van Cruyningen J.P., Puro M.J., Tailings Disposal Area Development at Inco Sudbury Operations, Canadian Mineral Processors 25th Anniversary Conference, Ottawa, Jan 20 - 23 1987.

Welch D,E,, Hope G.H., Ankenmann M.R., The Geotechnical Aspects of the Tailings Basin Development for the Page-Williams Gold Mine, Hemlo, Ontario, 40th Canadian Geotechnical Conference, Regina, Saskatchewan, Canada, October, 1987.

Williams D.J., Morris P.H., Effect of Placement Technique on the Properties of Slurried Fine Grained Coal Mine Tailings., Proc. Int. Conference on Mining & Industrial Waste Management, Johannesburg, 1987.

Williams D.J., Morris P.H., Effect of Placement Technique on the Properties of Slurred Fine Grained Coal Mine Tailings, Proc. Int. Conference on Mining & Industrial Waste Management, Johonnesburg, 1987.

Wood K., Design & Operation of Thickened Tailings Disposal System at Les Mines Selbaie, 86th Annual General Meeting of CIM, Ottawa, 15 - 19 April, 1984.

Yeomans T.J., Tailings Disposal at Kidd Creek Mines Ltd., 18th Canadian Mineral Processors Operators Conference, Ottawa, 28 November, 1985.

Pneumatic Placement of Mill Tailings at 25 Tons/Hour

John H. Archibald*

INTRODUCTION

It is the objective of this paper to present at least three successful techniques that were employed to overcome serious problems in the pneumatic transportation of fine ball mill tailings. Pneumatic transportation of mill tailings has been accomplished in the past, however only for tailings containing very large particles such as are found in coal mining operations and in hardrock mining operations where waste rock is blown back into the stope. Transporting tailings in the fine sand regime still remains a tricky problem especially if the tailings are very wet. This paper describes the results obtained from a research program conducted by Joy Manufacturing Company Ltd. in Cambridge, Ontario in 1980-81. The work did not lead to a commercially available product and therefore this paper is not an endorsement of Joy Mfg. Co. Ltd. The work did however lead to successful techniques for the pneumatic transportation of very fine ball mill tailings.In underground mining sand fill is used for filling up the space vacated by the mining of ore. The usual practice is to substitute ball mill tailings (waste rock that has been crushed and ground and contains very little ore minerals) for the sand. Some mine operators use sand from gravel pits, but this is rare. At any rate, the term "sandfill operation" generally refers to an operation that is using mill tailings directly from its ball mills and hydraulic classifiers or cyclones. Typically sandfill is pumped into a mine with an ordinary soft-rubber-lined centrifugal slurry pump. This means that the slurry of mill tailings would be about 50% water and 50% mill tailings by weight, otherwise the slurry pump could not handle the mixture. In order to place this water slurry in the mined space, or stope, a crude dam is usually built in the stope and chinked with burlap to prevent gross leakage while the sandfill is being slurried into the stope. The dam contains the slurry for a sufficient length of time to allow the solids to settle out of suspension. This part of a sandfill operation can be subject to a fair amount of operator error.

It is not uncommon in some mines to have one of these dams or bulkheads fail, releasing many tons of wet sand into the main haulage drift. This can knock out underground rail haulage for days and seriously effect mine profits. In addition, building bulkheads to contain the sandfill, so that it can drain, takes time and requires that the active mining of that particular stope be suspended for a few weeks. This leads to an overall inefficiency in the mine's operation.Ideally, the best way around these problems would be to send the fill underground as a water slurry, dewater the slurry in the haulage drift, and then in some efficient manner, transport and place the dewatered fill to the stope. This would eliminate the need for constructing bulkheads, shorten the time that mining operations would be banned from that stope and lead to a more effective use of labor and capital.This ideal method would be inherently safer and with the addition of Portland cement to the fill would lead to a much more robust surface to run equipment on during the mining and mucking operations.

SYSTEM OPERATION THEORY

The underground sandfill plant that was built for the research consisted of a high speed bowl centrifuge, a 100 Hp. 1,000 CFM @ 15 PSIG blower and a four inch diameter 300 foot pipeline. The centrifuge was fed a normal hydraulic mill tailings slurry (50% water 50%mill tailings). At speeds of about 2,200 RPM the centrifuge was capable of delivering over 40 tons/hour of a dewatered material

*President, Archibald Scientific, Inc. 2519 Kendall Ave., Madison, Wisconsin 53705, (608) 233-6910

known as underflow having a compostion of about 20% water and 80% mill tailings by weight all within the fine sand regime. The consistancy of this centrifuge underflow at 80% solids is very much like the consistancy of a fine wet beach sand just after a wave has receeded from a beach. This product, the centrifuge underflow, was the item to be pneumatically transported and placed in a stope by a machine that unofficially was called the "STOPEPAK" by the research team. The STOPEPAK consisted of a vertical walled hopper that led the underflow to a multi-pitched feed screw which injected the underflow into the beginning of the pipeline as shown in FIG.(1).

The injection point was somewhat critical to successful transportation of a given material. Adjustments to the injection point were provided for the different materials investigated. At the injection point the pressurized underflow, pressurized by virtue of the multi-pitched feed screw, was broken up with a high pressure (100PSIG) air blast from a nozzle to insure total mixing of the underflow with the transport air which was moving at about 425 feet per second within the three inch diameter throat.

After the high velocity steel throat injection section the airborne underflow entered a 3 to 4 foot long mixing section allowing a certain

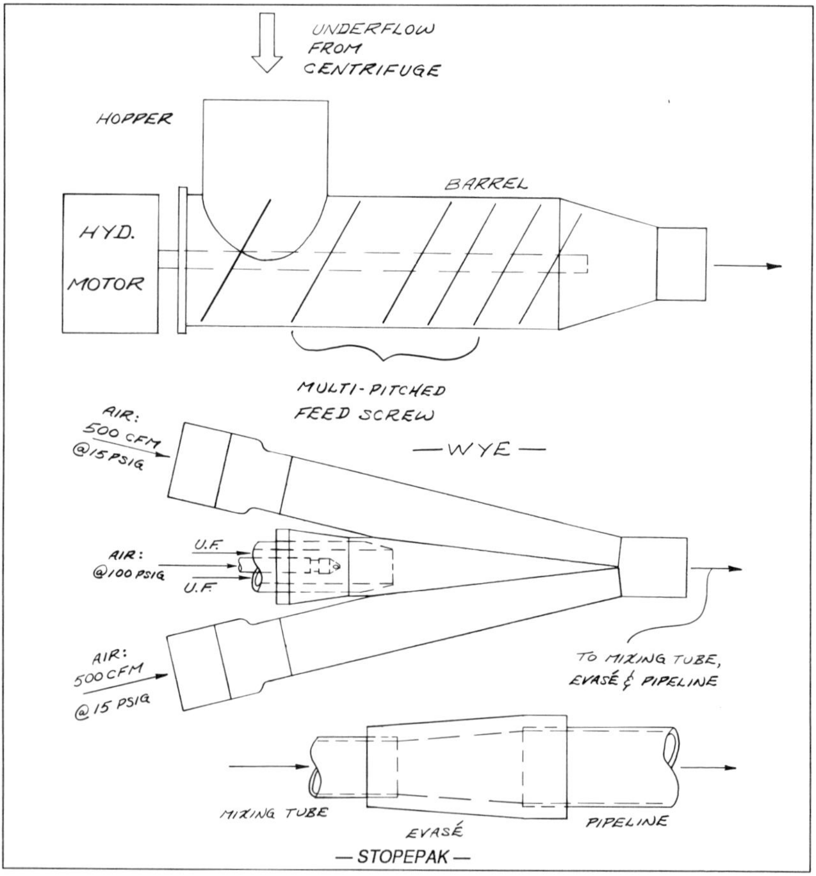

FIGURE 1

"MEAN FREE PATH" steadying and further mixing. It was instructive to visualize the airborne underflow at the injection point with a view leaning toward the kinectic theory of gases in that there was a lot of random motion and collisions of particles with the pipeline wall. In confined regions, for instance at the injection point, the mean free path of the particles was on the same order as the steel pipe throat diameter, i.e., about three inches. A calculation of the system's mean free path at the injection point appears in the appendix and is to the author's knowledge the first time this approach has been used in this field[1]. It was of paramount importance for successful injection to guarantee that the mean free path of the underflow at the injection point was less than the pipeline inside diameter. If the mean free path was longer than this, then the underflow particles, or "molecules", would not "see" the pipeline entrance and would build up in the wye and choke the flow off completely in a matter of seconds. In addition to maintaining a small mean free path to help the swiftly moving airborne underflow enter the pipeline proper, a gradual transition in section was made by using an evase as shown in FIG.(1), which was constructed of steel.

The pipeline itself, it was found, was best made from HD-UHMW polyethylene, (High Density Ultra High Molecular Weight). This material provided flexibility and the ability to negotiate 6 to 8 foot radius bends. Straight sections however could be of normal schedule 40 rigid grey PVC pipe, (Polyvinyl Chloride). The extremely low wall friction of the HD-UHMW PE pipe was by far the most important factor contributing to the success of the STOPEPAK transport system. If an identical steel pipeline had been substituted into the system the transport system would have failed due to a rapid buildup of underflow on the walls of the pipeline causing more drop out and the eventual choking off of the pipeline; all within the first two minutes of operation. The author has often wondered why ordinary positive displacement concrete pumping systems have not switched from steel pipelines to HD-UHMW PE pipelines. Wall friction is decidedly lower between concrete or underflow and HD-UHMW PE and this should lower the pumping pressures considerably which now stand at about 1,500 PSIG at the inlet to a conventional steel pipeline concrete pumping system. The STOPEPAK was capable of sustained operation at 30 tons per hour. The pipeline was 200 feet long and would carry the underflow at a velocity closely approaching the pipeline air velocity which was 240 feet per second. When the underflow left the end of the pipeline it would travel in free flight an additional 100 to 150 feet. Later, greater pipeline lengths were achieved, the maximum being 300 feet and with no indication of drop out. Early in the program a sight tube was designed that worked for many hours, allowed easy viewing, and above all allowed particle velocity and size to be determined at any point along the pipeline. This new design is illustrated in FIG.(2). The overall length of the sight tube section was

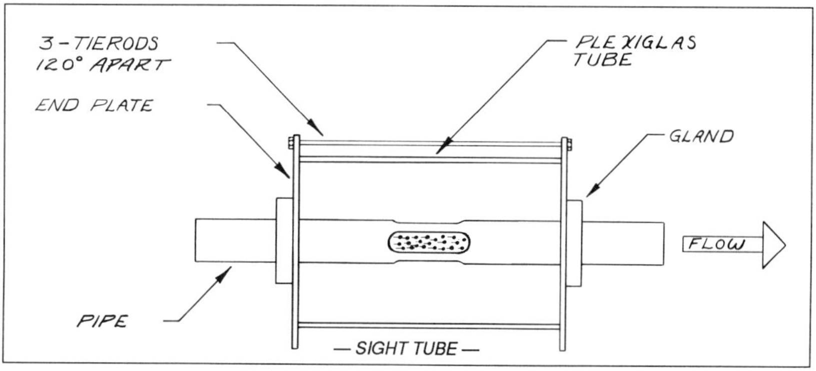

FIGURE 2

five feet with four viewing windows four inches long which were cut out of the PVC pipeline wall. The whole sight tube section could be housed in a light tight box and still permit normal system operation. The purpose of the light tight box was to allow the use of the Rayleigh[2] photographic method of velocity determination of the underflow particles inside the pipeline.

The Rayleigh photographic method of stopping high velocity objects on photographic film is well known but little used. The object or motion to be photographed is put in a darkened room or enclosure. Then, a camera is focussed on the object or its path and the shutter is held open. The shutter may remain open indefinitely if the enclosure or room is sufficiently dark. When the photograph is wanted a very high brilliance and short duration spark is initiated. The shutter is then closed and the film is developed.The author's improvement on the above technique was to substitute two xenon flash tubes for the single spark. The first xenon flash tube, when triggered, served as a reference. At a known time increment later, said increment being chosen electronically ahead of time, the second flash was initiated. This produced two images on the same piece of film which was then developed. Since a scale of length was photographed simultaneously with the high velocity object, and since a known path length had been traversed as evidenced in the resulting photograph; then it became a simple matter to determine sizes and velocities from these resulting photographs quite accurately. The associated timing circuit developed for this technique used modern integrated circuits. The method gave a permanent photographic record of conditions and velocities within the pipeline at any point along the pipeline without resorting to the hideous trouble and expense of ordinary high speed photography. The normal pipeline exit product was a fine spray of underflow material travelling at approximately transport air velocity. During normal operation of the STOPEPAK it was noticed that occasionally a very large globule of underflow would exit the pipeline. Two theories arose to explain this odd behavior.

The first theory called for a local drop out zone that would gradually grow with time. Then when it reached a certain size this drop out pile would be blown loose from the wall and exit the pipeline at irregular intervals. We had seen local drop outs before so this seemed at least plausible.

The second theory was that the flow in the pipeline was in some manner velocity modulated along its length. This would lead to a snowman effect, that is, particles would tend to grow as they coagulated with faster or slower moving particles that happened to be moving in the same region of space at the same time. The first theory was dropped becaused we usually associated any drop outs with total system choke off due to rapid build up on the pipeline walls as mentioned earlier.

We accepted the second theory and this meant that the snowmen would have to be formed in midair. This seemed plausible as well. By a very simple analysis shown in the appendix the author was able to arrive at a maximum probable size of these globules or, "snowmen", as they were called. The analysis agreed well with photographic evidence then available.

SUMMARY

The pneumatic transportation of very wet and fine underflow sandfill material is possible with properly designed machines when all of the many properties of this very strange material are taken into account.

The rheological properties for this substance are not published in any scientific form. This is very surprising because many hundreds of thousands of tons of mine tailings are transported in one way or another. The designer of equipment to handle this substance in its dewatered state (80% solids) is severely hampered by not knowing its behavior under different shear rates. The material is highly non-Newtonian and never seems to behave in a predictable manner. Ninety percent of the author's time was devoted to getting a feel for this strange material. At the time of the research there was absolutely no data available on any of its flow properties. Use was made of a crude flow net drawing to explain why it would not negotiate certain kinds of pipe bends and reducers in its bulk state, i.e., when it was fed into hoppers by gravity alone. The flow net gave insight to its properties, but was in no way quantitative. It is the kind of substance that needs a national research effort to unravel its many intricacies.

A very important factor was found by experiment, that this material was transportable only if very high velocities were maintained in the pipeline. These velocities were higher than what was then conventional, but the material was also unconventional.

The research was conducted on full sized equipment which minimized modelling expenses. Since the properties of the material were not well understood modelling would have been very difficult. In the end the results would need to be verified at the full scale in any event and it seemed inappropriate to do things twice since the full scale was not very large. Once we had a working system we did design others and they all seemed to work as long as we did not depart from the working design too far.

One question that often arises is that of the properties of the inplace material. With the injection of Portland cement we were able to get a very strong backfill material. To our knowledge its air content was typical of other nonpneumatically placed sandfill. Its strength was also about the same, and very uniform. Uniaxial compression tests were performed on cores of the cemented and uncemented material to verify its strength and for comparison purposes. Unfortunately, the author was not able to obtain this particular data for this paper.

BIBLIOGRAPHY

1). Van Nostrand's Scientific Encyclopedia, 3 rd. ed., D. Van Nostrand Co., Princeton, New Jersey, 1958, pg. 1038.

2). The Photographic Study of Rapid Events, W.D. Chesterman, Clarendon Press, Oxford, 1951, pg. 106.

APPENDIX

MEAN FREE PATH DETERMINATION

From the kinetic theory we have the expression for the mean free path of a gas molecule.

$$\lambda = \frac{1}{N\sigma} \tag{1}$$

where, λ, is the mean free path in feet, N, the number of "molecules" per cubic foot and, σ, is the collision cross sectional area in square feet. σ, the collision cross sectional area is further defined below in FIG.(3) in terms of, δ_p, the airborne underflow "molecular" diameter. This diameter, δ_p, was known from the photographic work and was about 0.060 inches, or 0.0050 feet (1.52 MM.). By definition, δ_p, is the mean airborne underflow "molecular" diameter and the cross sectional collision area can be found as,

$$\sigma = \pi(\delta_p)^2 \tag{2}$$

When, σ, from equation (2) is put into equation (1) we have the final expression for the mean free path length,

$$\lambda = \frac{1}{N\pi(\delta_p)^2} \tag{3}$$

The mean underflow "molecular" diameter, δ_p, in this equation was known from the photographic investigation. That leaves, N, the number of "molecules" per cubic foot to be determined. This may be gotten as follows. The weight rate of flow in the pipeline may be written,

$$W = \gamma AV \tag{4}$$

in which, W, is the weight rate of flow in pounds per second, γ, is the effective specific weight of the "underflow gas" in pounds per cubic foot, A, is the cross sectional area of the

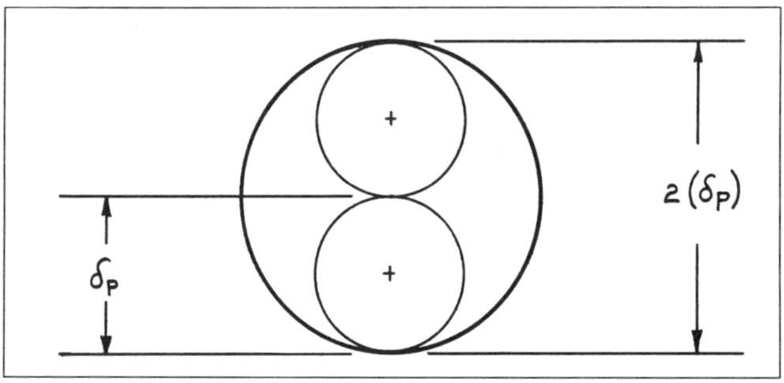

FIGURE 3

pipeline in square feet and, V, is the mean underflow gas velocity at the entrance to the pipeline in feet per second. Solving (4) for, γ and putting in the other known values;

$$\gamma = \frac{(30 \text{ tons/hr.})\left(\frac{1 \text{ hr.}}{3{,}600 \text{ sec.}}\right)\left(\frac{2{,}000\#}{1 \text{ ton}}\right)}{(\pi/4)(1/4)^2(425 \text{ ft./sec.})}$$

$$\gamma = 0.80\#/\text{ft.}^3 \ (12.82 \text{ Kg./M.}^3)$$

The dispersion factor, D.F., is defined as the ratio of the effective specific weight of the "underflow gas" to the specific weight of the bulk underflow material.

$$\text{D.F.} = \left(\frac{\gamma}{\Gamma}\right) \quad (5)$$

with, γ, being the effective weight of the "underflow gas" in pounds per cubic foot and, Γ, being the specific weight of the bulk underflow material. Γ, was easily measured and was about 150 pounds per cubic foot.

The total weight, W_T, of one cubic foot of the "underflow gas" is;

$$W_T = wN \quad (6)$$

where, W_T, is the total weight of one cubic foot of the "underflow gas" in pounds, w, is the weight per particle or "molecule" of the underflow gas in pounds per particle and, N, is the number of "molecules" in one cubic foot of the "underflow gas".

Noting that, w, the weight per "molecule" of "underflow gas" can be written as,

$$w = \Gamma V_P \quad (7)$$

with, w, being the weight per "molecule" of "underflow gas" in pounds, Γ, being the bulk specific weight of underflow in pounds per cubic foot and, V_P, being the volume of "underflow gas molecule" in cubic feet.

Further noting that the total weight, W_T, in (6) can also be described by,

$$W_T = \gamma V_T \quad (8)$$

where, γ, is as before and , V_T, is the volume of "underflow gas" considered (one cubic foot).

Equations (6) and (7) can be combined to give,

$$W_T = \Gamma V_P N \quad (9)$$

Whereupon equations (9) and (8) can be set equal to each other,

$$W_T = \Gamma V_P N = \gamma V_T \quad (10)$$

Solving (10) for, , leads to,

$$N = \left[\frac{\gamma}{\Gamma}\right]\left[\frac{V_T}{V_P}\right] = (\text{D.F.})\left[\frac{V_T}{V_P}\right] \quad (11)$$

Putting known expressions and values into (11) gives,

$$N = \left[\frac{(0.0053)}{\left(\frac{3.14}{6}\right)\left(\frac{0.060}{12}\right)^3}\right] =$$

$$N = 405 \text{ "Molecules"/ft.}^3$$

Finally, with (3) the mean free path length may be determined,

$$\lambda = \frac{1}{(405)(3.14)\left(\frac{0.060}{12}\right)} =$$

$$\lambda = 0.157\text{ft.} = 1.88 \text{ in. } (47.75\text{MM.})$$

It is seen from the above determination of, λ, that this value is less than the inside diameter of the entrance to the mixing section of the pipeline which was three inches. This means that the "molecule" could "see" the entrance point, and successfully start on its long journey down the pipeline.

MAXIMUM PROBABLE SIZE OF GLOBULES EXITING PIPELINE

Shown in FIG.(4) is the bottom half of an ideal maximum probable size underflow globule in the transported state known as the snowman. To facilitate a maximum size determination the interior is assumed full of underflow at the effective specific weight value. Acting against the bottom hemisphere's weight and to hold the bottom hemisphere to the top hemisphere is the uniformly distributed surface tension force σ_s.

The maximum weight that can be supported by the hemispherical water shell or cupped "membrane" is,

$$W = 2\sigma_s C \quad (12)$$

with, W, as the maximum supported weight in pounds, σ_s, as the surface tension of water in pounds per foot, and, $C = \pi D$, as the circumference of the hemisphere in feet. Then,

$$W = 2\sigma_s(\pi D) \quad (13)$$

where, D, is the diameter of the hemisphere in feet.

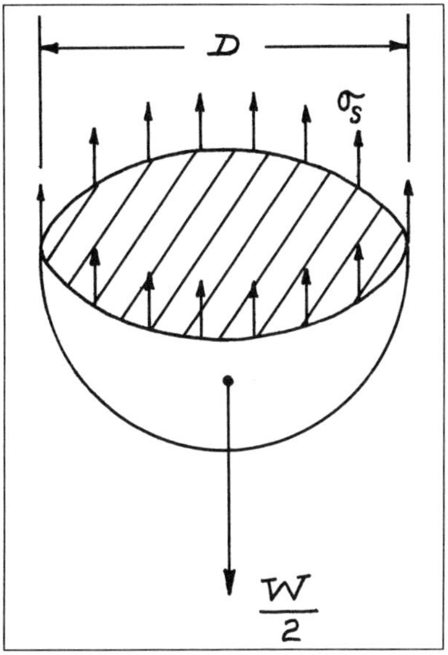

FIGURE 4

If the total sphere, or snowman, is filled to capacity and just at the point where its bottom half would "fall" out then we will have a certain maximum specific weight of the contents at this size given by,

$$\gamma = \frac{W}{V} \tag{14}$$

where, γ, is the maximum effective specific weight of the contents in pounds per cubic foot.Putting (13) in (14) yields,

$$\gamma = \frac{12\sigma_s}{D^2} \tag{15}$$

When the sphere bursts or splits in two we may assume that the new sphere formed by the action of the surface tension forces will have an identical overall effective specific weight since it is being transported successfully. Knowing the value for the surface tension of water equation (15) is solved for, D,

$$D = \left[\frac{12\sigma_s}{\gamma}\right]^{1/2} \tag{16}$$

giving numerically the value for the maximum globule or snowman diameter,

$$D = \frac{(12)(0.000773\#/ft.)^{1/2}}{(0.80\#/ft.^3)}$$

$$D = 0.1077 \text{ ft.} = 1.29 \text{ in.} (32.77MM.)$$

This value agrees remarkably with photographs taken at the pipeline exit. If in (16), Γ, is instead substituted for γ, then we get an estimate on the maximum size particle being transported at the bulk specific weight value,

$$D = \left[\frac{12\sigma_s}{\Gamma}\right]^{1/2} =$$

$$D = 0.0079 \text{ ft.} = 0.094 \text{ in.} (2.39 \text{ MM.})$$

Again, this value agrees well with photographs of the maximum size of these dense phase particles being transported within the pipeline.This is a further check on the validity of the assumption that the surface tension forces come into play and are responsible for two different kinds of transported particles. One kind of particle is transported at the bulk specific weight and of a certain maximum size and the other kind of transported particle possesses the effective specific weight but has a different maximum size and is totally independent from the other particle's constraints.

Slurry Backfilling of an Underground Coal Mine

Adrian Brown*, James Murphy**, and Henry Bartell***

The Wyoming Abandoned Mine Lands program is currently remediating the impacts resulting from mining conducted in Wyoming over the last century. An important part of that program has called for the control of subsidence over abandoned underground coal mines. This paper describes the planning, performance, and effectiveness of a subsidence control project conducted in Hanna, Wyoming, in which 800,000 tons of remotely placed hydraulic backfill was used as a supporting medium for a portion of an underground coal mine that underlies the main part of the town. This project is the first fully successful application of this technology in the west.

Introduction

The Town of Hanna, Wyoming is located approximately 100 miles west of Cheyenne, Wyoming (Figure 1). In the period 1889 to 1950 the underground mines of the Hanna coalfield were a major source of steaming coal for the Union Pacific Railroad. The exploitation of the thick seams of coal at Hanna left mined rooms under the town which were up to 35 feet high, 40 feet wide, and 500 feet long. Most of these rooms are now flooded. In the period since mining the town and its surroundings have been subjected to both sinkhole and continuous (trough) subsidence, resulting in some damage to buildings, and continuing risk to the inhabitants of the town. The present-day town, and the outlines of the mined locations, are indicated on Figure 2.

In 1984 the Wyoming Department of Environmental Quality (WDEQ) retained MidWest Mining Company (MWM) to perform an investigation of the subsidence risk in the town, and if the risk warranted it, to design a subsidence control program, and finally to act as project engineers during its implementation. The project was initiated in September, 1984, the investigation report was presented in February, 1985, and the backfilling program was performed between October, 1985 and October, 1986. Detailed information on the project is available in a series of reports prepared by MidWest Mining Company (1985, 1986, 1987). Funding was provided by the U.S. Department of Interior's Office of Surface Mining.

* President, MidWest Mining Company and Adrian Brown Consultants, Suite 302, 155 South Madison Street, Denver, Colorado, 80209.

** Engineer, Abandoned Mined Land Program, Wyoming Department of Environmental Quality, Herschler Building, Cheyenne, Wyoming, 82002.

*** Vice President, MidWest Mining Company and President, Henry Bartell and Associates, 390 Del Street, Lander, Wyoming, 82520.

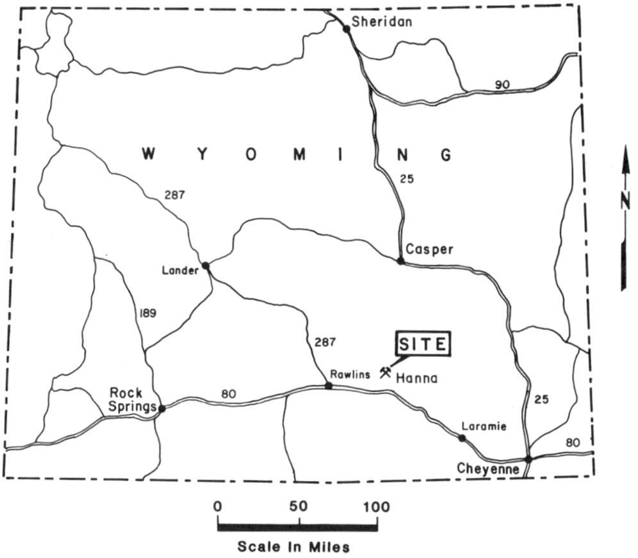

Figure 1. Location of the Site

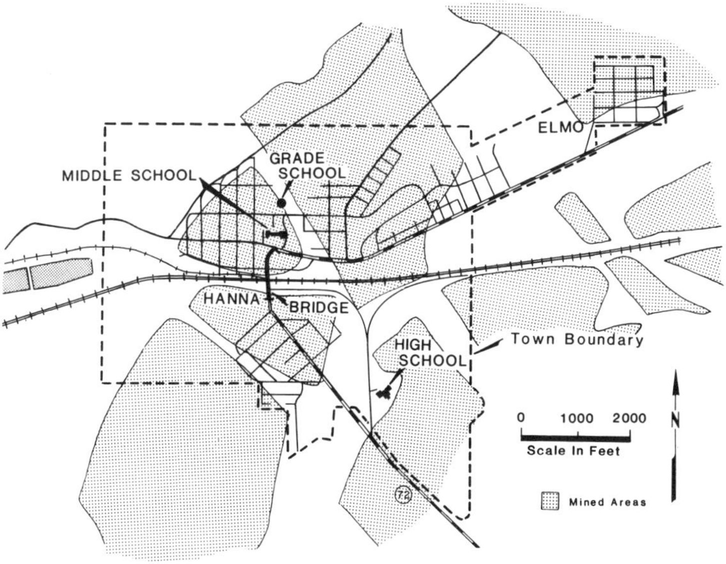

Figure 2. Town of Hanna and Location of Underground Mines

Investigation

The objective of the investigation was to identify the location, condition, and impact on overlying structures and services of the mined voids that were known to underlie the Town of Hanna. The investigation steps, and their results, are summarized below:

1. Collection of historical information on mining. Mining at Hanna was performed almost entirely by the Union Pacific Railroad Company (UPRR). Detailed records of the mining were kept, including geological information and mine plans, which UPRR made available for this project. A typical layout of the Hanna No. 4 Mine rooms beneath the town is presented in Figure 3, and a detail of the mining arrangement is provided in Figure 4.

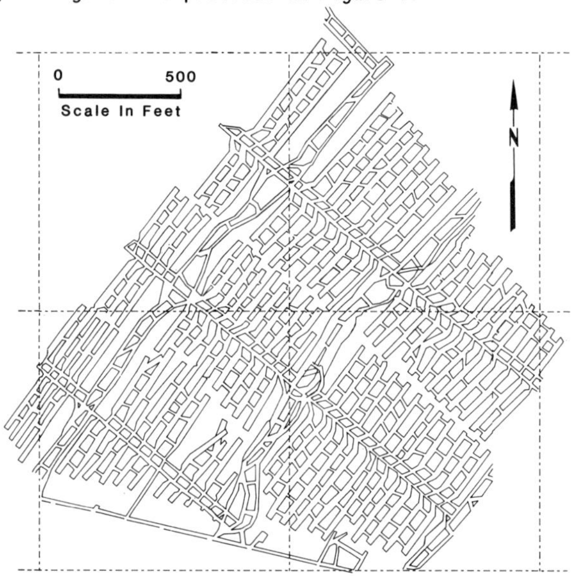

Figure 3. Typical Room Layout, Hanna No. 4 Mine

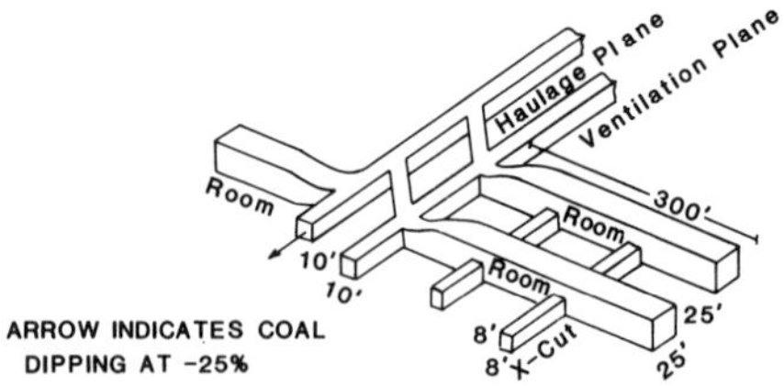

Figure 4. Typical Room Detail, Hanna No. 4 Mine

2. Collection of geological, geotechnical and other information relevant to the area. The Hanna Coal Basin is one of the deepest sedimentary basins in the United States, and contains considerable hydrocarbon resources. As a result, it has been extensively studied over the last century. Information on the basin was provided by the U.S. and Wyoming Geological Surveys and other sources, and incorporated into the study. Figure 5 shows the contours of the main coal seams that were mined in the Hanna area.

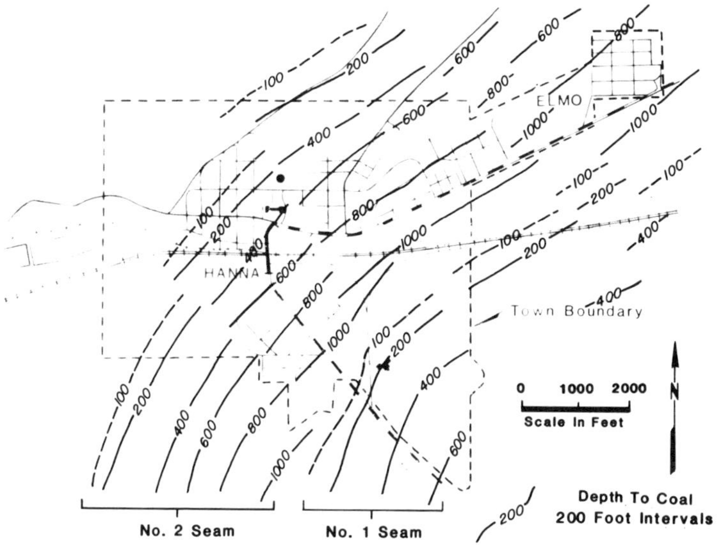

Figure 5. Contours of Uppermost Coal Seams

3. Site inspection. The town and adjacent areas were inspected for for signs of mining and mine-related subsidence. The information collected was used to validate the mine plans, and to calibrate the subsidence prediction model developed during the study. A plan of the areas in the vicinity of Hanna where subsidence of various types has been observed is presented in Figure 6.

4. Geotechnical field program. An extensive drilling and testing program was conducted in the town area. This investigation established that the mine maps are in general accurate, that the mined voids have generally remained as they were when mined, that the mine pillars beneath the town are in general relatively competent, and that the shale mine roof materials are presently relatively intact.

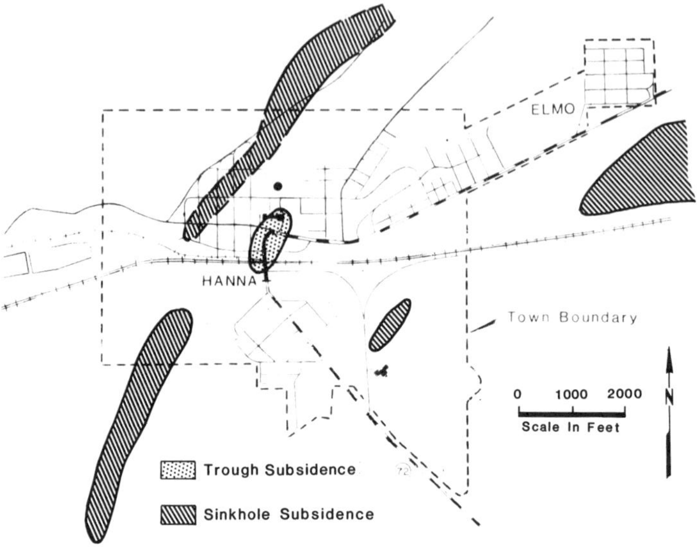

Figure 6. Location of Observed Subsidence

5. Geohydrology field program. A total of 18 wells were completed in the mined voids of the Hanna coalfield for the investigation of water levels and water quality. It was found that the quality of the water in most of the mines was poor: total dissolved solids ranged from 3,000 mg/l to 10,000 mg/l. Within any single mine, the water quality was essentially constant. A seven day pump test of the No. 4 mine voids indicated that the mined voids were interconnected, and while a very large volume of water was available from storage in the voids, there is very little recharge to, or seepage from, the mine.

6. Backfill source evaluation. Five potential borrow areas were evaluated, of which four were found to be suitable for the purpose. Each contained fine to medium grained sedimentary rocks, and the erosion products of those materials. The grainsize analyses of the materials identified are presented in Figure 7.

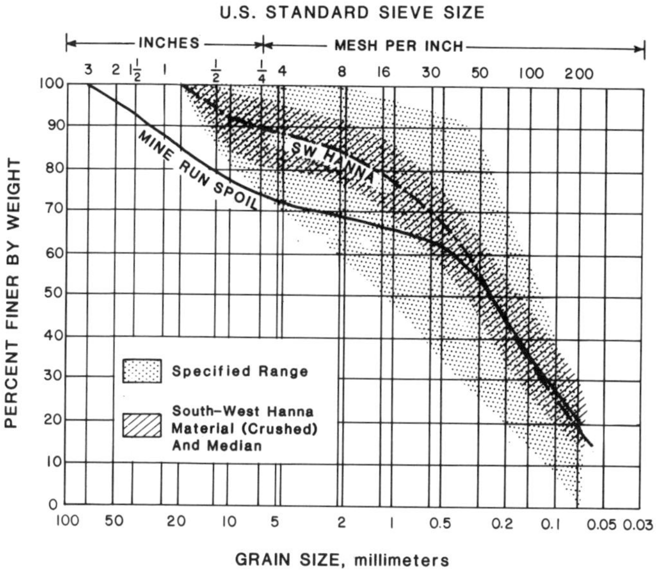

Figure 7. Grainsize Analyses of Backfill Materials

Subsidence Risk Assessment

Using the above information, a subsidence risk assessment for Hanna was performed. This evaluated the likelihood of subsidence over the mined voids, and the risk of significant impact of that subsidence. Two types of subsidence were evaluated in this study:

1. Sinkhole subsidence. This style of subsidence results from the collapse of the roof of a mined room, and the propagation of that collapse to the surface. Such subsidence creates deep, sharp sided subsidence craters, which have the capability to destroy any structures located on them. These features may form suddenly, creating the potential for a threat to human life. Analysis of the Hanna situation using the approach suggested in Piggoh and Eynon (1977) indicated that such subsidence would not be likely to propagate to the surface if the mined voids were more than 200 feet below the ground surface, and that the greatest risk was for voids whose roofs were within 100 feet of the surface. Thus any structure underlain by shallow workings can be impacted by sinkhole subsidence, so that any such inhabited location was considered to be at risk from this style of subsidence. Fortunately only the northwest area of the town is threatened in this way.

2. Trough or continuous subsidence. This style of subsidence results from the compression or failure of pillars in the mined area, which is reflected by differential subsidence and strains at the surface. In the Hanna context, analysis using the National Coal Board (1975) approach indicates that pillar failure is possible at depths of cover in excess of 250 feet, and would be more likely as the depth of cover increased. The surface strain resulting from these failures decreases with cover thickness. Analysis indicates that only rigid structures and linear features in excess of 100 feet long have the potential to be significantly impacted by such strains. Due to the depth of the coal and the location of the mine voids, only the Hanna-Elk Mountain Middle School and the highway bridge over the rail tracks (Figure 2) are located such as to be subject to damage as a result of trough subsidence; both of these structures show subsidence distress.

The subsidence risk situation identified in the town of Hanna is summarized on Figure 8. Based on the identified risk, it was decided by the WDEQ to undertake subsidence control measures in the northwest sector of the town, as this was the only area in the town where there was risk of sinkhole subsidence, and key structures and utility installations were threatened by continuous subsidence.

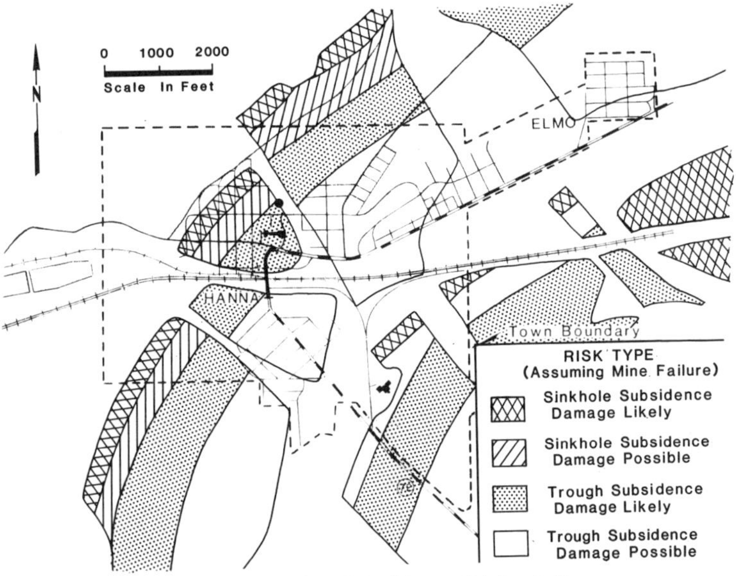

Figure 8. Mine Subsidence Risk at Hanna

Subsidence Control Approach

The subsidence control technology that was selected for the Hanna project was remote slurry backfilling, which was pioneered by the U.S. Bureau of Mines in the 1970's (Whaite and Allen, 1975). In this approach, subsidence control is obtained by hydraulic backfilling of the majority of the underground void with a granular material. The backfilling is achieved by pumping water from the mine, adding the backfill material, mixing, and pumping the resulting dilute slurry down access wells into the rooms to be filled. Once in the mined room, the solids settle out of the slurry and deposit in the mined voids; the water then returns to the point where it is pumped out of the mine and is used again. As the rooms fill up, the slurry is forced to move along the roof of the mine, thus keeping open the passageway along which the high velocity slurry moves, and continually filling any voids caused by consolidation of the backfill. Once the slurry stream reaches an area which is open, the deposition process resumes, filling that space. Tests in mines in Pennsylvania successfully filled up to 17 acres from a single injection hole.

The effectiveness of this approach to backfilling in the control of subsidence results from two mechanisms. With respect to sinkhole subsidence, the distance that a sinkhole can propagate upward is limited by the size of the mine void available to accept the material that falls from the roof of the cavity. Introducing backfill into the mined cavity reduces the void volume, and thus reduces the cover above the mine which is needed to protect against this type of subsidence. With respect to trough or continuous subsidence, the differential subsidence resulting from pillar failure is a function of the total subsidence that can occur in the mining panels. This in turn is a function of the available void space, as the pillars can only fail until the overburden load is supported by the backfill. Thus the backfilling of the voids reduces the "available" subsidence, and the maximum likely resulting strain that would occur in a surface structure. In addition the backfill may provide a measure of passive support for the pillar materials if they were to fail after emplacement of the backfill.

In summary, the slurry backfill approach derives its feasibility from the nature of flow in confined depositional systems. It derives its effectiveness from the ability of the backfill to support the roof of the mine after limited roof failure or pillar failure has occurred.

Subsidence Control Program

Review of previous slurry backfilling projects indicated that dilute mixtures in the order of two pounds of backfill per gallon of water (about 20% solids by weight of total slurry) were optimal for transporting the backfill material the required distances underground. For the backfill material selected, a slurry flow velocity of eight to twelve feet per second was computed to be optimal for pipeline transport. As it was desired to inject 600,000 tons of backfill over a six month period, the above parameters lead to the selection of a system utilizing two 8 inch diameter delivery systems, with injection rates between 3,300 and 5,000 tons of material per day.

The backfilling process comprised three principal activities: excavation of backfill material from a borrow area and delivery to the slurry plant; drilling and casing of holes into the underground voids to act a conduits for the injection of slurry to the voids; and mixing and pumping of the slurry to the underground voids via delivery pipes and the injection holes. In order to maximize the opportunity for a wide range of contractors to bid on the project, each component was performed in this project by a separate contractor. The layout of the project components that was actually used for the project is presented in Figure 9. Within the shaded backfill area a temporary network of pipelines was laid out to facilitate delivery of slurry from the mains to the injection holes. Details of each of the three main components of the project are presented in the next three sections of the paper.

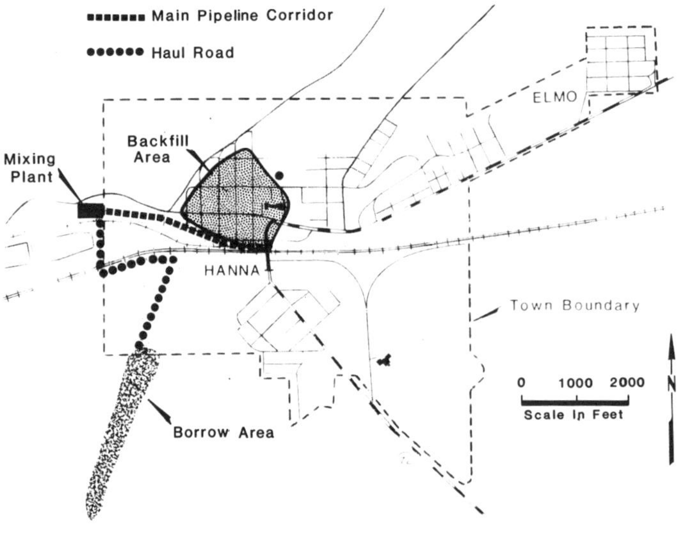

Figure 9. Layout of the Slurry Backfilling System at Hanna

Material Supply

The materials supply contractor was responsible for delivering a minimum of 100,000 tons per month of granular backfill material to a 12,000 ton stockpile in the slurry plant area, with a daily delivery requirement of between 3,300 and 5,000 tons. McMurry Bros. Co. of Casper, Wyoming was awarded this contract. The notice to proceed was issued on November 7, 1985.

The contractor was given wide latitude in selecting his source of material and plant site. The site selected was to the southwest of the center of Hanna, approximately one mile from the slurry plant (Figure 9). The original site was 25 acres in extent, to provide the 600,000 tons of borrow material initially required. The area was expanded to approximately 31 acres to allow borrowing of the additional 200,000 tons ultimately required to fill the voids.

Mobilization began on February 24, 1985. Plant construction, topsoil removal and stockpiling, site fencing, haulage road construction, and plant erection were completed by March 21st, using two 24 yard twin engine scrapers, a motor patrol and two D-9 bulldozers.

Production was started on March 24th. Two D-9 bulldozers with rippers were used for excavating and pushing the material to the crushing/screening plant. The material was sized using a cone crusher, and the material was screened to the specified 3/4 inch minus material using a shaker screen. Oversize material was recycled. The screened material was dumped onto a live stockpile and loaded into the haulage trucks with a Cat 920C loader. A haulage truck fleet of three to four 25 ton belly dump trucks were used throughout the project for transporting the crushed material to the 12,000 ton slurry plant stockpile. The material supplied was weighed on a 50 ton scale supplied by the contractor and manned by MWM personnel. The contractor also provided a Cat 824 wheel dozer at the stockpile for material movement activities.

The borrow area equipment was run on a two shift, seven day per week basis to maintain production rates. Truck haulage to the slurry plant was done on a 10 hour per day, seven days per week basis. Twenty-five persons were employed by the contractor at the job site, with additional maintenance and repair personnel being brought in for overhaul of major equipment. Daily production varied widely over the duration of the project, due to plant breakdowns, difficult material, and wet weather. Despite these variations however, supply of backfill material to the slurry contractor was significantly interrupted on only one day.

Two principal problems were encountered by the contractor. First, the excavation site was located in an area where shallow underground coal mining had taken place, and the abandoned workings had been on fire in the past. As a result, sinkhole subsidence had taken place in the area, leaving large craters, and subsurface voids, which were subsequently uncovered during the borrowing activities. While inadvertent exposure of the voids was a concern during the borrow activities, there were no accidents. Second, high winds were prevalent throughout the project which caused dusty conditions. A 2000 gallon water truck was in use much of the time for controlling dust, with water consumption averaging 10,000 gallons per day and peak consumption of 24,000 gallons per day. Very few complaints were received from residents about dust.

Delivery of backfill material was terminated on October 22, 1986. Based on individual truck weight receipts, 790,277 tons of material was delivered to the slurry plant. The contractor completed reclamation of the land on November 9, 1986, with reseeding, fertilizing and mulching of the entire disturbed area.

Injection and Observation Hole Drilling

Based on design slurry injection rates, injection wells were drilled at 12 1/4 inch diameter, and completed with 8 inch steel casing cemented to full depth. A total of 101 useful injection wells were drilled; an additional 14 wells failed to encounter voids and were plugged. The injection well drilling totaled 25,000 feet, 20,000 feet of which were cased for subsequent slurry injection. In addition, a total of 25 monitor wells were drilled for checking progress, effectiveness of backfilling, and quality of the emplaced backfill.

The drilling contract was awarded to Russell Drilling Company, of Harvey, North Dakota. Drilling of injection holes began on October 10, 1985, and was completed by the end of July, 1986. The injection and monitor well drilling was accomplished with up to three conventional rotary drill rigs. Truck mounted mud tanks with shale shakers were used in conjunction with the drill rigs and cuttings were collected in trailers for transport to the city dump.

Upon completion of backfilling operations, the wells were sealed to within three feet of the surface with cement grout, the casing was cut off three feet from the surface, and each well was permanently identified with its UIC Permit Number, and date of completion. The remainder of the drill hole was then filled in and compacted, with restoration of any pre-existing surface material. Reclamation of drill holes was undertaken sequentially after the injection into each hole was complete, and was completed by the end of November, 1986.

Slurry Mixing and Injection

Eby Mine Services of Brighton, Colorado was awarded the contract to perform slurry injection for the Hanna Mine Subsidence Control Project. The notice to proceed on the injection contractor was issued on November 7, 1985. The contractor had the responsibility of performing the detailed design of the slurry mixing plant and the pipeline delivery system; this design was completed by the end of 1985. Construction started during the first week in January 1986. By the end of March, 1986, the slurry injection system was essentially complete. System checkout, calibration and proof testing was completed by April 1, 1986. The general layout of the slurry mixing and pumping system is presented in Figure 10. The slurry injection system included a substantial instrumentation capability, including monitoring of backfill material input, water consumption rates, pump pressures, individual slurry pipeline flow rates, and wellhead pressure. This information was available in real time in the slurry plant control room, and was also recorded on strip charts as a permanent record.

The general operation of the plant was as follows. Water supply was provided by two 16 inch cased wells drilled into the mine haulageway, which had a combined capacity of about 4,200 gpm. The water was piped to the slurry plant through a single 14 inch steel pipeline. The backfill material loaded from the 12,000 ton capacity stockpile and introduced into the system through a hopper/feeder. This discharged onto a variable speed 48 inch inclined conveyor belt which emptied into a mixing tank where the slurrying water was added. A belt scale was located on the conveyor for weighing the material. From the mixing tank the slurry was passed into a constant speed slurry pump, and then into a variable speed slurry pump, and was then discharged to the delivery line. If slurry dilution was required, water could be introduced just downstream of the slurry tank. There were two essentially identical delivery systems from the slurry tank, which together were capable of a slurry flow rate of 4,200 gpm at delivery pressures up to 200 psi.

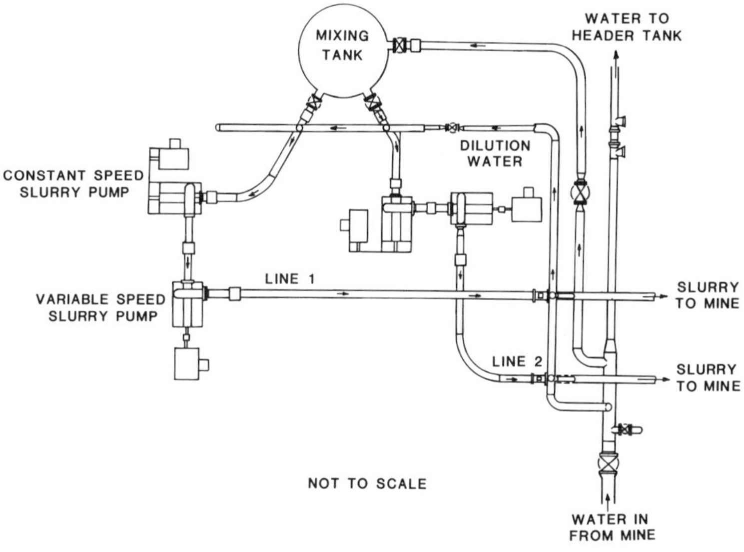

Figure 10. Layout of Slurry Mixing and Pumping System

The slurry was delivered to the general injection area in two main pipelines; delivery to specific well locations was achieved by temporary distribution lines. Each distribution line was connected by a wye fitting to two wellheads, with delivery being controlled by a diversion valve on each branch. After filling the first of the room, the second room would be started, while the unused branch of the line would be installed on a third injection well, and so on. In this way, a well was always available for filling or purging the system.

Slurry injection operations began on April 2nd, 1986. After a checkout and experimentation period, slurry injection evolved into a two shift per day, five day per week operation, with maintenance, line routing, etc. being accomplished on the weekends. For most of the project, operations were carried on by a three man crew during each shift - a plant operator, a loader operator and a laborer. Four employees were also required for general maintenance, pipeline relocation, changing wellheads, etc. Supervision for the project was provided by a project manager and a shift foreman.

A total of 790,000 tons of backfill material was injected into the mined rooms indicated in Figure 11, with production rates in excess of 100,000 tons per month. Essentially all of the area targeted for backfilling was treated, with the exception of a block of five rooms, which remained unfilled for safety reasons. Injection ended on October 24th, 1986

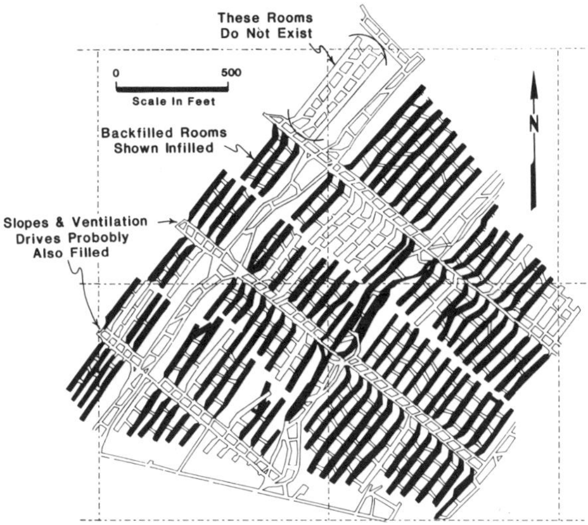

Figure 11. Rooms Treated with Backfill

During the backfilling operation it was established that backfilling could be accomplished slurry densities up to 7.2 pounds of material per gallon of water (55% by weight), without causing operational problems. This density was used in the latter part of the project.

In general, the plant was well designed and functional. As a result, daily operations were usually routine. However, the necessity of handling hard packed material as well as rocks and concrete from the stockpile bank occasionally created some difficulty. Major maintenance centered around overhaul of the slurry pumps, as it was found that under the operating conditions impellers and volute liners had a life averaging about 100,000 tons. Complete replacement of these parts was usually accomplished in two to three hours. Extensive replacement of the plant piping and some sections of slurry pipeline were necessary after 675,000 tons of material had been injected.

Effectiveness of backfilling

The Hanna Mine Subsidence Control Project, as the name implies, was carried out for the purpose of preventing surface subsidence due to prior underground mining activity. In addition to limiting roof collapse underground, the project was designed to stabilize an area under the town in which structural damage had occurred. The selection of hydraulically emplaced backfill was predicated on the need to fill large voids which were in general flooded and inaccessible. This section assesses the effectiveness of the backfilling, from a volumetric and support standpoint.

As originally planned, the project called for the emplacement of 600,000 tons of granular backfill material. This quantity was derived from the known geometry of the underground workings, assuming an eighty percent fill factor. In previous projects of this type, injection to each room was terminated when the pressure at the surface went positive, and the rooms "refused" to take slurry, which was a signal that the rooms were full. In the Hanna case, refusal rarely occurred. This is considered to be due to a combination of the open, large rooms, the slope of the coal seam, and the relatively fine material that actually reached the rooms. As a result of this failure to reach refusal, it became necessary to develop a criterion for ceasing to backfill each room. The initial criterion chosen was 120% of the volume of the room being backfilled. Based on testing information, this was reduced to 100% of room volume at about the half way point of the project, and to 80% in certain of the non-critical rooms filled last.

As a result of the ease with which the material could be emplaced underground, the project substantially exceeded the goals which had been established, and 790,000 tons of material was emplaced. In general, the individual rooms were well filled and greatly exceeded expectations. However a few rooms were not backfilled, because they were "cut off" from the return water flow system, and became pressurized when injected. The unfilled rooms are indicated on Figure 11. In addition, a few rooms refused to take the quantity of material which was planned.

To test room filling effectiveness, several rooms were evaluated in detail. This was done by drilling a number of observation holes into the test rooms, and monitoring the extent of filling of the rooms using these holes. A typical result of these evaluations is shown on Figure 12, which shows essentially complete filling, supportive of the conclusion that backfilling of the rooms and voids is largely complete.

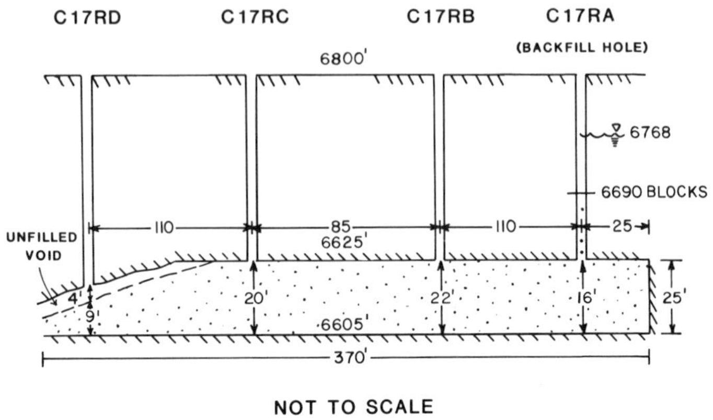

Figure 12. Schematic of Testing of Filling of Room C17R

In addition to the volumetric filling tests, checks were made in the laboratory and in the field to check the load-carrying capacity of the backfill, which is critical to the effectiveness of the program. The laboratory testing demonstrated that the backfill material is a poorly graded sand-silt mixture, with fair to excellent drainage characteristics. In addition, the compression of the material under load was checked using a consolidation test procedure which indicated about 15% consolidation as load increased from essentially zero to 10 tons per square foot (140 psi). These results support the expectation that the material would be effective as a support medium in the mined rooms.

The field tests involved penetration tests, conducted using drilling equipment. The in-situ tests consisted of pushing a drill rod with an 8.3 square inch blank end into the backfill material through an observation hole. A total of 11 rooms were tested in the program. Figure 13 shows the penetration curves from the test data in the same room that is presented in Figure 12. The data indicates that the backfill has a significant resistance to penetration, which is very great near the injection hole, and reduces as the distance to the injection hole becomes greater. The majority of the penetration tests revealed penetration resistances greater than half the maximum overburden load within five feet of the top of the backfill, and greater than the maximum overburden load within ten feet of the top of the backfill. It is considered that these results indicate that the backfill will provide effective support of future loads resulting from roof or pillar failure.

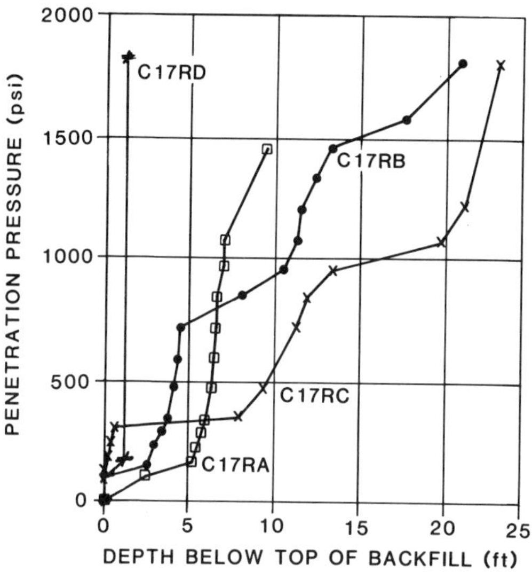

Figure 13. Penetration Tests of Backfill - Room C17R

Before, during, and after the performance of the backfilling, three other monitoring activities were undertaken to evaluate any short term impact of the activities. First, a survey of the structures in the backfill area was undertaken to provide a baseline for assessing any claims for damage that might result from the project. No claims were made for any subsidence-induced damage. Second, a network of 40 subsidence monitoring hubs was installed in the area to be treated. Periodic surveys were then taken during the project to detect any relative movement. No significant differences were noted throughout the duration of the project. Third, the water level in the mine workings was monitored during the project, to ensure that the level neither dropped to an extent that might cause subsidence, nor rose to a level that would flood low-lying areas of Hanna. This monitoring indicated that levels were appropriately controlled during the project.

Problems and solutions

The Hanna Mine Subsidence Project emplaced nearly 800,000 tons of backfill material underground without significant problems. However some operational difficulties were encountered and overcome during the project. This section is devoted to a discussion of these difficulties and solutions, and is included to allow subsequent similar projects to avoid them.

1. Injection Monitoring. Monitoring of progress while backfilling was limited to "sounding" monitor wells and injection wells. Though useful and necessary for determining the effectiveness of backfilling, "sounding" is less satisfactory than visual observation. Three attempts were made to use television during the project without success. The blackbody conditions underground absorbed the light from the camera and limited the depth of field to less than 20 feet. No improvement was noted when additional lighting was provided, even in dry voids.

2. Premature Plugging. It became evident within the first week of backfilling that flow to any injection hole had to be continuous; any cessation in flow generally caused the hole to block. It was found that the best means of preventing sand from flowing back into a well during periods when backfilling was not taking place was to continue injecting clear water at a high rate.

3. Location of Voids. The mine mapping, while generally accurate, indicated development in excess of actual room length in some instances. As it was considered desirable to locate the injection wells within 20 feet of the blind end of the room, on occasions drilling did not intersect the mined void, and the injection hole had to be located further into the room. In a few instances, rooms were not found where shown on the map.

4. Collapsed Rooms. Many of the rooms in the area of the Middle School were found to have collapsed, leaving only small void heights. However backfilling of these rooms was routine, with the great majority of the rooms accepting the expected backfill tonnage. It is considered that this success indicates that hydraulic backfilling can be employed even when rooms have partially collapsed, thus preventing further subsidence.

5. Surface Collapse. Several injection wells required surface casing for successful completion. A near-surface clay formation that blanketed some of the project area had a tendency to slough in the presence of water. When the drill hole intersected the void, drilling fluid drained down the hole, the pressure in the drill hole dropped, and occasionally the surface material would collapse into the void, leaving a surface slump on the order of fifteen feet in diameter and up to twenty feet deep. To combat this condition, over 1000 feet of surface casing was installed.

6. Lumpy Material. Presence of lumps in the slurry contractor's stockpile was a continuing problem. As a result of lack of space on the stockpile, the dump trucks and loaders were forced to drive over the stockpile materials, thus compacting them and generating lumps. The problem was not adequately rectified during the project: impact lump breakers were installed at the hopper and belt discharge, and hand labor was employed to remove lumps from the belt.

7. Injection Well Eruptions. Late in the project, a potentially dangerous condition developed in a panel towards the northwest of

the filled area where eruptions of slurry would occasionally occur through the injection wells. These eruptions reached heights in excess of two hundred feet in the air. On one occasion, pipeline whipping due to horizontal discharge of material caused property damage. These eruptions were considered to be caused by the isolation of sections of the mine, and the compression of air in these isolated portions. The best solution for preventing violent eruptions lies in keeping the conduit to the mine pool open. This entails consistently carrying out the backfilling in an up-dip direction in the mine.

8. Excess Volume Requirements. During the project, 32% more material was emplaced than was originally estimated. This is believed to be a result of better than expected backfilling (close to 100% of the room voids), together with losses of backfill from the area of interest. It is known that the fines fraction did not readily settle out and it is likely that some material flowed out from the entrances to the rooms and down the haulageways. A possible solution to the problem of material not being emplaced in the area desired is to place gravel bulkheads in the main haulageways and other potential loss pathways. However in this project placement of these bulkheads was considered to be too difficult and expensive to be warranted.

Conclusion

The Hanna slurry backfill project indicates that large volumes of mined voids can be remotely backfilled to reduce subsidence risk. The use of a dilute slurry emplacement process produced a very large volume of emplaced material for each injection location, and replaced the voids left by self-consolidation of the materials to produce essentially complete backfilling of the voids. The characteristics of the emplaced materials have been shown to be adequate to support the overlying materials with only slight further subsidence. At a total cost of about $8 per ton emplaced (1987 dollars), this technique is considered by the authors to be an effective subsidence control strategy for large abandoned underground coal mine.

Bibliography

MidWest Mining Company, 1985. Hanna Subsidence Control Project 7A - Phase I Report. Prepared for the Wyoming Department of Environmental Quality, Land Quality Division, Abandoned Mine Land Program, by MidWest Mining Company, Casper, February (five volumes).

MidWest Mining Company, 1986. Hanna Subsidence Project 7A - Land Use Recommendations to Minimize Subsidence Risk in Hanna, Wyoming. Prepared for the Wyoming Department of Environmental Quality, Land Quality Division, Abandoned Mine Land Program, by MidWest Mining Company, Hanna, May.

MidWest Mining Company, 1987. Hanna Subsidence Project 7A - Slurry Backfill Project Completion Report. Prepared for the Wyoming Department of Environmental Quality, Land Quality Division, Abandoned Mine Land Program, by MidWest Mining Company, Hanna, February (three volumes).

National Coal Board, 1975. Subsidence Engineer's Handbook, London.

Piggot, R.J, and Eynon, P., 1977. "Ground Movements Arising from the Presence of Shallow Abandoned Mine Workings" in Geddes J.D., ed., Large Ground Movements and Structures, New York, Halstead Press, pp. 749-780.

Whaite, R.H., and Allen, A.S., 1975. Pumped-Slurry Backfilling of Inaccessible Mine Workings for Subsidence Control. U.S. Bureau of Mines Information Circular 8667, USGPO, Washington.

Effects of Slurry Backfilling - Lessons Learned

Robert R. Faddick,[1] Member, ASCE, and William R. Eby[2]

ABSTRACT: In 1986 the underground coal mine voids in the town of Hanna were slurry backfilled as part of the Abandoned Mine Lands Program for the Department of Environmental Quality in the state of Wyoming. The purpose of this hydraulic fill was to alleviate subsidence throughout much of the town. To minimize disturbance to the population, over 100 injection boreholes were drilled to depths down to 150 m into which some 782,000 tonnes of sand were slurried. What made this project unique technically, were the large tonnage of injected material, the use of twin slurry pipelines operated simultaneously from a common source, and the highly instrumented operation of the slurry plant system. The project was completed successfully, but in a relative sense, because it was not without its share of difficulties. Some of these technical difficulties, pertaining to the fill material used, the operation of the pipeline system, and sequence of backfilling, are discussed with perhaps the hope that they may be interpreted as lessons learned for future slurry backfilling operations.

Introduction

Backfilling is almost as old as mining itself. Probably the earliest recorded backfilling event in the U.S. was in 1864 in Pennsylvania (Lightfoot 1951). Presumably European and Asian practices are even older. The reasons for backfilling include an attempt to improve the structural strength and integrity of the mine during mining activities, to prevent subsidence of the overburden, to control the elements of fire, gas and water intruding into the areas of mining activity, and to reduce ventilation requirements. A wide assortment of materials has been used ranging from the mine waste itself, to ordinary

---

[1]Prof. of Civ. Engrg., Engrg Dept., Colorado School of Mines, Golden, CO 80401.
[2]V.P., Eby Mine Services, Brighton, CO 80601.

sand, flyash, garbage, and, of course, water. While there are many similarities in the backfilling process, there are usually enough site-specific distinctions to prevent identical comparisons.

The backfilling process involves several components. These are:

a) Preparation Unit
b) Horizontal Pipeline to Borehole c) Vertical Pipeline to Borehole
d) Underground Void
   i) elbow at base of borehole to transfer flow to horizontal pipes in underground tunnels where access is possible.
   ii) termination of vertical pipe in borehole at top of void. Each is discussed in detail.

a) Preparation Unit - this includes a tank or hopper for mixing the solids phase and the fluid phase. The fluid phase may be air or water resulting in pneumatic transport or hydraulic transport, respectively. The prime mover is usually a centrifugal blower or pump. For slurries, positive-displacement pumps such as concrete pumps, pressurized lock hoppers, ram pumps, piston and plunger pumps have also been used. On occasion the feed tank or hopper is directly above the borehole and gravity drives the mixture down the borehole.

At this point it should be clarified that fluid transport principles are essentially identical whether the fluid be air or water. At first glance pneumatic and hydraulic transport schemes may appear to be distant cousins, but in reality, are fraternal twins. There are basically two differences in properties between air and water: viscosity and density. The reduced density and viscosity of air relative to water mean that the solids phase can only be suspended by a higher air velocity, i.e. a higher flowrate. This in turn creates higher power requirements and greater potential abrasive wear for pneumotransport over that of hydrotransport. Also, the solids loading (concentration) or mass flowrate is considerably less with air than with water because of the higher particle kinetic energy in airflow.

In general terms, abrasive wear is proportional to the square of the velocity and power is proportional to the cube of velocity. Thus, pneumotransport systems usually entail higher operating costs than hydrotransport systems. Otherwise the solids loading or concentration, headloss, and flow regimes evolve in the same manner regardless of the fluid medium and produce identical phenomena. The same principles of the conservation of mass, momentum and energy apply to both systems. Upon relatively rare occasions of elevated temperatures and pressures, compressibility effects of air must be considered.

Although there are exceptions, a general rule of thumb is that pneumotransport backfilling is used where water is scarce, the underground void is dry, or where water would deteriorate or contaminate the underground environment. Hydrotransport backfilling is used where water is available and the underground void is either dry or inundated.

b) Horizontal Pipeline to Borehole - this is required with pressure systems and some gravity systems. If high velocities are required and the solids are abrasive, the pipe must be made of wear-resistant materials. For both pneumotransport and hydrotransport systems, the operating velocity required to keep the heaviest and largest particles in suspension controls the minimum flowrate in the entire system, including both horizontal and vertical pipelines. This is because the forces of gravity and pressure are perpendicular to each other in horizontal pipeflow but are co-linear in vertical pipeflow. In the latter case, the velocity and concentration profiles are essentially uniform with no stratification of the larger and heavier solids.

c) Vertical Pipeline in Borehole - this pipe is usually vertical although sloped pipes have also been used. The pipe may be as long as 1100 m with diameters up to 300 mm or 400 mm. Smaller diameters are more common because of minimal backfill requirements, the need to use several boreholes to expedite backfilling, and to offset the cost of drilling and casing larger diameter holes at great depth.

An open system means the pipe at the top of the borehole is open to atmosphere via a vented pipe or open valve. This allows zero gauge or atmospheric pressure to exist at the top of the borehole. Such systems are often less stable in operation resulting in pressure-flow fluctuations, vibration, and excessive wear. Normally the system is closed and the pressure in the upper part of the pipe is negative (partial vacuum). This does not create any particularly adverse problems and in fact usually creates a more stable flow.

d) Underground Void - Most mine backfilling systems are controlled, that is the vertical borehole pipe connects to a horizontal pipe and is discharged to a particular location in the mine. There, personnel direct and monitor the backfilling process where access to the voids is possible.

Some backfilling systems are done in a blind fashion, that is, the vertical borehole terminates in a void where no access by personnel is possible. Then the backfilling is performed with limited control from the surface.

An intermediate system, perhaps considered as semi- controlled, is possible where in the first case the horizontal pipe discharges through a barrier or bulkhead to extinguish an underground fire. In the second case a remotely-controlled nozzle can be rotated to backfill in a radial pattern.

Theory

This paper deals with blind filling, in particular, how the entire system operates to achieve as complete a filling of the underground void as possible.

Basically there are two cases to consider: a dry void and an inundated void.

Dry Void

A vertical pipeline discharging slurry into a dry void provides kinetic energy equivalent to the product of mass flowrate and velocity squared. The pressure of the flow stream in the void is close to zero. If the kinetic energy is low, a cone builds up with an angle of repose of the sand and water flows off the cone with little sediment transport. If the kinetic energy and flow of the slurry are sufficiently high, the cone angle decreases as the finer solids are carried away in open channel flow transport under the principles of sediment transport. If the kinetic energy and flow are increased further, the cone will develop a small crater in the top. In the extreme case the kinetic energy and the flow can be increased to the point where the bottom of the void is scoured clean and an cone (like a volcano) is produced (Stewart and Heslep 1974). The basic assumption underlying the above description is that the water has somewhere to go, flowing away from the borehole carrying solids with it. If the water and solids are confined, the sediment transportation is restrained as the system changes from open channel flow to pressurized conduit flow with a noticeable decrease in velocity and sediment transport.

If air is the fluid medium, the velocity will be high, even with dense-phase flows. Consequently the solids will readily blow away from the borehole as long as the air has an escape route and cannot put a back pressure on the void.

The pneumotransport method is more efficient in filling a dry void because the sand blows away from the borehole and deposits by layering with no cone buildup. However, the pneumatic backfilling system is more energy intensive and abrasive.

Inundated Void

When the void is filled completely with water, the backfilling process reverts from open channel flow (with a free surface acted upon by atmospheric pressure) to pressurized conduit flow. The pressure is exerted by the elevation of the water table. Again, assuming that the inflow from the borehole pipe has somewhere to go, the velocity of flow capable of maintaining solids in suspension and transporting them away from the borehole is now reduced considerably. This is because the entire void cross section is flowing at a considerably reduced velocity.

In a dry void the slurry channels its way down the cone creating local velocities high enough to transport the sediment. In an inundated void the entire void becomes the channel and local velocities reduce considerably. To achieve significant sediment transport in an inundated void requires higher kinetic energy and flow than in dry voids given a constant void geometry and unrestrained water discharge.

The sediment transport mechanism in both cases is identical. The local velocity must be several fold that of the settling velocity of the coarsest solid. For a given discharge in the borehole, the same flowrate will occur in the void but the entire cross section of an inundated void is put into play resulting in a much reduced velocity.

Other Factors

Particle size and distribution are the keys to slurry transport. Ideally the distribution should be wide to permit good particle packing resulting in high solids concentration for a given unit of pumping power. Low values of specific energy (kw-hr/tonne-km transported) are desirable both for efficient pipeline transport and to reduce abrasive wear on the pipe-pump componentry. Once the high-concentration slurry reaches the vertical borehole, contrasting schools of thought prevail.

One school of thought believes that the slurry stream should have a high velocity (>5 m/s) and low solids concentration (<20% by weight), so that the particles possess high kinetic energy and will migrate further in either a dry or inundated void. Thus more solids are pumped into fewer boreholes. This school also prefers a narrow size distribution and a non-cohesive material so that the solids will roll over one another more readily in the transport process and migrate further into the void.

The other school believes that a slower stream velocity (<4 m/s) and a higher solids concentration (>40% by weight) will give a better packing density underground although more boreholes may be required.

The geometry of the underground mine is equally important. Numerous crosscuts and drifts free of man-made barriers or roof and pillar falls, allow good circulation of the slurry. The dip of the voids can be important, particularly in a dry mine. Opinions are varied as to the preference of backfilling down-dip or up-dip. The latter course seems to be more popular, the idea being that greater compaction of backfilling is achieved. Nevertheless, the operation works against gravity and usually requires more boreholes to fill the voids.

The Hanna Project

The details of the project are given elsewhere by the authors (Faddick and Eby 1986). Only a brief summary is provided here. The oral presentation is accompanied by a series of slides to acquaint the audience with the scope of the project.

Details

Owner: State of Wyoming, Dept. of Environmental Quality, Land Quality Div., Abandoned Mine Lands Program
Contractor: Eby Mine Services (EMS), Brighton, CO
Period: Dec. 1985-Nov. 1986; Injections: April-Oct., 1986
Mine: multi-entry room and pillar coal mine; 14 deg. dip to SE
Depth: 13 m in NW part of town, 150 m in SE part of town
Voids: 7.6 m wide, 7.3 m high, 60 to 120 m long, 100 boreholes
Water: Mine totally inundated. Water drawn from mine by two 60 m-deep well pumps (300 L/s total capacity)
Backfill: sand, 20 mm top size by maximum 25% passing 75 microns
Plant: 1 mix tank, twin 200 mm diameter slurry pipes, each up to 2100 m-long. (steel and plastic), single 350 mm-diam. steel water line. Gate valves for diverting from one borehole to another during backfilling. Two sets of two Warman 200x150 mm centrifugal slurry pumps in series with Ni-hard impellers, variable frequency and belt drives. Sand, water, and slurry all metered separately. Central control console capable of automatic or manual control of entire system.
Slurry: Contract Specs: 15-23% solids by weight, 2.4 to 4.6 m/s slurry velocity, 180 tph, 4550 tpd, 91,000 tons/month, 545,000 tons to be injected.
Actual Operation: up to 50% solids by weight, 2.4-5.5 m/s slurry velocity, up to 532 tph. Solids exceeded 20 mm top size on occasion.
Injected 782,000 tonnes.

Problems Encountered

Over-all, the project was completed successfully in that the contract amount of material was backfilled on time, and an additional 236,000 tonnes were injected prior to the winter season.

Several problems were encountered that perhaps could have been eliminated with better initial planning and careful thought. The problems are listed as they occurred chronologically along with any solutions that worked or at least made the situation tolerable.

1. The project was divided into three contracts: borehole drilling, backfill supply, and backfill injection. The purpose was to allow small local contractors to compete although it was recognized that more supervision would be required. The end result was three different contractors, scheduling complications, project-interface disputes, delays, and increased costs. A single contract would have been more cost- and time-effective for the State.
2. The backfill injection specifications were too rigid for the solids concentration range, allowing only 15 to 23% by weight. Field demonstrations were alluded to in the specifications but not allowed. It took months of persuasion by the backfilling contractor (EMS) to convince the State to allow higher solids concentration to be demonstrated. After successful demonstrations, concentrations up to 50% by weight were pumped. This allowed the project to finish on time before winter and to permit injection of an additional 236,000 tonnes beyond the original contract amount. The reason for the under-estimation was apparently the large number of crosscuts between rooms overlooked in the calculation of void volumes.
3. No surface compaction tests were made with the backfill material. Such tests would have disclosed the cementitious properties due to calcium carbonate in the material. This led to the formation of large, hard, agglomerates (chunks up to 30 cm-long). These chunks wreaked havoc on the loading equipment and the pumps, leading to considerable down time and substantial maintenance expense. There was difficulty in assessing blame between the contractors and owner.

4. Early in the project, the boreholes were taking more tonnage than estimated. Consequently, most of the holes were not filled to refusal for fear that the project would exceed its budget and backfill supply.

5. An unexpected phenomenon occurred early in the project: when the slurry flow to a borehole was stopped (such as when lumps blocked the lump breakers on the inclined conveyor belt and the solids feed had to be stopped), the borehole would plug and would not accept either water or slurry flow upon restart. On several occasions the hole had to be drilled clear and water or slurry introduced quickly as the drill was removed. This was possible by passing the drill through a 150-mm vent pipe attached above each 200-mm diameter injection borehole elbow. It was surmised that the combination of high water table, the fineness of the sand in the early stages of the project, the high velocity and low concentration that kept the sand in suspension, all contributed to an underground slurry pressure head. This caused a backflow into the borehole and a sand plug to develop once water or slurry flow ceased. It became apparent that when stopping slurry flow into a borehole, the borehole would have to be flushed continuously with water to keep it clean from sand backflow. EMS made a valved connection in each slurry pipeline downstream of the slurry pumps to the water line to allow water well pumping directly to the boreholes. This technique worked satisfactorily.

Conclusions: Lessons Learned

1. Backfilling underground voids does not lend itself to multiple contracts and contractors.
2. Backfilling projects should be assigned a range of estimated injection tonnages so that contractors can size their slurry plant for flexibility and also prepare flexible schedules.
3. Neither the contractor nor the owner possess infinite knowledge of the underground voids and how to fill them. Field demonstrations must be allowed and encouraged, with appropriate time dedicated to this task for observation and down-hole monitoring. If void conditions vary, such as the water table level, several demonstrations may be necessary to assess the variances.
4. The progression of filling boreholes and direction (up-dip vs. down-dip) should be mutually agreed upon by owner and contractor based on observations of the backfilling effectiveness during the field demonstration.

5. The backfill material must be studied thoroughly, including its compaction density, cohesion, and cementitious characteristics.
6. The purpose of a backfill project is to control subsidence by filling the mine voids as completely as practicable. Considering the expense of drilling boreholes and setting up a slurry plant, the borehole should be filled to refusal. Even then the voids will not be filled completely. The incremental cost of backfill material is usually small compared to the fixed plant costs.
7. It is interesting to compare backfilling specifications in the western and eastern US coal regions. Wyoming specifications will not tolerate any positive pressures at the borehole collar, whereas Pennsylvania will allow up to 350 kPa in non-critical urban areas as defined by the State Engineer. Yet, knowing the water table level in the hole, and the slurry hydraulic gradient, the borehole pressure or partial vacuum can be calculated and field verified. With a flexible slurry flow operation obtained with variable speed pumps, an operator can fill a hole to zero-pressure refusal with good control. With a reasonably wide particle size distribution he can divert a slurry of medium solids concentration to another borehole without interrupting the backfilling process.
8. The slurry plant should be capable of bypassing water to a borehole in the event it has to be kept flushed during slurry plant shutdowns.

Appendix I. References

Faddick, R.R., and Eby W. R. (1986). "Slurry Backfilling of Mine Voids in Hanna, WY," proc. 1986 National Symposium on Hydraulic Mining, Hydrology, Sedimentology, and Reclamation, Univ. of KY, Lexington, KY, Dec. 8-11.

Stewart, J. D. and Heslep, M. E. (1974). Dow Chemical patent #3,817,039, "Method of Filling Subterranean Voids with a Particulate Material," June 18.

Lightfoot, W. E. (1951). "Hydraulic Filling in Metal Mines," Special Report 12, Division of Mines, San Francisco, CA, Dec.

CONSTRUCTION METHOD FOR IMPROVING UNDERWATER SAND FILLS

W.E. Hodge*

ABSTRACT: A novel construction method is suggested for improving the engineering characteristics of underwater sand fills. The idea of pumping water out of submerged sand fills concurrently with fill placement, in order to generate inward seepage flows through the sand mass, is presented as a possible means of producing higher densities, steeper side slopes and enhanced erosion resistance. Two separate model test programs designed to investigate this method are described in detail. The second series carried out at the Hydraulics laboratories of the National Research Council of Canada was undertaken to verify the initial test series performed in Ireland. It is concluded that this method achieves the desired affects under laboratory conditions and at the model scale used. A working hypothesis is offered as a basis for further research into the forces which effect sand deposition underwater. A reason for the very flat slopes typically observed under field conditions is also suggested.

INTRODUCTION

Dredged sand is frequently used as a material for constructing underwater earthfills. A noteworthy example of this is in the recent offshore hydrocarbon searches in the Arctic waters of the Beaufort Sea in North America. These exploration programs have in most cases been conducted from man-made surface piercing islands constructed using dredged sand, or from mobile drilling platforms ballasted down on top of submerged sand fill berms.

In many cases sand is used because it is the only readily available granular material close to the site. Unfortunately, the ready availability and relatively cheap cost of dredged sand is offset to a large extent by the poor engineering characteristics of the resulting sand fills. As well as being poor compared with gravel or rockfill, the properties of underwater sand fill cannot be predicted at the design stage. One does not know what sort of strength and deformation characteristics will be attained until the fill is in place, after which time these parameters can be estimated from insitu testing. In other words, a typical dredged sand fill is not an engineered earthfill and you take what you get; consequently, it is not possible to optimize the design and minimize costs. This may be an acceptable situation for temporary islands or construction facilities in shallow waters, but it is not acceptable for permanent structural support or for situations where the sand fill may be subjected to significant static or dynamic loading.

Dredged sand, when placed through water, has three undesirable characteristics which compare unfavourably with coarser grained fills such as gravels or quarried rock:

i) Sand fills can be of variable density, containing large zones of very loose material.

* Phoenix Engineering Ltd., Vancouver, British Columbia, Canada

ii) Sand mounds normally have small side slope angles.
iii) Sand is easily eroded by currents or wave action.

Each of these characteristics has a serious impact on the acceptability of a sand fill mound placed conventionally through water:

a) Inadequate or unreliable packing density is generally the most serious disadvantage. Under static loading, low densities result in poor strength and deformation behaviour. A submerged sand fill is, however, rarely subjected to only purely static stresses. Natural causes such as wave action, earthquakes, and in the Arctic, drifting ice can induce dynamic stresses. Cyclic stresses imposed on submerged sand fills may be sufficient to cause liquefaction, or at lesser intensities, can lead to large deformations. Adequate density is necessary therefore, both in providing resistance to failure and in limiting the amount of structural movement.

b) Side slopes as flat as 3° to 5° are common in underwater sand fills. The volume of fill required to produce a given sized island or berm increases dramatically with flattening of side slopes. In the particular case of Arctic work, the sand volume required for conventional fills in deeper water may exceed the dredging capacity of the fleet working within the construction window which is dictated by the ice-free season. Flat slopes also have an environmental ramification, inasmuch as the larger the sand volume needed, the greater the extent of natural seabed disturbance, when both the area buried beneath the fill and the area removed at the borrow pit are included.

c) Erosion is not so much of a problem under current construction conditions where the sand fills normally end up at very flat slope angles. Erosion would become much more of a consideration if it proved possible to place sand at steeper side slopes.

The question addressed here is whether there is a simple practical method whereby a dredged sand dumped through water can be treated in such a manner as to produce a fill with acceptable engineering characteristics, which could be predetermined with some degree of reliability at the design stage. The purpose of the research work outlined in this paper was to attempt to assess one such method, i.e., pumping water out of the submerged sand fill during placement.

The research was carried out in several phases. Initially, the idea of dewatering during placement was checked out in rough fashion in the author's backyard using simple apparatus. These results were sufficiently encouraging to warrant more detailed work. In September 1983, the research was taken to University College Cork, Ireland, where suitable soil mechanics and hydraulics laboratory facilities were made available. The "Cork" testing gave qualititive, and to a limited degree, quantitative support to the broad concept. The Canadian Government was next approached with a request for financial assistance to do a field test. The National Research Council of Canada (NRC) insisted on larger scale verification testing in their Hydraulics laboratories in Ottawa prior to committing funds to a field trial. NRC also required that the verification testing be supervised by an independent geotechnical consulting firm. The author's company, Phoenix Engineering Ltd. (Phoenix), designed the test program and testing equipment for the "Ottawa" work, and engaged Golder Associates Ltd. to monitor the research. The NRC model testing commenced in August 1985. Following the Ottawa program, Phoenix set about the development of the field equipment necessary to build a prototype sand fill at an offshore site.

PUMPING WATER FROM SUBMERGED SAND FILLS

For the various reasons outlined above, the use of dredged sand placed underwater is currently limited to specific cases of fairly small temporary structures which are not sensitive to deformation. Despite its abundance and low unit placement cost, its use cannot be contemplated at present for such structures as earthdams, causeways or breakwaters because of its poor performance. The objective of the work reported here was to test a simple technique in the hope that it might provide a sand fill with characteristics similar to those of coarser grained fills.

Sand is composed of cohesionless discrete particles, but so too is quarried rock. They achieve roughly the same density when placed underwater, but the fill behaviours are significantly different. The difference is obviously associated with particle size. This may be stated in terms of specific surface where the smaller the grain size, the greater is the ratio of the surface (water) forces to the body (gravitational) forces. Or, phrased differently, the smaller the grain size, the greater will be the influence of viscous forces associated with water movements on the individual particles of sand. At the beginning of this study these drag forces or seepage forces were considered to be the likely root cause of the undesirable characteristics of submerged sand fills. For that reason, it was also believed that if seepage forces could have such a profound detrimental effect on submerged fills, then it was also possible that reversing the direction of the same seepage forces might have an equally powerful beneficial effect.

The destructive nature of pore pressure and seepage forces depends entirely upon the direction in which these forces act, i.e., from inside the body of the deposit outwardly. This perversive natural tendency towards outward seepage is due to the fact that pore pressure is normally generated within compressible saturated soil structures when the mass is subjected to increased loading. If, however, seepage forces could be made to act inwardly on a submerged fill they should have a beneficial effect comparable in influence to the detrimental effect of outwardly flowing water. This is conceptually easy to do - by withdrawing water from within the fill in order to lower the pore water pressure to a level below that of the outside ambient water pressure. Water would then flow into the submerged fill rather than flow out of the fill. The procedure of dewatering existing subaerial slopes to improve stability of open cuts has been, of course, standard practice for many years. But it has not been used underwater, presumably because it appears absurd to dewater an underwater fill. The novelty of the technique proposed here, however, is not simply related to providing a steeper slope in an existing deposit underwater; the novelty lies in attempting, during the fill placement operation itself, to cure all three of the behavioural problems of a sand fill in a single operation.

The hypothesis which evolved from this reasoning and which was subsequently tested in the research documented here may be stated as follows: The act of dumping dredged sand underwater introduces energy into the previously placed fill. This energy is converted into pore pressures within the sand mound. The outward gradient potential from the sand mass into the surrounding water results in outward seepage. These seepage forces decrease slope stability, decrease the packing density of surficial particles as they are being deposited and also lessen resistance to erosion on the face. Negating excess pore pressures and reversing the direction of the seepage forces while the fill is being placed would lead to steepening of side slopes, improve packing density and also the sand's resistance to erosion. Figure 1 depicts the general concept. A mathematical model was constructed for the forces acting on a surficial particle and this is summarized in the Appendix.

The essence of the concept was that the seepage forces must be made to act inwardly concurrently with sand placement. Slope steepness must be attained during construction; it is obviously not possible to pull up flat slopes after they are placed. Also, it is at the moment when a sand particle is falling out of suspension and coming to rest on the surface of the fill that its packing density can most readily be improved. A surface particle is not burdened by the weight of fill above it, and consequently seepage forces are more likely to be able to pull it into the spaces between adjacent particles at that moment. When a particle is buried it can only move into closer packing at the expense of generating additional pore pressures, and this movement can only be forced by applying external effort. Finally, because inward seepage forces increase effective normal stress on surficial grains it was believed that this would also increase their resistance to erosion. As will be discussed later, the laboratory and model testing programs reported here led to some contradictions of this mechanistic understanding and forced some rethinking of the general concept.

INITIAL LABORATORY TESTING (CORK)

The initial laboratory testing was carried out at Cork during October and November 1983. The testing program described below was designed to determine if the direction and magnitude of seepage forces prevailing during sand desposition underwater could be correlated with both side slope steepness

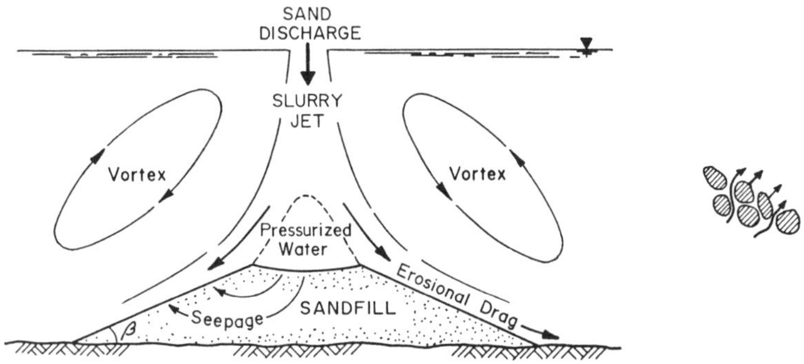

Destabilizing Water Forces Generated by Sand Placement.

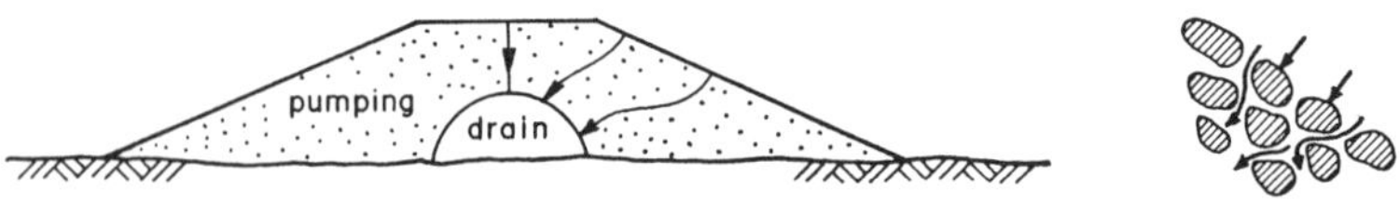

Beneficial Inward Seepage Forces Generated by Pumping.

FIG.1.- CONCEPTUAL MODEL USED AS BASIS FOR INITIAL HYPOTHESIS

and with packing density. The sand selected for use in the Cork tests was a fine beach sand of the gradation shown on Figure 2. The grains were mainly subangular quartz with some shell fragments. The natural angle of repose of this sand in air was 33.7°.

Model Islands

This test series consisted of building small submerged sand fills inside a glass-sided water tank to determine, in a simple visual way, if seepage flows could be used to increase or decrease the steepness of side slopes underwater. The apparatus shown on Figure 3 was constructed for that purpose. A total of 16 models, about 200 mm diameter, were examined. Four of the tests were carried out by pouring dry sand directly into the water; a sand-water slurry was used in the remaining 12 tests.

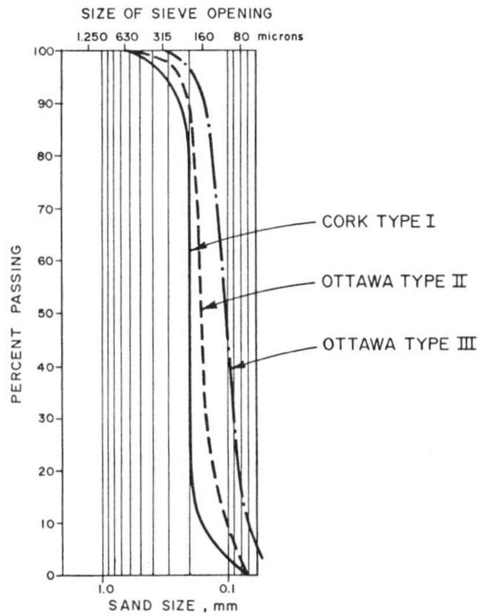

FIG.2.- GRADATION OF SANDS USED IN MODEL STUDIES

Prior to commencing sand placement, the base of the tank was covered with a layer of sand. A drainage element was embedded in this sand at the centre of the tank. The drain was dome shaped and was filtered against sand entry. The drain was connected by tubing to a back-pressure control vessel. By adjusting the water level at the brim of the back-pressure tank with respect to the water level within the modelling tank, a water pressure gradient could be exerted across the sand fill while it was being built.

The slurry was mixed by allowing dry sand to pour from a sand hopper into a small water reservoir. The slurry was siphoning into the modelling tank through a hose whose outlet was fixed in place just below water level at the centre of the tank. The slurry density and the rate of slurry discharge were controlled by the orifice size on the sand hopper, and by the difference in water levels between the mixing reservoir and the modelling tank. Four base line models were constructed without drainage, eight tests used inward seepage and four tests were subjected to outward seepage flows. In these latter tests only small incremental head differences could be exerted against the slopes without causing quicking of the top of the mound.

Figure 4 contains photographs of two sand fills which were formed with the aid of inward seepage flows. Figure 4a shows a mound built by discharging dry sand; the slope angle attained was 40°. Figure 4b shows a mound formed by a slurry discharge. The slurry imparted more energy to the water above the crown than did the dry sand; consequently, there was considerably more flattening of the upper sand surface as a result of the impinging jet. The asymmetrical shape of this particular model was caused by the jet being misaligned with the drain. The side slope closer to the drain is steeper (44°) than the far side slope (36°). This disparity is consistent with the hypothesis because the slope closer to the drain was subjected to a higher hydraulic gradient than the opposite slope.

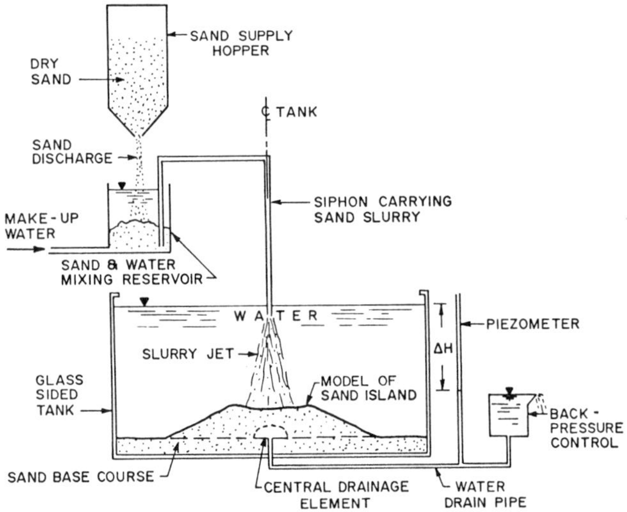

FIG.3.- MODEL ISLAND CONSTRUCTION APPARATUS USED AT CORK

In order to assess the results of the 16 tests on a comparative basis the plot shown in Figure 5 was made. Here, the side slope angles measured on the models are plotted against the ratio of the hydraulic head difference (H) across the sand fill and the radial distance (R) between the drain and the side slope. In a confined aquifer this ratio would be equivalent to hydraulic gradient, but due to the exposure of the top of the fills to open water, the actual gradient affecting the side slopes is less than indicated by the ratio of H/R. In this figure opposite slopes have been plotted separately. This data indicates a trend towards steeper side slopes with increasing magnitude of inward seepage. Flatter side slopes were measured when the direction of water flow through the sand was outwards. These test results appear to support the basic concept that side slope steepness could be modified during fill placement by imposing seepage forces on the growing mound.

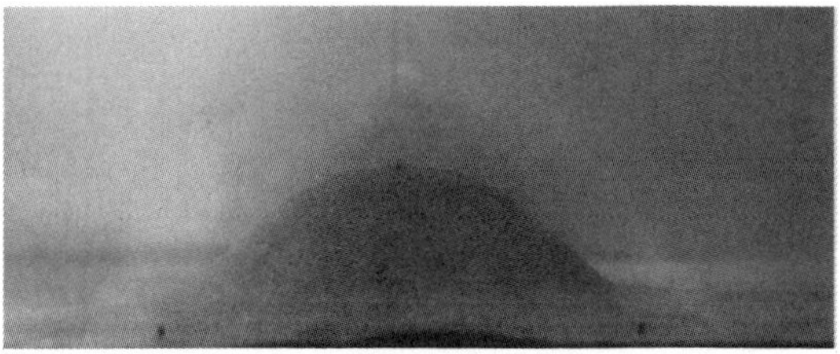

(a) Mound Formed by Dry Sand Pluviation into Water Tank.

(b) Assymetrical Mound Formation when Slurry Discharge Misaligned with Drain.

FIG.4.- SELECTED PHOTOGRAPHS OF MODEL ISLANDS TAKEN DURING CORK TESTING

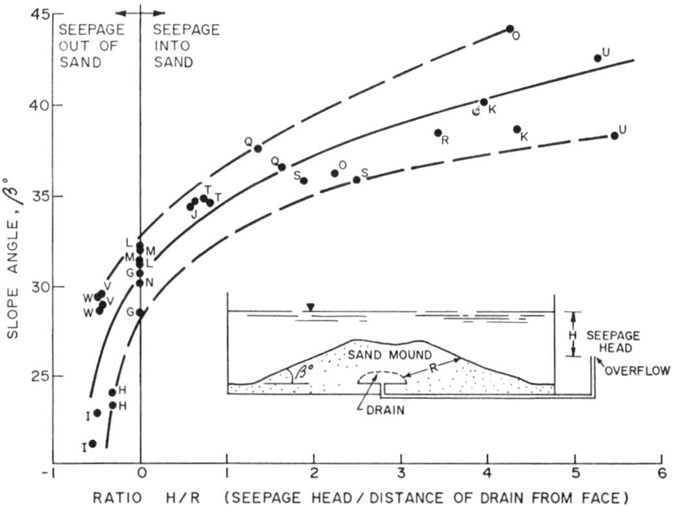

FIG.5.- RELATIONSHIP BETWEEN SLOPE ANGLE AND SEEPAGE BASED ON CORK MODEL ISLAND DATA

Density Tests

The apparatus shown schematically on Figure 6 was constructed in order to ascertain if the density of sand could be influenced by exerting seepage forces on the sand during the course of its deposition. Because seepage force per unit volume is directly proportional to hydraulic gradient, the test apparatus was designed so that the hydraulic gradient could be kept constant throughout each sand placement.

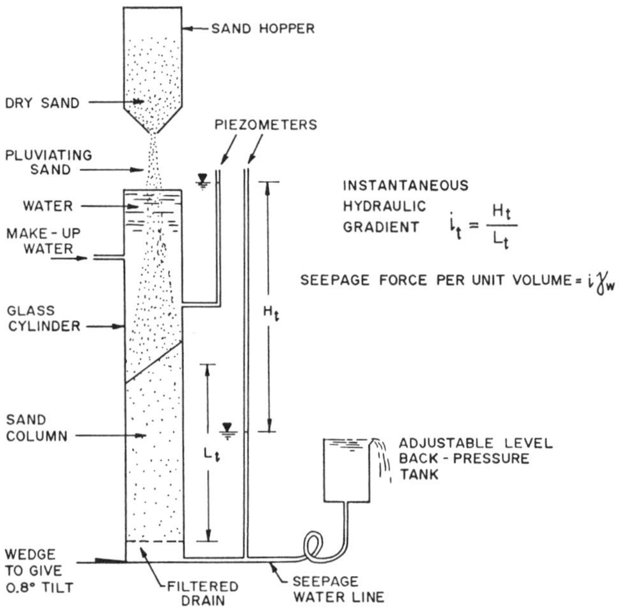

FIG.6.- DENSITY AND TILT TEST APPARATUS USED AT CORK

The test set-up was as follows: A glass cylinder measuring 850 mm long and 90 mm diameter was fitted with a filtered drain at its base. This cylinder was kept full of water. Each test consisted of discharging 1500 g of dry sand from an overhead hopper into the cylinder, under the influence of a constant seepage gradient. The seepage was developed by connecting the base drain to a back-pressure tank. By raising or lowering the back-pressure tank the hydraulic head differences acting across the sand column could be varied. The head difference was measured using two piezometers. The headwater peizometer measured the fluid density increase due to the sand in suspension; the tailwater piezometer accounted for hydraulic losses in the system. At any time (t) the hydraulic gradient (i) acting on the sand column is equal to the ratio of the difference in piezometer elevations (H_t) and the mean length of the sand column (L_t).

Each test was conducted by opening the orifice of the sand hopper and then continually adjusting the elevation of the back-pressure tank in order to maintain the appropriate differential head across the growing sand column. In this way, a sand deposit was formed which was at all times subjected to a predetermined constant hydraulic gradient, $i=H_t/L_t$. The resulting sand density was established by measuring the volume of sand in the cylinder at the end of the test (while the gradient was still imposed) and then weighing the entire sand mass after oven drying. Using this apparatus a total of 43 tests were carried out which produced data for three pluviation rates and several different seepage gradients. The sand discharge rates were 5.4 g/s, 2.8 g/s and 0.83 g/s and are referred to as fast, intermediate and slow, repectively. The hydraulic gradients ranged from 0.5 upwards to 10 downwards.

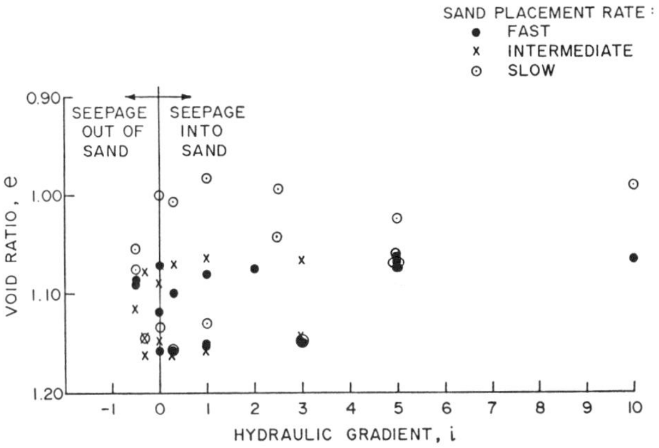

FIG.7.- RELATIONSHIP BETWEEN SAND DENSITY AND HYDRAULIC GRADIENT BASED ON CORK CYLINDER TEST DATA

The results of these tests are shown on Figure 7, where the dry density (in terms of void ratio) is plotted against hydraulic gradient. From this it may be seen that there is apparently no correlation between seepage gradient and sand density. This finding contradicts the initial concept. However, as will be discussed later, there is a redeeming feature in this data which led to a modified concept, and possibly to a more fundamental understanding of the behaviour of sand deposition.

Tilt Tests

The density test apparatus described above was intended to serve a double purpose. It had earlier been found that by tilting the glass cylinder slightly, the top of the depositing sand would form as an inclined surface. When sand discharges symmetrically into a cylinder, the sand surface formed during the course of placement is roughly horizontal. A slight tilting of the tube is sufficient to upset this balanced depositon. Any imbalance causes sand to build up on one side of the tube more than the other, thus producing a slope at the surface of the sand. This phenomenon is quite reproducible and yields a relatively planar surface.

In 26 of the density tests discussed above the cylinder was tilted 0.8° and the inclination of the resulting slope was measured while the gradient was still in effect. The slope steepness was found to vary, both with the rate of pluviation and with the seepage gradient being exerted across the sand column. These results are shown in Figure 8. For each of the pluviation rates it is possible to draw a curve through the data which suggests that, for the specific case of the very confined environment of the test cylinder, slope steepness is related to the rate of sand deposition. The results also indicate that up to a point, steeper side slopes are a function of hydraulic gradient. It should be noted here that slopes steeper than the angle of repose of the sand only remained stable while the inward seepage gradient was maintained. When the seepage was stopped the slopes ravelled back to the angle of repose.

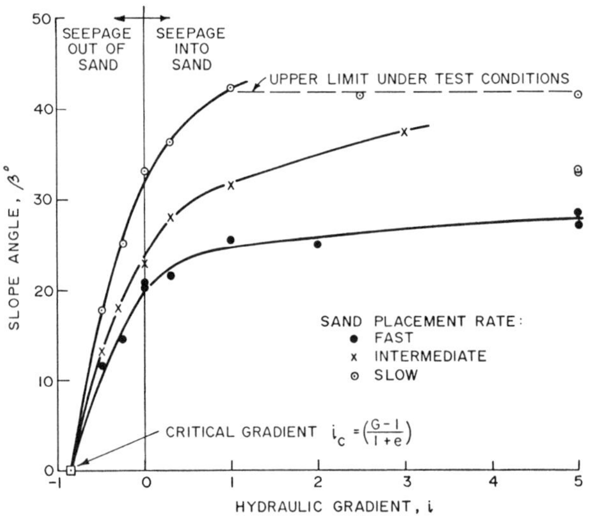

FIG.8.- RELATIONSHIP BETWEEN SLOPE ANGLE AND HYDRAULIC GRADIENT BASED ON CORK TILT TEST DATA

In these tests it was possible to introduce controlled upward seepage through the sand. This outward flow of water led to flatter slopes, to the extent that, as the gradient approached the critical (or quicking) gradient, the slope inclination approached the horizontal. The initial implication drawn from this data was that the flat side slopes normally associated with underwater sand fills is a result of excess pore pressures causing outward seepage forces. The flatter slopes observed for faster sand placement rates were attributed to higher pore pressures arising within the sand mass because of the greater energy associated with the more rapid deposition. This topic will be discussed further following presentation of the Ottawa test results.

Although the tilt test data relating slope steepness to hydraulic gradient is interesting in its own right, the more enlightening aspect of the data is related to sand density. During the course of the work it was noticed that there was a significant density difference between sands that were built on a steeply inclined surface and those that were deposited on a flatter surface. Consequently, the density data was plotted against the side slope angle measured at the surface of the sand at the end of each test. This plot is shown on Figure 9, where it may be seen that, in comparison with the random density relationship depicted on Figure 7, there is a discernible correlation between sand fill density and the slope of the sand surface. Examination of the stratification within the test columns indicated that the slope steepness remained virtually constant as the column of sand continued to build, therefore, the final surface slope was a good representation of the slope which prevailed throughout the test. Since the magnitude of the shear stress acting on surficial grains is a direct function of the surface slope inclination this data suggests the idea that the density of sand deposited underwater is to some degree dictated by the shear stress field into which the particles fall.

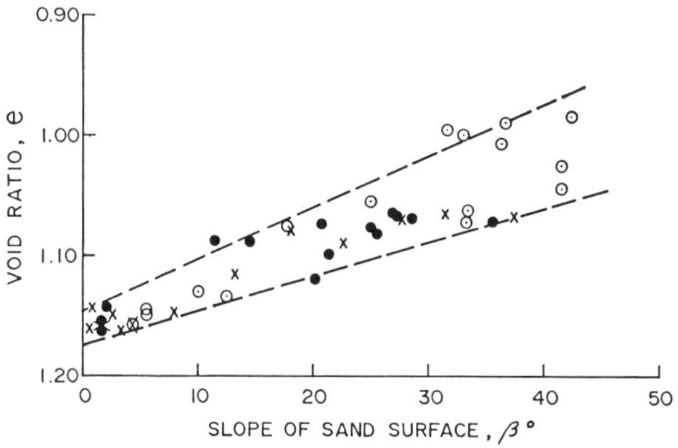

FIG.9.- RELATIONSHIP BETWEEN SAND DENSITY AND SLOPE ANGLE BASED ON CORK TILT TEST DATA

Tentative Conclusions

The simple small scale tests described above led to the following tentative conclusions:

The inclination of the side slopes forming a sand fill placed through water is influenced by the magnitude and direction of seepage forces acting on the sand mound at the time of sand deposition.

The density of sand packing is not influenced by seepage forces but appears to be related to the slope of the fill surface.

These statements are consistent with geotechnical theory. Since the sand/water interface is an equipotential surface, seepage forces act in a direction normal to the slope. Because seepage forces contribute to normal effective stresses they would be expected to influence slope stability. Sand fill density on the other hand, is not significantly improved by normal forces. Common practice is to compact granular soils by means of vibratory loading, which is essentially a matter of subjecting the sand structure to cyclic shear forces.

CONFIRMATORY MODEL TESTING (OTTAWA)

The model work outlined below was carried out during the summer of 1985 in the Hydraulics laboratories of NRC in Ottawa. Apart from NRC's funding requirement for "arms-length" verification testing of the initial Phoenix research, there were several other reasons for undertaking this work. It was necessary to establish that the steeper underwater slopes obtained in Cork could be reproduced using a different sand and a larger model size. In other words, it was necessary to free the initial conclusions from doubts associated with those small scale models and any peculiarities of the sand type used. The earlier model work used a central drain to cause inward seepage; this method could not be used in the field. It was necessary, therefore, to find out if similar beneficial results could be obtained using a ring of vertical wells such as might be used in a field application. Also, the fact that there appeared to be no correlation between density and seepage forces (normal stresses) warranted further investigation using a larger cylinder and a slurried sand fill. Finally, some testing was needed to see if sand erodibility could be modified using seepage forces. With these objectives in mind, Phoenix drew up a testing program which was subsequently approved by NRC. Phoenix designed and fabricated the testing equipment and apparatus necessary to pursue this program and then turned the testing over to Golder Associates' engineers who carried out the work.

The Ottawa program was divided into three separate series of tests to assess the effect of inward seepage forces on sand fills:

A. Construction of models to measure side slope angles and penetration resistance.
B. Direct measurement of sand density.
C. Wave flume tests to measure sand erodibility.

The sand fills tested at NRC were formed using a sand slurry placed through still water. Each sand fill was subjected to a constant value of hydraulic head differential during placement. The "islands" were totally submerged and did not break the water surface even during wave flume testing. Two types of sand were selected for testing and the gradations of those sands are shown on Figure 2. The silt in the Type III sand made visual observation very difficult and consequently the main body of work was done using Type II which is a clean subangular quartz sand.

Series A - Model Islands

The apparatus and testing arrangement is shown schematically on Figure 10. An existing 3.0 m wide and 2.7 m deep flume was divided into two compartments separated by a 0.8 m high wall. The smaller compartment was used for sand storage; the larger one was used to build the model islands. Both compartments were filled with water. Sand fill was pumped as a dense slurry (about 30% sand by volume) into the larger compartment using an air-activated, double-diaphragm pump and the suction/discharge arrangement shown on Figure 10. The pumping rate was about 2 l/s of fluid. The seepage apparatus consisted of a circular header to which vertical wells were attached. Three different header sizes were used with nominal ring diameters of 0.5 m, 0.7 m and 0.9 m. The wells consisted of 400 mm long, 13 mm tubing with the lower 150 mm perforated and covered by a 140 micron steel mesh. Two suction lines were attached to the 60 mm header pipe and connected to a back-pressure tank through a valve assembly. With this arrangement, water was drawn into the wells by the siphon effect created when the back-pressure tank was lowered beneath the water level of the main tank.

The model construction was carried out in the following manner: A 150 mm thick base course of sand was placed in the main tank and the surface was leveled. The seepage header and wells were lowered onto the sand surface and pushed into the sand bed to bury the perforated length of the wells. The tank was filled with water to a level of 230 mm below the top. The nozzle of the discharge pipe was fixed at 50 mm below water level in a position directly above the central axis of the header. Seepage flow was commenced by lowering the back-pressure tank to produce the desired differential head. Sand slurry was then pumped from the stockpile tank into the water above the seepage ring. As the submerged sand mound formed, the sand level at the line of wells increased leading to deeper burial of the perforations. As soon as the perforations were buried by about 100 mm, the header was raised by means of a hoist, until the sand cover above the top of the well screen was reduced to about 50 mm. During the period while the ring was being lifted, the slurry discharge was interrupted. This sequence of discharging slurry and then incrementally raising the seepage ring was continued until the surface of the sand mound was within about 0.3 m of the discharge nozzle. At this point, the sand filling was terminated but the seepage flow was maintained. Finished models were 1.6 m to 2.4 m in diameter and from 0.4 m to 6 m high. A total of 20 models were constructed in this way, seventeen using the clean 170 micron sand (Type II) and three models using the 105 micron silty sand (Type III). Of the 17 tests conducted with the clean sand, 11 (standardized) tests were performed using the medium size header ring, with all wells in place and constant suction. In the remaining six tests, procedural variations were introduced such as changing the header size, removing some wells or interrupting seepage flow.

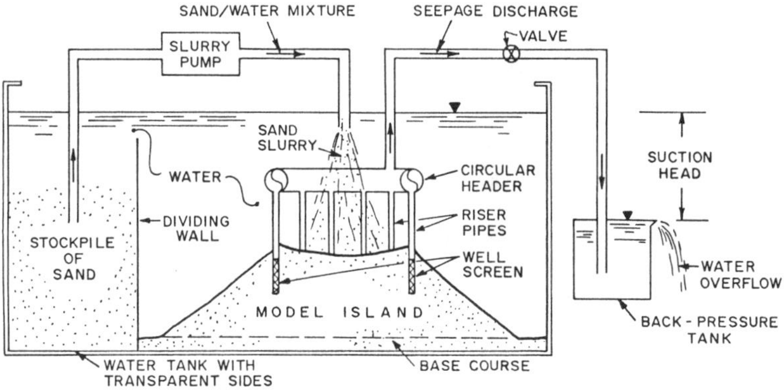

FIG.10.- MODEL ISLAND CONSTRUCTION APPARATUS USED AT OTTAWA

The photographs on Figure 11 show the Phoenix seepage ring and three stages of model building. In the early stage, sand can be seen accumulating in an annular ring coincident with the line of wells. The ripples on the sand bottom outside the fill indicate that there is significant energy being carried by the slurry jet. Both the inside and outside slopes of the annulus of sand are steep. The intermediate stage shows the sand slurry falling on a mound which is growing symmetrically beneath the discharge nozzle. At the end of construction the steep side slopes typical of this process are obvious.

Slope Angle

With the suction still maintained, physical measurements of the sand mounds were taken along two orthogonal sections. The results for a typical model are shown on Figure 12. In this case it may be seen that the sections of the slope close to the wells are steeper than 40°. Such unnaturally steep slopes could only stand while suction was being applied to the fill; once the seepage was cut off the outside slopes gradually ravelled back to about 29°. The silty sand responded in substantially the same way as the coarser, cleaner sand with sections of the slopes being steeper than 40°.

(a) Drainage Header and Wells

(b) Early Stage of Construction

(c) Slurry Discharge

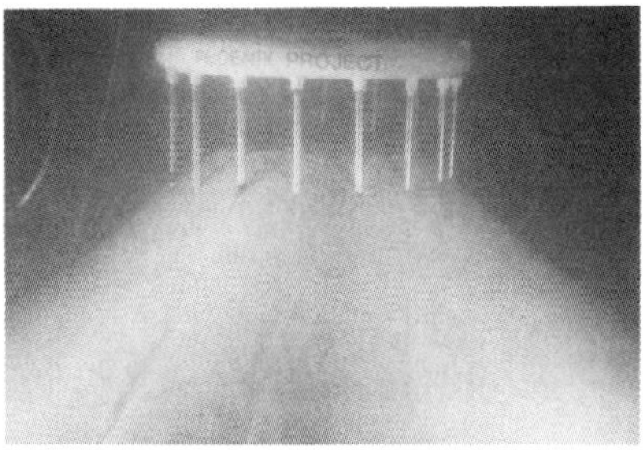

(d) Completed Underwater Model

FIG.11.- PHOTOGRAPHS OF MODEL ISLAND APPARATUS AND CONSTRUCTION PROCESS TAKEN DURING OTTAWA TESTING

The results of the 11 standardized tests are summarized on Figure 13, where average side slope steepness is plotted against the differential head applied across the seepage system. The observations made during this series of tests may be summarized as follows:

a) Side slope steepness during construction can be increased by applying a suction to the sand fill. This method works for both of the sand sizes tested.
b) Flat slopes such as observed in nature could not be produced in the laboratory at this scale.

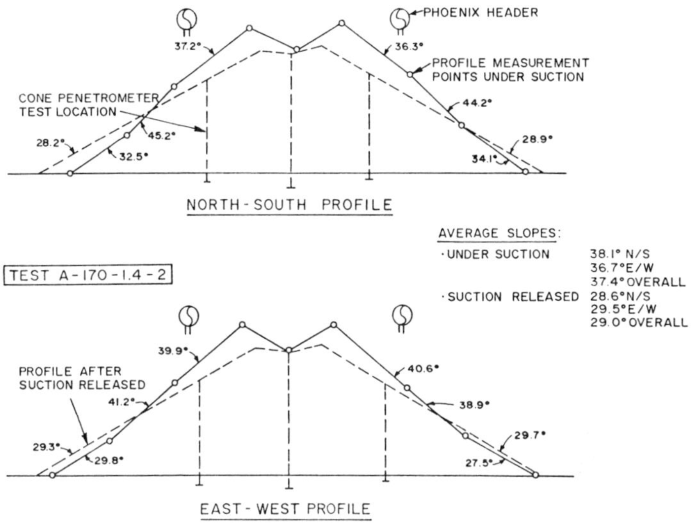

FIG.12- TYPICAL MODEL ISLAND GEOMETRY FROM OTTAWA TESTING

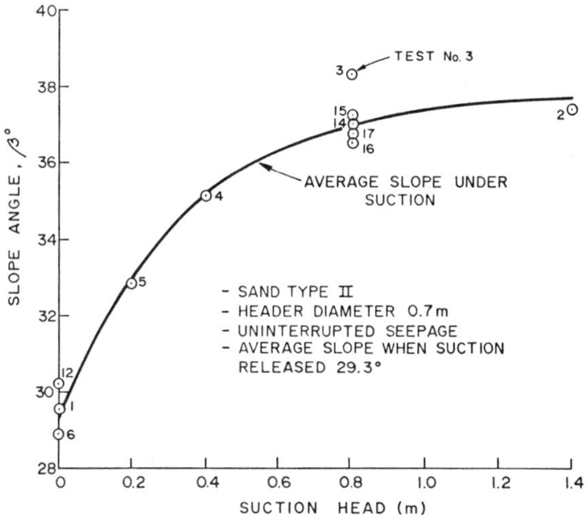

FIG.13.- RELATIONSHIP BETWEEN SLOPE ANGLE AND SUCTION HEAD BASED ON OTTAWA MODEL ISLAND DATA

c) The steepest slopes were closer to the top of the mound and nearer to the influence of the wells.
d) Beyond a head differential of about 0.8 m there was no appreciable improvement in slope steepness for this size model.

An incidential observation was made while hoisting the seepage ring. The force required to lift the header was very much greater when the suction was still active than when water flow was cut off. This additional force is due to the anchoring effect produced by the increased effective lateral stress acting around the wells.

Penetration Resistance

An attempt was made to evaluate the packing density of the sand fill in each of the models. The best method to determine the average density would have been to measure the sand volume accurately and then remove the entire volume of sand, then weigh it after drying. This was not done because of time and cost constraints. As an alternative it was decided to do miniature cone penetration tests. A right cylindrical probe with a 60° apex cone and 280 mm^2 cross sectional area was constructed for this purpose. Prior to doing the penetration tests the seepage flow was stopped and the side slopes allowed to stabilize. The probe was then pushed into the sand fill at five locations, as shown on Figure 12; one in the centre and four on the side slopes near the island rim.

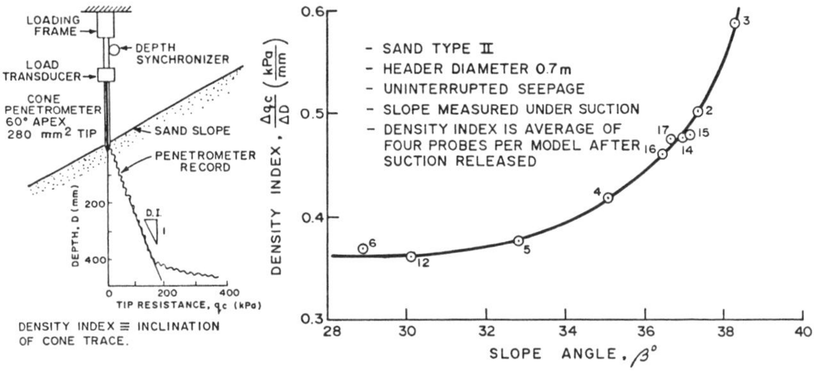

FIG.14.- RELATIONSHIP BETWEEN DENSITY INDEX AND SLOPE ANGLE BASED ON OTTAWA MODEL ISLAND DATA

The resistance to penetration was recorded using an electronic load cell. This method gave a qualitative indication of relative fill density for different models, but it could not be directly related to absolute values of density because of the lack of confinement close to the slopes. The sounding profiles showed a linear increase of resistance with depth and it was decided to use the slope of this profile i.e., the rate at which penetration resistance increased, as the measure of packing density.

The results of the probe measurements are summarized on Figure 14 for 10 of the 11 standardized test mounds shown on Figure 13. The probe malfunctioned during Test No. 1 and this data was therefore deleted. The penetration resistance for the average of the four individual soundings per model is plotted against the average side slope angle for that test. It should be noted that slope angles were measured while the seepage flow was occurring but the penetration tests were made after the seepage was cut off. The data points show a strong correlation between penetration resistance and slope angle. This suggests that sand density is a function of the construction slope angle as was previously suspected at Cork. A more surprising result is that the penetration tests carried out in the centre of the islands showed a decrease in resistance with increasing seepage force. This is directly contrary to what was expected and this aspect will be discussed later.

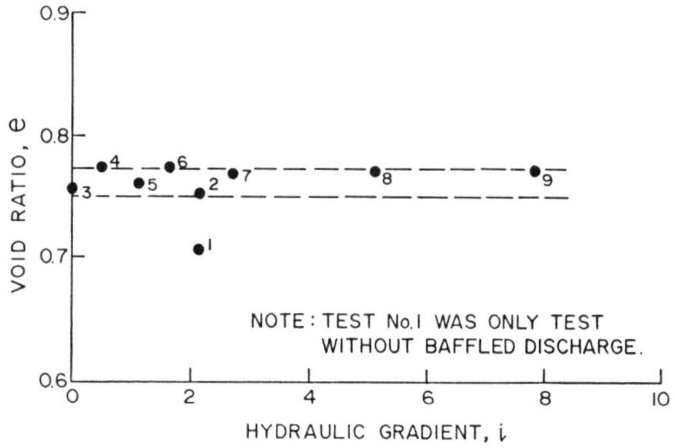

FIG.15.- RELATIONSHIP BETWEEN SAND DENSITY AND HYDRAULIC GRADIENT BASED ON OTTAWA DENSITY TEST DATA

Series B - Density Tests

The testing carried out at Cork had indicated that the density of sand fill was not improved by being placed under the influence of inward seepage flow. The following tests were performed in Ottawa at larger scale to see if the earlier results were valid and not just an artifact of the equipment and scale size. A 250 mm diameter, 370 mm high cylinder with a filtered drain in the base was used. This cylinder was submerged in the model tank and sand slurry was discharged into it in the same manner as was used to construct the model sand mounds. A baffle was used between the discharge nozzle and the top of the cylinder in all but one test. The sand was observed to form by depositing on a flat surface. For each of the ten tests performed, a fixed differential head was used throughout the filling process. After filling, the seepage flow was cut off and the cylinder was recovered from the tank. The surface was struck off and the known volume of sand was oven-dried and weighed. The results are summarized on Figure 15. It is evident from this data that placement of sand under an inward seepage force has, in itself, no effect upon sand fill density.

Series C - Wave Flume Experiments

This testing was undertaken to determine to what extent inward seepage would improve the resistance of sand fill to erosion. The test arrangement was designed to measure sediment transport during the passage of a wave train across a submerged sand mound, rather than the effects of wave run-up on a beach.

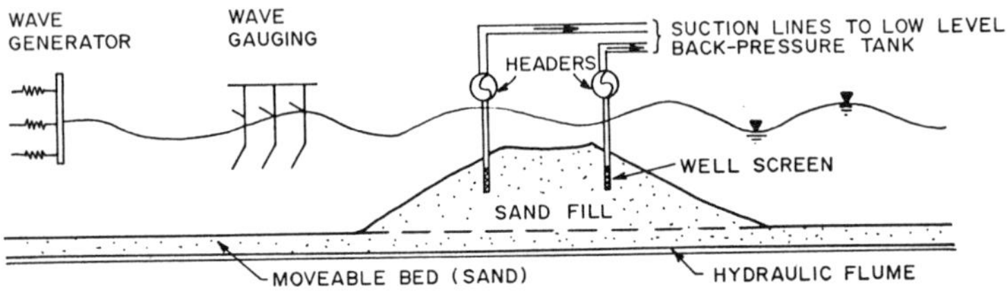

FIG.16.- WAVE FLUME APPARATUS USED AT OTTAWA

The general test arrangement is shown schematically on Figure 16. The central piece of equipment was the 1.3 m wide Ottawa Wave Flume, where it was possible to generate and monitor a wide spectrum of wave forms. Sand fills were built inside the flume under still water conditions using the same method as was used for the Series A testing. The fill sections in this case were of constant section across the width of the flume, rather than circular, as in Series A, i.e., an embankment shape rather than an island. In order to form this shape, the discharge nozzle was moved back and forth across the flume between two parallel lines of wells. The fill was brought to within 150 mm of the water surface. Following construction of the sand fills the suction (in cases where suction was used) was shut off to allow the fill to recline to its angle of respose ; this ensured a constant embankment shape irrespective of the suction imposed during construction. Transducers were then embedded in the fill at various depths to monitor pore pressures as a function of wave action. The wave climate suggested by NRC Hydraulics engineers and adopted by Phoenix, was a 110 mm high wave of 1.46 second period with a two hour storm duration. The experiments were recorded by a video camera which viewed the sand fill and wave passage through a window in the flume wall. A total of 12 tests were performed in this manner and the results are summarized in Table 1. Because some of the fills liquefied under wave loading the results will be discussed under two separate headings.

Liquefaction

The first two tests (No. 1 and No. 2) were built with the well system in place but without applying suction, either during construction or during wave loading. These tests were intended to provide base data against which to measure the amount of improvement that seepage forces would provide to the surficial sand fill. As it turned out these "untreated" fills liquefied in a most dramatic fashion within seconds of being subjected to the design wave. The video records of these events, where the upper third of the embankments turned to a fluid and collapsed is a remarkable illustration of sand liquefaction. Following the initial loss of mass the remaining fill actually could be seen to densify under the influence of the cyclic stresses imposed by the passing waves.

These video records together with the continuously monitored peizometric pressures provide a mass of data which may form the basis of a subsequent paper. Figure 17 is an example of these records.

TABLE 1. - NRC WAVE FLUME TESTING SUMMARY

Test No.	Header Spacing (m)	Suction Applied (m)		Comments
		Construction	Erosion	
1	0.80	0	0	Liquefied
2	0.80	0	0	Liquefied
3	0.80	1.0	1.0	Stable
4	0.80	0	0	Stable under 20,40,60, 80,100 mm sequence. Later liquified under 100 mm waves.
5	0.80	0.5	0.5	Stable
6	0.80	0	0	Stable under increasing wave heights.
7	0.35	1.0	1.0	Stable
8	0.35	1.0	1.0	Stable
9	0.35	1.0	0.1	Stable
10	0.35	1.0	0.5	Stable
11	0.80	1.0	0.5,0.4,0.3, 0.2,0.1,0.05, and 0.	Stable while suction incrementally reduced to zero.
12	0.35	1.0	0.5 and 0	Stable while suction reduced to zero.

Test No. 4 also was built and tested without the benefit of applying suction to the fill. In this case, however, the wave height was gradually increased from 20 mm to the full test height (100 mm) in increments of 20 mm in an attempt to determine the wave height which would initiate liquefaction. The fill did not liquefy under this sequence. After allowing the water to become calm, the fill was then subjected to the full design wave, but this time the wave was generated in a single increment and liquefaction did occur. Test No. 6 was conducted in the same manner as Test No. 4 with the wave height being gradually increased from 20 mm to 100 mm. Liquefaction did not occur and the test was carried to completion to measure erosion. Liquefaction did not occur in any of the tests where the fill was built under a suction head, even when the suction under wave loading was marginal, as in Test No. 9 (0.1 m), or gradually reduced to zero as in Tests No. 11 and No. 12. These results suggest that the Phoenix process produces a sand structure that is sufficiently dense to withstand failure under cyclic loading of a reasonably large magnitude.

Erosion

The amount of erosion that took place during the course of each test was measured by taking profiles of the sand in the vicinity of the mound, both before and after the passage of the wave train. The main test variables were the suction heads applied during construction and during wave loading, the spacing of the suction header manifolds, and the depth of burial of the well screens. The best set of comparative data is from Test No's. 7, 8, 9, 10, and 12 where the suction head used to build the mounds was 1 m and the header spacing was 0.35 m. This data is shown on Figure 18, where the erosion is quantified in terms of the volume of sand lost from the mound and in terms of the depth of sand scoured. These parameters are plotted against the suction head that was exerted during exposure to the test wave climate. It is apparent from both these measures that inward seepage has a significant effect on the reduction of erosion.

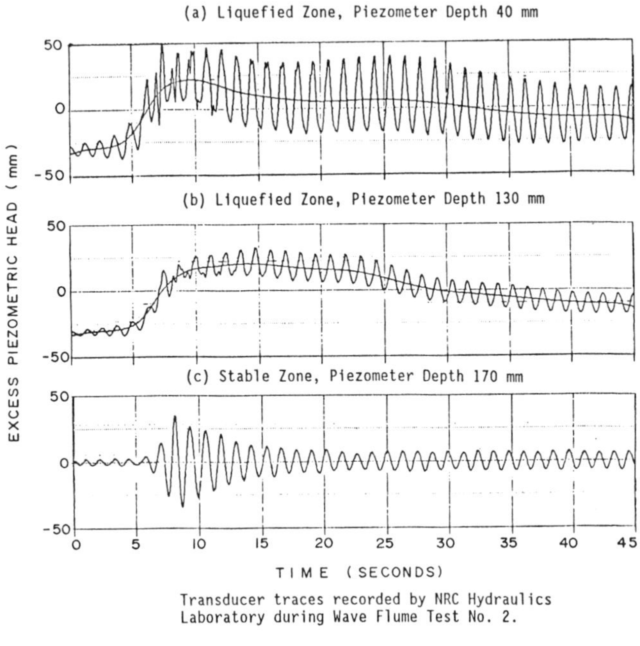

FIG.17.- SELECTED PORE PRESSURE RECORDS FROM OTTAWA WAVE FLUME TESTS

DISCUSSION AND INTERPRETATION OF TEST RESULTS

Slope Steepness

The fact that side slopes can be built more steeply underwater with the aid of inward seepage seems to have been established in the model studies. What has also been demonstrated by the models (excluding the Tilt Tests) is that very flat slopes, such as those recorded in the field, cannot be reproduced in the laboratory. This requires some explanation. During the early stages of this work, the author adopted a pseudo-static approach by considering that the side slopes achieved during fill placement underwater were simply a matter of slope stability, where the limiting slope was a direct function of the level of pore pressure within the sand mass. By this reasoning, very large pore pressures would lead to very flat slopes; the extreme case of horizontal slopes being consistent with pore pressures equivalent to the critical hydraulic gradient. In qualitative terms, it seemed reasonable to assume that high pore pressures would exist within a sand mass due to the dynamics of sand placement. It can be shown by such mechanisms as the Impulse-Momentum Theory that the energy introduced into the vicinity of a sand mound by the dredge discharge and the falling slurry is converted into water pressure within the mass. The higher pore water pressures beneath the slurry discharge would result in seepage flow away from the centre of the fill towards the outside slope, thus reducing stability and slope inclinations. In quantitative terms, the mathematical model summarized in the Appendix was useful in understanding the laboratory test results. This model should also be useful in accounting for the very flat slope angles observed in numerous civil-marine sand fill structures. But, in practical terms, it is not possible to justify the high pore pressure levels which would be necessary to cause slopes of the order of 5° in large fills. The energy required does not seem to be available in most instances - a point which was borne out by the small laboratory models themselves. As discussed below, the answer to this physical phenomenon is now believed to be related more to sand packing density than to slope stability.

Sand Density

The density testing to date consistently supports the conclusion that subjecting a sand to inward seepage forces does not in itself produce a denser packing. Seepage forces act in the direction of the water movement, i.e., parallel to the flow lines. The surface of the fill is an equipotential line and consequently the flow lines are normal to the sand/water interface as is shown on Figure 1. Therefore, the effect of seepage forces on a submerged sand is to increase the normal stresses on the sand without changing the shear stresses. Viewed thus, it is not altogether surprising that seepage forces do not improve packing density, since we know from experience that, although normal stresses are effective in consolidating cohesive deposits, it is necessary to introduce shear stress energy into a non-cohesive mass to decrease void ratio by physically moving grains into a closer packing arrangement.

If the general statement above is correct, why were improved densities recorded in some tests reported here? The answer appears to be surprisingly simple. When a sand grain settles on the side of the fill during construction, it comes to rest on a slope. Because it is on a slope, it is in a shear stress regime. The mere fact that the sand grain is on a sloping face means that it is subjected to a shear stress. The steeper the slope the greater the shear stress. The slope steepness is controlled by seepage force; the shear stress is proportional to slope steepness; and the density is a function of shear stress. Therefore, it is only indirectly that seepage forces improve density and this density increase is only to be expected in the vicinity of the slope.

It is important to note that the permanent improvement in densities obtainable close to a sloping surface is only achievable during the period of construction. In an existing fill, individual sand particles at depth are burdened down by the weight of the soil overlying them. In order to move into a closer packing, a considerable amount of energy is required to overcome the frictional and kinematic resistance to physical relocation. In fact, if this were attempted in a loose submerged sand, the energy input required to densify a deposit at depth could result in a liquefaction failure of the sand mass. This undesirable aspect of densifying an existing submerged sand is not present in the situation discussed here because closer packing is achieved on a continuing basis during sand placement. As each grain falls onto the surface of the fill it rolls into a packing that depends on the slope steepness. It is then held in place by its own frictional resistance to sliding. The ability of the grains to move into closer packing is not inhibited by any oveburden load. Also, there is no deleterious generation of entrapped pore pressures, since the particle is still at the fill surface, i.e., the drainage path is zero length. From this line of reasoning, it appears that the most efficient and effective time to improve the density of a sand fill is while it is being placed. To some extent this may be accomplished simply by depositing it on a sloping surface.

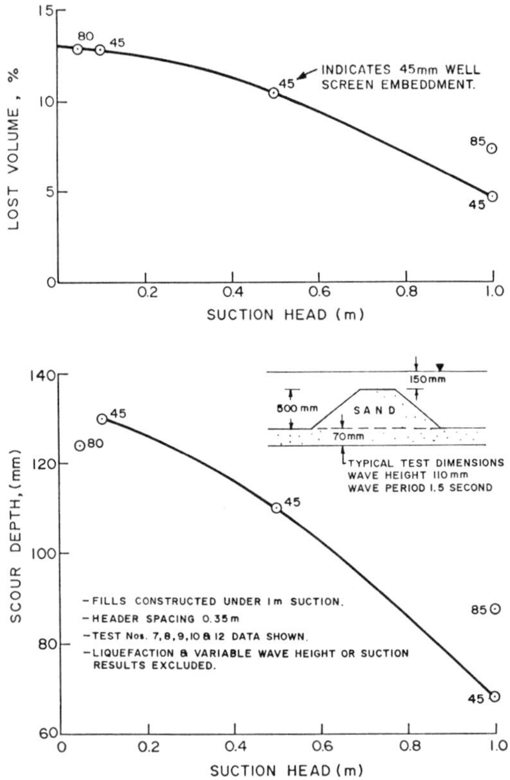

FIG.18.- RELATIONSHIP BETWEEN EROSION AND SUCTION HEAD BASED ON OTTAWA WAVE FLUME DATA

The cone penetration work carried out on the Ottawa models showed that when suction is used, the inside part of the fill was looser than the perimeter. The greater the suction and, consequently, the denser the outer ring, the looser was the interior. This apparent anomaly is consistent with the author's experience of some sand fills, where it seems that the loosest fills are formed when sand is dredged into a confined space within rigid (unyielding) side walls.

It can only be speculated that this may be because sand, when deposited within a confined area, tends to accumulate under a relatively flat surface where it would be unable to attain the closer packing brought about by sedimentation on a slope.

Erosion Resistance

The benefits of suction in increasing resistance to erosion is likely to be readily apparent to geotechnical engineers. The inward flow of water results in an increase in the normal effective stresses acting on the surficial sand grains. This effect can be visualized either as an increase in particle "weight" or as an "apparent cohesion" affect at the sand/water interface. This increases the resistance of surface particles to movements stimulated by the drag forces associated with external water movements.

CONCLUSIONS

Both the Cork and Ottawa results suggest that pumping water from within a submerged sand fill during fill placement has the following beneficial results:

i) It causes an increase in particle packing in the neighbourhood of the slopes. The density improvement achieved by this process is a permanent change in the sand structure.

ii) It increases the steepness of the side slopes of the fill. Side slopes steeper than the sand's angle of repose will stand while the suction is being applied, but will ravel back to about 30° when the seepage flow is stopped.

iii) It reduces erodibility of the sand surface. The enhanced erosion resistance remains in effect only while water is being pumped from within the sand mass.

The most enlightening results at both laboratories were unexpected and were by-products of tests done for other purposes. At Cork, a correlation between sand density and depositional slope angle or shear stress was found. At Ottawa, liquefaction of "untreated" sand fills under wave loading was observed. Although these results need considerably more verification before accepting them completely, they do provide the basis for a more credible and coherent explanation of the physical factors that affect submerged sand fills.

The starting hypothesis was that unsatisfactory sand fill characteristics could be attributed to high pore pressures generated within the fill as a result of the energy introduced into the system by the dumping process. This idea was changed during the course of the work to the following working hypothesis which will form the basis of the next stage of research and development:

a) Sand fills constructed by dumping through water form a loose mass which may initially stand at slopes approaching the angle of repose. Natural forces, such as wave loading or slope steepening by erosion, produce shear stresses within the mass which lead to liquefaction type failures of the fill. The very flat side slopes normally associated with these fills are a result of liquefaction of a loose sand mass.

b) The engineering characteristics of the fill can be improved by pumping water out of the fill during construction. Pumping causes a reduction in pore water pressure that serves two purposes, viz., it eliminates any excess pore pressure within the mass, then upon further depression, it leads to negative pore pressures which induce stabilizing, inwardly acting, seepage forces. The inward seepage forces cause steeper side slopes to build during fill placement. Sand deposited on these steeper side slopes naturally adopts a denser packing because of the shear stresses which exist in the vicinity of such slopes. The density improvement, which is a permanent change in the sand structure, is sufficient to resist liquefaction type failures under subsequent natural loading.

In practice, this method would be used to create a sand fill of adequate density to resist liquefaction of the mass and, incidentally, to produce side slopes of more acceptable steepness. Pumping would be continued until after slope protection material was placed, and possibly, later reactivated as an added measure of safety during severe storms.

Undoubtedly, there is need for further large scale model research work to verify and clarify the concepts presented here. However, the next stage of development planned will be to take the work out of the controlled environment of a laboratory and test it in the field. A relatively large test section will be constructed in a lake or reservoir. At this scale it will be possible to make a quantitative assessment of density using standard insitu techniques. The effect of variations in gradation and silt content will be encountered, although the testing to date suggests that these factors are not of fundamental importance other than in affecting the quantity of water which needs to be pumped. Some development work has already been done on the field equipment required to circulate water through a submerged fill during construction. The field trials will check the functional adequacy and ease of deployment of this system.

ACKNOWLEDGEMENTS

The author wishes to express his gratitude to the following:

Dennis Diggle of Foundex Explorations Ltd., for his moral support through the early research period and for his practical contributions to the subsequent development of the prototype equipment.

Professor Eamonn Dillon of the National University of Ireland, for his infectious engineering enthusiasm and for providing the laboratory facilities in which the Cork testing was carried out.

Nigel Skermer, for his help and comments during the course of the work and also for his criticism of this document.

Herb Hawson, for his promotional zeal and fund raising efforts.

The National Research Council of Canada for the use of their Hydraulics laboratory and facilities.

APPENDIX

MATHEMATICAL MODEL OF FORCE EQUILBRIUM FOR SURFACE PARTICLE RESTING ON THE SLOPE OF A SUBMERGED SAND MASS DURING FILL PLACEMENT AND PUMPING OPERATION.

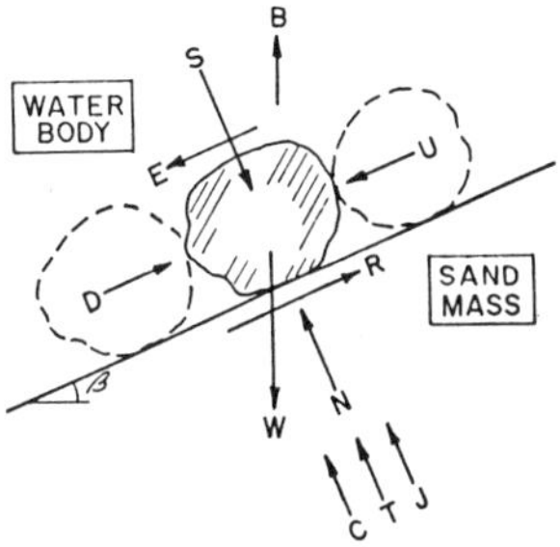

FORCE ON PARTICLE

- B bouyancy
- C excess pore pressure generated by consolidation
- D downslope particle support
- E erosional drag
- J excess pore pressure generated by jet deflection
- N normal intergranular
- R resultant shear
- S seepage from pumping
- T excess pore pressure generated by shear contraction
- U upslope particle push
- W weight

Assuming Limit Equilibrium of Spherical Particle

$$\left\{\cos\beta + (i_s - i_c - i_t)\left(\frac{1+e}{G-1}\right)\right\} \tan\phi = \sin\beta + \left(\frac{v}{v_t}\right)^2 J$$

where i_s is the applied inward seepage gradient

$$i_c = \frac{\Delta h}{\Delta t} \left\{\left(\frac{G-1}{1+e}\right) + i_s\right\} \frac{h}{C_v}$$

$$i_t = \frac{\Delta h}{\Delta t} \left\{\frac{e_{max} - e}{e_{max} + 1}\right\} / k$$

β	slope angle
e	void ratio of sand compatible with shear stress regime
G	specific gravity of sand
ϕ	angle of internal frictional resistance of sand
v	erosional velocity against particle
v_t	terminal velocity of particle in water
J	exposure factor to erosional drag
$\frac{\Delta h}{\Delta t}$	rate of sand placement
h	height of sand column subject to seepage gradient
Cv	coefficient of consolidation of sand mass
$e_{max,min}$	void ratios, loosest and densest packing density
k	coefficient of permeability

TAILINGS MANAGEMENT IN AUSTRALIA

Doug Cooper M.ASCE*

ABSTRACT: The disposal and management of tailings is assuming greater importance as stricter environmental controls are imposed on mining operations. Requirements to reduce operating costs and maximize water return are additional pressures. Despite the many advances in mineral processing and mining methods, the traditional means of creating an environment whereby peripheral embankments can be raised by upstream construction remain essentially the same as those in use at the turn of the century. The author discusses the development of those methods and the application to modern tailings management techniques.

INTRODUCTION

Mining commenced in Australia in the mid 1800's and today has grown to be one of the country's major industries. Australia has an abundance of virtually all the minerals and precious gems and, as such, has seen a rapid expansion in mining activity over the last 100 or so years. With the development of mines over virtually all parts of the country came mineral processing and the inevitable production of waste, including low grade (sub-economic) ore and tailings. This paper describes some of the history of tailings management in Australia, and suggests some future directions that the industry may take and the part that the civil/geotechnical engineer may play. As gold mining currently accounts for over 80% of the active mining operations throughout Australia with a large percentage occurring in Western Australia, the author will concentrate on this aspect of the industry.

HISTORICAL BACKGROUND

The earliest recorded production of gold mine tailings dates back to 1870 when small scale mining and processing commenced. Crushing and milling processes were crude and tailings tended to be coarse and free-draining. Small daily production tonnages resulted in low volumes of slurry and apparently few operational problems were encountered. The occasional spill of tailings was not of great concern. Most mining operations were carried out only when ample water was available and little effort was made to recycle water.

As mining methods improved with the introduction of efficient drill and blast techniques, the tonnages of material treated increased. Mining became a continuous operation not dependent upon seasonal rainfall. Permanent towns were developed in remote, arid areas and the need to conserve water from the mining operations became important. Tailings management became an essential part of the overall mining and ore treatment operations.

EARLY MANAGEMENT TECHNIQUES

The vast majority of mining operations in Australia were, and still are, located in the inland arid parts of the country where the topography tends to be virtually featureless. Little opportunity existed for valley type storages and, hence, most tailings storages were developed as turkey nest dams, ie the storages were raised with embankments on all sides.

* Principal, Australian Groundwater Consultants Pty Limited, 300 Albany Highway, Victoria Park, Western Australia, 6100, Australia.

There is little record of engineers being involved in the design, construction and management of tailings storages during these early years, however, despite this, there were some remarkable structures built.

It is believed that the skills in managing tailings were introduced by miners from South Africa and the USA where large scale mining preceded that in Australia by some years. Early mines were very selective and little waste material from which to construct embankments was produced. The generally arid weather conditions allowed the adoption of spigot deposition methods. Peripheral embankments were economically and safely raised using tailings.

Although records and photographs are scarce, remnants of early mining operations show that the storages were very stable. The author has been involved in the remining of several very old (turn of the century) dumps where the outer faces have been as steep as five vertically to one horizontally. Cutting into the faces has shown that very basic systems of slurry distribution have created a stable outer section to the dumps comprising free draining and well consolidated, coarse tailings.

Figure 1 shows the main gold mining areas in Western Australia.

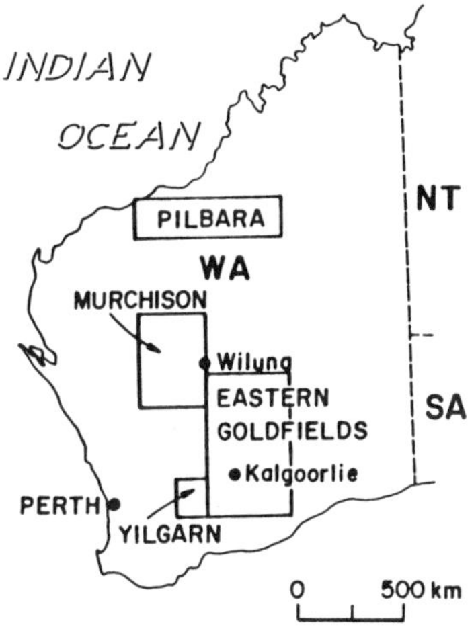

FIG 1. Goldfields of Western Australia

The management of the early storages followed two basic systems. Earlier, when the ore grind was relatively coarse, a single point discharge beaching system was used where the tailings slurry was brought up to the current surface in a single pipe and the flow directed across the surface using bags filled with dry tailings. By careful placement of sandbags, the slurry was switched from side to side over the surface of the storage. Perimeter embankments were constructed using the upstream technique in lifts of about 150 mm by excavating drying tailings using shovels. Released water was allowed to decant through a simple overflow pipe (wood stave in the earliest cases) protected by sandbags. The resultant surface of the storage was a gently sloping, concave surface with the high point at the discharge pipe (Figure 2).

The large Wiluna tailings dump is typical of those created in this manner. Over 6 Mt of tailings were placed in a mine life which covered a period of 30 years.

The climate at Wiluna is typical of inland Australia. Average annual rainfall is 200 mm whilst the annual evaporation is over 3 200 mm. Although very intense rainfall does occur, the total volume of storm inflow is generally small. Heavy rainfall in the form of short storms falls mainly in the summer months and the water collected is quickly evaporated. Figures 3a and 3b show the distribution of rainfall and evaporation across Western Australia. There is no threat of seismic activity.

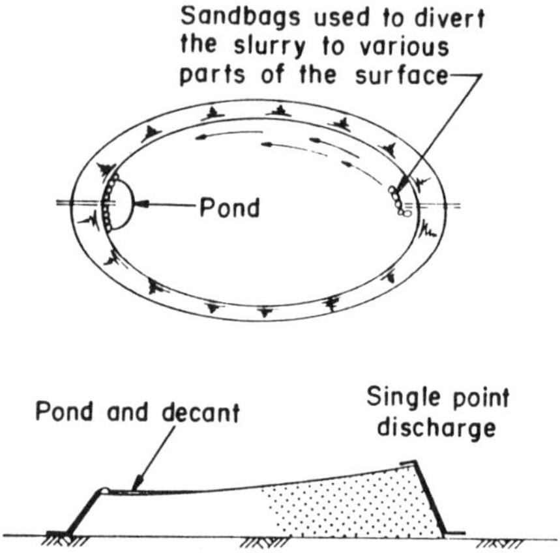

FIG 2. Plan and Section Through a Typical Earlier Tailings Storage

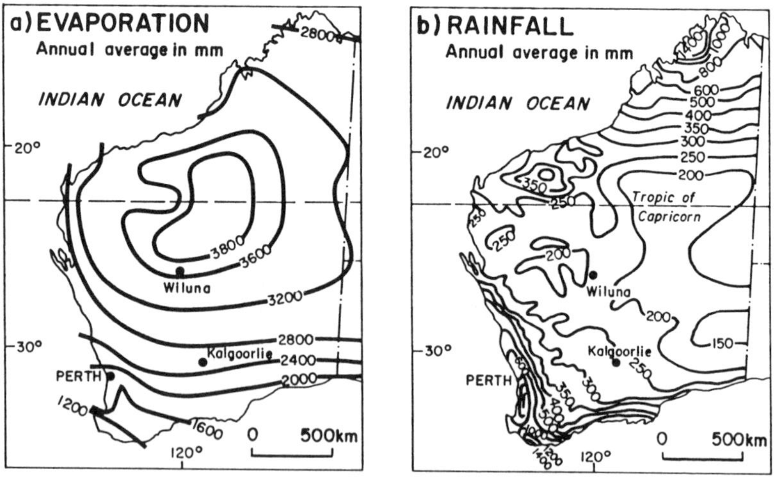

FIGS 3(a), (b). Climatic Conditions in Goldfields of Western Australia

As tonnages increased and the grinding of the ore improved resulting in finer average particle size, the single point discharge technique became more difficult to manage. It was still possible to safely raise the outer embankments in those areas close to the discharge point where the coarse sand tended to be deposited, however, the fineness of the material remote from the discharge required extended drying time and, hence, more surface area. Typical surface areas provided a ratio of 75 ha per million tonnes of tailings.

To improve the efficiency of segregation of the slurry, a ring main was introduced to evenly distribute the slurry around the full perimeter of the storage. By using several offtakes at any one time to distribute the flow, the coarse fraction of the tailings was deposited near to the edge of the storage and management of embankment construction was made considerably easier (Figure 4). Perimeter bunds were raised by men and shovels by progressively rotating the discharge points around the storage and allowing beach areas to dry to the extent that the tailings could be placed into 150 - 200 high lifts. The large tailings dumps at Kalgoorlie were usually constructed in this manner.

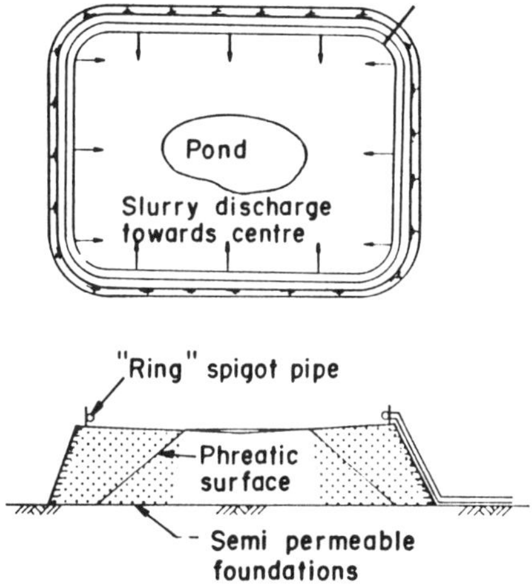

FIG 4. Plan and Section Through a Typical Storage Using Spigot Discharge from the Peripheral Embankment

As a result of the perimeter spigotting, water released from the tailings and incident rainfall ponded in the centre of the storage and, in most cases, was simply allowed to evaporate. For the first time, however, the operators of tailings storages were faced with the possible development of a phreatic surface and, hence, a threat to the stability of the outer embankments. Due to the generous surface areas and the generally low rainfall, few storages were actually threatened by high phreatic surfaces. Overall rates of rise of the storages of 0.5 to 1 m per year were common.

OPERATING CONSTRAINTS

In many mining areas of Australia, the projects are very concentrated. The Kalgoorlie area is a good example where a dozen or more gold projects are vying for the available land. Virtually all the ground is held in mineral claims and the proven non-ore bearing areas are becoming limited. There is, therefore, pressure to limit the plan areas of the tailings storages. A further reason for limiting the plan area is to maximize water return. Experience has shown that, during the hot months of the year (when daily maximum temperatures can reach 45°C), beach loss accounts for a large percentage of the total water losses. The larger the beach area, the larger the loss.

There is an inevitable tradeoff between ease of management, water return and overall economics (especially where the land occupied by the storage either has to be purchased from an adjoining tenement holder or compensation paid to the State for future rehabilitation). As the early operators discovered, the slower the rate of rise of the tailings beach, the more manageable is the storage. A lower moisture content provides for greater bearing capacity, easier excavation and embankment construction and, importantly, less shrinkage once placed. However, a compromise on storage plan area has to be reached to satisfy both the management of the tailings and the potential loss of water due to flow over long beaches to the decant point.

As a result of the desire to conserve water, several tailings storages currently under construction have been subdivided into smaller operating areas, each equipped with a separate decant. The Paddy's Flat Mine storage is an example of this approach (Figure 5). This storage is also equipped with an underdrainage system to maximize the recovery of water which would otherwise be lost to seepage.

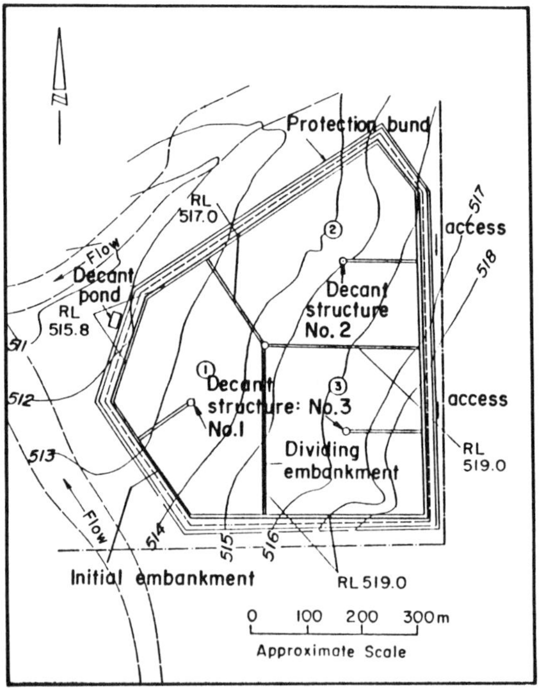

FIG 5. Plan of the Paddy's Flat Tailings Storage

CURRENT MANAGEMENT TECHNIQUES

The ore from the majority of Australia's gold mines is derived from open cut excavation and, hence, there are considerable volumes of waste rock removed (a typical stripping ratio could be around 1 ore to 12 waste). In many cases, however, the waste is not suitable for the construction of impermeable embankments. Mine operators are not keen to be placed in a position where the construction of the tailings storage can dictate mining sequence. The large capacity mine haulage equipment is not generally suitable for the placement of relatively narrow (5 m) crest width embankments. Overfilling to accommodate large equipment is an unnecessary use of resources. There are still, therefore, economic and practical reasons for the adoption of upstream construction of the peripheral embankments using tailings.

The spigot techniques developed during the operation of early mines has proved so successful that the basic methods are still in use 50 years hence. The great change which has occurred over the last few years has been the move to machine excavation which has completely replaced hand techniques. The cost of labour has forced the change which has been made possible by the efficiency of the modern hydraulic backhoe.

The changeover has not been without problems and these have been exacerbated by the trend to the mining and milling of highly oxidized ores. The geotechnical parameters of this material and the implications with respect to short and long term stability are discussed later. The need to conserve water has recently added a significant constraint to the management strategies.

Whilst hydrocyclones have been used in the past to separate coarse from fine tailings, there are currently few projects where the combination of site, tailings characteristics and mine operating requirements make this form of construction an attractive proposition.

Spigot discharge of tailings takes several different forms, however, the most common is to provide distribution pipework on the embankment crest with small diameter offtakes at regular intervals. With smaller projects, say those around 1 500 - 2 000 tpd, the spigot pipework is also the distribution pipework. The lack of flexibility in management that this causes can be overcome without too much difficulty.

On larger operations, such as the KALTAILS project located at Kalgoorlie 600 km east of Perth and where the production rate will exceed 18 000 tpd, the main tailings delivery pipework will be located at the toe of the starter embankment and the distribution or spigot pipework fed by a series of valved offtake riser pipes. The advantage of this approach is that the large diameter (400 mm) main delivery pipework can be permanently located in a bunded pipe trench which more easily satisfies environmental considerations. The offtake risers and the small diameter distribution pipework constructed using HDPE can be relocated without the need to use mechanical handling equipment. Embankment construction can therefore proceed more rapidly. This arrangement, which effectively duplicates pipework around the full perimeter of the storage, can be justified on the larger projects where the added flexibility more than compensates for the increase in capital costs (Figures 6 and 7).

The aim of spigot discharge is to reduce the velocity of the tailings slurry by spreading the total flow. It is common to spigot over a distance of 100 - 200 m for a typical 1 500 - 2 000 tpd operation. This would require, say, 10 to 20 outlets being used at any one time. The velocity of the slurry flow away from the discharge points would normally be around 1 m/s or less, depending on the percentage solids of the slurry.

A layer of slurry up to 200 mm deep is permitted to build up along the active beach before closing the spigots and moving to the next section of the storage. Regular rotation of the spigot outlets therefore leads to the even buildup of the beach area. This is important as it is not sufficient for small sections of sloping beach to be formed. Unless there is a general increase in level, there will always be the possibility of slurry flow parallel to the embankments, with the consequence that streamflow will develop. Beach areas are very susceptible to erosion.

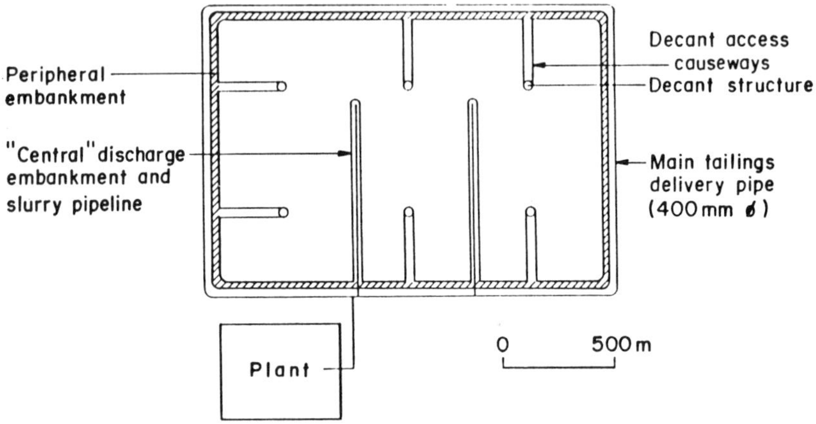

FIG 6. Plan of the KALTAILS Storage

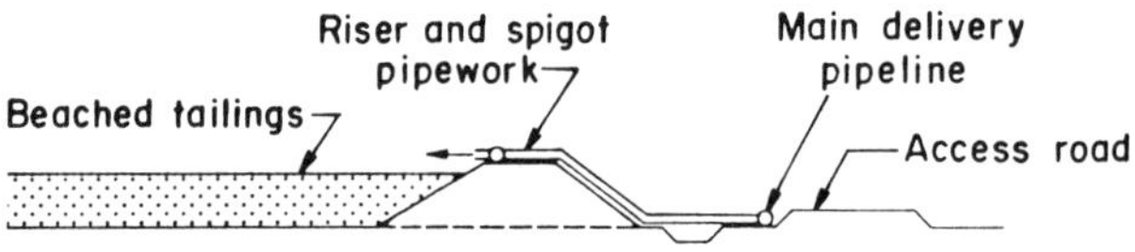

FIG 7. Typical Cross-section Through Peripheral Embankment

Embankment construction methods vary depending on the competency of the tailings being excavated. Typically, machines of up to 2 m^3 capacity are used, with reach being more of a criteria than breakout force. Specialist machines, purpose-built for the task, have not been introduced, however, such a move is not too far distant.

A typical "lift" of the peripheral embankment is around 1.5 to 2 m (Figure 8), although at some mines lifts in excess of this are attempted, often with less than satisfactory results. Most mines attempt to schedule embankment construction once a year to reduce mobilization costs, however, there are often false economics in this approach as the higher lifts require more fill, ie a single lift of, say, 3 m requires 50% more fill than three 1 m lifts and, hence, construction time is extended. Virtually all the work in raising the embankments is carried out by contract rather than by mine owned machinery.

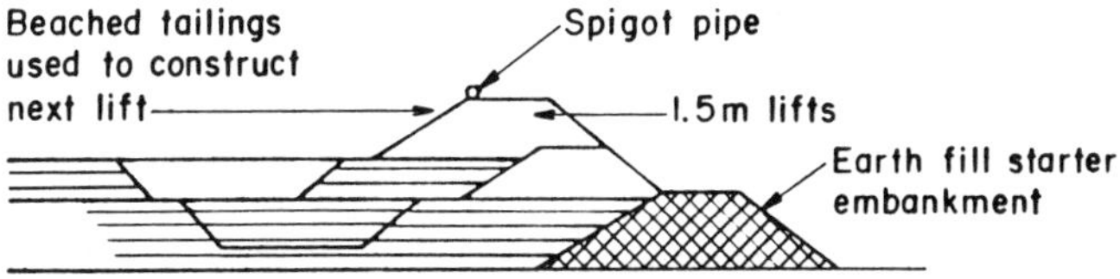

FIG 8. Construction of the Peripheral Embankment

GEOTECHNICAL ASPECTS

The successful management of a large tailings storage implicitly requires an understanding of the mechanics of slurry flow, the principals of drainage and consolidation and, importantly, the development of the strength of the tailings. In practice, typically, mine personnel do not have the background knowledge to enable them to be experts in the field of geotechnical engineering and, therefore, they largely rely upon experience and "gut feeling". However, in an increasing number of cases, outside consultants are being used to advise mines on the optimum means of efficiently and safely operating the storages.

The most relevant geotechnical parameters which influence tailings storage construction are (not necessarily in order of importance):

(i) bearing capacity;
(ii) moisture content/compactive effort relationship;
(iii) permeability;
(iv) shrinkage;
(v) consolidation;
(vi) rate of loss of moisture (from beach areas).

Whilst the majority of hard ore tailings are relatively permeable, the natural clay/silt content of tailings which result from the processing of oxidized ores tends to significantly reduce the natural permeability of the material. As a result, the tailings, even when beached by spigot methods, drain more slowly and hence tend to settle less rapidly. Initial void ratios are therefore higher and more fluid is retained. Figure 9 shows typical grading curves for the sulphide and oxide gold ores of Western Australia. Table 1 provides typical values for the other geotechnical parameters.

Largely due to the increase in the percentage of fine silt sized particles, oxide tailings take significantly longer to reach a moisture content where they can be excavated to raise embankments. Beach areas also take longer to reach a stable condition such that loading will not induce a bearing capacity failure.

An important geotechnical parameter in the design of stable and secure upstream constructed embankments is that of shrinkage coefficient. Piping failures are probably the single most common initial failure mode in the breakdown of the peripheral embankments of tailings storages. The majority of embankment "failures" can be traced to the passage of a generally very small flow of water or tailings slurry followed by a fairly rapid increase in flow as the hole or holes erode. Shrinkage cracking, resulting in the formation of continuous or semi-continuous small passages through the embankment, is the most common cause of the initial "failure".

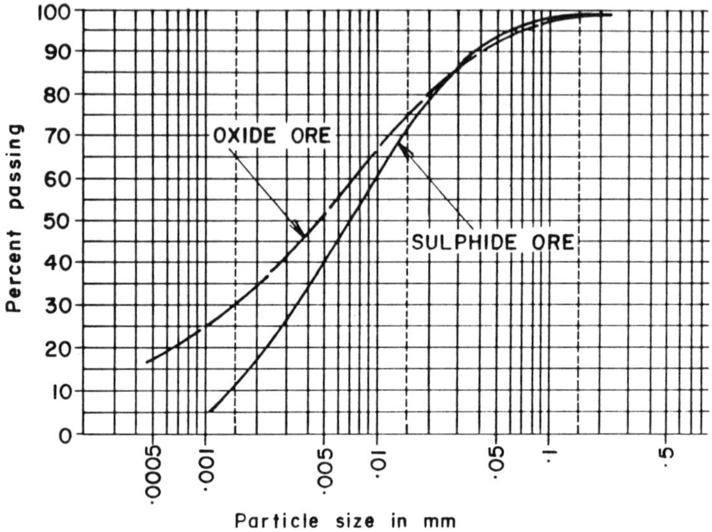

FIG 9. Typical Grading Curves for Oxide and Sulphide Tailings

TABLE 1. Typical Geotechnical Parameters

	Sulphide Ores Underground	Oxide Ores Open Cut
Initial settled dry density (t/m^3)	1.2 - 1.3	1.0 - 1.2
Final insitu dry density (t/m^3)	1.45 - 1.55	1.35 - 1.4
Cohesion c' (kPa)	5	10 - 15
Friction angle ϕ'	35 - 40^0	30 - 36^0
Shrinkage coefficient	4 - 6%	6 - 12%

All samples taken from the embankment end of the beach.

Often the failure path commences at the tailings beach which forms the foundation of the next lift of the embankment. If, for any reason, the surface of the old lift is cracked, a ready made passage can be available for water and tailings slurry to pass beneath the newly formed embankment.

Desiccation of the beach areas is, of course, desirable as this enhances the density and, hence, strength of the material. In most instances, desiccation cracks are filled with tailings and sealed during the following deposition cycle. Occasionally, however, either due to high water levels or badly managed deposition, slurry and water enters the cracks at a rapid rate building up a small hydraulic head and forcing a passage beneath the embankment.

Cracking within the embankment is caused by placing the excavated tailings at a moisture content well above optimum, thus preventing adequate compaction. The inclusion of slimes layers which have resulted from poor deposition techniques increases the shrinkage of the placed material.

The excavation of a trench to win borrow material for upstream construction exacerbates the problem of deep cracking as it allows a head of slurry to develop quickly. Such problems can be minimized by creating a series of excavations separated by a bridge of undisturbed beach rather than a continuous trench (Figure 10). Separate excavations allow the controlled filling of the areas adjacent to the embankment and prevent released water ponding in low points for long periods.

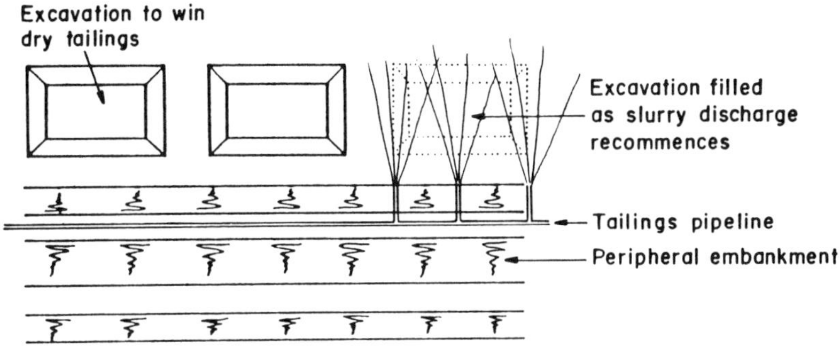

FIG 10. Plan showing the pattern of excavation to win dry tailings for embankment construction

The difficulty of handling oxide tailings can be traced to the fact that the material has a higher moisture content when initially beached than typical unoxidized ore tailings and, hence, takes longer to drain and air dry to a stable moisture content. Oxide tailings contain more silt and clay, and therefore behave more as a fine silt than do tailings derived from underground sources. A higher shrinkage coefficient makes moisture control during construction of embankments of even greater importance.

To partially overcome the problems presented by the oxide ores, the rate of rise of tailings storages has been reduced to permit greater drying time. In terms of plan area, the ratio of annual production to beach area has risen to around 35 - 40 ha per million tonnes per annum from 25 - 30 ha per million tonnes per annum for unoxidized ore tailings. The increase in total beach area results in larger water losses by evaporation, and it is becoming more common to subdivide storages into smaller operating units.

Underdrainage systems have been installed at several recently constructed tailings storages for the express purpose of gold recovery. Metallurgical work has shown that continued slow leaching of the tailings can liberate small but significant amounts of gold which are transported through the storage and collected by underdrainage. Underdrainage assists in the general dewatering of the stored tailings and promotes more rapid consolidation. In some instances, underdrainage is used as a means of reducing the hydraulic head acting on the foundations and, hence, seepage losses.

ENVIRONMENTAL CONSIDERATIONS

With the ever increasing emphasis on environmental matters, decommissioning of tailings storages has assumed great importance. It is therefore necessary to carefully consider the long-term consolidation, strength and chemical composition of the stored tailings. The long-term stability of tailings storages relies heavily on the control of incident rainfall, and surface drainage works play a large part in reducing erosion. As the majority of operators wish to be able to "decommission and walk away" from worked out mines, it is necessary to be able to predict long-term settlement of the tailings and design drainage works accordingly.

The calculation of settlement of very fine grained solids which commence consolidation from a relatively loose state under largely unknown drainage conditions is extremely difficult. Normal consolidation testing does not take into account the large strains and the consequent changes in permeability and new methods are being introduced. Finite element analyses incorporating large strains can provide theoretical solutions, however, the results will only be as accurate as the input parameters permit.

Rowe cell consolidation testing has been used recently in Australia to model large strain consolidation, the advantage of the Rowe cell being that the sample can commence the consolidation cycle from a much lower and more realistic density, ie closer to field density. The use of the Rowe cell to determine the parameters which are relevant to long-term consolidation is relatively new and it will be some years before field experience is available to test laboratory derived values (Rowe and Bardon, 1966).

The analysis of composite embankments, such as those which occur when tailings are used for the upstream face with earth or rockfill used on the downstream side, is complex. The simplistic approach often taken is that "if you can build it, it is safe". This is a fine philosophy providing that the phreatic surface is controlled at all times, ie the in-service conditions do not differ markedly from those pertaining when the embankments were under construction. The importance of pond control is stressed to all operators.

Cyanide is used in the processing of both sulphide and oxide ores, either in a CIL (carbon-in-leach) process or CIP (carbon-in-pulp) process. Cyanide is present in the tailings which leave the mill generally in a concentration of 50 - 100 ppm. At least part of this concentration is "complexed", ie not all is free cyanide. Some of the cyanide complexes are extremely stable, others appear less so. Free cyanide is fairly quickly volatilized by sunlight, and beach deposition provides the perfect conditions for cyanide loss. Whilst extensive testwork under Australian conditions has not been carried out, a study by Worsley Alumina in 1986 confirmed many of the findings of overseas researchers (Murphy and Robertson, 1979, Simovia, Snodgrass et al, 1984). It is of interest to note that most operators like to retrieve free water from the surface of the tailings storage as quickly as possible to retain at least some of the cyanide present. Cyanide is, in many operations, the single most expensive operating cost per tonne of ore treated.

Cyanide, generally in complexed forms, is present in seepage from the base of tailings storages, however, in most parts of Australia it is of little concern. This is primarily because local groundwaters are highly saline and are not used for either human or stock water. There is also strong evidence to suggest that the saline and generally acidic conditions, which typically exist in the weathered rock profile covering large areas of Australia, act to both destroy and/or adsorb cyanide in complex forms (Fuller, 1984).

Where underlying groundwaters are used for water supplies, additional care is taken to select tailings storage sites where the foundations are relatively impermeable. In a few isolated cases, a clay seal has been used to further reduce foundation permeability. Underdrains perform a dual function in such cases, both recovering water which seeps down through the tailings mass and reducing the hydraulic head acting on the foundations. The monitoring of groundwaters beneath tailings storages is becoming normal practice, however, real difficulties with obtaining reliable readings of cyanide levels persist, particularly where complex cyanide ions (cyanates) are present.

FUTURE DIRECTIONS

If anything, there is likely to be an increase in the scale of mining, especially gold mining, in Australia. Mining will be concentrated in the arid low rainfall zones unless significant discoveries are made in other areas. Hence, the traditional methods of tailings disposal will be used in the future with possibly some minor modifications. Strict controls on the activities of mining companies will require that the geotechnical engineer will play an increasingly important part in the design and management of tailings storages. Regular reporting to the authorities on the stability and general construction progress of tailings storages will be a requirement.

It is clear from current experience that more effort needs to go into the understanding of the parameters which govern the long-term consolidation and strength of tailings, especially for those derived from oxide ores. The continued expansion in the scale of mining will place even greater emphasis on the stability of storages and on aspects such as water recovery. Underdrainage will become more common both to increase water recovery and reduce seepage into groundwater systems.

Methods of raising embankments will become more sophisticated with specialist machinery possessing increased reach being employed. This will overcome some of the difficulties currently being experienced with deep excavations close to the peripheral embankments. The role of the geotechnical engineer will assume greater importance as the benefits of good management are recognized by the mining companies.

As the environmental controls tighten, mining companies will be required to install monitoring devices such as piezometers and settlement gauges. Regular surveys will be required to enable the ordered construction of the embankments to be undertaken. Seepage detection and monitoring will also be required, especially where cyanide or heavy metals are present in the tailings.

APPENDIX. REFERENCES

Fuller, W H (1984). "Cyanides in the Environment with Particular Attention to the Soil." In "Cyanide and the Environment"; Proceedings of a Conference on Cyanide and the Environment, Tucson, Arizona, 1984, Vol I. Colorado State University, 1985.

Murphy, K L, and Robertson, J L (1979). "Factors Affecting Natural Degradation of Free and Metal-Complexed Cyanides from Gold Mining Effluents." International Environmental Consultants, Canada. (Unpublished).

Rowe, P W, and Barden, L (1966). "A New Consolidation Cell." Geotechnique, Vol 16, No 2.

Simovic, L, Snodgrass, W S, Murphy, L L and Schmidt, J W (1984). "Development of amodel to Describe the Natural Degradation of Cyanide in Gold Mill Effluents." In "Cyanide and the Environment"; Proceedings of a Conference on Cyanide and The Environment, Tuscon, Arizona, 1984, Vol II. Colorado State University, 1985.

Worsley Alumina. "Report on Studies into Cyanide Decay in Process Residues." April 1986. (Unpublished).

Beaching and Permeability Properties of Tailings

Andy Fourie*

A laboratory technique for determining the beach profile of hydraulically placed tailings is described. Three different tailings were tested, and the results indicate that varying the percentage solids or the nature of the tailings liquor has little effect on the resulting normalised profile. An evaluation of the likelihood of a tailings deposit having anisotropic permeability characteristics is also presented.

Introduction

The most common method of disposing of the waste from any mining operation is to mix the pulverised rock waste with water to form a slurry, transport this slurry via pipeline to a specific disposal area and then deposit the tailings into the impoundment where it will eventually consolidate under the surcharge of subsequently placed material. When planning a facility of this kind, the designer is faced with choosing an impoundment area which may be developed to provide sufficient storage for the volume of tailings likely to be generated during the design life of the tailings dam. It is common to construct an elevation-storage diagram such as that shown in Figure 1 for a number of possible storage areas and then, depending on limitations on allowable tailings dam height, an appropriate ground surface area can be chosen.

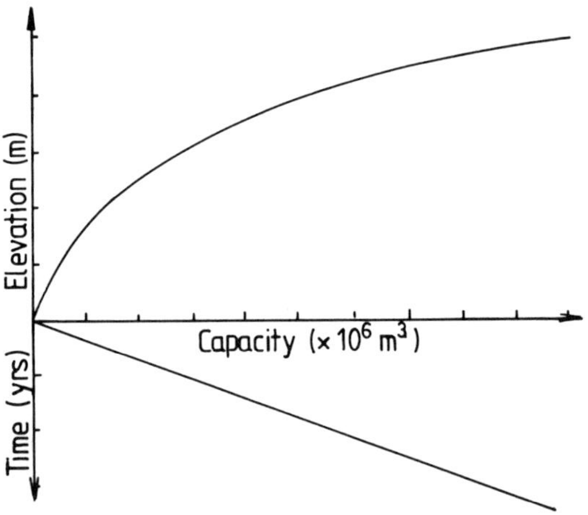

Figure 1 Typical storage curve

* Lecturer in Civil Engineering, University of Queensland, St. Lucia, Queensland 4067, Australia.

To construct a diagram such as that shown in Figure 1, information about the profile of the deposited tailings is necessary. This information, which is also required for the siting of decant facilities, e.g. Robinsky (1978) provides a relationship between the percent solids of copper tailings and the slope of the resulting tailings deposit. However, the depositional characteristics of a particular waste material may only be known for one particular set of conditions. The addition of a dispersant, for example, will have an effect on the resulting beach profile, and the tailings dam designer may have no precedent on which to base estimates of the likely new profile. In addition, the necessary field data may not be available when a new mining operation is to be commissioned. The development of a laboratory technique for estimation of the beaching characteristics of tailings is thus highly desirable.

Previous laboratory study

Blight, Thomson and Vorster (1985) describe the use of a small-scale laboratory flume which was found to provide an acceptably accurate method for the prediction of beach profiles of prototype tailings dams. A similar flume to that described by the above authors was constructed at the University of Queensland and is illustrated in Figure 2. The flume is made of perspex with occasional stiffeners to provide rigidity.

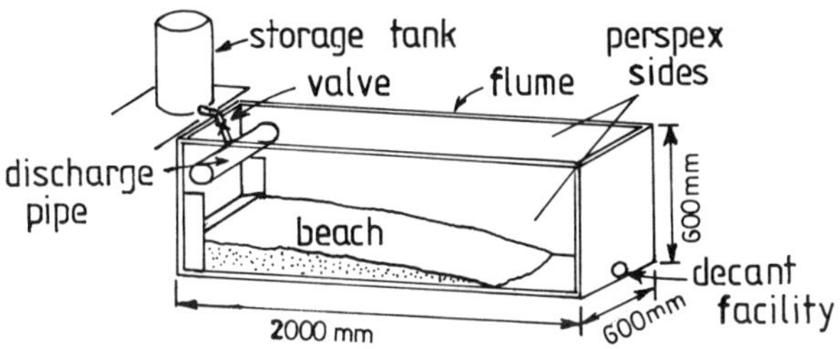

Figure 2 Laboratory flume

Tailings were mixed at the required percent solids in the elevated storage tank using an electrically-driven agitator. Deposition was begun by opening the valve leading to the horizontal discharge pipe. This pipe had four 1cm diameter holes along its underside in order to ensure a uniform thickness of deposited tailings across the width of the flume. Flow through the valve was regulated to provide deposition of approximately $0.12m^3$ of tailings slurry in a four-hour period. This rate of deposition applied to all tests described in this paper, unless otherwise indicated. At the deposition end of the flume, a simple energy dissipator was provided to prevent the formation of a splash pool at this location. This had the advantage of increasing the length of the undisturbed beach. At the far end of the flume a discharge facility was provided whereby the level of the resulting tailings liquor pond could be controlled.

Blight, Thomson and Vorster (1985) describe the results of tests with gold tailings using the laboratory flume. They found that if the height and length of the beach at any point are normalised by the total height and the length from the discharge point to the edge of the tailings pool, respectively, as shown in Figure 3, the results of the model tests superimpose virtually exactly on the normalised profile of an active full-size tailings dam. A relationship of the form

$$h/_Y = (1 - H/_X)^n$$

was postulated, where the variables are as defined in Figure 3. The exponent n is determined from the results of laboratory flume tests.

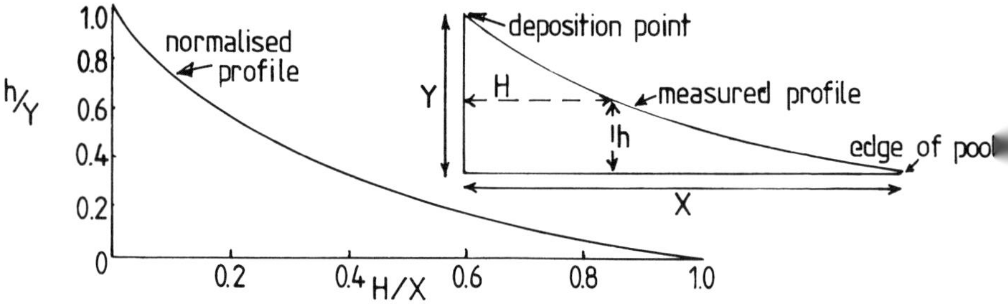

Figure 3 Normalising procedure for measured tailings beach profile

Previous work by Blight and Bentel (1983) on a number of full-scale tailings dams showed that a unique curve in terms of $h/_Y$ verus $H/_X$ existed for gold, copper, diamond and platinum tailings. The exponent n, therefore, although being empirically determined, appeared to constitute a material characteristic. The following section describes a series of laboratory tests on three different tailings using a laboratory flume similar to that described by Blight et al (1985). The objective of these tests was to validate the findings described above, as well as to evaluate the effect of changing the percent solids and the nature of the tailings liquor for a particular type of tailings.

Material tested

The three tailings tested were:

(i) bauxite tailings from North Queensland, Australia

(ii) nickel ore slurry from New Caledonia

(iii) coal tailings from South-East Queensland.

Results of particle-size analyses for the three materials are shown in Figure 4, and some typical properties are given in Table 1.

	Bauxite	Nickel	Coal
Liquid Limit	44	75	42
Plastic Limit	26	57	21
Specific Gravity	2.81	3.48	1.93

Table 1: Some engineering properties of tailings used in flume tests

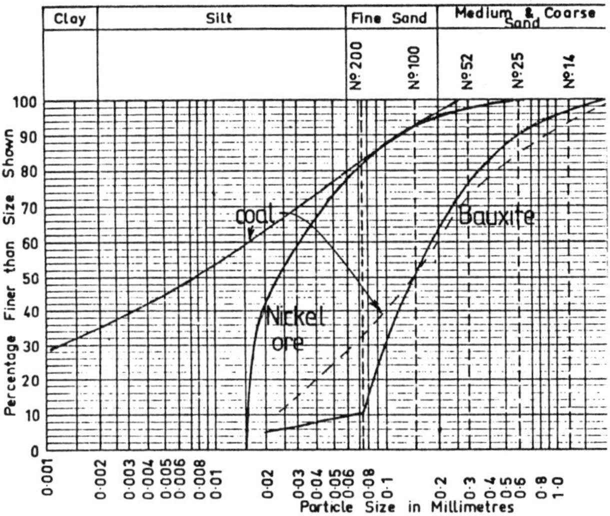

Figure 4. Particle size distribution of tailings used in laboratory study

Results of laboratory tests

(a) Bauxite tailings

The Bauxite tailings were produced at the Weipa mine in North Queensland. Two flume tests were carried out to determine the effects of varying the initial solids content on the resulting beach profile. Solids contents of 32% and 27% were used, resulting in a relatively thick consistency. The measured laboratory profiles are shown in Figure 5. During deposition very little bleeding of tailings water was evident. A thin film (1-2cm thick) of water developed in front of the advancing tailings profile. The perspex sides of the flume allowed visual observation of the beach formation process. It was evident that the tailings material closest to the base of the flume (i.e. the tailings deposited earliest) remained virtually stationery while newly deposited tailings flowed over it in layers which were approximately horizontal. As the thickness of tailings at any particular section

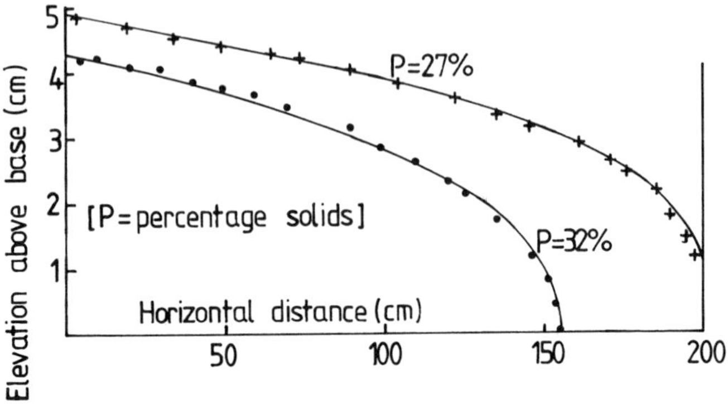

Figure 5 Measured profiles of bauxite tailings beaches

increased, so too did the thickness of material which remained stationary. This was true for all profiles along the length of the flume. The beach was therefore developed by the flow of thin horizontal layers on sheets.

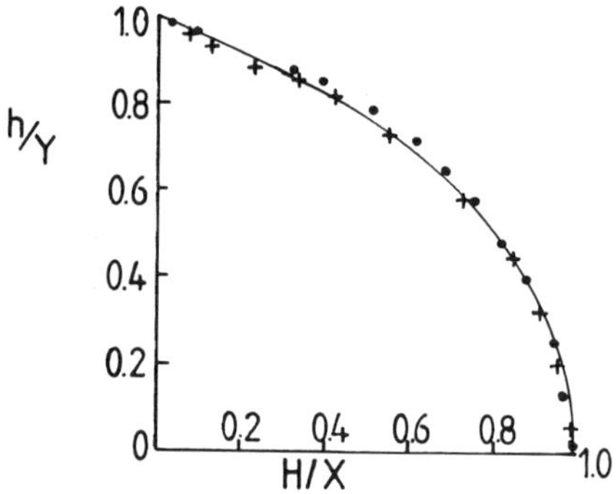

Figure 6. Normalised beach profiles bauxite tailings

The normalised curves are plotted in Figure 6, and although there is some variation in the results, the two curves are very similar in shape. A distinct difference which occurs from the results presented by Blight and Bentel (1983) is that their normalised curves for four different types of tailings were always concave upwards, whereas the results in Figure 6 are distinctly convex upwards. Plotting ln $(^{h}/_{Y})$ against ln $(1 - {}^{H}/_{X})$ produces a value of approximately 0.43 for the parameter n. Values of n obtained by Blight and Bentel were always greater than unity, reflecting the different shapes mentioned above. This is attributed to both the relatively high initial solids content and to the gap graded nature of the bauxite tailings. An additional factor may be the rate of deposition. This factor was not varied during the work described in this paper. It should be noted, however, that both convex and concave upwards profiles were found for a coal tailings using the same rate of deposition as mentioned earlier. These results are described in a subsequent section.

Despite the difference in the two curves shown in Figure 5, it appears that changing the initial percentage solids has not had a dramatic effect on the resulting normalised profile. Wates, Stevenson and Purchase (1987) produce field measurements which indicate that the average beach slope (given by ($^{Y}/X$) in Figure 3) increases as the relative density of the deposited slurry increases. The results shown in Figure 5 agree with this finding. However, when these two profiles are normalised the resulting curve as given in Figure 6, fits both sets of data equally well. These results therefore indicate that, within the range tested, the initial solids content affects the resulting average beach slope while having little or no influenece on the normalised beach profile.

(b) Nickel ore slurry

It is planned to transport nickel ore slurry to Townsville in North Queensland where it will be processed. Prior to processing the material is to be solar dried. The laboratory flume was used to investigate the optimum slope of the drying beds to ensure a uniform thickness of slurry after placement. Tests were run at initial solids contents of 42% and 35%, with the base of the flume inclined downwards from the deposition point at slopes of 4° and 1.6° respectively. There is thus a difference in these test conditions from those described above. This was unavoidable as the nickel ore slurry testing was carried out with the specific aim of evaluating the influence of the initial base inclination.

During the deposition process even less bleeding of tailings water was evident than occurred during the bauxite tailings tests. Once again a mudflow type flow regime developed, with horizontal movement being confined to discrete layers. The normalised curves for these two experiments are shown in Figure 7. There is exceptionally good agreement, and again the curves are convex upwards. The calculated value of the exponent n was 0.36. Once again, these results seem to indicate the parameter n is insensitive to changes in the initial solids content, although all the tests described to date were deposited as thick slurries.

Two other points arose from the tests using nickel ore slurry. For both the above tests, samples were taken at three positions along the model beach: at the deposition point, at the leading edge of the flow and midway between these two points. Particle size analyses on these samples showed no perceptible difference. Segregation definitely did not occur, as would be expected from both the thick consistency of the slurry and the relatively uniform particle size. As shown later for coal tailings, the length of the flume was quite adequate to allow particle segregation if the tailings slurry exhibited this tendency.

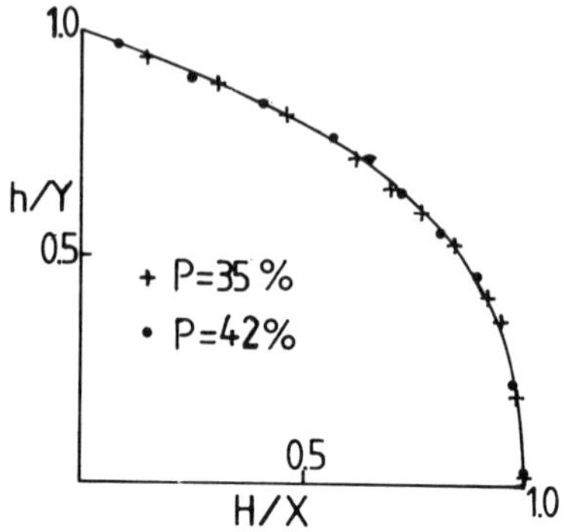

Figure 7 Normalised beach profiles nickel ore slurry

The test with an initial solids content of 35% was continued after the profile used in Figure 7 had been measured. Deposition continued until the slurry reached the far end of the flume, and a "backwater" effect was observed. The profiles shown in Figure 8 illustrate how the slope of the profile becomes significantly flatter after encountering an obstacle such as the end wall of the flume. The opposite observation would of course be expected for the condition where a tailings pond formed at the end of the beach. Upon encountering the pond edge, the slope of the tailings profile becomes steeper.

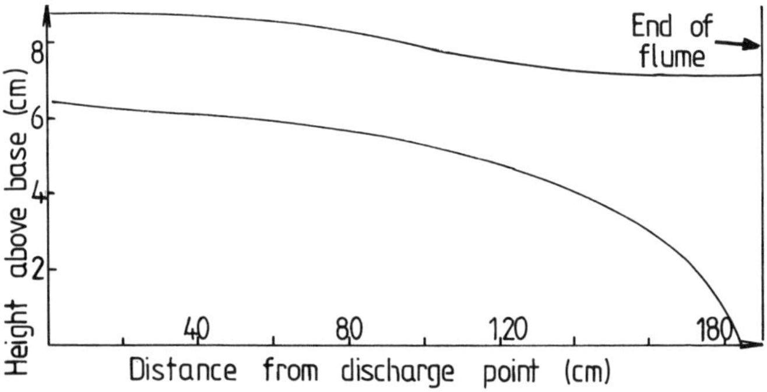

Figure 8 Change in slope of beach upon encountering end of flume

(c) Coal tailings

The tailings used in this series of tests was much more uniformly graded than the bauxite tailings and the nickel ore slurry. It was also substantially finer than either of the other two, having some 35%

finer than 2 μm. Four different tests were run using the coal tailings. The solids content and type of water used was varied, as detailed in Table 2.

TEST #	1	2	3	4
Initial solids content (%)	43	50	15	48
Dispersant added ?	YES	NO	NO	NO
Type of water	TAP	TAP	TAILINGS	TAILINGS

Table 2. Details of tests using coal tailings

Note: In Test #1, the dispersant used was a mixture of 33 g Sodium Hexametaphosphate to 7 g Anhydrous Sodium Carbonate, dissolved in sufficient water to produce a one-litre solution. This was added to the wet slurry in 1:10 proportions to produce a solids content of 43%.

During deposition a noticeably larger amount of bleeding of tailings water occurred than for either the bauxite or nickel ore slurries. This was obviously the case for test #3, with its markedly lower initial solids content, but was also observed in the three other tests.

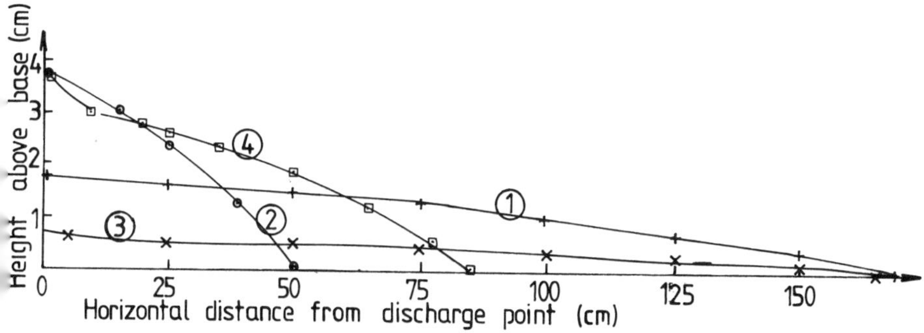

Figure 9 Measured beach profiles: coal tailings

The resulting beach profiles for the four tests are shown in Figure 9. The effect of altering the nature of the tailings water is demonstrated by the differences in curves 2 and 4, which had similar initial solids contents. Although there is a 2% difference in solids content it is unlikely that this is sufficient to cause the difference evident in Figure 9. Furthermore, curve 1, which had a solids content of 43% has a very shallow slope, which is almost comparable to curve 3, the result for a solids content of only 15%. This effect is attributed to the very high concentration of dispersant added to test sample #1. Comparing tests 3 and 4, it can be seen that decreasing the initial solids content has the effect of producing a much shallower profile.

The finding that the average beach slope increases as the initial percentage solids increases seems reasonable, as discussed by Wates, Stevenson and Purchase (1987). However, Blight (1987) has recently presented results for a fly ash slurry which indicate that the average beach slope does not increase indefinitely with increasing percent solids. There is some value of percent solids beyond which the average beach slope falls rapidly with increasing solids content. This value of solids content was also indicated as marking the transition from a "particle settling" to a "mud-flow" regime. In the tests described by Blight (1987), the maximum average beach slope occurred at a solids content of 40%. Blight (1987) postulates that the percentage of solids corresponding to the maximum beach slope decreases with increasing rate of deposition, and furthermore that some minimum value of beach slope will occur (about 35% solids content in the paper by Blight (1987)), whereafter beach slope will again increase wtih increasing solids content.

The results shown in Figure 9 are not inconsistent with these findings. Tests 1, 2 and 4 exhibited a "mud-flow" type flow regime, with the average beach slope increasing with increasing percent solids. The solids content used in these tests is therefore greater than the value corresponding to a minimum beach slope as detailed by Blight (1987). This minimum solids content would have to be markedly lower (< 43% as opposed to 55%) than the value for the fly ash slurry described by Blight (1987). A contributing factor to this difference may be the rate of deposition, as in the work described in this paper the full deposition process took 3-4 hours. The deposition rate in the tests described by Blight (1987) are not known.

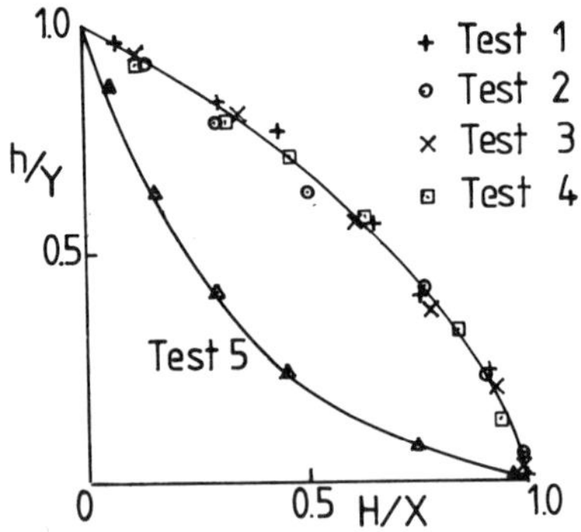

Figure 10 Normalised beach profiles: coal tailings

If the normalised plots are produced for the coal tailings using the procedure described previously, all the data appears to cluster around the same curve in terms of h/Y against H/X, as shown in Figure 10. While there is a degree of scatter in the data it is random, and the results indicate that the parameter n may be a soil characteristic. This hypothesis was further tested by carrying out an additional test using coal tailings with a much coarser grading (the particle size curve for this material is shown dotted in Figure 4). An initial solids content of 37% was used in this test, producing a normalised profile as shown in Figure 10. This curve is concave upwards and is significantly different from the coal tailings results for tests #1-4. Thus the parameter n is not unique to a particular tailings mineral but is influenced by the particle size distribution.

It has been found that the normalised beach profile formed by fine grained coal tailings (UCS classification CL) is similar in shape to that formed by bauxite tailings and nickel ore slurry, having a convex upwards profile. However, when a coarser coal tailings, classified as a silty sand, was tested, a concave upwards profile resulted. While the technique of using a small scale laboratory test to determine the beaching characteristics of tailings has been shown to be useful in examining the influence of parameters such as initial solids content and nature of tailings water, a number of questions still remain. Using a normalised plot of the beach profile as suggested by Blight and Bentel (1983) produces an exponent n which describes the curvature of the profile. The results presented in this paper indicate that, for a particular tailings, this exponent is insensitive to changes in both initial solids content and the nature of the tailings water, whereas it is very sensitive to changes in particle size distribution. This latter observation needs to be thoroughly investigated, using a range of gradings of the same tailings.

A possible explanation for the observations described above could be that for tailings which produced a value for n less than unity (resulting in convex upwards profiles), little or no segregation occurs. When n exceeds unity and a concave upwards curve results, it is an indication of segregation occurring, the reason being that coarser particles settle out rapidly and can thus produce a steeper beach slope. Finer particles which are transported further down the beach accordingly form shallower slopes. This hypothesis was checked by taking samples at four equal intervals along the model beach produced by test #3 (see Table 2), using the fine grained coal tailings. The resulting particle size analyses for these samples are shown in Figure 11, and it is clear that segregation is occurring with coarser particles settling out earlier. This observation therefore disputes the hypothesis discussed above, and future work will be directed at determining whether it is in fact realistic to obtian convex upwards profiles using slurries with low initial solids contents.

Permeability characteristics of hydraulically deposited tailings

This section describes the results of consolidation tests using both conventional and Rowe consolidation cells. Tests were conducted on the bauxite and coal tailings. As tailings is usually deposited hydraulically, it may be expected that the resulting sediment will have an anisotropic structure, with individual particles having a preferred orientation. The objective of the consolidation tests was to determine the differences in the vertical and horizontal coefficients of permeability (k_v and k_h). The likelihood of anisotropic permeability coefficients occurring in a tailings deposit has been recognised previously, and a series of chart solutions based on finite element

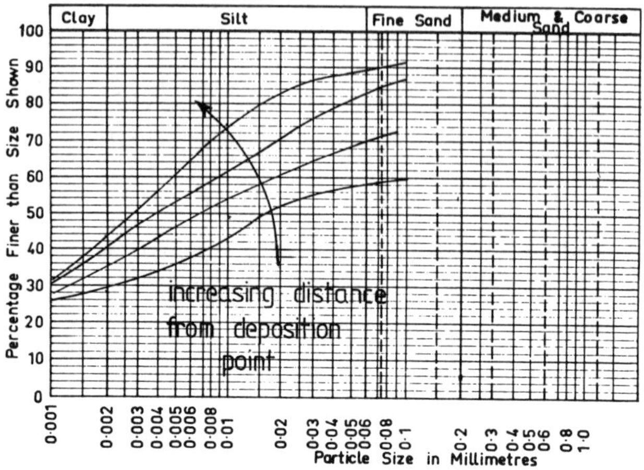

Figure 11 Variation of particle size distribution with distance from discharge points

analyses produced (Kealy and Busch, 1971). However, there is not a great deal of information available on measured values of k_v and k_h.

To try to reproduce the field condition reasonably accurately, slurries were made up at water contents of approximately 1.2 times the Liquid Limit of the tailings. The slurry was placed in the consolidation cell and incremental loads applied. To determine k_h, only radial drainage was allowed in a 150mm diameter Rowe cell (comprehensively described in Rowe and Barden, 1966). The determination of k_v also required the use of Rowe cells as the wet consistency of the slurries made placement and initial loading extremely difficult. For the bauxite tailings there was no discernible difference between the horizontal and vertical permeabilities, as shown in Table 3.

	TAILINGS		
	BAUXITE	COAL AT POINT OF DEPOSITION	AT END OF BEACH
k_v(m/sec)	3.15×10^{-6}	1.9×10^{-7}	6.5×10^{-8}
k_h(m/sec)	2.1×10^{-6}	1.8×10^{-6}	3.4×10^{-7}

Table 3. Coefficients of permeability from laboratory tests

The coarse nature of the bauxite tailings suggests the particles are relatively uniform in size, as for a sand, and not elongated or plate-like as occurs with clays. There is unlikely to be a preferred direction of drainage, and this is borne out by the results in Table 3.

Consolidation tests on coal tailings were carried out on samples from both the point of deposition and the end of the beach produced in the laboratory flume (equivalent to lines 1 and 4 in Figure 11). Each of the results presented in Table 3 is the average of three tests carried out at pressures in the range 10-25kPa. Therefore, although it must be acknowledged that a degree of uncertainty is unavoidable in defining parameters such as the coefficients of consolidation and permeability derived from oedometer tests, the results in Table 3 are believed to be representative of the field behaviour of the various tailings.

The significance of the results presented in Table 3 is not the order of magnitude difference between the vertical and horizontal permeabilities for the coal tailings, but rather the 5 to 10-fold variation in permeability along the length of the tailings beach. Kealy and Busch (1971) present the results of an extensive finite element study of tailings impoundments in which they show that the location of the seepage free surface is significantly affected by a variation in permeability along the tailings beach, whereas it is virtually insensitive to variations in the ratio of horizontal to vertical permeability. Cognisance should therefore be taken of the particle sorting which may occur along the length of a tailings beach, and the effect this has on relative permeability values. However, it should be noted that particle sorting is a complex process and dependent on a number of factors, such as particle size distribution, solids content of slurry, number and spacing of discharge outlets and particle specific gravity. (Abadjiro (1985) and Blight (1987)).

Conclusions

(i) A laboratory beach-modelling flume has been used to test the variation of beach profile with factors such as percentage solids and nature of tailings water. When the data are normalised by the full height and length of the laboratory beach as described in the text, it is found that for a particular type of tailings, the normalised profile is relatively insensitive to variations in percentage solids and tailings water. It is emphasised that additional testing is required to confirm this finding.

(ii) The normalised profile was found to be highly sensitive to a variation in relative particle grading for the coal tailings tested.

(iii) Flume tests on coal tailings at low percentage solids resulted in significant particle segregation.

(iv) Laboratory consolidation tests showed that approximately a 10-fold decrease in permeability was associated with this change in grading from the deposition point to the edge of the tailings beach.

(v) The ratio between horizontal and vertical coefficients of permeability was found to be about unity for poorly-graded coarse bauxite tailings. This value increased to approximately ten for the well-graded fine grained coal tailings.

Acknowledgements

The permission of Comalco Ltd. to publish the data on bauxite tailings, and the permission of Queensland Nickel Ltd. to publish the data on nickel ore slurry is gratefully acknoweldged. Assistance with the laboratory tests was provided by Messrs. E. Krees, G. Grahame, P. McMilland and S. Taylor, and Ms. H. McCullough.

References

Abadjiru, C.B. (1985) Estimation of the physical characteristics of deposited tailings in the tailings dam of non-ferrous metallurgy. Proc., 11th Int. Conf. on Soil Mech. and Found. Eng., San Francisco, Vol. 3, pp. 1231-1234.

Blight, G.E. The concept of the master profile for tailings dam beaches. Proc. Int. Conf. on Mining and Industrial Waste Management, Johannesburg, 1987, pp. 95-100.

Blight, G.E. and Bentel, G.M. The behaviour of mine tailings during hydraulic deposition. Jnl. S. Afr. Inst. Min. Metall., Vol. 83, No.4, pp. 73-86 (1983).

Blight, G.E., Thomson, R.R. and Vorster, K. Profiles of hydraulic-fill tailings beaches, and seepage through hydraulically sorted tailings, Jnl. S. Afr. Inst. Min. metall., Vol. 85, No.5, pp. 157-161 (1985).

Kealy, C., and Busch, R. Determining Seepage Characteristics of Mill-Tailings Dams by the Finite-Element Method, U.S. Bureau of Mines R17477 (1971).

Robinsky, E.I., Tailing disposal by the thickened discharge method for improved economy and environmental control. Proc. 2nd Int. Conf. on Tailings Disposal, Denver, 1978, pp. 75-92.

Rowe, P.W. and Barden, L. A New Consolidation Cell. Geotechnique, Vol. 16, No.2, pp. 162-170 (1966).

Wates, J.A., Stevenson, C., and Purchase, A.R. The effect of relative densities on beaching angles and segregation on gold and uranium tailings dams. Proc. Int. Conf. on Mining and Industrial Waste Management, Johannesburg, 1987, pp. 89-93.

SLURRY PIPELINES: Present and Future

Michael E. Henderson, Associate Member ASCE*

ABSTRACT: A review of recent slurry transportation research is made, which has resulted in improvements in mathematical models and their reliability. The Traynis model is discussed, with comments made on its use in designing slurry systems and its range of applicability. Research and operational experience at the Hydraulic Transport Research Facility are presented. The economics of slurry transportation are discussed, along with potential standard and novel applications.

Introduction

Hydraulic conveying of solid material has long been used to economically and efficiently transport materials from one location to another. The first documented application of slurry transportation was in 1891 in London, where a 660-yard long pipeline carried coal from barges on the Thames River to London. Since then, slurry pipelines have been used to transport coal (for example, the Consolidation Coal pipeline in Ohio, and the Black Mesa pipeline in Northern Arizona), iron ore concentrate (for instance, the La Perla pipeline in Mexico), wood chips, gravel, and many other types of material.

Interestingly, even though slurry transportation has been widely used for many years, little is known about the physical processes involved, or their mathematical formulation. Also, despite its long history of use, each new application of slurry transportation is often met with resistance from the owners or operators who regard the method as uncontrollable and undependable.

The purpose of this paper is to:

1) briefly describe the state-of-the-art in designing slurry pipelines,

2) describe a recent application in slurry transport,

3) discuss some operating experience in dealing with slurry systems, and

4) offer some novel applications for future projects.

* Engineering Manager, J. D. Welsh and Associates, Inc., 260 Coney Island Drive, Sparks, Nevada 89431

Slurry Pipeline Design

The fluid mechanics and mathematical modelling relating to hydraulic transportation of solids are not well known. Past attempts to correlate observations with input parameters have consisted of statistical approaches, dimensional analyses (most notable that of Durand, 1951-55), mechanistic modelling (Wilson, 1970; Eyler et al, 1982), and finite element modelling (Roco, 1984). Since each of these approaches generally has a constant or series of constants to correlate the fundamental approach with a set of data, the designer is cautioned to investigate the data set used to develop these constants to ensure that his particular design is within the limits of such data.

The best approaches to pipeline design are listed below, in descending order of certainty:

1. Build a full-scale pipeline.

2. Build an operational prototype.

3. Use a mathematical model to predict design and operational parameters.

Of these approaches, the first and second usually prove to be costly and may not be warranted, considering that most designs are within the development data set for the available mathematical models. It is the author's opinion that one of the better mathematical models available today is the Traynis formulation (as presented in Faddick, 1982), and it is proposed for design purposes.

The Traynis equation is given by:

$$i_m = i_w \left[(1 + C_v (\frac{p_s - p_l}{p_l}) \right] +$$

$$\left[\sqrt{gD} / (kC_d V) \times (p_s - p_{hm}) C_{vc} / p_l \right]$$

where:
i_m = hydraulic gradient of slurry
i_w = hydraulic gradient of clear water
C_v = total solids volumetric concentration
C_w = coarse solids volumetric concentration
p_s = solids density
p_l = liquid density
p_{hm} = density of heavy medium produced by fines
g = gravitational acceleration
D = inside pipe diameter
k = constant (for coarse coal = 1.9)
C_d = drag coefficient for coarse coal fraction = 0.75
V = mean slurry velocity

Figures 1 and 2 show the Traynis (Faddick, 1982) model predictions against data gathered by the U.S. Bureau of Mines at Hydraulic Transport Research Facility (HTRF) (Henderson, 1983-85). Although these figures were developed primarily from fine and coarse coal data, similar and equally reproducible results have been obtained from common sand and gravel slurries.

The Traynis Equation, like any other, should be used over the range of conditions for which it was developed. The author has experience in using this equation for pipe diameters from 2 to 42 inches, transporting material from sand-size mill tailings to coarse coal with a 3-inch top size, in concentration ranges of 10 to 55 percent by weight. Although the Traynis Equation was developed for primarily heavy or mixed media type slurries, the author has no experience in applying the correlation in such a manner.

The other primary concern in designing slurry pipelines is prediction of the settling velocity--where solids will drop from suspension and start sliding along the bottom of the pipeline. It is the author's opinion that no single adequate correlation exists to predict the settling velocity across the entire range of pipe sizes, material types (specific gravity), shapes, and other variables. Figure 3 shows the settling velocity for fine and coarse coal, under the conditions noted. For pipelines transporting sand-sized particles, the reader is referred to Makharadzye (Makharadzye, 1981). For coarse rock, or other materials having a specific gravity higher than that of coal, the designer should search the available data bases to determine if similar operational experience has been obtained elsewhere. As will be described in the operations section of this paper, design of the pumping system and the instrumentation system for a slurry pipeline has a large effect on the actual velocity at which solids will begin to settle out of the flow.

Hydraulic Transport Research Facility

In the 1970's, as a result of the 1973 Arab oil embargo, the Department of Energy (DOE) was given the mission of improving the productivity of the coal mining industry in the United States. The DOE proposed introduction of slurry transportation of coarse coal to dramatically reduce the transportation costs of raw or clean coal. To promote the application of coal slurry technology, the DOE built the Hydraulic Transport Research Facility in Pittsburgh, Pennsylvania. The HTRF consists of six-, twelve-, and eighteen-inch steel pipelines in horizontal, inclined, and vertical lengths of pipeline and was used to investigate the conditions surrounding coal slurry transportation. The facility was operated from 1980 through 1983 under the direction of the DOE and, later, the U.S. Bureau of Mines. During its period of operation, the HTRF gathered data for coarse and fine clean and run-of-mine coal, sand, rock, and tunnel muck. The data collected are presented in the reports referenced in this paper.

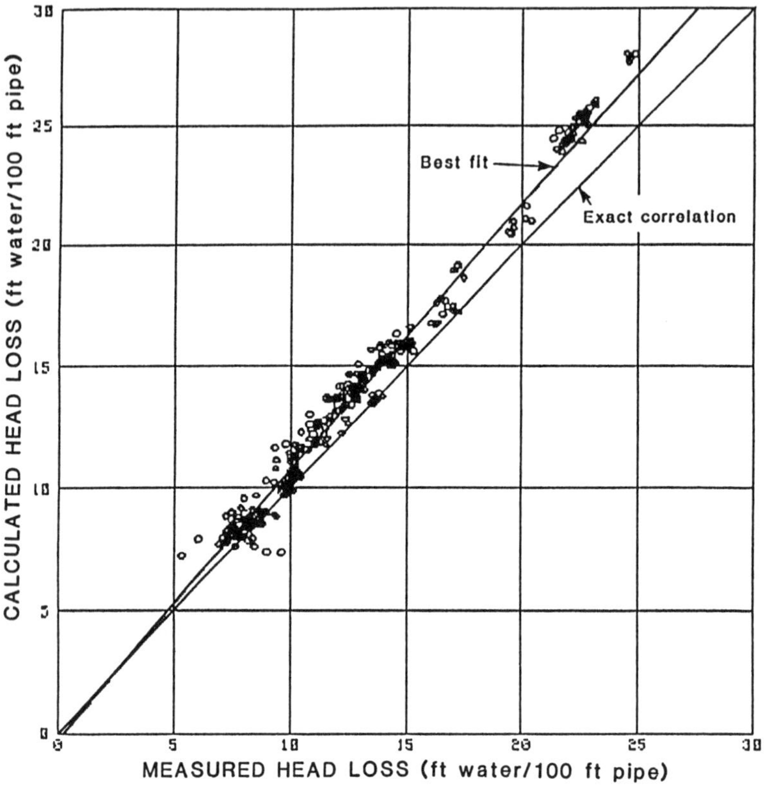

(HTRF data)

FIGURE 1

Measured head loss compared to that calculated from the Traynis equation (6-inch data). (Set no. 2)

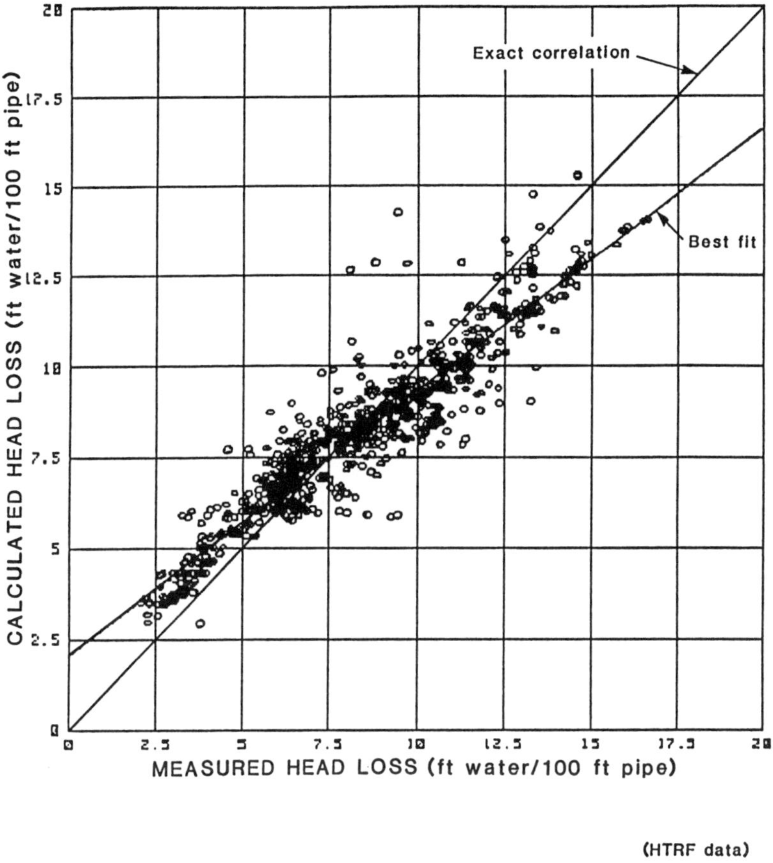

FIGURE 2

Measured head loss compared to that calculated from the Traynis equation (12-inch data).

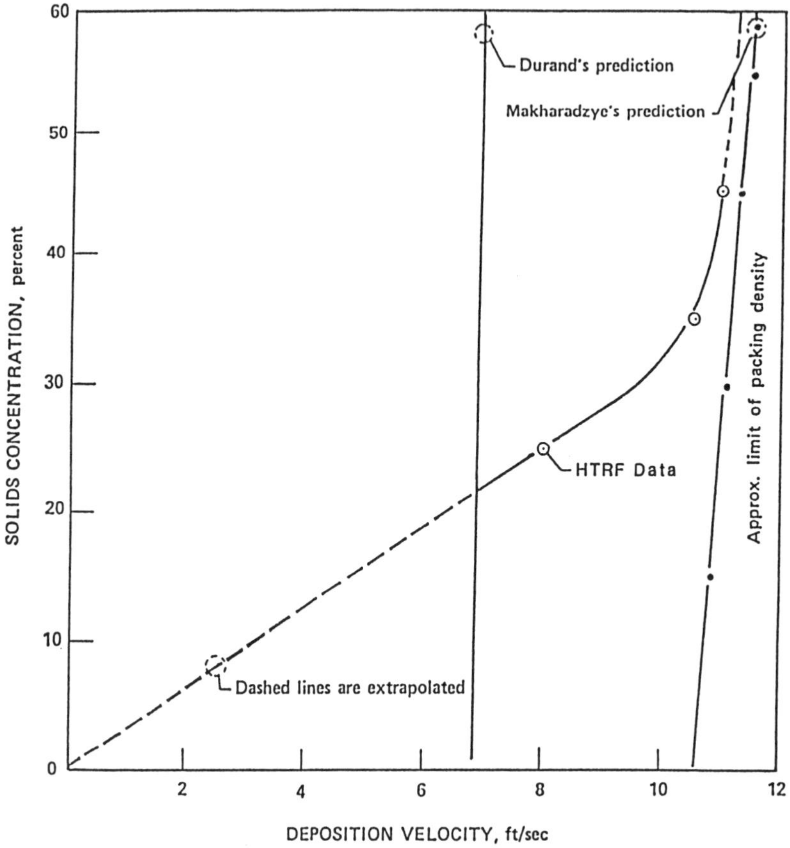

FIGURE 3

12-inch (300 mm) pipeline deposition velocity.

Experience gained at the HTRF regarding pipeline wear indicated that no appreciable wear had occurred for the several hundred thousand tons of coal transported through the pipelines. Although wear is very transport-, material-, and system-dependent, HTRF experience indicates that, for this application, severe wear should not become a problem. Visual observations made through a plexiglass section in the six-inch pipeline indicated that the coarser fraction of the coal was being transported in an annulus in the pipeline center, with few coal particles actually impacting the pipe wall--explaining the lack of noticeable wear.

Operating experience with settling velocities indicated that the coal behaved as described in Figure 3 when the system was operated in the standard, no-operator-feedback mode. However, when the HTRF pumping system was retro-fitted with a computerized controller, the system was able to operate down to velocities of approximately one foot per second without plugging. This is primarily attributed to the increase in pump pressure at the point of incipient bedload formation and to rapid system surging caused by pump speed changes driven by the controller.

The operational experience gained at the HTRF indicated that very coarse material could be transported for substantially lower energy requirements than was previously predicted. Figure 4 (Elsibal, 1982) shows the expected economies for various modes of coarse coal transportation. As can be seen, for short to medium length pipelines, coarse coal offers extremely attractive capital economics. Additionally, coarse material transport has the added advantage of avoiding problems inherent with fine material dewatering. Coarse material can usually be dewatered by gravity methods, resulting in substantially lower capital and operating costs.

Novel Applications

As described above, slurry transportation is seen as an economical and safe way of transporting material over almost any distance. Its application is limited where materials will be transported over a sufficiently short time period, because the high initial capital costs will not be offset by lower operating costs over the short period. Applications in which slurry transportation could significantly reduce project costs and improve efficiencies are:

1. Transportation of low or high level radioactive wastes to interim or final disposal areas. -- Slurry transportation is seen as a reasonable way to move such materials, due to the hazards associated with dust and conventional vehicle accidents with these operations.

2. Tunnel muck excavation. -- Conventional tunnelling machines or methods are often limited by removal of the waste generated by the mining operation. Slurry transportation has the promise of substantially reducing the costs of this materials handling and improving underground safety, compared to conventional conveyor-haulage methods.

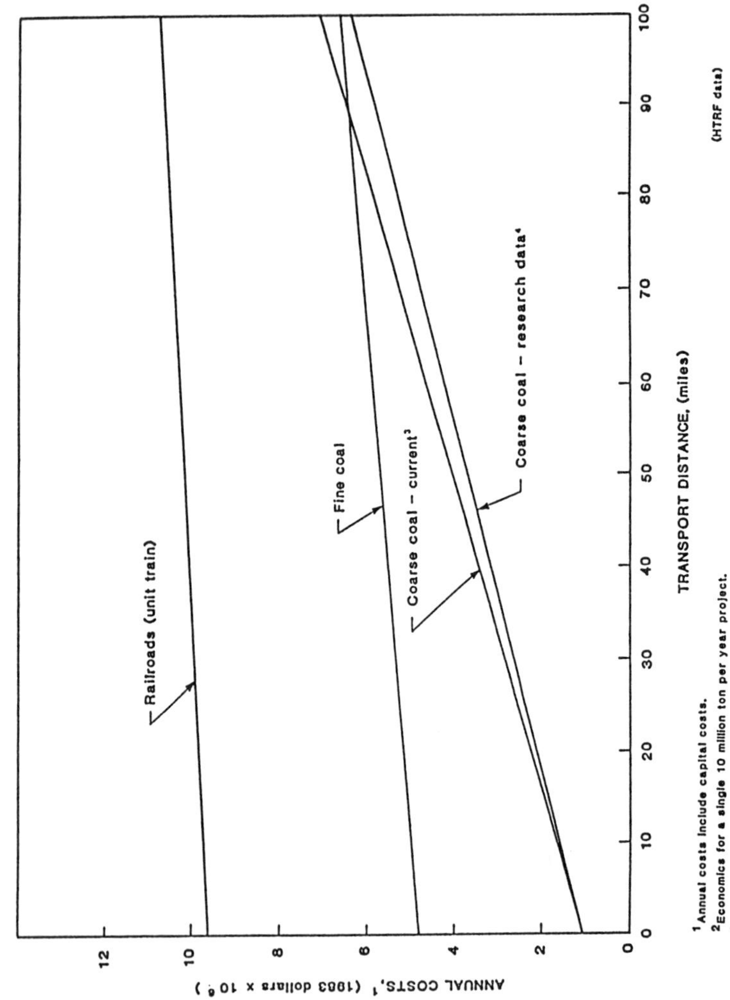

[1] Annual costs include capital costs.
[2] Economics for a single 10 million ton per year project.
[3] Currently available engineering design.
[4] Using results of the HTRF data.

FIGURE 4
Relative economics of coal transportation.

3. Hydraulic monitoring and transportation of large volumes of fine to medium grained materials. -- This could be applicable in large open pit mining operations where the material being moved is in a fairly loose state. Hydraulic mining gained a bad reputation due to poor mining practices in the late 1800's; however, using existing technology, most of the concerns can be negated in today's environment.

Summary

Hydraulic transportation of solids can substantially reduce the cost of transporting material in most operations. This paper has proposed a mathematical model which can be adequately used to design slurry transport systems, has described two slurry pipeline operations, and has suggested a few novel applications for such material transport systems.

This paper was designed to provide an overview of slurry transportation systems. For additional information, the reader is referred to the list of references following.

REFERENCES

Durand, R. "Transport Hydraulique De Graviers et Galets en Conduite." (Hydraulic Transport of Gravel and Sand in Pipes). La Houille Blanche. Special Report. 1951.

Durand, R., "The Hydraulic Transportation of Solids in Pipes: Experimental Studies for the Arrighi Power Plan Ash." La Houille Blanche. Vol. 6. 1951.

Durand, R. And E. Condolios, "The Hydraulic Transport of Coal and Solid Materials in Pipes." Proceeding of a Colloquium on the Hydraulic Transport of Coal. National Coal Board of England. 1952.

Durand, R., "Basic Relationship of the Transportation of Solids in Pipes-Experimental Research." International Association of Hydraulic Research, 5th Congress. 1953.

Durand, R., "Flow of Mixtures in Vertical Pipes, Etc." La Houille Blanche. Vol. 1. 1953.

Durand, R. "Seance Du Mercredi Matin 26 Novembre 1952." (Meeting on Wednesday Morning on November 26, 1952) La Houille Blanche. Special Report. 1953.

Durand, R., and E. Condolios. "Measurement of Pressure Losses Occurring in Clear Water Storage Pipes." La Houille Blanche. Vol. 9. 1954.

Durand, R., Confidential Report to the French Department of Ports and Roads. 1955.

Elsibal, N. G., P. E. Snoek and J. D. Pitts, "How Particle Size Affects the Cost of Short Distance Slurry Pipelines, Oil and Gas Journal. July 26, 1982.

Eyler, L. L., N. J. Lombardo and J. S. Barnhart, "FY 1980 and 1981 Progress Report - Hydrotransport Plugging Study." Battelle. 1982.

Faddick, R. R., "Technical Aspects of Shiploading Coal Slurries." Proceedings, Seventh International Technical Conference on Slurry Transportation, March 23-26, 1982. Slurry Transport Association. 1982.

Henderson, M. E. U. S. Bureau of Mines Open-File Reports :

______, 1983. Initial Experimental Studies on the Hydraulic Transport of Coarse Coal by Pipeline. Presented at the 8th International Technical Conference on Slurry Transportation, March 15-18, 1983, in San Francisco, CA.

______, 1984. Numerical Models for Predicting Head Loss in Coarse Particle Hydraulic Transport. Presented to the Graduate Faculty of School of Engineering, University of Pittsburgh in partial fulfillment of Master of Science in Civil Engineering.

______, 1984. Coarse Coal Hydrotransport Testing at the Hydraulic Transport Research Facility; 2-inch by 0 Clean Coal, 12-inch pipeline (Dec., 1984).

______, 1985. The Use of Bureau of Mines Facilities and Onboard Expertise to Expedite Development of Improved Rapid Underground Excavation Systems. Presented to the U.S. Defense Nuclear Agency.

______, 1985. Coarse Coal Hydrotransport Testing at the Hydraulic Transport Research Facility; 2-inch by 0 Clean Coal, 12-inch pipeline: R203 (June, 1985).

______, 1985. Coarse Coal Hydrotransport Testing at the Hydraulic Transport Research Facility; 2-inch by 0 Clean Coal, 12-inch pipeline: R203 (July, 1985).

______, 1985. Coarse Coal Hydrotransport Testing at the Hydraulic Transport Research Facility; 2-1/2-inch by 1-1/2-inch Clean Coal, 6-inch pipeline: R104 (July, 1985).

______, 1985. Hydraulic Transport Research Facility: Coal Degradation Report.

Makharadzye, L., "Determination of the Critical Velocity of a Slurry in a Pressurized Pipe." From Translations of Foreign Hydrotransport Technology Vol. 2 - Russian Technology. Terraspace, 1981.

Roco, M. C. "A Finite Volume Approach for Transport Flow Phenomena." The 5th International Symposium on Finite Elements and Flow Problems. January 1984.

Wilson, K. C., "Slip Points of Beds in Solid-Liquid Pipeline Flow". Journal of Hydraulic Division, American Society of Civil Engineers. 1970.

BEHAVIOR AND PROPERTIES OF HYDRAULIC FILLS

by
Robert L. Schiffman[1]
Steven G. Vick[2]
Robert E. Gibson[3]

ABSTRACT

This paper will review the theories of sedimentation and consolidation as these theories can be used to predict the development of pore-water pressures, settlement and undrained shear strength of hydraulic fills. Some commentary is made with respect to the compressibility, permeability and undrained shear strength of hydraulic fill material. A review of certain effects of large strain consolidation is presented. Finally a recent design case history is described.

INTRODUCTION

Deposition of hydraulic fill is analogous to the genesis of a soil deposit. That is the process, from filling, through self-weight and loading consolidation, can go through most, if not all, of the processes of soil formation. The difference between a natural and a hydraulically filled clay deposit is a matter of age, with the age of the clay fill measured in years and the age of natural clays (with the exception of deltaic deposits) measured in thousands of years or longer.

The genesis of developing soil deposits has been described by Imai (1981) as consisting of three stages as shown in Figure 1 and defined as follows:

> "In the first stage, no settling takes place, but flocculation yields flocs. In the second stage, the flocs gradually settle and form a layer of sediment, which undergoes consolidation and reduction of water content. The boundary between the upper settling zone and the sediment is the birth place of new sediment. While the sediment grows, the settling zone becomes thinner and finally vanishes. In the last stage, all of the sediment thus formed undergoes self-weight consolidation and finally approaches an equilibrium state."

---

[1]Professor of Civil Engineering, University of Colorado, Boulder, Colorado, U.S.A.

[2]Consulting Geotechnical Engineer, Indian Hills, Colorado, U.S.A.

[3]Director, Golder Associates, Maidenhead, U.K., and Adjoint Professor, University of Colorado, Boulder, Colorado, U.S.A.

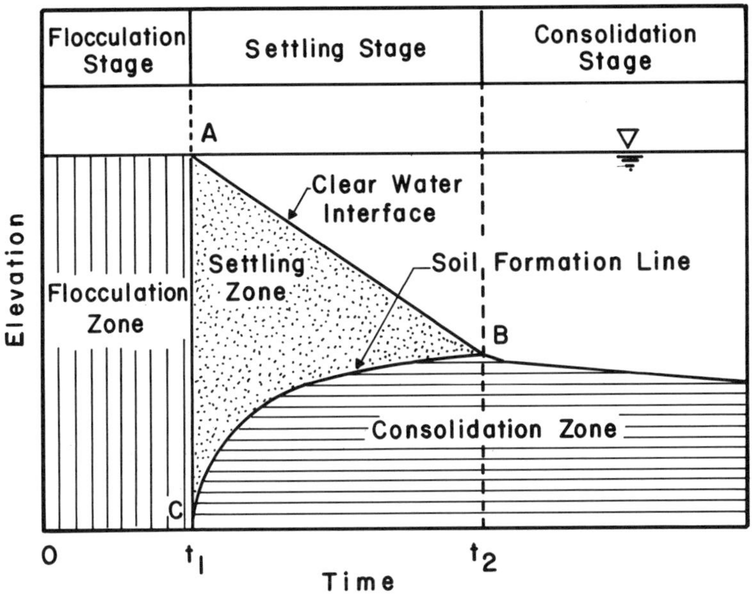

Figure 1
FLOCCULATION/SEDIMENTATION/CONSOLIDATION
(Imai, 1981)

Sedimentation

The theoretical background of sedimentation was established by the works of Kynch (1952) and Richardson and Zaki (1954). Richardson and Zaki theoretically examined the forces acting on the solid particles in dispersed conditions and showed that fluidization and sedimentation derive from the same principles. Although ignoring the transient process, i.e., the time dependence, they recognized that the drag forces depend only on the relative velocity of the two phases, which are equal in fluidization and sedimentation.

The first step away from this traditional approach to settling was by Kynch (1952). Kynch realized that the settling process of uniform dispersions is a highly transient process, and that "some" continuity equations need to be considered for the problem to be well-defined. Although representing a fundamental understanding of the settling process, the theory of hindered settling developed by Kynch focuses its attention on the continuity of the solid phase, i.e., the problem is simplified. Effective stresses in the sediment which is formed at the bottom of the dispersion are ignored, so that the velocity of the solid particles is a function of porosity only. This leads to Kynch's "flux-plot", which has been, and remains, the common basis for analysis and design of sedimentation processes in various disciplines such as chemical, environmental, coastal and geotechnical engineering.

Within the context of geotechnical engineering, experimental studies of sedimentation are reported by Been (1980), Imai (1980, 1981) and McRoberts and Nixon (1976). Relaxing Kynch's assumption of equi-sized particles on the basis that no segregation of particle sizes occurs at high solid concentrations, McRoberts and Nixon applied Kynch's theory to dispersions of natural silts and fine sand.

A more detailed discussion of the interdisciplinary applications of sedimentation is presented elsewhere (Schiffman, et al., 1985).

Consolidation

In this discussion of consolidation we will assume that the consolidating medium is fully saturated with an incompressible viscous fluid. Furthermore, it will be assumed that the consolidating medium consists of a system of incompressible particles which change position under the action of an imposed disturbance. We call this deformable part of the soil the "soil skeleton." Furthermore, the theory of consolidation is based upon the laws of continuum mechanics, so that the skeleton and water have distinctly defined and mathematically continuous properties. If one examines the theory of consolidation of a water saturated clay, one must rationally address the mechanical behavior of the soil skeleton, for this is the behavior which, by the effective stress principle, is the seat of the deformations and shear strength (Terzaghi, 1943).

In terms of the mechanical properties of the soil skeleton there are two distinct types of consolidation, which are:

- Primary (hydraulic) consolidation, which arises when the effective stress-strain properties are independent of time. As a consequence, the rate of generation of strain is related to the hydraulic conductivity (permeability) of the soil skeleton. This process can be called "hydraulic consolidation."
- Secondary consolidation ("secular effect") which arises when the soil skeleton is intrinsically time-dependent; i.e., the soil skeleton has a definable viscosity which results in a stress-strain-time relationship (Buisman, 1936). As a consequence, the progress of consolidation is related not only to the hydraulic conductivity (permeability) of the soil skeleton but also to its capacity to creep. Both these processes will occur simultaneously, but one or the other may dominate at different times.

Other mechanical properties can be attributed to the soil skeleton such as plastic yielding and combinations of elastic, inelastic, viscous and plastic response, but these are not pertinent to this discussion which will not discuss the application of the secular effect to hydraulic fill behavior. The evidence appears to indicate that the most important feature of the behavior of hydraulic fills is hydraulic (primary) consolidation. This effect along with sedimentation seems to provide the most significant magnitudes of deformation and shear strength of the fill.

The consolidation component of this paper will be exclusively related to one-dimensional nonlinear finite strain theory (Gibson, et al., 1967). The theory has been verified for soft or very soft disturbed soils (such as hydraulic fills) by centrifuge studies reported by Bloomquist and Townsend (1984), Croce, et al., (1984), Leung, et al., (1984), Mikasa and Takada (1984), Scully, et al., (1984), Townsend and Hernadez (1985), Townsend et al., (1985), and McVay, et al., (1986). Laboratory-derived properties and new testing techniques have been developed by Imai (1979, 1980, 1981), (Imai, et al., 1984), Znidarcic, et al., (1984), Znidarcic, et al., (1986), Carrier (1985), Carrier and Berkman (1984), and Martinez, et al., (1987).

The theory has been used in practice in a number of ways including capacity estimates for dredge-fill and tailings impoundments, settlement estimates for design of reclamation covers, and pore pressure and undrained shear strength estimates for stability analyses. For example, a detailed discussion of pore pressure considerations in stability of mine tailings impoundments has been presented by Vick (1983) and discussed by Morgenstern (1985). Bromwell (1984a, 1985) has applied the theory to dredged fills, Florida phosphatic clays and metal tailings. Cargill (1984), and Krizek and Somogyi (1984) have applied the theory to dredged fill impoundments. Caldwell, et al., (1984) and Ferguson, et al., (1985a, 1985b) have used the theory to size impoundments for gold and silver mine tailings, while DeMello and Carrier (1987) have provided conceptual designs for the disposal of bauxite tailings.

In all, as discussed by Schiffman, et al., (1984) and Templeton, et al., (1985), this theory has been successfully used to analyze the consolidation behavior of soft natural marine soils and a variety of hydraulic fills ranging from very soft phosphatic clays to metal tailings, as well as dredged fills.

Sedimentation/Consolidation

Almost all of the literature cited above recognizes the simultaneous presence of sedimentation and consolidation in a variety of industrial operations. These include batch and continuous separation processes, gravity thickening of sludges, mine waste disposal and dredging operations, to cite a few. However, little work has been done towards linking sedimentation and consolidation in a single framework. Some centrifuge modeling evidence has been reported by Bloomquist and Townsend (1984) and Townsend, et al., (1985). Been (1980) showed the relationship between the theories of Kynch (1952) and Gibson, et al, (1967).

A unified theoretical basis for sedimentation and consolidation processes of solid-water mixtures has been developed (Pane, 1985; Pane and Schiffman, 1985). In this theory the behavior of the mixture as a whole is governed by the same material properties during both processes. The link is provided by a modified effective stress principle (Schiffman, et al, 1984) and by extending the concept of the permeability to the dispersed state (Pane, 1985). We note here that the concept of hindered settling and the qualitative linkage between sedimentation and consolidation have long been recognized by environmental (sanitary) engineers (Mohlman, 1934).

This paper will touch upon the following subjects related to the genesis of hydraulic fills in place. First, the theories of sedimentation and consolidation will be reviewed. Second, some commentary will be provided concerning the significant properties of hydraulic fills. Third a review of the theoretical effects of consolidation will be presented. Finally, a recent design case history will be presented.

THEORIES OF SEDIMENTATION AND CONSOLIDATION

In this section we will review the theories of sedimentation and consolidation which are pertinent to this paper. However, before that can be accomplished we must carefully define the coordinate systems which are used in the theories.

Coordinate Systems

The mechanics of solids provides a relatively simple measure of deformation which is the relative movement of adjacent points under the action of an applied stress. However, in fluid mechanics the behavior of a flowing fluid can be examined in two ways (Lamb, 1932). The first is to establish a fixed point in space and observe the behavior of the fluid as it moves past this point. The distance of this fixed point from some datum is its Eulerian coordinate x. Another way to describe the behavior of a fluid is to select any particle and observe its behavior as in the course of time it moves from its initial position, which is located at a distance a from some datum. This is called a Lagrangian system and we will call a the "initial spatial coordinate."

Soil mechanics addresses both solid particles and fluid which are moving with respect to each other, as shown in Figure 2. In this figure the two types of coordinate system which are useful in describing the consolidation process are shown. Figure 2a shows an Eulerian system. In this system there is a box fixed in space through which both solids and fluids flow. During the process of consolidation we will fix our attention on what happens in this box, i.e., observe there the current magnitude and rate of variation of various qualities such as the pore-water pressures, void ratio, etc.

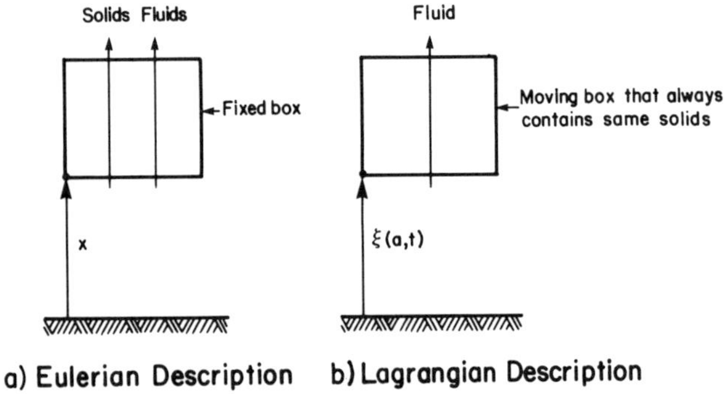

Figure 2
EULERIAN AND LAGRANGIAN COORDINATES

In Figure 2b we also have an elementary box, which in this case always contains the same solids. Clearly, if the solids are in motion this box will be moving also. This is the Lagrangian system in which only the fluid flows through the moving box.

In describing this motion we recognize that there are two measures of distance. First is the initial spatial (Lagrangian) coordinate a, which locates the box before any motion has commenced. Second is a "locator variable" ξ, a function of a and t, which is the distance of the box from some datum during the motion.

We are free, therefore, to describe the process of consolidation in the Eulerian manner, where the time and spatial variation of the various field quantities are expressed as functions of (x,t). Alternatively, if the Lagrangian system is adopted, the description is given in terms of (a,t) or (ξ,t). Each system has its own inherent advantages and disadvantages.

As shown in Figure 3 we can relate the Eulerian and Lagrangian coordinate systems to the behavior of a piezometer which is either fixed in space (Eulerian) or embedded in an elementary moving box. In the Eulerian system we focus our attention on an element of space or "box" $ABCD$ which is located at a distance x from a fixed datum plane. The points A, B, C, and D are labelled points of space.

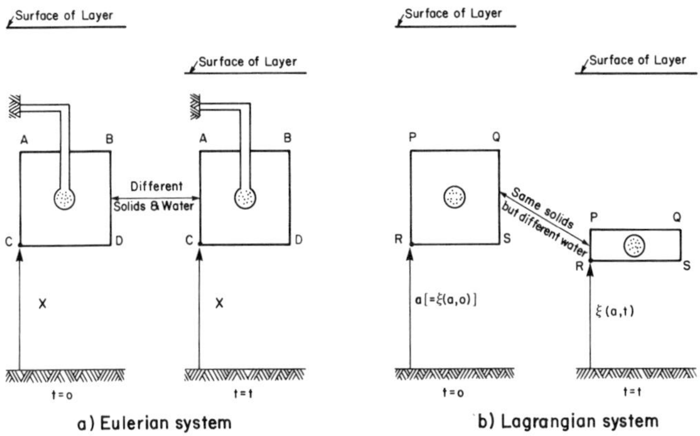

Figure 3
COORDINATES AND PIEZOMETER LOCATIONS

We noticed that in the Eulerian coordinate system a piezometer fixed in space is measuring the pore-water pressure as both the fluid and solids pass through the box at velocities v_w and v_s respectively. In the Lagrangian system the points P, Q, R and S are labelled grains in the skeleton. Hence the box $PQRS$ moves as consolidation takes place and it always contains the same solids. The piezometer also moves, as it is surrounded by these solids, and it measures the pore-water pressures at different locations at different times. Although the governing equations can be formulated in either Eulerian (x,t) or Lagrangian coordinates (a,t), we shall continue our discussion of small strain consolidation in terms of an Eulerian system. The consideration of the Lagrangian system will be used in discussing large strains.

Small Strain Consolidation

The governing equation is based upon the balance laws and constitutive relationships for each phase of the mixture of soil particles and pore fluid. The two fundamental balance equations express the conservation of linear momentum (equilibrium) and the conservation of mass (continuity). This is a quasi-static theory and no inertial forces are present. The required balance laws (equilibrium and continuity) must also reflect the balance of each constituent phase, soil skeleton and fluid. Since the effective stress equation provides a relationship between total stress (the bulk mixture), effective stress (the soil skeleton) and the pore-water pressure, the balance laws can be expressed in terms of any two of the three entities. Thus the required balance laws are:

- The equilibrium of the bulk (fluid and soil skeleton) mixture.

$$\frac{\partial \sigma}{\partial x} \pm \gamma_t (x) = 0 \tag{1}$$

where σ is the total stress, γ_t is the total unit weight which is a function of the Eulerian coordinate, x and the $\pm$ sign indicates whether the coordinate x is measured against/with gravity.

- The fluid equilibrium equation.

$$\frac{\partial u_w}{\partial x} - \frac{\partial u}{\partial x} \pm \gamma_w = 0 \tag{2}$$

which upon integration becomes

$$u = u_w \pm \gamma_w x + C(t) \tag{3}$$

where u_w is the pore-water pressure, u is the excess pore-water pressure defined as the increment (+ or -) with respect to static pore-water pressure, γ_w is at unit weight of water, and C is a constant or, at most, an arbitrary function of time.

- The continuity of the fluid phase.

The equation of continuity of the fluid assumes only that the pore-water is incompressible.

$$\frac{\partial (n v_w)}{\partial x} + \frac{\partial n}{\partial t} = 0 \tag{4}$$

where n is the porosity and v_w is the absolute seepage velocity of the pore-water.

- The continuity of the soil skeleton.

As explained earlier, both soil particles and pore-water are flowing through the element ($ABCD$) shown in Figure 3a. Then a continuity equation is required for the movement of solids through the element fixed in space.

$$\frac{\partial}{\partial x}[(1-n)v_s] + \frac{\partial}{\partial t}(1-n) = 0 \tag{5}$$

where v_s is the absolute velocity of the solids.

- The continuity of the mixture.

Adding the component continuity equations (4) and (5), leads to

$$\frac{\partial}{\partial x}[(1-n)v_s + nv_w] = 0 \tag{6}$$

Integrating equation (6) with respect to x results in

$$(1-n)v_s + nv_w = f(t) \tag{7}$$

where $f(t)$ is an arbitrary function of time to be determined later.

The fluid flow relationship for soil is usually Darcy's law (Darcy, 1856) which provides a proportionality between the velocity of water and the gradient of the total head. A more realistic representation of the effect of flow recognizes that during consolidation both solids and water are moving relative to each other. Thus a proper representation of Darcy's law is the Darcy-Gersevanov relationship (Gersevanov, 1934)

$$n(v_w - v_s) = -\frac{k}{\gamma_w}\frac{\partial u}{\partial x} \tag{8}$$

where k is the coefficient of permeability (hydraulic conductivity).

Equation (8) must reduce to Darcy's law when $v_s = 0$, i.e., the soil skeleton has a rigid structure. Furthermore, when no relative motion occurs the total head gradient must be zero.

Up to this point we have made no assumptions regarding the stress-strain relationships or the magnitude of strain. As such the governing equation of consolidation is

$$\frac{\partial}{\partial x}\left[\frac{k}{(1+e)\gamma_w}\frac{\partial u}{\partial x}\right] + f(t)\frac{\partial}{\partial x}\left[\frac{1}{1+e}\right] = \frac{1}{(1+e)^2}\frac{\partial e}{\partial t} \tag{9}$$

Since we are dealing only with primary consolidation and monotonic consolidation, the void ratio e is a single-valued function of the effective stress σ'. Thus

$$\frac{\partial e}{\partial t} = \frac{de}{d\sigma'}\frac{\partial \sigma'}{\partial t} = -a_v \frac{\partial \sigma'}{\partial t} \tag{10}$$

where the compressibility a_v must be a single-valued function of the void ratio. Then the governing equation becomes

$$(1+e)\frac{\partial}{\partial x}\left[\frac{k}{(1+e)\gamma_w}\frac{\partial u}{\partial x} + \frac{f(t)}{1+e}\right] = \frac{a_v}{1+e}\cdot\left[\frac{\partial u}{\partial t} - \frac{\partial \sigma}{\partial t}\right]. \tag{11}$$

Equation (11) accounts for non-linearity of properties since both k and a_v may be functions of the void ratio e. However, we note that we have one governing equation with dependent variables e and u and an unknown function of time $f(t)/(1+e)$ which accounts for the self-weight of the hydraulic fill, and is by itself related to the velocity of solids v_s and the hydraulic conductivity k. Were we to assume small strains and constant properties a_v and k, then the governing equation for an accumulating soil deposit which considers self-weight consolidation would be (Gibson, 1958)

$$c_v \frac{\partial^2 u}{\partial x^2} = \frac{\partial u}{\partial t} - \gamma' \frac{dh}{dt} \tag{12}$$

where the constant coefficient of consolidation c_v is

$$c_v = \frac{k}{m_v \gamma_w} = \frac{k(1+e_0)}{a_v \gamma_w} \tag{13}$$

in which γ' is the effective unit weight of the soil (a constant), m_v is the coefficient of volume change, e_0 is the constant initial void ratio and h is the height to the top of the clay layer (against gravity) at any time.

Equation (12) can also be expressed as

$$c_v \frac{\partial^2 u_w}{\partial x^2} = \frac{\partial u_w}{\partial t} - \left[\gamma_w \frac{dH}{dt} + \gamma' \frac{dh}{dt}\right] \tag{14}$$

where u_w is the pore-water pressure and H is the height of the water (against gravity) where $(H \geq h)$ at all times. In the following discussion of nonlinear finite strain theory we will compare our results to the small strain, linear relationships given by equations (12) or (14).

Large Strain, Nonlinear Consolidation

The development of the governing relationship for large strain, nonlinear consolidation (Gibson, et al., 1967) is best accomplished using Lagrangian coordinates (a,t) as shown in Figures 2 and 3. We find that

$$\frac{\partial}{\partial a}\left[\frac{(1+e_0)k(e)}{\gamma_w(1+e)}\frac{\partial u}{\partial a}\right]=\frac{a_v(e)}{1+e_0}\left[\frac{\partial u}{\partial t}-\frac{\partial \sigma}{\partial t}\right] \tag{15}$$

where e_0 is the initial void ratio, which is generally a function of the initial thickness a and in hydraulic fills may or may not be a constant. Clearly, however, the void ratio e is not a constant as time progresses. Thus this governing equation contains two dependent variables and without oversimplifying cannot be used in problems where large strains, such as exist in hydraulic fills, are important.

However, an unambiguous derivation of the governing equation can be developed in which the void ratio e is the dependent variable.

$$\frac{\partial}{\partial a}\left[\frac{(1+e_0)k(e)}{\gamma_w(1+e)a_v(e)}\frac{\partial e}{\partial a}\right]\pm f(e)\frac{\partial e}{\partial a}=\frac{1}{1+e_0}\frac{\partial e}{\partial t} \tag{16a}$$

where

$$f(e)=-\left(\frac{\gamma_s}{\gamma_w}-1\right)\frac{d}{de}\left[\frac{k(e)}{1+e}\right] \tag{16b}$$

where $f(e)$ is the term in the governing equation accounting for the self-weight of the hydraulic fill, and γ_s is unit weight of solids. This is **not** the same $f(t)$ used in equations (9) through (11).

A solution in terms of (a,t) is not sufficient since a is the initial position of a particle. What we must find is the position of a particle during the process of consolidation. This can be accomplished with reference to the phase diagram shown in Figure 4. If the current layer thickness is designated h_ξ then from the relationship shown in Figure 4,

$$h_\xi(t)=\int_0^{h_a}\frac{1+e(a',t)}{1+e(a',0)}da' \tag{17}$$

where h_a is the initial layer thickness.

The above relationships are quite adequate for hydraulic fill layers which start off with a fixed, known thickness and are subsequently loaded. However a major problem is that of a layer whose thickness is accumulating by a known deposition. In these cases we can use a reduced coordinate system which was first developed by

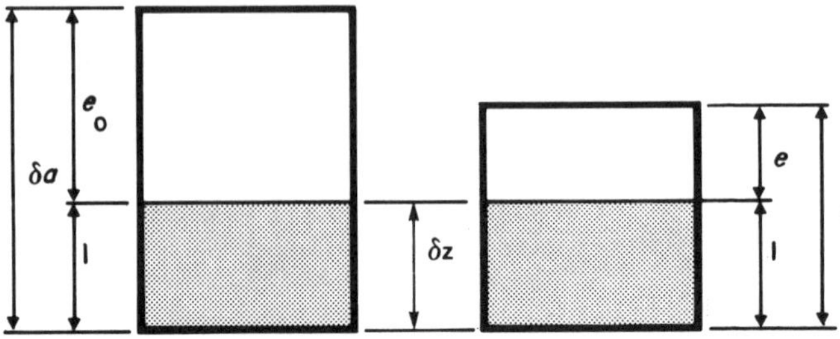

Figure 4
CHANGE IN VOID RATIO

Terzaghi (1923, 1925), Ortenblad (1930) and later independently by McNabb (1960). We define z, as shown in Figure 4, as the volume per unit area of solids lying between the datum plane and a given Lagrangian coordinate point a at time t. While z is a volume of solids, it has the units of length and can be used as a coordinate. Thus the governing equation can be expressed as

$$\frac{\partial}{\partial z}\left[g(e)\frac{\partial e}{\partial z}\right] \pm f(e)\frac{\partial e}{\partial z} = \frac{\partial e}{\partial t} \tag{17a}$$

where

$$g(e) = -\frac{k(e)}{\gamma_w (1+e) a_v(e)} \tag{17b}$$

The advantage of using the reduced coordinate z is that it represents the volume of solids, so that knowing the volume of solids in a hydraulic fill system the consolidation during accumulation can be calculated. Also from Figure 4 the relationship between z and the locator variable ξ can be developed. Then

$$h_\xi(t) = \int_0^{h_z(t)} \left[1 + e(z',t)\right] dz' \tag{18}$$

represents the actual height of the accumulating layer in terms of the momentary volume of solids.

Effective Stress Principle

Geotechnical engineering is built upon the principle of effective stress. This principle states that there is a state of stress, called the effective stress σ', which is responsible for the deformation of the porous mineral skeleton. Given the assumption that the porous medium is a two-phase system consisting of a deformable mineral skeleton filled with an incompressible liquid (water) the effective stress principle can be formulated as

$$\sigma = \sigma' + u_w \tag{19}$$

where σ is the total stress applied to the system and u_w is the pore-water pressure.

The effective stress principle governs the deformation of a soil; i.e., a deformable porous medium where the mineral skeleton has a definable stress-strain relationship. This is the material in the consolidation zone shown in Figure 1. In the settling zone, above the consolidation zone shown in Figure 1, the material is not a porous solid in the classical sense. Measurements in a sedimentation column (Michaels and Bolger, 1962; Been, 1980; Been and Sills, 1981) have shown that at the top of the settling zone the total stress and the pore-water pressure are equal. This would indicate that the particles have not aggregated and thus the effective stresses are zero. At the soil formation line (*CB* in Figure 1), particle aggregation has been established and the effective stresses are fully developed. It appears (Michaels and Bolger, 1962; Been, 1980) that there is also a thin transition zone separating the setting and consolidation zones where the effective stresses are non-zero but do not follow equations (16).

These observations have led us to reformulate the effective stress equation as follows (Schiffman, et al., 1984; Pane, 1985; Pane and Schiffman, 1985)

$$\sigma = \beta(e)\sigma' + u_w \tag{20}$$

where the interaction coefficient β is a monotonic function of the void ratio. A reasonable form of the $\beta(e)$ function is shown in Figure 5. The interaction coefficient β is equal to zero for values of the void ratio greater than e_m. In this case the particles (or aggregates of particles) are so distant that their interaction or contact is negligible and the mixture behaves as a heavy fluid. For values of the void ratio less than e_s there is full particle-to-particle contact and the interaction coefficient is equal to unity, such that the conventional effective stress principle fully applies and the mixture behaves as soil. Two kinds of (β,e) constitutive relationships are shown in Figure 5. In the general case (b) the interaction coefficient is a continuous function of the void ratio. In case (a) an abrupt change in β occurs at e_m.

Linked Sedimentation and Consolidation

There are basically three types of hydraulic fills, whose behavior varies with the coarseness and plasticity of the fill material. We can categorize these in terms of the interaction coefficient β described above. On the coarser end where the fill is either sand, silt, or a combination thereof, the coefficient β is either zero or close to it. Then Kynch's (1952) theory applies as shown by McRoberts and Nixon (1976). On the fine and plastic end where clays are being deposited, β may vary from zero to unity going through sedimentation to consolidation as shown in Figure 1. It has been

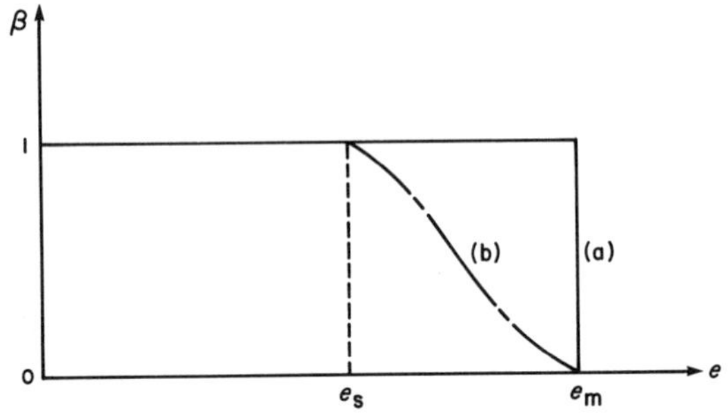

Figure 5
FORMS OF INTERACTION COEFFICIENT β
(Schiffman, et al., 1984)

shown that for phosphatic waste clays with an initial solids content of 17% or below ($e \geq 12.0$) both sedimentation and consolidation phases occur (Bloomquist and Townsend, 1984; Townsend, et al., 1985).

All of these phenomena can be accommodated by a linked theory of sedimentation and consolidation (Pane, 1985, Schiffman, et al., 1985).

$$\frac{\partial}{\partial z}\left\{ a(e)\left[\sigma' \frac{d\beta}{de} - \frac{\beta(e)}{a_v(e)} \right] \frac{\partial e}{\partial z} \right\} \pm f(e)\frac{\partial e}{\partial z} = \frac{\partial e}{\partial t} \tag{21a}$$

where

$$a(e) = -\frac{k(e)}{\gamma_w(1+e)} \tag{21b}$$

The governing equations (21) assumes that β is a continuous function of the void ratio. We choose as a first approximation a condition in which β is a step function at the soil formation void ratio, e_m. That is, there is no transition zone between consolidation and sedimentation. Under this condition when $\beta = 0$, ($e > e_m$) the governing equation is

$$f(e)\frac{\partial e}{\partial z} = \frac{\partial e}{\partial t} \tag{22}$$

which is Kynch's (1952) equation in terms of the reduced z coordinate. When $\beta = 1$, ($e < e_m$) then equations (21) reduce to equations (17) which govern nonlinear finite strain consolidation.

HYDRAULIC FILL PROPERTIES

There are four properties that govern the behavior of a hydraulic fill. These are compressibility, permeability, the interaction coefficient β, and the undrained shear strength. The first three of these properties relate to the sedimentation/consolidation behavior while the fourth property relates to the stability of hydraulic fill structures. We will not discuss the $\beta(e)$ function here for reasons of space and ignorance of the relation of β to the pore-water chemistry, composition of the soil skeleton, in many cases the initial void ratio of the mixture, and even time.

Compressibility

The measurement of compressibility (void ratio-effective stress) of a hydraulic fill is in principle a simple process. Using the oedometer as an example it would appear that all one has to do is to load a contained sample and record the height (void ratio) of the sample in the drained state. For relatively stiff fill materials a standard oedometer test is sufficient. However, serious problems develop when the fill material is very compressible and is tested at high void ratios. These are primarily the test mechanism which will maintain an intact sample and the time of testing since very soft fill material take an inordinate amount of time to consolidate. In order to provide a base line for consideration we will concern ourselves with initial void ratios well above 7 to 10 and times to consolidate in a step-load environment that is of the order of months, for example the Florida phosphatic clays (Bromwell and Carrier, 1979).

From a growing realization that the step-load oedometer test is awkward and time consuming, there is considerable interest in the development of quicker tests to accurately assess compressibility properties. It is not our concern here to discuss the various types of test hardware. For this we refer the reader to several papers which adequately cover the subject (Znidarcic, et al., 1984; Znidarcic, et al., 1986; Olson, 1985; Martinez, et al., 1987; Yong and Townsend, 1985). These papers at least provide sufficient references to review the actual test design and procedure.

One feature of compressibility testing which deserves further consideration is shown in Figure 6 (Scully, et al., 1984) which provides a series of compressibility measurements on a Florida phosphatic clay. There is an appearance of a "preconsolidation" effect for samples with high initial void ratios although all the samples were initially in a thoroughly disturbed state and hence should be considered normally consolidated. We attribute this effect to self-weight consolidation and/or sedimentation which can take place during sample preparation before the compressibility test has been run.

Permeability

Just as our discussion of compressibility was not concerned with the design of devices, this discussion of permeability is concerned only with the principles of testing which affect the measured permeability. Figure 7 shows a set of permeability test results performed on a Florida phosphatic clay (Scully, et al., 1984) companion to the data shown in Figure 6. We see here the effects of sedimentation and/or self-weight consolidation which we believe occurs before the test has been performed.

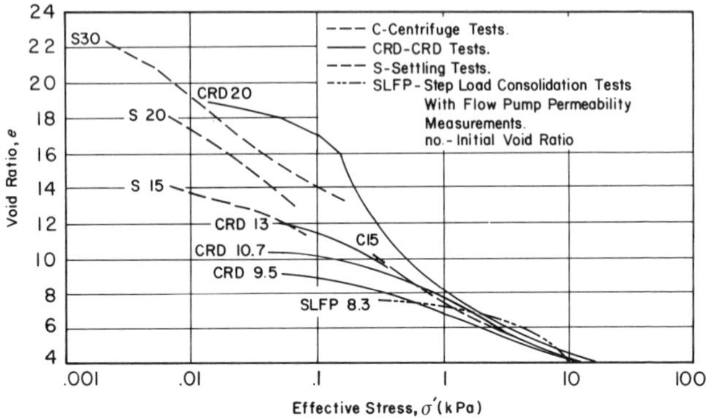

Figure 6
COMPRESSIBILITY RELATIONS
(Scully, et al., 1984)

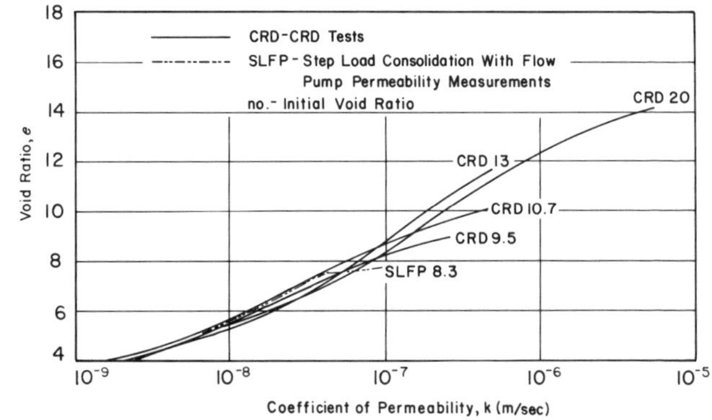

Figure 7
PERMEABILITY RELATIONS
(Scully, et al., 1984)

Defining the permeability as an empirical Darcy-Gersevanov relationship

$$k = -\frac{n\,(v_w - v_s)}{\frac{\partial u}{\partial \xi}}\,\gamma_w \tag{23}$$

a series of tests were performed on a remoulded kaolin with void ratios which varied between approximately 6 and 140. These data are shown in Figure 8 (Pane, 1985). The test data at the low void ratios (<6) was performed by a flow pump technique (Olsen, 1966; Pane, et al., 1983) for here the interaction coefficient β was clearly unity. At the very high void ratios where β is clearly zero (sedimentation), tests were performed by a fluidization experiment (Pane, 1985) which maintained the velocity of solids at zero. It is noted that the permeability between low (e=9) and high (e >50) void ratios is interpolated by an inversion process using the velocity of the soil-water interface at early settling times.

Certain conclusions can be drawn from the experimental data shown in Figure 8. First, the variation of the coefficient of permeability is enormous. It increases almost eight orders of magnitude for void ratio changes of less than two orders of magnitude. Second, at very large void ratios the coefficient of permeability is approximately a linear function of the void ratio. Third, in the intermediate range of void ratio (shown by the dotted line in Figure 8) there are no known techniques for a direct measurement of the permeability. In fact, even the inversion process briefly mentioned above is subject to interpretation. In this region interaction between aggregates is starting to occur and the mixture consists of aggregates which behave as a soil (β = 1) and particles which are subject to hindered settling (β = 0).

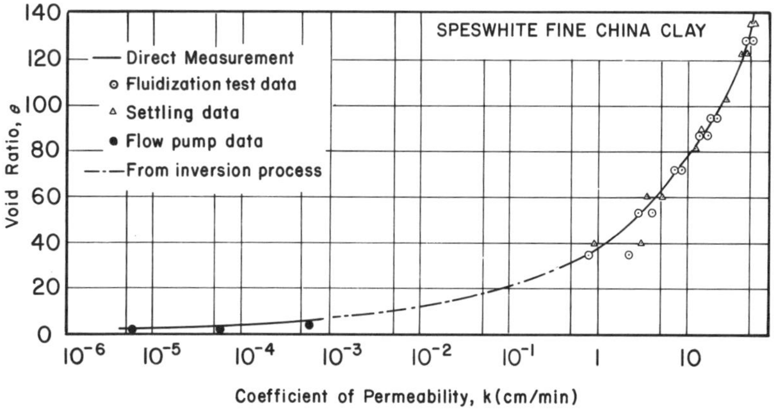

Figure 8
VOID RATIO-PERMEABILITY RELATIONSHIP
(Pane, 1985)

Another major feature in the direct permeability measurement of high void ratio soils (β = 1) is the phenomenon of seepage-induced consolidation (Pane, et al., 1983). Here the process of measuring permeability will induce seepage stresses causing consolidation. Thus the test will cause the sample to consolidate and change the permeability. An example of this is shown in Figure 9. The remedy here is simple. Permeability tests for soft soils, such as hydraulic fills should be performed at very low gradients, preferably less than one.

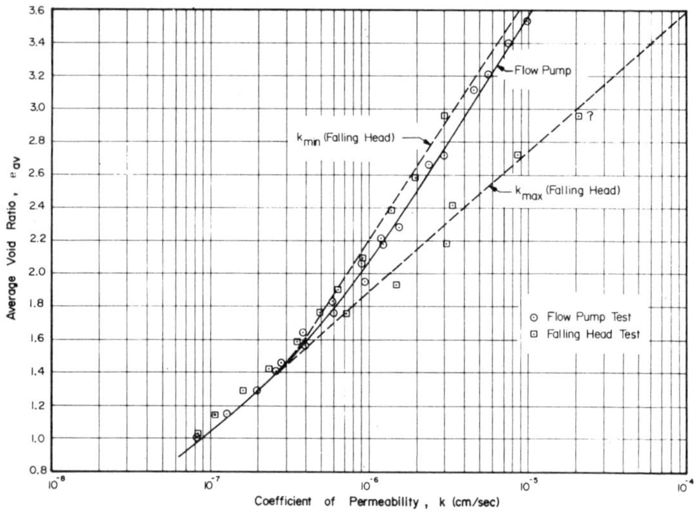

Figure 9
VOID RATIO-PERMEABILITY RELATIONSHIPS
(Pane, et al., 1983)

Coefficient of Consolidation

The definition of the coefficient of consolidation c_v is given by equation (13). Normal practice is to measure c_v and m_v by conventional methods and then calculate k via equation (13). Figure 10 shows a comparison of three procedures and a validation by centrifuge tests. The curve marked "calculated c_v" was developed by calculating c_v from directly measured compressibility and permeability data, and using conventional consolidation theory. The curve marked "measured c_v" was obtained by the normal laboratory procedures (square root of time method) and conventional consolidation theory. Both values of c_v used were taken at the average void ratio. The finite strain curve used directly measured, nonlinear void ratio-effective stress and void ratio-permeability curves as inputs to a nonlinear finite strain consolidation analysis given by equations (17) and (18) above. The process was verified by centrifuge experiments.

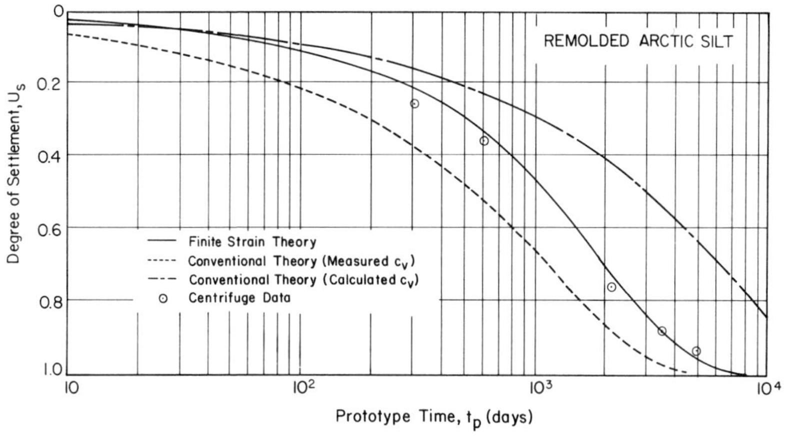

Figure 10
PROGRESS OF SETTLEMENT

We see in Figure 10 that the use of c_v and conventional consolidation theory falls far from the mark. As a consequence we firmly believe that the coefficient of consolidation as a measure of the progress of settlement of soft hydraulic fills will provide erroneous results. Aside from mysticism and lucky guessing the only appropriate means of accurately calculating the progress of consolidation of soft hydraulic fills is by means of directly measured compressibility and permeability data and nonlinear, finite strain consolidation theory.

Undrained Shear Strength

Shear strength under undrained conditions is that most often relevant to hydraulic fills due to the inability of cohesive materials to experience dissipation of pore pressures during shearing, and also due to interest in evaluating the potential for flow failures in both cohesive and cohesionless fill types. Like sedimentation and consolidation, undrained shear strength properties are influenced by techniques used to measure them. Strength of a hydraulic fill, however, is of concern for conditions considerably more diverse and less easily modeled than its sedimentation/consolidation genesis. Hence, both the methods of measurement and definitions of undrained strength vary according to the behavior mode being considered. These include:

- Strength behavior of the slurry during deposition.
- Strength behavior during and subsequent to consolidation.

- Strength behavior after strains sufficient to trigger rapid increase in pore pressure, which may be due to either static or cyclic loading. This refers to the onset of liquefaction flow failure.

Undrained shear strength during deposition is of interest where it is desired to predict the slopes which a thickened slurry will achieve during discharge. This condition has been studied by Palmer and Krizek (1987), who characterized undrained shear strength of bauxite red mud slurry by the yield stress τ_y. Capillary and rotational viscometer measurements shown in Figure 11 illustrate that τ_y is quite sensitive to slurry solids content, and that considerable scatter in the data occur. Used in a two-dimensional limit equilibrium model, it was found that laboratory measurements of τ_y consistently overestimated flow distance in field tests.

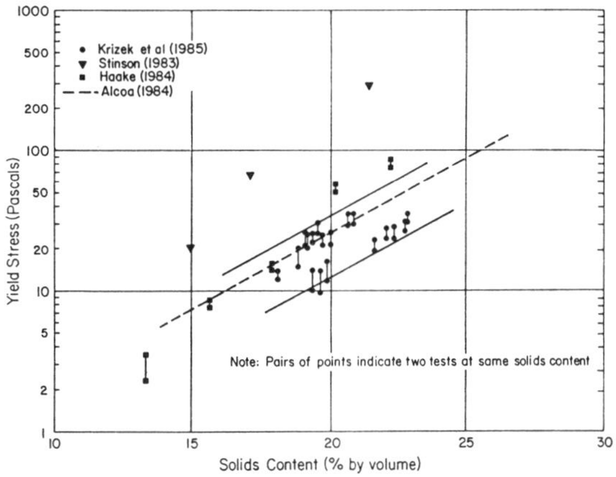

Figure 11
YIELD STRESS VS. SOLIDS CONTENT FOR BAUXITE RED MUDS
(Palmer and Krizek, 1987)

The more common hydraulic fill condition requires estimating undrained shear strength during the consolidation process or after it is complete. Here conventional geotechnical concepts of undrained shear strength apply, and for those hydraulic fill materials having significant plasticity, undrained shear strength can be evaluated according to normalized soil properties approaches developed for natural clays by Ladd and Foott (1974). It is convenient to express undrained shear strength according to the ratio s_u/σ_c'. This allows changes in undrained shear strength over time to be tracked within a deposit according to the consolidation-induced pore pressures that occur.

For typical hydraulic fills, normally consolidated conditions apply. For such natural soil deposits, Bishop and Henkel (1962) show a correlation with plasticity index (in percent) according to the form:

$$s_u/\sigma_c' = 0.11 + 0.0037\,(PI) \tag{24}$$

For hydraulic fills composed of normally consolidated phosphatic clays Bromwell and Raden (1979) cite $s_u/\sigma_c' = 0.22$ from $\overline{CK_0U}$ direct simple shear tests, a value similar to that for many natural clays of high plasticity. For a deposit of cycloned copper slimes tailings having a *PI* of approximately 6, Bromwell (1984b) found $s_u/\sigma_{v0}' = 0.26$ from field vane data. CIU triaxial test data indicated s_u/σ_{1c}' of 0.30 to 0.33, with similar values of 0.32 to 0.34 from CAU tests. Vick (1983) reported that cycloned, non-plastic lead-zinc slimes did not exhibit normalized soil behavior at higher OCR, but that s_u/σ_c' could be taken as approximately 0.20-0.22 with sufficient accuracy for engineering purposes at $OCR = 1$.

Those portions of coarser hydraulic fills deposited on above-water beaches may consist predominantly of cohesionless sands and silts at low to moderate relative densities typically in the range of 30-50% (Vick, 1983). Such materials may exhibit a peak undrained shear strength followed by a post-peak reduction at small strains, as reported by GEI (1982) for cycloned copper sand tailings and by Highter and Tobin (1980) for garnet, zinc, and iron ore sand tailings. In laboratory investigations on natural sands at relative densities typical of many hydraulic fills, Kramer and Seed (1988) found this effect to be related to the magnitude of confining stress and initial consolidation shear stresses. The influence of anisotropic consolidation was considered to be so strong as to limit the usefulness of ordinary limiting-equilibrium methods of stability analysis.

The final class of undrained strength behavior is for conditions of incipient or actual flow failure, and laboratory investigations have been taken several approaches. Using the previously discussed yield stress concept in a Bingham-plastic model to predict the distance of flow slide travel, Bryant, et al., (1984) measured normalized τ_y/σ_1' and plastic viscosity η_p for eight essentially non-plastic silty sand tailings materials using a vaned coaxial viscometer. Test results were found to be strongly influenced by small differences in water content, and τ_y could not be related to index properties of the materials. Undrained shear strengths back calculated from actual flow failures did not agree with laboratory values, with the difference attributed to intertial effects and viscous resistance. Another approach is the "steady-state" concept which defines undrained shear strength at the "steady-state of deformation" which occurs after very large strains (Poulos, 1981). This approach has been used to evaluate the potential for flow failures in bauxite red muds (Poulos, et al., 1985). However, the undrained steady-state strength s_{us} is sensitive to very small changes in void ratio which makes it difficult to evaluate in laboratory tests and produces extreme variations in soft hydraulic fill deposits during consolidation where void ratio varies markedly with time and location. In an attempt to overcome difficulties in laboratory strength determinations, Seed (1987) has proposed to base the residual undrained shear strength, s_r, on values back calculated from actual flow failures and related to the corrected SPT blowcount $(N_1)_{60}$ as shown on Figure 12. This procedure assumes that the same characteristics that influence undrained shear strength during flow failure also influence blowcount and that the prior blowcount in a deposit reflects its behavior as it undergoes flow failure.

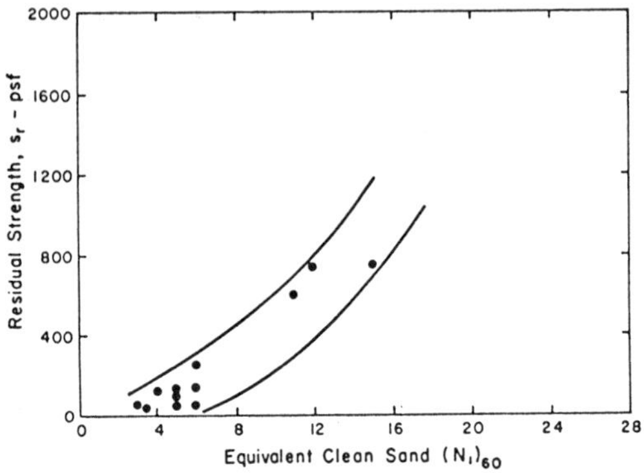

Figure 12
TENTATIVE RELATIONSHIP BETWEEN RESIDUAL UNDRAINED SHEAR STRENGTH AND SPT N-VALUES FOR SANDS
(Seed, 1987)

From these discussions it is clear that concepts of undrained shear strength vary considerably for hydraulic fill applications, and understanding of strength behavior lags behind that of sedimentation and consolidation processes. Only for cohesive materials during and after consolidation are undrained strength estimates reasonably straight forward using procedures based on behavior of normally consolidated natural clay deposits. For cohesionless sands and silts undrained shear strength may be subject to large reductions from peak values. Although several approaches have been used to estimate the reduced values, much remains to be understood about this and related aspects of undrained shear strengths characterization for hydraulic fills in engineering practice.

SEDIMENTATION EFFECTS

As discussed previously there are both sedimentation and consolidation effects in the settling process of hydraulic fills. Normal engineering practice tends to ignore sedimentation aspects of the process. However, in many cases, as shown in Figure 13, the sedimentation aspects may be quite important, and in the laboratory case shown, accounts for a major part of the settlement. Furthermore, the time of sedimentation is not insignificant (Pane, 1985).

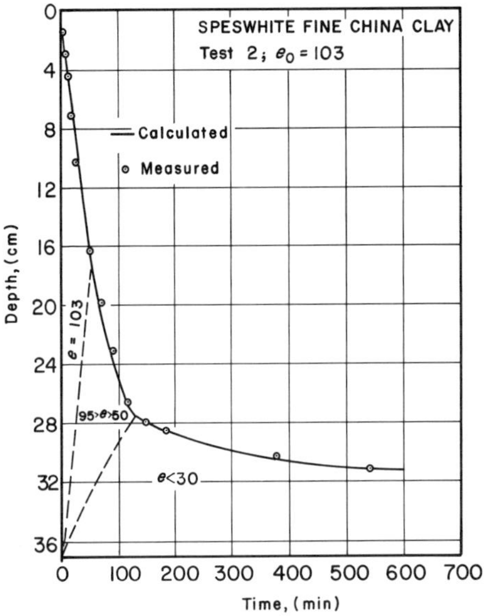

Figure 13
SEDIMENTATION/CONSOLIDATION
(Pane, 1985)

FINITE STRAIN CONSOLIDATION EFFECTS

In this section we will reiterate two effects related to nonlinear finite strain consolidation which are of some importance in field situations (Schiffman, et al., 1984). The first of these concerns the settlement due to loading of a soft marine soil. The consolidation properties of the soil are equivalent to a soft clay hydraulic fill. When we surcharge this deposit we have the choice of two variables which provide the degree of consolidation. The first of these we call the degree of settlement U_s defined as the ratio of the current settlement to the ultimate consolidation settlement,

$$U_s(t) = \frac{S(t)}{S(\infty)} \tag{25}$$

where S is the settlement of the layer. The second relationship indicating the progress of consolidation we call the degree of excess pore-water pressure dissipation U_e. We define this as

$$U_e(t) = 1 - \frac{u_{av}(t)}{u_{av}(0)} \tag{26}$$

where u_{av} is the average excess pore-water pressure. If the conventional theory of consolidation is used, both U_s and U_e are identical and are called the degree of consolidation. However if nonlinear finite strain consolidations used, we find that U_e and U_s are quite different as shown in Figure 14. What we see here is that the pore-water pressure dissipates at a slower rate than settlement progresses. Thus even after the settlement of a fill is substantially progressed the fill may still contain significant pore-water pressures. Thus, the stability of a hydraulic fill cannot be judged on the basis of the progress of its settlement. This phenomenon has been observed in many centrifuge tests where a clay slurry is brought to a state of normal consolidation. Even when the displacement of the test package has ceased under a high g-level, significant pore-water pressures exist, and the shear strength is substantially lower if the excess pore-water pressures had fully dissipated.

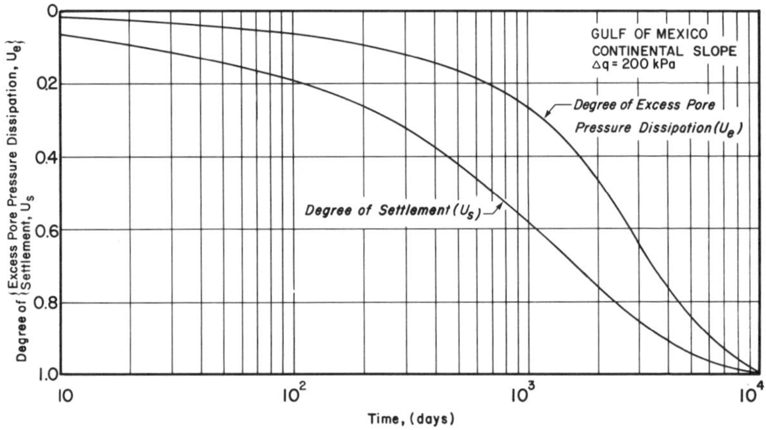

Figure 14
PROGRESS OF CONSOLIDATION
(Schiffman, et al., 1984)

A second example concerns the reconstruction of the events that led to the failure of a copper tailings impoundment built by upstream construction methods. At issue were the pore-water pressures which existed at the time of failure. Consolidation properties and the amount of material deposited were known. It was also known that failure occurred 3390 days after the start of construction. The impoundment at that time had reached a height of 49.5 meters above its dry permeable base. Furthermore the void ratio e_0 of the slimes at deposition was 1.45 and the average void ratio

e_{av} of the slimes at 3390 days was 0.8. During the development of the impoundment a surcharge was placed on the surface beginning at 3030 days and continuing at a constant rate of loading of 0.47 kPa/day for 360 days. Given field monitoring of the actual height and the average void ratios as functions of time, the material height function $h_z(t)$ necessary as input to the governing equation could be determined directly, according to the relationship

$$h_z(t) = \frac{h(t)}{1 + e_{av}(t)} = \frac{h_a(t)}{1 + e_0} \tag{27}$$

This procedure provides benchmarks that must be matched for the validation of the analysis. These benchmarks may be the measured current height, void ratio and pore pressure profiles, etc., depending on the availability of field data.

Figure 15 presents the history of the Lagrangian (no consolidation) calculated and measured heights of the slime impoundment. The calculated height is 51 meters, which represents a 3% difference with the measured benchmark height. The calculated average void ratio at 3390 days is 0.82.

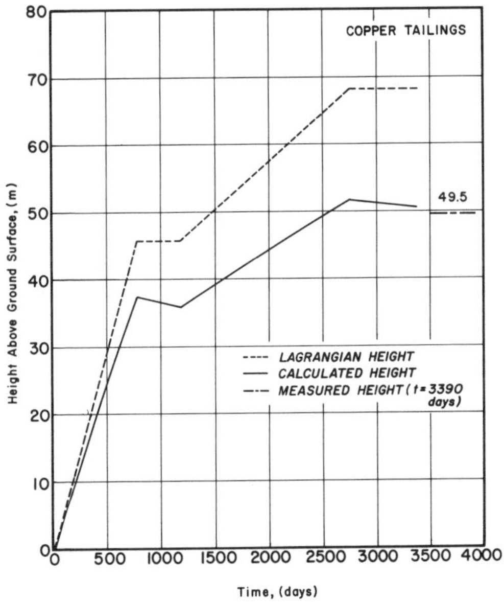

Figure 15
HEIGHT-TIME RELATIONSHIPS
(Schiffman, et al., 1984)

In order to compare the nonlinear finite strain approach to a companion linear small strain analysis, pore-water pressure profiles were calculated at 2760 and 3390 days. Figures 16 present the static u_0 and the pore-water pressure profiles u_w for the slimes. The earliest date (2760 days) represents the end of filling, some 270 days before the commencement of surcharging. The second date represents the end of surcharging. It is noted that at 3390 days the water level is above the slimes, thus providing a non-zero static pore-water pressure at the interface between the slimes and the surcharge.

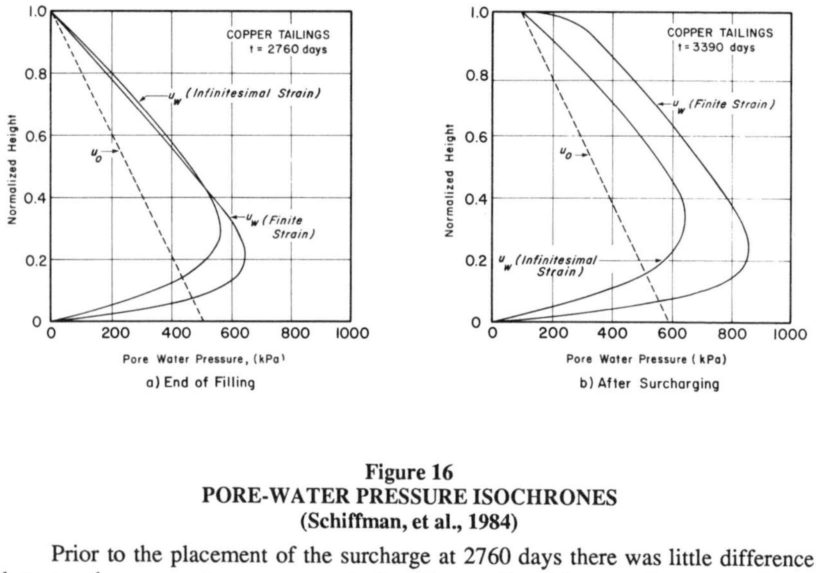

a) End of Filling

b) After Surcharging

Figure 16
PORE-WATER PRESSURE ISOCHRONES
(Schiffman, et al., 1984)

Prior to the placement of the surcharge at 2760 days there was little difference between the pore-water pressures calculated by small and large strain analyses in the upper 75% of the slime pond. This observation should not be generalized because there are numerous other examples (Schiffman, et al., 1984) where there is no congruence between small and large strain analysis of filling procedures. We frankly do not have any hypothesis as to why the end of filling pore-water pressure isochrones shown in Figure 16a are so similar. We observe further the effect of the surcharge, shown in Figure 16b. Here the use of conventional, small strain theory seriously underestimates the pore-water pressures compared to finite strain theory. This observation is fully consistent with the first example in this section.

USE OF FINITE STRAIN METHODS FOR UNDRAINED STRENGTH DETERMINATION

For hydraulic fills of cohesive, fine-grained materials a problem of engineering interest is the magnitude of undrained shear strength that the fill will achieve, accounting for both its rate of increase with time as deposition proceeds and its distribution with depth. It has been previously shown that cohesive fills during and after consolidation are among the few cases for which undrained shear strengths can be predicted with reasonable confidence according to the variation in s_u with effective stress and, hence, pore pressure. It has also been shown that conventional consolidation theory can yield excess pore-water pressures in soft deposits that are too low and therefore undrained shear strengths that may be unconservative. The use of non-linear finite strain consolidation approaches in conjunction with normalized undrained strength parameters provide a useful method to address this problem.

The so-called "thickened discharge" method of tailings disposal was first proposed by Shields (1974) and subsequently developed by Robinsky (1979). It consists of thickening a mill tailings slurry using flocculants and mechanical thickening devices to a density such that the material will achieve a flat slope upon discharge. The need for sedimentation and impoundment of water in the tailings disposal area are thereby avoided, and the tailings can be deposited in a gently-sloping conical pile. Among the proposed applications of this method is to deposit the thickened tailings cone on the surface of an existing impoundment, as depicted on Figure 17. If the existing tailings are soft and fine-grained, undrained shear strength is of interest in determining whether the existing materials can support the overlying thickened deposit.

This condition was investigated for an operating lead-zinc mine where it has been proposed to deposit thickened slimes tailings to a height of 15 m upon a diked impoundment of existing slimes 15 m deep. The slimes are clayey silt with 98% passing the no. 200 sieve and about 18% clay (0.001 mm). The material has a *PI* of 8, and the clay particles are indicated to be predominantly kaolinite. By thickening the material to a void ratio of 2, it has been estimated that pile slopes of 3% can be achieved.

As the conical pile is deposited, the pile toe will advance laterally over the existing slimes at a rate governed by the mill tailings discharge. For purposes of consolidation analysis, this can be idealized as a one-dimensional accreting layer at any point as shown on Figure 18. At the end of deposition, no loading is imposed at the pile toe and here the undrained shear strength profile with depth in the existing slimes is governed by initial static pore-water pressures. Beneath the pile center, excess pore-water pressures can be calculated from the Lagrangian accretion rate of 4.5 m/yr. over a 3.3 year loading period to yield the vertical distribution of undrained shear strength at the end of deposition. Undrained strength s_u profiles at intermediate locations can then be interpolated.

For the case described oedometer tests had been performed on the slimes, but samples were not available for more detailed testing. Two samples of copper slimes with similar gradation and index properties, for which detailed void ratio-permeability and void ratio-effective stress relationships had been previously determined, were selected for comparison. As shown on Figure 19, these curves were found to reasonably bracket the available oedometer data and were used in the finite strain consolidation analyses. It was also determined from index property correlations and testing on similar materials that $s_u/\sigma'_{vc} = 0.25$ would reasonably represent the undrained shear strength of the normally consolidated thickened discharge slimes and the existing slimes deposit.

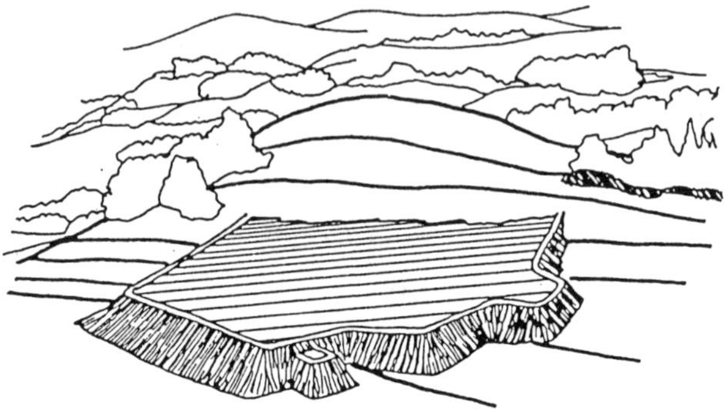

a) Initial condition-diked slimes impoundment

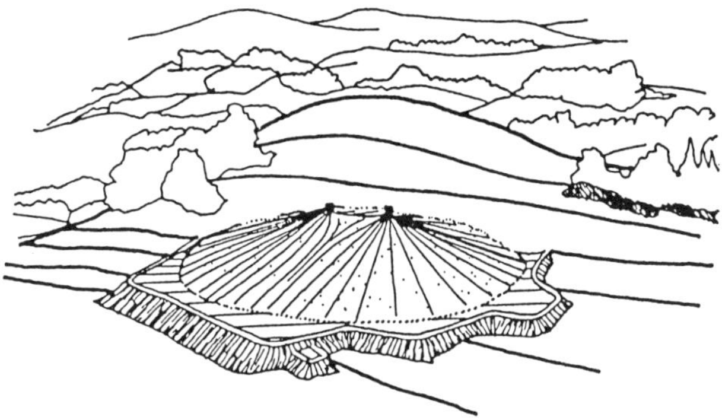

b) Final condition-thickened discharge slimes pile on existing slimes

Figure 17
THICKENED-DISCHARGE IMPOUNDMENT CONFIGURATION
(Robinsky, 1979)

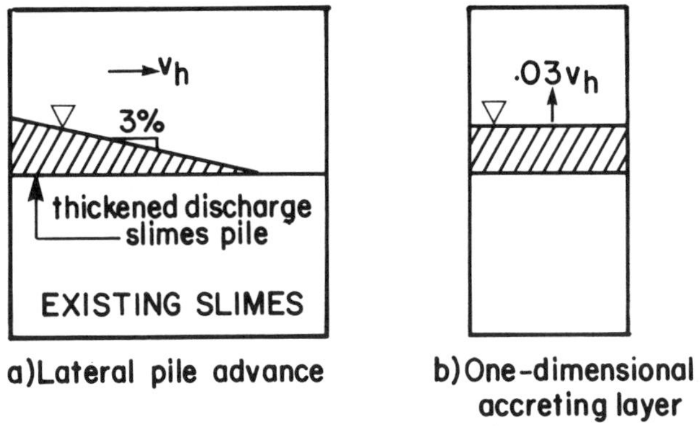

Figure 18
IDEALIZED LOADING CONDITIONS

Finite strain analyses were performed beneath the pile center using the program ACCUM. The program addresses one-dimensional nonlinear finite strain consolidation, which is reasonable in this case. An impervious lower boundary was also assumed.

Resulting vertical profiles of static and excess pore-water pressures at the end of accretion are shown on Figure 20a. It can be seen that total (static plus excess) pore-water pressures in the existing slimes are about 50 to 60 percent higher than static values alone. It was also found that the two bracketing curves for void ratio-permeability and void ratio-effective stress produced no significant difference in predicted pore pressures.

Predicted distributions of undrained shear strength are shown on Figure 20b. The profile at the pile toe is entirely within the existing slimes, which are also represented by the lower 15 m beneath the center of the pile. In both cases a nominal $s_u = 2.5\ kPa$ at the slimes surface was attributed to desiccation and evaporation effects. The two curves provide for comparisons of the consolidation-induced s_u increase that occurs in the existing slimes beneath the center of the pile. Significant increases occur only within the upper portion of the existing deposit by the end of pile accretion. Undrained shear strength s_u increases by about one-third at mid-height of the existing slimes, and by about half this amount at the bottom of the existing deposit.

These results are sensitive to the particular accretion rates used, which follow directly from pile slopes and tailings discharge rates, and cannot be generalized to other conditions. They illustrate, however, that finite strain consolidation approaches have useful application to strength prediction problems for certain classes of hydraulic fill deposition. In some cases, it may be possible to obtain meaningful results from index property data of the kind available during early stages of investigations, and this situation should improve as a more comprehensive information base is collected from a wider variety of hydraulic fill materials.

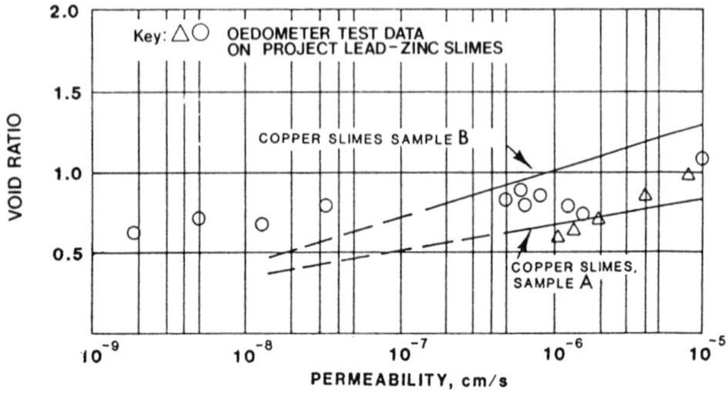

a) Void ratio-permeability relations

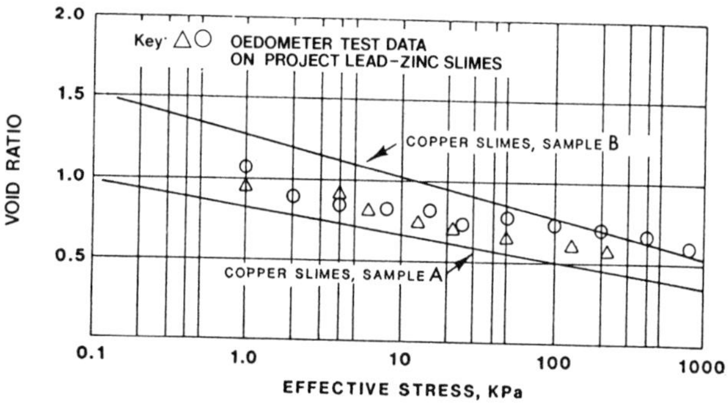

b) Void ratio-effective stress relations

Figure 19
CONSOLIDATION PROPERTIES

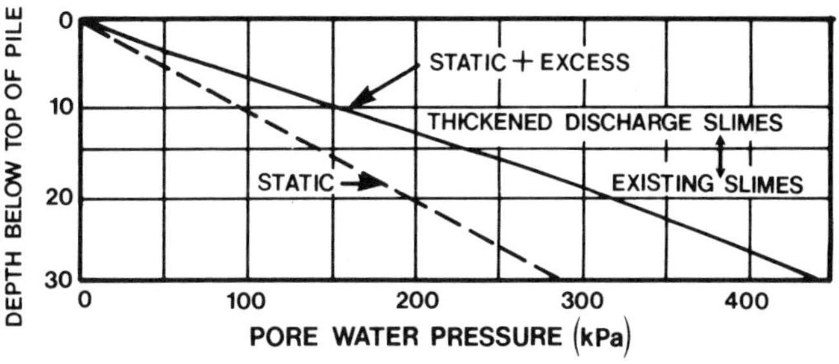

a) Pore-water pressures beneath center of pile at end of deposition

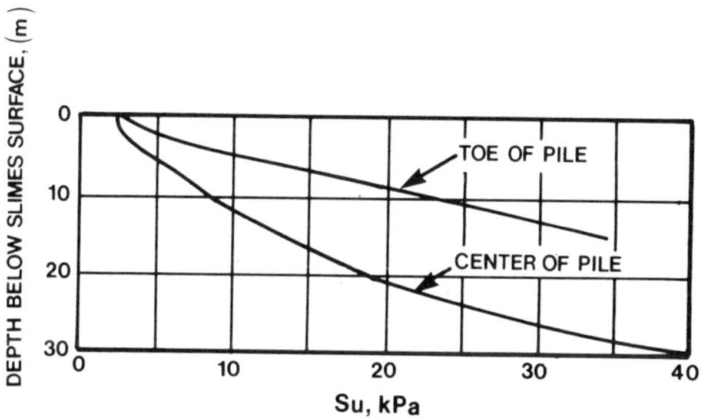

b) Undrained shear strength s_u profiles with depth

Figure 20
PORE-WATER PRESSURE AND UNDRAINED SHEAR STRENGTH PROFILES

CONCLUDING COMMENTS

We conclude with the following five comments. First, given the recent developments in sedimentation/consolidation theory, it is possible to predict field behavior of these processes with a reasonable degree of confidence provided the compressibility and permeability properties are known with sufficient accuracy. Second, we need to continue developments of devices and techniques which will accurately measure compressibility and permeability at high void ratios (effective stresses below 1 kPa). Third, the sedimentation/consolidation model described in this paper provides pore-water pressure information useful in estimating undrained shear strengths for certain classes of stability problems and for certain types of hydraulic fills. Fourth, the use of a constant, or even variable value of the coefficient of consolidation in solving consolidation problems related to soft hydraulic fills may provide unreliable pore-water pressure calculations. Finally, in examining problems involving more than one-dimension it is important to develop nonlinear finite strain consolidation models that account for multi-dimensionality. This implies that there is a need for practical and accurate two- and three-dimensional constitutive models.

ACKNOWLEDGEMENTS

We are pleased to acknowledge the partial financial support provided by the U.S. National Science Foundation, and the assistance of Ms. Dunja Peric and Mr. D. R. Gillette. We are further pleased to acknowledge the firm of Bromwell and Carrier, Inc. of Lakeland, Florida for making their data available to us.

REFERENCES

Been, K., (1980), "Stress Strain Behaviour of a Cohesive Soil Deposited Under Water," Ph.D. Dissertation, University of Oxford, Oxford, United Kingdom.

Been, K., and Sills, G.C., (1981), "Self-Weight Consolidation of Soft Soils: An Experimental and Theoretical Study," Geotechnique, 31, pp. 519-535.

Bishop, A.W., and Henkel, D.J., (1962), *The Measurement of Soil Properties in the Triaxial Test,* Second Edition, Edward Arnold, Ltd., London, United Kingdom.

Bloomquist, D.G., and Townsend, F.C., (1984), "Centrifugal Modeling of Phosphatic Clay Consolidation," *Sedimentation/Consolidation Models,* (Edited by R.N. Yong and F.C. Townsend), ASCE, pp. 565-580.

Bromwell, L.G., (1984a), "Consolidation of Mining Wastes," *Sedimentation/Consolidation Models,* (Edited by R.N. Yong and F.C. Townsend), ASCE, pp. 275-295.

Bromwell, L.G., (1984b), "Stability Investigation of Tailings Dam 1, Tyrone Mine, Tyrone, New Mexico," Report to Phelps Dodge Corporation, Tyrone Branch, Tyrone, New Mexico, Report Prepared by Bromwell and Carrier, Inc., Lakeland, Florida.

Bromwell, L.G., (1985), "Applications of Consolidation Theory to Disposal Planning for Florida Phosphatic Clays," *Flocculation, Sedimentation and Consolidation,* (Edited by B.M. Moudgil and P. Somasundaran), The Engineering Foundation, pp. 457-470.

Bromwell, L.G., and Carrier, W.D. III, (1979), "Consolidation of Fine-Grained Mining Wastes," Proceedings, Sixth Pan-American Conference on Soil Mechanics and Foundation Engineering, *1,* pp. 293-304.

Bromwell, L., and Raden, D., (1979), "Disposal of Phosphate Mining Wastes," *Current Geotechnical Practice in Mine Waste Disposal,* ASCE, pp. 88-112.

Bryant, S., Duncan, J.M., and Seed, H.B., (1984), "Application of Tailings Flow Analysis to Field Conditions," University of California, Berkeley, U.S. Bureau of Mines, Open File Report 53-84.

Buisman, A.S.K., (1936), "Results of Long Duration Settlement Tests," Proceedings, International Conference on Soil Mechanics and Foundation Engineering, *1,* pp. 103-106.

Caldwell, J.A., Ferguson, K., Schiffman, R.L., and Van Zyl, D., (1984), "Application of Finite Strain Consolidation Theory for Engineering Design and Environmental Planning of Mine Tailings Impoundments," *Sedimentation/Consolidation Models,* (Edited by R.N. Yong and F.C. Townsend), ASCE, pp. 581-606.

Cargill, K.W., (1984), "Prediction of Consolidation of Very Soft Soil," Journal of Geotechnical Engineering, ASCE, *110,* No. 6, pp. 775-795, Discussions and Closure, *111,* No. 11, pp. 1356-1364.

Carrier, W.D., III, (1985), "Consolidation Parameters Derived from Index Tests," Geotechnique, *35,* pp. 211-213.

Carrier, W. D., and Beckman, J.F., (1984), "Correlations Between Index Tests and Properties of Remoulded Clays," Geotechnique, *34,* pp. 211-228, Discussion and Reply, *35,* pp. 223-229.

Croce, P., Pane, V., Znidarcic, D., Ko, H-Y., Olsen, H.W., and Schiffman, R.L., (1985), "Evaluation of Consolidation Theories by Centrifugal Modelling," *Application of Centrifuge Modelling to Geotechnical Design,* Proceedings of a Symposium on the Application of Centrifuge Modelling to Geotechnical Design, (Edited by W.H. Craig), A.A. Balkema, Rotterdam, The Netherlands, pp. 381-402.

Darcy, H.P.G., (1856), *Les Fontaines Publiques de la Ville de Dijon,* Delmont, Paris, France.

De Mello, L.G.F.S., and Carrier, D.W., (1987), "Conceptual Design of Disposal of Bauxite Tailings," Proceedings, Eighth Pan-American Conference on Soil Mechanics and Foundation Engineering, *3,* pp. 51-58.

Ferguson, K.A., Hutchison, I.P.G., and Schiffman, R.L., (1985a), "Water Balance Approach to Prediction of Seepage from Mine Tailings Impoundments, Part I, General Water Balance Approach and Some Typical Modeling Results," *Seepage and Leakage from Dams and Impoundments,* (Edited by R.L. Volpe and W.E. Kelly), ASCE, pp. 250-262.

Ferguson, K.A., Hutchison, I.P.G., and Schiffman, R.L., (1985b), "Water Balance Approach to Prediction of Seepage from Mine Tailings Impoundments, Part II, Theoretical Aspects of Water Balance Approach to Seepage Modeling and Detailed Case History Results," *Seepage and Leakage from Dams and Impoundments,* (Edited by R.L. Volpe and W.E. Kelly), ASCE, pp. 263-285.

GEI (Geotechnical Engineers, Inc.), (1982), "Liquefaction Induced by Cyclic Loading," Report to the National Science Foundation, Project 80696.

Gersevanov, N.M., (1934), "Dinamika Mekhaniki Grunkov," (Foundations of Soil Mechanics), *2,* Gosstroiisdat, U.S.S.R.

Gibson, R.E., (1958), "The Progress of Consolidation in a Clay Layer Increasing in Thickness with Time," Geotechnique, *8,* pp. 171-182.

Gibson, R.E., England, G.L., and Hussey, M.J.L., (1967), "The Theory of One-Dimensional Consolidation of Saturated Clays, I. Finite Non-Linear Consolidation of Thin Homogeneous Layers," Geotechnique, *17,* pp. 261-273.

Highter, W., and Tobin, R., (1980), "Flow Slides and the Undrained Brittleness Index of Some Mine Tailings," *Engineering Geology, 16,* Elsevier Press, pp. 71-82.

Imai, G., (1979), "Development of a New Consolidation Test Procedure Using Seepage Force," Soils and Foundations, *19,* No. 3, pp. 45-60.

Imai, G., (1980), "Settling Behavior of Clay Suspension," Soils and Foundations, *20,* No. 2, pp. 61-77.

Imai, G., (1981), "Experimental Studies on Sedimentation Mechanism and Sediment Formation of Clay Materials," Soils and Foundations, *21,* No. 1, pp. 7-20.

Imai, G., Yano, K., and Aoki, S., (1984), "Applicability of Hydraulic Consolidation Test for Very Soft Clayey Soils," Soils and Foundations, *24,* No. 2, pp. 29-42.

Kramer, S.L., and Seed, H.B., (1988), "Initiation of Soil Liquefaction Under Static Loading Conditions," Journal of Geotechnical Engineering, ASCE, *114,* Paper No. 22334, pp. 412-430.

Krizek, R.J., and Somogyi, F., (1984), "Perspectives on Modeling Consolidation of Dredged Materials," *Sedimentation/Consolidation Models,* (Edited by R.N. Yong and F.C. Townsend), ASCE, pp. 296-332.

Kynch, E.J., (1952), "A Theory of Sedimentation," Transactions of the Faraday Society, *48,* pp. 166-176.

Ladd, C.C., and Foott, R., (1974), "New Design Procedure of Stability of Soft Clays," Journal of the Geotechnical Engineering Division, ASCE, Proceedings Paper 10664, *100,* pp. 763-786.

Lamb, H., (1932), *Hydrodynamics,* (Sixth Edition), Dover Publications, New York, New York.

Leung, P.K., Schiffman, R.L., Ko, H-Y., and Pane, V., (1984), "Centrifuge Modeling of Shallow Foundations on Soft Soil," Proceedings, Sixteenth Annual Offshore Technology Conference, Paper OTC 4808, *3,* pp. 275-282.

McNabb, A., (1960), "A Mathematical Treatment of One-Dimensional Soil Consolidation," Quarterly of Applied Mathematics, *17,* pp. 337-347.

McRoberts, E.C., and Nixon, J.F., (1976), "A Theory of Soil Sedimentation," Canadian Geotechnical Journal, *13,* pp. 294-310.

McVay, M., Townsend, F.C., and Bloomquist, D., (1986), "Quiescent Consolidation of Phosphatic Waste Clay," Journal of Geotechnical Engineering, ASCE, *112,* Paper No. 21047, pp. 1033-1049.

Martinez, R.E., Bloomquist, D., and Townsend, F.C., (1987), "Automated Large Deformation Consolidometer," Proceedings, Eighth Pan-American Conference on Soil Mechanics and Foundation Engineering, *2,* pp. 65-76.

Michaels, A.S., and Bolger, J.C., (1962), "Settling Rates and Sediment Volumes of Flocculated Kaolin Suspensions," Industrial and Engineering Chemistry Fundamentals, *1,* No. 1, pp. 24-33.

Mikasa, M., and Takada, N., (1984), "Selfweight Consolidation of Very Soft Clay by Centrifuge," *Sedimentation/Consolidation Models,* (Edited by R.N. Yong and F.C. Townsend), ASCE, pp. 121-140.

Mohlman, F.W., (1934), "The Sludge Index," Sewage Work Journal, *6,* pp. 119-122.

Morgenstern, N.R., (1985), "Geotechnical Aspects of Environmental Control," Proceedings, Eleventh International Conference on Soil Mechanics and Foundation Engineering, A.A. Balkema, Rotterdam, The Netherlands, *1,* pp. 155-185.

Olsen, H.W., (1966), "Darcy's Law in Saturated Kaolinite," Water Resources Research, *2,* pp. 287-295.

Olson, R.E., (1985), "State of the Art: Consolidation Testing," *Consolidation of Soils: Testing and Evaluation,* (Edited by R.N. Yong and F.C. Townsend), ASTM STP 892, pp. 7-68, Discussion, pp. 68-70.

Ortenblad, A., (1930), "Mathematical Theory of the Process of Consolidation of Mud Deposits," Journal of Mathematics and Physics, *9,* No. 2, pp. 73-149.

Palmer, B., and Krizek, R.J., (1987), "Thickened Slurry Disposal Method for Process Tailings," *Geotechnical Practice for Waste Disposal '87* (Edited by R.D. Woods), Geotechnical Special Publication No. 13, ASCE, pp. 728-743.

Pane, V., (1985), "Sedimentation and Consolidation of Clays," Ph.D. Thesis, Department of Civil Engineering, University of Colorado, Boulder, Colorado.

Pane, V., and Schiffman, R.L., (1985), "A Note on Sedimentation and Consolidation," Geotechnique, *35,* pp. 69-72.

Pane, V., Croce, P., Znidarcic, D., Ko, H.Y., Olsen, H.W., and Schiffman, R.L., (1983), "Effects of Consolidation on Permeability Measurements for Soft Clay," Geotechnique, *33,* pp. 67-72.

Poulos, S.J., (1981), "The Steady State of Deformation," Journal of the Geotechnical Engineering Division, ASCE, Proceedings Paper 16241, *107,* pp. 553-562.

Poulos, S.J., Robinsky, E.I., and Keller, T.O., (1985), "Liquefaction Resistance of Thickened Tailings," Journal of Geotechnical Engineering, ASCE, *111,* Paper No. 20213, pp. 1380-1394.

Richardson, J.F., and Zaki, W.N., (1954), "Sedimentation and Fluidisation: Part I.," Transactions, Institution of Chemical Engineers, *32,* pp. 35-53.

Robinsky, E.I., (1979), "Tailings Disposal by the Thickened Discharge Method for Improved Economy and Environmental Control," Proceedings, Second International Tailing Symposium (Edited by G. Argall), Miller Freeman, San Francisco, California, pp. 75-95.

Schiffman, R.L., Pane, V., and Gibson, R.E., (1984), "The Theory of One-Dimensional Consolidation of Saturated Clays, IV. An Overview of Nonlinear Finite Strain Sedimentation and Consolidation," *Sedimentation/Consolidation Models,* (Edited by R.N. Yong and F.C. Townsend), ASCE, pp. 1-29

Schiffman, R.L., Pane V., and Sunara, V., (1985), "Sedimentation and Consolidation," *Flocculation, Sedimentation and Consolidation,* (Edited by B.M. Moudgil and P. Somasundaran), The Engineering Foundation, pp. 57-121.

Scully, R.W., Schiffman, R.L., Olsen, H.W., and Ko, H-Y., (1984), "Validation of Consolidation Properties of Phosphatic Clay at Very High Void Ratios," *Sedimentation/Consolidation Models,* (Edited by R.N. Yong and F.C. Townsend), ASCE, pp. 158-181.

Seed, H.B., (1987), "Design Problems in Soil Liquefaction," Journal of Geotechnical Engineering, ASCE, *113,* Paper No. 21704, pp. 827-845.

Shields, D., (1974), "Innovations in Tailings Disposal," Proceedings, First Symposium on Mine and Preparation Plant Refuse Disposal, National Coal Association, pp. 86-90.

Templeton, J.S., III, Murff, J.D., Goodwin, R.H., and Klejbuk, L.W., (1985), "Evaluating Soils and Hazards in the Mississippi Canyon," Proceedings, Seventeenth Annual Offshore Technology Conference, Paper OTC 4964, *3,* pp. 63-72.

Terzaghi, K., (1923), "Die Berechnung der Durchlassigkeitsziffer des Tones aus dem Verlauf der Hydrodynamischen Spannungserscheinungen," Sitzungsberichte, Akademie der Wissenchaften in Wien, Mathematisch-naturwissenschaftliche Klasse, Part IIa, *132,* No. 3/4, pp. 125-138.

Terzaghi, K., (1925) *Erdbaumechanik auf bodenphysikalischer grundlage,* Deuticke, Wien, Austria.

Terzaghi, K., (1943), *Theoretical Soil Mechanics,* John Wiley and Sons, Inc., New York, New York.

Townsend, F.C., and Hernandez, J., (1985), "Predictions of Phosphatic Clay Consolidation by Numerical and Centrifugal Models," *Flocculation, Sedimentation and Consolidation,* (Edited by B.M. Moudgil and P. Somasundaran), The Engineering Foundation, pp. 449-455.

Townsend, F.C., Bloomquist, D.G., and McClimans, S.A., (1985), "Evaluation of Reclamation Schemes by Centrifugal Modelling," *Flocculation, Sedimentation and Consolidation,* (Edited by B.M. Moudgil and P. Somasundaran), The Engineering Foundation, pp. 471-487.

Vick, S.G., (1983), *Planning, Design, and Analysis of Tailings Dams,* John Wiley & Sons, New York, New York.

Yong, R.N., and Townsend, F.C., (ed.) (1985), *Consolidation of Soils: Testing and Evaluation,* ASTM Special Technical Publication 892, ASTM, Philadelphia, Pennsylvania.

Znidarcic, D., Croce, P., Pane, V., Ko, H-Y., Olsen, H.W., and Schiffman, R.L., (1984), "The Theory of One-Dimensional Consolidation of Saturated Clays, III. Existing Test Procedures and Analysis," Geotechnical Testing Journal, GTJODJ, *7,* No. 3, pp. 123-133, Discussion and Closure, *8,* No. 3, pp. 143-144.

Znidarcic, D., Schiffman, R.L., Croce, P., Pane, V., Ko, H-Y., and Olsen, H.W., (1986), "The Theory of One-Dimensional Consolidation of Saturated Clays: V. Constant Rate of Deformation Testing and Analysis," Geotechnique, *36,* pp. 227-237.

REPORT ON SESSION 2 - MATERIAL PROPERTIES

By W. David Carrier, III[a], F ASCE

BASIC STATISTICS OF SUBMITTED PAPERS

There were twelve papers submitted to this session; ten of these were accepted and their authors and titles are presented in Table 1. The papers cover six hydraulic fill types, as shown in Table 2: coal refuse, copper mill tailings, general, landfill, oil sand mine tailings, and uranium mill tailings. The latter was the most popular topic, with a total of three papers. There are a total of 22 different authors and co-authors, predominantly from the industrial sector. All of the latter are with consulting firms; none is with an operating company. The authors are affiliated with organizations in the following countries: United States (13), Canada (6), Australia (2), and Venezuela (1).

ANALYSIS OF REFERENCES

I analyzed the references cited by all of the papers, in order to determine if there were preferred sources of information. I found that there were several general references which tended to repeat (such as Castro et al., 1982, Jamiolkowski et al., 1985, Lambe and Whitman, 1969, NAVFAC DM-7.1, 1982, and Robertson and Campanella, 1984). However, when I considered only those references specifically related to the material properties of hydraulic fills, and eliminated "self-references", there were just four authors which were cited more than once: Blight, Department of Energy (11 remedial action plans for uranium tailings), McClelland, and Vick. And of these, only the book by Vick (1983) was cited in more than one paper; i.e., twice. It is a pity that Vick's book was published just as mining took a nosedive in the United States, and as a result, it has not received as widespread distribution as it should. I can personally recommend it as an excellent reference.

Interestingly, there were no references to the US Army Corps of Engineers, in particular the Dredged Material Research Program (DMRP). This despite an expenditure of $100 million and the publication of at least 100 government reports over the last 15 years.

[a]Bromwell & Carrier, Inc., Lakeland, Florida

There were also no references to the work at the University of Colorado or to that of our firm, which makes me wonder why Schiffman is giving the state-of-the-art and I am giving this report!

Finally, there were no references to the classic paper by Gilboy (1934), which elucidated the mechanics of a hydraulic fill dam and reduced its design to just three parameters: the friction angle of the shell, the ratio of the unit weights of the shell and the core, and the permeability of the core. In our firm, we continue to make use of the principles described in the Gilboy paper in the design of retention dikes used for disposal and reclamation of certain mine tailings.

Table 1

Authors and Papers Submitted to Material Properties Session

Authors	Title
Chen, P.K. Keshian, B., Jr. Guros, F. B.	Settlement of Uranium Mill Tailings
Chen, H. W. van Zyl, D. J. A.	Shear Strength and Volume Change Behavior of Copper Tailings Under Saturated Conditions
Cowherd, D. C. Perlea, V. G.	In-Situ Testing of Fine Coal Refuse by Nuclear Method
Keshian, B., Jr. Rager, R. E.	Geotechnical Properties of Hydraulically Placed Mill Uranium Tailings
Kuerbis, R. Negussey Vaid, Y. P.	Effect of Gradation and Fines Content on the Undrained Response of Sand
Larson, N. B. Keshian, B., Jr.	Prediction of Strains in Earthen Covers
McNeilan, T. W. Skaggs, R. L.	In-Place Properties of 7,000,000 cu yd Hydraulic Landfill
Plewes, H. D. McRoberts, E. C. Chan, W. K.	In-Situ Density Measurements in Sand Tailings Using Geophysical Nuclear Logging Techniques
Vasquez-Herrera, A. Dobry, R. Ng, T. T.	Pore Pressure Buildup and Liquefaction Failure of Anisotropically Consolidated Sand Due to Cyclic Straining
Williams, D. J. Morris, P. H.	Properties of Slurried Coal Tailings

Table 2

Hydraulic Fill Type and Author Affiliation

Hydraulic Fill Type	No. of Papers	Author Affiliation University	Industry	Total
Coal Refuse	2	2	2	4
Copper Mill Tailings	2	3	2	5
General	1	2	1	3
Landfill	1		2	2
Oil Sand Mine Tailings	1		3	3
Uranium Mill Tailings	3		5	5
Totals	10	7	15	22

REVIEW OF PAPERS

In the following sections, I have briefly summarized the conclusions of each paper in this session and have offered my own comments. When appropriate, I have also summarized the index properties of each material.

It is well to remember that every water-borne sedimentary deposit started its geologic life as a hydraulic fill. Hence, the study of the material properties of man-made hydraulic fills is also important in the understanding of natural deposits that cover vast portions of this planet.

COAL REFUSE

1. "In-Situ Testing of Fine Coal Refuse by Nuclear Method," by D. C. Cowherd and V. G. Perlea

Summary

Table 3

Index Properties of Coal Refuse

	D_{50} (mm)	-No. 200 (%)	-2μ (%)	G	LL	PI
Coarse Fraction	0.05-20	0-55	0-20	1.5-2.8	NP-35	0-13
Fine Fraction	0.003-0.07	15-95	0-42	1.3-2.5	NP-51	0-15

The authors present the properties of coal refuse from Indiana, Kentucky, Ohio, West Virginia, and Virginia. Because the specific gravity of coal refuse is extremely variable, it is somewhat complicated to evaluate relative density based on unit weight alone. The authors describe a nuclear sounding method that measures both total unit weight (with gamma rays) and water content (with fast neutrons). The measurements are made in situ by means of an access tube installed into the refuse. Given γ_t and w, then e, γ_d, and G can be calculated at every measurement point, assuming saturated conditions. The minimum void ratio is measured on recovered samples and is based either on ASTM D 4253 or ASTM D 1557 (Modified Proctor), whichever is lesser. The maximum void ratio is based on ASTM D 4254; the authors recommend Method A of the latter.

With the minimum, maximum, and in situ void ratios, it is then straightforward to calculate relative density, D_R, vs. depth in the coal refuse: typical values range from 30 to 100%. The calculated profiles of relative density may then be used in studies of liquefaction potential of the deposit.

Discussion

In order to obtain accurate measurements, the authors recommend that the nuclear gauge be calibrated in the specific material under consideration. Presumably, this procedure is covered in more detail in the references cited in the paper. Also, the access tube must have an effect on the calibration, and presumably, again, the authors have taken this into account. The access tube may be left in place and used for repeated tests at the same location, thereby permitting detailed evaluation of self-weight consolidation.

2. "Properties of Slurried Coal Tailings," by D. J. Williams and P. H. Morris

Summary

Table 4

Index Properties of Coal Tailings

D_{50} (mm)	-No. 200 (%)	G	γ_d (kN/m^3)	LL	PI	pH	Conductivity (μS/cm)	Clay Minerals
0.005-0.01	67-90	1.73-1.81	7.7-9.2	38-50	24-25	7.4-7.9	1.2-2.9	K, I, M

The authors present the properties of coal tailings from New South Wales and Queensland, and describe in detail the processes of beaching, sedimentation, consolidation, crusting, and rehabilitation. They report that the hydraulic beaching exponent, n, of coal tailings is 1.5. They conclude that sedimentation is affected by the initial solids content of the slurry, the proportion of fines, the type of clay mineral, and the water chemistry. The authors present oedometer test data for an undisturbed sample and a reconstituted sample of coal tailings from the New Hope Colliery and demonstrate that the results are significantly different (see Fig. 2 in paper). They also show that the permeability, k, and the coefficient of compressibility, c_v, of the undisturbed sample are as much as two orders of magnitude greater than the respective properties of the reconstituted sample (Figs. 6 and 7 in paper). From these measurements, the authors conclude "...that carefully obtained field specimens must be obtained for oedometer testing if representative parameters are to be determined." The authors also propose a permeability relationship for the undisturbed sample, of the form:

$$k = a(e - c)^b \quad (1)$$

Approximate values of a, b, and c are 3.4×10^{-6} m/s, 5.4, and 0.5, respectively. This relationship agrees with the undisturbed data for a void ratio between 0.9 and 0.7; but below 0.7, it predicts a permeability that is much less than that of the reconstituted sample at an equivalent void ratio. Finally, the authors present undrained shear strength, S_u, vs. water content, w (Fig. 9 in paper). For normally consolidated coal tailings, the relationship is approximately given by:

$$S_u = 10^{2.16 - 0.021w} \quad (2)$$

where w is in % and S_u is in kPa.

Discussion

The authors' observation that compressibility of a clayey material is changed by remolding is well known. This effect is very important for soils with low activity, and less so for soils with high activity, such as phosphatic clay (c.f., Carrier and Beckman, 1984). From the point of view of laboratory testing, the significance of this effect is that a reconstituted sample must be remolded at a high water content, similar to the initial slurry deposition conditions. We recommend a remolding water content corresponding to at least 3 times the liquidity index. Provided there are no irreversible chemical changes occurring in the disposal pond, such a reconstituted sample will yield very accurate consolidation parameters. Hence, undisturbed samples are, in fact, rarely required for oedometer testing.

This change in compressibility due to remolding is also of great practical import. We design disposal systems in which a dredge removes tailings that have been partially dewatered by sedimentation and self-weight consolidation. When the clay is remolded by the dredge pump and re-deposited, without adding water, the compressibility curve shifts downward. As a result, more tailings can be stored in the available volume than could otherwise be accommodated.

The reason that the permeability is less for the reconstituted sample is not because it was reconstituted, but because it was tested at a lower void ratio than the undisturbed sample. We have reviewed permeability data from many, many sources (Carrier and Beckman; Olsen, 1985) and have found that liquidity index (or void ratio for a given clay) is the primary controlling factor; permeability does not appear to be affected by activity, by over-consolidation ratio, or by remolding. In my opinion, the authors' permeability data for the undisturbed sample and the data for the reconstituted sample are continuous, rather than separate. I recommend a permeability relationship for both samples of the form:

$$k = 7.0 \times 10^{-8} \frac{e^{5.2}}{1 + e} \text{ m/s} \tag{3}$$

Whereas Equ. (1) predicts a permeability for the undisturbed sample that is much less than for the reconstituted sample when the void ratio is less than 0.7 (k approaches 0 as e approaches 0.5), Equ. (3) is consistent with both sets of data.

COPPER MILL TAILINGS

3. "Effect of Gradation and Fines Content on the Undisturbed Response of Sand," by R. Kuerbis, D. Negussey, and Y. P. Vaid

Summary

The authors report the results of undrained monotonic and cyclic triaxial tests on specially prepared sand samples. Copper mill tailings sand was obtained from the Brenda Mine in British Columbia as a source material: the sand was separated by sieve size and recombined to form pre-set gradations. Varying amounts of cohesionless silt, obtained from a glacial lake deposit, were then added to the sand in order to create a suite of materials in which each gradation variable could be controlled separately. Each of the triaxial samples were prepared at low density, using a special slurry deposition method. The authors report a high degree of sample homogeneity with this method, compared with water pluviation, which results in segregation.

With uniform or well-graded clean sands, the authors demonstrate that triaxial compression behavior is dilative under monotonic loading, whereas the extension behavior is contractive. As a result, the authors conclude that the steady state concept for liquefaction analysis (c.f. Castro, et al., 1982) does not appear to be applicable, at least with these sands. Uniform sand tends to be more contractive in extension and less dilative in compression compared to a well-graded sand.

Considering only uniform clean sands at the same relative density, as D_{50} decreases, the compression behavior is more dilative and the extension behavior is more contractive.

As the silt content of the sand is raised, up to 20%, the compression behavior is somewhat more dilative and the extension behavior is somewhat less contractive. The authors attribute the comparatively small change in behavior to a constant sand skeleton void ratio: the silt simply occupies the interstices between the sand grains without affecting the undrained friction angle or the pore pressure response.

Under cyclic loading, the undrained shear strength decreases with increasing silt content, when the relative density of the sample is based on the total void ratio. When the relative density is based on the sand skeleton void ratio, the cyclic shear strength actually increases slightly as the silt content is increased, up to 20%. The authors conclude that the cyclic shear strength of a silty hydraulic fill sand could be determined on clean sand at the corresponding sand skeleton void ratio. The resulting values would be slightly conservative.

Discussion

The authors' conclusion regarding the lack of a unique steady state line is very significant and undoubtedly will generate discussion by other investigators in the soil dynamics field.

The concept of a sand skeleton void ratio is very important. It is defined as:

$$e_S = \frac{V_V + V_F}{V_S} \tag{4}$$

where e_S = sand skeleton void ratio
V_V = volume of voids (water when saturated)
V_F = volume of fines
V_S = volume of sand

The authors present an expression for evaluating e_S (see Equ. (1) in paper), but we have simplified it somewhat as follows:

$$e_S = \frac{e_T\ (SFR + 1) + 1}{SFR} \tag{5}$$

where e_T = total void ratio (conventional definition)
SFR = sand-to-fines ratio, by dry weight

Hence, if the silt content is 13.3%, SFR = 86.7/13.3 = 6.52 and e_S = $(7.52e_T + 1)/6.52$. If e_T = 0.547 (see Fig. 9 in paper), e_S = 0.784.

The silty sand is gap-graded, and this is why, of course, that e_S controls the behavior of the sand. But it is important to recognize that gap-grading can occur in a hydraulic fill and to be aware of the consequences. Furthermore, as the percentage of fines is increased, and particularly if the fines are plastic, a transition occurs where the sand particles no longer control the behavior but are suspended in a matrix of fines. The controlling void ratio is then that of the fines, which we define as:

$$e_F = \frac{V_V}{V_F} \tag{6}$$

Thus, $$e_F = e_T(SFR + 1) = e_S SFR - 1 \tag{7}$$

I will return to this concept at the end of this report.

4. "Shear Strength and Volume Change Behavior of Copper Tailings Under Saturated Conditions," by H. W. Chen and D. J. A. van Zyl

Summary

Table 5

Index Properties of Copper Mill Tailings

D_{50} (mm)	-No. 200 (%)	-2μ (%)	G	$\gamma_{d\,max}$ (kN/m^3) Standard Proctor	Modified Proctor	PI
0.1	37	5	2.74	18.5	19.6	NP

The authors describe the results of 73 static triaxial compression tests. The test types included both CID (strain-controlled) and CIU (strain-controlled and load-controlled). The silty sand used in all the tests was obtained from an Arizona copper mill tailings dam, taken from the beach surface near the point of deposition and far from the pond. The index properties of the test material is summarized in Table 5 above.

A major objective of this study was to determine the effect of lubricated end platens in triaxial testing. The authors describe the specially manufactured platens in detail. Because the tests were run to a relatively large axial strain (20 to 35%), the diameter of the platens were made about 30% larger than the diameter of the test specimen. In addition, a small pin was installed at the center of each platen, in order to prevent the soil sample from sliding sideways during triaxial testing.

The authors conclude that lubricated platens do not significantly affect the CID and CIU shear strength. On the other hand, lubricated platens are very important with respect to volume change or pore pressure generation. In particular, the authors find that the critical void ratio is greater with lubricated platens.

The authors conclude that CID, as well as CIU, tests can be used to estimate the residual or steady state shear strength of loose saturated tailings. This result is contrary to measurements made by Castro (1969) and Casagrande (1976), on Ottawa sand, in which the steady state CIU strength (load-controlled) was about one-fifth the ultimate CID strength (strain-controlled). The authors attribute the differing results to the significant differences in particle shape and size distribution: the Ottawa sand is round and uniform; whereas the copper mill tailings sand is angular and well-graded.

The authors recommend that $(\bar{\sigma}_1/\bar{\sigma}_3)_{max}$ be used as the failure criterion for CIU tests on loose tailings sand. The authors also report that Rowe's stress dilatancy theory (Rowe, 1962) can be used to correct CID tests run at low confining stress.

Finally, the authors conclude that CIU strain-controlled and load-controlled tests are essentially the same in all respects, with one exception: the deviator stresses at steady state $(\bar{\sigma}_{dr})$ are consistently higher for strain-controlled tests when $\bar{\sigma}_{dr} \leq 200$ kPa.

Discussion

Although the authors conclude that the critical void ratio is higher for lubricated platens, there is considerable scatter in the data (see Fig. 4 in paper).

The authors state that all of the test specimens were compacted in ten layers using a small drop hammer. It is possible that this procedure produces a soil which is over-consolidated at low confining stress and normally consolidated at high confining stress. Obviously, this would complicate the interpretation of the test results.

Although the authors conclude that CIU and CID tests can be used interchangeably to determine the steady state line, the opposite results obtained by Castro (1969) and Casagrande (1976) suggest that this may not be a general rule. Additional tests on other materials are needed to help resolve the puzzle.

GENERAL

5. "Pore Pressure Buildup and Liquefaction Failure of Anisotropically Consolidated Sand Due to Cyclic Straining," by A. Vasquez-Herra, R. Dobry, and T. T. Ng

Summary

Table 6

Index Properties of an Alluvial Sand

D_{60}/D_{10}	-No. 200 (%)	G
2.80	13	2.64

The authors report on the results of 21 strain-controlled cyclic torsional triaxial tests, anisotropically consolidated, undrained, with pore pressure measurements: CyT-$\overline{CAU}$. All of the tests were performed on a silty sand obtained from a natural alluvial deposit. The samples were prepared in a loose condition, such that all of the triaxial tests were contractive.

The authors found a unique steady state failure envelope of angle $\bar{\alpha}_{us}$ = 29°, independent of test type (monotonic or cyclic) or by the consolidation stress ratio, $K_c = \bar{\sigma}_{1c}/\bar{\sigma}_{3c}$. The authors also report a unique triggering of flow failure strength envelope of angle $\bar{\alpha}_{up}$ = 23°. For the stress path of the CyT-$\overline{CAU}$ test, the value of excess pore pressure that will trigger a flow failure, u_t, is given by:

$$r_{ut} = \frac{u_t}{\bar{\sigma}_{3c}} \approx \frac{1}{2}\left(K_c + 1 - \frac{K_c - 1}{\tan\bar{\alpha}_{up}}\right) \tag{8}$$

Since K_c could typically equal 2 in a tailings dam slope, r_{ut} could be about 0.3. The authors conclude that the common assumption that initial liquefaction occurs when $r_{ut} \approx 1$ can be unconservative.

Using multiple regression analysis, the authors have curve-fit the test data and found that the number of cycles to flow failure triggering, n_t, is given by:

$$n_t = \frac{r_{ut}}{\alpha\,(\gamma_{cy} - \gamma_t)^{\beta}\,(2 - r_{ut})} \tag{9}$$

where γ_{cy} = cyclic shear torsional strain (%)
γ_t = threshold strain = 10^{-2}%
$\alpha = 4.78235 - 1.91176K_c$
$\beta = 2.95686 - 0.7843K_c$

The authors show that the number of cycles, n_t, for a given level of cyclic straining, γ_{cy}, is less for a steep slope (large K_c) than for a shallow slope (small K_c). In other words, a steeper slope will fail earlier than a flatter slope.

The authors have also curve-fit the pore pressure buildup, r_u, during triaxial cyclic straining:

$$\frac{r_u}{r_{ut}} = \frac{A\left(\frac{n}{n_t}\right)}{1 + B\,\frac{n}{n_t}} + C\left(\frac{n}{n_t}\right)^{8} \tag{10}$$

where n = number of cycles (up to n_t), at a given γ_{cy}
$A = 5.27917 - 2.133K_c$
$B = 2.182$
$C = 0.6779\,(K_c - 1)$

The authors show that the rate of pore pressure buildup in a steep slope is minimum at the beginning of cyclic straining and maximum just before failure. The opposite behavior occurs in shallow slopes.

The authors conclude that their empirical cyclic pore pressure model can be incorporated into a computer program, in order to perform a dynamic effective stress analysis of a hydraulic fill embankment during earthquake loading.

Discussion

Having had some experience in developing empirical curve-fits myself, I am very appreciative of the authors' efforts. Empirical curve-fits play an important role in science: it is a well-established tradition that first a Brahe collects accurate data, then a Kepler fits a reasonable curve to the data, and finally a Newton discovers the physical law that not only explains the original data and the curve, but predicts behavior not yet even observed. The authors have accomplished the first two valuable steps; the third leap remains to be made. For example, the coefficients in Equs. (9) and (10) must be influenced by particle shape and size distribution, but these effects cannot yet be predicted a priori.

Although the authors conclude that a dynamic effective stress analysis can be made for the slope stability of a hydraulic fill embankment, I am not personally convinced that even a static analysis can be made (yet): rather, I think that an undrained strength analysis is required. The authors' conclusions will likely lead to a lively discussion.

LANDFILL

6. "In-Place Properties of 7,000,000 cu yd Hydraulic Fill," by T. W. McNeilan and R. L. Skaggs

Summary

Table 7

Index Properties of a Landfill

	D_{50} (mm)	-No. 200 (%)	-2μ (%)	LL	PI	Organic Content (%)
Silty Sand	0.1-0.18	13-32		NP	NP	
Clayey Silt						
ML				30- 50	2-20	
MH	0.0065-0.02	65-97	8-25	70-110	30-60	1-9

The authors report on the creation of an 80-ha hydraulic landfill in the Port of Los Angeles. The fill consists of approximately equal portions of silty fine sand and clayey silt, inter-mixed and inter-layered. The fill is still settling, under self-weight consolidation, nearly five years after deposition. The authors predict that primary consolidation will be completed in five to ten more years and that the total settlement will range from 90 to 245 cm, depending on the local thickness of the landfill. They further estimate that the total settlement will be equal to about 15 to 20% of the cumulative thickness of the clayey silt layers at any given location.

The authors present the results of an extensive field and laboratory investigation of the fill. Undisturbed samples were obtained by means of an all-terrain drilling rig, with 1.7-metre wide, low contact pressure tires. Loose sand samples were frozen in the field and X-rayed in the laboratory. The field testing included cone penetration tests, vane shear tests, and piezometer installations. The laboratory program included index tests (see Table 6 above); UU, CIU, CID, and semi-consolidated undrained triaxial tests; and incremental and controlled-rate-of-strain consolidation tests.

The authors conclude that the undrained shear strength at a given location and elevation in the fill varies within ± 20% as measured by different tests. However, local layering has a profound effect, because of changes in the liquid limit of the clayey silt. The undrained strength measured in the fill ranges from 4.8 to 52 kPa.

The authors conclude that the fill was underconsolidated at the time of their investigation. As a result, the in situ undrained shear strength is significantly less than measured in conventional CIU triaxial tests. Furthermore, the value of the cone bearing factor, N_k, varies with depth similar to an isochrone of effective stress in a double-drained clay layer: the authors find that N_k varies from about 5 to 14 and is lowest at mid-depth of a layer and greatest near the top and bottom (see Fig. 10 in paper).

Discussion

The authors find that the rate of self-weight settlement of the fill is controlled by the coefficient of consolidation of the highly plastic clayey silt (MH). They conclude that the interlayered seams of silty sand are, therefore, not continuous over significant areas of the site and that the seams should be ignored. However, in our research work concerning dewatering of phosphatic clay, we found that a seam of clean sand, that was specially constructed to be continuous in order to enhance drainage, actually had very little effect on the rate of self-weight consolidation. The reason is that the hydraulic gradient within the sand seam is too small compared to the typical lateral distance to an external drainage boundary. So, just because the self-weight settlement is slow does not, in itself, mean the sand seams are not continuous. In fact, these seams might significantly increase the rate of settlement due to a load of limited size, such as an embankment.

In the last sentence of this paper, the authors conclude that "...when hydraulic landfills will contain significant percentages of clayey materials, special fill placement techniques should be used if the landfill is to be used in the near future." But they offer no suggestions as to what those special techniques might be.

OIL SAND MINE TAILINGS

7. "In-Situ Density Measurements in Sand Tailings Using Geophysical Nuclear Logging Techniques," by H. D. Plewes, E. C. McRoberts, and W. K. Chan

Summary

The authors describe a nuclear logging device for measuring total unit weight at depths of 50 to 100 m. The tests were conducted in a tailings dam at an oil sand mine in Alberta, as part of an analysis of liquefaction potential. The technique was borrowed from the oil well logging industry because the great depths were considerably beyond current correlations for correction of SPT blow count values.

Two nuclear devices were tested: both utilized cesium 137 as a source, and both had dual detectors. By having two detectors, one a short distance from the source (15 cm), and the other a long distance (30 to 48 cm), it is possible to partially compensate for the effects of borehole rugosity and mudcake on the borehole wall. Repeat tests in the same boring indicate a precision of ± 0.3 kN/m^3. The authors emphasize that the nuclear logging device must be calibrated in the specific material being tested and that a smooth and stable borehole wall is essential.

Discussion

The authors describe in detail a core sampling method that was used to confirm the nuclear density measurements. However, they report that the nuclear densities consistently exceeded the core densities, by as much as 14%. They conclude that the core samples must have expanded due to stress release, geometry of the sampler, and freezing of the sample. The authors find that back-calculating total unit weight from the measured water content of a sample, assuming saturated conditions, agrees very well with the nuclear density log. From this, the authors conclude that the logging technique and the data interpretation procedures are validated. I, on the other hand, conclude that recovering a disturbed sample and measuring the water content is a convenient and accurate way to determine the total unit weight -- and obtain a sample, too. We have used this method for years in soft clayey materials and it is reassuring to learn that it works equally well in sandy materials.

At depths of 50 to 100 m, the measured N values were 120 to 250 blows per foot. The authors extrapolated the recommendations from Skempton (1986), and calculated corrected N values of 14 to 30. But they report that these latter values did not correlate well with the nuclear density logs. So, in the end, no conclusions are drawn regarding the liquefiability of the tailings dam.

URANIUM MILL TAILINGS

8. "Geotechnical Properties of Hydraulically Placed Mill Uranium Tailings," by B. Keshian, Jr., and R. E. Rager

Summary

Table 8

Index Properties of Uranium Mill Tailings

	D_{50} (mm)	-No. 200 (%)	-2μ (%)	G	LL	PI
Sand	0.015-0.25	<30	0-12	2.43-2.73	NP	NP
Sand-Slime	0.035-0.11	30-70	6-21	2.46-2.76	NP	NP
Slimes	0.006-0.016	>70	8-42	2.59-2.92	22-100	2-68

The authors report on the properties of 24 uranium mill tailings piles located in Arizona, Colorado, Idaho, New Mexico, North Dakota, Oregon, Pennsylvania, South Dakota, Texas, Utah, and Wyoming. Each of the piles is being reclaimed as part of the Uranium Mill Tailings Remedial Action (UMTRA) program presently being carried out by the US Department of Energy (DOE). The piles consist of mixtures of sand (<30% passing a No. 200 sieve, by definition), sand-slime (30-70% fines), and slimes (>70% fines). The index properties of the tailings components are summarized in Table 8 above, and the engineering properties in Table 9 below.

Table 9

Engineering Properties of Uranium Mill Tailings

	k (cm/s)	$\frac{C_c}{1 + e_o}$	c_v (cm^2/s)	C_α	$\bar{\phi}$ ($\bar{c}$ = o)
Sand	10^{-4}-10^{-2}	0.008-0.292	2.1×10^{-3}-2.0×10^{-1}	---	32-38
Sand-Slime	2×10^{-6}-5×10^{-4}	0.032-0.264	1.0×10^{-3}-5.0×10^{-2}	5.0×10^{-4}-1.5×10^{-2}	27-38
Slimes	10^{-7}-10^{-4}	0.090-0.295	2.5×10^{-4}-6.2×10^{-1}	3.0×10^{-3}-1.25×10^{-2}	21-37

The authors conclude that the physical properties of the various uranium mill tailings are remarkably similar, in spite of different milling processes. The two exceptions are the sites in Oregon and Texas, owing to differences in the ore bodies. The other 22 piles were all derived from the Colorado Plateau region.

The authors show that the permeability of the tailings is a function of the void ratio and the percent of fines. They conclude that k can be predicted within ± one order of magnitude. Based on over 500 piezocone holes, the authors have developed correlations between point stress and friction ratio for identification of tailings types. They conclude that the piezocone is a very important characterization tool.

Based on two test fills (described in more detail in the paper by Chen et al.), the authors conclude that $C_{c\ Field}$ is within ±20% of $C_{c\ Lab}$, but that $c_{v\ Field} = 2 - 8 \times c_{v\ Lab}$. In other words, the magnitude of settlement can be accurately calculated but the rate is generally much faster than predicted. They attribute the higher rate to micro layers of sand within the tailings. The authors conclude that a small test fill is a cost-effective tool.

Discussion

For anyone interested in the properties of uranium mill tailings, this paper is a primary source of information. It also cites eleven DOE remedial action plans which provide additional details for specific sites.

The plasticity chart (see Fig. 6 in paper) shows a wide range in the Atterberg limits for the slimes portion of the tailings. Even when the two exceptional sites are excluded, the classification of the slimes varies from ML to CL to MH to CH. It would have been helpful if the authors had reported the Atterberg limits of just the -No. 200 sieve fraction of the slimes. Then perhaps the similarity among the sites would be more obvious.

The plot of Standard Proctor maximum dry density vs. optimum moisture content for the sand and sand-slime portions (see Fig. 7 in paper) also shows a wide range: $\gamma_{d\ max} = 9.1\text{-}18.4\ kN/m^3$ and $w_{opt} = 9\text{-}60\%$. And the plot of permeability vs. dry density (see Fig. 9 in paper), frankly, looks like a buckshot pattern. The trend of the slimes data even seems to be in the wrong direction: the permeability decreases as the density decreases (void ratio increases).

The authors have assembled a valuable body of data. Now they need to process and normalize it to make it more meaningful and useful.

9. "Settlement of Uranium Mill Tailings," by P. K. Chen, B. Keshian, Jr., and F. B. Guros

Summary

The authors report on the results of two test fills constructed on top of uranium mill tailings at Ambrosia Lake, New Mexico. The standard reclamation design for the UMTRA consists of a compacted soil layer for containment of radon gas, topped with a riprap cover for erosion protection. The DOE requires that this earthen system be functional for 200 to 1000 years without maintenance. Hence, the

Table 10

Index and Engineering Properties of Uranium Mill Tailings

At Ambrosia Lake, New Mexico

	-No. 200 (%)	LL	PI	G	k (cm/s)	e_o	C_c	C_r	c_v (ft^2/yr)[a]
Sand	25	NP	NP	2.62	5.6×10^{-4}	0.71	0.10	0.016	1.0×10^{-1}
Sand-Slime	44	14	3	2.65	4.4×10^{-6}	0.71	0.12	0.011	1.1×10^{-2}
Slimes	87	43	21	2.65	NR	1.43	0.54	0.056	6.2×10^{-3}
All	47	19	7	2.64	2.8×10^{-4}				

[a]Units reproduced from Table 2 in paper; but values appear to be expressed in cm^2/s.

purpose of the test fills was to measure the amount and rate of settlement and to compare the results with the predicted values.

The field investigation consisted of 281 borings (including installation of monitor wells) and 126 piezocone soundings. Each test fill was 60 metres square, 6.1 to 8.2 metres high. Each was instrumented with settlement points, piezometers, etc.

The authors conducted a detailed parametric sensitivity analysis and they conclude that the two most important physical parameters are: OCR of the tailings and location of drainage layer(s). The authors conclude that the magnitude of settlement is accurately predicted based on laboratory values of C_c, C_r, and OCR. They conclude that the piezocone measurements significantly overestimate OCR, and hence are unconservative with respect to settlement.

The authors found that most of the consolidation of the tailings occurred during construction of the test fills. From this, they conclude that $c_{v\ Field}$ of the tailings is greater than $c_{v\ Lab}$. They also conclude that $c_{v\ Field}$ of the natural shale underlying the tailings must be more than 100 times greater than $c_{v\ Lab}$.

Discussion

The authors' conclusion regarding the accuracy of $C_{c\ Lab}$ is somewhat misleading. Although most of the tailings at this site are sand-slime, the authors found that $C_{c\ Field}$ was approximately equal to $C_{c\ Lab}$ of the slimes, not $C_{c\ Lab}$ of the sand-slime. Referring to Table 10 above, it can be seen that C_c of the slimes is about 4 times that of sand-slime. Consequently, the measured settlement must have been several times the initial estimate, and only by adjusting the value of C_c were the authors able to obtain agreement. Apparently, the slimes control the amount of settlement, and this is a very important conclusion.

The authors' conclusion regarding the piezocone overestimating OCR implies that N_k must be much greater than 17, perhaps as great as 40 or more. This is a significant observation and deserves additional study.

10. "Prediction of Strains in Earthen Covers," by N. B. Larson and B. Keshian, Jr.

Summary

The authors report on a detailed prediction of performance of a proposed earthen cover for the Falls City, Texas, uranium tailings pile. The radon gas barrier is 0.9 m thick, capped with a riprap layer 0.5 m thick. Portions of the existing tailings will be excavated and re-placed on top of undisturbed tailings up to 12 m deep, prior to construction of the cover. Hence, it was necessary to predict the movement of the earthen cover: vertical settlement, horizontal displacement, and horizontal strain. The latter is the most important, because of the requirement to prevent cracking of the radon gas barrier.

The authors performed piezocone soundings on a 60-m grid, supplemented by 12 borings. Detailed profiles were prepared for each piezocone location, based on material characterizations previously established (see Keshian and Rager above). Based on laboratory measurements of c_v, the authors calculated that the primary settlement would be complete 3 months after construction of the cover was finished. Based on experience with other test fills (see Chen et al. above), the authors conclude that primary will occur during the construction period of 3 months. From this, the authors conclude that only secondary settlements are important with respect to cracking of the earth cover.

The authors calculated a maximum secondary vertical settlement of about 30 cm, over the design life of the cover (1000 yr). Because the secondary settlement varies from point-to-point, a detailed graphical contour map was prepared. A special computer program was used to assist in the analysis: the first derivative of the settlement contours yielded the horizontal displacement; and the second derivative yielded the horizontal strain within the cover. Because the maximum calculated tensile strain (0.003%) was significantly less than the minimum cracking strain reported in the literature (0.08%), the authors conclude that the cover will perform satisfactorily. Just to be on the safe side, the cover will be compacted wet-of-optimum.

Discussion

Although the OCR does not appear to be important in this analysis (because primary settlement is ignored), it is worth noting that the authors used an N_k value of 17 and conclude that OCR values from the piezocone soundings were slightly underpredicted. This is in contrast to Chen et al., who conclude that the piezocone overestimates OCR.

As I indicated in my discussion of McNeilan and Skaggs above, sand layers within fine-grained material may increase the rate of settlement of an embankment of limited size, but may have very little effect when the load covers a large area. Hence, the authors reliance on test fills may be unconservative. In this particular case, even a slight delay could lead to large primary settlements after construction is completed.

Even if all the primary settlement occurs during construction, simple calculations indicate that the maximum amount will be in excess of 1 m. If nothing else, the earth-moving contract documents need to reflect the volume change; this probably seems obvious, but it is amazing, to me, at least, how many lawsuits revolve around pay quantities.

In Chen et al. above, the authors demonstrate that in order for a predicted primary settlement to agree with the measured value, it is necessary to assume that all of the tailings is slimes, rather than the complex mixture of sand, sand-slime, and slimes that is known to exist. Hence, it is not clear to me why the authors chose to distinguish among the tailings material is such great detail in order to calculate secondary settlement. In other words, if the calculations for primary settlement are basically crude, why bother with such a sophisticated analysis of secondary settlement? Furthermore, the procedure used by the authors to calculate horizontal strain is related to the grid size of the piezocone soundings: if the grid spacing were reduced, the calculated horizontal strain would be increased.

On the other hand, if C_α of slimes were used exclusively, presumably the amount of calculated secondary consolidation would be greater; but the horizontal strain would be related solely to changes in geometry, and, thus, would probably be less than the authors calculated.

It should also be noted that the secondary consolidation is projected over nearly 4 orders of magnitude in time, from 3 months to 1000 years. This probably sets a new record for extrapolation.

Finally, while it is true that compacting wet-of-optimum will reduce the potential for cracking while the earth fill remains wet, it will increase cracking if and when the fill dries out.

EPILOGUE

As a Reporter, I am entitled to add my own three cents' worth, if I so choose. Consequently, here are some extra thoughts and ideas:

FIRST CENT: NON-LINEAR SOIL PARAMETERS

Our forefathers created the classical soil parameters: $\bar{\phi}$, $\bar{c}$, A, B, S_u, $S_u/\bar{\sigma}$, k, C_c, c_v, C_α, E, ν. These are all linear approximations of reality. They were developed and refined because the first and second generation soils engineers were limited to certain closed-form solutions and tables of numerical solutions.

The microcomputer revolution has changed all that. Obtaining a non-linear solution that was impractical, or impossible, a few years ago is virtually trivial now for the third and fourth generation soils engineers. We know that the Mohr-Coulomb envelope is curved, that fluid flow is non-Darcian, and that the compression index is not constant; yet, we persist in measuring and reporting linear soil properties, sometimes "bending" them to fit an available computer program. We need to start all over again with non-linear soil parameters.

SECOND CENT: PLASTICITY CLASSIFICATIONS

The Unified Soil Classification System (ASTM D 2487) is based on the seminal work of Arthur Casagrande (1947, 1948), over forty years ago. The USCS recognizes just two levels of clay plasticity: CL and CH. The CL category applies to lean clays (low to medium plasticity) with a liquid limit less than 50. The CH category applies to fat clays (high plasticity) with a liquid limit greater than 50. The selection of 50 as the Papal Line of Demarcation between CL and CH was probably influenced by the types of clays found in those geographic areas that dominated the early history of soil mechanics: Boston Blue Clay, London Clay, and Norwegian "Quick" Clay all have liquid limits less than 50 and are consequently classified as CL. Perhaps CH was meant to cover the rest of the world.

In any case, the inadequacy of this system is demonstrated in Figure 1, which plots the results of Atterberg limits tests on samples of dredged material from various locations in the United States (Bartos, 1977). The CL clays occupy just a small portion of the left-hand corner of the plasticity chart; the CH clays cover the rest. But it is silly to classify a clay with a liquid limit of 250 the same as a clay with a liquid limit of 51. I propose new categories with the following liquid limits: CH from 50 to 100; CV (for very high plasticity) from 100 to 150; CU (ultra-high) from 150 to 200; and CS (super high) above 200. While we are at it, we can also reintroduce the CI (intermediate) classification proposed by Casagrande (1947) for clays with a liquid limit between 35 and 50.

Forty years ago, clay was dredged from the bottom of Boston Harbor in order to construct a large fill for the expansion of Logan Airport (Casagrande, 1949; Carrier and Bromwell, 1984). This is a landmark project in soil mechanics, because no data was then available regarding the consolidation of hydraulic clay fills And yet, Boston Blue Clay typically has a liquid limit of just 40. Today's engineer routinely deals with clays of much higher plasticity and, consequently, a more refined classification system is needed.

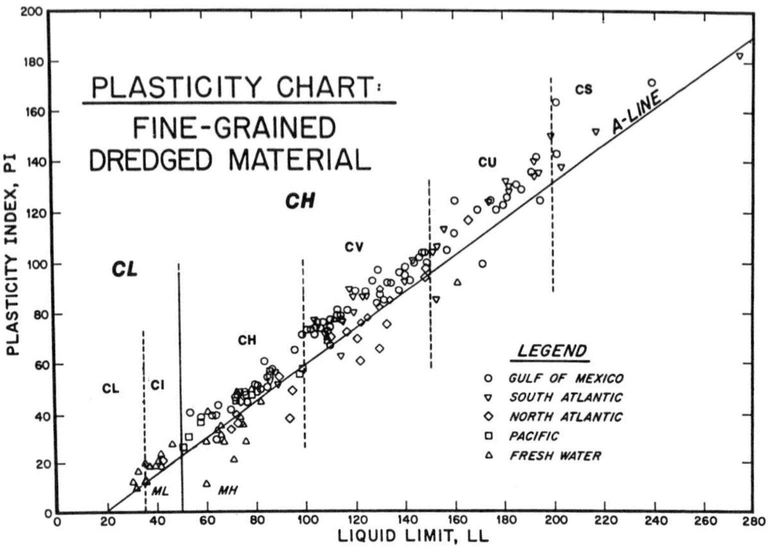

Fig. 1: Plasticity of fine-grained dredged material. The current ASTM clay classifications are shown as CL and CH. New proposed classifications are: CL (low plasticity), CI (intermediate), CH (high), CV (very high), CU (ultra-high), and CS (super high).

THIRD CENT: SAND-CLAY BEHAVIOR

Further on the subject of soil classification, the USCS distinguishes between a clayey sand and a sandy clay by the percentage of material coarser than a No. 200 sieve (74 microns): more than 50% and it is a sand; less than 50% and it is a clay. This definition is useful at "normal" water contents; but at high water contents (such as occur naturally in marine and fluvial deposits, or artificially in dredged materials, hydraulic fills, and mine tailings), it can be very misleading. For example, when four parts sand is mixed with one part clay in order to reclaim a phosphate mine in Florida, the soil is definitely a clayey sand, right? Wrong. There is so much water in the soil mix that it initially behaves like a pure clay with respect to undrained shear strength, consolidation, viscosity, etc. Not until a "normal" water content is reached does the soil behave like a sand. The clay is so plastic that the time required for dewatering can be years.

Scott and Cymerman (1984) have developed a ternary diagram which is very useful for distinguishing between sandy behavior and clayey behavior. As shown in Figure 2a, the three extreme points on the diagram represent sand, fines (silt and clay), and water. The boundaries separating different types of behavior are determined by tests run on a given soil; typical results are presented in Figure 2b. These sand-clay boundaries vary from soil to soil, depending on the plasticity and activity of the fine fraction of each soil. Eventually, enough data will be collected such that empirical guidelines can be established.

I am convinced that a number of tailings dams have failed because the engineer in each case considered the retained "slimes" to be sand-like when it actually behaved as if it were clay-like. Hence, the effect of water content on soil classification is very important.

Furthermore, when testing soil samples from a hydraulic fill, it is essential that the Atterberg limits be measured on the -No. 200 fraction, as well as on the -No. 40 fraction. Only the latter is specified in ASTM D 4318; but it is the plasticity, and the void ratio, of the fine fraction that can control the behavior of the fill. Unfortunately, none of the papers in this session reported Atterberg limits for the fine fraction. This is of fundamental importance.

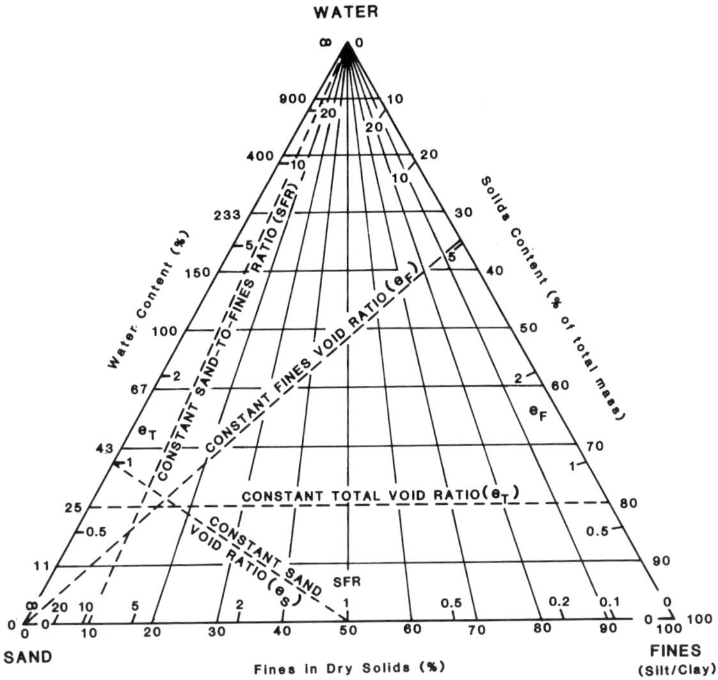

Fig. 2a: Ternary diagram for classification of soils, including effect of water content.

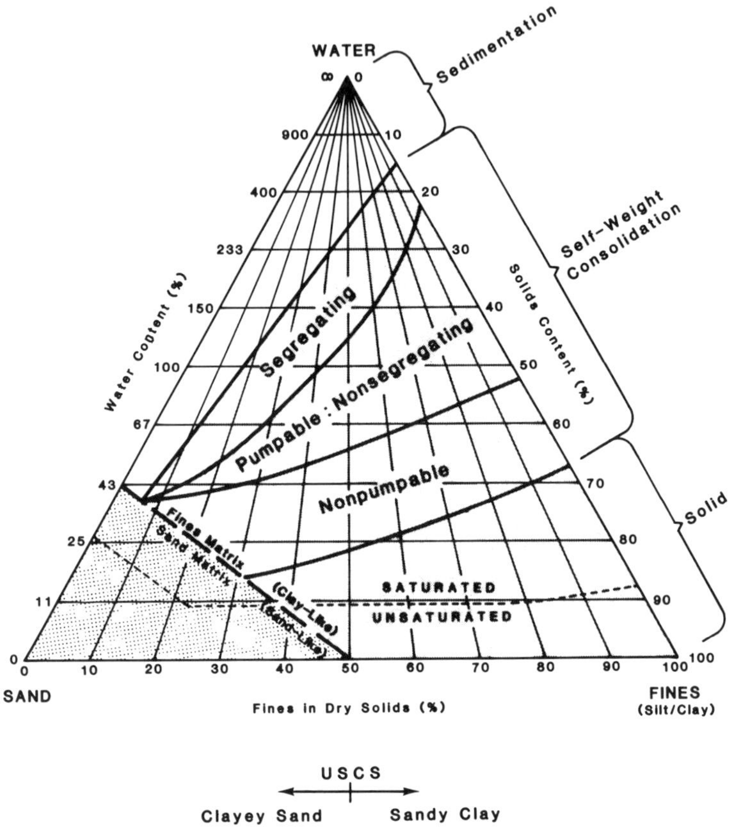

Fig. 2b: Example of sand-like and clay-like behavior (after Scott and Cymerman, 1984).

REFERENCES

ASTM D 1557 (1978) "Moisture-Density Relations of Soils and Soil-Aggregate Mixtures Using 10-lb (4.54-kg) Rammer and 18-in. (457-mm) Drop," Standard Test Methods.

ASTM D 2487 (1985) "Classification of Soils for Engineering Purposes," Standard Test Method.

ASTM D 4253 (1983) "Maximum Index Density of Soils Using a Vibratory Table," Standard Test Methods.

ASTM D 4254 (1983) "Minimum Index Density of Soils and Calculation of Relative Density," Standard Test Methods.

ASTM D 4318 (1984) "Liquid Limit, Plastic Limit, and Plasticity Index of Soils," Standard Test Method.

Bartos, M. J. (1977) "Classification and Engineering Properties of Dredged Materials," Dredged Material Research Program Technical Report, D-77-18, Army Corps of Engineers, Waterways Experiment Station, Vicksburg, Mississippi.

Carrier, W. D., III, and Beckman, J. F. (1984) "Correlations between index tests and the properties of remoulded clays," Geotechnique, Vol. 34, No. 2, pp. 211-228.

Carrier, W. D., III, and Bromwell, L. G. (1984) "Logan Airport Re-Visited," Sedimentation Consolidation Models, Predictions and Validation, Proceedings of a Symposium sponsored by ASCE, San Francisco, pp. 344-355.

Casagrande, A. (1947) "Classification and Identification of Soils," Proceedings, ASCE, Vol. 73, pp. 783-810.

Casagrande, A. (1948) "Classification and Identification of Soils," Transactions, ASCE, Vol. 113, pp. 901-991.

Casagrande, A. (1949) "Soil Mechanics in the Design and Construction of the Logan Airport," Journal of the Boston Society of Civil Engineers, re-printed in Contributions to Soil Mechanics, 1941-1953, BSCE (1953), pp. 176-205.

Casagrande, A. (1976) "Liquefaction and Cyclic Deformation of Sands--A Critical Review," Harvard Soil Mechanics Series No. 88, Harvard University, Cambridge.

Castro, G. (1969) "Liquefaction of Sands," Harvard Soil Mechanics Series No. 81, Harvard University, Cambridge, January, pp. 1-112.

Castro, G., Enos, J. L., France, J. W., and Poulos, S. J. (1982) "Liquefaction Induced by Cyclic Loading," Geotechnical Engineers, Inc., Winchester, MA. Prepared for the National Science Foundation, Washington, DC.

Gilboy, G. (1934) "Mechanics of Hydraulic-Fill Dams," originally published in the Journal of the Boston Society of Civil Engineers, Vol. XXI, No.3; reprinted in Contributions to Soil Mechanics, 1925-1940, Boston Society of Civil Engineers, 1940, pp. 127-147.

Jamiolkowski, M., Ladd, C. C., Germaine, J. T., and Lancellotta, R. (1985) "New Developments in Field and Laboratory Testing of Soils," Proceedings, 11th International Conference on Soil Mechanics and Foundation Engineering, San Francisco, Vol. 1, pp. 57-153.

Lambe, T. W. and Whitman, R. V. (1969) Soil Mechanics, John Wiley and Sons, New York.

NAVFAC DM-7.1 (1982) Soil Mechanics, Department of the Navy, Naval Facilities Engineering Command, Alexandria, VA.

Olsen, H. W. (1985) Personal communication.

Robertson, P. K. and Campanella, R. G. (1984) "Guidelines for Use and Interpretation of the Electric Cone Penetration Test," prepared for Hogentogler and Co., Inc., Gaithersburg, MD, at the University of British Columbia.

Rowe, P. W. (1962) "The Stress-Dilatancy Relation for Static Equilibrium of an Assembly of Particles in Contact," Proceedings of the Royal Society, London, Series A, Mathematical and Physical Sciences, Vol. 269, No. 1339, pp. 500-527.

Scott, J. D. and Cymerman, G. J. (1984) "Prediction of Viable Tailings Disposal Methods," Sedimentation Consolidation Models, Predictions and Validation, Proceedings of a Symposium sponsored by ASCE, San Francisco, pp. 522-544.

Skempton, A. W. (1986) "Standard Penetration Test Procedures and the Effects in Sands of Overburden Pressure, Relative Density, Particle Size, Aging, and Overconsolidation," Geotechnique, Vol. 36, No. 3, pp. 425-447.

Vick, S. G. (1983) Planning, Design, and Analysis of Tailings Dams, John Wiley and Sons, New York.

GEOTECHNICAL PROPERTIES OF HYDRAULICALLY PLACED URANIUM MILL TAILINGS

BERG KESHIAN, JR. (1) M.ASCE

RONALD E. RAGER (2) M.ASCE

ABSTRACT

The reclamation of 24 uranium mill tailings piles under the DOE Uranium Mill Tailings Remedial Action (UMTRA) program has provided an opportunity to compare the material properties of uranium tailings from different processing operations and different mineral belts. These results have shown that although the tailings have had different grinding processes and have come from different parts of the country, the material properties fit into a very close band of results.

Although mill tailings material characteristics have been presented for other mill tailings (Vick 1984), no such comparison exists for United States uranium mill tailings. This paper presents the compilation and comparison of data from uranium mill tailing piles data collected as part of the UMTRA program.

INTRODUCTION

The reclamation of uranium mill tailings piles requires the *in situ* characterization of their material properties in order to prepare a design that will meet design standards. For the UMTRA program, the EPA longevity standards under PL 95-604 (200-1,000 years) were practical. This strict longevity standard also requires that a surveillance program be implemented, but that no planned maintenance be incorporated into the design. The DOE technical assistance contractor (TAC) who prepares the remedial action plans (RAPs) will implement the standards, after concurrence by the state or tribe in which the pile occurs and the Nuclear Regulatory Commission (NRC) who issues a license as a final concurrence. Included in these RAP's is a site characterization report that contains all data obtained for a particular site.

Of the 24 sites that are part of the UMTRA program (see Figure 1), some will be relocated, and others stabilized-in-place (SIP). For those sites that will be relocated, the characterization has centered more on testing remoulded

(1) Senior Project Manager, Roy F. Weston, Inc., Albuquerque, New Mexico.

(2) Geotechnical Manger, UMTRA Program, Sergent, Hauskins and Beckwith, Albuquerque, New Mexico.

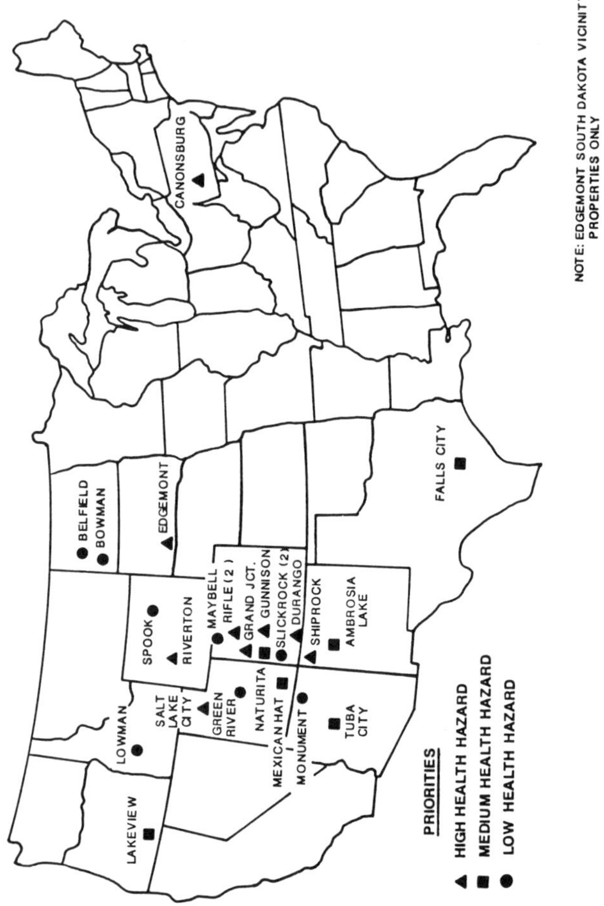

Figure 1. LOCATION OF UMTRA PROJECT SITES

samples of sand/slime material, based on the assumption that relatively good mixing will take place during relocation. Some sites are not really tailings, but contaminated materials while others (10 sites) have been characterized concentrating on the *in situ* properties, because the original intent was to stabilize-in-place, although the plan may now be to relocate the pile. Of these 10 sites, 5 had piezocone investigations, in addition to a limited geotechnical investigation. The following sections describe a brief background of each of these sites, the results of the characterizations of each site, and how they compare to each other.

SITE BACKGROUND

Each of the sites for which characterization data is to be presented was processed or deposited in a little different way. This section gives a brief description of each of the sites from which data is presented.

RIVERTON - The Riverton (RVT) tailings pile is located just south of Riverton, Wyoming, and covers 72 areas, ranging from 10 to 20 feet in thickness, between the Wind and Little Wind Rivers, and is made up of 702,000 cubic yards of material. The tailings were processed from 1958 to 1963, using both the acid and carbonate leach processes. Although the original recommendation was to stabilize the tailings in place, the present plan is for the tailings to be relocated to the UMETCO Mine, 40 miles east of Riverton in the Gas Hills of Wyoming.

AMBROSIA LAKE - The Ambrosia Lake (AMB) tailings pile is located 25 miles northwest of Grants, New Mexico, and covers 111 acres ranging from 10 to 25 feet in thickness, and is made up of over 2 million cubic yards of material. The tailings were processed from 1958 to 1963, using a finely ground alkaline pressure leach process with deposition into the tailings pile by perimeter and point discharge. Reclamation of the site is scheduled to begin in 1988, and will be consolidated into an 80-acre site using the stabilization-in-place alternative.

FALLS CITY - The Falls City (FCT) tailings pile is located 50 miles south of San Antonio, Texas, and covers 146 acres ranging from 10 to 40 feet above ground, as well as two piles 20 to 40 feet below grade, which were created by filling open pits. The tailings pile is actually seven separate piles which will be consolidated into one pile that will contain over 2 million cubic yards of material. The ore was processed from 1961 to 1973, using the acid leaching process, and deposition of the tailings was by pumped slurry, using spigots around the perimeter of the above grade tailings piles. The below grade pits were filled by point discharge.

GRAND JUNCTION - The Grand Junction (GRJ) tailing pile is located in the town of Grand Junction, Colorado, along the banks of the Colorado River. The pile covers 61 acres, ranging from 10 to 30 feet in height and contains a volume of approximately 1.7 million cubic yards of tailings. The ore was processed from 1951 to 1970, using the following 4 processes:

1. Pressure Carbonate leaching
2. Acid leaching using mechanical agitation
3. Alkaline carbonate leaching using mechanical agitation
4. Acid leaching using air agitation

The tailings were deposited will perimeter spigotting. The present plan is to relocate the tailings to a site named Cheney Reservoir, approximately 20 miles south of the present location.

MAYBELL - The Maybell site (MAY) is located 5 miles northeast of the town of Maybell, Colorado, in the northwest corner of the state. It covers approximately 80 acres, ranging from 5 to 60 feet in thickness, and contains a volume of approximately 2.6 million cubic yards of tailings. The ore that came from an adjacent open pit was processed from 1957 to 1964, using the acid leach process. The tailings were deposited as a pumped slurry with one main discharge point. As the starter dam was filled, tailings sand was used to raise the dam by the upstream method.

The planned method of stabilization for the pile is to consolidate the tailings into a 60-acre pile in place.

MEXICAN HAT - The Mexican Hat (HAT) tailings pile is located 5 miles south of the town of Mexican Hat, Utah, near the Arizona border. It covers approximately 70 acres, ranging from 10 to 50 feet in thickness, and is made up of two adjoining piles, and the tailings pile contains approximately 1.7 million cubic yards. The ore was processed from 1957 to 1965 using the acid leach process with an additional flotation process to recover copper.

The larger tailings dam was constructed as a typical valley fill with spigotting along one side to raise the dam by the upstream method.

The planned method of relocation is to consolidate the small 15-acre pile into the larger pile, resulting in a 55 acre pile. Construction is presently scheduled to start in the summer of 1988.

RIFLE - The Rifle (RFL) site is actually made up of two tailing piles, Old Rifle and New Rifle, which are approximately 6 miles apart. The Old Rifle tailings pile is 13 acres in size, ranging from 10 to 20 feet in thickness and contains approximately 400,000 cubic yards of tailings. The ore was processed from 1924 to 1932 and 1942 to 1946 for vanadium and from 1947 to 1958 for uranium and vanadium. The early processes used salt roasting, water leaching, and precipitation with sulfuric acid, while the 1947-1958 period used acid leaching and salt roasting.

The New Rifle site covers 35 acres, ranging from 20 to 35 feet high, and contains 2.5 million cubic yards of tailings. The ore was processed from 1958 to 1984, using the solvent extraction process, and was deposited by multiple discharge perimeter spigotting.

Both piles are located along the Colorado River in Rifle, Colorado, and are presently planned for relocation to a site named Estes Gulch, located 10 miles north of the town of Rifle.

SALT LAKE CITY - The Salt Lake City (SLC) tailings piles was located 5 miles south of Salt Lake City, along Highway I-15. The tailings pile covered 20 acres, were 5 to 15 feet high, and contained 1.5 million cubic yards of tailings. The ore was processed from 1951 to 1968, using the acid leach process, and was deposited using multiple discharge perimeter spigotting. The tailings pile has been relocated to Clive, Utah, 80 miles west of Salt Lake City.

TUBA CITY - The Tuba City (TUB) tailings pile is located 5 miles northeast of Tuba City, Arizona. The pile covers 40 acres, ranges in height from 5 to 45 feet, and contains approximately 1 million cubic yards of tailings. The tailings were processed from 1956 to 1966, using the acid leach and carbonate leach processes. The tailings were deposited as a pumped slurry with perimeter spigotting to raise the dam by the upstream method, once the starter dam was full. The present plan is to stabilize the pile in place with construction starting in early 1988.

DEPOSITIONAL CHARACTERISTICS

Tailings are typically described as being made up of three material types unique to mine wastes: sand, sand/slime, and slime. The three material types, the total percent of each, and the grain size distribution within each material type is a result of the grinding/process used, the ore body characteristics and the depositional method used. Typically tailings are pumped as a slurry to the tailings pond where one or several of the following methods will be used:

- point discharge into a dam
- multiple point spigotting along a valley fill
- cycloning with multiple point discharging
- spigotting along the perimeter of a starter dam

Each of these methods will produce a different grain size distribution of the materials across the tailing pile. Also, large tailings piles with a pond will generally have more slimes than a small pond due to less mixing as the spigots are moved.

However, before one can talk about the three material types, one must first decide how to classify a tailings sample as one of the three material types. For the UMTRA program, it was decided that the following distribution would be used:

Sand	-	Less than 30% passing No. 200 sieve
Sand/Slime	-	Between 30% and 70% passing No. 200 sieve
Slimes	-	Greater than 70% passing No. 200 sieve

This breakdown was chosen based on experience with other tailings as well as that of uranium tailings. Once the classification was chosen it was then possible to compare the material properties of the three classifications.

GRAIN SIZE

Figure 2 shows the grain size of each classification was plotted for each site. It can be seen that the grouping is very tight. This was true for all sites, although they are not presented. From the plot for each site and each material, an average grain size curve was determined for the three material types. These average curves were then plotted as shown on Figures 3, 4 and 5.

Figure 3 shows the range of grain-size curves for sand tailings from 12 sites. Most of the sites are very tightly grouped. Figure 4 shows the range of

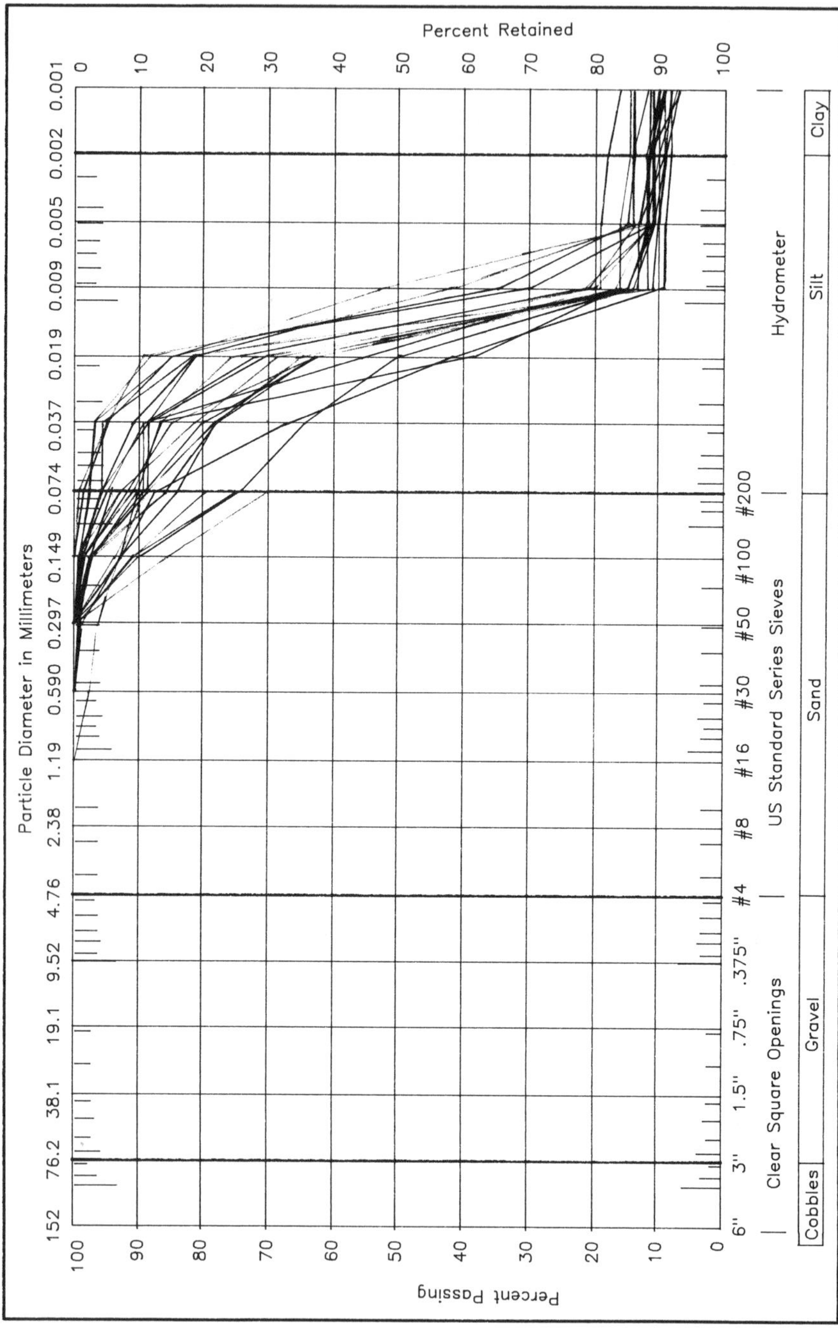

Figure 2. Typical Range of Grain Size Distributions for Slimes

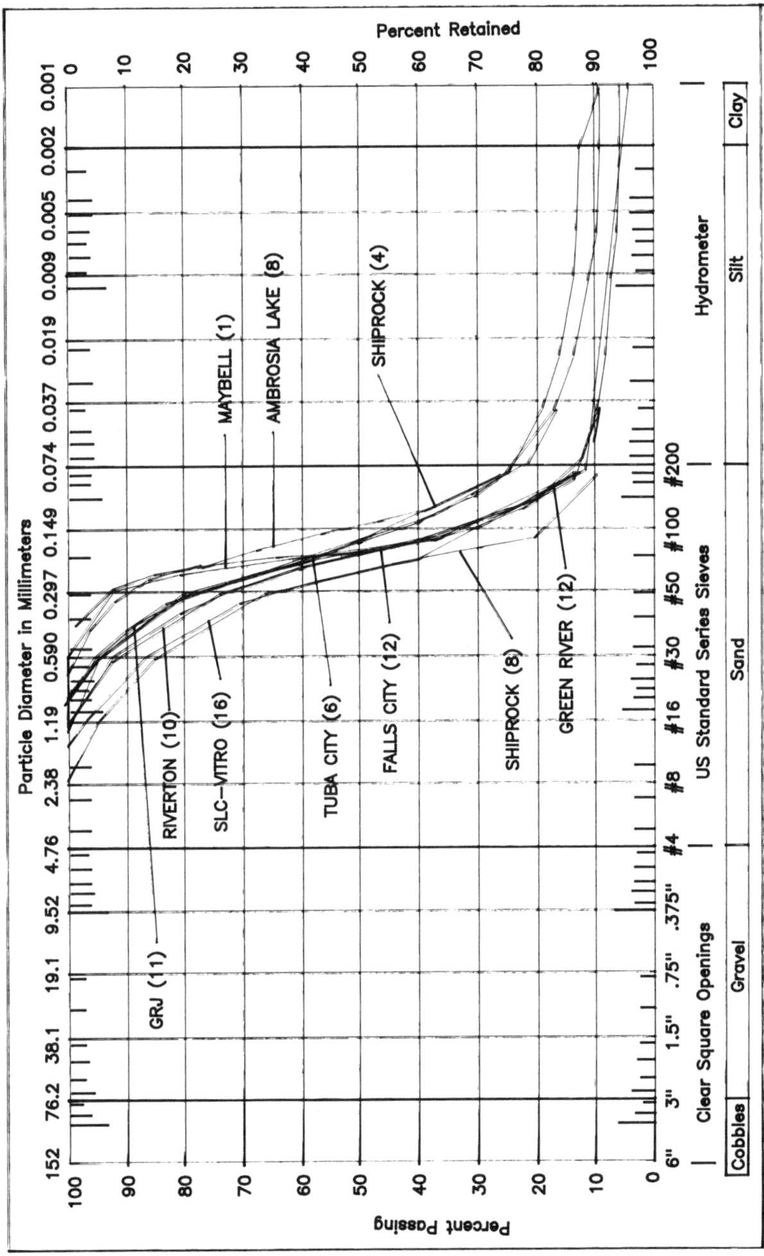

Figure 3. Grain Size Distribution – Sand

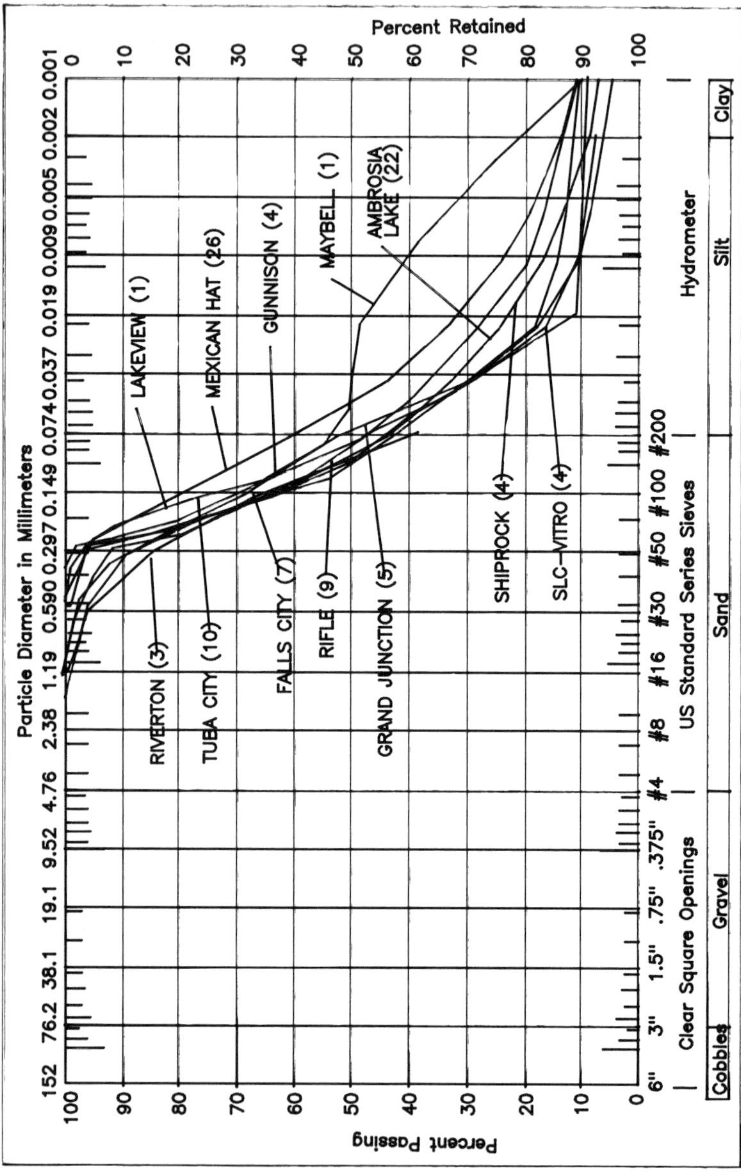

Figure 4. Grain Size Distribution – Sand/Slime

grain-size curves for sand/slime tailings, which are again tightly grouped with the exception of one sample from the Maybell, Colorado Site. Figure 5 shows the range of grain size curves for slime tailings. Once again, they were tightly grouped with the exception of one sample from Lakeview, Oregon, which showed almost four times as much clay as the samples from other sites, and the Tuba City, Arizona, where the samples had more silt and less sand and slimes from the other sites.

ATTERBERG LIMITS

Atterberg limits were performed on each classification of tailings. All sand samples and, for the most part, all sand/slime samples tested as nonplastic. Figure 6 shows the data points of all slime samples tested from the 10 sites. This plot shows that, with the exception of Maybell and Falls City, the PI is less than 30 and the liquid limit less than 60, classifying as CL or ML materials. The Falls City tailings are very different, classifying as a CH material with an average liquid limit of 80 and an average PI of 45. The Maybell material classifies as MH material with an average liquid limit of 90 and a PI of 30.

SPECIFIC GRAVITY

Specific gravity tests have been performed on samples from most sites on each material type. The results of these tests are shown on Table 1, which shows the specific gravity ranges from 2.43 for sand at Lakeview to 2.92 for slimes at Maybell. Excluding Lakeview and the slimes at Maybell, a value of 2.70 for all material types would represent material from most sites.

TABLE 1 SPECIFIC GRAVITY TEST RESULTS

Site	Material Type	Specific Gravity
Green River	Sand	2.66
Mexican Hat	Sand	2.70
	Sand/slime	2.76
	Slime	2.76
Ambrosia Lake	Sand	2.65
	Sand/slime	2.66
	Slime	2.67
Lakeview	Sand	2.43
	Sand/slime	2.46
	Slime	2.59
Rifle	Sand	2.73
	Sand/slime	2.73
	Slime	2.81
Maybell	Sand	2.68
	Sand/slime	2.71
	Slime	2.92

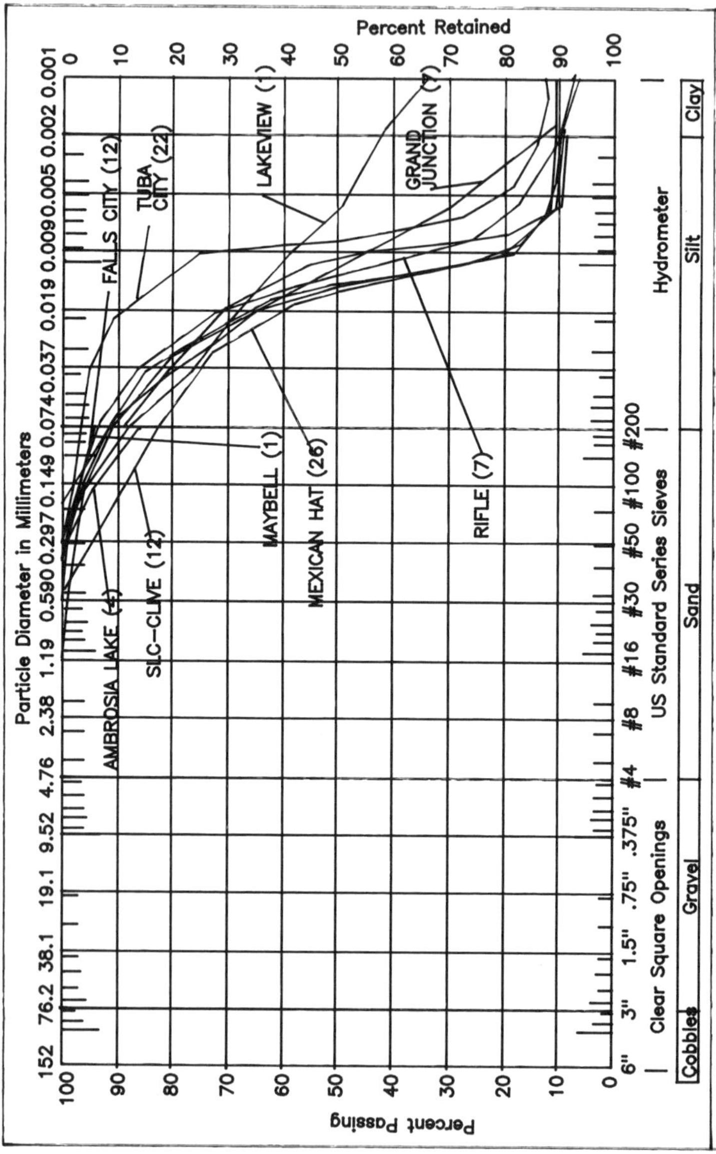

Figure 5. Grain Size Distribution — Slime

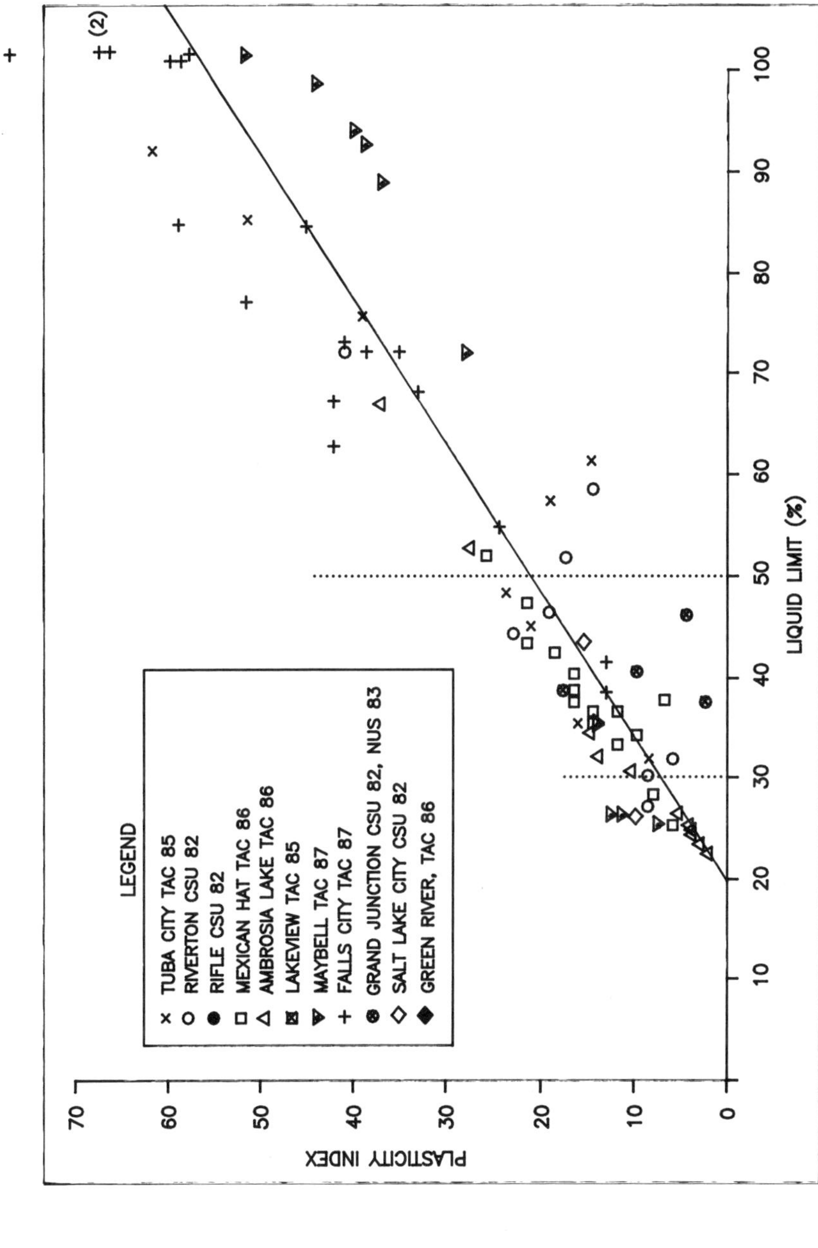

Figure 6. ATTERBERG LIMITS

MOISTURE-DENSITY RELATIONSHIP

Moisture-density relationships are determined for sand and sand/slime materials that will be moved and compacted as part of the consolidation process. Figure 7 shows a composite of all maximum dry densities versus optimum moisture contents data from standard proctor tests. These results show a range of maximum dry densities from 58 to 117 pcf. Most data is in the range of 90 to 110 pcf, with a optimum moisture content range of 15 to 25 percent.

PERMEABILITY

Permeability is one of the most difficult properties to achieve consistent results, even with a material that is very uniform. Figures 8 and 9 show the permeability data for the 10 sites which range over 6 orders of magnitude. For sand, the permeability ranges from 10^{-2} to 10^{-4} cm/sec; for sand/slime, the permeability ranges from $5x10^{-4}$ to $2x10^{-6}$ cm/sec, with most samples at 10^{-4} to 10^{-5} cm/sec ; and for slimes, the permeability ranges from 10^{-4} to 10^{-7} cm/sec. One major reason for the range of permeability is the dry density at which the permeability test was run. This is demonstrated on Figure 9, which shows permeability as a function of dry density. This figure shows that as the density increased, the permeability decreases.

Although not tested, permeability also varies between the horizontal and vertical directions by as much as two orders of magnitude, due to the layered depositional process.

COMPRESSIBILITY

Because of the loose deposition density and grading characteristics, tailings are more compressible than most natural soils of similar properties. The compressibility of sand tailings is not a major concern because the rate of consolidation is almost instantaneous. Values of the Compression Index (C_c) for the three material types are shown on Table 2, along with the dry density or void ratio (represented by C_c). These results show that for slimes, C_c ranges from .182 to 1.35, with an average of .515; for sand/slimes, C_c ranges from .061 to .66, with an average of .210; and for sand, C_c ranges from .015 to .29, with an average of .118.

Most tailings are normally consolidated, however, some tailings are over consolidated due to desiccation. The data presented is for normally consolidated tailings.

CONSOLIDATION

The time rate of consolidation for materials conforming to Terzaghi theory can be divided into the primary and secondary phases (Lambe and Whitman 1969). Primary consolidation governs the rate of pore pressure dissipation under constant load. This rate of consolidation is needed to determine how long primary settlement will take for calculations of differential settlement once construction is complete. Laboratory values of the coefficient of consolidation (c_v) are also shown on Table 2 for the three material types.

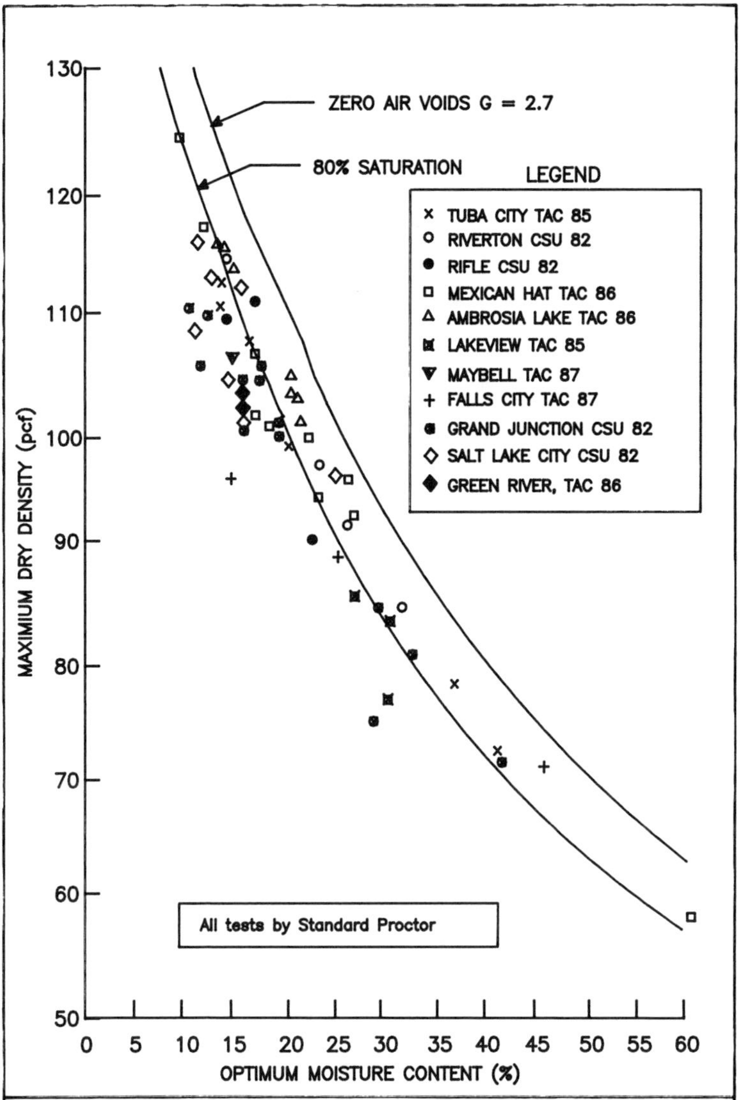

Figure 7. MOISTURE DENSITY RELATIONSHIPS

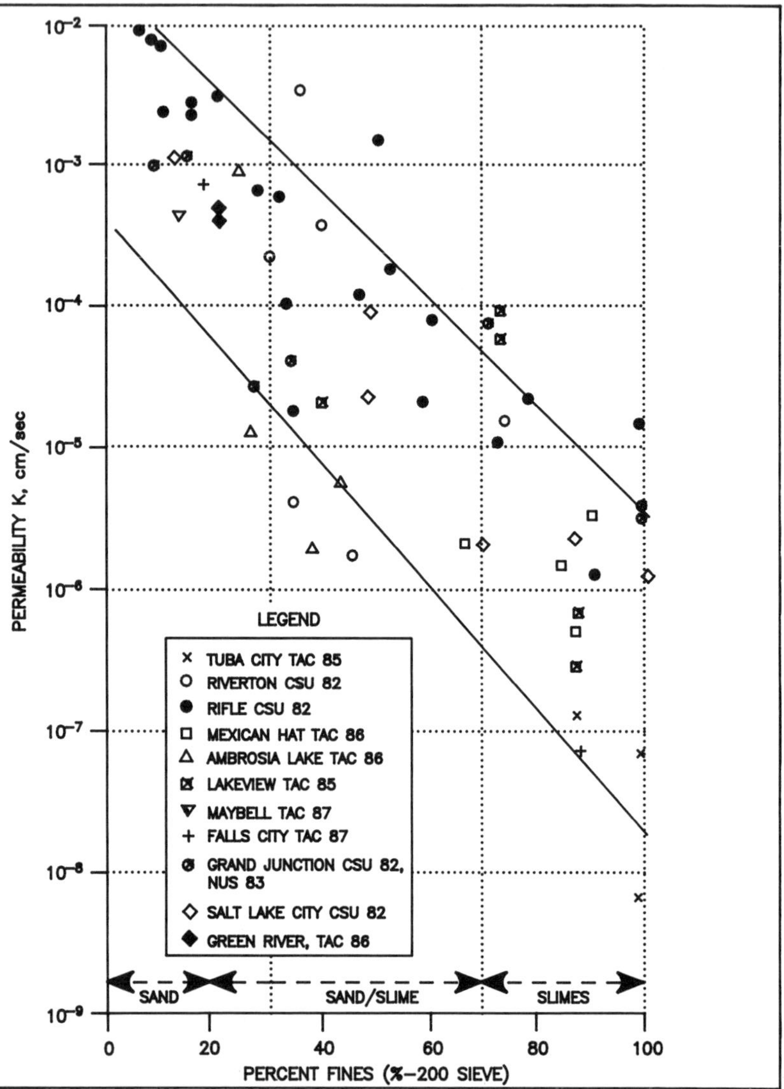

Figure 8. PERMEABILITY vs. PERCENT FINES

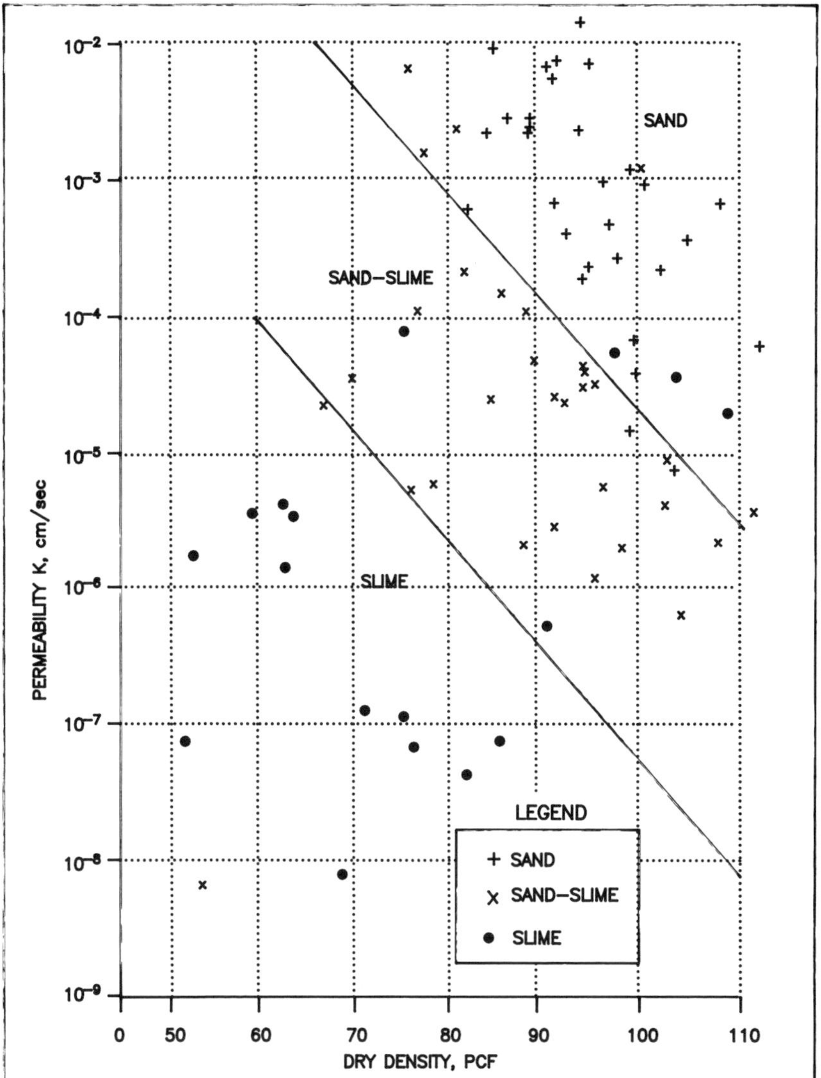

Figure 9. PERMEABILITY VS. DRY DENSITY

TABLE 2 CONSOLIDATION PROPERTIES

Site	Material Type	C_c	%-200	C_v (cm^2/sec)	γ_d(pcf)	C_α	e_o
Ambrosia Lake	Slimes	.256	70	2.4×10^{-3}	82.4	.0039	1.0
		.387	85	6.3×10^{-4}	64.0	.07	1.58
	Sand/slime	.14	44	3.3×10^{-3}	94.6	.0032	.756
		.088	44	9.2×10^{-3}	103.6	.0021	.591
		.065	46	5.7×10^{-3}	100.6	.0005	.644
		.104	32	2.5×10^{-3}	95.8	.0025	.746
		.070	--	1.0×10^{-3}	95.8	.010	.558
		.070	39	4.2×10^{-3}	105.4	.008	.558
Mexican Hat	Slime	.287	97	4.3×10^{-3}	81.7	.004	1.052
		.405	85	6.5×10^{-3}	67.4	.006	1.531
		.182	93	3.2×10^{-3}	84.0	.008	1.017
		.363	96	4.3×10^{-3}	74.9	.0095	1.268
		.322	88	3.6×10^{-3}	85.3	.003	1.013
Tuba City	Slimes	.4775	--	8.7×10^{-3}	75.8	.0125	--
Salt Lake City	Slime	.360	85	6.2×10^{-1}	78	.0022	1.17
	Remolded	.640	78	---	72	---	1.35
	Remolded Sand/Slime	.270	30	---	76	---	1.21
Falls City	Sand	.290	15	2.1×10^{-3}	81	.0053	1.10
	Remolded	.101	15	5.6×10^{-2}	88	.002	0.91
	Sand/Slime	.510	45	1.7×10^{-3}	72	.015	1.56
	Remolded	.303	45	5.0×10^{-2}	81	.010	1.09
	Slime	1.350	85	2.5×10^{-4}	51	.060	3.58
	Remolded	.460	85	3.1×10^{-2}	65	.020	1.56
Riverton	Sand/Slime	.66	47	3.2×10^{-2}	70	---	1.50
	Remolded Sand	.015	10	2.0×10^{-1}	84	---	1.00
Maybell	Sand/Slime	.452	52	2.5×10^{-3}	66.0	---	1.56
		.060	--	1.8×10^{-2}	90.3	---	.85
		.103	--	2.0×10^{-2}	97.9	---	.87
	Slime	1.197	--	5.4×10^{-4}	116.6	---	3.38
	Remolded	.872	96	1.5×10^{-3}	46.4	---	2.79
	Remolded Sand	1.243	--	4.4×10^{-4}	42.7	---	3.26
		.039	--	1.3×10^{-2}	92.0	---	0.83
Rifle	Sand	.170	5	2.1×10^{-2}	90.0	---	.73
	Remolded Sand/Slime	.127	40	2.0×10^{-3}	98.0	---	1.50
Grand Junction	Remolded Sand	.09	12	---	99.0	---	.70
	Remolded Sand/Slime	.13	32	---	95.0	---	.79

These results show that c_v for slimes ranges from 6.2 x 10^{-1} cm^2/sec to 2.5 x 10^{-4} cm^2/sec, with an average of 1.6 x 10^{-3} cm^2/sec; for sand/slimes, c_v ranges from 5.0 x 10^{-2} to 1.0 x 10^{-3} cm^2/sec, with an average of 7.9 x 10^{-3} cm^2/sec and for sand c_v ranges from 2.0 x 10^{-1} to 2.1 x 10^{-3} cm^2/sec with an average of 7.8 x 10^{-1} cm^2/sec.

The second part of settlement, important for determining the long term effect on the final contour of a reclaimed pile is secondary compression. For slimes and sand/slime materials, the secondary compression index varies, as shown on Table 2.

For sand/slime, long-term secondary consolidation (C_α) ranges from 0.015 to .0005, with an average of .004, and for slimes, C_α ranges from .0125 to .003, with an average of.005.

Although there is a wide value range for these material types, the average value shown is very representative of each material, with no sites showing any particular bias.

SHEAR STRENGTH

Although hydraulically deposited tailings are normally in a loose condition, the tailings exhibit a high-drained shear strength primarily due to of a high degree of particle angularity. Table 3 presents the results of consolidated undrained triaxial tests, with pore pressure measurements for each material type. For the most part, the samples exhibit little or no effective cohesion.

Shear strength does not seem to be affected by the dry density, although some samples show higher strengths at lower densities. However, this can be explained by a slight variability in sample preparation or material properties.

The important factor influencing shear strength for the tailings is the stress range over which the strength is measured. Although not graphically demonstrated, samples that were run as multi-phased, triaxial tests showed curvature at low stress levels. This accounts for the high cohesion in some samples, as well as the high shear strength in other samples. Taking into account these factors, the following range of effective friction angles is appropriate for each material type:

Sand	ϕ = 32° to 38°	use 35°
Sand/slime	ϕ = 27° to 38°	use 32°
Slime	ϕ = 21° to 37°	use 28°

These best estimate values, in most cases, are conservative, but provide a means of taking a quick look at their stability to see if there will be problems and to determine the amount of testing that will be required.

EFFECT OF PLACEMENT MOISTURE CONTENT

For the Salt Lake City tailings pile, the effects of placement-moisture content upon consolidation and shear strength of compacted tailings was

TABLE 3 SHEAR STRENGTH DATA

Site	Material	Phi	Cohesion (psf)	Dry Density (pcf)	Percent Moisture Content
Rifle	Sand/slime	48.9	0	77.4	44.1
		42.6	0	101.6	21.5
	Sand	38.3	0	93.8	24.1
		34.0	0	83.8	35.1
	Slime	41.7	146	71.8	52.6
Ambrosia Lake	Sand/slime	36.4	129	91.4	29.6
	Slime	30.0	0	65.4	55.3
Mexican Hat	Slime	39.1	83	88.9	34.3
		35.1	330	85.5	34.8
	Sand	36.6	250	85.0	34.5
		31.7	0	95.4	27.1
Riverton	Sand	36.0	144	115.5	7.7
	Sand/slime	27.5	317	101.2	48.1
Tuba City	Sand/Slime	33.0	0	99.0	27.2
		31.4	41	81.0	43.3
	Slime	20.0	165	69.5	53.1
		25.5	181	56.5	63.1
Lakeview	Sand/slime	36.7	372	68.6	30.0
		28.0	110	76.9	25.5
	Slime	25.5	34	74.7	30.2
Maybell	Sand/slime	30.4	170	63.5	67.8
	Slime	37.1	0	98.0	49.2
		33.8	65	81.0	41.0
Falls City	Sand	36.1	0	89.3	30.5
		26.9	27	83.4	32.3
		36.3	0	78.1	36.6
	Sand/slime	34.1	2	81.0	22.4
	Slime	27.0	0	86.3	52.0
Grand Junction	Sand/slime	30.2	400	64.0	37.2
		38.5	0	106.5	--
		44.	0	99.0	--
	Slime	30.2	100	60.6	62.3
		30.0	0	98.9	20.9
	Sand	43.0	0	109.6	--
Salt Lake City	Sand/slime	30.1	400	92.2	31.4
		31.3	400	91.7	30.1
		26.0	0	89.8	29.7
	Slime	31.0	360	--	--

Note: All tests were run consolidated undrained-with pore pressure measurements.

determined. Since tailings were to be mixed, transported, and placed at a disposal site, composite samples were prepared having a representative 40% passing the 200 sieve (sand/slime mixture). Individual test specimens were then prepared to 90% of Standard proctor, and precompaction moisture contents representing 70% to 100% of saturation.

Figure 10 depicts the results of a series of one-dimensional consolidation tests that were performed on samples prepared at moisture contents 69%, 80%, 88%, and 99% of saturation. The tests were performed unsaturated throughout the entire range of loading. The results indicate that the tailings are somewhat sensitive to compaction moisture content above 90% saturation. The influence is less pronounced below this value and the test results for less than 90% saturation approach the limit of error expected in the consolidation test. Nonetheless, for the 2600 psf loading, there is a two-fold increase in sample strain, when going from 70% to 90% saturation and nearly a five fold increase when going from 70% to 99% saturation (Table 4). This table was derived by reading the amount of strain obtained for the average embankment self-weight directly from the individual loading curves.

TABLE 4 SUMMARY OF UNSATURATED CONSOLIDATION TESTS UNDER 2600 PSF LOADING

Initial Dry Density (Pcf)	Percent Saturation After Compaction	Percent Strain
92.5	69	1.7
92.5	80	2.4
92.8	88	3.4
91.0	99	7.5

A series of triaxial compression tests were also performed on samples remoulded at between 65% and 96% compaction moisture and 90% of Standard proctor density. The tests were multiphased, consolidated, undrained triaxial compression tests with pore pressure measurements. The B parameters for these tests ranged from 0.96 to 0.98. Figure 11 depicts the variance of friction angle with compaction moisture content. While the results are widely scattered, it is apparent that there is a significant loss in the undrained shear strength for samples prepared above 80% saturation. The lowest strength obtained for the sand/slime mixture was for the sample prepared near saturation (that is, 98% saturated), and resulted in a 36.3 degree effective stress angle of internal friction.

The results of this series of laboratory tests indicate that the physical behavior of compacted uranium mill tailings, having a significant percentage of slimes, is similar to that of naturally occurring fine grained soils (Lambe and Whitman 1969). This is due to the large percentage of natural clay and silt minerals in the tailings grind that remain unaffected by the milling process. This behavior should be considered by investigators when testing compacted uranium mill tailings to obtain engineering design parameters.

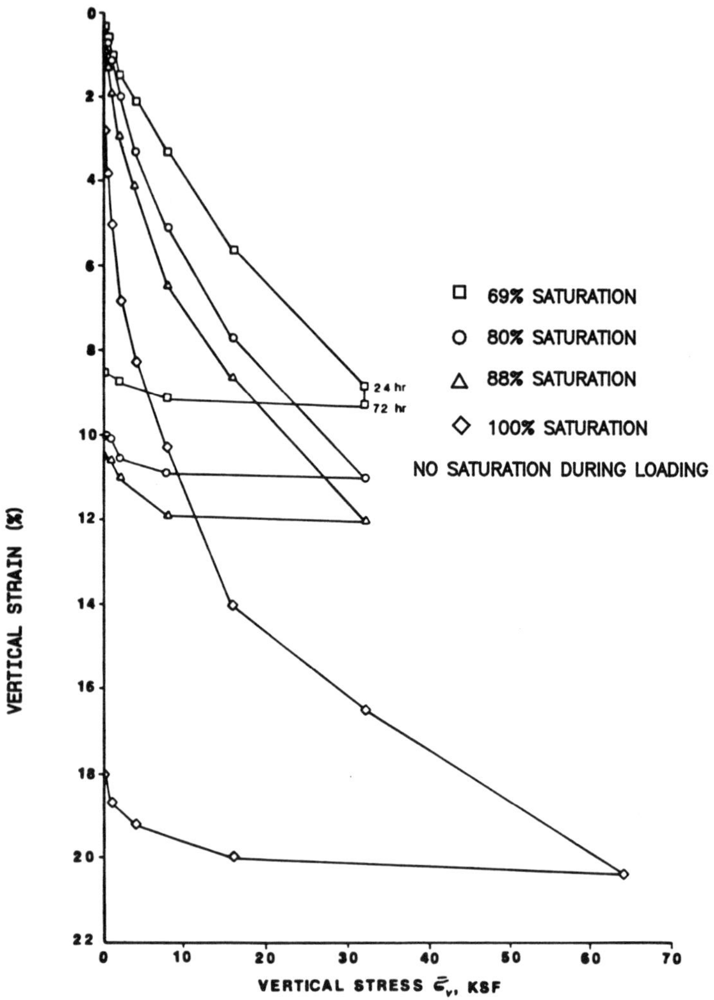

Figure 10. RESULTS OF UNSATURATED CONSOLIDATION TESTS AT VARYING INITIAL PERCENTAGE SATURATION

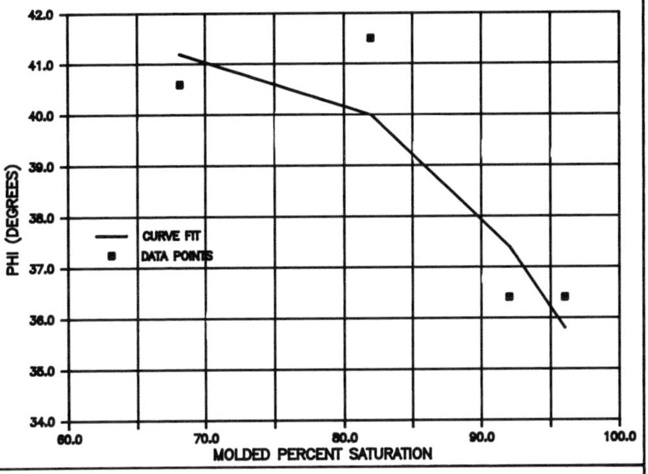

FIGURE 11-SHEAR STRENGTH vs. SATURATION

CHARACTERIZATION USING THE PIEZOCONE

The piezocone was used as the major characterization tool at the following five sites, supplemented by laboratory testing of *in situ* samples to develop correlations for the various material properties:

- Maybell, Colorado
- Mexican Hat, Utah
- Tuba City, Arizona
- Falls City, Texas
- Ambrosia Lake, New Mexico

Over 500 piezocone holes were pushed and over 100 samples were analyzed to develop the correlations. Figure 12 shows the curves that were developed for classifying material types. Again it was decided that it was more appropriate to use the three material types previously described than those developed by either ERTEC (1985) or Robinson and Campenella (1984). The figure shows that the Falls City and Maybell curves are lower than the other three sites, as was expected, based on the Atterberg limits for samples from these sites.

Additionally, a computer program was developed (Larson and Mitchell 1986) to determine undrained shear strength, overconsolidation ratio, blow count, and friction angle along with material classification. Other parameters calculated using data from literature, but not directly correlated, were modules of elasticity, relative density, and liquefaction susceptibility. An example of the output from the program is shown on Figure 13. The asterisk after the

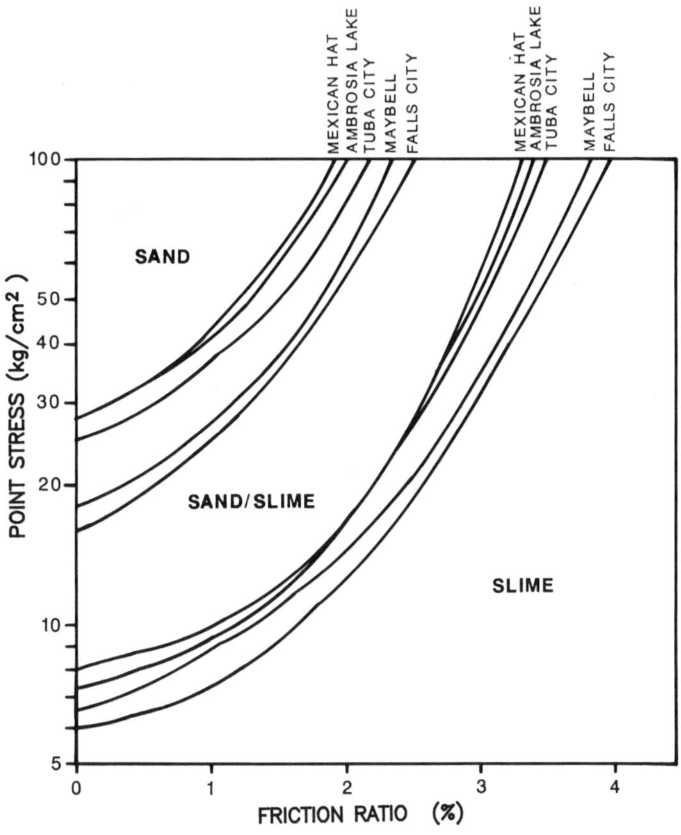

Figure 12. MATERIAL CLASSIFICATION USING PIEZOCONE

PEIZOCONE SOUNDING REPORT

SITE ID: TUB01 LOCATION ID: 004

AVERAGE SOIL WEIGHT (PCF): 110.000
GROUND WATER DEPTH (FT.): 50.000

LAYER	FROM (FT)	TO (FT)	THICKNESS (FT)	AVG. POINT STRESS (KG/CM2)	AVG. SLEEVE STRESS (KG/CM2)	AVG. PORE PRESSURE (KG/CM2)	AVG. FRICTION RATIO	MATERIAL TYPE	BLOW COUNTS PER FOOT	REL. DENSITY (%)	FRICTION ANGLE (DEG)	MOD. OF ELAST. (KG/CM2)	UNDRAINED SHEAR STRENGTH (PSF)	OCR
TAILINGS														
1	0.0	3.0	3.0	50.4	0.6	-0.1	1.3	SAND-SLIME*	16	63.	42.	177.	6054.	
2	3.1	3.6	0.6	32.9	0.6	0.0	1.9	SAND-SLIME*	10	51.	37.	115.	3950.	
3	3.7	8.0	4.3	56.7	0.5	0.2	0.9	SAND *	11	62.	42.	198.		
4	8.0	8.8	0.8	30.0	0.6	0.1	1.9	SAND-SLIME	10	40.	35.	105.	3601.	
5	8.8	14.2	5.4	58.0	0.5	0.1	0.9	SAND *	11	51.	37.	203.		
6	14.3	15.2	0.9	32.0	0.4	0.1	1.3	SAND-SLIME*	10	28.	33.	112.	3844.	
7	15.2	24.6	9.3	6.4	0.2	1.0	2.7	SLIME	4			23.	773.	1.1
% THICKNESS=		39.9	9.7	57.4	0.5	0.1	0.9	SAND	11	56.	39.	201.		
% THICKNESS=		21.8	5.3	42.2	0.6	-0.0	1.5	SAND-SLIME	13	52.	39.	148.	5070.	
% THICKNESS=		38.3	9.3	6.4	0.2	1.0	2.7	SLIME	4			23.	773.	1.1
FOUNDATION														
8	24.6	25.0	0.4	58.4	0.0	1.0	0.0	SAND *	11	33.	33.	204.		
% THICKNESS=		100.0	0.4	58.4	0.0	1.0	0.0	SAND	11	33.	33.	204.		

Figure 13. TYPICAL ENGINEERING PROPERTIES FOR EACH LAYER

material type classification means that the soil layer is susceptible to liquification.

It can be seen that the amount of data collected with the piezocone is much greater than one could hope to collect with conventional geotechnical investigations, due to the excessive drilling and laboratory testing costs. By doing a few good correlations of tailing properties, it is now possible to use the piezocone with a high degree of confidence.

FIELD COMPRESSIBILITY AND CONSOLIDATION

In order to determine how laboratory consolidation parameters compare to those that actually occur in the field, two - 200 ft square by 25 ft high test fills were constructed on the Ambrosia Lake tailings pile (Chen 1988). Piezometers and settlement measuring devices were installed, and measurements taken in order to back calculate C_c and c_v for slimes and sand/slimes.

For each instrumented profile, a pore pressure isopleth was calculated as shown on Figure 14. By calculating the area under the curve for each isopleth, the percent consolidation and the Terzaghi time factor "T" was calculated and the c_v was back calculated for each material in the profile. The results of the test fill showed that C_c could be calculated from the laboratory data to within $\pm$ 20% of the actual value, while back calculated values of c_v were two to eight times higher than those from laboratory tests.

These test fills showed that, because of layered deposition, characteristics that include micro layers of sand, the heterogeneity of the tailings, and the higher horizontal permeabilities, it is very difficult to accurately predict the consolidation rate of uranium tailings.

When the consolidated rate of tailings is important, and the schedule of a reclamation project depends on an accurate prediction of the settlement rate, a test fill is cost effective.

CONCLUSIONS

Based on the results of the data presented, it can be concluded that by using the definitions for sand, sand/slime and slime, there is very little difference in the material properties of uranium mill tailings, with only a few exceptions that can be explained as follows:

Lakeview - The Lakeview site has a low specific gravity and a higher than usual amount of clay. This difference is due to the difference in ore body. The Lakeview ore was mined from an area of high volcanics and the ore came from thermally altered intrusions in the basalt rock. There was also ash from volcanic eruptions mixed with the ore.

Falls City - The Falls City site had very high Atterberg limits. There is no explanation for this except for the variation in the ore body. Natural soils at the Falls City site also exhibit high plasticity. The slimes exhibit this mineralogic characteristic rather than being derived from this crushing process during milling.

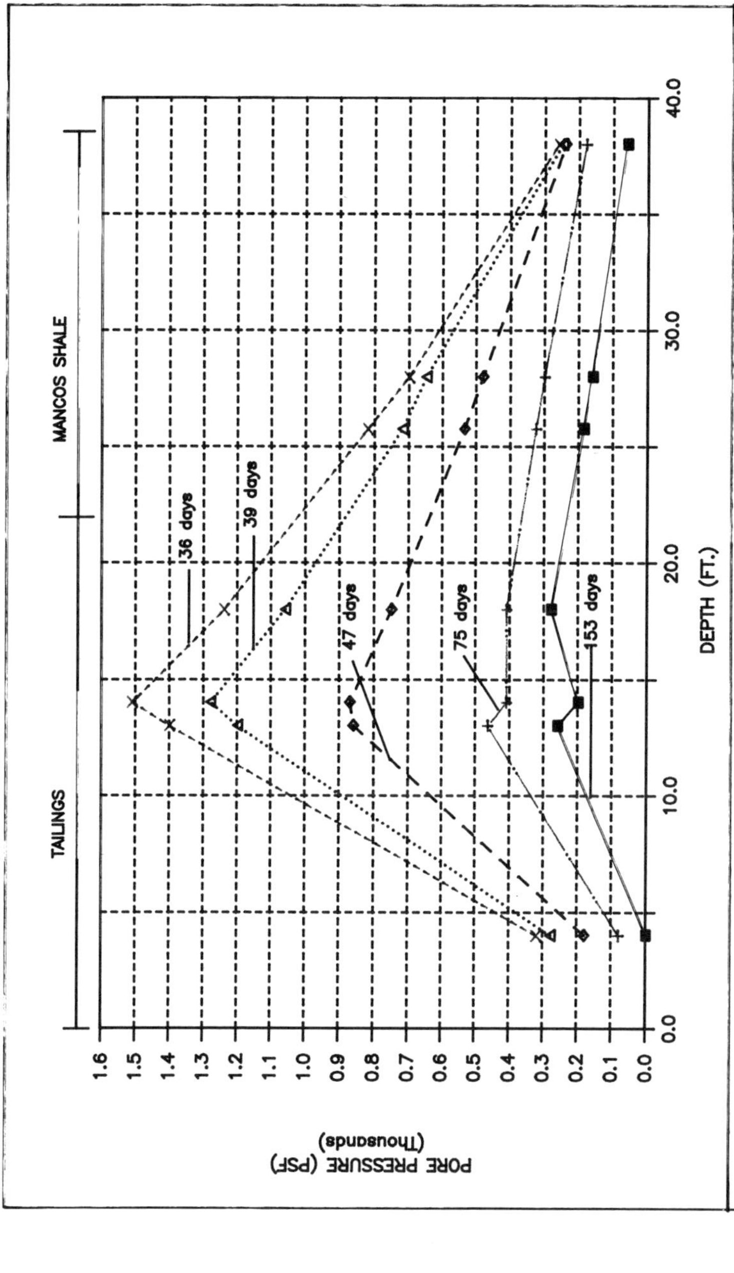

Figure 14-PORE PRESSURE ISOPLETHS AMBROSIA LAKE TEST FILL

It must also be noted that with the exception of these two sites, the sites examined are all in the Colorado Plateau region. Additionally, there was a lot of trucking of ore from one mine to several mills, and although the amount of each is not known, it does explain some of the similarity between sites. General mining areas can, however, be distinguished, including Utah (SLC,GRN), Wyoming (RVT), Colorado (GRJ, RFL, MAY), New Mexico (AMB) and Arizona (TUB, HAT). All are roll-front deposits, and all exhibit remarkably similar tailings physical properties, regardless of the milling process.

Also, the following additional conclusions can be reached from the data presented:

- Knowing the percent fines of the tailings, it is possible to predict the permeability to within two orders of magnitude.

- The coefficient of compression for the sand/slime tailings is approximately .2 and for slime tailings C_c is approximately 0.5.

- The coefficient of consolidation remains fairly constant in the 10^{-3} cm^2/sec range for normally consolidated tailings.

- The shear strength of tailings, although it has a broad range, can be estimated as follows:

Sand	$\phi = 35^{\circ}$	$c = 0$
Sand/slime	$\phi = 35^{\circ}$	$c = 0$
Slime	$\phi = 28^{\circ}$	$c = 0$

- The moisture content, or percent saturation of tailings when compacted, has an effect on the amount of consolidation and the shear strength at saturations above 90%.

- The piezocone is an effective tool for the characterization of tailings and is very cost effective. The use of a computer program is necessary to analyze the thousands of data points that are collected for a particular site. Reliable correlations have been developed for uranium tailings, which show a narrow band of variation for the three material types. Therefore, with minimal laboratory data, it is possible to use the piezocone as the major characterization tool.

- Results of test fills show that the amount of settlement that will occur in tailings can be predicted to within $\pm$ 20% of the actual. However, the rate at which settlement will take place can be under predicted by as much as an order of magnitude.

It is recommended that when settlement rates are important, that a test fill be run. Small test fills can be performed inexpensively, and the data and potential savings is more than worth the cost.

ACKNOWLEDGEMENTS

This study was completed as part of the DOE UMTRA Project, headquartered in Albuquerque, New Mexico, and was supported under DOE Contract NO. DE-AC04-8ZAL14086 to Jacobs Engineering Group Inc., the Technical Assistance Contractor (TAC). The TAC consists of Jacobs Engineering Group Inc.; Roy F. Weston, Inc.; and Sergent, Hauskins and Beckwith Geotechnical Engineers, Inc. The authors wish to express special thanks to Ned Larson and Barry Myers for valuable technical input, to Darcel Sandland for preparing the graphics, and to Lilly Rivera-Gregory for typing the paper.

REFERENCES

Chen, P. K., B. Keshian, and F. B. Guros. 1988. "Predicting Settlement of Hydraulically Placed Uranium Tailings", Hydraulic Fill Structures '88, Geotechnical Engineering Div., ASCE, August, 1988, Ft. Collins, Colorado.

DOE. 1987. "Remedial Action Plan and Site Conceptual Design for Stabilization of the Inactive Uranium Mill Tailings Site Near Falls City, Texas," Draft Report, November, 1987, UMTRA-DOE/AL-050520.0000, prepared by the U.S. Department of Energy, UMTRA Project Office, Albuquerque Operations Office, Albuquerque, New Mexico.

DOE. 1987. Remedial Action Plan and Site Design for Stabilization of the Inactive Uranium Mill Tailings Site at Riverton, Wyoming, Final Report, September, 1987, UMTRA-DOE/AL-050507.0000, prepared by the U.S. Department of Energy, UMTRA Project Office, Albuquerque Operations Office, Albuquerque, New Mexico.

DOE. 1987. Remedial Action Plan and Site Design for Stabilization of the Inactive Uranium Mill Tailings Site at Ambrosia Lake, New Mexico, Final Report, May, 1987, UMTRA-DOE/AL-050516.0000, prepared by the U.S. Department of Energy, UMTRA Project Office, Albuquerque Operations Office, Albuquerque, New Mexico.

DOE. 1987. "Remedial Action Plan and Site Conceptual Design for Stabilization of the Inactive Uranium Mill Tailings Site at Rifle, Colorado," Draft Report, May, 1987, prepared by the U.S. Department of Energy, UMTRA Project Office, Albuquerque Operations Office, Albuquerque, New Mexico.

DOE. 1987. Remedial Action Plan and Site Design for Stabilization of the Inactive Uranium Mill Tailings Site at Tuba City, Arizona, Final Report, May, 1987, UMTRA-DOE/AL-050518.0000, prepared by the U.S. Department of Energy, UMTRA Project Office, Albuquerque Operations Office, Albuquerque, New Mexico.

DOE. 1987. "Remedial Action Plan and Site Conceptual Design for Stabilization of the Inactive Uranium Mill Tailings Site at Green River, Wyoming," Draft Report, January, 1987, UMTRA-DOE/AL-050510.GRNO, prepared by the U.S. Department of Energy, UMTRA Project Office, Albuquerque Operations Office, Albuquerque, New Mexico.

DOE. 1987. "Remedial Action Plan and Site conceptual Design for Stabilization of the Inactive Uranium Mill Tailings Site at Maybell, Moffatt County, Colorado," Draft Report, prepared by the U.S. Department of Energy, UMTRA Project Office, Albuquerque Operations Office, Albuquerque, New Mexico.

DOE. 1986. "Remedial Action Plan and Site Conceptual Design for Stabilization of the Inactive Uranium Mill Tailings Site at Grand Junction, Colorado," Draft Report, June, 1986, UMTRA-DOE/AL-050505.000, prepared by the U.S. Department of Energy, UMTRA Project Office, Albuquerque Operations Office, Albuquerque, New Mexico.

DOE. 1986. Remedial Action Plan and Site Design for Stabilization of the Inactive Uranium Mill Tailings Site at Lakeview, Oregon, Final Report, April, 1986, UMTRA-DOE/AL-050510.0000, prepared by the U.S. Department of Energy, UMTRA Project Office, Albuquerque Operations Office, Albuquerque, New Mexico.

DOE. 1986. "Remedial Action Plan and Site Conceptual Design for Stabilization of the Inactive Uranium Mill Tailings Site at Mexican Hat, Utah," Draft Report, February, 1986, UMTRA-DOE/AL-050509.0000, prepared by the U.S. Department of Energy, UMTRA Project Office, Albuquerque Operations Office, Albuquerque, New Mexico.

DOE. 1984. Remedial Action Plan and Site Design for Stabilization of the Inactive Uranium Mill Tailings Site at Salt Lake City, Utah, Final Report, December, 1984, UMTRA-DOE/AL-0141.000, prepared by the U.S. Department of Energy, UMTRA Project Office, Albuquerque Operations Office, Albuquerque, New Mexico.

ERTEC. 1985. "The Electric Cone Penetrometer Test - A Users Guide to Contracting Services, Quality Assurance, Data Analysis", The Earth Technology Corporation, Huntington Beach, California.

Lambe, T. W., and Whitman, R. V. 1969. Soil Mechanics, New York, N.Y.: John Whiley and Sons.

Larson, N. and Mitchell, B. 1986. "Cone Penetration Use on Uranium Mill Tailings", Use of In situ Tests in Geotechnical Engineering Hydraulically Placed Tailings:, Hydraulic Fill structures '88, Geotechnical Engineering Div. Asce., June 1986, Blacksburg VA.

Robertson, P. K. and R. G. Campanella. 1984. "Guidelines for Use and Interpretation of the Electric Cone Penetration Test," prepared for Hogentogler and Co., Inc., Gaithersburg, MD, at the University of British Columbia.

UMTRCA (Public Law 95-604). 1978. "Uranium Mill Tailings Radiation Control Act of 1978 '42 USC 7901), November 8, 1978, 95th Congress of the United States of America, Washington, D.C.

Vick, Steven G. 1983. Planning Design and Analysis of Tailings Dams. New York, N.Y.: Whiley-Interscience.

In-Place Properties of a Hydraulic Landfill

Thomas W. McNeilan,* M. ASCE, and Ronald L. Skaggs,** A.M. ASCE

Abstract

Between Spring 1981 and Spring 1983, a 190-acre (75-hectare) hydraulic landfill was created on Terminal Island in the Port of Los Angeles. In the Spring of 1986, an extensive geotechnical investigation was conducted on the eastern 130 acres (51 hectare) of this landfill. In the area investigated, the hydraulic fill is typically about 40 ft (12 m) thick. The hydraulic fill consists of approximately equal portions of silty fine sand and clayey silt. Locally, the clayey silts are as much as 30 ft (9 m) thick.

The condition of the hydraulic fill was characterized using data from soil borings and laboratory testing, in situ piezocone penetration tests, in situ remote vane tests, and piezometers. This paper describes the subsurface conditions and summarizes the subsurface data collected. Conditions described include: stratigraphy, shear strength, compressibility, and groundwater/piezometric elevations. Predicted landfill settlements based on soil properties and stratigraphy are compared with historic topographic data.

Site Location, History, and Description

The project site is part of a man-made land addition to Terminal Island in the Port of Los Angeles. Originally Terminal Island was a barrier island formed from littoral and dune sands. The barrier sands graded seaward into finer sands and sandy silts and formed the floor of what is presently the outer harbor area. Placement of 11,000,000 cubic yards (8,400,000 cu m) of hydraulic fill between 1981 and Spring 1983 added 190 acres (75 hectare) of new land on the southern edge of Terminal Island (Figure 1). Because of the poor condition of the hydraulic fill, the 60-acre (24-hectare) western portion of the landfill was improved in 1984 by an excavation, mixing, and replacement process. The remainder of the site has remained vacant and unused since its creation.

The hydraulic fill material was obtained during a dredging program conducted to deepen the entrance channels and various basins of the Los Angeles Harbor. All material was dredged with a cutter-head suction dredge. Because the Port of Los Angeles is located in the ancestral Los Angeles River flood plain, the dredged materials consisted primarily of alluvial sands, silts, and clays that infilled

*Engineering Manager and **Project Engineer, McClelland Engineers, Inc., Ventura Division, 2140 Eastman, Ventura, CA 93003.

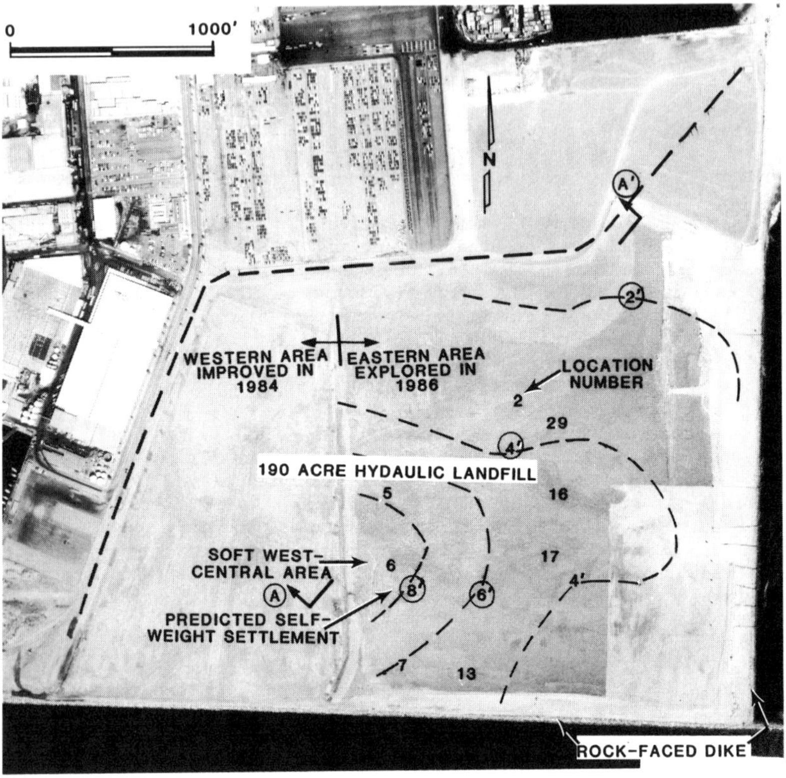

Figure 1. Air Photo of Site

the flood plain during geologically Recent times. Subsequent subsurface exploration has indicated that about one-half of the hydraulic fill consists of silty fine sands while the other half of the materials are predominantly clayey silts.

Landfill construction began by first constructing a rock-faced perimeter sand dike around the east and south boundaries of the landfill area. Following completion of the dike, the interior of the landfill was infilled. During deposition of the hydraulic landfill, the discharge pipes were moved around the northern perimeter of the landfill as well as around the perimeter dike. The soft condition of the fill prevented moving the discharge point into the interior of the fill. Because of its small particle size, much of the dredged material settled slowly. To contain the sediment-ladened water as long as possible, the contractor attempted to maintain the maximum pool within the landfill.

By Spring 1985, the surface of the unimproved landfill had settled 2 to 4 feet (0.6 to 1.2 m) and had developed a desiccation crust that was "walkable." The surface desiccation also resulted in 4 to 5 foot (1.2 to 1.5 m) deep, 1/2 to 3/4 foot (0.15 to 0.23 m) wide cracks. The innerconnected desiccation cracks resulted in 1 to 3 foot (0.3 to 1 m) diameter, polygonal-shaped, isolated columns of soil (Figure 2). In the west-central portion (Figure 1) of the unimproved area, the fill surface was still wet, undesiccated, and "unwalkable."

Figure 2. Desiccated Surface Crust

Throughout the summer of 1985, the shrinkage cracks became both wider and deeper as the crust hardened. The softer west-central portion of the fill improved significantly, but was still comparatively softer than the remainder of the site. About 1/2 to 1 foot (0.15 to 0.3 m) of additional surface settlement occurred between Spring 1985 and the end of the year.

From the end of 1985 through Spring 1986, the area experienced higher than average precipitation. By spring 1986, standing water was present in the surface shrinkage cracks over much of the site. After the last winter rain, the standing water evaporated and seeped into the underlying soils over a period of about one-half month to one month.

Through Spring and Summer 1986 and into the abnormally dry 1986-1987 winter season, the surface of the fill continued to desiccate. By Spring 1987, the desiccation cracks were typically about 1 foot (0.3 m) wide and as much as 5 to 7 feet (1.5 to 2.1 m) deep. The surface layer continued to dry through the dry summer of 1987 and a 2 to 3 inch (51 to 76 mm) thick layer of dust developed on the surface of the fill. This dust also partially infilled many of the desiccation cracks. Some of the desiccation polygons had broken apart and fallen into the cracks. By Fall 1987, total surface settlement experienced since the completion of dredging was typically on the order of 2-1/2 to 4-1/2 feet (0.8 to 1.4 m). In the softest west-central area of the landfill, the total surface settlement experienced since 1983 was as much as 6 feet (2 m).

Subsurface Investigation

In Spring 1986, a geotechnical study was conducted to define the conditions of the unimproved, eastern two-thirds of the hydraulic fill. The purpose of the study was to obtain data necessary to evaluate what methods could be used to economically improve the landfill to a usable state.

Subsurface explorations were conducted at 29 locations. The exploration program included: 15 sample borings, 20 in situ cone penetration tests (CPT), 4 in situ remote vane (RV) test holes, and groups of 2 to 4 piezometers at 4 locations. Most explorations penetrated through the fill into the natural materials underlying the landfill.

Soil Borings. All-terrain drilling equipment, shown in Figure 2, was required to access the soft, unstable site. The rig was adapted for the very soft conditions by replacing its standard 24-inch (0.6 m) wide tires with 66-inch (1.7 m) wide, low contact pressure terra tires. Using this equipment, all areas of the site including the very soft area in the west-central portion of the site could be accessed. Borings were drilled using wet rotary techniques with hollow-stemmed, 8-inch (20.3 mm) augers. Soil samples were obtained with a standard penetrometer test (SPT) sampler, 2-7/8-in. (73 mm) ID, 3-in. (76 mm) OD Shelby tubes, and a 2-3/8-in. (60 mm) ID, 3-in. (76 mm) OD liner sampler. Samples were obtained using both driven SPT procedures and pushed sampling techniques.

Selected Shelby tube samples of loose sands were drained, sealed using expandable brass and rubber plugs, and frozen in the field. These samples were transported to the lab in styrofoam packing boxes to minimize disturbance. At the lab, x-rays were taken to evaluate stratigraphy and sample disturbance and help choose the best samples for testing.

In Situ Testing. Cone penetration tests were conducted using a standard 10-square-centimeter, 60-degree apex conical cone [Doyle, et al., 1971]. The cone was instrumented with a friction sleeve and included a porous stone at the cone tip. The cone was advanced at a rate of 2 centimeters per second. The in situ remote vane tests were conducted using an electromechanical downhole remote vane system [Doyle, et al., 1971]. The vane was pushed into the soil using the drill rig and torque was applied to the vane using an internal battery-powered motor. The procedure was repeated at about 1 to 2 foot (0.3 to 0.6 m) depth intervals. Piezometer installations consisted of stand-pipe piezometers with screened intervals. The screened intervals were surrounded with sand packs, which were sealed top and bottom with Bentonite pellets.

Laboratory Testing Program. Subsequent to the subsurface exploration program, a laboratory testing program was developed to determine the classification and engineering properties of the soils. The testing included:

- Classification Tests - liquid and plastic limits, natural water content, organic content, and grain size analysis by hydrometer, percent passing the No. 200 sieve, and sieve analysis tests.
- Soil Condition Tests - unit weight determination, maximum-minimum density, and permeability tests.
- Static Strength Tests - pocket penetrometer, torvane, miniature vane, and unconsolidated-undrained, consolidated-undrained, semi-consolidated undrained, and consolidated-drained triaxial tests.
- Consolidation Tests - incremental and controlled rate of strain (CRS) consolidation tests.
- Chemical Analysis - salinity, pH, soluble chloride, and soluble sulfate.

Most testing was performed in compliance with applicable ASTM test procedures.

Subsurface Conditions

Hydraulic Fill Thickness. The top of the hydraulic fill was originally placed up to about El +22 ft (+6.7 m) (all elevations are relative to MLLW datum). The hydraulic fill increases in thickness from about 30 to 35 feet (9 to 11 m) at the northern perimeter of the site, to 40 to 45 feet (12 to 14 m) thick along the southern boundary. The fill is underlain primarily by dense to very dense silty fine sand deposits.

Fill Material Characteristics. The hydraulic fill is primarily intermixed and interlayered, silty fine sand (SM) and highly plastic clayey silt (MH) with lesser quantities of sandy silt and silt (ML). The particle size and plasticity data for the two predominant material types are shown in Figures 3 and 4. Although mappable zones exist where one or the other of these materials predominates, these two material types are intermixed and interlayered throughout the fill.

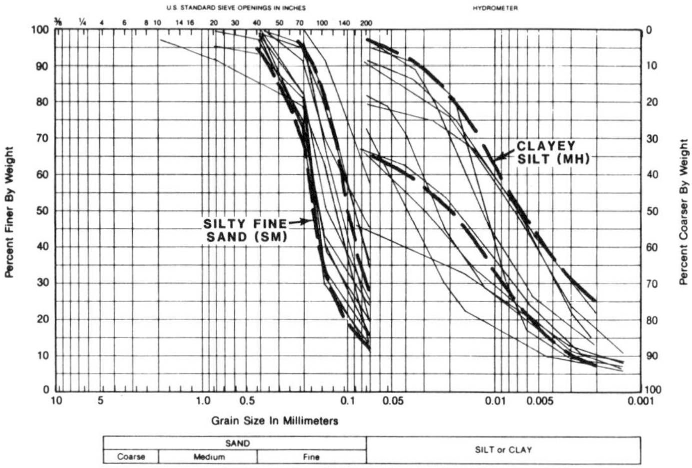

Figure 3. Grain Size Envelopes

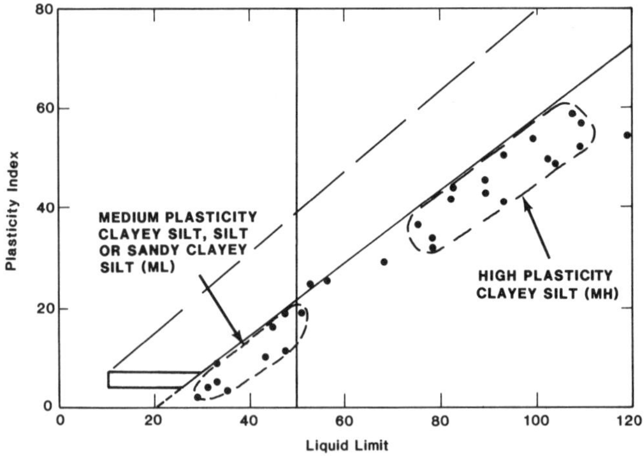

Figure 4. Plasticity Chart

The silty fine sand is nonplastic, poorly graded, and contains in excess of 12 percent fines. It is typically loose to medium dense, and occasionally dense. Particles larger than about 0.2 to 0.25 mm are almost exclusively shell fragments. Frequently the silty sand layers contain pockets and seams of clayey silt. The poorly graded silty fine sand has a coefficient of permeability of 10^{-3} centimeters per second.

The clayey silt plots below the A-line of the plasticity chart and is predominantly highly plastic with a liquid limit of 70 to 110. A smaller portion of the clayey silt is of medium plasticity with a Liquid Limit of 30 to 50. Below the desiccated crust, the clayey silt is very soft to firm. The clayey silt contains 8 to 25 percent clay-size particles (smaller than 2-micron size) and the organic content ranges from 1 to 9 percent. The layers of medium plasticity clayey silt also contain a significant sand-size fraction as well as frequent pockets and seams of sand and occasional stiff clay balls.

Stratigraphy. Although the fill is heterogeneous, a few generalities can be made in describing the stratigraphy. The highest percentage of sand is present around the eastern and southern perimeters of the fill where the dikes were constructed of predominantly sandy material. Sand also tends to predominate near the northern landfill perimeter where the hydraulic fill was discharged into the landfill area.

The surface of the fill consists predominantly of clayey silt. This surface layer is 5 to 15 feet (1.5 to 4.5 m) thick over most of the site. Locally, however, in the softer, west-central portion of the landfill, the clayey silt extends down to over 30 feet (9 m) deep.

As stated previously, a desiccated crust exists over the landfill. It was not until late in 1986 and early 1987, however, that a surface crust started to form in the soft west-central area of the site. As recently as Fall 1985, it was not possible to walk across this area of the site. The surface clayey silt layer is weakest where the fill is thickest. This correlation between thickness and strength is due to the underconsolidated condition of the thicker clayey silt layers.

Over most of the site, the surface clayey silt is underlain by a relatively persistent layer of medium dense to dense, silty fine sand. The sand layer typically extends down to about El 0. The bottom half of the landfill is composed of highly layered loose to medium dense, silty fine sand and soft clayey silt.

Figure 5 shows a diagonal cross-section across the landfill. The transition from northeast to southwest along this cross-section roughly corresponds to the direction of flow from the discharge pipe to the outlet weir during a significant portion of the filling.

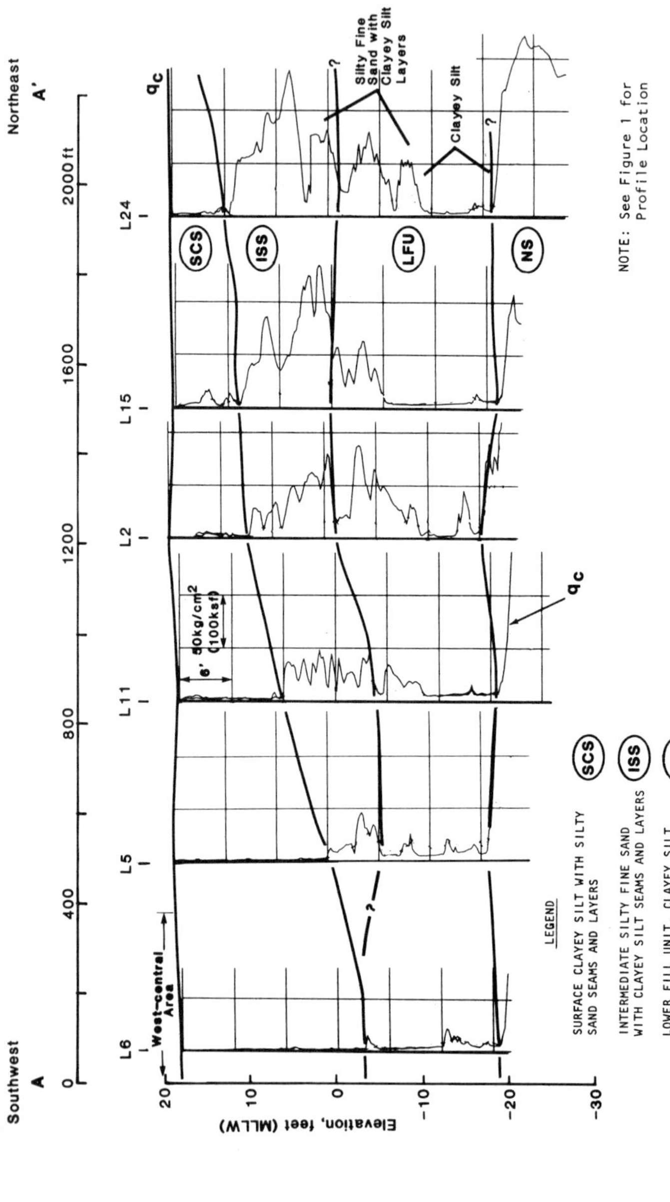

Figure 5. Generalized Subsurface Profile

Piezometer Surfaces. Because of the interlayered nature and fine-grained texture of much of the fill, the hydraulic fill contains various static and perched water levels. Near the base of the fill and along the southern and eastern perimeters of the fill, silty sand layers appear to respond to tidal fluctuations of the ocean. Elsewhere in the fill, however, water depths measured in the piezometers (Figure 6) indicate the piezometric surfaces are above the tidal range. Since Spring 1986, the piezometric surfaces in thick clayey silt layers have dropped an average of 3 feet (1 m).

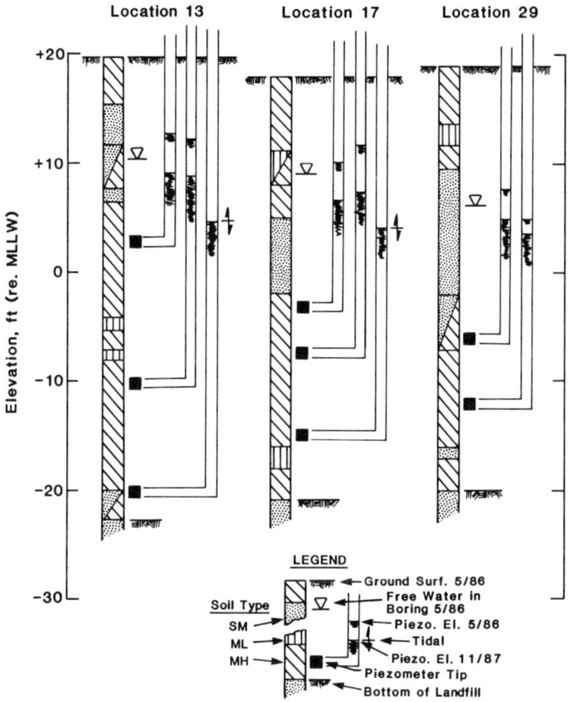

Figure 6. Piezometric Elevation Data

Soil Properties of Granular Layers

Relative Density. The dry density of the silty fine sand samples typically ranged from 85 to 105 pcf (1.36 to 1.68 g/cm³). In comparison, the measured minimum dry density ranges from 80 to 84 pcf (1.28 to 1.34 g/cm³) and the measured maximum dry density ranges from 105 to 109 pcf (1.68 to 1.75 g/cm³). Because the sands in many of the layers are below the critical void ratio, the calculated unit weights were determined using the sample volume measured in the field immediately

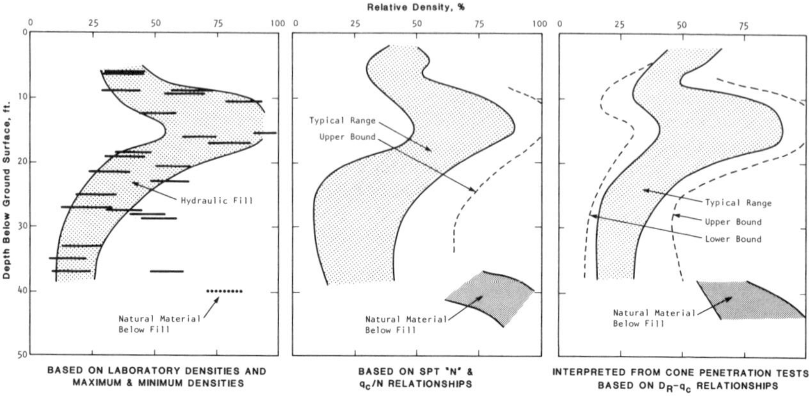

Figure 7. Relative Density Envelopes

upon recovery of the samples. As shown in Figure 7, the sands are typically medium dense to dense with relative density increasing with depth down to about El 0. Below El 0, the relative density of the sand layers decreases with depth, being typically medium dense down to about El -10 ft (-3 m) and loose to medium dense at the base of the landfill. The lower relative densities were measured in sand layers deposited below water.

The relative density also was estimated using relationships for standard penetration test blow count (N value) and relative density. We also used empirical relationships [Robertson and Campenella, 1984; Kasin, et al., 1986] to compare the SPT values with the N values inferred from cone penetration tests (CPT). Direct comparison of N values and the cone tip resistance, q_c, suggest an average q_c/N ratio of 4 for the hydraulically-placed silty fine sands. Several empirical correlations [Schmertmann, 1978; Mitchell and Lunne, 1978; Villiet and Mitchell, 1981; Baldi, 1982] also were used to directly estimate in situ relative density from the cone tip resistance. As can be seen from Figure 7, the relative densities interpreted from SPT N values and the CPT compare favorably with the relative densities measured in the laboratory.

Angle of Internal Friction. The angle of internal friction, ϕ', of silty sand layers were measured with triaxial tests and interpreted from the cone penetration test data using several empirical correlations [Drugunoglu and Mitchell, 1975; Robertson and Campenella, 1984]. The angle of internal friction trends within the granular layers are similar to the trends observed in relative density. The highest values of ϕ' were 34 to 38 degrees, measured between El +5 ft (+1.5 m) and El 0. The lowest values of ϕ' were measured near the base of the hydraulic fill where the angle of internal friction was as low as 28 to 32 degrees. The values of the angle of internal friction derived from the laboratory testing and CPT interpretations are comparable with the values that would be inferred from published relationships [Burmister, 1948] of relative density and ϕ' for similar material.

Soil Properties of Cohesive Materials

Compressibility and Drainage Characteristics. The measured consolidation and permeability properties for the highly plastic clayey silt and the medium plasticity clayey silt are tabulated below:

Measured Property	High Plasticity (MH) (LL = 70-110)	Medium Plasticity (ML) (LL = 30-50)
Compression Index, C_c	0.6 to 0.9	0.15 to 0.25
Compression Ratio, $C_c/(1 + e_o)$	0.2 ±0.04	0.1 ±0.02
Coefficient of Consolidation, c_v (cm²/day)	25 to 40	100 to 200
Coefficient of Permeability, k (cm/sec)	10^{-8} to 10^{-7}	10^{-7} to 10^{-6}

The values of C_c in the preceding table compare favorably with other published values. Similarly, the range of c_v compares favorably with published trend lines [NAVFAC, 1982] for virgin compression of undisturbed samples. Comparison of the vertical coefficients of consolidation measured in consolidation tests with the horizontal coefficients of consolidation, c_h, measured during the consolidation phases of triaxial tests, suggest that the horizontal coefficient of consolidation is probably about a factor of 2 greater than the vertical coefficient of consolidation. As will be discussed subsequently, historic settlement values backcalculated from the above data compare favorably with measured historic fill settlement.

Because the hydraulic fill contains thick clayey silt layers, substantial areas within the hydraulic fill are underconsolidated relative to the loads imposed by the overlying fill materials. This observation is substantiated by several types of data, including:

- Water levels measured in piezometers set within the thicker layers of clayey silt are significantly higher than the water levels measured in piezometers set in the basal sand (for example, Locations 13 and 17 in Figure 6) or in piezometers set in sands near the perimeter of the landfill.
- Consolidation tests on samples recovered from the west-central area, as well as other areas with substantial thicknesses of clayey silt, indicate that the samples are underconsolidated relative to the effective overburden pressure computed based on equilibrium piezometric conditions with respect to the surrounding ocean. Computations of the effective preconsolidation pressure in the consolidation tests on underconsolidated samples compare favorably with effective vertical stresses computed based on the piezometric elevations measured within the same layers.

- Liquidity index and S_u/p_o ratio data have been shown to provide qualitative estimates of consolidation state [McClelland, 1967; Lambe and Whitman, 1969; Wroth, 1979]. Although these indices are highly variable across the landfill site, a general trend is observable. The liquidity index increases and the S_u/p_o ratio decreases with distance from the top or bottom of a clayey silt layer. Both types of empirical data tend to suggest that when the sample depth is greater than about 3 to 4 feet (0.9 to 1.2 m) from a bounding sand layer, the sample is underconsolidated.

Computations using the laboratory consolidation parameters suggest that the midportion of clayey silt layers in excess of about 8 to 10 feet (2.4 to 3 m) thick were still consolidating 3 years after the landfill was completed. Based on these analyses, at the middle of 15 to 18 foot (4.6 to 5.5 m) thick clayey silt layers which are common throughout the fill, the degree of consolidation would be about 75 percent after 3 years.

Shear Strength. The undrained shear strength of the clayey silts was evaluated from: (1) miniature vane shear tests, (2) unconsolidated-undrained triaxial compression tests, (3) consolidated-undrained triaxial compression tests, and (4) in situ remote vane tests. Also, empirical relationships [Schmertmann, 1978; Azzouz, et al., 1983; Jamiolkowski, et al., 1985] were used to evaluate undrained shear strength from cone penetration test results where the pore pressure ratio parameter or the friction ratio indicated essentially undrained behavior. These results are summarized in Figure 8.

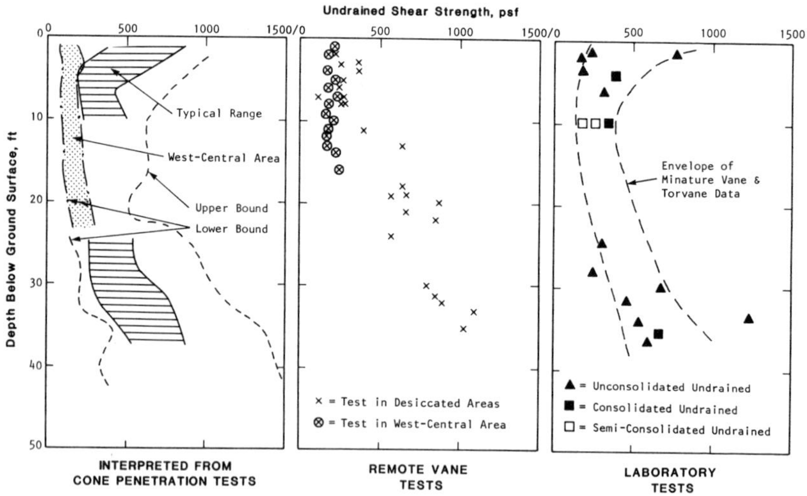

Figure 8. Undrained Shear Strength Envelopes

The variation in measured undrained shear strength between different tests at the same location and elevation typically is within ±20 percent. Where differences are greater than this, it is most likely due to variations in soil type, arising from the interlayered and intermixed characteristics of the deposit. Also we note that there is typically a variation of two in undrained shear strength of normally consolidated cohesive soils due to a variation in liquid limit from 30 to 110 [Skempton, 1957]. Thus, local layering can have a profound effect on shear strength. Localized differences in strengths measured by different types of tests also may be accentuated by the silty nature of the cohesive soils and the locally underconsolidated condition.

Effect of Underconsolidation. As stated previously where the clayey silt layers are thicker than about 8 to 10 feet (2.4 to 3 m), the data indicates the layers were underconsolidated at the time of our field investigation. To evaluate the effects of underconsolidation on shear strength, we ran special three-stage, semi-consolidated, undrained triaxial tests. In the first two stages of these tests, the specimen was allowed to partially consolidate to 33 and 67 percent of the time required for 100 percent consolidation, t_{33} and t_{67}, respectively, and then was sheared. In the third stage, the specimen was allowed to fully consolidate to t_{100} and then was sheared. The effect of underconsolidation on undrained shear strength is shown in Figure 9. These tests suggest that if only one-third the time required for the soil at this point in the layer to be fully consolidated has elapsed, then the strength at this depth will be only 40 to 45 percent of the strength after the soil has fully consolidated. For these conditions, we would expect results of a conventional laboratory consolidated drained or consolidated undrained test to exceed the in situ strength.

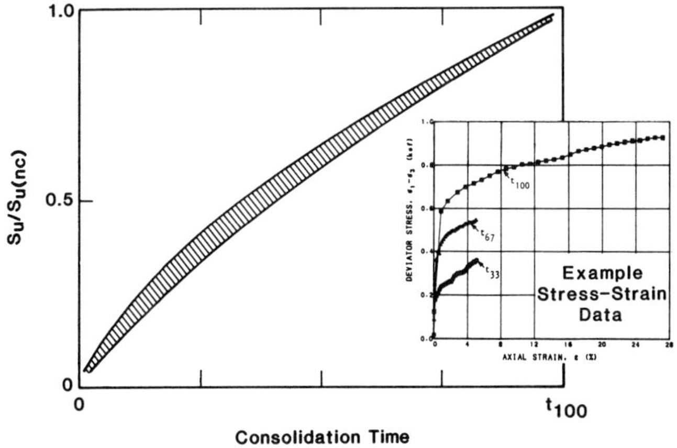

Figure 9. Effect of Underconsolidation on Undrained Shear Strength

In the west-central area, the surface clayey silt is over 20 to 30 feet (6 to 9 m) thick. Consolidation rate analyses suggest that 10 to 15 years, after fill placement, will be required before 90 to 95 percent consolidation is reached throughout the stratum. Since 3 years had lapsed when the field investigation was conducted, the time rate of consolidation analyses suggest the material at the center of the stratum had reached a t_{20} to t_{30} stage. Figure 9 suggests that for this condition, the current shear strength near the middle of the layer is probably about 30 to 50 percent of the ultimate shear strength. Shear strength envelopes on Figure 8 show that the shear strength of the upper soils in the west-central area are less than one-half of the typical range. Thus, the measured shear strengths in the west-central area appear to be consistent with the shear strength predicted for the apparent state of consolidation.

Cone Bearing Factor N_k. The bearing capacity factor, N_k, which is used to interpret undrained shear strength, S_u, from the cone tip resistance, q_c, is known to vary with overconsolidation as well as many other factors [Schmertmann, 1975]. Typically, all other parameters being consistent, N_k increases as the overconsolidation of the soil increases. Data to evaluate the effect of this latter condition are not prevalent in the literature; however, our data base of experience includes a number of sites located offshore from the mouths of active river deltas where underconsolidated conditions exist. At these sites, the value of N_k in underconsolidated layers has been less than two-thirds of the value in normally consolidated layers. We have evaluated the effect of consolidation state on N_k at this site by comparing the CPT data with both the in situ remote vane and unconsolidated-undrained triaxial compression test results.

Comparisons of the N_k values computed from the CPT resistance and the remote vane data are shown in Figure 10. Upon examining the pattern of N_k versus penetration, it is apparent that N_k is largest near the top and bottom of clayey silt layers and is lowest at mid-layer. For example, where the surface clayey silt is 18 feet (5.5 m) thick, the crescent shape of the variation of N_k within the layer mimics the theoretical shape of the isochrone of effective stress in a partially consolidated clay layer. These data are supportive of other data suggesting the thick clayey silt layers are underconsolidated.

Historic and Predicted Settlement

Since hydraulic fill placement was completed in spring 1983, the fill has continually compressed and settled under its own weight. Figure 11 shows the settlements versus time at four locations. Since July 1983 (three months after completion of fill placement), the site has settled 2-1/2 to 6 feet (0.8 to 2 m). The settlement data indicate that the site is still settling, albeit at a decreased rate. The settlement since April 1986 ranges from 1/2 to 1 foot (0.15 to 0.3 m). The greatest total and recent settlement has occurred in the west-central area where the surface clayey silt is thickest and softest.

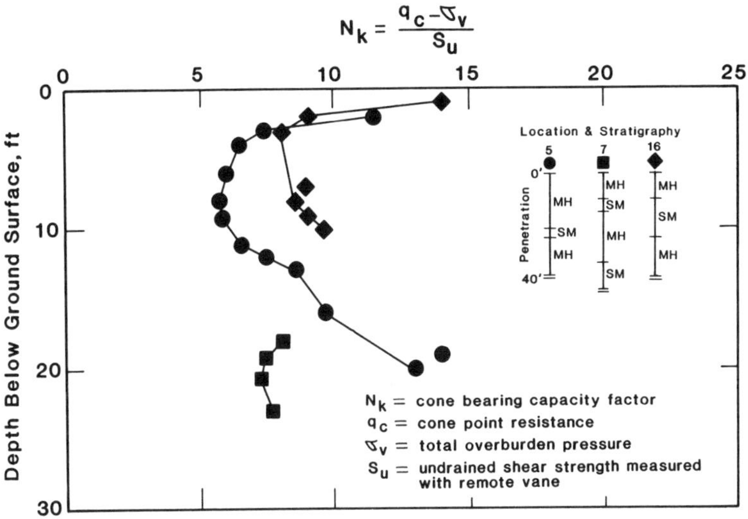

Figure 10. Effect of Underconsolidation on Cone Bearing Factor N_k

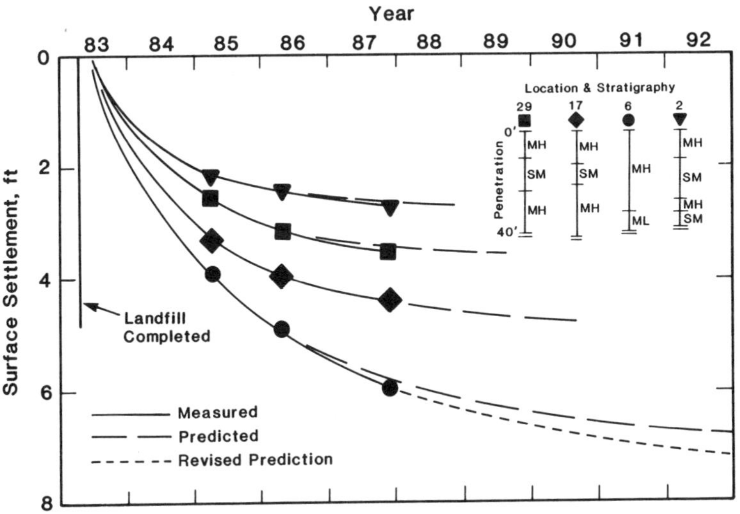

Figure 11. Historic and Predicted Settlement

Extrapolation of the historic settlement, shown in Figure 11, suggests that outside of the west-central area where the surface clayey silt is thickest, an additional site settlement of about 1/2 foot (0.15 m) will be realized over the next 2 to 3 years. Additional long-term settlement, under the current loads will be comparatively small. In the west-central area where the surface clayey silt is thickest and softest, extrapolation of the historic settlement suggests an additional foot of settlement over the next 2 to 3 years. Further settlement in the following 5 to 10 years could approach 1/2 to 1 foot (0.15 to 0.3 m).

In the Spring 1986, the future self-weight settlement of the landfill was predicted using the interpreted consolidation properties discussed previously. These analyses used the Terzaghi & Peck (1967) factors for pore pressure dissipation of hydraulically placed fill. The time rate data was used to evaluate: (1) the percentage of consolidation that had occurred in the thicker layers by March 1986, and (2) the time until 95 percent consolidation would be completed. The additional expected settlement then was computed from the percent consolidation to date and the settlement experienced to date.

Results of computations based on the c_v for the medium plasticity clayey silt (LL = 30 to 50) do not agree with the observed historic settlement. Predicted settlements based on the range of c_v for the highly plastic clayey silt compare favorably with the additional settlement realized since Spring 1986 (Figure 11). Because the settlement is sensitive to layer thickness, we evaluated the effect of interlayered silty sand on our settlement rates. These computations suggest that historic settlement rates in the interior of the fill are only explainable when the presence of small seams of silty sand are ignored. This implies that these sand seams frequently are not continuous over significant areas of the site. This implication is consistent with piezometric elevations measured in clayey silt layers containing silty sand seams.

Based on the measured settlement data (Figure 11), it appears that in many areas the hydraulic fill is continuing to settle nearly 5 years after being completed. Where clayey silt layers exceed 15 foot (4.6 m) thickness, the annual settlement is still in excess of 1/2 foot (0.15 m) per year. The data suggests that the ultimate settlement will be about 3 to 5 feet (0.9 to 1.5 m) over most of the landfill but will approach 8 feet (2.4 m) in the west-central area (Figure 1). Analyses of this data suggests that the ultimate post-completion, landfill self-weight settlement will be equal to about 15 to 20 percent of the cumulative thickness of clayey silt layers.

Concluding Comments

The project site is a good analog for estimating the geotechnical characteristics of uncontrolled hydraulic fills containing significant percentages of cohesive materials. Geotechnical studies of uncontrolled hydraulic fills containing significant clayey layers should recognize that:

- Thick deposits of clayey soils are likely to be underconsolidated.
- In underconsolidated layers, the in situ strengths may be less than the strength measured with laboratory tests on consolidated specimens.
- In underconsolidated materials, the value of the cone bearing factor N_k is less than the value normally used for normally consolidated materials.
- Because of these factors, geotechnical exploration programs should utilize multiple types of data and not rely on a single type of data.
- Where significant thicknesses of clayey material are present, time rate of settlement predictions need to recognize that thin sand layers not in communication with granular strata outside or beneath the landfill will not act as drainage layers.

The case history discussed in this paper shows that when hydraulic landfills will contain significant percentages of clayey materials, special fill placement techniques should be used if the landfill is to be used in the near future.

Acknowledgment

The data referenced in this paper was obtained under contract to the Port of Los Angeles (POLA). The authors wish to express their appreciation to the staff of the POLA for their support in this study.

Appendix - References

Azzouz, A.S., Baligh, M.M. and Ladd, C.C. (1983), "Cone Penetration and Engineering Properties of the Soft Orinoco Clay." Proceedings of the BOSS 83, Cambridge, Mass.

Baldi, G., Belotti, R., Ghionna, V., Jamiolkowski, M., and Pasqualini, E. (1982), "Design Parameters for Sand from CPT." Proceedings of the European Symposium on Penetration Testing, ESOPT II, Amsterdam.

Burmister, D. (1948), "The Importance and Practical Use of Relative Density in Soil Mechanics," Proceedings, American Society of Testing and Materials, Vol. 48, pp. 1249-1268.

Doyle, E.H., McClelland, B. and Ferguson, G.H. (1971), "Wire-Line Vane Probe for Deep Penetration Measurements of Ocean Sediment Strength," Proceedings of the Third Annual Offshore Technology Conference, Houston, Texas.

Durgunoglu, H.T. and Mitchell, J.K. (1975), "Static Penetration Resistance of Soils: I - Analyses, II - Evaluation of Theory and Implications for Practice," Proceedings, Specialty Conference on In Situ Measurement of Soil Properties, ASCE, Raleigh, North Carolina, Vol. I, pp. 151-189.

Jamiolkowski, M., Ladd, CC., Germaine, J.T. and Lancellotta, R. (1985), "New Developments in Field and Laboratory Testing of Soils," Proceedings of the 11th International Conference on Soil Mechanics and Foundation Engineering, San Francisco, California, Vol. I.

Kasim, A.G., Ming-Yau Chu and Jensen, C.N. (1986), "Field Correlation of Cone and Standard Penetration Tests," Journal of Geotechnical Engineering, American Society of Civil Engineers, New York, New York, Vol. 112, No. 3.

Lambe, T.W. and Whitman, R.V. (1969), Soil Mechanics, John Wiley and Sons, New York, New York.

McClelland, B. (1967), "Progress of Consolidation in Delta Front and Prodelta Clays of the Mississippi River," Marine Geotechnique, University of Illinois Press, pp. 22-40.

Mitchell, J.K. and Lunne, T.A. (1978), "Cone Resistance as Measure of Sand Strength," Journal, Geotechnical Engineering Division, ASCE, Vol. 104, No. GT-7, pp. 995-1012.

NAVFAC DM-7.1 (1982), Soil Mechanics, Department of the Navy, Naval Facilities Engineering Command, Alexandria, Virginia.

Robertson, P.K. and Campanella, R.G. (1984), Guidelines for Use and Interpretation of the Electronic Cone Penetration Test, Hogentogler and Company, Inc., Gaithersburg, Maryland.

Schmertmann, J.H. (1978), Guidelines for Cone Penetrometer Test Performance and Design, Federal Highway Administration, FHWA TS-78-209.

Schmertmann, J.H. (1975), "Measurement of In Situ Shear Strength," Proceedings, Specialty Conference on In Situ Measurement of Soil Properties, ASCE, Raleigh, North Carolina, Vol. II, pp. 57-139.

Skempton, A.W. (1954) "Discussion of the Structure of Inorganic Soil," ASCE Proceedings, Vol. 80, pp. 478-19 to 478-22.

Skempton, A.W. (1957) "Discussion on the Planning and Design of the New Hong Kong Airport," Proceedings of the Institute of Civil Engineers, Vol. 7, pp. 305-307.

Terzaghi, K. and Peck, R.B. (1967), Soil Mechanics in Engineering Practice, John Wiley and Sons, Inc., New York.

Villiet, W.C.B. and Mitchell, J.K. (1981), "Cone Resistance, Relative Density and Friction Angle," Proceedings, Cone Penetration Testing and Experience, ASCE National Convention, St. Louis, Missouri, October 26-30, pp. 178-208.

Wroth, C.P. (1984), "The Interpretation of In Situ Soil Tests," Geotechnique 34, No. 4, pp. 449-489, December, 1984.

Wroth, C.P. (1979), "Correlations of Some Engineering Properties of Soil," Proceedings, Second International Conference on Behavior of Offshore Structures, Vol. 1, pp. 121-132.

IN-SITU TESTING OF FINE COAL REFUSE BY NUCLEAR METHOD

By David C. Cowherd[1], M.ASCE and Vlad G. Perlea[2], M.ASCE

ABSTRACT: Fine coal refuse deposited as slurry may be liquefied under earthquake action. As part of an evaluation of liquefaction potential, the in situ density of fine refuse should be determined. An appropriate method for field measurements of density and moisture content using nuclear gauges is presented along with details on the use of this method in coal refuse impoundments and retaining structures based on many years of experience. The nuclear technique also permits an estimate of the average specific gravity of the mass of solids involved in the measurements, if the fine refuse is saturated. The importance of the variation in specific gravity for a correct estimation of relative density and using the nuclear gauges to determine specific gravity is emphasized and illustrated by case histories.

INTRODUCTION

Millions of tons of coal refuse are disposed in the United States each year. Most of the disposal is accomplished by constructing an embankment from the coarse portion of the refuse and pumping the fine portion (slurry) into the impoundment. Often, due to space constraints, it is necessary to construct the embankment in an upstream method. This method consists of building successive stages of coarse refuse over the previously disposed slurry.

The design of this type of embankment is complicated by the fact that some of the deposited fine materials may be susceptible to liquefaction under earthquake loading, and there is a lack of sufficient information about liquefaction potential of coal refuse. Failures by liquefaction due to earthquake shaking of coal refuse disposal facilities have not been reported, but tailings dams of similar materials hydraulically deposited (e.g. El Cobre, Chile, 1965, Mochikoshi, Japan, 1978) have failed.

Essentially, coal refuse resulting from the coal mining and cleaning process is of two types (Almes 1978, Cowherd 1977):

Coarse Coal Refuse - Coarse refuse is produced during the mining and separating process and generally consists of

---

[1] President and Chief Geotechnical Engineer, Bowser-Morner Associates, Inc., P.O. Box 51, 420 Davis Avenue, Dayton, Ohio 45401

[2] Senior Engineer, Bowser-Morner Associates, Inc., P.O. Box 51, 420 Davis Avenue, Dayton, Ohio 45401.

shales excavated along with the coal and later separated from the coals by a flotation process. Most coarse refuse is a well graded sand and gravel with some silt particles. Because of the coal screening process at the preparation plant, the coarse refuse is generally the +28 mesh material (greater than 0.6 mm). (Some process plants use the +100 mesh material [greater than 0.15 mm] in the coal mix and the coarse refuse is the +100 mesh material for these plants). The top size is generally about 76 mm (3 inches).

<u>Fine Refuse</u> - Fine coal refuse is collected from the preparation plant as a result of the coal washing operations. It is usually pumped to the disposal areas as a slurry. The particle sizes commonly vary from about the -28 mesh (sometimes the -100 mesh) to less than 0.0005 millimeters. It has a grain size comparable to a silty sand to a silty clay material.

The volume of these waste materials will vary between 10% and 50% of the usable coal production. The ratio of coarse to fine refuse will vary, but is usually about 50-50 by volume.

Ranges of particle sizes and Atterberg limits for coal refuse are shown on Figure 1 and in Table 1 (according to Almes 1978, Bishop & Rose 1976, D'Appolonia Cons. Engrs 1975, Hall et al 1976, Samarasinghe & Huang 1980, Saxena et al 1984, and authors' experience).

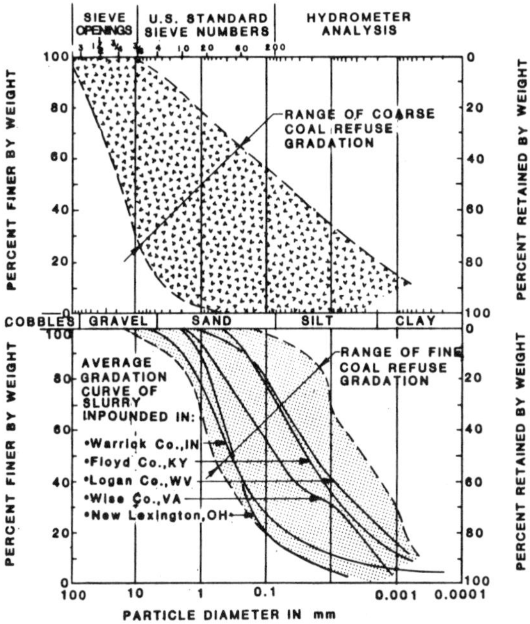

Fig. 1 - Range of Gradation for Coal Refuse Materials

TABLE 1. - ATTERBERG LIMITS OF COAL REFUSE MATERIALS

Reference (1)	Coarse Coal Refuse Liquid Limit (2)	Plasticity Index (3)	Specific Gravity (4)	Fine Coal Refuse Liquid Limit (5)	Plasticity Index (6)	Specific Gravity (7)
D'Appolonia Cons. (1975)	25-35	<12	1.5-2.8	30-40	<15	1.4-2.3
Bishop and Rose (1976)	25-35	2-12	1.8-2.6	--	--	--
Hall et al (1976)	21-32	1-7	1.7-2.7	20-43	2-14	1.4-2.5
Samarasinghe and Huang (1980)	Non-Plastic to 32	Non-Plastic to 10	1.6-2.7	20-51	0-14	1.3-2.5
Saxena et al (1984)	27	12	2.1-2.6	--	--	--
Materials Described in This Paper	--	--	--	23-38	5-8	1.4-2.1
Total Range	Non-Plastic to 35	0-13	1.5-2.8	Non-Plastic to 51	0-15	1.3-2.5

It should be noted that coal refuse materials are not included in the Unified Soil Classification System, even though current practice is to classify and test them in the same manner as other soil materials (D'Appolonia Cons. Engrs. 1975). Also, several samples were found to be non-plastic and Atterberg limits were omitted in most reports.

Ellison and Cho (1976) showed that coal refuse containing fines may have the same susceptibility to liquefaction as clean sandy soils. A fine coal refuse with 17% of particles less than 0.005 mm has been found to be liquefiable by dynamic laboratory tests. This material liquefied in laboratory dynamic tests at about 80% of the cyclic loading that caused very fine uniform sands, with a similar gradation, to liquefy. Although at this time the supporting data are lacking, the factors that control liquefaction susceptibility of conventional sands (discussed by Seed and Idriss, 1971) may be considered as affecting liquefiability of coal refuse, too. Among these factors, the relative density and specific gravity are supposed to be the most significant in these hydraulically placed deposits, which are uncompacted and essentially saturated a long time after deposition.

Coarse refuse is extensively used as a construction material in embankment structures for impoundment facilities and proper compaction and suitable drainage may decrease the liquefaction potential of this material to an acceptable level. On the other hand, the fine coal refuse hydraulically deposited and uncompacted should generally be considered as potentially liquefiable.

Although it is seldom used as a construction material in retaining embankments, the fine coal refuse can affect or even control the stability of disposal facilities when encountered in the foundation. Slurry may become a part of the foundation for an embankment in one of the following ways:

- Foundation of Impoundment Structures - Slurry may be found under a new embankment placed upstream of an old disposal area (see example in Fig. 2a).

- Foundation of Stages in the Upstream Method of Construction - Slurry deposited in a reservoir formed by a previously downstream developed embankment (see example in Fig. 2b).

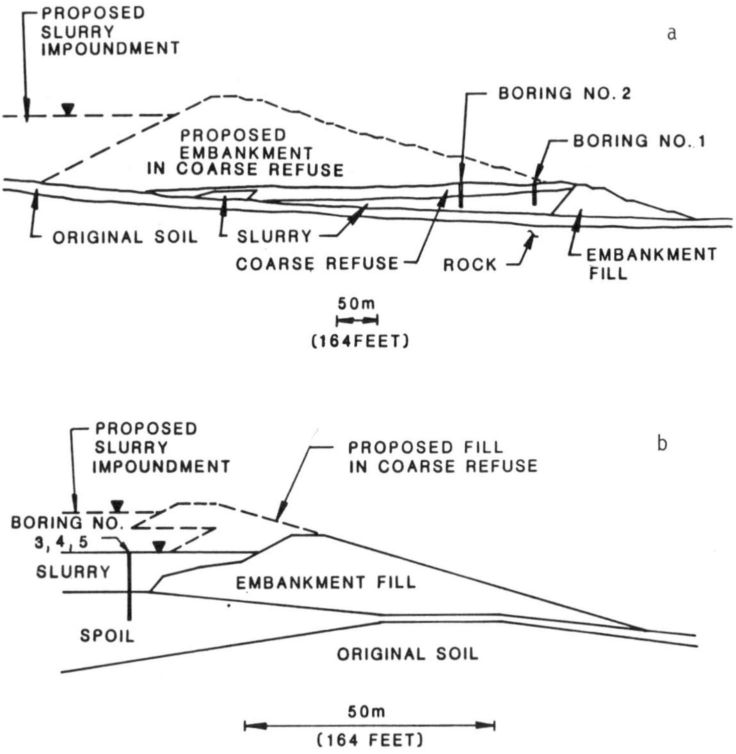

Fig. 2 - Examples of Coal Refuse Disposal = (a) In Wise County, Virginia; (b) In Warrick County, Indiana

DENSITY EVALUATION

Laboratory Tests for Relative Density Calculation

Maximum and minimum densities for coal refuse are commonly determined using the methods developed for soils, i.e., Standard Test Methods ASTM D 4253-83 and ASTM D 4254-83. The method for maximum index density determination may be applied to soils in which 100% by dry weight of soil particles pass a 75 mm (3 inch) sieve and which may contain up to 15% passing a No. 200 (0.074 mm) sieve, and 30% retained on a 37.5 mm (1.5 inch) sieve.

Many fine coal refuses may be, however, finer than these criteria. Accordingly, the use of ASTM D 4253-83 for all slurry materials is not appropriate. The authors consider that for these finer materials the use of modified Proctor maximum dry density as the maximum index density is appropriate (ASTM D 1557). For slurry having less than 15% particles passing the No. 200 sieve the ASTM D 4253 wet method may be used, but in parallel with the modified Proctor. The greatest value obtained by the two methods should be used as the maximum density for calculating relative density.

The methods for determining minimum density (ASTM D 4254-83) are also applicable to soils that may contain up to 15% by dry weight of soil particles passing a No. 200 (0.074 mm) sieve. The authors recommend method A of this standard for finer coal refuse, provided they still have cohesionless characteristics (by the lack of another appropriate method).

Two equations may be used to calculate the relative density of a cohesionless material.

$$Dr = \frac{\gamma_{dmax}\,(\gamma_d - \gamma_{dmin})}{\gamma_d\,(\gamma_{dmax} - \gamma_{dmin})} \times 100 \quad \text{..........(1)}$$

$$Dr = \frac{e_{max} - e}{e_{max} - e_{min}} \times 100 \quad \text{..................(2)}$$

Where γ_{dmax} and e_{min} = maximum dry unit weight and minimum void ratio obtained by ASTM D 4253 Method or Modified Proctor Test respectively; γ_{dmin} and e_{max} = minimum dry unit weight and maximum void ratio obtained by ASTM D 4254 A Method; and γ_d and e are the actual dry unit weight and void ratio of the slurry in the field.

The use of Eq. 1 supposes the same specific gravity of all investigated material as the sample used for maximum and minimum indexes determination. This is rarely the case in coal refuse deposits.

Void ratios used in Eq. 2 are less dependent on specific gravity than dry unit weights. Generally, minimum and maximum void ratios are functions of gradation parameters only and may be representative for great parts of deposited slurry, even if the specific gravity varies in a large range.

Table 2 lists maximum and minimum unit weight indexes of some slurries studied, as well as of some typical soils for comparison. When the specific gravity of a sample used for limit unit weights determination is not well known (due especially to very small samples used in specific gravity determination as compared with the relatively large sample used for limit unit weights determination) it is recommended that the obtained values of unit weight indexes be compared with the typical ones, in terms of void ratio.

TABLE 2. - DENSITY INDEXES OF COHESIONLESS MATERIALS

Type of Material	Dry Unit Weights (KN/m^3) Maximum γ_{dmax}	Dry Unit Weights (KN/m^3) Minimum γ_{dmin}	Specific Gravity G_s	Void Ratios Minimum e_{min}	Void Ratios Maximum e_{max}
(1)	(2)	(3)	(4)	(5)	(6)
Fine Coal Refuse:					
. Warrick County, IN	16.8	10.6	2.10	0.23	0.94
. Floyd County, KY	14.5	8.6	1.95	0.32	1.23
. Logan County, WV	13.5	7.5	1.76	0.28	1.31
. Wise County, VA	13.8	9.9	1.70[a]	0.21	0.68
. New Lexington, OH	10.9	7.2	1.45[a]	0.30	0.99
Values From Literature					
Fine Coal Refuse (Ellison & Cho 1976)	12.9	8.0	1.64	0.25	1.00
Various Sands (Castro & Paulos 1977):					
. Subrounded to Subangular, Well-Graded	22.8	19.0	2.84	0.22	0.46
. Subangular, Well-Graded	19.5	15.4	2.71	0.36	0.81
. Subrounded to Subangular, Poorly-Graded	18.2	15.2	2.69	0.45	0.74
. Subrounded, Poorly-Graded	17.3	14.1	2.65	0.50	0.84
. Angular, Poorly-Graded, Containing Platey-Shaped Grains	11.5	8.8	2.50	1.14	1.79
Ranges for Soils According to Hilf (Winterkorn & Fang 1975):					
. Well-Graded Gravel (GW)	20.7-22.9	17.1-19.3	2.65	0.13-0.25	0.34-0.52

[a]Estimated Values

TABLE 2. -- CONTINUED

Type of Material	Dry Unit Weights (KN/m^3) Maximum γ_{dmax}	Minimum γ_{dmin}	Specific Gravity G_s	Void Ratios Minimum e_{min}	Maximum e_{max}
(1)	(2)	(3)	(4)	(5)	(6)
. Poorly-Graded Gravel-Sand-Silt Mixture (GP-GM)	19.6- 22.2	15.9- 19.2	2.65 2.65	0.17- 0.32	0.36- 0.64
. Well-Graded Fine Sand (SW-SM)	18.2- 19.3	12.6- 15.4	2.65 2.65	0.34- 0.42	0.71- 1.07
. Poorly-Graded Fine Sand (SP-SM)	16.2- 19.5	12.4- 16.2	2.65	0.33- 0.60	0.60- 1.08
. Silty Sand (SM)	14.1- 21.1	11.0- 16.5	2.65	0.23- 0.83	0.57- 1.36

Nuclear Method For In-Situ Density Measurement

Direct measurements of density of cohesionless materials under the water table is difficult and subject to many sources of error. Among the indirect methods, the nuclear technique is one of the most accurate and economical method currently in existence (Cowherd 1963). The dry density (or unit weight) determination is made by separately measuring the in situ density and moisture content. Gamma rays are used for the measurement of total unit weight and fast neutrons are used for the measurement of moisture content.

Nuclear gauges have been used for many years to determine the in-place density and moisture content of soils as a compaction verification technique. Their use is described under the ASTM Standards D 2922-78 and D 3017-78. These standards cover surface nuclear gauges, but gauges used below the ground surface operate on the same principle. The source and detector are contained in opposite ends of a probe (Fig. 3). The gamma rays or neutrons are emitted into the

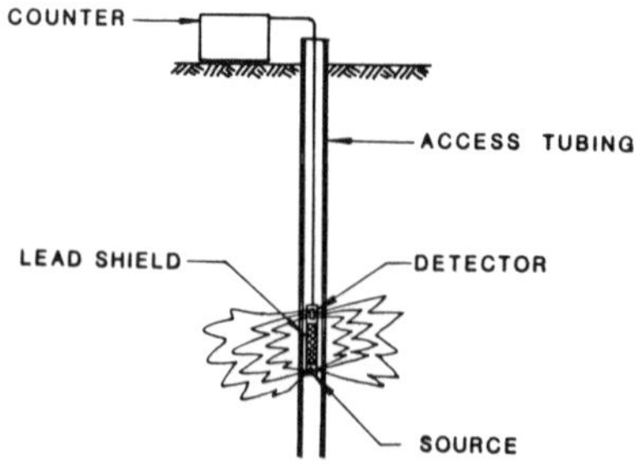

Fig. 3 Nuclear Gauge for Use in Deep Borings

material being tested from the source and are backscattered to the detector tube. The entire probe is lowered into an access tubing to the desired depth and the count is transmitted to a scaler at the ground surface. Such gauges require the installation of an access tube into the soil so the gauge can be lowered to the desired depth.

A calibration curve is prepared to bracket the expected densities in the slurry impoundment. In addition, a calibration curve is prepared for moisture content. By calibrating to the particular material in question the problem of different coals is removed and it is possible to determine the actual moisture content in-situ and obtain the dry unit weight as well as the wet unit weight (Cowherd, 1986).

An important advantage of using the nuclear method in coal refuse deposits is the non-sensitivity of calibration curves as relates to specific gravity, shape, size, and gradation of particles. This was demonstrated for soils (Fig. 4), when points corresponding to Ottawa (very uniform and roughly sypherical) silica sand, uniform creek fine sand, Ohio River well-graded sand, silica gravel both rounded and angular (crushed), granite, and clay, proved to lie on the same calibration curve (Cowherd, 1963). The sphere of influence is approximately 8-inches in diameter, so the method can identify rather thin abnormal layers from the density point of view.

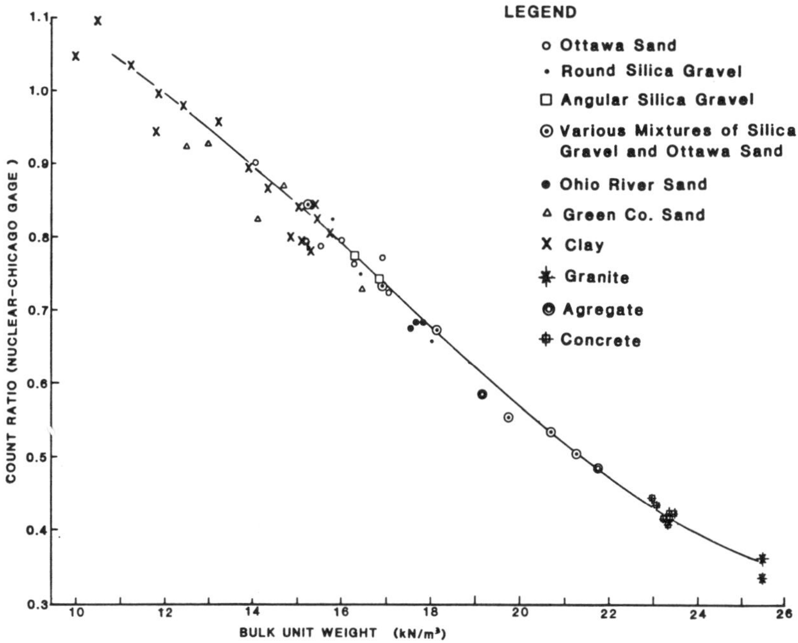

Fig. 4 Experimental Calibration Curve for Nuclear Moisture Measurements (Based on Cowherd, 1963)

Density calibration curves are sensitive to moisture content. Although the effect of water in voids may be theoretically predicted, it is preferable that a calibration curve be derived for the approximate moisture content existing in the field. Different calibration curves have been found for materials having different chemical composition, as the absorbtion coefficients at low gamma ray energies are different. However, points corresponding to coal (specific gravity, G_S = 1.30 to 1.45), fine coal refuse (G_S = 1.45 to 1.95), and coarse coal refuse (G_S = 1.90 to 2.20) from even different geographical regions have been generally been found to lie on the same calibration curve (Cowherd, 1986).

The moisture calibration curve may be affected by hydrogen atoms that are present in forms other than moisture content. Obviously, there are hydrocarbons within the fine coal refuse. However, by calibrating to the particular material in question this problem is removed.

Taking the above into account, it is obvious that a unique pair of calibration curves for a given deposit may be considered, irrespective of variations in specific gravity and gradation of particles.

Fig. 5 shows such calibration curves for two studied sites: curves Nos. 1 and 3 are for a slurry impoundment in Logan County, West Virginia, and curves Nos. 2 and 4 are for a slurry impoundment in Floyd County, Kentucky. The calibration curves yield, directly from nuclear measurements, the bulk unit weight (γ), and the moisture content defined by the ratio of the weight of water to the total volume of solids and voids (W_N).

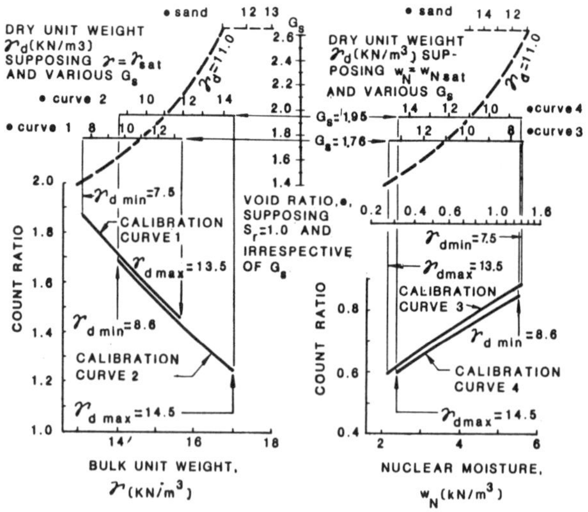

Fig. 5 - Examples of Calibration Curves (10 KN/m^3 = 63.7 lb/ft^3)

The dry unit weight, γ_d, in each measurement point may be obtained by the formula:

$$\gamma_d = \gamma - W_N \text{ ....(3)}$$

Theoretically, in saturated media, any of the two nuclear measurements may be used independently to obtain directly the dry unit weight, if the specific gravity is known:

$$\gamma_d = (\gamma_{sat} - \gamma_w) / (1 - 1/G_s) \text{ ....(4)}$$

Where γ_{sat} = the saturated unit weight, and γ_w = the unit weight of water, and

$$\gamma_d = (\gamma_w - W_{Nsat}) G_s \text{ ....(5)}$$

where W_{Nsat} = the saturated nuclear moisture.

The pair of nuclear measurements at a given point also permits an estimate of the average specific gravity of the mass of solids involved in the measurement. For this, γ_d in Eqs. 4 and 5 is eliminated.

$$G_s = \frac{\gamma_{sat} - W_{Nsat}}{\gamma_w - W_{Nsat}} \text{ ....(6)}$$

Three alternative procedures may be used to estimate the in situ relative density, D_r, in saturated deposits.

Procedure 1. - The values of total unit weight, γ, and nuclear moisture content, W_N, are used with Eq. 3 to obtain dry unit weight, γ_d. Knowing the maximum and minimum dry unit weights from appropriate laboratory tests, D_r may be computed on the basis of Eq. 1. An accurate use of this procedure implies that the specific gravity, G_s, in all measurement points is the same as that of the sample used in the laboratory. This generally is not true.

Procedure 2. - Both γ_d and G_s are computed from nuclear measurements, using Eqs. 3 and 6. From these values the void ratio, e, may be calculated for every point. Using maximum and minimum void ratios determined in the laboratory and Eq. 2, D_r may be evaluated. Usually the limit values of void ratio are more stable when specific gravity varies than limit values of unit weight. On the other hand, the obtained values may be easier compared with values from literature corresponding to soils having similar size, shape, and gradaton of particles. It is also recommended that G_s logs be checked by some direct determination on disturbed samples. This procedure is usually much more convenient for the study of slurry deposits than the first one.

Procedure 3. - When values of γ_d and G_s are not required, void ratio may be obtained directly from the nuclear moisture measurement only, from the following equation:

$$e = \frac{W_{Nsat}}{\gamma_w - W_{Nsat}} \quad \ldots\ldots\ldots\ldots\ldots\ldots\ldots(7)$$

This value is used with Eq. 2 for D_r determination. In fact, when G_s is determined by nuclear measurements only, e-values do not depend on γ and procedures 2 and 3 are identical regarding e and D_r determinations.

An important source of error may be the incomplete saturation of the deposit. This can be estimated taking into account the general formula:

$$G_s = \frac{\gamma - W_N}{\gamma_w - W_N/S_r} \quad \ldots\ldots\ldots\ldots\ldots(8)$$

and

$$e = \frac{W_N/S_r}{\gamma_w - W_N/S_r} \quad \ldots\ldots\ldots\ldots(9)$$

Where $S_r = W/W_{sat} = W_N/W_{Nsat}$ = degree of saturation.

For example, when $S_r = 0.9$ instead of 1.0 as considered when using Eqs. 6 and/or 7, the error in determination of e varies between 15% and 30%, on the unsafe side, for the range of nuclear moisture of 15 to 35 lb/ft^3. However, when $S_r \leq 0.9$ the material may in most cases be considered non-liquefiable (Finn 1981, Sherif et al 1977).

Coal Refuse Disposal in Wise County, Virginia. - This project consisted of the redesign of a combined refuse embankment to make it a coarse refuse embankment impounding slurry. There was slurry covered by coarse refuse of variable thickness already in place. The section of the existing embankment, about 64 m (210 feet) high, is shown in Fig. 2a. The construction had consisted of placement of a starter embankment, slurry upstream of the embankment, and placement of coarse refuse with additional slurry being placed within the embankment. The overall embankment consisted, therefore, of both coarse refuse and slurry. The configuration of the final design section is also shown on Fig. 2a. The Owner wanted to place an additional 119 m (390 feet) of coarse refuse over the existing embankment and impound slurry upstream of this structure. While the new structure itself is downstream construction it is necessary to construct a portion of the structure over slurry which was already in place.

Nuclear density and moisture determinations were used to evaluate the liquefaction potential of the slurry which was already in place. This was done and several borings were made into the refuse and slurry to assess the density and moisture content full depth.

Jumps from $D_r \simeq 0$ to $D_r > 100\%$ at 0.3 m (1-foot) intervals are possible in principle, but not credible in a free deposited slurry. Fig. 6 clearly shows that such jumps are artificially induced when Eq. 1 is used. Unit weight logs, with frequent values around minimum density and some values greater than the maximum one are evidently due to variation in G_s. The similarity in logs of γ_d and G_s is striking. Using procedure 2 or 3 (that is Eq. 2), no D_r values smaller than 30% or greater than 100% were obtained. All values greater than 100% (possible but not probable) obtained by the first procedure, were eliminated.

More credible results were obtained using Eq. 2. Large densities at the upper part of the slurry deposit are expected, due to a quicker consolidation in the vicinity of covering coarse refuse which represents both an overload and a drainage layer. A zone of loose material is signaled at the base of the deposit; where Eq. 1 gave D_r = 60% to 75%, the actual values (by Eq. 2) are in fact D_r = 30% to 40%. Some remaining anomalies may be due to inserted layers of coarse refuse; such layers can be recognized by their specific gravity, which usually differs significantly from that of the slurry.

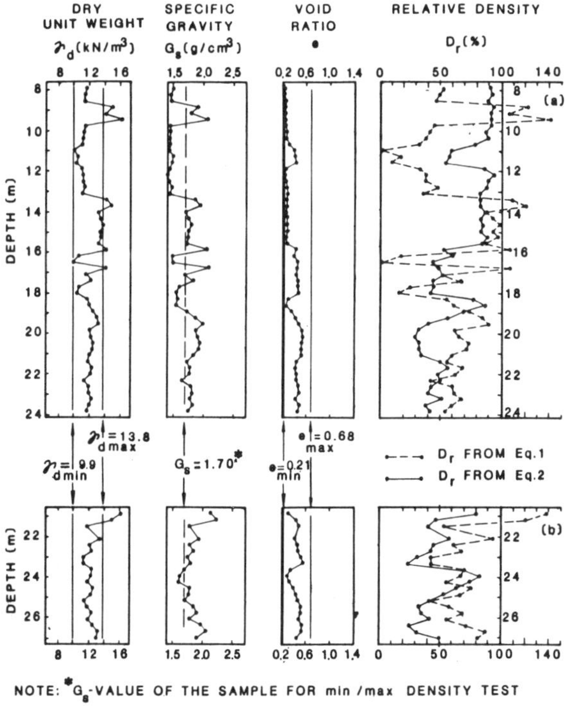

Fig. 6 - Results of Nuclear Measurements: (a) Boring No. 1; (b) Boring No. 2

It must be emphasized that in using Eq. 1 not only the scatter in D_r values is greater than the actual, due to variations in G_s, but also the average values are erroneous if the specific gravity of the sample used for laboratory limit index density tests is much different from the average value of the deposit. Such results may seem to compromise the nuclear method, which in fact, gave good results, but which could be wrongly interpreted.

Subsequent to the D_r determination, the liquefaction potential assessment was made based on laboratory dynamic tests on samples with various densities and the structure is currently being constructed. The nuclear method provided an economical tool for assessing the potential for liquefaction of slurry materials for a relatively large disposal project. A much higher cost would have been incurred if this project were evaluated with conventional methods.

Slurry Pond in Warrick County, Indiana. - An existing embankment constructed of earth fill and coarse refuse forms a pond in which slurry is deposited. The storage capacity of the impoundment is to be increased by an additional 6 m (20 feet) to the existing embankment in two 3-m (10-foot) lifts. Dikes of both stages will be constructed using upstream construction techniques from coarse refuse founded on slurry (Fig. 2b). Nuclear tests were used in several borings located in the pond as far from the embankment as possible. The results obtained in two of these borings are shown in Fig. 7.

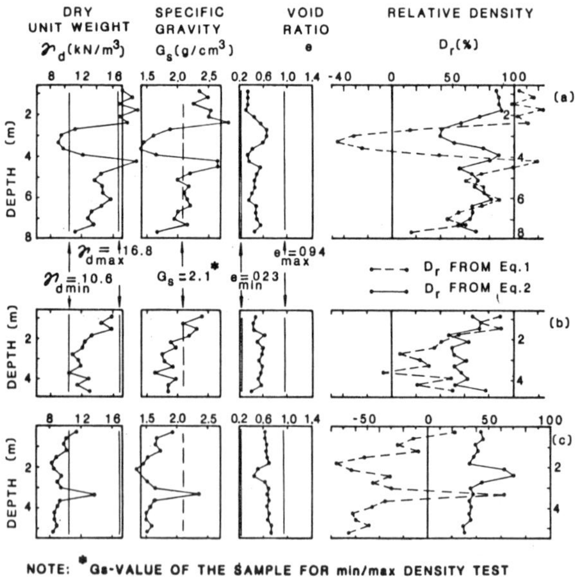

Fig. 7 - Results of Nuclear Measurements: (a) Boring No. 3; (b) Boring No. 4; (c) Boring No. 5

Very low values of γ_d were observed at depths of 3 to 3.6 m in boring 3. The relative density is in fact about 40% (not less than zero as found using Eq. 1). These low values are due to actual G_s values much lower than the average value. The use of Eq. 1 with boring 5 would show very loose material (D_r less than zero) at any depth except at 0.3 m and 3.4 m. In fact, at these elevations the relative density is around 40%, this being the mean value for the whole deposit at this location. The relative density is rather uniform (D_r = 30% to 50%) except for a layer at 2.1 m to 2.4 m where D_r = 60% to 70%; no such denser layer would be signalized by the γ_d values alone. The anomaly observed at 3.4 m is not in density, but in specific gravity only.

CONCLUSIONS

Fine coal refuse in disposal facilities may be susceptible to liquefaction under earthquake loading. One of the most important parameters controlling its liquefiability, the in situ density, is difficult to measure by direct methods.

Nuclear measurements of density and moisture content give continuous logs of dry density in a relatively quick and economic manner; the presently available equipment ensures the possibility to investigate deep deposits, down to tens of meters.

The accuracy of relative density determination is affected to a great extent by the large variability in specific gravity which characterizes coal refuse deposits. The specific gravity in the zone where dry density has been measured differs usually from the specific gravity of the sample used in laboratory for maximum and minimum density indexes determinations. It is practically impossible to obtain continuous logs of specific gravity by classic procedures. Nuclear measurements, however, gives this possibility when the deposit under investigation may be considered fully saturated (Eq. 6). Some case histories illustrate the need of taking into account the specific gravity variability and the advantage of using nuclear measurements in investigating the liquefaction potential of coal refuse deposits.

REFERENCES

Almes, R.G. (1978). "An Overview of Coal Tailings Disposal in the Eastern United States", Tailings Disposal Today, Volume 2: Proceedings of the Second International Tailings Symposium, Denver, Colorado, pp. 13-53.

Bishop, C.S., and Rose, J.G. (1976). "Physical and Engineering Characteristics of Coal Preparation Plant Refuse", Proceedings of the Seventh Ohio River Valley Soils Seminar, Lexington, Kentucky, pp. 4/1-11.

Castro, G. and Poulos, S.J. (1977). "Factors Affecting Liquefaction and Cyclic Mobility", Journal of the Geotechnical Engineering Division, ASCE, Volume 103, No. GT6, pp. 501-516.

REFERENCES -- CONTINUED

Cowherd, D.C. (1963). "A Study of the More Important Factors Affecting Nuclear Density Gauges", Thesis presented to the University of Kentucky in Lexington, Kentucky, in partial fulfillment of the requirements for the degree of Master of Science.

Cowherd, D.C. (1977). "Geotechnical Characteristics of Coal Mine Waste", Proceedings of the Conference on Geotechnical Practice for Disposal of Solid Waste Materials", ASCE, Ann Arbor, Michigan pp. 384-406.

Cowherd, D.C. (1986). "Liquefaction Study of Coal Refuse by Nuclear Gauges", Proceedings of the ASCE Specialty Conference on "Use of In Situ Tests in Geotechnical Engineering", Blacksburg, Virginia, pp. 487-501.

D'Appolonia Consulting Engineers, Inc. (1975). "Engineering and Design Manual Coal Refuse Disposal Facilities", Mining Enforcement and Safety Administration, 790 p.

Ellison, R.D. and Cho, Y.Y. (1976). "Liquefaction Considerations for Fine Refuse", Proceedings of the Seventh Annual Ohio River Valley Soils Seminar, Lexington, Kentucky, pp. 16/1-9.

Finn, W.D.L (1981). "Liquiefaction Potential: Developments Since 1976, State of the Art Report", International Conference on Recent Advances in Geotechnical Earthquake Engineering and Soil Dynamics, St. Louis, Missouri, pp. 655-681.

Hall, G., Robinson, M.K., and Smith, R.E. (1976). "West Virginia Experiences With Review of Coal Refuse Disposal Facilities", Proceedings of the Seventh Ohio River Valley Soils Seminar, Lexington, Kentucky, pp. 14/1-5.

Samarasinghe, M. and Huang, Y.H., (1980. "A Literature Review of Coal Refuse Dikes", Prepared for the Institute for Mining and Minerals Research, University of Kentucky, Lexington, Kentucky, 73 p.

Saxena, S.K., Lourie, D.E., and Rao, J.S. (1984). "Compaction Criteria for Eastern Coal Waste Embankments", Journal of the Geotechnical Engineering Division, ASCE, Volume 110, No. 2, pp. 262-284.

Seed, H.B. and Idriss, I.M. (1971). "Simplified Procedure for Evaluating Soil Liquefaction Potential", Journal of the Soil Mechanics and Foundation Division, ASCE, Volume 97, No. SM9, pp. 1249-1273.

Sherif, M., Ishidashi, I., and Tsuchiya, C. (1977). "Saturation Effects on Initial Soil Liquefaction", Journal of the Geotechnical Engineering Division, ASCE, Volume 103, No. GT8, pp. 914-917.

REFERENCES -- CONTINUED

Winterkorn, H.F. and Fang, H.Y. (Editors) (1975). "Foundation Engineering Handbook", Van Nostrand Reinhold Company, New York, 751 p.

NOTATIONS

The following symbols are used in this paper:

D = particle apparent diameter in hydrometer analysis or sieve openings,
D_{10} = effective size,
D_{50} = mean diameter,
D_r = relative density,
e = void ratio,
e_{max} = maximum void ratio,
e_{min} = minimum void ratio,
G_s = specific gravity,
S_r = degree of saturation,
W = (gravimetric) moisture content,
W_N = nuclear moisture content (weight of water to the total volume),
W_{Nsat} = saturated nuclear moisture ($W_{Nsat} = n \gamma_w$, where n is the porosity),
γ = bulk unit weight,
γ_d = dry unit weight,
γ_{dmax} = maximum dry unit weight,
γ_{dmin} = minimum dry unit weight,
γ_{sat} = saturated unit weight; and
γ_w = unit weight of water.

DOWNHOLE NUCLEAR DENSITY LOGGING IN SAND TAILINGS

H.D. Plewes, E.C. McRoberts and W.K. Chan*

Downhole nuclear density logging offers a direct method of obtaining in-situ density which is of great importance in liquefaction assessments. The basis for nuclear density logging techniques are reviewed. A range of technical considerations influencing the use and interpretation of logging equipment are discussed. Field calibration by comparison with densities obtained from bulk in-situ samples are presented.

INTRODUCTION

In general, any assessment of liquefaction potential of hydraulically placed fills begins with a definition of the in-situ conditions: effective stresses and density. Available liquefaction theories, such as the steady state or cyclic mobility approaches, and empirical correlations with observed performance all require a measure of either absolute or relative density. The most common techniques are the Standard Penetration Test (SPT) and Cone Penetration Test (CPT) where a range of empirical procedures are followed to obtain relative density. Extensive calibrations, usually in test chambers, are required to obtain reliable results.

Direct determination of density offers an alternative but difficult approach. It is possible to obtain density from carefully executed undisturbed coring of pre-frozen or grouted sand. These methods can be time consuming and costly. Moreover, only limited sampling is economically feasible compared with other techniques.

Recently, the authors had occasion to investigate the liquefaction susceptibility of a deep sand hydraulic fill at depths of 50 to 100 m and at calculated vertical effective stresses of 750 to 1500 kPa. At these depths, the use of existing SPT or CPT correlations would have required significant extrapolations of known correlations beyond the normal 100 to 300 kPa working effective stress range. Using the best extrapolations obtained in the literature, raw SPT values of 120 to 250 were corrected to 14 to 30. These corrections were considered to be highly unreliable. Coupled with likely variations in minimum and maximum density, any determination of absolute density was suspect. The CPT approach was also considered but, as it was felt that calibration tests at chamber pressures of 750-1500 kPa would be required, this technique was also set aside.

As the steady state method was being considered, it was determined that a direct density determination was most relevant.

* Hardy BBT Limited P.O. Box 746 Edmonton, AB Can T5J 2L4

Attempts to sample the sand in an undisturbed condition using conventional thin walled tube samplers were unsuccessful due to the high in-situ stress levels. The possible application of resistivity techniques were discounted due to concerns about resistivity variation within the hydraulic fill.

It was judged that downhole nuclear density techniques likely offered the best operative solution. While nuclear density determinations are often used in geotechnical practice, particularly for compaction control, their use in down-hole applications is more limited. This may partly be due to the lower energy sources available in the type of equipment discussed by Mittal and Morgenstern (1975) which also has the disadvantage of requiring a carefully installed access tube. However, higher energy source tools common in the oil well logging industry offer a more versatile technique. As a result of the author's experience in other oil sands and coal geophysical applications, it was felt that reliable results could be obtained. As well, Klohn (1984) has reported a limited test series in sand tailings using a downhole density logging tool. The results were considered only partially successful, although a good trend between nuclear density data and undisturbed samples was apparent. Tjelta et al (1987) have also reported good correlation between in-situ density measurements using a nuclear density cone penetrometer and densities computed from water contents of North Sea marine sediments.

DOWNHOLE NUCLEAR DENSITY LOGGING

General Theory

The technique of downhole nuclear density logging is based on the physical phenomenon of gamma ray scattering as a function of the bulk density of the surrounding formation (Pickell and Heacock, 1960). Figure 1 illustrates the configuration of the density tool within the borehole cavity. A focused stream of gamma rays or photons is irridated by the gamma ray source into the formation. A fraction of the incident photons will be absorbed or scattered rather than travel unimpeded through the formation medium. The rate of decrease of the gamma ray intensity per unit thickness of the medium is called the linear absorption coefficient, u, expressed as

$$\frac{dI}{dx} = - u \qquad (1)$$

or

$$I = I_o e^{-ux} \qquad (2)$$

where I_o = incident gamma ray intensity, counts per unit time

I = resultant gamma ray intensity, counts per unit time

u = linear absorption coefficient, cm^{-1}

x = thickness or distance travelled through the medium, cm.

Three primary interactions may occur between the gamma rays and the formation matter: Compton scattering, photoelectric absorption and pair production. The proportion in which these interactions occur

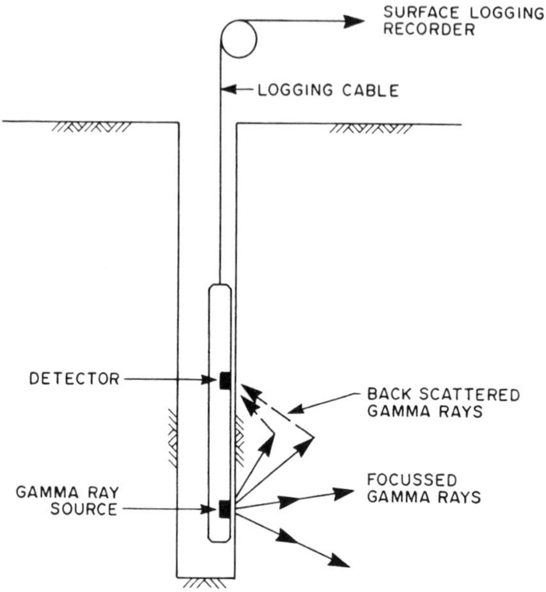

FIGURE 1 Configuration of Nuclear Density Tool in Borehole

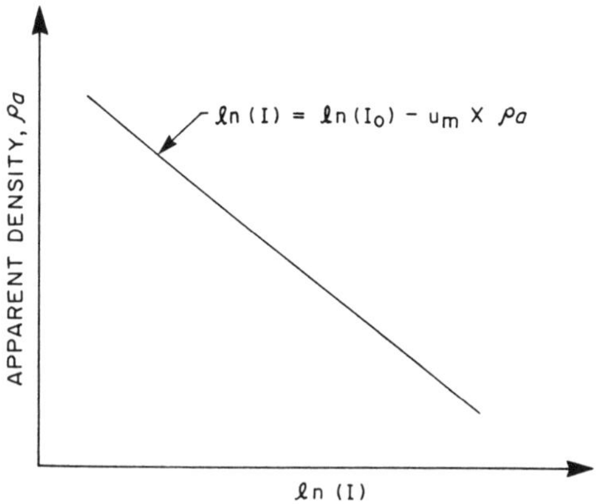

FIGURE 2 Relationship Between Apparent Density and Detector Gamma Ray Count

depends on the level of gamma ray energy and the atomic number, Z, of the elemental constituents of the formation matter. Cesium 137 radiation energy sources emitting 0.66 million electron volts (MeV) gamma rays are commonly adopted for industry density tools. For the range of atomic numbers (Z = 1 to 25) of common geologic elements, about 95 percent or greater of gamma ray interactions are by Compton scattering using the Cesium 137 source (Killeen, 1982).

Compton scattering is the process whereby a photon collides with an atomic electron and a new photon of lower energy is created that moves in a different direction than the incident photon. The difference in energies between the incident and new "scattered" photons is imparted to the "recoil" electrons. The frequency of occurrence of the Compton scattering phenomenon is inversely proportioned to the electron density where for a single element

$$\rho_{electron} = N(Z/A)\,\rho_b \tag{3}$$

where $\rho_{electron}$ = electron density, number of e^- per cm^3

N = Avogardro's number (6.03×10^{23} atoms per gram atomic weight)

Z = atomic number of an atom (number of electrons in the atom)

A = mass number (number of the protons and nuetrons in the nucleus)

ρ_b = bulk density of the element or molecular substance (g/cc)

The value of the ratio Z/A for most common elements occurring in soil is very close to 0.5 (Table 1). Hence, the extent of Compton scattering is chiefly proportional to the formation bulk density alone.

An apparent density, ρ_a, may be defined for a soil formation comprised of a number of constituent molecular substances such that

$$\rho_a = \sum_{i=1}^{m} 2(\Sigma\, Z/M)\, n_i\, \rho_{bi} \tag{4}$$

where ρ_a = apparent density, g/cc

m = number of constituent molecular substances

M = molecular weight

Z = sum of atomic numbers of the atoms in a molecule

n_i = porosity or volumetric proportion of each constituent molecular substance.

The weighted ratio of Σ Z/M for most molecular substances in common soils is also very close to 0.5 (Table 2) so that ρ_a is nearly identical to formation bulk density, ρ_b.

It is convenient to express the linear absorption coefficient in Equation (1) as

$$u = u_m \rho_a \quad (5)$$

where u_m = mass absorption coefficient, cm^2/g.

Therefore from Equation (2)

$$I = I_o e^{-u_m x \rho_a} \quad (6)$$

The variables in Equation (6) can be expressed for an actual density tool as follows:

$$I = I_o^{-u_m X \rho_a} \quad (7)$$

where I = radiation intensity at a detector, counts per second
I_o = Cs 137 source radiation intensity, counts per second
u_m = formation mass absorption coefficient, cm^2/g
X = distance from source to detector, cm.

Taking the natural log of both sides of Equation (7) yields

$$\ln (I) = \ln (I_o) - u_m X \rho_a \quad (8)$$

I_o and X are fixed parameters of the density tool. Data for u_m (Table 3) indicates that it is a satisfactory approximation to consider the mass absorption coefficient of common geologic materials to be constant. Hence, the measured detector gamma ray intensity is a logarithmic function of the apparent density as shown in Figure 2.

Instrument Design and Operation

The body of density logging instruments contain shielding, a Cesium 137 gamma ray energy source (0.66 MeV), gamma ray detectors and the associated electronics. The gamma ray source and detectors are inset into solid tungsten housings, the high density of which makes a very effective gamma ray shield. The shielding around the gamma ray source is designed to focus the flux of photons directly into the formation adjacent to the tool face. The shielding around the detectors allows only the gamma rays entering a predetermined limited volume confronting the detectors to be counted. Gamma rays scattered from the formation into the borehole fluid as well as gamma rays emanating from the source and travelling directly through the tool to the detectors are highly attenuated by the tungsten housings. With this shielding, a negligible borehole size effect may be observed for boreholes up to 25 cm diameter.

In operation, the instrument is lowered into the borehole with the caliper arm fully retracted (Figure 3). The electrical motor driven caliper arm is activated at the bottom of the borehole and the logging run is begun in the upward direction. The extended caliper arm applies a constant nominal net horizontal force of 650 N (150 lbs.), thereby maintaining intimate contact between the tool body and the borehole wall. Drilling fluid and soft mudcake are also squeezed out of the critical space between the natural formation and the source and detectors. Logging speeds of 10 metres per minute or slower have been adopted in the logging industry.

Calibration

The density tool is calibrated in pure laboratory mediums of precisely known composition and bulk density. These laboratory mediums comprise a wide range of bulk densitites which typically include water, magnesium, aluminum, limestone, sandstone and dolomite (Table 3). The response measured in counts per second by the density tool depends upon the number of electrons per unit volume of the formation which is directly proportional to the apparent density, ρ_a, defined in Equation (4).

Zone of Influence

The radial depth of density measurement is nominally one-half the spacing between the detector and the source. The actual zone of influence increases in mediums of very low electron density and decrease in mediums of very high electron density.

The spacing of the detectors is an important consideration in understanding the response of the density tool. For short spaced detectors (10 - 15 cm from the source), the density response is sensitive to the formation density near the borehole wall. The response may be influenced by local changes in formation density due to stress relief, mudcake against the borehole walls and contributions due to borehole roughness. Longer spaced detectors (greater than 30 cm from the source) are less sensitive to conditions near the borehole wall and are more sensitive to the density of the undisturbed formation.

The trade-off between the shorter and longer spaced detectors is the resolution of vertical density between vertical beds. The true density of any layer may only be established if the layer thickness is nominally greater than the spacing between the detector and the source. With consideration of increasing local borehole effects with shorter detector spacings, accurate density measurements in layers less than 15 cm thick is tentative.

Density Compensation

Density compensation is the process whereby the signals from a short spaced detector (SSD) and a long spaced detector (LSD) in a dual detector density tool (Figure 3) are combined to derive a corrected value for the formation density. The correction compensates to a great extent for the effects of mudcake on the borehole wall and to a more limited extent for borehole rugosity. Whereas the effects of mudcake are quantifiable, the influence of borehole rugosity is not.

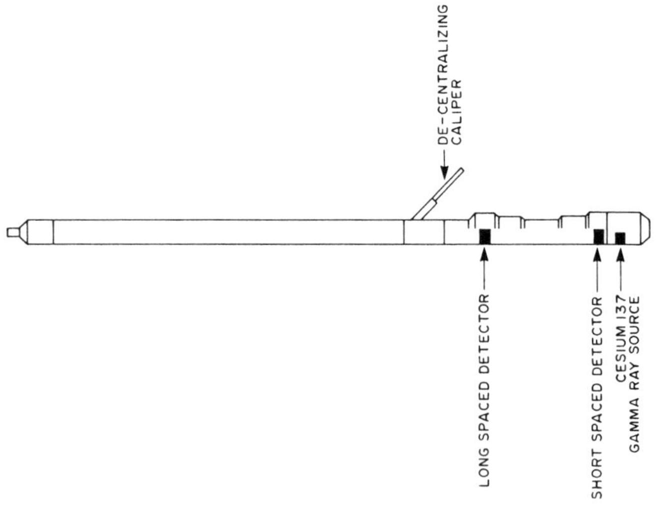

FIGURE 3 Dual Detector Compensated Nuclear Density Tool

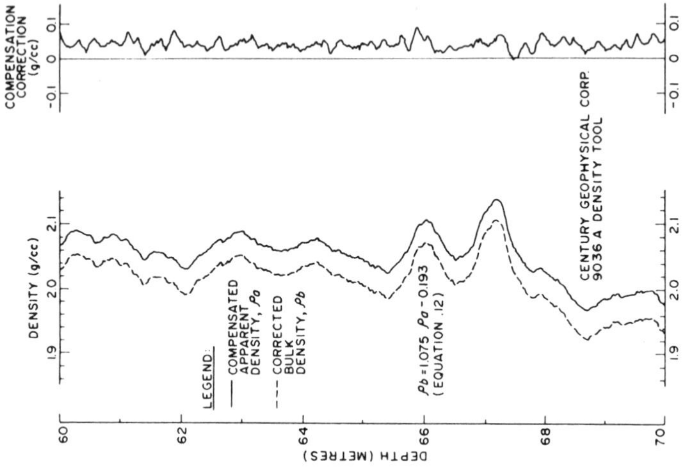

FIGURE 4 Compensated Apparent Density and Corrected Bulk Density Logs for Borehole H87-102

It is therefore very important to utilize drilling techniques which obtain a smooth and stable borehole wall.

A typical log for a dual detector compensated density logging tool is shown in Figure 4. The magnitude of correction between the compensated apparent density and the uncorrected LSD density (positive or negative) is provided as an additional trace with the compensated density log. The magnitude of the corrections are generally less than 0.05 g/cc for smooth borehole conditions.

Source Strength

The strength of Cesium 137 gamma ray sources used in the logging industry typically range between 0.15 Curies and 5 Curies, and sources up to 20 Curies are used in some oil field logging tools. Density measurements using source strengths less than 0.15 Curies can be influenced by natural gamma ray radiation within the formation. A 0.15 Curie source usually yields acceptable results for most applications and types of formations. As for the work described later in this paper, higher source strength is advantageous where high resolution of apparent density is important. This is particularly the case for compensated density tools where high resolution of the SSD and LSD responses is critical to the calculation of the corrected apparent formation density.

TECHNICAL CONSIDERATIONS

There is an extensive body of experience in the petroleum and mining industries on the use of downhole geophysical equipment. While much of this experience is applicable to geotechnical requirements, care must be taken in any off-the-shelf interpretation of logging test results. This section presents a review of some of the factors that should be considered.

Borehole Stability

A typical density tool receives about 90 percent of its reflected signal within 15 cm of the borehole wall. It is therefore necessary to ensure that the soil immediately adjacent to the borehole wall is at an in-situ condition.

The creation of a borehole cavity results in a redistribution of stress immediately surrounding the borehole. For small boreholes at relatively shallow depths, the provision of small borehole wall support from drilling fluid pressure results in a stable borehole and elastic equilibrium. At greater depths, boreholes can be stable although there may be a squeezing action and the development of a plastic zone. This plastic zone while failed in the geomechanical sense can still provide wall support. However, soil within the plastic zone will have undergone density change in response to the new plastic state of equilibrium.

In order to obtain representative results, the borehole wall must remain in the elastic state. This can be achieved by analyzing the stability of a borehole in a frictional material with consideration of the in-situ effective stress conditions prior to drilling and the applied mud fluid pressures. The use of high density muds may be required to ensure borehole stability. Techniques are available to compensate for the development of drilling mud cake on the borehole walls as discussed earlier.

The phreatic surface within the hydraulic fill investigated by the authors was nominally 45 m below the ground surface. Using a drilling mud of fluid density between 1.0 and 1.1 g/cc, it was possible to maintain borehole wall support pressures in the range of 450 and 600 kPa at the depths of investigation between 50 and 100 m. These support pressures were calculated to be sufficient to maintain elastic equilibrium in the formation around the borehole. Theoretically, the support pressures were actually high enough to cause elastic compression of the formation at the shallower depths. The changes in in-situ formation bulk density caused by elastic deformation (compression or expansion) of the soil within 15 cm of the borehole wall was estimated to be less than 1 percent or nominally 0.02 g/cc. This magnitude of density change in comparison to the precision of the nuclear density tool used in the investigation (discussed below) was considered acceptable.

Borehole Roughness

The roughness of the borehole wall strongly influences the measurement of density by the short spaced detector as discussed above. The density compensation process can correct for minor borehole roughnesses but the corrections cannot be explicitly quantified. Density measurements in zones of borehole washouts or irregular accumulations of mud cake (mud rings) are unreliable. It is imperative that smooth borehole wall conditions be obtained for proper performance of density logging techniques.

The high borehole wall support pressures achieved at the depths under investigation enabled a high level of borehole smoothness to be achieved with careful drilling procedures. Local roughnesses of the borehole wall of less than 0.2 cm were maintained throughout 95 percent of the depths of investigation.

Density compensation corrections were generally less than 0.05 g/cc, representing for the most part the influence of mud cake on the borehole wall. For comparison, density compensation corrections for local borehole roughnesses of 0.5 and 1.0 cm typically increased to 0.08 g/cc and 0.12 g/cc, respectively. Extreme density compensation corrections of 0.30 g/cc or greater were observed for borehole roughnesses of 5 cm caused by washouts. Assuming uniform conditions of borehole mud cake, it is considered that density compensations for borehole roughnesses greater than 0.5 cm are too extreme for the precision and sensitivity of density measurements required for liquefaction assessments.

Formation Properties

The density log presented in Figure 4 is a measure of the apparent density, ρ_a, which is directly proportional to the electron density of the formation. From Equation (4), the apparent density is identical to the true bulk density of the formation where the combined weighted ratio of $\Sigma Z/M$ of the formation medium is equal to 0.5. Where the weighted ratio $\Sigma Z/M$ deviates from 0.5, corrections to the apparent density must be undertaken to yield the true bulk density.

The requirements for these corrections normally stems from the effect of the presence of hydrogen in the formation fluids. The ratio of Z/A for hydrogen (0.992) is almost twice that for other elements in Table 1. The combined ratio of $\Sigma Z/M$ for pure water is therefore 0.555 (Table 2). Residual hydrocarbons as in the form of oil ($\Sigma Z/M = 0.570$) or gas ($\Sigma Z/M = 0.619$) also deviate significantly.

The hydraulic fill investigated is comprised of oil sand tailings. The tailings are fine grained sand with fines content (less than 0.0073 cm) generally ranging between 5 and 20 percent by weight. The composition of the sand grains and fines fraction is predominantly silica quartz (SiO_2). Incorporating the values of $\Sigma Z/M$ and specific gravity for SiO_2 and pure water from Table 2, the apparent bulk density (g/cc) from Equation (4) may be expressed for water saturated sand tailings as

$$\rho_a = 2.645 - 1.535\ n \qquad (9)$$

where n = total pore volume porosity

The true bulk density (g/cc)of the sand tailings is

$$\rho_b = 2.65 - 1.65\ n \qquad (10)$$

Substituting Equation (9) into Equation (10) yields the following relationship between the true bulk density as a function of the apparent density

$$\rho_b = 1.075\ \rho_a - 0.193 \qquad (11)$$

Since the density log is calibrated to measure ρ_a, then

$$\rho_b = 1.075\ \rho_{log} - 0.193 \qquad (12)$$

where ρ_{log} = apparent log density (g/cc).

The corrected bulk density from Equation (12) is compared with the apparent log density for Borehole H87-102 in Figure 4.

Table 4 lists the calculated weighted average of $\Sigma Z/M$ for the sand tailings at apparent log densities ranging between 1.90 and 2.10 g/cc. The weighted value of $\Sigma Z/M$ varies between 0.508 and 0.514.

The formation fluids are comprised of water combined with solvents used in the extraction of bitumen from the oil sand. It was recognized that the presence of the solvents could alter the $\Sigma Z/M$ ratio for the formation fluids from the assumed value of 0.555 for pure water. Field tests presented in the next sections show that the difference in $\Sigma Z/M$ for the formation fluids and pure water is small.

Bitumen unremoved from the oil sand by extraction processes are present within the pore volume of the sand tailings. The overall bitumen content amounts to less than 5 percent of the total pore volume. However, infrequent thin zones of tailings sand up to 15 cm thick are present in which the bitumen may comprise 25 to 50 percent of the pore volume. Considering a localized thin zone of saturated tailings sand with bitumen (fluid density = 1.0 g/cc) comprising 50 percent of the pore volume, the apparent density is given as:

$$\rho_a = 2.645 - 1.520\ n \tag{13}$$

where n = total pore volume porosity.

The corresponding true bulk density of the sand tailings is

$$\rho_b = 2.65 - 1.65\ n \tag{14}$$

Substituting Equation (13) into Equation (14) yields

$$\rho_b = 1.086\ \rho_a - 0.221 \tag{15}$$

or

$$\rho_b = 1.086\ \rho_{log} - 0.221 \tag{16}$$

Over the range of apparent log densities between 1.90 and 2.1 g/cc, the difference is predicted bulk density by Equation (12) and Equation (16) in less than 0.008 g/cc. The effect of the presence of bitumen in the sand tailings is small even for the extreme case considered.

FIELD OBSERVATIONS

In-situ density measurements in the oil sand tailings were undertaken using a widely available compensated nuclear density logging tool. The tool selected was the Century Geophysical Corporation 9036A density tool utilizing a 2.0 Curie Cesium 137 source and dual detectors spaced at 15 and 30 cm.

Evaluation of (1) the reliability of the density tool in terms of accuracy, and (2) the procedures of log interpretation discussed above were undertaken by a series of sampling and logging programs described below.

Repeatability of Density Logs

The precision of repeatability of the density tool were evaluated by conducting multiple "repeat" logs of sections of boreholes. Figure 5 shows the results of five repeat logs performed using the 9036A density tool between the depths of 80 and 95 m in Borehole H87-96. The five logs show a remarkably good degree of repeatability. The variation of apparent density measured at any elevation is

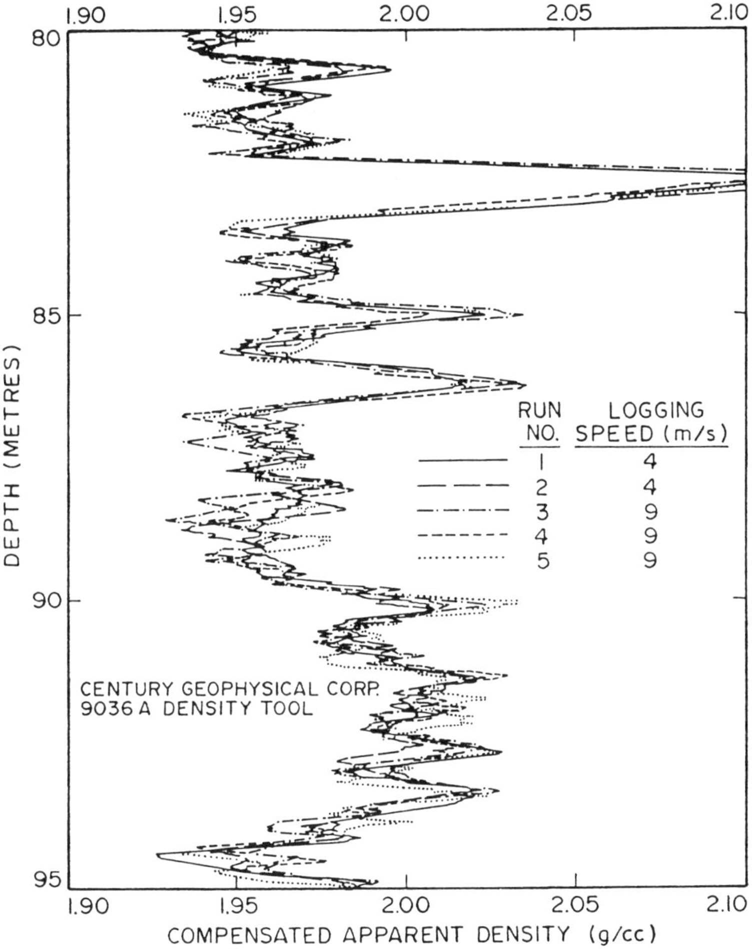

FIGURE 5 Multiple Repeat Density Logs in Borehole H87-96

generally less than 0.03 g/cc or roughly 1.5 percent of the magnitude of the apparent density. For any single density log, the range of repeatibility may fall between plus or minus 0.03 g/cc. Hence, the precision of density log measurements is taken as $\pm$ 0.03 g/cc.

Logging run Nos. 1 and 2 were conducted at logging speeds of 4 m/s, less than one-half of the standard logging speed of 9 m/s used for logging run Nos. 3, 4 and 5. As seen in Figure 5, no difference in precision of repeatibility of apparent density was found at the lower logging speed.

Density Response of Pore Fluid

The combination of solvents and water in the pore fluid could elicit an influence on the measured apparent density of the sand tailings. To assess this possibility the 9036A probe was dropped through a 3 m depth of pond water retained in the interior of the sand tailings dyke under investigation. The apparent density response is shown in Figure 6. The apparent density of the pond fluid ranges between 1.08 and 1.12 g/cc. For the true fluid density of 1.0 g/cc, the $\Sigma Z/M$ ratio for the pond fluid is back calculated to range between 0.54 and 0.56. The difference in $\Sigma Z/M$ from the value of 0.555 for pure water is small. The pore fluid in the sand tailings was therefore treated as pure water for purposes of log interpretation.

Core Sampling of Deep Beach Tailings

To obtain representative samples from the depths of interest (i.e. 50 to 100 m) for density measurement and other laboratory testing, unique procedures were developed to retain intact undisturbed core samples of the tailings sand. The procedures followed are described briefly below:

(1) The coring was undertaken using a 3 m long modified wireline Christiansen double core barrel. The standard core barrel shoe was replaced with a specially designed adjustable 6.85 cm inside diameter (ID) shoe. By extending or withdrawing the inner barrel, this shoe could be adjusted to extend slightly ahead of the face of the coring bit. A 0.15 cm extension of the shoe ahead of the coring bit was found to obtain optimal core recovery.

(2) A 7.30 cm ID PVC core liner was inserted into the inner core barrel for each coring run. This liner was used in conjunction with a steel spring core catcher behind the coring bit to retain the sand samples during retrieval and confine the samples during handling.

(3) Nominal coring runs of 1.0 to 1.5 m were performed. To prevent the tendency for the circulating drilling fluid to wash away the cohesionless beach sands ahead of the core barrel, a low pump pressure was maintained during coring.

(4) At the ground surface, the PVC liner was cut and sealed at both ends. The core sample was placed in dry ice and frozen. Visual

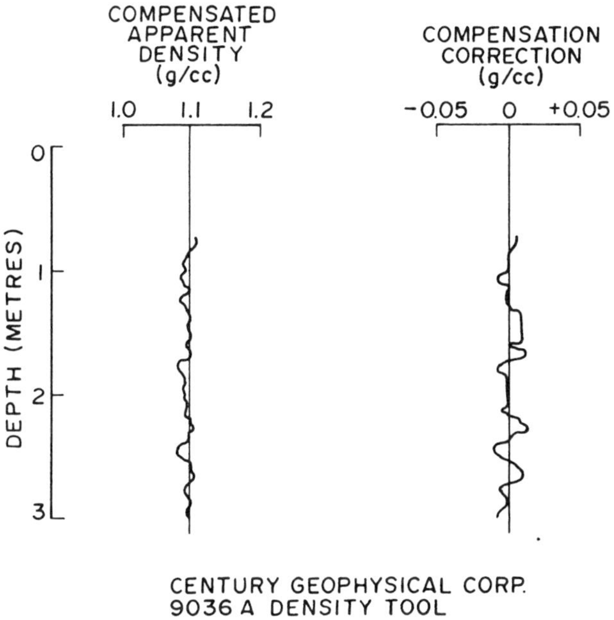

FIGURE 6 Compensated Apparent Density Log in Pond Water

examination of the frozen cores indicated no major distruptive influence due to ice lenses.

(5) Using the above procedures, good core recovery (75-100%) was typically achieved. In some of the cores, the top of the core run exhibited evidence of washing and disturbance of sand structure, but the remainder of the core run was visually undisturbed as bedding features and other structures from the deposition process of the sand tailings were preserved.

The bottom 45 cm of sand tailings was cut from the frozen core samples and measurements of frozen core density, water content and saturation recorded. The in-situ bulk densities calculated from the density log exceed the laboratory core densities in all cases by up to 14 percent. The saturation of the core samples generally ranged between 85 and 90 percent. Correction to the laboratory bulk density of the core samples for incomplete saturation did not satisfactorily improve the correlation.

It is considered the laboratory bulk densities of the core samples have to be corrected for the following influences:

a) Volume expansion change due to in-situ stress release,
b) Volume change resulting as the sample passes through the 6.85 cm I.D. core shoe into the 7.30 cm I.D. core liner, and
c) Volume change due to freezing.

It was not possible to satisfactorily subdivide and quantify the influence of each of these effects on the density of the core samples.

The in-situ bulk densities of the core samples were alternatively back-calculated using the measured water content and assuming 100 percent in-situ saturation of the sand tailings. This procedure to calculate in-situ density was considered valid on the following basis:

1) All core samples were recovered from saturated sand tailings below the phreatic surface.

2) The contact of the core samples with the drilling fluid was minimized by using the PVC liners in the inner core barrel and extending the "core shoe" of the inner barrel slightly a head of the coring bit.

3) The consistency of the core samples retrieved from the core barrel was very stiff similar to a cohesive material. This apparent cohesion is caused by negative undrained pore pressures developed in the sand due to release of the high in-situ stress levels. The sustained negative pore pressures suggests the amount of water imbibed into the core samples during the coring procedures is minimal. If any water were imbibed into the core samples, the calculated bulk densities would conservatively underestimate the in-situ bulk density.

The in-situ bulk density calculated by this method correlates well

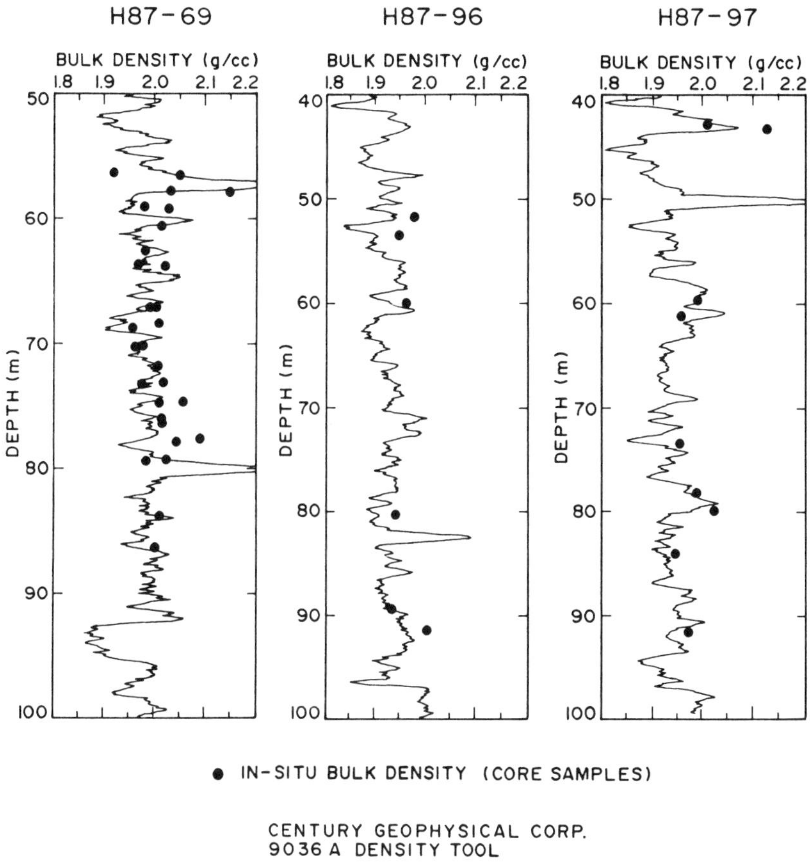

FIGURE 7 Comparison of Corrected Bulk Density Logs and Bulk Density of Core Samples for Boreholes H87-69, H87-96, H87-97.

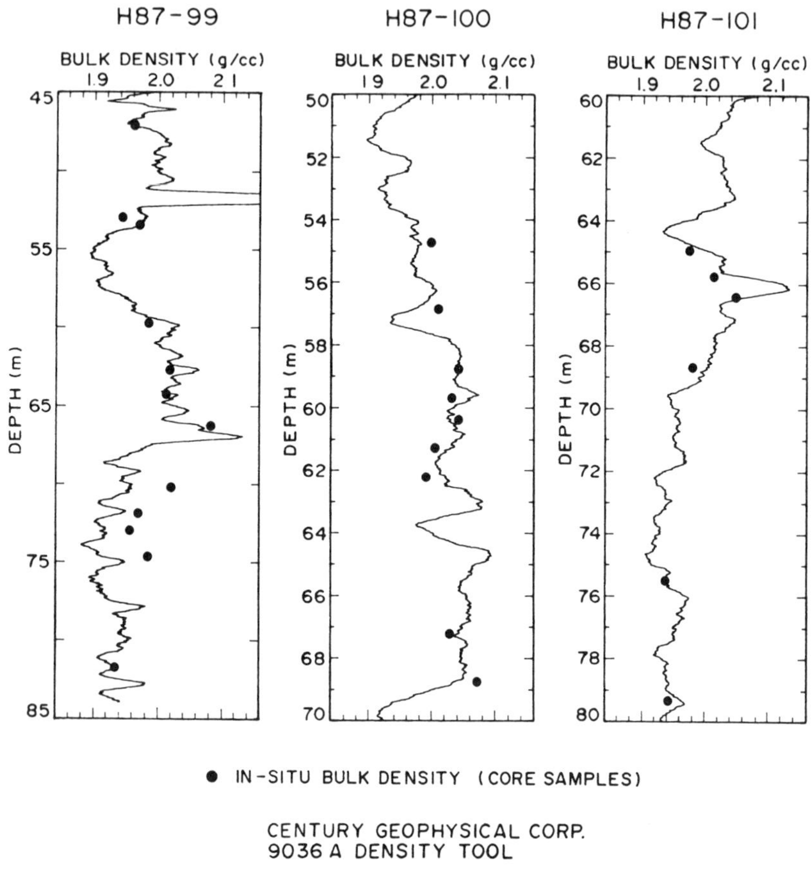

FIGURE 8 Comparison of Corrected Bulk Density Logs and Bulk Density of Core Samples for Boreholes H87-99, H87-100, H87-101.

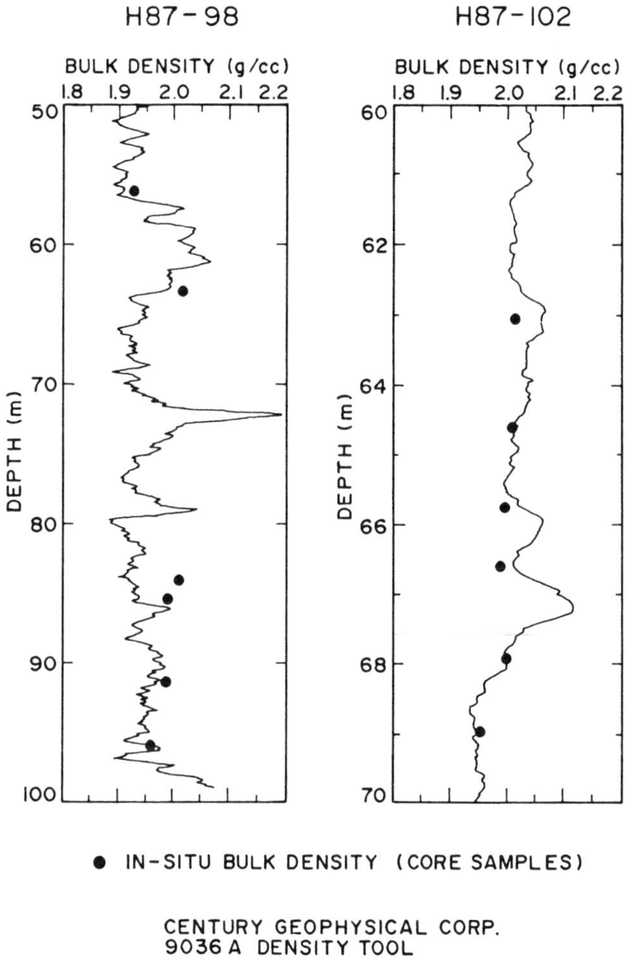

FIGURE 9 Comparison of Corrected Bulk Density Logs and Bulk Density of Core Samples for Boreholes H87-98, H87-102.

with the bulk density determined from the density logs. Figures 7, 8 and 9 show the comparison of corrected density versus the density logs in eight test holes. A total of 69 of 83 core samples (83 percent) in all boreholes fall within the range of precision of ± 0.03 g/cc for the density logs. This high degree of correlation is considered to validate the accuracy of the density logging technique and the suitability of the log interpretation procedures used.

CONCLUSIONS

Repeatability experiments indicate that the precision of the Cesium 137 gamma ray source downhole compensated nuclear logging equipment used in the reported investigation is ± 0.03 g/cc (bulk density). Comparisons with bulk densities determined from in-situ samples showed that 83 percent of individual comparisons fell within the ± 0.03 g/cc precision limit.

Off-the-shelf nuclear density equipment can be used for in-situ density determination for geotechnical purposes. However, appropriate interpretation techniques must be undertaken and calibrated to the material investigated in order to achieve optimum results. Development of drilling techniques to achieve a smooth and stable borehole wall are also an essential element in successful use of such equipment.

ACKNOWLEDGEMENTS

The authors wish to thank Suncor Inc. - Oil Sands Group for permission to publish the data presented in this paper. We also wish to acknowledge the cooperation of Elgin Exploration Company Ltd. of Calgary, Alberta in modifying their core barrels for the sampling program described in this paper. Their drilling expertise was also invaluable for providing optimal borehole conditions for density logging. The assistance of staff of Century Geophysical Corporation was also very generously provided.

APPENDIX I - REFERENCES

Dresser Atlas (1982). "Well Logging and Interpretation Techniques".

Killeen, P.G. (1982). "Gamma-Ray Logging and Interpretation" in Developments in Geophysical Exploration Methods - 3, Applied Science Publishers, New York, pp. 95-150.

Klohn, E.J. (1984). "The Brenda Mines Cycloned-Sand Tailings Dam", Int. Conf. on Case Histories in Geotechnical Engoineering, Vol. II, pp. 953-977.

Mittal, H.K. and Morgenstern, N.R. (1975). "Parameters for the Design of Tailings Dams", Can. Geot. Jour., Vol. 12, No. 2, pp. 235-261.

Pickell, J.J. and Heacock, J.G. (1960). "Density Logging", Geophysics, Vol. XXV, No. 4, 1960, pp. 891-904.

Skempton, A.W. (1986). "Standard Penetration Test Procedures and the Effects in Sands of Overburden Pressure, Relative Density, Particle Size, Aging and Overconsolidation", Geotechnique, Vol. 36, No. 3, pp. 425-447.

Tjelta, T.I., Tiegges, A.W.W., Smits, F.P., Geise, J.M., and Lunne, T.A. (1987). "In Situ Measurements by Nuclear Backscatter for an Offshore Soil Investigation", Norwegian Geotechnical Institute Publication No. 169, Oslo.

TABLE 1

Z/A RATIOS OF COMMON GEOLOGIC ELEMENTS

Element	Symbol	Z	Z/A
Hydrogen	H	1	0.992
Oxygen	O	8	0.500
Nitrogen	N	7	0.500
Carbon	C	6	0.499
Calcium	Ca	20	0.499
Sulfur	S	16	0.499
Silicon	Si	14	0.498
Magnesium	Mg	12	0.495
Potassium	K	20	0.487
Aluminum	Al	13	0.482
Sodium	Na	11	0.479
Chlorine	Cl	17	0.479
Iron	Fe	26	0.466

TABLE 2

Σ Z/M RATIOS OF COMMON GEOLOGIC COMPOUNDS

Material	Formula	Actual Bulk Density (g/cm^3)	Apparent Density (g/cm^3)	Σ Z/M
Quartz	SiO_2	2.65	2.645	0.499
Calcite	$CaCo_3$	2.71	2.71	0.500
Dolomite	$CaCo_3.MgCo_3$	2.87	2.864	0.499
Halite	NaCl	2.165	2.074	0.479
Water	H_2O	1.00	1.11	0.555
Oil	$C_n(CH_2)$	0.80	0.91	0.570
Gas	C_nH_{2n+2}	0.20	0.25	0.619

TABLE 3

MASS ABSORPTION COEFFICIENTS

Material	Mass Absorption Coefficient (cm^2/g) At 0.51 MeV Energy Level
Igneous Rock	0.087
Limestone	0.087
Shale	0.085
Sandstone	0.086
Water	0.096
Crude Oil	0.095

TABLE 4

Σ Z/M RATIOS FOR TAILINGS SAND

Apparent Log Density (g/cm^3)	True Bulk Density (g/cm^3)	Σ Z/M
2.10	2.065	0.508
2.00	1.957	0.511
1.90	1.850	0.514

SETTLEMENT OF URANIUM MILL TAILINGS

By P. K. Chen,[1] Berg Keshian,[2] Jr., and Frank B. Guros,[1] Members, ASCE

ABSTRACT

Two test embankments were constructed on top of an old tailings deposit near Ambrosia Lake, New Mexico to determine settlement characteristics of hydraulically-deposited uranium mill tailings. Before construction of the embankments, properties of in situ tailings and foundation soils were determined using data from boreholes, piezocone soundings, and laboratory tests. These properties were used to estimate post-construction settlement of a planned disposal embankment to be constructed on the tailings. However, excessive uncertainty existed in the following: a) field settlement rates of saturated and unsaturated tailings, b) degree of preconsolidation of the upper 15 feet of tailings, and c) the ability of an underlying silty sand foundation layer to facilitate drainage. Thus, assurance could not be provided that differential settlements of the radon barrier and erosion protection layers would be within allowable limits should the planned disposal embankment be constructed in a single-stage.

The two test embankments were constructed in single stages within a four-week period to heights of 20 and 27 feet. Settlements and pore pressures for the in situ tailings and foundation soils were measured. After these measurements were obtained, one-dimensional consolidation calculations were made using actual loading measured by total stress cells. The calculations resulted in values for key material properties that gave settlement and pore pressure responses in agreement with test embankment data. Further review of measurements and calculation results indicated that laboratory test results can accurately determine the primary embankment settlement magnitude but underestimate the settlement rates for both unsaturated and saturated tailings. Piezocone data provided valuable stratigraphic data, but resulted in over-estimation of preconsolidation for desiccated tailings relative to laboratory test data.

---

[1] Morrison-Knudsen Engineers, 180 Howard St., San Francisco, CA 94105

[2] Roy F. Weston, Inc., 5301 Central Avenue NE, Suite 1000, Albuquerque, New Mexico

Post-construction settlement estimates for the planned embankment were revised using results from the test embankment analyses. Revised estimates indicate that differential settlements of erosion protection and radon barrier layers will be within allowable limits following single-stage construction.

INTRODUCTION

The U.S. Department of Energy (DOE) is directing the Uranium Mill Tailings Remedial Action (UMTRA) Project, which involves construction of isolation systems for abandoned uranium mill tailings at 24 sites in 10 states. At several of these sites, tailings disposal embankments will be constructed either entirely or partially on existing hydraulically-deposited tailings. Design criteria established for these embankments by the U.S. Environmental Protection Agency require stabilization for 1000 years, to the extent reasonably achievable, or for 200 years in any case. While the completed embankments will be monitored after construction, designs do not generally rely on maintenance to ensure adequate performance. Reliance on controls that do not require maintenance and repair is preferable; e.g., a non-erodible rock riprap cover is preferable over covers that require maintenance and repair. A cover system comprising a compacted soil layer (radon barrier) and an overlying erosion protection layer is typically designed to isolate and stabilize the tailings. Differential settlements can affect the performance of this cover.

The results of a study are presented herein for estimating settlement of a disposal embankment to be constructed on an existing uranium mill tailings deposit at the Ambrosia Lake (Phillips/United Nuclear) UMTRA site in New Mexico (Figure 1). The existing tailings deposit was placed hydraulically at a millsite which ceased active milling in 1963. Ore was milled primarily using an alkaline leach process. The resulting tailings deposit is approximately 2000 feet square and projects 15 to 20 feet above the surrounding ground surface.

Reliable prediction of post-construction settlement of the embankment is necessary, particularly because compliance with standards must be demonstrated by adequacy of design prior to construction rather than by post-construction observations. The disposal embankment design includes placement of compacted tailings and a protective cover on top of the existing tailings deposit. To ensure adequate cover performance tailings settlement must be nearly complete before the cover is placed. A more costly dual-stage construction scheme (i.e., preloading) could be required if settlements caused by single-stage construction are estimated to be excessive. Therefore, reliable settlement estimates for single-stage construction were needed. However, site conditions and knowledge of material properties did not allow sufficiently reliable post-construction settlement estimates by conventional methods. Instead, the following approach for estimating post-construction settlement was adopted:

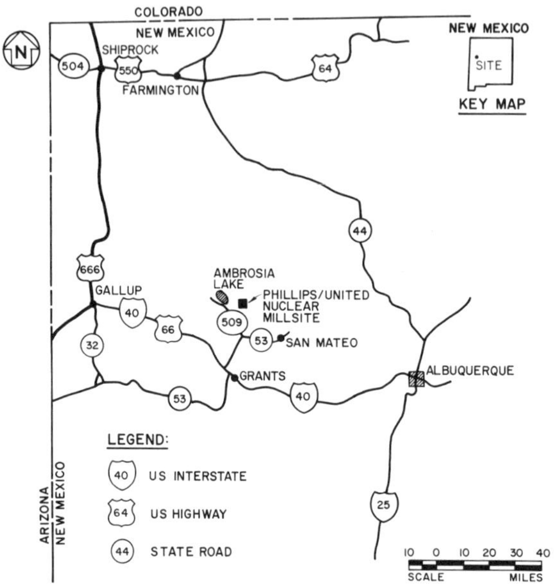

FIG. 1. LOCATION MAP

1. Conduct site characterization and material properties studies;

2. Perform preliminary analyses to estimate post-construction consolidation settlements and perform sensitivity calculations to identify key parameters;

3. Construct test embankments, monitor loading, and measure pore pressure and settlement of underlying tailings and foundation soils;

4. Backcalculate values for key soil parameters using test embankment data and compare with values used for preliminary analyses, obtained from boreholes, piezocone soundings and laboratory tests;

5. Use parameter values that accurately model test embankment performance to predict post-construction settlement of the disposal embankment.

SITE CHARACTERIZATION

Site characterization activities were conducted to determine site stratigraphy and material properties. Approximately 281 boreholes (including monitor wells), and 126 piezocone soundings were completed. Both disturbed and undisturbed samples were obtained, along with Standard Penetration Test (SPT) values. Laboratory test results are summarized in Table 1. Site stratigraphy is generally described as tailings overlying silty sand (alluvium), clay (weathered Mancos Shale), and shale bedrock, in descending order. At some locations, the silty sand alluvium layer is not present. The stratigraphy at the two locations where test embankments were later constructed is shown in Figure 2. The stratigraphy at the north test embankment location represents the general site stratigraphy. The east test embankment site is a location where post-construction settlements should be the greatest, due to the perched ground water in the tailings and the absence of the alluvium layer.

The shallowest continuously saturated aquifer beneath the site lies below the shale bedrock surface. A localized perched ground-water zone results from seepage recharge from a depression on the existing tailings deposit which collects surface runoff. The extent of water collected in the depression as well as the extent of the perched ground-water zone vary in size seasonally. The maximum extent of the perched ground-water zone in the tailings is approximately 900 by 400 feet. Beneath this perched ground water, the saturated tailings directly overlie the clay (weathered Mancos Shale) layer. The remainder of the tailings deposit comprises materials that are nearly saturated. These materials overlie the silty sand alluvium layer, which has saturation values on the order of 60 percent. Thus, the silty sand alluvium provides drainage for water percolating through the tailings, while the clay layer acts as a barrier to vertical flow.

Both piezocone soundings and grain size tests provide data for determining the distribution of various material types both within the tailings and in the underlying foundation soils. From the data it was concluded that the tailings deposit consisting of sands, sand-slimes and slimes is heterogeneous. The tailings deposit consists mostly of sand-slimes, defined for this study as tailings with 30-70% finer by weight passing a No. 200 sieve. Tailings sands (less than 30% passing the No. 200 sieve) are located around the perimeter of the pile. Tailings slimes (greater than 70% passing the No. 200 sieve) occur throughout the pile.

A laboratory testing program was conducted to determine engineering properties of the materials. The results of the tests performed are summarized in Tables 1 and 2. Table 2 presents material properties used in the settlement analyses.

For settlement estimates, virgin compression index (C_c), recompression index (C_r), coefficient of consolidation (c_v) and coefficient of recompression (c_{vs}), and over-consolidation ratio

TABLE I. - AVERAGE VALUES FOR INDEX PROPERTIES AND OTHER PARAMETERS

MATERIAL TYPE AND U.S.C.S. CLASSIFICATION[a]	SIEVE ANALYSIS % BY WEIGHT[b]		ATTERBERG LIMITS[c]		G_s	IN SITU		COMPACTION[d]		k[e]
	SAND	FINES	LL	PI		w (%)	γ_d (pcf)	w_{OPT} %	$\gamma_{d(MAX)}$ (pcf)	cm/sec
(1)	(2)	(3)	(4)	(5)	(6)	(7)	(8)	(9)	(10)	(11)
TAILINGS (ALL DATA) SM, CH, CL	53	47	19	7	2.64	27.0	91.1	—	—	2.8×10^{-4} f
TAILINGS SLIMES (SL) CH, CL	13	87	43	21	2.65	53.5	68.0	18.5	106.0	f
TAILINGS SAND-SLIMES (S-SL) SM	56	44	14	3	2.65	22.4	96.7	14.8	111.0	4.4×10^{-6}
TAILINGS SANDS (S) SM	75	25	NP	NP	2.62	13.4	95.6	—	—	5.6×10^{-4}
ALLUVIUM SM [CH, CL]	52	48	24	13	2.64	15.6	99.6	—	—	6.6×10^{-4} g
CLAY (WEATHERED MANCOS SHALE) CH, CL	27	73	48	24	2.67	17.9	106.5	—	—	4.0×10^{-8}

a. "[ ]" INDICATES RELATIVELY MINOR AMOUNT OF SOIL TYPES WITHIN LAYER.
b. SAND IS MATERIAL WITH SIZES SMALLER THAN NO. 4 SIEVE BUT RETAINED ON NO. 200 SIEVE. FINES ARE SIZES SMALLER THAN NO. 200 SIEVE.
c. ASSUMES LL=0 AND PI=0 FOR NON-PLASTIC (NP) TEST RESULTS. (APPROXIMATION ONLY)
d. ASTM D698: w= MOISTURE CONTENT, γ=DRY UNIT WEIGHT
e. PERMEABILITY TEST RESULTS.
f. ONLY TAILINGS SANDS AND SAND-SLIMES WERE TESTED
g. AVERAGE OF 3 SM SAMPLES SHOWN. IN ADDITION, ONE CH SAMPLE HAD $k = 2.3 \times 10^{-6}$ cm/sec TEST RESULT.

TABLE 2.—MATERIAL PROPERTIES FOR PRELIMINARY SETTLEMENT ANALYSIS

MATERIAL/ AND U.S.C.S. CLASSIFICATION[a] (1)	TOTAL UNIT WEIGHT (pcf) (2)	MOISTURE CONTENT (%) (3)	SATURATION, %		VOID RATIO (6)	C_c (7)	C_r (8)	C_v (9)	C_{vs} (10)	C_α (11)	OCR (12)
			AVERAGE (4)	ASSUMED FOR ANALYSIS (5)							
TAILINGS:											
SLIMES (SL) CH, CL	104.4	53.5	98	100	1.43	0.535	0.056	6.2×10^{-3}	a	3.1×10^{-3}	b
SAND-SLIMES (S-SL) SM	118.4	22.4	88	100	0.710	0.120	0.011	1.1×10^{-2}	a	1.6×10^{-3}	b
SANDS (S) SM	108.4	13.4	50	100	0.710	0.10 c	0.016 d	1.0×10^{-1}	a	1.6×10^{-3} e	b
ALLUVIUM SM (CH, CL)	116.8	15.9	62	DRAINAGE LAYER	0.634	0.053	0.010	1.6[f]	1.6[f]	7.2×10^{-4}	3[g]
CLAY (WEATHERED MANCOS SHALE) CH, CL	122.3	17.4	78	100	0.600	0.174	0.054	2.9×10^{-4} 1.8×10^{-3} h	2.9×10^{-4} 7.8×10^{-3} h	3.0×10^{-3}	3[g]

a. LABORATORY RESULTS UNAVAILABLE. SEE TABLE 3 FOR VALUES SELECTED FOR SENSITIVITY ANALYSIS.
b. SEE FIGURE 3.
c. ESTIMATED FROM PUBLISHED DATA FOR TAILINGS SANDS (VICK, 1983).
d. ESTIMATED FROM PIEZOCONE DATA.
e. NO DATA. ASSUME VALUE FOR S-SL.
f. VALUE SHOWN IS 100 TIMES GREATER THAN LABORATORY RESULTS TO MODEL ASSUMED DRAINAGE.
g. LABORATORY TEST RESULTS.
h. FIRST SET OF VALUES GIVES C_v FROM LABORATORY RESULTS AND ASSUMES $C_{vs} = C_v$.
SECOND SET OF VALUES IS ESTIMATED FROM PI DATA (NAVFAC DM 7.1, 1982).

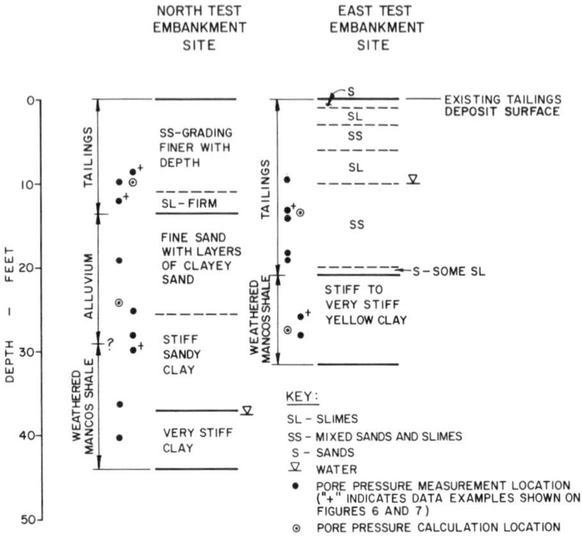

FIG. 2. TAILINGS AND FOUNDATION SOIL STRATIGRAPHY AT TEST EMBANKMENT SITES

(OCR) are the primary material properties used. Estimates of C_c, C_r c_v, and c_{vs} were based on laboratory consolidation testing of thin-walled tube undisturbed samples. Piezocone sounding data and laboratory consolidation test results were used to provide separate estimates of OCR.

The OCR values for saturated laboratory test samples were based on consolidation test data corrected for sample disturbance according to a method proposed by Ladd, 1973. OCR values were determined using in situ undrained shear strength (s_u) calculated from piezocone data (Robertson and Campanella, 1984) using a cone factor N_k equal to 17. Normally-consolidated s_u was estimated based on tailings PI data and a relationship with effective overburden stress (Jamiolkowski, et al., 1985). Finally, OCR was calculated from a relationship with the ratio of in situ s_u and normally-consolidated s_u (Ladd, et al., 1977, and Jamiolkowski, et al., 1985). A comparison of OCR values from laboratory test data and piezocone sounding data is shown on Figure 3.

PRELIMINARY SETTLEMENT ESTIMATES

Preliminary settlement estimates were made to evaluate the possible total magnitude of both primary and secondary consolidation as well as post-construction settlement. When

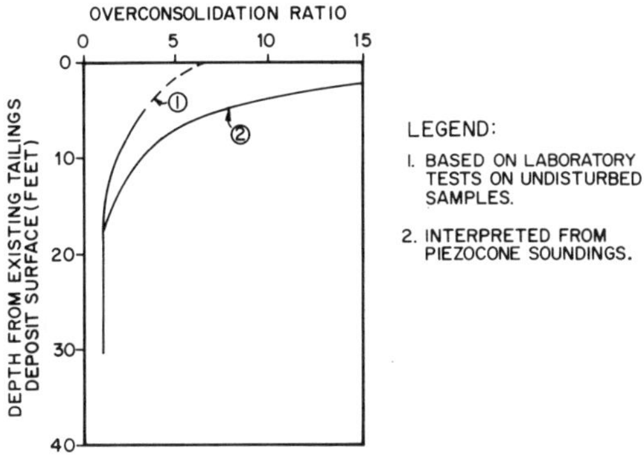

FIG. 3. OVERCONSOLIDATION RATIO VERSUS DEPTH FOR TAILINGS

specific test data were lacking, material properties were selected based on available site characterization data and correlations from literature sources. (Table 2). To evaluate effects of uncertainty in potential key material properties, sensitivity of settlement to variations in the following properties was evaluated:

1. OCR of tailings

2. Tailings c_{vs}

3. c_v and c_{vs} of foundation clay.

A range of values was selected for potential key parameters and used to estimate settlements for the five cases shown in Table 3.

Calculations were made with "CONSOL" (Wong and Duncan 1984) using one-dimensional Terzaghi consolidation theory. A total load of 25 feet of compacted tailings and cover was assumed, since embankment design was not finalized at the time. Three increments of loading were assumed, spread over a one-year period. Twelve locations on the tailings deposit, accounting for variations in stratigraphy, were selected for settlement analysis. Locations were selected to include areas that would exhibit typical, minimum, and maximum settlements. The results of these settlement analyses were also used to assist in the selection of the locations of the two test embankments.

Typical plots of estimated settlement versus time are shown on Figures 4 and 5. The results of calculations indicate that post-construction settlements are significantly affected by 1) variation of tailings OCR and 2) the presence of a silty sand layer which provides drainage for the tailings.

TABLE 3. Sensitivity Analysis for Preliminary Settlement Estimates

Case No. (1)	Tailings OCR (2)	Tailings c_{vs} (3)	Foundation Clay c_v [a] (4)
1	Laboratory Results	$c_{vs} = c_v$	Laboratory c_v
2	Piezocone Results	$c_{vs} = c_v$	Laboratory c_v
3	Laboratory Results	$c_{vs} = 10c_v$	Laboratory c_v
4	Laboratory Results	$c_{vs} = c_v$	c_v from PI Data (NAVFAC,1982)
5	Piezocone Results	$c_{vs} = 10\ c_v$	c_v from PI Data (NAVFAC,1982)

[a] $c_{vs} = c_v$ assumed

The results also indicate that the variation of c_{vs} of the tailings does not significantly affect the post-construction settlements.

Uncertainty in the behavior of the unsaturated tailings deposit compounded the difficulty of predicting settlement. Additionally, post- construction settlements could be excessive if the silty sand layer did not act as a drainage layer during consolidation. Available permeability data for this layer was insufficient to provide the assurance of adequate drainage. Thus, it was decided to construct test embankments and determine the field response to obtain data that would ensure reliable prediction of settlement for the planned disposal embankment.

TEST EMBANKMENT CONSTRUCTION AND INSTRUMENTATION

Two test embankments were constructed to observe consolidation behavior of the tailings and foundation soils. Construction and location of the test embankments were planned to meet the following conditions:

1. Adequate height to model the planned disposal embankment.

2. Adequate extent to facilitate primarily one-dimensional drainage.

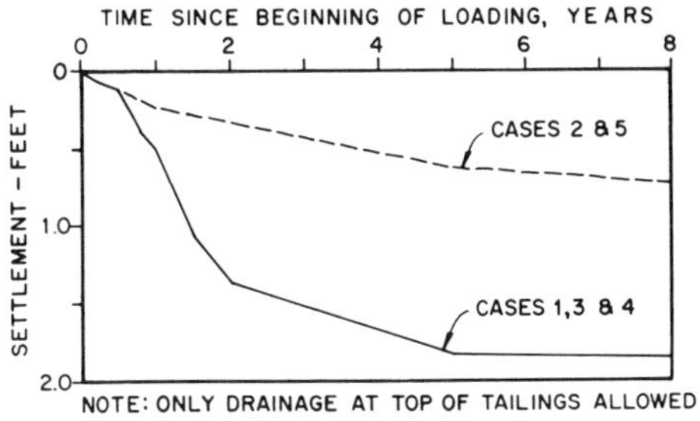

FIG. 4. PRELIMINARY SETTLEMENT CALCULATION RESULTS FOR EAST TEST EMBANKMENT SITE

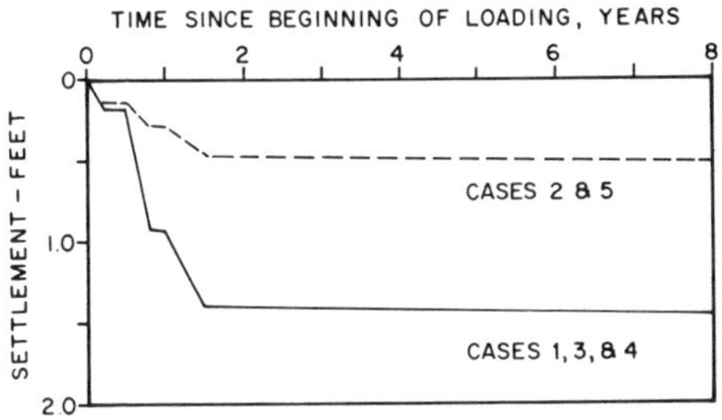

FIG. 5. PRELIMINARY SETTLEMENT CALCULATION RESULTS FOR NORTH TEST EMBANKMENT SITE

3. Check locations where both typical and maximum post-construction settlements were expected.

4. Provide drainage at the base of the embankments to facilitate analysis.

Additionally, soil borings and samplings were made to determine whether anomolous subsurface conditions existed at either location.

Both embankments were constructed with bases measuring 200 by 200 feet, with 2(H):1(V) sideslopes. The maximum height of the north test embankment of 27 feet was reached after 27 days. The east test embankment was constructed to a height of 20 feet over a 25 day period.

Instrumentation: In order to effectively monitor the results of the test embankment, an instrumentation network of pneumatic piezometers, settlement plates and total stress cells was installed. Beneath each test embankment a total of nine pneumatic piezometers, three settlement plates and two total stress cells was installed. Multiple piezometers were installed in each layer of tailings, alluvium, and weathered Mancos Shale. Each piezometer was installed within an individual auger hole using 1-2 foot fluid collection zone created with a sand pack and then grouting the hole to the surface. Settlement plates and total stress cells were placed at the base of each embankment.

All instrumentation was installed at least one week before construction of each test embankment in order to give the instruments time to equilibrate. Readings were taken on a daily basis for the first 60 days following installation. Once construction was complete and readings began to stabilize, the readings were spaced further apart.

Readings were taken for each date, and the following information was obtained and tabulated:

- Date of installation of each piece of instrumentation.
- Serial number of each instrument.
- Calibration factor(s) for each instrument, supplied with each.
- Readings for each instrument.
- Elapsed time since initial reading.
- Results of analysis for each reading.

The data were plotted on a daily basis during construction to ensure that the pore pressures did not become so great as to cause failure. Data obtained from the instruments are given below in Figures 6 and 7.

TEST EMBANKMENT RESULTS AND INTERPRETATION

Key aspects of the geotechnical behavior of the tailings and foundations soils were identified from the instrumentation data. These observations were evaluated qualitatively and compared with prior analytical predictions that incorporated previous best estimates of soil parameter values. Analyses were also used to backcalculate empirical values of key soil parameters from the field data. The objective of this program was to develop an understanding of the characteristic field behavior, and to reliably predict post-construction settlement of the final disposal embankment. The program was primarily directed towards determining the consolidation characteristics of the existing tailings and the underlying foundation soils. Analysis of the data is described below.

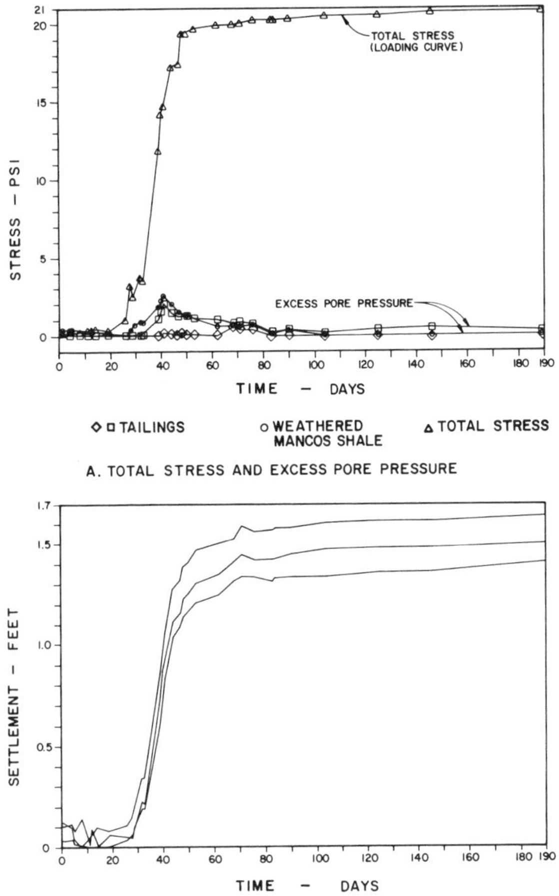

A. TOTAL STRESS AND EXCESS PORE PRESSURE

B. SETTLEMENT

FIG. 6. TOTAL STRESS, EXCESS PORE PRESSURE, AND SETTLEMENT MEASUREMENTS FOR NORTH TEST EMBANKMENT

Embankment Loading Rates: Based on the construction records and test embankment geometry, estimated load versus time plots were developed for both test embankments. The estimated loading rates from construction records agreed adequately with the actual total stress measurements. Therefore, the total stress measurements were used for the analysis. The sensitivity of calculated settlements and pore-pressures to step-loading increments was evaluated to establish a suitable step-loading approximation for the analyses.

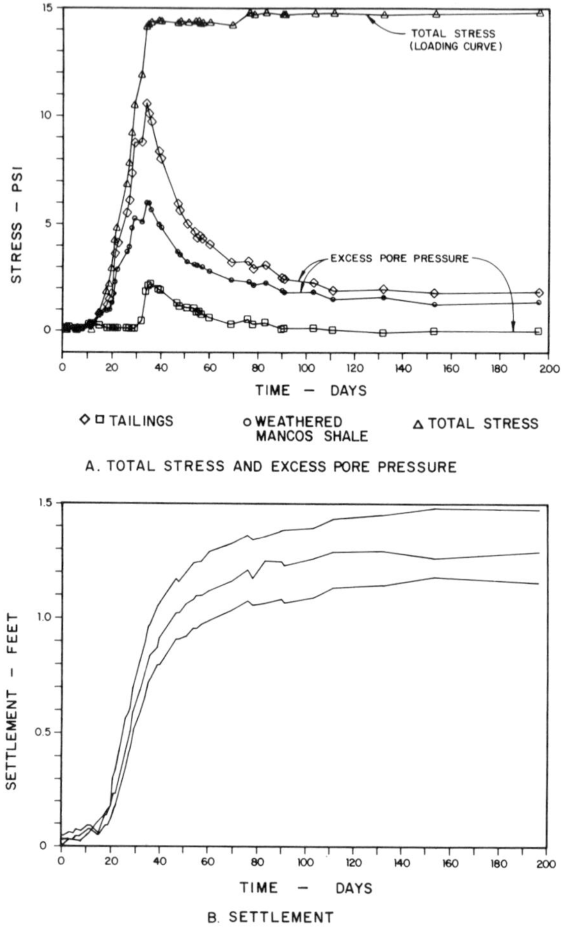

FIG. 7. TOTAL STRESS, EXCESS PORE PRESSURE, AND SETTLEMENT MEASUREMENTS FOR EAST TEST EMBANKMENT

Geotechnical Material Properties for Embankment Locations: Geotechnical data, including field borings, samples and laboratory test data, were obtained for the tailings and foundation soils beneath each test embankment. These data were analyzed and compared with the data obtained previously. It was concluded that anomalous material properties did not exist at the test embankment locations.

Settlement and Pore Pressure Measurements: Settlement and examples of data for pore pressure response with time for each test embankment are presented on Figures 6 and 7. Since variations between readings from several individual pore pressure gages were not great within either the tailings, silty sand alluvium or clay layer, average values were used for analysis. Most consolidation settlement took place during the construction period for tailings and foundation layers beneath both test embankments. The dissipation rate of pore water pressure beneath the east test embankment was slower than that of the north test embankment. This was expected because of the saturated tailings and the absence of the silty sand alluvium layer to provide drainage for the tailings beneath the east test embankment. Pore pressure data indicated that the layer of relatively dry material placed at the bottom of each test embankment provided adequate drainage.

Analysis: The settlement and pore pressure measurements from the test embankments were compared with the results of settlement and pore pressure calculations made using "CONSOL" (Wong and Duncan, 1984) for both test embankments. The primary goal was to derive a set of backcalculated soil parameters which would provide agreement between the calculated and measured settlements and pore pressures. Then, these soil parameters could be used to predict the post-construction settlement of the final disposal embankment.

The analytical approach was as follows:

- Soil profiles were developed based on the results of the field drilling and laboratory classification tests.

- Tailings beneath each embankment were assumed to be entirely slimes.

- Comparison of test embankment settlements with preliminary settlement estimates indicated that the tailings over-consolidation ratio (OCR) would be best predicted by the OCR curve obtained from the laboratory results.

- Only C_c for the tailings layer was adjusted to match total settlements between calculated and measured values. C_c and C_r values for the silty sand alluvium and clay (weathered Mancos Shale) layers were not adjusted because most of the settlement of these layers should take place in the recompression range. Therefore, the magnitude of settlement for these layers would be relatively small compared with tailings settlement.

- Values of c_v and c_{vs} were adjusted for tailings, silty sand alluvium, and weathered Mancos Shale until the calculated dissipation of pore pressure with time reasonably matched field measurements. For this study, it was also assumed that $c_{vs} = c_v$.

RESULTS

Calculated settlements and pore pressures are compared with field measurements in Figures 8 and 9. The following significant findings are based on the observations and analyses:

General:

- o The ultimate settlement per unit embankment height is essentially identical for both test embankments.

- o Although the majority of tailings are sand-slimes, backcalculated C_c values for tailings are closer to the corresponding laboratory C_c values for the slimes rather than those of the sand-slimes.

- o The backcalculated c_v values for tailings, and c_{vs} values for alluvium and clay are significantly greater than the corresponding c_v values derived from consolidation test results. In addition, backcalculated c_v values from the north test embankment are relatively greater than those from the east test embankment.

North Test Embankment:

- o Results confirm that the silty sand alluvium layer provided drainage during the consolidation process. Insignificant excess pore pressures were measured in this layer during construction.

- o The nearly-saturated tailings and foundation materials consolidated faster than the fully-saturated tailings and foundation materials below the perched ground water table under the east test embankment.

East Test Embankment:

- o A preliminary backcalculated calculation indicated that the pore pressure dissipation rate would be very slow during and after placement of embankment load based on 1) the assumption of upward single drainage for tailings and clay (weathered Mancos Shale) and 2) backcalculated c_v values from the north test embankment. Comparing the analysis results to the observed embankment behavior, this model apparently does not represent the actual condition.

  Although the tailings directly overlie the clay (weathered Mancos Shale), it is believed that the presence of clayey sand alluvium near the western edge of the embankment may provide lateral drainage for the tailings. Thus, a drainage layer of silty sand alluvium was assumed to be located between the tailings and the weathered Mancos Shale in the backcalculated analytical model. Backcalculated c_v values were obtained until a

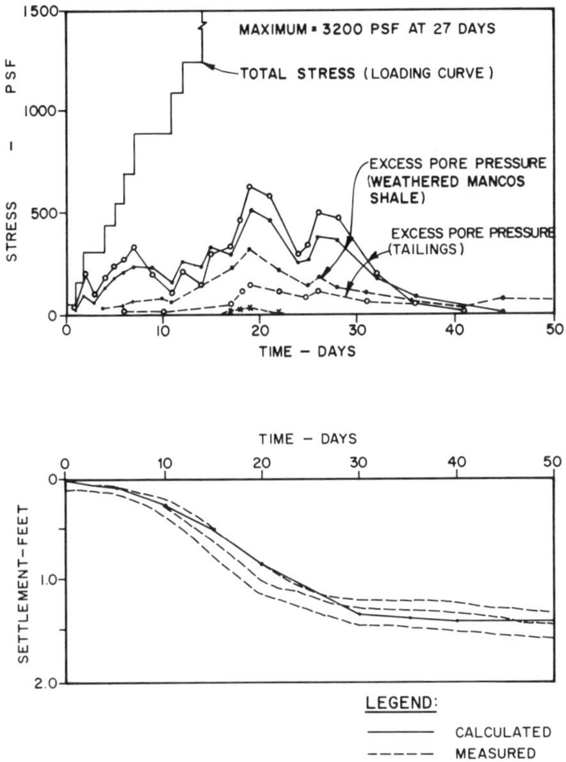

FIG. 8. COMPARISON OF CALCULATED AND MEASURED EXCESS PORE PRESSURES AND SETTLEMENTS-NORTH TEST EMBANKMENT

reasonable match of pore pressure fluctuations for tailings and the weathered Mancos Shale was achieved.

Compressibility and coefficient of consolidation were determined directly from test embankment data. Compressibility was determined from settlement data assuming laboratory test results for 1) tailings OCR and 2) relative compressibilities between material types. Coefficient of consolidation was calculated from plots of pore pressure versus depth for several points in time. Initial pore pressure distributions for each material were assumed to be distributions at the end of construction. Subsequent pore pressure dissipation was compared to the assumed initial distribution versus depth to calculate percent consolidation. Results of this direct determination are compared with preliminary and

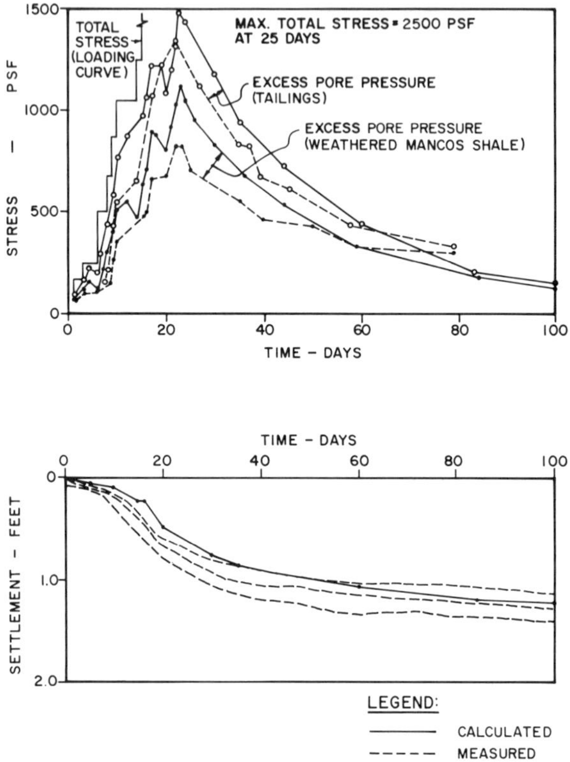

FIG. 9. COMPARISON OF CALCULATED AND MEASURED EXCESS PORE PRESSURES AND SETTLEMENTS-EAST TEST EMBANKMENT

backcalculated material properties in Table 4. The direct determination served as a verification of values obtained from backcalculated results by settlement calculation.

DESIGN PARAMETERS

Design parameters were based on backcalculated values from the test embankments for use in final settlement analysis of the disposal tailings embankment. East test embankment values were used at locations where perched ground water exists in the tailings; north test embankment values were used at the remaining locations.

TABLE 4. - COMPARISON OF KEY MATERIAL PROPERTIES FROM PRELIMINARY SETTLEMENT ANALYSIS AND TEST EMBANKMENT RESULTS

MATERIAL PROPERTY (1)	PRELIMINARY SETTLEMENT ANALYSIS (2)	BACKFIGURED VALUES FROM CALCULATION OF TEST EMBANKMENT SETTLEMENT		VALUES ESTIMATED DIRECTLY FROM TEST EMBANKMENT DATA[a]	
		NORTH EMBANKMENT (3)	EAST EMBANKMENT (4)	NORTH EMBANKMENT (5)	EAST EMBANKMENT (6)
TAILINGS					
C_c	0.535	0.460	0.400	0.503	0.460
C_r	0.056	0.056	0.056	0.121	0.109
C_v (FT^2/DAY)	0.58	4.7	2.1	3.1	4.6
ALLUVIUM					
C_r	0.010	b	b	0.009	---
C_{vs} (FT^2/DAY)	149	c	---	---	---
WEATHERED MANCOS SHALE:					
C_r	0.054	b	b	0.045	0.049
C_{vs} (FT^2/DAY)	0.03-0.73	16.5	2.55	5.3	1.7 (295)[d]

[a]ESTIMATES BASED ON THE FOLLOWING ASSUMPTIONS:
1. COMPRESSIBILITY OF TAILINGS, ALLUVIUM, AND WEATHERED MANCOS SHALE RELATIVE TO EACH OTHER WAS OBTAINED FROM LABORATORY RESULTS.
2. PORE PRESSURES AT END OF CONSTRUCTION ASSUMED AS INITIAL PORE PRESSURE DISTRIBUTION. SUBSEQUENT PORE PRESSURE DISTRIBUTIONS WERE COMPARED TO ASSUMED INITIAL DISTRIBUTION TO ESTIMATE PERCENT CONSOLIDATION.

[b]PRELIMINARY VALUE ASSUMED CONSTANT TO FACILITATE ANALYSIS.

[c]ASSUMED DRAINAGE LAYER. PORE PRESSURE BUILDUP NOT PERMITTED.

[d]ESTIMATE SHOWN IS FOR SINGLE DRAINAGE OF ENTIRE SECTION, FOR COMPARISON PURPOSES ONLY. BACKFIGURED VALUES AND VALUES ESTIMATED FROM TEST EMBANKMENT DATA ASSUME A DRAINAGE LAYER BETWEEN TAILINGS AND WEATHERED MANCOS SHALE.

CONCLUSIONS

The conclusions summarized below are based on comparison of properties from an initial investigation and from the interpretation of data from the two test embankments.

1. Total settlement of tailings, as affected by C_c, C_r and near-surface OCR, was accurately predicted based on laboratory test results. Interpretation of piezocone sounding data gave a near-surface OCR versus depth plot that showed significantly greater OCR than laboratory test results.

2. The settlement rate within the perched ground water zone was not accurately predicted using one-dimensional, single-drainage conditions as indicated by stratigraphy. It was necessary to assume an intermediate drainage layer for calculation results to agree with observations.

3. The settlement rates for nearly-saturated tailings and tailings in the limited perched ground-water zone are greater than rates predicted using the results of one-dimensional laboratory tests on saturated samples.

4. The silty sand alluvium layer acts as an effective drainage layer beneath the tailings.

5. In order for calculations to agree with observations, c_v for the weathered Mancos Shale had to be two orders of magnitude or more greater than c_v from lab results.

The test embankment program successfully demonstrated the viability of constructing the tailings disposal embankment in a single stage. In this case a test embankment, with a carefully designed and implemented instrumentation program, was an effective tool for geotechnical engineering design. Data and analysis demonstrated that settlement rate for a hydraulically-placed fill can be significantly faster than the rate predicted using laboratory test data as a result of heterogeneous placement of the material.

ACKNOWLEDGEMENTS

Permission of the Department of Energy to publish the contents of this paper is gratefully acknowledged. The writers also acknowledge with appreciation the contributions of Ned Larson of Jacobs Engineering, Inc., and Ed Tse, Reza Baradaran and Peter Chantalitanon of Morrison-Knudsen Engineers, Inc. Special thanks are due to G. R. Thiers, T. R. Wathen and E. S. Smith, Morrison-Knudsen Engineers, Inc., who reviewed the manuscript.

APPENDIX I. REFERENCES

1. Robertson, P.K., and Campanella, R.G., "Guidelines for use and interpretation of the Electronic cone Penetration test," Department of Civil Engineering, University of British Columbia, February, 1984.

2. Wong, K.S., and Duncan, J.M., "CONSOL - A computer program for 1-D consolidation Analysis of Layered Soil Masses, Microcomputer Version," Report "UCB/GT/84-06, Department of Civil Engineering, University of California, Berkeley, April, 1984.

3. Jamiolkowski, M., Ladd, C.C., Germaine, J.T., and Lancellotta, R., "New Developments in Field and Laboratory Testing of Soils," Proceedings, 11th International Conference on Soil Mechanics and Foundation Engineering., San Francisco, Vol. 1, p. 57, 1985.

4. Ladd, C.C., and Foott, R., "New Design Procedure for Stability of Soft Clays," Journal of Geotechnical Engineering, ASCE, Vol. 100, GT7, 1974, pp. 763-786.

5. Ladd, C.C., "Estimating Settlements of Structures Supported on Cohesive Soils", originally prepared for special summer program 1.345, "Soft Ground Construction", then revised for ASCE sponsored presentation, Massachusetts Institute of Technology, Cambridge, Massachusetts.

6. Soil Mechanics, Design Manual 7.1, NAFAC, DM-7.1, United States Department of the Navy, Alexandria, Virginia, May., 1982.

APPENDIX II. METRIC CONVERSION

1 ft	=	0.305 m
1 ft^2/day	=	0.093 m^2/day
1 ft^2/yr	=	0.093 m^2/yr
1 mile	=	1.61 km
1 pcf	=	0.157 kN/m^3
1 psf	=	0.04788 kPa
1 psi	=	6.89 kPa

EFFECT OF GRADATION AND FINES CONTENT
ON THE UNDRAINED RESPONSE OF SAND

R. Kuerbis[1], D. Negussey[2], and Y.P. Vaid[3], M.ASCE

Abstract

A systematic study of the effect of grain size, gradation and silt content on the monotonic and cyclic undrained loading response of sand is presented. The objective of the study is to gain an improved understanding of the behaviour of fluvial and hydraulic fill sands. A slurry method of deposition is developed to prepare homogeneous specimens of well-graded and silty sands, and to simulate hydraulic fill or fluvial deposition processes. Loose silty sands are shown to possess similar cyclic strengths regardless of silt content, even though they have widely different void ratios. The concept of sand skeleton void ratio is introduced in order to explain the observed behaviour of silty sands with varying fines content. Improved gradation tends to make clean loose sands more dilatant in compression and less contractive in extension.

Introduction

Hydraulic fill placement is a practical method of construction of earth and earth supported structures. The fill material used is invariably granular but may vary in average particle size, gradation and fines content. Deposition densities tend to be very loose rendering the material prone to liquefaction under static or cyclic loading. There are several documented case histories of catastrophic hydraulic fill failures. Loose fluvial sands also possess physical characteristics and liquefaction susceptibility similar to those for hydraulic fills.

In fundamental laboratory studies of undrained behaviour, water pluviation technique for sample reconstitution has been considered to simulate fluvial and hydraulic fill sand fabrics (8). This technique, however, results in segregation when used with silty or well graded sands. Consequently, laboratory studies of sand liquefaction have been confined mostly to clean uniform sands.

[1,2,3]Graduate Student, NSERC Fellow, and Professor, respectively, Dept. of Civil Engineering, University of British Columbia, Vancouver, Canada.

This paper presents a systematic study of undrained response of sand with the objective of gaining an improved understanding of the behaviour and performance of natural fluvial and hydraulic fill sands. Both, the effects of gradation and fines content on undrained behaviour are investigated. This was identified as a basic research need in the NRC Report (7). Selective sieving of a tailings sand was employed to obtain uniform and well graded sands. Silty sands were simulated by adding various percentages of a cohesionless silt to the processed sand. Use of identical parent material in various reconstituted sands eliminates possible influence of differing mineralogy among various sand gradations.

Fundamental studies of material behaviour require tests on homogeneous specimens. Consequently, a slurry method of deposition was developed that yields homogeneous specimens of well-graded as well as silty sands. This technique of placement simulates closely the deposition through water of hydraulic fills and fluvial sands. Homogeneity of laboratory test specimens is mandatory in order to determine element properties. It is not implied herein that hydraulic fills or fluvial sands may be homogeneous over large thicknesses.

The behaviour of each material is studied under undrained monotonic conditions. The behaviour of silty sands is studied in addition under cyclic loading. The test results are examined in an attempt to show a systematic variation in sand behaviour with change in gradation and silt content.

Material Tested

Brenda mine tailings sand was used as the parent sand. This is a coarse to fine grained angular sand derived from crushed granodiorite rock. The mineral composition of Brenda mine tailings sand is about 60% feldspar, 30% quartz. Of the remaining trace minerals, approximately half is mica.

The silt (Kamloops silt) is obtained from a glacial lake deposit. It is medium grained inorganic silt with a mineralogy similar to that of the tailings sand. The grain size distribution of the silt is shown in Fig. 1.

Oven dried tailings sand passing and retained on successive standard sieves was stockpiled in separate containers. Oven dry silt was pulverized and stored separately. Portions of the processed tailings sand were weighed and mixed in pre-set proportions to form uniform and well-graded sand samples. Various amounts of silt were added to the well-graded sand mix to form samples of different silt or fines content.

Gradations of processed clean sands tested are shown in Fig. 1. The designation 20/40 along the gradation curve, for example, implies sand passing #20 but retained on #40 sieve. Gradation curves for the silty sands are illustrated in Fig. 2. Some physical properties of the test materials are listed in Table 1.

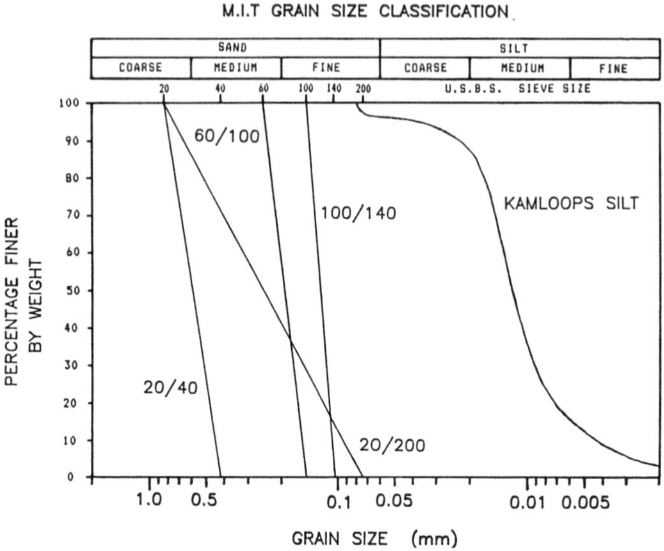

Fig. 1. Grain size curves for clean sands and silt tested.

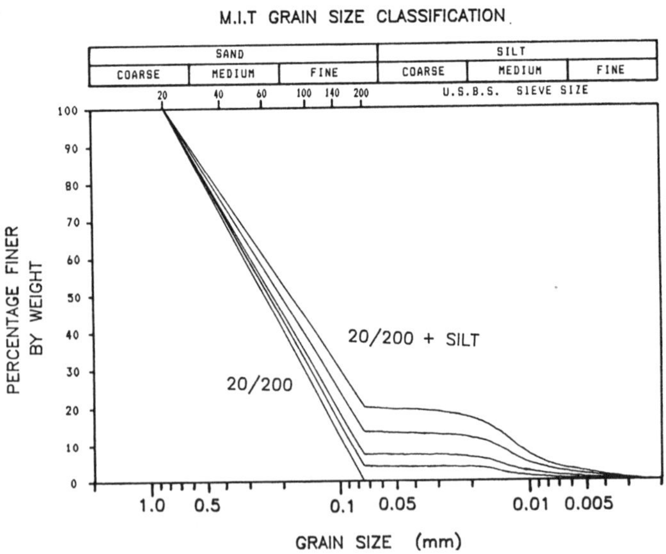

Fig. 2. Grain size curves for silty sands tested.

Table 1

		ASTM		Slurry		
		e_{max}	e_{min}	e_{max}	D_{50} mm	C_u
Sand	20/40	1.137	0.773	1.158	0.60	1.41
	60/100	1.185	0.790	1.132	0.19	1.29
	100/140	1.195	0.777	1.150	0.13	1.19
	20/200	0.864	0.518	0.890	0.25	3.44
Silt				1.447	0.012	3.09

The maximum void ratio of silty 20/200 sand deposited by slurry deposition depends upon silt content. It may be seen to decrease approximately linearly as silt content increases from zero to about 20% (Fig. 3). In contrast, maximum void ratio determined by the ASTM method is approximately equal to that of the clean sand and is essentially independent of silt content up to about 20%. Figure 3 also shows ASTM minimum void ratios as function of silt content.

Sample Preparation - Slurry Deposition Method

The details of the sample preparation method are described in (5). The technique consists essentially of mixing soil and deaired water to form a lean thoroughly mixed saturated slurry. Deposition of the soil from the slurry then occurs through the slurry water with a minimum of particle movement. This results in virtually no particle segregation even in silty sands, thus yielding very homogeneous specimens. The triaxial specimens tested were 6.3 cm dia. x 12.6 cm high.

The homogeneity of the slurry deposited triaxial samples was checked by comparing grain size distributions of horizontal slices. Horizontal slices were cut after the untested specimen had been solidified using the gelatin technique (2). Figure 4 shows that there is virtually no difference between grain size distribution of slices either for well-graded 20/200 clean or silty sand specimens, implying high degree of specimen homogeneity. A similar test on a water pluviated specimen of 20/200 sand clearly shows considerable segregation (Fig. 5.).

Test Results

Clean Sands

Monotonic undrained loading behaviour of uniform sand 60/100 is shown in Fig. 6 and that of well-graded sand 20/200 in Fig. 7. Both sands have approximately same mean grain size and thus differences in behaviour could be attributed solely to gradation differences. The behaviour of each sand is illustrated at three levels of isotropic consolidation stress on loosest specimens obtained by slurry deposition

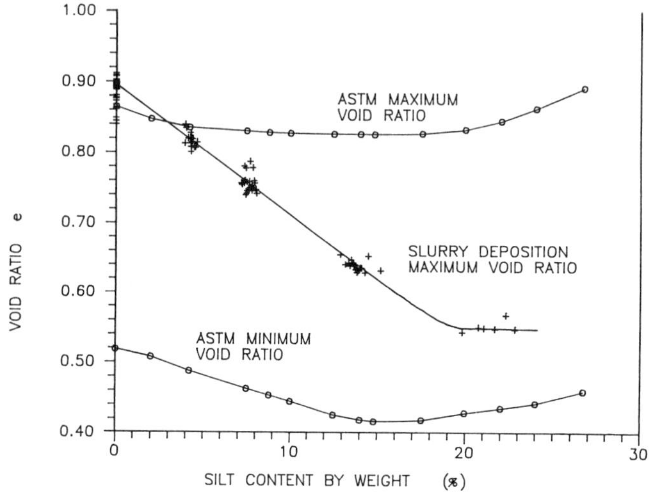

Fig. 3. Comparison of ASTM dry maximum and minimum void ratios and slurry deposition maximum void ratios of silty well-graded 20/200 sand.

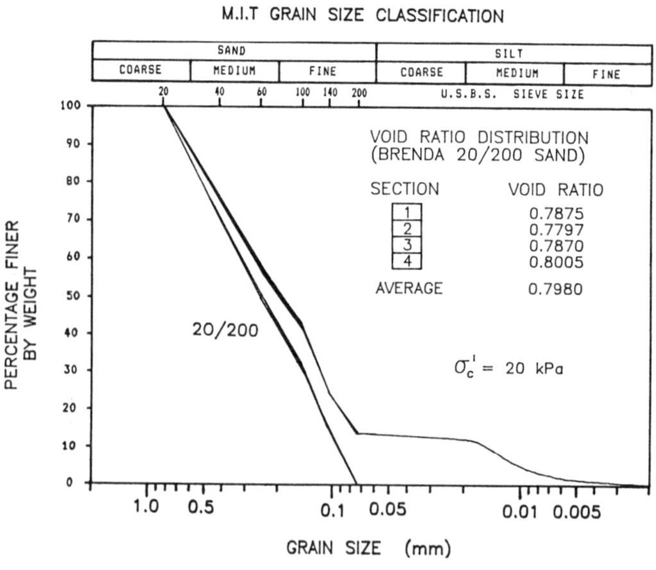

Fig. 4. Uniformity of slurry deposition method sand samples.

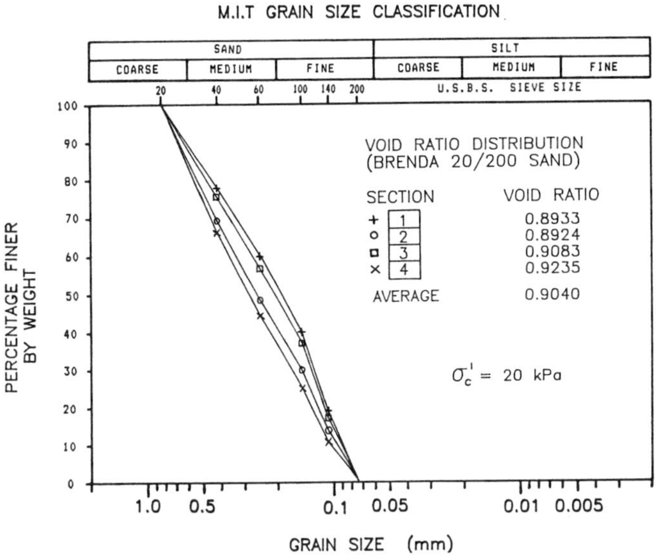

Fig. 5. Uniformity of a water pluviated sand sample.

In addition to the conventional compression loading, extension loading response is also shown.

Both sands show contrastingly different behaviour in compression and extension. At essentially identical initial stress and density states, compression behaviour is dilative as opposed to strongly contractive response in extension. These differences in behaviour are attributed to the presence of inherent anisotropy in pluviated sands (5). Such sand behaviour would have serious implications in the use of steady state concepts of design (1), wherein a unique steady state line obtained from compression tests is the key assumption.

Compression behaviour appears to be similar for both uniform and well-graded sand as is the extension behaviour. Uniform sand, however is more contractive in extension than the well-graded sand despite higher relative density at equal pressures. In terms of absolute density, however, uniform sand is looser and may account for its greater contractive tendency. As expected, increasing confining pressure induces increasing contractiveness or reduced dilative behaviour regardless of gradation.

The influence of average particle size D_{50} on monotonic behaviour is shown in Fig. 8. Considering behaviour of uniform sands only, decreasing grain size tends to make sand more dilative in compression and more contractive in extension. The relative position of the stress strain curves in each mode could partly be influenced by membrane penetration effects, which affect coarser sands more severely. Additional

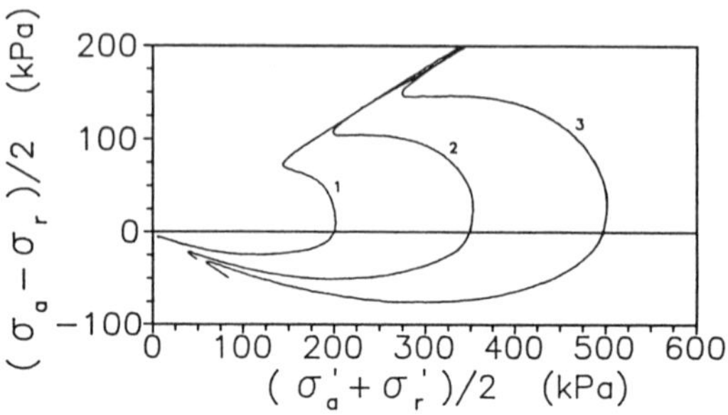

SAMPLE SET	σ'_{3c} (kPa)	COMPRESSION e_c	COMPRESSION Dr_c (%)	EXTENSION e_c	EXTENSION Dr_c (%)
1	200	1.027	40	1.025	40
2	350	1.044	35	1.056	33
3	500	1.038	37	1.033	38

$\phi_{Cpt} = 34^\circ$ $\phi_{Ept} = 34^\circ$

$\phi_{Cmax} = 36^\circ$ $\phi_{Emax} = 36^\circ$

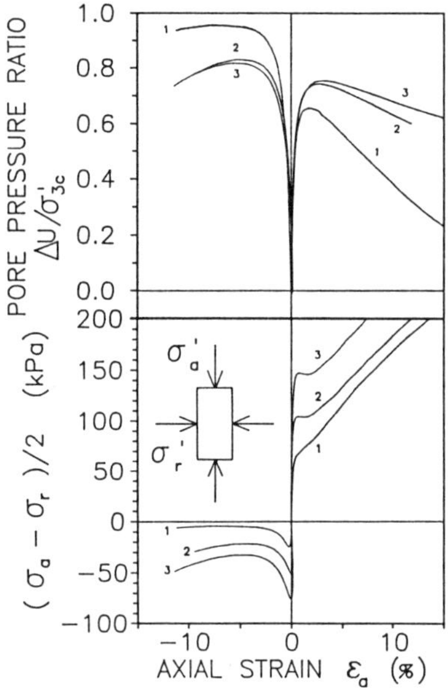

Fig. 6. Undrained monotonic triaxial test results for 60/100 sand.

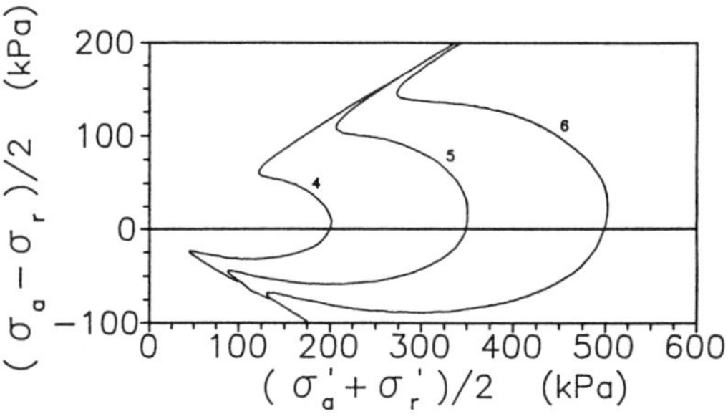

SAMPLE SET	σ'_{3c} (kPa)	COMPRESSION e_c	COMPRESSION Dr_c (%)	EXTENSION e_c	EXTENSION Dr_c (%)
4	200	0.767	28	0.768	28
5	350	0.763	29	0.765	29
6	500	0.770	27	0.771	27

$\phi_{Cpt} = 33.9°$ $\phi_{Ept} = 31.4°$

$\phi_{Cmax} = 36°$ $\phi_{Emax} = 33.6°$

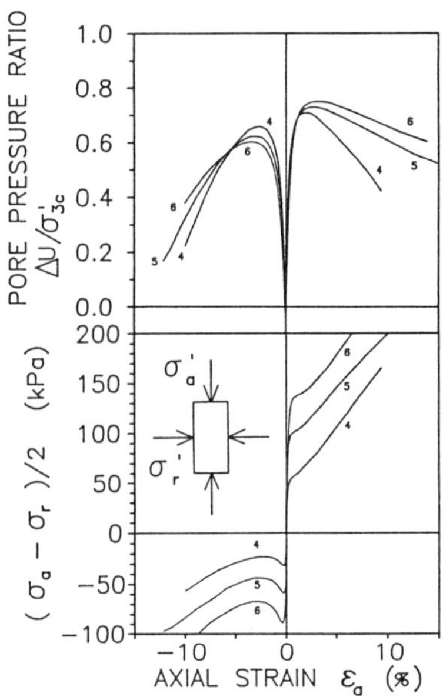

Fig. 7. Undrained monotonic triaxial test results for 20/200 sand.

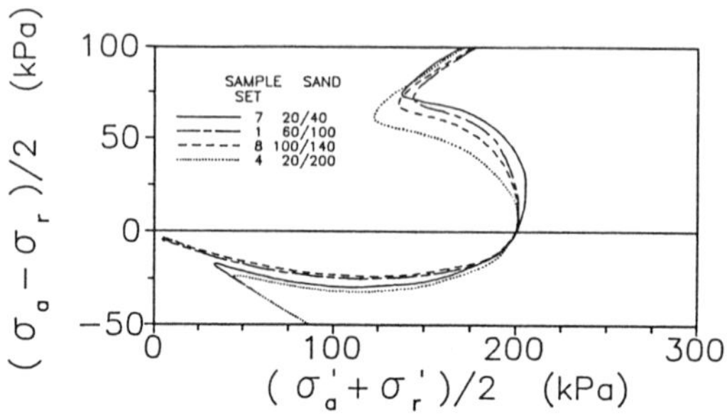

SAMPLE SET	SAND	COMPRESSION e_c	COMPRESSION Dr_c (%)	EXTENSION e_c	EXTENSION Dr_c (%)
7	20/40	1.009	35	1.011	35
1	60/100	1.027	40	1.025	40
8	100/140	1.016	43	1.006	45
4	20/200	0.767	28	0.768	28

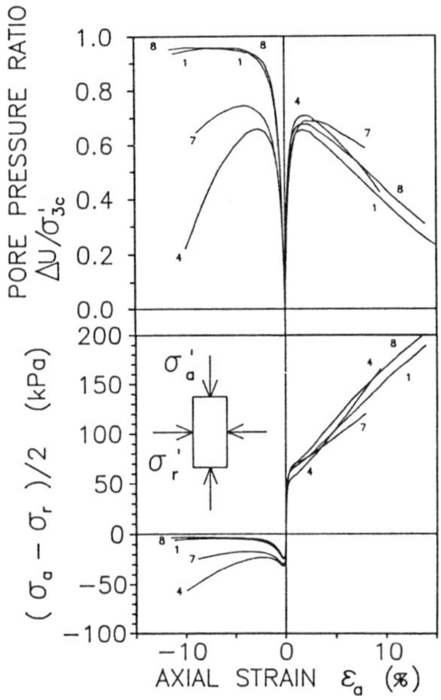

Fig. 8. Undrained monotonic triaxial test results for various sand gradations.

differences could be due to slight differences in relative densities of specimens.

Silty Sands

Monotonic Behaviour - Undrained stress-strain response and effective stress paths for silty sands are shown in Fig. 9 for specimens consolidated to 350 kPa. Test specimens were consolidated to the desired isotropic effective stress following loosest deposition by the slurry method. Since the loosest void ratio following slurry deposition decreases substantially with silt content (Fig. 3), the response shown in Fig. 9 represents very diverse void ratios or relative densities referred to the ASTM maximum/minimum dry state void ratios.

Figure 9 shows that the compression behaviour is dilative and that there is a definite trend towards increasing dilative behaviour with increase in silt content. In contrast, the behaviour in extension is contractive with some tendency towards reduced contractiveness with increasing silt content. This difference between compression and extension behaviour of silty sand is similar to that for clean sands discussed previously.

It has been suggested that relative density is not a suitable index for characterizing behaviour of silty sands (4). The results in Fig. 9, wherein silty sand at 98% relative density is seen to be contractive give further support to this contention. Void ratio instead of relative density has been used for comparing behaviour of silty materials (4).

The friction angle ϕ'_{PT} mobilized at phase transformation (3) is shown plotted against silt content in Fig. 10. It is interesting to note that this angle is essentially invariant and does not depend either on silt content or the mode of deformation. ϕ'_{PT} has been shown to be a unique property of a granular material (6). The results in Fig. 10 together with those in Fig. 9 showing minor changes in behaviour with increasing silt content, thus suggest that the silt in silty sands may simply be occupying void spaces in the sand skeleton and the measured behaviour is controlled essentially by the sand skeleton void ratio. This void ratio $e_{skeleton}$ can be calculated using the following expression

$$e_{skeleton} = [V_T G_s \rho_w - (M-M_{silt})] / (M-M_{silt}) \qquad (1)$$

in which V_T = total volume of the specimen, G_s = specific gravity of sand, ρ_w = density of water, M = total mass of the specimen and M_{silt} = mass of silt in the specimen. Sand skeleton void ratios (and relative densities) are shown in Fig. 9. It may readily be seen that various silt contents give rise to sand skeleton void ratios that are not too different, in contrast to the actual void ratio of the total soil. This may explain no radical differences in the behaviour of sand with various amount of silt.

Sand skeleton maximum void ratios of the as deposited silty sand by the slurry technique are illustrated in Fig. 11. An essentially constant value of skeleton void ratio that is independent of silt

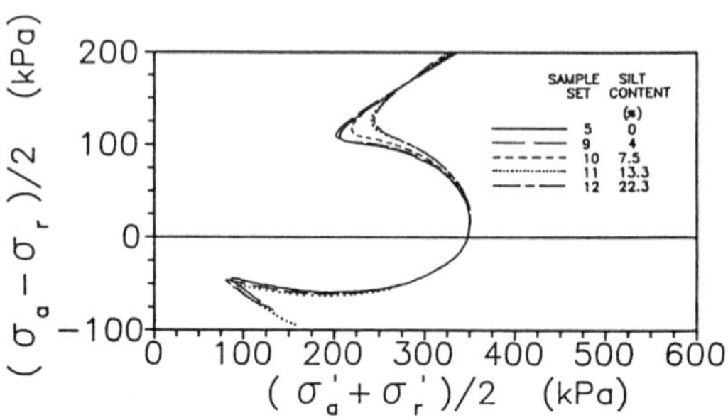

SAMPLE SET	SILT CONTENT (%)	TOTAL e_c	TOTAL Dr_c (%)	SAND SKELETON e_c	SAND SKELETON Dr_c (%)
5	0	0.764	29	0.764	29
9	4	0.728	31	0.802	18
10	7.5	0.669	44	0.805	17
11	13.3	0.547	69	0.784	23
12	22.3	0.446	98	0.863	0

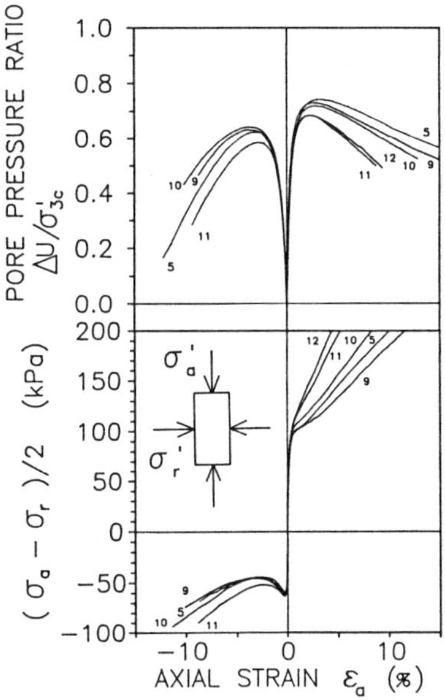

Fig. 9. Undrained monotonic triaxial test results for silty 20/200 sand.

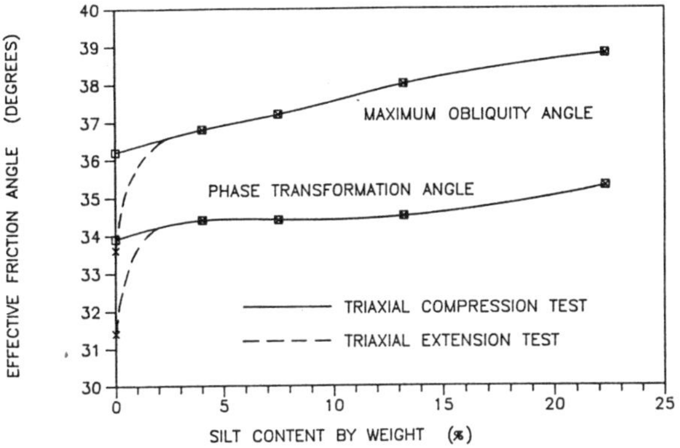

Fig. 10. Variation of silty 20/200 sand undrained friction angles with silt content.

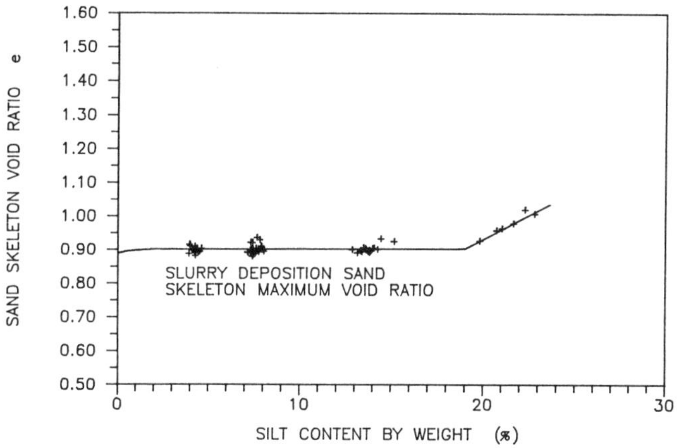

Fig. 11. Variation of maximum slurry deposition sand skeleton void ratio with silt content.

content up to about 20% provides definite evidence in support of silt occupying position within sand skeleton void space. This threshold silt content will of course vary with gradation and mean particle size of the sand.

Cyclic Loading Behaviour - Cyclic loading behaviour was studied at a fixed confining pressure of 350 kPa. The cyclic resistance curves showing cyclic stress ratio $\sigma_d/2\sigma'_{3c}$ to cause liquefaction ($2^{1/2}$% single amplitude strain in 20 stress cycles) versus void ratio e_c are shown in Fig. 12. It is seen that silty sands at loosest state have similar cyclic strength regardless of silt content, although there is a large variation in void ratio. The resistance curves are seen to essentially shift horizontally toward lower void ratios as silt content increases. At each silt content, there appears to be an approximately similar increase in cyclic resistance for a given decrease in void ratio. Over the range of overlapping void ratios of various silty sands, the cyclic resistance at a given void ratio decreases with increase in silt content. Such a comparison is not possible at all void ratios, since certain void ratios are not accessible to some silty sands.

An alternative comparison of cyclic resistance of silty sands as a function of ASTM as well as sand skeleton relative density is shown in Fig. 13. Sand skeleton relative density is calculated using $e_{skeleton}$ and e_{max} and e_{min}. The relative behaviour at different silt contents in terms of ASTM relative density of silty sand has the form similar to that in terms of void ratio in Fig. 12, i.e. higher the silt content, lower is the cyclic resistance for a given relative density. However, comparison in terms of sand skeleton relative density brings all resistance curves very close to each other. These curves show a slight increase in cyclic resistance with increase in silt content at a given value of sand skeleton void ratio. It would thus appear that the cyclic strength of a silty hydraulic fill sand could be determined by running tests on clean sand at identical sand skeleton void ratio of

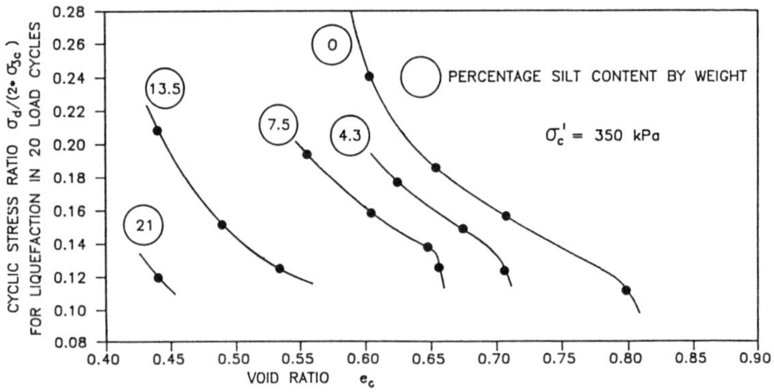

Fig. 12. Variation of silty 20/200 sand cyclic strength with silt content and void ratio.

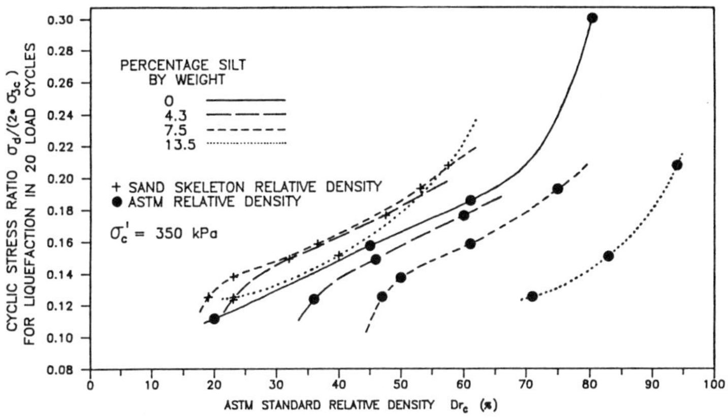

Fig. 13. Variation of silty 20/200 sand cyclic strength with relative density.

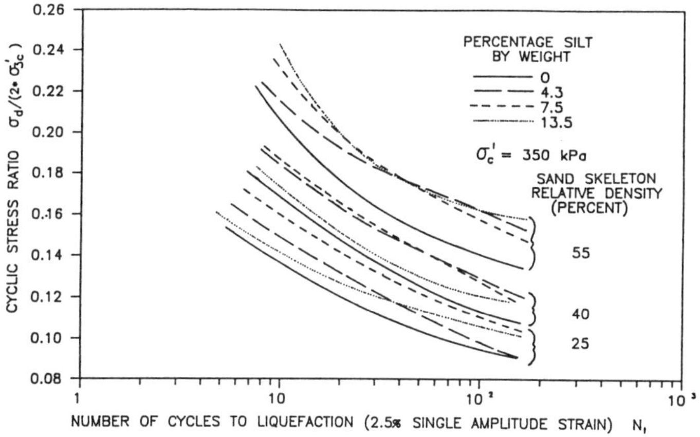

Fig. 14. Summary of silty 20/200 sand cyclic strength variation with silt content and sand skeleton relative density.

silty sand. This resistance estimate would tend to be on the conservative side.

Figure 14 illustrates a range of cyclic loading behaviour of silty sands at various fixed sand skeleton relative densities. It may be noted that at each sand skeleton density, increase in silt content leads to some increase in cyclic stress ratio to cause liquefaction in a fixed number of cycles.

Conclusions

From the results of 160 monotonic and cyclic undrained tests on well-graded, uniform and silty sands constituted of the same parent materials, the following conclusions can be drawn:

1. Water pluviated sand at a given density and confining stress may be dilative in triaxial compression yet contractive in extension. Well graded sand shows less of a difference between compression and extension mode of loading than poorly graded sand.

2. In uniform sands, decreasing grain size tends to make compression response more dilative, but extension response more contractive.

3. Increase in silt content up to 20% in the 20/200 sand tested makes compression behaviour somewhat more dilative and extension behaviour less contractive. The rather small change in behaviour with increasing silt can be explained by the fact that sand skeleton void ratio remains virtually unaltered with the addition of silt. Silt merely acts as a filler of sand skeleton voids and hence behaves as an inert component.

4. Silty 20/200 sands in the loosest slurry deposited state possess similar cyclic strength despite wide variation in void ratio. However, when compared at constant sand skeleton void ratio, cyclic strength improves slightly with increase in silt content.

Acknowledgements

This research was supported by a grant from the Natural Science and Engineering Research Council of Canada. Mrs. Kelly Lamb is thanked for her careful typing of the manuscript.

References

1. Castro, G., Poulos, S.J., France, J.W. and Enos, J.L. "Liquefaction Induced by Cyclic Loading," Report submitted to National Science Foundation, March, 1982.

2. Emery, J.J., Finn, W.D.L. and Lee, K.W. "Uniformity of Saturated Sand Samples," ASTM STP 523, 1973, pp. 182-194.

3. Ishihara, K., Tatsuoka, F. and Yasuda, S. "Undrained Deformation and Liquefection of Sand under Cyclic Stresses," Soils and Foundations, Vol. 15, No. 1, 1975, pp. 29-44.

4. Ishihara, K., Troncoso, J., Kawase, Y. and Takahashi, Y. "Cyclic Strength Characteristics of Tailings Materials," Soils and Foundations, Vol. 20, No. 4, Dec. 1980, pp. 127-142.

5. Kuerbis, R.H. "Effect of Gradation and Fines Content on the Undrained Response of Sand," thesis prestented to The University of British Columbia, 1987, in partial fulfillment of the requirements for the degree of Master of Applied Science.

6. Negussey, D., Wijewickreme, W.K.D. and Vaid, Y.P. "Constant Volume Friction Angle of Granular Materials," Canadian Geotechnical Journal, Vol. 25, No. 1, 1988.

7. N.R.C. "Liquefaction of Soils During Earthquakes," Report No. CETS-EE-001, Committee on Earthquake Engineering, National Academy Press, Washington, D.C., 1985.

8. Oda, M., Koishikawa, I. and Higuehi, T. "Experimental Study of Anisotropic Shear Strength of Sand by Plane Strain Tests, Soils and Foundations, Vol. 18, No. 1, 1978, pp. 25-38.

Pore Pressure Buildup and Liquefaction Failure of Anisotropically Consolidated Sand Due to Cyclic Straining

A. Vasquez–Herrera[1], R. Dobry[2], M. ASCE and T.–T. Ng[3]

Abstract

Hydraulic fill dams and mine tailings are very susceptible to liquefaction flow failure during strong seismic shaking. These earth structures are typically composed of very loose saturated silts and sands deposited without compaction by the hydraulic fill method. When subjected to earthquake ground shaking which is strong or long enough the soil might lose its strength, thereby creating large displacements or even failure of the earth structure.

The problem of seismic liquefaction flow failure of such loose sands is discussed in the paper assuming a fully undrained failure mechanism. A series of cyclic laboratory tests were performed to verify the influence of several parameters on the liquefaction resistance of a very loose, anisotropically consolidated, contractive silty sand. The experiments were of the strain–controlled cyclic torsional triaxial type ($C_yT - \overline{CAU}$), which provides in one test information on both: i) the initial stage of pore pressure buildup up to flow failure triggering, and ii) the subsequent stage of flow failure or steady state deformation after triggering. It was found that as the amplitude of cyclic shear strain, γ_{cy}, increases, the number of cycles to trigger failure, n_t, decreases for a given value of the anisotropic consolidation stress ratio, $K_c = \overline{\sigma}_{1c}/\overline{\sigma}_{3c}$; and that for a given value of γ_{cy}, n_t decreases with increasing K_c. The data also shows that for a given γ_{cy} and K_c, n_t increases as $\overline{\sigma}_{3c}$ increases. A simple pore pressure model is developed and fitted to these results, capable of predicting for this sand the excess pore pressure ($r_u = u/\overline{\sigma}_{3c}$) versus number of cycles, n, for any combination of K_c and γ_{cy}.

In all tests, flow failure triggering occurred at values of pore pressure ratio, r_u, significantly less than unity. From this, it is also concluded that it is possible for undrained liquefaction flow failure of a very loose sand to be triggered by an earthquake, before initial liquefaction ($r_u \simeq 1$) can occur in equivalent free field conditions.

Introduction

Saturated hydraulic fill embankments and tailing dams containing loose cohesionless soils have often experienced liquefaction flow failure due to earthquakes (Dobry and Alvarez, 1967; Seed et al., 1975; Seed, 1979; Marcuson et

[1] Research Engineer, INTEVEP, Los Teques, Venezuela.

[2] Professor of Civil Engineering, Rensselaer Polytechnic Institute, Troy, New York 12180.

[3] Graduate Student, Rensselaer Polytechnic Institute, Troy, New York 12180.

al., 1979; Lingyao et al., 1980; Troncoso, 1985). A number of methods for the evaluation of these seismically–induced flow failures have been proposed by different authors, involving the use of a variety of field testing techniques, static and cyclic laboratory tests, and empirical correlations based on available case histories (Seed et al., 1975; Seed, 1979; Lingyao et al., 1980; Finn, 1982; Poulos et al., 1984; Vasquez–Herrera et al., 1985; Dobry et al., 1985; Seed, 1986). Essentially all methods recognize that the development of this type of failure in the field involves two distinct stages: i) the stage of pore water pressure buildup and triggering of the failure, controlled by the intensity and duration of the **cyclic shear stresses and strains induced by the earthquake**, and ii) the flow failure itself, which is essentially a static slope failure controlled by the relation between the **static shear stresses** and the **steady–state or residual strength** of the soil after triggering. The differences between the various evaluation methods lie in the different relative emphasis they assign to the triggering and failure stages, their diverse assumptions affirming or denying the undrained character of the phenomenon, and the various experimental techniques they use to determine the triggering characteristics and the residual strength of the soil (National Research Council, 1985). A very important issue is the level of pore pressure buildup needed for triggering, and the applicability of cyclic laboratory test results representative of free field conditions, to a slope or embankment. Seed et al. (1975) and Seed (1986) suggested that a "pore–pressure ratio $\simeq$ 100%" is needed for flow failure triggering, similar to the free field, while the authors (Dobry et al., 1984, 1985; Vasquez–Herrera et al., 1985) have concluded that a pore pressure ratio significantly less than 100% is usually required, suggesting that predictions based in free field conditions can be unconservative.

An element of soil along a potential sliding surface within a hydraulic fill embankment dam is sketched in Figure 1, along with the stresses acting on the element. In general, the soil is in anisotropically consolidated state prior to the earthquake ($\bar{\sigma}_{1c} > \bar{\sigma}_{3c}$). The potential sliding surface is typically non– horizontal, whereas the seismic shear stresses and strains act predominantly on horizontal planes. Furthermore, the static shear stresses along the potential sliding surface remain more or less unchanged during the earthquake. This poses a major conceptual and technical difficulty which has prevented until now a better understanding of the problem: that is, how to simulate in a laboratory specimen a situation where the buildup of pore water pressures prior to flow failure is controlled by the cyclic (seismic) shear stresses and strains acting on a given plane, while the flow failure itself is driven by the static shear stresses acting on a different plane, and remaining essentially unaltered until flow failure starts.

During earthquake shaking, this representative soil element is subjected to an irregular sequence of stress/strain cycles of varying amplitude. Both seismic stresses and strains depend on the ground accelerations, the embankment geometry and the densities and stiffnesses of the soils involved. Therefore, earthquake loading is neither stress nor strain–controlled. Both cyclic stress and strain–controlled tests have been used to approximate the more complicated field situation in flow failure susceptibility evaluations. The authors prefer strain–controlled tests, as there is extensive experimental evidence showing that cyclic shear strain is a more fundamental parameter than cyclic stress in determining pore pressure buildup . As a result, the relation between cyclic strain and pore pressure has been shown to be very robust (Dobry et al., 1982; Dyvik et al., 1984; National Research Council, 1985; Vasquez–Herrera, 1988).

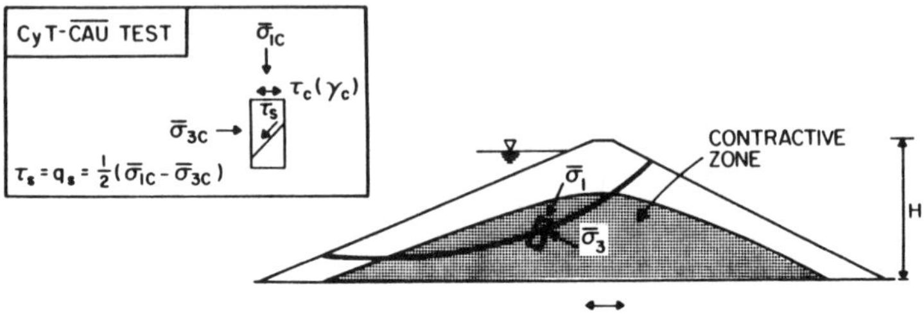

Figure 1. Possible Contractive Zone in Hydraulic Fill Dam and Simulation of Earthquake Loading by the CyT-$\overline{CAU}$ test.

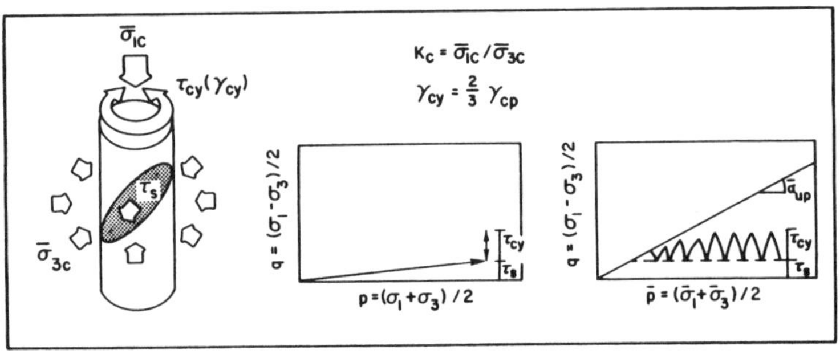

Figure 2. Stress Conditions in Strain Controlled CyT-$\overline{CAU}$ test.

The authors have proposed elsewhere a new strain–controlled laboratory technique to evaluate seismically induced flow failure in situ (Mohamad, 1984; Dobry et al., 1984; Vasquez–Herrera et al., 1985). This technique simulates both the triggering and flow failure stages of the process as part of one test. As illustrated in Figures 1 and 2, it consists of cyclic torsional strain–controlled undrained loading of anisotropically consolidated specimens ($C_yT - \overline{CAU}$ tests). In a typical $C_yT - \overline{CAU}$ experiment, a solid cylinder specimen, placed within a triaxial cell, is first anisotropically consolidated under principal stresses $\overline{\sigma}_{1c}$ and $\overline{\sigma}_c$, with $K_c = \overline{\sigma}_{1c}/\overline{\sigma}_{3c}$; these induce driving static shear stresses in the sample, represented by the maximum value $\tau_s = 0.5\ (\overline{\sigma}_{1c} - \overline{\sigma}_{3c})$ acting on 45° planes. Then, the drainage valves are closed, and the specimen is cyclically strained in torsion so as to produce a constant peripheral amplitude of cyclic shear strain equal to γ_{cp}. Although the cyclic shear strains are not uniform in a solid specimen, a study by Pierce (1985) showed that $\gamma_{cy} \simeq 2/3\ \gamma_{cp}$ can be taken as the representative equivalent cyclic shear strain, causing about the same amount of pore water pressure than if the cyclic strains had been uniform. Typical results of a $C_yT - \overline{CAU}$ test are shown in Figures 3 and 4.

During cycling, $q = \tau_s$ is constant at the end of each cycle, as sketched in Figure 2, and the corresponding effective stress path is the horizontal dashed line in the figure. Triggering occurs near the intersection between this dashed line and the "triggering envelope" of angle $\overline{\alpha}_{up}$ (point T in Figure 3). The number of cycles to triggering depends on the value of K_c, which determines the starting point of the stress path, on $\overline{\alpha}_{up}$, and on γ_{cy} which controls the rate of pore pressure buildup (see Figure 4). After triggering, the shear stress q decreases to the steady–state strength of the soil, S_{us} (point S in Figure 3). The value of S_{us} depends on the static consolidation stresses and the void ratio of the soil, but is completely independent on γ_{cy} and other cyclic loading parameters. Once triggering occurs, the necessary and sufficient condition for failure to happen is that $\tau_s > S_{us}$, or in other words, that the corresponding factor of safety $FS = S_{us}/\tau_s < 1$. Therefore, the test simulates in first approximation the seismic liquefaction flow failure of a representative soil element within the hydraulic fill structure, under the assumption of completely undrained loading during both the shaking and the subsequent failure (Mechanism "A", National Research Council, 1985).

The authors have presented elsewhere a systematic experimental study on the influence of $\overline{\sigma}_{3c}$, γ_{cy} and silt content on the pore pressure buildup, triggering and flow failure characteristics of a silty sand, for $K_c = 2$ (Vasquez–Herrera et al., 1985; Dobry et al., 1985). In the work reported herein, that study is extended to investigate the influence of K_c, and a model is proposed for evaluating pore pressure buildup and triggering for an arbitrary K_c.

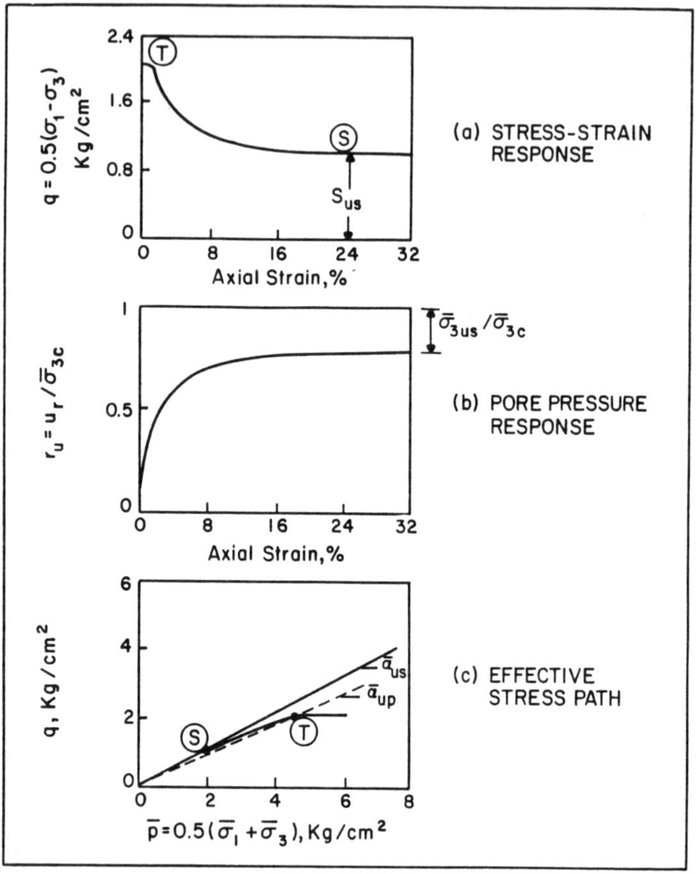

Figure 3. Cyclic Torsional CyT-$\overline{CAU}$ Test, $\bar{\sigma}_{3c}$ = 4.0 kg/cm^2
K_c = 2.0, γ_{cy} = 0.05%.

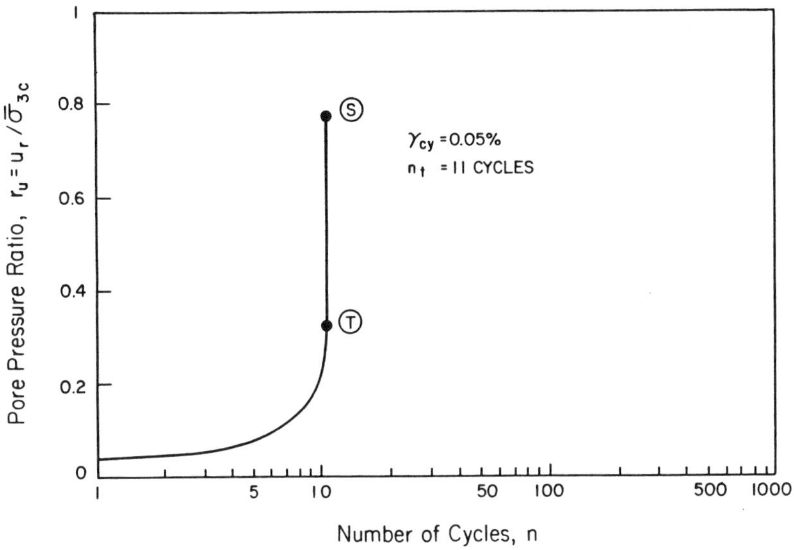

Figure 4. Pore Pressure Buildup Curve During the CyT-$\overline{CAU}$ test Shown in Figure 3.

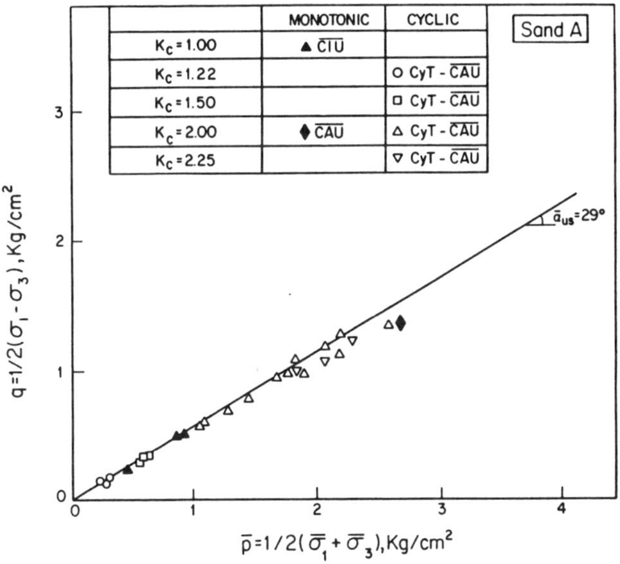

Figure 5. Steady State Strength Envelope from Monotonic and Cyclic Tests.

Experimental Investigation

A total of 21 $C_yT - \overline{CAU}$ tests were conducted on contractive specimens of a very fine and uniform silty sand from an alluvial deposit, with a non–plastic fines content of 13%, a uniformity coefficient $C_u = 2.8$ and a specific gravity $G_s = 2.64$. This is the same soil labelled "Sand A" in Vasquez–Herrera et al. (1985) and Dobry et al. (1985). The consolidation stresses and densities were chosen so as to ensure that the samples were fully contractive. The void ratios after consolidation ranged between 0.73 and 0.83, while $\overline{\sigma}_{3c}$ ranged from 2.7 kg/cm^2 to 4.8 kg/cm^2. In these $C_yT - \overline{CAU}$ tests, both γ_{cy} and K_c were varied, with $0.02 \leq \gamma_c \leq 0.15\%$, and with $K_c = 1.22$, 1.5, 2.0 and 2.25. In addition to these cyclic experiments, results are reported herein for monotonic $\overline{CIU}$ tests and one monotonic $\overline{CAU}$ test, on isotropically and anisotropically consolidated specimens, respectively.

The results of all these tests are summarized in Figs. 5 to 13, and are discussed in the following paragraphs.

In order for liquefaction flow failure to occur in a soil element tested undrained, two conditions must be satisfied. First, the soil must be contractive; that is, the minor principal effective consolidation stress, $\overline{\sigma}_{3c}$, must be greater than the undrained minor principal steady state stress, $\overline{\sigma}_{3us}$. Second, the driving shear stress $q = \tau_s$ must be greater than the undrained steady–state shear strength S_{us}. The in–situ driving shear stresses within the slope are those which would be used in a static equilibrium limit analysis, and thus are not necessarily equal to those created by the placement or consolidation of the soil (Castro et al., 1982; Castro, 1987). Specifically, for horizontal level ground in the free field the driving shear stresses are zero; therefore, liquefaction flow failure cannot occur. Both of the previously mentioned conditions can be illustrated using steady state lines, which represent the locus of those states for which steady state deformation is taking place (Castro, 1969, 1975).

The steady state strength line in $\overline{p} - \overline{q}$ space, is shown in Figure 5 for the silty sand used in this study. A unique steady state failure envelope of angle $\overline{\alpha}_{us} = 29^\circ$ is defined for both monotonic and cyclic tests. Furthermore, the variation in the K_c used in the cyclic experiments does not appear to influence $\overline{\alpha}_{us}$. The corresponding steady state lines in $e - \overline{\sigma}_3$ and $e - q$ spaces are presented in Figure 6. Here again it can be seen that the steady state line is unaffected by the type of test (monotonic or cyclic) or by the value of K_c. This reinforces the findings of Castro (1975) and Poulos (1981) that the steady state lines are unique and stress–path independent. Similar results have also been found by Dobry et al. (1984), Mohamad (1984) and Vasquez–Herrera et al. (1985).

If a soil element representative of the hydraulic fill mass is under pre–earthquake stress conditions that place it to the right of both $e - \overline{\sigma}_3$ and $e - S_{us}$ steady state lines, then the element is in unstable condition. However, flow failure will not be triggered unless the earthquake is strong and/or long enough, as indicated by Figures 3 and 4.

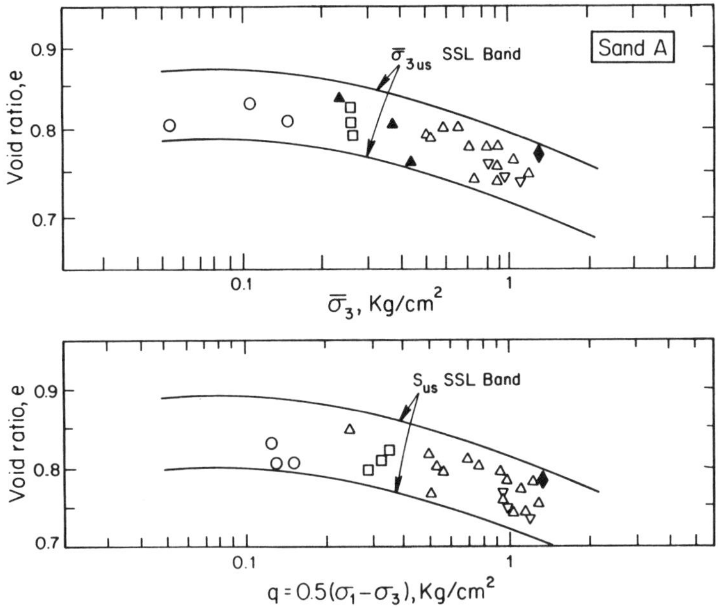

Figure 6. Steady State Lines from Monotonic and Cyclic Tests.

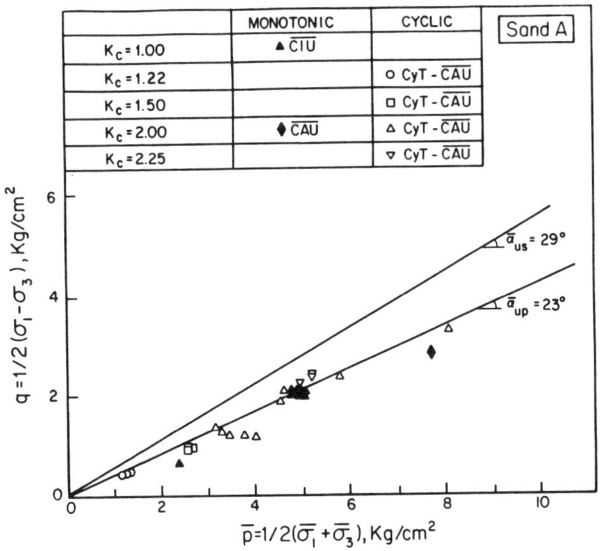

Figure 7. Triggering of Flow Failure Strength Envelope from Monotonic and Cyclic Tests.

Figure 7 summarizes the results obtained in all cyclic and monotonic tests, corresponding to the time of flow failure triggering. The graph shows that for this sand there is an approximately unique triggering strength envelope of angle $\bar{\alpha}_{UP} = 23^{\circ}$. It can be seen that K_c does not influence this $\bar{\alpha}_{UP} = 23^{\circ}$, which is also common between the cyclic $C_yT-\overline{CAU}$ and the monotonic $\overline{CIU}$ and $\overline{CAU}$ tests. For the particular stress path induced by the cyclic $C_yT - \overline{CAU}$ test, Mohamad (1984) and Dobry et al. (1985) have shown that the excess pore pressure necessary to trigger flow failure in the soil element, u_t, is given by

$$r_{ut} = \frac{u_t}{\bar{\sigma}_{3c}} \simeq \frac{1}{2}[K_c + 1 - (K_c - 1)/\tan \bar{\alpha}_{UP}] \qquad (1)$$

Both the above equation and Figure 7 clearly indicate that, for anisotropically consolidated contractive sands, triggering invariably occurs at values $r_u < 1$, and that $r_u \simeq 1$ can only occur if $K_c \simeq 1$. In particular, for $K_c = 2$ (a common value in the field) and $\bar{\alpha}_{UP} = 23^{\circ}$, $r_{ut} \simeq 0.32$. Therefore, the concept of "initial liquefaction" ($r_u \approx 1$) typically used in level ground liquefaction studies (Seed, 1979, National Research Council, 1985) can be unconservative when applied to embankments and slopes, since smaller values of pore pressure might be enough to trigger liquefaction flow failure.

Number of Straining Cycles to Trigger Failure

The most interesting data obtained in the experimental study relate to the pore pressure buildup and number of cycles during the first stage of the $C_yT - \overline{CAU}$ test. Unlike the S_{us} values and other characteristics of steady state failure, which can be determined from either cyclic or monotonic tests; pore pressure buildup and number of cycles to failure can be obtained only from cyclic experiments such as the $C_yT - \overline{CAU}$ test conducted by the authors.

The triggering relations between n_t and γ_{cy} obtained in the experiments for different K_c are presented in Figures 8–11. The trend for n_t to increase as γ_{cy} decreases is very clear; also, n_t becomes very large as γ_{cy} gets close to the threshold strain, which for this sand has been measured to be $\gamma_{cy} \simeq 10^{-2}\%$ (Vasquez–Herrera et al., 1985). The threshold strain is defined as that cyclic strain under which there is no sliding at the contacts between sand grains, and therefore there is no pore water pressure buildup, independently of the number of cycles. This existance of the threshold strain in Figures 8 – 11 is important as it establishes a lower bound for the cyclic shear strains that can cause pore pressure buildup (Dobry et al., 1981, 1982; Dyvik et al., 1984).

Further comparisons among Figures 8 through 11 show that the value of K_c plays an important role in the triggering relationship; for a given cyclic shear strain, γ_{cy}, the test with higher K_c fails in less cycles (smaller n_t). For example,

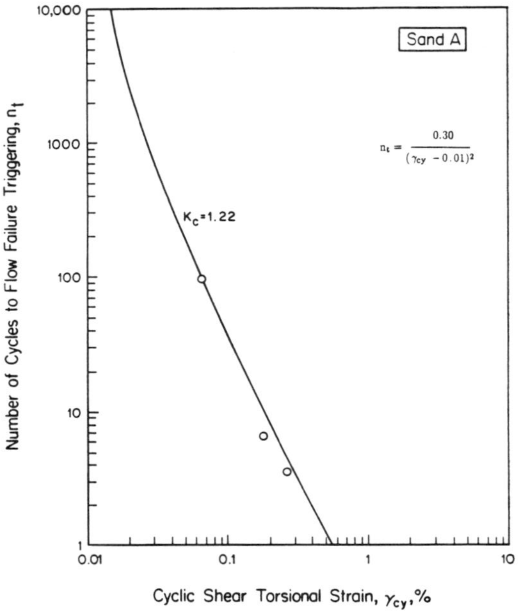

Figure 8. Number of Cycles of Shear Strain Needed to Trigger Liquefaction Flow Failure, $\bar{\sigma}_{3c}$=3.72 kg/cm^2, K_c=1.22, CyT-$\overline{CAU}$ Tests.

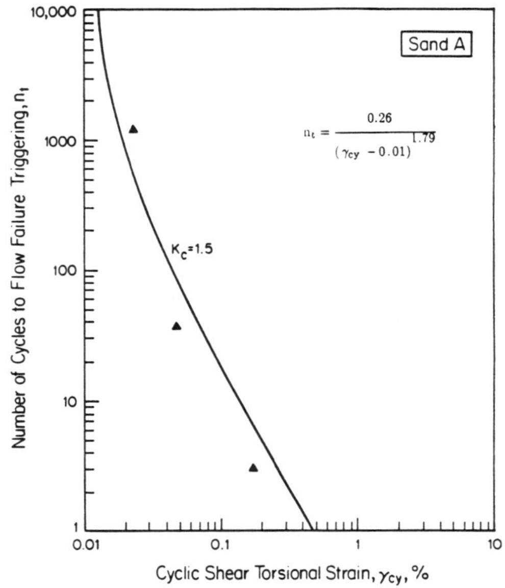

Figure 9. Number of Cycles of Shear Strain Needed to Trigger Liquefaction Flow Failure, $\bar{\sigma}_{3c}$=3.72 kg/cm^2, K_c=1.5, CyT-$\overline{CAU}$ Tests.

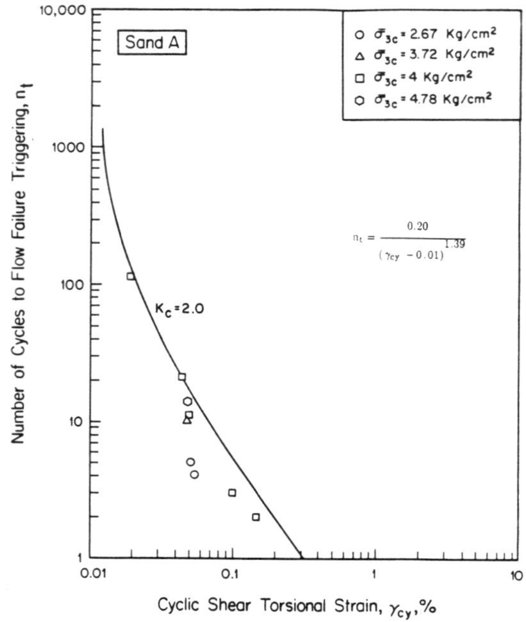

Figure 10. Number of Cycles of Shear Strain Needed to Trigger Liquefaction Flow Failure, CyT-$\overline{CAU}$ Tests.

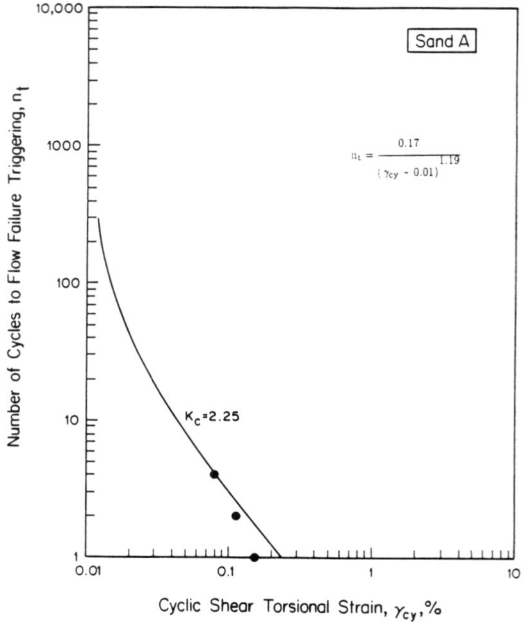

Figure 11. Number of Cycles of Shear Strain Needed to Trigger Liquefaction Flow Failure, $\bar{\sigma}_{3c}$=3.72 kg/cm^2, CyT-$\overline{CAU}$ Tests.

for $\gamma_{cy} = 0.2\%$ it takes about 8 cycles to reach failure if $K_c = 1.22$, whereas it is only $n_t = 2$ cycles if $K_c = 2$. Figure 12 depicts the triggering relationships for all values of K_c used for the tests; the lines in Figure 8–12 were fitted by multiple regression using the procedure described by Vasquez– Herrera (1988), and have the equation:.

$$n_t = \frac{r_{ut}}{\alpha(\gamma_{cy}-\gamma_t)^{\beta}(2-r_{ut})} \qquad (2)$$

where γ_t is the threshold strain = $10^{-2}\%$
$\alpha = 4.78 - 1.91\ K_c$
$\beta = 2.96 - 0.78\ K_c$
r_{ut} is given by Eq. (1)

The scatter in the values of n_t in Figure 10 for $K_c = 2$ is due mainly to the wide range of confining pressures used in that case. The line shown in Fig. 10 is for $\overline{\sigma}_{3c} = 3.72$ kg/cm², which is the consolidation pressure used on all tests with $K_c = 1.22$, 1.5 and 2.25, and in one of the tests with $K_c = 2.0$. The rest of the tests with $K_c = 2.0$ were done with $\overline{\sigma}_{3c}$ ranging from 2.7 to 4.8 kg/cm². As illustrated in Figure 13, the effect of confining pressure is to increase the number of cycles to failure as the magnitude of $\overline{\sigma}_{3c}$ increases.

Again, Figure 12 strongly suggests that the number of undrained cycles needed to trigger flow failure in a contractive sand, for a given intensity of cyclic loading (γ_{cy}), is significantly less if the slope is steeper (larger K_c). Furthermore, the figure also suggests that – under the assumption that the loading is undrained– the use of the number of cycles to initial liquefaction corresponding to free field condition ($K_c \simeq 1$) may be unconservative, and should not be used to predict flow failure in slopes or embankments involving contractive sands. A similar conclusion has been reached from tests on other sands using the same $C_yT\text{–}\overline{CAU}$ torsional technique (Vasquez–Herrera, 1988).

From a study of stress–controlled cyclic triaxial tests, using Ottawa sand, Vaid and Chern (1983) and Mohamad and Dobry (1986) also arrived to the conclusion that a high K_c can have a detrimental effect when **contractive** sand is used. However, in **dilative** Ottawa sand the effect of K_c was found to be always beneficial.

Pore Pressure Buildup During Cyclic Straining

In addition to the triggering relationships between γ_{cy} and n_t already discussed, another significant aspect of liquefaction triggering evaluation is the rate of pore pressure buildup. This is important because in practical engineering applications it may be necessary to predict the value of pore pressure for an element as it develops during cyclic loading, including the important case of variable γ_{cy} (irregular cyclic straining).

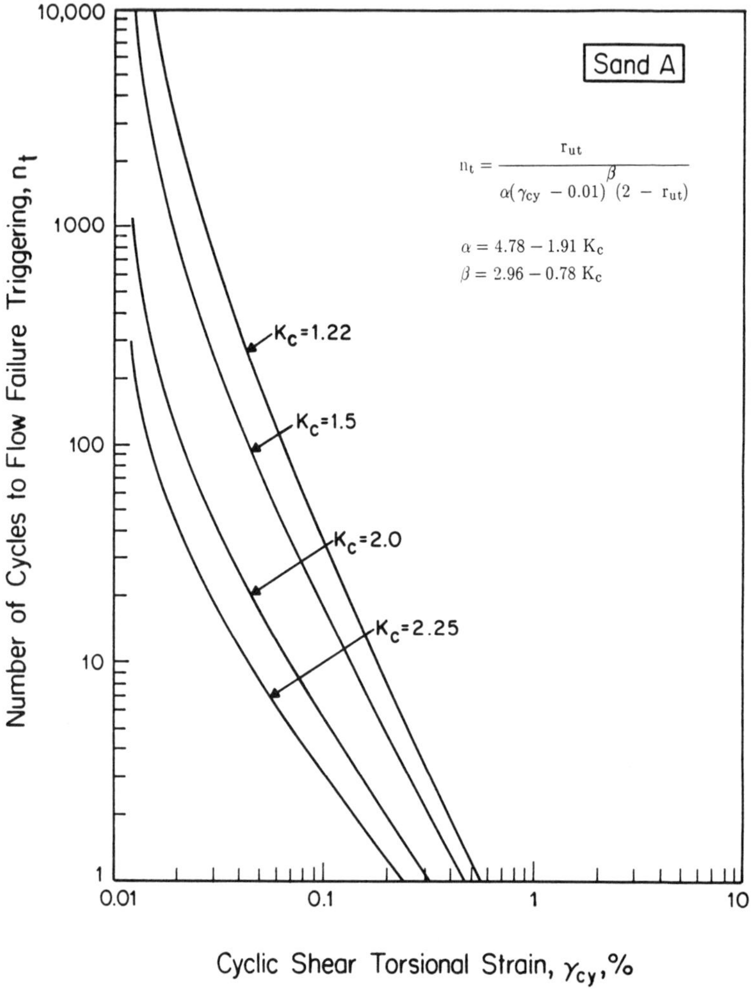

Figure 12. Triggering Relationships for $\bar{\sigma}_{3c}$ = 3.72 kg/cm^2 and for Various Values of K_c, CyT-$\overline{CAU}$ Tests.

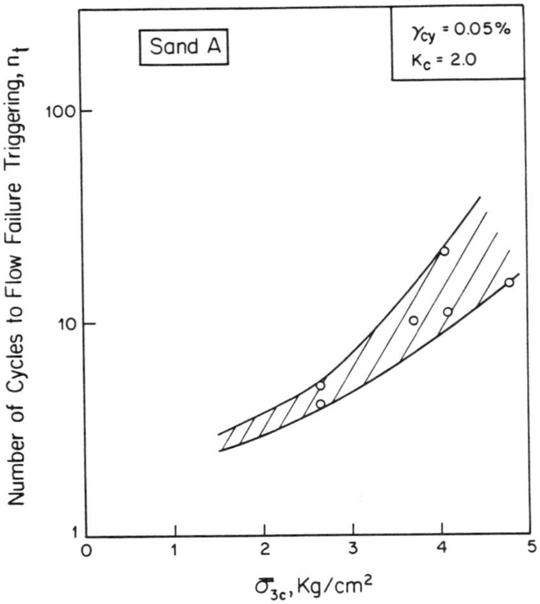

Figure 13. Influence of $\bar{\sigma}_{3c}$ on Number of Cycles to Trigger Flow Failure, CyT-$\overline{CAU}$ Tests.

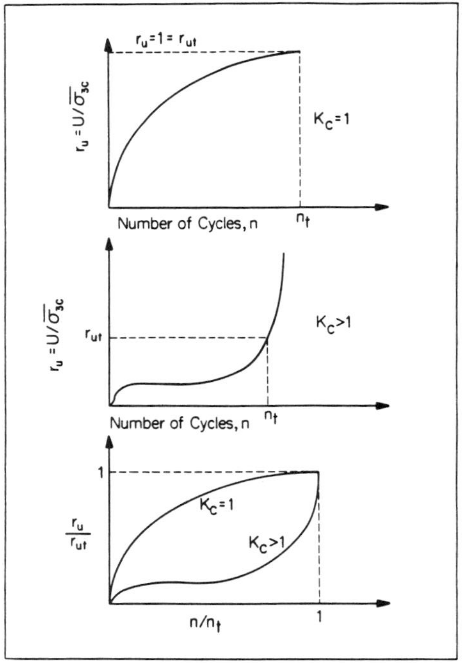

Figure 14. Pore Pressure Generation in Strain Controlled Cyclic Torsional Tests.

In strain–controlled cyclic torsional triaxial tests on isotropically consolidated ($K_c = 1$) contractive specimens, the generation of pore pressure versus number of cycles has a very specific shape, with the pore pressure ratio r_u plotting concave downward versus number of cycles n in arithmetic scale, as sketched in Figure 14(a). For the tests reported herein on Sand A, the authors found that this is no longer true if $K_c > 1$. This is illustrated in Figures 14(b), 15 and 16. At the beginning of the test the pore pressure curve is concave downwards, but there is always an inflection point beyond which the curve changes to concave upward until triggering. This means that K_c, in addition to its influence on the values of r_{ut} and n_t needed to trigger the failure, already discussed, also affects the rate of pore pressure buildup. In order to compare different tests for a specified K_c, the authors normalized all measured pore pressures by the pore pressure at triggering, that is, they established the ratio $u/u_t = r_u/r_{ut}$, and they also normalized the number of cycles n by establishing the ratio n/n_t. In this way, the curves of r_u versus n shown in Figures 14(a) and (b) become curves of r_u/r_{ut} versus n/n_t, and tests with different values of K_c and n_t can now be compared, as sketched in Figure 14(c). This normalization procedure worked very well for the data set reported herein, and for every given K_c it defined a unique line independent of γ_{cy} and of the specific values of r_{ut} and n_t. This is illustrated by Figure 15, which includes all $C_yT\text{–}\overline{CAU}$ results from the tests with $K_c = 2.0$. The uniqueness of the normalized line is very good indeed, especially when considering that the data points in the figure correspond to tests with cyclic strains ranging from 0.02% to 0.15%, numbers of cycles to failure from 2 to 121, and consolidation pressures from 2.7 kg/cm^2 to 4.8 kg/cm^2.

Unique curves, similar to that of Figure 15, were also obtained for the other values of K_c used in this research, and the corresponding lines are presented in Figure 16. This plot clearly shows the strong influence K_c has on the rate of pore pressure buildup. For a high K_c this rate is very slow at the beginning, with very little pore pressure increase until close to triggering; then a sharp inflection point occurs beyond which the rate becomes much faster until failure occurs. On the other hand, for a low K_c approaching the isotropic ($K_c = 1$) condition, the relation is close to a smooth concave downward curve with a mild inflection point near triggering.

On the basis of the triggering information summarized in Figure 12, plus the pore pressure curves of Figure 16, the authors were able to model the pore pressure buildup of the soil. The characteristics of the model and the procedure used to fit the model parameters to the experimental data are described by Vasquez–Herrera (1988). The corresponding expression for the lines of r_u/r_{ut} versus n/n_t, plotted in Figures 15–16, is:

$$\left(\frac{r_u}{r_{ut}}\right) = \frac{A\left(\frac{n}{n_t}\right)}{1+B\left(\frac{n}{n_t}\right)} + C\left(\frac{n}{n_t}\right)^8 \qquad (3)$$

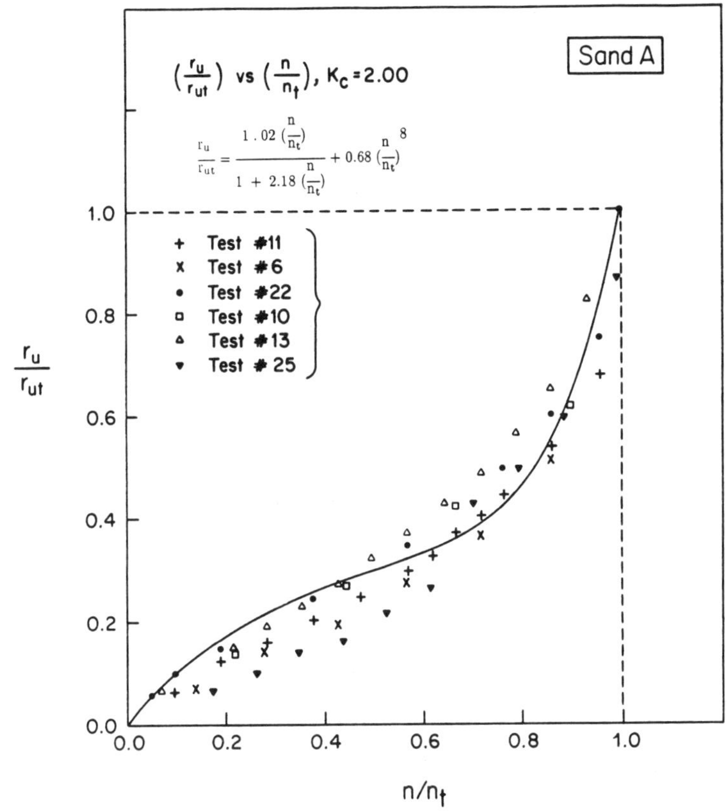

Figure 15. Normalized Pore Pressure Buildup Curve for CyT-$\overline{CAU}$ Tests, K_c = 2.0.

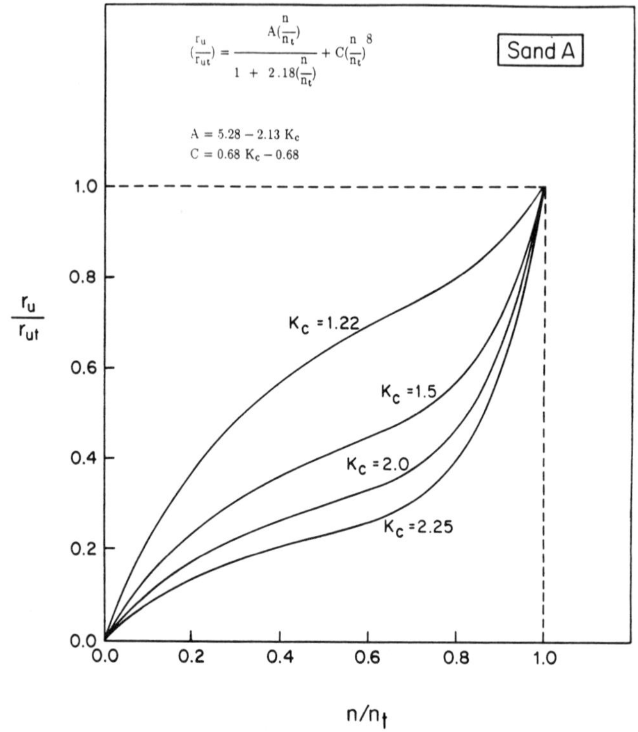

Figure 16. Normalized Pore Pressure Buildup Curves for Various Values of K_c, CyT-$\overline{\text{CAU}}$ Tests.

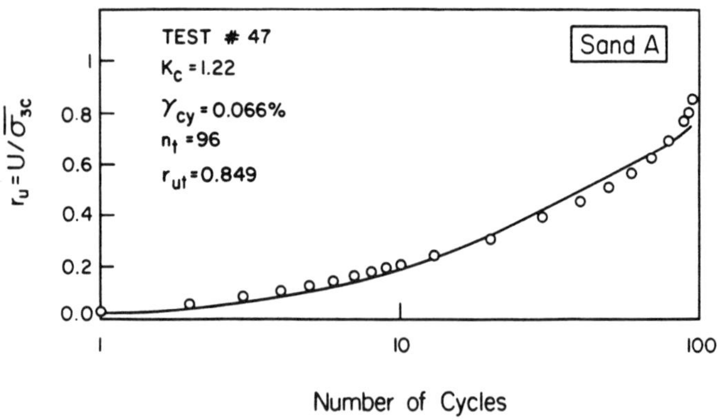

Figure 17. Comparison Between Predicted Versus Actual Values of Pore Pressure in a CyT-$\overline{\text{CAU}}$ Test with $\bar{\sigma}_{3c}$ = 3.72 kg/cm^2.

where $A = 5.28 - 2.13\ K_c$
$B = 2.18$
$C = 0.68\ K_c - 0.68$

with the normalizing parameters r_{ut} and n_t being obtained from Eqs. (1) and (2). The parameters A and C in Eq. (3) are functions of K_c, chosen so that for small K_c the first term of Eq. (3) dominates, whereas for larger K_c the second term becomes the dominant one.

A comparison between the pore pressure buildup measured in for one of the $C_yT - \overline{CAU}$ tests, and that predicted by combining Equations 1, 2 and 3, is shown in Figure 17, with good agreement.

Conclusions

The paper presents results of a number of undrained strain–controlled cyclic–torsional tests on anisotropically consolidated specimens of a silty sand ($C_yT - \overline{CAU}$ tests). The sand was tested loose and was in all cases contractive during cyclic straining. Selected monotonic $\overline{CIU}$ and $\overline{CAU}$ tests were also conducted. In the cyclic experiments, the degree of anisotropic consolidation $K_c = \bar{\sigma}_{1c}/\bar{\sigma}_{3c}$, the consolidation pressure $\bar{\sigma}_{3c}$, and the cyclic torsional strain γ_{cy}, were systematically varied between experiments to evaluate the influence of these parameters on the results. Data were obtained on: i) the rate of buildup of pore pressure ratio, $r_u = u/\bar{\sigma}_{3c}$, versus number of cycles of strain n, before flow failure triggering; ii) the conditions for triggering, including the values of $r_u = r_{ut}$ and $n = n_t$, as well as the effective stresses at triggering time; and iii) the effective stresses during flow failure after triggering, including the steady–state characteristics $\bar{\sigma}_{3us}$ and S_{us}.

Main conclusions extracted from the experimental results are:

1) The steady–state characteristics are the same for monotonically or cyclically induced flow failure, and are independent of K_c. For the sand tested, unique steady state lines could be defined, with a strength envelope $\bar{\alpha}_{us} = 29^\circ$.

2) In first approximation, the effective stress conditions at time of triggering are the same for monotonically or cyclically induced flow failure. For this sand and the tests performed, a unique triggering envelope could be defined, with $\bar{\alpha}_{up} = 23^\circ$.

3) For a given K_c and $\bar{\sigma}_{3c}$, n_t decreases as γ_{cy} increases.

4) For a given γ_{cy}, n_t decreases as K_c increases, indicating that, other things being equal, a steeper slope of contractive sand cycled undrained could fail in fewer cycles than a flatter slope. These results also suggest that the use of results corresponding to free field conditions ($K_c \simeq 1$) may be unconservative if applied to a slope or embankment.

5) For a given value of K_c and γ_{cy}, n_t increases as $\overline{\sigma}_{3c}$ increases.

6) The shape of the curve of pore pressure buildup versus number of cycles before triggering depends strongly on K_c, and thus on the steepness of the slope. For low K_c, the rate of pore pressure buildup is maximum at the beginning of cyclic straining, while the opposite is true for large K_c.

In addition, a model was successfully fitted to the test results, which allows to predict pore pressure buildup and triggering in the sand tested for any arbitrary combination of K_c and γ_{cy}. This model can be readily adapted to the case of irregular cyclic straining such as occurring during earthquakes. The main trends and results presented herein have also been shown to occur in other contractive sands. Therefore, the conclusions of this paper seem to provide a realistic representation of the behavior of anisotropically consolidated, contractive sand subjected to undrained seismic loading.

Acknowledgements

Panos Dakoulas, Ramli Mohamad and Mladen Vucetic provided useful help and advice on the laboratory tests reported in this paper. The work was supported by a grant from the Oil Research and Development Institute of Venezuela (INTEVEP). All this help and support is gratefully acknowledged.

References

Casagrande, A. (1936). "Characteristics of Cohesionless Soils Affecting the Stability of Slopes and Earth Fills," Journal of the Boston Society of Civil Engineering, reprinted in Contributions to Soil Mechanics, 1925 to 1940, Boston Society of Civil Engineers, October 1940, 257–276.

Castro, G. (1969). "Liquefaction of Sands," Ph.D. Thesis, Harvard University.

Castro, G. (1975). "Liquefaction and Cyclic Mobility of Saturated Sands," Journal of the Geotechnical Engineering Division, ASCE, **101**, GT6, 551–569.

Castro, G., Poulos, S.J., France, J.W. and Enos, J.L. (1982). "Liquefaction Induced by Cyclic Loading," Report to National Science Foundation, Geotechnical Engineers, Inc., Winchester, Massachusetts.

Castro, G. (1987). "On the Behavior of Soils During Earthquakes – Liquefaction," Proceedings 3rd International Conference on Soil Dynamics and Earthquake Engineering, Princeton, 1987.

Dobry, R. and Alvarez, L. (1967). "Seismic Failures of Chilean Tailings Dams," Journal of the Soil Mechanics and Foundations Division, Journal of the Soil Mechanics and Foundations Division, ASCE, **93**, SM6, 237–260.

Dobry, R., Stokoe, K.H. II, Ladd, R.S. and Youd, T.L. (1981). "Liquefaction Susceptibility from S–Wave Velocity," Proceedings Session on In Situ Testing to Evaluate Liquefaction Susceptibility, ASCE National Convention, St. Louis, Missouri.

Dobry, R., Ladd, R.S., Yokel, F.Y., Chung, R.M. and Powell, D. (1982). "Prediction of Pore Water Pressure Buildup and Liquefaction of Sands During Earthquakes by the Cyclic Strain Method," National Bureau of Standards, Building Science Series 138, U.S. Dept. of Commerce, July.

Dobry, R., Mohamad, R., Dakoulas, P. and Gazetas, G. (1984). "Liquefaction Evaluation of Earth Dams – A New Approach," Proceedings 8th World Conference on Earthquake Engineering, San Francisco, July.

Dobry, R., Vasquez–Herrera, A., Mohamad, R. and Vucetic M. (1985). "Liquefaction Flow Failure of Silty Sand by Torsional Cyclic Tests," in Advances in the Art of Testing Soils under Cyclic Loading, ASCE National Convention, Detroit, October.

Dyvik, R., Dobry, R., Thomas, G.E. and Pierce, W.G. (1984). "Influence of Consolidation Shear Stresses and Relative Density on Threshold Strain and Pore Pressure During Cyclic Straining of Saturated Sands," Miscellaneous Paper GL–84–15, Department of the Army, U.S. Army Corps of Engineers, Washington, D.C.

Finn, W.D.L. (1982). "Fundamental Aspects of Response of Tailings Dams to Earthquakes," Proceeding Session on Dynamic Stability of Tailings Dams," ASCE National Convention, New Orleans, 46–72, October.

Lingyao, L., Kuiefen, L. and Dongpin, B. (1980). "Earthquake Damage of Baihe Earth Dam and Liquefaction Characteristics of Sand and Gravel Materials," Proceedings 7th World Conference on Earthquake Engineering, Istanbul, Turkey, 171–178, September.

Marcuson, W.F. III, Ballard, R.F. Jr. and Ledbetter, R.H. (1979). "Liquefaction Failure of Tailings Dams Resulting from the Near Izu Oshima Earthquake, 14 and 15 January 1978," Proceedings 6th Pan American Conference on Soil Mechanics and Foundation Engineering, Lima, Peru, **2**, 69–80.

Mohamad, R. (1984). "Evaluation of Seismically–Induced Liquefaction Flow Failure of Earth Dams," Ph.D. Thesis, Rensselaer Polytechnic Institute, Troy, New York.

Mohamad R. and Dobry, R. (1986). "Undrained Monotonic and Cyclic Triaxial Strength of Sand," Journal of the Geotechnical Engineering Division, ASCE **112(2)**, 941–958, October.

National Research Council (1985). "Liquefaction of Soils During Earthquakes," National Academy Press, Washington, D.C.

Pierce, W.G. (1985). "Constitutive Relations for Saturated Sand Under Undrained Cyclic Loading," Ph.D. Thesis, Rensselaer Polytechnic Institute, Troy, N.Y.

Poulos, S.J. (1981). "The Steady State of Deformation," Journal of the Geotechnical Engineering Division, ASCE, **107**, No. GT5, 553–562, May.

Poulos, S.J., Castro, G. and France, J.W. (1984). "Procedure for Liquefaction Evaluation," Journal of the Geotechnical Engineering Division, ASCE, **111**(6), 772–792, June.

Seed, H.B. (1979). "Considerations in the Earthquake–Resistant Design of Earth Dams," Geotechnique, **29(3)**, 215–263.

Seed, H.B. (1986). "Design Problems in Soil Liquefaction," Journal of Geotechnical Engineering, ASCE, **113(8)**, 827–845, August.

Seed, H.B., Lee, K.L. and Idriss, I.M. (1969). "Analysis of Sheffield Dam Failure," Journal of the Soil Mechanics and Foundation Division, ASCE **93**, No. SM4, 1453–1490, July.

Seed, H.B., Lee, K.L., Idriss, I.M. and Makdisi, F.I. (1975). "The Slides in the San Fernando Dams During the Earthquake of February 9, 1971," Journal of the Geotechnical Engineering Division, ASCE, **101**, No. GT7, 651–688, July.

Troncoso, J.H. (1985). "Critical State of Tailing Silty Sands for Earthquake Loading," Proceedings II International Conference Soil Dynamics and Earthquake Engineering, QE II.

Vaid, Y.P. and Chern, J.C. (1983). "Mechanism of Deformation During Cyclic Undrained Loading of Saturated Sands," Soil Dynamics and Earthquake Engineering, **2(3)**, 171–177.

Vasquez–Herrera, A. (1988). "The Behavior of Undrained Contractive Sands and its Effects on Seismic Liquefaction Flow Failure of Earth Structures," Ph.D. Thesis, Rensselaer Polytechnic Institute, Troy, N.Y.

Vasquez–Herrera, A., Mohamad, R., Vucetic, M. and Dobry, R. (1985). "Liquefaction Flow Failure Characteristics of Some Silty Sands," Proceedings II International Conference on Soil Dynamics and Earthquake Engineering, QEII.

PREDICTION OF STRAINS IN EARTHEN COVERS

Ned B. Larson[1], Member ASCE
Berg Keshian, Jr.[2], Member ASCE

ABSTRACT

Cleanup and isolation of hydraulically placed uranium mill tailings and other contaminated materials at 24 uranium mill tailings sites are to be accomplished by the U.S. Department of Energy (DOE) Uranium Mill Tailings Remedial Action (UMTRA) Project through a series of remedial action activities that will culminate in the licensing of a remediated site for surveillance and maintenance by the U.S. Nuclear Regulatory Commission (NRC).

Generally, two methods are used to dispose of the tailings and contaminated materials: (1) excavation, transportation to another site, compaction, and placement of a long-term earthen cover system; and (2) placement of off-pile contaminated materials over the existing tailings pile and covering the entire pile with an earthen cover system. The second method is more difficult since it involves the characterization and prediction of how the undisturbed, hydraulically placed tailings will behave when subjected to the changes in stress condition as a result of reshaping and covering.

This paper discusses the Falls City, Texas, site and the methods used to characterize the tailings as well as the prediction techniques used to predict strain and possible cover cracking as a result of differential settlement in order to predict their performance for final design.

---

1 Deputy Manager of Engineering, UMTRA Project Jacobs Engineering Group Inc., Albuquerque, New Mexico

2 Senior Project Manager, Roy F. Weston, Inc., Albuquerque, New Mexico

INTRODUCTION

The Falls City, Texas, tailings pile is approximately 40 miles south of San Antonio, Texas. The uranium mill was constructed in the early 1960s, began operations in 1961, and continued operating until the shutdown in 1973 due to economic problems. Figure 1 shows the vicinity map for the site. Figure 2 shows the location of the seven existing tailings piles that will be consolidated into one pile at the site. Tailings piles 1, 2, and 7 were placed on the ground surface while piles 3, 4, 5, and pond 6 were placed in old open pits from which the ore was removed.

Due to groundwater considerations, it was determined that the tailings should be removed from pits 3, 4, 5, and 6 and placed on and around piles 1, 2, and 7. The final shape of the pile is shown on Figure 3 and a cross section of the final pile is shown on Figure 4. As can be seen on Figure 4, the remolded tailings would then be placed over approximately 40 feet of undisturbed tailings in some areas, and placed directly on the foundation materials and adjacent to the tailings in others. This raises concern regarding the differential settlement and ultimately strain in the cover caused by the differences in stiffness of the hydraulically placed tailings and the engineered compacted tailings.

Significant quantities of data were necessary to adequately calculate the settlements across the final reclaimed pile. The piezocone was chosen as the prime characterization tool and was supplemented with borehole and laboratory data in order to evaluate the tailings in their in-situ state while still maintaining reasonable exploration and laboratory testing costs. A grid of 200-foot centers was laid out and a piezocone sounding was performed at each grid point. Twelve boreholes were placed within three feet of the respective piezocone sounding locations so that correlations between the piezocone soundings and laboratory tests could be made. Not only would conventional drilling have been approximately eight to 10 times more expensive than the piezocone, but it would not have produced the continuous profile that the piezocone provided.

CHARACTERIZATION

The results of the piezocone correlations are discussed in detail by Larson and Mitchell (1986) and are briefly summarized here. The parameters that were correlated were the material classification, undrained shear strength, over consolidation ratio, blow count, and friction angle. Parameters that were determined from literature and not specifically correlated were modulus of elasticity, relative density, and liquefaction susceptibi-

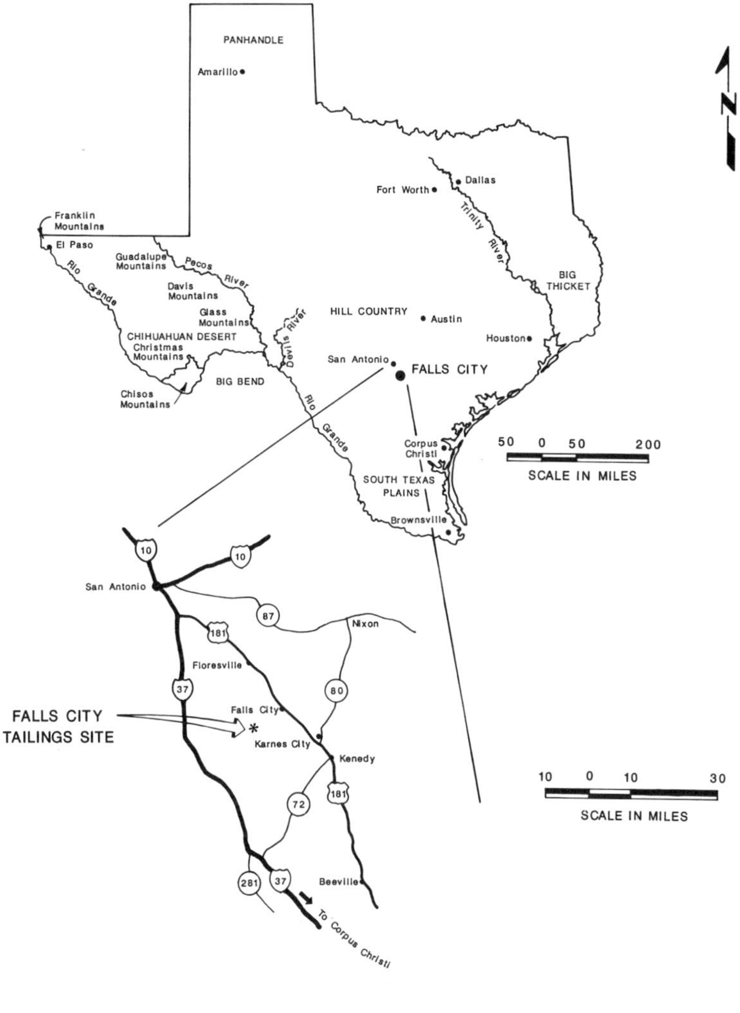

FIGURE 1
LOCATION OF THE FALLS CITY TAILINGS SITE
NEAR FALLS CITY, TEXAS

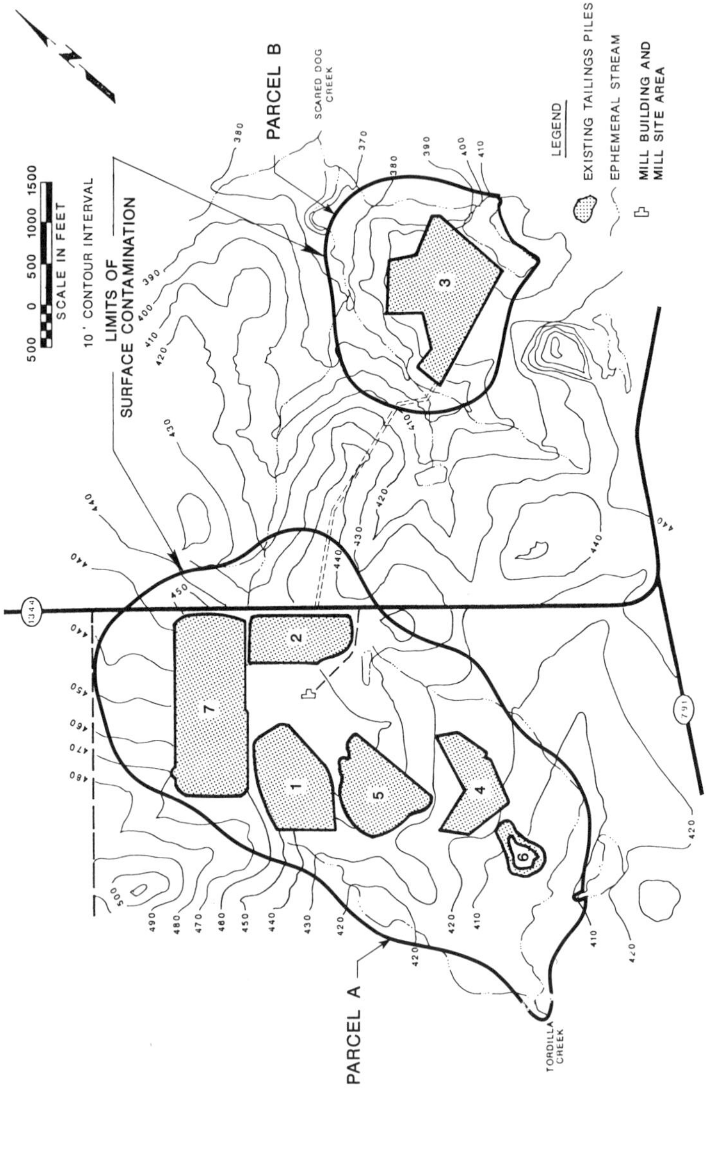

FIGURE 2
PRESENT CONDITION OF THE FALLS CITY TAILINGS SITE
NEAR FALLS CITY, TEXAS

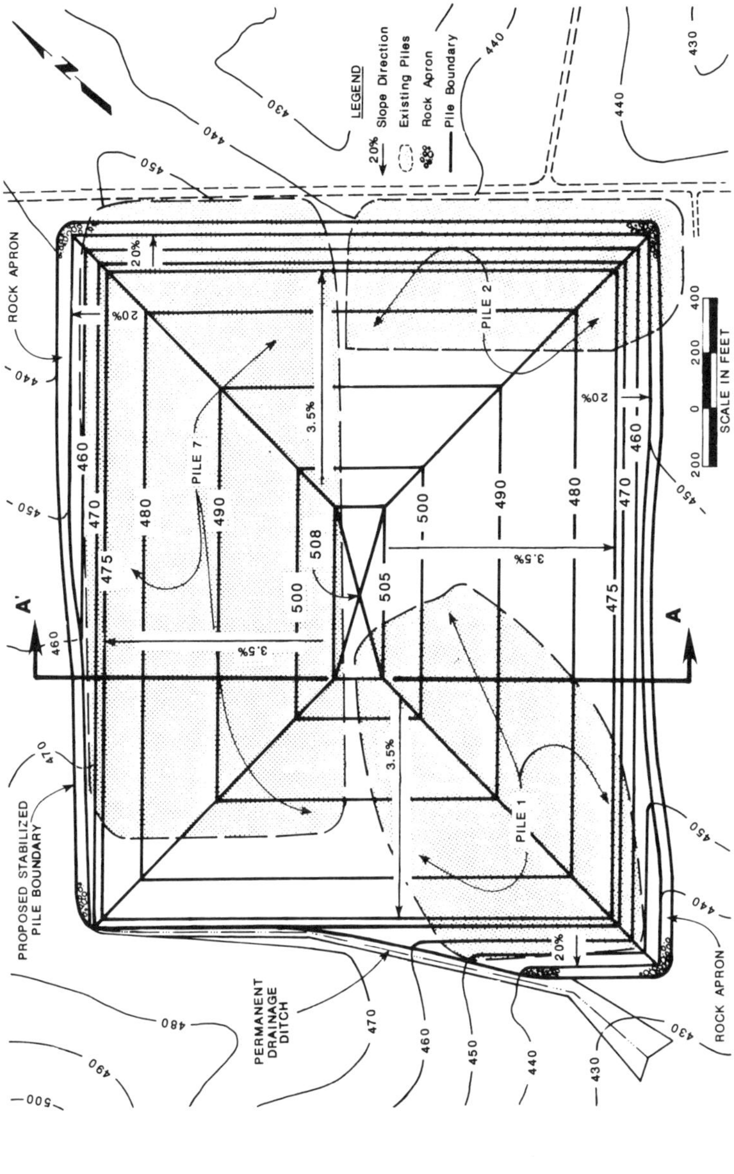

FIGURE 3
PROPOSED STABILIZED PILE CONFIGURATION FOR THE FALLS CITY TAILNGS SITE
NEAR FALLS CITY, TEXAS

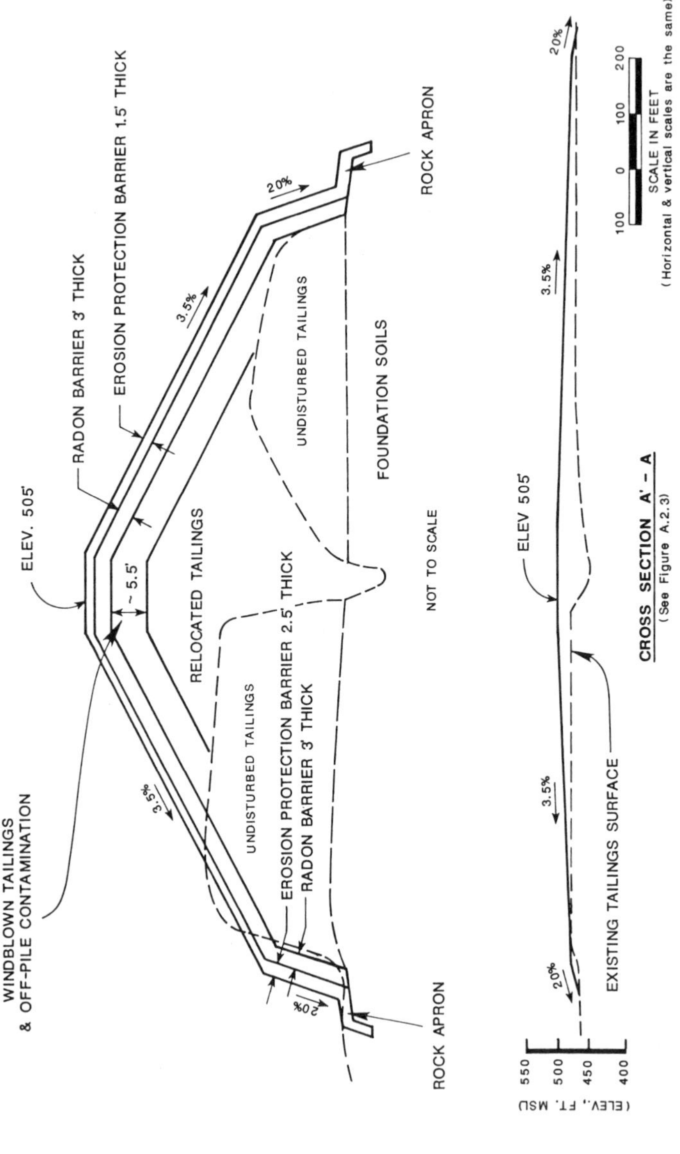

FIGURE 4
TYPICAL PILE CROSS SECTIONS, STABILIZATION ON SITE AT THE FALLS CITY TAILING SITE
NEAR FALLS CITY, TEXAS

lity. Each of the correlated parameters is discussed below.

Material Classification

The basis used for material classification was taken from work done by Robertson (1985) and Douglas and Olsen (1981). Although their work reduced the classification into several soil types, it was found that the tailings need only be divided into three different material types: sand, sand-slime, and slime. The division for each of these types was based on the following percent passing the 200 sieve divisions.

Sand	0 to 30 percent passing #200 sieve
Sand-slime	30 to 70 percent passing #200 sieve
Slime	70 to 100 percent passing #200 sieve

Figure 5 shows the results of the correlation between point stress and friction ratio when compared laboratory data for the three tailings types.

Blow Counts

The basis for the blow count correlations came from work done by Robertson and Campanella (1984). The point stress to blow count ratio for the sands, sand-slimes, and slimes averaged 4.5, 3.0, and 1.5 for each of the materials, respectively.

Undrained Shear Strength

The undrained shear strength was determined from the following equation:

$$S_u = q_c - \sigma_v / N_k$$

where:

S_u = undrained shear strength.
q_o = point stress.
σ_v = total overburden stress.
N_k = cone factor.

The N_k factor average from 8 laboratory UU tests was 17 which correlates well with published data.

Friction Angle

The friction angle was modified from Robertson and Campanella (1984). The results of 10 unconsolidated undrained triaxial tests with pore pressure measurements were compared to the values that were predicted using the

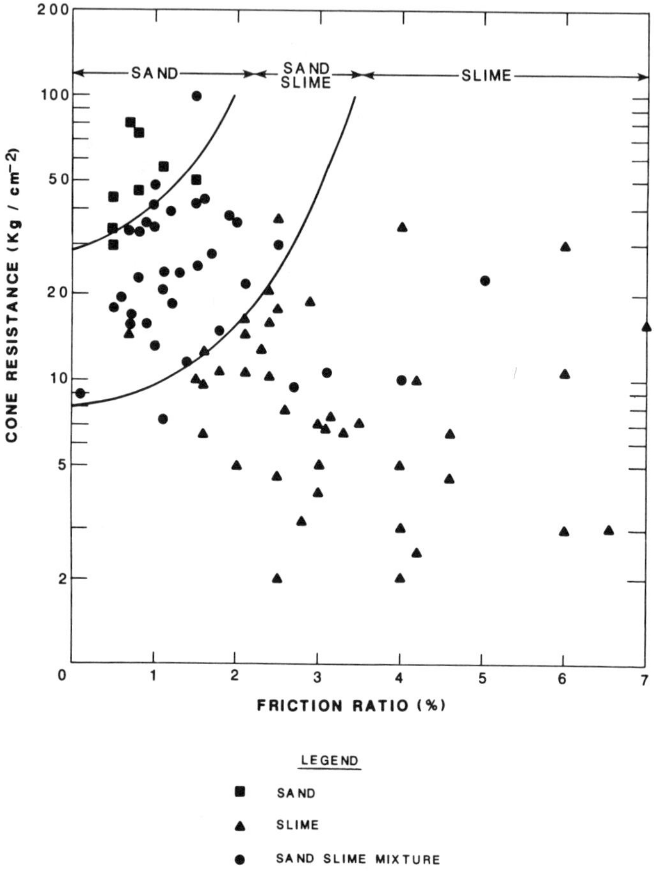

FIGURE 5
PIEZOCONE CLASSIFICATION CHART USED FOR TAILINGS

method shown by Robertson and Campanella and it was found that their method generally overpredicted the triaxial friction angle for the tailings. To compensate for this overprediction, the friction angle predicted by the piezocone was multiplied by a factor of 0.9.

Over-Consolidation Ratio

The over-consolidation ratio (OCR) calculation was modified slightly from the method presented by Schmertman (1978). This method uses the ratio of the undrained shear strength determined from the point stress (S_u) and the total overburden stress at that point (σ_v) and divides it by the normally consolidated S_u/σ_v ratio, which was determined to be 0.33. The OCR could then be calculated from the following equation:

$$OCR = (S_u/\sigma_v) / 0.33)^{1.25}$$

where

S_u = undrained shear strength
σ_v = total overburden stress
0.33= Normally consolidated S_u/σ_v ratio

In general, this relationship slightly underpredicted the over-consolidation ratio; however, since there was not enough data to justify modifying it, the over-consolidation ratio was used as calculated since the relationship would yield slightly conservative results.

ANALYSIS

These parameters were calculated for all layers of the profile by using a specifically developed computer program (Larson and Mitchell, 1986) to analyze all data gathered in the field. A typical profile of a piezocone sounding is shown in Figure 6. The computer was programed to classify every single data point gathered in the field and, based on the resulting classification, a corresponding color was placed in a bar code on the edge of the screen. This method is simulated in Figure 6 except that the different material types were separated into different columns. This simulation was necessary since the different colors for the materials could not be reproduced in the printing process. By moving a horizontal line up and down on the screen using a "mouse" peripheral, different layer horizons could be identified and the engineering parameters could be calculated and used in subsequent analyses. Table 1 shows a typical set of results for a piezocone sounding.

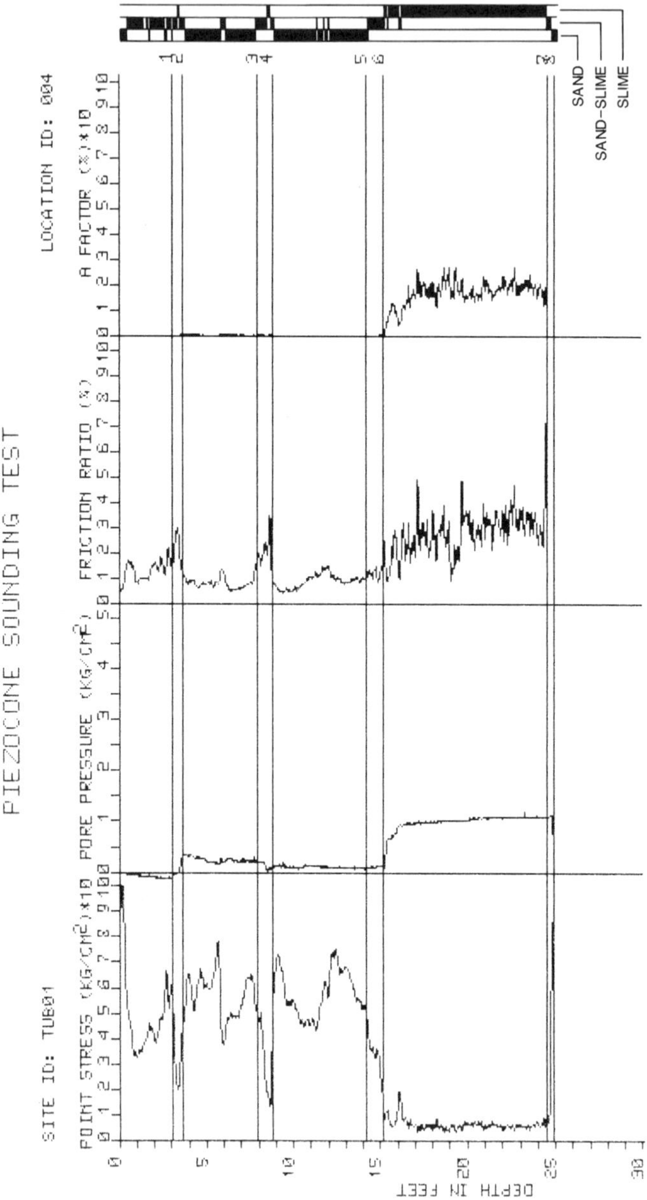

FIGURE 6
TYPICAL PIEZOCONE PROFILE AND COMPUTER GRAPHICS USED TO AID IN THE IDENTIFICATION OF SOIL LAYERS FROM PIEZOCONE DATA

Table 1. Typical engineering parameters for each layer in Figure 6

Site ID FCT-01 Location ID 004
Average soil weight (pcf) 110.0
Groundwater Depth (ft) 50.0

Layer	From (ft)	To (ft)	Thick-ness (ft)	Avg. point stress (KG/CM2)	Avg. sleeve stress (KG/CM2)	Avg. pore pressure (KG/CM2)	Avg. friction ratio	Material Type	Blow counts per ft.	Rel. density (%)	Fric. angle (DEG)	Mod. of elast. (KG/CM2)	Undrained shear strength	OCR
TAILINGS														
1	0.0	3.0	3.0	50.4	0.6	0.1	1.3	Sand-slime	16	63.	42.	177.	6054.	
2	3.1	3.6	0.6	32.9	0.6	0.0	1.9	Sand-slime	10	51.	37.	115.	3950.	
3	3.7	8.0	4.3	56.7	0.5	0.2	0.9	Sand	11	62.	42.	198.		
4	8.0	8.8	0.8	30.0	0.6	0.1	1.9	Sand-slime	10	40.	35.	105.	3601.	
5	8.8	14.2	5.4	58.0	0.5	0.1	0.9	Sand	11	51.	37.	203.		
6	14.3	15.2	0.9	32.0	0.4	0.1	1.3	Sand-slime	10	28.	33.	112.	3844.	
7	15.2	24.6	9.3	6.4	0.2	1.0	2.7	Slime	4			23.	773.	1.1
% Thickness =		39.9	9.7	57.4	0.5	0.1	0.9	Sand	11	56.	39.	201.		
% Thickness =		21.8	5.3	42.2	0.6	0.0	1.5	Sand-slime	13	52.	39.	148.	5070.	
% Thickness =		38.3	9.3	6.4	0.2	1.0	2.7	Slime	4			23.	773.	1.1
FOUNDATION														
8	24.6	25.0	0.4	58.4	0.0	1.0	0.0	Sand*	11	33.	33.	204.		
% Thickness =		100.0	0.4	58.4	0.0	1.0	0.0	Sand	11	33.	33.	204.		

SETTLEMENT ANALYSIS

Once a profile for a grid point was determined from the piezocone, a settlement analysis could be performed on that profile. Both primary and secondary settlements must be determined. Primary settlement begins during the construction of the embankment. As long as primary settlement is completed near the end of embankment construction, it has a minimal impact on the final pile. Since hydraulically placed tailings piles are generally highly layered, a computer program capable of modeling multiple layers and different drainage patterns was used. The program CONSOL (Wong and Duncan, 1985) was capable of modeling the tailings layers and the rate at which the pile was to be built. As the pile is built, the tailings will settle and additional material can be placed. If the tailings experience all primary settlement by the end of the recontouring and placement phase of construction, the tailings piles can be final graded and the cover system placed. Because of this, the only impact that primary settlement has on the pile design is the time required for primary settlement; however, the amount of primary settlement is not of significant importance.

The profiles of the piezocone soundings were checked; those profiles with the thickest amount of slimes and the largest loadings were analyzed for primary settlement. At Falls City, the soil profiles with the thickest slime layers were also the ones directly under the areas to receive the most fill for the final shaping. These profiles were analyzed; the time required for primary consolidation is only three months after completion of construction. Since experience from test fills on hydraulically placed uranium tailings (Chen, et al., 1988) has shown that the time for primary consolidation is generally overpredicted by 5 to 10 times using laboratory determined Cv values, it was concluded that primary consolidation at the site would not be a problem. If the analysis had shown that the time for primary consolidation was significant, then the construction sequence would have had to been broken up so the the final cover would not be placed until primary consolidation was completed and test fills would have been recommended to verify the predictions.

Secondary settlement tends to be more critical to the final performance of the remediated tailings pile and can have an adverse effect on the pile. However, unlike primary settlement, it is more difficult to control. If secondary settlement is excessive, there are few things that can be done to mitigate its effects. Because of this, special attention needs to be given to secondary settlement. The amount of secondary settlement was calculated using a combination of piezocone data and laboratory data. The following equation was used to calculate the amount of secondary settlement.

$$S=C * H * \log (t_2/t_1)$$

where

S = secondary settlement.
C = Strain secondary compression coefficient .
H = thickness of layer to be analyzed.
t_1= Time of the beginning of secondary settlement.
t_2= Time at the end of secondary settlement.

The secondary compression coefficient was determined from laboratory data; the height of each layer was determined from the piezocone profiles. Since it takes approximately three months for primary consolidation to be completed, t_1 was set to three months; t_2 was 1000 years (the desired design life of the embankment). Using the above equation, the secondary settlement was calculated for every soil profile made inside the final embankment footprint.

Secondary settlement in itself is not necessarily all that serious; however, differential secondary settlement is very critical. Unlike primary settlement, it is not possible to simply analyze the profile that is the most critical because it is the difference in settlement between two profiles that is critical. It is for this reason that the secondary settlement was analyzed at every point on the grid and then compared to the settlements for the surrounding profiles. The results of the settlements were analyzed so that the strains could be calculated and compared to those needed to cause cracking of the cover.

STRAIN CALCULATION

The method used to calculate the final strain in the cover was generally based on approaches to analyze the effects from subsidence due to mining or pumping of underground fluids (e.g., water or oil) (Brauner, 1976; Lee and Shen, 1969). Figure 7 shows the relationship between vertical settlement, horizontal displacement, and horizontal strains. As a trough is formed due to settlement of the underlying soils, subsidence is also accompanied by horizontal displacements. These horizontal displacements are larger than those normally expected from the curvature caused by the subsidence. The magnitude of the horizontal displacements are proportional to the slope of the subsidence profile or the first derivative of the trough created by the subsidence profile. Likewise with the horizontal strain, the point of maximum

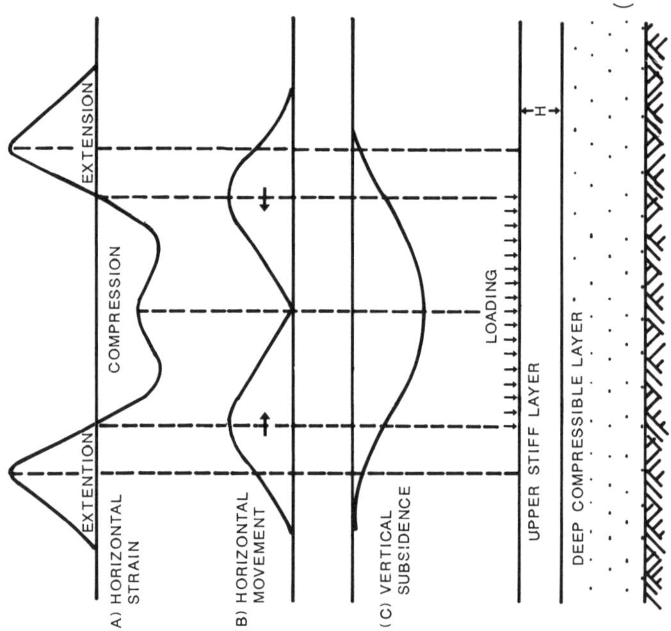

(From Lee & Shen)

FIGURE 7
IDEALIZED SURFACE MOVEMENTS AT A SUBSIDING AREA

horizontal strain is the point of steepest slope on the horizontal displacement curve or the second derivative of the subsidence profile. The shapes of the curves and relative positions of the movement curves agree with both observed data and theories.

In order to evaluate the strains, the shape of the settlement trough must be calculated by determining the settlement in the tailings caused by the added load of the shaped fill. Once the shape of the settled area is determined, then the horizontal displacements in the cover can be calculated by taking the first derivative of the curve. After that, the second derivative of the curve is calculated; it is this curve that represents the horizontal strain caused by settlement. If a cross section is drawn through a series of holes suspected of being the most critical and the deformed shape of the surface is plotted, then the derivative can be obtained using graphical methods. In some cases, it is very difficult to determine which is the most critical surface. Therefore, many sections have to be analyzed. Even then, the most critical section may not have been identified. Since the effects of differential settlement are critical, a method of analysis was devised such that the entire cover could be checked, thereby eliminating the need to guess which section was the most critical.

With the use of computers and advanced analytical software, the methods used to evaluate tensile strains for cover cracking can be improved over graphical methods used in the past. The computer program used to analyze the tensile strains was CPS (Contour Plotting System) from Radian Corporation and was run on a Micro VAX.

All northing and easting coordinates and settlements were entered into the computer and were contoured and plotted. (Figure 8 shows the results of the secondary settlement contours.) Next, the horizontal displacement was calculated by taking the first derivative of the settlement data and contour plotted as shown in Figure 9. Figure 10 shows the second derivative, which represents the strains in the cover due to differential settlement. The positive values represent tensile strains; the negative values represent compressive strains. The contour plot of the strains can then be closely examined and values of maximum strain can be located. This method allows the entire cover to be effectively analyzed and evaluated rather than simply taking a few sample sections and then extrapolating their interpretation to the whole pile.

FIGURE 8
SETTLEMENT CONTOURS

FIGURE 9
HORIZONTAL DISPLACEMENT

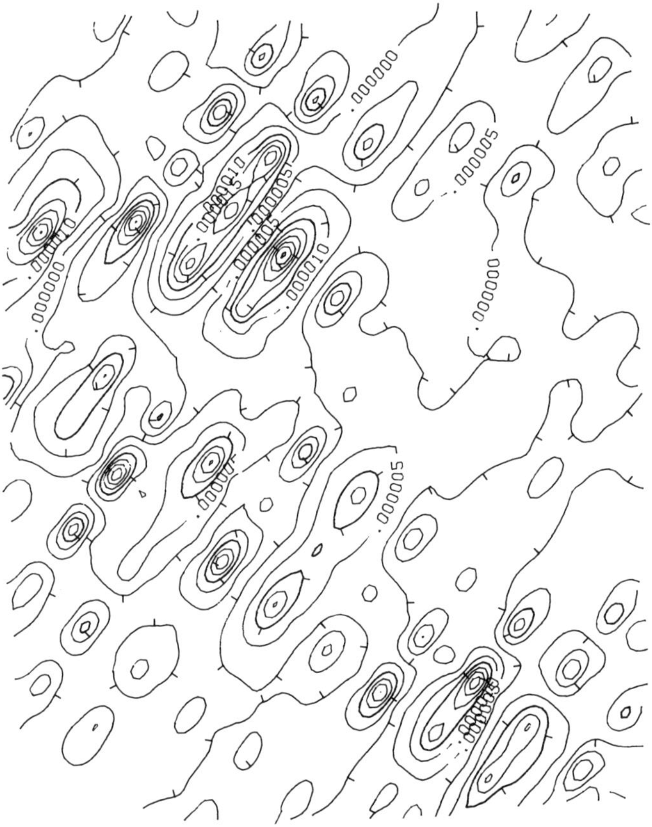

Units are feet/feet

FIGURE 10
HORIZONTAL STRAIN

EVALUATING STRAINS

Once all strains have been calculated, they must be evaluated and analyzed to determine if they will have harmful effects on the final pile configuration. Table 2 shows the results from several researchers (Moore and Hor, 1984; Ajaz and Parry, 1975; Carrigan 1972; Leondards and Narain, 1963) who have investigated the amount of tensile strain that compacted soil can experience before cracking. These results were also compared to many years of observations made on the effects of tensile strains caused by subsidence from underground mining. Table 2 also shows the allowable tensile strains for concrete and masonry walls. Although these structures are considered to be much more rigid that compacted soils, a comparison of the allowable strains shows that they are not significantly different.

By comparing the maximum strain calculated on Figure 10 (0.003 percent), it was concluded that cracking of the final cover was not a problem for the life of the embankment since this strain is much less than any listed on Table 2. As an added assurance, the cover will be constructed above optimum moisture content, therefore increasing the cover's ability to withstand tensile strains.

If cracking from secondary settlement had been a problem, then few steps could be taken to eliminate its effects on the embankment. The only reasonable thing that can be done to reduce secondary consolidation is: (1) to reduce the height of the in-situ tailings, and (2) to replace the material with compacted material that has a lower secondary coefficient of consolidation.

CONCLUSIONS

Although hydraulically placed tailings are often highly variable in both material type and behavior, they can be effectively and economically characterized using the piezocone as a logging tool. By using both published and site-specific data, the degree of confidence in the results can be increased. Not only can the material classification be obtained from the piezocone for engineering purposes, but other parameters such as friction angle, blow counts, undrained shear strength, and over-consolidation ratios can be calculated for settlement and stability. The piezocone also enabled significant amounts of data to be collected quickly and economically.

By using the methods discussed in this paper, the long-term performance, settlements, and strains of the cover can be calculated and evaluated. Any potential

Table 2. Tensile strains at cracking

Compaction water content relative to optimum water content %	Material type	Tensile strain at cracking %	Reference
-3 to +4 (OMC = 25.9%)	Limestone caly	0.13 to 0.31	Leonards & Narain (1963) (bending tests)
-3 to 0 (OMC = 27.5%)	Kaolin	0.21 to >0.55	Carrigan (1972) (model tests)
-3 to +7 (OMC = 24.6%)	Gault clay	0.5 to 1.7	Ajaz and Parry (1975) (bending tests)
-3 to +3 (OMC = 13%)	Balderhead clay	0.14 to 1.6	
-6 to +2	Gault clay	0.08 to 0.34	Ajaz and Parry (1975) (tension tests)
-1 to +5	Balderhead clay	0.08 to 0.48	
N/A	Plaster	0.1	Voight and Pariseau (1970)
N/A	Reinforced concrete	0.25 to 0.40	Voight and Pariseau (1970)

problems can be identified during the design and steps can be taken to mitigate the problems before construction.

ACKNOWLEDGEMENTS

This study was completed as part of the DOE UMTRA Project, headquartered in Albuquerque, New Mexico, and was supported under DOE Contract No. DE-AC04-82AL14086 to Jacobs Engineering Group Inc., the Technical Assistance Contractor (TAC). The TAC consists of Jacobs Engineering Group Inc.; Roy F. Weston, Inc.; and Sergent Hauskins and Beckwith Geotechnical Engineers, Inc. The authors wish to express special thanks to Ron Rager and Jack Caldwell for valuable technical input; to Dianna Muller for help in manipulating and plotting the data on the computer; and to Pam Preisen for editing assistance.

REFERENCES

Ajaz, A. and Parry, R.H.G. (1975). Stress-Strain Behaviour of Two Compacted Clays in Tension and Compression. Geotechnique. Vol. 25, No. 3, pp. 495-512.

Brauner, G., 1971, "Subsidence Due to Underground Mining, 1. Theory and Practice in Predicting Surface Defermation," U.S. Bureaue of Mines IC 8571, Washington, D.C.

Carrigan, L.F. (1972). Tensile Cracking in Clay Cores of Earth and Rockfill Dams. Thesis (M.Eng.Sc.) University of Melbourne.

Chen, P.K., B. Keshian, and F.B. Guros, 1988, "Predicting Settlement of Hydraulically Placed Uranium Tailings", Hydraulic Fill Structures '88", Geotechnical Engineering Div., ASCE, August, 1988, Ft. Collins, CO.

Douglas, B.J. & R.S. Olsen 1981. "Soil Classification Using Electric Cone Penetrometer," Symposium on Cone Penetration Testing and Performance, Geotechnical Engineering Div., ASCE, October 1981, St. Louis, MO.

Lee, K.L. and C.K. Shen, 1969. "Horizontal Movements Related to Subsidence," Journal of Soil Mecahnics and Foundations, ASCE, New York, New York.

Leonards, G.A. and Narain, J. (1963). Flexibility of Clays and Cracking of Earth Dams. Jnl. Soil Mech. and Found. Div. ASCE. Vol. 89, No. SM2, pp. 47-98.

Moore, P.J. and A.Y.T. Hor, 1984. "Cracking Behaviour of Compacted Clay," Fourth Australia-New Zealand Conference on Geomechanisms.

Robertson, Peter K. 1985. "In-situ Testing and its Applications to Foundation Engineering," Invited Colloquium Paper at the 38th Geotechnical Conference, Edmonton, Alberta, Canada.

Robertson, P.K. & R.G. Campanella, 1984. "Guidelines for Use and Interpretation of the Electric Cone Penetration Test," prepared for Hogentogler and Co., Inc., Gaithersburg, MD, at the University of British Columbia.

Voight, B. and W. Pariseau, 1970, "State of the Art of Predictive Subsidence Engineering", Journal of Soil Mechanics and Foundation Engineering, ASCE, SM2, New York, New York.

Wong, K.S. and J.M. Duncan, 1985. "CONSOL: A Computer Program for 1-D Consolidation Analysis of Layered Soil Masses," Department of Civil Engineering, Virginia Technical Institute, Blacksburg, Virginia.

Seepage Effects on Sedimentation of Fly Ash Slurry

Richard E. Wardwell, M.ASCE; R. Craig Findlay, and Matthew M. Muzzy*

ABSTRACT: This paper describes the geotechnical and hydraulic characteristics of hydraulically filled fly ash slurry which has been placed in an unlined lagoon. The disposition of the fly ash slurry is influenced by both gravitational effects (as modeled by Kynch theory of hindred settling and self-weight consolidation) and by seepage forces as the carrier fluid infiltrates and drains from the deposit. The seepage flow affects both the engineering design of the impoundment and the environmental impacts of the lagoon on the geohydrologic regime. This paper summarizes the field observations and laboratory testing conducted on the fly ash. Based on these pilot test results, a predictive analogue is presented for discussion purposes. Comparisons are made between these results, and recommendations provided for further research work needed to develop a sedimentation model which incorporates seepage forces.

INTRODUCTION

Hydraulically placed waste materials are prevalent in many industrial and environmental disposal applications. These include mine and mill tailings, dredged sediments, FGD sludges, hazardous waste lagoons and other industrial waste ponds. The design of these disposal facilities is related to the storage capacity, stability, and environmental impacts associated with the hydraulic wastes. In turn, the environmental impacts are influenced, to a large extent, by the geohydrologic conditions at a site and the changes in these conditions made by the disposal facility.

The geotechnical engineering characteristics of hydraulic wastes are instrumental in estimating this behavior. Specifically, the evaluation of vertical strains which occur during sedimentation is critical in the design of a storage facility. Some studies have been conducted to help understand the sedimentation and consolidation characteristics under "no flow" conditions. However, research is needed to evaluate the same process when the sedimentation is influenced by seepage forces. Flow occurs with an unlined pond or when providing supplemental underdrainage during remediation of a

* Senior Geotechnical Engineers, E.C. Jordan Company, Portland, Maine 04112.

facility. Estimations of the vertical strains determine, to a large extent, the feasibility of using a selected remediation technique.

This paper summarizes the geotechnical studies conducted on existing fly ash ponds. Cost effective disposal of fly ash is of paramount importance to coal-fired power generation which produces over 38 million tons of fly ash annually in the United States. The geotechnical engineering behavior of the ash is instrumental in determining the economics of waste management and the resulting environmental impact associated with leachate migration and reclamation stability of the deposit. As a demonstration of this, the work presented herein evaluates the design capacity and environmental impacts of fly ash ponds as influenced by seepage rates during sedimentation and consolidation. As will be discussed, fly ash is predominantly a fine grain, granular material lacking plasticity. The studies have therefore focused on the sedimentation process, and, specifically, the influence of seepage forces on the void ratio, density, and hydraulic conductivity (permeability) of the deposit material during sedimentation. A simplistic approach is presented to incorporate seepage forces into sedimentation theories and to provide a transition from sedimentation to the consolidation analysis.

The behavior and predictive techniques presented in this paper can be used to develop cost-effective design for waste disposal of hydraulically transported materials. While the studies here are related to a fly ash deposit, this work could be applied to other hydraulic wastes including mine and mill tailings, and hazardous waste sediments and lagoons. The results from this work provide the basis for further development to include the influence of seepage forces on existing consolidation theories for fine grained materials.

A brief description of the site is presented in the following section followed by a summary of the field and laboratory tests. A discussion of the results and a suggested method to incorporate seepage forces into sedimentation theories follows. The suggested analogue is preliminary and incomplete by nature. It is presented herein to encourage further thought and research development.

SITE DESCRIPTION

Fly ash from a generating station's electrostatic precipitators is collected by a hydrovactor system. The slurried fly ash is sluiced to one of two unlined fly ash settling ponds, shown on Figure 1, for temporary storage and dewatering. The ponds are used on an alternating basis, with one pond being excavated while the other pond is in service. This cycle takes approximately 36 months, with each pond receiving hydraulic waste for a period of about 18 months. A typical profile through the ponds is shown in Figure 2. Fifteen to twenty-five feet of stratified fly ash and occasional layers of

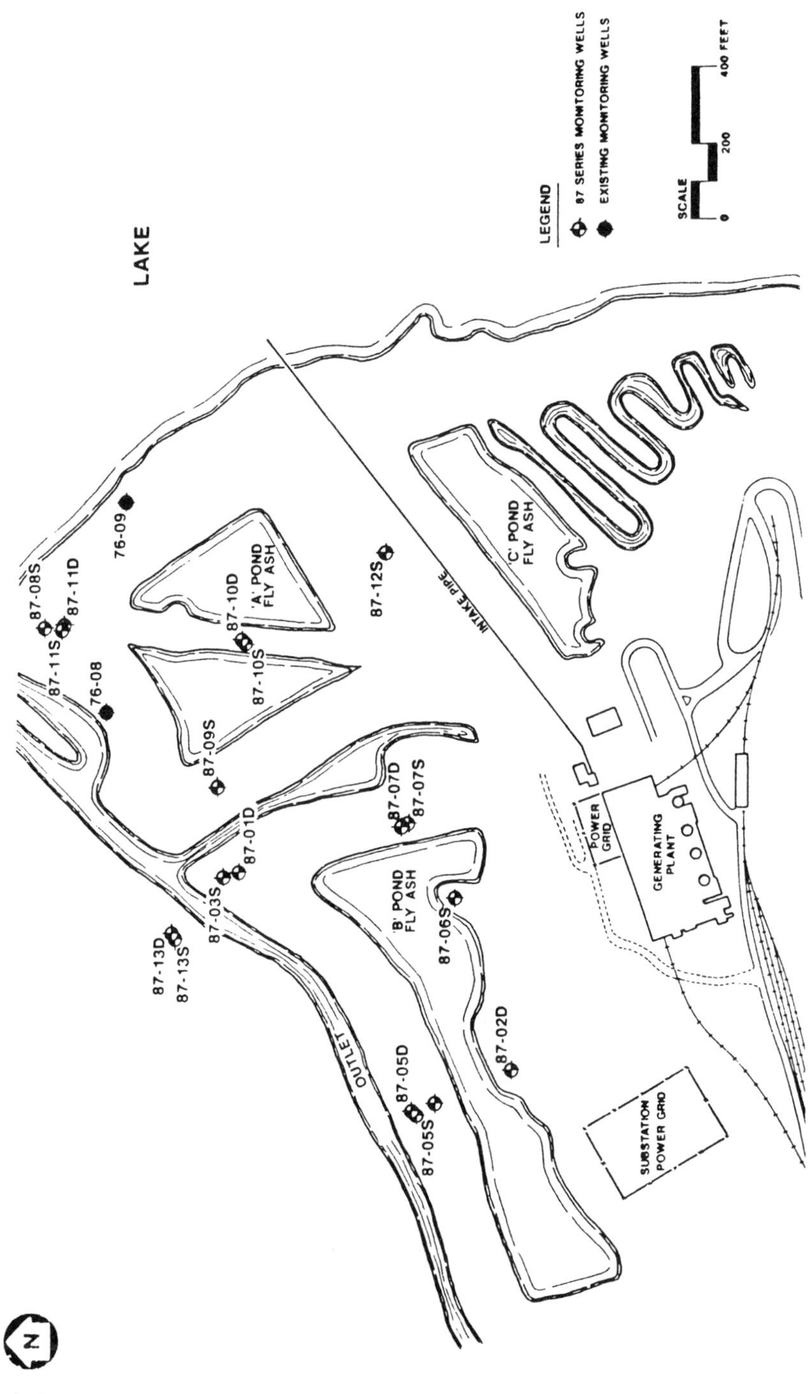

FIGURE 1
SITE PLAN

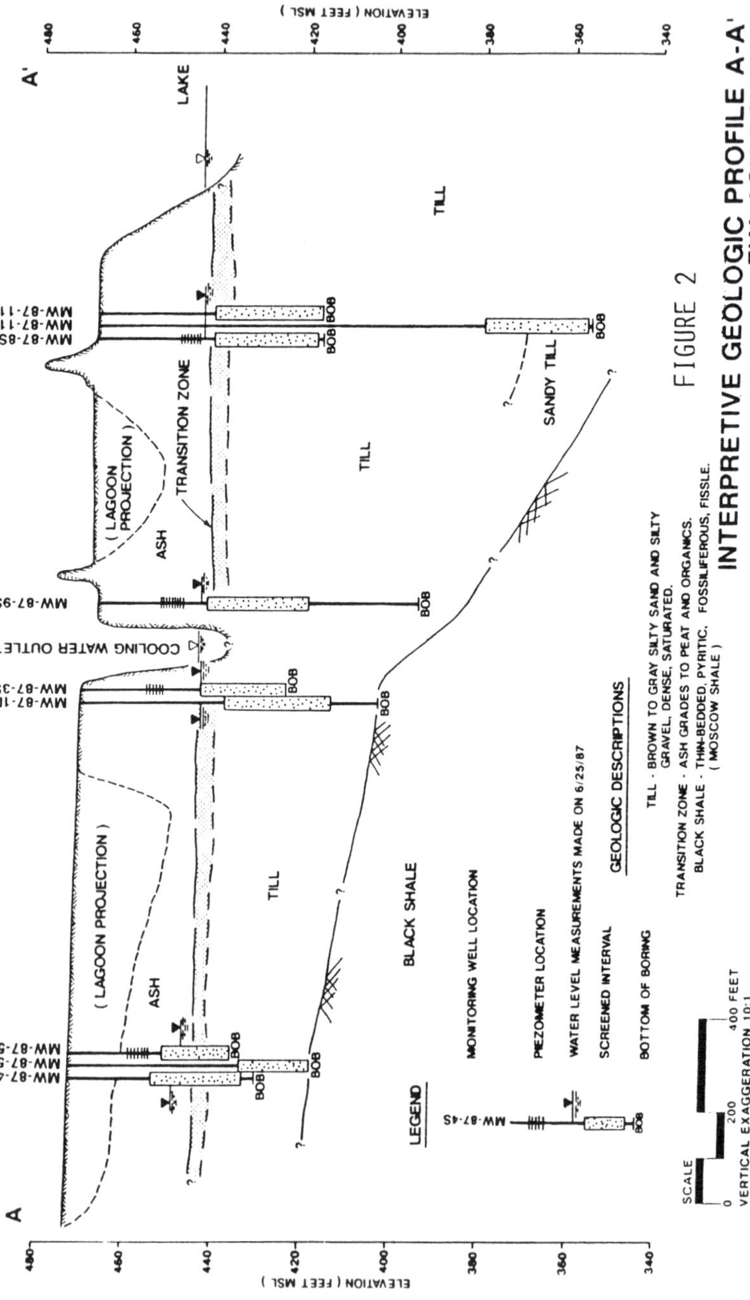

FIGURE 2

INTERPRETIVE GEOLOGIC PROFILE A-A'
FLY ASH PONDS

bottom ash are placed into the ponds directly on the underlying peat and deltaic alluvium gravelly sand and silt.

The ash is placed into the ponds at a solids content of about 5.5% or a wet/dry density of approximately 64.6/0.06 pcf. Water is decanted from the pond surface and recycled to the plant. Once filling is completed, the hydraulic spigot is switched to the other pond, and the existing surface water allowed to dissipate. The fly ash is then excavated with traditional earth moving equipment. The stability of the deposit in supporting this equipment indicates the possibility that seepage forces from pond have caused additional sedimentation of the material.

GEOTECHNICAL TESTING

Fly Ash Classification

Fly ash is generally considered to be composed of very small granular particles. The majority of these particles are glassy spheres, with the remainder being crystalline matter and carbon (DiGioia, 1979, Chae and Synder, 1977). The fly ash in the dewatering ponds is a gray to black, sandy silt with a greasy texture. The grain-size distribution is presented in Figure 3. As shown, more than 90% of the ash is composed of non-plastic uniform silt-sized particles, with organic contents ranging from 1.2% to 11.7%. The ash is classified as ML by USCS. The characteristics of this fly ash are similar to those determined for other ashes (Chae & Snyder; Srinivasan, et.al.; Cunningham, et.al.; DiGioria, et.al; Seals, et.al.; 1977).

The non-plastic nature of the ash is associated with the spherical shape, small surface area, and uniform silt size of the individual particles. The specific gravity of the particles is approximately 2.60. Based on this specific gravity, the maximum modified proctor dry density is estimated to be 110 pcf at an optimum water content of less than 10% (Chae et.al.; 1977). These values are consistent with the granular nature of the ash. The specific gravity is in the high range of reported values varying from 2.3 to 2.6 with an average of about 2.4 (DiGioia & Nuzzo, 1972). The high specific gravity would indicate a predominance of spherical particles and a lack of crystalline structure. Based on this identification testing alone, it would be anticipated that the resulting density would be more influenced by rapid sedimentation than by long term consolidation. As such, the resulting densities may be greatly influenced by hydraulic gradients and resulting seepage forces.

Field Testing

A field exploration program was conducted in July, 1987 on the fly ash surface in the bottom on Pond A. At the time of the testing, this surface was approximately 10 feet below full pond elevation. Approximately 9 feet of ash remained in the bottom of the lagoon.

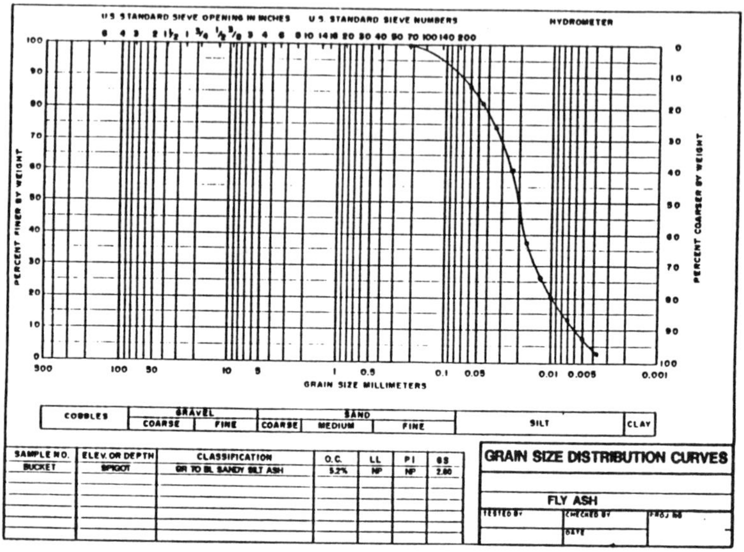

FIGURE 3
FLY ASH CLASSIFICATION RESULTS

TABLE 1
TEST PIT DATA

Sample		Dry Density (pcf)	Water Content (%)
Test Pit No.	Depth (ft.)		
1	0-1	-	50.2
	2	-	62.6
	4	-	50.6
	5	-	17.2
	6	-	68.1
3	0-5 (1)	49.0	37.9
	5.0 (1)	60.9	45.1
4	2	-	44.2
	4	-	41.7
	5.2	-	24.0

(1) average of four samples

The field testing included a geophysical survey of the area, geologic logging and sample collection from backhoe test pits, and double-ring infiltration testing at selected locations. A summary of the water content, and density testing from the test pit samples is presented in Table 1. The average dry density from 4 deep tests was approximately 61 pcf.

A bucket of the slurry being discharged into Pond B was obtained. As mentioned, the slurry is placed at a solids content of about 5.5%. The solid portion settled in the bucket to an average dry density of 76.7 pcf at a saturated water content of 42.9%. Some vibration compaction probably occurred as the bucket was transported from the field to the laboratory.

The results of the infiltration tests are presented in Table 2. Based on this, it is estimated that average permeability is approximately 3x10E-04 cm/sec.

SEDIMENTATION AND PERMEABILITY TESTING

Test #1: Sedimentation Followed by Flow

Samples of the fly ash were tested in the laboratory for sedimentation and permeability characteristics. Initially, a sample of slurried ash was placed in a 6.5 inch diameter, 5.3 foot high plexiglass cylinder and allowed to settle under hydrostatic condition. The height of the sample was measured once the slurried had cleared sufficiently to observe the top layer. After the initial sedimentation, an outlet at the bottom of the cylinder was opened to promote flow. At this time, a constant head was maintained on the sample, and measurements taken on the flow rate and sample height. This would be indicative of flow from a lined lagoon which has subsequently drained. Table 3 presents the resulting void ratio, dry density, settling velocity, and effective stress at the bottom of the cylinder. A plot of the sample void ratio versus time is presented in Figure 4. This plot is indicative of the self-weight consolidation and subsequent underdrainage of the slurried ash. After "no-flow" sedimentation, the sample had achieved a void ratio of about 1.73 and dry density of 59.5 pcf. The density measured in the field test pits of 61 pcf compares favorably with the laboratory observed values.

Once this sedimentation was completed, flow was initiated. Figure 4 shows the complete sedimentation process, while Figure 5 shows the decrease in void ratio once flow was initiated. During flow, the hydraulic gradient increased from 5.8 to 6.3 due to sample compression. At the same time, the flow rate decreased due to changes in permeability of the sample. At the completion of the test, the void ratio had decreased to 1.52 with an equivalent increase in dry density to 64.3 pcf. This corresponds to an addition strain of about 8% when flow is initiated after the sedimentation process.

TABLE 2
FIELD INFILTRATION TEST RESULTS

Test No.	Depth (ft.)	% No.200 Sieve	Organic Content (%)	Water Content (%)	Infiltration (X10E-4 cm/sec)
1	0.0	79.8	4.3	37.6	9.3
	0.3	96.3	3.8	43.2	3.4
2	0.0	99.0	1.2	39.1	1.7
	0.3	-	-	-	1.1
3	0.0	93.8	11.7	52.6	1.1
	0.3	94.0	10.0	42.2	3.5
4	0.0	97.4	6.3	47.3	1.9

TABLE 3
TEST #1 – SEDIMENTATION FOLLOWED BY FLOW

TEST 1 - SEDIMENTATION W/O FLOW, FOLLOWED BY CONSTANT HEAD FLOW

SAMPLE DRY	5812.90 Gm	STOKEIAN VELOCITY:	2.082 CM/MIN
CLYNDR AREA	213.83 Cm2	MEAN GRN SIZE:	0.0200 MM
SP. GRV.:	2.60	NO FLOW EFF STRESS:	34.25 PSF
VOL SOLID	2235.731 CC		
CYL HT:	161.50 CM		
STONE HT:	2.54 CM		
HEAD:	165.20 CM		
HYDRO K:	0.01386 AT 20 DEG C		

ELAPSED FLOW TIME (MIN)	FLOW RATE (CC/SEC)	VOID RATIO (e)	VERTICAL SAMPLE PERM'BLTY (CM/SEC)	k/dL (1/SEC)	DRY SAMPLE DENSITY (LB/FT3)	COMBINED SEP&SED VELOCITY (CM/MIN)	FLOW V.STRESS @BOTTOM (LB/FT2)
0.00	0.000	5.67	NA	0.00E+00	24.34	2.082	NA
0.00	0.000	3.43	NA	0.00E+00	36.64	2.082	NA
0.00	0.000	2.09	NA	0.00E+00	52.52	2.082	NA
0.00	0.000	1.96	NA	0.00E+00	54.72	2.082	NA
0.00	0.000	1.73	NA	0.00E+00	59.52	2.082	NA
0.00	0.000	1.73	NA	0.00E+00	59.52	2.082	NA
0.00	0.000	1.73	NA	0.00E+00	59.52	2.082	NA
0.00	0.000	1.73	NA	0.00E+00	59.52	2.082	NA
0.25	2.000	1.69	1.59E-03	5.66E-05	60.37	2.976	58.52
0.50	1.000	1.67	7.90E-04	2.83E-05	60.80	2.531	45.90
1.00	0.700	1.66	5.51E-04	1.98E-05	61.02	2.397	42.30
2.00	0.433	1.64	3.39E-04	1.23E-05	61.46	2.278	39.19
4.00	0.300	1.61	2.32E-04	8.49E-06	62.14	2.219	37.67
8.00	0.200	1.58	1.53E-04	5.66E-06	62.83	2.174	36.53
16.00	0.156	1.55	1.18E-04	4.42E-06	63.53	2.154	36.04
30.00	0.113	1.54	8.52E-05	3.20E-06	63.77	2.135	35.55
60.00	0.091	1.53	6.79E-05	2.56E-06	64.01	2.124	35.29
120.00	0.088	1.52	6.60E-05	2.50E-06	64.26	2.123	35.27
180.00	0.089	1.52	6.68E-05	2.53E-06	64.26	2.124	35.28
240.00	0.101	1.52	7.54E-05	2.85E-06	64.26	2.129	35.41

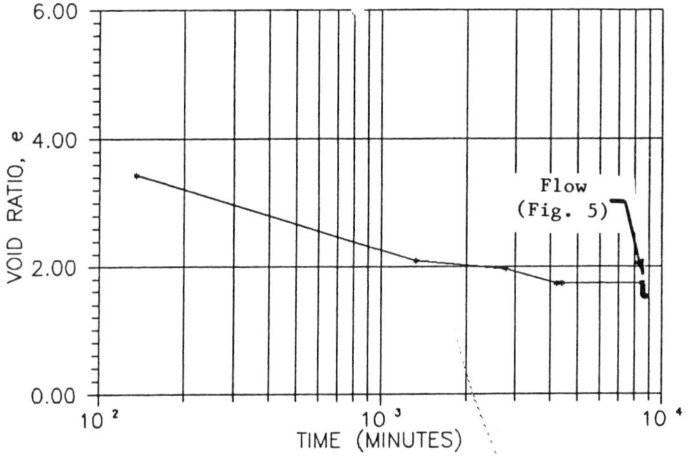

FIGURE 4
TEST #1 - VOID RATIO vs. TIME

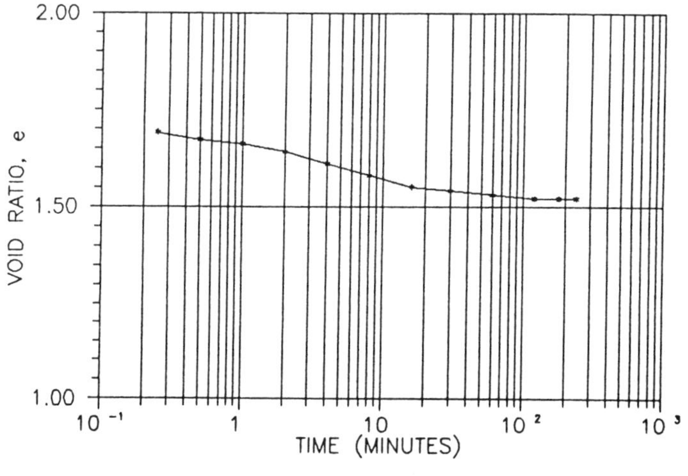

FIGURE 5
TEST #1 - VOID RATIO vs. TIME, FLOW AFTER SEDIMENTATION

Test #2: Sedimentation with Flow

To determine the influence of hydraulic gradients and, seepage forces on sedimentation, another test was run using virtually the same head loss through the sample, but allowing flow to occur as soon as the material was placed as a slurry. This situation represents sedimentation of a hydraulic fill when placed into an unlined or drained lagoon.

Data from Test #2 is presented in Table 4. A plot of void ratio versus time is presented in Figure 6. Measurement of sample height was not possible during the early stages of Test #2 because the supernatant was cloudy above the specimen. Because of this, values for void ratio and density were not possible until the latter stages of the test. Figure 6 shows the results after one hour of testing. More complete and accurate data would have been possible if the actual sample densities and pore water pressures were measured during the test. In future testing, densities should be obtained by direct measurements, or better yet, with nuclear density scans, and pore pressure measurement made with transducers (Bean & Sills, 1981).

For the equivalent hydraulic gradient, the void ratio decreased to 1.38 when seepage occurred during disposition. This void ratio is equivalent to a density of 68.1 pcf. This is 13% more strain than the self-weight sedimentation and 6% more strain than the "flow" test which occurred after the initial deposition.

Flow and Permeability

Flow rates were also measured during both tests as presented in Tables 3 & 4. A plot of these rates with time is presented in Figure 7. As may be seen, the relative decrease in flow and the final stabilized flow rate is approximately the same for both tests. However, the first test started at a lower flow value due to the structure and fabric established during the initial sedimentation process.

From the laboratory results, permeability was calculated from the measured flow rates. The results of this are presented in Tables 3 & 4. Calculated values were not possible during the early part of Test #2 due to the inability to accurately measure sample height. To evaluate the relative difference in values, permeability per unit length of sample, K/dl, was calculated. Results from this are presented in Table 3 & 4 and plotted in Figure 8. These results from this are similar to the flow diagrams.

TABLE 4
TEST #2 - SEDIMENTATION WITH FLOW

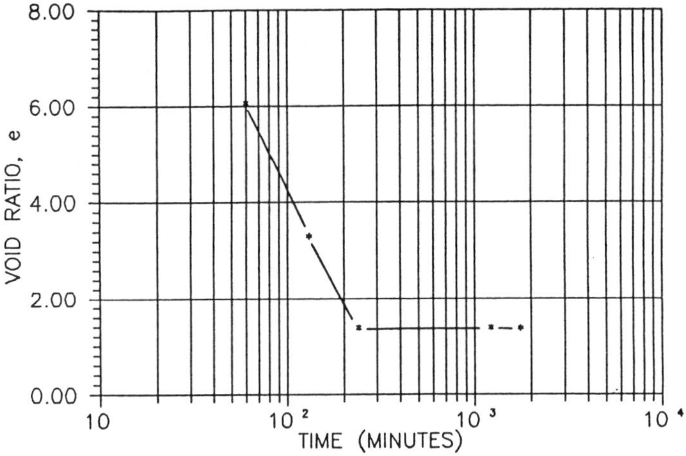

TEST 2 - SEDIMENTATION W/ CONSTANT HEAD FLOW

SAMPLE DRY WEIGHT:	5812.30	Gm
CLYNDR AREA:	213.83	Cm2
SP. GRV.:	2.60	
VOL SOLIDS:	2235.500	CC
CYL HT:	152.30	CM
STONE HT:	2.54	CM
HEAD:	165.20	CM
MEAN GRN SIZE:	0.0200	MM
HYDRO K:	0.01386	AT 20 DEG C

ELAPSED ELAPSED TIME (MIN)	FLOW RATE (CC/SEC)	VOID RATIO (e)	VERTICAL SAMPLE PERM'BLTY (CM/SEC)	K/dL (1/SEC)	DRY SAMPLE DENSITY (LB/FT3)	COMBINED SEP&SED VELOCITY (CM/MIN)	FLOW V.STRESS @BOTTOM (LB/FT2)
0.00	0.000	NA	NA	0.00E+00	NA	NA	NA
0.25	4.267	NA	NA	1.21E-04	NA	NA	NA
0.50	2.000	NA	NA	5.66E-05	NA	NA	NA
1.00	1.467	NA	NA	4.15E-05	NA	NA	NA
2.00	1.233	NA	NA	3.49E-05	NA	NA	NA
4.00	0.917	NA	NA	2.60E-05	NA	NA	NA
8.00	0.771	NA	NA	2.18E-05	NA	NA	NA
15.00	0.386	NA	NA	1.09E-05	NA	NA	NA
30.00	0.313	NA	NA	8.87E-06	NA	NA	NA
60.00	0.241	6.06	5.03E-04	6.81E-06	22.98	2.161	36.20
131.00	0.208	3.30	2.66E-04	5.90E-06	37.69	2.158	36.14
240.00	0.118	1.39	8.38E-05	3.35E-06	67.85	2.139	35.66
1223.00	0.053	1.39	3.74E-05	1.50E-06	67.85	2.108	34.88
1758.00	0.072	1.38	5.09E-05	2.05E-06	68.12	2.117	35.11

FIGURE 4
TEST #2 - VOID RATIO vs. TIME

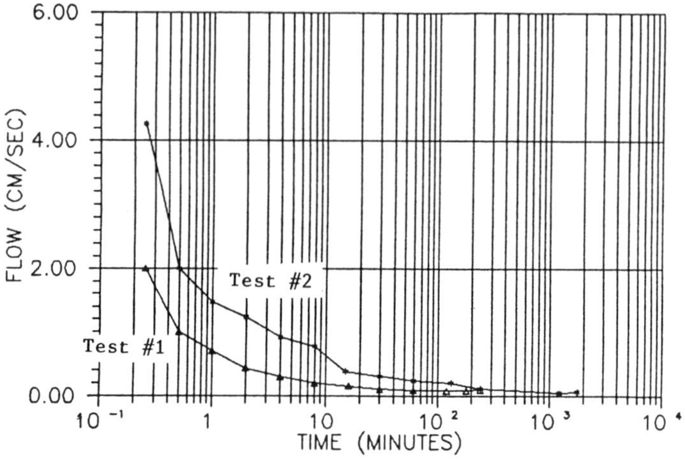

FIGURE 7
FLOW vs. TIME

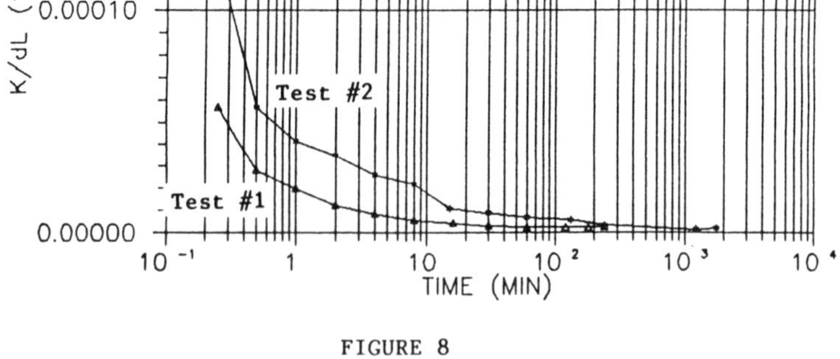

FIGURE 8
PERMEABILITY vs. TIME

EVALUATION

Discussion of Results

General: Settling of hydraulic fills is characterized by phases (Yong, 1984). When introduced into a disposal area, the slurry is influenced by gravitational (Stokesian) and interactive forces until the concentration of solids is sufficient to account for proximal hindrances and the slurry acts as a dispersion. A model to predict sedimentation during the hindered phase was developed by Kynch (1952). Kynch theory uses the assumption that the speed of a particle in a dispersion is determined by local particle density, and, as such, the relationship between the two can be reduced from observations on the fall of top of the dispersion. As the particle density increases, particle-to-particle contacts happen, structure develops, and self-weight consolidation compression occurs. At this point, the deposit begins to act as a soil rather than a dispersion.

For the fly ash, fairly high densities and stability were achieved even under the "no flow" conditions. Due to the granular nature of the fly ash, compression is influenced primarily by dispersion settlement, as model by Kynch, and, to a lesser degree, by consolidation. Regardless, it has been demonstrated that the consolidation and hindered settling are derived from the same basic principles and verified for the transition zone between a dispersion and soil (Been & Sills, 1981).

However, based on the preliminary values presented herein, it is apparent that sedimentation is influenced to a large degree by seepage flow. For the test on the fly ash, approximately 8% of vertical strain occurred even after the initial sedimentation occurred under "no flow" conditions. When the same flow conditions occurred during the sedimentation process, the sample settled an additional 6%. Since virtually the same flow conditions were present in both situations, it is apparent that the velocity of water affects the sedimentation of a slurry. The movement of water through the grains of a loose, normally consolidated deposit, impacts enough drag forces on the grains to cause additional reorientation and resulting densification of the deposit. Consideration of these strains is critical when evaluating the behavior of reclaimed deposits if supplemental drainage is provided as a stabilizing and dewatering technique.

Seepage Effects: To help evaluate the effects of seepage rates on compressibility, comparisons were made between the finalized void ratio and the weighted average seepage flow (equivalent flow) which occurred during the sedimentation process. A plot of this is presented in Figure 9 and a plot of permeability per unit sample length and void ratio is presented in Figure 10. As may be seen by these figures, the flow of water apparently impacts additional drag forces and particle reorientation. During sedimentation of a dispersion, the water velocity caused by seepage impacts additional

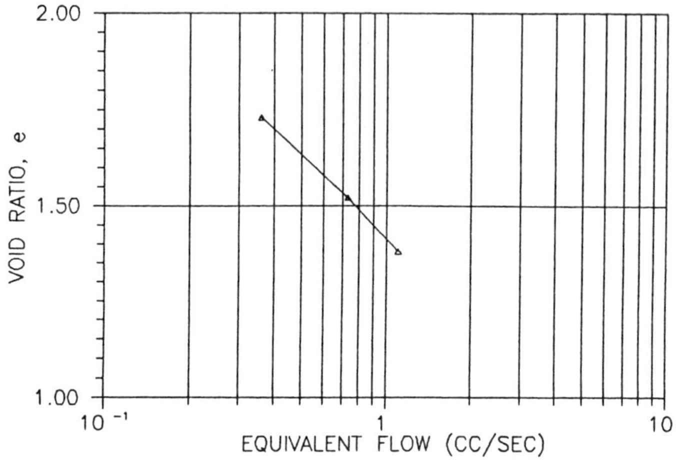

FIGURE 9
VOID RATIO vs. EQUIVALENT FLOW

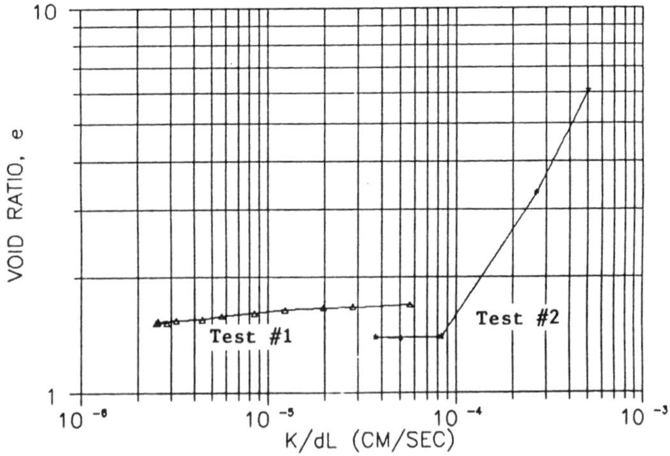

FIGURE 10
VOID RATIO vs. PERMEABILITY

momentum to the particle grains. The dissipation of additional momentum would increase the likelihood that each particle would approach its minimum energy state. With this, soil grains would try to pack into their minimum void ratio. As a result, higher densities are achieved.

Since momentum is based on velocity, it is probable that the resulting void ratio would be directly related to the seepage velocities which occur during sedimentation. A procedure has been developed to evaluate the effects of seepage forces on the consolidation of loose slurries (Imai, 1979). However, the effects of seepage is more important during sedimentation and prior to the development of significant intergranular stresses. A technique needs to be developed for estimating seepage effects during sedimentation and extended to include consolidation. The goal is to develop one unified approach which may be used to predict the ultimate settlement of hydraulic fills.

Predictive Analogue

Existing Models: Theories are available to predict hindered and consolidation behavior associated with the sedimentation and ultimate compression of slurried materials (Kynch, 1952; Imai, 1979; Been & Sills, 1981; Yong, 1984). Further work needs to be done to include seepage effects in the combined sedimentation and consolidation theories developed by Been & Sills (1981) and extended by Pane and Schiffman (1985). The resulting development of a practical analogue would be useful in the design of hydraulic fills such as tailing impoundments and in the remediation of hazardous and other waste materials such as dredged sediments, lagoon sludges, slurried ash, and other saturated fine grained materials. In addition, it would be useful, in some cases, in evaluating natural soils deposited in a water environment. Preconsolidation pressure is observed in many marine and lacustrine deposits, which, based on their geologic history, would be normally consolidated. This preconsolidation pressure may, in fact, be influenced by drainage forces which occur during deposition. Any developed technique would therefore be useful in better estimating the stress history of such a deposit.

The existing theories include relationships between solids concentration and settling velocity (Yong, 1984). Kynch (1952) stated that the velocity is related to particle density by the equation:

$$V_{(c)} \frac{\partial c}{\partial \zeta} + \frac{\partial c}{\partial t} = 0 \quad \text{(Eq. 1)}$$

where c is the solids concentration, and V is particle velocity in direction, ζ, during time, t. This relationship has been connected to a general consolidation theory by an interaction coefficient relating effective stress to void ratio (Pane & Schiffman, 1985). One model considers the upward diffusion of water and downward convection of solids in the analyses of sedimentation (Yong and Elmonayeri, 1984). In adapting this model for seepage forces, the net fluid flux would be the difference between the upward flux presented in the reference and the downward water movement caused by seepage. Using similar terms as designated by Yong and Elmonayeri (1984), the net flow would be:

$$q_{fs} = -K \frac{\partial \Psi}{\partial \zeta} + Kp \frac{\partial \Psi}{\partial \zeta} \qquad \text{(Eq. 2)}$$

where Kp is permeability and $\partial\Psi$ is the potential gradient in the $\partial\zeta$ direction. While this could be incorporated into the continuity equation, the resulting relationship would be cumbersome to evaluate for most practical application.

Model Criteria: A technique needs to be developed to incorporate seepage effects into the sedimentation process. To be a useful tool, this model should be:

- relatively simple and practical to use,
- reasonably accurate in predicting field strains and within the precision necessary for the proposed application,
- sensitive to various boundary conditions,
- provide a continuous predictive mechanism which incorporates both hindered, and traditional consolidation settlement.

None of the previous relationships between solids concentration and settling velocity consider particle momentum caused by seepage forces. Since momentum is influenced by velocity, it is probable that the analogue should consider seepage velocity rather than either hydraulic gradient, seepage force, or some other parameter.

SUGGESTED MODELS

Basic Approach: In considering an approach to this problem, it was felt that it would be desirable to express seepage effects as an equivalent effective stress. In this manner, continuity could be maintained between dispersion and consolidation strains. As will be discussed in the next section, more extensive work is needed in the development of such a model. It is not the intent of this paper to propose a predictive model, which fully evaluates the effects of seepage on the sedimentation process. However, at this time, it is useful to document some thoughts which occurred during field observations and the laboratory measurements made on the settling process of a hydraulic fill.

Model Development: Since fly ash is predominantly spherical in nature, Stokes law was considered where the relationship between sphere diameter and free fall velocity in a fluid is expressed as:

$$D = C \quad (V_{sk})^{\frac{1}{2}} \qquad \text{(Eq. 3)}$$

where V_{sk} is the Stokesian velocity of particle flow due to gravity and C is an constant depending upon the fluid viscosity. The mean particle size, D, falls though a hydrostatic fluid at velocity, Vsk. When flow is occurring, the particle is falling at the combined gravitational and velocity effects or:

$$V_T = V_{sk} + V_f \qquad \text{(Eq. 4)}$$

where V_T is the total velocity and V_f is the seepage velocity of fluid flow. The seepage velocity may be expressed in Darcian terms as:

$$V_f = \frac{K}{\phi} \frac{dh}{d\ell} \qquad \text{(Eq. 5)}$$

where K is the permeability, ϕ is the porosity, and dh/dl is the hydraulic gradient.

The combined velocity, expressed in Eq. 4, causes a particle to have more momentum than a particle in only gravitational free fall. To account for this, the "equivalent" effective diameter is increased to account for additional seepage velocities. The larger effective diameter may be expressed as:

$$D' = C \quad (V_{sk} + V_f)^{\frac{1}{2}} \qquad \text{(Eq. 6)}$$

where D' is the "equivalent" effective diameter. This "equivalent" effective diameter would have the effect of increasing the "equivalent" mass which, in turn, would increase the "equivalent" effective stress during the settling process. To incorporate this into a model, the "equivalent" particle mass, Mp' is expressed as:

$$Mp' = G_s V'_p$$

$$Mp' = G_s \frac{\pi D'^3}{6} \qquad \text{(Eq. 7)}$$

where Gs is specific gravity of ash, Vp' is the volume of the "equivalent" sphere, or $\pi D'^3/6$. The "equivalent" dry density of the soil mass with fluid flow, expressed as a ratio of density with only free particle flow, γ_d, is:

$$\gamma'_d = \frac{G_s(1/(1+e_f))\ \gamma_w}{G_s(1/(1+e))\ \gamma_w} \gamma_d$$

$$\gamma'_d = \frac{1+e}{1+e_f} \gamma_d \qquad \text{(Eq. 8)}$$

The volume of solids may be expressed as the product of the sphere volume and number of particles, n, or $n\pi D^3/6$. With this, the void ratio is:

$$e = \frac{1-(1/6)(n\pi D^3)}{(1/6)(n\pi D^3)}$$

and

$$e = \frac{6}{n\pi D^3} - 1 \qquad \text{(Eq. 9)}$$

Repeating for the "equivalent" soil and incorporating Eq. 9 into Eq. 8 yields

$$\gamma'_d = \frac{6(1/(n\pi D^3))}{6(/(n\pi D'^3))}$$

$$\gamma'_d = \frac{D'^3}{D^3} \gamma_d \qquad \text{(Eq. 10)}$$

For seepage effects on sedimentation, the "equivalent" effective stress, σ'_v, would then be an increase over the traditional effective stress, σ_v, by a factor of D'^3/D^3. The traditional effective stress, σ'_v, is calculated on the basis of Terzaghi infinite strain, or other theory, and using the appropriate pore water pressures. Seepage effects may be incorporated into the consolidation process using an approach developed by Imai (1979).

Application: To apply this model, only flow rates and boundary conditions are needed to estimate the resulting densities and effective stresses in a soil. However, in practical applications, the seepage rates are not constant because both the hydraulic gradients change as the settled layers build and the permeability changes as the sample densifies. The first problem with layer build-up can be addressed by using discrete time increments and integrating over the period of sedimentation. The second problem can be addressed by evaluating the changes in permeability with changes in void ratio for a loose sample as presented in Figure 10. Layer build-up combined with adjustments made for changes in permeability will provide a predictive mechanism which can be used to evaluate changes in effective stress.

As mentioned, void ratios could not be measured during the early part of Test #2 due to cloudy nature of fluid. If measurements could be made, then the seepage velocities and equivalent stresses could be calculated. Void ratio vs. a weighted average equivalent effective stress could be plotted in a manner similar to the void ratio vs seepage flow relationship plotted in Figure 9. The slope of this plot could then be used in a fashion similar to using Cc in compressibility estimations from consolidation. By using small discrete time steps the weighted average flow rate can be ignored and the seepage rate during testing plotted directly against void ratio.

In summary, it may be possible to determine the settling characteristics of a hydraulic fill by:

- o dividing the problem into discrete time steps,
- o estimating the hydraulic gradient based on the site-specific field conditions,
- o calculating the seepage velocities for the initial layer,
- o predicting the resulting density, effective stresses and strains caused by these velocities,
- o repeating for the subsequent layers,
- o recalculating the density of the first layer using traditional Terzaghi or finite strain theory for the changes in effective stresses with subsequent deposition.

LIMITATIONS AND FURTHER WORK

This paper summarizes field and laboratory results of geotechnical testing conducted on a slurried fly ash. The limitations of the data include:

- o restriction to only silt-size, granular type ash not highly influenced by flocculation,
- o inability to observe sample height and lack of instrumentation to actually measure densities and pore water pressures,
- o ignoring the influence of particle segregation during testing,
- o evaluating the combined effects of seepage and gravitational influences,
- o constraints with narrow range of seepage velocities observed.

This paper also presents some thoughts on a model relating to the evaluation of seepage effects on sedimentation of a slurried fly ash. Predictive models need to be developed to provide a mechanism for estimating strains for hydraulically placed materials. These models will be useful in the design of waste impoundments and in evaluating the feasibility of various remediation techniques for hazardous waste sludge lagoons and sediment ponds.

An approach which might be useful in predicting seepage strains has been presented. It is realized that the development of this analogue is far from complete, and based only on some preliminary laboratory testing and field observations. While further studies are being conducted, the results to date have been summarized herein to encourage further discussion, testing, and thought experiments. Further work should be conducted to:

- verify the effects of a more broad range of seepage velocities at various hydraulic gradients,
- continue the development of a technique for continuity between seepage sedimentation and traditional consolidation prediction theories both with and without seepage forces,
- extend the testing and modelling to include the time rate of strain with seepage,
- separate out the influence of gravity by conducting experiments on single diameter particles using a long horizontal tube,
- verify predictive models with comparison made between laboratory test results and field observations.

References

Been, K. and Sills, G.C. (1981), "Self-weight Consolidation of Soft Soils: An Experimental and Theoretical Study," Geotechnique, 31, pp. 519-535.

Chae, and Snyder, (1977), "Laboratory Compactions of Fly Ash," Geotechnical Practice for Disposal of Solid Waste Materials, ASCE Specially Conference of the Geotechnical Engineering Division, Ann Arbor, June 13-15.

Cunningham, Lukas, and Anderson (1977), "Impoundment of Fly Ash and Slag-A Case Study," Geotechnical Practice for Disposal of Solid Waste Materials ASCE Specially Conference of the Geotechnical Engineering Division, Ann Arbor, June 13-15.

DiGioia, McLaren, and Taylor (1979), Fly Ash Structural Fill Handbook, EPRI, Palo Alto.

DiGioia, Muyers, and Niece (1977) "Design and Construction of Bituminous Fly Ash Disposal Sites," Geotechnical Practice for Disposal of Solid Waste Materials ASCE Specially Conference of the Geotechnical Engineering Division, Ann Arbor, June 13-15.

DiGioia, and Nuzzo, (1972) "Fly Ash as a Structural Fill" Journal of the Power Division, ASCE, June.

Imai, G., (1979) "Development of a New Consolidation Test Procedure Using Seepage Forces," Soils & Foundations (Japan), 19, pp. 45-60.

Kynch, G.J., (1952) "A Theory of Sedimentation," Transactions of Faraday Society, 48, pp. 166-176.

Pane, V. and Schiffman, R.L. (1985), "A Note on Sedimentation and Consolidation," Geotechnique, 35, May, pp. 69-72.

Srinivasan, Beckwith, and Burke, (1977) "Geotechnical Investigations of Power Plant Waste," Geotechnical Practice for Disposal of Solid Waste Materials ASCE Specially Conference of the Geotechnical Engineering Division, Ann Arbor, June 13-15.

Yong (1984) "Particle Interaction and Stability of Suspended Solids" Sedimentation/Consolidation Models-Predictions and Validations, ASCE, San Francisco, October 1.

Yong, and Elmonayeri (1984) "Convection-Diffusion Analysis of Sedimentation in Initially Dilute Solids-Suspensions", Sedimentation/Consolidation Models- Predictions and Validations, ASCE, San Francisco, October 1.

PROPERTIES OF SLURRIED COAL TAILINGS

By David John Williams[1] and Peter Henri Morris[2]

ABSTRACT: The processes of beaching, sedimentation, consolidation, crusting and rehabilitation which fine grained coal mine tailings initially deposited as a slurry undergo are discussed. The relevant engineering properties at different stages of the life cycle of the tailings are given and measured values for them are compared with values for other coal tailings and similar materials. Ongoing research is aimed at building a framework from which the behaviour of coal mine tailings deposited as a slurry may be predicted.

INTRODUCTION

Tailings which typically comprise 20 per cent of the waste from run-of-mine coal, leave the washery as a very watery slurry with a solids concentration of only a few per cent by weight and are piped to a thickener. In the thickener, flocculants are added to aid sedimentation. Flocculants are expensive and their consumption is a significant cost in the production of coal. The thickener underflow is a slurry with a solids concentration of typically 30 to 40 per cent. The overflow water is essentially clear and suitable for re-cycling to the washery. In many mines, the thickened slurry is pumped to a tailings dam or pond for disposal. Some of the older mines have a less elaborate thickening process and pump tailings slurry with a solids concentration of about 15 per cent to a tailings dam or pond. In other mines, particularly where tailings disposal areas are limited, the thickened underflow is further dewatered in a centrifuge or band press filter. Such facilities are expensive to purchase and expensive to operate due to the large quantities of flocculant which they consume. They use about 0.05 per cent flocculant by weight or about 5 to 10 times the concentration required in a thickener. Flocculation is sensitive to the pH balance of the slurry feed and requires neutral or slightly acidic conditions to be effective. This is particularly the case in centrifuges and band press filters so that both the flocculant demand and pH balance must be continually monitored and the chemical additives varied to ensure optimum operation.

In the past, the philosophy behind the disposal of coal mine tailings was one of containment of the slurry, with little or no attention given to dewatering of the tailings nor to their eventual rehabilitation. In cases where water was in short supply, it was re-cycled to the washery from the pond. Thickeners were first introduced to minimise the amount of water discharged with the tailings. With the increasing quantities of tailings being produced, the diminishing available disposal areas and the tougher legislation governing tailings containment and

---

[1] Lecturer, Department of Civil Engineering, University of Queensland, St Lucia, Queensland 4067, Australia.

[2] Research Student, Department of Civil Engineering, University of Queensland, St Lucia, Queensland 4067, Australia.

rehabilitation, coal mine operators are now being forced to move towards the engineered disposal of tailings. This necessitates an understanding of the physical processes involved and requires the determination of material properties to facilitate prediction of the progress of these processes. It also necessitates the investigation of alternative disposal techniques.

Field and associated laboratory investigations have been underway over the last 2 years, centred on a number of mature coal mines in the Bundamba District of the West Moreton Coalfields near Ipswich in South East Queensland, Australia, and some of the more recently developed coal mines of the Bowen Basin in central Queensland. The processes of beaching, sedimentation, consolidation, crusting and rehabilitation of tailings have been studied. Material properties which have been studied include plasticity characteristics, particle size distribution, particle relative density (specific gravity), vane shear strength and consolidation characteristics.

PROCESSES INVOLVED

Beaching. - Blight and Bentel (1983), in extensive studies of the beaching profiles of metalliferous and diamond mine tailings, were able to reduce all observed beach profiles for a particular tailings material to a single nondimensional master profile of the form

$$\frac{h}{Y} = \left(1 - \frac{H}{X}\right)^n \qquad (1)$$

The variables h, H, X, Y and n are defined in Appendix II and Y/X represents the average gradient along the beach. The exponent n is characteristic of the tailings material and was found by Blight and Bentel (1983) to have values of 1.4, 1.5, 2.0 and 4.0 for copper, diamond, platinum and gold tailings, respectively. They also found from measurements that the average particle size (represented by D_{50}, the size through which 50 per cent of the material passes) at a distance H from the point of deposition is approximately inversely proportional to H. The shear strength of beached tailings was found to decrease from the point of deposition by between 50 and 70 per cent over the length of the beach to ponded water. Other parameters affecting beaching such as slurry concentration and the particle relative density of the solid particles were not investigated.

In an earlier paper (Williams and Morris, 1987b), the authors measured the field beaching profile of a coal tailings beach at Aberdare Colliery in the West Moreton Coalfields of South East Queensland. The coal tailings is deposited at an average solids concentration of 15 per cent by weight. It typically has a plasticity index I_p of 18 per cent and a liquid limit w_L of 43 per cent, and comprises about 50 per cent silt sized particles, 30 per cent clay sized, and 20 per cent fine to medium sand sized particles. The average particle relative density G_s of the all-in tailings is 1.76 and that of the silt and clay sized fraction is 1.84. For the Bundamba District, Norman (1960) published mean densities of 1.28 for coal and 2.10 for mineral matter. Based on these values, the estimated average coal content of the all-in tailings is 42 per cent and that of the fines is 32 per cent.

Tailings beaches form by means of channels which continuously sweep across the width of the beach. In the immediate vicinity of the channel, erosion of the tailings takes place, particularly close to the deposition point where the slurry flow is concentrated and its velocity high. In fitting a dimensionless beach profile, the first 10m from the deposition point was omitted. The effective length of beach was 70m to the ponded water. Beyond that, the slope of the slurry surface steepens underwater. The n value obtained for the coal tailings beach was 1.5 and the average slope about 1 in 100.

The beaching process gives rise to the hydraulic sorting of particles. With increasing distance down the beach the particle size distributions become finer grained, although the distribution near the ponded water is still coarser than the all-in distribution. As a result of the variable densities of the constituents of the tailings, the hydraulic sorting is not simply related to particle size. Analyses presented in Williams and Morris (1987b) show that hydraulic sorting is dominated by the denser mineral particles over the first third of the beach from the deposition point and by the lighter coal particles over the remainder of the beach to the ponded water. It is proposed to further investigate the processes of beaching and hydraulic sorting of coal tailings in a small scale laboratory model similar to that used by Blight et al (1985).

Sedimentation. - The process of sedimentation is a function of the initial solids concentration, the particle size distribution, the chemical composition of the tailings and tailings water, and the concentration of any flocculant and acidic additives. The initial solids concentration is of prime importance as it dictates the degree of interference between settling particles. As the solids concentration and degree of interference increase, the slurry takes on soil-like behaviour. With further expulsion of water from the slurry, solid particles come into contact, consolidation takes over from sedimentation, and the slurry develops a measurable shear strength.

Fig. 1 demonstrates the marked effect of initial solids concentration on the rate and total amount of sedimentation of coal tailings from the Moura Coal Mine at the southern end of the Bowen Basin in Queensland. The tests were carried out in 1 L graduated measuring cylinders on slurries with 15 per cent and 30 per cent solids by weight. The tailings samples were taken from the deposition point at Moura and the tailings water decanted off and stored. The solid particles were then air-dried and measured quantities re-mixed with the tailings water to form the required concentrations. A suspension was formed in the measuring cylinder by repeatedly inverting it, after which the water/slurry interface was monitored with time. To quantify the effect of solids concentration on sedimentation, the times for 50 per cent and 90 per cent sedimentation, s_{50} and s_{90} respectively, may be compared. This is done in Table 1. Also given in Table 1 are the s_{50} and s_{90} values for sedimentation tests on a variety of coal tailings prepared in the same way as the Moura test at 15 per cent solids concentration.

The Moura and German Creek tailings from the Bowen Basin, together with the New Hope tailings from South East Queensland give very similar sedimentation results, indicating similar particle size distribution and chemistry. The Goonyella tailings from the Bowen Basin displays a much slower sedimentation rate and Riverside tailings, from nearby in the Bowen Basin, is slower still to settle out.

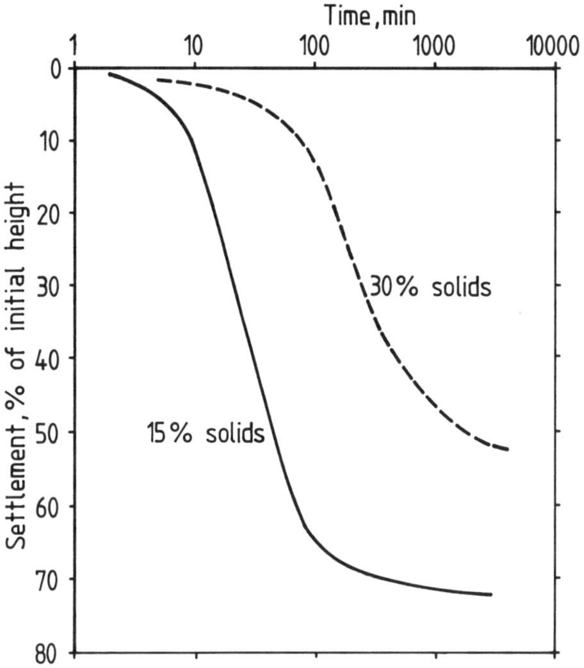

FIG. 1. - Effect of Initial Solids Concentration on Sedimentation of Moura Coal Tailings

TABLE 1. - Sedimentation Times for Different Initial Concentrations and for Different Coal Tailings in Tailings Water

Coal tailings	15% solids		30% solids	
	s_{50},min	s_{90},min	s_{50},min	s_{90},min
(1)	(2)	(3)	(4)	(45)
Moura	27	100	220	1300
German Creek	32	90		
New Hope	25	80		
Goonyella	115	770		
Riverside	250	1100		

From a study of the limited data available on the different coal tailings, presented in Table 2, some explanations for their different semimentation behaviour become apparent. Comparing the data for Moura and Riverside tailings, which have the same dominant clay mineral, the very much faster sedimentation of the Moura tailings is attributable to its much lower fines content, which reduces the influence of the clay mineral type. This is reflected by the very much larger conductivity for the Moura tailings. Comparing the data for Moura and New Hope tailings, the higher fines content of New Hope tailings is balanced by the less reactive dominant clay mineral kaolinite. A comparison of the G_s values for New Hope, Coonyella and Riverside tailings suggests that the sedimentation behaviour of Goonyella tailings would be intermediate between that of the other two, as was observed. It appears that the proportion of fines is the best indicator of sedimentation behaviour, but the dominant clay mineral present is also of importance.

TABLE 2. - Typical Characteristics of Different Coal Tailings

Coal tailings (1)	I_p % (2)	w_L % (3)	% fines[a] (4)	D_{50} mm (5)	G_s (6)	pH (7)	Conductivity $mS^b.cm^{-1}$ (8)	Dominant clay Mineral (9)
Moura			67			7.6	2.9	illite/montmorillonite
German Creek						7.9		
New Hope	24	38	80	0.01	1.73			kaolinite
Goonyella					1.77			
Riverside	25	50	90	0.005	1.81	7.4	1.2	illite/montmorillonite

[a]Silt and clay sized particles

[b]Micro-Seimens per centimetre

The influence of the chemistry of the tailings water on the sedimentation process may be demonstrated by mixing the air-dried tailings in distilled rather than tailings water. This was done for all five tailings at 15 per cent solids and the results are summarised in Table 3. Comparing the results in Tables 1 and 3, all of the tailings except for Goonyella tailings settle out much more slowly when mixed in distilled water, reflecting the presence of flocculants in these tailings waters. The sedimentation of Goonyella tailings appears not very sensitive to the water used, which suggests that natural dispersants and added flocculants are roughly in balance in the Goonyella tailings water or that it is alkaline, rending flocculants ineffective.

TABLE 3. - Sedimentation Times for Different Coal Tailings in Distilled Water at 15 per cent Solids

Coal tailings (1)	s_{50},min (2)	s_{90},min (3)
Moura	96	820
German Creek	290	1800
New Hope	50	140
Goonyella	105	520
Riverside	8500	40000

The effect of adding flocculants may be demonstrated by adding hydrated lime to the tailings slurry. This has been done for Moura and Riverside tailings mixed in distilled water at 10 per cent solids. The results are presented in Table 4. With the addition of 3 per cent or more hydrated lime, a maximum sedimentation rate is achieved for the Riverside tailings and a similar sedimentation rate is achieved for Moura tailings with 3 per cent hydrated lime. A flocculant dosage of 3 per cent would be prohibitively expensive in practice and it is not suggested that it be adopted; however, the exercise serves to highlight the effect of flocculants. For a difficult tailings material such as that at Riverside Coal Mine, the potential increase in the sedimentation rate with the addition of flocculant is about an order of magnitude. For a more well behaved tailings material such as that at Moura Coal Mine, the potential increase in the sedimentation rate with the addition of flocculant is about 2-fold. Clearly, the slow sedimentation of clay-rich coal tailings necessitates the addition of substantial quantities of flocculant, and the sedimentation of such tailings will be extremely slow at high initial solids concentrations.

TABLE 4. - Sedimentation Times for Moura and Riverside Tailings in Distilled Water at 10 per cent Solids with Different percentages of Hydrated Lime

Coal tailings	0% lime		1% lime		3% lime		5% lime	
	s_{50},min	s_{90},min	s_{50},min	s_{90},min	s_{50},min	s_{90},min	s_{50},min	s_{90},min
(1)	(2)	(3)	(4)	(5)	(6)	(7)	(8)	(9)
Moura	48	65			17	42		
Riverside	195	920	73	230	17	90	19	120

Consolidation. - It is difficult to delineate where sedimentation ceases and consolidation begins and a gradual transition may well mask the change-over. It is also difficult to monitor consolidation in isolation in the field and it is usually tested in the laboratory using an oedometer. In a conventional odeometer it is not possible to test a slurry because a finite level of stress must be applied to the sample to start the test. Longworth and McKenzie (de Ambrosis and Seddon, 1986) have modified the conventional oedometer to allow an initial stress application of only 1kPa. However, a further problem arises due to the relatively high initial permeability of the slurry, which necessitates the rapid acquisition of data by computer.

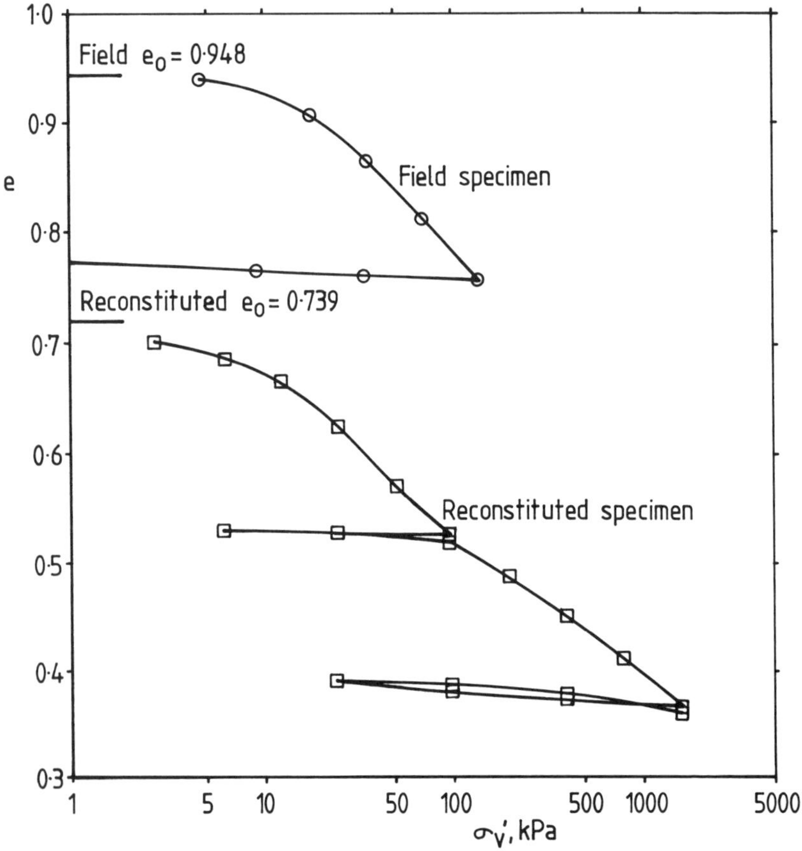

FIG. 2. - Oedometer e versus $\log_{10}\sigma_v'$ Curves for New Hope Colliery Tailings

Work by the authors has highlighted yet another problem. It has been found that a carefully obtained field sample of coal tailings generates a very different set of results in an oedometer test to that generated by a reconstituted test specimen of the same material. This results from the very loose packing which forms on the sedimentation of solid particles from a slurry. As the coal tailings dry, the cardhouse structure apparently undergoes some cementation which results in the structure being largely retained on subsequent wetting up of the tailings. The structure is destroyed on reconstituting the tailings, resulting in a relatively dense packing with particles aligned. Oedometer tests were performed on an 'undisturbed' field specimen and on a reconstituted specimen of New Hope Colliery tailings. The field specimen was allowed to wet up in the oedometer prior to the first increment of load and the reconstituted specimen was prepared at a moisture content just above its liquid limit so that it was just handleable in a conventional oedometer. The results of the tests in the form of voids ratio e versus $\log_{10}$ vertical effective stress σ_v' curves are presented on Fig. 2. The field specimen was readily handleable at a voids ratio of 0.945 (moisture content w of 54.0 per cent), while the reconstituted specimen was only just handleable at an e of 0.739 (w of 42.4 per sent). It can be seen from Fig. 2 that while the initial voids ratios for the two specimens are substantially different, the slopes of the normally consolidated and overconsolidated lines for the two specimens are similar. The compression indices C_c are 0.152 and 0.186 for the field and reconstituted specimens, respectively, and the corresponding swell indices C_s are 0.00830 and 0.00810, respectively.

The parameters calculated from the two oedometer tests are summarised in Table 5. The symbols used in Table 5 are defined in Appendix II. These results highlight the dramatic effect of specimen preparation. The coefficient of consolidation c_v and coefficient of permeability k are each up to 2 orders of magnitude higher for the field specimen, and the coefficient of volume decrease m_v is up to 1 order of magnitude lower for the field specimen. These results demonstrate that carefully obtained field specimens must be obtained for oedometer testing if representative parameters are to be determined. Petersen et al (1982) found that field values of k for brown coal power station ash tailings were 2 to 3 orders of magnitude higher than laboratory values of k determined from reconstituted specimens.

Crusting. - The process of crusting takes place when the surface of the tailings is exposed to the atmosphere. The process involves a lowering through evaporation of the water table beneath the surface of the slurry, accompanied by settlement and the generation of pore water suctions. The pore water suctions will exist initially under near saturated conditions, but as evaporation continues, the near surface tailings will become unsaturated and shrinkage cracks will develop. There is some evidence to suggest that significant desaturation of the tailings commences at about its plastic limit: at a w in the range 20 to 25 per cent, so that only the upper 0.1 to 0.2m may be affected. Rainfall will result in a wetting up of the tailings surface and the infilling of the shrinkage cracks with slurry borne by the rainfall runoff. Runoff may also pond over the crusted surface, causing reversal of the crusting process. Accompanying crusting there is an increase in the shear strength of the near surface tailings. The magnitude of the shear strength achieved close to the surface indicates that the tailings material is capable of supporting a pore water suction in excess of 200kPa. A secondary effect of crusting is that the accompanying densification of the near surface tailings causes a dramatic reduction in permeability. This effect becomes more pronounced as the tailings desaturates, and restricts further dewatering of tailings at depth through the upward drainage of water.

TABLE 5. - Parameters Calculated from Oedometers Tests on New Hope Colliery Tailings

Stress increment kPa (1)	Final e (2)	Mean[a] c_v $m^2.s^{-1}x\ 10^{-7}$ (3)	m_v MPa^{-1} (4)	k $m.s^{-1}x\ 10^{-9}$ (5)
(a) Field Specimen				
0 - 5	0.940	-[b]	0.561	-
5 - 10	0.930	-[b]	0.965	-
10 - 20	0.907	58.8	1.21	69.9
20 - 40	0.865	25.7	1.11	28.0
40 - 80	0.812	15.3	0.719	10.8
80 - 160	0.756	26.2	0.396	10.2
160 - 0	0.779	-	-	-
(b) Reconstituted Specimen				
0 - 2.75	0.703	0.764	7.62	5.71
2.75 - 6.25	0.688	0.833	2.45	2.00
6.25 - 12.5	0.668	1.37	1.95	2.62
12.5 - 25	0.628	1.39	1.92	2.66
25 - 50	0.574	1.48	1.35	1.96
50 - 100	0.527	3.51	0.602	2.07
100 - 6.25	0.530	-	-	-
6.25 - 25	0.529	-	0.0303	-
25 - 100	0.521	-	0.0684	-
100 - 200	0.489	4.87	0.215	1.03
200 - 400	0.452	6.42	0.127	0.800
400 - 800	0.409	10.6	0.0735	0.761
800 - 1600	0.369	11.0	0.0368	0.399
1600 - 25	0.391	-	-	-
25 - 100	0.388	-	0.0292	-
100 - 400	0.380	-	0.0197	-
400 - 1600	0.362	-	0.0105	-

[a]Mean of $(time)^{0.5}$, log (time) and velocity method calculations.

[b]Drainage too rapid to obtain settlement/time data manually.

The crusting process has been monitored in the field by means of sets of levelling plates, over a range of depths, at two locations in New Hope Colliery tailings. Consolidation during the monitoring period is considered to be minor. At one location, where the depth of tailings was about 4m, levelling commenced on 26th May 1986 and has continued at about monthly intervals since then. Levelling plates were installed at 0.1m intervals of depth from 0.1 to 1.7m depth in an area which had previously crusted, but which had recently been covered by ponded water. The results of the monitoring are presented on Fig. 3 as plots of settlement S versus depth z at different dates. Several important features are revealed by Fig. 3. Firstly, nearby ponded water inhibited settlement close to the surface until some time after the ponded water had itself evaporated. Secondly, settlement is consistently retarded at about 1.0m and 1.5m depth. This is considered to be related to remnant crusts as these depths correspond to higher levels of shear strength. A remnant crust, being of lower permeability, settles less than the more permeable material above and below. The remnant crusts become more distinct with time. The inhibiting effect on settlement of rainfall between measurements is also clear, and inundation of the tailings reverses the crusting process. Another feature is that the crusting process extends to considerable depth, eventually approaching the full depth of the tailings at this location of about 4.0m.

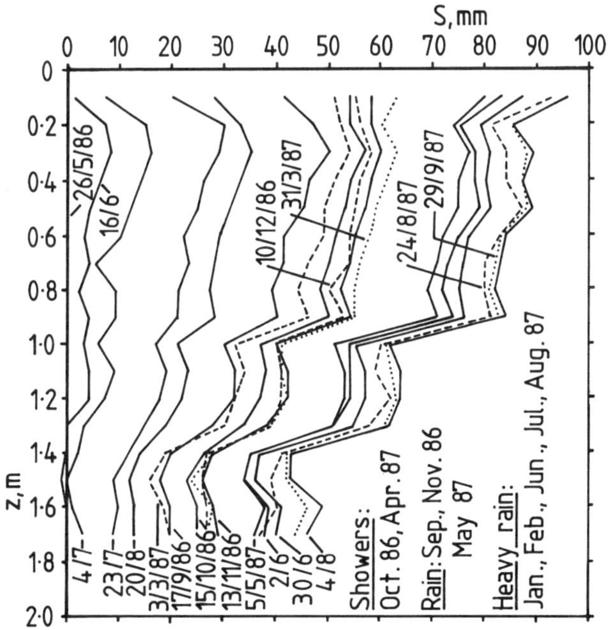

FIG. 3. - Settlement S versus Depth z for 4m Depth of Tailings at New Hope Colliery

Monitoring of the crusting process at a second location, where the depth of tailings is in excess of 10m, commenced on 14 April 1987 and has continued at about monthly intervals since then. In this case, levelling plates were installed at 0.1m intervals of depth from 0.2 to 1.9m depth, again in an area which had previously crusted and had recently been covered by ponded water. The results of the monitoring are presented on Fig. 4. The deeper tailings seem to experience a much less marked crusting gradient, with the crusting process immediately extending to great depth. The influence of nearby ponded water is less marked and there is little evidence of remnant crusts. Since this location is within the main body of the tailings deposit, intermediate crusts may not have had the opportunity to develop.

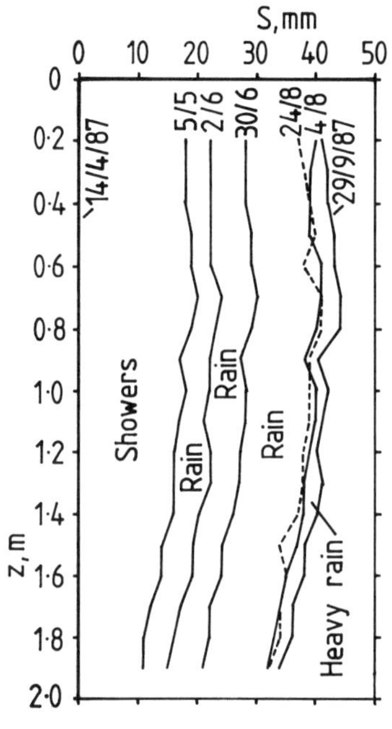

FIG. 4. - Settlement S versus Depth z for in excess of 10m Depth of Tailings at New Hope Colliery

Rehabilitation. - The process of rehabilitation of dis-used tailings disposal areas generally involves the placement of a layer of fill over the crusted surface of the tailings. The bearing capacity of old tailings is a function of the strength and thickness of the crust and the strength and thickness of the underlying softer tailings. The crust may be many times stronger than the underlying tailings, and whereas the strength of the crust increases sharply towards the surface, the strength of the underlying tailings may be relatively constant with depth. The profile of tailings strength with depth will be a function of the age of the tailings, the history of the tailings, its particle size distribution and its chemical composition. As the crusting process is at least partially reversible following rainfall, the bearing capacity of the tailings will vary with time in a complex way.

The bearing capacity of a tailings deposit may be estimated by idealising the strength profile as a two-layer case of a stiff layer over-lying a soft layer and applying the analysis of Brown and Meyerhof (1969). This analysis considers a stiff cohesive layer of finite thickness over-lying a semi-infinite soft cohesive soil. It is based on failure by punching shear through the stiff layer and makes use of a bearing capacity factor modified accordingly. The estimated bearing capacities and corresponding thicknesses of loose coarse reject or overburden (placed at a unit weight of $14kN.m^{-3}$) for various dis-used tailings deposits in the West Moreton Coalfields are given in Table 6. The symbols used in Table 6 are defined in Appendix II. In arriving at the results in Table 6 it has been assumed that the fill plan dimensions are large compared with the thickness of the stiff layer. The strength profiles on which the calculations are based were obtained by vane shear testing using a vane 110m long by 55mm in diameter, rotated by hand using a torque wrench. The vane shear strengths used were the peak values obtained, uncorrected for the plasticity index of the tailings. It can be seen from Table 6 that the bearing capacity of coal tailings deposited as a slurry increases with time since the end of deposition, although at a diminishing rate. The thickness and strength of the crust varies greatly between deposits, but the strength of the underlying soft layer shows a consistent increase in strength with age of the deposit and this appears to govern the bearing capacity of the deposit. Except for very young deposits, dis-used coal tailings could support a substantial thickness of loose fill.

TABLE 6. - Estimated Bearing Capacities of Various Dis-Used Tailings Deposits in the West Moreton Coalfields

Coal tailings (1)	Age years (2)	c_2/c_1 (3)	T m (4)	Bearing capacity kPa (5)	Equivalent loose fill thickness m (6)
New Hope	1.6	10/40 = 0.25	0.9	63	4.5
New Hope	6.0	10/25 = 0.40	0.3	65	4.6
Aberdare	0.4	2/5 = 0.40	0.25	12	0.9
Aberdare	1.1	7/40 = 0.18	1.6	50	3.6
Rhondda No.1	16.0	15/30 = 0.50	0.8	95	6.8

In a previous paper (Williams and Morris, 1987a), the authors described the results of an instrumented trial fill on 2.3 year old New Hope Colliery tailings. A fill embankment 2m high and 10m long by 3m wide at the crest was placed on about 2m depth of crusted tailings. The applied stress was about 27.5kPa at the crest, which compared with an estimated bearing capacity of the tailings of 64.3kPa: a factor of safety of 2.3 against punching shear failure. The ratio c_2/c_1 was 0.23

and thickness of the crust was taken to be 0.3m. Surprisingly, little settlement of the tailings under the fill loading took place. A maximum settlement of 39mm or 1.8 per cent of the tailings thickness diminishing to a maximum settlement of 11mm at the toe of the fill was observed. During fill placement, the tailings heaved up to 5mm between the toe of the fill and 4.5m beyond the toe, before settling. Standpipes placed at a depth of about 1.8m beneath the surface of the tailings (about 0.4m below the watertable) recorded a time-lagged response to the placement of fill and subsequently indicated a slight mounding of the watertable in the tailings beneath the fill. This latter effect results from the fill preventing evaporation of moisture from the underlying tailings.

Settlement was also monitored with depth beneath the tailings surface. It was found that the crust essentially maintained its original thickness and that most of the settlement took place towards the base of the tailings. This may be indicative of minor lateral spread of the softer tailings at depth. The low levels of settlement observed are perhaps not too surprising when it is recalled that the surface of the tailings may have been under a pore water suction and hence effective stress of the order of 200kPa. Although the pore water suction would diminish rapidly with depth, the applied stress of 27.5kPa is small by comparison. It is proposed to construct a second trial fill in two stages, taking the second stage to failure.

Observations of tailings rehabilitation in the Hunter Valley reveal two quite different approaches. If fill is placed before the tailings have crusted, the fill will sink into the tailings forming a loose packing with tailings slurry in the voids. This mixture will then support substantial heights of fill without bearing capacity failure or gross deformation occurring. If on the other hand a crust is first allowed to form, it will prevent the fill from penetrating the tailings to any great extent until the crust fails. Failure is typically characterised by heaving of the tailings at the toe of the fill. Further placement of fill would mobilise a circular slip through the softer tailings underlying the crust.

SOME ENGINEERING PROPERTIES OF COAL TAILINGS

Particle Relative Density. - In earlier discussion it was pointed out that the particle relative density of the solid particles in coal tailings, which is partly related to particle size, has an influence on the process of beaching. It will also influence sedimentation. Particle relative density G_s is plotted against mean particle size D_{50} to a $\log_{10}$ scale for various coal tailings on Fig. 5. The Hunter Valley data shown on Fig. 5 is taken from de Ambrosis and Seddon (1986). The majority of the data fall within the range suggested by de Ambrosis and Seddon for Hunter Valley Coal Tailings.

Dry Unit Weight versus Moisture Content. - It is difficult to sample soft coal tailings without affecting its unit weight. However, estimates of the dry unit weight γ_d may be obtained by thin walled tube sampling and moisture content may be determined with reasonable reliability. This was done for Moura coal tailings by Hollingsworth Consultants (1986). If the tailings are assumed to be saturated, then G_s and e may also be estimated. The resulting ranges of G_s, e and γ_d

for Moura tailings are compared with those for Eastern U.S., Western U.S. and Great Britain coal tailings quoted by Vick (1983) in Table 7. The symbols used are defined in Appendix II. The Moura data is most comparable to that of the U.S.

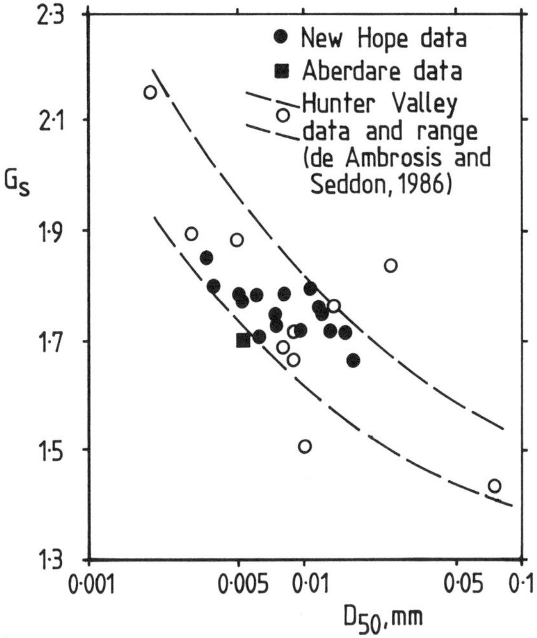

FIG. 5. - Particle Relative Density G_s versus Mean Particle Size D_{50} for Various Coal Tailings

TABLE 7. - **Comparison of G_s, e and γ_d Values for Coal Tailings Worldwide**

Coal tailings (1)	G_s (2)	e (3)	γ_d, kN.m^{-3} (4)
Moura, Australia	1.3 - 1.7	0.4 - 0.8	7.7 - 9.2
Eastern U.S.	1.5 - 1.8	0.8 - 1.1	7.1 - 8.6
Western U.S.	1.4 - 1.6	0.6 - 1.0	7.1 - 11.0
Great Britain	1.6 - 2.1	0.5 - 1.0	8.6 - 13.4

Permeability. - The void ratio e of tailings slurries reduces dramatically as they undergo sedimentation, consolidation and crusting. Coal tailings slurry may initially be deposited at an e value of up to 10, it will settle out to an e of about 3 and consolidate to an e of the order of 1. Crusting will reduce the e to as low as about 0.5. The permeability of tailings is a function of its particle size distribution, plasticity and e. Over the large range of e which coal tailings follow, the permeability may reduce by 3 or more orders of magnitude.

Based on the oedometer test results for New Hope Colliery tailings given in Table 5, plots of e versus k to a $\log_{10}$ scale have been drawn on Fig. 6 for the field and reconstituted specimens. Also shown on Fig. 6 is the approximate range of data obtained by de Ambrosis and Seddon (1986) for reconstituted Hunter Valley coal tailings. The Hunter Valley data straddles the reconstituted New Hope data, indicating similar behaviour for coal tailings from both the Ipswich and Hunter Valley regions.

For a particular location within a tailings deposit, the form of the relationship between k and e suggested by Blight et al (1985) becomes

$$k = a\,e^{b} \qquad (2)$$

In equation (2), a and b are characteristics of the tailings, in particular its particle size distribution, at the location in question. The same form of relationship was used by Al-Tabbaa and Wood (1986) to fit data obtained for kaolin. However, it is likely that as desaturation occurs on crusting, a limiting value of e (say c) will be approached and k will become very low. To provide for this, a relationship of the following form is postulated.

$$k = a(e - c)^{b} \qquad (3)$$

The value of c for coal tailings may be of the order of 0.5. A possible curve described by equation (3) for the New Hope field specimen data is shown on Fig. 6.

Comparing the data for coal tailings with permeability data for metalliferous slimes presented by Vick (1983), it is apparent that the New Hope field specimen has a similar permeability to that of metalliferous slimes at the same e. The permeability data for reconstituted coal tailings plot beyond the lower limit of the permeability data for metalliferous slimes. It is not known how the permeability data for metalliferous slimes was obtained, whether by field or laboratory testing. If it was by laboratory testing it is not known whether or not the slimes were reconstituted, or if so, what effect this would have. Vick (1983) also discussed the difference in magnitude between the horizontal and vertical permeabilities of a tailings deposit, the former being up to 2 orders of magnitude higher.

Coefficient of Consolidation. - The coefficient of consolidation c_v increases dramatically with decreasing e. Based on the oedometer test results for New Hope Colliery tailings given in Table 5, plots of e versus c_v to a $\log_{10}$ scale have been drawn on Fig. 7 for the field and reconstituted specimens. Also shown on Fig. 7 is the approximate range of data obtained by de Ambrosis and Seddon (1986) for reconstituted Hunter Valley coal tailings. The Hunter Valley data falls between the data for the field and reconstituted New Hope Colliery specimens.

Comparing the data for coal tailings and the c_v data for mattalliferous slimes presented by Vick (1983), it is apparent that the New Hope field specimen has a similar c_v to that of metalliferous tailings at the same e. The c_v data for the reconstituted New Hope specimen and for the Hunter Valley tailings plot beyond the lower limit of the c_v data for metalliferous slimes.

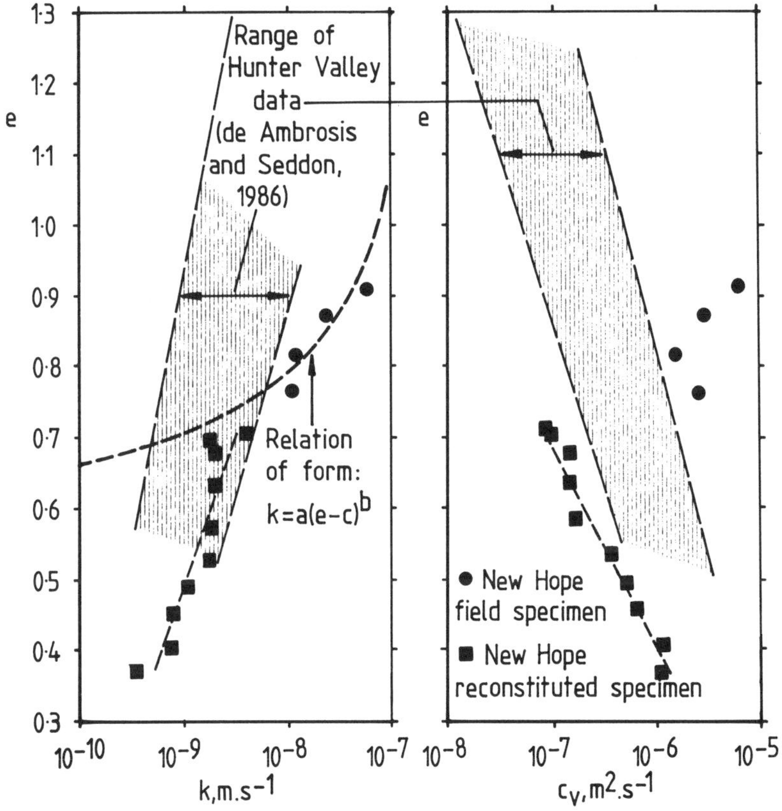

FIG. 6. - Void Ratio e versus Coefficient of Permeability k for Coal Tailings

FIG. 7. - Void Ratio e versus Coefficient of Consolidation c_v for Coal Tailings

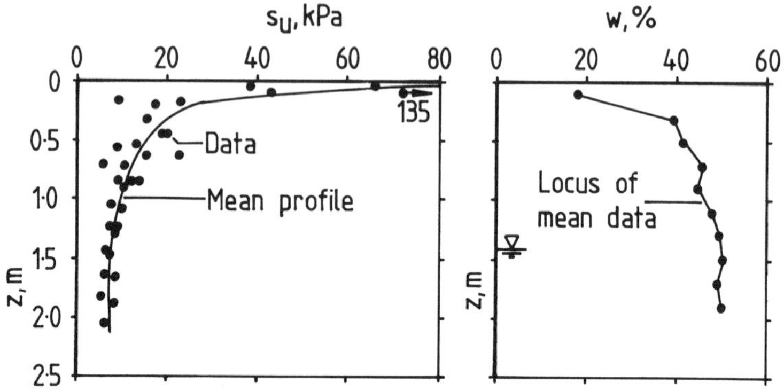

FIG. 8. - Typical Profiles of Undrained Shear Strength s_u and Moisture Content w for New Hope Colliery Tailings

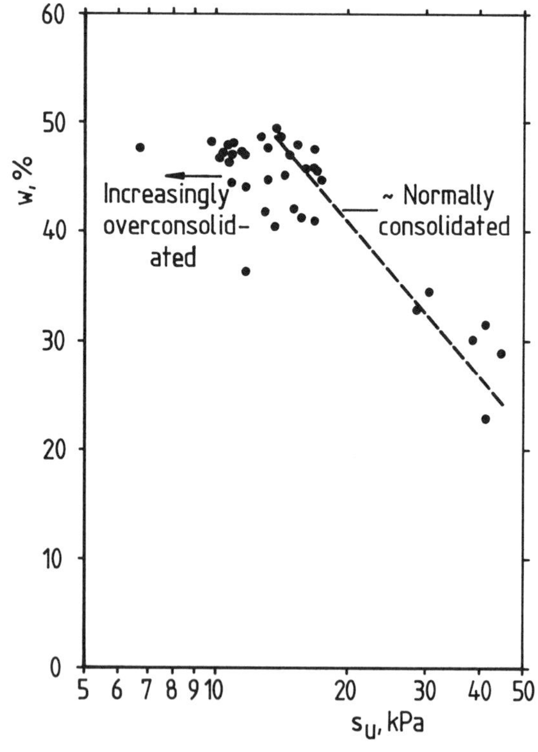

FIG. 9. - Water Content w versus Undrained Shear Strength s_u Correlation for New Hope Colliery Tailings

Shear Strength versus Moisture Content. - Moisture content is a measure of the shear strength of a soil, and for coal tailings which desaturate from the surface it is preferred over void ratio for the purposes of correlation with undrained shear strength. Typical undrained peak vane shear strength and moisture content profiles with depth for New Hope Colliery tailings are shown on Fig. 8. These indicate a distinct crust about 0.3m thick and over 4 times stronger than the underlying softer tailings. A correlation between moisture content and undrained shear strength to a $\log_{10}$ scale for New Hope tailings is shown on Fig. 9. The plot has a similar form to an e versus $\log_{10}$ effective stress plot from an oedometer test and may be interpreted in a similar way. A straight line representing normally consolidated behaviour may be drawn and data at high moisture contents plot to the left of this line. These data largely represent over-consolidated behaviour arising from past cycles of re-wetting and re-drying of the surficial crust and remnant crusts at depth. Since most of the tailings tested were above the watertable they would be under pore water suction, which would have the effect of steepening the slope of the normally consolidated line from that expected for no suction. As the moisture content reduces below the plastic limit (about 25 per cent), desaturation would set in and it would be expected that a limiting value of undrained shear strength would be approached. These features are currently the subject of detailed study.

CONCLUSIONS

The processes involved in the disposal and rehabilitation of coal mine tailings have been described, and some of the important engineering properties have been discussed. This provides the basis for a framework for the prediction of the behaviour of coal mine tailings deposited as a slurry and points to alternative disposal and rehabilitation strategies.

ACKNOWLEDGEMENTS

The co-operation of the managers of New Hope, Aberdare and Rhondda Collieries near Ipswich, Queensland; Moura, German Creek, Goonyella and Riverside Coal Mines in the Bowen Basin, Queensland; and of Coal and Allied Operations in the Hunter Valley, New South Wales, is gratefully acknowledged. Assistance from the State Electricity Commission of Victoria is also acknowledged. Technical assistance from Messrs Peter McMillan and Steve Taylor was invaluable.

APPENDIX I - REFERENCES

Al-Tabbaa, A. and Wood, D.M. (1986). 'Some measurements of the permeability of kaolin'. Report No. CUED/D-SOILS/TR183, University of Cambridge, England.

Ambrosis, L. de and Seddon, K. (1986). 'Disposal of coal washery waste'. Workshop Notes: Mine Tailings Disposal. Dept of Civ. Engrg., Univ. of Queensland, Brisbane, Australia.

Blight, G.E. and Bentel, G.M. (1983). 'The behaviour of mine tailings during hydraulic deposition'. J. Sth. African IMM, Apr. 1983, 73-86.

Blight, G.E., Thomson, R.R. and Vorster, K. (1985). 'Profiles of hydraulic-fill tailings beaches, and seepage through hydraulically sorted tailings'. J. Sth. African IMM, May 1985, 157-161.

Brown, J.D. and Meyerhof, G.G. (1969). 'Experimental study of bearing capacity in layered clays'. Proc. VII Int. Conf. on Soil Mech. and Found. Engrg., Mexico City, Vol. 2, 45-51.

Hollingsworth Consultants. (1986). 'Investigations for tailings dewatering at Moura 2C tailings dam'. Report No. 216200 to Thiess Dampier Mitsui.

Norman, A.J. (1960). Queensland coals, physical and chemical properties and classification. Queensland Coal Board, Brisbane.

Petersen, R.C., Schneider, P. and Raisbeck, D. (1982). 'Latrobe Valley ash properties investigation'. Report No. ID30, State Electricity Commission of Victoria.

Vick, S.G. (1983). Planning, design and analysis of tailings dams. John Wiley and Sons.

Williams, D.J. and Morris, P.H. (1987a). 'Bearing capacity and deformation characteristics of ponded fine grained coal mine tailings'. Proc. Nat. Conf. on Mining and Environment - A Professional Approach, Brisbane, Jul. 1987, 139-144, Melbourne, AusIMM.

Williams, D.J. and Morris, P.H. (1987b). 'Effect of placement techniques on the properties of slurried fine grained coal mine tailings'. Proc. Int. Conf. on Mining and Industrial Waste Management, Pretoria, Aug. 1987, 9 pages.

APPENDIX II - NOTATION

The following symbols are used in this paper:

a = permeability constant of proportionality

b = permeability exponent

c = limiting void ratio

C_c = compression index

C_s = swell index

c_1 = cohesion of crust

c_2 = cohesion of lower layer

c_v = coefficient of consolidation

D_{50} = size through which 50 per cent of particles pass

e = void ratio

G_s = particle relative density (or specific gravity)

H = distance down beach to a particular point

h = height of beach above pool at point H

I_p = plasticity index

k = coefficient of permeability

m_v = coefficient of volume decrease

n = hydraulic beaching exponent

S = surface settlement

s_{50} = time for 50 per cent sedimentation

s_{90} = time for 90 per cent sedimentation

s_u = undrained peak vane shear strength

T = thickness of crust

w = moisture content (relative to weight of solids)

w_L = liquid limit

X = total length of beach

Y = maximum height of beach above pool

z = depth below surface

γ_d = dry unit weight

σ_v' = vertical effective stress

SHEAR STRENGTH AND VOLUME-CHANGE BEHAVIOR OF COPPER TAILINGS UNDER SATURATED CONDITIONS

By H. W. Chen[1], A.M. ASCE and D. J. A. van Zyl[2], M. ASCE

ABSTRACT

A total of 73 static triaxial compression tests was performed to investigate the shear strength and volume- (pore-pressure) change behavior of a southern Arizona copper mine tailings material under saturated conditions. For the isotropically consolidated undrained (CIU) tests, both strain-controlled and load-controlled loading methods were used, while only strain-controlled loading was used for isotropically consolidated drained (CID) tests.

Selected groups of tests within each of the categories were performed using both conventional and lubricated end platens to evaluate the effects of end boundary conditions. Test-specimen densities and effective confining stresses were systematically varied in these tests. In order to study the residual shear strength and the shear-stress-induced pore-water pressure and volume change at steady state, test specimens were loaded to relatively large axial strains (20 to 35 percent) in this study.

This paper describes the material characteristics, test procedures, and results.

INTRODUCTION

The hypothesis of 'critical void ratio' was first postulated by Casagrande (1936) to relate flow-slide failures to the in situ densities of saturated cohesionless soils. Casagrande described the volume-change behavior of sands associated with shear deformation in a direct shear box. He noted that dense sand expands and loose sand contracts until a certain large horizontal deformation at which both dense and loose sand tend to approach the same 'critical void ratio.' At that point, under a constant shear stress, 'cohesionless soil can undergo any amount of deformation or actual flow without volume change.' Because of the limitations of the testing equipment, the 'critical void ratio' of a soil can not easily be reached in the laboratory. In view of this, Casagrande (1936) introduced the 'lower' critical void ratio. The 'lower' critical void ratio is obtained as follows: A series of

---

[1]Project Engineer, International Technology Corporation, Irvine, California.
[2]Principal, Welsh Engineering, Inc., Englewood, Colorado.

consolidated-drained triaxial tests using the same confining pressure is performed on specimens with different initial densities. The initial void ratio of a test specimen is plotted against the amount of volume change at maximum deviator stress. The void ratio at which the curve passes zero volume change is the 'lower' critical void ratio of the material. The 'lower' critical void ratio is lower in value than the critical void ratio estimated by using only the ultimate volume change of loose specimens. So it is considered to be conservative.

For the last two decades, considerable research has been conducted on the liquefaction potential of saturated loose sands and tailings materials under seismic loading. Most of these studies have used laboratory cyclic triaxial tests. Static-loading triaxial compression tests, especially load-controlled undrained tests, have also been used to study the liquefaction potential of saturated loose granular materials (Castro, 1969; Castro et al., 1982). However, most of the reported static-loading triaxial compression tests have been performed on uniformly graded Ottawa sand or on river sands which have only a small fraction of minus No. 200 sieve material and have relatively rounded particles. Metal mine tailings usually have large amounts of silt-sized slimes, while both sand- and silt-size particles in mine tailings are very angular in shape and have rough surface textures due to the processes of crushing, milling, and mineral extraction. The only published data on the liquefaction potential of silty sand tailings under undrained, monotonic load-controlled loading are found in Highter and Tobin (1980) and Highter and Vallee (1980). In their studies, three mine tailings materials, garnet, zinc, and iron ore, were investigated. Two of the three were uniform sands. Only one, the iron ore tailings, had about 20 percent by weight passing the No. 200 sieve. However, the steady-state line for the iron-ore tailings was not reported.

The primary objectives of the research reported here were:

- o To study the stress-strain and volume- (or pore-water pressure) change behavior of copper mine tailings in the laboratory by means of static-loading triaxial-compression tests.
- o To contribute to the relatively small amount of existing static-loading triaxial-test data on tailings materials as they are found in the literature and to provide information about the effects of different laboratory testing techniques on the test results.

The scope of work was to:

- o Evaluate the influence of initial void ratio and effective confining stress on specimen volume-change for CID tests.
- o Evaluate the influence of initial void ratio and effective confining stress on induced pore-water pressure and residual shear strength at relatively large (20 to 35 percent) axial strains for CIU tests.
- o Perform tests for comparison purposes by using both traditional frictional loading end platens and the rubber disc-silicone greased lubricated loading ends.

- o Conduct comparison tests between strain-controlled loading and load-controlled loading methods.

MATERIAL CHARACTERISTICS

The tailings material used in this investigation was collected from the tailings impoundment of the Cyprus Sierrita copper mine located about 65 km southwest of Tucson, Arizona. The material was collected from the beach surface at a location close to the deposition points and away from the pond.

The tailings material is a well graded, non plastic silty sand with about 37 percent by weight passing the No. 200 sieve. The grain-size distribution is shown in Figure 1; it has a D_{10} of 0.011 mm and its coefficient of uniformity is 12. The specific gravity of solids is 2.74. Both sand and silt particles are angular in shape with rough surface cleavages, as shown in the photomicrograph in Figure 2.

The tailing's maximum dry density obtained from the Standard Proctor compaction test (ASTM D 698) was 1,890 kg/m³ (118 lb/ft³) with an optimum moisture content of 11.4 percent. The Modified Proctor compaction test (ASTM D 1557) yielded a maximum dry density of 2,000 kg/m³ (125.1 lb/ft³) and an optimum moisture content of 9.9 percent.

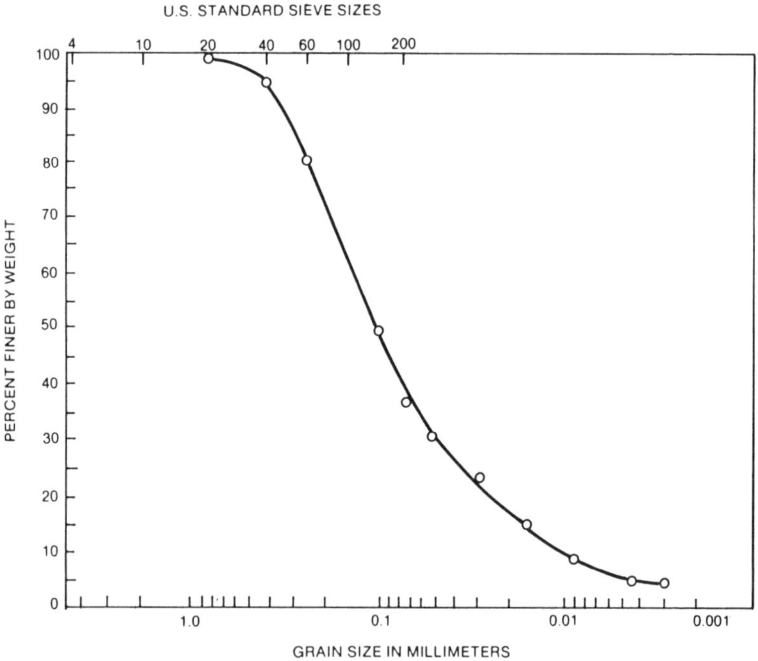

FIGURE 1
COPPER TAILINGS GRAIN SIZE DISTRIBUTION

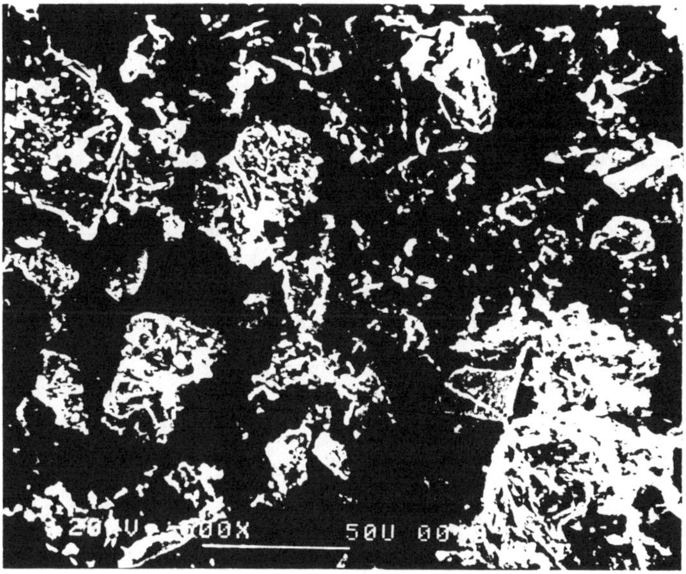

FIGURE 2
PHOTOMICROGRAPH OF COPPER TAILINGS MATERIAL

TESTING APPARATUS DESCRIPTION

The major testing apparatus used in this investigation included a Geonor triaxial test apparatus with modified volume measuring device, modified Soiltest loading frame, electronic instrumentation, and lubricated loading end platens. Detailed testing apparatus description has been presented by by Chen (1984).

Geonor Triaxial Test Apparatus

The Geonor triaxial test apparatus, as described by Andreson and Simons (1960), was used, with modification, to conduct the strain controlled static triaxial compression tests in this study. In addition to the components made by the manufacturer, two volume change measuring devices similar to the 'paraffin volume gage' described by Bishop and Henkel (1962) were incorporated in the Geonor system to separately obtain test-specimen and triaxial-cell volume-change information during back-pressure saturation and shear testing.

Modified Soiltest Loading Frame and Electronic Instrumentation

The Geonor triaxial test apparatus, without the strain-controlled loading press, was used together with a dead-weight loading yoke and frame to conduct all the load-controlled triaxial compression tests in this study.

A Soiltest consolidation test frame was modified to accommodate the triaxial cell. The loading yoke was hung on a cable that circles around a large pulley. The pulley was fixed to the top of the test frame by means of a ball-bearing assembly. The weight of the yoke was balanced by counter weights hung on the other end of the steel cable. The load-controlled test setup was similar to the one reported by Castro (1969).

For those tests in which liquefaction was expected to occur with large subsequent strain rate, an electronic data acquisition system including high output force transducer (load cell), linear variable differential transformer (LVDT), pressure transducer, and high speed strip chart recorder were used.

Lubricated End Platens

Lubricated end platens of plexiglass were manufactured. The diameter of the polished plexiglass end platens was about 30 percent larger than the diameter of the test specimen so that the radial expansion of the specimen at large vertical strain could be accommodated. A cavity inlaid with porous plastic was located at the center of both end platens for the purpose of connecting to drainage lines. The porous plastic was made of high-density polyethylene with an average pore size of 70 microns. A 0.76-mm- (0.03-inch) diameter steel pin was forced into the porous plastic disk and secured with about 6 mm sticking out. The steel pin penetrated the test specimen to prevent the specimen from sliding sideways during shear testing. The effects of the presence of the pin on the stress-strain behavior of soil specimens was assumed to be negligible in this study. Two 0.25-mm (0.01-inch) thick rubber discs were inserted between the specimen and loading ends. A thin layer of high-vacuum silicone grease was spread uniformly between the rubber discs and between the loading ends and the interior rubber disc to reduce friction. The porous plastic was left in direct contact with the test specimen by means of a hole located at the center of the rubber discs.

TRIAXIAL COMPRESSION TEST PROCEDURES

In this section, test procedures including sample preparation, volume-change measurements for partially saturated specimens, and the strain-controlled and load-controlled loading methods, are presented.

Sample Preparation

All test specimens, 3.57 cm (1.4 inches) in diameter and 8.0 cm (3.15 inches) high, were compacted in a split mold in 10 layers of about equal thickness using a small drop hammer. The number of blows on each layer was kept constant for a given specimen. A molding water content of 10 percent was used so that the compacted silty sand specimen had enough strength to survive handling during the test setup. To check the test-specimen uniformity, two specimens of different densities were each carefully sliced into five sections of about equal thickness using a razor blade. The volume of each oven-dried section was measured using the mercury displacement method. Since the molding water content was below 18 percent, which is the shrinkage limit of the material, the specimens did not change volume during the oven-drying process. It was found that the dry densities of the slices varied from 99.4 percent to 101.4 percent of the mean dry density and the moisture content varied from 96.2 percent to 102.5 percent of the mean moisture content.

Volume-Change Measurement for Partially Saturated Specimen

It was noticed that for specimens with initial densities lower than about 1,520 kg/m^3 (95 lb/ft^3), a decrease in sample volume took place during both the percolation and the back-pressure saturation process. Since void ratio is one of the primary parameters in this study, specimen volume change occurring at any stage of the test had to be monitored closely. Therefore, two twin burette volume change devices were employed, one connected to the cell to track specimen volume change during the saturation and percolation stages, and a second to the specimen to record volume change during consolidation and shear.

The use of a strain-gage band or a lateral-strain indicator to measure specimen volume change directly is not feasible because the process of mounting such devices to a small, loose specimen could cause disturbance or destroy the specimen. Therefore, an indirect method of measurement by monitoring the amount of cell water entering or exiting the triaxial cell was adopted.

To calibrate the triaxial cell, a metal dummy specimen was subjected to the same confining pressures that would be used for real test specimens in the percolation and back-pressure saturation stages of the test. Volume change of the cell was recorded by using the volume-change device for a period of time that also simulates the real test. Any volume-change occurring under these conditions was attributed to deformation of the cell itself. This 'calibration' cell volume-change was then subtracted from the actual test volume change data measured during a test on tailings. The net difference was the volume change of the tailings specimen. It was found that with the two volume-change devices graduated at 0.1 and 0.2 cm^3 per finest division, the overall accuracy of this method was 0.5 cm^3, which corresponded to a void-ratio change of about 0.012 for a specimen volume of 80 cm^3.

Strain-Controlled Axial Compression

When Skempton's B parameter reached 0.95 or greater, the sample was considered to be fully saturated. Volume change was then monitored as the volume of pore water expelled from the sample under the effective consolidation stress.

Based on the measurements of hydraulic conductivity and coefficient of consolidation, a strain rate of 0.17 percent per minute was selected for both the drained and undrained shear tests (Bishop and Henkel, 1962).

Specimen groups with initial densities of 1,310; 1,460; 1,540; 1,620; 1,730; and 1,790 kg/m^3 (82, 91, 96, 101, 108, and 112 lb/ft^3) were prepared and tested under the strain-controlled condition. Effective consolidation stresses of 39, 98, 392, and 883 kPa (5.7, 14.2, 57, and 128 psi) were used. The test was terminated when one or more of the following conditions were reached:

1. Steady-state strength was reached.

2. More than 25 percent vertical strain was reached.

3. Apparent failure plane or planes were formed.

4. Specimen tilted or part of the specimen slid out of the loading ends.

Load-Controlled Axial Compression

All of the load-controlled tests in this study were performed as consolidated undrained tests with lubricated end platens. A dead weight was added to the hanger and the loading yoke approximately every 5 minutes after the vertical deformation rate caused by the previous load increment dropped below 0.025 mm/min. (0.001 in./min.). Smaller weights (each weighing about 1/100 of the total load) were added when failure was anticipated.

The electronic data-acquisition system was used in this phase of the study. The axial load, sample vertical deformation, and pore pressure were read from the digital multimeter or from the slow-running (50 cm/hr) oscillograph chart until the point of incipient liquefaction. Then the strip chart speed was increased to 125 mm/sec in order to record the variation of the transducer output for a fraction of a second. The test was automatically terminated at about 35 percent of vertical strain when the loading hanger hit and was stopped by a wooden block stopper.

ANALYSES AND DISCUSSION OF TEST RESULTS

A total of 73 CID and CIU triaxial compression tests was conducted during this study. The various types of tests and the number of each type are shown in Table 1. The test results are summarized and discussed in this paper while the stress-strain and volume- (or pore-water pressure) change data for all 73 tests can be found in Chen (1984). In general, when lubricated end platens were used, the test-specimen shape of a right cylinder was maintained even at very large strains. This implies uniform stress and strain distribution in the specimen.

Table 1. Type and Number of Tests Conducted

	Strain-Controlled		Load-Controlled	
	LEP[a]	REP[b]	LEP[a]	REP[b]
CIU	24	8	12	0
CID	21	8	0	0

[a]Lubricated end platens.
[b]Regular end platens.

Typical Stress-Strain and Volume-Change Curves for CID Tests

The two types of stress-strain and volume-change curves typically observed during the CID tests are illustrated in Figure 3. Sample A was loose with a dry density of 1,420 kg/m³ (89 lb/ft³, 71 percent maximum Modified Proctor dry density) as prepared, and was consolidated under a low $\bar{\sigma}_c$ of 98 kPa (14.2 psi). The stress-strain curve rose at a mild slope and eventually flattened at large strains (>25 percent). The void-ratio curve shows that the sample contracted continuously and approached a constant volume at large strains.

Sample B was relatively dense with a dry density of 1,800 kg/m³ (113 lb/ft³, 90 percent maximum Modified Proctor dry density) as prepared, and was consolidated under a very low $\bar{\sigma}_c$ of 39 kPa (5.7 psi). The stress-strain curve started with a steep slope, reached an apparent peak at approximately 2.5 percent strain, and then dropped. Multiple failure planes were observed to develop right after the peak was reached. The volume-change curve showed that the sample contracted slightly at the beginning of axial straining then started to dilate. The test was stopped at 9.5 percent strain because the nonuniform state of stress and strain was clear. A dense sample under a high confining stress would behave like Sample A because the tendency to dilate will be inhibited by the high confining pressure.

Lubricated Versus Regular End Platens for CID Tests

Data obtained in this study indicate that, with otherwise identical conditions, the use of lubricated ends generates a slightly lower shear strength, about 1 to 2 degrees lower in terms of internal friction angle ϕ, as compared to the use of regular ends. A flatter initial slope on the stress-strain curve and an apparent tendency of more dilation in volume are also associated with using lubricated ends. In general, the results obtained here agree with those found in the literature where non-tailings materials were tested (Barden and McDermott, 1965; Bishop and Green, 1965; and Raju et al., 1972).

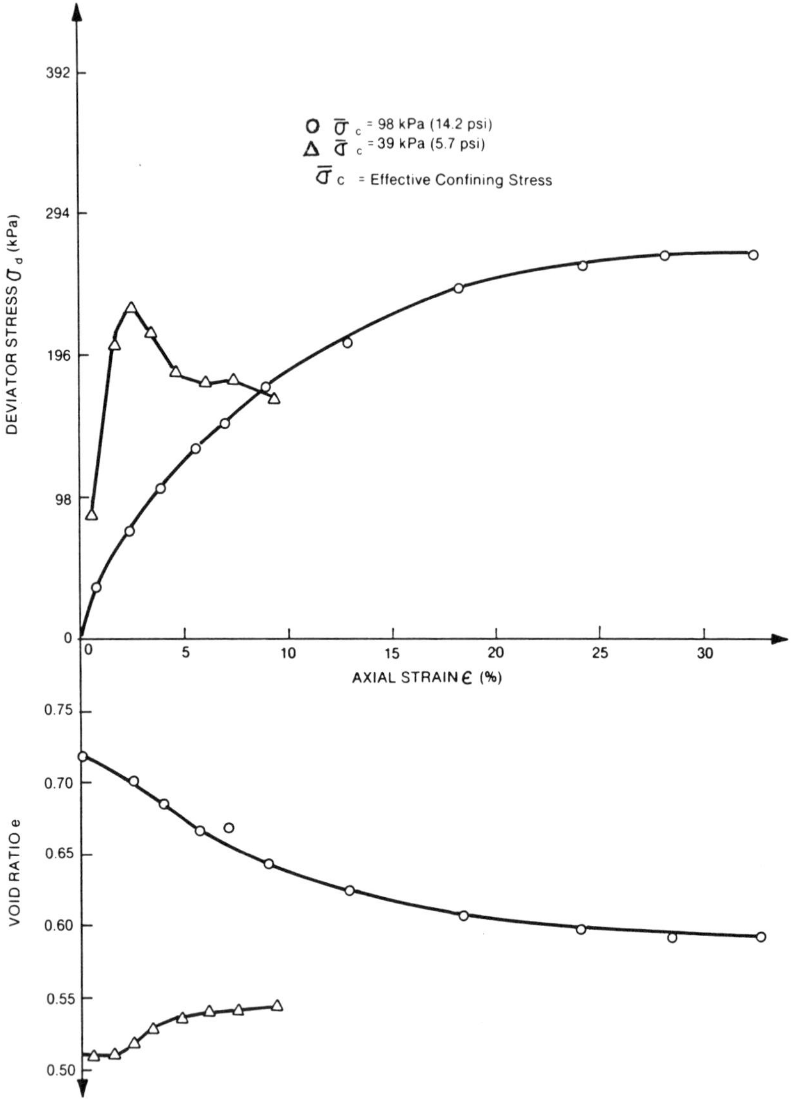

FIGURE 3
TYPICAL STRESS-STRAIN AND VOLUME CHANGE CURVES FOR CID TESTS USING LUBRICATED END PLATENS

Critical Void-Ratio Determination

For all of the CID tests in which the specimens experienced contraction in volume during shear, the ultimate void ratios, corresponding to ultimate axial strains ranging from 17 to 38 percent for different tests, were plotted against the effective confining stresses as shown in Figure 4. The range-of-data scatter for the critical void ratio is about 0.05. In general, however, the critical void ratios obtained from tests with lubricated ends are greater than the ones obtained from regular ends. This gain indicates that lubricated ends apply less restraint and allow more sample dilation. Lines obtained from linear regression analysis are fitted to the data. These lines were called 'e_s lines' by Castro (1969); the same terminology is used in this paper. The critical void ratios for specimens tested with lubricated ends corresponding to the effective confining stresses of 39, 98, 392, and 883 kPa are 0.61, 0.57, 0.49, and 0.44, respectively.

The 'lower' critical void ratio, as defined by Casagrande and Watson (1938) and modified by Taylor (1948), was obtained by plotting volumetric strain at peak deviator stress against void ratio after consolidation for various confining stresses, and interpolating and extrapolating the data to find the void ratio at zero volumetric strain. The 'lower' critical void ratios obtained for specimens tested with lubricated ends are 0.55, 0.52, 0.47, and 0.42 for the same series of effective confining stresses as above. The 'lower' critical void ratio is lower in value than the critical void ratio because the tendency of the material to continue to dilate after the peak deviator stress is usually not accounted for.

Strength Parameters Obtained from CID Tests

Four CID tests were performed on samples with about the same initial density of 1,810 kg/m³ (113 lb/ft³) under different effective confining stresses. Lubricated loading ends were used for all four tests. Assuming that the tailings possessed no cohesion and using the equation

$$\left(\frac{\bar{\sigma}_1}{\bar{\sigma}_3} \right)_{max} = \tan^2 (45 \bar{\phi}/2)$$

effective internal friction angles ($\bar{\phi}$) of 48, 43, 36, and 34 degrees were obtained for specimens under effective consolidation pressures of 39, 98, 392, and 883 kPa, respectively. When Mohr circles were drawn for these results, a curved failure envelope, especially for lower $\bar{\sigma}_c$, resulted. The high friction angles at low consolidation stresses and subsequently the curved failure envelope for this series of dense specimens may be due to:

1. Strong sample dilatancy during shear under low confining stresses.

2. Breakage of individual grains under high confining stresses.

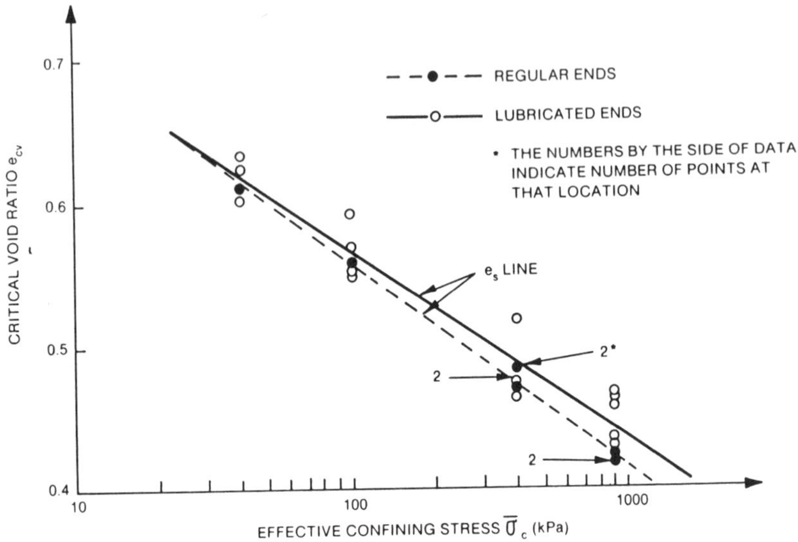

FIGURE 4
CRITICAL VOID RATIOS AND e_s
LINE OBTAINED FROM CID TESTS

No attempt was made during this study to determine the breakage of tailings grains. The volume-change behavior can be explained by the stress-dilatancy theory proposed by Rowe (1962).

Following Rowe (1962), the volume-change correction was made by obtaining a modified $(\bar{\sigma}_1)_{max}$, which is called $(\bar{\sigma}_1)_{mod}$, by dividing the $(\bar{\sigma}_1)_{max}$ by 1 plus the slope of the volumetric strain axial-strain curve and using the $(\bar{\sigma}_1)_{mod}$, together with $\bar{\sigma}_3$ to plot the 'corrected' Mohr circle. The resulting failure envelope is shown as the broken line in Figure 5. It appears to be linear and passes through the origin with a $\bar{\phi}_\mu$ of 34 degrees (the effective friction angle corrected for dilatancy). The uncorrected curved failure envelope is shown in the same figure as a solid curve. Note that the curve is not tangent to the corrected Mohr circles at low $\bar{\sigma}_c$, and it is a straight line at higher $\bar{\sigma}_c$ with a slope of approximately 34 degrees.

A third failure envelope is shown in Figure 5 as a broken line with dots. This straight line is an extrapolation of the uncorrected, curved failure envelope for $\bar{\sigma}_c$ less than 392 kPa (57 psi). This

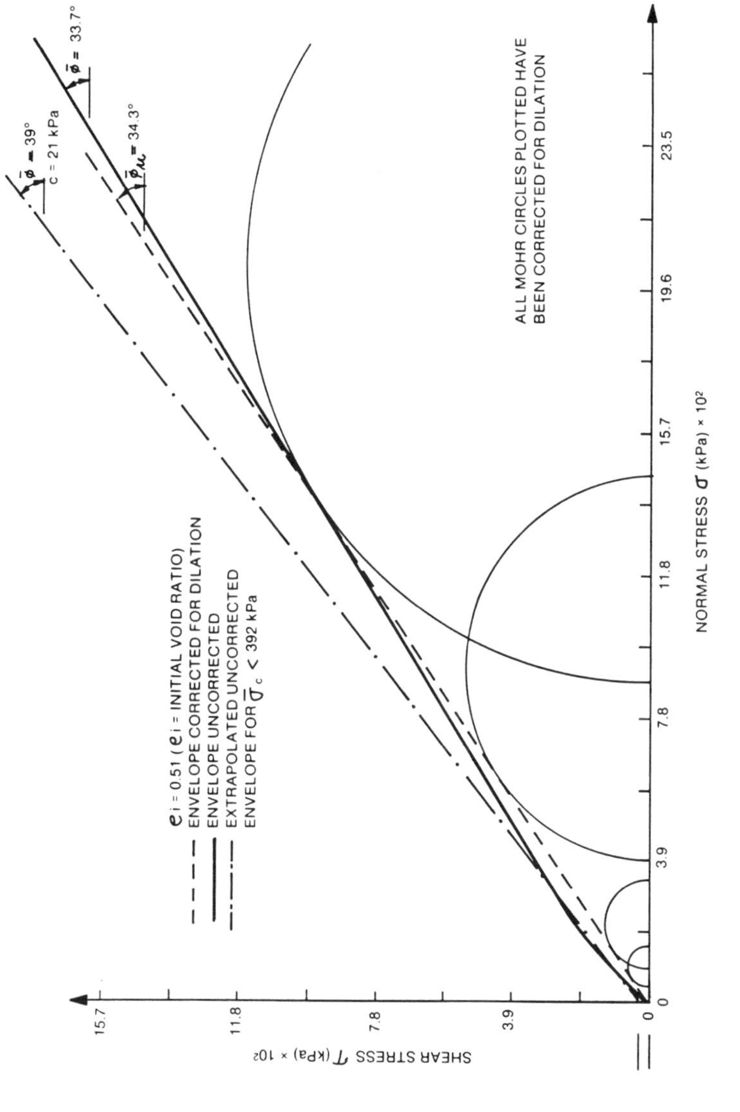

FIGURE 5
FAILURE ENVELOPES FOR CID TESTS

envelope does not go through the origin ($\bar{c}$ = 21 kPa or 3 psi) and has a slope of 39 degrees.

The corrected straight-line failure envelope, which goes through the origin, suggests that the Rowe stress-dilatancy equation works well for this tailings material and that the effects of particle breakage under higher $\bar{\sigma}_c$, as performed in this study, may be small. However, if it is assumed that the field conditions, which include the effective confining stresses, are closely simulated in the laboratory, the shear-strength parameters obtained from the laboratory tests should be used without the stress-dilatancy correction. The curved failure envelope can be approximated as straight lines in the range of applicable effective confining stresses.

Typical Stress-Strain and Pore-Pressure Curves for CIU Tests

Three typical stress-strain and pore-pressure-change curves observed in this study for strain-controlled consolidated-undrained tests with lubricated ends are illustrated in Figure 6. Specimen A with an initial void ratio of 0.886 was consolidated under 883 kPa (128 psi) to a void ratio of 0.559. The stress-strain curve rises sharply to a peak, levels off, and then approaches a steady state before indicating failure at about 16 percent strain when failure planes form. The steady state or residual strength (σ_{dr}) of the sample is about 90 percent of the peak strength. The shear-induced pore pressure also rises sharply until the maximum deviator stress is reached. The slope of the pore-pressure curve decreases after that point and approaches a steady state with a considerable amount of residual effective confining stress ($\bar{\sigma}_{3f}$).

Specimen B, with an initial void ratio of 0.794, was consolidated under 392 kPa (57 psi) to a void ratio of 0.606. The shapes of the stress-strain and pore-pressure curves are basically the same as those of Specimen A, except that the initial slopes of both curves are milder. No failure planes were observed during the test even though an axial strain of about 30 percent was reached before the test was terminated. The stress-strain and pore-pressure-strain curves for Specimen B both approached steady-state conditions after a strain of about 20 percent. The residual strength (σ_{dr}) was about 60 percent of the peak strength. The residual effective confining stress ($\bar{\sigma}_{3f}$) was also smaller compared to that of Specimen A.

Specimen C with an initial void ratio of 0.805 was consolidated under 98 kPa (14.2 psi) and was looser than Specimen B after consolidation. The stress-strain and pore-pressure-strain curves both approached steady state after a strain of about 15 percent. The residual shear strength was only about 25 percent of the peak strength and the induced pore pressure rose to almost the same value as the effective consolidation stress ($\bar{\sigma}_c$), resulting in a very low effective confining stress ($\bar{\sigma}_c$) at steady state.

The behavior of tailings samples under strain-controlled undrained loading was found similar to that of sands under load-controlled undrained loadings as reported by Castro (1969).

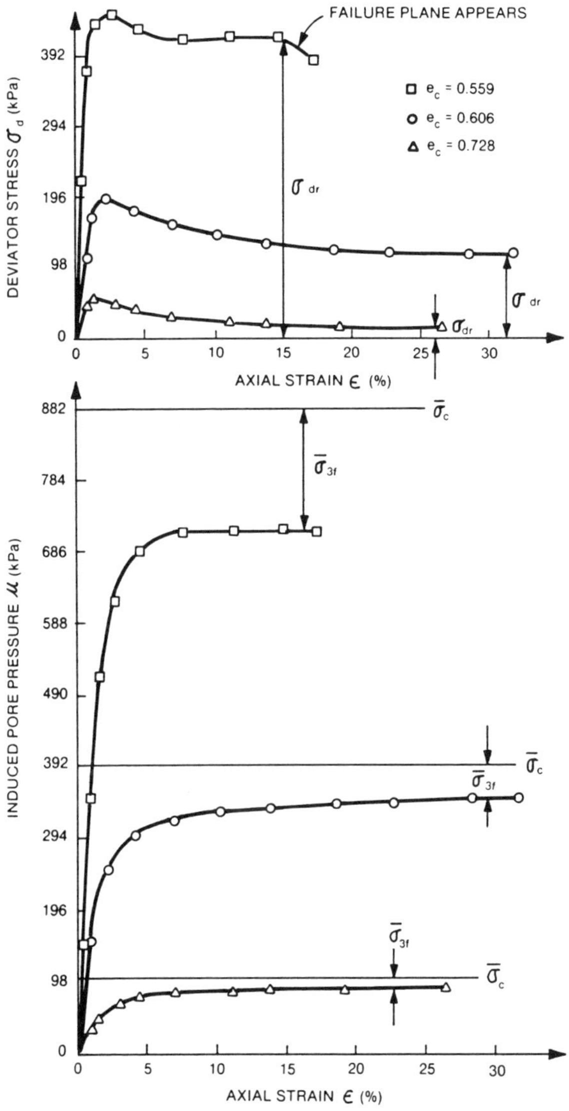

FIGURE 6
TYPICAL STRESS-STRAIN AND PORE PRESSURE CURVES FOR STRAIN-CONTROLLED CIU TESTS

Lubricated Versus Regular End Platens for CIU Tests

Data obtained in this study show that the use of lubricated ends in CIU tests results in lower induced pore pressure than using regular ends. However, the resultant specimen shear strength using lubricated ends can be either slightly higher or lower (about 5 percent) than that obtained using regular ends. These results do not agree with the test results reported by Lee (1978), which indicate that for medium dense Sacramento river sand, the shear strength is about 20 percent higher when lubricated ends are used than when regular ends are used.

The test results obtained in this study indicate that for the range of sample densities and effective confining stresses investigated, both for CIU and CID tests, the differences in shear strength caused by using different loading end platens are relatively small. In view of this, if only the shear-strength parameters are of interest for a tailings material, the benefits of using lubricated ends may not out weigh the extra testing difficulties involved. However, if the volume-change behavior or the critical void ratio is of interest in a study, the advantages of using lubricated ends are obvious. The use of lubricated ends either eliminates or delays the formation of failure planes for relatively dense materials and enables the tests to be carried out to a larger axial strain, or until the steady state is approached. It allows more dilative response and reduces the buildup of shear-induced pore pressures. Therefore, the use of lubricated ends is recommended for tests in which residual shear strength or critical void ratios are of interest.

Failure Criteria and Strength Parameters for CIU Tests

Two failure criteria are generally adopted in triaxial testing of soils, i.e., the maximum deviator stress, $(\bar{\sigma}_d)_{max}$ and the maximum principal effective stress ratio, $(\bar{\sigma}_1/\bar{\sigma}_3)_{max}$. For drained tests, the two criteria coincide. For undrained tests on saturated samples with contractive volume changes, pore pressure builds up under shear due to the tendency of sample contraction and $\bar{\sigma}_3$ decreases with increasing strain. The $(\bar{\sigma}_1/\bar{\sigma}_3)_{max}$ max may occur after the $(\bar{\sigma}_d)_{max}$ is attained. If saturated samples with dilative volume change during shear are tested undrained, negative pore pressures develop because of the dilative tendency of the material. In that case, $\bar{\sigma}_3$ increases with increasing strain and $(\bar{\sigma}_1/\bar{\sigma}_3)_{max}$ may occur before the $(\bar{\sigma}_d)_{max}$ is reached.

For most of the CIU tests performed in this study on loose tailings specimens, the total Mohr failure envelope for peak deviator stress plotted higher than the envelope for residual deviator stress, as shown in Figure 7a. However, when the effective envelope was plotted, the opposite occurred as shown in Figure 7b. The position switch of the envelopes is explained as follows:

> For loose samples, the induced pore pressure at steady (residual) state is higher than the pore pressure at peak deviator stress because the specimens try to contract more during

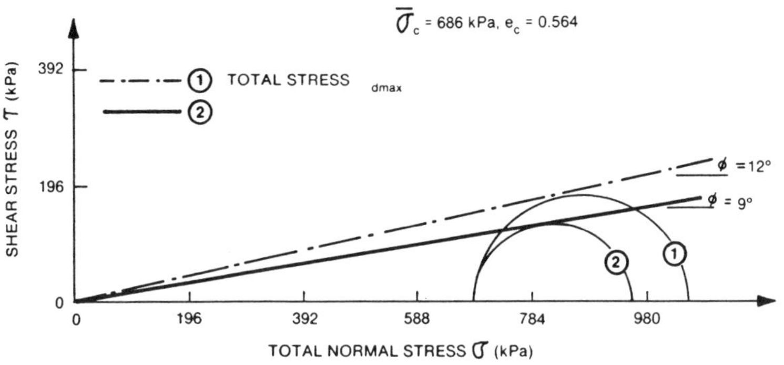

FIGURE 7a
CIU TEST SHEAR STRENGTH PARAMETERS USING TOTAL STRESS ANALYSIS

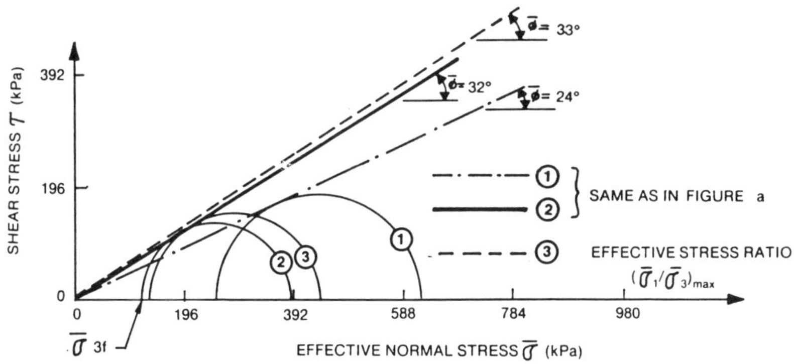

FIGURE 7b
CIU TEST SHEAR STRENGTH PARAMETERS OBTAINED BY USING EFFECTIVE STRESS ANALYSIS

steady state. Therefore, the effective stresses are lower at steady state than at peak deviator stress, and the Mohr circles for the former condition plot closer to the origin than those of the latter. This results in a failure envelope with greater slope, provided that there is no cohesion.

The failure criteria $(\bar{\sigma}_1/\bar{\sigma}_3)_{max}$ was also used to plot the third Mohr circle in Figure 7b. It shows that the $(\bar{\sigma}_1/\bar{\sigma}_3)_{max}$ criterion generates a $\bar{\phi}$ of 33 degrees while the effective residual deviator stress σ_{dr} generates a $\bar{\phi}$ of 32 degrees and the effective $\bar{\sigma}_{dr}$ generates a much lower $\bar{\phi}$ of 32 degrees and the effective $(\bar{\sigma}_d)_{max}$ generates a much lower $\bar{\phi}$ of 24 degrees. The same trend was noticed for all the CIU tests performed in this study except for medium dense samples in which dilative volume response occurred and the use of the effective $(\bar{\sigma}_d)_{max}$ and $\bar{\sigma}_{dr}$ resulted in comparable $\bar{\phi}$.

From the above discussion, use of the $(\bar{\sigma}_1/\bar{\sigma}_3)_{max}$ failure criterion is suggested for shear strength parameter determination for CIU tests performed on loose tailings material.

Steady-State Line (SSL) for CIU Tests

Steady-state line and steady-state strength analyses suggested by Castro et al. (1982) were performed on the CIU test data obtained from this research. This was done for two purposes:

1. To estimate the liquefaction potential of the tailings material,

2. To compare the effect of different testing techniques, e.g., the use of load- and strain-controlled loading methods and lubricated versus regular end platens, on the position of the steady-state line.

Steady state is the state of continuous yielding at constant void ratio and constant shear strength.

In Figures 8a and 8b, three average steady-state lines (SSL) are plotted for strain-controlled tests with lubricated and regular ends and for load-controlled tests with lubricated ends for comparison purposes. In Figure 8a, the void ratio after consolidation, e_c, is plotted against the logarithmic deviator stress at steady state $(\bar{\sigma}_{dr})$, while in Figure 8b, e_c is plotted against the logarithmic effective confining stress at steady state $(\bar{\sigma}_{3f})$. Each of these average steady-state lines is the result of a least-squares regression analysis.

The steady-state line for $\bar{\sigma}_{dr}$ may be used to estimate the steady-state shear strength for Cyprus Sierrita copper tailings of different densities under various confining pressures. The steady-state line for

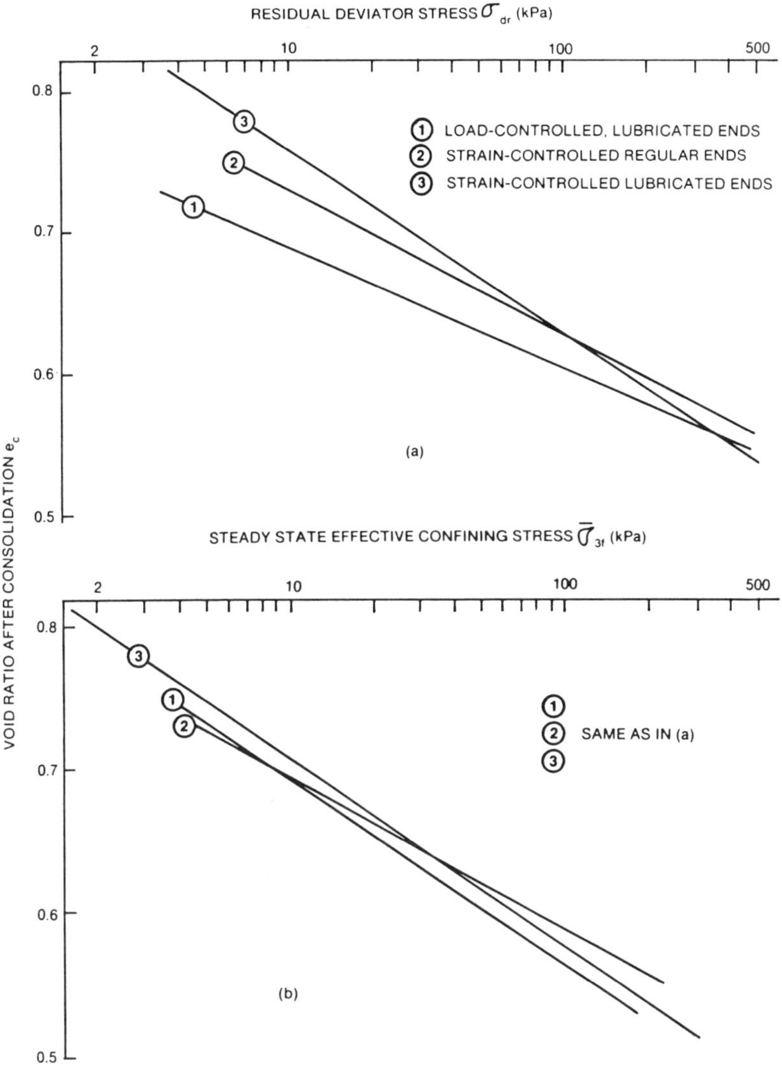

FIGURE 8
AVERAGE STEADY STATE LINES FOR DIFFERENT END PLATENS AND LOADING METHODS

$\bar{\sigma}_{3f}$ can be used to estimate the liquefaction potential of this tailings material.

It is evident from Figure 8b that for a given e_c, the three lines yield comparable value of $\bar{\sigma}_{3f}$. The differences among the three steady-state lines are acceptable in view of the original data scattering. However, for residual deviator stress, $\bar{\sigma}_{dr}$, as shown in Figure 8a, the steady-state line for load-controlled tests with lubricated ends is consistently below the other two lines for $\bar{\sigma}_{dr} \leq$ 200 kPa (29 psi). The discrepancy may arise from two possible sources:

1. The sample failure modes are different for load-controlled and strain-controlled tests. Casagrande (1976) speculated that a 'flow structure' may be formed in test specimens at steady state under a very high strain rate for load-controlled tests.

2. The resolution of the strip chart recorder was not high enough for the load-controlled test in which liquefaction occurs.

The steady-state line obtained from the strain-controlled CIU tests with lubricated ends is designated as No. 1 in Figure 9. In the same figure, the SSL obtained from the load-controlled CIU tests is designated as No. 2 and the e_s line obtained from CID tests with lubricated ends is designated as No. 3. The three lines are parallel to each other in this figure and are almost overlapping for the effective confining stresses between 25 and 200 kPa (3.6 and 29 psi, respectively). This indicates that at the same void ratio the test specimens have the same ultimate strength whether tested under undrained or drained conditions. This conclusion contradicts the results obtained from tests performed on Ottawa sand reported by Castro (1969) and Casagrande (1976). Castro's results indicate that the steady-state shear strength obtained from load-controlled CIU tests is about one fifth of the ultimate strength obtained from strain- controlled CID tests. Compared to Ottawa sand, the copper tailings are highly angular in particle shape, rough in surface texture and have a much higher percentage passing the No. 200 sieve. The different results obtained for Ottawa sand and copper tailings can be attributed to these property differences.

CONCLUSIONS

The conclusions reached from analyzing laboratory test data collected during this study are as follows:

1. For copper tailings samples, the use of lubricated end platens did not significantly affect the shear-strength parameters or the locations of the steady-state lines as compared to the results obtained by using regular friction loading ends. However, lubricated ends did create more uniform stresses and strains as evidenced by the deformed shapes of test specimens. The use of lubricated ends allowed more specimen volume dilatancy. Both the

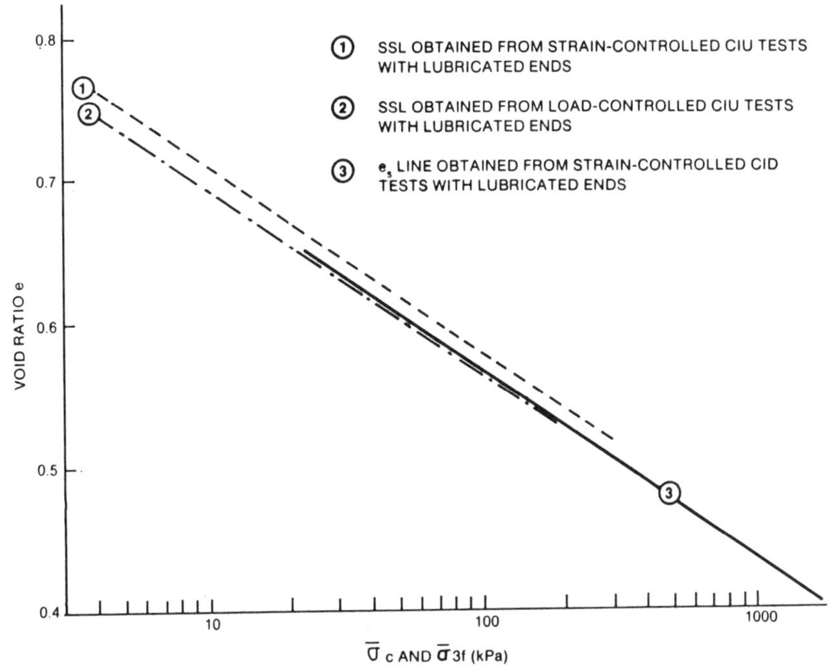

FIGURE 9
STEADY STATE LINES AND e_s LINE

critical void ratios and 'lower' critical void ratios were greater in tests where lubricated ends were used.

2. It is suggested that $(\bar{\sigma}_1/\bar{\sigma}_3)_{max}$ be used as the failure criterion to establish shear strength parameters for CIU tests performed on loose tailings material.

3. When CID tests are performed on tailings materials using low confining stresses, the specimens will dilate strongly, resulting in high internal friction angles with apparent cohesion. If needed, the effect of dilation on shear-strength parameters can be corrected by using Rowe's stress dilatancy theory (Rowe, 1962).

In general, the strain-controlled and load-controlled loading methods generated comparable test results for CIU tests performed on 'identical' test samples. The differences might be within the range of testing errors. These include comparable peak deviator stresses, comparable percentage of axial strains which correspond to the peak deviator stresses, comparable shear-induced pore-water pressures corresponding to both peak and residual (steady-state) deviator stresses, and comparable positions of steady-state lines (zones) for effective confining stresses at steady state ($\bar{\sigma}_{3f}$). There was one exception, the deviator stresses at steady state ($\bar{\sigma}_{dr}$) are consistently lower in value for load-controlled tests than for strain-controlled tests for $\bar{\sigma}_{dr} \lesssim 200$ kPa (29 psi). Due to the low resolutions of part of the data recorded during liquefaction in load-controlled tests, there is greater concern over the accuracies of the calculated $\bar{\sigma}_{dr}$ values for the range below 10 kPa (1.5 psi).

The critical void-ratio line (e_s line) obtained from CID tests for copper tailings material overlays in part the steady-state line obtained from CIU tests. This test result indicates that for the overlapped range of $\bar{\sigma}_c$ and $\bar{\sigma}_{3f}$, at the same void ratio, the test specimens have the same residual or steady-state shear strength whether tested under undrained or drained conditions. This suggests that CID tests, as well as CIU tests, can be performed to estimate the residual shear strength for loose saturated tailings should the large unidirectional deformation occur.

REFERENCES

Andreson, A., and Simons, N. E. (1960). 'Norwegian triaxial equipment and technique,' Proceedings of the ASCE Research Conference on the Shear Strength of Cohesive Soils, Boulder, Colorado, 695-710.

Barden, L., and McDermott, J. W. (1965), 'The use of free ends in triaxial testing of clays,' Journal of the Soil Mechanics and Foundations Division, ASCE, 91, SM6, 1-24.

Bishop, A. W., and Green, G. E. (1965), 'The influence of end restraint on compressive strength of cohesionless soil,' Geotechnique, 15(3), 243-266.

Bishop, A. W., and Henkel, D. J. (1962). 'The measurement of soil properties in triaxial tests,' 2nd Edition. E. Arnold, London, 228.

Casagrande, A. (1936). 'Characteristics of cohesionless soils affecting the stability of slopes and earth fills,' Contributions to Soil Mechanics, 1925-1940. Boston Society of Civil Engineers, October 1940, 257-276. Originally published in the Journal of Boston Society of Civil Engineers. January 1936.

Casagrande, A. (1976). 'Liquefaction and cyclic deformation of sands--a critical review,' Harvard Soil Mechanics Series No. 88, Harvard University. Cambridge, Massachusetts. 1976.

Casagrande, A., and Watson, J. D. (1938). Appendix B II of: 'Compaction tests and critical density investigation of cohesionless materials for Franklin Falls Dam,' report for the U.S. Engineer Corps. Boston. 1938.

Castro, G. (1969). 'Liquefaction of sands.' Harvard Soil Mechanics Series No. 81, Harvard University, Cambridge, Massachusetts. January 1969. 1-112.

Castro, G., Enos, J. L., France, J. W., and Poulos, S. J. (1982). 'Liquefaction induced by cyclic loading,' Geotechnical Engineers, Inc. Winchester, Massachusetts. Prepared for the National Science Foundation. Washington, D.C. March 1982.

Chen, H. W. (1984). 'Stress-strain and volume change characteristics of tailings materials,' dissertation submitted to the University of Arizona, Tucson, Arizona, in partial fulfillment of the requirements for the degree Doctor of Philosophy, 1984.

Highter, W. H., and Tobin, R. F. (1980), 'Flow slides and the undrained brittleness index of some mine tailings,' Quarterly Journal of Engineering Geology, 16, 71-82.

Highter, W. H., and Vallee, R. P. (1980), 'The liquefaction of different mine tailings under stress-controlled loading,' Quarterly Journal of Engineering Geology, 16, 147-150.

Lee, K. L. (1978). 'End restraint effect on undrained static triaxial strength of sand,' Journal of the Soil Mechanics and Foundations Division, ASCE, 104, GT6, 687-704.

Raju, V. S., Sadasivan, S. K., and Venkataraman, M. (1972), 'Use of lubricated end platens in triaxial tests on sands,' Soils and Foundations, 12(4), December 1972, 35-43.

Rowe, P. W. (1962). 'The stress-dilatancy relation for static equilibrium of an assembly of particles in contact,' Procedure of the Royal Society. London, England. Series A, Mathematical and Physical Sciences. 269(1339), October 1962, 500-527.

Taylor, D. W. (1948). 'Fundamentals of soil mechanics,' John Wiley and Sons, New York, New York,1948.

STRENGTH FOR STATIC AND DYNAMIC STABILITY ANALYSIS

Steve J. Poulos, M. ASCE[1]

ABSTRACT

One of the larger sources of error in stability analysis often occurs due to improper selection of shear strength. A review of case histories indicates that most reported failures have occurred in soils that have a large ratio of peak to steady state strength in the field. Therefore, use of the peak strength for stability analysis may result in a metastable embankment, even though the designer may believe it is stable. Progressive failure, e.g., due to earthquake shaking, can then lead to failure.

Designs should be based on a minimum factor of safety of one on the <u>in situ steady state strength</u> for the loading condition involved to avoid the metastable condition. Factors of safety on peak strength should be high enough that the factor of safety on the steady state strength will not be less than one, unless adequate field evidence and experience justify the less conservative design.

INTRODUCTION

Various standards have been established, based on a great deal of empirical experience, for the factors of safety that are acceptable for design of embankments and their foundations. In nearly all cases, the factor of safety standards are based on a determination of the peak shear strengths of the soils involved.

One exception is the case of highly dilative clays and clay-shales. The drained steady state strength often is used for judging their stability (Skempton, 1964). Usually a factor of safety slightly greater than one is sought. A more recent exception is the recommendation to use undrained steady state strength and a factor of safety greater than one for design of embankments of loose, saturated sands (Poulos, Castro, and France, 1985).

The purpose of this paper is to point out that the <u>steady state strengths</u> of soils ultimately <u>control</u> stability of permanent structures, for both static and earthquake loading, in all cases. The steady state strength (Poulos, 1971, 1981) is the strength at very large strains, i.e., the strength when the soil is in a steady state of deformation. The in situ peak strength is of temporary nature and usually should not be relied upon for the long term. Past failures

---

1 Principal, GEI Consultants, Inc., 1021 Main Street, Winchester, Massachusetts 01890

will be reviewed to demonstrate the need to focus attention on the steady state strength.

REVIEW OF PAST FAILURES

In Table 1 a number of stability failures that have occurred are listed in a format that reflects the basic factors that control the strength of the soils that probably caused the failures. These factors are the soil composition and structure, the state of the soil in situ, and the loading.

The references for the failures in Table 1 are:

Point bars	U.S. Army (1956)
Waco	Beene (1967)
San Luis	Smart (1984)
Vaiont	Hendron and Patton (1986)
Calif Slopes	Gould (1960)
London clay	Skempton (1964)
Panama canal	Casagrande (1949)
Silo, Canada	Terzaghi and Peck (1967)
Fort Peck	Casagrande (1964)
Calaveras	Hazen (1920)
Lower San Fernando	GEI (1988)
Sheffield	Seed, Lee and Idriss (1969)
Highway, Canada	Lo and Stermac (1965)
Great Salt Lake	Casagrande (1960)
Sensitive clay	Ladd (1972)
N J Turnpike	Poulos (1979)

Table 1 shows that most reported failures occur mainly due to undrained shear of sand, silt, or clay deposits that were contractive (loose or soft) and due to drained shear of dilative (heavily overconsolidated) clays.

Few failures have been reported for the following cases:

Drained shear	- Contractive	- Sand, Silt, Clay
	- Dilative	- Sand, many Silts
Undrained shear	- Dilative	- Sand, Silt, Clay

Failures can occur in all of these cases, but apparently few have actually occurred.

A review of the data for the cases in Table 1 indicates that there is a common thread among most of the failures. In each case the stress strain curve of the soil that controlled the failure would display a substantial peak (Poulos, 1979). The author estimates that the peak strength of the soil in the principal failure zones may have been about two to ten times greater than the steady state strength. Whether the failure was essentially drained, as in the London clays, or undrained, as for the Lower San Fernando Dam and the silo failures, the stress strain curves would be expected to have the general shape shown in Fig. 1.

TABLE 1.—CLASSIFICATION OF SOME FAILURES BY SOIL TYPE AND LOADING

Cause of Failure	Drained Loading				Undrained Loading			
	Contractive		Dilative		Contractive		Dilative	
	Sand	Clay	Sand	Clay	Sand	Clay	Sand	Clay
A. Long term Steady seepage	N	N	N	Waco Vaiont	NA	NA	NA	NA
B. Toe erosion	N	N	N	N	Point bars	Quick clays, silts	N	N
C. Construction	N	N	N	N	Fort Peck Calaveras	Highway, Canada Sensitive clay Silo, Canada Great Salt Lake N.J. Turnpike	N	N
D. Rapid drawdown	NA	NA	NA	NA	N	N	N	San Luis
E. Earthquake	Footings, Chile	N	N	N	Lower San Fern. Sheffield	Quick clays	N	N
F. Excavation	N	N	N	Panama Canal London clay	N	N	N	N
G. Water pressure	Aberfan	None	Müren Debris flows	Calif. slopes	Aberfan	N	N	NA

NOTES: (a) N means no example found. (b) NA means not applicable. (c) Sand here refers to a bulky-grained soil such that grain orientation has a negligible effect on φ_s. Silts may fall in this category if the percentage of platey grains is negligible. (d) Clay refers to platey-grained soils with plasticity such that grain orientation affects φ_s. Clayey silts and some clayey sands may be included. (e) Category G of the causes of failure refers to cases in which water is forced in slowly from the outside, which is a "drained" condition. It can cause dilative soils to dilate and/or the effective stress to decrease so that failure eventually occurs. (f) See text for references to failures listed in this table.

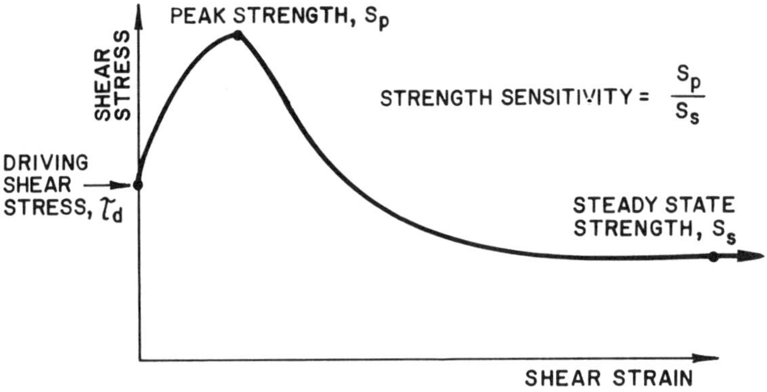

FIG. 1 STRESS STRAIN CURVE SHOWING SUBSTANTIAL PEAK

In the blank spaces in Table 1, where no failures are entered, the stress strain curves normally display no peak, or the peak would be only slightly greater (0% to 50%) than the steady state strength, such as shown in Fig. 2. Loose sands sheared drained and dense sands sheared undrained show stress-strain curves with no peak, curve A in Fig. 2, even at very large strains. Dilative sands and contractive clays sheared drained generally display a small peak, such as curve B in Fig. 2.

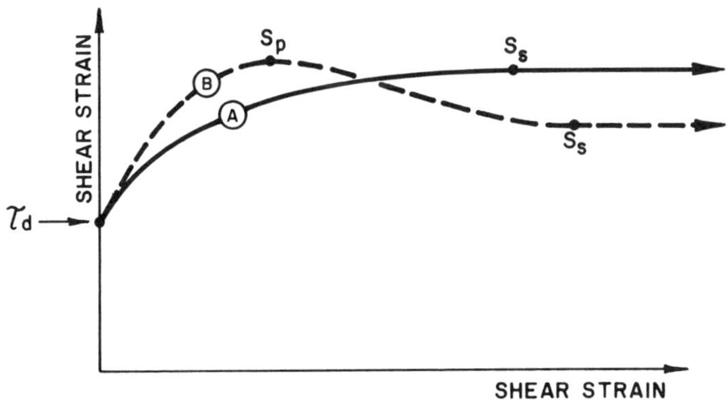

FIG. 2 STRESS STRAIN CURVE SHOWING NO PEAK OR SMALL PEAK

[Dilative clays sheared undrained represent a special case not covered in this paper. If they are sheared through their undisturbed structure where grain orientation has not developed, the stress strain curve may show a small peak. But if time is allowed for internal redistribution, the local negative pore pressures will draw in water to allow local volume increase and gradual grain orientation, even though the overall volume of the mass remains essentially constant (undrained). The resistance eventually can reach the drained steady state strength. If grain orientation had previously existed in the plane of shear, the material in that zone may be close to the steady state "structure." The stress strain curve would show a small peak. The peak is similar to that observed when reversing direction in a direct shear test after the original structure of the soil had been disturbed by previous reversals.]

The topic of this conference is hydraulic fill structures. Such embankments often exist or are placed in a relatively loose state and they remain saturated over large zones. Therefore, it is appropriate to note in Table 1 that most of the publicized failures have been in deposits of saturated soils that were contractive, i.e., had a tendency to decrease in volume when subjected to pure shear, and were loaded in undrained shear. The balance of this paper will be focused on the failure of contractive sands sheared undrained, although the main points are applicable for all soils.

STABLE, UNSTABLE AND METASTABLE CONDITIONS

Consider the slope shown in Fig. 3. Along a potential failure surface there exist driving shear stresses, τ_d, (Poulos, Castro and France, 1985) due to the weight of the mass and due to any other loading that may be permanent. The driving shear stresses can be estimated using common methods of static stability analysis, such as described in manuals of the Corps of Engineers (U.S. Army, 1970).

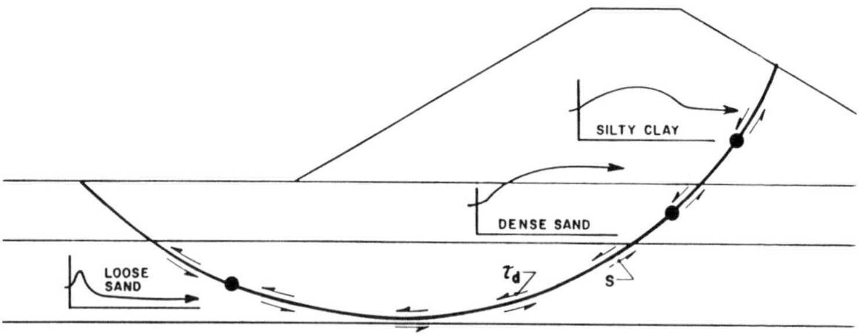

FIG. 3 TRIAL FAILURE SURFACE SHOWING STRESS STRAIN CURVES

Driving shear stresses are stresses that tend to drive a stability failure if the resisting forces are too low. The in situ stresses beneath level ground are not driving shear stresses. A static stability analysis for such a case would result in zero driving shear stress because there are no unbalanced forces to cause a stability failure. The oscillating stresses that occur during an earthquake also are not driving shear stresses. These rapidly alternating seismic stresses can cause strain and, therefore, can trigger a stability failure, but the driving shear stresses, which are unidirectional, are the stresses that cause the failure if the shear resistance drops. The average driving shear stress is statically determinate, although the distribution of shear stress is dependent on the stiffness of the soil layers, until failure occurs in each soil.

The shapes of the stress strain curves for the soils along the failure surface may vary considerably, as shown in Fig 3. The factor of safety along a trial failure surface is given by:

$$F = \frac{\int S \, \Delta L}{\int \tau_d \, \Delta L} \qquad (1)$$

where: F = Factor of safety
S = Shear strength
L = Arc length along trial failure surface
τ_d = Driving shear stress

Factors of safety are applied when designing embankments to take into account the uncertainty of the measured strengths and computed stresses and lack of knowledge about the composition and distribution of soils. Larger factors of safety are frequently recommended if the consequences of failure are large (Casagrande, 1964). If the peak strength is used when calculating the factor of safety, one must also take into account progressive failure (Poulos, 1971), as will be illustrated further below.

Figure 4 shows the definitions of the stable, unstable, and metastable conditions. It is assumed that the shape of the stress strain curve in Fig. 4 is the correct shape <u>in the ground</u> for the loading condition being considered. [To obtain the in situ stress strain curve, one must test the in situ soil and one must attempt to reproduce the in situ structure, state, and loading, be it drained or undrained, after the embankment has been built, when the design loading conditions are to be imposed.] If the weighted average of the driving shear stresses along the failure surface is greater than the weighted average of the peak strengths along that surface, the factor of safety on peak strength, F_p, is less than one and failure occurs. The mass is unstable. But if the average driving shear stress is less than the average steady state strength, the factor of safety on steady state strength, F_s, is greater than one and failure cannot occur. The mass is stable.

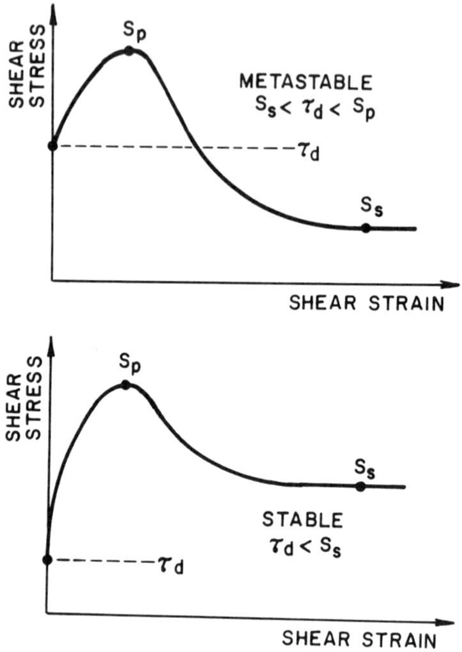

UNSTABLE $\tau_d > S_p$

FIG. 4 DEFINITIONS OF STABILITY CONDITIONS

When the driving shear stress lies between the peak and steady state strengths, the mass is metastable. In that case, any loading, be it rapid or slow, that causes the strain to be great enough to exceed the strain at peak in situ may trigger failure. The embankment is poised for failure when it is designed to be metastable.

Many of the failures listed in Table 1 for dilative clays and for contractive sands, silts, and clays represent cases in which the masses most likely were metastable with respect to the loadings that caused failure. A few may have been in the unstable range.

Sometimes all of the soils along a trial failure surface may have stress strain curves without a peak. As shown in Fig. 5, the metastable condition does not then exist. The mass would be either unstable or stable.

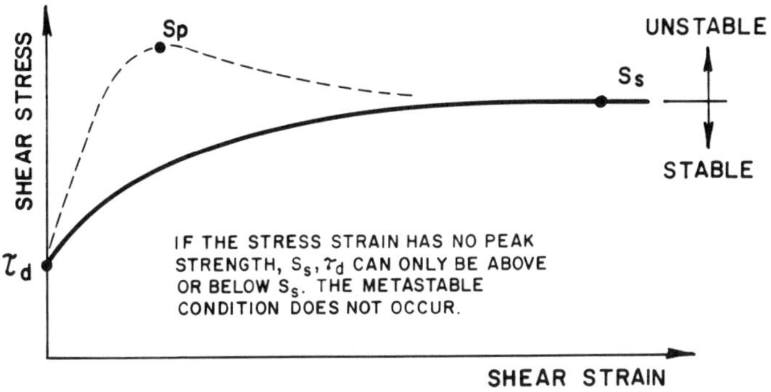

FIG.5 STABILITY CONDITIONS FOR STRESS STRAIN CURVE WITH NO PEAK

There may be a tendency in geotechnical practice to assume that the shapes of the stress strain curves for many soils is as shown in Fig. 5. If there is a peak, but it is not taken into account, then one may be designing an embankment with a factor of safety in the metastable range without realizing it.

It is vital to know the steady state strengths along the failure surface if one is to have confidence that a given design will be stable. If one measures the steady state strength and adjusts the design so that F_s is greater than one, then a stability failure would not occur unless the weakest layer had not been discovered or an error was made in the testing or computations. When all progressive failure has taken place, the steady state strength is still present to resist shear stress.

Note that the symbol S is being used to refer to a shear strength, which is the shear stress on the failure plane at failure and which is defined by:

$$S = q \cos \phi \qquad (2)$$

where: ϕ = slope of the strength envelope. Use peak points for ϕ_p and steady state points for ϕ_s.

q = maximum shear stress in sample, i.e., shear stress on 45 degree plane in triaxial cell or $(\sigma_1 - \sigma_3)/2$.

S is measured directly in rotation shear, direct shear and simple shear tests.

Any given sample of soil in situ subjected to a particular loading condition has two well defined strengths, the peak strength, S_p, and the steady state strength, S_s. The peak strength is the maximum shear stress on the failure plane that occurs during the loading, as shown in Fig. 1. The steady state strength is defined as the strength that exists after the shear loading has eliminated all effects of the original structure of the sample and after any orientation of grains that might affect strength has occurred. At the steady state of deformation, the sample is being sheared at constant volume, constant effective stress, and at constant strain rate (Poulos, 1981).

The peak strength and the peak strength envelope are functions of the soil composition, the structure of the sample, the state (meaning the void ratio, effective normal stress, and shear stress at start of loading), and the loading method. Thus there are numerous peak strengths and peak strength envelopes for any given soil in situ, whether sheared drained or undrained. These strengths can be lost by progressive failure in situ.

In contrast, the slope of the steady state strength envelope, ϕ_s, is a function only of the soil composition (Poulos, 1971). For this reason the steady state strength could be referred to as a "fundamental" strength. This strength cannot be lost by progressive failure. The steady state strength at a given point in the ground is a function only of the soil composition and whether the test is carried out drained or undrained.

The peak and steady state strengths for a given sample can be determined for drained or undrained shear. The notation S_{dp} and S_{up} is used to refer to the drained and undrained peak strengths, respectively. The notations S_{ds} and S_{us} are used to refer to the drained and undrained steady state strengths. The drained steady state strength, S_{ds}, at a given point in the ground is only a function of ϕ_s and the effective stress on the failure plane at failure. The undrained steady state strength, S_{us}, (and the corresponding pore pressure buildup) is a function only of ϕ_s and the in situ void ratio.

STRAINS IN SITU AND PROGRESSIVE FAILURE

For evaluating stability, it is important to be aware of the strains required to reach both the peak and the steady state strengths and the rate of decay of resistance as the peak strain is exceeded. These characteristics should be considered if one desires to depend in practice on mobilizing any strength greater than the steady state strength. In Fig. 6 the stress strain curves for highly contractive specimens of a sand and a clayey silt are compared. Note the large differences between the strains at peak, the rate of decay after peak, and the strains needed to reach the steady state.

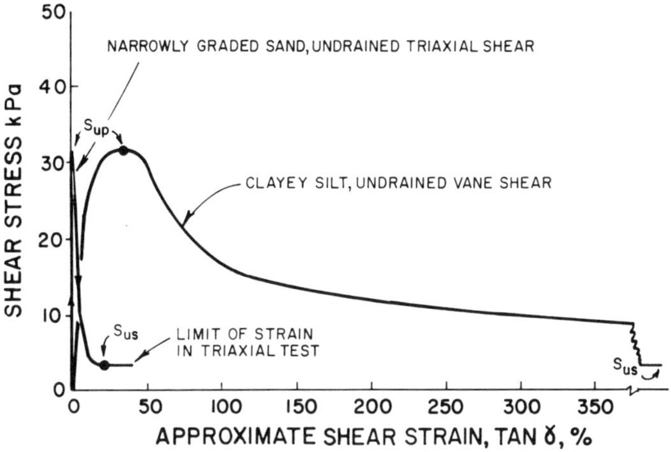

FIG. 6 COMPARISON OF STRESS STRAIN CURVES FOR A SAND AND CLAYEY SILT

The strains shown in Fig. 6 for the clayey silt are so large as to be unreachable in the triaxial or direct shear test. The peak strain for the clayey silt would not be reached in such a test unless very special procedures were used. Furthermore, the steady state strength could not even be closely approached in such a test. Therefore, it was necessary to use the vane test to estimate the stress strain curve shown for the clayey silt.

For comparison, what strains are reachable in the field? Consider an embankment that is being constructed, or one that is subjected to an earthquake, or is subject to fluctuating ground water levels that cause strain in the mass. If a maximum strain of 5% occurs at some location, the overall movements in the field may be large. For example, if the main zone of shear is 30 feet wide and 5% shear strain occurs (where the 5%-value is equal to tan γ times 100), then 1.5 feet of movement will occur. But if the 5% strain is confined to a 6-inch-thick layer, the movement will be only 0.025 feet (1/4 inch)--a hardly noticeable deformation. For many clays failure does not occur until the strain considerably exceeds 5%.

On the other hand, if 1.5 feet of movement were confined to a 6-inch-thick layer, the shear strain in that layer would be 300%. In Fig. 6 it is seen that this strain exceeds the peak strain even for the clayey silt. Therefore, one would not be able to depend on the peak strength of this clayey silt in situ if the strains are sufficiently localized due to layering or other nonhomogeneities.

For contractive sands sheared undrained, the strain at peak is so small that it would not be a good idea to depend on the peak strength regardless of its value in the field. Instead one should design based on the steady state strength.

Common practice in geotechnical stability analysis is to use an average shear stress in the denominator of the factor of safety, Equation (2), and the peak strengths in the numerator. The peak strength is combined with an average shear stress. In other fields, such as structural analysis of concrete, the peak (compressive) strength is combined with a peak compressive stress, and a factor of safety of about two is used. Thus, in structural analysis, a factor of safety of two is applied at the most critical location. That compares with the typical practice of using a factor of safety of 1.5 as the average factor of safety in soils.

Bishop (1952) showed that the maximum driving shear stress along a critical failure surface in a homogeneous elastic embankment is twice as large as the average driving shear stress on that surface. Therefore, if one uses a factor of safety of 1.5 on the average driving shear stress, the factor of safety at the point of maximum stress is about 0.75. For design of concrete structures, a factor of safety of two is used at the most critical point and in geotechnical practice that same factor of safety is commonly only 0.75!

In spite of the above fact, most stability analyses that are made do not lead to failures. A principal reason for the absence of more numerous failures is that the peak strengths, which are the strengths normally used in the analyses, are often only 0 to 50% greater than the steady state strength. When that is the case and if a factor of safety of 1.5 is used on the peak strength, failure will not occur because the factor of safety on the steady state strength is greater than one. That is, the driving shear stress, obtained by dividing the peak strength by the factor of safety, would be smaller than the steady state strength. The mass would be stable.

However, if the peak strength is twice as large as the steady state strength, or greater, then the factor of safety on the steady state strength is less than one if conventional factors of safety are used.

Often failures have been credited to improper measurement of peak strength. And much attention has been paid to improve our ability to measure the peak strength properly. But when failures occur, quite often the major problem resides in the use of the peak rather than the steady state strength. One should assume that progressive strain in nature almost always can be relied upon, in time, to bring the strain beyond the peak strain if the mass is metastable.

Based on laboratory tests, we may be misled into relying on a peak strength and thinking that a strain at peak of, say, 20% is "large." But, as discussed previously, seemingly small movements in situ could result in very large strains - strains that are much larger than those that we commonly relate with laboratory testing. Those strains often are large enough in situ to exceed the strains at peak for most soils, so that the peak no longer is mobilizable in the ground. But in laboratory testing the strains usually discussed are those associated with the triaxial test, which can be misleading and unconservative.

The author feels that it should be assumed that the peak strength of a soil cannot be mobilized in situ for permanent structures unless proven otherwise by satisfactory empirical evidence. For example, it is clear that when cuts are made in London clay, the clay can mobilize part of its peak strength for limited time periods (Skempton, 1964). But slow progressive failure occurs as water flows to the most highly stressed zones. The resistance eventually reaches the drained steady state strength as the clay dilates towards the steady state line (Poulos, 1971).

Quite often the shape of stress strain curve shown in Fig. 5 is reported for saturated soft clays sheared undrained. If such a clay sample were sheared undrained to the large strains that can be reached in the ground, the resistance would drop to the steady state strength. The observed stress strain curve for such a clay represents only a part of the entire stress strain curve. One may be tempted to think that the strength observed in the test is fully mobilizable in situ, especially if a broad peak is displayed in triaxial shear. The conventional factor of safety of about 1.5 on that strength may then result in a metastable embankment.

STABILITY AND DEFORMATION PROBLEMS

The determination of the suitability of a given design, regardless of whether the structure is composed of soil, concrete, steel, wood, or plastic, requires that two criteria be satisfied. First, the strength must be sufficient to prevent collapse, or stability failure. Second, the deformations that occur under design conditions should be limited to those that will be neither uncomfortable nor result in undesirable cracking or other effects. The author refers to these two classes of design problems as <u>stability</u> and <u>deformation</u> problems, respectively (Poulos, 1971).

A well-known example of the relationship between stability and deformation problems was provided by Taylor (1948). He pointed out that footings on sand, so long as they are wide enough, have very high bearing capacity. The factor of safety for a typical design might be 3 to 10 on the drained peak strength of the sand. But the allowable pressure on the footing is controlled by the settlement of the footing or, more precisely, the differential settlement that can be tolerated between adjacent footings. When the footing is narrow, the bearing capacity drops rapidly, so that the allowable bearing pressure controls design. Thus, for footing design, the allowable loads may be controlled by the results of a stability (bearing capacity) analysis or by a deformation (settlement) analysis. Two analyses are needed.

For design of an embankment, one should first design it to be stable, and then check to make sure that the deformations will be small enough to be tolerable. The procedures used to analyze embankment stability (U.S. Army, 1970; Poulos, Castro and France, 1985) are quite different from the procedures used to estimate deformations and to evaluate their importance (Castro, 1986; Makdisi and Seed, 1978; GEI, 1988).

As noted earlier, seismic shear stresses, which fluctuate rapidly during the earthquake, are not driving shear stresses. It is the driving shear stresses that tend to deform the mass downslope during shaking. The seismic stresses cause deformation. If the mass is stable, the factor of safety on steady state strength, F_s, is greater than one and the shear deformation caused by the seismic stresses stops shortly after the earthquake, Fig. 7. Some consolidation occurs subsequently.

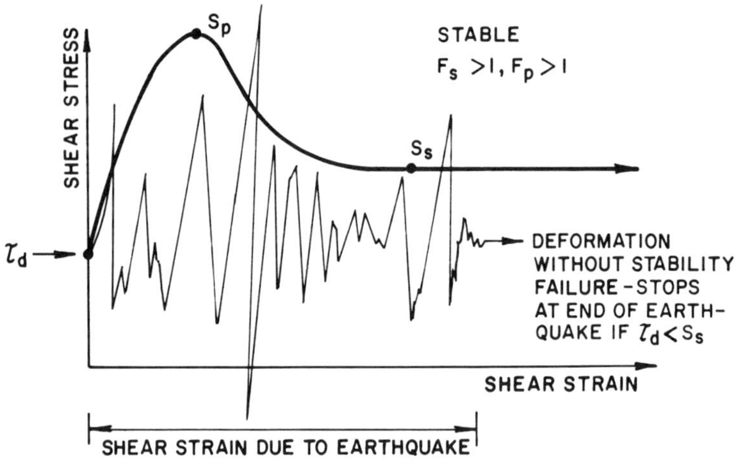

FIG. 7 DEFORMATION WITHOUT STABILITY FAILURE DUE TO EARTHQUAKE SHAKING

If the mass is metastable, $F_s < 1$ and $F_p > 1$, then the same seismic deformations can lead to a stability failure, Fig. 8. The strains due to the shaking superimposed on a driving shear stress that is greater than the steady state strength, may exceed the strain at peak and cause failure. In saturated sands such failures in undrained shear are observed as liquefaction or flow slides. Therefore, earthquake shaking can cause deformations without stability failure or, if the mass is metastable, the shaking can trigger a stability failure.

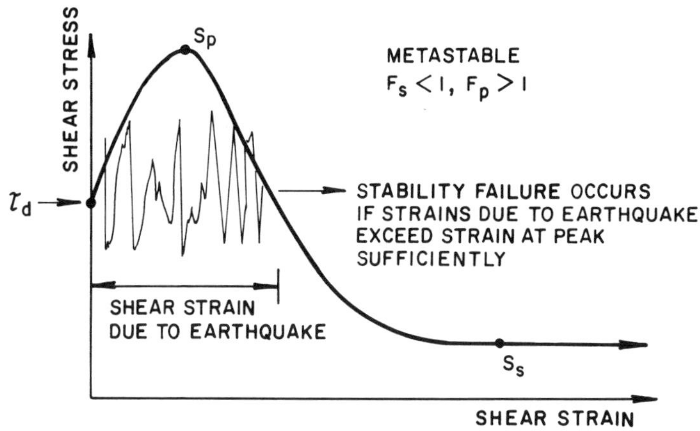

FIG. 8 STABILITY FAILURE TRIGGERED BY EARTHQUAKE SHAKING

The manner in which a deformation or stability problem arises in a slope due to an earthquake can be understood by analogy to a block sliding down an inclined plane. The static coefficient of friction, f_s, might be 20% to 30% greater than the kinetic coefficient of friction, f_k. If the plane is inclined steeper than f_s can support, the block is unstable and slides downslope without provocation. If the inclination is flatter than is consistent with f_k, then no amount of shaking will cause a precipitous downslope movement. However, those vibrations will cause continuous downslope movement so long as they continue.

But if the slope angle lies between the above two angles, the block is metastable. With sufficient shaking the block starts moving, the lower, kinetic friction, takes over, and the block moves to the bottom of the plane.

This analogy explains each type of event that may occur during an earthquake: Downslope movement without a stability failure, downslope movement leading to liquefaction, and the need to apply a certain intensity of shaking to a metastable mass to trigger liquefaction (flowslide).

The stress strain curve for the sliding block on a plane has a peak, just as does the stress strain curve for a highly contractive, saturated sand, silt or clay in undrained shear.

COMMENTS ON CURRENT FACTOR OF SAFETY STANDARDS

Several of the points made above can be illustrated by evaluating standard procedures for selecting factors of safety and shear strengths. In particular, one can understand those cases in which it is unconservative to use peak strengths. For this purpose, the 1970 edition of the Corps of Engineers Engineering Manual, Stability of Earth and Rockfill Dams (U.S. Army, 1970) will be used as a guide. Table 1 in that reference shows recommended factors of safety, and the text contains recommendations for shear strength selection. Two of the cases listed in the table will be discussed below[2].

Case I - End of Contruction The Corps' recommendations are as follows: Assure a factor of safety of 1.3 (1.4 for embankments higher than 50 feet), for both the upstream and downstream slopes. The factor of safety is based on the strength measured in Q tests, assuming the embankment is composed largely of impervious soils in which pore pressure can build up during construction. The Corps notes in the table that this recommendation does not apply to embankments on clay shale foundations.

Interpretation of Case I: The factor of safety on the Q strength is applied to the peak strength. Using the notation in this paper, $F_p = 1.3$ (or 1.4). This factor of safety is smaller than the factor of safety of 1.5 that is required by the Corps for the long-term steady seepage condition. The reason for the lower allowable factor of safety is that the end of construction condition usually does not represent a perfectly undrained condition. Therefore, the peak undrained strength, S_{up}, measured in the Q test should be a low side (conservative) estimate of the peak strength that will exist in situ.

The qualification by the Corps that their recommendations do not apply to embankments on clay shale foundations probably stems from recognition of the need to be very conservative with design on such foundations. Layers with the lowest drained steady state strength must be discovered and the pore pressures in the foundation understood prior to judging the safety of a clay-shale foundation.

Comments on Case I: The recommended factors of safety, $F_p = 1.3$ or 1.4, for Case I are, in the authors's opinion, satisfactory for most embankments that are compacted to at least 90% of the maximum density achieved using the test ASTM D1557 (modified compaction). Impervious soils that are compacted to such a percent compaction usually display pore pressure build up during undrained shear, but the undrained peak strength, S_{up}, is unlikely to be more than about 20% larger than the undrained steady state strength, S_{us}. Therefore, if the factor of safety F_p is 1.3, the factor of safety F_s will be greater than one. The recommendation by the Corps thus leads to a stable embankment in most cases, although the potential embankment deformations may have to be considered during design (Poulos, 1988).

---

2 The author reviewed an early draft of this table circa 1967 at the request of Dr. A. Casagrande and was in full agreement with the published recommendations.

If the clay used in the embankment were compacted to a low percent compaction and/or were highly plastic and tough, then the Q strength, which is supposed to represent the in situ value of S_{up}, could be considerably higher than the in situ undrained steady state strength, S_{us}. The recommended factor of safety of 1.3, being based on the peak strength, could leave the embankment in a metastable condition and the recommended factor of safety would be unconservative. The strains due to the construction loading should be assumed large enough to overcome the strain at peak. Therefore, one should design so that $F_s > 1$. Only if extenuating circumstances exist should one investigate the reasonableness of designing with a less conservative factor of safety than $F_s = 1$ for this case in which the steady state strength is considerably lower than the peak.

The recommendations given by the Corps for Case I did not cover hydraulic fill embankments, in which large zones of contractive saturated sandy materials often exist. In such materials liquefaction stability failures can occur. It is this condition that existed during construction of the Fort Peck dam before its partial failure in 1937 (Casagrande, 1964). The author's recommendation that F_s be greater than one is particularly important for this case. Use of a peak strength should not be considered, since in such soils the peak occurs at such low strain that it is risky to allow the driving shear stresses to exceed the undrained steady state strength at any time.

In the above paragraph, the author has assumed that the hydraulic fill is relatively loose, which is the case for many existing embankments. However, inexpensive techniques are available to compact hydraulic fills during placement. Such compaction is recommended as a standard practice to avoid zones of contractive saturated sands in embankments.

In summary, for Case I the Corps' recommendations seem appropriate to the author for most compacted embankments of impervious clays. But if the clays are soft or highly plastic and tough, whether they exist in the foundation or in the embankment, the more general recommendation that $F_s > 1$ should be applied. The manual does not cover the case of embankments composed of hydraulic fills. For these cases the undrained steady state strength should be estimated conservatively, and the factor of safety on that strength should be somewhat greater than one.

Case VII - Earthquake This case covers earthquake loading superimposed on the steady seepage condition. A factor of safety of 1.0 is recommended by the Corps. The strength to be used is the S strength or the average of the S and R strength, whichever is lower. The driving shear stress computed from the stability analysis for this case is augmented by adding a horizontal seismic shear stress that depends on the earthquake intensity at that location.

Interpretation of Case VII: The factor of safety being used is again F_p. The strength recommended is the S strength for dilative soils and the average of the R and S strength for contractive soils.

For dilative (overconsolidated) soils, the S strength is used because the negative pore pressures that develop in such soils during undrained shear in the laboratory may not occur in situ (Casagrande, 1959). In the field, gas can come out of solution during shear and the soils may not remain fully saturated.

For contractive (normally consolidated) soils, use of the average of the R and S strengths was originally intended to take into account the fact that the R strength measured in the laboratory on isotropically-consolidated specimens may be low. It had been shown that specimens that are consolidated anisotropically in the laboratory have higher S_{up}-values than specimens consolidated isotropically (Ladd, 1965). Therefore, the strength in situ could be higher than that measured using isotropic consolidation.

Use of the average between the R and S strengths is reasonable for taking into account the effect of anisotropic consolidation without complicating the testing procedures. However, before using this average, it is necessary to be sure that the peak R strength is measured on specimens that are at the same void ratio in the laboratory as they will be in the field when the loading is applied. This means that corrections would have to be made to the peak undrained strengths measured on compacted or undisturbed specimens to adjust the laboratory-measured strength to the void ratio that exists or will exist in situ at the time of loading.

Comments on Case VII: The case of earthquake loading of embankments composed of or on contractive, saturated sands is not covered in the Corps manual, nor is it covered in the most recent guidance on the subject known to the author (U.S. Army, 1983). The author's recommendations for this case are covered in Poulos, Castro and France (1985), Castro and Keller (1988), and Poulos (1988). The general recommendation that F_s be greater than one applies to earthquake loading of saturated sands. The peak strength should not be used.

The Corps' recommendation of using a horizontal seismic force and reducing the allowable factor of safety to 1.0 for earthquake loading is based on the idea that the sum of the driving shear stress and the seismic shear stress may exceed the peak undrained strength and that failure may then occur.

However, there are two quite distinct conditions that can arise during an earthquake:

(1) If $F_s > 1$, which means $\tau_d < S_s$ (Fig. 7), then the seismic shear stresses will cause only deformations, as explained under the previous heading. Even if the peak strength is exceeded during the earthquake, the duration is small and the movements will be limited. No horizontal seismic force need be introduced to make this stability analysis.

(2) If $F_s < 1$, which means $\tau_d > S_s$, then the earthquake can cause enough strain to overcome the strain at peak so that only the steady state strength can be mobilized. A liquefaction (flow slide) failure

can then be triggered. Again no horizontal seismic force need be introduced to evaluate stability.

The F_s-value should be known in all cases when evaluating the effect of an earthquake on an embankment. This information is particularly important for contractive saturated sands because the strain at peak often is small and the peak strength usually cannot be relied upon.

Even if $F_s > 1.0$, a design earthquake with a peak ground acceleration of about 0.2 or greater may cause deformations large enough to influence some design features. For this reason, deformations should be estimated in some cases even if $F_s > 1$ to check that the embankment can tolerate them adequately.

RECOMMENDATIONS

It is recommended that the steady state strength be measured and used to compute the factor of safety for stability analysis. Doing so will avoid relying on the peak strength, which can be destroyed by progressive strains in situ. Techniques for measuring the steady state strength should be improved and disseminated more widely.

If the stress strain curve in situ has a peak, a portion of that strength might be used, but only in those special cases where field evidence demonstrates how much of the peak strength can be relied upon in the ground.

Field and analytical investigations are needed to learn how to take into account the entire shape of stress strain curves in situ. The conditions under which it is satisfactory to permit driving shear stresses in situ that are higher than the steady state strength for a given loading condition should be understood and simple techniques developed to aid design of slopes.

The stability calculations to determine the driving shear stress in a mass are essentially the same whether stability due to a static or an earthquake loading is being considered. The computation of the driving shear stresses is done by conventional methods of stability analysis. If $F_s > 1$, the embankment is stable but deformations occur during the earthquake which, in themselves, may lead to failure. These deformations should be estimated in certain cases. The methods currently available are described by Newmark, 1965; Makdisi and Seed, 1979; Castro, 1986.

MEASUREMENT OF UNDRAINED STEADY STATE STRENGTH

The S_{us}-values for a given soil in situ are dependent only on void ratio. Furthermore, for sands, particularly narrowly graded sands, the undrained steady state strength is very sensitive to void ratio. Seemingly minor changes in gradation and void ratio can affect this strength (Poulos, Castro, and France, 1985). In situ, minor changes or variations in the way in which a sand is deposited can result in void ratio variations that are significant with respect to the undrained steady state strength. Therefore, multiple measurements of undrained steady state strength of sand samples that are quite close together and in the same layer can display wide variations. Although the S_{us}-values may seem at first to be difficult to measure, if one wishes to understand the stability of such materials, it is necessary to make the measurements.

Using techniques that have been developed over the past twenty years, one can measure the undrained steady state strength quite well. But much research and analysis of field cases is needed to improve the measurement for various soils and to estimate the strengths that are mobilizable in situ.

Variation in measured undrained steady state strengths that occur in sands should be studied so that the strengths that should be used in the analysis of stability can be verified. Ultimately one can base the procedures used to evaluate stability only on well-documented case histories. The standard will be set through a combination of understanding stress-strain behavior and progressive failure, combined with a careful evaluation of in situ failures and near-failures.

SUMMARY

The stability of slopes and their foundations is controlled ultimately by the steady state strength, not by the peak strength which is now used most often to define acceptable factors of safety in practice. This statement applies to stability evaluations for both static and earthquake loadings.

An embankment is said to be stable if the average driving shear stress is less than the average steady state strength on the critical surface. It is unstable if the driving shear stress exceeds the peak strength. The metastable state lies between the two. In this case, the driving shear stress is lower than the peak but higher than the steady state strength. Such an embankment can be caused to fail by straining it. In nature it should usually be assumed that sufficient progressive strain occurs to cause the peak strength to be destroyed.

Many of the failures reported in the literature occurred in cases for which the embankments probably were in a metastable condition prior to failure.

In some cases, the metastable state is acceptable in practice for a limited time. But for permanent structures, it is recommended that the factor of safety on the steady state strength, F_s, always be greater than one, unless there is proof from field behavior that a less conservative approach can be used.

The computations and strength selections required to understand stability due to static or earthquake loadings are the same. However, substantial earthquake loadings can cause large deformations which may be intolerable, even if a stability failure does not occur. For the design of slopes, it is important to separate the stability analysis from the deformation estimates and to evaluate each. For embankments that are designed with factors of safety on the undrained steady state strength from about 1.0 to 1.2, the deformations during an earthquake with peak ground acceleration greater than about 0.2 g could control portions of the design.

The selection of strength for a stability evaluation is separate and distinct from the analysis to compute the driving shear stresses. Computation of the driving shear stresses is the major focus of all conventional stability analysis techniques that have been published. But current techniques for selection of strength introduce a far greater chance for major errors than do the stress computations.

Measurement of the steady state strength, particularly the undrained steady state strength of contractive sands, requires care. Techniques for measurement of undrained steady state strength have been developed in detail in only one laboratory. In the past three years measurements have been made in a few other laboratories. It is recommended that research and practical developments be directed toward simplification and improvement of its measurement in many laboratories.

Continued efforts should be made to analyze both new and old case histories, taking into account the steady state strengths and recognizing that progressive failure will destroy in situ peak strengths if the mass exists in a metastable condition.

ACKNOWLEDGMENT

The author thanks Dr. Gonzalo Castro, Principal; Dr. Ronald C. Hirschfeld, Principal; and Mr. Michael Paster, Senior Geotechnical Engineer; of GEI Consultants, Inc., for their review and comments.

APPENDIX I.-REFERENCES

Beene, R. R. W. (1967). "Waco Dam Side," **Journal ASCE**, 93, (SM4), July, 35-44.

Bishop, A. W. (1952). **The Stability of Earth Dams**, Ph.D. Thesis, University of London, Imperial College.

Casagrande, A. (1949). Discussion of "Excavation of Slopes in the Panama Canal," by W. V. Binger and T. F. Thompson, Transactions ASCE, Vol. 114, Paper No. 2378, 870-874.

Casagrande, A. (1959). "Lectures on Soil Mechanics I and II," Harvard University, Cambridge. (Unpublished.)

Casagrande, A. (1960). "An Unsolved Problem of Embankment Stability on Soft Ground," **First Pan-American Conference on Soil Mechanics and Foundation Engineering**, Vol. II, Mexico City, 721-746.

Casagrande A. (1964). "The Role of the 'Calculated Risk' in Earthwork and Foundation Engineering," ASCE Terzaghi Lectures 1963-1972, 1974, Second Terzaghi Lecture.

Castro, G. (1986). "Seismic Stability of Earth and Earth Supported Structures - Liquefaction," State-of-the-art paper presented at the 8th International Symposium in Earthquake Engineering, India, December.

Castro, G. and Keller, T. O. (1988). "Re-Evaluation of the Lower San Fernando Dam," GEI Consultants, Winchester, Massachusetts. Submitted to U.S. Army Corps of Engineers, Waterways Experiment Station.

Geotechnical Engineers Inc. (1985). "Effect of Densification on Seismic Response of Foundation Soils, Jackson Lake Dam," Project 84416, Vols. 1 and 2 with Appendices, May 17, 27. (Unpublished.)

Geotechnical Engineers Inc. (1986). "Stability Evaluation Hinckley Dam Embankments, Final Report, Gregory B. Jarvis Power Project," submitted to New York Power Company, Geotechnical Engineers Inc., Winchester, Massachusetts, August.

Geotechnical Engineers Inc. (GEI) (1987). "Response to FERC Questions, Meeting of January 9, 1987, Hinckley Dam," Geotechnical Engineers Inc., Winchester, MA, February 5, 11 pp. plus Appendices A and B. (Unpublished.)

Gould, James P. (1960). "A Study of Shear Failure in Certain Tertiary Marine Sediments," Research Conference on Shear Strength of Cohesive Soils, ASCE, Boulder, June, 615-641.

Hazen, Allen (1920). "Hydraulic Fill Dams," Transactions ASCE, 83, 1713-1745.

Hendron, A. J. and Patton, F. D. (1986). "A Geotechnical Analysis of the Behavior of the Vaiont Slide," **Civil Engineering Practice**, Boston Society of Civil Engineers Section/ASCE, Fall, Vol. 1, No. 2.

Ladd, C. C. (1965). "Stress-Strain Behavior of Anisotropically Consolidated Clays During Undrained Shear," **Proceedings Sixth International Conf. on Soil Mechanics and Foundation Engineering**, Vol. I, 282-286.

Ladd, Charles C. (1972). "Test Embankment on Sensitive Clay," Proceedings ASCE Specialty Conference on Earth and Earth Supports Structures, Purdue, Vol. 1, Part 1, 101-128.

Lo, K. Y. and Stermac, A. G. (1965). "Failure of an Embankment Founded on Varved Clays," **Canadian Geotechnical Journal**, Vol. 2, No. 3. (See also Closure, Vol. 3, No. 1.)

Makdisi, F. I. and Seed, H. Bolton (1978). "Simplified Procedure for Estimating Dam and Embankment Earthquake-Induced Deformations," **Journal of the Geotechnical Engineering Division ASCE**, 104(GT7),849-867.

Newmark, Nathan M. (1965). "Effects of Earthquakes on Dams and Embankments," Fifth Rankine Lecture, **Geotechnique**, London, England, Vol. 15, No. 2, 139-159.

Poulos, Steve J. (1971). "The Stress-Strain Curves of Soils," Geotechnical Engineers Inc., Winchester, MA, January. (Unpublished.)

Poulos, Steve J. (1979). "Effects of Large Strains on Shear Strength Selection for Stability Analysis," **Lecture Series on Embankment Dams, Design and Construction**, American Society of Civil Engineers, Boston Society Section, Boston, December, 1-16.

Poulos, Steve J. (1981). "The Steady State of Deformation," **Journal of Geotechnical Engineering Division**, ASCE, 107(GT5), May, 5513-5562. Closure discussion in 108(GT8), August 1982, 1087-1091.

Poulos, Steve J. (approx. 1988). "Liquefaction and Related Phenomena," **Advanced Dam Engineering for Design, Construction and Rehabilitation**, Chapter 9, Robert Jansen, ed., Van Nostrand Reinhold, New York. (To be published.)

Poulos, Steve J.; Robinsky, Eli I.; and Keller, Thomas O. (1985). "Liquefaction Resistance of Thickened Tailings," **Journal of Geotechnical Engineering**, ASCE, 111(GT12), December, 1380-1394.

Poulos, Steve J.; Castro, Gonzalo; and France, John W. (1985). "Liquefaction Evaluation Procedure," **Journal of Geotechnical Engineering**, ASCE, 111(GT6), June, 772-795.

Seed, H. Bolton; Lee, Kenneth L.; and Idriss, I. M. (1969). "Analysis of Sheffield Dam Failure," **Journal of the Soil Mechanics and Foundations Division**, ASCE, 95(SM6), 1453-1490.

Skempton, A. W. (1964). "Long-Term Stability of Clay Slopes," **Geotechnique**, Vol. 14, No. 2, London, England, 77-101.

Smart, John (1984). "Drawdown Failure of Upstream Slope of San Luis Dam," Lecture at ASCE Geotech III, Session F2 on Unsatisfactory Performance of Dams, Atlanta, May. (Unpublished. Data in USBR files.)

Taylor, Donald W. (1948). **Fundamentals of Soil Mechanics**, Wiley, New York.

Terzaghi, Karl and Peck, Ralph B. (1967). **Soil Mechanics in Engineering Practice**, Wiley, New York, 513.

U.S. Army Corps of Engineers (1970). "The Stability of Earth and Rockfill Dams," **Engineering Manual**, EM1110-2-1902, 1-29.

U.S. Army Corps of Engineers (1983). Engineering and Design, Earthquake Design and Analysis for Corps of Engineers Project, Regulation No. 1110-2-18-6, May 16.

U.S. Army Waterways Experiment Station (1956). "A Review of Soils Studies," **Potamology Investigations Report 12-5**, Vicksburg.

APPENDIX II.-NOTATIONS

F Factor of safety

F_p Factor of safety based on peak strength

F_s	Factor of safety based on steady state strength
f_k	Kinetic friction coefficient
f_s	Static friction coefficient
L	Arc length
q	One half of principal stress difference
Q-test	Unconsolidated-undrained triaxial test
R-test	Consolidated-undrained triaxial test
S-test	Drained triaxial test
S	Shear strength (shear stress on failure plane at failure)
S_{dp}	Drained peak shear strength
S_{ds}	Steady state shear strength measured in drained shear
S_p	Peak shear strength
S_p/S_s	Strength sensitivity
S_s	Steady state shear strength
S_{up}	Undrained peak shear strength
S_{us}	Steady state shear strength measured in undrained shear
γ	Shear strain
γ_p	Shear strain required to reach peak strength
γ_s	Shear strain required to reach steady state of deformation
$(\sigma_1 - \sigma_3)$	Principal stress difference
τ_d	Driving shear stress, from stability analysis
ϕ_p	Slope of strength envelope, plotted in terms of effective stress, through points on Mohr diagram representing peak strengths. The value is usually different for drained, ϕ_{dp}, and undrained, ϕ_{up}, shear.
ϕ_s	Slope of strength envelope, plotted in terms of effective stress, through points on Mohr diagram representing steady state strengths. The value is the same for drained or undrained shear.

Evaluation of Seismic Behavior of Hydraulic Fill Structures

Jorge H. Troncoso*, M.ASCE

Abstract

Evaluation of seismic behavior of hydraulic fill structures, such as tailings dams, during the March 3, 1985 Central Chile Earthquake is presented. The observed behavior is compared with analytical studies using finite element methods together with information available from previous in-situ investigation of the dynamic and static properties of tailings deposits.

The analytical work shows a predictable importance of the variation in deposit properties with height of the dam, and the existence of critical zones where instrumentation should be concentrated to monitor stresses and pore water pressures. This information can be used to improve the design of future stages of the tailings dam construction. The importance of evaluating actual seismic behavior to reduce construction costs is stressed.

The failure of two tailings dams in the strong intensity zone of the March 3 earthquake is described and discussed. Other observed local damages are analyzed.

Introduction

Evaluation of actual seismic behavior of earth structures is necessary because of the many unknown factors that are generally associated with design and construction. This fact is particularly true in the case of hydraulic fill structures because of the characteristics particular to their deposition and sedimentation processes of construction. The design of tailings dams requires the use of all the information that can be gathered in an evaluation effort because these dams, built from mineral residues, introduce new variables which have to be added to the ones which pertain to every soil structure. As examples, aging effects together with structural effects induced by earthquake motions may produce very important changes in the properties which ultimately control stress-strain behavior and stability of a dam.

Recognizing the importance of evaluation techniques for improving safety and economy in the design of hydraulic fill structures, the purpose of this paper is to present experience on effects of recent earthquakes in Chile, in relation to actual seismic behavior of tailings dams, to evaluate changes in the properties of tailings soils

*Professor of Civil Engineering, The Catholic University of Chile, V. Mackenna 4860, Santiago, CHILE.

as a function of age and seismic history, and to develop criteria to optimize instrumentation. The main objectives behind these research efforts are the reduction of future construction costs through better understanding of fundamental soil properties, the reduction of instrumentation expenses, through more accurate knowledge of the critical zones of a dam where devices should be located, and finally the validation of theoretical analyses, through comparison with observations of the effects of earthquake motions on hydraulic fill structures of different types and ages.

Evaluation of the Structural Performance of Tailings Deposits

The evaluation stage of a structure comprises all the tasks devoted to verify design hypothesis and to probe the validity of predictions of static and dynamic performance during and after construction. This phase is important because its results may allow for optimizing the project. In fact, evaluation should allow to start with an economic initial design, based on most recent advances in the state of knowledge, which should also be flexible enough to permit adjustments suggested by backfeeding the results of observed behavior. The result is to achieve a final work of minimum cost compatible with adequate safety.

This methodology of integrated design and construction is readily applicable and obviously useful for tailings deposits. This is so because these deposits are gradually built over long periods of time and, in spite of the fact that the rate of disposal may be as large as tens of thousands of cubic meters per day (equivalent to construction of a 15 m high, 100 m long earth dam daily), the increase in height of the dam may be only a few meters per year. Under such conditions, enough time is available to control the mechanical behavior of the deposit, to analyze alternatives, and to implement convenient changes to improve safety and reduce operational costs. Earthquake resistant design specially may benefit from such methodology if seismic motions and their consequences are recorded in the first years of construction when the risks of potential failure are smaller.

Evolution of dynamic soil properties is another condition that justifies periodic measurements of representative parameters. Such records are specially useful in cases of tailings dams built by stages, or dams which are to be raised or loaded by newer deposits in future times.

Analytical Criteria for Evaluation of Seismic Performance

General. To program the evaluation tasks through the lifetime of a tailings dam, it is necessary to analyze the structural changes which are to occur from the first stages of construction up to the abandonment condition. In typical deposits, as shown in Fig. 1, these changes refer mainly to increases in height, natural period and depth of the impounded slimes plus the possible seismic loadings and aging effects. As a result of all these changes, the state of stress throughout the dam also varies continuously. When an earthquake occurs, the initial static state of stress determines the dynamic response of the structure.

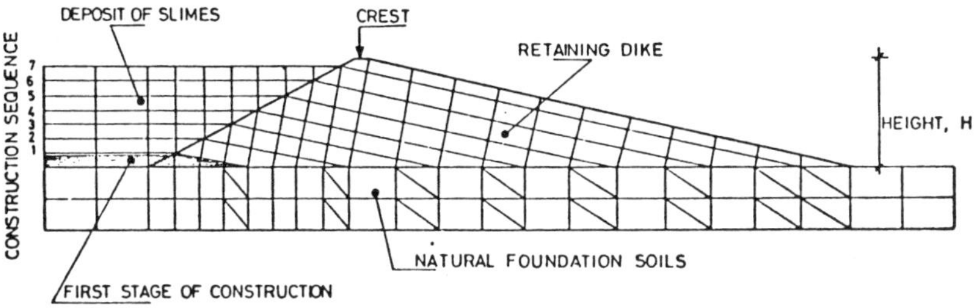

FIG. 1 FINITE ELEMENT MODEL FOR A TYPICAL HYDRAULIC FILL TAILINGS DAM BUILT BY DOWNSTREAM METHOD.

Static Analyses. The state of stress in different zones of a tailings dam may be analyzed using finite element methods. For instance, a distribution of principal stress ratio and of deviator stress are shown in Figs. 2 and 3 as obtained from the ISBILD program of the model of a typical deposit built by the downstream method as shown in Fig. 1. This deposit is formed by a retaining dike, built with single-cycloned compacted sands, and of impounded slimes composed of fine silty soils sedimented and consolidated from the slurry of the overflow of the cyclons. In this model the foundation soils are firm alluvial materials of low compressibility. These variables are important because they determine the excess static pore water pressure, Δu, which is the initial condition for any increment of load in the dam, since:

$$\Delta u_d = A\Delta(\sigma_1-\sigma_3) \tag{1}$$

where A is Skempton's pore water pressure parameter and σ_1,σ_3 are principal stresses. Such results indicate zones of high deviator stress concentrated in the lower parts of the upstream slope and of the central part of the dam. Because these particular zones are usually submerged below the phreatic surface, excess pore water pressures may also occur in these zones in a seismic event. High pore water pressures, which may take longer times to dissipate in the sands of tailings dams than in natural sand deposits, because of low permeability associated to very fine grinding and large silt content, indicate a very unfavorable initial condition for the occurrence of an earthquake, as shall be explained now.

The shear modulus, G, which is the property that controls shear stress-shear strain behavior of soils, is a stress dependant variable proportional to the square root of the effective mean normal stress; therefore, higher pore water pressures lead to smaller shear moduli. Once an earthquake starts, the lower shear moduli of soils in the low parts of the upstream slope of the dam should permit high shear strains with resulting high dynamic pore water pressures from the very beginning of the event. Because of this unfavorable initial state, a failure condition could be initiated in this zone and propagated towards the downstream slope.

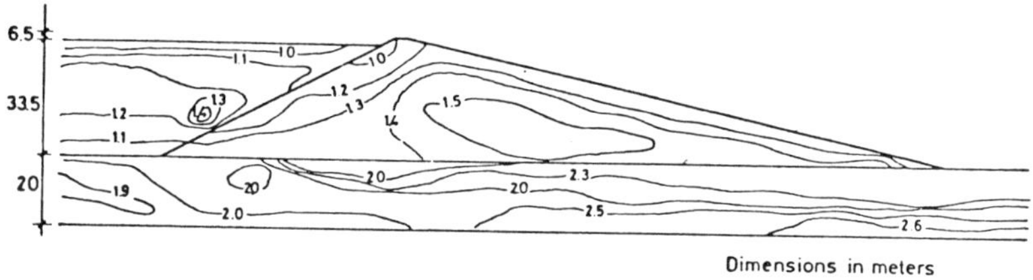

FIG. 2 STATIC PRINCIPAL STRESS RATIO, σ_1/σ_3, DISTRIBUTION IN TAILINGS DAM. (ISBILD)

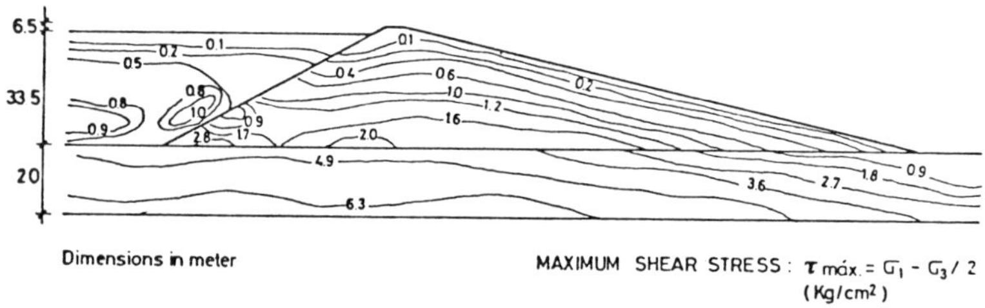

FIG. 3 STATIC DEVIATOR STRESS, $\sigma_1 - \sigma_3$, DISTRIBUTION IN TAILINGS DAM (ISBILD).

The principal stress ratio, σ_1/σ_3, is equal to 1 in near surface elements and increases to values close to 1.5 in the deeper central part of the downstream slope. This change from isotropic to anisotropic stress condition occurs for every layer of tailings from the instant of deposition, as succesive layers are deposited. The σ_3/σ_1 ratio less than 1 indicates that the stress path which corresponds to this method of construction tends toward a condition with the soils dilating in the direction of the minor principal stress.

The angle of rotation of the axis of principal stress, θ, varies inside the dam form a 30° to 40° value under the downstream slope to a 10° to 20° value in the central and upstream slope zones (Troncoso, 1986).

Similar analyses performed for the upstream and the centerline methods of construction, would lead to different stress distributions and, therefore, to different initial conditions to earthquake response and corresponding differences in failure potentials.

Dynamic Analyses. Evaluation procedures during construction of a tailings deposit are particularly useful when the effects of seismic events can be recorded. In order to analyze the importance of changes induced by growth in these hydraulic fill structures, it is interesting to compute the effects of different seismic ground motions in tailings dams at different stages of construction. To this purpose the theoretical responses of dams with the geometry of the model described in Fig. 1, but whith different heights, have been submitted to different actual seismic motions. The results shall now be compared to evaluate relations between dynamic characteristics of the structures and the characteristics of the earthquakes. The seismic input motions are the records obtained of the Chilean earthquake of March 3, 1985, at four stations founded on sites with the characteristics detailed in Table 1.

Table 1

Characteristics of Accelerograms of the March 3, 1985
Earthquake Magnitude M_S = 7.8
Epicenter 72° Long W, 33° 20' Lat.S

Station	Longitude	Latitude	Maximum Acceleration (gal)	Direction	Epicentral Distance (km)
Llo-Lleo	71° 36'	33° 41'	426	S80E	25
			669	N10E	
Almendral	71° 38'	33° 01'	293	N50E	64
			163	S40E	
Viña del Mar	71° 35°	33° 02'	228	N70W	62
			356	S20W	
UTFSM	71° 38'	33° 01'	164	S20E	63
			179	N70E	

The reason for using these records as input motions is because they were obtained at stations close to the epicentral area (Saragoni,1985). Also these stations are near the sites of several tailings dams investigated by the author.

The finite element model used for the dynamic analyses have the general geometry shown in Fig. 1 with downstream slope of 4.5(H):1(V), upstream slope of 2(H):1(V), heights of 20, 40 and 80 m, and crest widths of 3,6 and 12 m. The program used for computations was QUAD-4 (Idriss et al, 1973).

The dynamic properties of the tailings soils in the form of initial shear modulus, damping ratio and unit weight, and the curves of variation of shear modulus and damping ratio as functions of strain, were introduced in accordance with data available form previous research efforts (Troncoso, 1986). These data were the same for all models, assuming that the composition of the tailings soils was the same. The phreatic surface was fixed to follow the upper surface of the deposit of slimes, the boundary between the slimes and the retaining dike of sands and the bottom of the retaining dike.

The results obtained from these analyses of seismic behavior should correspond to the behavior of tailings dams which hypothetically would have been located at the sites of the accelerograph stations, at a stage of construction corresponding to that of the different heights analyzed. The longitudinal axis of all the dams was taken perpendicular to the direction of the recorded motion.

As expected, the same earthquake motion generates larger maximum shear stresses in higher dams, as shown in Fig. 4 and Fig. 5, where the responses of dams of 80, 40 and 20 m high are compared for the Llo-Lleo S 80 E and the El Almendral S 40 E ground motions, respectively. The analyses show how the maximum dynamic stresses are concentrated at the toe of the upstream slope, thereby confirming that this part of the structure constitutes a critical zone where failures may start.

If the distributions of maximum shear stress generated by different seismic motions are compared for dams of heights of 80 m and 20 m as in Tables 2 and 3, respectively, some interesting conclusions may be obtained. To make the comparison easier, in these Tables all the stresses are normalized with respect to the maximum shear stress recorded at the base of each dam under the central vertical axis during the strongest motion which is the Llo-Lleo N10E. (In Tables 2 and 3 the "central vertical axis "is the vertical axis by the crest and the "medium vertical axis" is the vertical axis by the central section of the downsteam slope). The earthquake motions are listed in the order of decreasing damage potential. Based on these results, it is possible to appreciate that the maximum shear stress decreases as elevation increases inside the dam, but at a lower rate in the higher dam. The magnitudes of shear stresses, which can be related to the damage potential that can be caused by a given earthquake, are different for dams of different height. For both dams, the most destructive ground motion is the Llo-Lleo N10E record, while the El Almendral S40E record has relatively smaller effects in the lower dam than in the higher one.

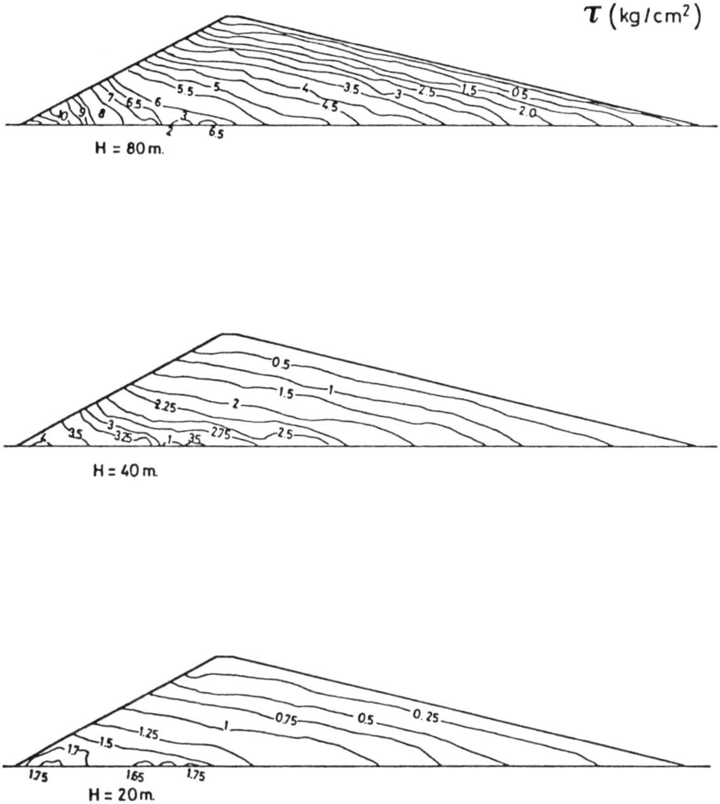

FIG. 4 MAXIMUM SEISMIC STRESSES IN TAILINGS DAMS OF DIFFERENT HEIGHTS. MARCH 3, 1987 CHILEAN EARTHQUAKE RECORDED AT LLO-LLEO S 80 E.

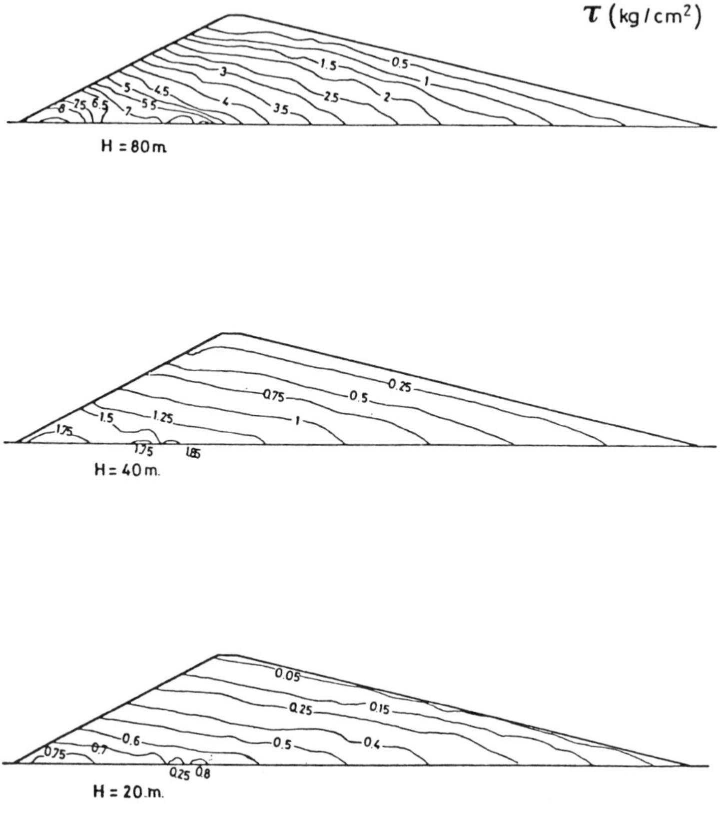

FIG. 5 MAXIMUM SEISMIC SHEAR STRESSES IN TAILINGS DAMS OF DIFFERENT HEIGHTS. MARCH 3, 1987 CHILEAN EARTHQUAKE RECORDED AT EL ALMENDRAL S40E.

Table 2

Relative Effects of Different Ground Motions
in an H = 80 m-high-Tailings Dam
Maximum Normalized Shear Stress
Chilean Earthquake of March 3, 1985

Ground Motion	Direction	Central Vertical Axis Elevation:				Medium Vertical Axis Elevation:	
		0	0.25 H	0.5 H	0.75 H	0	0.25 H
Llo-Lleo	N10E	1.00	0.89	0.69	0.35	0.58	0.39
Llo-Lleo	S80E	0.46	0.42	0.31	0.15	0.23	0.19
El Almendral	N50E	0.46	0.35	0.27	0.12	0.19	0.12
Viña del Mar	N70W	0.39	0.35	0.27	0.15	0.15	0.12
Viña del Mar	S20W	0.35	0.27	0.23	0.12	0.19	0.15
El Almendral	S40E	0.35	0.27	0.23	0.12	0.12	0.08
UTFSM	N70E	0.27	0.23	0.19	0.12	0.12	0.08
UTFSM	S20E	0.15	0.15	0.12	0.08	0.08	0.08

Table 3

Relative Effects of Different Ground Motions
in an H = 20 m-high-Tailings Dam
Maximum Normalized Shear Stress
Chilean Earthquake of March 3, 1985

Ground Motion	Direction	Central Vertical Axis Elevation:				Medium Vertical Axis Elevation:	
		0	0.25H	0.5H	0.75H	0	0.25H
Llo-Lleo	N10E	1.00	0.80	0.52	0.24	0.50	0.36
Llo-Lleo	S80E	0.50	0.44	0.38	0.14	0.20	0.16
Viña del Mar	S20W	0.44	0.36	0.24	0.10	0.20	0.16
El Almendral	N50E	0.40	0.32	0.22	0.12	0.20	0.14
Viña del Mar	N70W	0.34	0.26	0.18	0.10	0.16	0.12
UTFSM	N70E	0.26	0.22	0.14	0.08	0.14	0.08
UTFSM	S20E	0.24	0.20	0.14	0.08	0.14	0.10
El Almendral	S40E	0.24	0.20	0.14	0.06	0.10	0.10

Amplification of seismic effects are clearly shown in Fig. 6 where maximum shear stresses generated by different ground motions are compared for dams of 20 m and 80 m heights:

Seismic response of the tailings dams to the excitation of different earthquake records may also be analyzed by comparing the acceleration amplification at similar points inside the dams. Table 4 shows the average acceleration amplification in a column of elements located vertically under the crest and the acceleration amplification at the crest of the dam. It is observed that the crest acceleration is strongly amplified as the height of the dam increases and that the rate of amplification is very different for the different ground motions. These results should be attributed to the different frequency contents and time histories of the earthquake motions and the corresponding unequal responses of structures of different dynamic characteristics.

Table 4

Ratio of Maximum Accelerations (A) Related to Input Base Acceleration (A_b)

Earthquake Ground Motion	Height of Dam, H (m) 80 Crest	80 Average	40 Crest	40 Average	20 Crest	20 Average
Viña del Mar, N70W	2.51	1.73	1.20	1.09	1.03	1.01
Almendral, S40E	1.91	1.47	1.19	1.11	1.11	1.07
Llo-Lleo, S80E	1.54	1.15	1.23	1.08	1.04	1.03

If the Acceleration Amplification Ratio, A/A_b, is plotted as a function of the ratio of the predominant period of the ground motion to the fundamental period of the dam, as in Fig. 7, it is observed that for every ground motion record the amplification increases as the ratio of the periods increases. More data are needed to investigate resonance effects when the ratio of periods gets closer to unity. However, it is interesting to notice that one of the most destructive earthquakes, Llo-Lleo, corresponds to a relatively low rate of amplification as the ratio of periods is larger that 1.

These results are useful in predicting the seismic behavior of a tailings dam as it increases in height. They are also valuable in interpreting the results of observed behavior and in extending them to future stages in the growth of the structure.

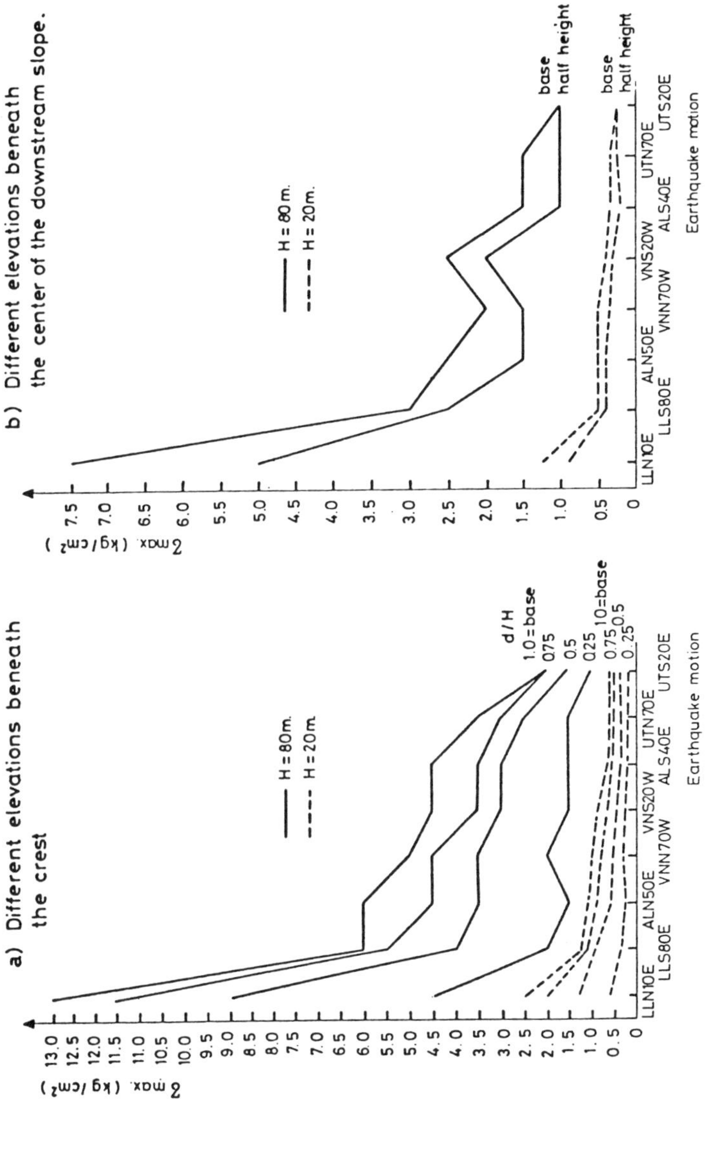

FIG. 6 MAXIMUM SHEAR STRESS, AT DIFFERENT ELEVATIONS INSIDE A TAILINGS DAM, GENERATED BY DIFFERENT EARTHQUAKE MOTIONS.

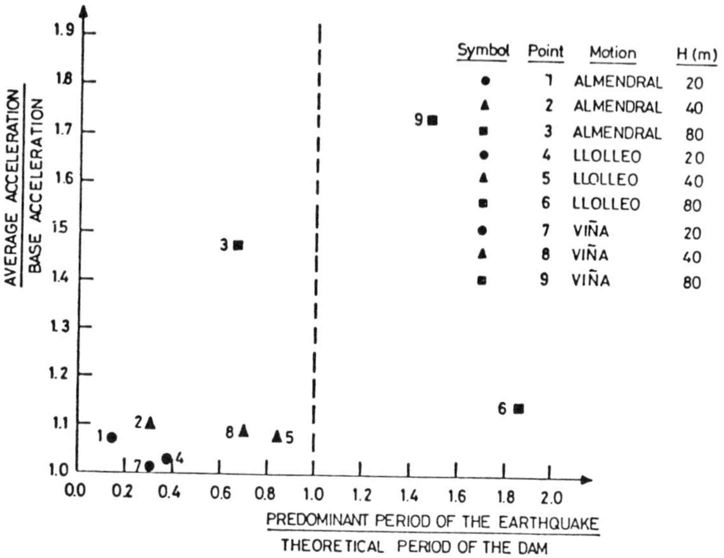

FIG. 7 SEISMIC ACCELERATION AMPLIFICATION

Observed Seismic Effects of the March 3, 1985 Earthquake in Tailings Dams

The March 3, 1985 M_s = 7.8 earthquake caused liquefaction and flow slide failures of two tailings dams in the central zone of Chile, one about 40 m and the other 20 m high. Other higher tailings dams responded satisfactorily to the very strong earthquake ground motions. Limited residual displacements and cracking were observed in dams of heights around 80 m, located at distances between 120 and 200 km from the epicenter.

A summary of these seismic effects observed by the author is as follows.

Flow Slides. A -20 m-high tailings dam, called Veta del Agua N° 1, located about 130 km from the epicenter failed in the mode of flow slide. The liquefied tailings advanced 8 km downstream following the course of El Sauce creek. This dam had been built by the centerline method and the downstream slope was very steep, in the ordef of 2(H):1(V) to

1.5(H):1(V); no drainage blanket had been provided. Because of these unfavorable conditions and the strong accelerations at the site (probably 0.3 g or even more) plus the fact that the dam was located on a hillside with 10% slope, this failure was predictable. In fact, a project to build a reinforcement berm to the downstream slope of the dam had been studied, but its construction had not been initiated at the time of the earthquake.

The density of silty sands which formed the retaining structure was lower than the critical density and, therefore, the development of high pore water pressures was to be expected under seismic loadings. A research project currently in progress will determine the steady-state strength of the silty sands and clayey silts of the deposit and the most probable mechanism of failure (Castro and Troncoso, 1988). Information collected at the site a few days after the earthquake revealed that the slimes did liquefy. One of the operators of the dam who witnessed the failure reported an explosive burst which broke the central and highest part of the retaining dam, projecting soils from the upper parts fordward and forming a breach which permitted the liquefied materials to flow downstream and to empty 40 percent of the deposit. Such type of failure is indicative of the development of very high pore water pressures inside the dam.

Another dam, 40 m high, called Cerro Negro N° 4, located at 140 km from the epicenter, which is also being investigated, failed during the earthquake apparently by a mechanism of flow slide. This dam was built by the downstream and centerline methods, with a 2.5(H):1(V) downstream slopes. Tailings sands were cycloned silty sands; no mechanical compaction was provided. After failure a deep gorge was formed, about 30 m wide at the dike crest, permitting the flow of water and soils which subsequently followed the course of Pitepeumo river for about 10 km. About 25 percent of the soils of the whole deposit were lost in this failure.

None of these dams, Veta del Agua nor Cerro Negro, had piezometers or other instruments installed. Therefore, there were no means to verify the location of the phreatic surface before or at the time of the earthquake. However, because no drainage blanket had been built in either site it is estimated that the depth of the phreatic surface might have been very shallow underneath the toe of the Cerro Negro dam and probably at the toe or slightly above it at Veta del Agua. Main differences between drainage conditions of both dams were less permeable alluvial foundation soils and steeper downstream slope in Veta del Agua and more permeable fluvial foundation soils in Cerro Negro. The critical zone, defined by the analytical studies at the lower part of the upstream slope of the retaining dikes, must have been saturated in both dams.

Volume Changes. An interesting case of seismic behavior has been reported by the author (Troncoso, 1986) in relation to a tailings dam, about 80 m high, located at 120 km form the epicenter of the March 3, 1985 earthquake. This dam is being built, by the downstream method, with cycloned sands compacted by bulldozer with a 4.5(H):1(V) downstream slope. The silty sands of this dam had been the subject of several research projects before the earthquake, including the determination of curves of critical relative density by monotonic triaxial tests, and the performance of standard penetration tests in boreholes. Pore water pressures had also been recorded during the occurrence of previous earthquakes, in addition to accelerations at the crest, by means of a unique system of seismic piezometers and one accelerograph, as reported elsewhere (Troncoso, 1983). The interruption of power, which occurred at the very begining of the March, 1985 earthquake, prevented to obtain records of accelerations and pore pressures in this event. However, the periodic readings of Casagrande piezometers did indicate that the location of the phreatic surface above foundation level was about 2 or 4 meter in the upstream zone and very close to zero in the central and downstream part of the retaining dike. Such favorable condition was due to provision of a drainage blanket underneath the downstream zone and to the high permeability of the alluvial and fluvial deposits which formed the natural foundation at the site. At noon the day after the earthquake a strong aftershock happended and lasted for about 30 seconds; at this time liquefaction of the slimes was observed by the author in the form of surface boils. Large deformations of the surface of the impounded slimes were also noticeable. The behavior of the dam was satisfactory with only small residual crest displacements and cracking parallel to the crest. Because of the strong movements observed in the earthquake and in the main aftershock, two borings were performed a few days after the earthquake, at the same locations of previous ones, in order to compare standard penetration resistances before and after the seismic events. The SPT results were normalized to account for different effective stresses and then used to estimate relative densities from the surface to the bottom of the downstream slope prism. The elapsed time between performances of both sets of SPT tests was four months. Little aging effects should be expected for such short period of time; therefore, the main reason for changes in SPT resistance had to be the change in volume of the sands. The results, presented in Fig. 8, indicate that the critical density of the silty sands of this dam were in the range of 50 to 60 percent, and that the sands which were looser than such critical density before the earthquake contracted, while those which were denser, dilated. These results fully confirmed laboratory results obtained previously for the same soils of this tailings dam.

Behavior of Abandoned Deposits. Most tailings dams in a 150 km radius from the epicenter suffered liquefaction of the impounded slimes even when some of them were as old as 40 years. Such behavior of the deposits of slimes was common to dams of different type: downstream or centerline construction or regular earth retaining dike dam. The retaining structures of some of these deposits showed evidence of longitudinal cracks, some as wide as one centimeter, and small displacements. One old dam which had failed in 1981, and where the stability of the remaining masses had not been improved, underwent liquefaction of some deep strata where the degree of saturation was very high because of water flow from the adjacent hill side. As a consequence, the downstream slope failed in a mechanism of sliding wedges.

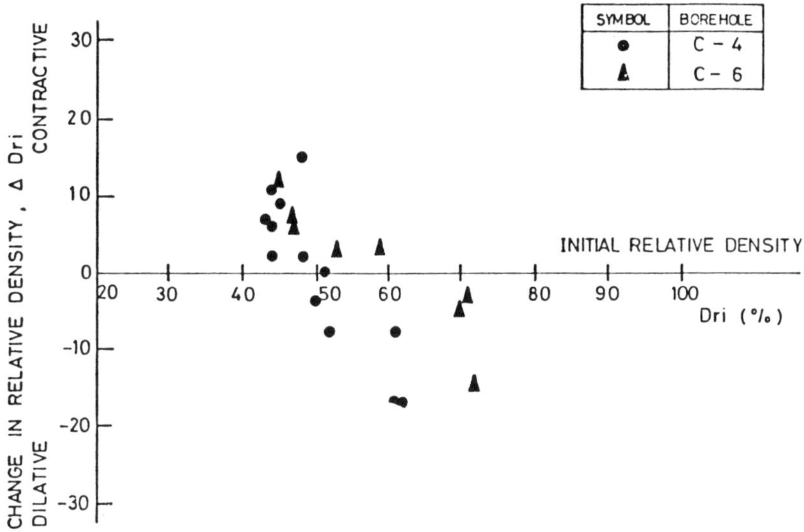

FIG. 8 SEISMIC BEHAVIOR OF TAILING SANDS AS DETERMINED BY IN-SITU TESTS (Troncoso, 1986)

Except for these damages most of the old dams performed satisfactorily in spite of the fact that many of them had been built with rudimentary methods of cycloning and without compaction. Because of these observations, research has been started to measure cyclic strength of undisturbed samples of deposits of different ages. Initial results indicate a clear tendency for an increase of cyclic strength as a function of age. Increases as large as 250 percent in 30 years have been found for tailings sands with 25 percent silts (Troncoso, Ishihara and Verdugo, 1988) and of 200 percent in 4 years for tailings sands with 33 to 44 percent clayey silts (Troncoso, 1986).

Such effects of aging may be explained by evolution of the structure of the soils, especially in the physico-chemical forces that act between the solid particles. The changes may be attributed to the following causes, particularly important because of the composition of tailings materials,:

a) chemical reactions between the mineral components of the particles, their double layers and the fluids;
b) decrease of void ratio because of consolidation or densification induced by seismic history;
c) changes of the dielectric constant of fluids in the pores
d) cementation and oxidation
e) leaching

The effects of aging and seismic history in evolution of hydraulic fill structures are important and must be considered in the design because as the lifetime increases so does the risk of occurrence of stronger earthquakes. In particular, design for abandonment conditions may benefit of any increase in strength which could be reliably predicted. In order to assess the validity of these predictions, evaluation techniques should include periodic testing of soil dynamic properties.

Conclusions

Evaluation of seismic behavior of hydraulic fill structures such as tailings dams has been analyzed both from a theoretical viewpoint and in light of the observed performance of these structures in the March 3, 1985 earthquake in Central Chile.

Analytically, it has been demonstrated that there are some critical zones of the dams where instruments should be installed. The lower part of the upstream slope of the retaining dike is one of these zones. Piezometers and load cells should be installed in such zones to measure pore water pressures and normal stresses and to evaluate shear stresses and effective normal stresses. Effective stress paths could then be monitored and verified.

Seismic response of a dam changes as the height of the dam increases and different amplification effects should be expected for different ground motions. Evaluation of response and amplification requires installation of accelerographs at the crest and at the toe of the dam. Recording of pore water pressures during and inmediately after earthquakes, with dynamic piezometers or with static piezometers connected to scanner recorders, is useful to verify and to improve predictions of liquefaction potential.

The effects of the March 3, 1985 earthquake in several tailings dams located within the affected area have indicated some important factors which have to be considered in the evaluation stage of a project. Seismic history and aging are important causes of evolution and changes in the mechanic and dynamic properties of the soils that should be monitored by periodic in-situ measurements of strength such as standard or static penetration, vane shear, shear wave velocity, and laboratory testing of undisturbed samples.

Acknowledgements

This paper is based on research projects sponsored by Fondo Nacional de Desarrollo Científico y Tecnológico de Chile (FONDECYT)and by the United Nations Industrial Development Organization (PNUD). The author would also like to acknowledge cooperation of Mr. E. Corominas, research engineer, and of Mr. R. Brito, research assistant, in the performance of these projects. Thanks are also extended to Dr. J. Vásquez, colleague form the Structural Engineering Department who kindly agreed to review the manuscript of this paper.

Appendix 1. References

Castro, G. and Troncoso, J.H. 1988 "Behavior of Tailings Dams During the March 3, 1985 Chilean Earthquake" NSF Research Project, in progress.

Idriss, I.M., Lysmer, J., Hwang, R., and Seed, H.B. 1973 "QUAD-4: A Computer Program for Evaluating the Seismic Response of Soil Structures by Variable Damping Finite Element Procedures". Report N° EERC 73-16, University of California.

Saragoni, R., González, P. and Fresard, M., 1986 "Analyses of the March 3, 1985 Earthquake Accelerograms", "The Earthquake of March 1985-Chile", Universidad de Chile and Acero Comercial S.A. Report pp. 43-83 (in spanish).

Troncoso, J.H. 1983 "Seismic Pore Water Pressures in Tailings Dams" Proceedings VII Panamerican Conference on Soil Mechanics and Foundation Engineering, Vancouver.

Troncoso, J.H., 1986 "Critical State and Seismic Response of Sands of Mineral Residues" Final Report, Fondecyt Research Project N° 85/122. December, (in spanish).

Troncoso, J.H., 1986 "Critical State of Tailing Silty Sands for Earthquake Loadings" Soil Dynamics and Earthquake Engineering, Vol. 5, N° 4.

Troncoso, J.H. 1986 "Liquefaction Potential of Tailings Sands as a Function of Age", Anales del Instituto de Ingenieros de Chile, Vol.98, N° 2, September (in spanish).

Troncoso, J.H., Ishihara, K. and Verdugo, R., 1988 "Aging Effects in Tailings Soils" submitted for publication in the IX World Conference on Earthquake Engineering, Tokyo.

THE DISASTER OF STAVA/NORTHERN ITALY

Giorgio Berti[1], Floriano Villa[2], Daria Dovera[3]
Rinaldo Genevois[4], Josef Brauns[5]

ABSTRACT

In July 1985, a failure of a pair of tailings impoundments occurred in Northern Italy. The paper describes the geological and geotechnical conditions of the site and discusses the particular methods of construction and operation of the basins in the course of their lifetime.

INTRODUCTION

On July 19, 1985 at 12.23 pm the tailings dams of the Prestavel Mine at Stava di Tesero (Trentino/Northern Italy, see Figure 1) failed and caused a disastrous mud

Figure 1. Location map

1 Prof. Avvocato, Catholic University of Milano, 20100 Milano/Italy
2 Prof. Dott. Geol., University of Venezia, 30100 Venezia/Italy
3 Dott. Geol., viale Bacchiglione 20, 20139 Milano/Italy
4 Prof. Dott. Geol., University of Ferrara, 44100 Ferrara/Italy
5 Prof. Dr.-Ing., University of Karlsruhe, D-7500 Karlsruhe 1/FRG

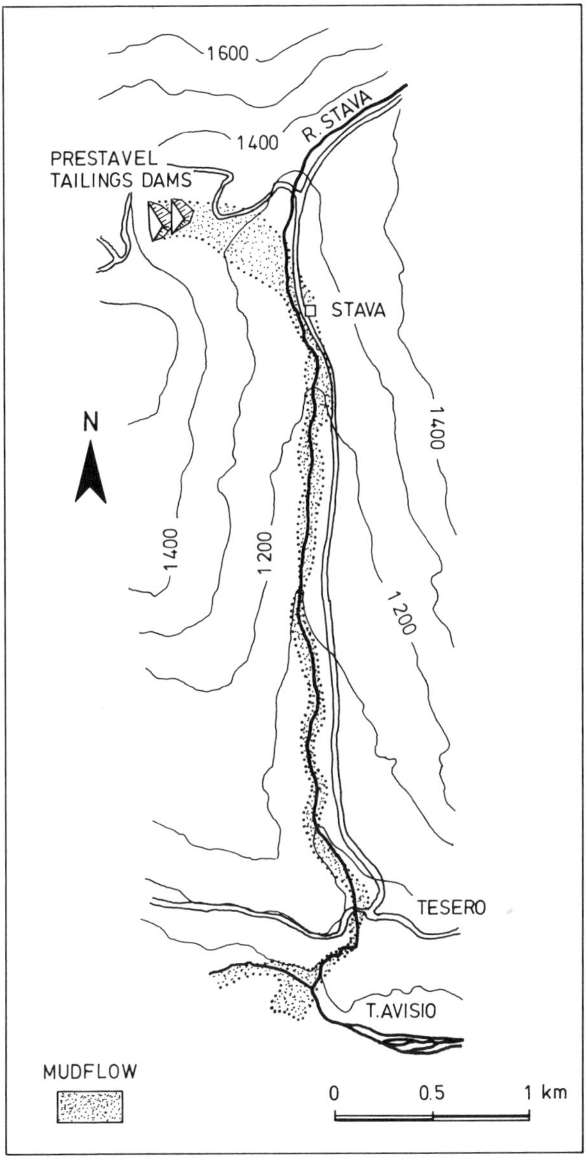

Figure 2. Position of tailings dams and the mud flow's course

flow over the valley downstream destroying everything (see Figure 2). A total quantity of 190,000 m^3 liquid mud was released in a sudden burst, washed away several 'alberghi' which were situated just downhill from the tailings dams and where many people (parents with their children spending their summer holidays in this alpine region) had gathered to take their lunch, and rushed down the narrow Stava valley to Tesero, a small town at a distance of about 4 km, in a few minutes. On its way from the dam site to the junction of the Stava river with the Avisio river near Tesero, the flood wave, varying its height depending on the morphology of the Stava valley, destroyed a large number of further houses. There was no warning time at all and 268 people died due to the extremely unfavourable time of the event. The consequences would have been even more serious, if this had occurred during the night, being an even more unfavourable time.

Legal action has been started in order to ascertain the penal and civil liabilities of engineers, mine's owners and public authorities. At present, the legal and administrative situation concerning tailings dams in Italy is complex. The authors want to give a short legal background on the matter before presenting and discussing technical aspects of the tragic event.

LEGISLATIVE ASPECTS IN ITALY

In Italy there are no specific regulations concerning tailings dams. Rules which may pertain can be interpreted and applied only by the judge on the basis of the principle of liability and of the rights to health or to environment. The split of jurisdiction among the different administrative authorities encourages each of them to reject all responsibilities. Therefore, it is absolutely necessary to create a technical and administrative branch having the specific task and duty to control tailings dams from the safety point of view.

As the Italian legal system includes no regulations applying wholly and consistently to tailings dams, we must refer to rules concerning other fields. Rules concerning dams could be applied to tailings dams, though they are non-exhaustive. They refer mainly to the construction and the maintenance of the dam system itself and not to its efficiency and functioning in respect to other related activities or to the impact on the territory and the surrounding social communities. Secondly, tailings dams have technical features (i.e. usually variable height and capacity of the impoundment) that do not allow the general application of regulations on dams, based on fixed capacity.

The same can be said as regards the regulations concerning mining activities that also affect tailings dams. Mining rules are applied by the public authorities not to limit, but to increase industrial production. In this background, tailings dams can without doubt be submitted to mining police rules, because they are considered as a kind of mining activity.

Finally tailings dams must conform to the rules and legal principles for woodlands and mountain areas. These rules are established to restrict any disturbance to the land, rather than regulating the industrial activities that will be established on it. Even with such a variety of rules, there are no local public authorities with specific duties to control industrial activities that can be applied to tailings dams.

The Italian laws relating to these points can be found in the list of references, numbers 3 to 8.

GENERAL FEATURES OF THE AREA

1. Geomorphological Aspects

The Prestavel Mine is situated in the North-Western part of the municipal territory of Tesero (see Figures 1 and 2), Trentino in Northern Italy. The location of the tailings impoundments, named Pozzole, is situated on the right side of the Stava River, at an altitude of about 1350 m. The impoundments have been constructed near a stream called Rio Porcellini, which has formed a side valley to the Stava River (see Figure 2). This stream is fed by several intermittent tributaries.

The construction of the tailings dams and impoundments changed the local morphological conditions substantially, as did finally the catastrophic failure of these deposits.

2. Geologic, Stratigraphical and Tectonic Aspects

The stratigraphical sequence of the area in question is as follows (upward sequence, see also Figure 3):

Lagorai porphyries	Permian
Werfen formation	Scythian
Morainic deposits	Quaternary
Fluvial and alluvial deposits	Quaternary
Talus	Quaternary

In the Northern portion of the area, the Lagorai porphyries outcrop. These are rhyolitic ignimbrites and reddish quartzlatites with a solid structure, characterized by widespread subvertical fractures. Many vein

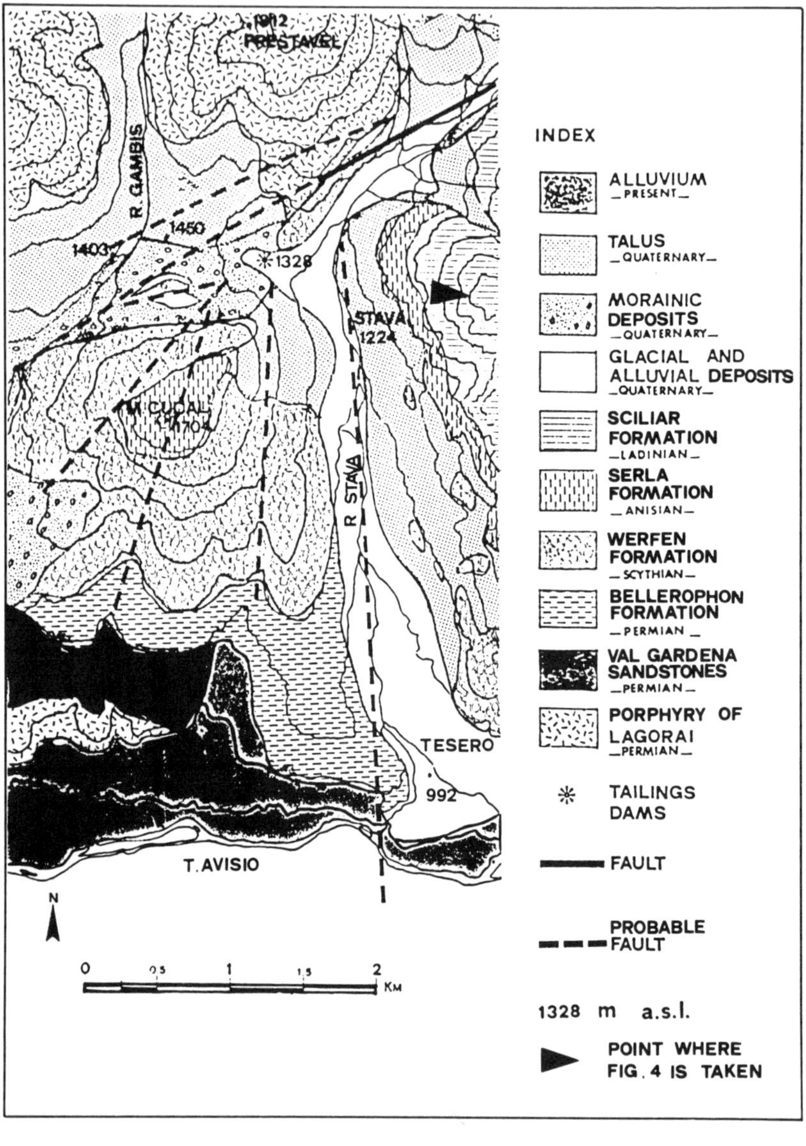

Figure 3. Geologic map of the area

elements are present in this formation along the fracture planes; some of these veins, enriched with fluorite, are the material of interest at the Prestavel mine. On the Southern side of the aforementioned lithological complex the Werfen Formation outcrops, formed by alternating layers of mainly calcareous rocks and siltstones. The contact of the two formations is due to tectonic action.

The Quaternary deposits, which are of glacial, fluvio-glacial and clastic origin, are widely distributed throughout the area. In particular, the fluvioglacial deposits constitute the foundations of the impoundments as revealed from the official geological investigations after the collapse. Deeper calcareous deposits in the area show various degrees of karst phenomena with a strong influence on the local hydrological conditions. According to cartographic and photographic documents, there were dolines underneath the dams and in their vicinity.

After the collapse, the structure of the Prestavel tailings dam site was reconstructed on the basis of photographs, other documents and direct in situ invest-igations. The main tectonic element is the Stava Line with an orientation WSW-ENE, along which the Lagorai porphyries and the Werfen formation are in contact. The Stava Line influenced the evolution of the impoundment area, clearly modifying the surface hydrology and, con-sequently, the geomorphological and hydrogeological features.

3. Hydrogeological Aspects

The hydrogeological features of the area are closely related to its tectonic structure. The watercourse of the Rio Porcellini, which exists as a surface stream uphill from the tailings impoundments, disappears before reaching the site. It surfaces as a spring to join the Rio Stava a few hundred meters further downstream, as can clearly be seen from Figure 4 (photograph taken from Mount Cornacci on the left bank of Rio Stava, as is indicated in Figure 3).

Figure 4 shows the morphological situation, which is related to the main Stava Line (see Figure 3 for com-parison) and a Rio Porcellini Line, which cross in the area of the tailings dam site to form the Pozzole depres-sion. The Rio Porcellini Line, from the hydrogeological point of view, has two effects. On the one hand, it drains water coming from the Mount Prestavel side (where the Prestavel Mine is located, see Figure 3) towards the Rio Stava; on the other hand, the Rio Porcellini Line blocks off the Stava Line and provides the area with a supply of water flowing along the line. This results in the presence of a number of springs, which also existed in the area where the two tailings dams and impoundments were built. In fact, during the enlargement of the tailings impoundments, the watercourses and springs have been partially covered and heavily surcharged. After the collapse of the dams and removal of the major portion of the tailings sediments, several springs have reappeared feeding water into the lower portion of the Rio Porcellini, just as they did before the existence of the tailings dams.

Figure 4. View of the site after the failure of the dams

(Photo taken from a point indicated in Figure 3)

STRUCTURE AND HISTORY OF THE PRESTAVEL TAILINGS DAMS

As the tailings disposal facility was to be built on a slope, an initial starter dam was constructed using the available slope material. The starter dam of the lower impoundment was built in 1961 with a maximum height of 9 m. Special measures were to be taken to "key" the starter dam into the swampy and unstable ground (Rossi 1973). The starter dam was covered with sand cycloned from the tailings using two cyclones moving along the dam and raising in steps of about 5 to 6 m. In general, the upstream method of construction was used over the period from 1961 until the late sixties, when efforts were started to create a new impoundment just upstream of the existing one on the same slope (see the common section of both dams and impoundments in Figure 5). The final height of the lower tailings dam was about 26 m.

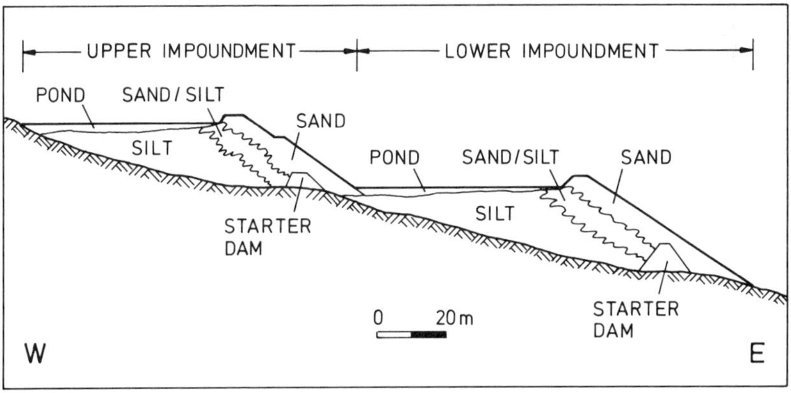

Figure 5. Schematic section of Prestavel tailings dams

(taken from final report of the official experts team, Colombo/Datei/Dolzani/Fuganti)

During construction and operation of the lower impoundment, the cyclone overflow was fed into the pond and the clear water was drawn off by a concrete pipe with a number of intake structures, installed underneath the starter dam and running along the bottom of the impoundment. These are shown in Figure 6, which provides a schematic plan of both basins in their final form and also gives the installations to operate the entire tailings disposal system.

The height of the starter dam of the upper impoundment

was only around 5 m (see Figure 5). In the following stages the 'centerline method' was apparently followed for a while resulting in placement of cycloned dam material partly on slimes of the lower basin. Later, the 'upstream method' of construction was applied again and with raising steps about 3 m in height. This procedure was followed until the dam had a total height of 14 m in 1975, with a downstream slope of 39° (1.2 H: 1.0 V).

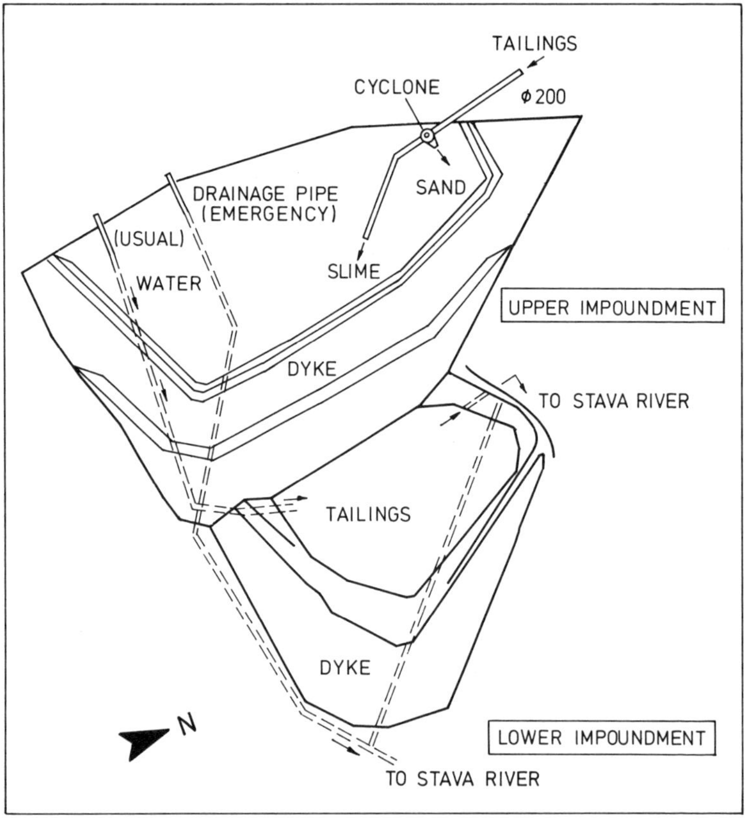

Figure 6. Schematic plan of the operational systems (Muramoto et al. 1985)

As is to be seen from Figure 6, the clear water was drawn off from the upper impoundment by decant pipes installed on the impoundment floor. These were provided with holes(at 0.30 m elevation increments),which were to be closed periodically by conical concrete plugs according to the requirements of clarifying the tailings effluent (see Figure 7).

A second identical decant pipe was subsequently installed (termed "usual" in Figure 6), which fed the clear water drawn off from the upper basin into the lower one for secondary purification. This procedure was implemented to improve the cleaning effect of the system, which seems to have been insufficient in that period. The initial decant pipe was subsequently used as a standby emergency outlet.

In connection with the planned extension of the mining activities in 1975, a first stability calculation was performed when the upper dam had reached a total height of 14 m. As a result of this analysis, a berm, 4 to 5 m in width, was provided and the slope of the downstream dam face was reduced to 34° (1.5 H: 1.0 V). Construction continued until 1978, when changes in mining conditions interrupted disposal activities for several years.

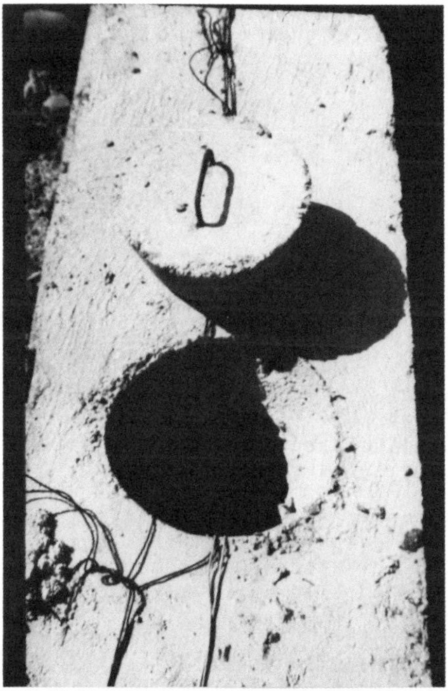

Figure 7. Intake of the decant pipe in the upper basin with a concrete plug

Regular operation was resumed in 1982 and continued until the catastrophic collapse in July 1985, when the upper dam had reached a maximum height of 29 m.

As regards the method of operation and construction, cycloning along the crest of the dam was abandoned in the final stages of the facilities. As can be seen from Figure 6, the sand was finally cycloned with one cyclone at the left edge of the upper impoundment. Raising of the outer dam was performed mechanically using slope material taken from the ground along the upstream edge of the upper reservoir.

FAILURE EVENTS AND POSSIBLE REASONS

Although the catastrophic failure of July 19, 1985 occurred in a sudden burst, a number of events were observed in the preceding 6 months:

1) In January 1985, an earth flow or local slide occurred in the lower portion of the upper dam on its right side (in the area where the decant pipes pass underneath the dam, see Figure 6) as a consequence of freezing water in or/and around the service pipe during heavy frost conditions.

2) Until March 1985, water was observed to seep from the zone of the slide of January.

3) In June 1985, the decant pipe underneath the lower basin impoundment (being in continuous service due to the operation of both impoundments in sequence) broke and all free water and liquid mud escaped from the pond through the broken pipe towards the Rio Stava. A wide and deep crater formed above the point of failure.

4) A bypass pipe had to be installed through the top portion of the lower dam (left end, see Figure 6) and the broken decant pipe had to be blocked in order to restore use of the system.

5) During the interruption of the operations after the collapse of the decant pipe in the lower pond, the water level in the upper impoundment was lowered as far as possible to allow excavation of slope material for the raising of the dam. (right portion)

6) July 15, 1985 (i.e. 4 days prior to the bursting of the dams); both ponds were filled with tailings fluid again and taken into operation under modified conditions (see above).

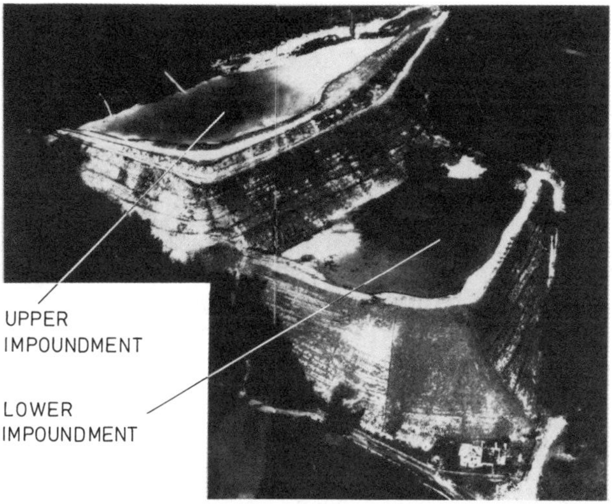

Figure 8. Latest view of the tailings impoundments of Prestavel (view from SE)

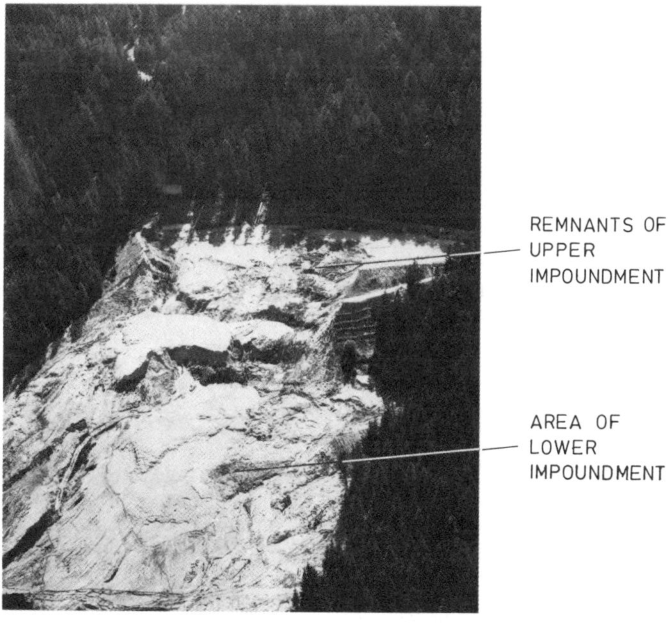

Figure 9. The site after the failure (view from E)

7) On July 19, 1985 (day of failure):

*11.30 and 12.15 pm: Two power failures from a power line running across the Rio Porcellini depression occurred at a distance of nearly 200 m (downhill) from the lower dam. These events were not understood at the time.

*12.23 pm: A sudden burst of the two impoundments occurred with the consequence of a catastrophic mud flow through the Stava Valley, reaching Tesero (distance about 4 km) within a period of 5 to 6 minutes.

Figure 8 gives the latest photograph taken of the two impoundments prior to the failure. Figure 9 gives an impression of the large extent of the burst, which started with a slide of the upper dam pushing away nearly the entire body of the lower one. Figure 9 clearly shows the remains of the sickle-shaped upper dam and the very minor remnants of the lower dam and impoundment.

As can be seen in Figure 9, roughly a third of the total volume of the sediments rushed away, forming a total volume of mud of nearly 200,000 m^3.

In view of the 268 victims and the enormous losses in property, a clear, rational and complete interpretation of the reasons and mechanisms of the worst tailings dam failure in European history is urgently required. Unfortunately, the site information from before the disaster is extremely poor.

An investigation program has been carried out and supervised by an official group of experts formed by the District Attorney. It would be worthwhile but impossible to adequately present here a critical review of the voluminous reports prepared on the subject. In order to demonstrate the great difficulty in establishing whether or not the instability of the dams could have been foreseen, a profile of the two dams on the sloping foundation is given in Figure 10. This section also indicates the various zones of soil incorporated and some relevant geotechnical data obtained from field and laboratory investigations.

A systematic stability analysis was made by the authors on this section accounting for reasonable variations of soil parameters and seepage conditions. The results of these investigations are presented in Figure 11 and reveal how the safety of the two dams decreased during the course of construction. As can be seen from the diagram, the differences between the calculated safety factors for favorable or unfavorable combinations of the relevant parameters are relatively large. For instance, in the latest stage of 1985, the minimum safety factor calculated was less than 0.75, whereas the maximum FS proved to be more than 1.25.

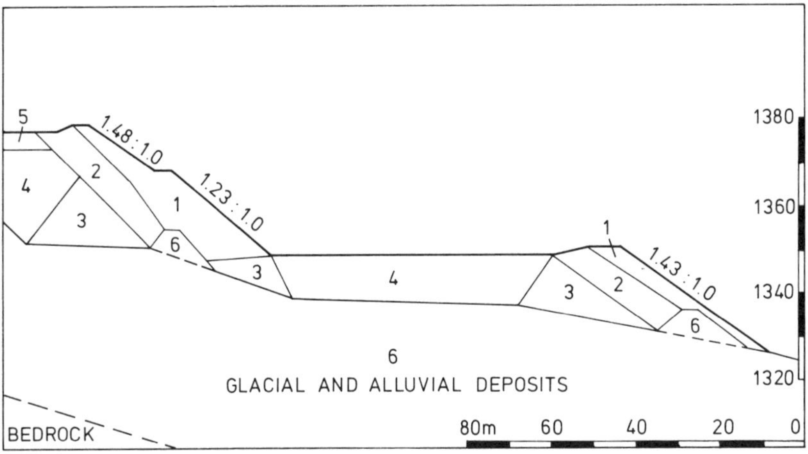

N.o	SOIL TYPE	SATURATED UNIT WEIGHT (KN/MC)	COHESION (KN/MQ)	FRICTION ANGLE (°)
1	SANDS	19.6-20.6	0	38 - 42
2	SANDS AND SILTS	17.6-19.6	0	32 - 36
3	NORMALLY CONSOLIDATED, PLASTIC SILTY CLAYS	16.7-17.6	30 - 40	0
4	UNCONSOLIDATED SOFT TO VERY SOFT SILTY CLAYS	15.7-16.7	20 - 30	0
5	UNCONSOLIDATED SILTY CLAYEY MUDS	14.7-15.7	8 - 20	0
6	NATURAL SOIL: GRAVEL SAND SILT MIXTURE	20.6-21.6	10-20-40	38 - 40

Figure 10. Schematic longitudinal section through the tailings dams indicating soil types and their properties (evaluated from final report of the official experts team)

In addition to this spread of analytical results, one has to keep in mind that the post-failure reconstruction of a profile like that given in Figure 10 is a difficult process involving many approximations and assumptions. This is mainly due to the very complex and heterogeneous structure of such tailings dams and deposits, which, in turn, depend mainly on the variable conditions of construction and operation of embankments of this sort.

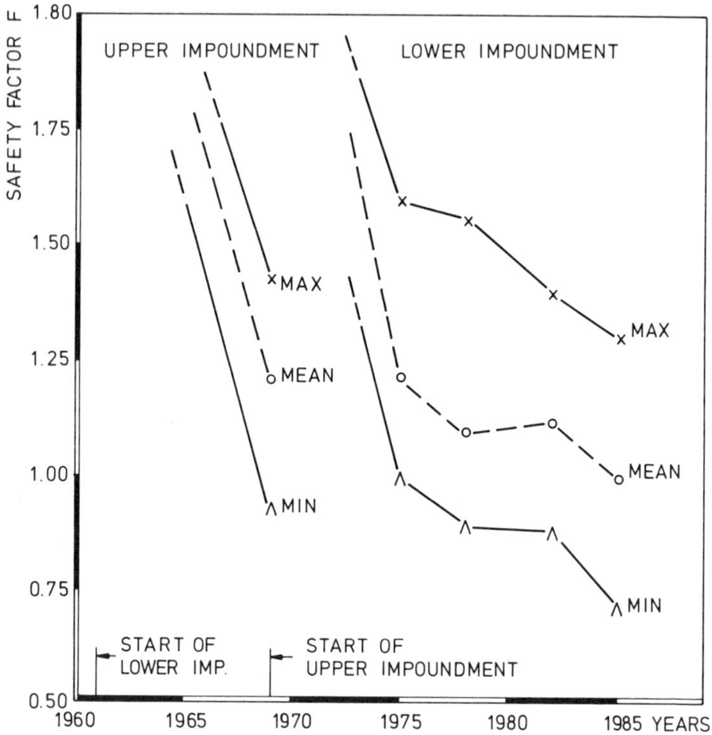

Figure 11. Evolution of overall safety of the tailings dams of Prestavel with time

Generally speaking, it must be concluded from the conditions, under which the tailings dams of Prestavel came to exist and were expanded to their final dimensions, that their reserve stability was always very low. This holds for both the lower and the upper dam.

A number of critical observations characterizing the situation can be briefly listed as follows:

1) The tailings dams were built without prior geological hydrogeological or geotechnical investigations, as may have happened in quite a number of other cases in the early sixties.

2) A rational design of the structures and systematic stability analyses were never made during the lifetime of the dams and construction and operation were purely empirical. Not even a grain size analysis of the tailings is to be found in the documents.

3) There was not a single instrument in the dams to observe movements or phreatic lines. Nor were any tests performed as to the quality and characteristics of the sediments in the tailings ponds, despite the application of the "upstream method" of construction, which relies on the characteristics of the sediments.

4) Sufficient clarification of the tailings fluid (clean water went to Rio Stava, not back into the mill) could only be achieved by using both impoundments in sequence (cascade) and with high water levels in both as shown by existing photographs. In the absence of sandy "beaches", the dam crest was in contact with water over large portions of its length in both the lower and upper dams. This is a dangerous situation, which is to be strictly avoided for dams built with the "upstream method".

5) No measures were taken to effectively divert the surface run-off or to drain springs.

Taking all these conditions together, it can be said that a high failure potential must have existed in these two dams as soon as they reached certain dimensions, and that they were prone to fail under any additional effect acting as a trigger.

A number of such possible triggering effects existed under the particular conditions described above:

1) Effects such as local liquefaction of sands (due to movements in embankment zones with marginal stability) or build-up of excess pore pressure due to additional loads on soft sediments may play a role in such cases (heights of raising steps with cycloned sand were around 5 m).

2) Sudden refilling of both basins with tailings fluid after repairs to the decant pipe in the lower basin may have led to water infiltration into the outer zone of the dams (cycloned sand) with the consequence of a slope failure.

3) Depending on the mode of repair of the collapsed decant pipe, the full reservoir pressure may have built up in the remaining open portion of the pipe, leading to infiltration and pressurisation of the lower portions of the sediments in the lower impoundment. Such an excess

uplift pressure would have substantially reduced the effective stresses and the overall safety of the entire system. (If such an effect had triggered off the event, the failure would have started in the area of the lower impoundment.)

4) A detrimental effect may have resulted from the excavation of slope material in the upper basin shortly before refilling it with tailings fluid towards the 15th July, 1985. This may have led to pressurisation of a pervious gravelly upper layer of the glacial and alluvial deposits underneath the two impoundments (see Figure 5). According to site investigations performed after the failure, a layer of this kind appears to exist; but unfortunately information on the local stratigraphy is quite incomplete. As regards the possibility of the pressurisation of the foundation from the reservoir, the sloping of the site plays an important role. With reference to Figure 5, one can easily imagine that, under certain stratigraphic conditions (a sequence of more or less pervious layers) in the sedimentary foundation, excess pore pressures may have developed due to infiltration from the upper pond, resulting in critical uplift pressures underneath the toe of the upper dam, which rests on impervious sediments of the lower basin. This area is, in fact, the one, which is most sensitive as regards the stability of the entire upper impoundment.

Pressurisation of the foundation is very likely to have played a decisive role in the Stava disaster, because the particular geological and hydrogeological conditions of the site favored such phenomena. Unfortunately the role of foundation pore pressures, caused either by the hydrogeological conditions and heavy rainfalls preceding the event, or by water infiltration from the upper pond, can only be explained by indirect means.

In the opinion of the authors severe excess pore pressures in sensitive parts of the foundation are the most likely trigger of the event and could alone explain the explosive power and extent of the failure (see Figure 9) and the dynamics of the mudflow through the Stava Valley. In this case, one would have to conclude that not only did severe omissions occur with regard to proper planning, construction, observation and maintenance of the tailings dams and reservoirs of Prestavel, but that the site chosen for the project was largely inadequate.

CONCLUDING REMARKS

The disaster of Stava in July 1985 was the most tragic failure of a tailings dam that ever happened in Europe. The paper deals with the geological, hydrogeological and geotechnical conditions of the site and discusses aspects of design, operation/construction, maintenance and control of the tailings disposal facilities. In view of the responsibilities to be allotted, it is difficult to make exact statements. However, for the sake of preventing similar events in the future, it is important to underline, that the failure of Stava happened to a tailings structure, the design and operation of which violated almost every principle of the upstream method of construction applied in this case. In addition to this, the site chosen for the project was risky with regard to the local geological and hydrogeological conditions; this fact was not taken into account at all.

Thus, the disaster of Stava was new and most strong proof of the fact that tailings impoundments are engineering structures, which must be planned, designed, constructed and observed just like other water retaining structures (dams) taking into consideration the geological and hydrogeological site conditions. The most important point is to realistically predict and to effectively control the pore pressures in and around the structure in a way, which exludes any risk of a failure under all varieties of circumstances, which may exist during its entire lifetime (including the period after abandonment). The authors agree that this is a very general concluding remark and, in fact, no news to the engineering community. It does, however, represent the most important point, which in actual cases is the most difficult one to verify.

ACKNOWLEDGEMENTS

Thanks are given to Associazioni Cristiane Lavoratori Italiani (ACLI)/Milano for the kind permission to publish this paper, which summarises studies made in order to provide a large number of bereaved with information on the background of the case. ACLI ran the family hostels, which were lost in Stava, and about 50 of the victims were ACLI members or their relatives.

APPENDIX

LIST OF REFERENCES

1. MURAMOTO, Y., UNO, T., TAKAHASHI, T. (1985)."Investigation of the Collapse of the Tailings Dams at Stava in Northern Italy." Metal Mining Agency of Japan, Confidential Report, Tokyo.

2. ROSSI, G. (1973). "I bacini di decantazione dei rifiuti degli impianti di trattamento dei minerali". Industria Mineraria, No. 10, 11.

3. D.M. LL.PP. n. 1992 del 24 marzo 1982. "Norme tecniche per la progettazione e la costruzione delle dighe di sbarramento." G.U. n. 212 del 4.8.1982.

4. D.P.R. n. 128 del 9 aprile 1959. "Norme di polizia delle miniere e delle cave". G.U. n. 87 del 11.4.1959.

5. D.P.R. n. 1363 del 1 nov. 1959. "Approvazione del regolamento per la compilazione dei progetti, la costruzione e l'esercizio delle dighe di ritenuta". G.U. n. 72 del 24.3.1960.

6. R.D. n. 1126 del 16 maggio 1926. "Approvazione del regolamento per l'applicazione del R.D. 30.12.1923, n. 3267." G.U. n. 154 del 6.7.1926.

7. R.D. n. 1443 del 29 luglio 1927, modificato dalla L. 7. novembre 1941. "Norme di carattere legislativo per disciplinare la ricerca e la coltivazione delle miniere del Regno." G.U. n. 194 del 23.8.1927 e G.U. n. 301 del 23.12.1941

8. R.D.L. n. 3267 del 30 dicembre 1923. "Riordinamento e riforma della legislazione in materia di boschi e di terreni montani". G.U. n. 117 del 17.5.1924.

Evaluation of Two Hydraulic Fill Dams in Alberta

By Neil H. Wade[1], M.ASCE, Tim W. Peterson[2], Lloyd J. Courage[3] and Robert A. Keys[4]

Abstract

In the foothills of the Rocky Mountains of Alberta eleven hydroelectric sites have been developed since 1911. Two of these developments include earthfill dams constructed with hydraulic fill. Minnewanka Dam, a zoned embankment containing a partly hydraulic and partly rolled impervious core, is 617 m long and 35 m high. The hydraulic fill portions are confined to the highest sections of the dam which infill natural gorges. Ghost Dam consists of two sections constructed with hydraulic fill, the 615 m long by 24 m high South Dam and part of the 225 m long by 21 m high North Dam. The results of dam safety evaluations on these structures, conducted in accordance with guidelines established by Provincial regulatory authorities, are summarized highlighting methods of construction, material properties, stability assessments and modifications required.

Introduction

In the Province of Alberta, the majority of existing hydroelectric developments are owned by private utility companies. Between 1911 and 1955, eleven hydroelectric developments were constructed in southwestern Alberta on the Bow River and its tributaries by Calgary Power Ltd., now known as TransAlta Utilities Corporation. Two of these developments, the Cascade and Ghost projects, respectively, incorporate zoned embankments with the impervious cores constructed by hydraulic fill placement techniques (Wade et al, 1985).

In the late seventies, the government of Alberta (1979) instituted Dam and Canal Safety Guidelines which require that the owner report on the condition of each major dam, identifying any deficiencies and providing recommendations for any repairs, modifications, etc. required to ensure project safety. The report for each project must contain an assessment of the overall

1 Asst. Chief Geot. Engr., Monenco Consultants Ltd., Calgary, Canada.
2 Superv. Geot. Engr., Monenco Consultants Ltd., Calgary, Canada.
3 Asst. Mgr. Hydraul. Res., Monenco Consultants Ltd., Calgary, Canada.
4 Mgr., Hydro Operations, TransAlta Utilities Corp., Calgary, Canada.

stability of all earthfill structures giving factors of safety for conditions of rapid drawdown and steady state seepage at full supply level (FSL) and probable maximum flood (PMF). Soil properties as well as the condition of both embankment and foundation must be verified by testing and instrumentation unless sufficient data exist to support the selected values. The Alberta guidelines are similar to those adopted by other regulatory agencies including the U.S. Corps of Engineers (1979) and British Institution of Civil Engineers (1978). The more stringent safety requirements have resulted in a situation where many existing dams no longer exhibit acceptable factors of safety.

The method of construction used, the properties of the hydraulic fill materials and the results of the safety evaluations on these embankments are outlined in this paper.

Cascade Development

Background

Lake Minnewanka was first developed as a storage reservoir in 1912 by the construction of a concrete gravity dam on the Cascade River downstream of the lake outlet. The lake level was thereby raised 5 m and outflows from the reservoir were regulated to augment winter flows for two existing hydro plants downstream on the Bow River. In 1942 the water level in the lake was raised an additional 20 m when Minnewanka Dam, sited just downstream of the original gravity dam, and a new concrete regulating dam in the gorge between Lake Minnewanka and Two Jack Lake were constructed.

The present development, located beside Highway No. 1 approximately 100 km west of Calgary in Banff National Park, utilizes water from the Cascade River, the upper reaches of the Ghost River and the Lake Minnewanka drainage area (Figure 1). A canal and diversion dike diverts the Ghost River into Lake Minnewanka, and Minnewanka Dam, constructed across the original Cascade River channel at the lake outlet, maintains water levels in the lake. Water from the lake is conveyed through a small regulating structure into a 5 km long canal leading to an intake structure which feeds two generating units. From the powerhouse, the tailrace outflows pass under the highway and the Canadian Pacific Railway before returning to the Cascade River just above its confluence with the Bow River.

Construction of Minnewanka Dam

A zoned earthfill embankment with a central impervious core comprised of partly hydraulic fill and partly rolled earthfill, Minnewanka Dam has a total length of 617 m and a maximum height of 35 m. The hydraulic fill sections (Figure 2), totalling about 140 m in length, are confined to the two highest parts of the dam where it crosses the Cascade riverbed and a deep gorge near the south end. In each area, hydraulic fill placement was accomplished by monitors mounted on rafts in the central core pool and directed at the dumped

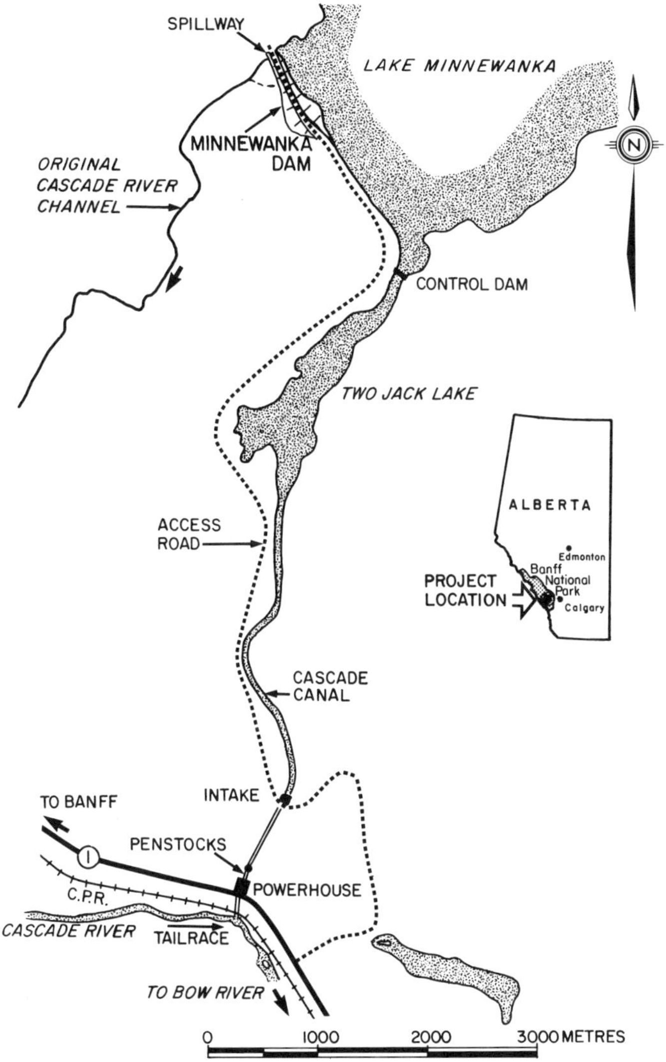

Fig. 1 Cascade Development Site Plan

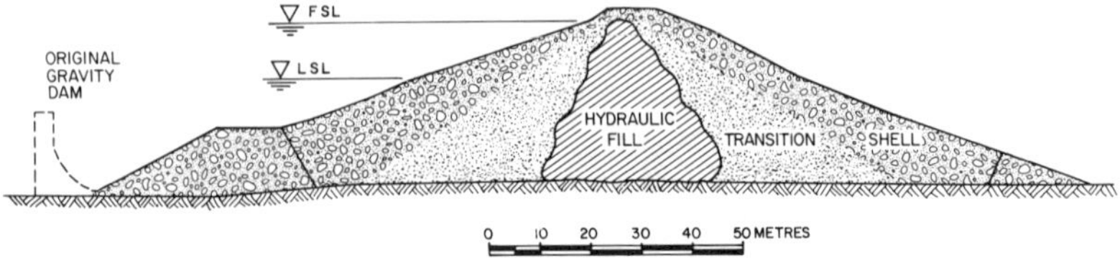

Fig. 2 Minnewanka Dam Hydraulic Fill Section

side slopes to wash fines towards the pool. As the thickness of sediments in the pool built up, additional silty sandy gravel was end dumped along the shoreline to maintain freeboard. The shells were raised simultaneously by placing gravel in lifts and compacting with rollers. Construction records indicate that foundation preparation in the Cascade riverbed, which was carried out during the winter, was an arduous task but this work was trivial compared with that required at the south gorge where it was necessary to excavate 16 m through gravel and boulders to expose bedrock.

Material Properties

As part of the safety evaluation conducted in 1980, a geotechnical investigation was undertaken at Minnewanka Dam to determine subsurface conditions and provide data for stability analyses (Monenco, 1980). The field investigation consisted of drilling a number of holes in the embankment to recover representative disturbed and undisturbed samples of the fill materials, perform insitu permeability and standard penetration (SPT) tests, and install standpipe piezometers. Test pits were excavated on the dam slopes to obtain samples of the granular shell materials.

A laboratory program, conducted to establish material classification and strength parameters, consisted of natural moisture content determinations, Atterberg Limits tests, grain size analyses, and direct shear and triaxial tests. Results of the testing are summarized in Table 1.

The field and laboratory test results were examined to identify zones of fill and foundation material with broadly similar geotechnical characteristics. Boundaries of zones delineated were found generally to correspond with those defined on the original construction drawings.

(i) Hydraulic Fill Core and Transition Material

The core material consists of a non-to-low plastic mixture of silt, sand and gravel with some clay size particles. As indicated on Figure 3, the gradation of the core materials varies somewhat as would be expected from the placement procedures. The proportions of the various grain size components are approximately as follows:

Gravel content = 0 - 23%
Sand content = 34 - 37%
Fines content = 66 - 40%

Atterberg Limits tests gave liquid and plastic limit values ranging from non-plastic to 25% and 21% respectively. In-place moisture contents for the core material varied from 12 to 24% with

Table 1 Summary of Material Properties

Zone	Dam	(%) M.C.	(%) LL	(%) PL	(kPa) c'	(Deg.) ø'	SPT (Blows/30 cm) Range	Avg.
Hydraulic Fill Core	Minnewanka	12-24	NP-25	NP-21	0	33-38(a)	8-18	14
	Ghost North	16-34	NP-41	NP-23	0	36.3-37.5(a)	16-37	-
	Ghost South	11-37	NP-30	NP-25	0	36.3-37.5(a)	0-22	11
Transition	Minnewanka	6-14	NP-17	NP-13	0	33-38(a)	22-40	30
	Ghost South	10-21	NP-26	NP-15	0	35(c)	16-25	20
Shell	Minnewanka	-	-	-	0	40-45(b)	25-52	38
	Ghost North	8±	-	-	0	38(c)	24-91	29
	Ghost South	-	-	-	0	38(c)	24-91	29

Legend

M.C. = moisture content
LL = liquid limit
PL = plastic limit
NP = non-plastic
c' = effective cohesion
ø' = effective friction angle
(a) from triaxial tests
(b) from direct shear tests
(c) after Peck, Hanson & Thornburn (1953)

the higher values occurring in the siltier materials. Four triaxial tests performed on samples of core material gave the following effective strength parameters:

$$c' = 0$$
$$\phi' = 33° \text{ to } 38°$$

SPT values for the insitu core material ranged from 8 to 18 blows per 30 cm. A typical penetration resistance profile for one hole at the interface between the core and transition materials is shown on Figure 4. In general the fine grained materials exhibited the lowest penetration resistance indicating the silts and sands are in a loose to medium dense state.

The transition zone consists of a low plastic silty sand and gravel mixture. The approximate range in particle size distribution, shown on Figure 3, is as follows:

Gravel content = 6 - 72%
Sand content = 42 - 24%
Fines content = 52 - 4%

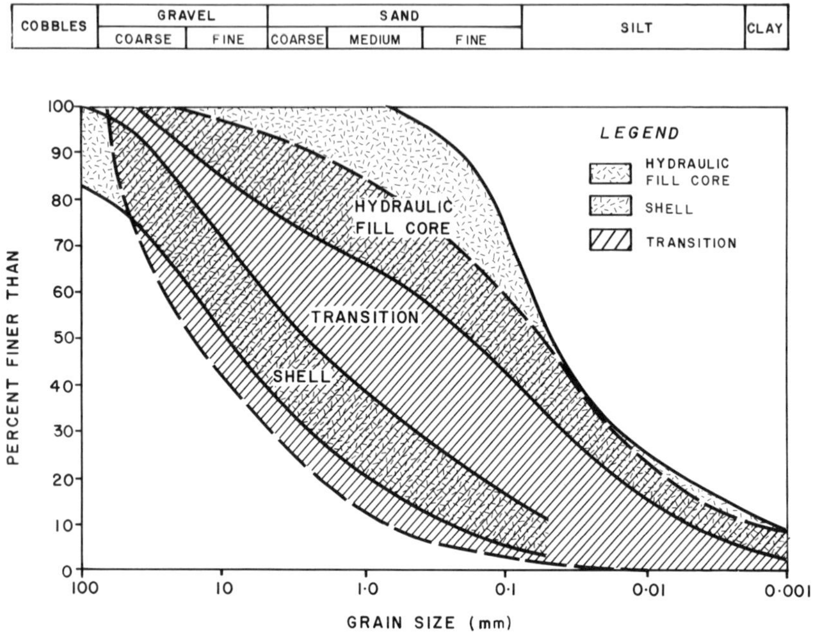

Fig. 3 Grain Size Distribution for Minnewanka Dam Materials

In-place moisture contents measured in the transition materials varied from 6 to 14% with an average of less than 10%. The approximate SPT blow counts for the transition zone ranged from about 22 to 40 blows per 30 cm as shown on Figure 4. Higher blow counts recorded were not considered representative since they were likely affected by the coarser gravel-sized particles, which in many cases resulted in less than 100% recovery in the split spoon sampler.

The boundary between the transition zone and shell could not readily be delineated from drilling. Constructon records indicate that the end dumped transition material comprised silty sandy gravel from which fines had been partly washed to form the core. It was conservatively assumed that the strength of the transition material was equivalent to the upper bound strength of the core material obtained from triaxial tests.

(ii) Shell Materials

Gradation tests on samples from test pits on the dam slopes show the material to consist of a sand and gravel mixture with a trace of silt and occasional cobbles, although a few boulders were encountered in the excavation which were not included in the sampling. Grain size tests (Figure 3) showed the following approximate proportions for shell material:

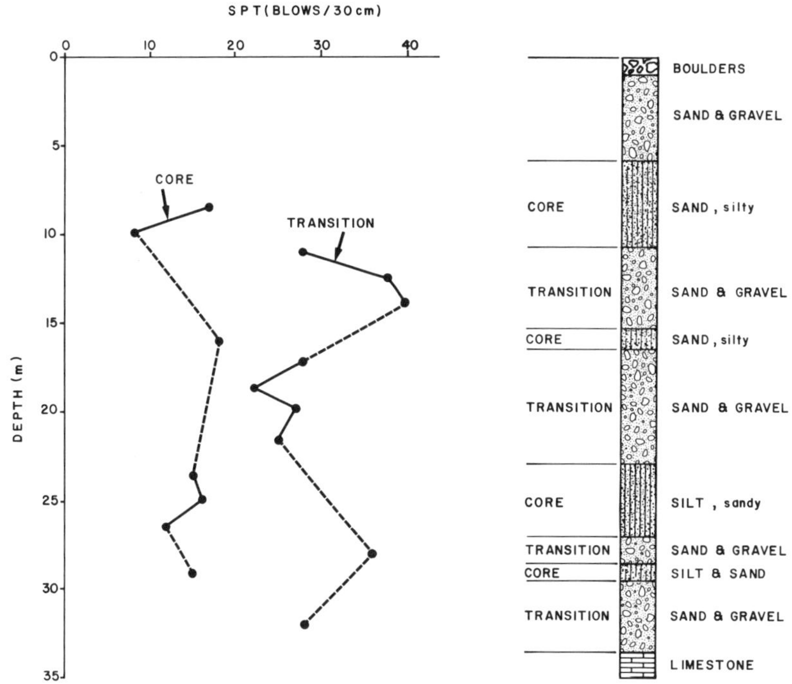

Fig. 4 Typical Log of Hydraulic Fill at Minnewanka Dam

Cobble content = 18 - 0%
Gravel content = 44 - 42%
Sand content = 33 - 46%
Fines content = 5 - 12%

Direct shear tests conducted on remoulded samples of shell material from the test pits gave peak strength parameters of c' = 0 and ϕ' = 40° to 45°.

(iii) Bedrock

Bedrock beneath the dam consists of hard, slightly weathered, massively bedded, dolomitic limestone with an unconfined compressive strength in excess of 200 MPa and a unit weight of 2647 kg/m^3.

Stability Analyses

The material parameters used in the stability analyses of the hydraulic fill sections of the dam are summarized in Table 2. Loading conditions considered in the analyses included steady-state seepage with the reservoir at full supply level (FSL), earthquake loading with the reservoir at FSL, and rapid drawdown of the reservoir. The analyses were carried out using the Morgenstern-Price and Simplified Bishop methods of analysis. Piezometric levels within the embankment at FSL were estimated by extrapolating observed piezometric readings at lower reservoir levels. The equivalent earthquake load which would produce a factor of safety of unity was determined from pseudostatic analyses to be 0.4 g (Monenco, 1980). This earthquake is well above the maximum credible earthquake (MCE) of about 0.1 g assumed for the project site. The rapid drawdown case assumed lowering of the reservoir from FSL to the spillway sill, a drop of 4.3 m, in 24 hours (Monenco, 1987). Results of the analyses, listed in Table 3, indicate the dam has adequate factors of safety for the loading conditions considered. It was concluded that liquefaction of the hydraulic fill materials under the assumed MCE was not likely and that the dam had adequate freeboard to accommodate the anticipated minor earthquake-induced settlements and deformations.

Table 2 Parameters Used in Stability Analyses - Minnewanka Dam

Material	Bulk Density (kg/m^3)	c'(KPa)	Ø'(deg)
Hydraulic Fill Core	2246	0	34
Transition	2246	0	38
Shell	2246	0	40
Bedrock	2647	42.5	45

Table 3 Minnewanka Dam - Stability Assessment

	Factor of Safety	
Loading Conditions	Allowable Min.	Calculated
1. Steady-state seepage @ FSL	1.5	2.1
2. Steady-state seepage @ PMF	1.1	1.3
3. 0.4 g Earthquake* @ FSL	-	1.0
4. Reservoir Rapid Drawdown	1.2	2.3

*Earthquake loading to produce a factor of safety = 1.0

Ghost Development

Background

The Ghost Hydroelectric Development is located on the Bow River approximately 50 km west of Calgary. It was originally constructed in 1929 with a Main Concrete Dam, North and South Earthfill sections, a Sluiceway and a Stoplog Spillway. The latter was replaced by the South Dike in 1974. This development is currently undergoing substantial upgrading to increase spillway capacity following dam safety evaluation recommendations.

The layout of the Ghost Development is shown on Figure 5. The main concrete dam is located in the old Bow River bed. North of the concrete dam and sluiceway the 225 metre long by 21 m high North Dam was constructed partly with a hydraulic fill and partly with a rolled fill core. To the south the 615 metre long 24 metre high South Dam was constructed entirely with hydraulic fill enclosed within gravel zones. Typical sections are illustrated on Figure 6.

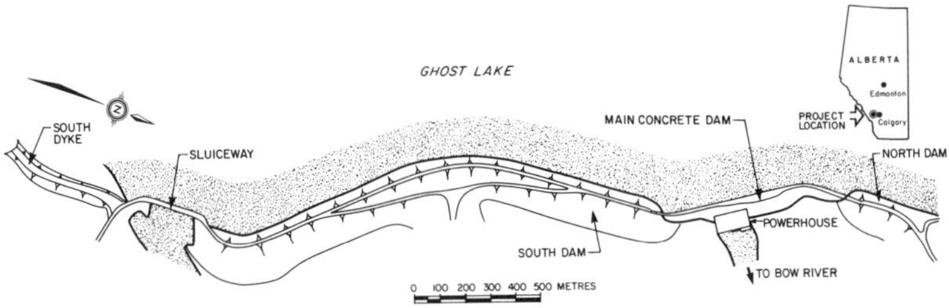

Fig. 5 Ghost Development Site Plan

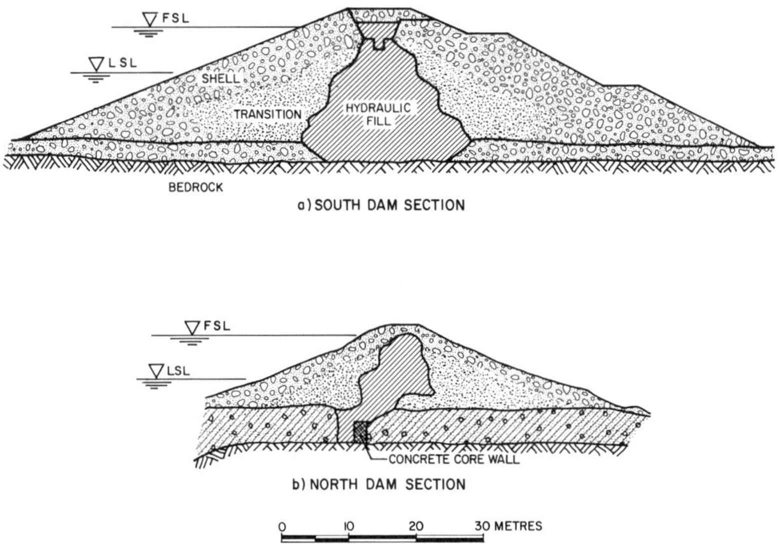

Fig. 6 Ghost Dam Hydraulic Fill Sections

Construction of Ghost Dam

Detailed construction reports were available for review during the Dam Safety Evaluation and provide a good record of the problems encountered and how they were solved, as indicated by the following excepts:

(i) North Dam

"The north earthfill connects the river section with the north bank of the river. The design and construction of this section presented probably the most difficult problem in connection with the whole structure. In the first place there was very little material suitable for an earthfill within an economical hauling distance of the north side of the river. In the second place Ghost Hill, which forms the left bank of the Ghost River, was composed largely of sand standing on a very steep slope and in its natural state was not suitable as (an abutment) for the dam.

Where the Ghost River joins the Bow, there is a sharp point of land at the north end of the dam. Through this point the distance from the Ghost to the Bow River is comparatively short. Bedrock here is at an elevation of about 3845 ft (1172 m) and along the rock contact springs occurred at a number of points. Overlying the rock, the material consisted of a mixture of gravel and sand and disintegrated shale. In addition to sloping off the Ghost Hill and blanketing with clay, the question of disposing of the springs had to be considered. It was finally decided to sluice off the face of the North Hill hydraulically, place a clay blanket tied into bedrock or overlying clay, following the contour of the hill up to above high water level of the reservoir, and to take care of the springs and whatever seepage might occur along the clay core contact by tile drains leading to the inspection tunnel (in the main concrete dam). This plan was carried out. Some of the core was sluiced into place, but the most of it was placed by trucks. The main body of the fill was built of gravel and sand, some of it sluiced into place but most of it deposited by trucks.

Much consideration was given to the North Hill problem before any plans were approved, and it was not until May 16, 1929, that operation commenced in earnest. The methods of construction finally adopted are briefly outlined as follows:

First - The face of the hill was stripped hydraulically, the gravel and boulders which accumulated at the toe being removed partly hydraulically and partly by dragline.

Second - A core trench at the toe of the hill was excavated by dragline down to rock or clay and connected to the north transition concrete section by a concrete core wall founded on bedrock and extending up to ground surface.

Third - Sluicing operations left the bank very steep and irregular. This was sloped off by means of a dragline working from the top and overcasting.

Fourth - A clay core tied into bedrock or overlying blue clay in the core trench was placed by truck, hauled from clay pits located on the north side of the Bow River within one-half mile (1 km) of the fill.

Fifth - A gravel berm was built up by trucks on the upstream side of the clay core or at the outer toe of the fill.

Sixth - (The area) between the gravel fill, the clay core, and the bank was filled in partly hydraulically and partly by trucks with a mixture of gravel, sand, and clay."

Total excavation and fill quantities for the North Dam are as follows (Abbott, 1930):

Hydraulic excavation	60,750 m^3
Machine excavation	14,850 m^3
Total	75,600 m^3

Gravel fill	111,000 m^3
Rolled core	40,750 m^3
Hydraulic fill	10,000 m^3
Total	161,750 m^3

(ii) South Dam

"Material from the clay borrow pit was transported by trains and dumped into sluice boxes. Here the material was broken down by a stream of water from monitors with a nozzle pressure of something like 60 psi(400 KPa) and sluiced into hog boxes and from there pumped by means of dredge pumps into the core pools of the dam.

The material as pumped from the hog boxes consisted of boulders, gravel, and clay, which when deposited, forms first the beaches consisting of the heavier and coarser materials, and then the core consisting of the finer material which is deposited in the deeper water.

One feature in connection with the hydraulic fill which caused some concern, was the consolidation of the core. If the hydraulic fill is placed too rapidly and consolidation does not keep pace, there is a great danger of slips. A number of hydraulic fill dams constructed in the United States were known to have failed in this manner, and great care was taken to investigate the core during construction. Tests were made periodically by determining to what depths a pointed rod would penetrate the core under various weights, thus enabling a comparison with other similar structures."

According to Abbott (1930), the width of the core at any height during construction "...was maintained to equal the remaining height of the dam as closely as possible. A record of the outline of the clay core was kept by examination of the fill on each lift. The edge was considered to be where there was no trace of gravel or any lumps of clay...The average lift of the dam was 3.4 ft (1 m) per week, and the maximum for one week was 5.5 ft (1.7 m). The maximum yardage pumped in 24 hours was 4,035 cubic yards (3087 m^3)." Total excavation and fill quantities for the South Dam were approximately as follows:

Stripping	14,000 m^3
Core Trench excavation	19,250 m^3
Total excavation	33,250 m^3
Gravel fill	480,000 m^3
Hydraulic fill	306,000 m^3
Total fill	786,000 m^3

Material Properties

Other than a pre-construction site investigation to prove the depth to bedrock, there is no record of exploration activities or laboratory testing being carried out until the dam safety evaluation in 1983 (Monenco, 1985). Early that year, an exploratory drilling program was conducted which included nine drillholes in the hydraulic fill sections of the dam. All holes were completed to bedrock and subsequently instrumented with piezometers. Undisturbed samples of core material were obtained using a piston sampler or Shelby Tube and several disturbed samples were obtained of all soil types encountered in the drillholes.

(i) Granular Shell

The shell material consists of well to poorly graded sand and gravel with some cobbles and up to 10 percent silt. It is dense to medium dense with rounded to subrounded particles. The thickness of the shell overlying the core could not be determined from the drill holes, however a shell thickness of two metres was exposed in one test pit. A drillhole located close to the upstream edge of the dam crest was expected to penetrate the core of the dam but because of high gravel content and mud loss between 1.5 m and 6.4 m, the hole was abandoned when a boulder was encountered at a depth of 6.4 m. A second hole drilled 4.5 m north of the first encountered mostly gravel to a depth of 9.5 m before penetrating stiff clayey, sandy silt of medium plasticity. Since the seepage flows measured in a weir immediately downstream of this area are much lower than would be expected if the dam core were composed of the granular material, it was concluded that the holes were located near the shell-core contact or just upstream of the main hydraulic fill core zone. The approximate grading of the shell and core materials is given on Figure 7 and the material properites are listed in Table 1.

The samples recovered from drilling and the excavation for the 1983 repair work at the south dam show that the shell material in all the earthfill structures is similar. Standard Penetration Test (SPT) results varied from a low of 12 to a high of 91 blows per 30 cm, the higher values reflecting the presence of gravel and cobbles rather than very dense conditions. After adjusting for the overburden pressure (Seed & Peacock, 1971), the minimum SPT value increased to 24 blows per 30 cm.

Based on the Gibbs and Holtz correlation (Peck et al, 1953; Bowles, 1977), strength parameters of $\phi' = 38°$ and $c' = 0$ were selected for use in the stability analysis. Published data for similar soils (USBR, 1977; Mitchell, 1977) are consistent with these values.

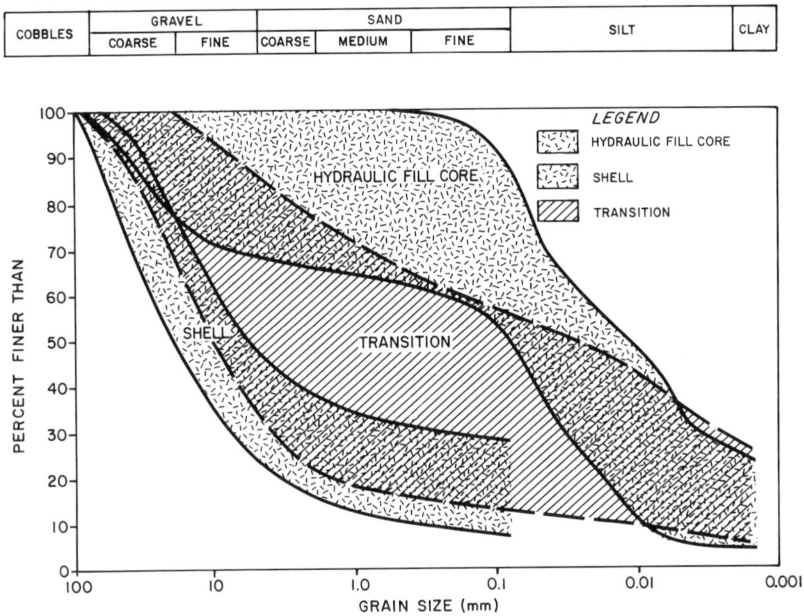

Fig. 7 Grain Size Distribution for Ghost Dam Materials

A 6 m thick layer of alluvium, consisting of an organic, medium dense to dense clayey silt and fine sand, was encountered beneath the shell at one location. Between the alluvium and bedrock is a deposit of medium dense to dense glacial till with a fines content varying from 16 to 37%, although a coarse, bouldery gravel overlying bedrock was observed in a pre-construction test pit (Allan, 1928).

(ii) Hydraulic Fill Core

The core material at the South dam consists of clayey silt and fine sand with little or no plasticity. The material is saturated and loose, especially near the top. The SPT blow count varied from 0 to 22 blows per 30 cm, depending on the gravel content. The top of the core, which was located as much as 4.6 m below the dam crest, was raised in April and May 1983 to within 1.8 m of the crest. The grain size curves for the core material, including the material used to raise the core, are shown on Figure 7.

Highly plastic, gravelly till-like material with high SPT blow counts was encountered beneath the core in several drillholes. Because of the gravelly nature and high moisture content of this material, it is thought to be transition material that migrated to the bottom of the core pool during construction.

At the North Dam, SPT values varied from 16 to 37 blows per 30 cm. It was assumed that the core materials for both dams would have the same drained shear strengths. Triaxial test results on core material from the South Dam gave effective friction angles (ϕ') of 36.3° and 37.5° with zero cohesion. Therefore strength parameters of ϕ' = 35° and c' = 0 were selected for stability computations on both dams.

(iii) Transition

SPT values for this material at the South Dam varied from 12 to 25 blows per 30 cm. After adjusting for the overburden pressure, the corrected minimum value increased to 16 blows per 30 cm. Using the Gibbs and Holtz correlation, strength parameters of ϕ' = 35° and c' = 0 were selected. Although the composition of the transition material varies considerably (Figure 7) the selected values were considered reasonable for the finer materials and conservative for the coarser fraction.

Due to the method of fill placement, the boundaries between the different zones are irregular and often indistinct. The transitional material between zones is similar to native till found locally but generally wetter and loose to medium dense. The fines are a clayey silt with little or no plasticity and the moisture content varies from 10 percent to 15 percent.

(iv) Bedrock

Bedrock beneath the North Dam and South Dam core zones is a soft clayey siltstone with clay filled joints and hardness between R0 and R2 (ISRM, 1981) corresponding to uniaxial compressive strengths ranging from 0.25 MPa to 25 MPa. Rust stained joints indicate some water movement and insitu permeability tests show that the bedrock ranges from moderately pervious to relatively tight with permeability values ranging from 10^{-3} to 10^{-6} cm/sec.

Structurally, the Cretaceous strata strike approximately NNW-SSE and dip upstream at between 20° and 30°. The rock sequence, which consists of feldspathic sandstones and argillaceous siltstones with localized accumulations of pebble conglomerate, is cyclically repeated within a series of near vertical thrust faults approximately parallel to the general strike direction. The closest fault occurs about 550 m upstream of the sluiceway.

Direct shear tests on representative samples indicated strength parameters for the siltstone sequence of c' = 12 kPa, ϕ' = 29° parallel to bedding and c' = 42 kPa, ϕ' = 45° across bedding.

Stability Analyses - Ghost Dam

The computer program "SLOPE II", originally developed at the University of Saskatchewan, was used for the slope stability analyses. The Simplified Bishop method, one of six methods included in this program, was used to analyse circular and composite potential slip surfaces. The search feature of the program was used to locate the potential slip surface with the minimum factor of safety. Parameters used in the stability analyses are summarized in Table 4.

Due to the relatively conservative cross-sections of the earthfill structures, all of the critical potential slip surfaces are located in the shell and/or transition material. The factors of safety, listed in Table 5, are considered to be adequate under all the loading conditions considered for the analyses.

Two earthquake conditions, the operating basis earthquake (OBE) and the maximum credible earthquake (MCE), are generally considered for dams (Seed and Peacock, 1971). Structures should remain elastic and suffer no damage with the OBE while damage can be tolerated with the MCE provided that the water retaining structures remain stable. The MCE can be regarded as the tectonic equivalent of the PMF. The OBE is assumed to produce a horizontal ground acceleration of 0.9 percent gravity at Ghost. This value is for an earthquake with an estimated return period of 100 years and is derived from a statistical analysis undertaken by the Pacific Geoscience Centre of 79 years of record. The MCE is the most severe earthquake that could be expected to occur based on geological and seismological evidence. The MCE was assumed to occur on the closest fault to the development which is capable of producing an earthquake. The potential earthquake source zone closest to Ghost is the southern British Columbia zone which has a maximum expected earthquake magnitude of 6.5 (Basham and Heidebrecht, 1984). The boundary of this zone lies just west of Ghost Development. The closest recorded seismic event was a magnitude 3.1 earthquake in 1974 at a distance of 68 km from Ghost. Since a magnitude 6.5 earthquake located at a distance of 61 km from Ghost would produce a horizontal earthquake acceleration of 10% gravity at Ghost (Hasegawa et al, 1981), this value was adopted for the MCE. This value is in agreement with the minimum value recommended by the USBR (1977) which also recommends use of a vertical acceleration of 5% gravity. Due to the cyclical nature of earthquakes no changes to uplift pressure in the rock foundations were assumed to occur.

For a loose, saturated, silty material like that comprising the core of the earthfill dams, excess pore water pressure would be generated by the cyclic shear stresses induced by an earthquake. This pressure could be high enough to cause liquefaction of the core. To simulate the effect of such earthquake induced pore water pressure, an excess pore water pressure over and above the steady-state water table at FSL was applied to the core of the North Dam and the underlying alluvium as well as the core and transition of the South Dam. The excess pore water pressure was computed by mul-

Table 4 Parameters Used in Stability Analyses - Ghost Dam

Material	Bulk Density (kg/m^3)	c'(KPa)	Ø'(deg)
Hydraulic Fill Core	2198	0	35
Transition	2406	0	35
Shell	2406	0	38
Siltstone bedrock (along bedding)	2500	12	29
Siltstone bedrock (across bedding)	2500	42	45

Table 5 Ghost Dam - Stability Assessment

	Factor of Safety		
	Min.	Calculated	
Loading Conditions	Allowable	South Dam	North
1. Steady-state seepage @ FSL	1.5	2.3	2.0
2. Steady-state seepage @ PMF	1.1	1.3	1.7
3. 0.2 g Earthquake @ FSL	-	0.9*	1.0
4. 0.009g Earthquake @ FSL	1.0	2.2	1.9
5. Reservoir Rapid Drawdown	1.2	1.5	1.5

* Potential failure restricted to upstream shell.

tiplying the earthquake-induced cyclic shear stress at depth h (Seed and Peacock, 1971) by the pore pressure parameter, A, as follows:

$$u = As \doteqdot 0.65Ab\,\gamma h \,\frac{a}{g} \quad \ldots\ldots\ldots\ldots\ldots\ldots \quad (1)$$

in which u = excess pore pressure; s = cyclic shear stress; b = stress reduction factor varying from 1 at ground surface to about 0.9 at a depth of 10 m; a = ground acceleration; and g = the acceleration of gravity. In the computations A was taken as unity for loose hydraulic fill and b was also assumed equal to unity.

The results of a pseudostatic sensitivity analysis for earthquake loads indicates that a maximum ground acceleration of approximately 20% gravity is required to reduce the factor of safety to unity for both the North and South Dams.

Drillholes in the core of the South Dam revealed exceedingly low SPT values in this hydraulic fill which indicated that much of this material could be in a metastable state. It was considered prudent to check the stability of the dam with the conservative assumption that the core had liquefied by using zero strength for the core material. Under these conditions the South Earthfill Dam has a minimum factor of safety of 1.4 and 1.6 for the upstream and downstream slopes respectively when subjected to a maximum ground acceleration of 10% gravity.

Performance of Minnewanka and Ghost Hydraulic Fills

Regular inspections of the earthfill dams at Minnewanka and Ghost indicate that all structures have performed satisfactorily since construction. Upstream and downstream slopes are in good condition apart from some minor riprap deterioration on the Ghost North Dam.

Seepage at Minnewanka and Ghost North is considered normal and that at Ghost South has been substantially reduced since the core was raised in 1983. Due to evidence of leakage over the top of the Ghost South Dam core at that time, the top 5.8 m of the dam was removed and the core was brought up to within 1 m of full supply level. Inspection of the hydraulic fill core material exposed during excavation showed that the core was well consolidated and in very good condition.

Conclusions

As a result of the dam safety evaluations, it was concluded that the hydraulic fill embankments at both the Cascade and Ghost developments were in good condition and can be expected to remain stable over the long term under all anticipated loading conditions.

Acknowedgements

The authors would like to thank TransAlta Utilities Corporation, owner of the projects, for permission to publish this paper.

APPENDIX - REFERENCES

Abbott, W.H. (1930) "Hydraulic Fill on the Ghost Power Development", The Engineering Journal, Vo. XIII, No. 6; June.

Allan, J.A. (1928). "Geological Report on Ghost Water Power Project, Radnor, Alberta", report prepared for Calgary Power Ltd., Calgary; Sept.

Basham, P.W. and Heidebrecht, A.C. (1984) "Engineering Applications of New Probabilistic Seismic Ground-Motion Maps of Canada", National Research Council, Div. of Building Research, Ottawa; Paper No. 1135.

Bowles, J.E. (1977). "Foundation Analysis and Design", McGraw-Hill, New York.

Government of Alberta (1979), "Dam and Canal Safety Guidelines," Alberta Dept. of Environment, Dam Safety Branch, Edmonton; Publ. No. D.S. 3.

Hasegawa, H.S., Basham, P.W. and Berry, M.J. (1981), "Attenuation Relations for Strong Seismic Ground Motion in Canada", Bull, Seis. Soc. Am., 71, No. 6, Dec.

Institution of Civil Engineers (1978), "Floods and Reservoir Safety - An Engineering Guide," Institution of Civil Engineers, London.

ISRM (1981). "Rock Characterization, Testing and Monitoring - ISRM Suggested Methods", pub. for the Commission on Testing Methods, Int'l Soc. for Rock Mech.; Ed. E.T. Brown; Pergamon Press.

Mitchell, J.K. (1977). "Soil Property Correlations", Sym. on Recent Developments in the Understanding and Characteristics of Soil Properties, sponsored by Woodward-Clyde Consultants, Oakland, California; July.

Monenco (1980). "Dam Safety Evaluation for the Cascade Development", report prepared by Montreal Engineering Company Ltd. for Calgary Power Ltd., Calgary; Vols. I & II; October.

Monenco (1985). "Dam Safety Evaluation for Ghost Development", report prepared by Monenco Limited for TransAlta Utilities Corporation, Calgary; Vols. I and II; October.

Monenco (1987). "Dam Safety Report, Cascade Development, Review I", report prepared by Monenco Limited for TransAlta Utilities Corporation, Calgary; April.

Peck, R.B., Hanson, W.E. & Thornburn, T.H. (1953) "Foundation Engineering", Wiley, New York.

Seed, H.B. and Peacock, R. (1971). "Test Procedures for Measuring Soil Liquefaction Characteristics", J. SMFE, ASCE, Aug.

USBR (1977). "Design of Small Dams", Bureau of Reclamation, U.S. Dept. of the Interior, Washington, D.C.

U.S. Corps of Engineers (1979), "Engineering and Design, National Program for Inspection of Non-Federal Dams," Dept. of the Army, Office of the Chief of Engineers, Washington, D.C.

Wade, N.H., Courage, L.J. & Pildysh, M. (1985). "Dam Safety Evaluation on a Number of Canadian Dams", Proc. ICOLD, Lausanne, Switzerland.

A METHOD TO EVALUATE THE DYNAMIC STABILITY OF HYDRAULIC FILL STRUCTURES

by

Shobha K. Bhatia[1], M.ASCE, Joseph P. Schwab[2]
and Isao Ishibashi[3], M.ASCE

ABSTRACT

An effective stress method of analysis is presented for evaluating the liquefaction resistance of saturated granular soils having an initial static shear strain/stress. Inclusion of static shear is of considerable importance in successfully predicting the liquefaction potential of hydraulic fill dams in which significant in-situ shear stresses/strains exist prior to an earthquake. State-of-the-art torsional simple shear test results for combined static and cyclic loading were used in the formulation of the stress-strain and pore pressure relationships utilized in the analysis. Effective stress analysis of a hydraulic fill dam revealed the following: 1) the areas of high pore pressure generation coincided with those areas having the largest initial static shear strain, 2) the adverse effects of high pore pressures in the localized regions of high static strain is mitigated somewhat by dissipation between regions of high and low pore pressures, 3) consequently the overall decrease in cyclic shear strength of the structure is significantly less than that predicted from the laboratory tests, and 4) liquefaction initiates beneath the sloped section of the structure near the toe and radiates outward, the extent of which is a function of the amplitude of the in-situ static strains and earthquake accelerations.

INTRODUCTION

The catastrophic failure of the Lower San Fernando Dam during the 1971 earthquake focused the attention of many regulatory agencies and the design profession on reevaluating existing conventional design tools (Seed et al., 1978). Attempts were made to adopt the more recently developed dynamic analysis procedures for evaluating the dynamic stability of hydraulic fill or tailing structures.

Hydraulic fill dams around the world have suffered extensive damage from earthquakes. The state-of-the-art paper by Seed et al. (1978) describes the failures of some hydraulic fill dams in USA,

---

1. Associate Professor, Syracuse University, Syracuse, NY 13244-1190
2. Senior Engineer, Niagara Mohawk Power Corp., Syracuse, NY 13202
3. Professor, Old Dominion University, Norfolk, Virginia 23508

Japan, Chile, Mexico and U.S.S.R. as well as noting that many dams have performed satisfactory when subjected to moderate size earthquakes. Seed et al. (1978) concluded: "Hydraulic fill dams have been found to be vulnerable to failures under unfavorable conditions and one of the particular unfavorable conditions is the shaking produced by strong earthquake". It was also pointed out that our efforts should be concentrated in analyzing dams which are likely to present problems because they incorporate large bodies of cohesionless material which if saturated, may loose most of their strength. It has been repeatedly shown by many investigators, such as Akiba and Semba (1941), and Seed et al. (1973), that a primary cause of damage in hydraulic fill dams is the build-up of pore water pressure in the dam and the corresponding loss of strength that occurs as a result of these pore pressures. Therefore, for these types of structures, dynamic analysis techniques should be used to account for the loss of strength of cohesionless soil during an earthquake.

TYPES OF DYNAMIC ANALYSIS

Many analytical techniques, ranging from very crude to extremely complex are available for seismic stability of hydraulic fill structures, but most suffer from either theoretical limitations or practical drawbacks.

In a hierarchy of complexity, these methods include empirical evaluation, simplified liquefaction analysis, pseudo-static methods in conventional or modified form and full dynamic analysis (Vick, 1983). On most simplified levels, it is instructive to make a preliminary assessment of seismic stability on the basis of the observed performance of hydraulic fill structures during earthquakes. Seed et al. (1978) for conventional water dams constructed by hydraulic fill method related performance to estimated maximum acceleration. They showed that many hydraulic fill dams survived at accelerations less than 0.2 g. However, this assessment alone is not sufficient for most dams, especially those which might be subjected to larger earthquakes.

The simplified liquefaction analyses of the type proposed by Seed and Idriss (1971) and Seed et al. (1983) have been used to evaluate a factor of safety against liquefaction for upstream tailings embankments and hydraulic fill deposits. The main problem with this approach stem from the application of a one-dimensional level-ground method of analysis to the two-dimensional configuration of a hydraulic fill dam (Idriss and Seed, 1967; Vick, 1983). In the modified pseudo-static method, pore pressures induced by cyclic shear are accounted for in a more explicit way. One procedure is to first consolidate a sample in the laboratory, then subject it to the number of cycles of undrained shear corresponding to a magnitude and duration of the design earthquake, and finally to shear the sample in a static undrained mode.

A strength envelope results that is similar to a conventional envelope from static consolidated undrained tests, except this envelope incorporates the pore pressures due to seismic loading. Lee and Roth

(1977) and Ramanujam et al. (1978) used such an envelope in the analysis of hydraulic fill dams.

A complete dynamic analysis represents the state-of-the-art technique for seismic analysis and attempts to account for explicitly and individually all the factors needed to realistically model the behavior of a dam during an earthquake. Typical dynamic analyses for conventional dams have been described by Seed (1979a), Lee and Roth (1977), Finn (1982) and others. A typical dynamic analysis incorporates the following steps:

1) Pre-earthquake static stresses in the dam are determined - This is usually performed by using a static finite element analysis to determine the effective normal stresses and shear stresses along the potential failure planes. These stresses are necessary for proper interpretation of laboratory cyclic test data. The best procedure for the static analysis is to use a finite element program such as ISBILD (Ozawa et al. 1974) or FEADAM 84 (Duncan et al. 1984) that incorporates the hyperbolic stress strain relations of Wong and Duncan (1974). An assessment of the quality of the results from this program has been given by Finn (1980).

2) The dynamic response of the dam and foundation to base rock motion is determined - From the base motion, the response of the dam is determined by dynamic finite element modeling, using either equivalent linear or nonlinear procedures. The calculated history of shear stresses at every element are converted to an equivalent series of uniform cyclic shear. These cyclic shear stresses are used to interpret laboratory test data. Finn (1980) has critically evaluated and discussed the advantages and disadvantages of existing programs (Quad-4, Lush and Flush) which are generally used to evaluate dynamic shear stress history at particular elements.

3) The dynamic soil behavior is determined - This is accomplished by performing cyclic tests (triaxial, simple shear or torsional shear) under initial consolidation stress conditions corresponding to those calculated by the static stress analysis. Failure in laboratory samples may be defined in terms of initial liquefaction or with reference to a particular level of axial strain. It has been shown by Ishibashi et al. (1985), that the laboratory test interpretation of data from cyclic triaxial, simple shear and torsional shear has been very different. Also the influence of static shear stresses on the cyclic behavior of sand are not fully understood.

4) Evaluate dam stability - Stability of the dam may be addressed in terms of either stresses or strains occurring locally within the dam. The stability of the embankment is evaluated by comparing stresses induced by the earthquake to those required to cause liquefaction or to exceed prescribed limits of local strains.

The location and extent within the dam for which induced stresses exceed those required to cause liquefaction or produce unacceptable levels of strains determine the extent of liquefaction or unacceptable performance. Additionally a stability analysis can then be performed to indicate the overall factors of safety against sliding in the zones of liquefaction or the limits of excessive deformation (Seed et al. 1973).

At present, most programs available for evaluating the dynamic response of dams to base rock motion are based on the equivalent linear method of analysis. In this method, a non-linear hysteretic material is approximated by a lumped elastic model. The major difficulty with this method is that it does not allow the direct computation of permanent deformations. The concept of strain potential in this method, although quite a plausible basis for judgement, violates compatibility requirements and makes the resulting deformation field, largely arbitrary. This difficulty is not removed by the Serff et al. (1976) procedure for converting strain potential to compatible deformation. In addition, the equivalent methods are total stress method and do not directly account for the effect of increasing pore water pressure on soil stiffness. It has been shown by Finn et al. (1978) for a level sandy site subjected to seismic loading that the total stress analysis tends to overestimate the dynamic response when pore pressures exceed 30% of the effective overburden pressure.

In order to alleviate some of these limitations, in recent years a number of programs have been developed for two dimensional nonlinear dynamic effective stress analysis such as TARA-3 (Finn et al. 1986). The program TARA-3 is based on an effective stress method of analysis, and in this program the nonlinear hysteretic behavior of soil is modeled by a hyperbolic skeleton curve with loading and unloading defined by the Masing Criterion (Masing, 1926).

The recently developed programs such as TARA-3, perhaps present the most suitable tools to evaluate the dynamic stability of dams. However, the constitutive relationships used in these programs require special soil tests to determine constants. In addition, the predictive capability of these programs have yet to be proven for hydraulic fill dams. Consequently, for some time in the future, designers will still rely on the dynamic analysis methods provided by Seed et al. (1978).

STATIC AND DYNAMIC SOIL PROPERTIES

Since seismic deformations are not given by dynamic analysis based on the concept of equivalent linear material, these deformations are estimated from the static and dynamic stresses together with the strain data from cyclic triaxial tests. Correspondence between the soil elements in the field and the triaxial sample is established by ensuring that the same initial static stresses exist on failure planes in the field and on failure planes in the triaxial samples. Because in most of the methods of analysis, seismic motions are assumed to be shear waves propagating vertically upward, horizontal planes within the dam are treated as potential failure planes. The shear stresses (static and cyclic) are assumed to be the most significant in generating deformations. The stresses on a typical soil element under the slope in a hydraulic fill dam are shown in Figure 1. This figure includes a representation of the corresponding Mohr's stress circle. These soil elements are subjected to an initial static shear stress (τ_o) and a cyclic shear stress (τ_c) on the horizonal and vertical planes.

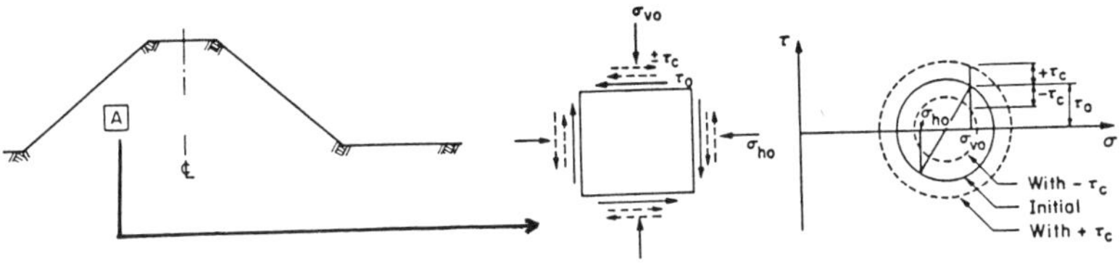

Figure 1. Ideal In-Situ Stress Condition with Static and Cyclic Shear under the Slope

Various investigators (Lee and Seed, 1967; Seed and Lee, 1969; Seed et al., 1973) have used the cyclic triaxial device to investigate and measure cyclic response of soil. In this type of device, they simulated a state of initial static shear stress before and during earthquake loading by first anisotropically consolidating the soil samples with an effective stress ratio (kc $= \frac{\sigma'_{10}}{\sigma'_{30}}$) greater than one. The samples are then subjected to a cyclic deviatoric stress. Further, it is assumed that the potential plane of failure is inclined at (45° + ϕ/2) to the horizontal plane.

Typical cyclic triaxial test results for a soil sample from a hydraulic fill dam are shown in Figure 2 and gives the cyclic shear stress τ_c required to cause 5% strain in 10 cycles of loading for various initial static stress conditions specified by α* value (Seed et al.1978). Based on their experimental results from cyclic triaxial tests, Seed and his co-workers concluded that the larger the initial static shear stress on a plane with a given normal effective stress, the greater the maximum number of cycles of deviatoric stress required to induce a certain amount of strain in a fixed number of cycles. They also observed that in the case of no shear stress reversal ($\tau_c < \tau_o$), the sample tended to deform progressively in each successive cycle, but pore pressure generation did not increase sufficiently to cause initial liquefaction. On the other hand, in the case of partial or total shear stress reversal ($\tau_o < \tau_c$), pore pressures may build up to the value of initial confining stress. These conclusions were derived from experiments conducted on medium-dense and dense sand samples.

The cyclic response of a soil with initial shear stress is commonly expressed in terms of either the cyclic stress ratio τ_c/σ_{vo}' or the peak stress ratio τ_p/σ_{vo}'. The peak stress ratio includes the static

---

*$\alpha = \bar{\tau}_{fc}/\bar{\sigma}_{fc}$

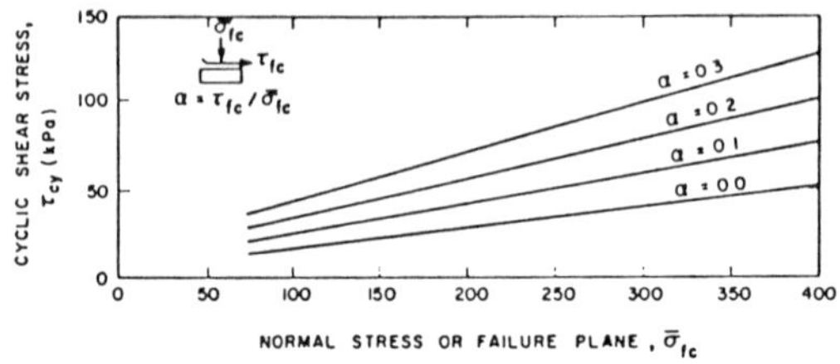

Figure 2. Typical Test Data from Cyclic Loading Test for Medium Dense to Dense Sands (after Seed et al. 1978).

component in the ratio defining liquefaction and as such gives the total response of a soil under cyclic loading.

Vaid and Chern (1983) observed from their extensive experimental data on loose, medium-dense, and dense clean sand that resistance of a soil to liquefaction or cyclic straining increases with initial shear stress. For dense sand this increase is directly proportional to the increase in the static shear stress. However, a loose sand (Dr $\simeq$ 45%) with high initial static shear stress may experience a decrease in liquefaction resistance when compared to the response with no initial shear stress. Typical results from Vaid and Chern (1983) for clean sand are shown in Figure 3. Review of these figures shows that for all ranges of soil density the effect of static shear on resistance to liquefaction is extremely sensitive to the amplitude of strain used to define liquefaction.

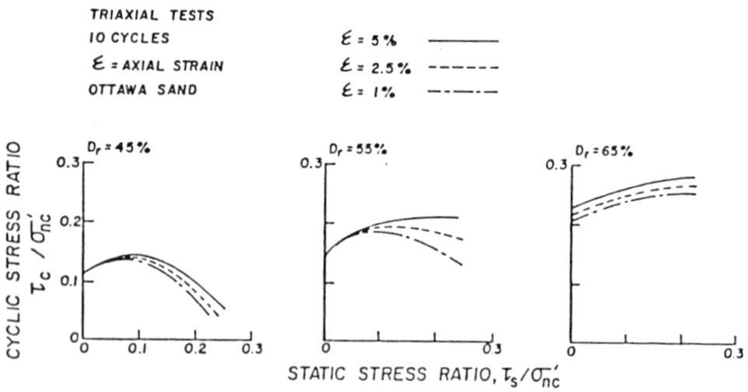

Figure 3. Effect of Initial Static Shear Stress on Cyclic Stress Ratio Required to Cause Various Levels of Strain (after Vaid & Chern, 1983).

Recent studies by Vaid and Chern (1985), using cyclic triaxial device have improved the basic understanding of soil behavior under static and cyclic shearing. However there still remains major concern related to the applicability of these results for a hydraulic fill dam analysis for the following reasons

(a) In a cyclic triaxial device, the effective octahedral confining stress changes during the application of the cyclic deviatoric stress which does not occur in situ. The continuous change of σ_{oct} induces uneven cyclic strains which causes the pore pressure build-up pattern to be non-uniform, and

(b) There is no rotation of principal stress axes in the cyclic triaxial test, while they do rotate in the ideal in-situ case. The effect of this discrepancy is unknown but considered to be minor (Ishibashi et al. 1985).

It has been shown that cyclic simple shear and cyclic torsional shear devices are better suited to duplicate stress conditions developed in a soil element during an earthquake. Many investigators (Yoshimi and Oh-Oka, 1975; Vaid and Finn, 1977; Ishihara and Takatsu, 1977) have used these devices. These studies have qualitatively demonstrated the importance of initial static shear stress and stress reversal, but have failed to establish a unified conclusion (Ishibashi et al., 1985).

The reason that previous researchers have not been successful in finding conclusive results lie in the useage of stress parameters for defining liquefaction or accumulated strains (Ishibashi et al., 1985). Many researchers now realize that the liquefaction potential can be analyzed more consistently by strain method than by the conventional stress method (Dobry et al., 1982; Seed and Arango, 1983; and Finn and Bhatia, 1981). Bhatia et al., (1988) studied the influence of static shear strain by performing strain controlled undrained tests.

MATERIAL PROPERTIES AND TESTING PROCEDURES

Standard Ottawa sand (ASTM, C-109) was used in this investigation. The material properties of the sand are summarized in Table 1. Two different densities (loose ($Dr \simeq 40\%$)) and dense ($Dr \simeq 80\%$) were prepared by the dry-tamping method. After saturating the samples, 20 psi (138 kN/m^2) confining pressure was applied. The samples were tested under strain controlled condition using the torsional shear apparatus. The details of this apparatus and testing procedure are discussed by Ishibashi et al., (1985). Samples were statically sheared with a given initial shear strain in a drained condition. Then the specimens were subjected to a number of uniform sinusoidal cyclic shear strain (with or without strain reversal) under undrained conditions. During the test, pore pressures build up, shear stresses and shear strains were monitored.

Table 1. Material Properties of Ottawa Sand

Designation	D_{10}, mm	D_{50}, mm	D_{60}, mm	G_s	e_{max}	e_{min}
ASTM, C-109	0.20	0.36	0.40	2.65	0.76	0.50

EXPERIMENTAL RESULTS

Pore Pressure Response

The effect of initial static strain on pore pressure development is illustrated in Figure 4 for cyclic strains up to 0.33%. This figure in general shows that for lower levels of static strain ($\gamma_o \leq 0.11\%$), higher pore pressures result for all amplitudes of cyclic strain. At higher levels of static strain (γ_o up to 0.3%), pore pressure build-up is either equal to or slightly less than that without static strain. This indicates that there may be a limit to the static strain amplitude which influences pore pressure build up and liquefaction. In this study, this limit appears to be approximately γ_o = 0.3%.

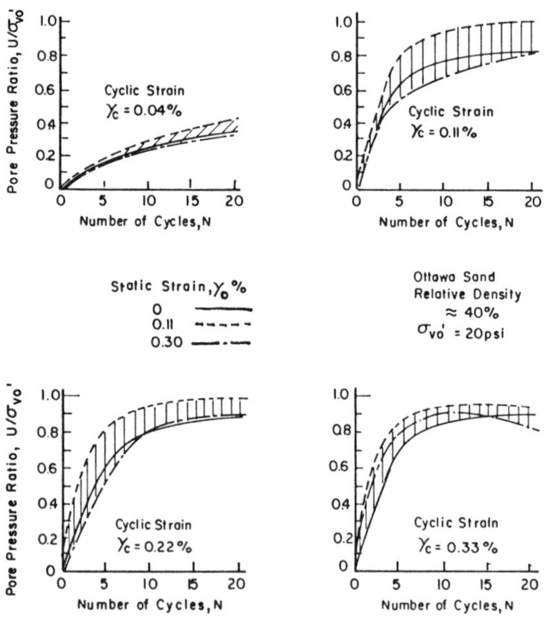

Figure 4. Pore Water Pressure Ratio Versus Cycles for Static and Cyclic Loading. (Torsional Shear Device)

There is some rationale for this behavior in terms of the soil particle arrangement that can occur during static and cyclic loading. Under a high static loading, considerable rearrangement and densification occurs such that the soil becomes stiffer and exhibits a reduced tendency for further particle rearrangement under cyclic loading. Consequently, smaller build up in pore pressures occurs, and slightly higher resistance to liquefaction is exhibited by the soil. On the other hand, for lower values of static shear strain, the potential still exist for considerable rearrangement of the particles under cyclic loading. The static strain in this case tends to assist the cyclic strain in overcoming the friction and interlocking resistance of the soil. It is this additive effect which results in a greater rearrangement potential and a larger build-up in pore pressures and explains why the largest pore pressure does not occur without static strain. Although soils without static strain exhibit the greatest potential for particle rearrangement, the lack of static strain to help overcome the friction and interlocking resistance of the soil usually results in slightly lower or comparable pore pressure levels compared to combined static and cyclic response. Additional tests are required to confirm this unusual pattern of behavior.

Stress-Strain Response

The stress-strain behavior with static shear strain continued to reflect the highly nonlinear hysteretic characteristics observed for cyclic loading only. The addition of static strain resulted in a duplication of some of the trends identified previously, and produced significant new behavior which is shown in Figure 5. As can be seen from this figure, in general for a given cyclic strain amplitude, the shape and the size of the stress-strain curve appears essentially unchanged by static shear particularly after the first cycle. The only noticeable difference is merely a shifting of the loops along the strain axis.

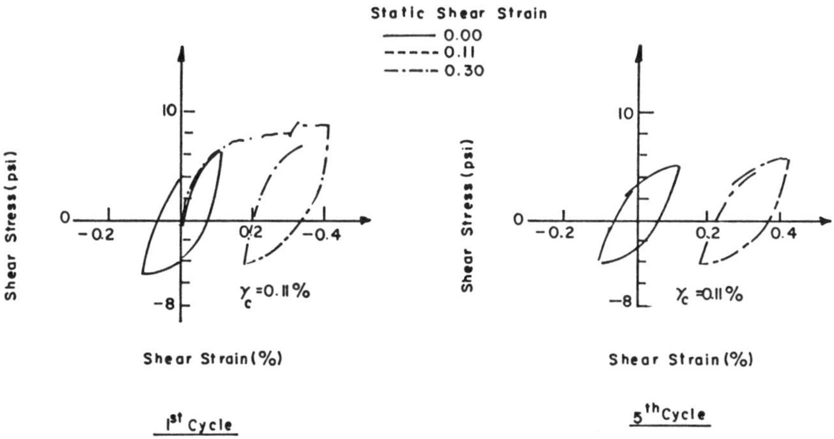

Figure 5. Undrained Stress-Strain Response With and Without Static Shear Strain.

ANALYSIS

Porewater Pressure Model

Prediction of the cyclic behavior of a soil requires a constitutive model for describing the stress-strain response and pore water pressure generation during cyclic loading. Development of constitutive models can be quite complicated due to the nonlinear behavior exhibited by soils, and often an understanding of the physical process occurring in the soil is lost. Therefore, in addition to accurately predicting cyclic behavior, the constitutive models should be simple enough so as not to lose their physical meaning.

The damage parameter concept formulated by Finn and Bhatia (1980), (1981), Bhatia and Nanthikesan (1987a, 1987b), has been found to be particularly suitable for predicting pore pressure generation behavior of sands. A primary advantage of this approach is that a single curve in terms of damage parameter can be used for a wide range of strains and stresses. The concept of damage parameter originated from the intrinsic time concept of endochronic theory (Valanis, 1971). Bazant and Krizek (1976) used endochronic variables to describe the volumetric strain response during cyclic loading. In both cases, the volumetric strains were then incorporated into a porewater pressure model which gave the increments of pore water pressure in terms of the volumetric strain and the bulk moduli of the soil skeleton and water. Both approaches were based on very limited test data.

It can be seen from Figure 4, that the pore water pressure ratio is a function of cyclic and static shear strain amplitudes, and number of cycles, N. This can be expressed as:

$$u/\sigma'_{vo} = f(\gamma, N) \qquad (1)$$

Alternative to N in describing the strain history of the sample is the length, ξ, of the strain path. The variable ξ is a monotonically increasing and continuous variable. A definition which equates an increment in the length of strain path with an increment in deviatoric strain is:

$$d\xi = \tfrac{1}{2}\left|(d\varepsilon_{ij} \cdot d\varepsilon_{ij})\right|^{\frac{1}{2}} \qquad (2)$$

and for simple shear condition and reduces to:

$$d\xi = \tfrac{1}{2}\,|\,d\gamma\,| \qquad (3)$$

The pore water pressure ratio in a constant strain test may then be defined as:

$$u/\sigma'_{vo} = g(\gamma, \xi). \qquad (4)$$

Some of the data in Figure 4 is shown in Figure 6 with u/σ_{vo}' plotted against the strain path ξ. In calculating the strain path, both static and cyclic strains were considered. As can be seen from Figure 6, for a given cyclic strain amplitude (γ_c), different curves result for each value of γ_o. Ideally, if the influence of static shear strain

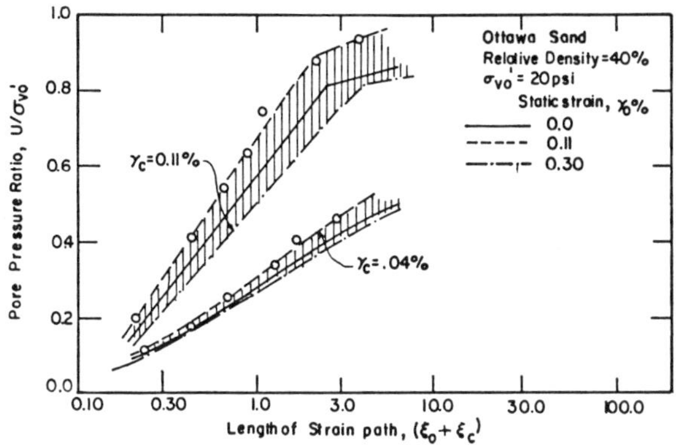

Figure 6. Pore Water Pressure Ratio vs. Length of Strain Path.

had been the same and with no experimental scatter, for a given γ_c there should have been one unique curve. Based on this consideration, for each level of cyclic shear strain, an average curve has been used in the analysis. This will result in pore pressure response that increases with static strain amplitude which is not completely in accordance with Figure 4. However, the trend of larger pore pressures with static shear compared without static shear is duplicated and this is the major finding of this study. Further testing is necessary to verify the increasing and then decreasing pore pressure response with static strain evident in Figure 4. Before generalizing the constant cyclic shear strain, the explicit dependence on cyclic shear strain amplitude, γ_c, must be removed. Ideally, u/σ_{vo}' should be expressible as a monotonically increasing function of a single variable, K, which could be unambiguously defined for static strain, constant cyclic strain and irregular strain histories. It has been shown by Finn and Bhatia (1980) that the porewater pressure ratio u/σ_{vo}' may be expressed as a function of K if a transformation T exists such that for K = T ζ For the present data, such a transformation $T = e^{\lambda\gamma c}$ exists where the value of λ = 3.94 and ${}_c$ is expressed in percentage. The procedure used to determine λ is explained by Finn and Bhatia (1981). Using this value of λ each data point $(u/\sigma_{vo}'$, γ, ζ) was converted to a data point $(u/\sigma_{vo}'$, K) using the transformation.

$$K = \zeta\, e^{3.94\gamma c} \qquad (5)$$

The new data points are shown plotted in Figure 7 against the natural logarithm of K. A nonlinear least square curve fitting method was used to determine the curve shown in Figure 7 describing the relationship between u/σ_{vo}' and K. The equation of this curve is:

$$u/\sigma'_{vo} = G(K) = A/B \text{ Ln } (1+ B\, K) \qquad (6)$$

with A = 0.42 and B = 1.41

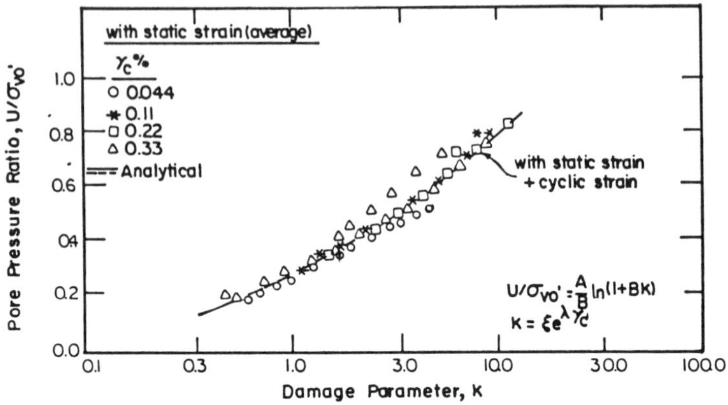

Figure 7. Pore Water Pressure Ratio vs. Damage Parameter for Combined Cyclic and Static Loading.

The accuracy with which the basic test data in Figure 4 is represented by equation (6) was checked using the inverse transformation of T, and comparing with the original data. This verification procedure showed that for the range of shear strain used in the testing program, the porewater function G(K) gives good representation. For other soil densities this procedure will also result in equation 6, however different values will be obtained for A, B and λ.

Stress-Strain Model

To describe the undrained stress-strain relationship for sand under undrained condition, an analytical model must be capable of duplicating the signifciant strain softening which occurs as pore pressure increases during cyclic loading. Bhatia and Nanthikesan (1987a, 1987b) used the stress-strain relationships based on endochronic formulation for undrained behavior of clean sands. They found the relationship were very complex and that the required constants were difficult to determine. For this reason, a nonlinear hysteretic model using hyperbolic skeleton curves and unloading-reloading response defined by the Masing criterion (Masing, 1926) was developed. This representation is similar to one described by Lee and Finn (1978) except this model uses a different scaling factor instead of the Masing factor of 2. The static response predicted by this model is based on the experimental static strength curve. The primary objective of this model is to incorporate the effect of static strain on the cyclic response of sand. The effect of static strain is to initially produce a significant increase in stresses at low cyclic strain amplitudes, while for higher amplitudes it results in only slightly higher stresses. The analytical model developed quite successfully duplicates this behavior. The details of this model are discussed in detail by Bhatia, Schwab and Ishibashi (1988). A typical predicted and experimental stress-strain relationship is shown in Figure 8.

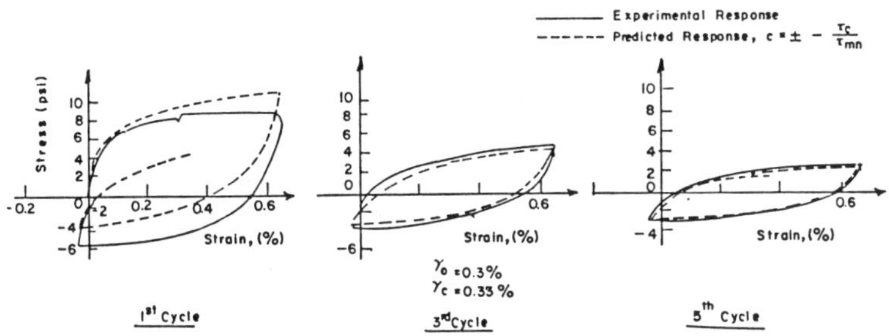

Figure 8. Experimental and Predicted Undrained Stress-Strain Response for Static Strain Amplitude γ_o = 0.3% and Cyclic Strain Amplitude, γ_c = 0.33%.

EFFECTIVE STRESS MODEL

If an effective stress model is to be used for analyzing tailings or hydraulic fill dams inclusion of the effects of static shear on the response of the soil is necessary. The effective stress model developed is a fully coupled dynamic model capable of predicting the liquefaction resistance during an earthquake for a hydraulic fill dam having in-situ static shear strain/stress. To simplify the analysis, the response under two dimensional loading is treated similarly to that for the one dimensional case (i.e. as a horizontally layered soil), except the pore pressure model has been extended to include the effect of static shear strain. This reduces the problem to that of a lumped mass shear beam subjected to horizontally propagating shear waves from the base. The material properties of the soil are assumed to vary only in the vertical direction. Finn et al., (1977) showed that the dynamic response of a level soil deposit could be reasonably predicted by representing the soil as a lumped mass shear beam shaken by vertically propagating shear waves. Inherent in the representation of the soil as a lumped mass shear beam is the assumption of an essentially horizontal deposit having no initial shear stress. Therefore, usage of this method requires major modification to incorporate the effect of static shear stresses or strains. This was accomplished by using the two dimensional pore pressure model and stress-strain relationships discussed in previous section. The effective stress analysis includes the effect of pore pressure on soil stiffness. This is accomplished by means of a pore pressure generation model which continuosly calculates the variation in pore pressure during an earthquake. Inclusion of dissipation due to pressure gradients between soil layers is also included.

The response predicted by the developed program for a soil layer subjected to El Centro earthquake motion was compared with the results given by DESRA-2 (Finn et al., 1977) and Shake (Schnabel et al; 1972). The results compare reasonably well (Bhatia et al., 1988).

ANALYSIS OF A HYDRAULIC-FILL DAM

In order to demonstrate the applicability of the developed program, a 50 ft high earth dam has been analyzed. Due to the sloping sides, static shear stresses/strains will vary considerably throughout the cross-section of the dam. The finite element program ISBILD, which incorporates the hyperbolic stress strain relations of Wong and Duncan (1974), was used to calculate the static strains throughout the cross-section of the dam. The program analyzed the dam as constructed in 5-10 feet lifts using soil properties given in Table 2. The variation of static shear strains throughout the dam cross section is shown in Figure 9. As can be seen from this figure, the static shear strains vary considerably throughout the cross-section of the dam, and the maximum strains occur in the lower sloped section of the dam. The static strains computed at two locations (1 and 2) are given in Table 3 for five foot increments. In evaluating the potential for failure due to an earthquake, localized areas susceptible to liquefaction must be identified. For the structure shown in Figure 9, the dynamic response of the centerline (location 1) and along the sloped surface (at location 2) were analyzed using the developed program. These two locations represent the condition of minimum and maximum static strains determined from the static analysis. Of particular interest is the influence that the larger static strains at location 2 has on pore pressure build-up.

Table 2. Program Soil Properties Used in Analysis.

Description		1	2	3	4	5	6	7	8	9	10
Thickness (ft)		5	5	5	5	5	5	5	5	5	5
Density (lb/ft^3)		125	125	125	125	125	125	125	125	125	125
Initial shear modulus, Gmo (#/ft^2)		621	981	1162	1317	1456	1583	1701	1811	1914	2012
Initial shear strength, mo (#/ft^2)		426	673	797	903	999	1086	1167	1242	1313	1380
Permeability, K (ft/sec)		.01	.01	.01	.01	.01	.01	.01	.01	.01	.01
Hardening constants	H_1	.0185	.0185	.0185	.0185	.0185	.0185	.0185	.0185	.0185	.0185
	H_2	1.001	1.001	1.001	1.001	1.001	1.001	1.001	1.001	1.001	1.001
	H_3	7	7	7	7	7	7	7	7	7	7
	H_4	4	4	4	4	4	4	4	4	4	4
Bulk Modulus, C_b		.0322	.0322	.0322	.0322	.0322	.0322	.0322	.0322	.0322	.0322
Damping constants,	Alpha	0	0	0	0	0	0	0	0	0	0
	Beta	.005	.005	.005	.005	.005	.005	.005	.005	.005	.005
Pore Pr. constants,	A	0.42	0.42	0.42	0.42	0.42	0.42	0.42	0.42	0.42	0.42
	B	1.41	1.41	1.41	1.41	1.41	1.41	1.41	1.41	1.41	1.41
	Lambda	3.94	3.94	3.94	3.94	3.94	3.94	3.94	3.94	3.94	3.94

The El Centro earthquake modified to a_{max} = 0.15 g was used in these analyses. Additionally, to simplify the analysis seepage pressures within the dam were omitted and the initial phreatic or free surface co ncided with the slope of the dam. The results of this analysis was compared with the response of a 50 foot horizontal soil deposit with similar properties subjected to the same motion. The only difference between the cases (dam and soil layer) is the presence of static strain on the dam. The results revealed the following:

1) At location 1, very small levels of pore pressure were generated with the maximum values (u/σ_{vo}' = 0.50) occurring in the layer with the largest static shear strain (layer 3). Pore pressure response for layers 3 and 6 also reveals that dissipation effect are quite significant.

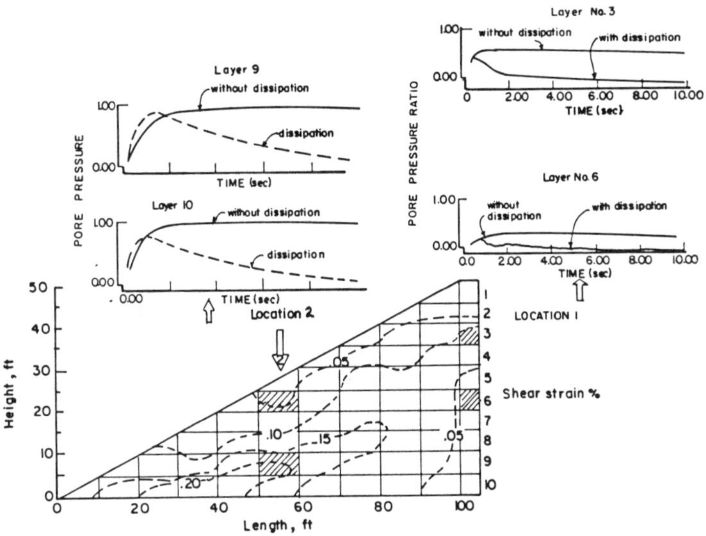

Figure 9. Static Shear Strain Contours in an Earth Dam

Table 3. Static Strains Within Earth Dam at Locations 1 and 2.

Layer	Location 1	Location 2
1	0	-
2	0	-
3	0.128	-
4	0.093	-
5	0.037	0.002
6	-0.037	0.031
7	-0.005	0.096
8	0.011	0.134
9	0.015	0.245
10	0.010	0.163

2) At location 2, significant levels of pore pressure occur only in those layers with significant amplitudes of static strain as evident from Figure 9. Near the surface where the static strains are small only nominal pore pressure buildup occurs. However, dissipation effects are quite significant due to the pressure gradients between layers of high and low pore pressures.

3) Comparison with the results for a horizontal soil deposit having identical soil properties reveals that the significant pore pressure build up evident within localized areas of the dam does not occur. For the horizontal deposit, the cyclic stresses and strains are approximately half of those generated in the dam (Figure 9). However, in both cases the cyclic fluctuations are the greatest at the base and diminish significantly near the surface.

Dynamic analysis was also performed on a typical centerline-type tailing dam constructed of uncompacted cycloned tailings sand at a relative density of 40%. In the center of the tailings dam is a starter impervious dike. On the downstream side of the dam, the upper 60 feet of material is unsaturated while the sand below the water table is assumed to be fully saturated. Vaid and Chern (1985) investigated the cyclic strength of various tailings sands and found the cyclic strength in many cases is very similar to clean Ottawa sand. However, tailings sand can liquefy over a wide range of densities compared to round Ottawa sand. The relative density, consolidation pressure and the overconsolidation ratio are found to be most important controlling factors for tailing sands whereas relative density alone is a good parameter for Ottawa sand. Therefore, the laboratory results on Ottawa sand are not directly applicable to tailings sands, however a qualitative response of tailings sand can be obtained from the constitutive relationships based on Ottawa sand.

For the 100 feet high tailings dam in Figure 10, static strains were calculated using the finite element computer program FEADAM84 Duncan et al., 1984). The dynamic analysis has been performed at

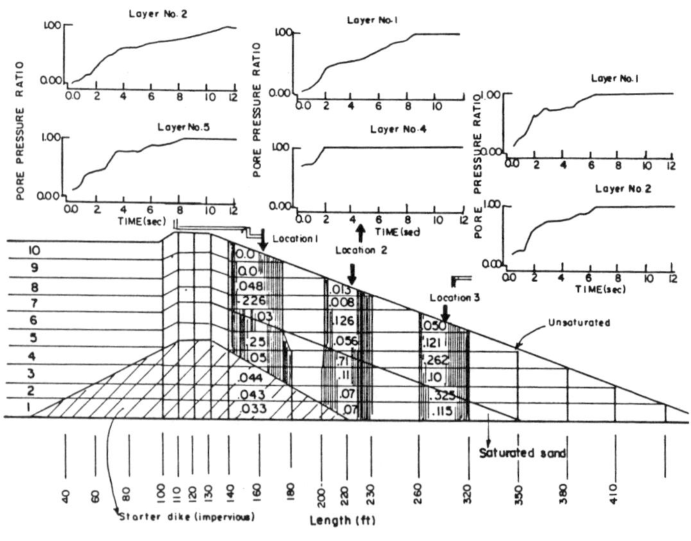

Figure 10. Typical Tailing Dam with Static Shear Strain Values.

three locations with varying magnitudes of static strains. The dam was subjected to the El Centro time history modified to a_{max} = 0.3 g and 12 second duration. In Figure 10, pore pressure generation versus time plots are shown for various elements within the dam. It can be observed that elements with higher initial shear strain develop higher pore water pressures.

CONCLUSIONS

The main objective of this investigation was to develop and test an analytical method for predicting the cyclic response of a soil deposit under combined static and cyclic loading. Development of the new method followed along three steps.

First, an extensive review of state of the art knowledge of the cyclic behavior of sand with and without static shear stresses/strains was undertaken. This review identified key factors influencing behavior, and was used in the development of the experimental program which focused on the response of loose sand in strain-controlled undrained conditions under combined static and cyclic loading. A comprehensive evaluation of the test results was performed to establish important trends and behavior patterns. This evaluation was conducted to devise a new pore pressure and stress-strain model for incorporating the effects of static strain. Development of these models and their formulation in an effective stress analysis comprised the third step. The effective stress analysis model presented herein is an extension of one dimensional analysis. The effect of static loading is reflected in the new pore pressure and stress strain models.

The effective stress analyses of two typical dams clearly shows that liquefaction resistance decreases as static strain increases. The rate of decrease and the affected zone is directly proportional to the static strain levels. It was also noticed that the omission of dissipation tends to under estimate shear strength in the critical zones and may produce overly conservative results. To date, dynamic analysis based on the equivalent linear method has a good record in practice, and it will be some time in the future before a two-dimensional dynamic analysis (with hyperbolic Masing stress strain relationship) is used in practice. The proposed method of analysis provides an alternate approach for dynamic analysis.

ACKNOWLEDGEMENTS

The support for this research was provided by the National Science Foundation under Grant CEE-8313 863 to Syracuse University. This support is gratefully acknowledged. Appreciation is expressed to Y. DeGouff who typed the paper and to R.D. Ziemer who drew the figures.

REFERENCES

1. Akiba, M., and Semba, H., 1941, The earthquake and its influence on reservoir in Akita prefecture, J. Agric. Engng. Soc. Japan, 13, No. 1.

2. Bhatia, S.K., Schwab, J. and Ishibashi, I., 1985, Evaluation of the Porewater Pressure for Seismic Stability Analysis of Natural Slopes, Discussion Eleventh International Conference on Soil Mechanics and Foundation Engineering, San Francisco, Calif., August.

3. Bhatia, S.K. and Nanthikesan, S., 1987a, The Development of Constitutive Relationship for Seismis Pore Pressure, 3rd International Conference on Soil Dynamics and Earthquake Engineering, Princeton, June 22-24.

4. Bhatia, S.K. and Nanthikesan, S., 1987b, Constitutive Relationships for Cohesionless Soil, paper submitted to Journal of Soil Dynamics and Earthquake Engineering.

5. Bhatia, S.K., Schwab, J.P. and Ishibashi, I., 1988, Influence of Static Shear on Cyclic Undrained/Drained Behavior of Sand, paper submitted to ASCE, Geotechnical Division.

6. Biot, M.A., 1941, General Theory of Three-Dimensional Consolidation, J. Appl. Phys., 12, pp. 155-64.

7. Finn, W.D.L., Lee, K.W. and Martin, G.R., 1977, An Effective Stress Model for Liquefaction, Journ. of the Geotech. Engr. Div., ASCE, Vol. 103, No. GT6, Proc. Paper 13008, June, pp. 517-533.

8. Finn, W.D.L., Martin, G.R. and Lee, M.K.W., 1978, Comparison of Dynamic Analyses for Saturated Sands, Proceedings, ASCE Geotechnical Engineering Division Specialty Conference, June 19-21, Pasadena, Calif., pp. 472-491.

9. Finn, W.D.L., 1980, Evaluation of Critical State and Hyperbolic Stress-Strain Models, Proceedings, NSF/NSERC North American Workshop on Plasticity and Generalized Stress-Strain in Geotechnical Engineering, McGill University, Montreal, May 28-30, Vol. 2.

10. Finn, W.D.L. and Bhatia, S.K., 1981, Prediction of Seismic Porewater Pressure, Proceeding of the 10th International Conference of Soil Mechanics and Foundation Engineering, Stockholm, Vol. 3, pp. 201-206.

11. Finn, W.D.L., 1982, Fundamental Aspects of Response of Tailing Dams to Earthquakes, Dynamic Staiblity of Tailing Dams, Preprint, 82-539, ASCE, pp. 46-72.

12. Finn, W.D.L., M. Yogendra, Kumar, N. Yoshida, and H. Yoshida, 1986, TARA-3: A Program for Nonlinear Static and Dynamic Effective Stress Analysis, Soil Dynamics Group, University of British Columbia, Vancouver, B.C.

13. Idriss, I., and Seed, H., 1967, Response of Earth Banks During Earthquakes, Journ. Soil Mech. and Fdn. Div., ASCE, Vol. 93, No. SM3, pp. 61-82.

14. Ishihara, K. and Takatsu, H., 1977, Pore Pressure Buildup in Initially Sheared Sand Subjected to Irregular Excitation, Proceeding of the Sixth World Conference on Earthquake Engineering, New Delhi, India, pp. 2163-2168.

15. Ishihara, K., J. Troncoso, Y. Kawase and Y. Takahashi, 1980, Cyclic Strength Characteristics of Tailing Materials, Soils and Foundation, Vol. 20, No. 4, Dec.

16. Ishibashi, I., Kawamura, M. and Bhatia, S.K., 1985, Effect of Initial Shearing on Cyclic Drained and Undrained Characteristics of Sand. Report submitted to NSF for Grant No. CEE-83140009 and CEE-8313863, May.

17. Lee, K.L. and Seed, H.B., 1967, Dynamic Strength of Anisotropically Consolidated Sand, Journal of the Soil Mechanics and Foundation Division, ASCE, Vol. 93, No. SM5, pp. 169-190.

18. Lee, K. and Roth, W., 1977, Seismic Stability Analysis of Hawkins Hydraulic Fill Dam, Journ. Geotechnical Div., ASCE, Vol. 103, No. GT6, pp. 627-644.

19. Masing, G., 1926, Eigenspannungen und verfestigung beim Messing, Proceeding 2nd International Congress of Applied Mechanics, Zurich.

20. Ozawa, Y. and Duncan, J.M., 1973, ISBILD: A Computer Program for Analysis of Static Stresses and Movement in Embankments, Report No. TE-73-4, College of Engineering, University of California, Berkeley, Calif.

21. Ramanujam, N., Holish, L., and Chen, W., 1978, Post Earthquake Stability Analysis of Earth Dams, Proc. Conf. on Earthquake Eng. and Soil Dynamics, ASCE, Vol. 2, pp. 762-776.

22. Seed, H. and Idriss, I., 1977, Simplified Procedure for Evaluating Soil Liquefaction Potential, Journal Soil Mech. and Fdn. Div., ASCE, Vol. 97, No. SM9, pp. 1249-1273.

23. Schnabel, P.B., Lysmer, J. and Seed, H.B., 1972, SHAKE: A Computer Program for Earthquake Response Analysis of Horizontally Layered Sites, Report No. EERC 72-12, EERC, University of California, Berkeley, Calif. Dec.

24. Seed, H.B., Lee, K.L., Idriss, I.M., and Makdisi, F., 1973, Analysis of the Slide in the San Fernando Dams During the Earthquake of Feb. 9, 1971, Report No. EERC 73-2, University of California, Berkeley.

25. Seed, H.B., 1973, Stability of earth and rockfill dams during earthquakes. Embankment-dam engineering, Hirschfeld, R.C. and Poulos, S.J., Eds., New York: John Wiley and Sons, 239-270.

26. Serff, N., Seed, H.B., Makdisi, F.I. and Chang, D.K., 1976, Earthquake Induced Deformation of Earthdam, Earthquake Engineering Research Center, Report No. EERC 76-4, University of California, Berkeley, Calif.

27. Seed, H.B., Makdisi, F.I. and De Alba, P., 1978, Performance of Earth Dams During Earthquake. J. Geotech. Engng. Div. 104, No. GT7, Proceeding Paper 13870.

28. Seed, H., 1979a, Soil Liquefaction and Cyclic Mobility Evaluation for Level Ground During Earthquakes, Journ. Geotech. Eng. Div., ASCE, Vol. 105, No. GT2, pp. 201-255.

29. Seed, H.B. and Arango, I., 1983, Evaluation of Liquefaction Potential Using Field Performance Data, Journal of the Geotechnical Engineering Division, ASCE, Vol. 109, No. 3, pp. 458-482.

30. Vaid, Y.P. and Finn, W.D.L., 1979, Static Shear and Liquefaction Potential, Journal of the Geotechnical Engineering Division, ASCE, Vol. 105, GT10, pp. 1233-1246.

31. Vaid, Y.P. and Chern, J.C., 1983, Effect of Static Shear on Resistance to Liquefaction, Soils and Foundations, Vol. 23, No. 1, pp. 47-60.

32. Valanis, K.C., 1971, A Theory of Viscoplasticity without a Yield Surface, Archiuum Mechaniki Stoswanej, Vol. 23, No. 4, pp. 517-533.

33. Vick, Steven, G., 1983, Planning Design and Analysis of Tailing Dams, Wiley Series in Geotechnical Engineering, John Wiley and Sons, 1983.

34. Wong, K.S. and Duncan, J.M., 1974, Hyperbolic Stress-Strain Parameters for Nonlinear Finite Element Analysis of Stresses and Movements in Soil Masses, Geotechnical Engineering Report. No. TE-74-3 to National Science Foundation, Office of Research Services, University of California, Berkeley, July.

35. Yoshimi, Y. and Oh-Oka, H., 1975, Influence of Degree of Shear Stress Reversal on the Liquefaction Potential of Saturated Sand. Soils and Foundations, Vol. 15, No. 3, pp. 27-40.

STABILITY OF HYDRAULIC SANDFILL
TAILINGS DAMS

By Robert C. Lo[1], Earle J. Klohn[2], Fellows, ASCE, and W.D.L. Finn, M.ASCE[3]

ABSTRACT

This paper reviews possible static and dynamic loading conditions which might induce instability in hydraulic sandfill tailings dams. It outlines a general approach used by the authors to evaluate tailings dam stability. For static loading, this entails determining the shearing strength appropriate to the loading condition. For dynamic loading, the approach follows a staged process, with each successive stage involving a more complex form of analysis. Two finite-element methods for dynamic analyses are discussed; an equivalent linear method based on total stress and a non-linear hysteretic method based on effective stress. A discussion of the performance of tailings dams constructed by the upstream method is also included.

INTRODUCTION

The compactive effort employed to densify hydraulic sandfill tailings dams is often less than that used for modern earthfill water storage dams. The reason for this situation is both economical and functional. Tailings are essentially the waste products of a mine. Their disposal is usually carried out by the most economical means possible, albeit with increasing government constraints due to safety and environmental concerns. Moreover, the function of a tailings dam is to retain mainly tailings together with an amount of free water dictated by the mining operation. Usually a wide tailings beach separates the dam from the free water in the pond. This lessens the requirement for water tightness of the dam itself, and permits larger dam deformations to be acceptable. The possibility of developing instability in a less compacted hydraulic sandfill tailings dam and/or its foundation is a major threat to the dam's safety.

This paper first reviews possible static and dynamic loading conditions which might induce slope instability in tailings dams. Next, it outlines a general approach used by the authors to evaluate tailings dam stability. For static loading, this entails determining the shearing strength appropriate to the loading condition. For dynamic loading, the approach follows a staged process, with each successive stage involving a more complex form of analysis. Two

1 Associate, Klohn Leonoff Ltd., Richmond, British Columbia, Canada
2 Chairman, Klohn Leonoff Ltd., Richmond, British Columbia, Canada
3 Professor, Department of Civil Engineering, University of British Columbia, Vancouver, British Columbia, Canada

finite-element methods for dynamic analyses are discussed; an equivalent linear method based on total stress and a non-linear hysteretic method based on effective stress. A discussion of the performance of tailings dams constructed by the upstream method is also included.

LOADINGS LIKELY TO INDUCE SLOPE INSTABILITY

The behaviour of a sandfill depends on density, stress state, degree of saturation and drainage conditions during loading. Under both static and dynamic loading, substantial loss of shearing strength may occur if the saturated sand is in a contractive state, and is sheared under undrained conditions. For saturated dilative sand, cyclic mobility may occur under undrained cyclic loading conditions.

Static and dynamic loading conditions with the potential to induce instability in tailings dams are discussed below:

Static Undrained Loading

Three types of undrained loading may arise during construction of a tailings dam.

(a) Addition of Dam Fill: The annual raising of the dam for storing the next year's tailings production may constitute the most important undrained loading, depending on the rate of construction.

(b) Foundation Straining: For dams overlying a weak foundation, foundation straining may also add to the undrained loading on the dam fill.

(c) Slope Oversteepening: During the life of the tailings dam, it is conceivable that dam slopes may accidentally be oversteepened locally by internal erosion due to seepage and/or surface erosion caused by concentrated runoff or pipeline breakage. Should the erosion occur rapidly, the resulting slope steepening process would constitute an undrained loading in the vicinity of the eroded area.

Dynamic Undrained Loading

Dynamic loading at a tailings damsite usually results from earthquakes and blasting associated with mining activities.

(a) Earthquake: The attributes of an earthquake relevant to dam stability are its magnitude and hypocentral distance from the site. Together with the regional geology and local site conditions, these determine the peak acceleration, duration, and frequency content of the earthquake input to the dam foundation. The response of the tailings dam to this earthquake input will determine the extent of pore pressure build-up within the dam and the amount of deformation or even slope failure that might occur.

(b) Blasting: Blasting may be carried out in the vicinity of a tailings damsite. However, multiple delays and blasting monitoring are used to control blasting operations, and the detrimental release of large explosive energy occurs rarely. In addition, waves generated by blasting are usually dominated by high frequency waves, which do not significantly affect major tailings dams which have much lower natural frequencies. In special circumstances, controlled blasting has been used to induce temporary liquefaction of small volumes of the damfill or foundation material in order to densify the loose soil.

STABILITY ASSESSMENT UNDER STATIC LOADING

In areas where the only threat of slope instability comes from static loading, dam stability can be assessed by carrying out limit-equilibrium stability analyses using the following three different shear strengths appropriate to the loading conditions:

(1) Drained strength (effective stress analysis);
(2) Peak undrained strength (effective stress analysis);
(3) Steady-state strength (total stress analysis).

Drained Shear Strength - Effective Stress Analysis

The stability analysis using drained shear strength is used to evaluate the long term stability condition when all excess pore pressure induced by dam construction is fully dissipated. It also can be applied to the construction condition, when practically no excess pore pressure is developed due to sufficiently slow rates of construction. Such an analysis generally demonstrates a very stable condition of the dam under drained condition.

Undrained Shear Strength

Casagrande (1936) in a classic paper discussed the significance of the critical void ratio of cohesionless soils. He pointed out that when saturated sands are sheared in an undrained state, negative pore pressures are developed at large strains if the void ratio is less than the critical value (dilative state), and positive pore pressures are developed if the void ratio is greater than the critical value (contractive state).

Hence for dense sands the undrained strength is higher than the drained strength due to the development of negative pore pressure. This higher undrained strength is, however, not used in stability analysis, for it is prudent not to rely on the temporary presence of negative pore pressure. For loose or medium dense sands, on the other hand, the development of positive pore pressure causes the undrained strength to be lower than the drained strength. Figure 1 illustrates the effective-stress vector curve of a contractive soil during the undrained shear stage in a consolidated-undrained shear test. Starting from an initial consolidated state, the soil passes through a peak state before approaching a steady-state failure condition.

The peak state may be absent for medium dense sands.

(a) Peak Strength - effective stress analysis:
The peak strength may be expressed in terms of effective stress friction angle, ϕ'_p (Vaid and Chern 1985) as shown in Figure 1, and can be used in effective-stress limit-equilibrium stability analyses. When peak strength is used in analyzing the slope stability, an adequate factor of safety is required to guard against the soil reaching the peak strength and passing on to the steady state strength. In checking the factor of safety, it is important to incorporate appropriate undrained excess pore pressure due to field loading in the stability analysis. Furthermore, construction pore pressure should be regularly monitored and controlled by adjustment of the rate of fill placement. Local slope oversteepening should also be avoided to prevent the initiation of a progressive failure.

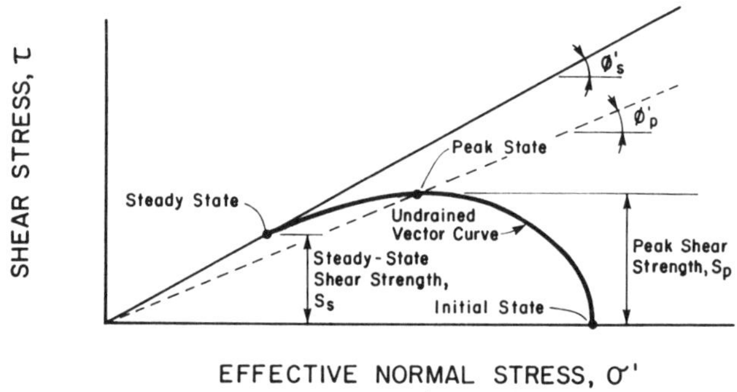

Figure 1. Effective - Stress Undrained Vector Curve (After Vaid and Chern 1985)

(b) Steady State Strength - total stress analysis:
Castro (1969) used consolidated-undrained triaxial tests to study the critical void ratio of sands and found that this ratio varies with the effective confining pressure. The line describing such a relation on a void ratio versus effective confining pressure plot (see Fig. 2) is called the steady-state line. From Fig. 2, the effective minor principal stress, σ'_{3s}, at the steady-state condition can be obtained for a given void ratio. The corresponding steady-state shear strength, s_s, can then be determined from the Mohr diagram (see Fig. 3) and expressed in the following form:

$$s_s = \frac{\sin \phi'_s \cos \phi'_s}{1-\sin \phi'_s} \quad \sigma'_{3s}$$

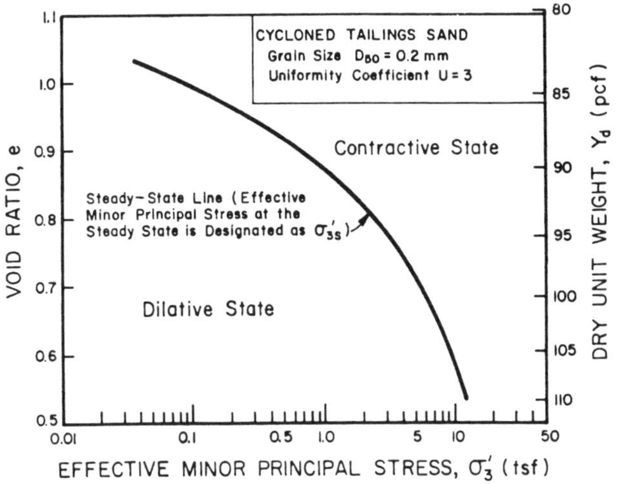

Figure 2. Steady-State Line in Semi-Logarithmic Plot (After Castro 1969 and Klohn and Lo 1983)

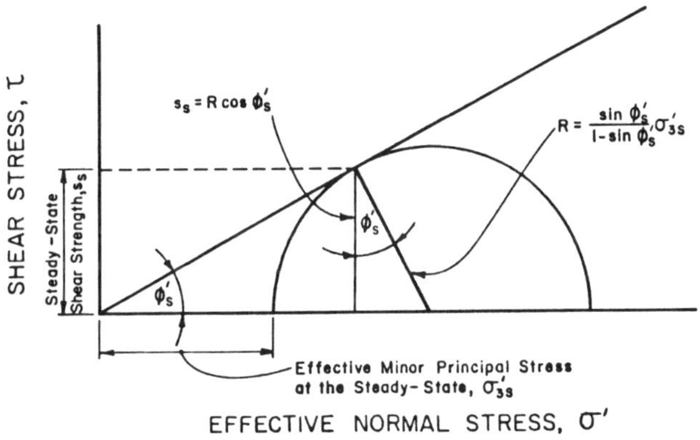

Figure 3. Effective Stress Conditions at Steady State (After Castro 1969 and Klohn and Lo 1983)

where ϕ'_s is the effective friction angle corresponding to the steady-state condition obtained from the consolidated-undrained tests with pore pressure measurements. The steady-state strength thus obtained is plotted against the void ratio in Fig. 4.

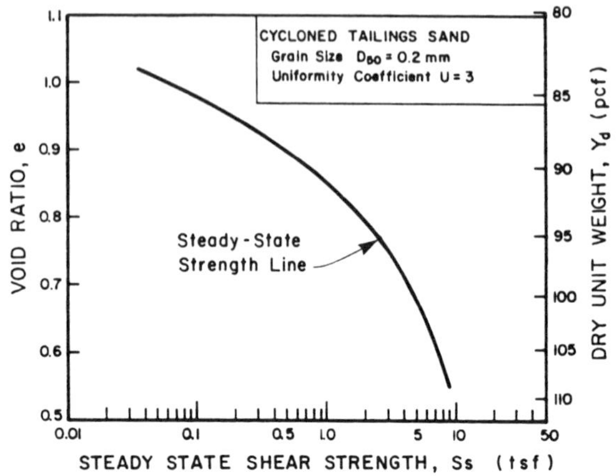

Figure 4. Steady-State Strength in Semi-Logarithmic Plot (After Castro 1969 and Klohn and Lo 1983)

The use of the steady-state strength in a total stress limit-equilibrium stability analysis represents the most conservative evaluation of the slope stability under static loading. It implies that the loading is applied rapidly allowing no time for drainage to occur. Moreover, the loading is of sufficient magnitude to drive the soil involved past the peak state. Failures of upstream slopes of the Tar Island tailings dike during construction (Mittal and Hardy 1977) represented such an extreme condition.

EARTHQUAKE EFFECTS ON SANDFILL TAILINGS DAM DESIGN

When a dam is subjected to earthquake forces there are three distinct aspects of the problem that must be considered over and above those considered for the conventional static analysis. These are: earthquake inertial forces, loss of strength due to strain or remolding, and loss of strength due to pore pressure development. Where the dam is used to retain tailings, a fourth factor, the potential and extent of liquefaction or loss of strength of the tailings materials stored in the pond, must also be considered. This occurrence places a sudden, additional thrust force on the tailings dam which does not occur when only water is stored behind the dam.

There are three main methods of tailings dam construction used in practice: upstream method in which often uncompacted tailings sands are constructed over spigotted tailings, and downstream and centerline methods where design measures such as good drainage and/or compaction are usually employed to construct sandfill over prepared foundations. Where the earthquake risk is very high, tailings sands are normally heavily compacted to produce a very dense sandfill. Such sandfill will not lose its strength to any significant extent during an earthquake. Consequently the stability of structures constructed in this way is normally assessed using pseudo-static stability analysis with an appropriate seismic coefficient as described by Seed (1979).

When the earthquake risk is low to moderate, dams constructed of medium dense sandfill are generally acceptable. However, any possible loss of strength due to straining and pore pressure build-up under earthquake loading should be investigated. Stability analyses of these dams should include an estimate of seismic pore pressure and potential dam deformations. Procedures for making these estimates are presented in later sections.

Upstream tailings dams are usually more susceptible to earthquake induced failures, because they are constructed over the top of the previously deposited tailings which are often in a loose, saturated state. The authors consider that the use of the upstream method of tailings dam design should be limited to areas of low seismicity. Past performance records of upstream tailings dams are discussed below.

EARTHQUAKE PERFORMANCE OF UPSTREAM TAILINGS DAMS

Conlin (1987) summarized available data on the earthquake performance of Chilean and Japanese upstream tailings dams in the literature. Safety criteria based on these data are shown on Fig. 5 together with similar criteria based on the earthquake behaviour of gently sloping ground (Youd and Perkins 1987), and hydraulic fill water dams (Seed et al. 1978) using attenuation relationships by Leps and Jansen (1984). On the basis of these criteria potential problems may exist for tailings dams, hydraulic-fill water dams, or near-level ground sites plotted above the appropriate curve. Conversely, for sites plotted below the respective curve satisfactory performance is expected. The limiting curve corresponding to nonoperating tailings impoundments plots higher than that for operating tailings impoundments. This reflects the more favourable conditions which exist in a non-operating tailings impoundment, where tailings materials tend to be normally consolidated and free water is usually absent. Figure 6 shows Conlin's safety criteria plot on which additional case histories from Chile, China and U.S.A. have been superimposed, relevant data of these additional cases are given in Table 1.

Many factors affecting the seismic behaviour of a tailings dam are not explicitly considered in plots like Figs. 5 and 6. T' ›se include such factors as the flatness of the slopes, the position of the saturation level within the dam, the characteristics and degree of consolidation (Vick 1983) of the tailings sands, the earthquake frequency content, and local site effects. For example, Chinese

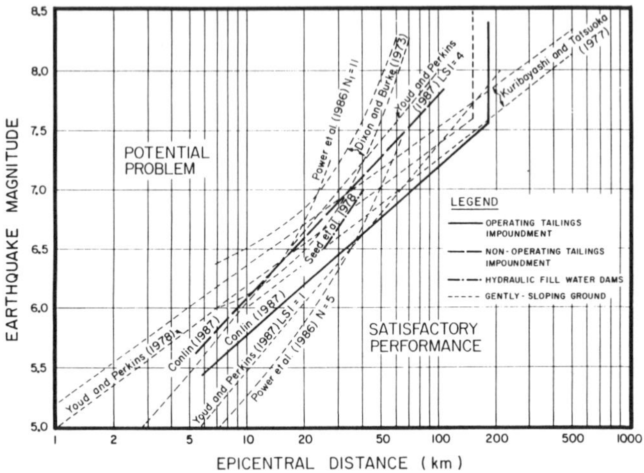

Figure 5. Earthquake Performance of Upstream Tailings Dams, Hydraulic Fill Water Dams and Gently Sloping Ground (After Conlin 1987, Seed et al. 1978, Leps and Jansen 1984, and Youd and Perkins 1987)

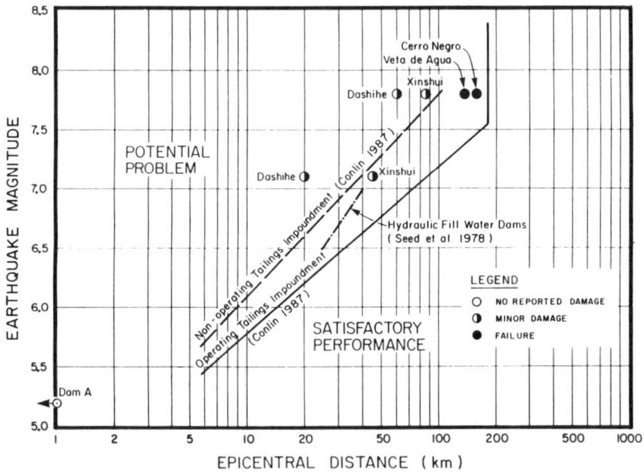

Figure 6. Susceptibility of Upstream Tailings Dams to Earthquake - Induced Damage (After Conlin 1987)

upstream tailings dams usually have much flatter downstream slopes than those in other countries, typically 1V on 4H or flatter. Therefore, there is considerable scattering of data points in plots like Figs. 5 and 6 (Conlin 1987). However, for a more restricted database involving dams sharing similar characteristics and sites sharing similar seismo-tectonic settings, case history data points would be expected to follow a tighter band in such plots.

TABLE 1

RELEVANT DATA ON UPSTREAM TAILINGS DAMS SHOWN ON FIGURE 6

DAM	COUNTRY	MINE	EARTHQUAKE	DAM PERFORMANCE	REFERENCE
Cerro Negro	Chile	Copper	March 3/85 M = 7.8	Flow slide failure	EERI 1986
Veta de Agua	Chile	Copper	March 3/85 M = 7.8	Flow slide failure	EERI 1986
Dashihe	China	Iron	July 28/76 M = 7.8 (Main shock) M = 7.1 (After shock)	Longitudinal cracks on D/S slope, sand boils and water spouts on tailings beach	MMI 1985
Xinshui	China	Iron	July 28/76 M = 7.8 (Main shock) M = 7.1 (After shock)	Seepage and settlement on D/S slope, sand boils and water spouts on tailings beach	MMI 1985
Dam A	U.S.A.	Copper	Sept 5/62 M = 5.2	No reported damage	Klohn Leonoff 1987

DYNAMIC STABILITY ASSESSMENT FOR DOWNSTREAM (CENTERLINE) TAILINGS DAMS

For tailings dams constructed by the downstream (or centerline) method involving medium dense sands, both the selection of methods of dynamic stability analysis and the interpretation of results obtained from these analyses require special attention. The authors have adopted the following staged approach (Lo et al. 1982, Klohn and Lo 1983, Klohn 1984) in assessing the dynamic stability of sandfill tailings dams. In ascending order of cost and complexity, these are:

1. Steady-State Strength Analysis;
2. Simplified Dynamic Analysis;
3. Finite-Element Dynamic Analysis.

Steady-State Strength Analysis

In this type of analysis the steady-state strength along a potential failure surface is used in a conventional static limit-equilibrium stability analysis as discussed in earlier sections. The inertial forces caused by the earthquake are not included as these are of very short duration and the dam is assumed capable of safely absorbing any movements resulting from those inertial forces.

The steady-state method is a quick and relatively inexpensive form of analysis and represents the worst possible conditions that could develop for the dam involving saturated cohesionless materials of low to medium density during an earthquake. At sites involving loose, saturated materials where the seismic risk is high, the method is considered to provide a reasonable assessment of the dam stability under earthquake loading. However, for medium-dense cohesionless materials at sites of low to moderate seismicity the extent of seismic loading is often insufficient to trigger a steady-state condition. Under these conditions the steady-state analysis is considered to provide an overly conservative assessment of the dam's seismic stability. If a dam is judged to be safe by steady-state analysis and the deformation associated with going to the steady-state can be accepted, more sophisticated analysis is not required. On the other hand, if the steady-state analysis indicates that a problem in stability might exist, then the next stage in the evaluation would be to run either a simplified dynamic analysis or a finite-element dynamic analysis to check whether the design earthquake is capable of triggering the process leading to the steady-state (Finn 1982, Lo et al. 1982, Klohn and Lo 1983, Poulos et al. 1985b, Mohamad et al. 1985, Castro 1987).

While the concept of steady-state strength analysis is sound and its application appears simple, there remains an important practical problem in assessing the steady-state strength applicable in the field. Poulos et al. (1985) cautioned that sampling, handling, and testing of samples may alter the void ratio of test specimens in the laboratory. Meticulous field and laboratory programs are required as well as adjustments to laboratory strengths to account for density differences between the soil in situ and the laboratory specimen

(Castro et al. 1985, Seed et al. 1987). Seed (1987), by evaluating flow slide case histories, arrived at a correlation between the field residual (steady-state) strength and the normalized standard penetration test blow counts. This field strength is generally significantly lower than that obtained from the laboratory procedure. Seed attributed this discrepancy to the effect of water content redistribution involved in the field behaviour of sand deposits. It is also possible that the full-scale flow slide may involve a flow structure as envisioned by Casagrande (1975) which may not be fully realized in the laboratory setup. Since the field residual strength represents the lowest available strength based on the back analysis of actual flow slides, it offers a lower bound shear strength for assessing slope stability.

Simplified Dynamic Analysis

Klohn et al. (1978) developed a pseudo-static method of dynamic analysis for dams which incorporates seismically induced porewater pressures. The method is incorporated in the computer program SEISLOP (Grigg and Lo 1979), and involves the following steps:

1. A design earthquake and its related parameters are selected for the site.

2. Static stress conditions on an assumed failure surface are evaluated using a conventional limit-equilibrium form of analysis.

3. Inertia forces are then applied to the dam, using a conventional limit-equilibrium analysis to determine the resulting cyclic shear stresses on the assumed failure surface. The inertial forces are calculated using appropriate seismic coefficients.

4. Using the design earthquake parameters, the physical properties of the cohesionless material, and cyclic triaxial test data, the pore pressure increases caused by the cyclic inertia forces are evaluated. These cyclic pore pressures are then added to the original pore pressures developed by seepage flow. Potential increase of pore pressure due to the additional pond force resulting from some loss of strength in tailings stored in the pond can also be accommodated.

5. Finally, the resulting increased pore pressures are used to calculate the reduced effective normal stresses on the failure surface for the condition immediately following the last cycle of the earthquake.

The SEISLOP simplified dynamic method of analysis requires more field and laboratory data than does the limit-equilibrium stability analysis using the steady-state strength, and also requires the selection of a design earthquake. However, the SEISLOP analysis program is simpler and less costly to run than the programs for finite-element analysis, while at the same time acknowledging the importance of energy input levels from the earthquake and the effects of seismically induced porewater pressures. It does, of course, also include many more approximations than does the finite-element method.

Finite Element Dynamic Analysis

Finite element dynamic analysis represents the current state-of-the-art for dynamic analysis of dams (National Research Council 1985). It involves complicated, finite-element, computer programs; requires the input of a broad spectrum of engineering skills and experience and are time consuming and expensive. The following comments highlight the necessary steps in conducting such an analysis:

1. Determine the static and dynamic properties of the soils comprising the dam and its foundation.

2. Subject representative samples of the soils to laboratory tests simulating the combined effects of the initial static stresses and the superimposed dynamic stresses and determine their effects in terms of the generation of porewater pressures and the development of strains.

3. Determine the static, pre-earthquake stresses within the dam.

4. Determine the time history of base excitation to which the dam and its foundation may be subjected by the design earthquake.

5. Determine the dynamic response for the dam and its foundation to the seismic loading from Step 4 above.

6. If Step 5 is conducted using a true non-linear effective stress method of analysis (Finn and Yogendrakumar 1987), all response parameters such as earthquake-induced pore pressure and deformation are obtained directly. If conventional equivalent linear total stress analysis (Seed 1979) is used, the relevant response parameters are deduced with the aid of the laboratory tests which simulate the field conditions.

7. Carry out a post-earthquake limit-equilibrium analysis including pore pressures developed during the earthquake. In making this post-earthquake analysis it is also assumed that the lower limiting value of the shear strength along a failure surface is defined by the steady-state strength regardless of the pore pressure values calculated from the dynamic finite element analysis.

Obviously, a complete and detailed dynamic analysis is a major undertaking which requires an extensive data base and specialized skills. For these reasons such analyses should be reserved for critical structures such as major dams or tailings dams where failure would pose a serious threat to life and property and where the staged assessment involving the first two steps has indicated that a more detailed review of the stability problem is required.

The total stress equivalent linear dynamic analysis has been used for the last two decades to assess the overall stability of dams under seismic loadings. Numerous case histories involving back analyses of failed hydraulic fill dams as well as design and review analyses for existing dams have been accumulated to date. This body of work provides an invaluable reference base for the designer. An effective stress non-linear hysteretic dynamic analysis, TARA-3, (Finn et al. 1986) has recently been applied to assess seismic performance of tailings dams (Finn et al. 1988, Scott et al. 1989). The theoretical framework for both analyses is briefly discussed in the following sections.

EQUIVALENT LINEAR ELASTIC TOTAL STRESS ANALYSIS

Dynamic response analysis as practiced widely today had its origin in the pioneering research of Seed and his co-workers at the University of California at Berkeley. The history and use of this method has been discussed by Finn (1987). Seed and Idriss (1969) modelled the non-linear hysteretic stress-strain properties of soils by using an equivalent linear elastic method of analysis. This model was incorporated into the computer program SHAKE (Schnabel et al. 1972) which is widely used to predict site specific ground motions involving level ground. The model was generalized to two and three dimensions in the finite element programs QUAD-4 (Idriss et al. 1973) and FLUSH (Lysmer et al. 1975). QUAD-4 is used primarily for the analysis of the seismic response of earth and rockfill dams; FLUSH is used for soil-structure interaction studies.

The fundamental assumption of the equivalent linear method of analysis is that the dynamic response of soil, a non-linear hysteretic material, may be approximated satisfactorily by a damped elastic model if the properties of that model are chosen appropriately. These properties are obtained by an iterative computational process.

In QUAD-4 analysis, the stress-strain properties of the soil are defined in each finite element by the Poisson ratios and shear strain dependent shear moduli and equivalent viscous damping ratios. Extensive data on shear strain dependent moduli and damping for incorporation in QUAD-4 have been given by Seed and Idriss (1970).

For dynamic analysis, initial values of moduli and damping ratios are selected corresponding to small shear strain values ($\gamma \cong 10^{-4}\%$) or to strain levels judged appropriate for the anticipated earthquake or cyclic loading, and an elastic analysis is carried out for the entire duration of the earthquake. An average or effective strain (usually assumed to be about 65% of the maximum value) is computed in each finite element and moduli and damping ratios are selected from the data base compatible with these average strains and are used for the next iteration. The procedure is repeated until no significant changes in moduli or damping ratios are considered necessary. The response determined during this last iteration is considered to be a reasonable approximation to some aspects of the non-linear response.

Since the final analysis with strain compatible soil properties is purely elastic, the permanent deformations caused by earthquake shaking cannot be computed by this type of analysis. The deformations of a dam are estimated using either acceleration data or shear stress data from the equivalent linear analysis. If the damfill does not suffer significant degradation in strength during shaking, the deformations are derived from the acceleration data. If significant strength loss occurs due to high porewater pressures, then the deformations are estimated from the shear stress output with the help of laboratory test data. The procedure is outlined in Fig. 7. A critical review of these procedures for estimating the deformation of dams using data from equivalent linear total stress analysis has been given by Finn (1987). Some of the main points of the review are presented below.

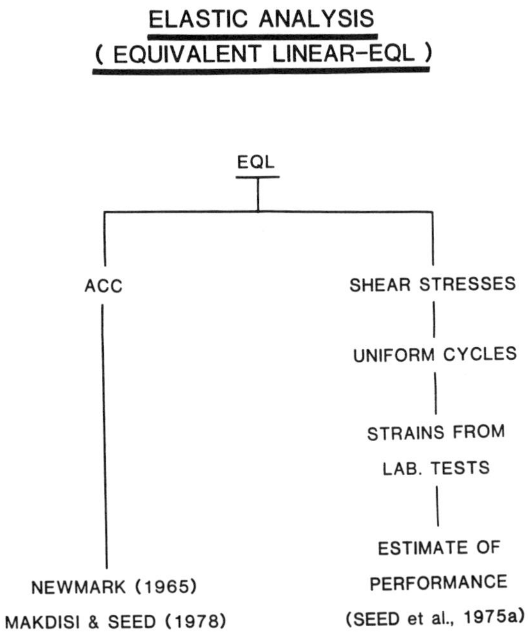

Figure 7. Procedures for Estimation of Dam Deformation by Equivalent Linear Elastic Analysis (After Finn 1987)

Deformations from Acceleration Data - Various potential failure surfaces in the dam are analysed with varying seismic coefficients to find the lowest yield acceleration, a_y, required to cause failure along the critical failure surface (Fig. 8). The average acceleration time-history of the sliding block is obtained from the QUAD-4 analysis using the technique described by Makdisi and Seed (1978). The yield acceleration is deducted from the average acceleration time history and the net acceleration shown shaded in Fig. 8 is available to generate permanent displacements. The analysis is conducted on the equivalent model of a horizontal sliding block on a plane with only one way motions allowed (Fig. 8).

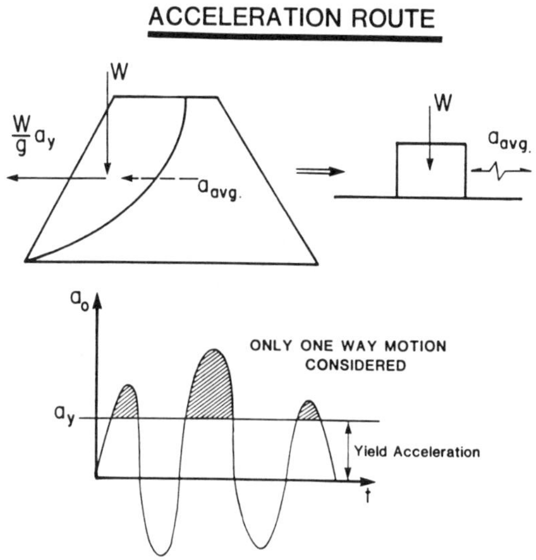

Figure 8. Estimation of Dam Deformation by Newmark's Concept of Yield Acceleration (After Finn 1987)

This type of analysis was pioneered by Newmark (1965). The version described above takes into account the flexibility of the dam when calculating the average acceleration acting on the potentially sliding block.

Makdisi and Seed's procedure (1978) for determining the average acceleration of the potentially sliding block is based on the method described by Chopra (1966). This involves calculating the resultant force acting on the base of the sliding block and dividing this force by the mass of the block to get the average acceleration. Sometimes in practice the average acceleration is computed by averaging the accelerations at nodal points of the finite element mesh within the sliding block.

Makdisi and Seed (1978), on the basis of extensive dynamic analysis studies, have presented charts that allow estimates of the order of displacements in well-designed earth and rockfill dams up to 70 m in height to be made conveniently without resorting to dynamic analysis.

The Makdisi-Seed method is particularly appropriate when there is no significant degradation in strength and stiffness of the fill in the dam during earthquake shaking. In these circumstances deformations tend to be localized along potential failure surfaces. If there is significant strength degradation, the deformation pattern will probably be more diffused. In this case, the stress route to the estimation of strains and deformations is, in theory at least, more applicable.

Deformations from Stress Data - When the stress route to deformations is followed (Seed 1979), the shear stress history in a finite element is first converted to an equivalent number of uniform stress cycles (Seed et al. 1975). The strain potential in the finite element is considered to be the strain developed in a laboratory specimen consolidated to the same condition as the finite element and subjected to the equivalent uniform cycles of loading. These strains are incompatible because they are determined from isolated element in laboratory tests. They are nevertheless an indication of probable performance of the dam during an earthquake. Further analyses are required (Lee 1974 and Serff et al. 1976) to harmonize these incompatible strains.

Both methods of deformation analysis provide the opportunity to compare the computed performance of a dam under study with that of other dams which have been shaken by earthquakes. This provides an additional basis for the exercise of design judgement.

NON-LINEAR HYSTERETIC EFFECTIVE STRESS ANALYSIS

In TARA-3 analysis (Finn et al. 1986), response in shear is assumed to be non-linear and hysteretic with unloading and reloading stress-strain paths defined by the Masing criterion (Masing 1926). The reponse of the soil to uniform all round pressure is assumed to be non-linearly elastic and dependent on the mean normal effective stress. Porewater pressures during shaking are computed using the Martin-Finn-Seed porewater pressure model (Martin et al. 1975) modified to take into account the effects of initial static shear stress. Moduli and strength are continuously modified during analysis to reflect changes in the effective stress regime. A detailed description of the constitutive relations in TARA-3 is given by Finn (1985), and Finn and Yogendrakumar (1987).

For a non-linear dynamic analysis of a dam, a static analysis is first carried out to determine the stress and strain states throughout the dam. Each element of the dam exists in a particular state of stress and strain prior to the earthquake. The response to seismic forces at any point in the dam depends on this initial stress-strain state. However, this is ignored in current engineering practice and all dynamic response calculations are conducted assuming that the initial stress-strain state is zero for all elements even for those which carry high shear stresses. TARA-3 has the capability to start the dynamic analysis either from the origin as in current practice or from the stress-strain condition just before the earthquake as shown in Figure 9. The latter case leads to accumulating permanent deformations in the direction of the smallest residual resistance to deformation (Finn 1987).

As shaking proceeds, two phenomena occur; porewater pressures develop in saturated portions of the dam, and volumetric strains and associated settlements develop in the unsaturated regions. The program takes into account the effects of the porewater pressures on moduli and strength during dynamic analysis and computes the additional permanent deformations due to gravity acting on the softening soil.

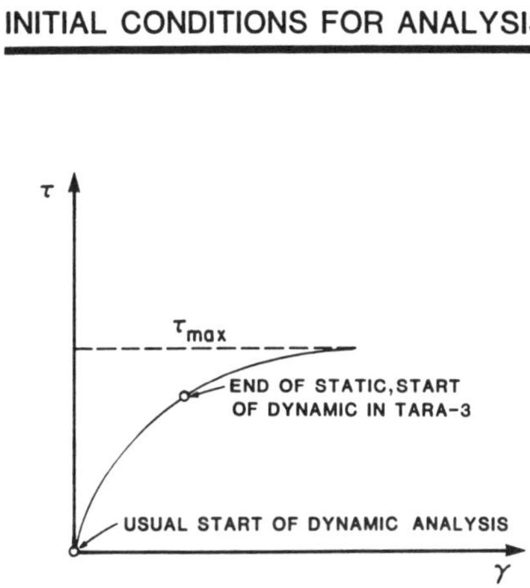

Figure 9. Initial States for Equivalent Linear and Non-Linear (TARA-3) Methods of Analysis (After Finn 1987)

The shear modulus, G, is proportional to the mean normal effective stress σ_m', with a minimum value equal to the atmospheric pressure, raised to some power n, often taken as 1/2,

$$G \;\alpha\; (\sigma_m')^{1/2}$$

or

$$G \;\alpha\; (\sigma_m - u)^{1/2}$$

where σ_m is the mean normal total stress and u is the porewater pressure. It is clear that G decreases as u increases. The decrease in G affects both the dynamic and gravitational response of the dam.

The hysteretic stress-strain response of soil leads to accumulating permanent deformations due to cyclic loading. The deformation increases with each cycle of loading due to ratchetting effect as shown in Fig. 10.

RESIDUAL DYNAMIC DISPLACEMENTS

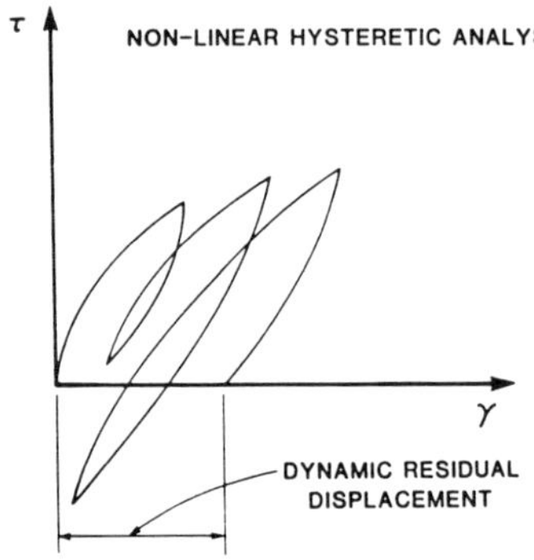

Figure 10. Residual Dynamic Displacement Due to Non-Linear Hysteretic Response to Seismic Loading (After Finn 1987)

The final deformed shape of the dam results from the sum of permanent deformations due to the hysteretic dynamic stress-strain response, constant volume deformations in saturated portions of the embankment, volumetric strains in unsaturated portions and deformations due to consolidation as the seismic porewater pressures dissipate. The final post-earthquake deformed shapes of two example embankments analyzed by TARA-3 program are shown in Figs. 11 and 12. Figure 11 shows the deformed shape of a saturated homogeneous embankment constructed of cohesionless materials. The figure depicts the classical spreading of an embankment due to high seismically induced pore water pressure. Figure 12 shows the deformed shape of an embankment with a central clay core and cohesionless upstream and downstream shells. High pore pressure is generated in the saturated upstream shell due to earthquake shaking. Consequently large deformations occur upstream, and the core is deformed towards the upstream side.

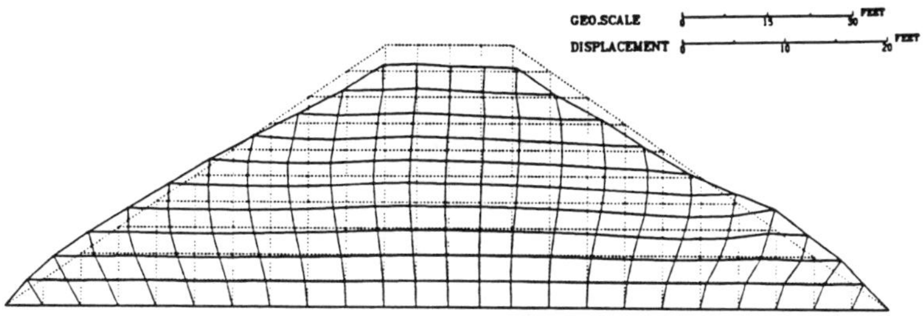

Figure 11. Deformed Shape of a Saturated Homogeneous Embankment After Earthquake (After Finn et al. 1987)

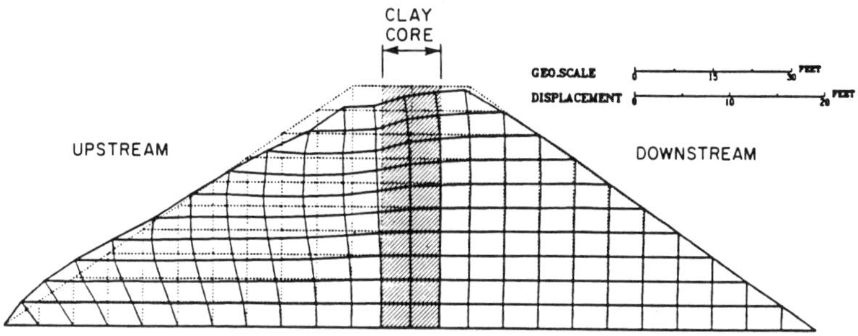

Figure 12. Deformed Shape of an Embankment with a Central Core After Earthquake (After Finn et al. 1987)

The non-linear effective stress analysis offers substantial advancement in analytical capability based on theoretical considerations. However, the use of such analysis in practical application requires further calibration against actual dam performance in past earthquakes. The TARA-3 computer program has been validated extensively in studies of seismic tests on centrifugal models. The response parameters computed by TARA-3 agree with those recorded in a variety of model tests within limits acceptable for engineering purposes.

CONCLUSIONS

This paper outlined a general staged approach used by the authors in assessing the stability of tailings dams subject to static and dynamic loadings. For static loading, it is essentially to carry out limit-equilibrium stability analyses using one of the following three shear strengths most appropriate to a given loading condition: (1) drained strength, (2) peak undrained strength, and (3) steady-state strength. For dynamic loading, it progresses from (1) simple limit-equilibrium stability analysis using steady-state strength to (2) simplified SEISLOP dynamic analysis to (3) finite-element dynamic analysis. Salient features of two finite-element dynamic analyses: total stress equivalent linear method and effective stress non-linear hysteretic method are presented. It is believed that downstream (centerline) constructed hydraulic sandfill dams can be engineered to resist even the most severe earthquake loadings. However, for most applications an optimum design can be achieved by using an appropriate stability analysis to assess the dam stability according to the project requirements.

REFERENCES

Casagrande, A. (1936). "Characteristics of Cohesionless Soils Affecting the Stability of Slopes and Earth Fills". Contributions to Soil Mechanics, 1925-1940, Boston Society of Civil Engineers, Boston, Mass., 1940. pp. 257-276. Originally published in the Journal of the Boston Society of Civil Engineers, January.

Casagrande, A. (1975). "Liquefaction and Cyclic Deformation of Sands - A Critical Review". Proceedings, Fifth Pan-American Conference on Soil Mechanics and Foundation Engineering, Buenos Aires, Argentina, November.

Castro, G. (1969). "Liquefaction of Sands". Harvard Soil Mechanics Series No. 81, Harvard University, Cambridge Mass., January.

Castro, G., Poulos, S.J. and Leathers, F.D. (1985). "Re-examination of Slide of Lower San Fernando Dam". J. Geotech. Engrg. Div., ASCE, 111(9), 1093-1107.

Castro, G. (1987). "On the Behaviour of Soils During Earthquakes - Liquefaction." Proc. Third Int'l Conf. on Soil Dynamics and Earthquake Engineering, June, Princeton, N.J. Soil Dynamics and Liquefaction Volume, Edited by Cakmak A.S., Elsevier.

Chopra, A.K. (1966). "Earthquake Effects on Dams." Thesis presented to the University of California, at Berkeley, Calif. in partial fulfillment of the requirements for the degree of Doctor of Philosophy.

Conlin, B.H. (1987). "A Review of the Performance of Mine Tailings Impoundments Under Earthquake Loading Conditions". Proceedings, Vancouver Geotechnical Society Seminar on Earthquake Geotechnique, May.

Dixon, S.J., and Burke, J.W. (1973). "Liquefaction Case History". J. Soil Mech. Found. Div., ASCE, 99(11), 921-937.

EERI (1986). "The Chile Earthquake of March 3, 1985." Earthquake Spectra, 2(2), 249-291.

Finn, W.D. Liam (1982), "Fundamental Aspects of Response of Tailings Dams to Earthquakes". Proceedings, ASCE Annual Fall Convention, Technical Session on Dynamic Stability of Tailings Dams, October, New Orleans, 46-72.

Finn, W.D. Liam (1985). "Dynamic Effective Stress Response of Soil Structures; Theory and Centrifugal Model Studies". Proc. Fifth Int'l. Conf. on Num. Methods in Geomechanics, Nagoya, Japan, Vol. 1, 35-36.

Finn, W.D. Liam, Yogendrakumar, M., Yoshida, N. and Yoshida, H. (1986). "TARA-3: A Program for Nonlinear Static and Dynamic Effective Stress Analysis". Soil Dynamics Group, University of British Columbia, Vancouver, British Columbia.

Finn, W.D.L. (1987). "Seismic Response of Fill Dams". International Symposium on Earthquakes and Dams, Chairman's Opening Address, Beijing. May.

Finn, W.D.L. and Yogendrakumar, M. (1987). "Centrifugal Modelling and Analysis of Soil-Structure Interaction". Proceedings, Fifth Canadian Conference on Earthquake Engineering, July, Ottawa, 453-460.

Finn, W.D.L., Yogendrakumar, M. and Nichols, A. (1987). "Seismic Response Analysis of Level Sites, Embankments and Soil-Structure Systems". Proceedings, Vancouver Geotechnical Society Seminar on Earthquake Geotechnique, May.

Finn, W.D.L., Yogendrakumar, M., Lo, R.C., Yoshida, N. (1988). "Direct Computation of Permanent Seismic Deformation." Proc. of 9th World Conference on Earthquake Engineering.

Grigg, R.F. and Lo, R.C. (1979). "Simplified Seismic Stability Analysis: Program Documentation." Preprint 3746, ASCE Convention, Atlanta, Georgia, Oct. 23-25, 19 pp.

Idriss, I.M., Lysmer, J., Hwang, R., and Seed, H.B. (1973). "QUAD-4: A Computer Program for Evaluating the Seismic Response of Soil-Structures by Variable Damping Finite Element Procedures". Report No. EERC 73-16, Earthquake Engineering Research Center, University of California, Berkeley, California, July.

Klohn, E.J. (1984). "The Brenda Mine's Cycloned - Sand Tailings Dam". Proc. of the Case Histories in Geotechnical Engineering Conf., May, St. Louis, Missouri, Vol. II, 953-977.

Klohn, E.J., Maartman, C.H., Lo, R.C., and Finn, W.D.L. (1978). "Simplified Seismic Analysis for Tailings Dams", Proceedings ASCE Geotech. Engng. Div. Speciality Conference on Earthquake Engineering and Soil Dynamics, Pasadena, California, June 19-21, Vol. I, 540-556.

Klohn, E.J. and Lo, R.C. (1983). "Earthquake Resistant Design of Tailings Dams". Proceedings, Fourth Canadian Conference on Earthquake Engineering, June, Vancouver, 614-625.

Klohn Leonoff (1987). Project files.

Kuribayashi, E., Iwasaki, T. and Tatsuoka, F. (1977). "A History of Soil Liquefaction in Japan." Proc. 6th World Conference on Earthquake Engineering, India, Vol. 3, 2448-2454.

Lee, K.L. (1974). "Seismic Permanent Deformations in Earth Dams". Report No. UCLA-ENG-74-97, School of Engineering and Applied Science, Univ. of Calif., Los Angeles.

Leps, T.M. and Jansen, R.B. (1984). "Proposed Guidelines for Dam Seismic Stability Assessment". USCOLD News, Nov.

Lo, R.C., Klohn, E.J., Maartman, C.H. and Finn, W.D.L. (1982). "Dynamic Analyses of Sandfill Tailings Dams". Proceedings, ASCE Annual Fall Convention, Technical Session on Dynamic Stability of Tailings Dams, October, New Orleans, 73-91.

Lysmer, J., Udaka, T., Tsou, C.F. and Seed, H.B. (1975). "FLUSH: A Computer Program for Approximate 3-D Analysis of Soil-Structure Interaction Problems". Report No. EERC 75-30, Earthquake Engineering Research Center, University of California, Berkeley.

Makdisi, F.I. and Seed, H.B. (1978). "Simplified Procedure for Estimating Dam and Embankment Earthquake-Induced Deformations". J. Geotech. Engrg Div., ASCE, 104(7), 849-867.

Martin, G.R., Finn, W.D. Liam and Seed, H.B. (1975). "Fundamentals of Liquefaction Under Cyclic Loading". J. Geotech. Engrg. Div., ASCE, 101 (5), 324-438.

Masing, G. (1926). "Eigenspannungen und Verfestigung beim Messing". Proc. 2nd Int'l. Congress of Applied Mechanics, Zurich, Switzerland.

Mittal, H.K. and Hardy, R.M. (1977). "Geotechnical Aspects of a Tar Sand Tailings Dyke". Proceedings of ASCE Conf. on Geotechnical Practice for Disposal of Solid Waste Materials, Michigan, June 13-15, 327-347.

MMI (1985). Ministry of Metallurgical Industry, China, Personal Communication.

Mohamad, R., Dakoulas, P., Gazetas, G., Dobry, R. (1985). "Liquefaction Flow Failure Evaluation of Earth Dams". Proc. 11th Int'l Conf. on Soil Mech. and Found. Engrg. Vol. 4, 1865-1868.

National Research Council (1985). "Liquefaction of Soils During Earthquakes". National Academy Press, Washington, D.C.

Newmark, N.M. (1965). "Effects of Earthquakes on Dams and Embankments". Geotechnique, 15(2), 139-160.

Poulos, S.J., Castro, G. and France, J.W. (1985). "Liquefaction Evaluation Procedure". J. Geotech. Engrg. Div., ASCE, 111(6), 772-792.

Poulos, S.J., Robinsky, E.I., and Keller, T.O. (1985b). "Liquefaction Resistance of Thickened Tailings", J. Geotechn. Engrg. Div., ASCE, 111(12), 1380-1394.

Power, M.S., Berger, V., Youngs, R.R., Coppersmith, K.J. and Streiff, D.W. (1986). "Evaluation of Liquefaction Opportunity and Liquefaction Potential in the San Diego, California Urban Area." Woodward-Clyde Consultants Report to the U.S. Geol. Survey.

Schnabel, P.B., Lysmer, J. and Seed, H.B. (1972). "SHAKE: A Computer Program for Earthquake Response Analysis of Horizontally Layered Sites". Report No. EERC 72-12, Earthquake Engineering Research Center, University of California, Berkeley, California, December.

Scott, M.D., Lo, R.C., Klohn, E.J., Finn, W.D.L. and Yogendrakumar, M. (1989). "Non-linear Dynamic Analysis of L-L Tailings Dam", Submitted to Twelfth Conf. of Int'l. Soc. of Soil Mech. and Found. Engrg., Rio de Janeiro.

Seed, H.B. and Idriss, I.M. (1969). "Influence of Soil Conditions on Ground Motion During Earthquakes". J. Soil Mech. Found. Engrg. Div., ASCE, 95(1), 99-137.

Seed, H.B. and Idriss, I.M. (1970). "Soil Moduli and Damping Factors for Dynamic Response Analysis". Report No. EERC 70-10, Earthquake Engineering Research Center, University of California, Berkeley, December.

Seed, H.B., Lee, K.L., Idriss, I.M. and Makdisi, F.I. (1975a). "The Slides in the San Fernando Dams During the Earthquake of February 9, 1971." J. Geotech. Engrg. Div., ASCE, 101(7), 651-688.

Seed, H.B., Idriss, I.M., Makdisi, F. and Banerjee, N. (1975b). "Representation of Irregular Stress Time Histories by Equivalent Uniform Stress Series in Liquefaction Analyses". Report No. EERC 75-29, Earthquake Engineering Research Center, University of California, Berkeley, October.

Seed, H.B., Makdisi, F.I. and De Alba, P. (1978). "Performance of Earth Dams During Earthquakes". J. Geotech. Engrg. Div., ASCE, 104(7), 967-994.

Seed, H.B. (1979). "Considerations in the Earthquake-Resistant Design of Earth and Rockfill Dams". Nineteenth Rankine Lecture, Geotechnique 29(3), 215-263.

Seed, H.B. (1987). "Design Problems in Soil Liquefaction". J. Geotech. Engrg. Div., ASCE, 113(8), 827-845.

Seed, R.B., Jong, H.L. and Nicholson, P.G. (1987). "Laboratory Evaluation of Undrained Cyclic and Residual Strengths of Lower San Fernando Dam Soils". Research Report No. SU/GT/87-01, Dept. of Civil Engrg. Stanford University, June.

Serff, N., Seed, H.B., Makdisi, F.I., and Chang, C.K. (1976). "Earthquake Induced Deformations of Earth Dams". Report No. EERC 76-4, Earthquake Engineering Research Center, University of California, Berkeley.

Vaid, Y.P. and Chern, J.C. (1985). "Cyclic and Monotonic Undrained Response of Saturated Sands". Proc. of ASCE Convention, Session on Advances in the Art of Testing Soils Under Cyclic Conditions, Oct., Edited by Khosla, V.

Vick, S.G. (1983). "Planning, Design and Analysis of Tailings Dams." John Wiley & Sons.

Youd, T.L. and Perkins, D.M. (1978). "Mapping of Liquefaction Induced Ground Failure Potential." J. Geotech. Engrg. Div.. ASCE, 104(4), 433-446.

Youd, T.L. and Perkins, D.M. (1987). "Mapping of Liquefaction Severity Index". J. Geotech. Engrg. Div., ASCE, 113(11), 1374-1392.

Liquefaction of Hydraulically Placed Sand Fills

by K. Been*, M.ASCE, J.H.A. Crooks*, M.ASCE,
B.H. Conlin** and D. Horsfield*

Recent events in the Canadian Beaufort Sea have raised the question as to whether underwater hydraulic placement of clean sand fill can result in a contractive state which would lead to a flow slide as a result of gravity stresses in the sand mass. A review of reported liquefaction failures of underwater sand fills subjected to monotonic loading was undertaken to determine what evidence there is regarding the state of a sand deposited underwater. There is a remarkable lack of conclusive data regarding the state of underwater fills that failed. The mechanisms by which flow slides can occur in a sand that is in a dilatant state is also discussed to investigate whether the flow slide are indicative of the initial state of a sand. It is shown that high pore pressures can occur in dilatant sands, which may explain the flow of initially dilatant sands.

Introduction

Undensified hydraulically placed sand is an economically attractive fill for use in coastal areas at wharf structures and as reclaimed land, underground backfill in mines, retaining structures and foundation materials in mine tailings impoundments, and recently for the construction of sand islands and berms in the Beaufort Sea.

The potential for slope failures due to liquefaction in hydraulic sand fills is frequently considered a major drawback to the use of this technique for construction. This is particularly true in seismically active areas and there are many reported cases of liquefaction of hydraulic sand fills resulting from cyclic loading (i.e. earthquake or wave loading). However, there are also reported cases of failures which are not associated with cycling loading; these are generally attributed to strain softening of contractive sands which are at states above the steady state line (Castro and Poulos, 1977). The authors have consistently observed that sands deposited underwater in the laboratory lie below the steady state line and do not exhibit strain softening behaviour. Also, Terzaghi and Peck (1967) state..."A clean sand deposited underwater is stable, although it may be loose, because the grains roll down into stable positions. In a sand capable of spontaneous liquefaction some agent must interfere with this process". There is some difficulty with the terminology adopted by Terzaghi and Peck, specifically their use of "stable" and "spontaneous liquefaction". Based on the discussion on terminology provided below, their statement is interpreted to mean that a clean fill deposited underwater will be in a dilatant rather than contractive state. To achieve a contractive state, there

*Golder Associates, 7017 Farrell Road S.E., Calgary, Alberta T2H 0T3
**Golder Associates, 224 West 8th Avenue, Vancouver, B.C. V5Y 1N5

must be some agent which interferes with the deposition process. This interpretation would be consistent with the authors' laboratory observations that sands pluviated through water are dilatant. The interfering process could include partial saturation leading to an apparent cohesion for fills placed above the water table, the simultaneous deposition of sand with silt which infills the voids and prevents the sand grains from reaching a stable structure, or high upward seepage gradients during deposition.

The question as to whether or not a sand which is placed through water can lie above the steady state line and exhibit strain softening behaviour under non-cyclic loading conditions, is important in the design and construction of artificial exploration islands in the Canadian Beaufort Sea. These structures are typically constructed using undensified hydraulic sand fills. About thirty surface piercing and caisson retained islands have been constructed since 1970 in water depths ranging from 5 - 45 m using a wide variety of materials and construction techniques. Problems associated with mass failures have been encountered in six cases, which represents a significant failure rate. In five of these cases, the failures were the result of failure of the foundation clay during construction (1 case), erosion after construction (1 case), wave induced excess porewater pressures and/or erosion during construction (2 cases) and dynamic ice loading during operation (1 case). Incomplete dissipation of excess porewater pressure in the sand masses generated during construction may have been a contributing factor in some cases. None of these failures are attributed to overstressing of strain softening sands which were initially at states above the steady state line. In the sixth case, the Nerlerk berm, there is disagreement as to the cause of the observed failures which exhibited the characteristics of flow slides. Sladen et al (1985, 1987) consider that the Nerlerk failures were the result of static overstressing of a contractive hydraulic sand fill which existed in a state above the steady state line. Been et al (1987) consider that the sand was in a slightly dilatant state and that the failures were triggered by deformations in the underlying foundation clay and/or residual excess porewater pressures in the sand.

The Nerlerk berm failures have raised some important questions for the Beaufort Sea operators. Based on back analyses of the failures, Sladen et al (1985) concluded that the sand was in a very loose condition above the steady state line. Cone penetration test data interpreted by Been et al (1987), laboratory test observations and the statement by Terzaghi and Peck suggest that the sand would be dilatant, in a state slightly below the steady state line. There are two possible explanations for this apparent contradiction.

Either: The CPT interpretation based on large laboratory calibration chamber testing is grossly in error. In this case the Nerlerk sand could have been in a very loose contractive state and the flow failures could have been caused by static overstressing within the sand fill itself.

Or: The CPT interpretation is correct and the sand was in a dilatant state. In this case, sands that lie below the steady state line can still be involved in this type of failure.

The issue of errors associated with interpretation of state in sands from CPT data is discussed by Jefferies (1988) and Been et al (1988). They concluded that these errors would not explain the different between the above two explanations of the Nerlerk berm failures. These arguments are not repeated in this paper. Instead, the objective of this study was to develop further information on the subject by studying case histories involving failures of hydraulic sands fills for evidence of whether the sands were in a contractive or dilatant state, the trigger mechanisms associated with the failure and the mechanics of flow of the failure mass. Cases involving cyclic loading were not included since it is generally accepted that both contractive and dilatant sand masses can fail due to cyclic loading.

Terminology

There is frequently confusion regarding the term liquefaction and the associated behaviour of sands. For the purposes of this paper, the following terminology will be used:

Liquefaction is a condition in which a sand loses a large percentage of its shear resistance due to high porewater pressures which reduce the effective stresses to low values. Under certain boundary conditions the sand can flow in a manner resembling a liquid until the shear stresses acting on the sand mass and its shear resistance become equal. Increased porewater pressures in sands can be generated in several ways, such as cyclic loading, seepage or shear strains.

Liquefaction is thus used as a general term to describe a certain phenomenon which can occur in sands. Confusion exists in the literature regarding the term liquefaction because it has also been used to describe specific causes of liquefaction. Liquefaction causes deformations which in the extreme are evidenced by the sand flowing.

A flow slide is the term that will be used in this paper for a failure in which the sand has flowed like a liquid. It is thus a morphological term, and does not imply knowledge of trigger mechanisms or conditions before failure.

A contractive sand is at a state above the steady state line (Castro, 1969). This term is preferred to "loose", as loose is commonly associated with relative density and there is a wide range of relative densities that can be above or below the steady state line (SSL) depending on other conditions such as material gradation and confining stress. Contractive sands subjected to shear stress will exhibit high positive porewater pressure.

A dilatant sand is at a state below the SSL. This term is preferred to "dense" for the same reasons described in the preceding paragraph. However, it should be noted that sands below the steady state line may develop positive pore pressures (under cyclic loading, for example) or contractive strains before dilation occurs. Dilation refers to the rate of change of volumetric strain, not the absolute value of volumetric strain.

A retrogressive failure is a slope failure in which the lower part of the slope fails first. Following the initial slip, the slide develops progressively upslope, slice by slice, as the material behind the back scarp of the failure loses support.

Hydraulic sand fills are placed by pumping sand into the fill in a slurry form. Three types of hydraulic fill can be distinguished - those placed entirely underwater, those discharged into the water from air, and those placed above water. The latter two type are different from fills placed underwater in that there is the potential for air entrainment, and therefore the fill may not be fully saturated at the time of deposition.

Case Histories

A computerized search based on the keywords "sand", "hydraulic fill", "liquefaction" and combinations thereof was carried out and produced remarkably few relevant case records. These are described below:

Nerlerk

The Nerlerk berm was constructed in 1982/83 using dredged underwater hydraulic sand fill placed as a foundation pad for a caisson-type oil exploration structure. When the berm reached a height of about 24 m above seabed (20 m below water level), several large failures occurred as construction progressed. These failures are described by Sladen et al (1985). Figure 1 shows a typical prefailure section at Nerlerk. Prefailure slopes were typically between 3:1 and 12:1, while post failure slopes range from about 12:1 and flatter. In plan, the failures had bowl-shaped crests and a tendency to a narrower necked portion below the

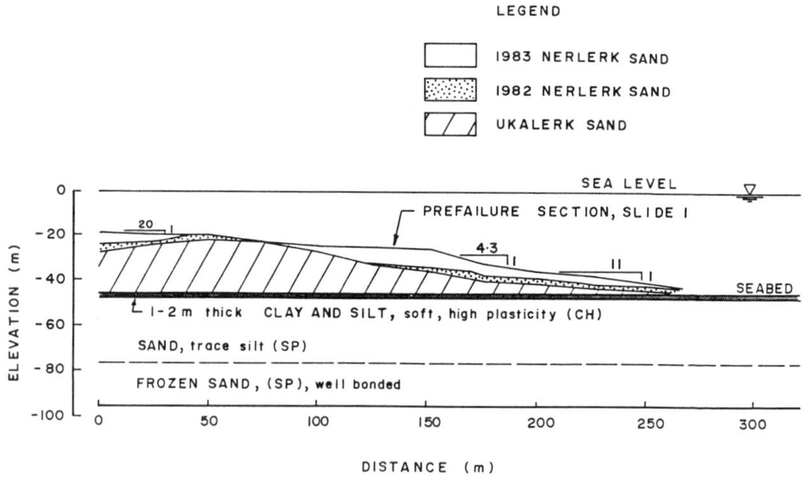

Figure 1: Section Through Nerlerk Berm Before First Slide

bowl features which is typical of subaqueous flow slides in sands. The Nerlerk case history is notable in that detailed information such as bathymetry, material type and cone penetration resistance is available. Unfortunately, as the berm was underwater, additional information and visual observations were impossible.

As discussed previously, there is disagreement as to the nature of the Nerlerk berm failures. In summary, Sladen et al consider that the Nerlerk flow slides were the result of the upper hydraulic sand fill being in a contractive state before failure and were triggered by static overstressing of the sand. On the other hand, Been et al (1987) consider that the sand was in a dilatant state and that the flow slides were triggered by deformations in the underlying clay and possibly residual positive pore pressures remaining from construction.

The Nerlerk berm case history may be summarized as follows:

- Method of deposition: Entirely under water
- Fill densities: D_r = 30 - 50%, CPT interpretation indicates a dilatant state, back analyses indicate contractive state.
- Trigger mechanisms: Underlying soft clay present, strength unknown. Possible excess porewater pressures in sand fill.
- Flow slide mechanism: Inferred to be retrogressive by Sladen et al.

Fort Peck Dam

A large slide occurred in the upstream shell of the Fort Peck Dam near the end of construction in 1938 (Casagrande, 1965). This dam was constructed by the hydraulic fill method using river sands and fine grained alluvial soils. The foundation consisted of alluvial sands, gravels and clays, underlain by Bearpaw shales containing bentonitic layers. At the time of failure the dam was about 60 m high with an average slope of 4:1. The failure occurred over a 500 m section, and was preceded by bulging. At some time after these initial strains, a flow slide developed, with very large displacements (up to 450 m) and very flat (20:1) final slopes. The post failure appearance was that of intact blocks in a mass of thoroughly disturbed material.

The Fort Peck Dam slide was investigated by a 9-man review board, whose members held diverse views of the cause of failure. A majority of the board concluded that the slide was due to shear failure of the shale foundation and that "the extent to which the slide progressed upstream may have been due, in some degree, to a partial liquefaction of the material in the slide". The minority view was that "liquefaction was triggered by shear failure in the shale, and that the great magnitude of the failure was principally due to liquefaction". From the point of view of this paper, it is noted that the failure occurred in the shale, which then triggered a flow slide in the overlying sand. The sand was in a quick condition after the slide but the failure was not initiated by the contractive sand.

In summary, for the Fort Peck Dam case record:

- Method of deposition: Hydraulic placement, but the authors do not have data on water levels during placement of the fill or the underlying alluvial sands.
- Fill densities: No reliable data
- Trigger mechanism: Underlying shales
- Flow slide mechanism: Complex

Wharf on Willamette River

This flow slide is well documented by Cornforth et al (1973), as it was in fact an instrumented test section adjacent to several previous flow slides. Figure 2 shows conditions before and after the slide, which was induced by dredge excavation at the toe (to a slope of about 34°). As shown, the stratigraphy consisted of natural alluvium, overlain by about 15 m of hydraulic sand fill and 6 m of bulldozed sand fill at the top of the slope. At the onset of failure, bubbles and an aerated silt scum formed on the water surface. Failure progressed up the slope to the waterline in the form of curved slices which enlarged back into the slope as the failure progressed. Most of the sliding occurred in 20 minutes. Instrumentation confirmed that sliding occurred at the contact between the hydraulically placed sand and the overlying bulldozed sand. Based on this case record, it appears that the hydraulically placed sand was more stable than the overlying bull-dozed sand. The bulldozed sand was likely partly saturated as evidenced by the bubbles during failure and thus could have been in a contractive state.

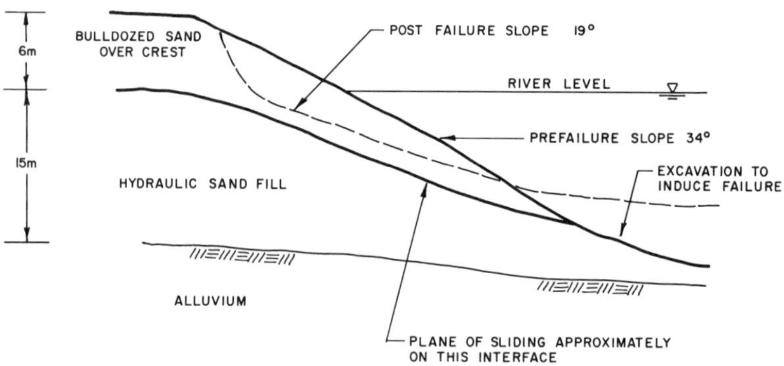

Figure 2. Sketch of Pre- and Post-Failures Slopes at Willamette River Wharf Test Section, Descibed by Cornforth et al (1972).

In the context of this paper, the following are noted:

- Method of deposition: Bulldozed fill flowed, hydraulic fill stable
- Fill densities: D_r = 20 - 40%, SPT 'N' value = 3
- Trigger mechanism: Undercut toe to steep angle (> ϕ')
- Flow slide mechanism: Retrogressive, well documented.

Helsinki Harbour

This flow slide has been reported by Andresen and Bjerrum (1967). The documentation of this event was partly destroyed during the war and the currently available information is therefore incomplete. In 1936, a harbour extension was being constructed in Helsinki by pumping sand into a water filled basin formed by outer banks of blasted rock fill. The water depth was 2 - 8 m, and the seabed consisted of a 5 - 10 m thick deposit of loose silty sand and clay. At the time of hydraulic filling, a 25 m section of the rock fill dyke was incomplete, leaving an opening. A small slide took place at the opening of the basin, but a decision was nevertheless made to place a few more loads of sand. When the next load was pumped in, however, a large flow slide was released with about half the sand from the basin flowing through the relatively small opening and coming to rest 30 - 60 m away. Figure 3 is a plan sketch of this failure. Despite the very soft foundation soils, it appeared that the slide had occurred entirely within the sand. The initial small failure provides evidence of a local instability in a critical part of the fill

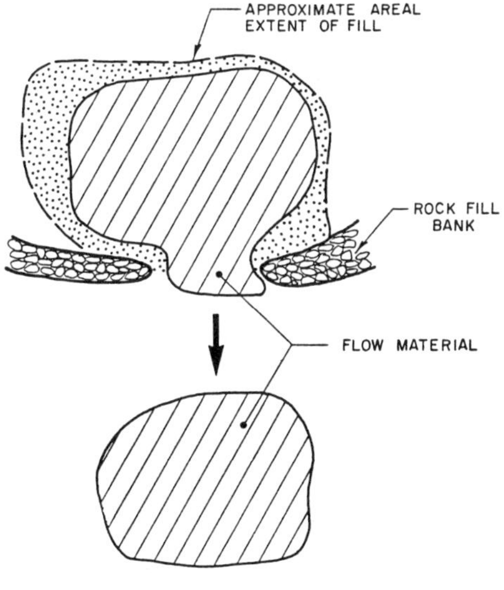

Figure 3. Plan Sketch of the Helsinki Harbour Slide in 1936 (After Andresen and Bjerrum, 1967)

due to failure of the underlying soft clays or high construction pore pressures. Andresen and Bjerrum classify this failure as a retrogressive flow slide. The very soft clay (reported unit weight of 1500 kN/m^3) provides a plausible triggering mechanism for the retrogressive slide.

For the Helsinki harbour case record:

- Method of deposition: Hydraulic, water level unknown
- Fill densities: No data
- Trigger mechanism: Probably local instability of underlying soft materials.
- Flow slide mechanism: Inferred to be retrogressive (Andresen and Bjerrum, 1967).

Dock Wall, Chicago

Peck and Kaun (1948) provide an interesting description of a flow slide that occurred in fine sand during reconstruction of a dock wall inEast Chicago in 1946. The sand, with D_{50} = 230 microns and uniformity coefficient of 1.7, was hydraulically placed with a dredge. The flow slide occurred through a narrow 0.8 m opening created by removal of two sheet piles. About 500 m^3 of material was lost in 30 minutes before the gap could be closed. SPTs in the fill prior to construction suggested it was in a very loose state with a single blow of the hammer driving the sampler between 1 and 2 feet in many parts of the deposit. Peck and Kaun conclude that the extent of flow was a direct result of the looseness of the sand.

For this case record:

- Method of deposition: Hydraulic, water level unknown
- Fill density: Very loose based on SPT, but no data regarding state (dilatant or contractive)
- Trigger mechanism: Removal of sheet pile
- Flow slide mechanism: Retrogressive, but not specifically documented

Tar Sand Tailings Dyke

Mittal and Hardy (1977) discuss the performance of the tar sand tailings dykes at the GCOS plant in Alberta. They describe several instabilities related to high rates of construction (i.e. high excess pore pressures). These failures were manifest as classical circular arc failures. However, a failure that occurred in August 1974 is of particular interest. At this site, the tailings dyke was raised using upstream construction methods which required placing of compacted sand in cells over uncompacted beach sands. The beaches were formed by simply pumping the tailings into the pond during winter. On August 23, 1974 failure occurred when a compacted mat 13 m high had been constructed over loose beach sand. This beach sand had been placed by discharging sand into shallow (less than 6 m) water and above water for the final metre or so. The construction rate for the 11 days before the slump was well above average and Mittal and Hardy concluded that dissipation of porewater pressures in the underlying beach sands was not rapid enough to maintain stability. About 4.5 m of subsidence occurred, without significant lateral movements and no observed toe uplift. The compacted mat material remained intact and then broke into slabs. Mittal and Hardy hypothesize that a layer of sand about 4.5 m thick below the mat liquefied and flowed into the pond. Figure 4 is a sketch of this proposed failure mechanism, which is somewhat surprising in that very little lateral displacement of the compacted mat occurred.

In summary, for the tar sands tailings dyke:

- Method of deposition: Tailings beach deposition into shallow water and above water.
- Fill density: D_r = 30 - 40%, N = 38 - 56 but no data regarding state (dilatant or contractive)
- Trigger mechanism: High pore pressures due to rapid construction(?)
- Flow slide mechanism: Confined flow(?)

Discussion

Six case histories of flow slides (including Nerlerk) involving hydraulic fills have been briefly described above. Relevant observations regarding these fills are summarized on Table 1.

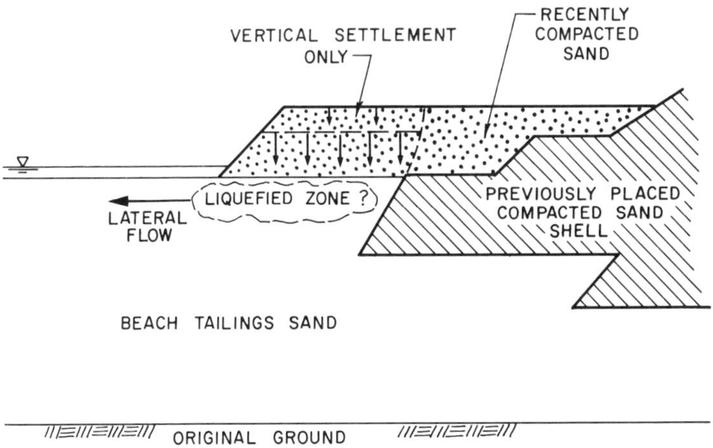

Figure 4: Sketch of Inferred Failure Mode for Tar Sands Tailings Dyke Described by Mittal and Hardy (1977)

TABLE 1

SUMMARY OF CASE RECORDS OF STATIC LIQUEFACTION FAILURE

CASE RECORD	FORT PECK DAM (1938)	WHARF, WILLAMETTE RIVER (1967)	NERLERK (1983)	HELSINKI HARBOUR (1936)	DOCK WALL CHICAGO (1946)	TAR SAND TAILING (1974)
FAILURE DESCRIPTION						
Height of Fill (m)	60	25	24	8	7	13
Average Slope	4:1	4.6:1	7:1	3.1	2.6:1	4.6:1
Foundation Soils	Bearpaw shale Alluvial sand	Organic silt	NC Clay	Soft clay	Dense sand	Tailings
Type of Slide	Spontaneous	Retrogressive	Uncertain	Retrogressive	Retrogressive	Confined flow(?)
Time	10 min.	20 min.	Unknown	Unknown	30 min.	Unknown
Trigger Mechanism	Failure in foundation shale, for several hours before flow	Toe undercutting to induce slide	Controversial, clay or contractive sand	Additional placement after first toe failure	Removal of sheet piles	Rapid construction (excess pore pressures?)
SAND FILL						
D_{50} (mins)		0.35 - 0.55	0.23 - 0.28	Details not reported	0.22	0.15
% Fines (passing #200 sieve)		0	Not reported, ≈ 5% (?)		0	up to 15 - 20
Uniformity Coefficient		2.5	1.7		1.7	5.7
Grain Shape					Angular	Angular
Relative Density %		20 - 40	30 - 50		Very loose	30 - 40
SPT N Value		3	-		<1	38 - 56
CPT q_c Value (bars)			40 - 60			40
Placement	Hydraulic Fill and Alluvial sands	Bulldozed over underlying hydraulic fill	Underwater Hydraulic	Hydraulic fill in or into shallow water	Hydraulic, water level not reported	"Beach" sand foundations for cell into shallow water.
					Very steep slope at toe after sheet pile removal.	Liquefaction and flowing of "beach" sand forming foundation for compacted mat.

The purpose of this review was to assess flow slides in hydraulic sand fills, in order to determine firstly if these materials flow during failure, and if so does this necessarily mean that the sand state was above the steady state line before failure or are there alternative explanations for the failure. Table 2 is a list of the relevant conclusions in this regard. There is little doubt that flow slides can occur in underwater fills. However, there is also no data to support any inference regarding whether the material in the flow slide was initially above or below the steady state line. The only evidence remaining is from the trigger and flow slide mechanisms.

In all cases there is a potential trigger mechanism that would account for the failure. At the Fort Peck dam it was undoubtedly the Bearpaw shales that failed first; the Willamette River wharf was undercut at the toe and the Chicago dock failed after removal of a sheet pile retaining wall. At the Helsinki harbour, the occurrence of a toe failure before the flow slide suggests that the underlying foundations were the trigger. The tar sand tailings failure appeared to be related to rate of construction and slow dissipation of pore pressures. The Nerlerk berm was underlain by soft clays, but there is disagreement about whether or not these triggered the failure.

The flow slide mechanism information is interesting in that all the slides which occurred underwater, and were therefore invisible, are classified as retrogressive. The most reliable information is the Willamette Wharf, which was an instrumented test section. The Fort Peck Dam failure is difficult to classify, as it appears there were blocks of intact material "floating" in a large liquefied mass of sand. The flow mechanism for the tar sands dyke hypothesized by Mittal and Hardy is difficult to understand based on the available evidence and has loosely been defined in this paper as "confined flow".

In summary, there is a trigger mechanism to cause initial failure in all the flow slide cases examined and there is a total lack of direct evidence regarding sand states (contractive or dilatant). Therefore, this review provides no basis to conclude that the occurrence of a flow slide is a reliable indicator of the initial state of the sand mass. If the authors' interpretation of the Nerlerk berm failures is correct, then it remains to examine whether flow failures are possible in dilatant sands. While this will not prove that the sand states before flow were dilatant, it would suggest that the fact that a flow slide occurs cannot be used as evidence that the state of the sand before flow occurs is above the steady state.

Flow Slide Mechanism in Dilatant Sands

Flow slides have been described mechanistically by workers such as Koppejan et al (1948), Terzaghi (1956) and Andresen and Bjerrum (1967). They describe a mechanism where the toe of the slope slides away, relieving the lateral stresses on the steeper scarp above the slide material. As a result of this unloading, the material has a tendency to swell and negative pore pressures occur. Water is drawn into the slope as the negative excess pore pressures dissipate, until limit equilibrium is again reached and the next slice slides away from the slope. A positive excess pore pressure may also develop due to straining of the

TABLE 2
SUMMARY OF REVIEW CONCLUSIONS

	Nerlerk	Fort Peck Dam	Willamette Wharf	Helsinki Harbour	Chicago	Tar Sand
Method of deposition	Underwater hydraulic	Hydraulic and alluvial	Bulldozed/ hydraulic	Hydraulic	Hydraulic	Tailings beach
Possiblity of "interfering agent" referred to by Terzaghi and Peck	No data	No data	Evidence of air entrapment	No data	No data	Possibly; high silt content and air
Fill state (above or below SSL)	CPT → below Back analyses → above	No data	No data	No data	No data	No data
Trigger mechanism	Controversial, clay or fill placement on contractive sand	Shale foundation	Toe undercutting to 34°	Local instability	Shale pile removal	High pore pressures(?)
Flow slide mechanism	Retrogressive	Complex	Retrogressive	Retrogressive	Retrogressive	Retrogressive

steepened slope. This mechanism continues until denser or stronger materials are encountered or the slope becomes very flat. However, it is frequently argued that provided the sand is in a dilatant state, any flow will be arrested by the negative pore pressures that may be generated and therefore that this mechanism will not occur in dilatant sands. Further clarification is required.

In searching for further evidence of flow in dilatant sands, the authors have noted the following:

- There are numerous published instances of so called liquefaction failures due to seismic or dynamic loading (Seed 1987). When these seismic failures occur, they frequently result in flow slides. However, the structures are generally sound prior to the extreme seismic loadings, and it is reasonable to assume that lesser cyclic loads had been experienced before failure. This suggests that the materials were not at states well above the SSL before the earthquakes, even though flow slides subsequently occurred. In addition, it is sometimes noted that delayed failures occur some time after the seismic event (for example, the Mochi-Koshi Tailings dam slide described by Ishihara, 1984 and Marcuson et al, 1974). These delayed failures are attributed to time dependent redistribution of excess pore pressures, again indicating an initial dilatant state.

- In the laboratory, dilatant sands do develop contractive strains and positive pore pressures. These may be induced by cyclic loading. More significantly, monotonic loading will induce initial contractive strains or positive pore pressures before dilation occurs. Figure 5 shows a typical undrained test on a dilatant material and it is apparent in this case that the excess pore pressure reaches about 50% of the initial confining stress before dilation occurs.

Based on the above, it is concluded that flow slides of dilatant sands initially below the steady state line do occur in practice and that laboratory testing provides support for the mechanism described above in that positive pore pressures are generated before dilation occurs.

There is much diversity in the nature of flow slides, with the distances of flow varying from a few metres to several hundred metres or more. While the distance of flow will be related to many factors, it is reasonable to expect that the initial state of the sand should be important. The steady state approach (Castro and Poulos 1977) and the collapse surface approach (Sladen et al 1985) require that a sand lies above the SSL before flow occurs, and that sands initially below the SSL will not develop large strains under monotonic loads. However, based on the data reviewed in this paper, there appears to be an intermediate range of sand states which are not sufficiently loose to be above the steady state line, but will still flow if failure is induced by some other mechanism. While it is clearly not possible to quantify the threshold state for flow slides to occur in general, because other factors come into play, some guidance can be obtained from laboratory data.

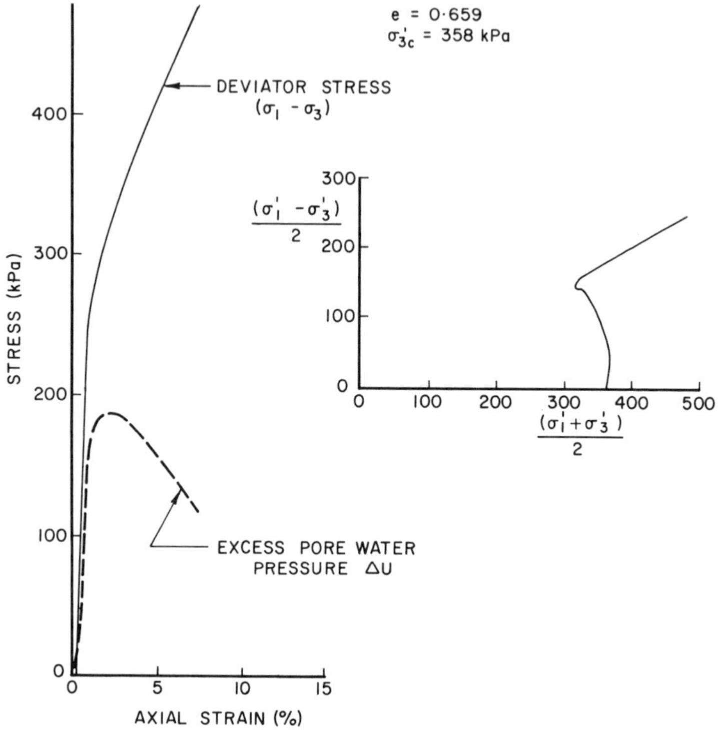

Figure 5. Typical Undrained Triaxial Test Result for a Weakly Dilatant Sand, with 5% Fines.

Figure 6 shows the definition of state parameter (Been and Jefferies, 1985) which is used by the authors as a description of "density" of sands. The state parameter, ψ, is simply the distance on an e-log p' state diagram of a material from the steady state line (Castro 1969) at the same stress level.

Under monotonic loading, a sand will contract or dilate towards the SSL. Thus, a material with a positive ψ lies above the SSL and will be contractive. Conversely, a material below the SSL with negative ψ will be dilatant. However, cyclic loads may induce cyclic mobility and high positive pore pressures in negative ψ materials. Figure 7 shows that significant positive pore pressures can also be induced in negative ψ materials under monotonic loading. The change in mean normal stress, $\Delta p'$, at the peak pore pressure normalized with respect to the initial mean stress, p'_o, is plotted against the value of ψ for 14 sands. The change in mean effective stress is preferred to the peak pore pressures in this plot, which is for triaxial test data. In the triaxial test there are two components of pore pressure: the first component is due to the mean total stress increase and the second is a direct result of the shear strains. The change in mean effective stress which is plotted on

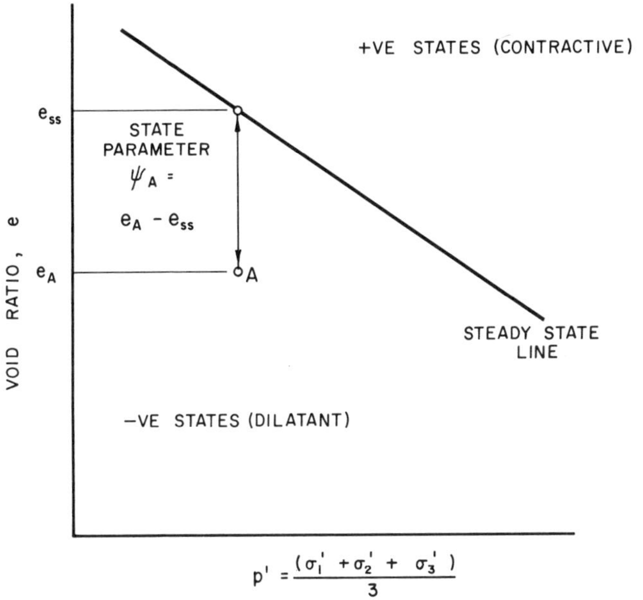

Figure 6. Definition of State Parameter

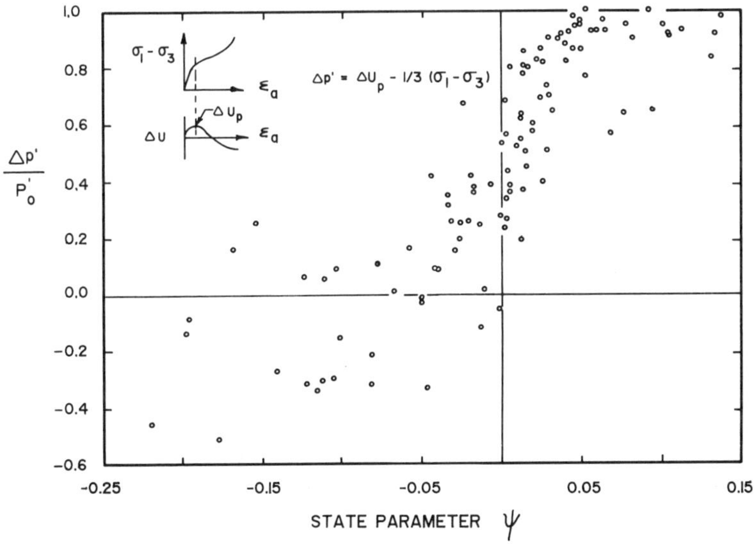

Figure 7. Change in Mean Normal Effective Stress at Peak Porewater Pressure: Undrained Triaxial Tests on 14 Sands

Figure 7 represents only the shear induced pore pressures. It is notable that even samples with ψ = -0.05 (i.e. the material void ratio is 0.05 more dilatant than the steady state) may show significant mean stress decrease before dilation occurs.

Thus when a slightly dilatant sand is subjected to strain as a result of deformation of an adjacent material, relatively high positive porewater pressures develop initially in the sand. Therefore the shearing resistance of the sand is reduced and failure within the sand can occur. The shearing resistance as further strains occur will depend on the initial density, volume changes which occur in the mass, and the pore pressures. While many other factors such as acceleration, permeability, kinetic energy, large strain dilation/contraction, stress changes and the like also come into play, this rationale provides clarification of the occurrence of flow slides by the mechanism described previously in what are normally considered dilatant sands.

In summary, two important aspects of sand behaviour allow retrogressive failures in dilatant sands to occur by the mechanism documented by Andresen and Bjerrum (1967) described previously:

- Significant positive shear induced pore pressures are initially generated in mildly dilatant sands. Dilation is by definition the rate of change of volumetric strain, but it is the actual value of volumetric strain or pore pressure that determines instantaneous shearing resistance.

- Negative pore pressures that do develop in sand slopes as retrogressive failures occur are dissipated by a reverse drainage process, i.e. water is drawn into the slope.

The above discussion represents a balance between positive and negative pore pressure generation and dissipation. Clearly the process is complex and dependent on slope geometry and boundary conditions in addition to sand state and permeability. It is not possible to conclude that a sand above a certain state (i.e. the steady state line) can flow, and that a sand will be stable below this state.

Conclusions

The purpose of this paper was to review the field evidence regarding failures in underwater fills. The Nerlerk CPT data suggested the sand was dilatant, but flow slides occurred. This leads to questions regarding the CPT interpretation, or whether dilatant sands might flow under certain conditions.

There is a complete lack of evidence regarding the state of sands (above or below the steady state line) prior to the case record failures reviewed in this paper. The laboratory observation that sands placed underwater will generally lie below the steady line cannot be substantiated in the field without involving CPT or SPT interpretations. However, the Willamette River Wharf test section does suggest that the underwater placed fill was in a denser state than overlying bulldozed material.

The cases reviewed do indicate that in all cases there was a significant trigger mechanism to initiate the failure, independent of the sand state. Thus it may be concluded that "spontaneous liquefaction" triggered by "slight disturbances" (Terzaghi and Peck, 1967) were not observed. Slight disturbances are not sufficient to trigger liquefaction in clean underwater sand fills.

Given the absence of evidence regarding the state of sand fills, but the fact that significant trigger mechanisms appear to be necessary to initiate failure, a mechanism by which retrogressive flow slides occur in dilatant sands was explored. It is concluded that such a mechanism exists. The laboratory observations and CPT interpretations regarding the likely state of underwater fills are therefore not refuted by the evidence.

Acknowledgements

The authors gratefully acknowledge the support of their colleagues at Golder Associates for the opportunity to write this paper. Fruitful discussions were also held with M. Jefferies of Gulf Canada during this study. The reviewers provided many thoughtful comments that helped clarify the presentation of this paper. The assistance of K. Koch and R. Kroeker in preparing the manuscript is also appreciated.

References

Andresen, A. and Bjerrum, L. (1967). Slides in subaqueous slopes in loose sand and silt. In Marine Geotechnique, edited by A.F. Richards, University of Illinois Press, Chicago, pp. 221-239.

Been, K. and Jefferies, M.G. (1985). A state parameter for sands. Geotechnique, 35 (2), pp. 99-112.

Been, K., Conlin, B.H., Crooks, J.H.A., Fitzpatrick, S.W., Jefferies, M.G., Rogers, B.T. and Shinde, S. (1987). Discussion on "Back analyses of the Nerlerk berm liquefaction Slides". Can. Geotechnical Journal, 24 (1), pp. 170-179.

Been, K., Crooks, J.H.A. and Rothenburg, L. (1988). A critical appraisal of CPT calibration chamber tests. Proc. 1st International Symposium on Penetration Testing, ISOPT-1, Orlando, Florida, pp. 651 - 660.

Casagrande, A. (1965). Role of the "Calculated Risk" in earthwork and foundation engineering. J. of the Soil Mechanics and Foundation Engineering Division, Proc. ASCE, Vol. 91, SM 5, pp. 1-40.

Castro, G. (1969). Liquefaction of Sands. Ph.D. Thesis, Harvard University. (Also Harvard Soil Mechanics Series, No. 81).

Castro, G. and Poulos, S.J. (1977). Factors affecting liquefaction and cyclic mobility. J. of Geotechnical Engineering Division, Proc. ASCE, Vol. 103, GT 6, pp. 501-516.

Cornforth, D.H., Worth, E.G. and Wright, W.L. (1972). Observations and analysis of a flow slide in sand fill. In Field Instrumentation in Geotechnical Engineering, British Geotechnical Society Symposium, Butterworths, London, pp. 136-157.

Ishihara, K. (1984). Post earthquake failure of a tailings dam due to liquefaction of the pond deposit. Proc. Intl. Conf. on Case Histories in Geotechnical Engineering, St. Louis, Mo.

Jefferies, M.G. (1988). Verification of q_c- ψ function in sand strata. Proc. 1st International Symposium on Penetration Testing, ISOPT-1, Orlando, Florida, pp. 793-804.

Koppejan, A.W., van Wamelen, B.M., and Weinberg, L.J.H. (1948). Coastal flow slides in the Dutch province of Zealand. Proc. 2nd Intl. Conf. on Soil Mechanics and Foundation Engineering, Rotterdam, Vol. V., pp. 89-96.

Marcuson, W.F. III, Ballard, R.F. Jr. and Ledbetter, R.H. (1979). Liquefaction failure of tailings dams resulting from the Near Izu Oshima Earthquake, 14 and 16 January 1978. Proc. 6th Pan American Conf. on soil Mechanics and Foundation Engineering, Lima.

Mittal, H.K. and Hardy, R.M. (1977). Geotechnical aspects of a tar sand tailings dyke. Proc. Conf. on Geotechnical Practice for Disposal of Solid Waste Materials, Ann Arbor, Michigan, pp. 372-347.

Peck, R.B. and Kaun, W.V. (1984). Description of a flow slide in loose sand. Proc. 2nd. Intl. Conf. on Soil Mechanics and Foundation Engineering, Rotterdam, Vol. V, pp. 31-33.

Seed, H.B. (1987). Design Problems in Soil Liquefaction. J. of Geotechnical Engineering, Proc. ASCE, Vol. 11, GT8, pp. 827 - 845.

Sladen, J.A., D'Hollander, R.D., Krahn, J. and Mitchell, D.E. (1985). Back analysis of the Nerlerk berm liquefaction slides. Canadian Geotechnical J., Vol. 22, 4, pp. 579-588.

Sladen, J.A., D'Hollander, R.D., Krahn, J. and Mitchell, D.E. (1987). Reply to discussion on "Back analysis of the Nerlerk berm liquefaction slides". Canadian Geotechnical J., Vol. 24, 1, pp. 179-185.

Terzaghi, K. (1956). Varieties of submarine slope failures. Proc. 8th Texas Conf. on Soil Mechanics and Foundation Engineering, Austin, pp. 1-41.

Terzaghi, K. and Peck, R.B. (1967). Soil Mechanics in Engineering Practice, John Wiley and Sons, New York (p. 108).

STABILITY ANALYSIS OF A SLURRY DEPOSITED DIKE

F. Masse[1], D. K. Atmatzidis[2] M. ASCE, and R. J. Krizek[3] M. ASCE

ABSTRACT--The disposal of various process tailings often involves impounding the wastes in diked containment areas. To optimize the filling procedure, it is sometimes desirable to build a dike from the waste material to support a pipeline which serves to distribute the slurry more uniformly over the entire surface of the disposal area. Studied here in terms of both a rotational failure and a flow analysis is the stability of such a dike for the disposal of bauxite residue or red mud from aluminum production. Representative values for the undrained shear strength of the existing sedimented mud were obtained by means of vane shear and dilatometer tests in the field, and data on the strength-moisture content relationship and drying characteristics of thickened muds were obtained from an extensive laboratory investigation. In general, the flow analysis has been found to be less critical than the rotational failure analysis, but both indicate that very low dike slope angles will be required for stability. One technique for building a stable dike is to superpose slurry layers incrementally by allowing adequate time for each layer to desiccate and gain strength before depositing a successive layer, and calculations have been made to suggest a reasonable compromise between layer thickness and drying time to accomplish this goal.

INTRODUCTION

The extraction of alumina from bauxite by the Bayer process results in the production of large volumes of tailings (known as "red mud"), which contain sodium hydroxide, aluminum oxide, iron oxide, and other undissolved solids. These tailings are often deposited in diked containment areas at a solids content (weight of dry solids divided by total weight, expressed as a percentage) ranging from about 20% to more than 40%. One operating procedure, sometimes termed drystacking, utilizes a pipeline with numerous discharge points along the perimeter of the containment area; a thickened slurry (usually at a solids content greater than 40%) is released intermittently from each discharge point to form a "tongue", which is then allowed to settle, dewater, desiccate, and thereby improve in strength before the next layer of waste is applied (Palmer and Krizek, 1987). As a result, the deposited solids form a layered system which, in general, has low dry density and strength.

---

[1]Former Graduate Student, Department of Civil Engineering, Northwestern University, Evanston, Illinois.

[2]Professor of Civil Engineering, University of Patras, Patras, Greece.

[3]Professor and Chairman of Civil Engineering, Northwestern University, Evanston, Illinois.

At a specific disposal site in Point Comfort, Texas, the disposal area was very large in areal extent and discharing slurry from the perimeter dikes resulted in deposits sloping toward the center of the containment area. Hence, optimum use of the available disposal volume could not be made because a valley would form at the center, and improvements in the disposal operation were desired. Toward this end, the construction of an internal dike, consisting of layers of sedimented "red mud" capable of supporting a pipeline, was considered as a possible solution. Presented herein are the results of an investigation which was conducted to evaluate the stability and identify the problems associated with the construction of such an internal dike in an existing and operating red mud disposal facility. Although this investigation was initially prompted by the particular conditions that prevailed at the Point Comfort area, the results obtained and the conclusions advanced can serve equally well as basic information to evaluate comparable situations for other materials.

SITE AND MATERIAL CHARACTERISTICS

The disposal area at Point Comfort, Texas has been operated for about two decades by pumping into it dilute suspensions of red mud at solids contents on the order of 15% to 20%. The material was usually deposited in layers of 8 to 10 cm in thickness and allowed to settle before adding another layer. The plan of the diked containment area and a profile along the centerline of the proposed internal dike (as conditions existed in the summer of 1985) are shown in Figure 1. Based on the available survey information, it can be observed that (a) depending on location, the proposed internal dike would have a height ranging from 3.1 to 8.2 meters and it would rest on a "foundation" of previously deposited tailings with a thickness ranging from 11.8 to 6.7 meters, and (b) since the sum of the dike height and the previously deposited tailings thickness is constant, small dike heights would be associated with large tailings thicknesses.

An earlier geotechnical investigation of the disposal site (Bromwell Engineering, 1983) yielded information on the variation of the solids content and strength of the previously deposited tailings. This investigation was conducted from a floating platform and the depth to which these tailings were sampled ranged from 1.2 to 5.4 meters, the maximum depth to which the piston tube sampler could be pushed. Samples were taken at distances from the containment dikes ranging from 46 to 500 meters. The average solids content was found to be about 48%, and there was no consistent variation of solids content with depth or distance from the dike. Isotropically consolidated undrained triaxial compression tests with pore pressure measurements yielded three different values for the effective angle of internal friction (29.8^0, 31.6^0, and 42.2^0), and this is a further indication of property variations within the deposited bauxite residue. Further details regarding these tests can be obtained from the report by Bromwell Engineering (1983).

Based in large part on previous experience and results and observations reported by Bromwell Engineering (1983), it was concluded that the existing bauxite residue was very soft and that obtaining and transporting undisturbed samples for laboratory strength tests would be very difficult. Accordingly, a series of in situ tests (vane shear and dilatometer) were conducted to evaluate the "undisturbed" shear strength of the deposit. As shown in Figure 1a, the majority of these tests were

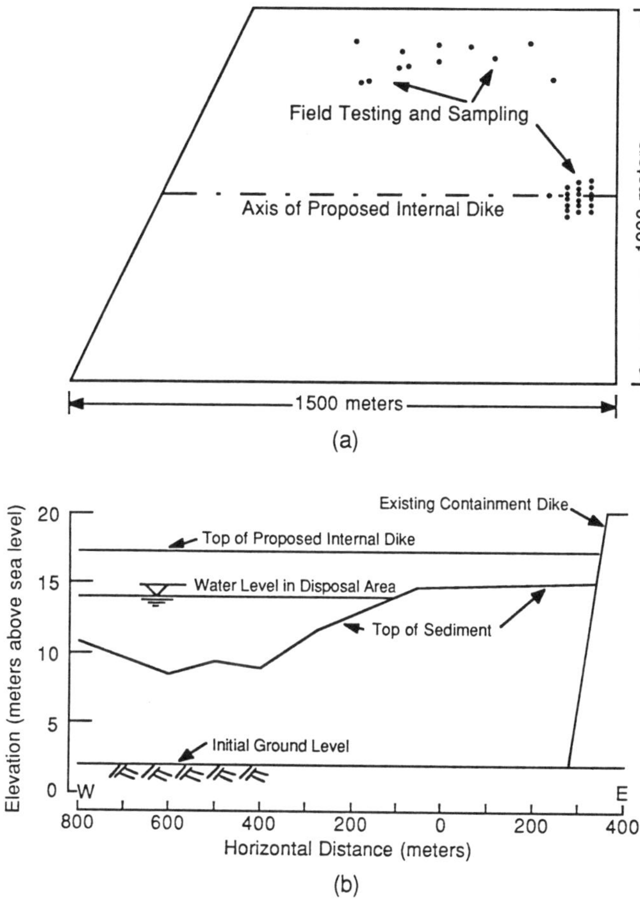

Figure 1. Plan View of Disposal Area and Profile Along Axis of Proposed Internal Dike

located at the east end of the centerline of the disposal area, where the proposed internal dike was to be started; however, some tests were conducted in the northern part of the lake to assess the homogeneity of the deposit.

The measured solids content of the deposit to a distance of about 150 meters from the north and east containment dikes had an average value of about 53%. This average, as well as most of the individual values obtained, are somewhat higher than the values reported by Bromwell Engineering (1983). This can be attributed to the fact that the deposit was submerged when Bromwell Engineering sampled it in 1983, while its surface was above the water table (see Figure 1b) when the Northwestern University team sampled it in July 1985 (primarily to determine in situ density); hence, the deposit probably consolidated somewhat under the influence of the increased effective stresses over the period of time between the two sampling operations. Accordingly, it was decided that a value of 50% would be representative of the solids content of the previously deposited bauxite residue, as well as the material deposited to form the new internal dike, and this led to the selection of a value of 14.4 kN/m^3 for the total unit weight of the material to be used in subsequent stability analyses.

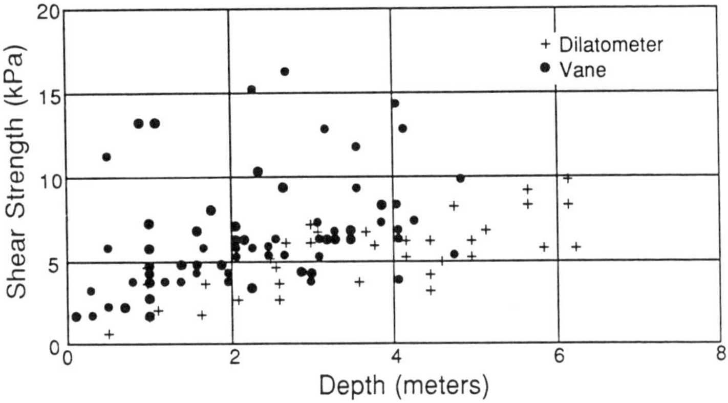

Figure 2. Variation of Shear Strength with Depth at 127 Meters from East Containment Dike

The results of the vane shear and dilatometer tests indicated that the undrained shear strength of the deposited bauxite residue varies, in general, with depth and with distance from the containment dike. Typical results obtained for a distance of 127 meters from the east dike and depths up to 6 meters from the surface of the deposit are shown in Figure 2. Similar results were obtained for distances of 47, 92, and 150 meters from the east containment dike. Although the undrained shear strength generally increased with increasing depth, the deposit was not homogeneous and there existed hard layers with high shear strength between soft layers. This is attributed to an observed "varving phenomenon" that occurred during deposition and intermittent desiccation of a given layer before placement of a subsequent layer. The undrained shear strength decreased with increasing distance from the containment dike, but there is evidence that this trend is much less pronounced for distances over 130 meters from the dike. This is undoubtedly due to the selective deposition of the larger particles in the dilute slurry closer to the points of discharge (namely, near the peripheral dikes). Average strength values for depths up to six meters are summarized in Table 1.

Table 1. Average Undrained Shear Strength (in kiloPascals) of Bauxite Residue in Disposal Area

Depth (meters)	Distance from Containment Dike (meters)			
	47	92	127	150
1	7.2	5.3	3.0	3.0
2	8.7	4.3	3.5	3.3
3	10.3	5.9	4.5	3.7
4	12.0	7.6	5.1	5.7
5	12.7	7.5	6.2	7.0
6	—	9.5	7.9	9.0

Parris (1986) conducted an extensive series of laboratory and field drying tests on samples of the bauxite residue slurry that was pumped into this containment site, and summarized in Figure 3 is the resulting correlation between the average moisture content and the undrained vane shear strength. It can be observed that the strength increases drastically as the moisture content decreases; for example, drying from a moisture content of 80% to one of 55% increases the undrained shear strength by about an order of magnitude (from 1 kPa to 10 kPa). However, the drying rate of a sedimented layer of red mud slurry depends on the initial thickness of the layer, and Figure 4 gives a summary of results determined by Parris (1986); all samples for these tests were prepared by sedimenting appropriate quantities of slurry with an initial moisture content of at least 120%, and drying them under measured environmental conditions.

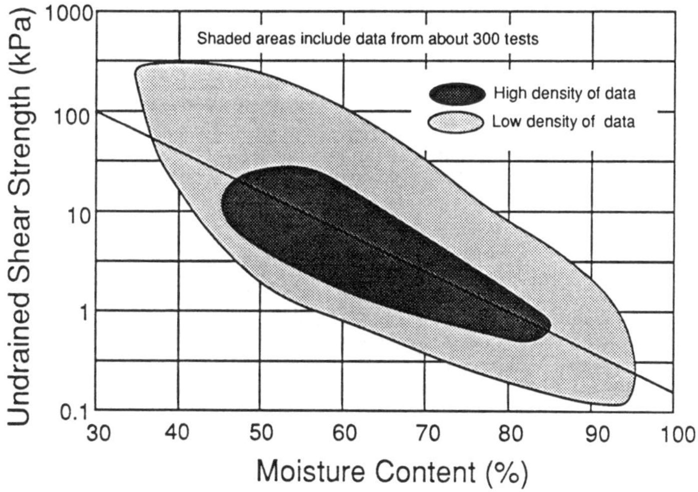

Figure 3. Variation of Undrained Shear Strength with Moisture Content of Sedimented Bauxite Residue

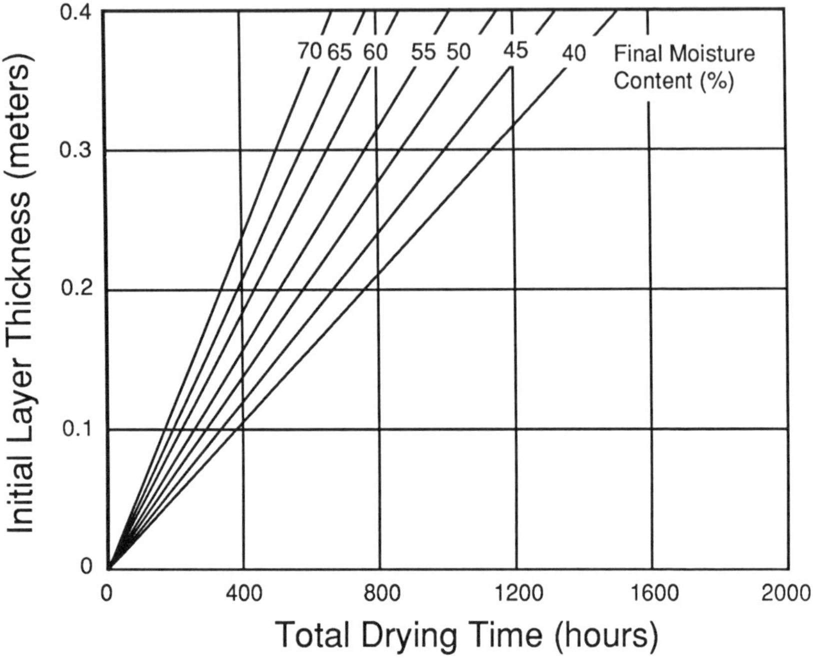

Figure 4. Drying Characteristics of Sedimented Bauxite Residue (after Parris, 1986)

STABILITY ANALYSIS

Considering the available topographic and strength data, it was decided to define eight different cross-sections along the axis of the proposed internal dike (at distances ranging from 10 to 850 meters from the east containment dike) in order to incorporate the variations in dike height, depth of existing deposit, and shear strength in the stability analyses to be performed. Three types of strength-versus-depth profile were selected for the underlying sediments, corresponding to the values presented in Table 1 for cross-sections at distances of 47, 92, and 127 meters from the east containment dike; for depths larger than 6 meters, the strength was considered to be equal to that measured at 6 meters. The undrained shear strength of the dike materials was taken as either 5 kPa or 3 kPa, corresponding to the measured strength of deposited material at depths less than one meter. For an effective stress analysis, the angle of internal friction was set equal to 31^0 (an average of 29.8^0 and 31.6^0, rounded upward). The geometric and strength data selected for each cross-section are summarized in Table 2.

Table 2. Data for Selected Profiles along Axis of Proposed Internal Dike and Minimum Strength Required for Flow Stability

Distance from Containment Dike (meters)	Strength Profile* below Internal Dike	Depth of Slurry (meters)	Height of Internal Dike (meters)	Lateral Extent of Internal Dike (meters)	Minimum Strength of Dike Material (kiloPascals)
10	1	11.8	3.1	178	2.0
100	2	11.8	3.1	178	2.0
160	3	11.8	3.1	178	2.0
280	3	11.5	3.4	194	2.2
580	3	9.4	5.5	314	3.5
670	3	7.9	7.0	402	4.5
850	3	6.7	8.2	472	5.0

* Profiles 1, 2, and 3 correspond to cross-sections at distances of 47, 92, and 127 meters from the east containment dike, respectively, as shown in Table 1.

Rotational Failure

The stability of the proposed dike at the end of construction was evaluated by using the available undrained shear strength data. Computations were performed according to the Simplified Bishop Method (Bishop, 1955) using an appropriate computer code; it was decided to use the Simplified Bishop Method because it yields safety factors which are generally similar to those obtained by more sophisticated methods, such as those of Spencer (1967) and Morgenstern and Price (1965), and because there is evidence that this method is easier to program and cheaper to use (Fredlung and Krahn, 1977; La Rochelle and Marsal, 1985). Stability computations were performed by first assuming a value for the inclination of the slopes of the dike and then determining the corresponding minimum factor of safety. The results obtained are summarized in Figure 5. It is seen that very small angles are required for the slope of the proposed dike to ensure stability against rotational failure, and this can clearly be attributed to the very low undrained shear strength of both the underlying sediment and the dike materials. More specifically, a dike slope angle of 3^0 does not ensure stability along the entire length of the dike, while a dike slope angle of 2^0 appears to result in a stable dike for either 5 kPa or 3 kPa undrained shear strength of the dike material. As should be expected, the lower factors of safety are obtained for the cross-sections with the largest dike height.

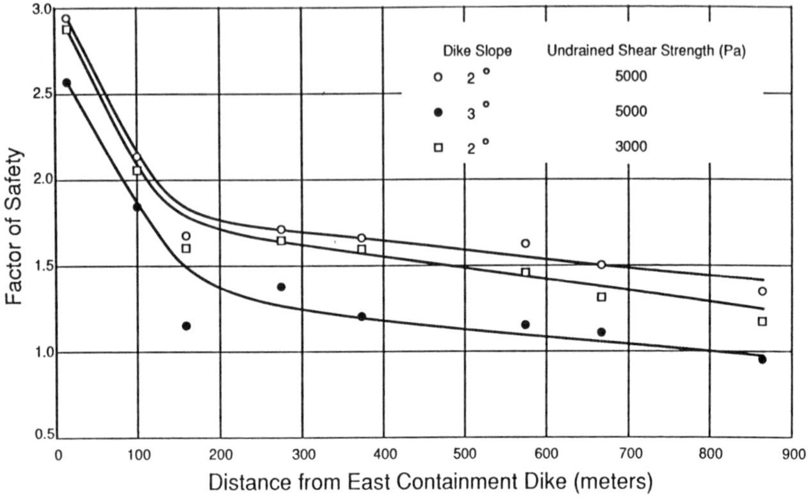

Figure 5. Factors of Safety for Rotational Failure Analyses of Proposed Internal Dike

An observation of importance can be made by considering the stability of the proposed dike up to a distance of about 160 meters from the east containment dike. Throughout this distance, the height of the proposed dike remains constant (3.1 meters), but the strength of the underlying sediment decreases with increasing distance from the containment dike; as shown in Table 1, the average undrained shear strength of the sediment over a depth of about 12 meters is reduced from about 11.6 kPa to about 6.5 kPa and this results in a drastic reduction of the factor of safety (from about 2.9 to about 1.7 for dike material with a strength of 5 kPa and from about 2.6 to about 1.2 for dike material with a strength of 3 kPa). Accordingly, it can be concluded that the single most important factor controlling the slope of the dike is the strength of the underlying sediment. For example, a safety factor of about 1.3 would be obtained if a dike with slopes of 5^0 were founded on the strongest available sediment within the containment facility (Type 1 profile); for the same factor of safety, the slopes could be increased to about 20^0 if the underlying sediment had an undrained shear strength of 100 kPa. However, the former (slopes of 5^0) would require an unrealistic improvement in the strength of the bauxite residue within the containment area, while the latter (slopes of 20^0) would probably be applicable only to a new containment facility located on soils with adequate strength.

To evaluate the long-term stability of the proposed dike, an effective stress analysis was conducted by setting the effective angle of internal friction equal to 31^0, the effective cohesion equal to zero, and varying the inclination of the dike slopes up to 5^0. The factors of safety obtained were very large (in general on the order of ten) and indicated that the long-term stability of the proposed dike is not critical if the slopes of the dike are small.

Flow Analysis

To supplement the rotational failure analysis, the stability of the proposed internal dike was evaluated by applying the mechanics of mud flow mobilization to a semi-infinite slope. As illustrated in Figure 6, the equilibrium equation for a semi-infinite slope of cohesive material can be written as

$$\frac{1}{2}\gamma H^2 + W \sin i = \frac{1}{2}\gamma H^2 + \frac{1}{2}\gamma HL \sin i = \tau L \qquad (1)$$

where τ is the undrained shear strength along the failure surface, L is the flow length, H is the initial thickness of the flow, γ is the total unit weight of the material, and i is the field slope. Accordingly, a stable angle, α, can be computed from Equation (1) as

$$\tan\alpha = \frac{H}{L} = \frac{2\tau}{\gamma(H + L \sin i)} \qquad (2)$$

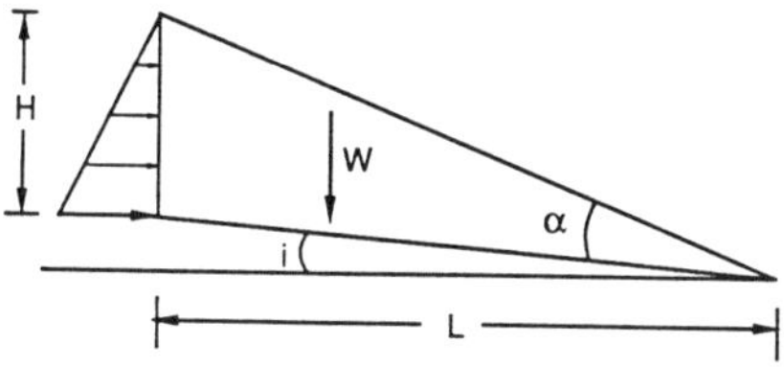

Figure 6. Geometry for Semi-infinite Slope Analysis

In his investigations of the mechanics of mudflows, Vallejo (1979, 1981) described the mudflow geometry as shown in Figure 7 and considered the material to vary from hard clay fragments to a soft clayey matrix or soil slurry. Assuming that hydrostatic pressure conditions exist within the mudflow at equilibrium, the equilibrium conditions can be written as

$$\frac{1}{2}\gamma_m(h + dh)^2 - \frac{1}{2}\gamma_m h^2 + \gamma_m h \sin\beta \, dx = S dx \qquad (3)$$

where S and γ_m are the undrained shear strength and the total unit weight, respectively, of the mud. According to Vallejo (1981), when the free surface of the mudflow and the bed on which it moves both have low angles of inclination, it can safely be accepted that $dh/dx = \alpha - \beta$ and $\sin\beta = \beta$, whereby Equation (3), for a safety factor of unity, is simplified to

$$\gamma_m h\alpha = S \qquad (4)$$

Accordingly, the factor of safety, FS, against mudflow mobilization can be expressed as

$$FS = \frac{S}{\gamma_m h\alpha} . \qquad (5)$$

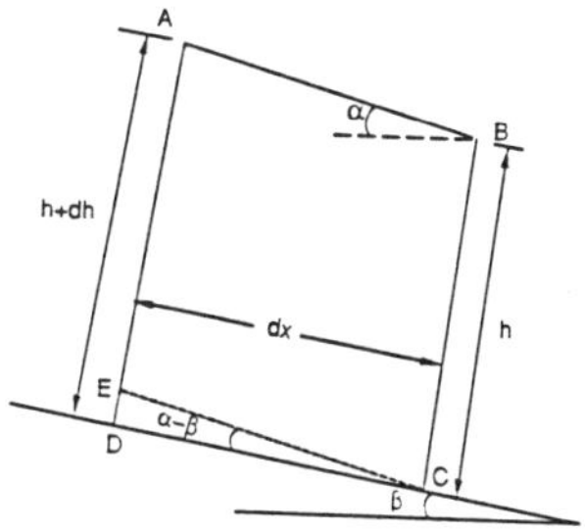

Figure 7. Geometry of Element in Mudflow Analysis

Since the rotational failure analysis indicated that very low slope angles are required to ensure a stable dike, the flow analysis was conducted by specifying a slope of 2^0 and determining the strength of the dike material required to avert mudflow mobilization. Factors of safety corresponding to the semi-infinite slope assumption were computed for two cases: (a) by using different inclinations of the field slope and (b) by setting the field slope equal to zero and varying the angle of the failure plane within the dike. Results obtained for the minimum undrained shear strength of the dike material required to yield a factor of safety equal to 1.3 (the lowest obtained for the rotational failure analysis of the dike) are summarized in Table 2. It can be observed that the resulting shear strength values are smaller or equal to those used for the rotational failure analysis and, therefore, it can be concluded that instability due to mudflow mobilization is not critical when compared to possible rotational failure.

DIKE CONSTRUCTION

The foregoing rotational and mudflow stability analyses are valid for the "end of construction" period and are based on the assumed final geometry of the proposed dike. However, since the dike is to be constructed of bauxite residue which is piped to the site and discharged into the containment area, it follows that the proposed dike will consist of a number of layers with varying thickness and lateral extent, and these layers must be superimposed on each other in a manner that will ensure stability during construction and will yield a dike with the desired final shape. Hence, the specification of construction procedures must include (a) a selection of the total number of layers required and an estimation of the shape (thickness and lateral spread) of each layer, (b) an evaluation of the stability of individual layers, (c) an evaluation of the stability of the partially completed dike at the time of placement of each layer, and (d) the formulation of a time schedule for the sequential application of the required number of layers. Although performing and reporting these operations for all of the selected dike cross-sections is beyond the scope of this presentation, the procedure will be illustrated by studying one dike cross-section, namely, that at a distance of about 10 meters from the east containment dike.

Geometry and Stability of Individual Layers

The geometry of the dike at the selected cross-section is characterized by a total height of about 3.1 meters and a slope of 2^0; this requires that the first layer of thickened slurry to be deposited would have a lateral spread of about 88 meters in both directions, while the spread of subsequent layers should decrease as the dike rises to full height. Evaluation of the stability of each layer requires the use of a relationship between layer thickness, flow length, field slope (slope of the bottom of the layer), and material properties (unit weight of slurry and shear strength at the interface). By considering the equilibrium of a small element of slurry, as shown in Figure 8, and

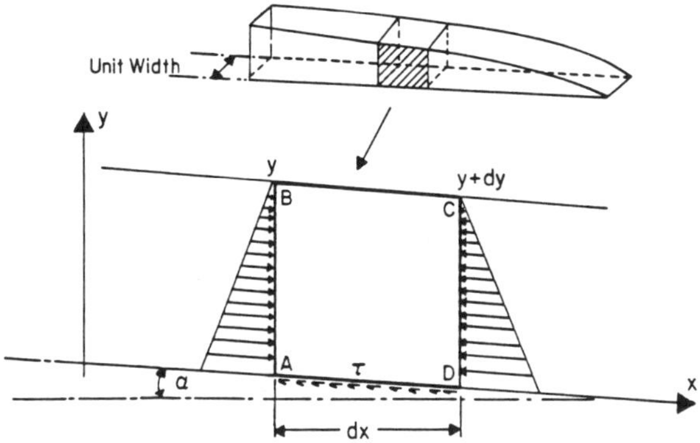

Figure 8. Equilibrium of Element in Mudflow Analysis

assuming that (a) flow is unidirectional, (b) the field slope is zero and the slurry thickness is constant perpendicular to the direction of flow, and (c) the pressure within the slurry is hydrostatic, Palmer and Krizek (1987) developed the following expression to delineate the mud profile:

$$x_2 - x_1 = \frac{y_2 - y_1}{S} + \frac{\tau}{\gamma S^2} \ln \frac{\gamma S y_2 - \tau}{\gamma S y_2 - \tau} \qquad (6)$$

where S is the field slope, γ is the total unit weight of the slurry, τ is the shear strength at the slurry-field slope interface, and $\ln$ designates the natural logarithm. This relationship has been shown to be successful in predicting the geometry and spreading of the same bauxite residue slurry considered in this investigation.

The stability of seventeen layers of deposited slurry was analyzed by use of Equation (6). According to information presented by Parris (1986) and Palmer and Krizek (1987), the shear strength of the slurry was not allowed to exceed 50 Pa, since the slurry would then have to be pumped at a solids content that could not be accommodated by the existing pumping system. The field slope for the first (bottom) layer was assumed to be equal to zero; for each subsequent layer, the field slope was set equal to the slope of the top of the preceding layer. The results obtained, in terms of final stable geometry for each layer and required shear strength of the slurry, are presented in Figure 9. It can be observed that the final total thickness of the layers is slightly greater than the required total thickness. This "overdesign" is recommended because there is evidence that the deposited material undergoes some volume reduction upon drying (Parris, 1986). However, the drying behavior and consolidation of the bauxite residue have not been adequately documented to allow their quantitative consideration in the design of this multi-layered dike.

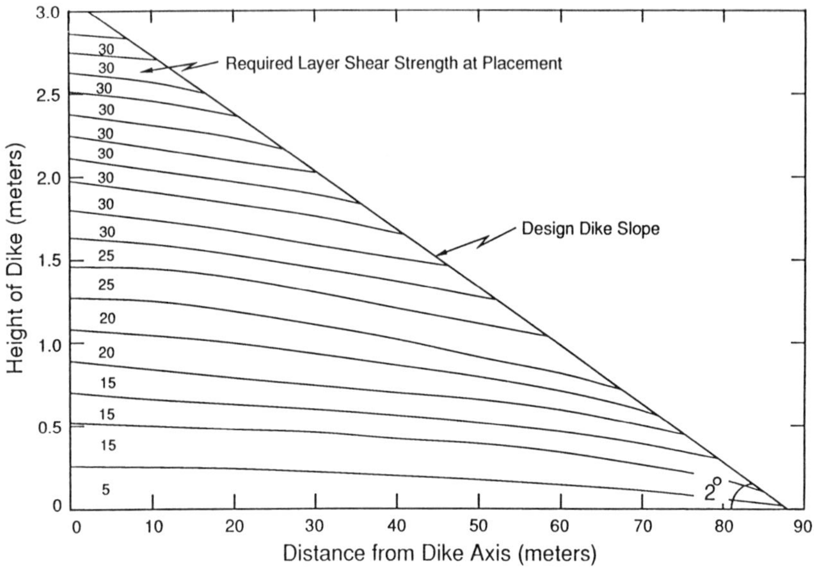

Figure 9. Superposition of Stable Layers to Form Dike (Strengths are expressed in Pascals)

Stability of Dike During Construction

Rotational and flow analyses were conducted after the placement of each layer to evaluate the stability of the partially completed dike. Since the geometry of the dike and the strength of the underlying sediments were known, the scope of this analysis was to determine the minimum required strength of the material in the previously deposited dike layers to yield an acceptable safety factor. Rotational failure analyses were performed by specifying that the maximum strength attained by any layer should not exceed 5 kPa and the results obtained are shown in Table 3. It can be observed that, when a layer is placed, it is sufficient for the immediately underlying layer to have reached a shear strength of 100 Pa. However, the strength of a layer must increase rather rapidly as the number of layers placed on top of it increases; on the average, the strength of a layer should increase to 1 kPa, 3 kPa, and 5 kPa as it becomes the third, fourth, and sixth layer, respectively, from the top.

Table 3. Strengths Required (in Pascals) before Adding New Layer during Dike Construction

New Layers \ Previous Layers	L1	L2	L3	L4	L5	L6	L7	L8	L9	L10	L11	L12	L13	L14	L15	L16	Safety Factor
L2	100																2.83
L3	1000	100															2.43
L4	3000	1000	100														2.25
L5	3000	3000	1000	100													1.73
L6	5000	5000	1000	1000	100												2.05
L7	5000	5000	3000	3000	1000	1000											1.68
L8	5000	5000	5000	5000	3000	3000	100										2.29
L9	5000	5000	5000	5000	3000	3000	1000	1000									1.90
L10	5000	5000	5000	5000	5000	3000	3000	3000	100								3.03
L11	5000	5000	5000	5000	5000	5000	5000	3000	1000	100							2.53
L12	5000	5000	5000	5000	5000	5000	5000	5000	3000	1000	100						2.14
L13	5000	5000	5000	5000	5000	5000	5000	5000	5000	1000	1000	100					2.31
L14	5000	5000	5000	5000	5000	5000	5000	5000	5000	1000	1000	1000	100				2.95
L15	5000	5000	5000	5000	5000	5000	5000	5000	5000	5000	1000	1000	1000	100			1.76
L16	5000	5000	5000	5000	5000	5000	5000	5000	5000	5000	3000	3000	3000	1000	100		1.66
L17	5000	5000	5000	5000	5000	5000	5000	5000	5000	5000	5000	3000	3000	1000	1000	100	1.68

The intent of the flow analysis is to determine the required shear strength at the interface between the dike and the existing sediments to ensure the stability of the dike as each layer is placed. The results indicate that the required strength at the interface increases with an increasing number of layers; for example, for 5, 10, and 15 layers, the required strength at the interface is 200 Pa, 800 Pa, and 1500 Pa, respectively, and the corresponding factors of safety are 1.51, 1.42, and 1.32. The highest value for the required shear strength was obtained for the case where all seventeen layers were in place; this value was found to be about 1.8 kPa, which is significantly lower than the 5 kPa that must be attained to ensure safety against failure.

Time Factor

The time required to complete construction of the proposed internal dike depends on the time required for each layer to attain the levels of undrained shear strength that will ensure stability of the partially completed dike. The four target strengths that almost all layers should attain are 0.1 kPa, 1 kPa, 3 kPa, and 5 kPa, and, according to Figure 3, these correspond to average moisture contents of 95%, 80%, 65%, and 60%, respectively. As shown in Figure 9, the layers forming the proposed dike have a maximum thickness ranging from 0.10 meters to 0.25 meters. The data presented by Parris (1986) and summarized in Figure 4 can be used to estimate the time required for any layer to achieve the required target strength or moisture content. For example, a layer with a maximum thickness of 0.20 meters will reach a strength of 5 kPa (moisture content of 60%) in about 20 days. Since information is not available for moisture contents of 95% and 80% (strengths of 0.1 kPa and 1 kPa, respectively), it may be assumed that the former can be achieved in not more than five to seven days and the latter can be reached within ten days to two weeks. Accordingly, estimates were made of the critical waiting period before each layer can be placed, and this time was found to range from seven days to fifteen days. The total time required for the construction of all seventeen layers was found to be 143 days in this example. However, any such estimate must be viewed with caution because environmental conditions obviously play an important role in the drying process, and a basic assumption in this analysis is that these conditions are similar to those experienced by Parris (1986).

CONCLUSIONS

The stability and construction of an internal dike, composed of bauxite residue deposited in layers of appropriate thickness and lateral spread, within an existing containment area of bauxite residue (red mud) was investigated. Based on the results obtained, the following conclusions can be advanced.

1. The construction of an internal dike using bauxite residue appears to be feasible, but is very limited in practicality; the low strength (3 kPa to 12 kPa) of the existing bauxite residue in the containment area, as well as the very low strength (considerably less than 5 kPa) of the dike material itself, dictate very small (2^0 to 3^0) angles of inclination for the dike slopes.

2. Rotational failure, rather than mudflow mobilization, is the mechanism which controls the stability of the dike; long term stability, in terms of effective stresses, is not critical in the rotational failure analysis.

3. A stable final shape can be achieved by incrementally superposing layers of thickened slurry; however, adequate time must be allowed for each layer to dry and improve in strength (up to 5 kPa) to ensure its equilibrium and the stability of the dike during construction.

4. Layers with a thickness of about 0.2 meters appear to offer a reasonable compromise between the rate of dike construction and the rate of strength gain of the sedimented bauxite residue due to desiccation.

5. Because the desiccation process is a significant factor in the time needed for construction of an internal dike composed of the waste material, environmental conditions will be very important; in fact, such a project is not even feasible under weather conditions where net evaporation does not occur.

ACKNOWLEDGEMENTS

This study was initiated as part of an ALCOA research project for which John Stinson was the project monitor. The help of Bertrand Palmer throughout the project, in general, and this study, in particular, is gratefully acknowledged and appreciated. And finally, thanks are extended to one of the reviewers, whose many constructive comments aided measureably in improving this presentation.

REFERENCES

Bishop, A. W. (1955), "The Use of Slip Circle in the Stability Analysis of Slopes," Geotechnique, Volume 5, Number 1, pp. 7-17.

Bromwell Engineering (1983), Geotechnical Evaluation of Alumina Residue Pond No. 4," Technical Report, Lakewood, Colorado.

Fredlung, D. G., and Krahn, J. (1977), "Comparison of Slope Stability Methods," Canadian Geotechnical Journal, Volume 14, pp. 429-439.

La Rochelle, P., and Marsal, R. J. (1981), "Slope Stability: General Report," Proceedings of the Tenth International Conference on Soil Mechanics and Foundation Engineering, Stockholm, Sweden, Volume 4, pp. 485-507.

Morgenstern, N., and Price, V. E. (1965), "The Analysis of the Stability of General Slip Surfaces," Geotechnique, Volume 15, Number 1, pp. 79-93.

Palmer, B., and Krizek, R. J. (1987), "Thickened Slurry Disposal Method for Process Tailings," Proceedings of the Symposium on Geotechnical Practice for Waste Disposal, American Society of Civil Engineers, Ann Arbor, Michigan, pp. 728-743.

Parris, T. E. (1986), Field Testing of Red Mud to Determine the Drying Characteristics, Master of Science Thesis, Department of Civil Engineering, Northwestern University, Evanston, Illinois, 300 pp.

Spencer, E. (1967), "A Method of Analysis of the Stability of Embankments Assuming Parallel Inter-Slices Forces," Geotechnique, Volume 17, Number 1, pp. 11-26.

Vallejo, L. E. (1979), "An Explanation for Mudflows," Geotechnique, Volume 29, Number 3, pp. 351-354.

Vallejo, L. E. (1981), "Stability Analysis of Mudflows on Natural Slopes," Proceedings of the Tenth International Conference on Soil Mechanics and Foundation Engineering, Stockholm, Sweden, Volume 3, pp. 541-544.

DENSIFICATION AND IMPROVEMENT OF HYDRAULIC FILLS

James K. Mitchell, F.ASCE[1]

Abstract

Hydraulic fills are very young sediments that almost invariably require improvement by densification or other means prior to their use. As placed, cohesionless fills have low relative density, low strength, and high liquefaction potential. Fine-grained hydraulic fills have a high water content, high compressibility, and low strength.

Almost all of the available soil stabilization and ground modification methods can be used for the improvement of hydraulic fills. The characteristics of these methods for application to hydraulic fills are described herein. Densification of cohesionless hydraulic fills during placement is likely to be more effective and economical than deep densification later. Installation of underdrains prior to placement can expedite the consolidation of fine-grained hydraulic fills.

Significant stiffening and strength increases may develop in cohesionless hydraulic fills during the first several weeks following their placement. Several examples of this as well as three recent case histories illustrating the application of different improvement methods are presented.

Introduction

Hydraulic fills of various soil types come from several sources. Some are incorporated in earth structures, some are used for land reclamation and development, and others are waste materials from such activities as dredging and mineral processing for which disposal space may be at a premium. Granular hydraulic fills are characterized by low relative density, low strength, and high liquefaction potential. Fine-grained hydraulic fills are characterized by very high water content, high compressibility, and low strength. Because of these unsatisfactory initial conditions, densification and improvement of most hydraulic fills is required either at the time of placement or subsequently for (1) safe support of structures and facilities, (2) prevention of liquefaction, (3) insurance against flow slides, and/or (4) maximizing disposal site capacity.

[1] Prof. of Civil Engineering, University of California, Berkeley, CA 94720.

The purposes of this paper are to review methods for stabilization and improvement of hydraulic fills and to illustrate the use of some of them by case history. Both cohesionless and cohesive hydraulic fills are considered, with emphasis on their special characteristics, their initial state, selection of improvement method, and the level of improvement. Different ground improvement methods are not described in detail, as they are amply covered in a number of other recent readily available references; e.g., ASCE (1978), Mitchell (1981), ASCE (1987).

Types of Hydraulic Fills and Their Initial State

Hydraulic fills are generally formed from either the discharge of slurried material that has been pumped through pipes or by dumping the material through water. The fill may be composed of a narrowly graded material or contain a range of particle sizes, depending upon the source material and the methods used. Similarly, the fill may be homogeneous throughout or exhibit considerable heterogeneity over short distances.

Hydraulic fills were classified according to the nature of the borrow material by Whitman (1970) as indicated in Table 1. This classification is useful in the selection of methods for densification and improvement.

Table 1

CLASSIFICATION OF HYDRAULIC FILL MATERIALS
(Adapted from Whitman, 1970)

Type of Borrow Material	Characteristics of Fill
Fairly clean sand (less than 15% finer than No. 200 sieve)	Reasonably uniform fill of medium density
Silty or clayey sand	Heterogeneous fill of large void ratio
Stiff cohesive soil	Clay balls with sand and clay matrix
Soft cohesive soil	Laminated normally consolidated or underconsolidated clay

The initial state and properties of different types of hydraulic fills have been described by Whitman (1970), and the engineering properties of fine-grained dredged materials have been discussed by Johnson et al. (1977). They are summarized briefly here.

<u>Clean Sands (less than 15% fines)</u>. The average relative density of fairly clean sand hydraulic fills after placement is typically 50 to 60 percent. However, it may be as low as 20 percent or as high as 80 percent locally. Variability in density and gradation throughout the fill will depend on such factors as distance from discharge points,

the presence of ponded water during placement, and variability of the source material.

Silty or Clayey Sands. These materials are usually placed so as to retain the fines. Otherwise, too much material will be lost, and there will be pollution from runoff. The resulting material is variable and compressible.

Stiff, Cohesive Soils. A stiff, cohesive borrow material produces a hydraulic fill of clay balls in a matrix of other soil. The compressibility may be controlled by the deformation of the clay balls; whereas, the consolidation rate is determined by the characteristics of the surrounding matrix. Large, unfilled voids between clay balls increase potential settlements.

Soft, Cohesive Soil Fills. Slurried, fine-grained hydraulic fills are placed in confined areas. The dry solids content when the material is pumped in is typically in the range of seven to 25 percent. Sedimentation occurs during the first few weeks after placement. After decantation of the free water, the water content should be about 1.2 to 1.3 times the liquid limit at the surface, decreasing to about the liquid limit at a depth of 10 ft. (3 m) (Johnson et al., 1977).

Methods for Densification and Improvement

Almost all of the available ground improvement and modification methods may be used for the improvement of hydraulic fills. The selection of the appropriate ones for a project depends on detailed evaluation of the following factors:

1. Intended use of the treated ground, level of improvement needed
2. Area and depth of fill to be treated
3. Hydraulic fill material type
4. Availability of required special materials, equipment, and skills
5. Environmental considerations and constraints
6. Prior experience
7. Time available
8. Cost

Hydraulic fills may be viewed as freshly deposited, very young water-borne sediments. As a result, their densification and improvement is likely to be easier from the standpoint of structure breakdown and particle rearrangement than is the case for most natural sediments of comparable composition and density. This is because time-dependent strength gain and interparticle cementation may not yet be as well developed as for older sediments. On the other hand the work may be more difficult to execute owing to the poor equipment trafficability and support properties of the soft fill.

Suitable methods and their characteristics relative to treatment of hydraulic fills are summarized in Table 2. Detailed descriptions of these methods are given in readily available literature; e.g., ASCE

Table 2

SUMMARY OF METHODS FOR IMPROVEMENT OF HYDRAULIC FILLS

	METHOD	PRINCIPLE	MOST SUITABLE HYDRAULIC FILL TYPES	MAXIMUM EFFECTIVE TREATMENT DEPTH	SPECIAL MATERIALS REQUIRED	SPECIAL EQUIPMENT REQUIRED	PROPERTIES OF TREATED MATERIAL	SPECIAL ADVANTAGES AND LIMITATIONS	RELATIVE COST
DENSIFICATION DURING PLACEMENT	SURFACE COMPACTION	Compaction in layers as fill is placed	Fairly clean sands (< 15% passing No. 200 sieve)	N/A	None	Rollers, Bulldozers	High (85%+) relative density, good uniformity	Simple, low cost, work in the dry	Low
	UNDERWATER SURFACE COMPACTION	Tamping of fill surface during placement	Fairly clean sands (< 15% passing No. 200 sieve)	N/A	None	Barge mounted tampers, underwater compactors	High relative density, good uniformity	Simple, may interfere with fill placement operations	Low to Moderate
	INTERNAL DEWATERING BY PUMPING	Inward seepage forces give density increase, allow steeper side slopes, reduce surface erosion	Clean Sand	?	None	Suction probe, suction pumps	Increased density	In early stage of development	?
	SURFACE DRYING	Drying shrinkage causes density and strength increase, compressibility decreases	Soft, cohesive soils	1 m (?)	None	None	Higher strength, lower compressibility than when placed	Limited depth of effectiveness; very slow placement and improvement rates	Low

Table 2 - continued

SUMMARY OF METHODS FOR IMPROVEMENT OF HYDRAULIC FILLS

	METHOD	PRINCIPLE	MOST SUITABLE HYDRAULIC FILL TYPES	MAXIMUM EFFECTIVE TREATMENT DEPTH	SPECIAL MATERIALS REQUIRED	SPECIAL EQUIPMENT REQUIRED	PROPERTIES OF TREATED MATERIAL	SPECIAL ADVANTAGES AND LIMITATIONS	RELATIVE COST
DEEP DENSIFICATION BY VIBRATION	BLASTING	Shock waves and vibrations cause liquefaction and displacement, with settlement to higher density	Clean sands (< 15% passing No. 200 sieve)	> 30 m	Explosives, hole casings	Jetting or drilling machine	Relative densities to 70-80%; may get variable density; time-dependent strength gain	Rapid, inexpensive, can treat any size area; effective at large depth; variable properties; little improvement near surface; large ground vibrations	Low
	VIBRATORY PROBE	Densification by vibration; liquefaction induced settlement under water	Clean sands (< 15% passing No. 200 sieve)	20 m (Ineffective near surface)	None	Vibratory pile driver, or rod probe	Relative densities of up to 80%. Ineffective in some sands	Rapid, simple, good underwater; soft underlayers may damp vibrations; difficult to penetrate stiff overlayers	Moderate
	DYNAMIC COMPACTION	Repeated application of high intensity impacts at surface	Cohesionless soils, waste fills	15 m	None	Weights to more than 100 tons, high capacity crane with long boom	Can obtain good improvement and reasonable uniformity, improvement decreases with depth	Simple, rapid, suitable for some soils with fines; usable above and below water; economical for large areas; vibrations in adjacent structures	Low
DEEP DENSIFICATION AND REINFORCEMENT	VIBRO-COMPACTION	Densification by vibration and compaction of backfill material	Clean sand (< 20% passing No. 200 sieve)	30 m	Granular backfill, water supply	Vibroflot, crane, pumps	High relative densities, good uniformity, increased K_o	Uniformity, control, high level improvement	Moderate
	COMPACTION PILES	Densification by displacement of pile volume and by vibration during driving	Clean sands, silty or clayey sand	> 20 m	Pile material (often sand or soil plus cement mixture)	Pile driver, special sand pile equipment	High densities, good uniformity, piles act as reinforcement	Useful in soils with fines; uniform compaction; easy to check results; slow; limited improvement in upper 1-2 m	Moderate to high
	VIBRO-REPLACEMENT STONE AND SAND COLUMNS	Hole in soft, fine-grained soil backfilled with densely compacted gravel or sand	Soft silty or clayey sands, cohesive soils	20 m	Gravel or crushed rock backfill	Vibroflot crane	Increased bearing capacity, reduced settlements, columns act as reinforcements	Faster than precompression; sand pockets densified by vibration; drainage through columns; limited bearing capacity	Moderate to high

Table 2 - continued

SUMMARY OF METHODS FOR IMPROVEMENT OF HYDRAULIC FILLS

	METHOD	PRINCIPLE	MOST SUITABLE HYDRAULIC FILL TYPES	MAXIMUM EFFECTIVE TREATMENT DEPTH	SPECIAL MATERIALS REQUIRED	SPECIAL EQUIPMENT REQUIRED	PROPERTIES OF TREATED MATERIAL	SPECIAL ADVANTAGES AND LIMITATIONS	RELATIVE COST
PRECOMPRESSION	PRELOADING	Load is applied sufficiently in advance of construction so that compression of soft soil is completed prior to development of the site	Cohesive soils	---	Earth fill or other material for loading the site; gravel drainage blanket, vertical drains	Earth moving equipment; vacuum drainage systems sometimes used; settlement markers, piezometers	Reduced water content and void ratio, increased strength	Easy, theory well developed, uniformity; requires long time (vertical drains can be used to reduce consolidation time)	Low (Moderate if vertical drains are required)
	SURCHARGE FILLS	Fill in excess of that required permanently is applied to achieve a given amount of settlement in a shorter time; excess fill then removed	Cohesive soils	---	Earth fill or other material for loading the site; sand or gravel for drainage blanket	Earth moving equipment; settlement markers, piezometers, inclinometers	Reduced water content, void ratio and compressibility; increased strength	Faster than preloading without surcharge, theory well developed; extra material handling; can use vertical drains to reduce consolidation time	Moderate
	ELECTRO-OSMOSIS	DC current causes water flow from anode towards cathode where it is removed	Soft silts and silty clays	---	Anodes (usually rebars or aluminum), cathodes (well points or rebars)	DC power supply, wiring, metering systems	Reduced water content and compressibility, increased strength, electrochemical hardening	No fill loading required, can use in confined area, relatively fast; non-uniform properties between electrodes, no good in highly conductive soils	High
	SEEPAGE CONSOLIDATION	Seepage forces downward through hydraulic fill to underdrains cause consolidation	Soft clay and silt	a few m.	Underdrains beneath fill	Pumps	Reduced water content, compressibility and void ratio, increased strength	No fill loading required, variable properties with depth, requires long time	Moderate

Table 2 - continued

SUMMARY OF METHODS FOR IMPROVEMENT OF HYDRAULIC FILLS

	METHOD	PRINCIPLE	MOST SUITABLE HYDRAULIC FILL TYPES	MAXIMUM EFFECTIVE TREATMENT DEPTH	SPECIAL MATERIALS REQUIRED	SPECIAL EQUIPMENT REQUIRED	PROPERTIES OF TREATED MATERIAL	SPECIAL ADVANTAGES AND LIMITATIONS	RELATIVE COST
ADMIXTURES	STRUCTURAL FILLS	Structural fill distributes loads to underlying soft soils by acting as a mat foundation	Use over soft, cohesive hydraulic fills	---	Sand, gravel, flyash, bottom ash, slag expanded aggregate, shell incinerator ash, cement	Mixing and compaction equipment	Soft subgrade protected by structural load-bearing fill	High strength, good load distribution to underlying soft soils	Low to high
	MIX-IN-PLACE PILES & WALLS	Lime, cement, or asphalt introduced through rotating auger or special in-place mixer	All soft or loose hydraulic fills	> 20 m	Cement, lime, asphalt, or chemical stabilizer	Drill rig, rotary cutting and mixing head, additive proportioning equipment	Solidified soil piles or walls of relatively high strength	Uses native soil, difficult quality assurance	High
INJECTION AND GROUTING	PARTICULATE GROUTING	Penetration grouting-fill soil pores with soil, cement, and/or clay	Clean sand and gravel	Unlimited	Grout, water	Mixers, tanks, pumps, hoses	Impervious, high strength with cement grout, eliminate liquefaction danger	Low cost grouts, high strength, limited to coarse-grained soils, hard to evaluate	Lowest of the grout systems
	CHEMICAL GROUTING	Solutions of two or more chemicals react in soil pores to form a gel or a solid precipitate	Medium silts and coarser	Unlimited	Grout, water	Mixers, tanks, pumps, hoses	Impervious, low to high strength, eliminate liquefaction danger	Low viscosity, controllable gel time, good water shut-off; high cost, hard to evaluate	High to very high
	COMPACTION GROUT	Highly viscous grout acts as radial hydraulic jack when pumped in under high pressure	Heterogeneous, compressible hydraulic fills; clay balls in matrix	Unlimited	Soil, cement, water	Batching equipment, high pressure pumps, hoses	Grout bulbs within compressed soil matrix	Good for voids and soft spots	Low material, high injection
	JET GROUTING	High speed jets at depth excavate, inject, and mix stabilizer with soil to form columns or panels	Sands, silts, clays	Tens of m.	Water, stabilizing chemicals	Special jet nozzle, pumps, pipes and hoses	Solidified columns and walls	Precision in locating treated zones, slow	Expensive

(1978), Mitchell (1981), ASCE (1987). In Table 2 the different methods are organized into the following categories:

Densification during placement
Deep densification by vibration
Deep densification and reinforcement
Precompression
Admixtures
Injection and Grouting

Qualitative, comparative costs for the different methods are given in Table 2. More specific values for some of the methods, from Welsh (1987), are given in Table 3. The values assume a treatment depth of about 30 ft (10 m).

Table 3

COSTS OF SOME GROUND IMPROVEMENT METHODS
(Data from Welsh, 1987)

METHOD	Based on Volume of Treated Ground		Based on Length	
	$/yd³	$/m³	$/ft	$/m
Dynamic Compaction	0.5-2	0.6-2.6		
Vibro-compaction	1-5	1.3-6.5	5-12	15-40
Vibro-replacement stone columns	3-9	3.9-11.8	10-16	30-50
Compaction Grouting	20-150	26-200		
Slurry Grouting	30-60	40-80		
Chemical Grouting	120-400	160-525		
Jet Grouting	100-300	130-400	25-100	80-330
Surface Compaction	2-5	2.5-6.5		
Remove and Replace	8-15	10-20		
Freezing	200-500	260-650		

Several general observations may be made about each of the categories of improvement methods.

Densification During Placement. Surface compaction of fairly clean sand hydraulic fills during placement is likely to be easier, more effective, and less expensive than deep densification later. Vibratory rollers and heavy construction equipment working on the beach surrounding the hydraulic fill discharge pipes can maintain optimum surface profiles for fill spreading and deposition, and compact concurrently to high relative densities; e.g. 85-90 percent.

Underwater surface compaction has been little used, but is possible by several means, including surface pounders, tamping plates attached to the end of piles, and underwater vibrating plate compactors operated by barge mounted equipment. For example, a Dynapac unit was developed in Sweden (CM&E, 1976) that weighs 1.8 tons and develops a 9.6 ton vertical force on a 3 ft x 3½ ft (0.9 m x 1.1 m) plate at a frequency of 24 Hz.

Underwater roller compactors are a reasonable expectation for densification of large volume hydraulic fills in the future. A crawler-tractor type unit was designed and patented by the late Arthur Casagrande for use on rockfill and thick soil layers in water depths to more than 300 ft (100 m).

A new technique for underwater densification of clean sand during placement has been described by Hodge (1988). Pumping from within the fill creates inward seepage forces which produce a higher density, allow steeper side slopes, and reduce surface erosion.

Surface drying for consolidation and strengthening of fine-grained hydraulic fills, while inexpensive, takes a very long time and is relatively ineffective. A drying crust inhibits evaporation and acts like a mulching layer. The final crust thickness depends on underdrainage, vegetation, and climate. Months to years may be needed to form a crust of adequate strength and thickness to support workers and equipment. Stabilization by desiccation can be improved somewhat by surface sloping, surface trenching, and mechanical agitation. Transpiration by vegetation may be a very effective means for drying the near surface zone (Johnson et al., 1977). However, the improvement of soft, cohesive hydraulic fills to a level suitable for most subsequent land uses will generally require consolidation to considerably greater depths than can be reached by surface drying.

<u>Deep Densification by Vibration</u>. Vibratory methods are very well suited for the deep densification of clean sand hydraulic fills. Deep dynamic compaction is likely to be the simplest and most economical method for densification of layers up to 30 or 40 ft (10 to 12 m) thick. The current state of practice using dynamic compaction is described in detail by Lukas (1986).

Vibratory probe densification of clean sand hydraulic fills, using probes of various types suspended from vibratory pile driving hammers, has been very effective in several cases, with maximum relative densities to more than 80 percent being achieved. Two recent cases have been described by Leycure and Schroeder (1987) which illustrate both the magnitude of improvement and the distribution of relative density with depth. Fig. 1, from their paper, shows this distribution of relative density with depth after vibratory probe densification of a clean medium sand at two sites at the Port of Portland. Maximum density was developed at a depth corresponding to an effective overburden pressure of about 1150 psf (55 kPa).

The observed density distribution with depth is explained by Leycure and Schroeder (1987) on the basis that at shallow depths the confinement needed for maximum densification is not available. At large depth the higher confining pressure increasingly resists the breakdown of the sand structure that is needed to start the densification process.

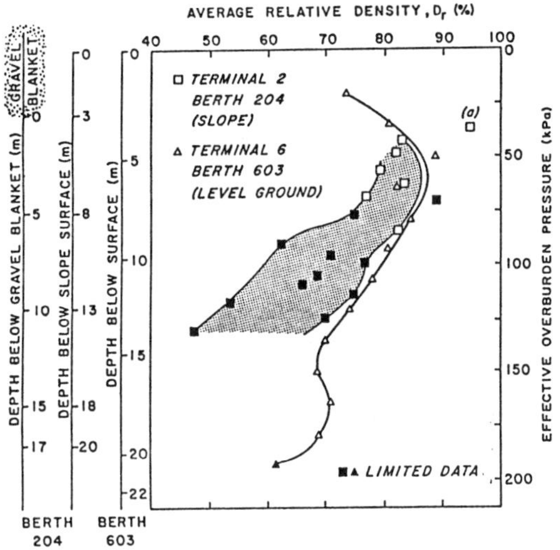

Figure 1. Relative Density Distribution as a Function of Effective Overburden Pressure and Depth after Densification of a Clean Sand Hydraulic Fill by Vibratory Probe Compaction (from Leycure and Schroeder, 1987)

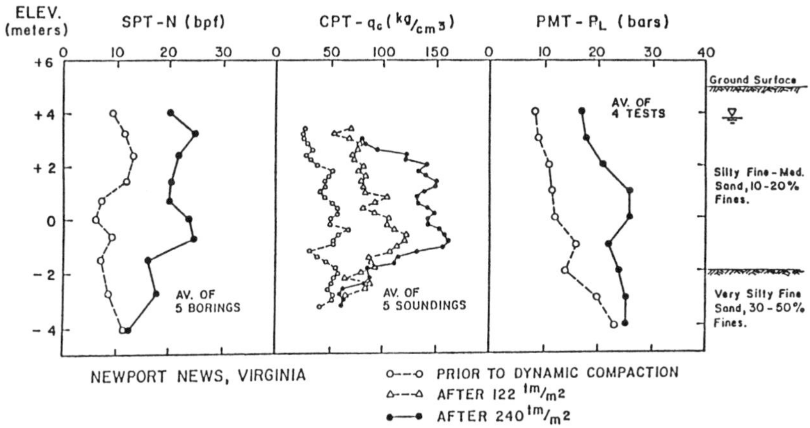

Figure 2. Variation of CPT Resistance, SPT Resistance, and Pressuremeter Limit Pressure with Depth after Densification by Dynamic Compaction (from Mayne et al., 1984)

A similar variation of properties with depth has been found for densification by dynamic compaction. Fig. 2 is an example. The same confining pressure effects are important as for vibratory probe densification. In addition, the amplitude of vibrations caused by the impacting weight diminishes with depth below the ground surface, and this further restricts the effective depth of densification by dynamic compaction.

Blasting may be an effective and economical means for densifying clean sand hydraulic fills, and may have application for special problems. It can be used at depths greater than reachable by other deep densification methods. Successful application to depths of 45 m (150 ft) at the Jebba Hydroelectric Development was described by Solymar (1984). Blasting was used to densify the hydraulic fill sand to a depth of 23 m (75 ft) inside the Molikpaq mobile arctic caisson. Headroom constraints beneath the deck of the caisson prevented the utilization of the methods for deep densification.

Deep Densification and Reinforcement. Vibro-compaction of cohesionless soils using a vibroflot and a backfill differs from vibratory probe compaction in that not only is the soil densified by vibration, but a densely compacted column of backfill is produced as well. Thus, the densified sand is further strengthened by the reinforcing action of the compacted sand or gravel piles.

Compaction piles and vibro-replacement stone columns are more suitable for hydraulic fills of silty or clayey sand and cohesive soil hydraulic fills. The importance of the reinforcing effect of the compaction piles and stone columns increases relative to the ground improvement due to densification by vibration as the fines content of the hydraulic fill increases.

An additional contribution to increased strength and liquefaction resistances as a result of deep densification and reinforcement of fairly cohesionless soils by vibratory and vibro-displacement methods is caused by an increase in the effective horizontal earth pressure coefficient, K_o.

Precompression. Precompression using preloading and surcharge fills has probably been the most widely used technique for the improvement of cohesive soil hydraulic fills. The use of sand drains for acceleration of the consolidation process has been almost completely supplanted by the use of prefabricated band-shaped drains. The theory and technology for precompression are both well-developed and comprehensively described in readily accessible literature; they are not discussed further here.

Electro-osmosis is a feasible means for consolidation of soft, fine-grained hydraulic fills. Except in special cases, however, the economics may not be favorable. One such case might be when time is not critical and wind power could be used for generation of the needed electrical power.

The potential for densification of fine-grained dredge spoil by seepage consolidation to underdrains was analyzed by Johnson et al. (1977). The effective stress profiles in a soil layer corresponding to different drainage and drying conditions are illustrated in Fig. 3. It may be seen that downward seepage forces may be used to develop significant effective stress for consolidation. Comparative costs were also developed by Johnson et al. (1977), and values, in 1976 dollars, are given in Table 4.

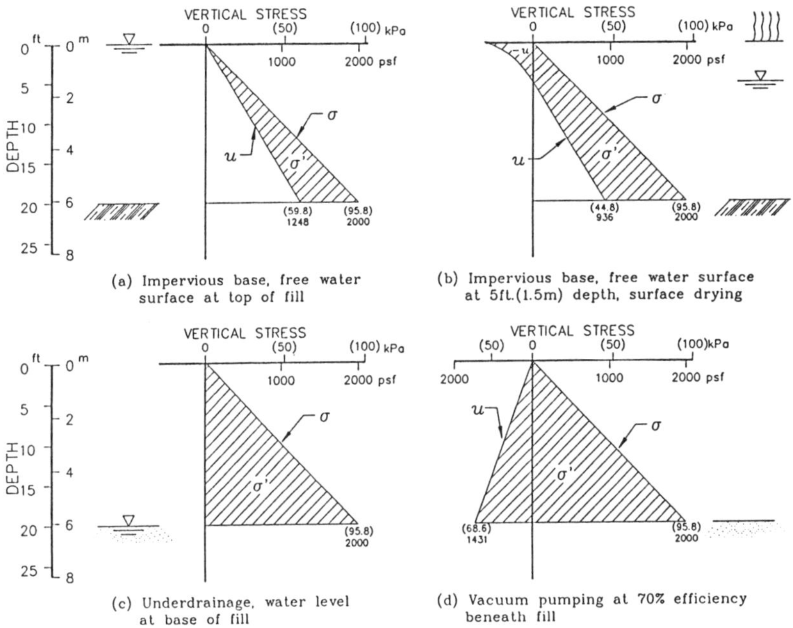

Figure 3. Vertical Stress Profiles at the End of Consolidation for Different Drainage Conditions in a Fine-Grained Hydraulic Fill. Total Unit Weight of Fill is 100 pcf (15.7 kN/m^3)

Table 4

APPROXIMATE PRECOMPRESSION COSTS FOR FINE-GRAINED HYDRAULIC FILLS
(Values from Johnson et al., 1977)

	Cost, in 1976 Dollars	
	Per Acre	Per Hectare
Surcharge Loading	8,000-16,000	20,000-40,000
Surcharge fill with vertical drains	15,000-30,000	37,000-75,000
Water surcharge and upper sand drainage blanket	13,000	32,000
Vacuum drainage from top	12,000	30,000
Underdrainage layer with collectors	3,000	7,500
Underdrainage with vacuum pumping	10,000	25,000
Seepage consolidation, underdrainage layer with collectors	3,000	7,500
Surface drying, with thin layer placement and trenching	2,000	5,000

It is evident from the data in Table 4 that seepage consolidation is very economical compared to other techniques, provided, of course, that the underdrain system is installed prior to hydraulic filling. It may be noted also that the use of vertical drains essentially doubles the cost of surcharge loading.

A unique method for improving the liquefaction resistance of a sand hydraulic fill beneath existing oil storage tanks in Kawasaki, Japan involved dewatering to increase the effective confining pressure (Ryan, 1988). The zone of soil to be strengthened by this form of precompression was surrounded by a slurry wall. The water level inside the encapsulated zone was then lowered until the increase in vertical effective stress was sufficient to give the hydraulic fill the required liquefaction resistance.

Admixtures. Structural fills produced by stabilizing the upper few feet of soil using cement, lime, or chemical admixtures can be used for support over soft hydraulic fills. Geotextiles may be used for the reinforcement of near surface layers.

The technology, available in Europe and Japan for a number of years, for construction of deep mix-in-place piles and walls was introduced into the U.S.A. in 1987 for foundation stabilization at Jackson Lake Dam in Wyoming. Equipment is available for construction of piles up to about a meter (3 ft) in diameter to depths up to 30 m (100 ft). Adjacent piles can be used to form walls or cells for encapsulation of liquefiable soil. The high strength of the piles themselves means that they can be used for direct support of structures over soft hydraulic fill.

Injection and Grouting. Permeation grouting using particulate and chemical grouts can be used to eliminate the liquefaction potential of cohesionless soils. Because of its high cost, however, permeation grouting is not likely to be used extensively in hydraulic fills. Jet grouting gives results similar to those obtained using mix-in-place piles and walls. The high cost of jet grouting, Table 3, precludes its large scale use in hydraulic fills except for local support of structures.

On the other hand, compaction grouting can be used to advantage in silty or clayey sand fills of large void ratio. Compaction grouting may have its greatest utility for filling voids in hydraulic fills composed of clay balls in a loose sand and clay matrix.

Time Effects in Hydraulic Fills

Many sands will develop significant stiffening and strength increase during the first several weeks following their deposition or densification (Mitchell and Solymar, 1984; Dowding and Hryciw, 1986; Schmertmann et al., 1986). As an example Fig. 4 shows the cone penetration resistance at several locations in a 10 m (30 ft) thick clean sand hydraulic fill placed in the Niger River during construction of the Jebba Dam. The fill was dredged from the river bottom and pumped in suspension to the discharge points. The left CPT record of each pair in Fig. 4 pertains to a CPT measurement between 4 and 10 days after fill placement; whereas, the right CPT record of the pair pertains to a CPT measurement between 50 and 80 days after fill placement. During the several week aging period the penetration resistance about doubled. Deep blasting, to depths of 45 m (150 ft), for densification of this river sand caused immediate densification, as evidenced by settlement of the ground surface. Significant increases in penetration resistance were not measured, however, until several weeks later, by which time the development of interparticle bonding had produced a strength increase (Mitchell and Solymar, 1984).

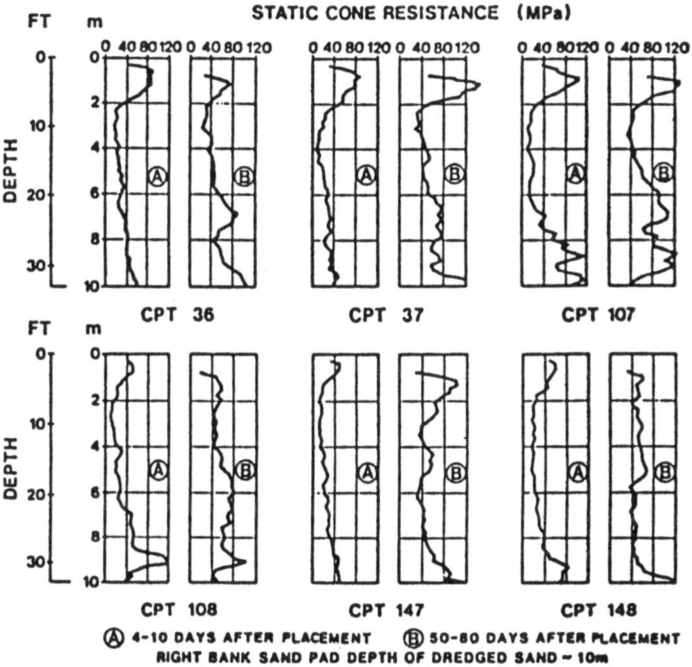

Figure 4. Effect of Time on the Cone Penetration Resistance of Hydraulic Fill at the Jebba Dam

The development of increased penetration resistance of a saturated sand densified by blasting under controlled conditions in a large laboratory chamber test is shown in Fig. 5 (Dowding and Hryciw, 1986). The significant increase in sand strength was attributed to both the time-dependent development of cementation between sand grains and the effects of bubbles from the explosive gases.

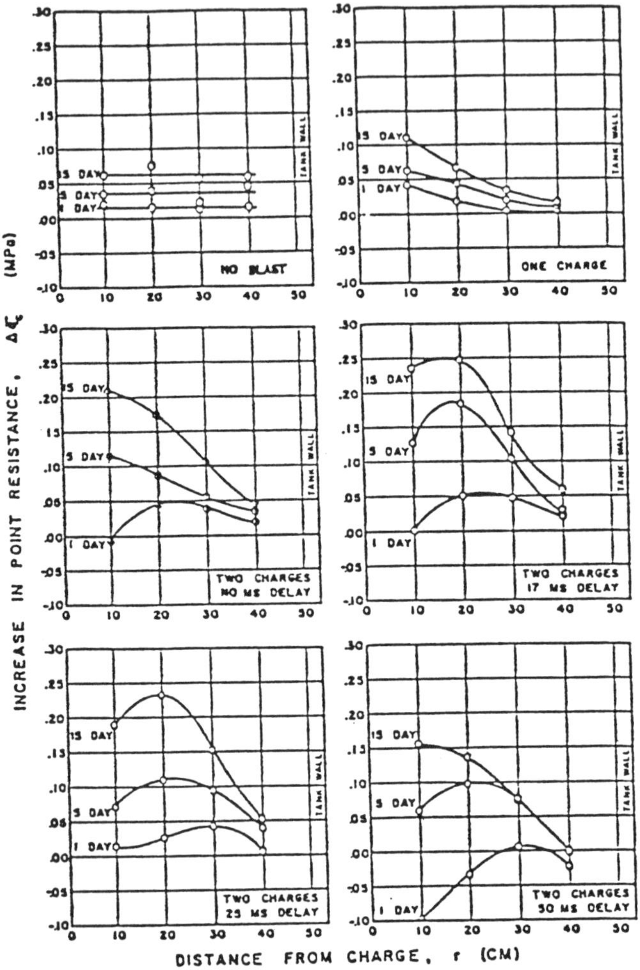

Figure 5. Time-Dependent Increase in Penetration Resistance of Sand Densified by Blasting in a Laboratory Chamber Test (from Dowding and Hryciw, 1986)

An excellent illustration of the time-dependent strength increase following densification of a medium to coarse sand hydraulic fill by dynamic compaction is given by Dumas and Beaton (1988). The clean sand fill was placed into the sea by end dumping. The average grain size of the sand is 0.79 mm, and the uniformity coefficient is 3.06. Increases in CPT resistance of up to 100 percent of the immediate post-compaction values were measured during the first 18 days following dynamic compaction.

Results of CPT tests at a point where the hydraulic fill thickness was 15 m (50 ft) and the dynamic compaction energy was 330 tm/m^2 are shown in Fig. 6. The very low values of penetration resistance prior to treatment by dynamic compaction imply that the sand fill must have been in a very loose state initially. The greater time-dependent strength increase near the surface and diminishing effect with depth are noted by Dumas and Beaton (1988). As this diminishing strength increase parallels the attenuation of dynamic stresses with depth during dynamic compaction, the authors suggest that the time-dependent strength increase may be related to the level of dynamic stress increase. It may be dependent on the static stresses as well, as are the time-independent strength increase and densification that result from ground improvement using a particular type and magnitude of dynamic input energy.

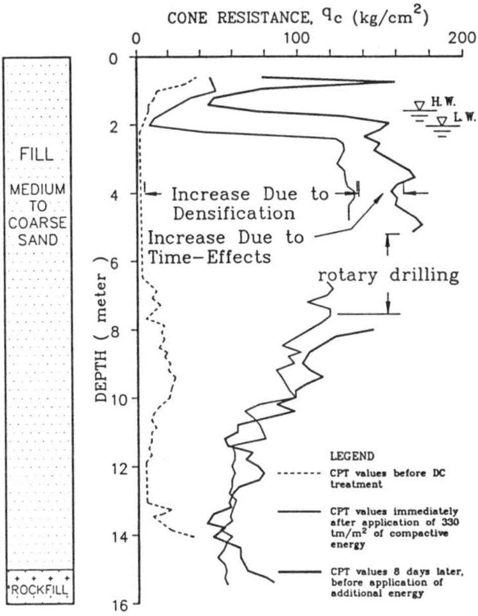

Figure 6. Results of CPT Testing at Impact Point No. 40-34 at the New Harbor Facility of Port Canada, Pointe Noire, Sept-Iles, Quebec Showing Strength Gains from Dynamic Compaction and Time Effects (redrawn from Dumas and Beaton, 1988)

Time-dependent strength gains in cohesionless hydraulic fills are of practical importance in the following ways:

1. The properties of freshly deposited or densified hydraulic fills of cohesionless soils are likely to improve with time even though there is no increase in density. This improvement can be assessed by penetration tests; however,

there presently exist no analytical means for predicting the magnitude and rate of improvement.

2. Densification of hydraulic fills of cohesionless soils is easiest during or shortly after their placement.
3. The quality assurance testing of hydraulic fills should be done at specified time periods after densification.

Case Histories

Three recent case histories illustrate several of the methods and considerations related to the densification and improvement of hydraulic fills that are discussed in the preceding sections. Two are concerned with the densification of predominantly cohesionless soils, and the other involves the improvement of a fine-grained hydraulic fill.

Harbor Bay Business Park. Approximately 10 million cubic yards of a silty fine sand hydraulic fill varying in thickness from about 20 to 27 ft (6.1 to 8.2 m) was pumped from barges onto a 320 acre (130 hectare) site adjacent to the Oakland International Airport in Alameda County, California during 1967 and 1968. Prior to filling a dike was constructed around the site to contain the hydraulic fill. Most of the dike was constructed by excavating and dumping soil from either side of the dike with a dragline. The lower part of the dike fill was dumped through water; the upper part was placed in the dry.

Bulldozers mounted on wide tracks were used to prevent inclusions of fines within the pumped sand fill. The dozers worked the surface of the sand fill behind the pipe discharge, pushing fines, when encountered, to disposal areas. While this procedure was successful in eliminating clay balls and significant lenses of fines from the sand fill, occasional layers of soft clay, consolidated from deposited fines, are present at the base of the sand.

The natural soil underlying the hydraulic fill is loose to medium dense silty and clayey sand except near the south corner of the site where up to 15 ft (4.6 m) of Bay mud is encountered. This silty and clayey sand has an average grain size of about 0.2 to 0.4 mm and 20 to 35 percent fines. The silty fine sand hydraulic fill has an average grain size of about 0.3 mm, a uniformity coefficient of about 6, and a fines content of about 8 to 16 percent.

Cone penetrometer tip resistance values in the upper part of the hydraulic sand fill are typically greater than 10 MPa, with values of 30 to 40 MPa measured in some places. These high values correspond to a dense to very dense material. Significant density increases could have been developed by the bulldozers working on the surface during placement of the fill. Additional strengthening could have resulted from time effects, from capillary stresses in the upper part of the fill, and from the precipitation of salts as brackish Bay water evaporated from the surface.

The lower part of the fill is less dense. CPT tip resistance values are generally less than 8 MPa, with a few values as low as 1 MPa. It is likely that these low values are associated with lenses of fine material. Some of the dike sand is also loose to medium dense, with CPT tip resistances from as low as 1 MPa to 4 MPa.

A program of dynamic compaction has been underway since 1985 along the dike to densify and strengthen the loose fill sands and underlying natural sand. The purpose of this densification is to prevent lateral spreading during a major seismic event. The densification is being done in strips, with the width of each strip varying from 75 ft to 100 ft (23 to 30 m). A plan of the project site is shown in Fig. 7, and a typical profile is shown in Fig. 8.

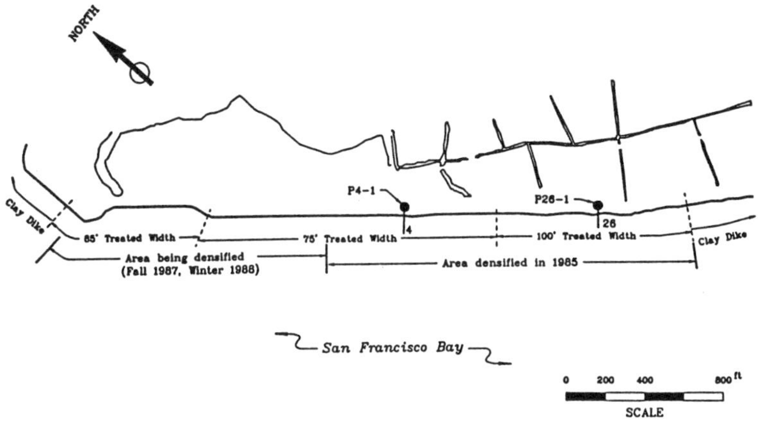

Figure 7. Plan of the Harbor Bay Business Park Site showing location of areas densified by Dynamic Compaction

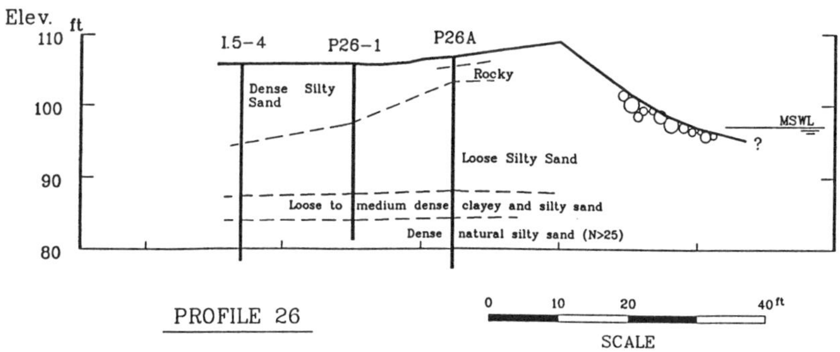

Figure 8. Typical profile through the dike area of Harbor Bay Business Park where hydraulic sand fill is being densified by Dynamic Compaction

For the densification that was done during 1985, a 20 metric ton 5 ft square pounder was dropped several times in each of several passes from a height of 30 m (100 ft). A total energy of 80 ton-meters per cubic meter of soil densified was used. This corresponded to 500 ton-meters per square meter of ground surface. The print pattern for a 90 ft (27 m) wide strip is shown in Fig. 9.

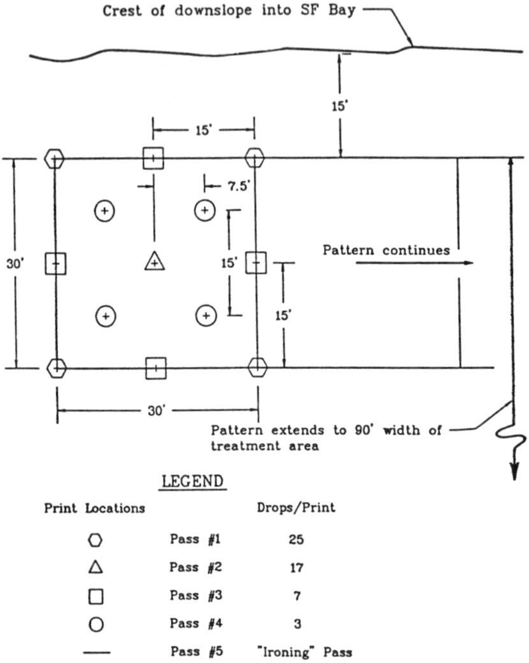

Figure 9. Dynamic Compaction print pattern used for densification of hydraulic fill along the dike at Harbor Bay Business Park; 1985 densification program.

The program presently underway (Fall 1987, Winter 1988) is similar, but the pounder weighs 18 metric tons and is dropped from a height of 24 m (80 ft). The greatest print spacing is 25 ft (7.6 m) instead of 30 ft (9.1m) as used in the 1985 densification. The specification calls for a mean penetration resistance after densification of 10 MPa in each CPT profile. No more than 10 percent of any sand layer shall have a tip resistance less than 8 MPa, and the mean value in any sand layer shall not be less than 9 MPa.

Cone penetration results for two locations from the 1985 densification program are shown in Fig. 10. CPT records are shown for the hydraulic fill prior to densification, shortly after the indicated pass of dynamic compaction, and five months later, after completion of the densification program, for CPT location P23-1 and 2-1/2 months later for CPT location P4-1. It may be seen from Fig. 10 that

(1) The penetration resistance was significantly reduced in the upper 7 ft (2-1/2 m) at CPT P26-1 as a result of dynamic compaction.

This layer, which was initially dense and strong, was loosened by the falling weight.

(2) Dynamic compaction resulted in a significant increase in penetration resistance in the hydraulic fill at depths greater than 7 ft (2-1/2 m) at CPT P26-1 and at all depths of fill except for a thin layer at about 4 to 5 ft (1½ m) depth at CPT P4-1.

(3) The improvement extended through the full depth of the hydraulic fill. Experience has shown that the maximum effective depth for densification by dynamic compaction is of the order of $0.5\sqrt{WH}$, where W is the pounder weight in tons and H is the height of drop in meters. For the above values of W and H this gives an effective depth of 12.2 m (40 ft), or twice the required depth of treatment. This meant that essentially all of the fill layer was above the depth where attenuation of the impact energy becomes large.

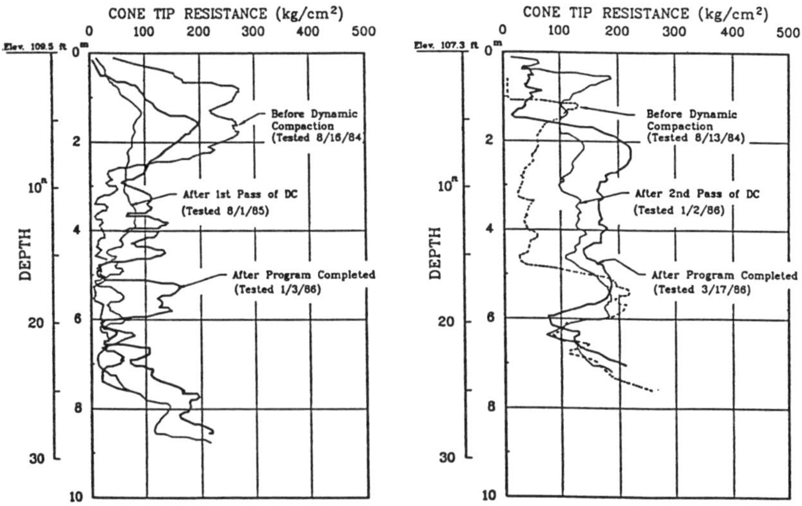

a. CPT No. P26-1 b. CPT No. P4-1

Figure 10. Cone penetration resistance profiles before and after dynamic compaction of silty fine sand hydraulic fill. (Courtesy of Hallenbeck and Associates)

Both open standpipe and porous tip piezometers were used to measure pore water pressures during the densification process. Sharp increases in pore water pressure were recorded immediately after the impact of the falling weight. As would be expected, the magnitude of this increase decreased with increasing distance from the impact point. In a number of cases the measured immediate pore pressure head exceeded the ground surface elevation, which meant that the impact had caused liquefaction. This was evidenced also by the drainage of water at the surface through cracks and fissures and by the formation of sand boils in the bottom of impact craters.

Dynamic compaction has been an effective and economical method for developing the required improvement of the silty fine sand at the Harbor Bay Business Park.

Terminal Island, Port of Los Angeles. Improvement of a 190 acre (75 hectare) hydraulic fill at Terminal Island in the Port of Los Angeles to a level suitable for its development posed quite different problems than existed at the Harbor Bay Business Park site. The hydraulic fill at Terminal Island is about 40 ft (12 m) thick and consists of silty fine sand in the lower portion and clayey silt in the upper part. In local areas the clayey silt is up to 30 ft (9 m) thick.

Dredging of Los Angeles Harbor to deepen the entrance channel and various basins provided 11,000,000 yd^3 (8,400,000 m^3) of hydraulic fill. The fill material consists of geologically recent alluvial sands, silts, and clays. It was placed between the existing shoreline of Terminal Island and a rock-faced perimeter dike around the east and south boundaries of the fill area. The location of the project is shown on the area map, Fig. 11. The initial water depth was about 20 ft (6 m), and the fill placement was done between 1981 and Spring 1983.

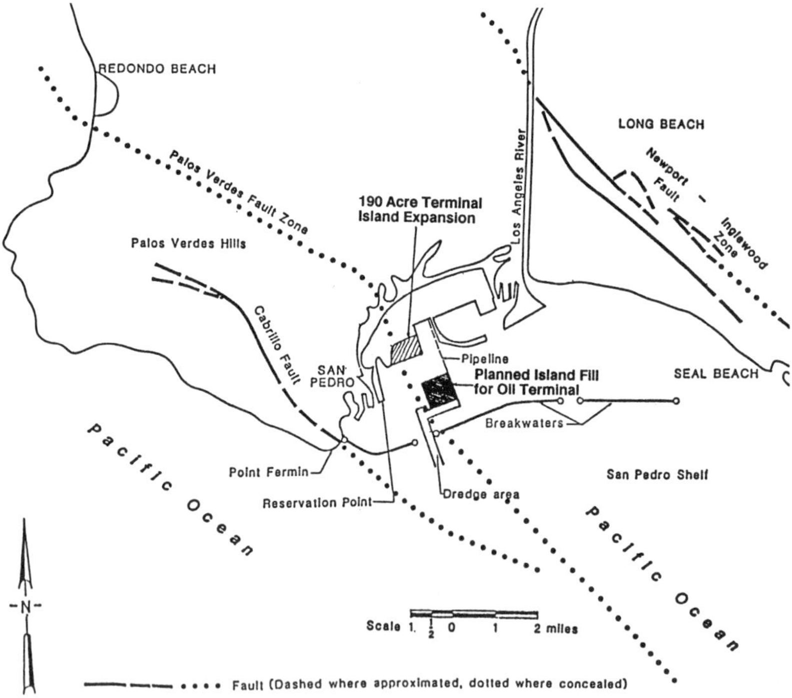

Figure 11. Area map showing location of hydraulic fills for Terminal Island expansion and Proposed Oil Terminal Terminal, Port of Los Angeles

The dredging sequence was such that the silty sand and clayey silt materials form separate strata within the vertical profile. Owing to the soft condition of the fill, the discharge point of the pipeline from the dredge could not be extended into the central portion of the area. As a result, the upper clayey silt stratum thickens towards the central portion of the site. The history, profile, and properties of this hydraulic fill are described and discussed by McNeilan and Skaggs (1988) in a separate paper to this conference. A generalized profile and grain size envelopes for the silty fine sand and clayey silt are shown in Figs. 12 and 13, respectively.

The problems to be overcome for satisfactory development of this site include large past and anticipated future surface settlements, very low bearing capacity, and the development of a blocky surface caused by desiccation cracking. The cracks, which created a surface of polygonal blocks about 3 ft (1 m) across, extended to depths of 5 to 7 ft (1.5 - 2.1 m) and were about 1 ft (0.3 m) wide by Spring 1987. Recorded and predicted settlements due to self weight of the fill for four locations are shown in Fig. 14 (McNeilan and Skaggs, 1988).

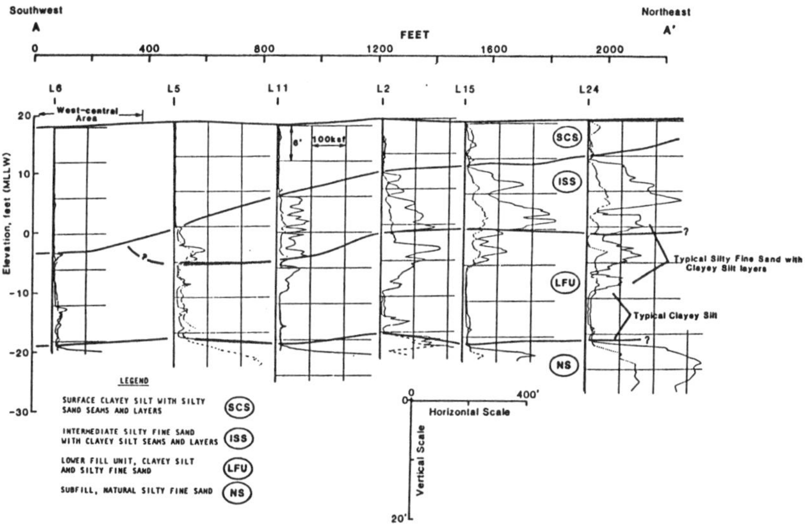

Figure 12. Generalized Soil profile through the Terminal Island hydraulic fill, Port of Los Angeles (from McNeilan and Skaggs, 1988)

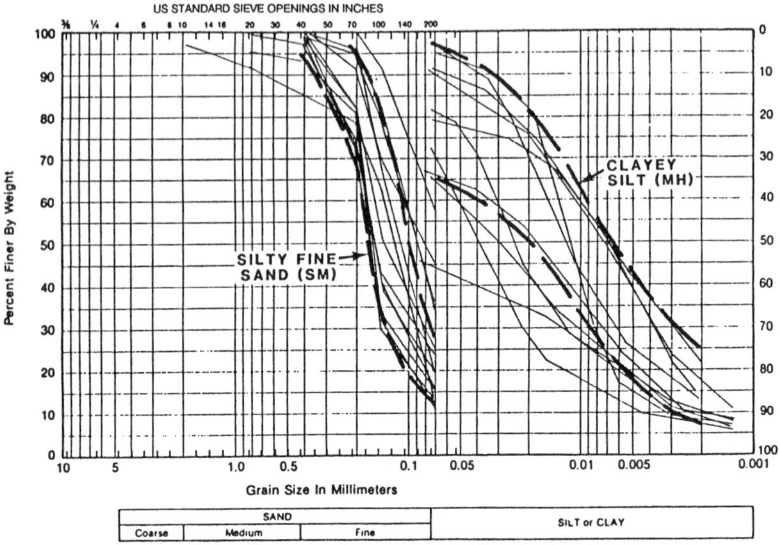

Figure 13. Grain Size envelopes for Terminal Island hydraulic fill, Port of Los Angeles (from McNeilan and Skaggs, 1988)

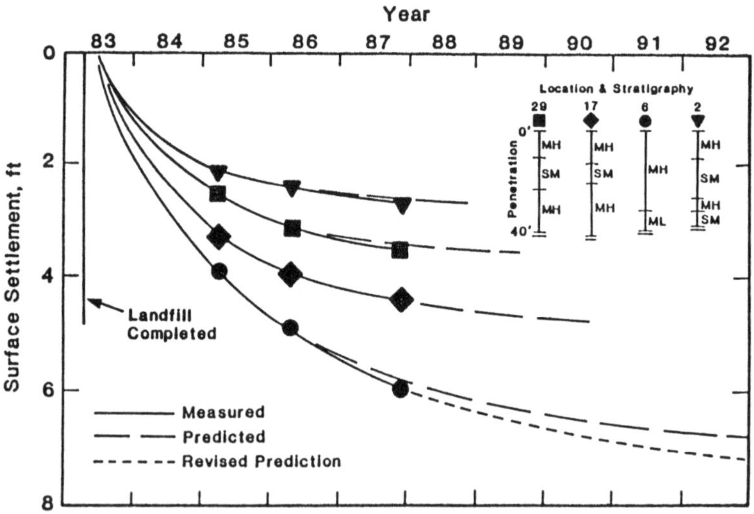

Figure 14. Recorded and predicted surface settlements due to self weight of the hydraulic fill at Terminal Island, Port of Los Angeles (from McNeilan and Skaggs, 1988)

An attempt was made to improve the 60 acre (25 hectare) western portion of the site in 1984 by excavating the surface layer, mixing it with sand, and replacing it. A mud wave developed during this process, and considerably more sand fill was required than had been anticipated. This area has been used as a storage yard for shipping containers.

The 122 acres (50 hectares) that are currently (1988) being improved are to be used initially for storage of bulk goods, such as coke, containers, and automobiles. In the future a new ship channel is to be dredged, and pile supported docks may be constructed outside the south perimeter dike. Terminal facilities will extend onto the site. Whereas the initial storage uses will require only vehicular access to the site, and substantial settlements can be tolerated, the ultimate development will include streets and utilities as well as settlement sensitive structures.

The undrained shear strength in the upper silty clay layer is of the order of only 200 to 400 psf (10 to 20 kPa). The resulting low bearing capacity means that fill and stockpile heights and side slopes must be carefully controlled to avoid foundation stability failures.

The geotechnical consultants, McClelland Engineers, Inc., evaluated all ground improvement techniques that might be suitable for this site. Consideration of the fine-grained character of much of the fill, the intended uses of the site and schedule for their implementation, and costs has led to the following program.

The large surface desiccation cracks have been filled with sand by spreading with a bulldozer. A vibratory roller was used to density the sand within the cracks to a relative compaction of 90 percent according to the ASTM D-1557 standard.

A 20 ft (6 m) high preload fill has been placed over the northern 30 acres (12 hectares) of the site. This will provide a preload bearing pressure of about 2000 psf (100 kPa), or 1 to 1.5 times the service load level to be applied in the future. Prefabricated band shaped vertical drains (wick drains) have been installed beneath the preload fill to achieve consolidation in about six months. The drain spacing is 6 ft (2 m), and 1×10^6 ft (3×10^5 m) of drain have been installed. Two to six feet (0.6 to 1.8 m) of settlement are anticipated under the preload fill, with the actual value depending mainly on the thickness of the clayey silt stratum.

After consolidation of the northern 30 acres has been completed, that portion of the preload remaining above the desired final grade (El. +19 ft) will be moved to a different portion of the site where the process will be repeated. This staged precompression program will produce an improved soil profile that will be suitable for the intended immediate uses. In the future when heavier and more settlement sensitive structures and facilities are to be constructed, either additional ground improvement or deep foundations at specific locations may be needed.

The ground improvement program that has been developed for the Terminal Island hydraulic fill includes several desirable features. The preloading method is known to be reliable and effective. It is simple in concept and execution, and the theory is well developed. It is an economical means for obtaining a modest, but uniform level of improvement over a large area. The initial preload section serves also as a test fill in which the effects of such parameters as drain spacing can be investigated and predictions of settlement rates and amounts and strength gains can be evaluated.

Proposed Oil Terminal, Port of Los Angeles. The third project also concerns a hydraulic fill in the Port of Los Angeles; however, the soil has yet to be placed. This means that there are opportunities to both place the hydraulic fill selectively and to improve it during or immediately after placement.

A 110 acre (45 hectare) artificial island fill to be used as an oil terminal is to be placed in Los Angeles Harbor at the location shown in Fig. 11. A deep water wharf is to be constructed along the south side of the island. The final ground elevation will range from +17 to +20 ft (+5.2 to 6.1 m). The water depth at the site is 33 to 40 ft (10 to 12.2 m).

Most of the soil to be used for the island will be a hydraulic fill obtained by harbor dredging. About 70 percent of the 11,300,000 cu yd (8,400,000 m^3) available from this source is fine sand (SP) or silty sand (SM), and the remainder is silt (ML) or sandy silt (ML).

The fill will be retained by containment dikes on all four sides of the island. Selective placement of the coarser and finer grained hydraulic fill materials is planned so that the better materials are deposited at the locations of structures and facilities. It has been assumed that the hydraulic fill will have a relative density averaging about 50 percent after placement. Tests and analyses have indicated that the sandy materials will require densification to a relative density of 75 percent or greater to prevent liquefaction and damage to facilities in the event of a large earthquake.

The design of the soil densification program has proceeded on the basis that the hydraulic fill will be placed from the harbor bottom to elevation +8 ft (+2.4 m) and then densified full depth. The designers recommend that the best and most cost effective means for this densification be determined by field test. Once the fill is above the water level it is to be compacted in lifts as it is placed. This procedure for handling the material above water can be expected to produce the maximum densification at least cost.

Several ground improvement methods have been considered for the deep densification of the sandy soils. Vibrocompaction, vibro-replacement stone columns, and dynamic compaction emerged as the three best candidates. Compaction grouting was also considered a suitable technique; however, cost considerations resulted in its elimination. The relatively high content of fines in the sandy soils leads to concern about the effectiveness of both vibrocompaction and dynamic compaction. Furthermore, the effective treatment depth for dynamic compaction using the largest readily available weights and cranes, 30 to 40 tons dropped from about 100 ft (30 m), is only about 30 ft (10 m). A thickness of fill greater than this must be densified.

It has been concluded, on the basis of these considerations, that of the three candidate methods only vibroreplacement stone columns will be capable of achieving the required site improvement for the range of materials to be included in the fill. The estimated cost for treatment by this method is about $10 per sq. ft. ($100 per sq. m.) or about $450,000 per acre ($1,125,000 per hectare). Because of this high cost, field tests are proposed to determine its effectiveness as well as that of other candidate methods.

This would appear to be a project where densification during placement could be both effective and economical. The fill is to be placed under controlled conditions, and the water depths are not excessive. Accordingly, underwater surface compaction, as discussed earlier in this paper, might be a suitable technique for consideration in addition to those discussed above.

Conclusion

Cohesionless hydraulic fills have low relative density, low strength, and high liquefaction potential when placed. Fine-grained hydraulic fills have a very high water content, high compressibility, and low strength. Consequently, densification and improvement of hydraulic fills are required before they can be used for most purposes.

Almost all of the available ground improvement and modification methods can be used for the improvement of hydraulic fills. The characteristics of these methods are summarized in Table 2. Densification by vibration has been the most commonly used technique for improvement of cohesionless hydraulic fills, whereas, precompression has been used the most for fine-grained hydraulic fills. Surface compaction of clean sand hydraulic fills during placement is usually easier, more effective, and less expensive than deep densification later. Underwater compaction during placement appears not to have been much used but warrants consideration for new projects. Seepage consolidation of fine-grained hydraulic fills can be both effective and economical provided underdrains can be installed prior to placement of the fill.

Hydraulic fills are very young sediments. Significant stiffening and strength increase may develop in cohesionless hydraulic fills during the first several weeks following their placement or densification. Several examples have been shown in this paper.

Three recent case histories have been described that illustrate some of the methods discussed. In any case the most suitable treatment method will depend upon the soil type, especially the fines content, the thickness of material to be treated, the level of improvement required, the time available, and costs.

Acknowledgments

The author is indebted to several individuals for supplying information and data used in the case histories section of this paper. These individuals and their organizations are:

Dr. Hei-Yip Lee and Curtis N. Jensen - Hallenbeck and Associates, Emeryville, CA

Thomas W. McNeilan and Ronald L. Skaggs - McClelland Engineers, Inc., Ventura, CA

Dr. Donald W. Quigley - Harding Lawson Associates, Tustin, CA

Several of the drawings were prepared by Dar-Jen Tseng, Graduate Student in Geotechnical Engineering at the University of California, Berkeley.

References

ASCE (1978) "Soil Improvement, History, Capabilities and Outlook", Report by the Committee on Placement and Improvement of Soils.

ASCE (1987) "Soil Improvement - A Ten Year Update", Geotechnical Special Publication No. 12.

CM&E (1976) Methods Memo, p. 132, April.

Dowding, C. H. and Hryciw, R. D. (1986) "A Laboratory Study of Blast Densification of Saturated Sand", Journal of Geotechnical Engineering, ASCE, Vol. 112, No. 2, pp. 187-199.

Dumas, J. C. and Beaton, N. F. (1988) Discussion to "Practical Problems from Surprising Soil Behavior", by J. K. Mitchell, Journal of Geotechnical Engineering, ASCE, in press.

Hodge, W. E. (1988) "Underwater Sand Fills", Hydraulic Fill Structures: ASCE Geotechnical Division Specialty Conference, Fort Collins, CO.

Johnson, S. J., Cunny, R. W., Perry, E. B. and Devay, L. (1977) "State-of-the-Art Applicability of Conventional Densification Techniques to Increase Disposal Area Storage Capacity", Technical Report D-77-4. Soils and Pavements Laboratory, U.S. Army Engineer Waterways Experiment Station, Vicksburg, Miss.

Leycure, P. and Schroeder, W. L. (1987) "Slope Effects on Probe Densification of Sands", Soil Improvement - A Ten Year Update, ASCE Geotechnical Special Publication No. 12, pp. 197-214.

Lukas, R. G. (1986) "Dynamic Compaction for Highway Construction, Vol. 1: Design and Construction Guidelines", Federal Highway Administration Report No. FHWA/RD-86/133, 230 pp.

Mayne, P. W., Jones, J. S. and Dumas, J. C. (1984) "Ground Response to Dynamic Compaction", Journ. of the Geotechnical Division, ASCE, Vol. 110, No. 6, pp. 757-774.

McNeilan, T. W. and Skaggs, R. L. (1988) "In-Place Properties of 7,000,000 cu yd Hydraulic Landfill", Hydraulic Fill Structures: ASCE Geotechnical Division Specialty Conference, Ft. Collins, CO.

Mitchell, J. K. (1981) "Soil Improvement: State-of-the-Art", Proc. Tenth International Conference on Soil Mechanics and Foundation Engineering, Stockholm, Vol. 4, pp. 509-565.

Mitchell, J. K. and Solymar, Z. V. (1984) "Time-Dependent Strength Gain in Freshly Deposited or Densified Sand", Journ. of Geotechnical Engineering, ASCE, Vol. 110, No. 11, pp. 1559-1576.

Ryan, C. R. (1988) "The Fringes of Technology in Geotechnical Construction", Lecture to the Geotechnical Group of the San Francisco Section, ASCE, January 14, 1988.

Schmertmann, J., Baker, W., Gupta, R. and Kessler, K. (1986) "CPT/DMT QC of Ground Modification at a Power Plant", in Use of In Situ Tests in Geotechnical Engineering, Geotechnical Special Publication No. 6, ASCE, pp. 985-1001.

Solymar, Z. V. (1984) "Compaction of Alluvial Sands by Deep Blasting", Canadian Geotechnical Journal, Vol. 21, No. 2, pp. 305-321.

Welsh, J. P. (1987) GKN Hayward Baker Ground Modification Seminar Notes.

Whitman, R. V. (1970) "Hydraulic Fills to Support Structural Loads", Journ. Soil Mechanics and Foundations Division, ASCE, Vol. 96, No. SM1, pp. 23-47.

Hydraulic Fill Embankments Utilizing Geosynthetics

Robert M. Koerner* and Bruce L. Uibel**

Abstract

The use of geosynthetic materials (geotextiles or geogrids for reinforcement and vertical strip drains for drainage) is revolutionizing the stabilization of soft foundation soils. Two separate classes are considered in this paper; one type being a large areal fill, the other being a long linear fill. While many similarities exist between the two, there are also decided differences. These aspects are described on the basis of their various design elements. Case histories are noted with particular emphasis on a 2400 m (8000 ft) long linear embankment under construction in Wilmington Harbor, Wilmington, Delaware. This particular project utilizes a 330 kN/m (1900 lb/in) geotextile with strip drains at 3 m (10 ft) centers. The dredged embankment is being placed directly on the fabric to a height of approximately 9 m (30 ft). It is the largest geosynthetic stabilized project of this type constructed to date.

Introduction

Fine grained saturated soils exist near most river estuaries and harbor areas around the world. Unfortunately, these are precisely the areas where industralization is the heaviest. Heavy buildings, factories, freight yards, stockpiles, storage tanks, railroads, heavy equipment, etc., etc., are all incompatible with such foundation conditions. The traditional options given to the owners of such facilities have been, (a) deep foundations placed through the unsuitable soils, (b) excavation and replacement with suitable soil, (c) stabilization with injected additives, or (d) wait until natural consolidation occurs. All of these methods have a certain degree of applicability, but suffer from being either expensive or time consuming.

The newly emerging alternate method (which is the focus of this paper) is to deploy a high strength geotextile or geogrid over the site, place a drainage layer/working blanket above it, install vertical strip drains to the bottom of the soft layer, and then complete the surcharge fill up to final grade. The sand blanket and surcharge layers can be placed by conventional earthmoving equipment or by dredging. As will be illustrated later, the entire process can be accomplished under water if the need arises.

*H. L. Bowman Professor of Civil Engineering and Director Geosynthetic Research Institute, Drexel University, Philadelphia, PA 19104
**Former-Chief, Geotechnical Section, U. S. Army Corps of Engineers, Philadelphia District, Philadelphia, PA 19106

Clearly, the geotextile or geogrid acts as a reinforcement material since the foundation soils are often less than 5 to 10 kPa ($\simeq$ 1.0 lb/in^2) which will hardly support the weight of an individual. The vertical strip drains (often called prefabricated vertical drains or wick drains) are geosynthetic composites and are used to drain the excess pore water from the foundation soil as it is mobilized by the surcharge fill. While not used to date on a large scale, sheet drains made from geosynthetics could be used for drainage layers if adequate granular soils at the site are not available. The final surcharge fill is usually from local soils.

Regarding configurations of these projects, there are two somewhat different variations. One is a large areal fill where the length and width of the site are approximately equal. In such cases there is no clearly defined principal stress direction and the strength of the geotextile or geogrid must be equally balanced in all directions. This, of course goes for seams in both directions as well as the material itself. Actual situations in this category are often industrial or office site development projects. The second variation is one in which the length of the fill is much larger than the width, i.e., a linear fill. In these cases the major principal stress direction can be identified and the geosynthetic reinforcement can be properly aligned. Seams can often be avoided, or placed in opposite, hence lower stress, directions. Situations in this category are roadway embankments and containment dikes. The Wilmington Harbor South Project which will be described later is in this category.

Types of Geosynthetics Used

Geosynthetics represent a rapidly growing technical area which draws from polymer, textile and geotechnical engineering disciplines in attempting to solve a wide range of soil-related problems.(1) Emphasis here is only on those products used in soft soil stabilization projects.

Geotextiles - Geotextiles consist of polymeric fibers made into permeable fabrics by conventional weaving or by nonwoven techniques. There are an extremely large number of geotextiles which are commercially available. Their tensile strength is usually assessed for design by a wide width tensile test which has been standardized, i.e. ASTM D-4595. The resulting strength varies from very low values up to 700 kN/m (4000 lb/in) and higher. The fabrics are generally seamed from one panel to another by conventional sewing but it must be realized that the strength of a sewn seam is always less than that of the fabric itself. See Figure 1 where it can be seen that a limit is reached whereby additional fabric strength accomplishes very little. Thus, unless they can be avoided, seams become limiting constraints.

The design of the geotextile reinforcement utilizes adaptations of conventional geotechnical design(2) as shown in the sketches of Figure 2. Each of these

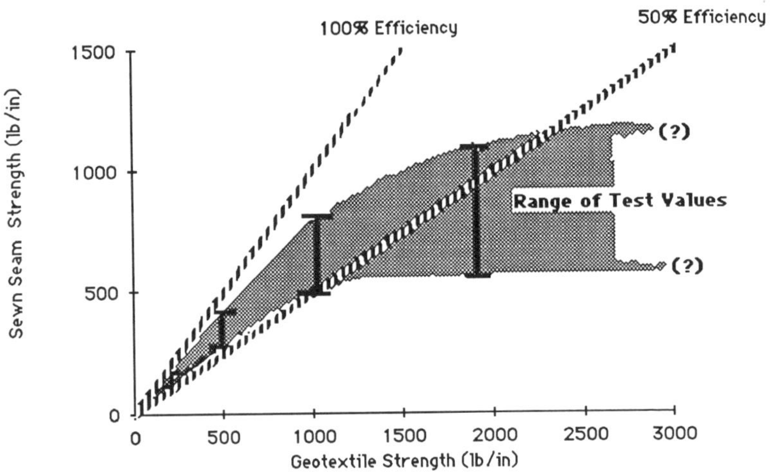

Fig. 1. - Performance Behavior of Geotextile Seams as Determined from Wide Width Tensile Tests and Proposed Seam Tests

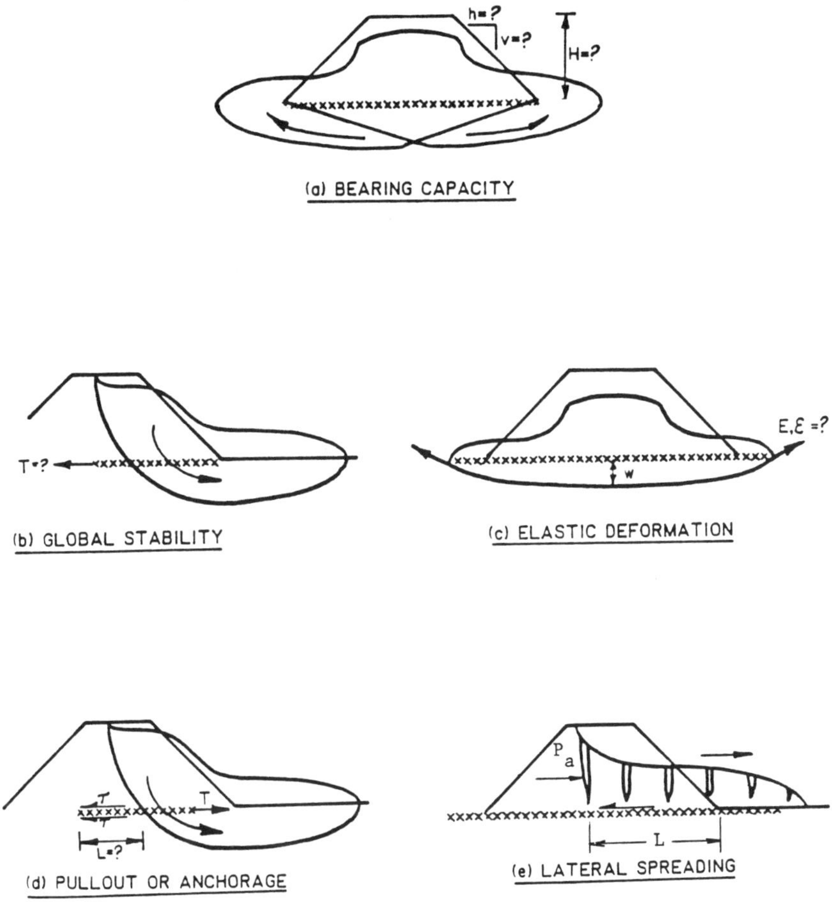

Fig. 2. - Geotextile Design Models for Use in Soft Soil Stabilization, after Koerner, et al.(2)

separate analyses results in a required fabric property or geometric property. Soil properties of unit weight, strength and consolidation characteristics must obviously be known in order to obtain a confident design. The option to these design models based on plastic equilibrium is finite element methods which has recently been reviewed by Rowe and Soderman.[3]

Geogrids - Geogrids are relatively stiff, gridlike materials with large apertures made from a variety of manufacturing methods, see Table 1. They reinforce a soil mass quite differently from geotextiles which mobilize strength from surface friction. Geogrids primarily resist tensile stress by soil bearing against their transverse ribs. The stress is transferred through the interconnecting junctions to the longitudinal ribs thereby resisting the applied stress. Thus junction strength is important and a major variable in these products.[4] Note should be made that friction also occurs along the geogrid's surface, but is a much smaller component than the bearing stress as just described.

Strip Drains - Rapid consolidation of fine grained saturated soils using radial drainage techniques has progressed from sand drains, to a number of intermediate variations, to the current use of strip drains. Also called "wick drains" and "prefabricated vertical drains", these geosynthetics generally consist of a fluted core to carry the expelled water and a geotextile sheath which acts as a filter. As seen in Table 2, however, there are a number of available products. Design procedures are adaptions of conventional consolidation theory using radial flow of the expelled pore water.[5,6] The required flow rate of the strip drains and some considerations regarding their deformed configurations have also been described.[7]

Sheet Drains - Large panels of geosynthetic drains could possibly be used as blanket drains if granular soil is not locally available. Many products are available in the drainage composite and geonet categories, see Table 3. These panels would be placed on top of the geotextile or geogrid reinforcement and be covered by a lightweight geotextile acting as a filter. Local soil (not necessarily granular) would then be placed above in order to provide an adequate working blanket. The problem envisioned with this type of configuration is that the soft underlying soil will extrude through the holes in the geotextile/geocomposite drain and not allow for hydraulic communication with the strip drain. This would require predug holes in the soil cover and backfilling them with sand or gravel. Thus flow communication from the strip drain to the geocomposite drain could be assured. It is, however,

Table 1 - Physical Properties of Geogrids

MANUFACTURER/AGENT	PRODUCT NAME	STRUCTURE	POLYMER	JUNCTION METHOD	Aperature Size MD/XMD (in)	Thickness (mils) @ rib/@ junction
ITW/Signode	TNX 250	intersecting strips	polyester	ultrasonic	2.5/2.5	30/60
	TNX-800	intersecting strips	polyester	ultrasonic	-	-
	TNX 5001	intersecting strips	polyester	ultrasonic	1/3.5	30/60
	TNX 8001	intersecting strips	polyester	ultrasonic	-	-
	CTX 120	intersecting strips	polypropylene	ultrasonic	2.5/2.5	26/54
	CTX 1255	intersecting strips	polypropylene	ultrasonic	1/2.5	22/72
Huesker Synthetic	HaTeGrid 80/30-20	woven	Polyester	entangled	.8/.8	-
	HaTeGrid 40/40-20	woven	Polyester	entangled	.8/.8	-
	HaTelit 30/13	woven	Polyester	entangled	1.2/1.2	-
Mirafi	Miragrid 5T	woven and knit	polyester-latex coated	knit stitch	1.2/1.3	80/85
ICI (Mirafi)	Paragrid 100/25S	intersecting strips	polyester-PP coated	melt bonded	5.5/2/5	100/170
Tenax	TT2	punched sht drawn	polyethylene	unitized	4.3/0.6	-
	TT3	punched sht drawn	polyethylene	unitized	4.9/.60	-
	TT4	punched sht drawn	polyethylene	unitized	5.3/.6	-
	MS3	punched sht drawn	polypropylene	unitized	-	-
	MS5	punched sht drawn	polypropylene	unitized	-	-
Tensar	SR1/UX1100	punched sht drawn	polyethylene	unitized	3.9/.6	20/80
	SR2/UX1200	punched sht drawn	polyethylene	unitized	3.9/.6	55/180
	SR3/UX1300	punched sht drawn	polyethylene	unitized	3.9/.6	70/230
	UX1400	punched sht drawn	polyethylene	unitized	5.7/.6	30/110
	UX1500	punched sht drawn	polyethylene	unitized	5.7/.6	50/170
	UX1600	punched sht drawn	polyethylene	unitized	5.4/.6	70/230
	WV/UX2100	punched sht drawn	polyethylene	unitized	1.1/.2	-/60
	SF/UX3109	punched sht drawn	polyethylene	unitized	4.3/1	-/80
	SB/UX4000	punched sht drawn	polyethylene	unitized	3.9/.6	-/80
	SSO/BX1000	punched sht drawn	PP/HDPE Copolymer	unitized	1/1.3	21/80
	SS1/BX1100	punched sht drawn	polypropylene	unitized	1/1.3	30/110
	SS2/BX1200	punched sht drawn	polypropylene	unitized	1/1.3	40/150
	SS3/BX1300	punched sht drawn	polypropylene	unitized	1.8/2.5	50/170
	GS1/BX2000	punched sht drawn	polypropylene	unitized	1.3/1.3	-/110
	GM1/BX3110	punched sht drawn	polypropylene	unitized	2.0/2.0	62/210

Table 2 - Properties of Geocomposite Strip Drains

Manufacturer or Agent	Product Name	Core Characteristics				Filter Type (cover)
		Material	Shape	Number of Channels	Size (mm)	
SGI	Kjellman	Paper	Straight channels	10	100 x 3.5	Paper
SGI	Geodrain	Plastic	Straight channels	27	92 x 4.0	Paper
Geosystems, Inc.	Alidrain	Plastic	Studded-2 sides	-	100 x 6.0	Geotextile
	Alidrain "S"		Studded-1 side	-	-	
	Aliwick		Straight channels	19	1	
Japanese (Harguim)	Castle	Plastic	Straight channels	36	94 x 2.6	Unitized
Japanese (Fukuzawa)	Bando	Vinyl chloride	Straight channels	48	96 x 2.9	Unitized
Dutch	Mebra Drain	Polyethylene	Straight channels	38	-	Geotextile
	Ester				95 x 3.3	Polyester
	Pap				95 x 3.2	Paper
	Prop				95 x 3.4	Polypropylene
	Sol				95 x 2.0	Polypropylene
-	Colbond	Geotextile	Needled, nonwoven	-	100 x 6.0	Geotextile
-	PVC	PVC	Straight channels	11	100 x 1.5	Unitized
ICE	Amerdrain	Polyethylene	Straight channels	8	92 x 10.0	Geotextile
Burcan	Hitek	Polyethylene	Waffle	-	100 x 6.0	Geotextile
Recosol	Desol	Polyolefine	Straight channels	24	92 x 2.0	Unitized

Table 3(a) - Properties of Available Geocomposites

Manufacturer/Agent	Product Name	Core Structure	Core Polymer	Geotextile	Roll Size (ft) width/length	Thickness (mils)	Crush Strength (psi)	Transmissivity (gal/min/ft) @ 1.45 psi	@ 14.5 psi
American Wick Drain	Amerdrain 480 mat	Nippled core	polyethylene	polypropylene	4/104	375	86	18	16
BASF Corp	Enkamat 7010	Open core	nylon 6	-	3.2/492	400	-	-	-
	Enkamat 7020	Open core	nylon 6	-	3.2/330	800	-	-	-
	Enkamat 9010	Open core	nylon 6	polyester	3.0/99	400	-	-	-
	Enkamat 9120	Open core	nylon 6	polyester	3.0/99	800	-	-	-
Burcan Industries	Hitek 6c	cuspated core	polyethylene	polypropylene	3.0/450	255	69	26	2.1
	Hitek 20c	cuspated core	polyethylene	polypropylene	3.6/125	785	35	9.6	9.6
	Hitek 40c	cuspated core	polyethylene	polypropylene	3.5/80	1575	17.5	22	-
Exxon	Tiger Drain	cuspated core	polyethylene	polypropylene	4/100	600	38	9	8
Geotech Systems	insul. drain board	EPS panel	-	-	4.0/4.0	1000	5.5	2.3	1.25
Huesker Synthetic	Ha Te-Drainmatte		polypropylene	PES	13/328	260	-	-	-
JDR Enterprises	J-Drain 100	extruded rib	polyethylene	polypropylene		250		7.2	5.8
	J-Drain 200	extruded rib	polyethylene	polypropylene		250		3	1.8
Mirafi	Miradrain 6000	extruded rib	polystyrene	polypropylene	4/8,25,50	380	104	15	-
	Miradrain 9000	extruded rib	polystyrene	polypropylene	4/8,25,50	380	125	15	-
	Miradrain 4000	extruded rib	polystyrene	polypropylene	4.0/8.0	750	30	5	-
Monsanto	Hydraway Drain	raised cyl. tubes	polyethylene	polypropylene	12,18/400	1000	95	20.3	19.6
Nilex	Nudrain A		polyethylene	polypropylene	1.6/49,98	1575	18.8	28.5	-
	Nudrain C		polyethylene	polypropylene	3.6/98	787	34.7	24.1	-
NW Fabrics	Permadrain	cuspated	polyethylene	polypropylene	any size	-	28	-	-
Pro Drain Systems	PDS 20	cuspated	polyethylene	polypropylene	3.7/10-500	750	-	9.6	9.6
	PDS 40	cuspated	polyethylene	polypropylene	3.3/10-250	1500	-	22	22
Tensar	DC 1100	extruded rib	polyethylene	polypropylene	5.3/100	230	-	5.5	4.5
	DC 1200	extruded rib	polyethylene	polypropylene	5.3/100	240	-	4	3

Note: Transmissivity values at 1 = 1.0

Table 3(b) - Properties of Available Geonets (all polytheylene)

Manufacturer/Agent	Product Name	Structure	Roll Size (ft) width/length	Thickness (mils)	Crush Strength (psi)	Transmissivity (gal/min/ft) @ i = 1.0	
						@ 1.45 psi	@ 14.5 psi
Carthage Mills	FX-2000 Geo-Net	extruded ribs	7.5/300	200	138	12.7	12.5
	FX-2500 Geo-Net	extruded ribs	7.5/300	250	138	16.6	14.8
	FX-3000 Geo-Net	extruded ribs	7.5/220	300	138	20.8	20.8
Conwed Plastics	XB8110	extruded	6.9/300	250	130	16.4	14.5
	XB8210	extruded	6.9/300	160	130	9.2	8.7
	XB8310	extruded	6.9/300	200	130	13	12.6
	XB8410	extruded	6.9/220	300	130	20.8	20.8
	XB8315CN	extruded	6.9/300	200	130	12.6	12.6
	LXE-9000	thermoformed net	4/100	600	40	37.2	37.2
Fluid Systems Inc.	(Tex-Net) TN-1001	extruded	7.5/300	250	130	5.25	4.38
	TN-3001	extruded	7.5/300	200	130	4.38	4
	TN-4001	extruded	7.5/300	300	130	7.8	6.5
	TN-3001CN	extruded	7.5/300	200	130	4.38	3.72
	(Poly-Net) PN-1000	foamed & extruded	6.75/300	250			
	PN-2000	extruded	6.75/300	160			
	PN-3000	extruded	6.75/300	200			
	PN-4000	foamed & extruded	6.75/300	300			
Geo-synthetics	GSI Net 100	foamed & extruded	-	250			
	GSI Net 200	extruded	-	160			
	GSI Net 300	extruded	-	200			
Gundle	Gundnet XL-1	extruded ribs	6.2/100	250		6.5	6
	Gundnet XL-3	extruded ribs	6.2/100	200		6	5.5
Low Brothers & Co	Lotrak 8	extruded mesh	6.6/164	120			
	Lotrak 30	extruded mesh	6.6/164	200			
	Lotrak 70	extruded mesh	6.6/164	290			
Tenax	CE 1	extruded ribs	4.8/66	250			
	CE 2	extruded ribs	12.4/82	200			
	CE 3	extruded ribs	7.4/82	160			
Tensar	DN1-NS1100	extruded ribs	5.2/98	220		6	5
	DN3-NS1300	extruded ribs	6.2/98	150		2.2	1.8
	- NS1400	extruded ribs	6.2/98	200		6	5

an extra and costly additional step.

Past Projects

Clearly, the Corps of Engineers has been the leading force behind the use of high strength geotextiles to reinforce very soft foundation soils.(8-11) Quite often the mission was to construct permanent linear dikes for dredged soil containment. As such, the necessity for high strength seams could be somewhat avoided by placing the warp (strong) direction transverse to the dike's alignment in the direction of the major principle stress. This allows for the weft (weaker) direction to be seamed and placed in alignment with the minor principal stress. Reference 12, lists twelve projects of this type. Foundation soil strengths for these projects were from 1 to 8 kPa (0.1 to 1.2 lb/in^2). The geotextile used had wide width tensile strengths from 160 to 500 kN/m (900 to 2900 lb/in). They were all relatively heavy woven fabrics made from polypropylene or polyester. The embankments placed above the geotextiles varied in height from 2 to 7 m (6.6 to 23 ft.). Post construction consolidation settlements varied from 1 to 5 m (3.3 to 16.4 ft.). To our knowledge all have been successful with only one known problem, that is, at one site a propagating tensile failure occurred in the fabric between instrumentation holes as surcharge was near full height. This illustrates the importance of reducing fabric strength to allow for holes such as those made by installation of instrumentation devices or strip drains.(2)

An example of a recently completed areal fill on dredged soil using a high strength geotextile was at Seagirt, Maryland for the Maryland Port Authority.(13) This 45 ha (113 acre) site contains 6 to 11 m (20 to 35 ft) of dredged soil at water contents 50% to 150% above the liquid limit. The goal was to prepare the site for ground surface loads of approximately 30 kPa (600 lb/ft^2) within six months! This goal was achieved by deploying a 1000 g/m^2 (30 oz/yd^2) geotextile having wide width strength of approximately 210 kN/m (1200 lb/in). Required seam strengths were 105 kN/m (600 lb/in) which was the limiting design constraint. A single lift of granular soil (the drainage blanket/working platform) 0.76 m (2.5 ft.) thick was placed from which strip drains were installed at 1.5 m (5 ft.) centers to the bottom of the in-situ soil. The final operation was to place a 2.4 m (8 ft.) surcharge fill over the entire site. Settlement plates and piezometers were the main control instruments from which the surcharge fill placement rate and dwell time was controlled. The project was very successful in that rapid consolidation occurred to the anticipated amount. The area is currently paved and used for heavy truck storage, loading and unloading.

Wilmington Harbor South Project(14-16)

(a) Overview

The Wilmington Harbor South Disposal Area project is a linear fill and is located near the confluence of the Christina and Delaware Rivers, approximately 47 km (29 miles) south of Philadelphia, PA, see Figure 3. The project will provide capacity for dredged material from Wilmington Harbor maintenance dredging. This effort involves the

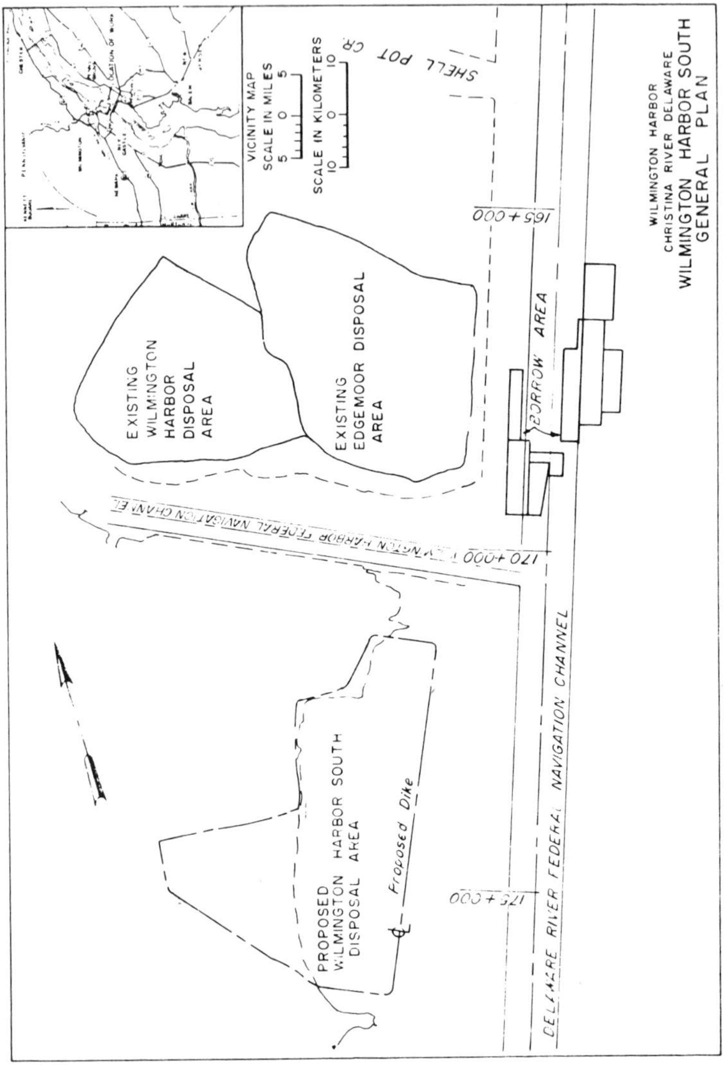

Fig. 3. - General Plan of Wilmington Harbor South Showing Limits of Disposal Area

removal of approximately 770,000 m^3 (1,000,000 yd^3) of shoal material from the harbor every year.

This new disposal area will provide capacity for approximately 9 million m^3 (12 million yd^3) of dredged material from maintenance dredging operations in Wilmington Harbor. This is the equivalent of approximately 24 years of use of the area providing reasonable disposal area management practices are followed.

(b) Design Features

The major feature of the disposal area project is the long linear containment structure shown on Figure 3. It will consist of a 2400 m (8000 ft) long "U-shaped" embankment constructed of dredged granular material from a borrow area located in the Delaware River off Cherry Island, just upstream of the site. The embankment ties into low lying ground at its upstream and downstream ends. The upstream or northern end of the dike extends eastward from a point on the low ground approximately 210 m (700 ft) south of the Wilmington Harbor South Jetty to the -0.61 m (-2 ft) contour, Christiana River Datum (CRD). From this point, the embankment turns south, generally following the -0.61 m contour for about 1372 m (4500 ft), where it turns to its southern terminus on fast land, just north of a railroad pier at Pigeon Point.

The foundation conditions at the site were determined during seven phases of subsurface investigations beginning in 1956 and extending through 1987. The dike alignment extends over weak, highly compressible silts and clays of the Holocene deposits which form these tidal flats and shallows of the Delaware River. Sands and gravels, primarily of Pleistocene Age, are encountered below the soft silts and clays at depths averaging 27 m (90 ft.). The oldest unconsolidated sediments at the site are of the lower Cretateous Age Potomac formation. The stratigraphy of the disposal area site is shown on Figure 4.

The borrow area contains sands and gravels with varying amounts of cobbles and boulders deposited in a Pleistocene channel located on the west side of the Delaware River approximately 1.6 km (1.0 mi.) upstream of the new disposal area. These deposits are covered with relatively thin Holecene silts and clays and are removed prior to the excavation and placement of the granular borrow materials. These blanketing materials are being disposed of in the Corps' Penns Neck Disposal Area near Pennsville, New Jersey. To date, the quality of the borrow has been consistently good. However, the contractor is reporting difficulty in attaining the depths of excavation called for in the plans, which could result in an underrun in the granular borrow available for dike construction.

The soil samples selected for testing were representative of the materials encountered within the major zones of the foundation and borrow area. The adopted design parameters for the various fill and foundation materials used in the slope stability, bearing capacity and settlement analyses are given in Table 4. It is noted that the characteristics of the Potomac formation were not quantified because

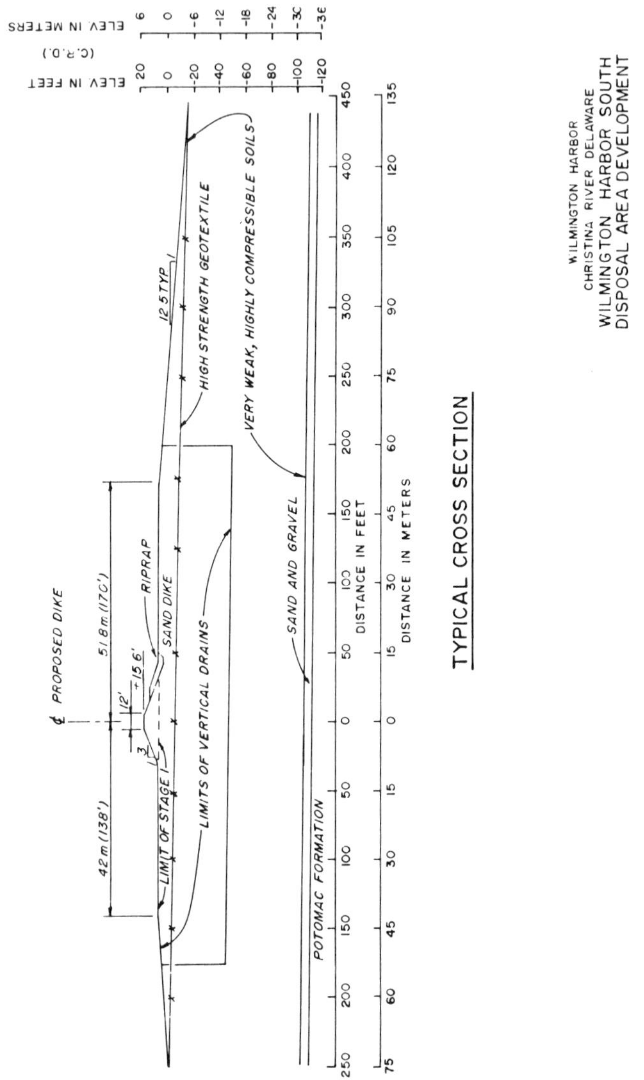

Fig. 4. - Wilmington Harbor South Disposal Area Development Showing Site Stratigraphy and General Dike Cross Section

Table 4 - Adopted Design Parameters

Material	Unit Weights Moist	Sat.	Shear Characteristics Q		R		S		Compression Constants C_c	C_v	E_s
	g/cc (PCF)	g/cc (PCF)	φ °	C kPa (PSI)	φ °	C kPa (PSF)	φ °	C kPa (PSF)		cm/day (in/day)	kPa (PSI)
Riprap	2.00 (125)	2.08 (130)	40	0	40	0	40	0	--	--	--
Sand Dike Stage I	1.60 (100)	1.87 (117)	30	0	30	0	30	0	--	--	13,790 (2000)
Stage II	1.60 (100)	1.87 (117)	30	0	30	0	30	0	--	--	27,580 (4000)
Stratum No. 1 Stage I	1.41 (88)	1.52 (95)	0	4.79 (100)	15	0	30	0	0.897	4.923 (0.763)	--
Stage II	1.41 (88)	1.52 (95)	0	7.18 (150)	15	0	30	0	0.897	4.923 (0.763)	--
Stratum 2A	1.33 (83)	1.52 (95)	0	9.58 (200)	15	9.58 (200)	30	0	0.897	4.923 (0.763)	--
Stratum 2B	1.33	1.52	0	14.36	15	9.58	30	0	0.897	4.923	--
Sand Layer	1.44 (90)	1.95 (122)	31	0	31	0	31	0	--	--	--
Dredged Material	1.46 (91)	1.51 (94)	0	0	0	0	0	0	--	--	--

of the very stiff consistancy of these soils. This formation was considered the "firm base" in the design analyses. The majority of the design testing was done at the Corps'South Atlantic Division Laboratory.

The poor foundation conditions encountered, the limited quantity of granular borrow available and environmental considerations led to the adoption of a wide-bermed embankment to enclose this disposal area, see Figure 4. This design assured stability and minimized displacement of foundation materials (mud waves) while at the same time being economical in its use of embankment materials. The concept behind this design involves "floating" the dike on the soft foundation with the use of a high strength geotextile for tensile reinforcement. The chosen geotextile employed on this project is woven of high-tenacity polyester yarn and was specifically designed for this application. The fabric's specifications are given in Table 5. Following placement of the high strength geotextile, the dike is constructed of granular materials placed in two stages. The first stage averages approximately 3 m (10 ft) in thickness by 180 m (600 ft) in width and forms the wide berm of the dike section. This first stage construction consists of the five separate hydraulic fills shown in section in Figure 5. When constructing the embankment, the outer two fills, designated 1 and 2, are placed concurrently, thus putting the central section of the fabric in tension. This provides added support for fill segments 3, 4 and 5, thereby preventing excess displacement of the foundation materials. Fill segments 1 and 2 proceed approximately 30 to 46 m (100 to 150 ft) in advance of fill segment 3 while segments 4 and 5 trail 3 by about 15 m (50 ft). The loads on the geotextile are balanced by keeping the total weights of segments 1 and 2 approximately equal and their leading edges even with each other as filling progresses. The same thing is done with segments 4 and 5, although this latter requirement is not considered as essential as the first on this particular project. The first stage reaches an average top of fill elevation of 2.1 m (7 ft) C.R.D. at the completion of its initial placement.

Table 5 - High Strength Geotextile Specifications

Tensile Strength (Warp and Fill Directions).........	260 kN/m (1500 lbs/in.)
Tensile Stress Strain Secant Modulus from........... 0 to 5% Strain	3,300 kn/m (19,000 lbs/in.)
Elongation at Failure............................	10 to 35%
Soil-Geotextile Friction Angle.....................	30 minimum
EOS (O_{95} value)..................................	0.30 mm (max.) (0.0117 in.)
Seam Strength......................................	140 kN/m (800 lbs/in.)

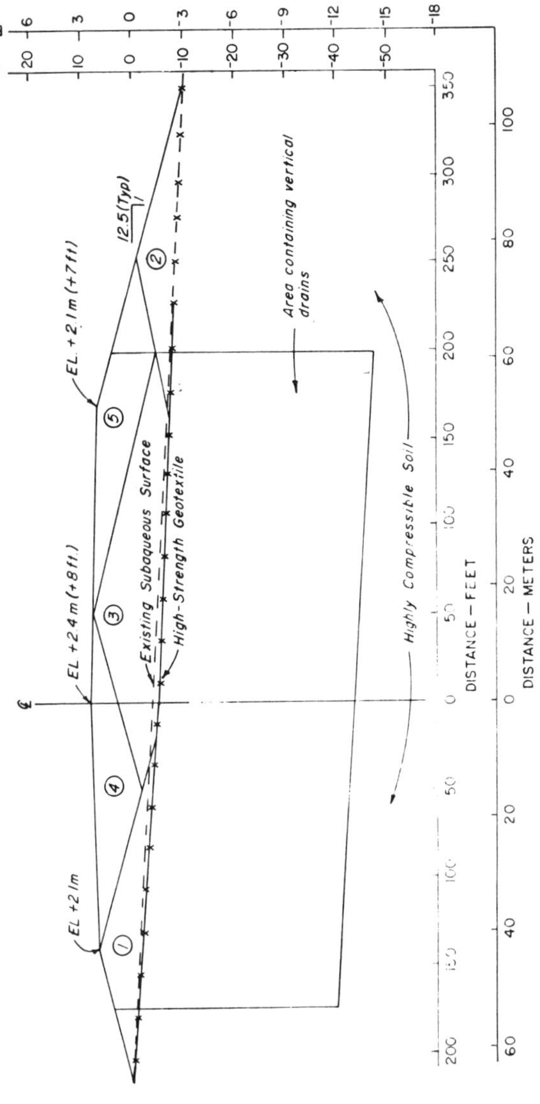

Fig. 5. - Typical Stage I Embankment

Stage I construction also includes the installation of strip (wick) drains through the granular fill to a depth of 12.2 m (40 ft) below the geotextile. These strip drains are essential to the design in that they increase the rate of consolidation and strength gain in this critical portion of the foundation. Strip drain installation follows the completion of the Stage I fill at a given location. The drains are placed in a triangular pattern at a spacing of 3 m (10 ft) as shown in Figure 6. These drains are expected to provide the drainage needed to allow the average settlement of 1 m (3 ft) expected during the one year hiatus between the Stage I and Stage II embankment construction. This settlement and the attendent strength gain in the foundation materials will allow construction of the second stage embankment section to El 4.6 m (15 ft).

The final operation in the Stage I construction is the installation of instrumentation consisting of piezometers for measurement of pore pressures at various levels in the foundation and Boros point anchors for determining the elevation of the geotextile at selected locations as foundation consolidation progresses. Slope inclinometers are also installed at this time.

Figure 7 shows the Stage II portion of the embankment consisting of a 3.3 m (11 ft) high dike with 3H:1V side slopes and a 3.7 m (12 ft) top width. The embankment employs a 75 cm (2.5 ft) thick layer of riprap extending from El.0 to El.4.6 m (0 to 15 ft) C.R.D. A layer of light weight geotextile is used as a filter between the granular embankment material and the riprap. Another feature incorporated into the design is the 30 cm (1 ft) thick layer of impervious soil on the inner slope of the Stage II dike. This is intended primarily to reduce erosion damage prior to filling, but also fills a secondary function as a growing medium.

(c) Construction Methods

The contractor chose to fabricate the high strength geotextile tensile reinforcement from panels measuring 3.7 and 5.2 m (12 and 17 ft) in width. These panels are continuous for their full length in the warp direction since the contract specifications allowed no seams in the warp direction. The fabric supplier assembled each 16.5 m (54 ft) wide section from two 5.2 m wide and two 3.7 m wide panels in a very long loft, see Figure 8.

The seamed sections are then rolled onto axles for shipment to the actual construction site. Upon arrival at the site, the sections are loaded onto barges, attached to wheels, laid down in position and then field seamed, see Figures 9 and 10. This results in an accordian folded mat of the proper length and width for the next segment of the geotextile mat to be placed.

When field seaming is completed, the geotextile mat is placed on the subsoil beneath the river's surface. The normal means of accomplishing this is to move the inter connected laying barges forward along the centerline of the embankment, letting the fabric feed off the barge over rollers and down into place on the river bottom. When starting off a new leg of the embankment, this process is

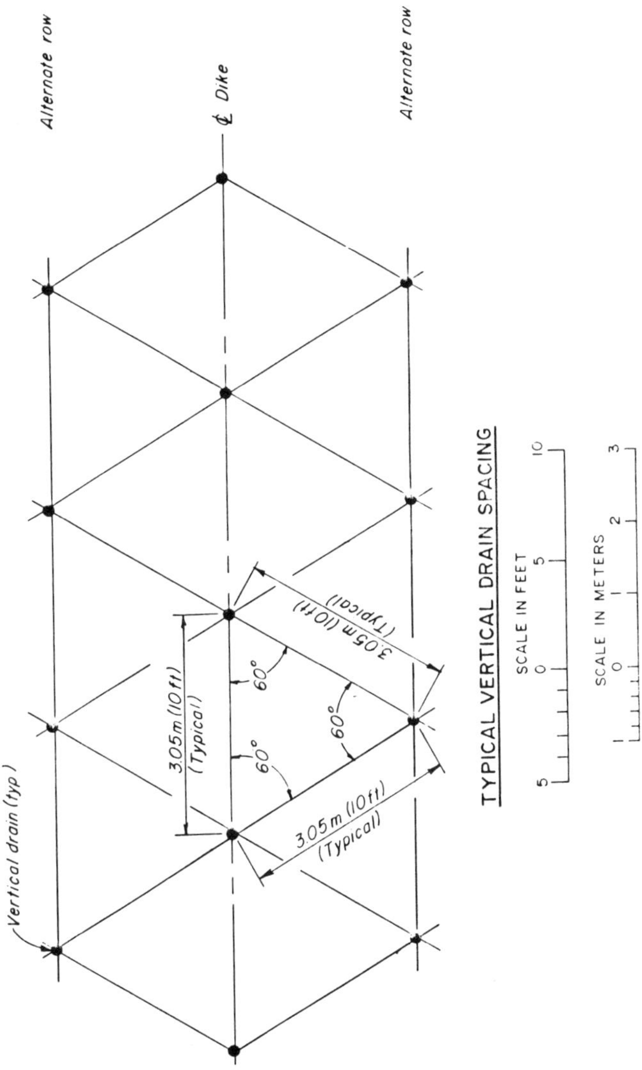

Fig. 6. - Typical Vertical Drain Spacing

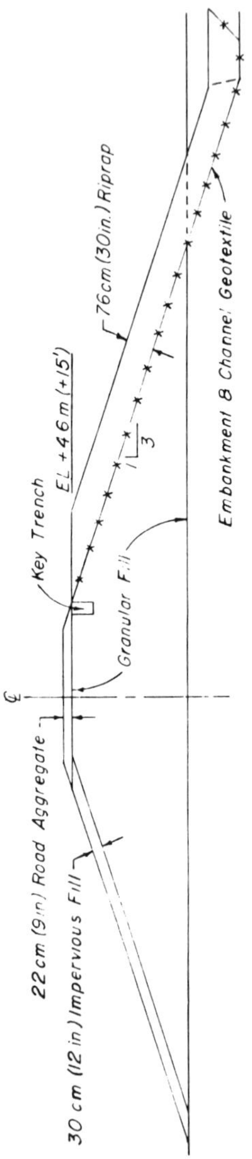

Fig. 7. - Typical Stage II Embankment

Fig. 8. - Loft Fabrication of Geotextile Panels into Sections

Fig. 9. - Unrolling of Geotextile Sections on Barge at Field Site

Fig. 10.- Sewing of Geotextile Section on Barge at Field Site

Fig. 11. - Areal View of Site at Approximately 50% Geotextile Placement Stage I Fill Placement with Dredge Supply Lines and Laying Barge in View

reversed with the fabric being pulled from the barge which is temporarily anchored.

The hydraulic placement of granular fill is begun once the high strength geotextile is in place. The prime contractor has chosen to branch five (5) lines off of a primary supply line to allow the selective placement of the granular material on the five fill segments (recall Figure 5). These lines are plainly evident in Figure 11.

The vertical drains are installed following completion of the Stage I fill as explained earlier. The subcontractor's wide tracked installation crane is rigged to install the drains to depths of 18.3 m (60 ft). Two views of this unit are shown in Figure 12. The mandrel point and flat anchor system employed by the drain installer have proven very satisfactory to date.

(d) Lessons Learned

A minimum seam strength of 140 kN/m (800 lbs/in) was specified for the high strength geotextile. The value was deemed necessary based upon the use of the fill slopes and low ground pressure equipment which were specified in the contract. During the early part of the construction period and extending, to a much lesser degree, to the present, considerable difficulty was encountered in meeting this requirement for all field seams. Some of the same difficulty was ecountered with the factory seams during a short period last summer. Following considerable design review and not a little soul searching, the Corp's allowed the sections with understrength seams to be deployed, but placed constraints on the placement of construction equipment on the fill. This combination of lower seam strength and control of equipment placement provided the same factor of safety as used in the original design.

The proper alignment of the fabric seams has proven a sporatic, but persistant problem since the start of Stage I construction. Early in our job experience (see Figure 13), this problem resulted primarily from our unfamiliarity with the way the material would act to the combined effects of river current and the rise and fall of the tide. The geometry of the bottom configuration and the barge orientation resulted in slightly different variation on the same theme later in the construction. All these problems have been corrected by improved fabric anchoring systems or barge manuevering to a degree sufficient to allow the construction to continue without major, costly correction to equipment or placement methodology.

The fill slopes specified in the design have been met in the actual construction. This required some experimentation on the part of the contractor in the placement of pipe discharges, timing of pumping operations to make best use of the tide stage, and the use of land based equipment to achieve the desired results. The underwater placement of the hydraulic fill can be difficult to control and constant attention to this placement is necessary to avoid overstress of the geotextile due to oversteepening of the transverse or longitudinal slopes.

Fig. 12. - Photographs of Wide Track Crane Installing Strip (Wick) Drains After Placement of Stage I Fill

Fig. 13. - Photograph of Geotextile as Placed at Western Terminus of Project Showing Difficulty of Maintaining Seam Alignment

Allowances made in the design for reduction in strength of the geotextile due to holes made during the strip drain installation were based on the estimated size of the hole and the intact fabric remaining along a given line on the fabric. A reduction of 5% was provided for this purpose and has proven to be adequate. The success of this assumption as well as others made during the design process can best be judged by the satisfactory performance of the embankment during the early, critical stages of the construction just completed.

The foundation displacement under the Stage I loading has agreed closely with the design assumptions based on the data obtained during strip drain installation, placement of Boros point anchors and borings made specifically for the purpose of locating the top of the geotextile for fill payment purposes.

The performance of the completed portions of the Stage I embankment as measured by the instrumentation has been as anticipated. Usually, there is little evidence of settlement of the fill until the strip drains are installed. This settlement becomes evident almost immediately after drain installation and, to date, has continued at a uniform rate at practically every station at which it is measured. Figure 14(a) shows a plot typical of the Boros point anchor readings for those installed through mid-December 1987. Piezometer readings have indicated the effectiveness of the strip drains in reducing the amount of pore pressure in the upper 12.2 m (40 ft) of the foundation materials as shown in the typical plot of Figure 14(b).

It appears that the embankment will continue to behave as anticipated and, based on the limited amount of information available at this time, the foundation will consolidate and increase in strength as expected. To verify this, slope indicators have been installed and show only minor movements to date. There is some possibility of overdesign of the embankment system since the actual stresses in the high strength geotextile under actual field loadings have not been measured. There are plans to make such measurements during the next season's construction which will be reported at a later date.

Conclusions

This paper has attempted to provide a synthesis of using geosynthetics to reinforce and then stabilize high water content fine grained soils. The technique has been successfully utilized on a number of major projects. Some of the projects have incorporated prefabricated vertical strip drains to further enhance the consolidation process.

Two examples were presented in this paper; one a large areal fill in Baltimore, Maryland, the other a large linear fill in Wilmington, Delaware. This latter project was explained in detail due to its size, scope, newness, and feedback of design information. In this latter regard, settlement plates, piezometers and inclinometers are being used to assess and control fill placement as consolidation proceeds.

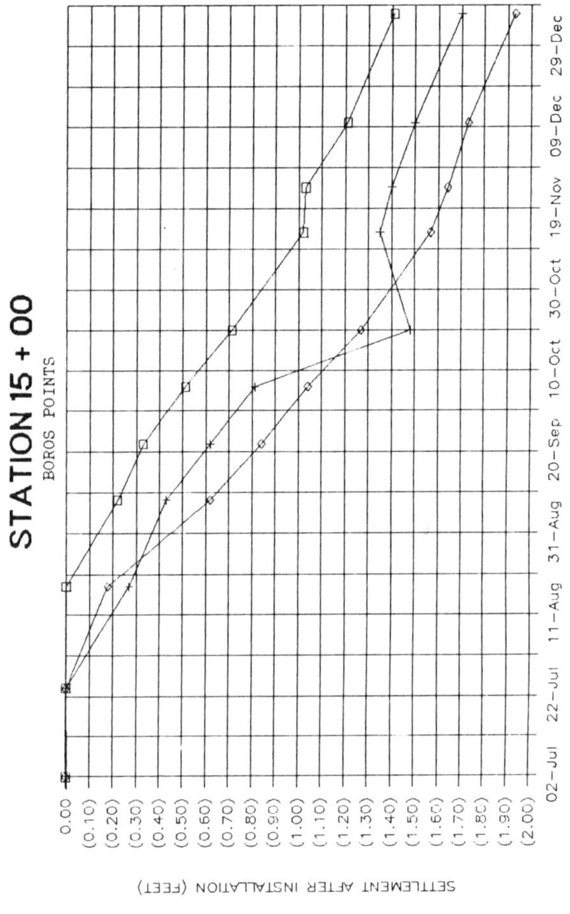

Fig. 14(a) - Typical Plot of Results from Boros Anchor Points

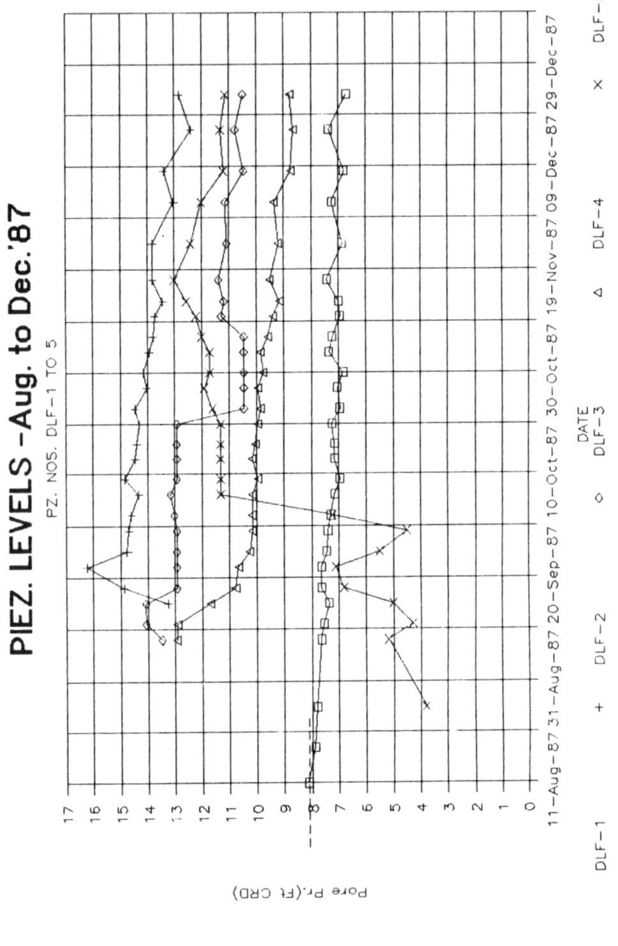

Fig. 14(b) - Typical Plot of Results from Piezometers

Acknowledgements

The Wilmington Harbor project was designed by the Geotechnical Section of the Philadelphia District of the U. S. Army Corps of Engineers. The work was done under the supervision of V. Calvarese, Chief, Design Branch and B. Uibel, Chief, Geotechnical Section. Lead design engineers for the geotechnical design were S. Fritzinger, J. McKenzie and R. Pincioti. Mr. R. Lazor served as the project geologist.

The prime contractor for the work is the American Dredging Company whose original bid price for the work was $19,315,638. Modifications to the contract since award have increased this to approximately $24 million. The contract period is expected to extend through January 1990. Wellman Quline, Inc. are subcontractors for the high strength geotextile.

References

1. Koerner, R. M., Designing with Geosynthetics, Prentice Hall Publ. Co., Inc., Englewood Cliffs, NJ, 1986.

2. Koerner, R. M., Hwu, B-L and Wayne, M. H., "Soft Soil Stabilizatin Designs Using Geosynthetics," Jour. Geotex. and Geomemb., Elsiever Publ. Co., (to appear, 1988).

3. Rowe, R. K. and Soderman, K. L., "Stabilization of Very Soft Soils Using High Strength Geosynthetics: The Role of Finite Element Analysis," Jour. Geotex. and Geomemb., Elsiever Publ. Co., (to appear, 1988).

4. Koerner, R. M., Swan, R. H. and Carroll, R. G., Jr., "Junction Strength of Geogrids," Geotech. Fabrics Rept., IFAI, St. Paul, MN (accepted, Oct. 1987).

5. Holtz, R. D., "Preloading with Prefabricated Vertical Strip Drains," Jour. Geotex. and Geomemb., Elsiever Publ. Co., (to appear, 1988).

6. Rixner, J. J., Kramer, S. R. and Smith, A. D., "Prefabricated Vertical Drains, Vol. 1, Engineering Guidelines," FHwA Report RD-86-168, Washington, DC, 1986.

7. Lawrence, C. A. and Koerner, R. M., "Flow Behavior of Kinked Strip Drains," Proc. Symp. Geosynthetics for Soil Improvement, ASCE-NAGS Conv., Nashville, TN, May, 1988.

8. Fowler, J., Analysis of Fabric-Reinforced Embankment Test Section at Pinto Pass, Mobile, Alabama," Ph.D. Thesis, Oklahoma State Univ., 1979.

9. Haliburton, T. A., Fowler, J. and Langan, J. P., "Design and Construction of a Fabric Reinforced Test Section at Pinto Pass, Mobile, Alabama, Trans. Res. Rec. 79, Washington, DC, 1980.

10. Fowler, J., "Theoretical Design Considerations for Fabric Reinforced Embankments," Proc. 2nd Intl. Conf. Geotex., Las Vegas, NV, 1982, IFAI, pp. 665-676.

11. Fowler, J., Peter, J. and Franks, L., "Influence of Reinforcement Modulus on Design and Construction of Mohicanville Dike No. 2," Proc. 3rd Intl. Conf. Geotex., Vienna, Austria, 1986, pp. 267-271.

12. Fowler, J. and Koerner, R. M., "Stabilization of Very Soft Soils Using Geosynthetics," Proc. Geosynthetic '87, New Orleans, LA, IFAI Publ., 1987, pp. 289-300.

13. Koerner, R. M., Fowler, J. and Lawrence, C. A., "Soft Soil Stabilization Study for Wilmington Harbor South Dredge Material Disposal Area," U. S. Army Engineer Waterways Experiment Station, Misc. Paper GL-86-38, Dec. 1986.

14. U. S. Army Corps of Engineers - Philadelphia District, Wilmington Harbor South Disposal Area Development, Wilmington Harbor, Christina River, Delaware, June 1985.

15. U. S. Army Corps of Engineers - Philadelphia District, Navigation Study Detailed Project Report and Environmental Assessment, Wilmington Harbor, New Castle County, Delaware, September, 1985.

16. U. S. Army Corps of Engineers - Philadelphia District, Final Environmental Impact Statement, Wilmington Harbor, Federal Navigation Project, Dredged Material Disposal Area, Wilmington Harbor, Delaware, October, 1985.

IMPROVEMENT OF HYDRAULIC FILLS BY GROUND MODIFICATION

by

Joseph P. Welsh, P.E.[1], M.ASCE

Abstract

Some of the Ground Modification^{sm} techniques that can be utilized for new or remedial construction over hydraulic fills are reviewed including deep dynamic compaction, vibro-compaction, vibro-replacement and compaction grouting. Case histories are included of new and remedial work where hydraulic fills have been improved by Ground Modification techniques.

Introduction

Ground Modification can be broadly defined as the in-place, controlled manufacture of ground materials to form part of a geotechnical construction system. Rarely can untreated hydraulic fills be directly built on, as objectionable settlements will result. Ground Modification or soil improvement techniques are performing an ever expanding role in treating these fills to a predetermined performance criteria. Each project will differ depending upon the grain size distribution and uniformity of the hydraulic fill, the allowable settlement, and actual loading conditions. A general overview is presented on some soil improvement techniques and factors that have to be considered in determining their ability to improve hydraulic fills. A series of case histories is included highlighting the use of various soil improvement techniques to treat hydraulic fills.

Deep Dynamic Compaction

Deep dynamic compaction is the utilization of heavy weights dropped from heights to densify soils. This soil improvement technique has been used extensively in the United States since the late 1970's and in Europe for well over a decade. As pointed out by Lucas (1986), deep dynamic compaction can improve all soils except for cohesive soils below the groundwater table. One restriction of this technique is that the depth of influence is limited as shown in the following:

$$D = N\sqrt{Wh}$$

D = depth of influence, in meters
N = site specific factor
W = falling weight, metric tonnes
h = height of drop, in meters

[1]Vice President, GKN Hayward Baker, 1875 Mayfield Road, Odenton, Maryland 21113

"N" was originally suggested to be unity by Menard (1975). This was modified by Leonards (1980) to 0.5 and recently suggested to be between 0.3 and 0.7 by Lucas (1986).

The maximum improvement is normally obtained between one-third and two-thirds of the depth of influence and the nature of the soil beneath the depth of influence will determine the amount of improvement that can be obtained in the lower portion of the zone of influence. A dense underlying strata, such as rock, glacial till, etc., will reflect the impact energy and higher density will be realized in the lower zone. The upper few feet has to be redensified either by an ironing pass (100 percent area coverage with low impact energy) or conventional rollers. One drawback with deep dynamic compaction is that one has little control of the in-situ moisture content of the soil. Therefore, if the soil is wet of optimum, maximum densification will not happen unless aided by natural or artificial drainage. Lucas (1986) indicates that the groundwater will have to be maintained at least six feet (2 m) below the bottom of the falling weight. Dewatering or placing fill prior to densification may be required to achieve this. He further points out that multi-drops at a particular location improves the densification in the zone of influence but will not increase the depth of influence. Generally, 3 to 10 drops per location on a 2-pass primary-secondary grid spacing will produce the most effective and economical results.

Deep dynamic compaction can be employed on sites larger than 50,000 square feet (4,645 m^2) in the range of \$1.00 per square foot (0.09m^2) of surface area if the weight to be dropped is between 10 and 20 tons (13.6 and 18.1 tonnes), from heights up to 100 feet (30 m) plus mobilization/demobilization costs ranging between \$15,000 and \$35,000. Heavier weights up to 40 tons (36.2 tonnes) can be dropped freefall, but this doubles the unit cost. Weights up to 100 tons have been dropped with special rigs, but mobilization/demobilization costs exceed \$250,000. Therefore, deep dynamic compaction of all soils, including hydraulic fills, is limited to an economical depth of 40 feet (12.2 m).

Vibro Techniques

In the mid 1930's, the use of in-place vibrators to densify soil was patented in Germany and has been employed in the United States since the early 1940's. Vibro-compaction is a method of deep densification of insitu granular soils by rearranging the loose cohesionless sand grains into a denser array (Figure 1).

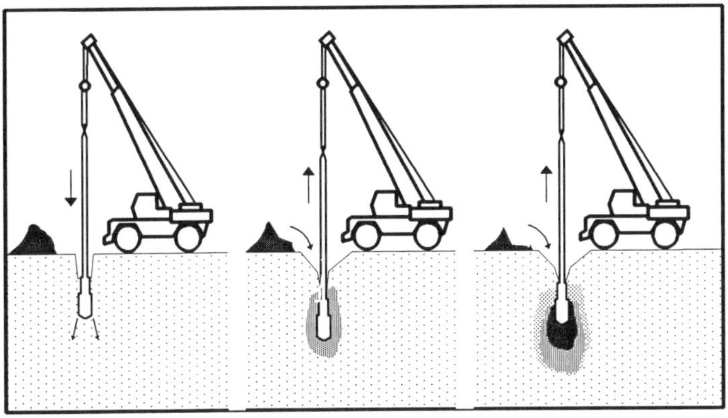

Figure 1 - Vibro-Compaction

The key factory in utilizing vibro-compaction for hydraulic fills is the gradation of the materials. It is soil character sensitive, proving ineffective where more than about 15 percent fines are present in the sand and where more than 3 percent of the fines are active clays. Depths in excess of 100 feet (30.5 m) have been treated and relative densities of up to 85 percent have been obtained.

As the vibrator liquefies and densifies the soil, it will effectively lower the whole site. Alternatively, backfill sand can be added to compensate for the volume reduction due to densification.

Vibro-replacement, using a stone backfill, reinforces cohesive soils and densifies granular soils with more than 15 percent finer than the No. 200 sieve as well as granular soils of uniform gradation. One of the latest developments in this technique is to use a dry bottom-feed technique to place the stone at the tip of the vibrator and minimize the environmental concerns caused by the jetting water employed to assist in inserting the vibrator. The dry top-feed vibro technique can be used without the jetting effect for depths up to 25 feet (7.6 m). This method will produce smaller diameter stone columns, while still achieving the soil improvement results desired. However, the use of stone raises the cost of densification from about $4.00 per ft^3 ($0.76\ m^3$) for vibro-compaction to between $5.00 and $9.00 per ft^3 ($0.76\ m^3$) for vibro-replacement.

Grouting Techniques

Grouting is the injection of a pumpable material into soil or rock to change the physical characteristics of the formations. There are four basic types of grouting, each with a different purpose and objective (Figure 2).

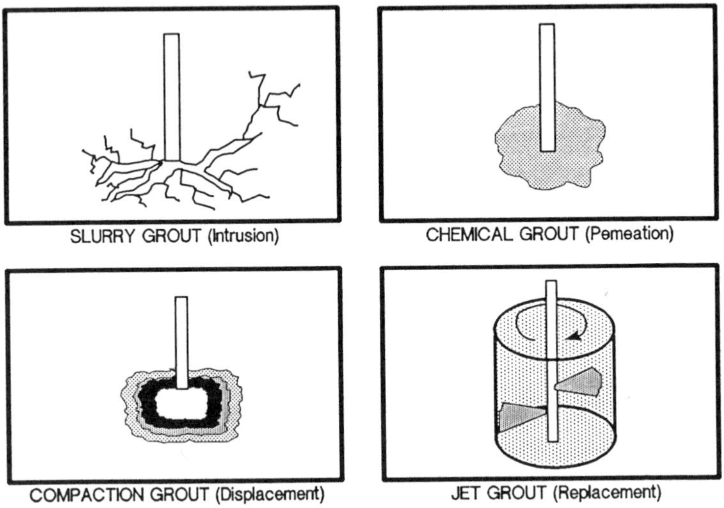

Figure 2 - Types of Grouting

Chemical grouting is the permeation of specially designed fluids into sands with less than 20 percent passing the No. 200 sieve, to increase their strength and/or reduce their permeability. Slurry grouting is the intrusion of particular grouts into fissures of rock or into the open voids of soils for the same purposes. Jet grouting in hydraulic fill would be done by mixing a high pressure grout with the soil. The grain size limitations and economics normally preclude the consideration of chemical and slurry grouting for treating hydraulic fills. The cost of jet grouting would normally preclude this technique for treating hydraulic fills. However, the ability of compaction grouting to densify hydraulic fills bears consideration.

Compaction grouting is one of the few soil improvement techniques with its origin in the United States. Warner (1982) gives an overall history of this technique, which involves pumping a soil-cement with a slump less than 2 inches (51 mm) under relatively high pressures. No in-depth research has been performed to determine the limitations on soils that can be treated by this technique, but as the low slump grout is exerting a relatively instantaneous load to densify the surrounding soil, it stands to reason that this method works most effectively in free-draining granular formations with its efficiency decreasing as the soil becomes more cohesive. Saturation and the soil's ability to dissipate the induced pore pressures play key roles in the effectiveness of this technique for many soils. When used for remedial treatment, the grout pipes are normally spaced closer than a 10 foot (3 m) grid; however, for new construction, wider spacings have been utilized depending upon the results anticipated to be obtained (Schmertmann, 1986; Welsh, 1986). Additional pipes can be installed if a relative density over 75 percent is required (Salley, et.al. 1987).

Compaction by explosives has been utilized in hydraulic fills to improve the relative density up to 75 percent; however, this technique is effective only in loose sands with less than 15 percent passing the 200 sieve and has the disadvantage that the use of explosives in construction normally is followed by damage or nuisance liability problems from adjacent property owners.

Case Histories

Redding Cogeneration Facility, Redding, CA

This 76-acre complex centers around a single 23 megawatt turbine generator to be fueled by wood chip waste from milling operations conducted on-site. Stock pile fuel will be delivered by conveyors to two boilers where raw material will be burned and converted to power by the turbine generator.

The entire 76-acre site and adjacent properties had been extensively mined in the early 1900's by the use of large floating dredges for gold recovery which consisted of separation of the coarse-grained gravels and cobbles from the finer grained sands, silts and clays. This separation caused by the dredging operation resulted in deposited mounds of tailing materials up to 30 feet (9.1 m) in height. The area selected for the 23 megawatt turbine generator was the largest mound of coarsest grain material available.

The initial geotechnical exploration disclosed 16 to 17 feet (4.9 to 5.2 m) of medium dense, gravelly sands underlain by a slightly sandier material with a looser consistency. This transition appeared to coincide with the groundwater level. Subsequent field testing indicated that the upper material also had loose zones. The consultants' concern for building on this site was purely from a seismic standpoint; even though the Redding area is a relatively aseismic zone, the concern was great enough to warrant an investigation into possible soil improvement techniques.

Deep dynamic compaction was selected as the most appropriate method to improve the soils. The initial depth of loose material approached 40 feet (12.2 m) in some areas of the structure. Since the most economical deep dynamic compaction program would treat only the upper 25-30 feet (7.6-9.1 m), the top 6 feet (2 m) of material was removed prior to the start of the work. An 18-ton (16.3 tonnes) weight was dropped from 75 feet (22.9 m) with up to 10 drops per location on a 16 foot (4.9 m) grid pattern. Treatment effectiveness was mainly evaluated by monitoring the depths of the craters generated and the overall depression of the site at the conclusion of the deep dynamic compaction. Post-densification testing was deemed too costly and too unreliable to be warranted due to the large gravel and cobble content of the soil.

Submarine Facility, Barrow-in-Furness, England

As reported by Bell, et.al., (1986) over 800,000 cubic meters of a hydraulic sandy fill was densified adjacent to an existing facility for the construction of a submarine hall. The fill was obtained by near shore dredging and placed in water depths ranging from 5 to 22 meters with a mean depth of about 10 meters. Beneath the hydraulic fill was a stiff boulder clay succeeded by various densities of fine to coarse sand, another boulder clay strata and then a slightly weathered mudstone with siltstone and gypsum veins. The design loading for the construction hall floor slab, 2 meters in thickness and 58 x 269 meters in plan, was up to 225 kN/meters^2. The foundation for the construction hall was up to 300 kN/meters^2 and raft foundations for the office tower carried loads up to 200 kN/meters^2. In addition to densifying the hydraulic fill for structural purposes, there was also a major concern for liquefaction based on an earthquake in 1865 with an intensity between VI to VII on the modified Mercalli scale.

Due to its economics, deep dynamic compaction was considered for this application; however, the proximity of adjacent structures and the depth of treatment required, ruled out this technique. Vibro systems were considered next and it was determined that by carefully monitoring the source of the hydraulic fill, it would be possible to prevent material with more than 12 percent passing the No. 200 sieve from being deposited. Thus, the more economical vibro-compaction could be utilized.

Over 3,000 grain size analyses were performed on samples obtained from borings and from bucket samples obtained by divers during placement of the fill. The densification of the dock fill was performed by using 165 HP GKN Keller S-Type electrical powered vibrators operating at 30 Hz. Up to 8 vibrators supported from crawlers cranes were employed and, in 5 months, densified almost 1.5 million cubic meters of soil. The maximum depth of treatment was 28.5 meters. The vibrator spacing varied between 2.8 and 3.2 meters on a triangular grid pattern and each probing was based on a 200 amp current reading.

As stated in Bell, et.al., (1986), the method of analysis which was adopted in assessing liquefaction resistance is a semi-empirical approach developed by the University of California at Berkeley, and detailed by Seed & Idriss (1981). Essentially the method allows the standard penetration tests' 'N' value to be related with the cyclic stress ratio causing liquefaction for any given depth. This stress ratio is then compared with the likely stress ratio produced by an earthquake of relevant magnitude, to provide an assessment of liquefaction potential. A design earthquake corresponding with 0.25 g maximum ground acceleration with a 10 cycle limit was adopted. The

Seed & Idriss method was modified in two ways for use at Barrow. Seed & Idriss differentiated between two types of liquefaction - susceptible sands based on D_{50} values less than 0.15 mm and greater than 0.25 mm. In view of the wide variation in grain size at Barrow, it was considered prudent to extend the method to include coarser grain sizes, although it was recognized that the larger sized materials would perform better than the finer materials in the event of an earthquake owing to greater permeability. In addition, rather than measuring SPT 'N' values after treatment, it was decided to employ electric static cone bearing resistance q_c (kg/cm^2) to assess improvement in density. Consequently q_c/N ratios were developed to allow conversion from SPT to q_c values, which Seed & Idriss employed. These assumptions led to liquefaction criteria for the dock fill in terms of cone resistance, depth and soil grading as shown in Figure 3.

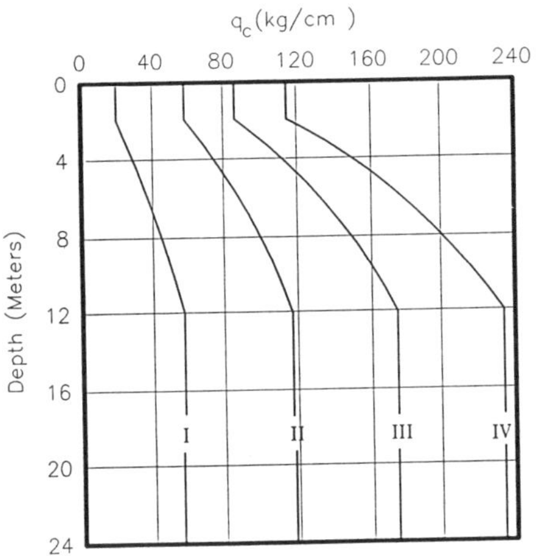

REF	q_c/N	D_{50}(mm)	DESCRIPTION
I	2.5	< 0.15	Fine silty sands
II	4	0.25–1.0	Fine-medium sand
III	6	1.0–5.0	Coarse sand with gravel
IV	8	> 5.0	Sandy gravels

NOTE: Normally type II will include particle size D_{50} down to 0.15 mm

Figure 3 - Liquefaction Resistance Criteria for 0.25G

As can be seen by Figure 4, the average cone penetration reading exceeded 200 kilograms per cm^2 for 15 meters. However, the q_c values dropped greatly at the area of demarcation between the before grades and the new hydraulic fill, strongly suggesting that the silt strata was not adequately removed or was redeposited prior to placement of the hydraulic fill. Accordingly, a 14 meter surcharge was placed over the construction hall area. As of 1987, no report has been issued giving the actual settlement readings.

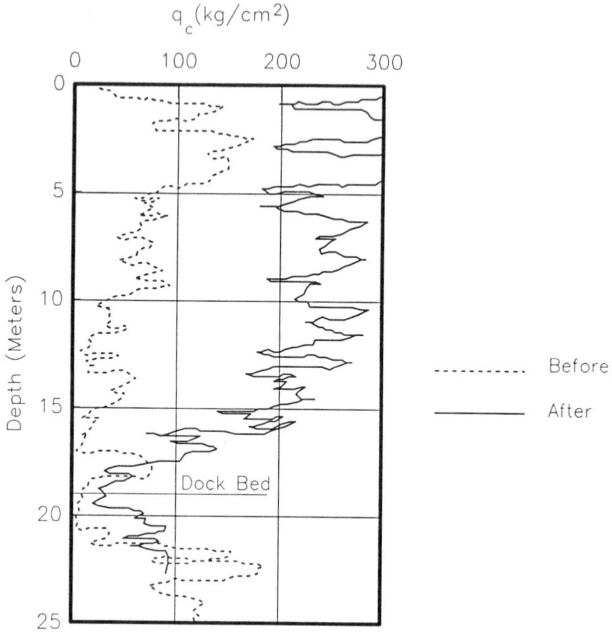

Figure 4 - Typical Pre and Post Treatment Cone Test Results

Interstate Route 10 through McElroy Swamp, Louisiana

As reported by Arman (1986) and Arman and Griffin (1988), a 4.3 mile section of Route I-10 between Baton Rouge and New Orleans, Louisiana, had experienced excessive pavement distress in the form of differential subsidences of 2 to 8 inches (51 to 203 mm) along its length. This section of I-10 crosses a fresh water swamp. It was constructed during the period of September 1967 to July 1970 by removing approximately 15 to 20 feet (4.5 to 6 m) of soft, alluvial, organic clays overlying competent Pleistocene deposits and by replacing the soft clays with hydraulically placed sands pumped from sand bars of the Mississippi River, a distance of approximately 15 miles. The specifications for the sand required that it contain no more than three percent organic material, with 75 percent to 100 percent passing the No. 40 screen and 0 percent to 10 percent passing the No. 200 screen. The solid-to-water ratio of the pumped slurry was reported to be 14 percent to 18 percent by volume (Starring, 1972). Following the pumping of the sand, a sand surcharge of approximately 200 psf was applied for a period of three years to eliminate primary consolidation. Thus, a sand embankment about 15 to 25 foot (4.5 to 7.6 m) thick was formed to provide a foundation for this section of the highway. The highway consisted of two roadways each with two lane pavements, 24 foot (7.3 m) wide, with paved shoulders 10 foot (3 m) wide on the outside and 6 foot (1.8 m) wide on the inside and a 52 foot (15.8 m) depressed median strip between the roadways (Figure 5).

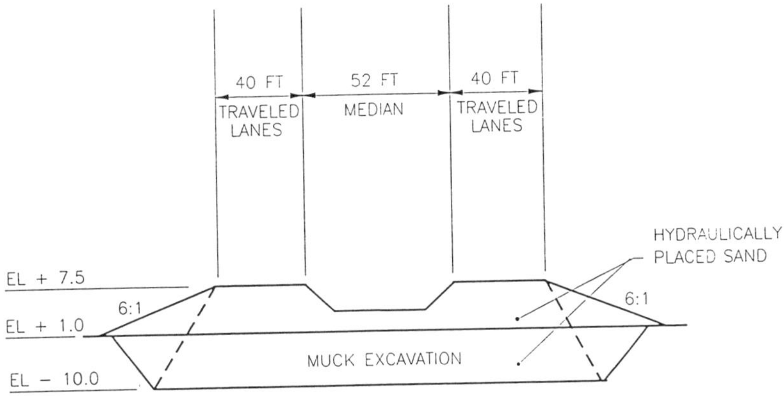

Figure 5 - I-10 Profile

Soil borings showed that the underlying material consisted basically of medium dense, fine silty sand from 0 to 6 feet (0 to 1.8 m) underlain by a loose fine silty sand to a depth of 9 feet (2.7 m). From 9 to 18 feet (2.7 to 5.5 m) the embankment is composed of very loose fine silty sand of extremely low density. The underlying material from 18 to 21 feet (5.5 to 6.4 m) is classified as very soft to stiff silty clay. The standard penetration tests varied from 35 blows in the upper 6 feet (1.8 m) of sand embankment to 6 blows between 6 and 9 feet (1.8 and 2.7 m) and 0 to 2 blows down through the hydraulic fill (Figure 6).

Figure 7 shows similar results for cone penetrometer tests. Grain size analysis show that the sand is uniform in size with about 95 percent retained on the 200 screen and negligible percentages of clays and colloids. An analysis of the clay beneath the hydraulic fill shows that in the worst case, assuming a 10 foot (3 m) thick layer of soft clay, single drainage and 0 tons per square foot original overburden, the settlement of the soft clay should have been completed in approximately 5 years; however, subsidence is still being experienced after 15 years and it is therefore assumed that a majority of the settlement is taking place in the hydraulic fill. The upper 6 feet (1.8 m) apparently has been densified by traffic loads.

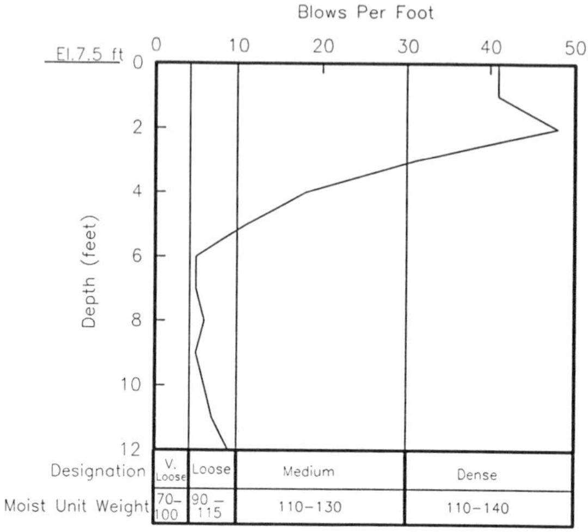

Figure 6 - Average Post-Treatment Standard Penetration Test Results

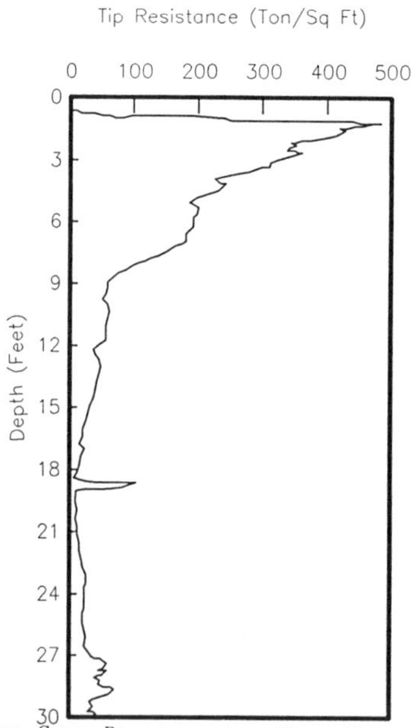

Figure 7 - Average Cone Penetrometer Test Results

The problem faced by the Louisiana Department of Transportation, was to abandon the existing roadway and build an elevated highway at an approximate cost of $56 million or remove the pavement and the embankment and reconstruct both for about $52 million; however, either solution would cause extensive economic impact on Baton Rouge and New Orleans. A solution to densify the hydraulic fill sand while maintaining traffic is the best alternative.

Deep dynamic compaction, vibro-compaction, vibro-replacement and compaction by explosives were considered for densifying the hydraulic sand fill. While they all would adequately perform at various economic considerations, the highway itself, would have to be rerouted and new pavement placed. With this in mind, it was elected by the Louisiana Department of Transportation to perform a grouting test program to try to densify the sand in-place. The techniques investigated were compaction grouting and lime/flyash injection, either individually or in combination.

The initial tests were performed in February and March 1988 in the median strip of the highway. The first phase of the grouting test involved three test pads injected with lime/flyash at various spacings and then injected with compaction grouting. One additional test pad three miles away was dewatered and then injected with lime/flyash; the takes here were higher, however no acceptable improvement was observed with the cone penetration testing. Preliminary results pointed out the effectiveness of the compaction grouting. The second phase of the test program was altered to perform only compaction grouting under the highway pavement. The purpose of the phase of the test was as follows:

1. to determine the improvement in densification that can be obtained by compaction grouting as measured mainly by before and after cone penetration results.

2. to determine the optimum spacings that can be utilized for the compaction grout pipes.

3. working under the highway paving and the hydraulic fill densified by the vibration of traffic, determine what is the maximum pressure than can be utilized to densify the lower hydraulic fill without causing objectionable heave.

4. to determine if all, or a majority of, the pipes can be placed from the shoulder area, thus minimizing traffic disruption

The complete results of this grouting test program will be presented in a future report.

Conclusion

Ground Modification techniques offer both a technical and economical solution to improving hydraulic fills, with each site being an additional learning experience to be shared with the engineering profession. In preparing this publication, other case histories were uncovered but for various reasons were unpublishable. Many valuable lessons are buried in engineer's and contractor's records and thus require these lessons to be relearned or prohibit soil improvement techniques to be utilized because no past published information is available. Unfortunately, in our society the sharing of experiences by publishing carries little reward for the non-academic except to the ego.

References

Arman, A., and Griffin, P.M., (1988), "An Investigation of the Failure of a Hydraulically Placed Embankment," Planned to be published by Transportation Research Board, 1988.

Arman, A. (1986), "An Investigation of Pavement Distress on Interstate Route 10 through McElroy Swamp," Louisiana State University, Baton Rouge, Louisiana, Research Project No. 86-16T(B), Louisiana Department of Transportation, Dated January 14, 1986.

Bell, A.L., Slocombe, B.C., Nesbitt, A.M. and Finey, J.T. (1986), "Vibro-Compaction Densification of a Deep Hydraulic Fill," Proceedings of Conference Building on Marginal and Derelict Land, Institution of Civil Engineers, 1-7 Great George Street, London, SW1 P3A, Held in Glasgow, Scotland, May 1986, pp. 567-573.

Leonards, G.A., Cutter, W.A. and Holtz, R.D. (1980), "Dynamic Compaction of Granular Soils," Journal of the Geotechnical Engineering Division, ASCE, Vol. 106, No. GT1, Proc. Paper 15144, January 1980, pp. 35-44.

Lukas, R.G. (1986), "Dynamic Compaction for Highway Construction," Vol. 1, Design and Construction Guidelines, Federal Highway Administration, Office of Research and Development, U.S. D.O.T., Washington, D.C., Report No. FHWA/RD-86/133, July 1986.

Menard, L. and Broise, Y. (1975), "Theoretical and Practical Aspects of Dynamic Consolidation," Geotechnique, Vol. 15, No. 1, March 1975, pp. 3-18.

Salley, J.R., Foreman, B., Henry, J.F., and Baker, W.H. (1987), "Compaction Grout Test Program - West Pinopolis Dam," Soil Improvement A Ten Year Update, ASCE Geotechnical Special Publication No. 12, Edited by Joseph P. Welsh, April 1987, pp. 245-264.

Seed, H.B. and Idriss, I.M. (1981), "Evaluation of Liquefaction Potential of Sand Deposits Based on Observations of Performances in Previous Earthquakes," Preprint 81-544, ASCE Fall Convention, St. Louis, Missouri, October 1981.

Schmertmann, J., Baker, W., Gupta, R., and Kessler, K. (1986), "CPT/DMT QC of Ground Modification at a Power Plant," Use of In Situ Tests in Geotechnical Engineering, ASCE Geotechnical Special Publication No. 6, New York, New York, USA, June 1986, pp. 985-1001.

Starring, J.W., (1972), "Sand Fill Pumped 15 Miles for Interstate Construction," Civil Engineering, ASCE, Vol. 41, No. 2, New York, New York, 1972.

Warner, J. (1982), "Compaction Grouting - The First Thirty Years," ASCE Specialty Conference on Grouting in Geotechnical Engineering, February 1982, New Orleans, Louisiana, pp. 694-707.

Welsh, J. (1986), "In Situ Testing for Ground Modification Techniques," Use of In Situ Tests in Geotechnical Engineering, ASCE Geotechnical Special Publication No. 6, New York, New York, USA, June 1986, pp. 322-335.

Dredged Fill Performance: South Blakeley Island

Marian E. Poindexter,* M. ASCE and James E. Walker, Jr.,** A.M. ASCE

Abstract

Recent developments in finite strain consolidation theory and analysis procedures have provided improved capabilities for predicting the long-term performance of confined disposal facilities for hydraulic fills. Such predictions are necessary for efficient utilization of both existing and planned containment facilities.

A computer model developed at the US Army Engineer Waterways Experiment Station has proven useful in analyzing dredged material containment facilities. This model, entitled Primary Consolidation and Desiccation of Dredged Fill (PCDDF), utilizes a finite difference numerical technique to solve the governing equation of finite strain consolidation theory; also included in the model is the capability for considering desiccation of the dredged material surface. The model is available in both mainframe and microcomputer versions.

The use of finite strain analyses in determining settlement rates has application in mineral processing operations and mining waste storage ponds, as well as in dredged material containment areas. Use of consolidation models such as PCDDF improve the estimation of hydraulic fill storage capacity and useful life. PCDDF can also be used to evaluate the cost effectiveness of various dewatering techniques in terms of accelerating desiccation and increasing settlements within an hydraulic fill. An evaluation of performance at the South Blakeley Island Disposal Site, Mobile Harbor, Alabama is presented in this paper.

Introduction

Quantities of sediment approaching 500 million cubic yards are dredged annually from the navigable waterways of the United States. After these sediments are dredged, they must be placed in an environmentally acceptable manner at designated disposal sites, many of which are diked, hydraulically filled containment areas located in an upland environment. Port development, coupled with stringent

---

*Research Civil Engineer, Environmental Engineering Division, US Army Engineer Waterways Experiment Station, P.O.Box 631, Vicksburg, MS 39180-0631.

**Civil Engineer, Operations Division, US Army Engineer District, Mobile, P.O.Box 2288, Mobile, AL 36628-0001.

environmental and economic constraints, has decreased significantly the land available for use as dredged material containment areas. Therefore, a critical need for maximum utilization of both existing and planned containment areas has been created in recent years.

Primary responsibility for maintaining navigation in this country belongs to the US Army Corps of Engineers pursuant to the Rivers and Harbors Act of 1899. Associated with the responsibility for dredging is the requirement for obtaining, approving, and/or utilizing disposal sites for placement of the dredged material. In order to more effectively carry out these responsibilities, the Corps of Engineers has conducted a number of major research projects to provide guidance for, and to quantify environmental effects of, dredging and dredged material disposal (Lazor et al. 1984). These efforts have culminated, in part, in development of analytical models and field management techniques which can be applied to confined hydraulic fills to increase their long-term storage capacity and thus improve their overall performance.

Guidance for designing, operating, and managing the Corps of Engineers hydraulic fill structures has evolved over the past ten years and presently includes several state-of-the-art laboratory testing and data analysis procedures (Palermo, Montgomery, and Poindexter 1978; Palermo 1986; Office, Chief of Engineers 1987; Poindexter 1987; Rollings, Poindexter, and Sharp 1988). One significant aspect of this guidance is the method of geotechnical analysis used in hydraulic fill performance predictions. Recent developments in finite strain consolidation theory and geotechnical laboratory testing procedures (Gibson, England, and Hussey 1967; Gibson, Schiffman, and Cargill 1981; Cargill 1985; Townsend 1987; Poindexter 1987) have provided improved capabilities for predicting the long-term performance of hydraulic fill structures. The use of this guidance has application in mineral processing operations and mining waste storage ponds, as well as in dredged material containment areas.

Procedures developed by the US Army Engineer Waterways Experiment Station for analysis of hydraulic fill structures have been successfully applied to a number of dredged material containment areas. Most recently the procedures have been used to analyze the South Blakeley Island containment area in Mobile, Alabama as reported in this document. Evaluation of the South Blakeley Island hydraulic fill included predictions of surface elevation over time, comparisons of predictions with field observations, comparisons of various levels of dewatering efforts, and discussion of the cost-effectiveness of dewatering.

Project Description

The Port of Mobile, Alabama has grown steadily in recent years and has become a major U.S. port in the Gulf of Mexico, passing approximately 40 million tons (36 million metric tons) of cargo through its facilities annually. It is located at the confluence of the Mobile River and Mobile Bay in the southwestern portion of the State of Alabama; its location at the ocean terminus of the

Tennessee-Tombigbee Waterway enhances the already thriving port. With current deepening of the Mobile ship channel to a depth of 55 ft (17 m) and subsequent port development, almost 200 million cubic yards (153 million m^3) of material, in addition to the annual maintenance dredging volume of 6.5 million cu yd (4.6 million m^3), is expected to be generated.

The combination of large quantities of dredged material and a growing metropolitan/industrial complex has caused a severe shortage of land available for new dredged material containment areas. Nearly all remaining undeveloped land is comprised of the Mobile River Delta and other pristine wetlands which are located adjacent to the Mobile River and ship channel. In an effort to minimize the impact of dredging and disposal of sediment upon the environment, the Mobile District has initiated a management plan for the Upper Mobile Harbor. This plan includes operation and management of the South Blakeley Island disposal site, in conjunction with management of three additional Upper Mobile Harbor disposal sites, to maximize the storage life of these hydraulic fill structures (see Figure 1).

The existing South Blakeley Island disposal site is a 200-acre (81 ha) diked facility which is located near the intersection of the Mobile River and Mobile Bay. This site, which has been in use for approximately 20 years, has a present dike height of 38 ft (11.6 m) above Mean Lower Low Water (MLLW). The land adjacent to the north, south, and west sides of the site is at approximately EL +10

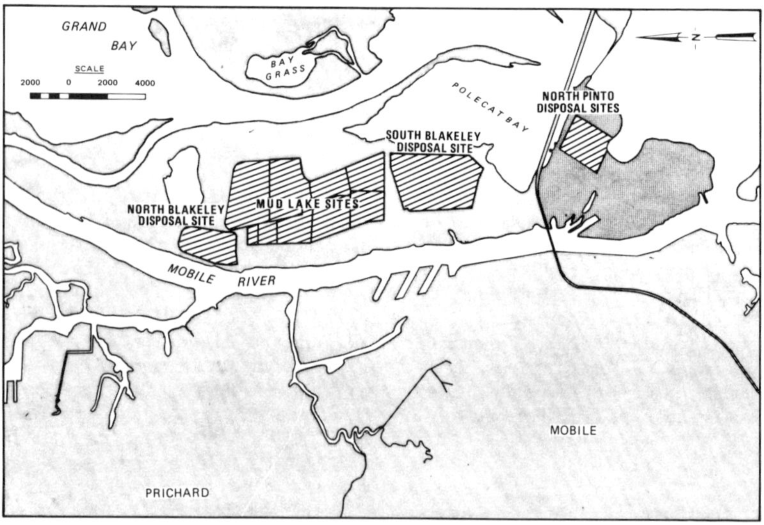

Figure 1. Vicinity map for Mobile Harbor, Alabama showing hydraulic fill disposal sites

MLLW. The east side of the facility is adjacent to Polecat Bay with the area immediately adjacent to the toe of the dike at EL +1 MLLW. The material in the containment area was at an elevation between EL +20 and EL +25 MLLW before the 1985-1986 disposal operation.

The disposal site has a generally rectangular shape. One spur dike extends across the site as shown in Figure 2. The hydraulic pipeline inflow location has traditionally been alternated between the two locations shown in this figure, while the outflow weirs are located along the eastern dike. The weirs used for a particular disposal operation depend upon the inflow pipeline location. When the inflow is located south of the spur dike, the weirs north of the spur dike are used for water removal during and immediately after disposal; for disposal inflow north of the spur dike, the southern weirs are used.

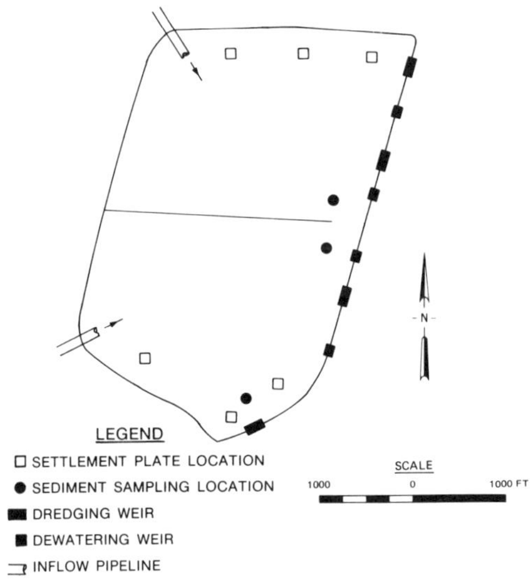

Figure 2. Configuration of South Blakeley Island disposal site

Two types of weirs are utilized at this site. The dredging weirs are typical rectangular box weirs which are used during the disposal operation to remove excess ponded water from the site; these weirs are 20 ft (6 m) by 4 ft (1.2 m), have a weir crest length of 20 ft (6 m), and are drained by multiple 30-in.(0.75 m) diameter discharge pipes. Smaller dewatering weirs have also been installed at the South Blakeley Island site to facilitate continued dewatering after completion of disposal operations; these smaller weirs can more easily be removed and reinstalled at a lower elevation, if/when the dredged material surface elevation falls below the minimum weir crest elevation. The four dewatering weirs are each composed of 4-ft (1.2 m) by 4-ft (1.2 m) vertical box culvert with one 30-in. (0.75 m) diameter discharge pipe which transports water from the weir structure. All weirs have adjustable crest elevations to facilitate removal of surface water from the site.

South Blakeley Island disposal site receives approximately 1 million cu yd (765,000 m^3) of dredged material from the navigation channel at approximately two year intervals. This sediment, which is predominantly fine-grained, is hydraulically pumped into the site at an average slurry concentration of approximately 145 g/l. Thus the material is subject to a considerable amount of settlement. To measure the amount of settlement which occurred in this disposal site, six settlement plates were installed. The locations of these plates are shown in Figure 2. Because the field monitoring efforts were not initiated before the last disposal operation, access to the site was limited at the time of settlement plate installation and the plates had to be located near the perimeter of the containment area. These plates were installed on the surface of the drying dredged material, a fact which further reduced their ultimate usefulness.

The last placement of hydraulic fill into the South Blakeley Island site occurred between October 1985 and January 1986. At that time, approximately one million cu yd (765,000 m^3) of sediment were dredged from the channel and placed into the site. This material initially occupied a volume of approximately 1.6 million cu yd (1.2 million m^3) in the disposal site, corresponding to a void ratio of 9.6. The lift thickness varied somewhat within the disposal site as a result of irregularities in the surface topography caused by previous construction borrowing and dike raising activities, but generally averaged 5.5 ft (1.7 m).

Management Practices

Several management practices were identified during the Dredged Material Research Program (Palermo et al. 1978) which assist in maximizing the storage capacity of dredged material containment areas. The practices include: (a) proper relative locations of inlets and outlets, (b) thin lift placement (<3-5 ft(1-2 m)), (c) progressive surface trenching, (d) removal of dried material, and (e) compartmentalization of sites larger than 300 acres (120 ha). These techniques were utilized at the South Blakeley Island disposal site to the maximum extent possible and are described below.

At the South Blakeley Island disposal site, the inflow pipeline was typically placed along the western side of the site, while the weir structures were located along the east dike as shown in Figure 2. By so locating the inflow and outflow points, the dredged material surface had a very gentle slope from the west dike to the weirs on the east. This aided in initial drainage of the site. Minor adjustments in inflow pipeline location were made as necessary to fill any scour holes or to otherwise correct major irregularities in the dredged material surface resulting from previous disposal operations and/or borrowing activities for dike raising.

After the disposal operation was completed in late January 1986, dewatering of the disposal site was begun. First, the weir crests were lowered on the four dredging weirs, thus decanting the ponded surface water. By using all dredging weirs, the depth of withdrawal at each weir was minimized; this in turn minimized the resuspension of sediment and ensured good effluent water quality. Removal of all surface water was accomplished within about 30 days. At that time, the dredged material surface was exposed to evaporative drying. From approximately March through July 1986, dewatering efforts consisted of periodic lowering of the weirs' crest elevations as the dredged material surface subsided.

In August 1986, the first active dewatering program began. This activity consisted of constructing a perimeter trench around the entire site. The trench was constructed by a dragline working from the dikes. Material was removed from the site using the maximum practical reach of the dragline and was placed on the inside slope of the confining dikes. The dewatering weirs were opened in August to facilitate the dewatering efforts.

The second phase dewatering program continued the dewatering activities at the South Blakeley Island site from October 1986 through March 1987. This effort involved continuation and deepening of the perimeter trench and initiation of interior trenching in the north cell of the disposal site. The perimeter trenching was accomplished by two draglines, while the interior trenching was performed by a backhoe operating on double mats.

The third phase of the dewatering effort was initiated in July 1987. In addition to one dragline and a backhoe, this phase utilized a rubber-tired vehicle which pulled a rotary ditcher. In Figure 5, the ditching equipment is shown operating in the South Blakeley Island site. This rubber-tired vehicle with rotary ditcher was used to construct and deepen the interior trenches in both the north and south portions of the South Blakeley Island site. The dragline, operating on double mats, continued with perimeter trench deepening and construction of sumps in front of the weirs. The backhoe was used for perimeter trenching. Operating on single mats, the backhoe was also used to clear any blockages which occurred at trench intersections.

The final phase of dewatering at South Blakeley Island, including some additional trench deepening, will begin during the spring of 1988. It also includes an earthwork contract for excavation and hauling of dried dredged material from the site. The material will be scraped from the surface and windrowed between the trenches. It will then be removed from the site by self-loading and/or top loading scrapers and used to raise and upgrade the confining dikes at the site. This final activity is scheduled for completion during September 1988.

The interior trenches, illustrated in Figure 6, extend from west to east in the South Blakeley Island site and were constructed at 100-ft (30 m) intervals. This spacing was selected since it was

Figure 5. Trenching equipment operating in South Blakeley Island disposal site

Figure 6. Trenching pattern used at the South Blakeley Island site showing perimeter trench in foreground

the minimum spacing which could be used while allowing adequate room for maneuvering of self-loading and top-loading scrapers between trenches. The trenches constructed by the rotary ditcher were trapezodial in shape and had a bottom width of 14 to 18 in. (0.3 to 0.5 m) with a side slope of 30 to 32 degrees. These trenches were initially constructed to a depth of about one ft (0.3 m) or the depth to which the trench would stand open, whichever was the lesser. With each progressive deepening operation, the trenches were deepened approximately one ft (0.3 m), until a trench depth of 4 ft (1.2 m) was obtained.

It was found empirically during this trenching operation that there is an optimum moisture content for trenching dredged material, particularly when using a rotary trencher. In the early stages of trenching, the material is typically too wet to maintain a regularly shaped trench of a given depth. During the intermediate stages, the material is wet enough to be easily thrown clear of the trencher and the trench, and it is strong enough to maintain the trench as cut. Later when the material has dried further, some operational problems are encountered as the material tends to adhere to the housing of the trencher, forming large clumps of material which eventually fall beside or into the trench and block drainage.

Settlement Predictions

To predict the performance of an hydraulic fill structure, the geotechnical properties of the hydraulic fill material must be determined, an appropriate procedure for analysis must be selected, and the adequacy of the predictions should be verified against field performance data. For analysis of the South Blakeley Island disposal site, dredged material samples were collected and tested; an analysis procedure which has been successfully used at other sites was selected; and field settlement data were collected for comparison to the model predictions.

Material Properties. Classification and consolidation tests must be performed to determine the necessary geotechnical properties of a dredged material hydraulic fill. Routine laboratory classification testing provides the specific gravity and Atterberg Limits, while the void ratio - effective stress ($e - \sigma'$) and void ratio - permeability ($e - k$) relationships must be determined from one or more of four laboratory consolidation tests available to the Corps of Engineers (Poindexter 1987, 1988). The laboratory consolidation tests used by the Corps of Engineers for dredged material testing are: a modified fixed ring consolidation (oedometer) test, the self-weight consolidation test, the large strain, controlled rate of strain consolidation test, and a developmental automated slurry consolidation test. A comparison of these test procedures has been presented by Poindexter (1987, 1988).

Laboratory testing on the Mobile Harbor dredged material was conducted at the Waterways Experiment Station. The material was classified according to the Unified Soil Classification System (USCS) as a CL-ML, with a liquid limit of 96.0, a plastic limit of 28.0, and a specific gravity of 2.70. Both the self-weight consolidation test and the modified fixed ring consolidometer were used to determine the compressibility and permeability characteristics of the material; results of these tests are shown in Figures 3 and 4.

Analysis Procedure. Because many soft, fine-grained materials may eventually undergo 50-percent strain or more, Terzaghi's conventional small strain theory is not technically applicable to analyses of hydraulic fill structures. A more appropriate approach involves use of a finite strain consolidation theory. The most general and least restrictive of the many one-dimensional primary consolidation formulations is the finite strain theory developed by Gibson, England, and Hussey (1967).

The governing equation for finite strain consolidation theory is based on the continuity of fluid flow in a differential soil element, Darcy's law, and the effective stress principle, similar to the conventional consolidation theory. However, finite strain theory can additionally consider vertical equilibrium of the soil mass, place no restriction on the form of the stress-strain relationship, allow for variable coefficient of permeability, and accommodate any degree of strain. The governing equation is:

$$\left(\frac{\gamma_s}{\gamma_w} - 1\right) \frac{d}{de}\left[\frac{k(e)}{1 + e}\right] \frac{\partial e}{\partial z} + \frac{\partial}{\partial z}\left[\frac{k(e)}{\gamma_w\,(1 + e)} \frac{d\sigma'}{de} \frac{\partial e}{\partial z}\right] + \frac{\partial e}{\partial t} = 0$$

where:
γ_s = unit of weight of solids
γ_w = unit of weight of water
e = void ratio
$k(e)$ = soil permeability as a function of void ratio
z = vertical material coordinate measured against gravity
σ' = effective stress as a function of void ratio
t = time

This equation is well suited for the prediction of consolidation in thick deposits of very soft hydraulic fills since it accounts for the large strains and nonlinear soil properties inherent in these materials.

Since the dredged material surface is allowed to undergo evaporative drying in many containment areas, it is essential to incorporate predictions of desiccation settlement in evaluations of disposal site capacity. An empirical description of the desiccation process has been developed in terms of water balance in the upper portion of dredged material layers (Cargill 1985). Procedures for calculation of soil evaporation rates and depths of influence have also been developed. Site-specific climatic conditions are incorporated in the analysis procedures. The predictive model developed utilizes void ratios instead of water contents in order to be compatible with the consolidation model.

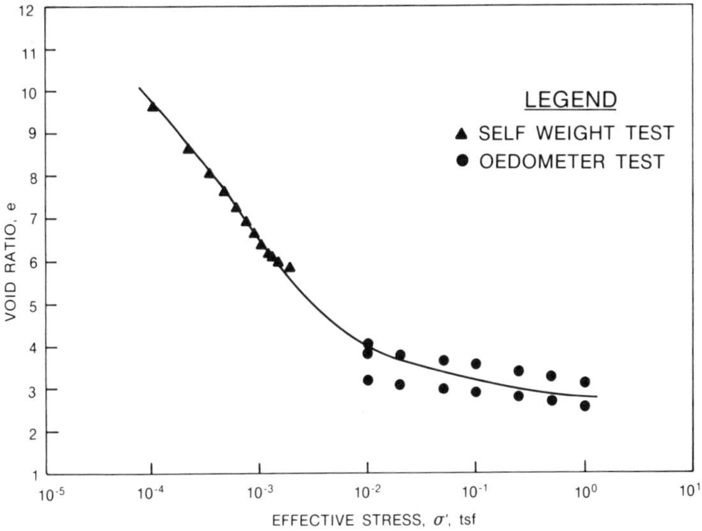

Figure 3. Compressibility data for Mobile Harbor sediment

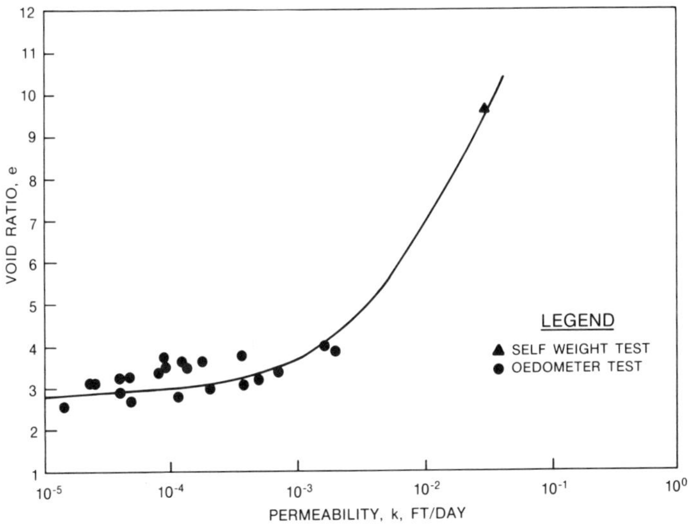

Figure 4. Permeability data for Mobile Harbor sediment

Both the finite strain consolidation model and the empirical desiccation model have been programmed for computer solution (Cargill 1982, 1985). The program Primary Consolidation and Desiccation of Dredged Fill (PCDFF) incorporates an explicit finite difference mathematical approximation to describe the consolidation process. Monthly adjustments in the top boundary condition and location are made to account for the amount of desiccation which has occurred. In addition to material settlement which comes from a calculation of void ratio distribution, the program also calculates the distribution of stresses and pore pressures throughout the dredged material layer. Any sequence of material deposition can be considered by the program.

PCDDF has been incorporated into the Automated Dredging and Disposal Activities Management System (ADDAMS), a collection of dredging related computer programs assembled by the Corps of Engineers (Walski et al. 1984). This makes the program more accessible to potential users and will allow its use in conjunction with other dredging and dredged material containment area design and analysis programs.

The model PCDDF was selected for use at the South Blakeley Island site since it has been used successfully to predict the performance of numerous hydraulic fills. For example, Cargill (1985) has reported very good comparisons of predictions with field data for dredged material disposal sites in Norfolk, VA, Charleston, SC, and Canaveral, FL. Sites in Savannah, GA and Wilmington, DE have also been analyzed and accurate predictions reported (Benson 1987). The model PCDDF has recently been used by Poindexter to predict the consolidation settlement as well as the void ratio, effective stress, and excess pore water pressure profiles for various phosphatic clay waste ponds as a part of a consolidation prediction symposium (Townsend 1987); results from PCDDF compared well with other state-of-the-art consolidation models.

Prediction. Within PCDDF, various levels of disposal site management for dewatering are simulated through the use of empirical coefficients. The coefficient for site drainage efficiency was varied for the South Blakeley Island disposal site prediction to simulate various levels of dewatering effort within the site. This coefficient can be varied from a value of 0.0 for an inactive dewatering scenario to 1.0 for a well-drained site which has an active dewatering program.

For a 5.5 ft (1.7 m) lift of dredged material placed in the South Blakeley Island disposal site, the variation caused in the predicted surface elevation as a result of variation of the drainage efficiency coefficient is shown in Figure 7. As illustrated in this figure, predictions using values of the drainage efficiency coefficient between 0.5 and 1.0 ultimately result in equal amounts of settlement, although the rate of settlement varies significantly during the first two to three years after hydraulic fill placement. The rate of settlement becomes significant if subsequent disposal operations are scheduled to occur before the curves converge. This

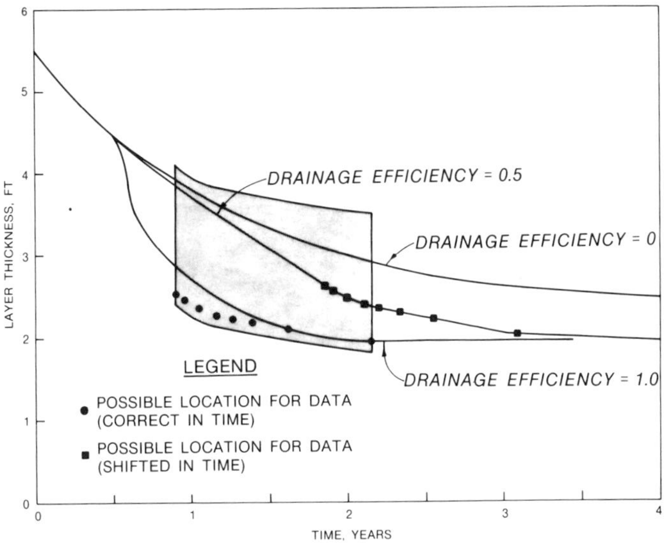

Figure 7. Predicted layer thickness over time for various drainage efficiency factors

figure indicates that a more aggressive trenching/dewatering program is warranted when frequent disposal operations will occur at a site, if capacity of the site is to be maximized. At present only a qualitative correlation exists between the value used for the coefficient of drainage efficiency and the intensity of the field dewatering efforts.

At the South Blakeley Island site field surveys of settlement plates within the site have been conducted periodically since September 1986 in an attempt to verify the adequacy of the predictive model while also establishing a stronger correlation between field activities and model parameters. Unfortunately, several factors combined to hamper this effort. Because the plan to monitor settlement in the South Blakeley Island site was not initiated prior to the 1985-1986 disposal operation, no pre-disposal surveys of the interior of the containment area were conducted to establish an average surface elevation. Also, the settlement plates were not installed before disposal; therefore they had to be placed on the surface of the new deposit, thus precluding the ability to determine the layer thickness at a given time. Since they were placed subsequent to disposal and only a thin surface crust had developed at the time of installation, the settlement plates could only be placed near the perimeter of the site; thus they do not necessarily indicate conditions representative of the majority of the site. Additionally, surveys of the plates were not conducted during the early stages of consolidation, resulting in an incomplete field data curve.

As a result of these problems, the field data obtained at South Blakeley Island can only be located relative to each other and with respect to time. The data cannot be definitely located with respect to thickness of the dredged material deposit, thus making it impossible to determine which drainage efficiency factor best models field behavior at this site. Therefore, all possible locations for the field data are shown by the shaded area in Figure 7. If a complete field data curve had been obtained, it could more properly be located relative to the various predictions. It should be noted that when the data are positioned as shown by the circles in Figure 7 the data for the last nine months of the field monitoring effort fall almost identically on the curve for a drainage efficiency coefficient of 1.0. Alternately, if the data are shifted one year to the right in time, the entire set of field data perfectly matches the predicted curve for a drainage efficiency factor of 0.5 at its point of maximum curvature, possibly indicating that the mechanism for modeling dewatering is correct but the predicted rate is too slow.

Collection of additional field data from the South Blakeley Island hydraulic fill would permit more correct interpretation of the data and possible improvement of the predictive technique. While additional and more comprehensive field monitoring is needed at this site to conclusively determine how well the model PCDDF and the field data compare, it should be remembered that PCDDF has been used successfully and accurately to predict hydraulic fill settlement at a number of other locations.

Storage Capacity Evaluation

Containment areas intended for use in conjunction with recurring disposal operations must be sized for long-term storage capacity over the service life of the facility. Storage capacity is defined as the total volume available to hold additional dredged material and is equal to the total unoccupied volume minus the volume associated with ponding and freeboard requirements. The estimation of long-term storage capacity is an important consideration for long-term planning and design of new containment areas or evaluation of the remaining service life of existing sites. The disposal site capacity will vary with the quantity of material placed, the level of dewatering efforts, and the time available for consolidation between disposal operations.

The computer model PCDDF was used to evaluate the storage capacity and to determine the remaining disposal life of the South Blakeley Island disposal site for both inactive and active dewatering scenarios. The material consolidation properties determined for the Mobile Harbor material were used in the analysis. The desiccation parameters were determined by Walker (1987). In order to conduct a conservative evaluation, the coefficient for drainage efficiency was taken as 0.5.

At the South Blakeley Island disposal site, approximately 1 million cu yd (765,000 m^3) of dredged material have historically been placed in the disposal site every two years. Considering the size of this site and the initial density of the hydraulic fill, placement of the anticipated quantity of material results in an initial dredged material layer thickness of 5.5 ft (1.7 m) (Walker 1987). The site had a maximum permissible dike height of 18 ft (5.5 m) above the existing dredged material surface before the 1985-1986 disposal operation. This corresponds to a height of additional dredged material of 14 ft (4.3 m) when the required 2 ft (0.6 m) of ponding depth and 2 ft (0.6 m) of freeboard are considered.

Results of the filling simulations are presented in Figure 8. As can be seen, the storage capacity will be exhausted after four additional disposal operations using the inactive dewatering scenario. With inactive dewatering, the remaining storage capacity of the site would be 4 million cu yd (3.1 million m^3) of channel sediment; this capacity would be exhausted in six years, eg. in 1991. Alternately, an active dewatering effort including a progressive trenching program would extend the disposal life of the South Blakeley Island site to 1993; this corresponds to a remaining disposal life of eight years and an additional dredged material volume of 5 million cu yd (3.8 million m^3).

It should be noted that the Mobile District's Upper Mobile Harbor dredged material management plan specifies that less than 1 million cu yd (765,000 m^3) of material will be placed in the South Blakeley site biannually. With implementation of this plan, the disposal life of the site can be expected to increase as the biannual volume of material placed in South Blakeley Island is decreased.

Cost Effectiveness

Quantifying absolutely the cost effectiveness of dewatering efforts is a difficult and complex task. The environmental impact of disposal site expansion and/or replacement can be almost impossible to quantify, although environmental impacts should certainly be minimized. A simple approach for analyzing the cost effectiveness of dewatering efforts is to compare the cost associated with raising the existing dikes at the disposal site. The latter approach was used to evaluate the cost effectiveness of dewatering efforts at the South Blakeley Island disposal site.

The total cost for dewatering one lift of dredged material in the South Blakeley Island disposal site, as reflected by the three dewatering activities presented earlier, was approximately $400,000. Based upon model predictions, if an active dewatering program is established after each disposal operation, then the South Blakeley Island site will be able to contain material from five disposal operations, instead of the four lifts which would deplete its capacity if an inactive dewatering program were followed. Therefore, by incurring expenses of $400,000 for dewatering each of four lifts of material, an additional lift of material can be placed into

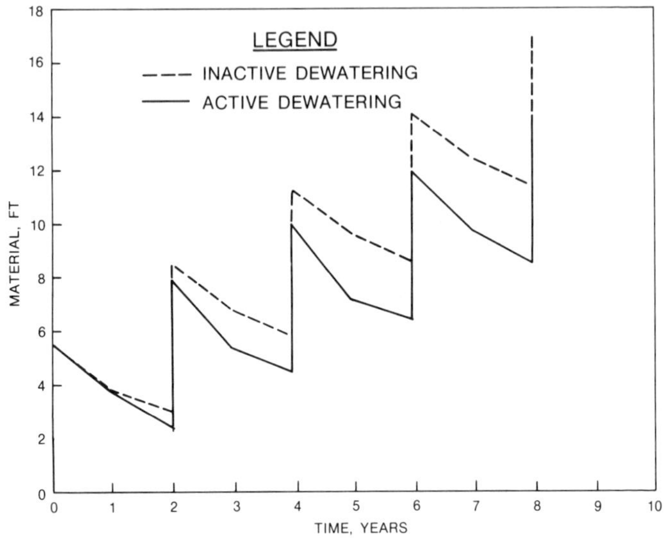

Figure 8. Projected filling sequence for South Blakeley Island disposal site

the South Blakeley Island site (with its existing dikes); this represents a total expenditure for dewatering of $1,600,000.

In contrast, if an inactive dewatering program were followed, the dikes at South Blakeley would have to be raised five feet before a fifth lift of dredged material could be placed in the site. Recent costs for raising the existing dikes by five feet at the South Blakeley site totaled $2,037,000.

In comparison, the dewatering efforts are somewhat less expensive ($1.6 million as compared to $2 million) than the dike raising activities required to provide the same additional capacity. An additional factor which favors dewatering is that by following this approach, the capability for raising the dikes in the future is still present and represents additional storage volume that would otherwise be nonexistent. Also when dewatering is accomplished, additional storage capacity can be obtained by removing the dried material for dike raising activities. Obviously, there are other factors which should be considered in a thorough evaluation of the cost effectiveness of dewatering dredged material. Such factors as the time value of money, the reduced cost of dike raising where dried material from within the site is used, the increase in storage capacity resulting from removal of the dried material, the expense associated with the environmental process of obtaining new sites, etc. should be evaluated.

Conclusion

Procedures have been developed for maximizing and predicting the settlement which will occur within an hydraulic fill structure. These procedures have been applied to the South Blakeley Island dredged material disposal site in Mobile, AL where a progressive trenching program effectively dewatered the recently deposited dredged material. It was found that additional comprehensive field data will be needed for thorough model verification and to more closely correlate empirical model coefficients with field operations. A re-evaluation of the remaining storage capacity of the South Blakeley Island containment facility should be conducted as the Upper Mobile Harbor dredged material disposal management plan is implemented and the future quantities of material to be placed in this site are determined.

Acknowledgment

Procedures for testing and analysis of dredged material hydraulic fills were developed at the US Army Engineer Waterways Experiment Station as a part of the Dredged Material Research Program and the Dredging Operations Technical Support Program. Field and laboratory data from the South Blakeley Island disposal site were collected, and made available by personnel of the US Army Engineer District, Mobile. Permission was granted by the Chief of Engineers to publish this information.

Appendix. - References

Benson, R. E., "Long Term Storage Capacity of Confined Disposal Sites: A Comparison of Predictions with Field Behavior," Environmental Effects of Dredging Technical Note, December 1987, US Army Engineer Waterways Experiment Station, Vicksburg, Miss.

Cargill, K. W., "Consolidation of Soft Layers by Finite Strain Analysis," Miscellaneous Paper GL-82-3, March 1982, U. S. Army Engineer Waterways Experiment Station, CE, Vicksburg, Miss.

Cargill, K. W., "Mathematical Model the Consolidation/Desiccation Process in Dredged Material," Technical Report D-85-4, July 1985, U. S. Army Engineer Waterways Experiment Station, CE, Vicksburg, Miss.

Gibson, R. E., England, G. L., and Hussey, M. J. L., "The Theory of One-Dimensional Consolidation of Saturated Clays. I. Finite Non-Linear Consolidation of Thin Homogeneous Layers," 1967. Geotechnique, Vol 17, No. 3, pp 261-273.

Gibson, R. E., Schiffman, R. L., and Cargill, K. E. 1981. "The Theory of One-Dimensional Consolidation of Saturated Clays, II. Finite Non-Linear Consolidation of Thick Homogeneous Layers," Canadian Geotechnical Journal, 18, pp. 280-293.

Haliburton, T. A., "Guidelines for Dewatering/Densifying Confined Dredged Material," Technical Report DS-78-11, September 1978, U. S. Army Engineer Waterways Experiment Station, CE, Vicksburg, Miss.

Lazor, R. E., Calhoun, C. C., and Patin, T. R., The Corps' Environmental Effects of Dredging Programs," November 1984, Proceedings of the ASCE Specialty Conference Dredging 84, Clearwater Beach, Fla.

Montgomery, R. L., Ford, A. W., Poindexter, M. E., and Bartos, M. J. "Guidelines for Dredged Material Disposal Area Reuse Management," Technical Report DS-78-12, December 1978, U. S. Army Engineer Waterways Experiment Station, CE, Vicksburg, Miss.

Palermo, M. R., "Development of a Modified Elutriate Test for Estimating The Quality of Effluent from Confined Dredged Material Disposal Areas," Technical Report D-86-4, August 1986, U. S. Army Engineer Waterways Experiment Station, CE, Vicksburg Miss.

Palermo, M. R., Montgomery, R. L., and Poindexter, M. E., "Guidelines for Designing, Operating, and Managing Dredged Material Containment Areas," Technical Report DS-78-10, December 1978, U. S. Army Engineer Waterways Experiment Station, CE, Vicksburg, Miss.

Poindexter, M. E., "Consolidation Properties of Dredged Material," May 1987, Proceedings of the University of Florida and Florida Institute of Phosphate Research Symposium on Consolidation and Disposal of Phosphatic and Other Waste Clays, Lakeland, Fla.

Poindexter, M. E., "Subaqueous Dredged Material Mounds: Effect on Disposal Site Capacity," May 1988, dissertation submitted in partial fulfillment of requirements for PhD, Texas A&M University, College Station, Texas.

Rollings, R. S., Poindexter, M. E., and Sharp, K. G. "Heavy Load Pavements on Soft Soils," August 1988, Proceedings of the 1988 Australian Road Research Board Conference, Canberra, A.C.T., Australia.

Townsend, F. C. "Summary of Numerical Predictions," May 1987, Proceedings of the University of Florida and Florida Institute of Phosphate Research Symposium on Consolidation and Disposal of Phosphatic and Other Waste Clays, Lakeland, Fla.

Walker, J. E., Jr., "Finite Strain Analysis of Settlements at South Blakeley Disposal Site, Mobile Harbor, Alabama," August 1987, independent study report prepared in partial fulfillment of requirements for Masters Degree, Colorado State University, Fort Collins, Colorado.

Walski, T. M., Hayes, D. F., Pranger, S. A., and Schaefer, T. E., "Automated Dredging and Disposal Activities Management System (ADDAMS)," November 1984, Proceedings of the ASCE Specialty Conference Dredging '84, Clearwater Beach, Fla.

AIR-JETTING DENSIFICATION OF SAND TAILINGS

Ronald L. Halley, Jr., M. ASCE, & Michael L. Jacobs, A.M. ASCE[1]

Abstract

Air-jetting densification of medium graded tailings upstream of the crest of an earthfill tailings embankment is discussed. The tailings embankment is located in an area of moderate to significant seismic activity. The in situ density of the tailings beach had averaged 1.3 g/cm^3, inadequate to resist liquefaction when subjected to the design earthquake loading of 0.2g horizontal ground acceleration. After air-jetting induced liquefaction, all but one of the measuring stations recorded densities exceeding 1.5 g/cm^3, the target density. This permitted utilization of the tailings beach to enlarge the existing embankment with modified upstream construction methods and resulted in significant savings of both construction time and cost.

Introduction

Rosario Dominicana is currently operating the Pueblo Viejo mine in the Dominican Republic, extracting gold and silver. The mine utilizes a Counter-Current-Decant (CCD)/Merrill-Crowe process plant. This process requires that the ore be crushed into a relatively fine material. The ore is processed at an average rate of 9,300 tonnes per day and the resulting tailings are spigotted along the upstream face of the Mejita tailings embankment, a zoned earthfill structure consisting primarily of compacted clay and presently 73 meters high. The embankment crest length is over 1,100 meters.

The embankment has been raised four times since the initial stage was completed in 1974, and only the second stage enlargement was anticipated during original design. These first two stages were designed by TAMS (20), and the third stage (second crest raising) was designed by Pincock, Allen, & Holt in 1980 (15). All subsequent stages have been designed by Hydro-Triad, Ltd. and the most recent addition was completed in 1986 (8).

At that time a final addition to the crest was anticipated, which will ultimately raise the embankment to a height of nearly 80 meters. In preparation for the final crest raising, a rockfill zone on the crest and upstream face of the existing embankment was constructed. This upstream construction method widened the crest from approximately 8 meters to 26.5 meters, but relied upon the tailings beach to laterally support the rock zone. Under static loadings the tailings are stable and the rock zone provides a solid foundation for the final stage. The Dominican Republic, however, is located in a region of moderate to significant seismicity. Under dynamic loadings, the tailings in their normal state may liquefy. Therefore, the upstream stability of the

[1] Respectively, President, and Water Resource Engineer, Hydro-Triad, Ltd., 1310 Wadsworth Blvd., Lakewood, Colorado 80215

proposed embankment increase was unacceptable since the tailings beach laterally supports the upstream rock zone. Stability analyses indicated, however, that a final stage could be constructed upon the rock zone with appropriate factors-of-safety if a zone of tailings could be made relatively immune from liquefaction.

To resist potential liquefaction, loose, cohesionless materials such as the tailings are often densified. Various densification methods have been utilized in the past, including dynamic compaction, vibroflotation, and use of explosives. A densification program was pursued at the Pueblo Viejo mine. The objective of the program was to render the tailings beach relatively immune from liquefaction under the design loadings, thereby permitting construction of the final embankment stage to be founded upon the recently constructed upstream rock zone and saving considerable time and cost over downstream construction alternatives.

Site Description

Tailings beach.- The tailings have been spigotted in a 45 percent solids by weight slurry in a pattern designed to create a tailings beach, and the beach slopes away from the embankment at about a one percent slope. Due to this deposition pattern, coarser material tends to settle near the embankment and the finer grained materials are carried out into the impoundment.

In preparation for the final crest raising, a rock zone was constructed on the crest and upstream slope of the embankment, widening the crest to 26.5 meters. At the time the rock zone was constructed, however, the tailings surface was encroaching upon the crest, leaving little of the upstream slope exposed. Therefore, to make full use of the available horizontal distance of the exposed embankment's upstream face, the upstream slope of the rock zone was constructed nearly vertical. Stability of the rock zone is provided by the tailings beach, which laterally supports the rock zone. The crest width was originally 8 meters wide, but was increased to a width of 26.5 meters, the crest elevation was increased 4.5 meters, and the downstream slope was held constant.

The in situ density of the beach sand tailings average 1.3 g/cm^3, but in the surface zones the density can be as low as 1.1 g/cm^3. The saturation level is typically one meter or so below the surface. The beach sands have approximately 45 percent fines and the coarse fraction rarely exceeds 2 mm. A gradation curve for an average tailings sample extracted from the beach sands is shown on Figure 1, and the soil properties are listed in Table 1 (1).

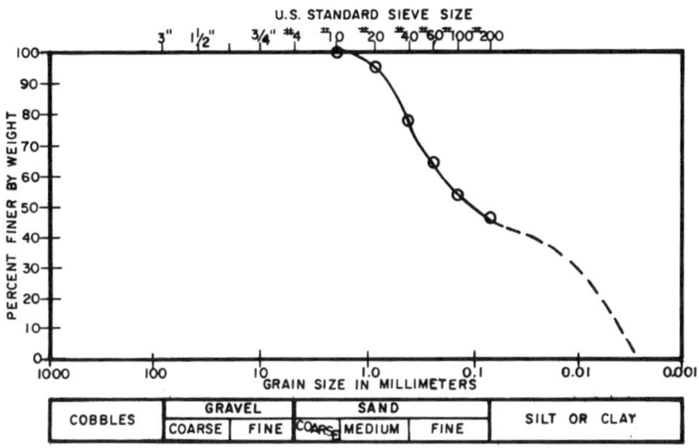

FIGURE 1 - BEACH SAND TAILINGS GRADATION CURVE

TABLE 1 - SOIL PROPERTIES OF THE BEACH SAND TAILINGS

Soil Property	Value
Percent Fines (-200 U.S. Sieve Material)	45 %
Minimum Void Ratio	0.38
Maximum Void Ratio	1.53
Specific Gravity	2.70
Liquid Limit	23
Plasticity Index	0 - 4
Standard Proctor Maximum Dry Density	2.1 g/cm^3
Optimum Water Content	13 %

As evidenced by the soil properties, the beach tailings contain little or no cohesive materials.

Seismicity.- A study was prepared for the client with the assistance of Dr. Steven Vick to assess the seismic hazard at the Pueblo Viejo mine site (6). Table 2 shows expected Probable Ground Acceleration (PGA) at the site as determined by probabilistic analysis.

TABLE 2 - EXPECTED PGA AT PUEBLO VIEJO MINE FOR DIFFERENT PROBABILITIES OF EXCEEDANCE

Annual Probability of Exceedance	Return Period Years	Expected PGA (g)	Contributing Magnitude
0.010	100	0.14	6.3
0.005	200	0.18	6.4
0.002	500	0.27	6.6
0.001	1000	0.35	6.8

The values in Table 2 are generally for bedrock acceleration and do not account for the dynamic attenuation of softer sediments which may be present in the embankment foundation or the response of the structure itself to these ground motions. A PGA of 0.2, corresponding to a return interval exceeding 200 years, was selected for dynamic design.

Liquefaction Potential.- The tailings, in their loose state, will likely liquefy under the design seismic loadings due to loss of grain to grain contact and thusly load transfer to the pore water. The loss of the basic soil or grain structure is due to the repeated shear strains imposed upon the tailings mass by propagation of earthquake created shear waves from the abutments and/or foundation. Should this occur, the upstream rock zone, which is laterally supported by the tailings, could slide out into the impoundment.

The potential for liquefaction of the tailings is dependent upon several variables, including: the saturation level; the probable dynamic loading due to earthquakes in and around the Dominican Republic; the in situ density or void ratio; the silt and clay content (gradation); and the cohesiveness (or lack thereof) of the material.

Unsaturated sands, although they may settle to a denser state under dynamic loadings, will not liquefy. After the completion of mining and project reclamation, the upper 12 meters of tailings along the embankment will be unsaturated and will contribute to the stability of the embankment. The tailings beneath this upper zone will remain permanently saturated and be subject to liquefaction under significant dynamic loadings.

Since liquefaction potential is affected by many variables, it is difficult to predict the probability of liquefaction by Standard Penetration Test (SPT) blow counts or other similar means, although considerable effort has been expended in this direction (16,19). Therefore, cyclic triaxial shear tests were performed on tailings samples of varying densities to assess the density needed to effectively prohibit liquefaction under the design loadings. Undisturbed samples were not obtained since it is very difficult to obtain undisturbed samples of loosely compacted sands, such as the tailings.

For the given seismic loadings, tailings gradation, and level of cohesiveness, cyclic triaxial shear testing indicated that a minimum density of 1.5 g/cm^3 was necessary to resist liquefaction under the expected dynamic loadings with appropriate factors-of-safety. Testing also confirmed that tailings at the existing density of 1.3 g/cm^3 or less would likely liquefy.

Densification Methodologies

For upstream stability of the final embankment stage, the possibility of tailings liquefaction had to be minimized. To accomplish this it was necessary to increase the average density of the tailings beach from 1.3 g/cm^3 to 1.5 g/cm^3 or more. Originally, densification by blast induced liquefaction was selected as the most cost effective means of achieving a stable tailings beach.

Densification by blasting was discussed by Lyman (13) as early as 1941, and further work on the subject has been described by several others, including Charlie (2,3), Dowding (4), Hall (5), Kummeneje (12), and Prugh (17). In 1981, Klohn (11) described an experimental field program for densification of loose sand tailings and states "In situ densification of loose, saturated, granular soils, using blasting techniques, provides an effective and economical means of improving the properties of the (tailings) deposit." Ivanov (10) provided the most detailed treatment of the subject in 1967, which was translated from Russian to English in 1972. He described Russian experience with underwater, surface, and deep densification on several projects. The literature supports the effectiveness and economic attractiveness of in situ densification by blast induced soil liquefaction.

Test Program.- A blasting densification test program was initiated at the mine site to determine if it was possible to stabilize the beach tailings by blast induced liquefaction. The initial results from the testing program were positive; the tailings would indeed liquefy and, after a period of settlement, the tailings assumed a denser state. A test blast in progress is shown in the following photograph. The small stream shown flowing on the right side of the photograph eventually began flowing across the blast test pad once the area settled.

BLASTING LIQUEFACTION TEST

During blasting the vibration levels were recorded. The vibrations resulting from the individual 5 kg charges were not high; however, any blasting would have to occur up against the embankment. Initial conservative vibration limits imposed on the embankment would have been exceeded had blasting densification occurred adjacent to the embankment. Therefore, although the blasting densification program was successful in achieving the tailings densities necessary to prohibit liquefaction, the risks involved with blasting immediately upstream of a 73 meter high embankment were significant.

Air-jetting.- It was during the test blasting program that air-jetting was conceived. To monitor pore water pressure response during the test blasts, piezometers were required. Five centimeter diameter PVC pipe with the submerged end screened were installed to a depth of approximately 7 meters. To install the piezometers it was necessary to jet them into the tailings with either water or air, and air was more convenient. While air-jetting the piezometers into the tailings, an area approximately 1.5 meters in diameter quickly liquefied, and liquefaction continued as long as air was injected. Since the goal of the blasting program was to achieve liquefaction, a proven densification method, and air-jetting would achieve the identical result without the concomitant effects of blasting, an air-jetting test program was pursued.

Densification Mechanism.- Air-jetting induced liquefaction is caused by compressed air injected into the tailings. The air is injected at the prescribed depth. The pressure of the air is immediately transferred to the pore water, causing liquefaction. Liquefaction is continuous as long as compressed air is injected. This action causes the sand structure to break down and lose strength. Under the weight of the overburden not present during initial deposition, and increased drainage, the tailings adopt a more dense and stable structure. Immediately after air-jetting and liquefaction the soil exhibits less density than it had originally, possibly due to entrapped air, but within 24 hours the density increases are significant. Incremental density increases were measured as late as three weeks later.

Added benefits of air-jetting over blasting include: no extreme ground or embankment motion; ease of application; the ability to maintain liquefaction; the capability of densifying tailings at the submerged upstream surface of the embankment; relatively inexpensive with minimal cost for consumables; expedient; and air-jetting is not as dangerous.

For the above reasons air-jetting was selected as the means to increase the density of the tailings beach, enabling the final stage to be founded upon the upstream rock zone (7).

Application

Initially, air-jetting liquefaction was accomplished by lowering a 5 cm diameter steel pipe into the tailings and pumping 24 cubic meters per minute of compressed air through it from an air compressor on the bank. Different sized nozzles were attached to the bottom of the pipe to accelerate air flow and keep the opening free of tailings. A 2 cm opening provided the best results. The pipe was suspended from a crane set up on the bank. Once the pipe was suspended above the point of application, the compressed air would be applied and the pipe lowered into the tailings.

Normally the pipe would settle to full depth under its own weight. Periodically a hardened desiccated layer would be encountered and the pipe would stall. By raising and dropping the pipe short distances in the liquefied soil column, the pipe would break through the denser layer.

Once the pipe reached the designated depth, air flow was maintained for two minutes. Surface boiling would be significant for an area approximately 1.5 meters in diameter. Water was also noted flowing from piezometers 8 to 10 meters away and 1.5 meters above the surface.

The liquefied area immediately after air-jetting had the appearance of a thick "gravy" but no significant settlements were evident. After 24 hours, however, settlements up to 50 cm were common and formed the shape of a crater. Much of the density increases occurred during the initial 24 hours, but density increases were measured as late as two weeks after initial liquefaction.

The tailings beach supporting the recently added rock zone is approximately 900 meters long. The entire embankment is about 1,100 meters long but 200 meters of that contains the water quality cycling (polishing) pond. Densified tailings are not necessary to support the rock zone in the polishing pond. The embankment layout and its relationship to the causeway and the densified tailings beach is shown on Figure 2.

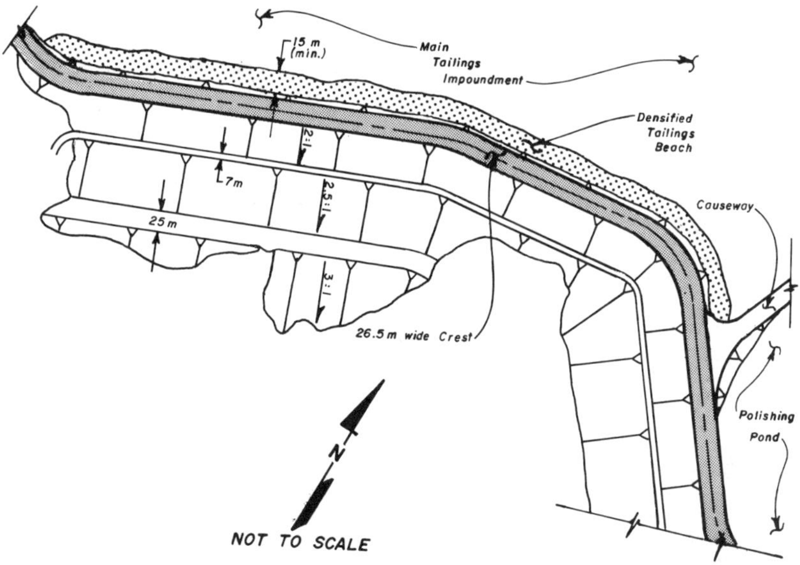

FIGURE 2 - MEJITA EMBANKMENT AND TAILINGS BEACH

To satisfy stability requirements, a 15 meter wide strip of tailings beach needed to be densified. Therefore, the area under consideration for the densification program was 13,500 square meters. An area of this size could not be densified economically by working one square meter at a time (considering overlap). To speed the process, a "liquefaction manifold" was developed.

Liquefaction manifold.- The liquefaction manifold was constructed to liquefy four separate areas simultaneously. The rectangular manifold frame was 2.6 meters by 2.2 meters and the length of each of the four legs was adjustable. The minimum length of each leg was 1.8 meters and the maximum length used in the program was approximately 14 meters.

The entire manifold was constructed of 5 centimeter inside diameter steel drill pipe. The upper frame was constructed such that it could be suspended by a crane set up on the embankment. The manifold was equipped with two air inlets at the top, permitting two separate air compressor hookups. A nipple was also included and affixed with an air hose , thereby allowing internal air pressure within the manifold to be monitored from the embankment. A 2 cm nozzle was attached to the bottom of each leg.

A larger pipe (10 cm diameter) was welded to one of the manifold legs and used to lower 5 cm diameter PVC pipe into the tailings during liquefaction. These PVC pipes were sealed on the bottom and used as casing for the gamma-gamma density logger (described in the next section), permitting tailings density measurements at a later date.

The manifold also had a water hose attached, which was connected to a pump on the embankment. Water was needed to fill the PVC pipes used to measure density, preventing them from floating out of the liquefied tailings column.

Several combinations of air compressors and air pressures were used initially before the most efficient combination was determined. The best results were achieved when between 15 and 17 cubic meters per minute of compressed air per leg were supplied to the manifold and the air pressure was maintained at or above 175 kPa.

To ensure complete coverage of densification in the tailings beach, an overlapping pattern was utilized. The manifold was set down at 2.5 meter intervals during the first insertion and was placed in-between during the second run. The width of application was restricted only by the reach of the crane.

Density Verification

A means of measuring in situ density changes was necessary to monitor effectiveness of the air-jetting liquefaction program. Densities of sand deposits are historically difficult to measure because slight disturbances significantly alter measured densities. Freezing samples in place prior to removal was considered but was discarded as being costly and impractical to implement. Equipment capable of measuring in situ samples with relatively little disturbance does exist but the cost was prohibitive and access onto the tailings beach would have been difficult. Therefore, geophysical means of soil density measurements were pursued. A technique commonly used in oil exploration, "gamma-gamma" formation density logging, was selected.

The density measurements were made using a Mt. Sopris Instruments Model 1000-C bore hole logging unit equipped with the special GG375 gamma-gamma probe assembly (14), permitting bulk density readings. Calibration was achieved by obtaining data on tailings of known density and lateral extent. Since the 1000-C was not equipped to ignore bore hole effects, a bore hole of constant diameter and material type was also necessary at every measuring station in the tailings to provide repeatable results. Five centimeter diameter water-filled PVC pipe was used for this purpose. The PVC pipe was installed during air-jetting to avoid disturbing the densified soil.

Field conditions were duplicated as closely as possible during instrument calibration. The operating principle of the 1000-C gamma-gamma logger, Compton scattering, assumes a formation of infinite lateral extent (18,21), but surface conditions, depth of bore hole, and boundary conditions also influence bulk density readings.

After several trials using different sized containers, a 30 cm diameter PVC pipe approximately one meter high was selected. This size diameter pipe virtually eliminated horizontal boundary influences during calibration readings. The readings were taken in the center of the tailings to avoid top and bottom effects, and the 30 cm diameter PVC pipe was sufficiently large in lateral extent to obtain accurate measurements. Five centimeter diameter PVC pipe was centered within the tailings material during calibration to duplicate field conditions.

The data are plotted on semi-logarithmic paper since the bulk density (and dry density once calibrated) is directly proportional to the inverse of the log of the counts per second (CPS) registered by the instrument. The calibration data was plotted and is shown on Figure 3.

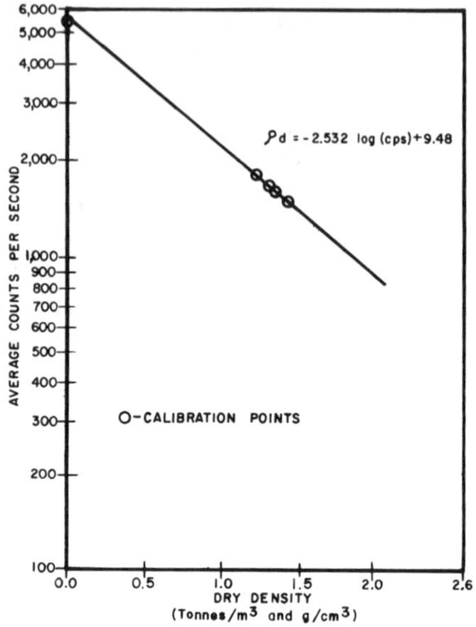

FIGURE 3 - GAMMA-GAMMA CALIBRATION CURVE

Densification Program Results

Air-jetting along the embankment began in November, 1986 and was completed in four months. A crew of approximately 10 people was necessary to execute the program. Work progressed slowly at first until most of the operational problems were resolved. With the knowledge available now the entire project could have been completed in about two months.

Two photographs of the liquefaction manifold in operation are included on the following pages.

LIQUEFACTION MANIFOLD DURING INITIAL APPLICATION

The first photograph depicts the initial liquefaction attempt with the manifold and the second shows the same manifold as the program neared completion. The first photograph indicates a relatively level tailings surface with desication cracks. Very few hoses were connected to the manifold initially, but in the second photograph the manifold was connected to several hoses. This reflected the need for additional air compressors and monitoring internal air pressure. Large settlements are also evident in the background of the second photograph.

LIQUEFACTION MANIFOLD NEARING PROJECT COMPLETION

A final photograph is included which clearly shows the effects of the completed densification program. Surface settlements exceeding 50 cm were recorded at all locations. In fact, much of the tailings settled below the phreatic surface, which had been approximately one meter below the original surface prior to air-jetting.

FINAL DENSIFIED TAILINGS BEACH

Since only final density results were needed, initial density readings prior to liquefaction were done only during the test program. Densities as low as 1.25 g/cm^3 were recorded, primarily in the upper tailings layers not having the benefit of overburden pressures. At greater depth, initial densities were found to be as high as 1.45 g/cm^3. This corresponded to studies done prior to 1984 by Hydro-Triad, Ltd. which showed density of tailings deeper within the impoundment, based on storage volume and tonnes of tailings processed by the mill, of approximately 1.45 g/cm^3. A later study completed in 1986 determined that average in situ densities within the impoundment had decreased to 1.35 g/cm^3, and the beach sands and other surface zones had densities even lower. In all cases, however, measured densities were below the 1.5 g/cm^3 needed to resist liquefaction under the design loadings.

Typical gamma-gamma down-hole surveys are shown on the following two figures. Figure 4 shows the surveys done on November 15, 1986 and on November 27, 1986 in a test hole near the embankment. The initial survey of November 15, 1986 was done prior to any air-jetting. The calculated density was 1.35 g/cm^3 identical to the latest average density calculated for the impoundment. The final density measurement made twelve days later indicated a density of 1.51 g/cm^3, a twelve percent increase.

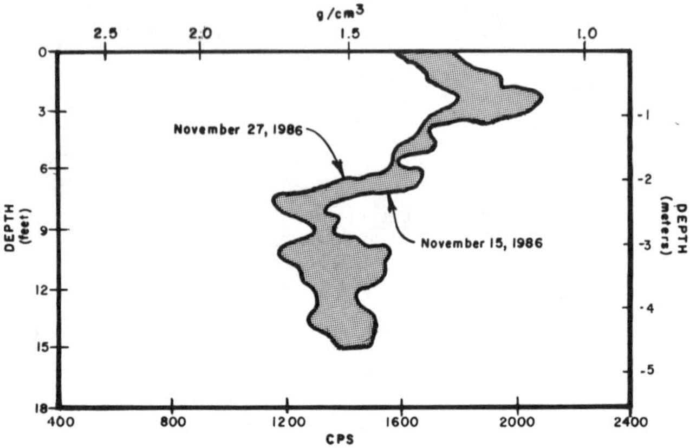

FIGURE 4 - GAMMA-GAMMA DENSITY SURVEY OF A TEST HOLE

Several intermediate measurements were also taken on this hole but are not shown for reasons of clarity. The largest density increase occurred during the first 24 hours, but the density continued to increase during the next 12 days. No further measurements were taken on this particular hole once the density exceeded 1.5 g/cm^3, the target density. Other measurements taken in the separate areas concur with these results. The upper tailings layers near the surface do not exhibit the degree of densification achieved by the lower zones due to the lack of overburden pressures during liquefaction. This program was concerned only with the density of the lower tailings layers.

Typical gamma-gamma measurements taken during the air-jetting program are shown on Figure 5. The first density measurement in this hole, which was located approximately midway across the embankment, was taken on January 27, 1987 four days after the area had been air-jetted.

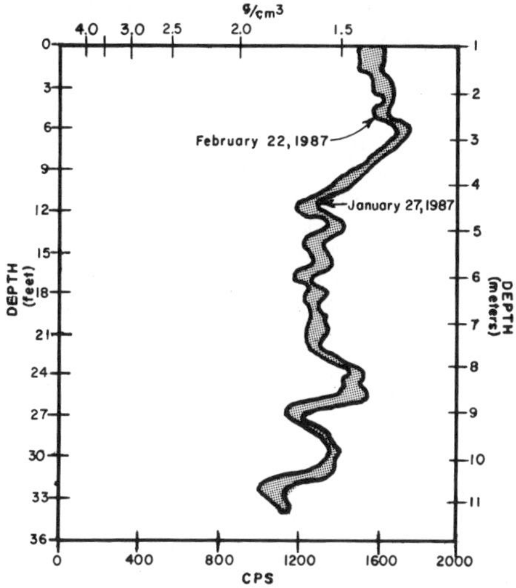

FIGURE 5 - TYPICAL GAMMA-GAMMA DENSITY SURVEY

The average density was 1.50 g/cm^3 (already at the target value). Subsequent measurements were taken in this hole, the latest on February 20, 1987. During the 24 day period, an increase in density of 4 percent was measured, raising the value to 1.56 g/cm^3. Again, the greatest percentage increase occurs immediately after air-jetting. Therefore, the initial density in the hole prior to air-jetting was probably much lower. These measured values were typical of all the readings along the embankment, but was expected due to the relative homogeneity of the material. A summary of final measured densities is given in Table 3 (9).

TABLE 3 - FINAL MEASURED DENSITIES

Measuring Station	Final Dry Density (g/cm^3)
1	1.4
2- 4	1.5
5-13	1.6
14	1.7
15-27	1.6
28-36	1.5
37-40	1.6
41	1.5

The PVC pipe density measuring stations were installed at 20 to 30 meter spacings across the entire length of the embankment. Every measuring station except one recorded densities at or exceeding 1.5 g/cm^3, the target density. The one inadequate density measurement was recorded in an area located adjacent to the causeway. Due to the spigotting arrangement used at Mejita, however, an abundance of finer grained particles is trapped in this area. Approximately 75 percent of the material near the causeway is finer than 0.074 millimeters, and the material exhibits more cohesion that the remainder of the tailings. Although this area did liquefy during air-jetting, it did not consolidate to a much denser state. The density in this area is adequate, however, since the abundance of fines and increased cohesion would resist liquefaction under the expected dynamic loadings.

Conclusion

The crest and upstream slope of the Mejita tailings embankment for the ultimate section was designed to be laterally supported by the tailings beach. The tailings will remain saturated 9 to 11 meters below the crest of the final embankment section after reclamation. Because the site is located within an area of moderate seismic activity, the possibility of tailings liquefaction existed. Should the beach tailings liquefy and lose strength, portions of the existing and ultimate crest would no longer have lateral support and could slide into the impoundment. This would cause the loss of all subsequent construction and possible breach of the embankment. To prevent this, a zone of the tailings were made relatively immune to liquefaction by increasing the in situ density. The density was increased by jetting air under pressure into the tailings, causing liquefaction, and then allowing the increased overburden pressure to consolidate the material to a denser state.

Every density reading in the Mejita tailings beach, except for one area near the causeway, indicates a dry density at or exceeding the target density of 1.5 g/cm^3. The target density was established by cyclic triaxial testing in the laboratory using molded samples of Mejita beach sands at known density.

Based on all available data, laboratory testing, and inspection, it can be stated with confidence that the air-jetting densification program was a success. The density of the beach sand tailings now exceed 1.5 g/cm^3, sufficient to resist liquefaction under the design loadings. This process proved to be an expedient, effective, and cost effective means of densifying loose, cohesionless, and saturated tailings sands. The enlarged tailings embankment will be stable under both static and dynamic loadings with appropriate factors-of-safety, and the air-jetting liquefaction procedure resulted in a cost savings to the client of more than $5 million (U. S.) over conventional downstream construction methods.

Appendix 1 - References

1. Bufete Nacional de Ingeneria, S.A., October, 1985, "Stage IV Tailings Dam Investigations," for Rosario Dominicana, S.A.

2. Charlie, W. A., Mansouri, T. A., and Ries, E. R., 1981, "Predicting Liquefaction Induced by Buried Charges," X-International Conference on Soil Mechanics and Foundations, Stockholm, Vol. 1, pp. 77-80.

3. Charlie, W. A., Veyera, G. E., Abi, S. R., and Patrone, H. D., 1983, "Blast Induced Soil Liquefaction," Symposium Proceedings, The Interaction of Non-Nuclear Munitions with Structures, U. S. Air Force Academy, Colorado, pp. 62-68.

4. Dowding, Charles H., and Hryciw, Roman D., February, 1986, "A Laboratory Study of Blast Densification of Saturated Sand," Journal of Geotechnical Engineering, ASCE, Vol. 112, No. 2, pp. 187-199.

5. Hall, C. E., June, 1982, "Compacting a Dam Foundation by Blasting," Soil Mechanics and Foundation Division, Vol. 88, No. SM3, pp. 33-51.

6. Hydro-Triad, Ltd., December, 1985, "The Seismicity of the Pueblo Viejo Mine Area, Final," Lakewood, Colorado, for Rosario Dominicana, S.A.

7. Hydro-Triad, Ltd., October 1986, "Tailings Densification Testing Report, Blasting and Air Jetting, Stage IV Enlargement Investigation, Mejita Oxide Tailings Embankment, Mina Pueblo Viejo," for Rosario Dominicana, S.A.

8. Hydro-Triad, Ltd., January, 1987, "Mejita Tailings Embankment, Stage IV Enlargement," for Rosario Dominicana, S.A.

9. Hydro-Triad, Ltd., April, 1987, "Mejita Beach Tailings Densification Summary Report," for Rosario Dominicana, S.A.

10. Ivanov, P. L., 1967, "Compaction of Non-Cohesive Soils by Explosions," translated from Russian and published for the U. S. Department of Interior, Bureau of Reclamation and the National Science Foundation, Washington, D. C. by the Indian National Scientific Documentation Centre, New Dehli.

11. Klohn, E. J., Garga, V. K., and Shukin, W., June, 1981, "Densification of Sand Tailings by Blasting," Proceedings of the Tenth International Conference on Soil Mechanics and Foundations, Stockholm, p. 725-730.

12. Kummeneje, O., and Eide, O., 1961, "Investigation of Loose Sand Deposits by Blasting," Proceedings, 5th International Conference of Soil Mechanics and Foundation Engineering, Vol. 1, pp. 491-497.

13. Lyman, A. K. B., May, 1941, "Compaction of Cohesionless Foundation Soils by Explosives," Proceedings, ASCE, Vol. 67, pp. 769-780.

14. Mount Sopris Instrument Company, 1977, "Portable Borehole Logger, Model 1000-C, Operation and Maintenance Manual," Delta, Colorado.

15. Pincock, Allen & Holt, Inc., August, 1980, "Engineering and Design Report, Stage III Embankment, Pueblo Viejo Tailings Project," Tucson, Arizona, U.S.A., for Rosario Dominicana, S.A.

16. Poulos, S.J., and Hed, A., "Density Measurements in a Hydraulic Fill," Evaluation of Relative Density and Its Role in Geotechnical Projects Involving Cohesionless Soils, ASTM STP 523, American Society of Testing and Materials, 1973, pp. 402-424.

17. Prugh, B. J., March, 1963, "Densification of Soils by Explosive Vibrations," Journal of the Construction Division, ASCE, Vol. 89 No. CO1, pp. 79-100.

18. Schlumberger, 1972, "Log Interpretation," Volume 1, Principles, New York, pp. 43-48.

19. Seed, H. B., and Lee, K. L., 1977, "Evaluation of Liquefaction Potential of Sand Deposits Based on Observations of Performance in Previous Earthquakes," Proceedings, Advisory Meeting on Earthquake Engineering and Landslides, Taipei, Republic of China, pp. 46-66.

20. Tippets-Abbett-McCarthy-Stratton (TAMS), January, 1974, "Design Recommendations, Tailings Retention Dam," New York, for Rosario Dominicana, S.A.

21. Tittman, J., and Wahl, J. S., "The Physical Foundations of Formation Density Logging," Proceedings of Conference, Nuclear Geophysics Conference, Cracow, Poland, September 24-October 1, 1962, Volume II, p. 339.

Mine Tailings, Dispersants and Flocculants

Robin Fell*

ABSTRACT A laboratory study has been carried out to assess the effect on settled density of adding dispersants and flocculants to high clay content tailings from ore treatment plants.

The study has shown that the settled density of the tailings is not improved by the addition of flocculants, but that the settled density of some tailings can be improved by adding small quantities of dispersants prior to deposition. This can also result in improved water recovery. The behaviour is dependent on the mineralogy of the tailings.

Introduction

In many mines it is necessary to dispose tailings from ore treatment plants into a storage dam without beaching the tailings to gain the benefits of dessication. In these dams the tailings often settle and consolidate to very low densities requiring large containment dams. The tailings also trap large quantities of water in their voids, reducing water recovery to the plant. This is a particular problem when the tailings contain a large percentage of clay size particles.

As part of a study to improve the deposited properties of mine tailings being carried out at the University of New South Wales School of Civil Engineering, tailings from seven mines were tested to assess whether the addition of dispersants and flocculants to the tailings prior to deposition would increase the settled density of the tailings. This paper presents the results of that study.

Five of the companies (Comalco, TDM Riverside, Mt Newman Mining, Hamersley Iron and Wambo Mining) are supporting a three year study coordinated by Australian Mineral Industries Research Association, which will concentrate largely on the effects of dessication but which will also use the results of this work to optimise deposition of tailings.

Tailings Used in the Study

The tailings used in the study were provided by the mining companies as being representative of their normal operating conditions. This usually involved taking sub-samples over a period of 1 to 2 weeks. The tailings were taken from the discharge pipe to the tailings disposal areas. All tailings were supplied as a slurry with the accompanying liquid used to transport them. All experiments used this water, and

* Professor, School of Civil Engineering, University of New South Wales, P.O. Box 1, Kensington, Sydney, Australia, 2033

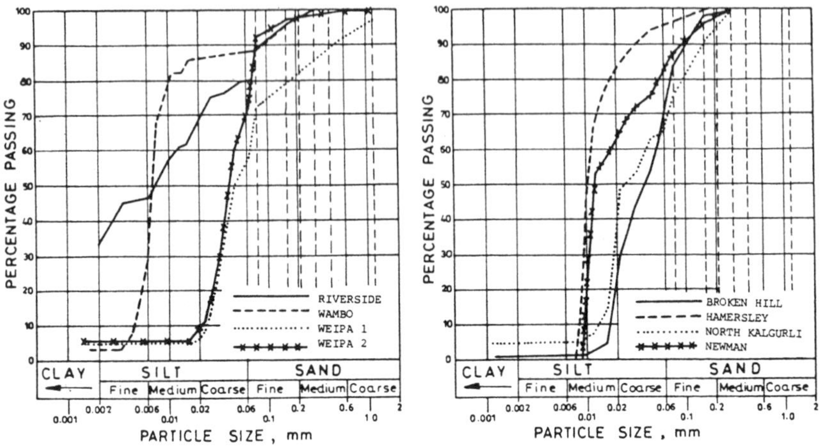

Figure 1. Particle Size Distributions — Without Dispersants

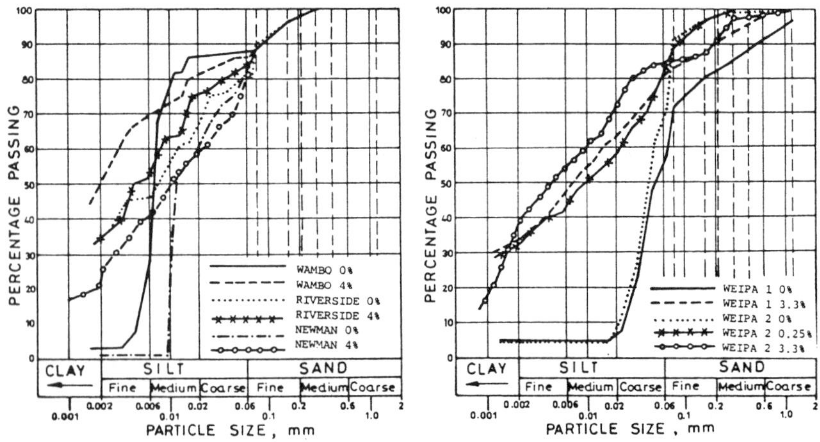

Note — 4% indicates 4% Calgon by weight of soil particles added

Figure 2. Particle Size Distributions — With and Without Dispersants

tailings were not allowed to dry at any time. Tables 1, 2 and 3 describe the tailings disposal systems, tailings chemistry and mineralogy, and soil mechanics properties. Figures 1 and 2 show particle size distribution for the tailings. These were determined by hydrometer testing with and without dispersant (Calgon).

As can be seen five of the tailings are high clay content tailings

Table 1. Details of Tailings Disposal

Tailings/ Mining Company	Type of Tailings	Tailings Production: Annual Tonnes	Tailings Production: Monthly Max Tonnes	Tailings Production: Discharge Solids Content-%	Type of Disposal System	Disposal Area Hectares	Climate: Rainfall mm/year	Climate: Evaporation mm/year	Degree of Dessication being Achieved
Weipa 1 Weipa 2 Comalco Aluminium Ltd	Bauxite benef-ication	(1) 1,200,000 (2) 1,300,000	200,000 210,000	(1) 20 (2) 20	(1) Discharged from spigots at 10m apart, the embankment to be built of recompacted dried tailings. (2) Discharge from a single discharge point, with an earthfill dam on a flat area.	(1) 160 (2) 110	2000	2300	(1) Near total dessication provided the cycle time is sufficiently long. (2) Very little dessication apart from near the discharge point.
Wambo Wambo Mining Corporation Ltd	Coal washery	100,000	17,000	20	(A) 1 to 3m deep ponds have been used in the past, with tailings being recovered and placed in mine spoil. (B) Presently discharging into underground workings which have a limited life.	(A) 4	700	1700	(A) Poor dessication after 3 years (only 0.3m crust). Tailings still semi fluid. (B) Not applicable.
Riverside TDM Riverside	Coal washery	550,000	60,000	31	Discharge from pipes spaced at about 300m apart into an earth and rockfill dam on a relatively flat area.	90	630	2200	Total dessication achieved within ≈ 100m of discharge pipe, but a relatively large water pond is maintained with no dessication in or near the pond.
Newman Mt Newman Mining Co Ltd	Iron ore benef-ication	240,000	50,000	30	Discharged from spigots at 10m apart into 3 areas in rotation, cycle time about 6 months. Tailings are being dessicated to form the embankments.	9 (at present)	300	3700	Total dessication being achieved within 50m of discharge, and near total beyond that.
Hamersley Hamersley Iron Pty Ltd	Iron ore concen-trator	1,200,000	190,000	27	Discharged from spigots 30m apart into a dam formed in a valley by 40m high earth and rockfill embankment.	26	440	3660	Up to 0.5m approx well dessicated crust over part of the disposal area. Very little dessication over the remainder.
North Kalgurli North Kalgurli Mines Ltd	Gold mining	580,000	50,000	55	Upstream method, with 2m starter dam. Tailings pile up to 18m high.	12	250	2400	Total dessication for the majority of tailings.
Broken Hill Zinc Corporation Limited	Lead-zinc mining	2,100,000	350,000	25	Upstream method up to 40m high.	35	280	3000 approx.	Total dessication for the majority of tailings.

Table 2. Mineralogy and Chemistry of Tailings Used in Study

Tailings	Mineralogy	Chemistry — % by Weight of Dried Tailings								pH Tailings Water
		S_iO_2	Al_2O_2	Fe_2O_3	CaO	MgO	Na_2O	K_2O	LOI(1)	
Weipa 1	Gibbsite(45%), Boehmite (18%), Kaolin(12%), Quartz(17%), Hematite (5%)	19	52	6	<0.1	<0.1	<0.1	<0.1	18	6.9
Weipa 2	Gibbsite(34%), Boehmite (35%), Kaolin(10%), Quartz(8%)	10	53	12	<0.1	<0.1	<0.1	0.1	19	6.4
Wambo	Na and Ca Montmorillonite, Kaolin, Coal, Quartz(2)	28	12	1	0.9	1.0	0.8	0.4	51	6.8 (1987 tailings)
Riverside	Illite, Ca Montmorillonite, Kaolin, Quartz(2)	29	13	0.5	<0.1	0.3	0.3	0.7	54	7.4
Mt Newman	Hematite(40%), Kaolinitic shale(50%), Quartz(6%)(2)	17	12	63	0.4	0.7	<0.1	0.4	5	7.4
Hamersley	Hematite(34%), Kaolinitic shales(52%), Limonite (8%), Goethite(19%)(2)	20	14	60	0.1	0.3	<0.1	0.1	7	7.1
North Kalgurli	Quartz, Kaolin, Gibbsite (2), Host rock, Dolerite, Basalt, Sulphide and Oxide ores	55	14	13	2.3	1.1	2.8	2.2	7	7.9
Broken Hill	Quartz(50%), Rhodenite (20%), Calcite(12%), Garnet(6%), Manganhedbergite (4%), Pyrrhotite(4%)	56	7	12	6.7	1.1	1.1	1.3	1	7.7

NOTES (1) Loss on ignition (LOI) at 1000°
(2) Mineralogy percentages are approximate and vary with production sources

Table 3. Soil Mechanics Properties of the Tailings As Supplied

Tailings	Water Content %	Dry Density t/m³	% Solids	Soil Particle Density t/m³	Atterberg Limits - %			Particle Size - %		
					Liquid Limit	Plastic Limit	Plasticity Index	Sand	Silt	Clay
Weipa 1	675	0.14	13	2.75	44	27	17	20	47	33
Weipa 2	362	0.25	22	2.85	43	26	17	15	45	40
Wambo	411	0.22	20	1.86	74	28	46	14	36	50
Riverside	250	0.32	29	1.74	44	28	16	16	50	34
Newman	192	0.46	34	3.70	33	22	11	20	60	20
Hamersley	169	0.50	37	3.50	30	21	9	5	55	40
North Kalgurli	71	0.94	59	2.81	28	21	7	Not available		
Broken Hill	462	0.20	18	3.05	Non plastic			See Figure 1		

NOTE (1) Particle size with dispersant added as per AS1289

derived from washing operations, but the clay mineralogy is quite variable. The other tailings (North Kalgurli and Broken Hill) are more conventional crushed rock tailings.

Because of their high clay content several of the mining companies add flocculants to assist settlement in the thickeners. Details are given in Table 4. The Wambo tailings are particularly dispersive naturally (because of the presence of sodium montmorillonite) and all coal is treated with gypsum prior to entering the washing cycle.

TABLE 4.—DETAILS OF MILL PROCESSING — FLOCCULANT TREATMENT

Tailings	Flocculants Added
Weipa 1	Nil
Weipa 2	Nil
Wambo	Gypsum — 29 kg/tonne Allied Colloids Magnafloc 156 and Catoleum 625 — 0.15 kg/tonne
Riverside	Allied Colloids Magnafloc 156 0.0025 to 0.008 kg/tonne Magnafloc 1697-0.02-0.03 kg/tonne
Mt Newman	Nil
Hamersley	Allied Colloids Magnafloc 155 — 0.01 to 0.02 kg/tonne
North Kalgurli	Flocculant is added, details not available
Broken Hill	Nil

Table 4. Details of Mill Processing — Flocculant Treatment

Flocculants and Dispersants Used in the Study

Several companies who supply flocculants and dispersants to the mining, ceramic, drilling and water treatment industries were approached for advice on which of their products would be most likely to be useful for treating the tailings. From these recommendations, literature provided by the companies, the use of chemicals already by the mining companies and consideration of the clay mineralogy, a range of flocculants and dispersants were selected for use in the study. Details are given in Tables 5 and 6. To these were added the inorganic salt flocculants, lime, gypsum and alum; and Calgon, which is commonly used as a dispersant in soils testing laboratories.

The Mechanism of Operation of Flocculants and Dispersants

Flocculants and dispersants operate in soil (tailings) suspensions by modifying the electric charge on the soil (tailings) particles, and/or by forming chains to link the particles together.

(a) Polyelectrolyte flocculants

The mechanism by which polyectrolyte flocculants such as the Magna-

Table 5. Flocculants Used in the Study

Name	Manufacturer	Chemical Description	Manufacturers Recommended Dosage kg/tonne	Environmental Effects	Commercial Application
Polyelectrolytes					
Magnafloc 155	Allied Colloids	High molecular weight, anionic, polyacrylamide	0.002-0.2	pH 5.5-6.5 Low toxicity(1)	Mineral processing — thickening, clarification, sedimentation
Magnafloc E10	Allied Colloids	Very high molecular weight, slightly anionic, polyacrylamide	0.002-0.2	pH 3.0-4.0 Low toxicity(1)	Mineral processing — thickening, clarification, sedimentation
Magnafloc 333	Allied Colloids	Very high molecular weight, non ionic, polyacrylamide	Not stated	pH 3.0-4.0 Low toxicity(1)	Mineral processing — thickening, clarification, sedimentation
Zetag 92	Allied Colloids	Ultra high molecular weight, medium high cationic polyelectrolyte	0.002-0.2	pH 4.2-5.0 Low	Dewatering and in sewage and industrial sludges
Inorganic Salts					
Alum		Aluminium sulphate			
Lime		Calcium hydroxide			
Gypsum		Calcium sulphate			

NOTES (1) As claimed by manufacturer

Table 6. Dispersants Used in the Study

Name	Manufacturer	Chemical Description	Manufacturers Recommended Dosage kg/tonne	Environmental Effects	Commercial Application
Calgon		Sodium hexametaphosphate	40 for soils testing	Phosphate promotes algac growth	Commonly used dispersant in soils testing. Clothes washing.
Giessfix C30	Zschimmer & Schwarz	Polymetric phosphate silicate compounds	1-5	pH=12. Possible phosphate promotes algae growth	Ceramics.
Busperse 39	Buckman Laboratories	Sodium polyacrylate-anionic polyelectrolyte	1-2	pH=9. No hazard(1)	Drilling mud.
Busperse BL257	Buckman Laboratories	"Organic substances"	0.1-0.5	pH >12. No hazard(1)	Portland cement manufacture.
Dispex N40	Allied Colloids	Sodium salt of polycarboxylic acid	0.5-5	pH 7-7.5 Not	Paint manufacturer, ceramics, concrete workability, cement manufacture
Dolapix PC67	Zschimmer & Schwarz	"Synthetic polyectrolyte"	1-10	pH 8-9	Ceramics
Lignin		High molecular weight polyphenol	Not stated	pH ≈ 7	Drilling mud
Tannin Acid		Gallic and ellagic acids	Not stated		

NOTES (1) As claimed by manufacturer

floc range of products work is outlined in Moss and Dymond (1978) and in Gadiel (1979). The polyelectrolyte flocculants are high molecular weight polyacrylamides. Polyacrylamides are essentially non-ionic but are commonly modified to carry a positive (cationic) or negative (anionic) charge to assist in adsorption onto the electrically charged particles in the tailings, and to cause the long chain polymer molecule to extend and uncoil so they attract more soil particles. The polymer therefore acts as a bridge between soil particles and large flocs, even big enough to be visible to the eye, can be formed.

(b) Inorganic salts

These operate by exchange of high charge density cations such as Ca^{++} and Al^{+++} for low charge density cations such as Na^{+} in the clay minerals. This gives a reduced diffuse double layer thickness, and less tendency for dispersion ie. flocculation is encouraged. Mitchell (1976) discusses these effects in some detail.

(c) Dispersants

When dissolved in water dispersants ionise to produce cations and anions. The anions adsorb onto the tailings particle causing the particles to become negatively charged (or increase the negative charge already present) and hence lead to dispersion by repulsion of the charged surfaces.

Use of Flocculants and Dispersants

Polyelectrolytes are used extensively in the mineral processing industry eg. Lynberg (1981) describes the use of Superfloc 1202, a high molecular weight anionic polyelectrolyte, along with lime to pretreat coal tailings before belt filtration; Lawver and Carrier (1984) discuss the use of polyacrylamide flocculants in high montmorillonite content phosphatic clays; Ihle and Sparrow (1980) describe experiments on Kaolin clay suspensions where the effect of cationic and non-ionic polyacrylamide flocculants on settling time and sorptivity (which is related to hydraulic conductivity) is assessed; Smelley et al (1980) describe the use of non-ionic polyethylene oxide polymer to assist in dewatering of coal, potash, phosphate, talc, T_iO_2 and uranium tailings using a rotating screen.

Salts such as alum have long been used in water treatment, and lime and gypsum in reducing the dispersiveness of soils in dam construction.

As described in Table 4 these flocculants are in use in the processing of some of the tailings used in this study.

Dispersants are used in the drilling industry (Chilingarian and Borabutz 1983) and in soils testing. The author is not aware of their use in mineral processing.

Column Settling Tests

The study was carried out by observing the rate of settlement of tailings which had been mixed with the required amount of flocculant or dispersant.

Two sizes of column were used:

- — 300mm x 35mm or 45mm dia glass flasks (small columns)
- — 1900mm x 90mm dia perspex (large columns)

The flocculant and dispersant were mixed for 15 minutes with the tailings in a flask using a paddle mixer. Experiments carried out at the beginning of the study showed that the results were not sensitive to the duration or speed of mixing, and the method adopted was selected for convenience and to ensure thorough mixing.

A range of "dosage" of the flocculant or dispersant was determined from the manufacturers' recommendations and from the results of testing.

Results of Column Settling Tests

Tables 7.1 to 7.4 and Figures 3 to 6 present a selection of the results of the small column settling tests. In all tables the "optimum" dosage is that at which maximum settled dry density is achieved.

Table 8 presents a summary of the results of the large column tests. The total time for settlement can be up to 240 days, and intermediate values are also presented in the table. Figures 7 and 8 present a selection of the results.

The apparent rapid settlement after about 5000 minutes in some tests (eg. Weipa, Giessfix) occurs as the cloudiness due to finely dispersed particles clears.

All small columns, and most large columns were undrained at the base. Two drained large column tests were carried out with a resultant increase in density and significantly reduced settling time as shown in Table 8. This is to be expected since the vertical seepage resulting from draining means that the tailings are subjected to higher effective stress.

To gauge whether water returned from the tailings storage would contain residual quantities of dispersant which might upset the washing or thickener operation, analyses were carried out of the total carbon content of water on the surface of the tailings after 24 hours settling. Results of these tests are given in Table 9.

Other features which were determined by testing the tailings were

- the settled density was not affected by the initial water content. Weipa 2 tailings "diluted" to twice the water content of the as supplied tailings settled to the same density in the same time
- the tailings used in the study did not alter properties with time. Tests repeated 12 months after the initial tests gave almost identical settling characteristics.

Discussion of Results

(a) Untreated tailings

Table 10 summarizes the results of testing in small and large

Table 7.1. Summary of Results of Small Column Settling Tests Weipa 2 Tailings

Flocculant	Dosage Kg/Tonne (1)		Settled Dry Density t/m³	
	Range	Optimum	Range (2)	Maximum (6)
As supplied	Nil			0.41(7)
Magnafloc 155	0.01-0.07	0.02	0.40-0.41	0.41
Magnafloc 333	0.01-0.07	0.07	0.40-0.43	0.43
Zetag 92	0.01-0.07	0.03	0.41-0.42	0.44
Lime	20			0.42
Gypsum	20			0.44
Dispersant	**Dosage Kg/Tonne (1)**		**Settled Dry Density t/m³**	
	Range	Optimum	Range (2)	Maximum (4)
As supplied	Nil			0.42, 0.44, 0.45
Calgon	1-120	10	0.45-0.57	0.61, 0.68, 0.68
Giessfix C30	1.5-6	6(3)	0.50-1.03	0.27, 0.41, 1.04
Busperse 39	0.05-6	6	0.41-0.81	0.81, 0.81, 0.81
Busperse BL257	1.5-6	6	0.41-0.58	0.33, 0.57, 0.58
Dispex N40	0.3-11	3	0.41-0.43	0.45, 0.45, 0.45
Dolapix PC67	1.4-7	7(3)	0.44-0.86	0.36, 0.86, 0.86
Lignin	1-6	6(3)	0.40-0.49	0.49, 0.49, 0.49
Tannic Acid	1-6	1(3)	0.39-0.37	0.39, 0.39, 0.39

Table 7.2. Summary of Results of Small Column Settling Tests Wambo Tailings

Flocculant	Dosage Kg/Tonne		Settled Dry Density t/m³	
	Range	Optimum	Range (2)	Maximum (4)
As supplied	Nil		0.33-0.34	0.32, 0.34, 0.37
Magnafloc 155	0.02-0.06	0.035	0.33-0.34	0.32, 0.34 -
Magnafloc 156	0.02-0.06	0.045	0.35-0.36	0.35, 0.36, 0.41
Magnafloc E10	0.02-0.10	0.10	0.33-0.35	0.34, 0.35, 0.37
Lime	20	-	-	0.25, 0.26, 0.26
Gypsum	20	-	-	0.32, 0.34, 0.39
Alum	25	-	-	0.25, 0.26, 0.28
Gypsum plus Magnafloc 156	25 Gypsum + 0.1 Magnafloc	-	-	0.35, 0.36, 0.39
Dispersant	**Dosage Kg/Tonne**		**Settled Dry Density t/m³**	
	Range	Optimum (5)	Range (2)(5)	Maximum (4)
As supplied	Nil		0.33-0.34	0.32, 0.34, 0.37
Calgon	1-44	12	0.33-0.23	0.33, 0.36, 0.39
Giessfix C30	1.5-6	1.5(3)	0.32-0.30	0.31, 0.32, 0.37
Busperse 39	1.5-5	5(3)	0.35-0.36	0.32, 0.36, 0.44
Busperse BL257	1.5-3	1.5(3)	0.35	0.33, 0.35, 0.41
Dispex N40	0.03-1.2	0.03(3)	0.34	0.32, 0.34, 0.38
Dolapix PC67	1.4-6	6(3)	0.35-0.36	0.33, 0.36, 0.39
Lignin	1.3-7	7(3)	0.35-0.36	0.34, 0.36 -
Tannic Acid	1.3-6	6(3)	0.36	0.32, 0.36 -

NOTES
(1) Actual amounts varying by up to 5%
(2) Range is for minimum and maximum amounts of additive
(3) Limited range tested
(4) Figures are for 10, 20 and 100 days
(5) Water cloudy for 20 days for 1 kg/tonne and greater of Busperse 39, Busperse BL257, Dolapix and Giessfix
(6) Figures for 10, 20 days are identical
(7) Earlier tests gave this value. Later tests gave values below
(8) Figures are for 20 day settling

Table 7.3. Summary of Results of Small Column Settling Tests
Riverside Tailings

Flocculant	Dosage Kg/Tonne		Settled Dry Density t/m³	
	Range	Optimum	Range (2)(5)	Maximum (4)
As supplied	Nil		0.54-0.57	0.50, 0.54, 0.61
Magnafloc 155	0.03-0.08	0.03(3)	0.61	0.54, 0.61, 0.61
Lime	17-18	17	0.41-0.39	0.40, 0.41, -
Dispersant	**Dosage Kg/Tonne**		**Settled Dry Density t/m³**	
	Range	Optimum (5)	Range (2)(5)	Maximum (4)
As supplied	Nil		0.54-0.57	0.50, 0.54, 0.61
Calgon	1.5-7.5	7.5	0.72-0.65	0.68, 0.72, 0.76
Giessfix C30	1.4-9	9(3)	0.65-0.71	0.63, 0.71, 0.93
Busperse 39	0.5-30	6	0.59-0.77	0.69, 0.77, 0.79
Busperse BL257	1.7-7	7	0.59-0.69	0.63, 0.69, 0.70
Dolapix PC67	1.3-6	6(3)	0.59-0.74	0.65, 0.74, 0.81
Lignin	3-12	12(3)	0.63-0.74	0.65, 0.74, 0.74
Tannic Acid	0.8-3.6	0.8(3)	0.60-0.59	0.51, 0.60, 0.65

Table 7.4. Summary of Results of Small Column Settling Tests
Hamersley Tailings

Flocculant	Dosage Kg/Tonne (1)		Settled Dry Density t/m³	
	Range	Optimum	Range (2)	Maximum (4)
As supplied	Nil			0.82
Magnafloc 155	0.1-0.3	0.1(3)	0.88-0.89	0.89
Dispersant	**Dosage Kg/Tonne (1)**		**Settled Dry Density t/m³**	
	Range	Optimum	Range (2)	Maximum (4)
As supplied	Nil			0.82, 0.82, 0.86
Calgon	0.5-5	5	0.83-0.98	0.96, 0.98, 1.09
Busperse 39	0.05-5	5	0.82-1.25	1.00, 1.06, 1.25
Dispex N40	0.1			0.86, 0.88, 0.93

NOTES (1) Actual amounts varying by up to 5%
(2) Range is for minimum and maximum amounts of additive
(3) Limited range tested
(4) Figures are for 10, 20 and 100 days
(5) Figures are for 20 day settling
(6) Water cloudy for 10 days for 5 kg/tonne Busperse 39 and for >100 days, for 9 kg/tonne Giessfix C30

columns. It can be seen that the tests result in a wide range of settled densities. The voids ratio is a better guide of settling behaviour and shows the markedly different behaviour between the crushed rock tailings (North Kalgurli and Broken Hill) and the washery tailings.

As would be expected the small columns which have lower average effective stress in the tailings generally result in lower settled densities than the large columns (generally 80-95%) but the ratio is not constant. It was not practicable to determine the variation of density with depth in the large columns.

Table 8. Summary of Results of Large Column Settling Tests

Tailings	Dispersant Added	Final Condition		Intermediate Density	
		Density t/m³	Time days	50 days	100 days
Weipa 1	As supplied	0.54	20	-	-
	Drained	0.76	6		
Weipa 2	As supplied	0.57	90	0.54	-
	4 kg/tonne Calgon	0.67	100	0.64	0.67
	10kg/tonne Calgon	0.67	70	0.63	0.67
	5 kg/tonne Busperse 39	0.64	215	0.27	0.31
	Drained	0.71	24		
	Drained + 4 kg/tonne Calgon	0.76	35		
Wambo	As supplied	0.42	175	0.35	0.39
Riverside	As supplied	0.63	200	0.52	0.57
	4 kg/tonne Busperse 39	0.76	240	0.63	0.68
Newman	As supplied	1.07	35	-	-
Hamersley	As supplied	0.96	135	0.83	0.93
North Kalgurli	As supplied	1.35	30	-	-
Broken Hill	As supplied	1.46	<1	-	-

NOTES (1) All columns 1900mm high, and undrained, except for "DRAINED" tests which are 1300mm high and drained
(2) Water is cloudy for a long period when dispersants are added

Table 9. Residual Chemical Content After Adding Dispersants

Tailings	Dispersant	Quantity of Dispersant kg/tonne	Total Carbon Added ppm	Total Carbon of Tailings Water ppm
Weipa 2	As supplied	Nil	Nil	6
	Busperse 39	0.05	1	6
	Busperse 39	0.1	2	6
	Busperse 39	1.0	20	9
	Dispex N40	0.05	2	6
	Dispex N40	0.1	4.5	6.5
Riverside	As supplied	Nil	Nil	
	Busperse 39	0.5	13	24
	Busperse 39	1.0	26	28
	Busperse 39	5.0	130	800
	Giessfix C30	5.0	4.5	1400
Hamersley	As supplied	Nil	Nil	23
	Busperse 39	0.05	2	21
	Busperse 39	0.1	4	23
	Busperse 39	0.25	10	23
	Busperse 39	0.5	20	26
	Busperse 39	1.0	40	33
	Busperse 39	2.5	100	48
	Busperse 39	5.0	200	64
	Calgon	0.5	3	31
	Calgon	1.0	6	36
	Calgon	2.5	13	52
	Calgon	5.0	27	73
	Dispex N40	0.1	9	22

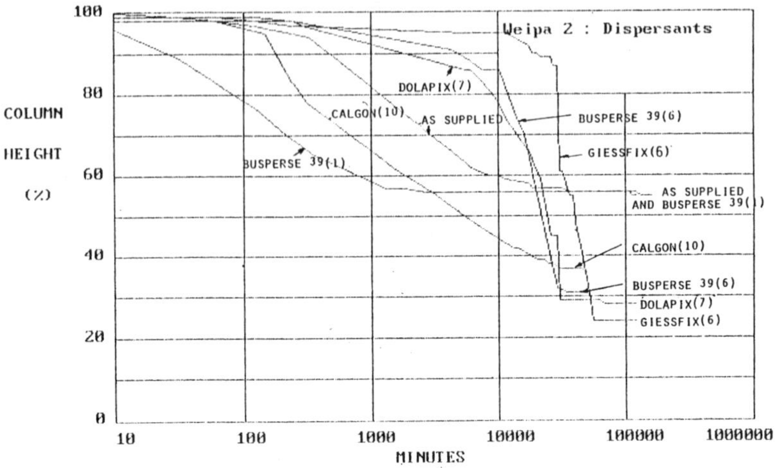

NOTE: Calgon(12) indicates 12kg/tonne Calgon added

Figure 3. Small Column Tests — Weipa with Dispersants

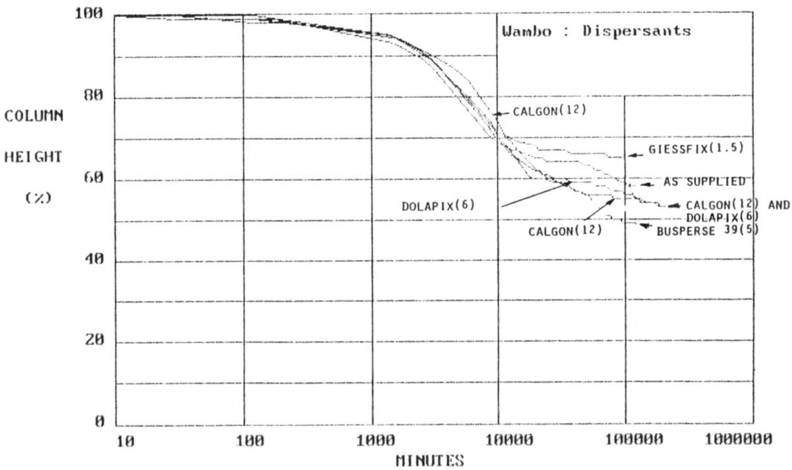

Figure 4. Small Column Tests — Wambo with Dispersants

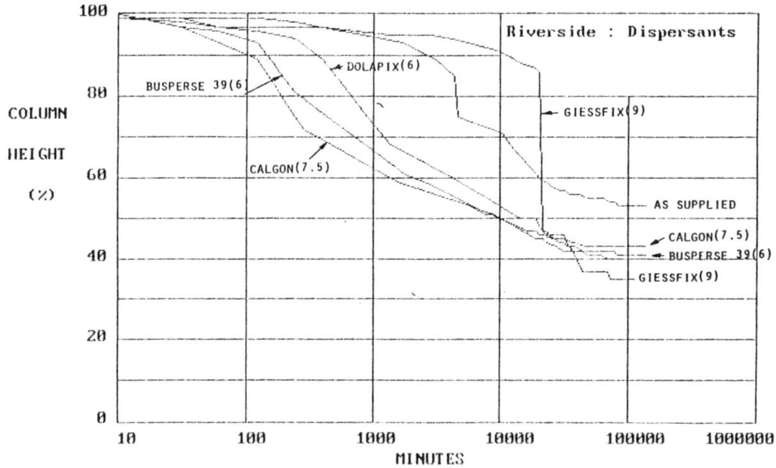

Figure 5. Small Column Tests — Riverside with Dispersants

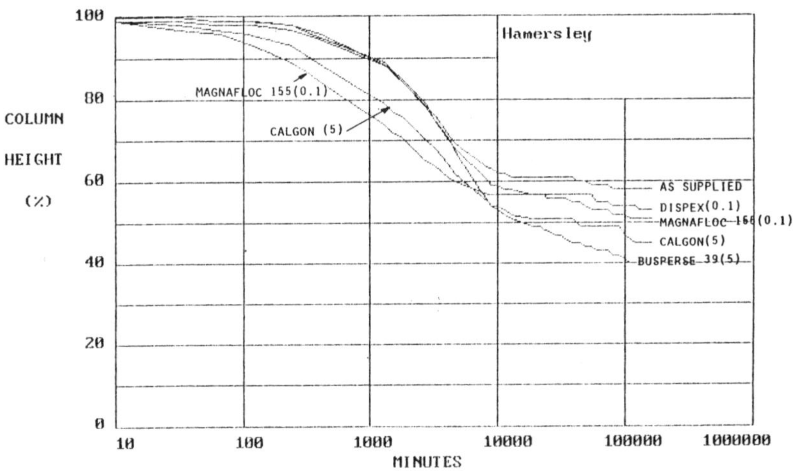

Figure 6. Small Column Tests — Hamersley

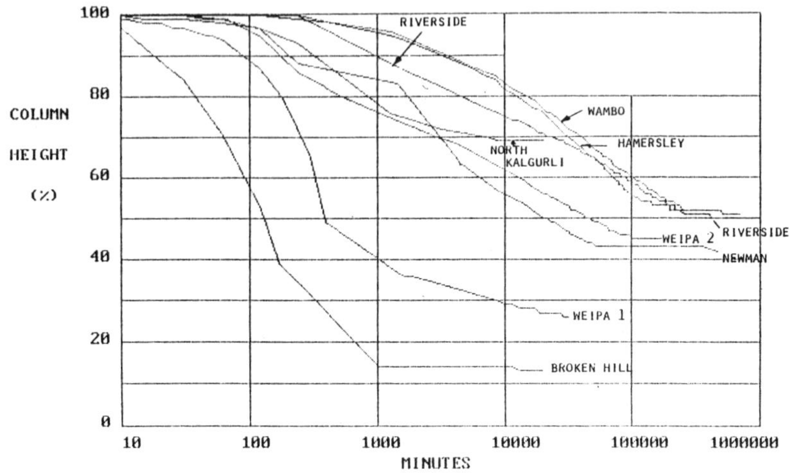

Figure 7. Large Column Tests — As Supplied

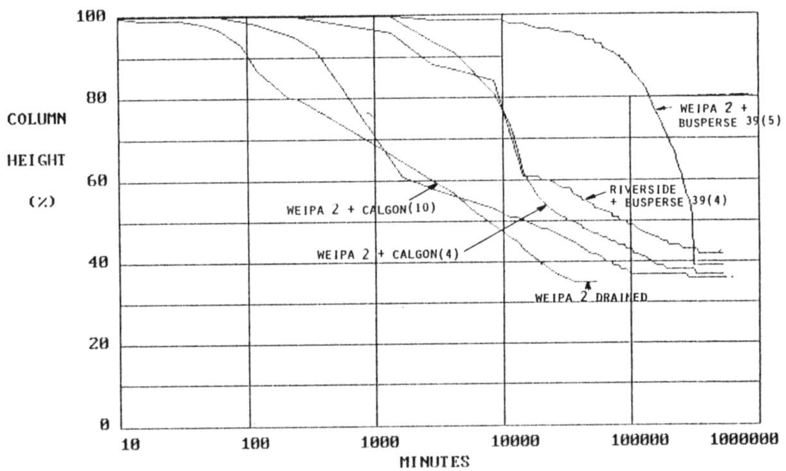

NOTE: Only "Weipa 2 Drained" is a Drained Test

Figure 8. Large Column Tests — With Dispersants and Drained

Table 10. Comparison between Results of Small and Large Column Settling Tests

Tailings	Dispersant Added	Final Condition				Dry Density	Voids Ratio
		Small Columns		Large Columns		Small Column / Large Column	Small Column / Large Column
		Dry Density t/m³	Voids Ratio	Dry Density t/m³	Voids Ratio	%	%
Weipa 1	As supplied	0.30	8.16	0.54	4.09	55	199
Weipa 2	As supplied	0.45	5.33	0.57	4.0	79	158
	10 kg/tonne Calgon	0.66	3.32	0.67	3.25	98	102
	6 kg/tonne Bus-perse 39	0.81	2.52	0.64	3.35	127	75
Wambo	As supplied	0.34(0.37)	4.47(4.03)	0.42	3.42	81(88)	130(118)
Riverside	As supplied	0.54(0.61)	2.22(1.85)	0.63	1.76	86(97)	126(105)
	7 kg/tonne Bus-perse 39	0.77(0.81)	1.26(1.15)	0.76	1.29	101(106)	98(89)
Newman	As supplied	0.96	2.85	1.07	2.46	90	116
Hamersley	As supplied	0.82	3.26	0.96	2.65	85	123
North Kalgurli	As supplied	1.29	1.18	1.35	1.08	95	109
Broken Hill	As supplied	1.82	0.68	1.46	1.09	125	62

NOTE: 1. Voids Ratio = Volume Voids/Volume Solids
2. Figures in brackets are for 100 day settling, other are for 20 days. For other than Wambo and Weipa, no settling occurs after 20 days.

Table 11 compares the results of the laboratory tests with densities measured in the mine tailings storages. It can be seen that there is not a consistent relationship between the field and laboratory values. This may be due to

— limited amount of field testing
— field testing being in the "slurries" remote from the discharge point, while laboratory tests are on tailings as discharged.

Table 11. Comparison of Laboratory and Field Settled Densities

MINE	SOIL PARTICLE DENSITY	SETTLED DRY DENSITY t/m³			SETTLED VOIDS RATIO		
	t/m³	Small Column	Large Column	Field (Depth) in Metres	Small Column	Large Column	Field (Depth) in Metres
Weipa	2.85	0.45	0.57 (0.71) (3)	0.64(0-0.35) 0.82(0.35-0.7) 0.84(0.7-1.05)	5.33	4.0	3.45(0-0.35) 2.48(0.35-0.7) 2.39(0.7-1,05)
Wambo	1.86	0.34	0.42	0.43min (1)	4.47	3.42	3.32max
Riverside	1.74	0.54	0.63	No data available	2.22	1.76	No data available
Newman	3.70	0.96	1.07	1.01(0-0.35) 1.19(0.35-0.7)	2.85	2.46	2.67(0-0.35) 2.11(0.35-0.7)
Hamersley	3.50	0.82	0.96	0.57(0-0.35) (2) 0.71(0.35-0.7) 0.90(0.07-1.05)	3.26	2.65	5.14(0-0.35) 3.92(0.35-0.7) 2.89(0.7-1.05)

(1) Minimum in pond which had been dessicating for 3 years — occurred at depth 0.6 and 1.5m.
(2) Sampling below water was difficult. Top value may be inaccurate.
(3) Drained column test.

(b) Effect of flocculants

The tests show that the addition of flocculants does not result in any significant change in settled density. In the case of Wambo, Riverside and Hamersley tailings, flocculants have already been added, but addition of further flocculant does not improve the settled density. The addition of lime and alum actually results in significantly reduced settled densities for Wambo and Riverside tailings.

(c) Effect of dispersants

Tables 12 and 13 summarize the effect of dispersants on small and large column settling tests. Table 14 summarizes cases when small quantities of dispersant resulted in more rapid settlement of tailings.

Table 12. Summary of the Effect of Adding Dispersants to Tailings in Small Column Settling Tests

Tailings	Dry Density Treated/Dry Density as Supplied - %					
	Calgon			Busperse 39		
	10 days	20 days	100 days	10 days	20 days	100 days
Weipa 1		160				
Weipa 2	145	154	151	193	184	180
Wambo	103	106	105	100	106	119
Riverside	136	133	125	138	143	133
Newman	104	109	112	111	117	131
Hamersley	117	119	126	122	129	145
North Kalgurli	100	100	100	102	102	102
Broken Hill	69	69	-	-	-	-

Table 13. Summary of the Effect of Adding Dispersants to Tailings in Large Column Settling Tests

Tailings	Dry Density Treated/Dry Density as Supplied - %	
	Calgon	Busperse 39
Weipa 2	117	112
Riverside		120

Table 14. Summary of Small Column Settling Tests where Small Quantities of Dispersant Result in Rapid Settling

Tailings	Dispersant	Quantity of Dispersant kg/Tonne	Time to Settle With Dispersant Hours	Time to Settle Without Dispersant Hours	Settled Dry Density With Dispersants / Settled Dry Density As Supplied	
					10 day	20 day
Riverside	Dispex N40	3	15	95	1.07	1.07
	Busperse 39	3-6	400(2)	400	1.09	1.09
Weipa 2	Busperse 39	0.9-1.4	15	95	1.40	1.35

NOTES: (1) Time is for settlement to 95% of total settlement
(2) 60 hours to settle to the same final density as untreated tailings

From these tables, Tables 7.1 to 7.4 and Figures 3 to 8, it can be seen that

- the addition of relatively large quantities of dispersants results in increased settled density (reduced voids ratio) for several of the tailings. The following dispersants are most effective:

Weipa 2 — Calgon, Busperse 39, Dolapix PC67
Riverside — Giessfix, Busperse 39, Dolapix PC67
Newman — Busperse 39, Calgon
Hamersley — Busperse 39, Calgon

- Wambo tailings are not significantly affected by dispersants and there is no effect on North Kalgurli and Broken Hill tailings
- the increase in settled density with addition of dispersants is significantly less in the large columns than for the small columns. The total time for settlement with dispersants in the large columns is very large
- the total time for settling is increased by the addition of large quantities of dispersant, but the time to reach the same density as the untreated tailings is not always increased
- as shown in Table 14 and Figures 3 and 5, small quantities of dispersant added to Weipa 2 and Riverside tailings results in more rapid settling along with some increase in settled density
- the tests reported in Table 9 indicate that addition of up to 0.5kg/tonne of Busperse results in little or no residual dispersant in the tailings water. Larger quantities (which also lead to cloudy water over the tailings) result in residual dispersant
- there is some concern that friction on the sides of the large columns may lead to "arching" effects, giving reduced settled densities. However none was observed, and in any case comparative testing "with" and "without" dispersant should be valid.

(d) Discussion in terms of clay mineralogy

The results of the testing can be explained in terms of the clay mineralogy of the tailings:

- Weipa 1 and 2 tailings are naturally flocculated clay minerals (Kaolin and amorphous clay minerals). The addition of small quantities of dispersant breaks up the aggregations of particles, allowing them to behave more as nearly spherical, high density particles which settle more quickly than the aggregated particles. Addition of larger quantities of dispersant gives even greater breakup of aggregated particles, and a wider grading which eventually settles more densely
- Newman and Hamersley tailings are also flocculated and behave to a certain extent similar to Weipa tailings. However the large quantity of iron oxide would possible be unaffected giving a lesser response
- Riverside tailings are a mixture of flocculated and dispersed clay minerals and behave intermediately between Weipa and Wambo
- Wambo tailings have a high percentage of sodium montmorillonite which is a naturally dispersive clay mineral, and hence the tailings are not much affected by adding dispersant
- the absence of clay in the Broken Hill tailings leads to it settling rapidly and the tailings being unaffected by flocculants and dispersants.

Engineering Implications

The tests carried out in this study indicate that:

- the addition of flocculants to tailings is unlikely to increase the settled density in the tailings dam. There are indications that the addition of lime to some tailings may even result in reduced settled densities
- the addition of dispersants to tailings which contain a high content of Kaolin and amorphous clay minerals can result in increased settled density. Tailings with no clay, or montmorillonite clays are not benefitted by adding dispersants
- the increases in density are relatively large 20-80% for shallow depths of tailings (300mm), but only 10-20% for 2m depths of tailings. Hence the potential benefit for large depths of deposition are likely to be relatively small. However this needs further investigation to assess the effects of arching in the large columns
- the maximum increases in density are achieved with relatively large dosages of dispersant, which result in an increased time for settlement, cloudy water, and a residual content of dispersant in the water, which may affect thickener operation if returned to the mill
- if the water was used elsewhere in the mine the residual chemical content need not be a problem
- the column tests are not necessarily a good guide to field performance, but should give a reasonable guide to relative performance, with and without dispersants and flocculants
- the addition of smaller amounts of Busperse 39 to Weipa 2 and Riverside tailings resulted in more rapid settling of the tailings, with some increase in settled density. The amount of residual chemical was small for Weipa 2 and apparently high for Riverside. The rapid settling behaviour should be of benefit where tailings are placed in layers, since water will bleed off more quickly, potentially resulting in better water recovery and less water to be removed by dessication
- there is some scope for improved water recovery as a result of adding a dispersant such as Busperse 39. Assuming that the 10 day small column values are indicative of tailings placed in thin layers, the increase in water recovery may be of the order of 5% of total water in the tailings for Weipa 2 and Newman, 10-15% for Riverside and Hamersley. Water recovery would be worsened for Wambo, with no change for North Kalgurli and Broken Hill. The 20 day values would indicate larger gains (25%) for Weipa 2, but evaporation during this time would be significant and more detailed analysis is necessary
- the column tests show that the settled density is independent of initial water content. This indicates that provided water can be recovered from the tailings dam, there is little benefit in thickening the tailings to the limits of the plant unless the resulting water content is lower than the settled water content. In practice water is lost in the storages by leakage and evaporation so

there probably are advantages in thickening

- effective underdrainage can result in significantly increased rates of settlement of the tailings
- the settled density represents the starting point from which drying by dessication can begin. Relatively small increases in density can lead to significantly shorter evaporation times. Hence there may be potential application for use of dispersants in sub-aerial deposition schemes, if the dispersants lead to higher settled densities.

Conclusions

It is concluded that there is scope for further worthwhile study of the effects of adding dispersants to the "flocculated" tailings — Weipa 2, Riverside, Newman and Hamersley. This would concentrate on assessing the effects of adding large quantities of dispersant on large depths of tailings using slurry consolidometers, and the effects on water recovery and dessication of using smaller amounts of dispersant resulting in more rapid settlement. The results of this study should be applicable to other high clay content tailings and to disposal of dredged clay fill.

Acknowledgements

The assistance of the seven mining companies who provided tailings for this study is acknowledged, as is the assistance of the Faculty of Engineering, UNSW. Individuals who have contributed to this work include Messrs. A. Harrington, L. O'Keeffe, G. Tamaddoni, F. Mizzi, A. Macken, R. Hogg and R. McKinnon. Ms P. Fiztgerald assisted in the study on flocculants.

References

Allied Colloids (1983). "Magnafloc Synthetic Flocculants".
Chilingarian, G.V., and Vorabutz, P. (1983). "Drilling and drilling fluids". Elsevier.
Gadiel, P.A. (1979). "A review of polyelectrolyte use in water treatment in Australia". Water, Vol 6 No 4, Dec. 1979, 10-12.
Ihle, S.W., and Sparrow, G.J. (1980). "The effect of flocculants on the dewatering time of a clay suspension". Proc. Australasian Inst. Min. Metall. No 274, June 1980, 37-40.
Lawver, J.E., and Carrier, W.D. (1984). "Modelling of phosphatic clay disposal systems". Minerals and Metallurgical Processing, Vol 1 No 3, Nov. 1984, 198-204.
Lynberg, T.M. (1981). "Use of double belt filter press in coal processing". Mining Congress Journal, July 1981, 53-56.
Mitchell, J.V. (1976). "Fundamentals of soil behaviour". Wiley.
Moss, N., and Dymond, B. (1978). "Flocculation: theory and application".
Smellay, A.G., Scheiner, B.J., and Zatko, J.R. (1980). "Dewatering of industrial clay wastes". U.S. Dept. Interior, Bureau of Mines Report of Investigations 8498.

Evaluation of a Coal Refuse Impoundment Built by the Upstream Method

Barry K. Thacker[1], C. Robert Ullrich[2], J. George Athanasakes[3], and George Smith[4]

Abstract

A case history of the Abner Fork Coal Refuse Impoundment located in Harlan County, Kentucky, is presented. The dam is being constructed using coarse coal refuse to create an impoundment where slurried fine coal refuse can be pumped for disposal. Because of the steep terrain at the site and the relatively high percentage of fine refuse being generated, the facility is being built by the upstream construction method. Because of the relatively rapid rate of construction of this facility, synthetic wick drains are being placed in the fine refuse to allow essentially complete pore pressure dissipation, as documented by field measurements with pneumatic piezometers. Also, cyclic triaxial testing and an associated seismic stability analysis have been performed. Based on the results of the analyses and monitoring, the coal refuse impoundment meets current, prudent, engineering criteria with regard to stability, even though a large portion of the dam is being built over hydraulically filled fine coal refuse.

Introduction

Coal contains impurities which cause sulfur dioxide and ash to be emitted when the coal is burned in a power plant; these impurities consist of shales and pyrites (coarse coal refuse) and

[1] Project Manager, Geologic Associates Division of EDGe, Inc., Knoxville, TN

[2] Associate Professor of Civil Engineering, University of Louisville, Louisville, KY

[3] Formerly Research Assistant, University of Louisville, Louisville, KY

[4] Project Engineer, Manalapan Mining Company, Brookside, KY

sand, silt and clay size tailings (fine coal refuse). To comply with government air quality regulations, coal is often processed to remove some of the impurities before the coal is transported to a power plant. The quantities of rejects (refuse) produced at a single coal processing plant can exceed one million tons per year, thereby presenting coal operators with the difficult task of disposing of these wastes in a cost effective, safe, and environmentally acceptable manner (10).

One method of disposing of coal refuse is in a slurry impoundment, which is created in the following steps:

(1) A dam is built with either coarse refuse or on-site borrow to create an initial impoundment.
(2) Fine refuse is pumped to the impoundment for disposal.
(3) Additional stages of the dam are built using coarse refuse from the plant to enlarge the impoundment and thus provide continued disposal capacity for fine refuse.

The additional stages of the dam may be constructed by the upstream, downstream, or centerline construction methods or a combination thereof. As shown in Figure 1, the three variations of slurry impoundment construction reference the direction in which the later stages of the dam progress in relation to the starter dam.

A potential failure surface typically passes entirely through the embankment in a dam built by either the downstream or centerline methods. Designers of these types of coal refuse impoundments may specify the degree of compaction of the coarse refuse and thereby have significant control of the engineering properties of the materials situated within the potential zone of failure. However, a potential failure surface within the facility built by the upstream method typically passes through the fine refuse that is placed by hydraulic means as well as through the embankment materials. Therefore, in upstream construction, the

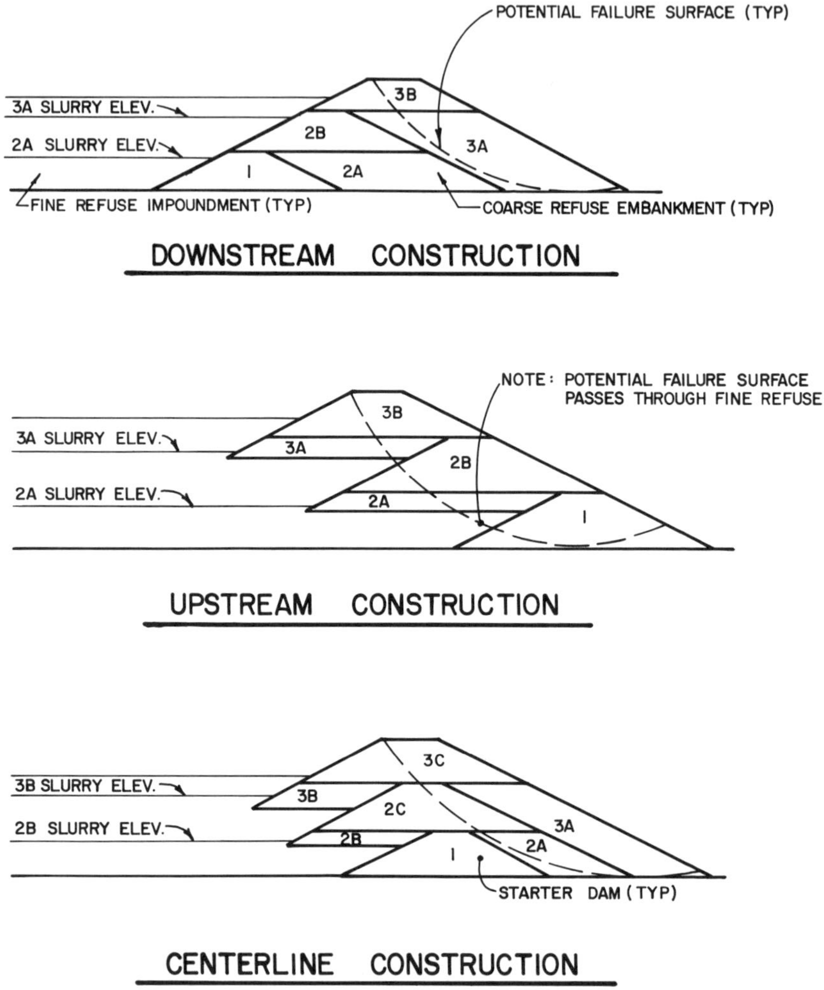

Figure 1. Types of Slurry Impoundments

stability of the facility is partially dependent on the shearing resistance developed in the fine refuse. Previously, designers had little control of the engineering properties of the fine refuse. They could only predict the degree of consolidation of the fine refuse at a given time and then design the facility accordingly. Because of this lack of control of the engineering characteristics of the fine refuse, many designers have preferred the use of the downstream method or the centerline method of construction for tailings impoundments.

The Highsplint Coal Preparation Plant is located in an area of Harlan County, Kentucky, where the valleys are narrow and the hillsides are steep. Moreover, the ratio by volume of coarse refuse to fine refuse produced at the plant is on the order of about 1.5 to 1. Prior to 1981, fine refuse generated at the plant was dewatered by mechanical means and mixed with the coarse refuse prior to disposal in a fill which was situated in a steep, narrow hollow. Because of the high percentage of clay size particles present, the fine refuse retained a significant quantity of water even after mechanical dewatering. The wet combined refuse was difficult to handle and nearly impossible to compact in the fill. At that time, plans were initiated to build a slurry impoundment so that the fine refuse could be placed hydraulically, thus avoiding the potentially unsafe disposal of the wet combined refuse in a fill.

Preliminary design calculations indicated that an inadequate quantity of coarse refuse was produced at the plant to build the slurry impoundment by the downstream or centerline construction methods within any of the hollows surrounding the plant. After an extensive feasibility study, the adjacent Abner Fork hollow was selected as the site for a slurry impoundment because of the presence of an existing freshwater impoundment. This existing facility could be used as a starter dam and an adequate quantity of coarse refuse would then be available to build the additional stages of the slurry impoundment in Abner Fork by the upstream method. Presented herein is a description of the design and construction monitoring of the Abner Fork Slurry Impoundment. Also, results of cyclic triaxial testing of the fine refuse and an associated seismic stability analysis are presented.

Design

The Abner Fork Slurry Impoundment was designed to be built in 16 stages as shown in Figure 2 (plan) and Figure 3 (cross-section). The facility was designed to enable slurried fine refuse to be pumped to an existing fresh water impoundment for disposal so that coarse refuse from the plant could be used as borrow to build subsequent stages of the embankment. Grain size distributions of the coarse refuse and fine refuse disposed at the Abner Fork site are shown in Figure 4. By discharging the fine refuse at the upstream face of a dam, a delta will form in which the coarser size particles are deposited at the upstream face and the finer size particles settle toward the rear of the impoundment. If the fine refuse consists of a large percentage of sand size particles and fine refuse is discharged evenly across the upstream face of the dam, then the foundation for the upstream construction will be predominantly a silty sand which consolidates relatively quickly.

However, as shown by the grain size distributions in Figure 4, the Abner Fork fine refuse consists almost exclusively of silt and clay size particles. In this case, the foundation for the upstream construction stages will be a clayey, sandy silt which consolidates relatively slowly. Moreover, the shearing resistance of the fine refuse is directly dependent on its degree of consolidation (i.e., pore pressure dissipation). Therefore, the stability analyses of the Abner Fork Slurry Impoundment had to consider the rate of pore pressure dissipation of the fine refuse and the rate of construction of the facility. A stability analysis of the Abner Fork Facility was performed using the strength parameters shown in Table 1 and construction pore water pressures, which were estimated by comparing the time rate of consolidation characteristics of the

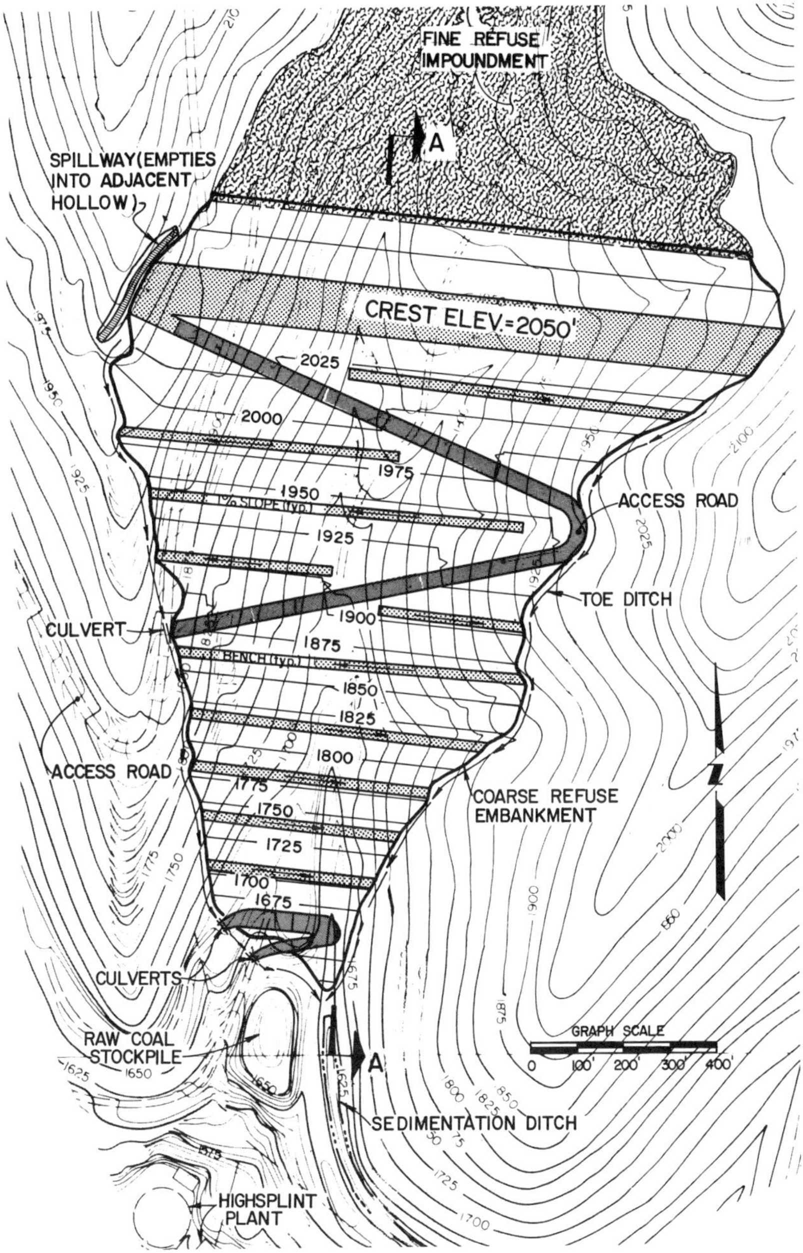

Figure 2. Plan of Abner Fork Slurry Impoundment

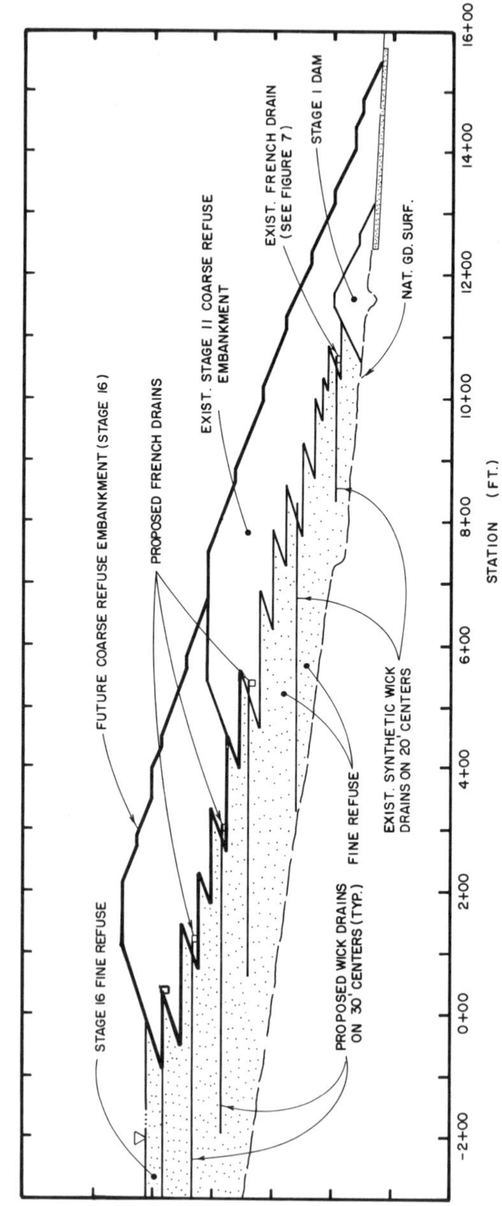

Figure 3. Cross-section A.A of the Abner Fork Slurry Impoundment

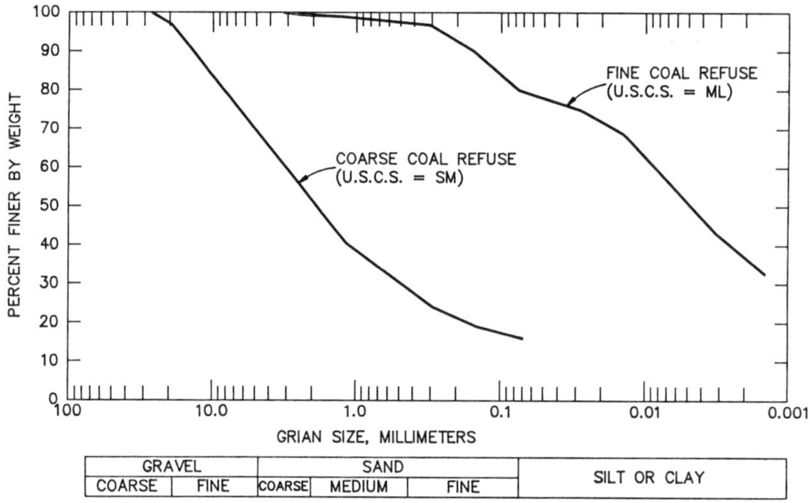

Figure 4. Particle Size Distributions

Table 1

Engineering Properties of
Abner Fork Coal Refuse

		Effective Stress Parameters	
Material	Unit Weight, pcf	$\bar{c}$, psi	$\bar{\phi}$,°
Coarse Refuse (90% Compaction-Standard Proctor)	110 (moist)	0	36
Fine Refuse (Hydraulically placed)	105 (saturated)	0	30

fine refuse to the predicted rate of construction of the embankment. Two conclusions were drawn from the results of the initial stability analyses:

1) Essentially complete pore pressure dissipation in the fine refuse was required to achieve the desired factors of safety for slope stability.
2) As originally designed, the predicted rate of construction resulted in only about 60 percent pore pressure dissipation.

Clearly, additional measures were required to allow essentially complete pore pressure dissipation within the fine refuse during construction. To speed consolidation of the fine refuse, synthetic wick drains were used to reduce the drainage distance within the fine refuse and thereby enable construction pore pressures to dissipate (7, 10). The required spacing of the wick drains, based on the time required for essentially complete pore pressure dissipation for the Abner Fork fine refuse, was estimated using the equation:

$$t = \frac{1.13}{c_v} H^2 \qquad (1)$$

Where:

t = Time required for 95% pore pressure dissipation

c_v = Coefficient of consolidation determined from consolidation test data = 0.0012 in^2/sec. for Abner Fork Fine Refuse

H = Drainage distance to nearest wick drain or free draining coarse refuse surface.

Construction Details and Monitoring

Prior to installation of the wick drains, the fine refuse was discharged evenly across the area upstream of the dam to form a delta. As shown in Figures 5 and 6, a pad of coarse refuse was

Figure 5. Fine Refuse Being Discharged at Upstream Face of Dam (Far Right) to Form a Delta with Coarse Refuse Being Placed and Compacted During Upstream Construction.

Figure 6. Synthetic Wick Drains Installed within Fine Refuse Impoundment

then placed over the fine refuse delta. The wicks were then pulled across the fines from within a boat or using a rope. Figure 6 shows the wick drains after they had been installed. These wick drains can be connected directly into the more pervious coarse refuse for drainage purposes, or preferably, they can be connected to a french drain installed within the upstream face of the dam. If the latter option is selected, the french drain can be connected to a pipe which extends through the dam to discharge downstream of the dam (refer to Figure 7).

Currently, Stage 11 of the subject facility, as shown on Figure 3, is nearly completed. Piezometers (both standpipe and pneumatic) have been installed within the dam and into the fine refuse. As shown in Figure 8, pore pressures have dissipated much more rapidly in stages of upstream construction where wick drains were installed than in cases where no wick drains were installed. Moreover, as shown in Figure 7, the fine refuse has currently consolidated to an average standard penetration test N-value of about 20.

Seismic Stability Analysis

In 1986, questions were raised by representatives of the Mine Safety and Health Administration regarding the seismic stability of tailings impoundments built by the upstream construction method. Initially, an attempt was made to evaluate the liquefaction potential of the Abner Fork Slurry Impoundment based on field performance data developed by Seed et al. (9) and Ishihara (5). However, because the referenced data were developed for materials with a specific gravity greater than 2.6 and considering that the specific gravity of fine coal refuse is about 1.8, it was determined that the previous data were not applicable to coal refuse impoundments. Consequently, a seismic stability analysis was performed on the Abner Fork impoundment based on site specific data using the method described by Klohn et al. (6).

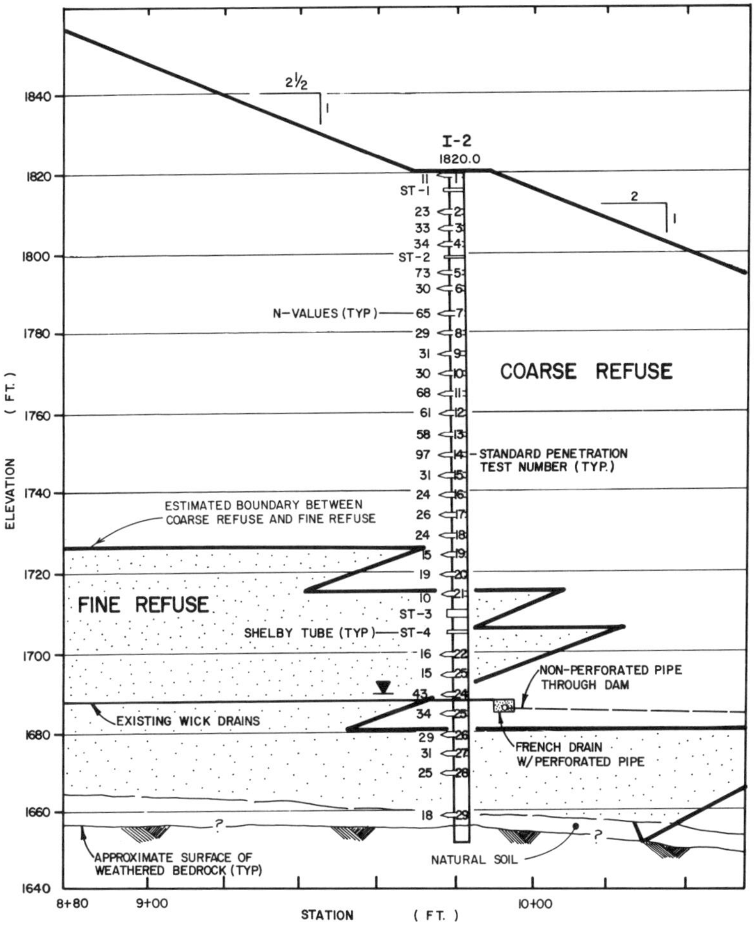

Figure 7. Detail of As-built Wick Drain/French Drain Dewatering System Including Subsurface Data

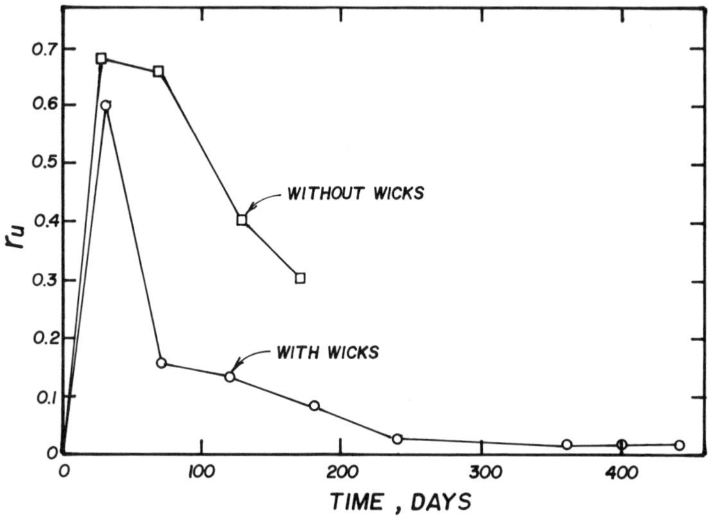

Figure 8. Measured Relationship Between Time and r_u Within Fine Refuse Deposits with and without Synthetic Wicks. (r_u Defined as the Pore Pressure Divided by the Total Overburden Pressure)

The Klohn method is an effective stress analysis using a conventional method of slices slope stability technique, incorporating estimates of excess pore pressures generated by earthquake shaking. The computer program STABR (3), which employs the Bishop Modified Method of Slices, was used to analyze the simplified cross-section shown in Figure 9. The failure surface was chosen to pass mainly through fine-grained coal refuse, in which excess pore pressures would be generated during an earthquake. The factor of safety for static conditions for the failure surface shown in Figure 9 was 1.90.

The computer program STABR calculated and output values of normal stress σ_v and shear stress τ on the base of each slice, as shown in the inset in Figure 9. The major principal stress σ_{1s} and the minor principal stress σ_{3s} were calculated for static conditions for each slice using the geometry of the Mohr circle shown in Figure 9. Effective principal stresses $\bar{\sigma}_{1s}$ and $\bar{\sigma}_{3s}$ were calculated by subtracting the static pore pressure u_s from the total principal stress values σ_{1s} and σ_{3s} for each slice. Finally, the consolidation ratio K_c was calculated for each slice using the equation:

$$K_c = \bar{\sigma}_{1s} / \bar{\sigma}_{3s} \quad (2)$$

For the Abner Fork impoundment, the average value of K_c was about 2 and the consolidation stress $\bar{\sigma}_{3s}$ along the failure surface ranged from 200 kPa to 500 kPa.

The pseudostatic approach was used to estimate dynamic stresses along the failure surface. In the STABR program, a horizontal pseudostatic force kW is applied at mid-height of each slice having weight W. The seismic coefficient k was chosen as follows: The 50-year, 95% probability earthquake was estimated to be magnitude 6.1 (1) and the distance from the site to the nearest major fault (Pine Mountain Fault) was estimated to be 5 km. Using the relationship between peak horizontal acceleration, and magnitude and distance, presented by Idriss (4), the peak

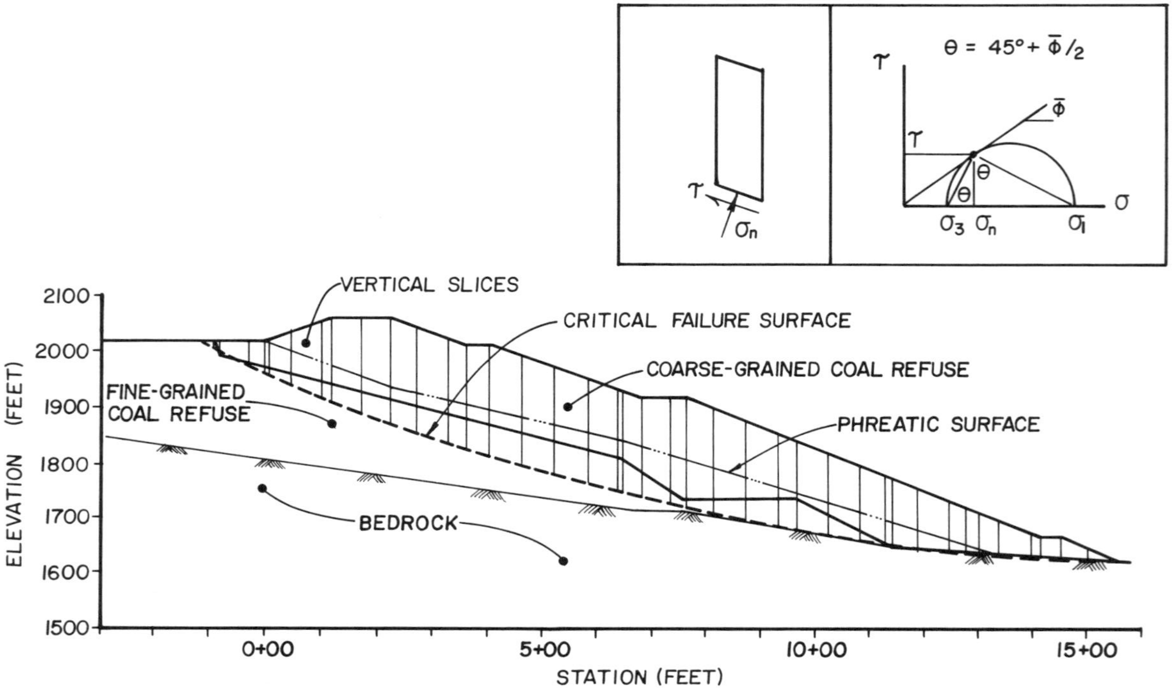

Figure 9. Simplified Cross-section and Critical Failure Surface

acceleration at the site was estimated to be 0.35 g. Since the estimated magnitude and peak acceleration of the design earthquake were similar to those of the 1943 El Centro earthquake, the analysis developed by Seed and Martin (8) was used to estimate seismic coefficient, number of force cycles, and frequency of force cycles. In the analysis, the shear wave velocity of the fine coal refuse was estimated to be 300 feet per second. The estimated seismic coefficient was 0.1 and the design earthquake was assumed to be composed of eight force cycles having a frequency of one cycle per second.

The computer program STABR was used to perform a pseudostatic slope stability analysis with k = 0.1 and to calculate σ_n and τ on the base of each slice. Values of dynamic principal stresses σ_{1d} and σ_{3d} were calculated for each slice using values of σ_n and τ. The cyclic stress ratio SR was calculated for each slice using the equation:

$$SR = \frac{(\sigma_{1d} - \sigma_{3d}) - (\sigma_{1s} - \sigma_{3s})}{2\bar{\sigma}_{3s}} \quad (3)$$

The average value of cyclic stress ratio for the cross-section shown in Figure 9 was 0.2.

In order to estimate excess pore pressures generated in the fine coal refuse during the design earthquake, a series of cyclic triaxial tests was performed on sedimented samples of the fine refuse using a CKC e/p Cyclic Loader (2). Four tests were performed on anisotropically consolidated samples with confining pressures ranging from 200 kPa to 500 kPa and K_c = 2. Samples were subjected to cyclic deviator stresses corresponding to SR = 0.2; eight cyclic stress cycles were applied to each sample at a frequency of one cycle per second and pore pressure buildup was measured, as shown in Figure 10. The maximum excess pore pressure measured in each test was normalized by dividing $\bar{u}$ by $\bar{\sigma}_{3s}$, and $(\bar{u}/\bar{\sigma}_{3s})$ max was plotted versus $\bar{\sigma}_{3s}$, as shown in Figure 11.

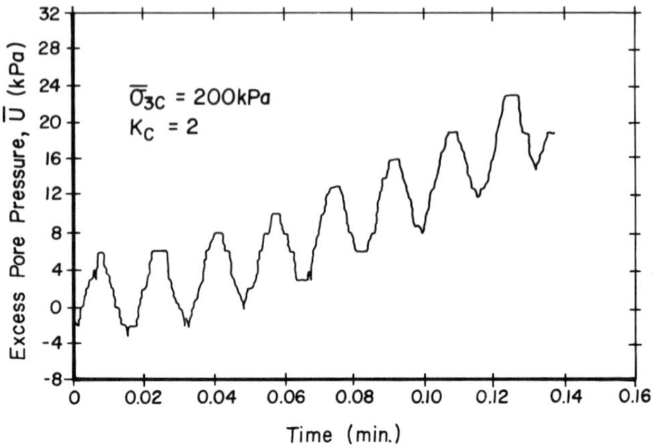

Figure 10. Excess Pore Pressure Versus Time in Cyclic Triaxial Test

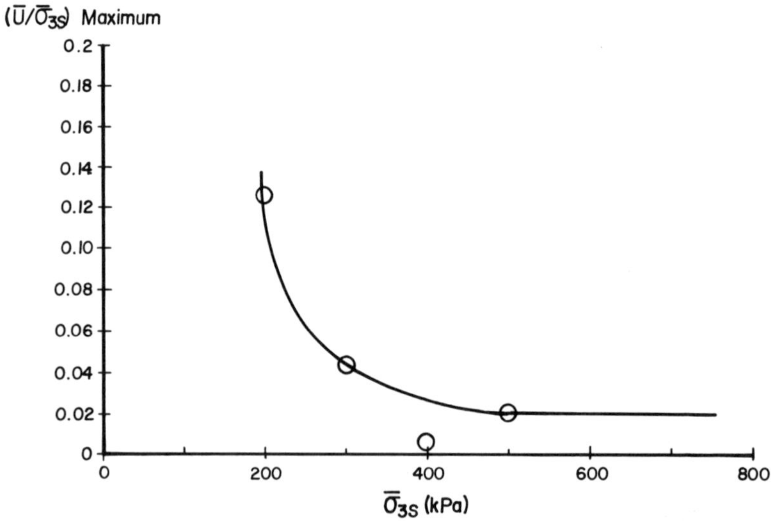

Figure 11. Relationship Between Maximum ($\bar{u}$ / $\bar{\sigma}_{3s}$) and $\bar{\sigma}_{3s}$

A final slope stability analysis was performed on the cross-section shown in Figure 9, in which a seismic coefficient of 0.1 was used and pore pressures were estimated by adding excess pore pressures from Figure 11 to static pore pressures indicated by the phreatic line in Figure 9. The calculated factor of safety was 1.29 in the seismic analysis.

The seismic stability analysis of the Abner Fork impoundment showed that in this case, inertia forces during the design earthquake are much more important than excess pore pressures in lowering the calculated factor of safety. A factor of safety of 1.31 was calculated for the case where a seismic coefficient of 0.1 was used and excess pore pressures were assumed to be zero as compared to a factor of safety of 1.29 when estimates of excess pore pressures were also included. The fine coal refuse in the Abner Fork impoundment apparently is not highly susceptible to pore pressure buildup because of its fine-grained nature and plasticity. Also, as shown in Figure 11, excess pore pressure buildup decreases with increasing consolidation pressure. Thus, the use of wick drains to speed consolidation of the fine coal refuse during construction reduces the risk of failure of the impoundment during an earthquake.

Conclusions

The results of design analyses and performance monitoring of the Abner Fork Slurry Impoundment lead the authors to the conclusions as follows:

1) In many areas of the Appalachian Coal Fields, cost considerations require that coal refuse disposal facilities be built by the upstream construction method.
2) Previously, designers could only predict the degree of consolidation and therefore the strength of the fine refuse at given times and then design coal refuse

disposal facilities built by the upstream method accordingly.

3) The use of wick drains as proposed herein provides designers with the ability to decrease the drainage distance within the deposit and thereby increase the rate of pore pressure dissipation and shearing resistance of the fine refuse.

4) Based on Klohn's method and the results of cyclic triaxial testing of the fine refuse, the Abner Fork Slurry Impoundment has a factor of safety of about 1.3 for the design earthquake condition.

5) The Abner Fork fine refuse is not highly susceptible to pore pressure build-up because of its fine grained nature, plasticity, and in-place density.

6) Inertia forces during the design earthquake are much more significant than excess pore pressure in lowering the calculated factor or safety.

7) The use of wick drains to speed consolidation and thereby increase the density of the fine refuse during construction reduces the risk of failure of the impoundment during an earthquake.

Acknowledgements

The authors wish to thank personnel from the Mine Safety and Health Administration for their assistance throughout this project.

References

1. Algermissen, S. T., et al., Probabilistic Estimates of Maximum Ground Acceleration and Velocity in Rock in the Contiguous United States, U.S. Geological Survey Open-File Report 82-1033, 1982.

2. Chan, C. K., "An Electropneumatic Cyclic Loading System," Geotechnical Testing Journal, ASTM, Vol. 4, No. 4, 1981, pp. 183-187.

3. Duncan, J. M., and K. S. Wong, STABR: A Computer Program for Slope Stability Analysis with Circular Slip Surfaces, Microcomputer Version, 2nd ed., User's Manual, Virginia Polytechnic Institute and State University, Department of Civil Engineering, 1985, 19 pp.

4. Idriss, I. M., "Evaluating Seismic Risk in Engineering Practice," Proceedings, 11th International Conference on Soil Mechanics and Foundation Engineering, San Francisco, Vol. 1, 1985, pp. 255-320.

5. Ishihara, K., "Post-Earthquake Failure of a Tailings Dam Due to Liquefaction of the Pond Deposit, " Proceedings, International Conference on Case Histories, in Geotechnical Engineering, Vol. 3, University of Missouri, Rolla, Missouri.

6. Klohn, E. J., C. H. Maartman, R. Lo, and W. D. L. Finn, "Simplified Seismic Analysis for Tailings Dams," in Earthquake Engineering and Soil Dynamics, ASCE, Vol. 1, 1987, pp. 540-556.

7. Morrison, A., "The Booming Business in Wick Drains," Civil Engineering, ASCE, Volume 53, No. 3, New York, NY, March, 1982.

8. Seed, H. B., and G. R. Martin, "The Seismic Coefficient in Earth Dam Design," Journal of the Soil Mechanics and Foundations Division, ASCE, Vol. 92, No. SM3, 1966, pp. 25-58.

9. Seed, H B., I. M. Idriss, and I. Arango, "Evaluation of Liquefaction Potential Using Field Performance Data," Journal of Geotechnical Engineering, ASCE 109 (3), 1983, pp. 458-482.

10. Thacker, B. K., and J. A. Schad, "Rapid Construction of a Combined Coal Refuse Embankment," Proceedings, 14th Ohio River Valley Soil Seminar, Louisville, KY, 1983, pp. 4-1 to 4-16.

Densification of an Existing Dam with Explosives

Gerald T. Handford*

In 1981, 1.4 million cubic meters of sand in an existing hydraulic fill tailings dam, were densified with explosives. The loose sand zones ranged in thickness from 5 to 18 meters, and were overlain by 11 to 16 meters of medium dense sand. In all cases, the loose zones lay below the phreatic surface. Penetration resistance increased by 40 to 90 percent over pre-blast values. The majority of tests exhibited a gain in penetration resistance after blasting. Time-dependent strength gain was observed where duplicate post-blast tests were performed.

Introduction

Syncrude Canada Ltd. is one of two open pit oilsand mining and upgrading operations in the world. Both sites are located near Fort McMurray in northern Alberta, Canada. Approximately 240,000 tonnes of tailings solids are produced daily at the Syncrude operation. The tailings are currently stored in an out of pit pond completely enclosed by a dam constructed from the coarse tailings fraction.

Current construction follows a modified form of upstream construction as illustrated in Figure 1. The compacted shell is constructed by sluicing the tailings into construction cells, typically 50 meters wide by 700 meters long, orientated parallel to the dam centerline. The sand is spread by dozers during the sluicing operation and attains an average relative density of 60 percent. In the winter, the tailings are discharged upstream off the compacted shell to form the beach. Beach sand deposited above water has an average relative density of 50 percent.

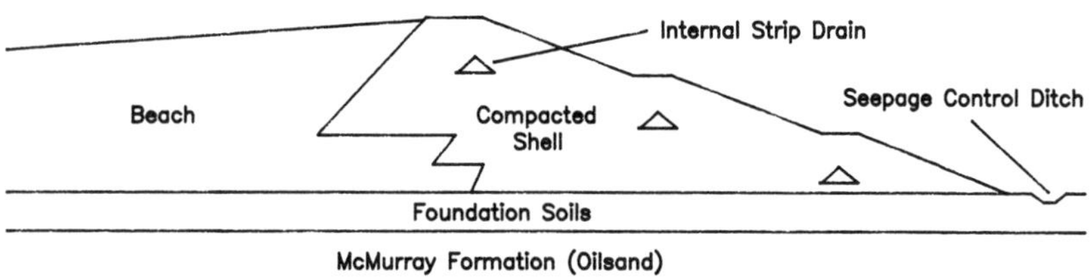

Figure 1. Typical Cross Section

* Supervisor, Tailings Engineering, Syncrude Canada Ltd., P.O. Bag 4009, Drop Point 0062, Fort McMurray, Alberta, Canada, T9H 3L1

The dam has been arbitrarily divided into 700 meter long cells, as illustrated in Figure 2. Loose sand zones in Cells: 18, 19, 27/28, and 29, were densified by blasting in 1981, to preclude liquefaction. Two of the loose zones originated from sluicing tailings into standing water (Cell 18 and 19), one from dumping borrow sand (Pf4) into an existing lake (Cell 27/28), and the last was in-situ Pf4 sand (Cell 29). In total, 1.4 million cubic meters of sand were densified in this program, at a cost of $0.22 per cubic meter. The total area requiring densification was 200,000 square meters.

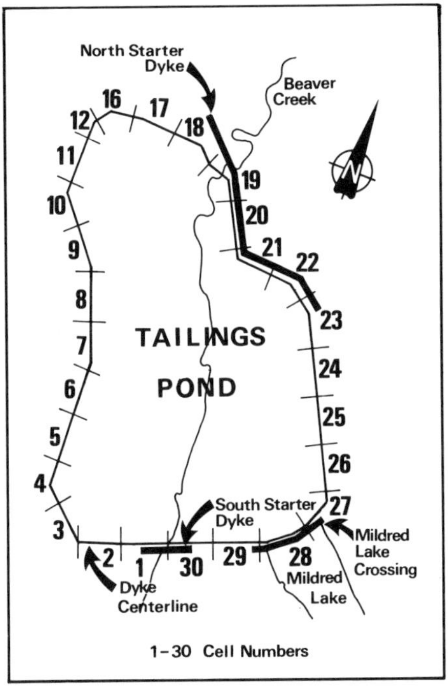

Figure 2. Plan View of Syncrude Tailings Dam

The effectiveness of blasting in densifying sands has been documented in the literature by Lymen (1942), Wild and Haslam (1962), Hall (1962), Prugh (1963), Mitchell (1970) and Solymar (1984) among others, and was chosen on the basis of cost effectiveness. This paper documents the explosive densification program, as there are some features different from the case histories published in the literature. These differences include: (1) the densification program was carried out beneath an existing tailings dam, (2) the dam lies directly on an overconsolidated clay exhibiting horizontal foundation displacement in response to dam construction, and (3) the majority of standard penetration tests (SPT) completed shortly after blasting showed an increase in penetration resistance.

Material Properties

The tailings sand is a fine uniformly graded quartz sand with angular to subangular grains. The Pf4 sand is a glacial outwash quartz sand and is uniformly graded with subangular to rounded grains. Typical grain size distribution curves are shown in Figure 3 for each sand. Mineralogy and index properties are available for tailings sand and are presented in Table 1.

TABLE 1. Mineralogy and Index Properties of Tailings Sand

PROPERTY	MEAN VALUE
Quartz content	95%
Feldspar content	2%
Amphibole content	1%
Pyrite content	1 - 2%
Muscovite content	1%
Coefficient of uniformity	2.3
Minimum void ratio (1)	.52
Maximum void ratio (2)	.84
Specific gravity	2.63

(1) Test method ASTM D4253-83 (1A)
(2) Test method ASTM D4254-83 (A)

Permeability is estimated to be 10^{-3} cm/sec for beached tailings sand and in-situ Pf4. For tailings sand in the compacted shell horizontal permeability is approximately 10^{-4} cm/sec and vertical permeability 10^{-5} cm/sec. From recent work, the saturation at the time of blasting is estimated to be 65 percent above the phreatic surface, 85 percent from 0 to 6 meters below the phreatic surface, and 95 percent at depths greater than 6 meters below the phreatic surface.

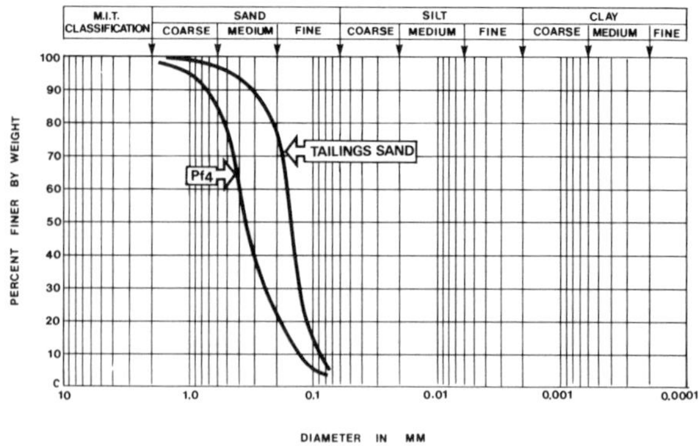

Figure 3. Grain Size Distribution Curves

Cell 18 Test Blast

A typical cross section through Cell 18 is shown in Figure 4. Cell 18 is underlain by an overconsolidated silty clay (Clearwater Formation). Foundation failures occurred in the Clearwater during construction of the starter dam. Subsequent construction of the tailings dam has resulted in horizontal displacement of 8 cm in the Clearwater in the area of the starter dam.

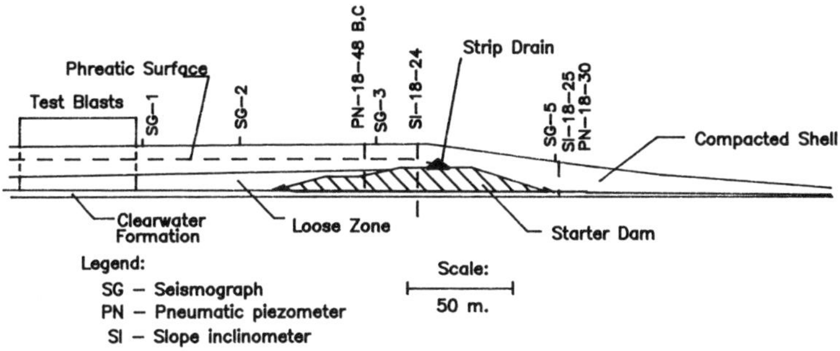

Figure 4. Typical Cross Section Cell 18

The topography in the tailings disposal area slopes slightly to the northeast with only a relatively small creek valley cutting through the area. To get the initial storage volume required for plant startup, 2.5 kilometers of starter dam was constructed with mobile equipment at the northeast corner of the disposal area. The initial tailings were discharged upstream off the starter dam. The surface runoff water and initial tailings water collected in the low areas against the starter dam resulting in the base of the tailings dam being constructed from subaqueously deposited tailings sand.

The zone of subaqueously deposited sand in Cell 18 was approximately 7 meters thick. At the time of densification this zone was overlain by 11 meters of compacted shell. Standard penetration tests, corrected for vertical effective stress, generally ranged between 10 and 20 blows per foot in the subaqueously deposited sand indicating a loose state. In contrast the penetration resistance in the compacted shell ranged from 60 to 90 blows per foot indicating that it was a medium dense sand.

Four small test blasts were carried out initially in Cell 18 to evaluate the effect of blasting on dam stability. The location of the test blastholes and monitoring instruments are illustrated in Figure 5. The first blast consisted of two holes, and was followed by blasts of 4, 7 and 6 holes. A 9 kg charge was used in all holes, with a 25 millisecond surface delay and a 10 meter spacing between holes. The peak particle velocity attenuation curve from the test blasts is plotted in Figure 6. The longitudinal peak particle acceleration is plotted in Figure 7 as a function of scaled distance.

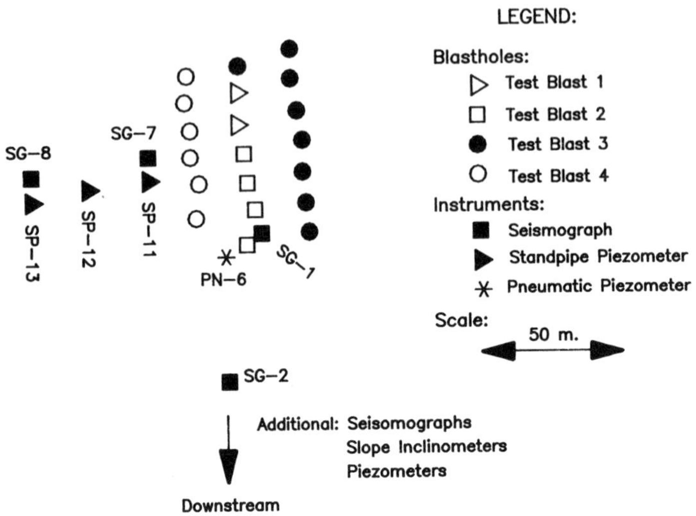

Figure 5. Cell 18 Test Blasts

Prior to the fourth blast, a series of standpipe piezometers were installed in the loose zone. The standpipes were located at distances of 15, 35 and 55 meters perpendicular to the line of blast holes as illustrated in Figure 5. The piezometric elevation for each standpipe vs. time after the blast is shown in Figure 8. The first observation of SP-11 (distance 15 meters) was 3.5 minutes after the blast, at which time water was flowing out the top of the standpipe (elevation 315.5). Piezometer SP-12 (distance 35 meters) shows an immediate increase in pore pressure of 1 meter due to densification of the loose sand and a further increase of 1.5 meters due to migration of excess pore pressures from the blast site. The pore pressure increase peaks 80 minutes after the blast. Piezometer SP-13 (distance 55 meters) does not show an immediate pore pressure response due to blasting, but shows a 1 meter increase 130 minutes after the blast. It took 33 hours for the blast induced pore pressures to dissipate completely.

In Figure 9, the piezometric profile perpendicular to the blast line is plotted for the pre-blast condition and at times of 10, 100 and 1000 minutes after the blast. The immediate pore pressure response as defined above, extends to a distance of 45 meters from the blast line (scaled distance 15 $m/^2 kg$). This distance defines the influence radius of the blast with respect to disturbance of the soil structure. Any pore pressure increase beyond this distance is due solely to migration of excess pore pressures from the blast site.

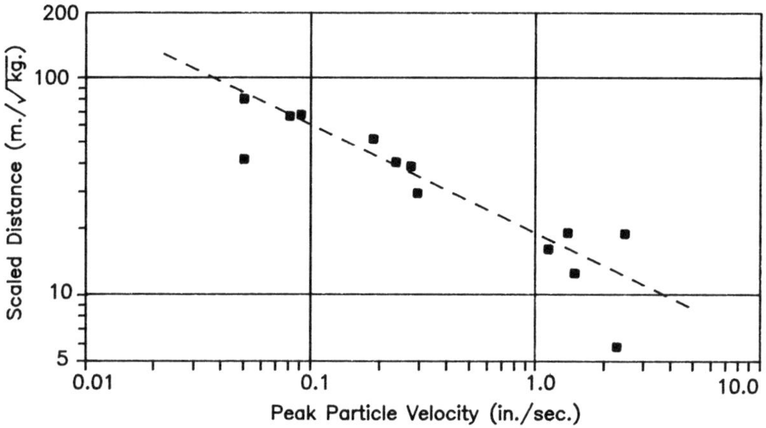

Figure 6. Peak Particle Velocity Attenuation Curve: Tailings Sand

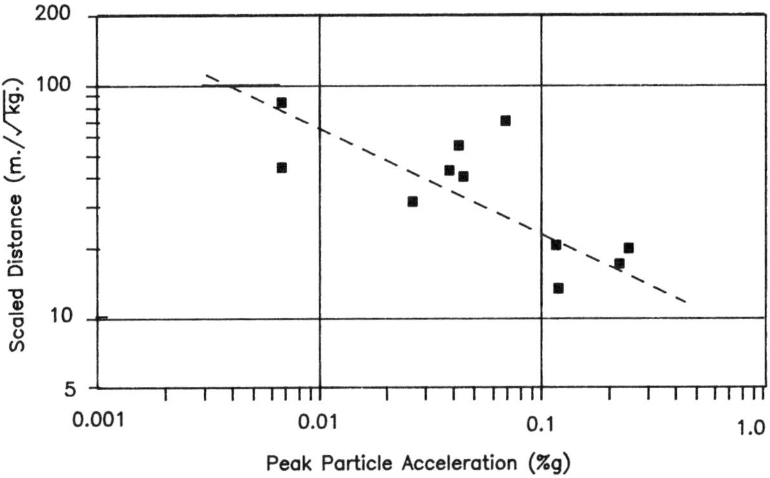

Figure 7. Peak Particle Acceleration: Tailings Sand

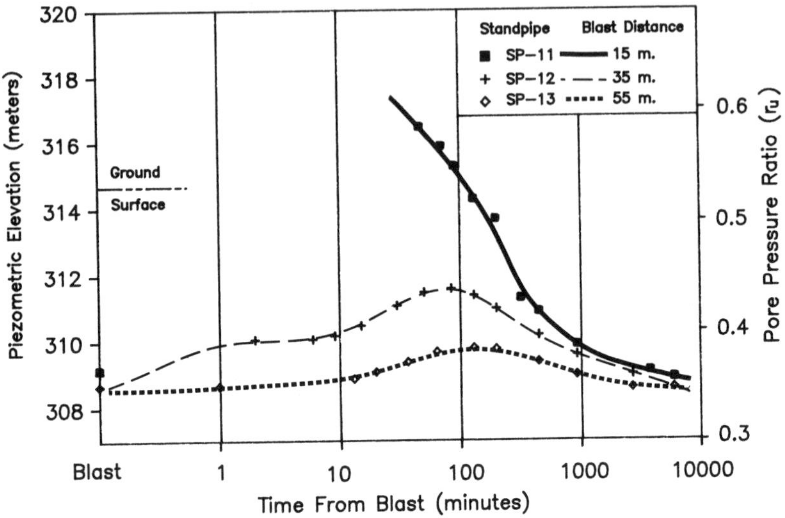

Figure 8. Piezometric Elevation vs. Time From Blast 4

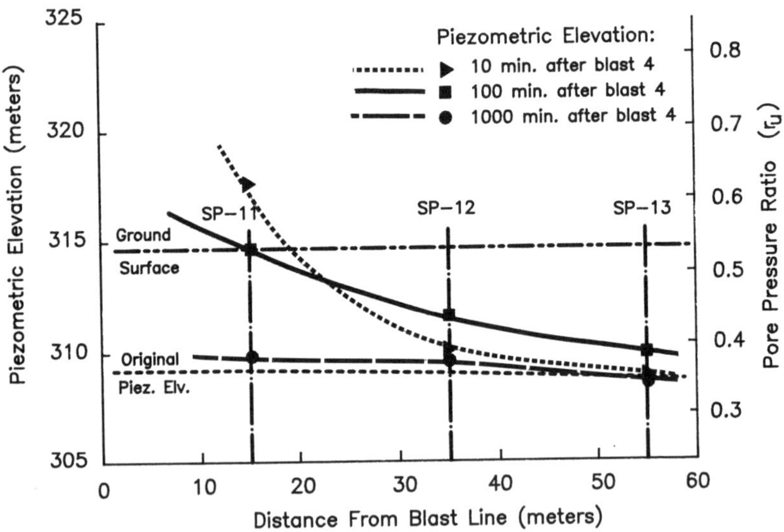

Figure 9. Piezometric Profile Perpendicular to Blast Line

Five slope inclinometers downstream of the blast line were monitored for foundation shear. The distance of the slope inclinometers from the blast line ranged from 100 to 200 meters. The maximum incremental displacement due to blasting was 1 millimeter, occurring in the Clearwater Formation. Six pneumatic piezometers in the same formation were also monitored. Three of the piezometers were located 200 meters downstream of the blast line and showed no response to blasting. Two piezometers located 110 meters from the blast line showed a temporary increase of approximately 0.2 meters. The final tip was adjacent to the blast area but was destroyed by the blast.

Having demonstrated that blasting would not induce a foundation or liquefaction failure if production blasts were restricted to a single line of holes orientated perpendicular to the dam centerline, work proceeded to evaluate the effectiveness of blasting in reducing liquefaction potential. The criterion used to evaluate liquefaction potential at the time was based on the standard penetration test. A joint field and laboratory program had been undertaken to correlate the standard penetration test with the difference between in-situ dry density and steady state dry density at the appropriate stress level. Based on that work a criterion of 21 blows per foot for tailings sand and 25 blows per foot for Pf4, after correcting for vertical effective stress, was established. While this was considered appropriate at the time there are several shortcomings with this approach, namely: (1) the subsequent work at Syncrude and elsewhere has demonstrated that horizontal stress varies significantly in hydraulic fill deposits and must be accounted for in interpreting penetration tests, (2) recent published works have documented the effects of aging on penetration resistance Mitchell and Solymar (1984), and (3) discovery of a potential source of error in steady state testing of very loose sands Sladen and Handford (1987). Alternative techniques to evaluate liquefaction potential are currently being evaluated at Syncrude.

Syncrude had previously used blasting to densify loose sand (Cell 21). Conditions in Cell 18, 19, 27/28 and 29 differed in that: (1) the latter loose zones were overlain by a thick layer of medium dense sand and (2) the depth to the loose zone was significantly greater. A second phase of the test program was implemented to evaluate the effect of the different boundary conditions on the effectiveness of explosive densification and to reduce densification costs.

In the second phase, eight test basts, consisting of two rows of holes using a 14 meter blasthole grid, were completed around previously drilled SPT holes. The charge weight was increased to 11.25 kg to account for the increased blast hole spacing. The loading factor averaged 0.8 grams of explosive per cubic meter of loose sand. Penetration tests were undertaken within days after each test blast, to evaluate the effectiveness of the blasting operation. Figure 10 illustrates typical before and after SPT profiles for four of these holes.

There are several points to note in the penetration profiles presented. Firstly, the smallest increase in penetration resistance

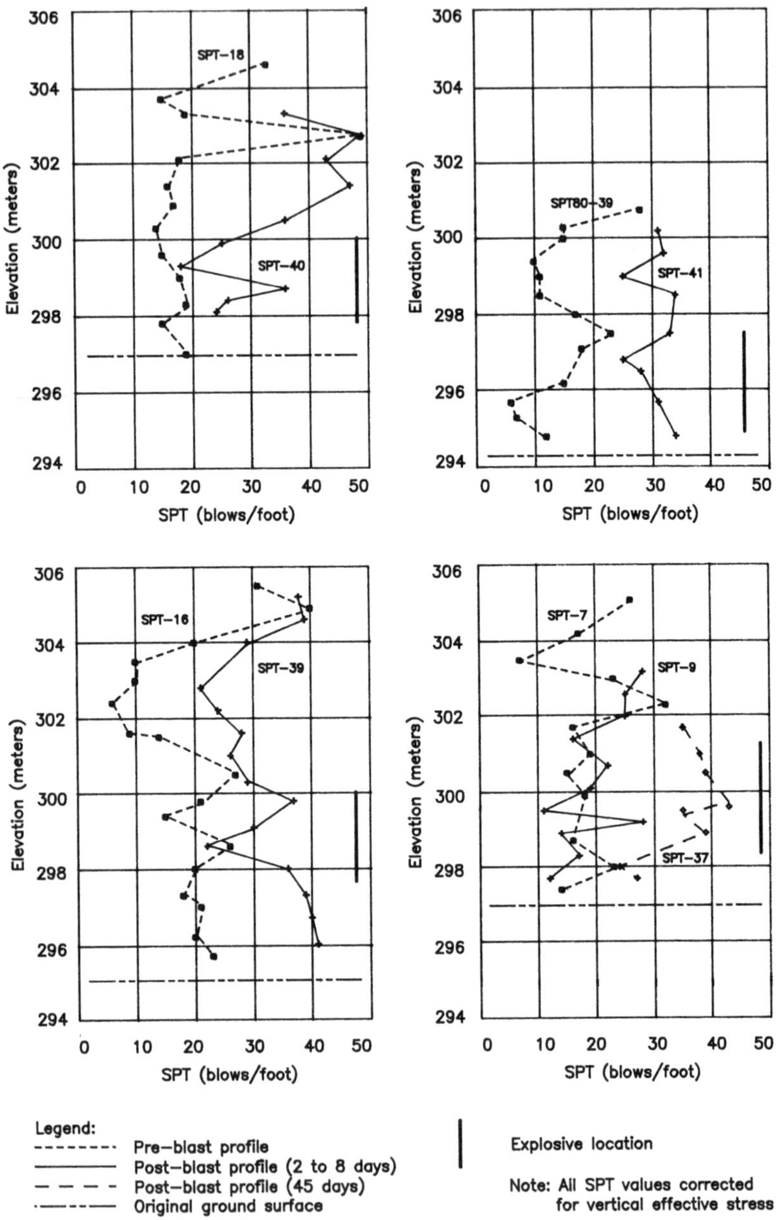

Figure 10. Typical Standard Test Profiles: Cell 18

seems to occur at the elevation of the center of the charge. Secondly, three of the four holes show an increase in penetration resistance within days after blasting. And finally, the fourth penetration profile shows an initial decrease in the first post-blast test, however a second penetration test at this location performed 45 days later, shows a sizeable increase in penetration resistance.

Blasting Details

The explosive used was 60% geogel, a CIL product used in the seismic industry. Geogel comes in plastic containers of small charges, that can be screwed together to form a charge of the desired weight. The blastholes were drilled with a 4.5 inch tricone bit and kept open with drilling mud. This method of drilling was found to maintain hole stability sufficiently long enough to place the charge with loading poles. The charges were installed to a depth of 1 meter above the bottom of the loose zone. The drilling mud was found to provide adequate stemming for the blastholes. The charge weights used, were a function of the depth to the bottom of the loose zone, as per Table 2. The loading factor varied from 0.4 g/m^3 in the deeper holes to 1.2 g/m^3 in the shallow holes.

Table 2 Charge Weight per Hole Using 60% Goegel

HOLE DEPTH (meters)	CHARGE WEIGHT (kg)
18 to 21	12
21 to 24	13
24 to 27	14
27 to 30	15
30 to 37	16

The drill pattern and firing sequence used are illustrated in Figure 11. Daily blasting was limited to two rows of holes with a minimum separation of 60 meters apart and a 25 millisecond surface delay between blastholes. The reason for this approach was twofold. Firstly, there was a desire to minimize the width of the liquified zone that would occur immediately following blasting. Secondly, it was thought that the repeated application of shock waves to the deposit would result in a better densification.

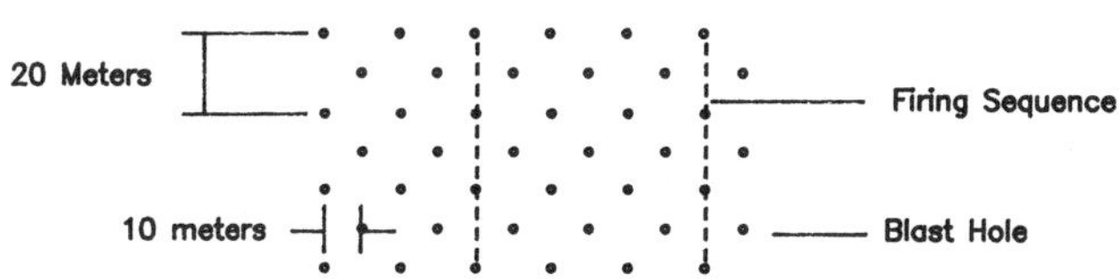

Figure 11. 10 x 20 Staggered Blast Pattern

Loose Zone Summary

The details of the loose zone in each cell are given in Table 3. The loose zone thickness was generally in the order of 5 to 7 meters, although in Cell 19 the thickness increased to 18 meters in the creek valley. The compacted shell thickness overlying the loose zones varied from 11 to 16 meters. In all cases the loose zones lay below the phreatic surface.

In Cell 27/28 and Cell 29, the loose zone extended from the centerline to the toe of the dam. A series of berms were cut in the downstream slope in these cells to allow drill rig access. In Cell 18 and Cell 19, the loose zone extended from the centerline to the downstream crest of the dam.

Table 3 Loose Zone Geometry Summary

Cell	Soil Type	Loose Zone Thickness (m)	Compacted Shell Thickness (m)	Maximum Loose Zone Depth (m)	Phreatic Surface Depth (m)	Surface Area (m^2)
18	TS	7	11	18	6	81,000
19	TS	6 - 18	16	34	6	50,000
27/28	Pf4	5	16*	21	10	45,000
29	Pf4	5	16*	21	10	29,000

* Less beneath the dam slope

Results

The SPT results are summarized in Table 4 for each loose zone, before and after explosive densification. An increase in penetration resistance of approximately 90 percent was achieved in the tailings sand, in holes drilled within weeks after blasting. At three of the SPT locations in Cell 18, additional penetration tests were completed a month and a half after completion of the densification program. These holes showed a further increase of 28 percent in penetration resistance over the first set of post-blast holes, and a 147 percent increase over the pre-blast holes. This increase is likely due to additional blasting in the vicinity of the holes, as well as the aging phenomena discussed by Mitchell (1984). In Figure 8 it was shown that the blast induced pore pressures dissipated completely within 33 hours thus it would not appear that the time-dependent gain in penetration resistance is related to dissipation of residual blast pore pressure.

The majority of post-blasting penetration holes did not show an initial reduction in penetration resistance after densification as has been noted elsewhere, Mitchell (1984) and Schmertmann (1987). This may be due in part to the significantly lower loading factor used in the Syncrude project than the 25-35 g/m^3 reported by Mitchell (1984).

The densification program achieved only a 40 percent increase in penetration resistance in the Pf4 sand in Cell 27/28 and Cell 29. The geometry of the loose zone, the loading factor and the timing of the post-blast penetratin tests in these cells was essentially the same as in Cell 18, suggesting that the difference in effectiveness is due to the change in soil type. This reinforces the necessity of determining the blast details for any particular site from a test program.

Table 4 SPT Penetration Results

Cell	Soil Type	Before n	Before N'	Before s	After n	After N'	After s	Percentage Increase
18	TS	121	15	5	65	28	9	87
					47	37	12	147*
19	TS	81	18	5	70	34	8	89
27/28	Pf4	30	19	5	18	26	7	37
29	Pf 4	26	21	7	15	30	7	43

* 1.5 months after densification
n number of tests
N' average corrected N value
s standard deviation

Conclusions

Explosive densification has been demonstrated to be effective in increasing penetration resistance at depths to 34 meters even though the loose zone was overlain by a medium dense sand layer of significant thickness. Whether this has successfully eliminated the risk of liquefaction cannot be assessed by penetration testing alone. It has also been demonstrated that explosive densification can be carried out beneath existing dams even when foundation conditions are less than favourable. Piezometric data gathered suggests that the influence radius of the blast with respect to disturbance of the soil structure, extends to a scaled distance of 15 $m/^2kg$. This distance would be expected to be a function of initial stress conditions and soil structure. Densification results in the Pf4 sand illustrate the influence of soil type and the necessity of conducting field trials for each different situation.

Acknowledgements

The initial explosive densification tests were conducted in 1979 by Farukh Khan and Keith Pedwell both of Syncrude Canada Ltd. at the time, and Vic Milligan of Golder Associates. Subsequent explosive densification was successfully carried out in Cell 21 in 1979 and 1980 by Keith Pedwell and Alan Fair. Their work provided an excellent starting point for the program carried out under the writer's direction in 1981. Acknowledgment is also due to Jeff Buckles, John Veldon, and Fred Payne of Syncrude Canada Ltd. who participated in the latter densification program.

References

1. Dowding, C.H. and Hryciw, R.D., "A Laboratory Study of Blast Densification of Saturated Sand", Journal of Geotechnical Engineering, ASCE, Vol. 112, No. 2, February, 1986, pp. 187-199.

2. Hall, C.E., "Compacting a Dam Foundation by Blasting", Journal of the Soil Mechanics and Foundation Division, ASCE, Vol. 88, SM3, Part I, Proc. Paper 3160, June, 1962, pp. 33-51.

3. Lymen, A.K.B., "Compaction of Cohesionless Foundation Soils by Explosives", Transactions, ASCE, Vol. 107, 1942, pp. 1330-1348.

4. Mitchell, J.K., "In-place Treatment of Foundation Soils", Journal of the Soil Mechanics and Foundation Division, ASCE, No. SM1, Proc. Paper 7035, January, 1970, pp. 73-110.

5. Mitchell, J.K. and Solymar, Z.V., "Time-dependent Strength Gain in Freshly Deposited or Densified Sand", Journal of Geotechnical Engineering ASCE, Vol. 110, No. 11, November, 1984, pp. 1559-1576.

6. Prugh, B.J., "Densification of Soils by Explosive Vibrations", Journal of the Construction Division, ASCE, Vol. 89, No. CO1, Proc. Paper 3458, March 1963, pp. 79-100.

7. Schmertmann, J.H., Discussion, Journal of the Geotechnical Engineering Division, ASCE, Vol. 113, No. 2, February, 1987, pp. 173-175.

8. Sladen, J.A. and Handford, G.T., "A Potential Systematic Error in Laboratory Testing of Very Loose Sands", Canadian Geotechnical Journal, CGS, Vol. 24, No. 3, August, 1987, pp. 462-466.

9. Solymar, Z.V., "Compaction of Alluvial Sands by Deep Blasting", Canadian Geotechnical Journal, CGS, Vol. 21, No. 2, May, 1984, pp. 305-321.

10. Wild, P.A., and Haslam, E.F., "Towers and Foundations for Project EHV", Journal of the Power Division, ASCE, Vol. 88, No. PO2, Proc. Paper 3193, July, 1962, pp. 69-111.

MODELING OF GEOTEXTILE REINFORCED MAT PLACED ON SOFT DREDGINGS

Joseph P. Martin,[1] M. ASCE, Robert B. Crawford,[2] A.M. ASCE, Robert H. Swan, Jr.,[3] A.M. ASCE and Anthony J. DePaul[4]

ABSTRACT

A former dredge spoil site adjacent to a deep channel was available for assembly of offshore drilling rigs, but it consisted of slurry-placed silty fine sand overlying deep, soft alluvial soils. A stable embankment or platform was needed. Placement of a high-strength geosynthetic on the subgrade would provide a separating blanket on which dredged borrow could be compacted, but the adequacy of support for a heavy crawler crane was problematical. To further improve bearing capacity, a "geomat" within the fill, consisting of compacted soil layers sandwiched between geotextiles was proposed. However, published work on the use of multiple reinforcing layers focuses on general shear failure, and assurance of safety against punching failure required further investigation.

A laboratory study was conducted to investigate the phenomenon of circular footing punching through reinforced soil into a soft foundation. It was found that significant increases in undrained bearing capacity were obtained with up to two layers of a high-strength fabric, but little additional benefit was obtained with more than three layers. The multilayer geomat appeared to provide both tensile reinforcement and soil confinement, bridging over a crater formed in the soft soil directlyunder the load. The efficiency in this regard increased with thinner interlayer spacing and higher geotextile burst strength. Stiffness was only marginally improved, however. This project demonstrated the value of a combination of densification and internal reinforcement to improve the bearing capacity of embankments of or on hydraulic fills.

INTRODUCTION

Construction of an offshore drilling rig required several acres for an assembly area slightly above but adjacent to a deep channel. One site on the Intracoastal Waterway was a swamp covered with loose silty fine sand dredgings. Various site stabilization alternatives were considered, including excavation and replacement, a structural platform supported on piles, and subgrade preloading accelerated with the use of wick (strip) drains. Each of these alternatives was rejected due to technical, economic or time constraints.

---

[1]Asst. Professor, Dept. of Civil Engrg, Drexel Univ., Phila., Pa.
[2]Res. Asst., Geosynthetics Research Inst., Drexel Univ., Phila., Pa.
[3]Staff Engr., GeoSyntec Inc., Boynton Beach, Fla.
[4]Staff Engr., A. DePaul Constr. Co. Inc, Phil., Pa.

High-quality construction materials are not readily available in quantity. However, the silty sand dredge spoils are of a composition suitable for construction of a broad embankment about 7' thick. The loose and saturated in-situ material behaves like an alluvial silt, with an unconfined compressive strength on the order of 250 psf. However, mechanical behavior improves greatly with densification and reduced water content. The main problem was to be able to place, dry and densify the material and to keep it in place without mixing with the soft existing subgrade.

Geosynthetics of high strength and modulus are increasingly used for separation, filtration and reinforcement functions under embankments on soft deposits (Koerner, 1986). At this site, a high-strength, pervious geosynthetic (geotextile and/or geogrid) could be readily placed on the existing soft deposits. Soil could then be placed mechanically or hydraulically and compacted without loss into the subgrade. The construction phase settlement and localized distortion during compaction would also elongate and thus tension the geosynthetic to mobilize its reinforcement value (Giroud and Noiray,1981; Williams, 1983; Koerner et al, 1987). Obtaining borrow from the dredge spoil on the margins of the platform would also provide wide, shallow drainage channels on the sides of the embankment. Analysis of slope stability on the three sides of the platform facing the swamp i.e., inland, would require evaluation of individual parameters such as soil-reinforcement friction (Bonaparte, 1987). However, it was recognized that cellular cofferdams would be necessary to stabilize the side of the platform along the navigation channel as dredging to provide a deep channel for supplies offloading and rig launching would be necessary.

A troublesome problem was support for a heavy crawler crane, as shown on Figure 1. This would weigh several hundred tons, and have tread lengths on the order of 16' long. With a shallow fill, attenuation of the crane contact pressure with depth would be limited, and the high-strength geosynthetic would respond more or less as a trampoline-like membrane supporting the crane a point load. Rupture of the underlying reinforcement is one potential mode of failure, especially if punching behavior creates a void beneath the crawler. If the crane were to dwell for an extended time at any particular location, creep deformations in both the geosynthetic and subgrade could result in the same effect.

Sole reliance on the bottom reinforcement would be risky in itself. Furthermore, even if deep bearing capacity was assured, severe surface rutting could occur with compression or local shear failure of the fill. With the main reinforcement/separator geosynthetic placed on a very soft base, thorough densification of the fill would only be possible in the upper few feet. Because crane operations required movement over almost all of the multi-acre embankment, its support dominated the entire design.

One proposed solution was reinforcement in the interior of the fill, as shown on Figure 1. Stiffening the upper two-thirds or so would distibute surface loads over a wider area of the softer fill and subgrade. The literature examined did not fully address these

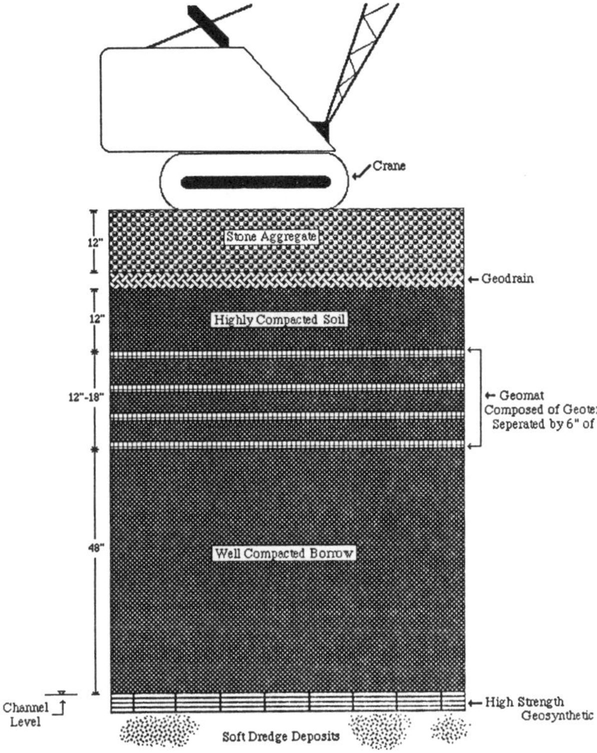

Fig.1 Schematic Cross-section of Reinforced Embankment on Dredgings

problems Studies of earth reinforcement beneath footings (Binquet and Lee, 1975; Akinmusuru and Akinbolade, 1981; Guido et al, 1985) dealt with reinforcement of homogeneous soil, and thus, general or local shear conditions. An experimental investigation was seen to be advisable to to develop sensitivity data and to assist in design of a prototype field study. The problem of multiple layer reinforcing over soft subgrades was also addressed by Sakti and Das (1986) in an experimental study done after the pilot one described herein.

PRELIMINARY ANALYSIS

There are a number of variables of interest in experiment or design regarding soil reinforcement, and several potential mechanisms of failure in fabric (planar) or strip reinforced soil have been identified, as illustrated in Figure 2 (adapted from Koerner, 1986).

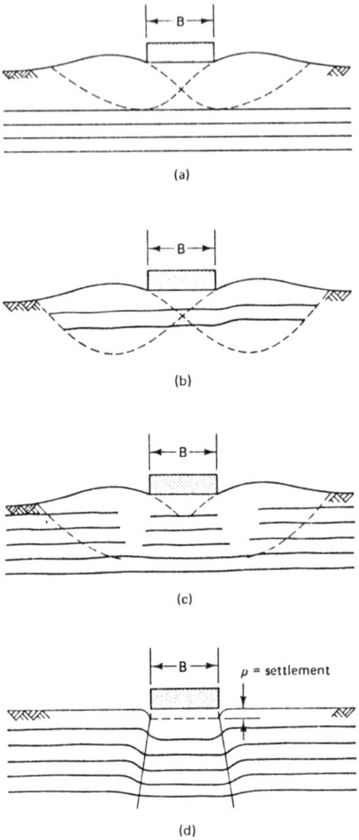

Fig.2 Mechanisms of Failure in Geosynthetic-Reinforced Homogeneous Soil: a) Shallow Shear, b)Anchorage Pullout, c)Reinforcement Tension, d)Reinforcement Creep. (Koerner, 1986)

To focus the laboratory effort, prior study of these variables and mechanisms was necessary. The major variables are:

-Soil and geotextile properties
-Soil-geotextile interaction (friction, etc.)
-Geometrical variables (spacing)
-Number of reinforcing layers

In this project, the soil was fixed, although its properties did vary with water content and unit weight. A single type of fabric was used, with the thought that if one parameter (strength, stiffness, soil-fabric friction, etc.) was shown to be limiting, another material could be used.

The first three modes of failure shown on Figure 2 were identified by Binquet and Lee (1975) and the fourth by Koerner (1986). The following symbols are used herein:

B = footing width
N = number of reinforcement layers
X_1 = depth to the uppermost layer
X = vertical spacing between layers

The case of $X_1/B > 0.67$ is the classical problem of a soft soil overlying a stiff stratum, with failure occuring by shear in the upper soil. This would be as disastrous as any other type of failure for a structural footing, but at this temporary construction site, it would be primarily an inconvenience unless equipment or personnel safety were threatened. After a localized failure, material could be replaced, as is the practice proposed by Giroud and Noiray (1981) for rut repairs on geotextile-reinforced gravel roads. In fact, it is useful to place the reinforcement as deep as possible in the embankment section. This would increase the thickness of fill above the geomat that could be highly compacted and thus improve shallowbearing capacity. Since the silty fine sand dredgings will not provide a clean, rapidly draining construction surface, an aggregate surfacing and subsurface drain would be required anyway, as also shown on Figure 1. Consequently, shallow failure is not a likely mode at this site.

The second type of failure shown on Figure 2 is for the case when $X_1/B < 0.67$ but the number of reinforcing layers provided is limited. In this case, the reinforcement anchorage length could be insufficient to resist pullout. This mode of failure is also unlikely at this site except near the edges of the fill. The reinforcement supports a moveable load, and thus must be continuous across the platform and extend well beyond the location of the crane at any particular time.

In the third type of failure, Figure 2c, the strength of the individual reinforcements is limiting. With a multitude of well-anchored reinforcements, the uppermost layer carries the highest load (Binquet and Lee, 1975). Tensile failure of the uppermost fabric would initiate a progressive failure. Even if rapid tensile failure doesn't occur, differential creep movements of the polymeric reinforcements would produce relaxations and progressive downward shifts of the load be expected. A long-term creep failure would result (Figure 2d).

These last two failure mechanisms do not lend themselves to a limit equilibrium expression for factor of safety. Variation in the reinforcement spacing will give different load-deformation results short of failure. Hence, a Bearing Capacity Ratio (BCR) is often used to indicate the improvement in foundation support provided by a particular reinforcement scheme or pattern. Typically, this is done by comparing centerline deflections to those obtained without any reinforcement. The reference bearing capacity is that of the footing on the "natural" soil in a particular condition (density, moisture content, etc.). Relative improvement due to reinforcing, additives, etc. is indicated by:

$BCR = q/q_o$

where:

q_o = pressure on the unreinforced soil at a given settlement
q = pressure on the reinforced soil at a corresponding settlement

This notation lends itself to iterative study of sensitivity to variables such as vertical and horizontal (for strip reinforcement) spacing, expressed in X/B ratios and also the increase in BCR for each additional layer, N. An extensive investigation of the effect of each of these variables on the BCR was done for strip reinforcing of a homogeneous sand by Akinmusuru and Akinbolade (1981) The influence of the number of geosynthetic layers on the BCR of a loose sand was also investigated by Guido et al (1985), with results of the form shown on Figure 3.

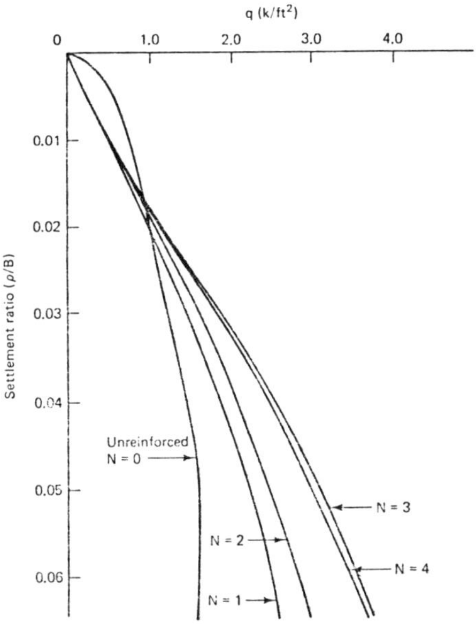

Fig.3 Test Results for Loose Sand Reinforcement (Guido et al, 1985)

The vertical reinforcement spacing, X/B, influences the confining stresses and shear deformations in the interlayer soils,and thus the unitary action of the reinforced soil zone. Akinmusuru and Akinbolade (1981) found that the BCR in the strip-reinforced sand tended to fall off when X/B exceeded about 0.5. The upper limit of improvement in BCR with additional reinforcement layers varied between the different studies. Akinmusuru and Akinbolade (1981) found minimal improvement beyond N=2 for closely spaced layers, whereas Binquet and Lee (1975) determined the maximum efficient N to be between 4 and 6. Obviously, this is a soil-structure interaction issue, depending upon the nature of the soil, the reinforcement, and the geometry. As illustrated on Figure 3, Guido et al (1985) found the maximum N to be 3 at high allowable settlement ratios but no benefit from reinforcement (BCR=1) was obtained at all at low settlements. This was attributed to the loose sand fill, such that considerable soil strain was necessary to mobilize any geotextile assistance. The classical rule of thumb in mechanics that the ratio of deformation moduli between the reinforcement and the reinforced material should be in the range of 10 to 20 certainly applies to homogeneous soil reinforcement as well.

SOFT SUBGRADE IMPLICATIONS

All of these investigations dealt used a homogeneous soil, with similar densities above, within, and below the reinforcement. For the present problem of a reinforced soil fill over a soft subgrade, there are two basic scenarios. In one, the load distribution through the reinforced embankment is so efficient (possibly assisted by the lowerseparation layer) that a more or less uniform pressure is applied to the soft subgrade. This mutes the effect of the soft soil,other than some increase in displacement at each load increment. The parametric sensitivities just described discussed above would be expected to apply. When the reinforcing layers are well spaced, the uppermost one is probably the primary tension member, not unlike a reinforced concrete beam under a negative moment. Closer spacing may provide a different distribution of loads betwen the reinforcements, but with the same practical effects.

However, if the subgrade deforms non-uniformly due to transversely varying vertical stresses, a "void" will tend to form in the soft soil directly below the applied load, with progressively increasing movement from beneath the load as it increases. The lowest reinforcement layer would undoubtedly be under the highest tension, bridging over the void. A punching failure would be the control on design, and the reinforcement deformations would be the reverse of the those shown for the long-term creep condition on Figure 2(d).

With a soft subgrade, the geomat structurally resembles either a beam on an elastic foundation or a membrane. Slippage in the soil between the fabrics would determine which is the case. A flexural stiffness, EI, could be backfigured, using the space betwen the lowest two layers in computing a nominal moment of inertia, I. With the membrane case, the contribution to the stiffness of the soil above and within the reinforcement would be limited. Flexural resistance would not actually exist. However, with closer spacing, the interlayer soil would be confined and contribute to E. The soil above the reinforcement would be laterally confined as indicated by the elastic curve resulting from positive moment response. Thus, the deflection at any load, as well as the final BCR, would appear to be a function of the interlayer spacing, X/B.

TEST MATERIALS

While the prototype site is in Louisiana, soil of similar geotechnical properties (gradation, plasticity, moisture-density) was obtained locally, i.e. silty dredge spoils from a site at the confluence of the Delaware and Schuylkill Rivers in Philadelphia. The soil was screened to remove particles coarser than the #16 sieve (1.18 mm) in order to more closely represent the Louisiana materials (Crawford et al, 1985).

The test soil, used for both the soft subgrade and the compacted platform fill, displayed the following properties:

Natural Water Content	28%
Natural Dry Unit Weight	95 lb/ft^3
Gradation	
% passing #40 sieve	55%
% passing #200 sieve	17%
% - 0.01mm	6%
Atterberg Limits	
Liquid Limit	27.5
Plastic Limit	19.5
Plasticity Index	8
Shrinkage Limit	16.0
Classification	
Unified	SM
AASHTO	A-2-4
USDA	Loamy Sand
Standard Proctor Results	
Max. Dry Unit Weight	113 lb/ft^3
Optimum Moisture Content	13.5%

It can be seen that the native material is very loose, but has sufficient cohesion from capillarity to allow preparation of unconfined compression specimens. The unconfined compressive strength was about 500 psf at w_c=22%, and decreased to 225 psf at w_c=26%. Intact specimens could not be fabricated beyond this moisture range.

The geosynthetic reinforcment that was used was AMOCO Propex 2006, a high-strength, high modulus geotextile made of a woven polypropylene slit film. This material is designed for separation of soil layers and soil reinforcement rather than the more traditional use of geotextiles as "filter cloth".Filtration is not as major concern, as similar material exists on either side of each layer. The following properties are manufacturer's data:

Effective Opening Size	20-50 microns
Puncture Strength (D-751)	120 lbs

Burst strength (Mullen test, ASTM-751)	600 psi
Tear Strength (ASTM D1117)	120 lbs
Grab Tensile Strength & Elongation (ASTM D1682)	273 lbs @ 15%

EXPERIMENTAL SETUP

The major component of the investigation was a 36" diameter, steel cannister as shown on Figure 4. The test stratigraphy was placed and

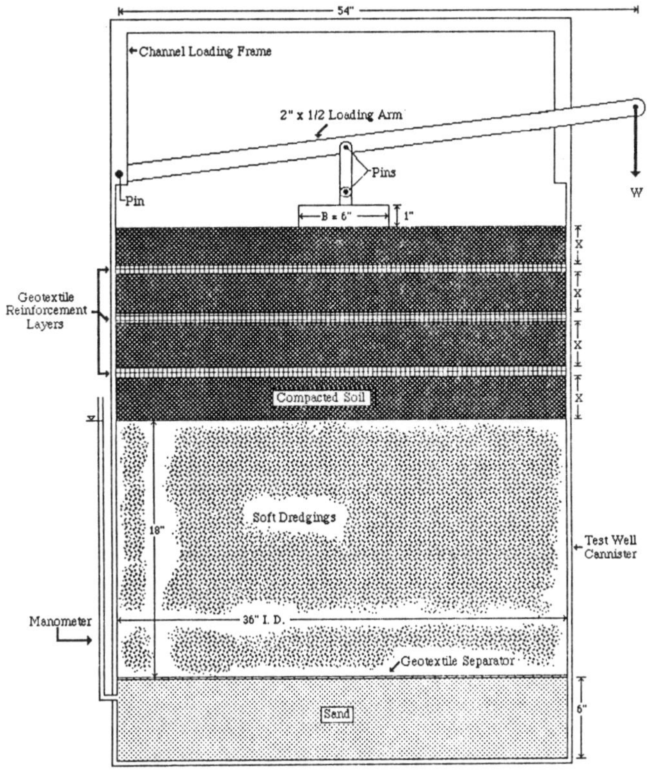

Fig.4 Experimental Apparatus

loaded through a bearing plate with a simple lever-arm apparatus. To provide a flexible base and a mechanism to control soil water content, a 6" sand cushion was placed at the bottom and covered with a geotextile separator. Above this, 18" of the dredge material was placed as a permanent test bed, but rebuilt after each test to destroy shear zones. At this test scale, it was not possible to deposit the soil as a slurry without causing separation of the silt and clay fractions, a feature not evident at the site. The relatively low plasticity soil did not appear to have a high sensitivity or well-developed structure. Therefore, bulk placement of the material near the natural water content and subsequent completion of saturation by low-gradient upflow from the sand layer was assumed to be a reasonable re-creation of the in-situ conditions.

With the permanent test fill in place, a total thickness of 18" of various compacted soil-geotextile cross-sections were placed on it in each iteration. A 6" diameter, 1/2" thick bearing plate bore directly on the surface of the soil. As indicated on Figure 4, the plate was loaded through a 54" steel bar, 2" deep and 1/2" thick, pinned at one end on the loading frame and overhanging the test well on the other. The loading arm bore on the bearing plate though a doubly pinned vertical bar, to assure unidirectional loading. The bearing plate was 18" from the loading arm pin, thereby giving a 3:1 multiplication on the load applied to the free end. This is similar to the technique used by both Akinmusuru and Akinbolade (1981) and Sakti and Das (1986).

TEST VARIABLES AND CONDITIONS

The two major variables tested were the number of reinforcement layers, N, and the interlayer spacing, X/B. The need for high confinement was demonstrated in the work cited above, such that the interlayer spacings were chosen to be X =0.25B and X=0.5B , or 1.5" and 3.0", respectively. In a scale-up to the prototype, this would not present a practical constraint on lift thickness. The depth to the first layer of reinforcing was identical to the interlayer spacing, as was the depth below the lowest reinforcement and the the compacted soil-soft soil interface. Details of the horizontal clearances are shown on Figure 5. The ratios of equivalent crane tread width to uppermost reinforcement depth (3:1 to 4:1) are not significantly different than in the expected field case. It is noted, however, that the normal (overburden) stresses on the reinforcements present in the field were not applied in this case; this would make the results conservative, especially with respect to horizontal strains of the fabrics.

The need for competent interlayer soil was also demonstrated in the loose sand tests reported by Guido et al (1985), but very high compaction of the soil below the geomat was difficult in a laboratory setting, and would also be thus in a field situation. Therefore, the soil within, below and above the reinforcing layers was compacted to 90% of maximum dry unit weight, and at the optimum water content.

UNREINFORCED REFERENCE TESTS

To provide the reference q_o, the bearing plate was first used directly on the soft soil. Results, averaging three trials (with variations in the range of 10% to 15%)are shown on Figure 6, with no detectable inflection in the load-deflection curve. This clearly indicates a punching-type movement. At each loading, almost all of the deflection occured in the first 10 to 15 minutes, followed by a very

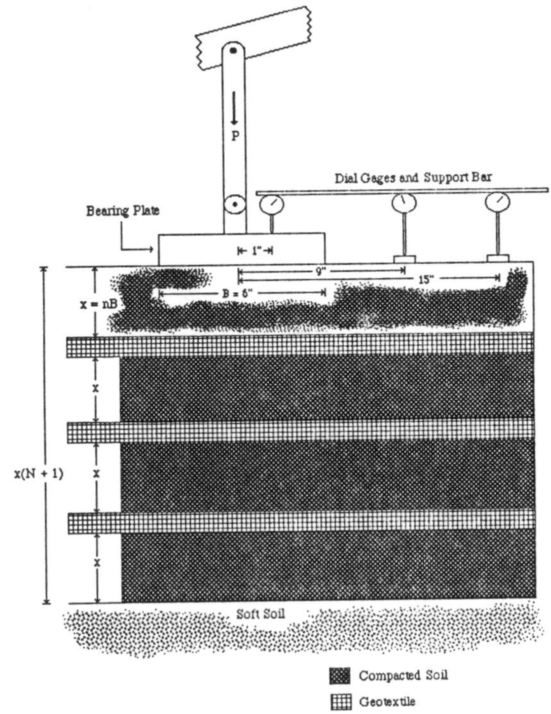

Fig.5 Interlayer Spacing Dimensions

slow consolidation-type movement. The soft soil foundation was rebuilt and re-leveled after each test.

Test failure was defined at 1" displacement (settlement ratio 0.17), and the reference bearing capacity (q_o) was 0.2 ksf. A bearing capacity factor (N_c) of 1.6 in a punching mode can be backfigured from the undrained expression $q_o = c \times N_c$.

The next tests demonstrated the need for reinforcement as well as separation of the compacted soil placed on the soft subgrade. 3.0" of the dredge material was carefully compacted directly on the soft soil and loaded with the 6" diameter plate. The classical curvilinear elastic-plastic load-deflection curve was observed, such that q at a settlement ratio of 0.17 increased to 0.36 ksf (BCR=1.8, N_c=2.8). However, at higher loads, the deflection curve became asymptotic to the N=0 line shown on Figure 6, i.e., a punching failure re-occured and no benefit from the compacted soil pad was derived at loads above 1.45 ksf.

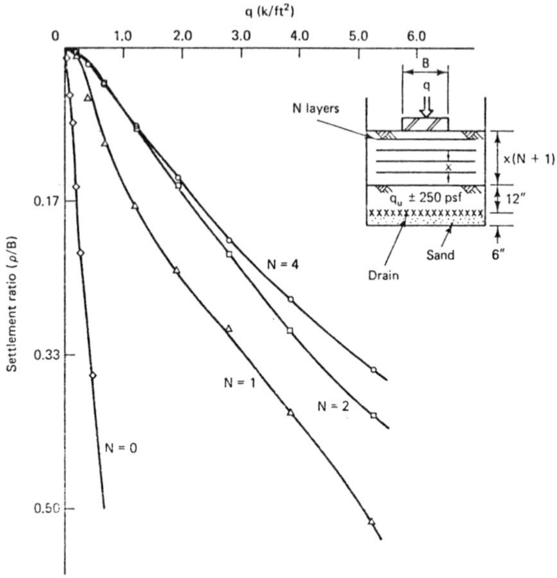

Fig.6 Unreinforced and X/B=0.25 Reinforcement Experimental Results

REINFORCING RESULTS

The results for the X/B=0.25 reinforcement spacing are also shown on Figure 6. For the single layer of reinforcing, placed 1.5" below the 6" plate, the bearing capacity at a 0.17 settlement ratio quadrupled to 1.3 ksf. This provided a BCR of 6.5. With the soft soil a depth of only 0.5B below the plate, a general shear failure surface would obviously be dominated by the strength of that material, such that the "failure" load of 1.3 ksf represents an equivalent N_c of 6.1. At larger deformations, which would clearly indicate performance failure in a practical sense, the single layer of reinforcement provided a still higher BCR. Shear failure above the reinforcing (Figure 2a) was clearly prevented.

With two layers of reinforcing enclosing the soil, the bearing capacity increased to 2.0 ksf at the 0.17 settlement ratio, representing a BCR of 10. The single-layer reinforcement appeared to provide a tensioned membrane type of support, as expected with the separation geosynthetic at the bottom of an embankment (Bonaparte, 1987), while the two-layer reinforcement acted in the manner of a beam of finite moment of inertia and flexural shear resistance due to restraint of the soil between the fabrics. This concept is supported by the minimal improvement at low settlements when a third and fourth layer were included in the geomat, as these tension members would be located at or near what could be construed as a neutral axis. However, significant improvement in capacity at very high deformations was observed with the additional reinforcement layers. This would be expected when high distortions indicate shearing of the interlayer soil, such that load would be carried by a number of individual reinforcements, in the manner shown on Figure 2c.

The results with the wider X/B=0.5 (3" spacing) tests were of the same pattern as those shown on Figure 6, but the bearing capacities at low settlement ratios were only two-thirds of the thin-spacing results i.e., lower BCR's. The single layer bearing capacity at the 0.17 settlement ratio was only 0.8 ksf, yielding an equivalent N_c=6.4 and BCR=4.0. The implication is that the condition of shallow general shear above the geomat (Figure 2a) was approached. With two reinforcement layers, the bearing capacity at a settlement ratio of 0.17 was 1.2 ksf. If the geomat did act as a beam, it would have been expected to have higher strength, due to the fifteenfold increase in the moment of inertia, I. A greatly reduced interlayer soil confinement and shear distortion resistance is indicated. Hence, the two fabric layers acted not as top and bottom beam reinforcing, but as independent tensioned reinforcements, as regarded in the Binquet and Lee (1975) analysis. The same pattern of minimal improvement with extra reinforcing layers at low deformations was observed in this set of experiments as well.

PUNCHING ANALYSIS

In many respects, these results indicate that bearing capacity analysis of reinforced embankments placed over soft material simply involves an adjustment of the previously developed soil reinforcement analysis. The load-deformation behavior shown in Figure 3 (Guido et al, 1985) is similar to the results shown in Figure 6. The 50% loss in bearing capacity as X/B was increased from 0.25 to 0.5 is more severely sensitive to spacing than that observed by Akinmusuru and Akinbolade (1981). Loss of BCR with higher spacing did not occur in those experiments until X/B exceeded about 0.6.

However, in the homogeneous soil case, the top reinforcement is clearly the one most heavily loaded in a more or less uniaxial condition (Figure 2c), while in punching resistance, the load on the lowest layer is also high. As vertical deformation increases, a transition from a situation analogous to the homogeneous soil case to a stiff-over-soft punching resistance by bridgeing appears to occur. The key distinction is that the properties of the reinforcement with respect to transverse loading resistance can be critical. When the test apparatus was dismantled and re-assembled after each iteration, a crater in the surface of the soft soil directly under the loading plate was observed. The fabric reinforcement layers more or less conformed to this shape for all values of N and both X/B ratios.

Sakti and Das (1986) conducted similar tests at the University of Texas at El Paso with reinforced soil placed over a subgrade of similar composition, strength and stiffness as that used in the Drexel investigation. However, the load was applied through a hydraulic system onto a square loading plate, and the reinforcement was a needlepunched nonwoven geotextile. Key properties of that reinforcement(Mirafi 140N) include:

Burst strength	210 psi
Puncture Strength	70 lbs
Grab Tensile Strength	120 lbs

It can be seen that the puncture and tensile strengths of the nonwoven geotextile are slightly more than one-half of those values used in the Drexel study, but that the burst strength, the resistance to failure under a distributed transverse load, was about one-third that of the slit-film woven material used at Drexel. The BCR's at corresponding settlement ratios were lower in the Texas tests than for the Drexel work by roughly the same ratio. In a situation where the reinforcement is poorly supported, it appears that failure is influenced by the resistance of the reinforcement to distributed transverse loading as a membrane. Bridging ability becomes limited as the reinforcement distorts. The data base for such a conclusion is admittedly limited, but the distinction between reinforcing of homogeneous soil and reinforcing over a soft subgrade is important.

SUMMARY AND CONCLUSIONS

Results have been presented for a physical model of the bearing capacity of a reinforced embankment placed over very soft silty fine sand. The problem addressed was improvement of punching failure resistance as opposed to general shear concerns. The pattern of results are similar to those published with respect to reinforcing homogeneous loose or dense sands, and also similar to another study of poorly supported embankment bearing capacity. Significant increases in bearing capacity at low to medium settlement ratios is provided by up to two layers of reinforcement. Inclusion of more layers of reinforcement improves capacity only at very high deformations, when it is clear that the soil has failed. Another critical variable is the interlayer spacing, such that as X/B increases, the improvement in bearing capacity decreases at all values of N.

The data supports, but does not prove, a theory that a soil-geotextile beam displaying flexural behavior is formed at low X/B ratios. This contrasts with the conventional view of independent membrane-like behavior of each reinforcement course. Perhaps this behavior occurs only with a yielding soft subgrade. Another influential material property is the bursting resistance of the reinforcement, which is clearly of major importance in punching failure situations.

In terms of the specific project, it is clearly unwise to linearly scale up test results for design, i.e., state an optimum X/B spacing. Field body forces and boundary conditions are not well represented in the laboratory. It would appear, however, that interlayer spacings of more than one foot would not provide sufficient soil confinement to allow the geomat to act as a unit. It would be advisable to provide three layers of reinforcement to assure soil restraint and provide tensile strength redundancy. Placement of internal reinforcement fairly high up in the embankment reducesthe intensity of load appliedto the foundation geosynthetic. This is more critical than increasing density below the geomat, as the lower grid or fabric protects the entire embankment, and is not repairable once placed in service.

References

Akinmusuru, J. O, and Akinbolade, J. A. "Stability of Loaded Footings on Reinforced Soil" Jour. Geotech. Eng, ASCE, Vol. 107, No. GT6, 1981, pp. 819-828

Binquet, J., and Lee, K. L."Bearing Capacity Analysis of Reinforced Earth Slabs" Jour. Geotech. Eng. Div., ASCE, Vol 101, No. GT12, 1975, pp. 1257-1276

Bonaparte, R. "Geosynthetic Earth Reinforcement" New Technology in Geotechnical Engineering, PennDOT Conf., Apr. 1987, Hershey, Pa.

Crawford, R. B., Swan, R. H., Jr., and Depaul, A.J. "Increasing Shallow Foundation Bearing Capacity using High Strength Geotextiles" Unpubl. Design Project, Drexel Univ., June, 1985

Giroud, J-P. and Noiray, L., "Geotextile-Reinforced Unpaved Road Design" Jour. Geotech. Eng., ASCE, Vol 107, No. GT9, 1981, pp 1233-1254

Guido, V. A., Biesiadecki, G. C., and Sullivan, M. J. "Bearing Capacity of a Geotextile-Reinforced Foundation" Proc., 11th ISSMFE, San Francisco, 1985, Vol. 3, pp. 1777-1787

Koerner, R. M., "Designing with Geosynthetics" Prentice-Hall Publ. Co., Englewoood Cliffs, N. J. 1986

Koerner, R. M., Hwu, B-L., and M. H. Wayne. "Soft Soil Stabilization Designs using Geosynthetics" Proc., Very Soft Soil Stabilization using High Strength Geosynthetics, Drexel Univ., Oct. 1987, pp. 26-57

Sakti, J. P., and Das, B. M., "Model Tests for Strip Foundation on Clay Reinforced with Geotextile Layers" Unpubl. Report, Univ. of Texas at El Paso, 1986

Williams, C. E., "Closure of an Industrial Waste Site Using a High-Strength Geotextile" Hazardous and Industrial Solid Waste Testing, Second Symposium, ASTM STP 805, R. A. Conway and W. P. Gulledge, Eds., 1983, pp. 283-295

PROPERTIES OF SEDIMENTED DOUBLE ALKALI FGD SLURRIES

R. J. Krizek,[1] S. W. Tatioussian,[2] and D. K. Atmatzidis,[3] Members ASCE

ABSTRACT--Some of the engineering properties that most strongly influence the disposal of double alkali flue gas desulfurization sludge in a settling pond and the subsequent reclamation of the site for some useful purpose are evaluated in an extensive series of laboratory sedimentation, permeability, and strength tests. Examined in this test program are the effects of codisposing of fly ash with the sludge, the addition of different stabilizing agents (lime and cement), changes in the concentration of solids, and differences in the curing time on the engineering properties of the sedimented slurries. An analysis of the sedimentation test results for slurries with various fly ash contents and a solids concentration of 15% indicated that the downward velocity of the suspension/clear liquid interface was about one foot per hour. When the solids concentration in the slurry was varied from 25% to 5%, the sedimentation rate nearly doubled. In addition, it was found that the addition of fly ash significantly increased the sedimentation rate. The low dry densities (between 20 pcf and 40 pcf) of the sediments suggest that large disposal areas will be needed to accommodate these sediments. The coefficient of permeability of these sedimented slurries is on the order of 10^{-4} cm/sec; the addition of 5% to 10% lime decreased the permeability by about one order of magnitude after approximately one month, but the addition of cement had a lesser influence. The shear strength (determined by a laboratory vane shear apparatus) of the sedimented materials was generally very low (less than 20 psf), but it was greatly influenced by curing time, especially for the first month. When the sediments were dried, the presence of 5% or 10% lime increased the strength by an order of magnitude or more, but the effect of adding cement, while significant, was less. Fly ash served as a good pozzolan that reacted with lime or cement and significantly improved the strength. Variations in the concentration of solids from 5% to 25% did not seem to have a significant effect on the strength and permeability of the sediments.

INTRODUCTION

The tightening of air pollution regulations due to the Clean Air Act of 1970 and 1977, the advent of the Resource Conservation and

---

[1]Professor and Chairman of Civil Engineering, Northwestern University, Evanston, Illinois

[2]Senior Project Engineer, Woodward Clyde Consultants, Santa Anna, California

[3]Professor of Civil Engineering, University of Patras, Patras, Greece

Recovery Act in 1976, and the shift to coal as a major fuel source have given rise to a variety of systems to remove sulfur dioxide from flue gases that are produced by the combustion of fossil fuels in utility and industrial boiler installations. Most of the flue gas desulfurization (FGD) systems use lime or limestone slurries to contact the gas stream and absorb the sulfur dioxide, and all produce waste material containing calcium sulfite and sulfate salts, unreacted lime or limestone, and fly ash. Some systems, the most common being the double alkali system, employ water soluable absorbers; a soluble salt, such as sodium hydroxide or carbonate is used as the reagent and the resultant sulfites and sulfates are precipitated with lime or limestone in a separate reactor. One of the problems associated with FGD systems is the disposal of the large quantities of waste that are produced.

The two basic types of FGD sludge disposal methods are dry landfilling and ponding or wet disposal. Processing of the waste before dry landfilling includes several variations of (a) interim sedimentation and natural dewatering, (b) mechanical dewatering, (c) blending with dry fly ash, and (d) blending with patented or unpatented additives as stabilizers. Wet disposal systems can be as simple as pumping the sludge to a settling pond and releasing the supernatant, but in some cases the bleed stream is first pumped to a clarifier or thickener and the underflow is then sent back to the pond for final disposal; alternatively, interim settling ponds can be used to reduce the water content, thereby reducing the size of the final disposal pond and the associated costs.

Wet disposal systems offer several advantages over dry landfilling systems; the primary benefits are (a) inherent simplicity of operation, (b) low operating cost, (c) flexibility to accept a wide variation in the bleed quantity and the solids content, (d) ease of operation under a wide range of conditions with minimum manpower, and (e) reduced adverse public reaction to problems of noise, dust, and traffic to and from the disposal area. One of the major disadvantages of wet disposal systems is the large land area required to accommodate the pumped slurry, which contains appreciable amounts of water. According to one estimate (Midkiff, 1979), the volume requirement for a wet disposal system can range from 0.2 to 10 acre-feet per megawatt-year. In addition, in areas of high rainfall accommodation must be made for the extra amount of water to prevent overflowing of the pond, while in arid regions the evaportation from a large pond may be undesirable if water is to be conserved. Further disadvantages of the wet disposal system, as compared to the dry disposal system, include (a) higher initial capital investment cost, (b) reduced flexibility with regard to the management of the disposal operation, (c) increased volume of waste, and (d) need for engineered containment dams and dikes.

Blending double alkali FGD sludge with dry materials, such as fly ash, aids considerably in improving the engineering properties of the landfilled waste (Krizek, Christopher, and Scherer, 1980; Krizek, Chu, and Atmatzidis, 1987). Since double alkali FGD sludge and fly ash are usually waste products from the same plant, their codisposal should be seriously considered and any advantages should be identified and exploited. The use of additives, such as lime, has been shown to further improve the engineering properties, and particularly the long term strength, of the landfilled double alkali sludge (Krizek, Chu, and Atmatzidis, 1987).

The choice of the most suitable disposal system for a given situation depends on many factors, such as capital cost versus operating expenses, distance of transportation, availability of suitable disposal area, and environmental constraints, and the issue becomes even more complicated when the use of dry admixtures and/or stabilizers is considered. Ponding or wet disposal of double alkali FGD sludge has not been accorded much attention heretofore and inadequate information is available to allow this disposal method to be properly evaluated as an alternative to dry landfilling. Accordingly, an experimental study was undertaken to obtain data which will be helpful in assessing the possibility of codisposing of double alkali FGD sludge and fly ash with or without stability additives in a settling pond and then reclaiming the disposal site for some useful purpose.

EXPERIMENTAL PROGRAM

Sedimentation tests were conducted on laboratory prepared slurries and the permeability, strength, moisture content, and dry density characteristics of the sedimented materials were measured to evaulate the effects of variations in the ash-to-cake ratio, solids concentration, amount and type of additive, and curing time.

Materials

The double alkali FGD waste (filter cake) and fly ash used in this investigation were both obtained from an electric utility plant located in the midwestern United States. The filter cake was produced at the final stages of the sulfur dioxide removal operation by dewatering the slurried sludge by means of a vacuum filter. In general, the resulting filter cake consists of small lumps with dimensions on the order of 0.5 to 1 cm. Although appearing to be relatively dry, the filter cake has a moisture content of 70% to 90% (defined as the weight of water divided by the weight of dry solids, expressed as a percentage) and a solids content of about 60% (defined as the weight of dry solids divided by the total wet weight, expressed as a percentage). The fly ash used in this study has a dark, steel gray color, which indicates that it is nonreactive (low content of calcium oxide), and a high content of unburnt carbon. The typical grain size distributions for the filter cake material and fly ash are shown in Figure 1 and indicate that these materials consist primarily of silt-size particles.

Lime and Type I Portland cement were used as additives or stabilizing agents. According to Krizek, Christopher, and Scherer (1980), the presence of lime or cement serves to increase the strength and decrease the permeability of FGD sludge mixed with fly ash and disposed by dry landfilling. This is due to pozzolanic and/or chemical reactions between the additive and the fly ash. The particular lime used was obtained in the form of pebbles and was ground to fine particle sizes.

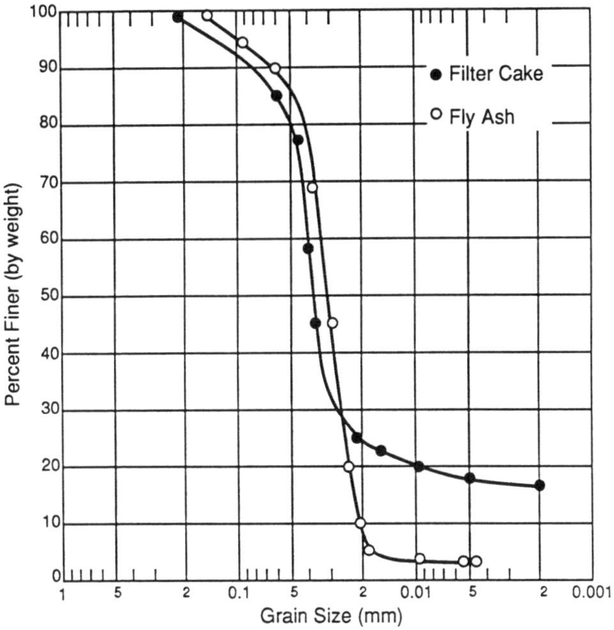

Figure 1. Typical Grain Size Distribution Curves for Filter Cake Material and Fly Ash

Testing Procedures

All tests were conducted in sedimentation cylinders which are shown schematically in Figure 2. An ordinary 1000 mℓ graduated cylinder, 18 inches long and 2.5 inches in diameter (Figure 2a), was used to conduct sedimentation tests only. The permeability and strength tests were performed in custom-built cylinders (Figure 2b) after completion of the corresponding sedimentation test in the same cylinder. Permeability tests were conducted on the sedimented material by allowing drainage through the porous stone and the valve at the bottom of the cylinder, and strength tests were conducted after the top section of the cylinder was removed to allow access to the sedimented material by a labortory vane shear device.

The samples for the sedimentation tests were prepared by (a) weighing the required amounts of each constituent (ash, cake, and additive), (b) transferring them to a mixing cup, (c) adding water until about two-thirds of the cup was filled, (d) mixing the components thoroughly for about three minutes, (e) pouring the mixture into the sedimentation cylinder, (f) cleaning the mixing cup of all solid particles and depositing the wash water in the test cylinder, (g) adding water to the cylinder until it contained 1000 mℓ of slurry, (h) capping the top (either by a rubber stopper or the palm of the hand) and carefully agitating the mixture by turning the cylinder back and forth for about one minute to enhance homogeneity, and (i) immediately thereafter setting the cylinder down. The approximate position of the solid/liquid interface was then measured as a function of the elapsed time.

Falling head permeability tests were conducted on the sedimented materials that accumulated in the bottom section of the custom-built cylinders. The supernatant was permitted to flow through the specimen until it reached a previously established position, and the elaspsed time, t, was measured. The coefficient of permeability, k, of the sediment was calculated by use of the equation

$$k = (L/t)\ell n(h_1/h_2) \qquad (1)$$

where L is the length of the specimen (height of sediment), and h_1 and h_2 are the initial and final heads, respectively; since the cross-sectional areas of the specimen and the inflow tube are the same,

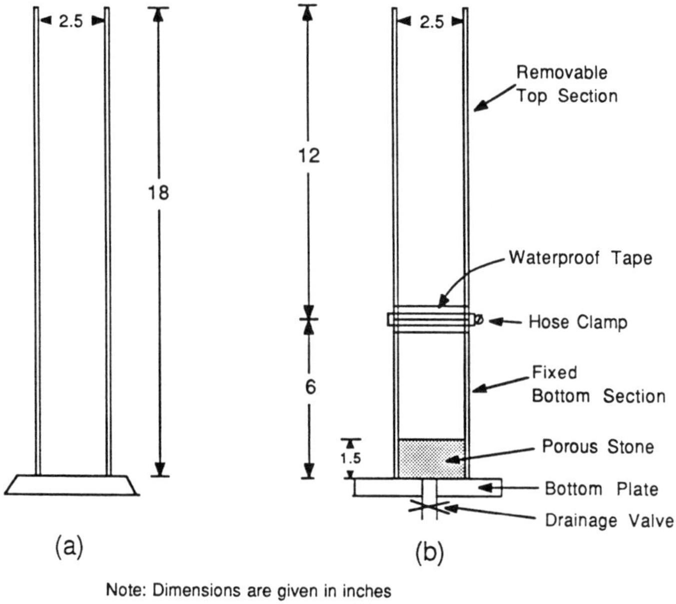

Figure 2. Schematic Diagrams of Test Cylinders

Equation (1) is a simplified version of the standard equation used to interpret data from a falling head permeability test.

A laboratory vane shear device was used to perform undrained shear strength tests on the sedimented materials in the custom-built cylinders. The test consists of inserting a four-bladed 1/2-inch diameter vane into the undisturbed sediments and rotating it to determine the torque required to cause a cylindrical surface and the associated ends to be sheared by the vane; this torque is then converted into an equivalent shear strength by the use of the equation

$$S_u = \frac{6}{7} \frac{M}{\pi D^3} \tag{2}$$

where S_u is the undrained shear strength, D is the diameter of the vane, and M is the torque required to develop the failure surface.

The effect of drying on the strength of the sedimented materials was evaluated by drying the specimens in stages to various moisture contents in a laboratory oven at a constant temperature of 60^0F. All specimens were retained in the containers in which they had sedimented (lower part of cylinder in Figure 2b), and the drying process was started after aging for seven days in a submerged condition. At the end of each drying period each specimen was sealed in a plastic bag and allowed to cure for six to eight hours in a humid room to enhance a more even distribution of moisture in the specimen.

Variables Investigated

Approximately one hundred specimens were prepared and tested to evaluate the sedimentation characteristics of the suspensions and the permeability and strength of the sedimented masses. The majority of tests were conducted on slurries with an initial solids content (ratio of weight of dry solids to total weight, expressed on a percentage) of 15% and the effect of the initial solids content on the measured properties was investigated by performing supplementary series of tests with slurries having solids contents of 5%, 10%, 20%, and 25%. Different amounts of fly ash (0%, 28%, 36%, 50% and 67% of the total weight of dry solids) were added to the double alkali filter cake to evaluate the possibility of codisposing these two wastes. Lime and cement in moderate quantities (0%, 5% and 10% of the total weight of dry solids) were used as additives or stabilizers. Finally, the effect of curing time on the permeability and strength of the sedimented material was quantified by allowing sediments to cure for up to ninety days.

RESULTS AND OBSERVATIONS

The results obtained from this experimental investigation are presented next and are placed in perspective with pertinent results available in the literature. Emphasis is given to comparisons between the properties of sedimented and dry landfilled double alkali FGD sludges because this will facilitate the evaluation of using a wet disposal system as an alternative to dry landfilling these materials.

Sedimentation Characteristics

The effects of variations in the initial solids content and the amount of fly ash added to the double alkali FGD filter cake on the sedimentation rate of the laboratory prepared suspensions are shown in

Figure 3. Sedimentation rates were computed for that period of each test in which the rate was, for all practical purposes, constant with time. In general, the bleed (rate of subsidence of the interface between suspension and clear supernatant water), as well as the rate

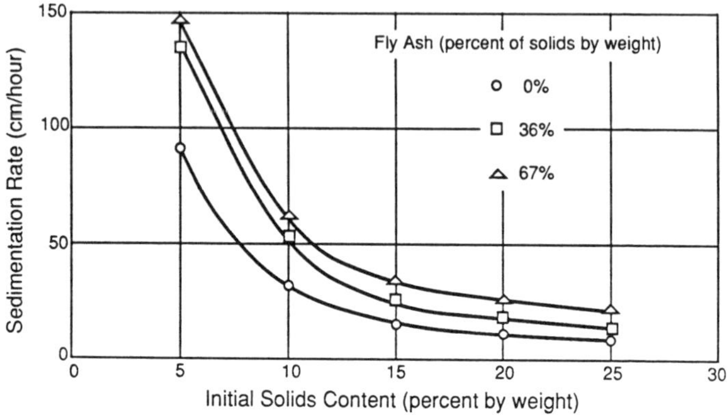

Figure 3. Sedimentation Rate of Double Alkali FGD Sludge Suspensions With or Without Fly Ash

of sediment accumulation at the bottom of the sedimentation cylinder, were practically constant for an initial period ranging from a few minutes to about one hour, depending on the initial solids concentration of the suspension. It can be observed that the sedimentation rate increases significantly as the solids concentration is decreased and that the addition of fly ash further increases the rate of sedimentation. For example, when the initial solids content of a plain filter cake slurry is increased from 10% to 20%, the sedimentation rate decreases from 40 cm/hour to 14 cm/hour; when the amount of fly ash added to the filter cake constitutes 67% of the total solids by weight, the sedimentation rates of suspensions with more than 10% solids by weight are approximately double the rates of similar suspensions which contain only filter cake.

The envelope of the results obtained from twenty-five different tests is presented in Figure 4 in terms of percentage of sedimented material as a function of time. These results were obtained for suspensions with an initial solids content of 15% with the fly ash content ranging from 0% to 67% of the solids by weight and an additive (lime or cement) content of 0%, 5%, or 10% of the solids by weight. It can be observed that, in most cases, about 80% of the sedimentation took place in less than an hour and the entire process was essentially completed in approximately two hours. Furthermore, variations in the ash-to-cake ratio and the addition of different amounts of lime or cement do not have a significant influence on the sedimentation rate.

Based on equations developed by Barnea and Mizrahi (1973) and nomographs prepared by Zanker (1980), theoretical sedimentation rates of about 11 cm/hour to 29 cm/hour were computed for particles with sizes

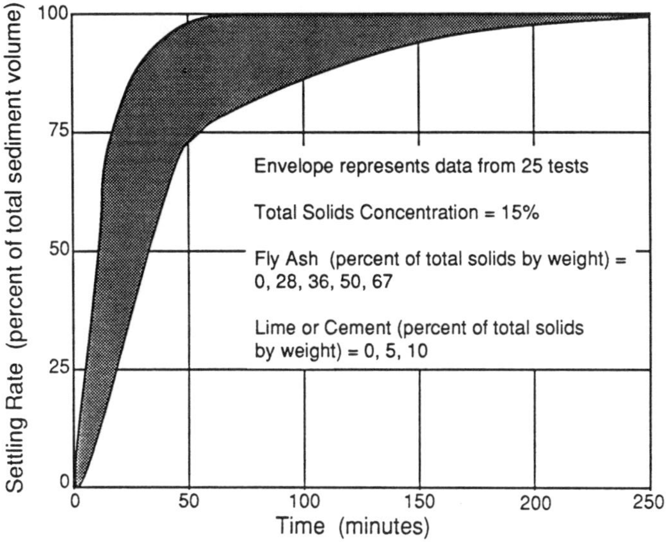

Figure 4. Settling Rate of Double Alkali FGD Sludge Suspensions Mixed with Fly Ash and Lime or Cement

ranging between 10 μm and 30 μm. Although these "sedimentation rates" represent the settling velocities of the solid particles and not the bleed rate of the suspension, the two can be considered essentially the same for suspensions of poorly graded particles, such as the suspensions tested in this investigation. Considering that the sedimentation rates measured for slurries with initial solids contents greater than 15% by weight ranged from 10 to 35 cm/hour, it can be concluded that theoretical methods are reasonably successful for predicting the sedimentation rates of double alkali FGD slurries. Based on the available data, first order appoximations can be advanced for the sedimentation time in a large basin or settling pond. For example, if the sedimentation rate of a slurry with 15% solids is taken to be 25 cm/hour , the time required for these solids to settle 5 meters would be about twenty hours.

Dry Density

The values obtained for the dry density of all sedimented materials are summarized in Figure 5. In general, the measured dry density was quite low, with values ranging from 20 pcf to 40 pcf and increasing with increasing total solids content and/or amount of fly ash added. The addition of various amounts of lime or cement did not cause any significant change in the dry density. The low densities and corresponding high void ratios may be due to the flocculated arrangement of the settled particles, as well as the microstructure (predominant rosette configuration) of the calcium sulfite produced by the double alkali system (Borden, 1977; Edwards, 1979; Krizek, Chu, and Atmatzidis, 1987). Approximately the same range of field dry densities (25 pcf to 35 pcf) was reported by (Krizek, 1979) for loosely dumped double

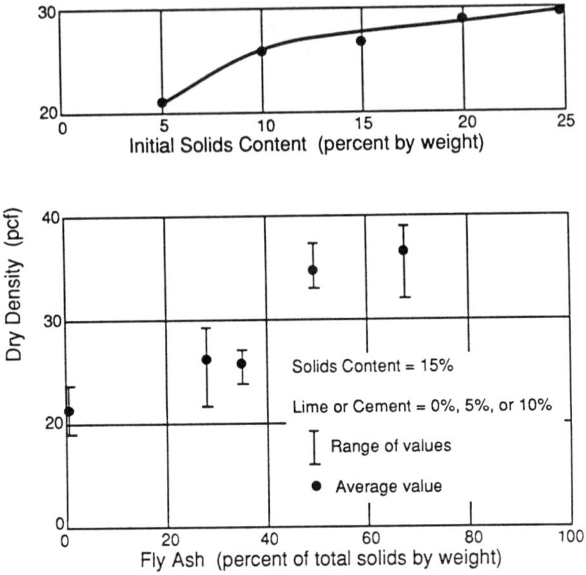

Figure 5. Dry Density of Sedimented Double Alkali FGD Slurry

alkali filter cake; these densities increased to approximately 40 pcf to 45 pcf after one or two passes of a dozer. This indicates that the area required for wet disposal might be as much as double the area required for dry disposal, and this constitutes one of the major disadvantages of wet disposal systems.

Permeability

The results of all falling head permeability tests conducted on sedimented slurries (void ratios ranging from 3.2 to 7.4) are summarized in Table 1 and indicate a coefficient of permeability on the order of 10^{-4} cm/sec. The permeability of the sedimented material measured within one day after completion of the sedimentation process does not appear to be significantly affected by the initial solids content of the slurry; variation of the solids content from 5% to 25% resulted in coefficients of permeability ranging between 1.6×10^{-4} cm/sec and 3.2×10^{-4} cm/sec with no trend or pattern. Similarly, the addition of fly ash in quantities up to 67% of the solids by weight appears generally to cause an increase in the coefficient of permeability, but the data in this case are also erratic. In contrast, the effect of curing time on permeability is more pronounced, especially when additives (lime or cement) are used. Although curing for seven days produced virtually no change in permeability, extending the curing period up to sixty days resulted in a decrease in the coefficient of permeability by up to one order of magnitude, and this can be attributed primarily to the development of pozzolanic reactions. Finally, it

Table 1. Permeability[1] of Sedimented Double Aklakli FGD Sludge[2] Mixed with Fly Ash and Lime or Cement

Fly Ash (% of solids by weight)	Curing Time (days)	Additive (% of solids by weight)				
			Lime		Cement	
		0	5	10	5	10
0	1	3.9	3.8	5.0	4.6	4.6
28	1	4.2	3.4	4.6	3.7	4.4
36	1	3.2[3]	3.2	4.4	4.4	2.1
50	1	3.7	3.0	3.3	3.4	5.4
67	1	4.7	2.5	2.1	2.9	2.9
36	1	3.2	3.2	4.4	4.4	2.1
36	7	5.6	2.8	2.8	6.0	6.5
36	28	1.8	0.6	0.6	3.5	3.1
36	60	1.6	0.3	0.3	1.6	1.6

[1]Values for the coefficient of permeability are in 10^{-4} cm/sec.

[2]All tests were conducted with slurries having an initial solids content of 15% by weight.

[3]Variations from 5% to 25% in the solids content of the slurry yielded permeability values of 2.4×10^{-4} ($\pm$33%) cm/sec.

should be observed that lime is more effective than cement in reducing the coefficient of permeability and increasing the amount of the additive from 5% to 10% has no beneficial effect in terms of permeability reduction.

Krizek, Chu, and Atmatzidis (1987) summarized accumulated permeability data for a variety of double alkali sludges for which specimens were compacted with 40% Standard Proctor compactive energy in conventional permeameters at their as-produced water content, thereby yielding dry densities of 0.6 g/cm^3 and 1.0 g/cm^3; the majority of the values obtained lie in the range from 10^{-6} cm/sec to 10^{-4} cm/sec, with most being larger than 10^{-5} cm/sec. Available data indicate that the permeability generally decreases as the dry density increases and is essentially independent of curing time. The addition of fly ash to double alkali sludges (Lin, 1979; Landivar, 1981; Weber, 1981) tends to increase their permeability several-fold, but values usually lie within the same order of magnitude, regardless of the type and amount of fly ash. The addition of lime to a double alkali-sludge-fly ash mixture (Lin, 1979; Landivar, 1981; Weber, 1981) generally decreases its permeability, but this decrease is a rather erratic function of the amount of lime added; whenever the permeability can be measured with reasonable confidence, it is usually on the order of 10^{-6} cm/sec to 10^{-5} cm/sec, regardless of the amount of lime in the mixture.

It becomes apparent that the permeability of sedimented materials and dry landfilled materials is esentially the same at the initial stages of disposal and that decreases in the permeability with time are more pronounced in the sedimented materials with additives, such as lime or cement. On the other hand, dry landfilled materials have the advantage of being able to be compacted, thereby increasing their density and hence reducing their permeability.

Strength

The undrained vane shear strength values obtained after seven days of curing for materials sedimented from slurries with an initial solids content ranging from 5% to 25% by weight show negligible variations. Accordingly, all strength testing was conducted on sedimented slurries with an initial solids content of 15% by weight, and the results obtained are presented in Table 2. It can be observed (see upper portion of table) that, with the exception of two specimens with high fly ash content and cement additive, the vane shear strength of sediments cured for seven days was generally less than 20 psf, which is indicative of extremely soft materials. This is consistent with the high water contents (usually in the range of 100% to 150%) and low dry densities (on the order of 20 pcf to 40 pcf) of these sediments. The relatively high (but still low in the absolute sense) vane shear strengths obtained from the specimens with 67% fly ash and the cement additive is undoubtedly due largely to the cementation bonds that developed as a result of chemical reactions between the fly ash and the cement, with the sludge acting basically as a filler.

Table 2. Vane Shear Strength[1] of Sedimented Double Alkali FGD Sludge Mixed with Fly Ash and Lime or Cement

Fly Ash (% of solids by weight)	Curing Time (days)	Additive (% of solids by weight)				
			Lime		Cement	
		0	5	10	5	10
0	7	1	2	2	4	6
28	7	1	2	2	16	15
36	7	2	3	3	14	56
50	7	2	3	3	3	20
67	7	3	10	5	585	875
100	7	1	-	-	-	-
0	21	1	-	-	-	-
36	21	250	125	160	200	625
100	21	100	-	-	-	-
36	14	80	-	-	-	-
36	60	600	-	-	-	-
36	90	650	-	-	-	-

[1]Values for vane shear strength are expressed in pounds per square foot.

Further data (see middle portion of table) illustrate to a limited degree the effects of curing time on strength. Seven specimens were prepared, five with a fly ash content of 36% (but containing different amounts of lime or cement), one of plain filter cake, and one of fly ash only, and vane shear tests were conducted after a curing period of twenty-one days. With the exception of the plain filter cake specimen, which did not demonstrate any variation in strength, all other specimens manifested a strength increase on the order of 50 to 100 times their 7-day strength. These improvements in the strength of the sediments may be attributed primarily to delayed reactions that took place and secondarily to the decrease in the moisture content of the additives. The chemical reactions are due mainly to the fly ash, which is known to be a very good pozzolan that reacts with lime and/or cement. Finally, supplementary tests on sediments of material containing 36% fly ash by weight, but no additives, yielded vane shear strengths on the order of 80 psf, 600 psf and 650 psf after curing periods of 14, 60 days and 90 days, respectively (see lower portion of table); the latter two values constitute an increase of 2.5 times over the 21-day strength and more than two orders of magnitude over the 7-day strength. Although the gain in strength is considerable over this extended period of time, it is important that significant gains were measured during the first to weeks.

All specimens subjected to drying manifested an increase in strength as the moisture content decreased. Typical results obtained for specimens containing 36% fly ash by dry weight are shown in Figure 6. In general, the largest increase in strength was obtained from specimens prepared with 10% lime, and this was followed in order by specimens with 5% lime, 10% cement, 5% cement, and no additive. When

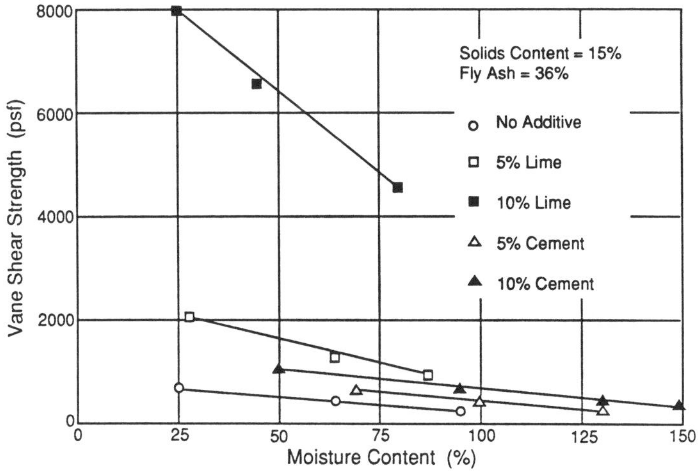

Figure 6. Effect of Drying on Shear Strength of Sedimented Double Alkali FGD Sludge Mixed with Fly Ash and Lime or Cement

the moisture content of specimens prepared with 10% lime and various amounts of fly ash was finally reduced to around 10% to 40%, the measured vane shear strength ranged from about 5600 to 8000 psf, the higher values corresponding to a fly ash content of 28% or 36%. Similarly, the presence of 5% lime yielded strengths ranging from about 2000 to 3000 psf for moisture contents between 20% and 35%, the higher values corresponding again to a fly ash content of 28%. The beneficial effect of lime becomes apparent when it is considered that, for the same final moisture contents, the strength of filter cake-fly ash mixtures with no additive was in the range of 750 to 1200 psf and the strength of plain filter cake was never more than 500 psf. Finally, it was noted that, in the absence of fly ash, the effect of adding lime or cement was drastically reduced; the vane shear strength attained for moisture contents in the range of 30% to 40% was about 1000 psf. The foregoing observations indicate that fly ash has a significant effect in improving (by a factor of 2 or 3) the strength characteristics of sedimented scrubber sludge and that the addition of lime, even in small quantities, can further increase the strength severalfold. Finally, there appears to be an optimum fly ash content which, based on the limited information available, is about 30%.

The unconsolidated-undrained shear strength of a saturated material is theoretically one-half of the unconfined compressive strength. Virtually all of the reported data on the unconfined compressive strength of double alkali sludges were generated at Northwestern University (Menoff, 1978; Weber, 1981; Chu, 1983) and have been summarized by Krizek, Chu, and Atmatzidis (1987). In general, specimens were compacted in the laboratory with 40% Standard Proctor energy at their as-produced water content, and the results can be considered indicative of the strength of dry landfilled and lightly compacted sludge. Since most of the sludge specimens tested in unconfined compression were not fully saturated, it is more appropriate to treat the resulting unconfined compressive strength merely as an empirical strength parameter; nevertheless, general comparisons can be made between trends observed for "dry landfilled" and "ponded" materials.

The initial unconfined compressive strength (up to one day of curing) ranges from about 200 to 1600 psf, with an average of about 800 psf at moisture contents ranging from about 50% to 90%. After 28 days of curing, the average strength increases somewhat to about 1000 to 1200 psf. The unconfined compressive strength of sludge-fly ash mixtures generally increases with increasing fly ash content up to a value of 40% to 50% fly ash (by dry weight) and can attain values on the order of 3000 psf. The addition of lime in small quantities results in strengths that are substantially higher than those of plain sludge or sludge-fly ash mixtures. For example, unconfined compressive strengths up to 5500 psf were measured for specimens with 5% lime and 20% fly ash. It becomes apparent that dry landfilled sludge and sedimented sludge which has been exposed to drying have similar quantitative and qualitative strength characteristics.

OPERATIONAL AND ENGINEERING IMPLICATIONS

The foregoing results and observations on the sedimentation rate, dry density, permeability, and strength of sedimented materials and comparisons with the properties of dry landfilled materials provide the basis for identifying some operational and engineering implications of using wet disposal systems as an alternative to dry landfilling for the disposal of double alkali FGD sludges. However, the selection of the proper disposal system can not be based solely on understanding the behavior and properties of the sludges; rather, a more comprehensive approach must be taken and should include all environmental aspects, operations, costs, availability of a suitable disposal area, and ultimate or long-term use of site.

The inherent simplicity in the operation of wet disposal systems provides a major advantage over dry disposal systems. The common factor in all wet disposal systems is that the sludge is transported and deposited in the final disposal site in a slurry state. Because of this, wet disposal systems (without additives) are usually simpler and less expensive to operate. However, the installation of retention dikes, pipelines, and pumps for wet disposal systems may lead to higher capital costs. Alternatively, a variety of systems may be used to transport and landfill the dewatered double alkali scrubber sludge, and the choice will depend on a number of site-specific factors; the filter cake can be transported to the disposal site via a conveyor belt system and deposited in place, or it could be transported in trucks and dumped directly into the landfill or into a distribution box which would then feed a conveyor belt or bucket chain to place the material.

Measured settling rates suggest that the slurry produced by wet disposal systems and deposited in a sedimentation pond will settle at a velocity that is compatible with the normal retention times of such basins. The low dry densities of the sedimented materials (on the order of 20 to 40 pcf) require disposal sites with large surface areas and this could constitute a major limitation of the wet disposal method when large tracts of land are not available or when land is at a premium. In contrast, the sludge in a dry disposal system could be placed with no compaction at a dry density near the upper limit of the aforementioned range, or it could be lightly compacted in 8-inch to 12-inch layers by one or two passes of a lightweight compactor or dozer with relatively low contact pressure to achieve about a 50% increase in dry density or a final dry density on the order of 60 pcf. This latter increase in density will also serve to increase the strength and decrease the permeability and compressibility of the material.

Another major problem of concern is the quantity and quality of the leachate that may enter the groundwater; this will obviously be governed in large part by the permeability of the soils immediately below the sedimented or landfilled materials and above the groundwater table, as well as by the permeability of the sludge itself. Permeability tests conducted on the double alkali FGD sediments indicated permeability coefficients on the order of 10^{-4} cm/sec which could be reduced by up to one order of magnitude if appropriate quantities of lime are added to the filter cake or the cake-fly ash mixture and the sedimented material is allowed to cure for about sixty days. By way of comparison, the coefficient of permeability for a variety of dry disposal double alkali sludges was measured to be in the range between 10^{-4} cm/sec and 10^{-6} cm/sec. The addition of lime was found to be effective in reducing the permeability of these sludges and it was suggested that, in some cases, coefficients of permeability on the order of 10^{-7} cm/sec may be obtained (Landivar, 1981; Weber, 1981); however, the general inability to consistently duplicate such low values of permeability suggests that they be viewed with caution (Krizek, Chu, and Atmatzidis, 1987).

The presence of a pool of standing water above the sludge deposited in a wet disposal system will increase the volume of leachate generated due to the availability of permeant and driving head. The actual quantity of leachate will vary with the nature of the site specific conditions, and an impervious liner may be necessary to contain the leachate within the area.

The shear strengths developed in the sedimented materials in the first three weeks or so were generally lower than the strengths developed in the dry landfilled materials. Although vane shear strengths as high as 8000 psf were measured in specimens prepared with 10% lime and dried to approximately 30% moisture content, long periods of time would be needed for the ponded materials to dry naturally to moisture contents between 30% and 40%. When ponded under normal conditions, the sedimented materials may eventually be capable of supporting lightweight structures; the actual bearing capacity of these sediments will depend on the amount and type of additive used.

Due to the loose nature of the sedimented materials, large settlements (on the order of inches or more) should be expected, but these settlements should occur soon after application of the loads. The actual amount of the settlement will depend, of course, on the thickness of the sedimented materials and the magnitude of the applied load.

Costs and problems associated with the ultimate abandonment and reclamation of wet (as well as dry) disposal areas are site specific, and even generalities are rather meaningless. The evidence to date, supported by results reported herein, indicates that at least some untreated sludges, especially those containing high sulfite/sulfate ratios, will remain structurally unstable unless treated with some effective stabilizer. Wet disposal systems utilizing additives, such as lime or cement, can improve the strength characteristics of the sediments and reduce environmental concerns. If properly applied and operated, a wet disposal system with additives can reduce leachate generation (by reducing the permeability and, to a certain extent, tying up the leachate salts) and improve structural stability so that future reclamation problems are reduced. It should be recognized, however, that both capital and operating costs increase for wet systems when additives are used. More extensive processing facilities increase capital costs, and the continued addition of additives increases operating costs. Whether or not wet disposal systems with additives have a cost advantage over dry disposal will depend on the specific situation.

In summary, it can be concluded that a properly reclaimed wet disposal site can be utilized for some productive purpose. If the materials are weak and compressible, possible uses (assuming appropriate cover) would include a recreational park, farmland, or wild life refuge. However, if stabilized with the proper additive, light industrial and residential construction may be feasible. The selection of a suitable combination of filter cake, fly ash, and additive will depend, in part, on the intended future utilization of the area.

CONCLUSIONS

The possible codisposal of double alkali FGD sludge and fly ash in a settling pond and then reclaiming the site for some useful purpose was evaluated in terms of specific engineering properties. Within the scope and limitations of this study, the following conclusions can be advanced:

1. Since fly ash and double alkali FGD sludge are usually waste products from the same plant, their codisposal in a sedimentation pond offers the advantages that the sedimentation rates are increased and the strength of the sediment is improved.

2. Sedimentation rates of about one foot per hour are typial of slurries with 15% solids content; the sedimentation rate is doubled when the initial solids content is decreased from 25% to 5%.

3. Sediments are generally in a very loose state (dry density in the range between 20 pcf and 40 pcf) and, as a result, large disposal areas are needed to contain these materials.

4. Although adding lime or cement in quantities up to 10% of the total solids serves to decrease the permeability of sedimented filter cake or cake-fly ash mixtures, the coefficient of permeability seldom attains values lower than about 10^{-5} cm/sec; this makes it necessary to furnish any disposal site with an appropriate "impervious" liner.

5. The shear strength of freshly deposited saturated sediments is generally very low (about 20 psf or less), but it can increase by one- to-two orders of magnitude with curing time, depending on the amount of fly ash and additive (lime or cement).

6. Exposure to conditions that cause drying of the sediments results in a further increase in strength; the highest shear strength values (about 8000 psf when the moisture content was reduced to about 30%) were obtained for sediments containing 10% lime and 36% fly ash.

REFERENCES

Barnea, E., and Mizrahi, J. (1973), "General Correlation for Fluidization and Sedimentation," Journal of Chemical Engineering, Volume 5, Number 2, pp. 171-189.

Borden, R H. (1977), Engineering Properties of Three Double Alkali Flue Gas Desulfurization Scrubber Sludges, M.S. Thesis, Department of Civil Engineering, Northwestern University, Evanston, Illinois.

Chu, S. C. (1983), Geotechnical Properties and Disposal Considerations for Flue Gas Desulfurization Sludges, Ph.D. Dissertation, Department of Civil Engineering, Northwestern University, Evanston, Illinois.

Edwards, L. O. (1979), Calcium Sulfite Crystal Sizing Studies, United States Environmental Protection Agency, Report Number EPA-600/7-79-192, 95 pp.

Krizek, R. J. (1979), "Landfill Management for Double Alkali Sulfur Dioxide Scrubber Sludge," Journal of the Energy Division, American Society of Civil Engineers, Volume 105, Number EY2, pp. 229-239.

Krizek, R. J., Christopher, B. R., and Scherer, S. D. (1980), "Stabilization of Double Alkali Scrubber Sludges," Journal of Civil Engineering Design, Volume 2, Number 3, pp. 279-304.

Krizek, R. J., Chu, S. C., and Atmatzidis, D. K. (1987), "Geotechnical Properties and Landfill Disposal of FGD Sludge," Proceedings of the Specialty Conference on Geotechnical Practice for Waste Disposal, American Society of Civil Engineers, Ann Arbor, Michigan, pp 625-639.

Landivar, H. E. (1981), Effect of Several Variables on the Permeability of Two FMC Double Alkali Filter Cakes, M. S. Thesis, Department of Civil Engineering, Northwestern University, Evanston, Illinois.

Lin, D. (1979), Character and Control of Leachate from FGD Double Alkali Solids Disposal, M. S. Thesis, Department of Civil Engineering, Northwestern University, Evanston, Illinois.

Menoff, S. D. (1978), Fly Ash Stabilization of Double Alkali Scrubber Sludges, M.S. Thesis, Department of Civil Engineering, Northwestern University, Evanston, Illinois.

Midkiff, L. (1979), "Disposal of Flue Gas Cleaning (FGC) Wastes," Power, Volume 123, Number 9, pp. 1-11.

Weber, R. H. (1981), Engineering Characteristics of Double Alkali Filter Cake Mechanically Mixed with Fly Ash and Lime, M. S. Thesis, Department of Civil Engineering, Northwestern University, Evanston, Illinois.

Zanker, A. (1980), "Nomographs to Determine Settling Velocities for Solid-liquid System," Chemical Engineering, pp. 147-150.

CEMENT STABILIZATION OF DEWATERED MINE TAILINGS AND ITS EFFECT ON VEHICLE MOBILITY

Edward A. Nowatzki[1], M.ASCE, and Andrew Mac G. Robertson[2], M.ASCE

ABSTRACT: Tailings impoundments are hydraulic fill structures typically encountered in mining operations. Tailings generated at the Green's Creek Mine in Alaska have specific gravities ranging from 3.2 to 4.2 with 95% of the material passing the #200 sieve (nonplastic silt). Because these tailings are virtually nondraining under gravity alone, conventional disposal techniques would result in an impoundment structure having a high risk of liquefaction since the mine is located in an active seismic area. Dewatering and stabilization of the tailings with additives prior to disposal would diminish potential for the risk of liquefaction and provide the mine with the added benefits of flexibility in shaping the geometry of the impoundment and of being able to operate vehicles over the stabilized impoundment during and after its construction.

The purpose of this study was to investigate the effect of degree of dewatering, tailings-to-cement ratio, and curing time on the stabilization of Green's Creek Mine tailings and to assess the improvement in trafficability that would result from such stabilization. Specimens were prepared from dewatered tailings at tailings-to-cement ratios of 80:1, 40:1, and 20:1. Direct shear tests and unconfined compression tests were performed on specimens after curing times of 2, 6, and 28 days. Coulomb strength parameters were obtained for the tailings under various combinations of these conditions and mobility analyses were performed for a typical 30-ton haul truck using a computer model developed by Karafiath and Nowatzki (1978). Vehicle performance curves were generated for all cases that did not yield a "NO GO" condition. The results were analyzed to determine which combination of degree of dewatering, cure time, and tailings-to-cement ratio provided acceptable vehicle performance.

Introduction

Drainage and control of water associated with tailings impoundments is critical to stability of the impoundment (especially in seismically active areas), the potential for pollution, and the method used in disposing of the tailings. Current designs for tailings disposal can generally be assigned to one of three groups:

----

[1]Associate Professor, Department of Civil Engineering and Engineering Mechanics, The University of Arizona, Tucson, AZ 85721

[2]Principal, Steffen Robertson and Kirsten, Inc., Vancouver, B.C.

1. Designs that seek to prevent seepage by the installation of an all-encompassing barrier or liner. Tailings in impoundments designed in this way generally remain saturated for long periods of time. Since such impoundments are particularly susceptible to liquefaction, this design is not preferred in seismically active areas.

2. Designs that allow for controlled seepage through engineered embankments and/or underdrain systems. Such designs, although preferable to the former, still produce tailings impoundments which, in the short term, may be susceptible to static instability or liquefaction under seismic loads. Impoundments designed in this way are generally more expensive than those constructed with an all-incompassing barrier or liner.

3. Designs that call for predisposal dewatering of the tailings in the plant to a degree that they can be handled by conventional "dry" materials methods. Such designs produce "dry" tailings impoundments that have relatively little free water (10%-40%). For this reason, long- and short-term pollution and instability problems are minimized. In addition, flexibility with respect to disposal procedures is greatly increased because of the enhanced trafficability of dewatered tailings. It is not surprising that this design is finding increasing application in the field of mine tailings disposal.

The following methods listed in approximate order of decreasing ease for removing water are currently available for the dewatering of tailings (Robertson et al., 1982).

(a) Gravity drainage - can be applied to freely draining tailings. The method is generally most effective when the material has a D_{50} greater than 0.2 mm and a D_{10} less than 0.075 mm.

(b) Cycloning plus gravity drainage - can be applied to well graded tailings. Cycloning dewaters the coarse fraction of the tailings while the fine tailings in the overflow are treated by one of the following methods.

(c) Centrifuge dewatering - can be used to dewater the total tailings, at least partially. The coarser solids are usually still "sloppy" and the very fine tailings in the effluent may need to be treated separately.

(d) Vacuum filtration - can be used to dewater even fine tailings in a relatively short period of time. Disk, drum, or belt filters are used to create high hydraulic gradients over relatively thin layers of "cake" through the application of vacuum. The effectiveness of the method is limited by cake cracking which may result in loss of vacuum.

(e) Pressure filtration - can be used to dewater even the finest of tailings. Belt or filter presses produce greater hydraulic gradients through the "cake" than can be produced by vacuum filtration alone. Positive controls increase the effectiveness and flexibility of this method.

"Dry" tailings produced by any of the methods listed above vary in consistency from a thick slurry to a granular material or cohesive cake with little free water. The drier the dewatered tailings, the greater the ease with which they can be handled and transported and the greater the stability of the disposal dump. Another matter of concern in "dry" tailings disposal is the trafficability of the dewatered product, i.e., the drier the dewatered tailings, the less likely will equipment operating on it encounter mobility problems during loading or following placement in the disposal area.

Despite these obvious advantages, there are a number of inherent difficulties in the development of a "dry" tailings deposit. These include:

(a) The cost of dewatering the tailings to the moisture contents at which the tailings will behave as a "dry" granular product as distinct from a thick slurry. Optimum moisture contents and dewatering costs are obtained by the selection of appropriate dewatering equipment.

(b) The handling of the tailings and their placement in the pile. Concerns include the tendency for the tailings to stick or set in the transportation containers and the ability to be able to travel over the tailings (trafficability) with rubber-tired equipment once they have been placed in the field. Improvement of handling and trafficability are achieved through adequate dewatering and the use of appropriate transport and placement equipment and techniques.

(c) The stability of the tailings once they have been placed in the tailings pile. Both static stability, in the slopes of the tailings pile, and dynamic stability, when subjected to dynamic earthquake forces, must be evaluated. Stability is achieved by constructing with low slope angles, low rates of rise, and by compacting or draining tailings to eliminate the potential for liquefaction.

This paper addresses problems related to the second of the inherent difficulties listed above. There is obviously a trade-off between the cost of dewatering tailings to the extent required to improve their trafficability versus the cost of stabilizing wetter tailings by additives. The purpose of this paper is to evaluate the effect of cement stabilization of "dry" (partially dewatered/nonsaturated) mine tailings on the mobility performance of a 30-ton haul truck typically used in open pit mining operations.

Description of the Site

The proposed Green's Creek Mine is located on Admiralty Island approximately 30 miles south of Juneau in southeastern Alaska (Figure 1). The proposed development comprises an underground mine, a mill, and a tailings impoundment. The site selected for the tailings impoundment is called the Upper Cannery Site. It is located approximately 5 miles by access road from the mill in a shallow section of the Tributary Creek Valley. The following paragraphs describe the major features of the site as they impact the impoundment design.

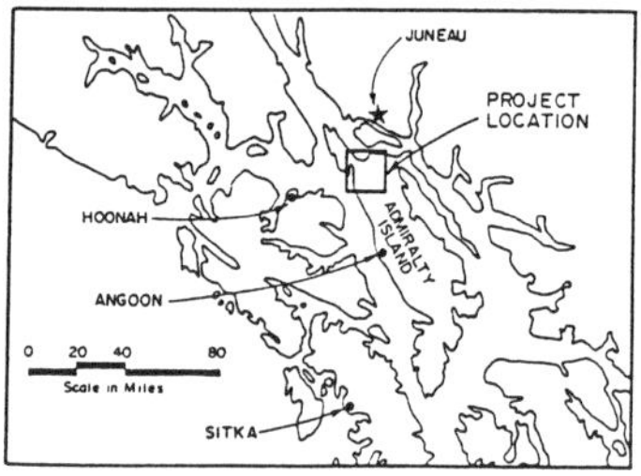

Fig. 1 Project Location

Topography - The site is located in the upper part of the Tributary Creek Valley. It is bounded on the north by a muskeg-filled saddle which forms the edge of the watershed. There is a steep mountain slope to the east, while on the west there is a low, discontinuous ridge which falls away to Hawk Inlet that leads to Chatham Strait and thence to the Pacific Ocean. The gaps in the discontinuous ridge are filled with sediment. The valley extends southward for about one mile to Zinc Creek.

Stratigraphic Sequence - As indicated by drillholes, seismic surveys, and vertical resistivity measurements performed across the valley and the western ridge, the stratigraphic sequence, listed in decreasing order, consists generally of the following units: peat, silty sand or gravel, hard to firm silty clay, silty gravelly sand, and bedrock consisting of a hard, bonded schist. The maximum depth to bedrock is approximately 65 feet except in a sediment filled trough along the western ridge where bedrock is at approximately 100 feet.

Seismicity - Southeastern Alaska is an area of high seismicity. A peak horizontal ground acceleration of 0.3 g was used for design. The seismic potential and the potential for liquefaction of the silty sand horizon had a major impact on the tailings impoundment design.

Groundwater Conditions - The water table in the muskeg-covered area is generally at or above the surface. In the ridges surrounding the valley, groundwater may occur at depths of up to 15 feet below the surface. The permeability of the silty clay underlying the peat and silty sand or gravel layers is relatively low, therefore infiltration and subsurface groundwater migration occur slowly. The water quality of the mill tailings effluent is expected to be superior to the quality of the natural groundwater. Therefore seepage control is required partly for discharge control, but mainly for the structural stability of the tailings impoundment.

Climatic Conditions - The Alaska Coastal Maritime Zone is characterized by cloudy skies, abundant precipitation, and moderate temperatures. Mean annual precipitation at the site is estimated at 60 inches with average monthly rainfall expected to vary from 3.5 inches in April to 8.5 inches in October. The maximum observed snowfall was 7 feet. The 10-year, 24-hour storm precipitation is approximately 5 inches.

Proposed Tailing Impoundment

The initial design for the impoundment called for direct discharge of the tailings into an area behind a large embankment constructed at the southern end of Tributary Creek Valley. The proposed tailings impoundment would provide for the disposal of more than 4 million tons of "wet" (saturated) tailings expected to be produced during the design life of the mine. This plan required the construction of a large embankment at a location underlain by deep deposits of soft, sensitive soils. To achieve embankment stability, it would be necessary either to remove the foundation soils by excavation or to improve their strength and reduce their liquefaction potential. A combination of excavation with replacement and stabilization by reverse osmosis consolidation was investigated and found to be economically unattractive. In addition, the construction of the large embankment needed for "wet" disposal would be very expensive. These factors taken together, required an unacceptably high capital outlay. Therefore, in order to reduce the initial capital cost associated with the conventional "wet" tailings impoundment construction, the smaller Upper Cannery Tailings Impoundment was proposed. At the same time, technical evaluations and cost effectiveness studies were undertaken to determine the feasibility of using dewatered tailings to construct a "dry" tailings impoundment. The following paragraphs describe the results of those studies and their influence on the final design of the tailings impoundment.

Factors Affecting the Decision to Use "Dry" Tailings - There were a number of factors that had to be considered in deciding whether or not to use "dry" tailings disposal for the Green's Creek Project. The more important considerations were:

(1) The mine owners considered using tailings as a backfill in the underground workings. However, because of their poor drainage characteristics, the tailings would not be suitable for use as a hydraulic fill. Therefore, before being transported underground for placement by mechanical means, the tailings would have to be dewatered. Since 50 to 60% of the tailings produced by the mill are to be used for mine filling, the incremental cost of dewatering the remainder of the tailings for transport to and placement on a "dry" tailings impoundment would be relatively low.

(2) The Tributary Creek Valley is underlain by soils that are prone to liquefaction under the seismic conditions of the area. This limits the embankment heights, side slopes, and construction rates that can be achieved safely. In addition, a large, saturated, tailings mass is potentially susceptible to liquefaction in the existing seismic environment.

(3) Excavation and/or stabilization measures required for construction of large embankments are costly and time-consuming.

(4) Good quality materials for construction of large embankments are scarce at the site. Moisture contents in the silty clays and silty sands are generally higher than those required for good compaction. Weather conditions do not favor moisture reduction by conventional spreading and drying techniques in the borrow pit. These factors increase the cost of large embankment construction.

<u>Advantages of Using Dewatered Tailings</u> - The "dry" tailings disposal scheme has the following advantages:

(1) The tailings do not require retaining embankments. The "dry" tailings impoundment can be constructed under controlled conditions to have low side slopes to insure stability.

(2) There is more flexibility in locating the tailings pile on the site to avoid unfavorable foundation conditions.

(3) Drainage from the impoundment is minimal. Therefore the heights of embankments for settling ponds can be limited to those required only to provide adequate retention time for waste water and runoff flows from the tailings pile, the mine, and the mill.

<u>Difficulties Associated with the Development of a "Dry" Tailings Impoundment</u> - There are a number of difficulties inherent to the development of a "dry" tailings deposit. These are discussed fully by Robertson et al. (1982) and include the following major problems:

(1) The cost of dewatering the tailings to the moisture contents at which they will no longer behave as a thick slurry.

(2) The handling and placement of the tailings and their trafficability by construction equipment once in place.

(3) The static and dynamic stability of the tailings pile. Stability is achieved by constructing at low rates of rise, by designing and building piles with low slope angles, and by providing drains to reduce and possibly eliminate the potential for liquefaction.

Description of the Proposed "Dry" Tailings Impoundment - A plan view of and sections through the "dry" tailings impoundment are shown in Figures 2 and 3, respectively. As indicated in the figures, the main components of the impoundment are:

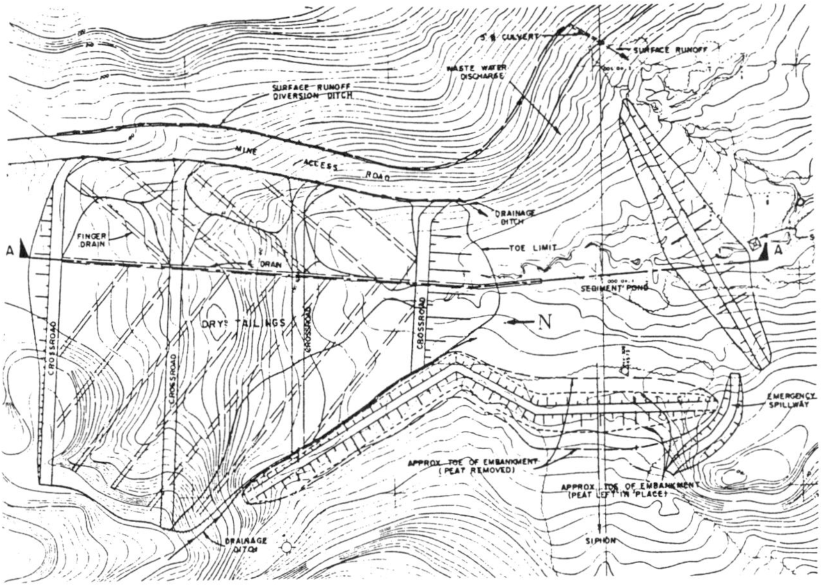

Fig. 2 Plan View of Project

(1) A centerline drain and a system of finger drains under the tailings pile.

(2) Cross roads constructed of waste rock from the mine or rock from a nearby quarry. These provide access to the tailings pile from the haul road adjacent to the impoundment on the east. They also serve as internal and surface drains for the "dry" tailings.

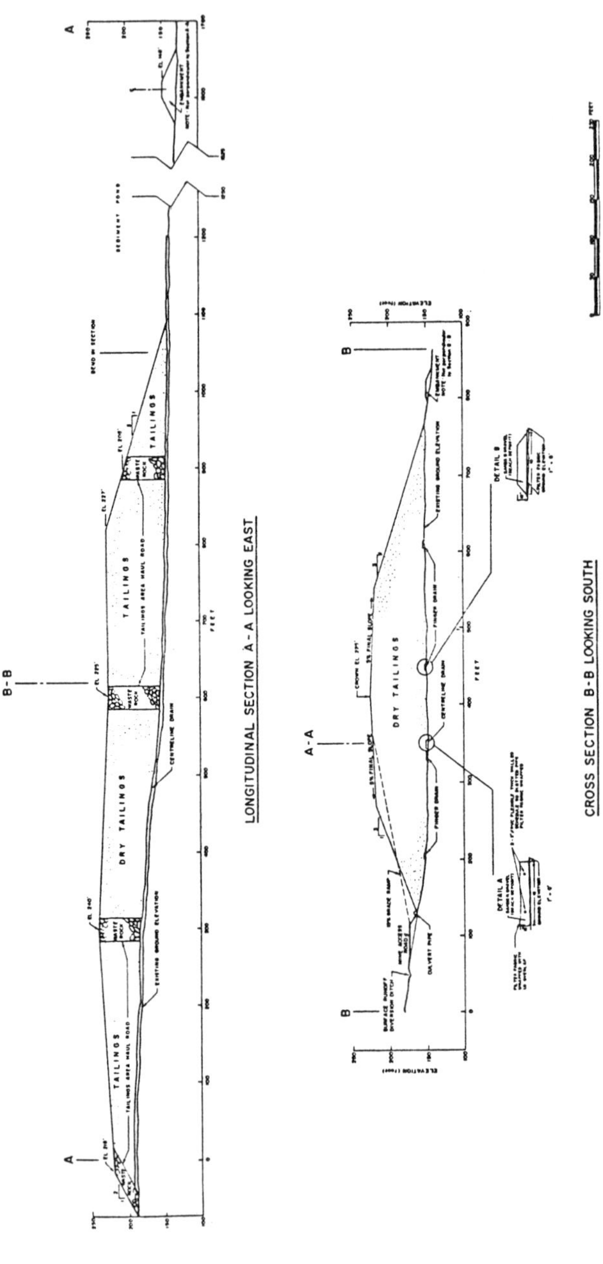

Fig. 3 Sections Through "Dry" Tailings Impoundment

(3) A diversion ditch upslope of the haul road around the tailings impoundment.

(4) The main (south) and saddle (west) embankments for the sediment pond, and the associated siphon and spillway discharges to Hawk Inlet.

(5) A seepage return structure immediately downstream of the main embankment.

These components constitute the major construction items for the project. The tailings pile shown in Figures 2 and 3 has a capacity of 23.4 x 10^6 cu ft. At an average compacted unit weight of 120 pcf, this would provide for an impoundment life of approximately 17 years at a production rate of 235 tons/day and 12 years at a production rate of 335 tons/day.

Trafficability Requirements for Impoundment Operations

In a project of this type, there are obviously a large number of potential technical problems that could impact the final design. These range from foundation and slope stability problems to considerations of groundwater quality, discharge and seepage, and construction problems peculiar to the environment of southeastern Alaska. One of these potential construction problems is the off-road mobility of construction equipment, especially on the "dry" tailings impoundment during its construction. Mobility of rubber-tire construction equipment on the tailings requires that:

(1) the tailings be dewatered to a moisture content below 16%. As indicated previously, this is not a restrictive requirement since 40 to 60% of the tailings will be dewatered anyway in order to make them suitable as mine backfill.

(2) the crest and sides of the impoundment be sloped to promote runoff and to prevent ponding, infiltration, and erosion (Refer to Fig. 3).

(3) stabilizing additives, such as cement, be used if needed.

(4) the tailings be compacted by a vibratory roller after placement but prior to moisture infiltration or setting of the cement additive.

(5) the tailings be compacted to approximately 93% of the Standard (ASTM D-698) maximum dry density which, in the case of Green's Creek Mine tailings, is about 120 pcf (specific gravity of solids = 3.23) at a moisture content of 16%.

(6) heavily trafficked surfaces be covered by one foot of rock waste as soon as vehicle mobility begins to degrade.

(7) internal drains and underdrains be installed to prevent the "dry" tailings from becoming saturated.

These requirements are based on a comprehensive testing program conducted as part of this project by Steffen, Robertson, and Kirsten and the results of mobility analyses performed specifically for the Green's Creek Mine tailings by the senior author. The details of these two investigations follow.

"Dry" Tailings Production and Impoundment Construction

Two processes were evaluated to determine their efficiencies in dewatering the Green's Creek tailings:

1. Filter presses - Testing resulted in the following filtration rates and dewatered moisture contents:

Filtration Rate ($lbs/ft^2/hr$)	Moisture Content (%)
41	10
51	11
64	13

2. Vacuum belt filters - Tests resulted in dewatered moisture contents of approximately 20%.

Moisture contents in the range attainable by vacuum belt filtration were considered too high for efficient handling of the tailings in the field or for adequate trafficability without stabilization. However, stabilization of these tailings would require substantially greater quantities of additives than tailings dewatered using filter presses. Therefore, for economic reasons, the use of filter presses with cement stabilization was chosen as the preferred method for dewatering the Green's Creek tailings.

Once the method for production of "dry" tailings had been determined, the following procedures for impoundment construction were developed.

1. Tailings dewatered to moisture contents of approximately 13-16% would be mixed with cement and placed in a "live" stockpile. Determination of the amount of cement required to stabilize the tailings will be discussed subsequently. The tailings would be loaded by front end loaders into 30-ton haul trucks and transported either to the mine for backfill or to the tailings impoundment for disposal.

2. As shown in Figure 4, trucks carrying the treated tailings to the impoundment would access an active disposal area or "panel" by driving onto the tailings pile along cross roads constructed of mine waste or quarry material. Under normal conditions the trucks should experience no difficulty in operating on the existing impoundment surface once the previously dumped tailings-cement mixture have been compacted and given time to cure.

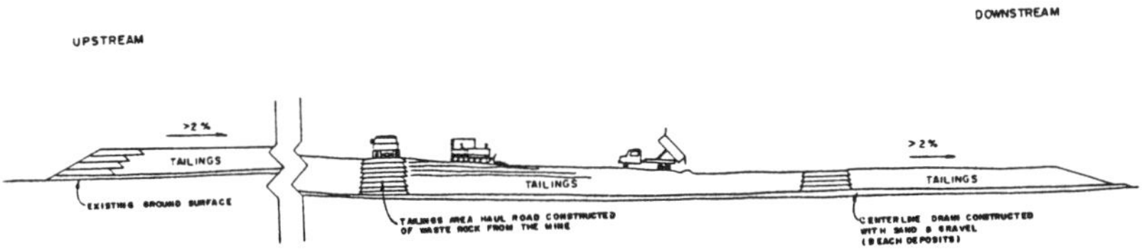

Fig. 4 Schematic of Construction Method

3. The newly-dumped treated tailings would be spread by a tracked dozer to form a layer about 1-1/2 feet thick. The tailings would then be compacted by a vibratory roller to an in-situ dry density of at least 120 pcf (93% of the ASTM D-698 maximum).

4. The pile would be raised in 1-1/2 foot layers with sufficient cross roads and tailings panel sites available to allow for flexibility in placement. Such flexibility would be required in order to allow the newly placed tailings to cure for 14 to 28 days prior to the resumption of disposal operations on that panel. Cross roads would be raised concurrently with the tailings to provide easy access from the roadways to all areas of the pile. A smooth grade of approximately 2% sloping toward the settling pond would be maintained on the impoundment surface to promote rapid surface water runoff (Refer to Figure 4). During periods of high precipitation the surface conditions could deteriorate to the point where wheeled vehicles would become immobile. In those areas, it would be necessary for trucks to deposit their loads adjacent to the cross roads for spreading by the dozers. In good weather or bad, traffic patterns would have to be planned to minimize repetitive operations in any one area that might result in a breakdown of the tailings surface. It may be necessary to place a one-foot thick waste rock-layer in such areas.

5. As sections of the pile approached the designed final height and profile, the slope of the top surface would be reversed by proper placement of tailings and covered with one foot of erosion-resistant fill. The surface would then be appropriately prepared, fertilized, and seeded to establish a vegetative cover.

Mobility Analyses

The mathematical model for pneumatic tire-soil interaction used in this study is presented in detail in a report by Karafiath (1974) and is described in the book by Karafiath and Nowatzki (1978). In general, the interaction between the soil and a vehicular footing, be it a rigid wheel, a tire, or a track comes into play primarily through the following three features of soil response and footing characteristics:

1. Geometry of the Vehicular Footing. For rigid wheels and tracks, the normal stresses at the vehicular footing-soil interface are limited only by the ability of the soil to carry the load in the plastic state of equilibrium. For tires, this limit is also related to the inflation pressure. In either case, the limit stresses are affected by the geometry of the contact area and by changes in the geometry due to deflection, either of the soil or the tire, or both.

2. Shear Stresses Acting as the Vehicular Footing-Soil Interface. The shear stresses generated at the interface by the torque or thrust applied to the running gear significantly affect the limit of the normal stresses that can be developed in the soil.

3. Balancing of Loads. The load on the running gear must be balanced by the interface stresses. Thus, when the interface normal stresses are affected by either the geometry of or shear stresses on the interface, balancing of the loads requires a change in the size of the contact area of the wheels, tracks, or tires. This is accomplished by sinkage.

In the computational scheme, the equations of plastic equilibrium are used in conjunction with the Coulomb failure criterion for soils and the geometric and stress boundary conditions peculiar to the tire-soil interaction problem being studied to determine the interface stresses just at the point of soil failure, i.e., at the point of limiting equilibrium. A finite difference representation of the governing differential equations is solved iteratively until the imposed boundary conditions are satisfied within a specified tolerance. The same mathematical model can be used to determine the performance characteristics of driven wheels, nondriven wheels, and braked wheels merely by modifying the boundary conditions.

The objective of the solution for mobility problems is to determine the performance characteristics of a vehicle. This is accomplished by calculating the net drawbar pull and torque for driven wheels and the motion resistance offered by nondriven wheels and summing them for a given vehicle's running gear assemblage to obtain the total vehicle characteristics. The solution also provides sinkage data and values of rimpull (applied tractive force) that can often be used directly with manufacturer's empirically derived performance curves to obtain velocity and gearing information. This type of information is particularly useful in comparative analyses for equipment selection (Nowatzki et al., 1982) or in cycle time analyses for heavy construction equipment (Nowatzki et al., 1983).

Basically, the computer program models a soil-structure interaction problem and requires the following input data:

1. Soil characteristics based on field loading and drainage (cohesion (c), internal friction angle (ϕ), unit weight (γ)).

2. Tire characteristics (radius, width, inflation pressure, deflection).

3. Loading conditions (individual wheel load at vehicle's rated loading, slip, grade).

The following sections describe how these data were obtained for the mobility analysis of vehicles operating on the Upper Cannery Tailings Impoundment.

Tailings Properties - The results of conventional laboratory tests performed on Green's Creek Mine tailings are summarized in Table 1. The values of the various properties listed in the table suggest that the tailings should behave as an inorganic silt. The trafficability of silts is extremely poor, even when they are not at or near saturation moisture contents. Therefore the "dry" tailings disposal scheme appeared to be unworkable unless the trafficability of the tailings could be improved. To this end, a study was conducted at SRK to determine the effect on the shear strength of each of the following treatments:

1. Increased compaction. As shown in Table 1, increased compaction to either 93% or 98 of the Standard (ASTM D-698) maximum dry density at optimum moisture content (12.5%) did not result in any significant change in the strength properties of unsaturated tailings. Therefore increased compaction was not a viable method for improving the trafficability of the tailings.

2. Addition of cement. The effect of cement stabilization on the shear strength of the tailings was evaluated by performing unconfined compression tests on specimens prepared in triplicate at tailings to cement ratios (T:C) of 80:1, 40:1, 20:1, and 10:1. The specimens were compacted to 93% of the Standard (ASTM D-698) maximum dry density of untreated tailings at a wet-of-optimum moisture content (16%) to insure total hydration of the cement. Specimens were cured for 28 days before being tested. The average values of the unconfined compressive strengths are presented in Table 2. As indicated in the table, there is a significant increase in the 28-day unconfined compressive strength with decrease in the T:C ratio, with

Table 1
Basic Engineering Properties of Untreated Tailings

Specific gravity of solids:
G_s = 3.23 - 3.24

Grain size distribution:

Grain size (mm)	% Finer
0.076	93-94
0.020	27-43
0.002	5-8

Compaction characteristics:

Compactive Effort ($ft\text{-}lb/ft^3$)	Dry Density (pcf)	Moisture Content (%)
12375 (ASTM D-698)	129 (max.)	12.5 (optimum)
5000 (moderate)	108	11
5000 (moderate)	118	16
1250 (light)	111	11
1250 (light)	113	16

Permeability:

Test Type	Material	Permeability (cm/sec)
Percolation	Hydraulic tailings	9.7×10^{-5}
Percolation	Dewatered tailings (lightly compacted)	9.1×10^{-5}
Percolution	Dewatered tailings (moderately compacted)	3.3×10^{-5}
Permeameter	Dewatered tailings (compacted to 90%- ASTM D-698)	1.8×10^{-5}

Relative density (ASTM D-4253):
Maximum: 125 pcf
Minimum: 84 pcf

Shear Strength Parameters (ASTM D-3080):

Degree of Compaction	Saturation Conditions	Direct Shear Strength Parameters C (psf)	ϕ (degrees)
Loose (W=11%)	unsaturated	67	41
Losse (W=16%)	unsaturated	166	39
93% Standard (W=12.5%)	unsaturated	141	41
98% Standard (W=12.5%)	unsaturated	125	36
93% Standard (W=12.5%)	saturated	62	36
98% Standard (W=12.5%)	saturated	62	32

Table 2

Average Values of Unconfined Compressive Strength (psf) for Tailings Treated with Portland Cement and Allowed to Cure Before Testing

Tailings: Cement Ratio	Unconfined Compressive Strength (psf) 2-day	6-day	14-day[a]	28-day
80:1	627	835	1166	1775
40:1	---	---	---	1740
20:1	---	---	---	8451
10:1	---	---	---	12864

(a)Estimated by equation shown in Figure 5.

the exception of the 80:1 and 40:1 specimens. In all cases, however, the unconfined strengths are much greater than those suggested in Table 1 for untreated tailings compacted to the same dry density at optimum moisture content.

3. Aging of cement-treated tailings. Since there was essentially no difference between the unconfined compressive strengths of specimens prepared at T:C ratios of 80:1 and 40:1, it was decided, for reasons of economy, to evaluate the effect of aging on the strength of tailings prepared at a T:C ratio of 80:1. Two sets of specimens were prepared in triplicate at that ratio under the same compaction and moisture conditions as were used for the 28-day specimens. One set of specimens was tested after 2 days, the other set after 6 days. The average values of unconfined compressive strength are presented in Table 2 along with the 28-day strengths. The increase in strength with time for the 80:1 T:C specimens is shown graphically in Figure 5.

As indicated previously, the purpose of the unconfined compression tests was to determine the increase in total shear strength that could be expected from an increase in the cohesion component following hydration of the cement. If it is assumed that the angle of internal friction (ϕ) is unaffected by the addition of the cement and remains approximately 36° for untreated tailings tested under saturated conditions and 40° for untreated tailings tested under unsaturated conditions (refer to Table 1), then an estimate can be made of the increase in cohesion due to the addition of cement as follows. Mohr circles are plotted for the unconfined ($\sigma_3 = 0$) compression tests performed on the 80:1 specimens after cure times of 2, 6, and 28 days. Failure envelopes are then drawn tangent to each circle at angles of 36° and 40°. The intercepts of the envelopes on the τ-axis provide an estimate of the cohesion values for the

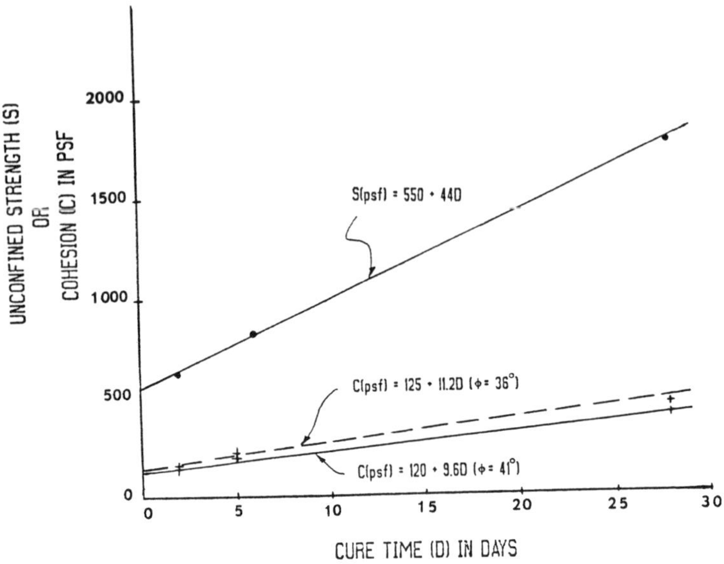

Fig. 5. Variation of Unconfined Compressive Strength and Estimated Cohesion with Curing Time (Tailings to Cement Ratio = 80:1)

the exception of the 80:1 and 40:1 specimens. In all cases, however, the unconfined strengths are much greater than those suggested in Table 1 for untreated tailings compacted to the same dry density at optimum moisture content.

3. Aging of cement-treated tailings. Since there was essentially no difference between the unconfined compressive strengths of specimens prepared at T:C ratios of 80:1 and 40:1, it was decided, for reasons of economy, to evaluate the effect of aging on the strength of tailings prepared at a T:C ratio of 80:1. Two sets of specimens were prepared in triplicate at that ratio under the same compaction and moisture conditions as were used for the 28-day specimens. One set of specimens was tested after 2 days, the other set after 6 days. The average values of unconfined compressive strength are presented in Table 2 along with the 28-day strengths. The increase in strength with time for the 80:1 T:C specimens is shown graphically in Figure 5.

As indicated previously, the purpose of the unconfined compression tests was to determine the increase in total shear strength that could be expected from an increase in the cohesion component following hydration of the cement. If it is assumed that the angle of internal friction (ϕ) is unaffected by the addition of the cement and remains approximately 36° for untreated tailings tested under saturated conditions and 40° for untreated tailings tested under unsaturated conditions (refer to Table 1), then an estimate can be made of the increase in cohesion due to the addition of cement as follows. Mohr circles are plotted for the unconfined (σ_3 = 0) compression tests performed on the 80:1 specimens after cure times of 2, 6, and 28 days. Failure envelopes are then drawn tangent to each circle at angles of 36° and 40°. The intercepts of the envelopes on the τ-axis provide an estimate of the cohesion values for the particular set of conditions plotted. A summary of results obtained in this way is presented in Table 3. The increase of cohesion with time is also shown graphically in Figure 5.

Table 3

Estimated Shear Strength Parameters for Cement-Treated Tailings (T:C = 80:1) Cured for Various Lengths of Time

Cure Time (days)	ϕ (deg)	C (psf)
2	41	150
2	36	160
6	41	200
6	36	215
14	41	254[a]
14	36	282[a]
28	41	400
28	36	450

(a)Estimated by equation shown in Fig. 5.

Vehicle and Tire Properties. Trucks of approximately 30-ton capacity are proposed for transporting the tailings from the mill site to the tailings impoundment. For the purposes of the mobility study, it was assumed that Terex 3305B 30-ton trucks outfitted with Michelin 18x25 tires (E-3 treads) would be used. The vehicle and tire characteristics required for input to the computer program are given in Table 4.

Results

The results of the mobility analyses performed for a fully loaded Terex 3305B haul truck outfitted with Michelin E-3 tires operating on the Upper Cannery Tailings Impoundment are summarized in Table 5. Results are presented for various degrees of tailings' treatment. The baseline performance for this study is that obtained for the untreated/unsaturated tailings.

Table 4

Vehicle and Tire Characteristics

Vehicle - Terex 3305B 30-ton truck

Front axle

Total weight (fully loaded) = 35780 lbs
2 wheels (nondriven)
Wheel load = 17890 lbs

Rear axle

Total weight (fully loaded) = 73720 lbs
4 wheels (driven)
Wheel load = 18430 lbs

Tires - Michelin E3 tread

Diameter	= 63.8 in
Free radius	= 31.9 in
Loaded radius	= 28.8 in @ 18410 lbs
Deflection	= 3.1 in
Section height	= 19.4 in
Section width	= 18.0 in
Inflation press	= 80 psi

If the net drawbar pull available for operations such as acceleration, climbing slopes, etc. is used as the indicator of the vehicle's performance, it can be seen from Table 5 that the fully loaded Terex 3305B should be mobile for all of the tailings' conditions considered, but not under all of the possible vehicle operating conditions. For example, on untreated/unsaturated tailings, the vehicle is expected to operate at slip values of 10% or less. Slip is defined as (1-v/u) x 100 where v = actual travel velocity and u = theoretical travel velocity. If, for some reason, the vehicle operator accelerates too quickly and the wheels begin to slip at a rate greater than 10%, the vehicle will bog down and become immobile, i.e., a "no go" condition will develop. Note that at 10% slip, the sinkage in this case is approximately four inches. By comparison, the vehicle is also likely to operate poorly in 80:1 (T:C) treated tailings even after a 14-day curing period if the slip approaches 40%. Note that in this case the net available drawbar pull is relatively low at 40% slip, even though it is maximum, while the sinkage is relatively high. The vehicle is expected to perform acceptably at lower slips, but the net drawbar pull will be lower. As expected, the conditions corresponding to 20:1 (T:C) treatment and a 28-day curing

Table 5

Summary of Results of Mobility Analyses

Tailings Condition	c (psf)	ϕ (deg)	γ (pcf)	Net Drawbar Pull (lbs)	Slip %	Sinkage (in)	Torque (max) (ft lbs)	Rimpull (lbs)	Mobility
Untreated/Unsat. (Shearbox Test)	140	40	108	25940 (Max)	10	3.8	78659	39134	Marginal to No Go
80:1 - 28 day s_{uc}=1775 psf	400	41	108	47544 (Max)	5	0.7	126080	52099	Acceptable
				37484	40	1.3	104860	44320	
80:1 - 28 day s_{uc}=1775 psf[a]	450	36	108	36113 (Max)	40	1.2	97269	41801	Acceptable
80:1 - 14 day s_{uc}=1166 psf([a])	254[a]	41	108	34702 (Max)	40	2.2	96473	42877	Poor to Acceptable
20:1 - 28 day s_{uc}=1166 psf	282[a]	36	108	29196 (Max)	40	3.9	85361	40605	Poor to Acceptable
20:1 - 28 day s_{uc}=8354 psf	2000	41	108	200650 (Max)	5	1.0	503820	206230	Acceptable
				168860	40	1.1	427320	174780	
20:1 - 28 day s_{uc}=8354 psf	2250	36	108	153130 (Max)	10	0.6	381610	157507	Acceptable
				136530	40	0.6	341560	140937	

(a) Estimated by equation shown in Fig. 5.

period result in the best mobility performance with the maximum value of net drawbar pull occurring at relatively low slips (5-10%) and with little decrease in net drawbar pull even at 40% slip. The price of this improved performance is the cost of the additional cement required to stabilize the tailings at a T:C ratio of 20:1.

Table 5 also lists values of "rimpull" for each tailings' condition considered. Rimpull is a mobility parameter frequently used by vehicle manufacturers as a quantitative measure of the driving force developed by a wheel as it acts upon a surface. The values of rimpull listed in the table are the maxima that can be applied before the vehicle becomes immobile. In all cases the rimpull values determined by the mathematical model are above those usually required for normal vehicle operations and suggest that the Terex 3305B will be mobile for all the conditions considered.

Conclusions and Recommendations

Based on the results of this study, it is concluded that Green's Creek tailings stabilized by the addition of 3% (by weight) cement, compacted to at least 93% of the ASTM D-698 maximum dry density, and allowed to cure for at least 14 days should be trafficable for fully loaded haul trucks of the type considered. For this reason, dewatering and cement stabilization appear to be feasible methods of pretreatment in disposing of tailings at the Green's Creek Mine. However, trafficable conditions may also be achieved without cement additions provided special operating methods are employed, such as dumping and pushing with a dozer. Such alternative placement and operating methods will be explored before a firm decision is made to resort to cement stabilization.

Despite the relatively good results obtained from this study, there are a number of factors that may make actual field conditions worse than those assumed in the analyses. These are:

1. The effect of repeated wheel loads in breaking down initially trafficable surfaces.

2. Freezing temperatures that may prevent or hinder curing and its associated strength gains.

3. Degradation of trafficability by saturation of impoundment surfaces, by ponding water or by alternate wetting and drying and freeze-thaw.

These factors require further study, preferably by full-scale field tests under actual operating conditions.

APPENDIX I. - References

Karafiath, L. L. (1974), Development of Mathematical Model for Pneumatic Tire-Soil Interaction, Technical Report No. 11900 (LL147), U.S. Army Tank Automotive Command, Mobility Systems Lab; Warren, Michigan.

Karafiath, L. L. and E. A. Nowatzki (1978), Soil Mechanics for Off-Road Vehicle Engineering, TransTech Publications; Clausthal, West Germany.

Nowatzki, E. A., L. L. Karafiath and R. L. Wade (1982), "The Use of Mobility Analyses for Selection of Heavy Mining Equipment," Proceedings of the Specialty Conference on Construction Equipment and Techniques for the Eighties, ASCE Construction Division; Purdue University, West Lafayette, Indiana.

Nowatzki, E. A., L. L. Karafiath and R. L. Wade (1983), "The Prediction of Cycle Time Performance of Haul Vehicles in Soft Coal Open Pit Mines," Proceedings of the Seventh Pan-American Conference on Soil Mechanics and Foundation Engineering, Vol. 2; Vancouver, British Columbia.

Robertson, A. MacG., J. W. Fisher and D. van Zyl (1982), "The Production and Handling of Dry Uranium and Other Tailings," Proceedings of the Symposium on Uranium Mill Tailings Management; Colorado State University, Fort Collins, Colorado.

ISLAND CONSTRUCTION IN THE CANADIAN BEAUFORT SEA

M.G. Jefferies, B.T. Rogers and H.R. Stewart
GULF CANADA RESOURCES LTD.
Calgary, Alberta, CANADA
AND

S.Shinde, D. James & S. Williams-Fitzpatrick
ESSO RESOURCES CANADA LTD.

ABSTRACT

Thirty six artificial islands were constructed in the shallow waters of the Canadian Beaufort Shelf during the period 1972-87. These islands extensively used hydraulically placed sand and a substantial body of experience was accumulated with the engineered construction of hydraulic fill structures. The Paper provides an overview of this experience.

INTRODUCTION

Canadian resource companies (exploration "operators") have used artificial islands as drilling platforms for hydrocarbon exploration in the shallow waters (<50m) of Arctic offshore since 1973. The Beaufort Sea is ice covered for 9 months of the year which constrains exploration by limiting the open water season for drillships. Islands are resistant to ice forces and offer the exploration operator the potential of a substantially longer drilling period than drillships (which may take 3 calendar years to complete a single well).

Islands are mainly constructed with hydraulically placed sandfills. Designs vary from simple surface piercing islands in one metre of water to monolithic steel caisson structures which utilize sand cores to provide ice resisting mass and are founded on zoned sandfill berms (Fig. 1). Thirty six islands have been constructed since the start of offshore exploration in 1973. Key details of these islands are summarized on Table 1 and their locations are shown on Fig. 2.

Islands were initially constructed in the shallow, near-shore areas but have been constructed in increasing water depths with time, (Fig. 3). To date, islands have been successfully constructed and withstood substantial design loadings in as much as 32 m of water but the one island attempted in 45 m of water (Nerlerk) had to be abandoned when it collapsed during construction.

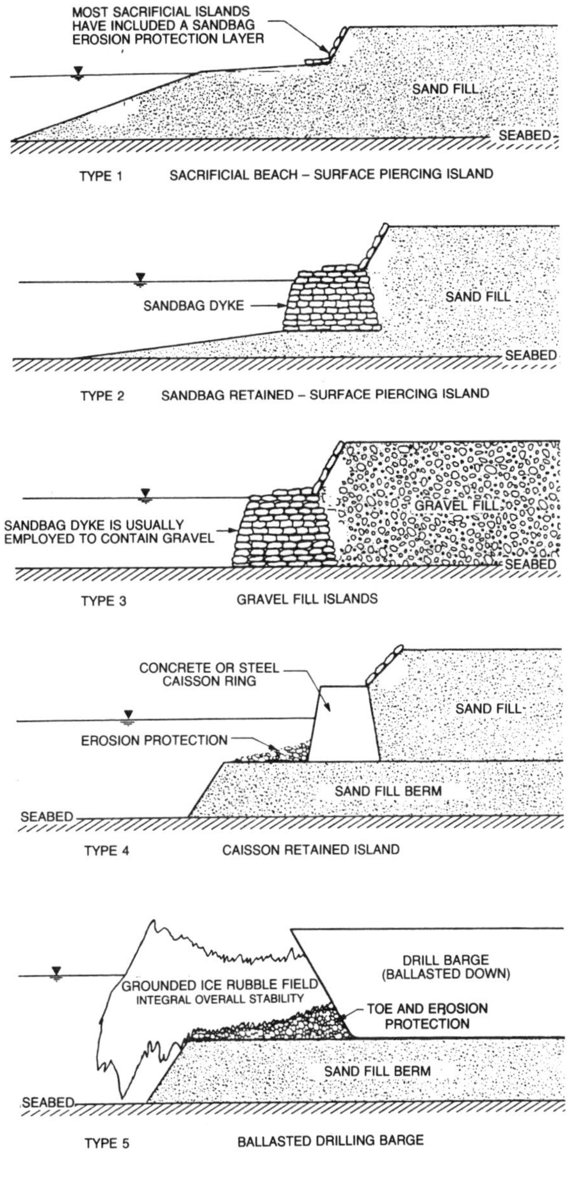

Figure 1

TYPES OF ISLANDS USED FOR EXPLORATION STRUCTURES

TABLE 1

SUMMARY OF ARTIFICIAL ISLAND CONSTRUCTION IN CANADIAN BEAUFORT SEA 1973 - 1987

	ISLAND NAME	EXPLORATION OPERATOR	YEAR COMP-LETED	WATER DEPTH m	FILL QUANTITY: m^3	ISLE TYPE	REF.
1	Tmmerk B-48	Esso Resources	1973	3.0	180,000	SBI	1,8
2	Adgo F-28	Esso Resources	1973	2.1	36,000	SRI	8
3	Pullen E-17	Esso Resources	1973	1.5	65,000	GHI	
4	Unark L-24	Sun Oil	1974	1.3	44,000	GHI	
5	Pelly B-35	Sun Oil	1974	2.0	35,000	BCI	
6	Netserk B-44	Esso Resources	1974	4.6	306,000	SRI	
7	Adgo P-25	Esso Resources	1974	1.5	27,000	SRI	
8	Adgo C-15	Esso Resources	1975	1.5	70,000	GHI	
9	Netserk F-40	Esso Resources	1975	7.0	291,000	SRI	
10	Sarpik B-35	Esso Resources	1975	4.3	118,000	GHI	
11	Ikkatok J-47	Esso Resources	1975	1.5	38,000	SRI	
12	Kugmallit H-59	Esso Resources	1976	5.3	160,000	SRI	
13	Adgo J-27	Esso Resources	1976	1.8	69,000	SRI	
14	Arnak L-30	Esso Resources	1976	8.5	1,070,000	SBI	
15	Kannerk G-42	Esso Resources	1976	8.5	1,070,000	SBI	
16	Isserk E-27	Esso Resources	1977	13.0	1,561,000	SBI	8
17	Issungnak 0-61	Esso Resources	1979	19.0	4,100,000	SBI	3,38,7
18	Issungnak 2-061	Esso Resources	1980	19.0	1,000,000	SBI	NOTE 1
19	Alerk P-23	Esso Resources	1981	10.5	2,360,000	SBI	61,58
20	N. Protection Island	Dome Petroleum	1981	4.6	2,000,000		NOTE 2
21	W. Atkinson L-23	Esso Resources	1981	7.5	955,000	SBI	
22	Tarsiut N-44	Gulf Canada	1981	21.0	1,800,000	CRI	13,10,11
23	Uviluk P-66	Dome Petroleum	1982	29.7	1,900,000	SSDC	65
24	Itioyok I-27	Esso Resources	1982	15.0	1,940,000	SBI	
25	Nerlerk B-67	Dome Petroleum	1983	45.1	4,000,000	SSDC	64,65, 66,67
26	Kogyuk N-67	Gulf Canada	1983	28.1	1,450,000	SSDC	42,65
27	Kadluk 0-07	Esso Resources	1983	14.0	436,000	CRI	37
28	Amerk P-09	Esso Resources	1984	26.0	1,162,000	CRI	6,5,61
29	Adgo H-29	Esso Resources	1984	3.0	75,000	SBI	
30	Nipterk I-19	Esso Resources	1984	11.7	931,000	SBI	
31	Tarsiut P-45	Gulf Canada	1984	26.0	343,000	CRI	4,12
32	Minuk I-53	Esso Resources	1985	14.7	1,986,000	SBI	
33	Amauligak I-65	Gulf Canada	1985	31.0	1,408,000	CRI	43,19
34	Kaubvik I-43	Esso Resources	1985	17.9	552,000	CRI	45
35	Arnak K-06	Esso Resources	1985	7.2	700,000	SBI	
36	Amauligak F-24	Gulf Canada	1987	32.0	2,000,000	SBI	69

LEGEND: SRI - Sandbag Retained Island
GHI - Island Constructed Using Dumped Gravel
SBI - Sacrificial Beach Island
CRI - Caisson Retained Island

NOTE: 1. Refurbishment of existing island.
2. Protection island was constructed in McKinley Bay as support facility, not an exploration island.

In 1983 Esso Resources Canada (ERCL) and Gulf Canada Resources (GCRL) initiated a co-operative study to synthesize the experience gained with hydraulic fill construction. This Paper, which presents an overview of the Canadian experience, is a summary of the findings of that joint project.

ISLAND TYPES

Sandbag Retained Islands

The sandbag retained island (SRI) was one of two types of islands constructed during the first 5 years of Beaufort exploration. A protective dike around the perimeter of the island consisting of a gravel berm below the water topped with sandbags to about a metre above mean sea level (msl) was used to contain the fill. The method is attractive in areas where fill is scarce and must be hauled some distance. The central core may be filled with either sand (7 cases) or gravel (12 cases). In most instances the fill is placed using clamshell grabs to transfer the fill from the transport barge to the island core. Some aspects of this type of islands are discussed in reference [1] with regard to the Netserk and Immerk projects.

Gravel Hauled Islands

Three gravel hauled islands (GHI) were constructed during the winter seasons from 1973 to 1976. Water depths ranged from 1.5 m to 4.3 m. To construct these islands, gravel was trucked from the near shore Ya-Ya lakes area over ice roads and then placed through slots cut into the offshore ice. Sandbags and filter cloth were placed during the summer season for erosion protection. However, Government discouraged use of the limited onshore gravel reserve in order to preserve the material for other uses. Since alternative gravel sources were too far away to be cost effective, this type of island construction ceased in the Canadian region of the Beaufort Sea.

Sacrificial Beach Islands

Sacrificial beach islands (SBI) consist of a drilling surface surrounded by long gradual beaches. As the name suggests, there is no expectation that the island will have other than a short life. The island geometry provides a working surface in the order of 100 m diameter that is resistant to wave action for the latter part of the open water season and to ice attack for approximately one year. This is achieved by using very flat perimeter beaches (typically 1V to 25H) where the island penetrates the waterline. Depending on the gradation of the beach sand sandbags may be used to protect the steep slope at the inshore end of the beach. Erosion of the beach is expected during the open water season, but the beach is sized for the probable storm scenario so that it is still adequate after erosion. In winter, the beaches cause the ice to fail in bending some distance from the work surfaces creating a protective ice rubble field around the island. This type of island has been discussed in the literature [1, 2] and such islands have been built in as much as 19 m of water (Issungnak) [3, 7].

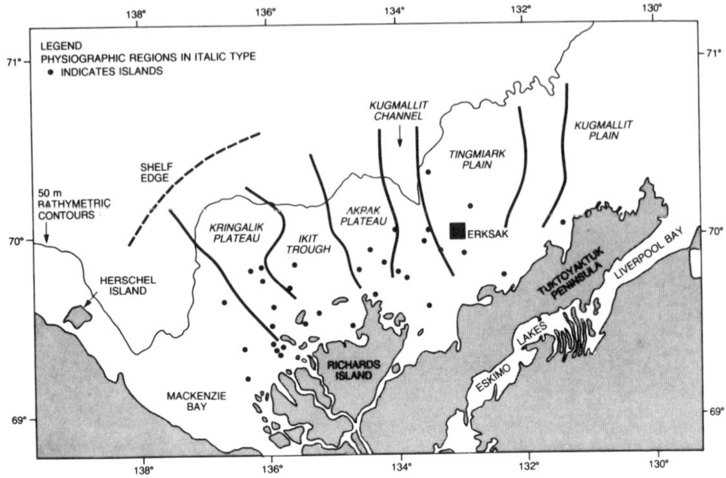

Figure 2
ISLAND LOCATIONS, PHYSIOGRAPHIC REGIONS AND ERKSAK BORROW PIT

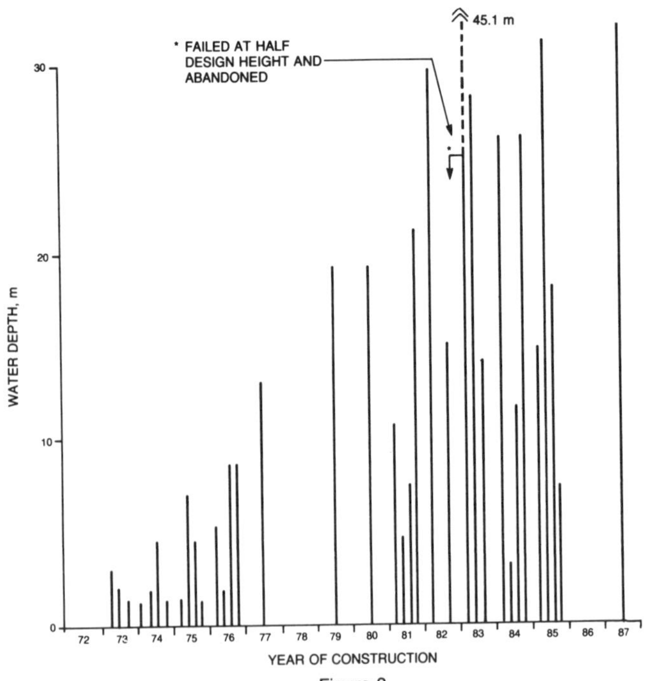

Figure 3
HISTORY OF ISLAND SIZE

Caisson Retained Islands

Caisson retained islands (CRI) use a caisson to substantially reduce the fill volume in the island but still retain sufficient sand mass in the core to resist ice load. As exploration moves to increasing water depths, the savings in sand volume become progressively larger. Moreover, when sand must be hauled over a long distance (eg: > 100 km at Tarsiut P-45,[4]) the advantage increases. The motivation for employing caisson islands was strictly economic - it was argued that halving the fill volume would halve the dredging cost [5]. Although this argument is not strictly correct, CRIs are significantly less expensive than SBIs in deeper (>20 m) water.

Caissons constructed to date are designed for exploration and are intended to be movable and used at several sites with differing water depths. This requirement for use in different water depths is met by placing the caisson on an underwater pad - called a berm - which provides the appropriate set-down depth regardless of water depth at a particular site. The berm is constructed with hydraulically placed granular material which is not compacted. The inner core of the caissons are also infilled with hydraulically placed sand.

There are three CRI systems that have been used in the Canadian Beaufort Sea to date, each substantially different for the other. The first CRI to be designed was that for Esso Resources Canada [5, 6, 7], Fig. 4. This caisson comprises an octagonal ring with each face of the ring being an independent unit. The units interlock and are held together with post-tensioned cables. The CRI is founded at 9 m and has 3 m of freeboard to island surface with wave or ice deflectors providing another 5 m of overtopping protection. The core is filled with sand or gravel which provided the drilling surface. The unit has been installed at three locations to date (Table 1).

The first CRI to be constructed was Tarsiut N-44 [9, 10, 11]. This CRI comprised four independent concrete caissons that were founded at 6.5 m while the berm extended down to the seabed at 21 m. The core was infilled with sand which provided the drilling surface. The units were only used at Tarsiut N-44, in 1981-3, and have since been abandoned.

The other CRI deployed in the Canadian Beaufort Sea is the Gulf Molikpaq [4, 12], Figure 5. Unlike the other two CRI's, the Molikpaq is a monolithic caisson with an integral deck which supports the drilling modules. To date, the Molikpaq has been deployed at 3 sites (Table 1).

Ballasted Barge Islands

Ballasted drilling barges are comparable in many respects to CRI's. Two such barges have been employed to date, one in 1974 at the Sun Oil Pelly B-35 Island, and one, the Canmar SSDC [13], which has a strengthened hull for extended arctic drilling. This latter unit has been used at two island sites in the Canadian Beaufort (Table 1). The SSDC used sandfill berms as it is founded at 9 m and was deployed in approximately 30 m of water at both Uviluk [14] and Kogyuk.

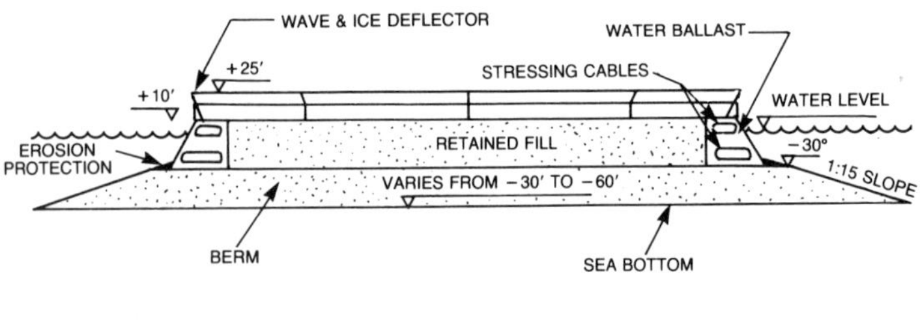

Figure 4

ERCL CAISSON ISLAND

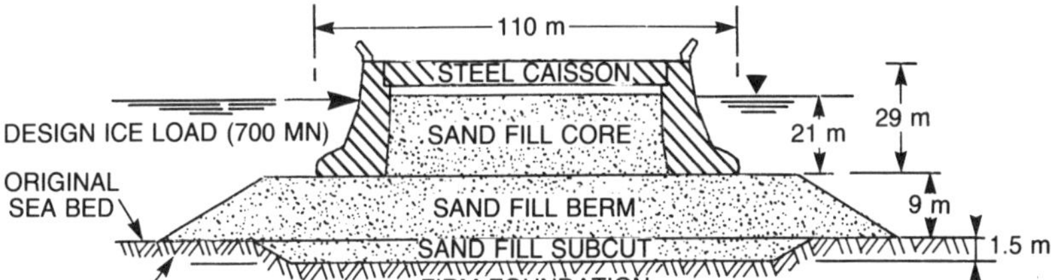

Figure 5
BEAUDRIL MOLIKPAQ

DESIGN

The purpose of design is to consider the loadings on the island and to develop the appropriate combination of island geometry and materials that can withstand the loads within a particular margin of safety at least cost. Conventionally, this is done by considering the factors independently and then making calculations on the overall system. Several iterations are usually required. For the present purpose it is sufficient to discuss each item separately without describing the optimization stage which is well understood.

Design Standards

The design and construction of offshore structures is usually carried out according to a code of practice (or its equivalent) that varies with the jurisdictional authority. The situation in Canada is rather different as no national code of practice has been published nor has any external code been approved for general use. The work reported in this Paper has been approved by the Canadian Oil and Gas Lands Administration (COGLA) on a case by case basis.

Exploration for hydrocarbons is a site specific activity of limited duration. In the Beaufort Sea most locations have only a single well, although a few sites have been considered for two or more wells. The required service life of the island is short, generally, less than two years. In this context, the advice of the British offshore code [15] is particularly pertinent:

> "...... If the intended service life of the structure is significantly less than thirty years, it may be appropriate to use an extreme design environmental condition having a recurrence period of less than 50 years."

This greater risk approach has become the norm for exploration structures in the Canadian Beaufort and is combined with monitoring programs to provide advance warning of possible overload or deviation between projected and actual performance [16].

Ice Loads

As noted in the introduction, ice cover is a major environmental feature of the Beaufort Sea. This ice may produce considerable horizontal loads on stationary structures. Ice loads used for design are influenced by two major factors: risk of exposure to thick ice; and, nature of the island.

In the shallow, near shore areas where the ice is landfast for much of the winter and extreme thicknesses of ice do not occur, design might be based on an ice sheet 2.0 m thick moving against the island/structure. The landfast ice extends (in an average winter) out from the shore to approximately the 20 m isobath. Beyond the 20 m isobath and extending as far as the shelf-edge a band of intermittently moving ice (called the shear zone) separates the near-stationary landfast ice from the constantly gyrating polar pack. Ice in the shear zone will normally be 'first-year' ice typically less than 2.0 m thick but there is a finite probability of incursions of older ice, called 'multi-year' ice in the form of floes that may be tens of miles in diameter with thicknesses typically between 3 to 6 m. Floes may also contain line features, known as 'pressure ridges' with substantially greater thickness of ice than the average ice floe thickness. Thus, the expected or design ice thickness to be resisted by the island depends very much on its location.

Ice loads may be computed from the design ice/structure interaction scenario using several different methods as there is not, as yet, an agreed practice. An excellent survey of the commonly used approaches is given in [17]. The important point to appreciate is that the expected load for any given thickness of ice depends on the structure itself with regard to both scale and failure mode.

Scale effects are significant in ice. The crushing pressure of thin ice in a laboratory tank against a model structure of a few centimetres width may exceed 5 MPa. Yet, the same type of ice may fail against a full size structure whose width is 100 m with a crushing pressure of less than 1 MPa even where geometric similitude has been preserved. Thus a two order of magnitude change of scale produces a five fold reduction in unit loading. Further data and discussion on this most important phenomenon may be found in [18].

The geometry of the island structure also has considerable influence on the ice load. If the ice/island interface is near vertical (eg; a CRI) then ice failure will tend to pure crushing whereby intact ice is fractured to a granulated material which is then extruded from the failure zone (see [19] for photographs of this process). On the other hand, if the ice/island interface slopes as in a SBI, the ice will tend to fail in bending with blocks of ice being pushed up the beach. The load to cause bending failure in ice is generally less than that required to cause crushing except in the case of very thick (> 12 m) sheet ice. Geometry also affects the ice/island interaction in the horizontal plane as, for example, in promoting splitting of ice floes (which lowers ice forces) or in the reduction in the importance of edge effects (which raises ice forces).

Overall, the evaluation of design ice load is a complicated process with scale effects, geometric effects, rate effects superimposed on uncertainty with regard to ice features. Although these factors might be expected to lead to considerable variation in opinion as to appropriate design loads, in fact, a consistent pattern emerges as may be seen on Table 2 which summarizes the design ice loading for four island projects. It illustrates that design load increases with distance offshore (water depth), and reflects the island type. Design ice loads are also comparable between operators. Probalistic studies suggests the load levels are associated with about a 4 per cent annual probability of being exceeded.

A final aspect of ice loads is the issue of load cycling. Several types of ice failure mode produce cycling of ice load and pure crushing of ice in particular produces systematic load cycling with an amplitude in the order of 50 per cent of the peak load [19].

TABLE 2

Project	Water Depth	Island Type	Design Ice Load
NIPTERK I-19	12 m	SBI	170 MN
AMERK P-09	26 m	CRI	400 MN
TARSIUT P-45	25 m	CRI	500 MN
AMAULIGAK F-24	32 m	CRI	700 MN

Frequencies of cycling are commonly in the order of 1 Hz while durations of such ice failure events may be of the order of 30 minutes. As will be readily appreciated, this type of loading has particular significance for hydraulically placed sandfills.

Sandfill Strength (Density)

The issue of sandfill density is one that attracts much attention in the context of hydraulic fills because sand strength is a strong function of sand density and hydraulically placed sandfills have a reputation of not being dense [20] unless compacted. The reality of underwater hydraulic fills is that they can never be made very dense even after extensive compaction. Because of this reality, comprehensive laboratory sand testing programs (approaching a total cost of $1M) were sponsored by the exploration operators to evaluate the drained, undrained and cyclic strength of Beaufort sands as functions of sand density and gradation.

A fundamental aspect of sand behaviour is dilatant response to stress. Dilatancy is the increase in volume with shear strain and is the mechanism by which sands may exhibit high friction angles [21]. Dilatancy is very desirable in engineered sandfills and is usually obtained by ensuring the sand is dense. However, it became apparent at an early stage of testing that loose clean Beaufort Sands (less than 5% silt) were dilatant, commonly at relative densities loose as low as 20%. But, there were many inconsistencies in sand behaviour as a function of relative density [22].

An alternative to relative density, the state parameter, was introduced [22] as a characterizing parameter for sandfill. The state parameter concept is an empirical adaption of critical state soil mechanics [23] and uses the distance from the steady (critical) state in e-p' space as a normalizing parameter for sand. It was found that both peak dilation rate and peak friction angle were strongly a function of the state parameter (ψ), regardless of sand gradation, mineralogy, silt content, etc. [22, 24]. Figure 6 shows both the definition of state parameter and its relation to peak friction angle and dilatancy. Although the state parameter approach is founded in

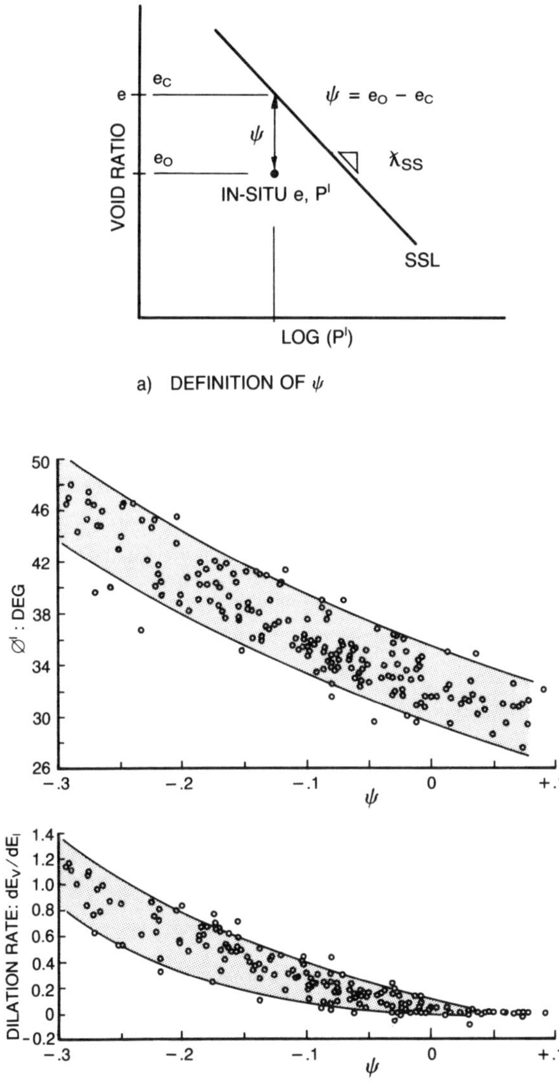

b) RELATIONSHIP BETWEEN Ø' & ψ FOR 20 SANDS

Figure 6

DEFINITION AND USE OF STATE PARAMETER

critical state soil mechanics it has the implicit operational form of stress-dilatancy theory [25]. These theories together provide theoretical assurance that the state parameter is a proper approach to the engineering of sandfills.

State is not the only parameter which influences the constitutive behaviour of sand. A second parameter, usually referred to as fabric, is equally important [26, 27]. The fabric parameter describes the relative orientation of sand particles and a whole range of fabrics can exist in sand at any particular density [28]. It has been shown, [26, 29] that fabric can change the shear stress - shear strain relationship by more than a factor of two and the peak friction angle by several degrees. However, the routine measurement of fabric has not been achieved and the avoidance of fabric in sand engineering necessarily introduces significant constraints on the precision of calculations.

Stability calculations of hydraulic fills placed considerable reliance on the 'steady state' concept [30]. This concept indicates that undrained strength of sand at large strain is a simple (logarithmic) function of initial void ratio regardless of whether the sand is monotonically or cyclically loaded. The steady state strength may be treated as an assured minimum value [31]. Stable behaviour of a sandfill may be obtained, according to this steady state concept, by ensuring the sand is dilatant, since a dilatant sand has a residual undrained strength greater than the peak drained (or partially drained strength).

Ice Resistance

As ice is a principal environmental load, providing adequate resistance to this loading is a major objective of island design. To date, design has employed upper-bound methods whereby a kinematically admissible failure mechanism is postulated and the work (force) required to drive this mechanism is compared to the ability of the mechanism to dissipate energy plasticly. Upper-bound approaches require consideration of a variety of mechanisms to find that with the least ratio of external work/plastic dissipation. Although an infinity of admissible mechanisms may exist, they can broadly be classed into two groups: those only involving the sandfill; and, those also involving foundation soils. The first group are often referred to as 'decapitation' failures by analogy to the top of the island moving as a block, shown as 'Mode 1' on Fig. 7. The second group are correspondingly referred to as 'basal' failures because of the importance to the foundation to this mechanism, shown as 'Mode 2' on Fig. 7.

'Mode 1' type failures usually display a logarithmic spiral velocity discontinuity with a factor of safety typically 15% less than that calculated by a planar sliding analysis. A factor of safety of 1.5 defined on the frictional peak stress ratio is usually sought for

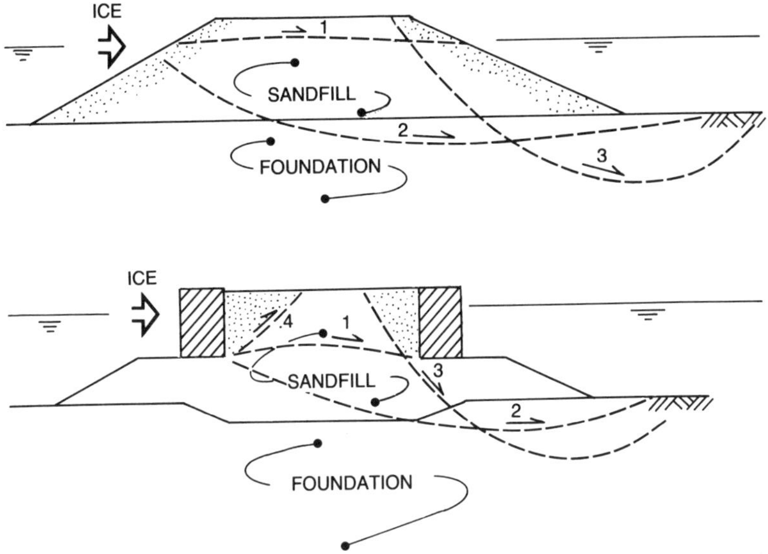

NOTE: 1 Decapitation failure by ice loading
2 Basal failure by ice loading
3 Slope or edge failure by self weight
4 Local passive failure

Figure 7
EXAMPLES OF KINEMATICALLY ADMISSABLE MECHANISMS FOR ISLAND DESIGN

'Mode 1' type failure. In the case of SBI platforms and other islands where the earthworks pierce the water surface, sufficient effective mass above the failure surface usually exists because of other design considerations. Experience has also shown that cyclic loading is not an issue with SBI type platforms.

Analysis of 'Mode 1' failures is quite complex in the case of CRI platforms because of the two issues of soil-structure interaction and cyclic loading. Although a factor of safety of 1.5 is still sought it is necessary to consider whether the caisson can enforce the unified action of 'Mode 1' or whether local failures ('Mode 3' on Fig. 7) might dominate. Both drained and steady-state analyses are usually carried out and may be supplemented with finite element analysis and physical model (centrifuge) tests [4].

Design action against 'Mode 1' failures of CRIs fall into two categories: i) structural and ii) site specific. Structural design of the caisson involves selecting the appropriate caisson geometry, set-down depth, structural system, etc. to meet the design load criteria. Discussion of these issues may be found in [32] and [34]. Once the caisson exists the question of its adequacy at a particular site arises. If the original design criteria are more stringent than the site specific ones then no further action is required. However, if the site specific criteria are more stringent than those assumed in the original design, then upgrading is required. There are three possible remedial measures:

- improve the sand properties;
- increase the effective stress regime;
- modify the sand.

Improvement of sand properties can be achieved either by using more permeable sand, which improves the time factor of the sand response to loading or by compacting the sand. More pervious sand was used at Kadluk O-07, Tarsiut P-45, Amerk P-09, Amauligak I-65, Kaubvik I-43 and Amauligak F-24. The core sand was compacted at Amauligak F-24.

Increasing the effective stress regime can be achieved by raising the set-down depth and/or installing a core dewatering system. Both actions reduce the internal phreatic surface with respect to the top of the core. Both actions were employed at Amauligak F-24 which used a 5 m raised set-down elevation and ten 200 mm dia. dewatering wells with a total of 225 hp installed pumping capacity.

Sand modification can be achieved either by structural grouting or freezing. To date neither technique has been used in any Canadian island project.

Basal failure, 'Mode 2', has not been found as critical as gravitational failure because the sliding plane is longer and flatter. However, concerns about basal sliding arose in the case of Tarsiut N-44, and Kadluk O-07 when pore pressures in the foundation clays continued to rise after construction had been completed [35, 36, 37, 38]. It was subsequently determined that the phenomenon was one of delayed peak pore pressure, not progressive failure.

Because of this phenomenon, design action for 'Mode 2' failures has continued to use short term (undrained) strengths. No allowance is made for consolidation induced strength gain. A factor of safety of 1.5 is usually taken as appropriate for 'Mode 2'.

The previous considerations relate to the overall action of the CRI. In addition local passive failure, 'Mode 4', is considered with the requirement that the factor of safety be at least equal to 'Mode 1'.

Gravitational Stability

Offshore islands impose substantial loads on the foundations because of self-weight. Although ice loading attracts attention because of the large horizontal forces involved, the fact remains that gravitational loads are at least three times greater than ice loads. Consideration of self-weight failure, 'Mode 3' on Fig. 7, is an important aspect of island design.

The approaches adopted by ERCL and GCRL have tended to diverge with regard to foundation treatment and its effect on design sideslope. The required construction effort and cost of the island/berm is strongly a function of sandfill volume, as halving fill volume may allow substantial reduction in the dredging fleet. But, the process is not one of proportionality because dredges tend to be charged on a per season basis. Typical trailer dredge production capabilities in the Canadian Beaufort Sea over an average haul of, say, 15 km, are 2.0 million m^3 sand/dredge/season. With such a production rate and a minimum commitment of one dredge it is quite clear from Figure 8 that it is irrelevant whether one builds a 1:5 or a 1:12 sloped island for island heights in the order of 10 m. The situation is much changed when island heights exceed 20 m. Again reference to Figure 8 shows

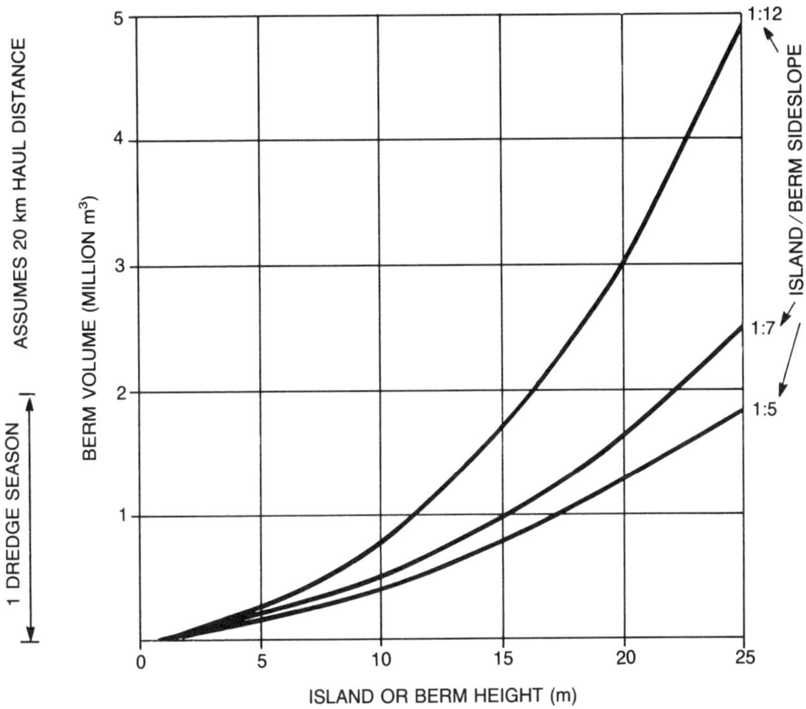

NOTE: Assumes 6000 m^3 hopper dredge @ 20 hrs/day
For 100 day open water season.

Figure 8

EFFECT OF SLOPE & HEIGHT ON ISLAND (BERM) VOLUME

that for these island heights a change from 1:8 to 1:12 slopes produces an incremental volume of approximately 1 million m^3. This requires the chartering of an additional dredge, or the lengthening of the construction period to two summer seasons. This marked difference in consequence of sideslope angle for project economics has dominated the issue of foundation treatment. Foundation treatment has been carried out at Amerk (ERCL), Kogyuk, Tarsiut N-44, Tarsiut P-45, Amauligak I-65 and Amauligak F-24 (all GCRL). Sideslope flattening was undertaken at Kaubvik I-43 (ERCL).

Design factors of safety have increased substantially with the increasing water depth where consequences of failure are very much more serious. In shallow water, where 'failure' amounts to slope slumping of no practical consequence to the function of the island, factors of safety less than 1.3 have been used. On the other hand, the stress concentration and monolithic nature of the Molikpaq caused factors of safety of 2.0 to be adopted against foundation failure by self-weight. Extensive monitoring of foundation performance has also been undertaken at CRI sites (Amerk P-09, Kaubvik I-43, Tarsiut N-44, Tarsiut P-45, Amauligak I-65, Amauligak F-24) to provide confirmation of predicted foundation performance.

Sand Permeability

The cyclic loading action of storm waves on island sandfills may fluidize the sand, which can then "flow", slump or become rapidly eroded. Cyclic inducing of excess pore water pressure has been analysed using one proposed method [48] for a typical 8:1 side slope berm (actually Amerk) for a typical design storm. The results are summarized on Fig. 9 in terms of the coefficient of consolidation, C_v, plotted against the predicted normalized residual porewater pressure ratio associated with the design storm.

As shown on Fig. 9, material type expressed in terms of C_v is the major factor in determining susceptibility to wave induced liquefaction. Wave induced excess pore pressures may be expected to be minimal provided C_v > 1000 cm^2/sec for the design Beaufort storm, regardless of sand density. This may be compared with a previous analysis [39] that concluded $K>10^{-4}$m/sec for the same reasons.

The coefficient of consolidation for sands is difficult to measure directly using standard laboratory equipment. The approach adopted was to measure the individual components, permeability and compressibility. Permeability, was measured either in triaxial test equipment or in a permeameter. Compressibility was measured in an oedometer. The range of C_v values for several Beaufort Sea sands are indicated on Fig. 9, while Fig. 10 shows the effect of grain size and silt content on C_v for Erksak sand.

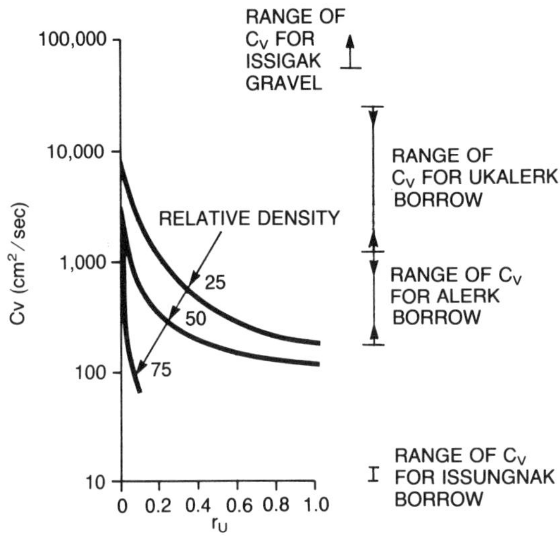

NOTE: Predicted by method in ref [48]

Figure 9
WAVE INDUCED EXCESS PORE PRESSURE

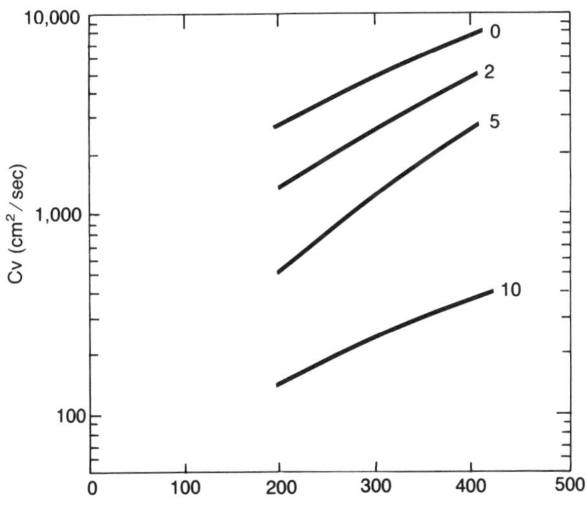

COEFFICIENT OF CONSOLIDATION

Figure 10
HYDRAULIC PROPERTIES OF ERKSAK SANDS

Specifications

The specification for sand quality used on the project must result in both an adequately dilatant and permeable sand in-situ. In fact, as will be subsequently shown, both objectives are complementary as the in-situ state of sand (which controls dilatancy) is linked with the permeability of the as-placed sand for a given construction method.

A specification for use during construction must be given in terms of variables easily observed and measured during construction. This conventionally implies sand gradation, which was adopted for the island fills reported here. It became recognized that a coarse sand could tolerate more silt content for any particular mechanical properties. Specification of sand quality therefore developed into a smooth range of median grain size/silt content combinations which were subdivided into material classes, Fig. 11. This figure shows a typical specification used for a zoned fill-quality berm.

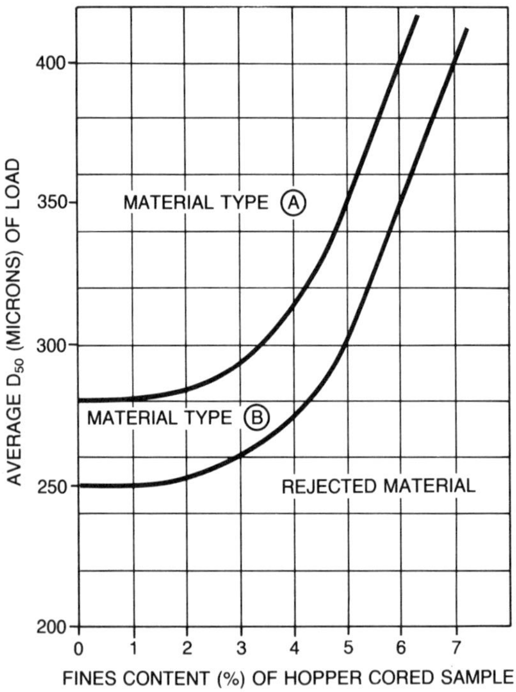

Figure 11
GRANULAR FILL SPECIFICATION

CONSTRUCTION

Once the desired berm/island geometry and sand properties have been defined, other engineering issues become important. These may be conveniently grouped into four topics:

i) sand availability and location with respect to project site
ii) construction technique including vessel selection and scheduling
iii) achieved in-situ sand density, island slopes and erosion protection
iv) potential for remedial works, if required.

The issues are all interrelated as, for example, both sand type and placement technique affect the achieved fill density. However, it is convenient to discuss each topic in turn before considering how they interelate.

Sand Availability and Location

Although sand is quite widespread in the Eastern portion of the Canadian Beaufort Sea, it is virtually absent in the Western part; as a rule of thumb, the 136°W longitude provides a divider as to those regions which may be expected to contain sand and those that do not (greater definition is contained in the physiographic studies [40]). Islands, however, have been constructed throughout the Beaufort Shelf.

Although many earlier projects used anchored dredges and local borrow, the trend since 1980 has been to use one particular borrow area and trailing suction hopper dredges. This borrow area is variously known as the Ukalerk/Erksak pit and its location is shown on Fig. 2.

The Erksak pit came into use with all three oil companies because it has shown an average grain size some 50% coarser than is typical for Beaufort Shelf sands. The coarsest, cleanest possible sand is desirable to achieve the maximum as placed density, steeper side slopes and, maximum resistance to wave induced liquefaction.

The in-situ gradation of the Ukalerk/Erksak pit is summarized on Fig. 12 in terms of median grain size, D_{50}, and silt content. Also shown for comparison are data from other, typical sand sources in the Beaufort Shelf. It can be seen from this figure that the Ukalerk/Erksak sand exhibits a median grain size of 300 - 400 microns and a silt content of <8% in the pit. Commonly at least half the silt content is lost during construction because of the washing action of the dredge.

Vessel Type and Selection

Dredges used to date may be classed into one of three types: anchored suction dredges, anchored cutter suction dredges and trailing suction hopper dredges. Figs. 13 and 14 illustrate a suction dredge and a hopper dredge, respectively.

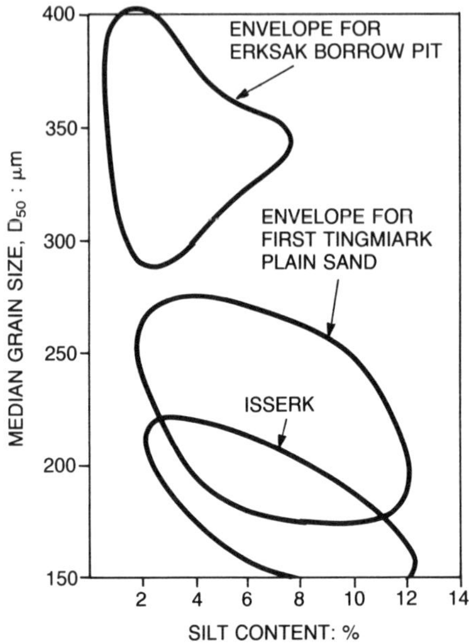

Figure 12
SUMMARY OF IN-SITU GRADATION OF BEAUFORT SEA CONSTRUCTION SANDS

An anchored suction or cutter suction dredge is moored during dredging operations and utilizes a large pump mounted on a barge or a vessel. The two types of dredge differ in that a mechanical cutter is provided on a cutter suction dredge to assist in fluidizing sand in the borrow pit; the suction dredge relies on the eroding action of the suction pipe sometimes supplemented with water-jetting. Sand, once fluidized is sucked up from the pit through the pumps and transported as a slurry in a pipeline to the project site. The pipeline may be either anchored itself at the project site providing a central spigot arrangement or alternatively, the pipeline may be connected to a construction barge having a means of directing the flow of sand. Anchored dredges require sand deposits at least 5 m thick to be effective, and these deposits should be within about 1 km of the project site. But, the delivered sand quality is not unduly sensitive to thin layers of overlying clay.

A trailing suction/hopper dredge is a vessel with the prime function of recovering material from the seabed and transporting it. A typical layout of such a vessel is shown on Fig. 14. Sand is recovered by the dredge pulling the drag arm across the seabed and pumping the

Figure 13
ILLUSTRATION OF SUCTION DREDGE

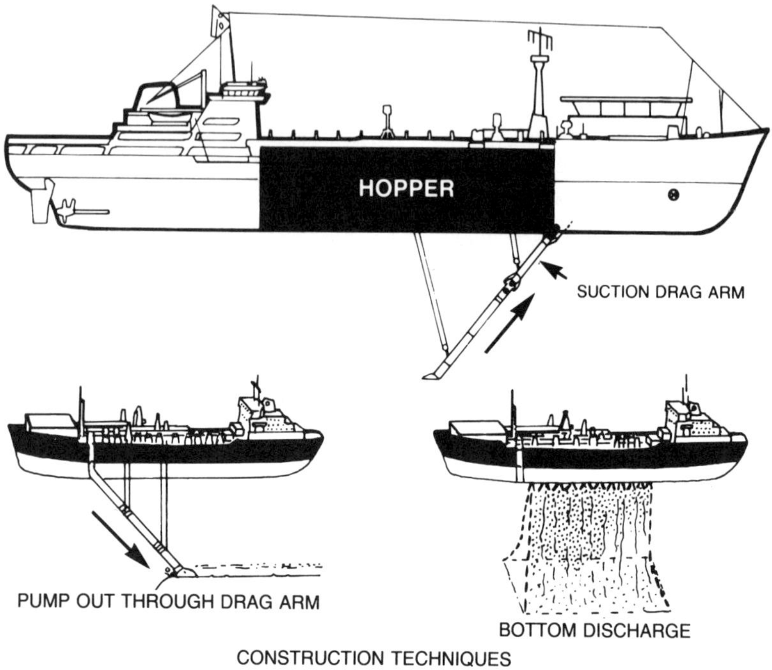

Figure 14
DETAILS OF TRAILER-HOPPER DREDGE

slurried sand into the hopper. Once the sand has been moved to the construction site it may be deposited in three ways. First, the dredge may unload the hopper using valves or doors in the base of the hopper - referred to as 'bottom discharging'. Second, the sand in the hopper may be slurried using the dredging pumps with supplementary water and discharged back down a drag arm. Thirdly, the sand may be slurried and bow discharged into a floating pipeline or through a nozzle/diffuser.

Trailer dredges operate most efficiently if sand recovery is continuous. The areal extent of the borrow should therefore typically be greater than 0.3 by 2 kms. Such sand deposits need not be thick and deposits as thin as 2 m have been worked in the Canadian Beaufort Sea. The sand deposits must be exposed at the seabed as even 0.1 m layer of overlying clay significantly degrades delivered sand quality. The clay can be stripped from potential borrow pits, but such stripping may severely affect project economics. Alternatively, abandoned islands have been used as borrow sources with the dredges either sailing around the island or anchored against the island slopes.

Sacrificial Beach Construction Method

Originally, SBI's were constructed using a stationary dredge using sand from the seabed close to the island. This approach was used in the construction of Issungnak [3]. A new construction method pioneered by ERCL at the Nipterk SBI used a trailing suction hopper dredge to haul fill to the site. Bottom dumping techniques are utilized to an elevation of approximately 8 m below mean sea level (bmsl) at which time draft restrictions prevent any further deposition. Bow and side discharging technique are then used to build up the center of the island. The drilling surface is constructed by bow discharging through a nozzle and then using land equipment to level the surface to the correct elevation. The beaches are extended by depositing material directly onto the desired area through a diffuser attached to the bow of the dredge. The development of the diffuser allows for the build up of the surface without the use of a pipeline, which is preferrable since pipeline use is weather sensitive because of hook-up requirements.

CRI Berm Construction - ERCL

ERCL carried out a comprehensive model study on the influence of sand placement technique on berm geometry. This work and the field trials that followed it are reported in detail in [41].

The ERCL berm design for their CRI comprises a berm rising from seabed to an elevation of 9 m bmsl. Sideslopes of 1:8 are desired if the foundation is sufficiently strong; this has not been the case at either Amerk or Kaubvik and berm sideslopes have been flattened to 1:15 in the lower zones so as not to overstress the clay. Excavation of soft clay was also carried out at Amerk.

The ERCL berm construction technique is based entirely on the bottom discharge technique. This technique - which provides excellent densities as subsequently documented - has the tendancy to produce rather flat sideslopes, less than 1V:20H if used indiscriminantly. The sequence developed by ERCL overcomes this difficulty by following steps:

i) the dredge is orientated radially to the island centre during discharge.

ii) a sequential ordering of discharge locations is followed with two discharges (approximately 12000 m^3) per location.

iii) the radius of the desired dump zones is a function of both water depth and berm crest diameter. In the case of the Amerk and Kaubvik islands, a dumping radius of 105 m was found effective in achieving the desired slopes.

iv) topping-off the berm is conducted with a much reduced dumping radius.

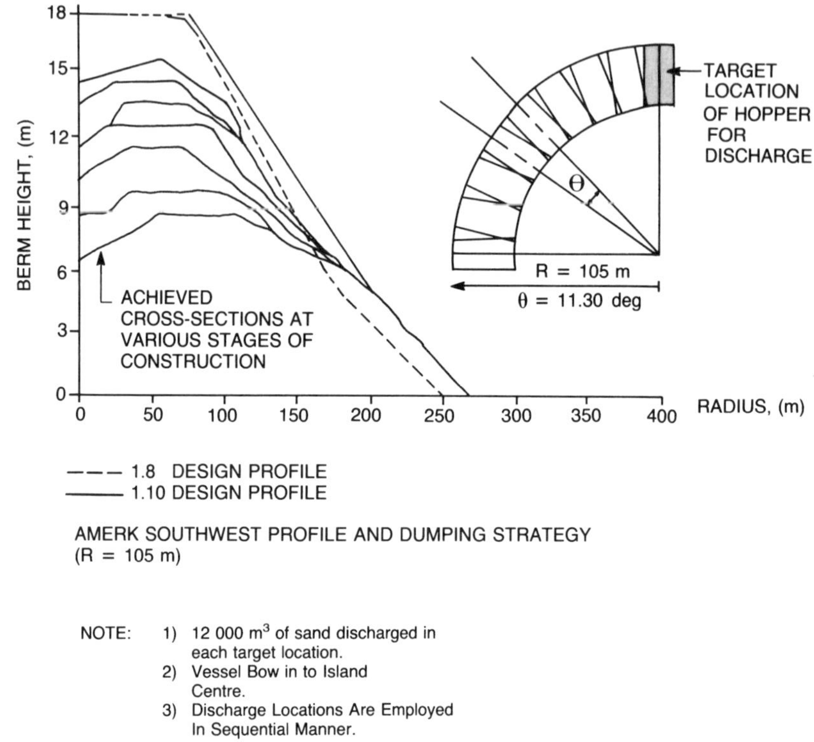

Figure 15
ERCL BERM CONSTRUCTION TECHNIQUE

The sequence of operations is shown on Figure 15 together with the achieved berm profile.

CRI Berm Construction - GCRL

The technique used by Gulf for four CRI berms (Kogyuk N-67, Tarsiut P-45, Amauligak I-65 and Amauligak F-24) amounts to an extension of the technique developed by Volker-Stevin [11] for the first caisson project, Tarsiut N-44. The original Gulf technique has been previously presented in some detail [42], but has subsequently received significant improvements [43]. The Gulf technique, in both original and upgraded form, involves sand discharge by both pump-out and bottom valve methods. Only trailing hopper dredges are employed.

The technique is based on constructing the berm in several "lifts". Each lift starts with the construction for a perimeter bund by pump-out discharge, the drag arm being positioned about 2.5 m above the sand surface and within a narrow horizontal zone dictated by the desired berm geometry. Drag arm positioning is detailed on Fig. 16. Perimeter bunds provide a means of obtaining steep slopes. Once a bund has been constructed, the centre zone is infilled by bottom valve discharge.

Subsequent modifications to the basic technique have arisen for two reasons. First, on both the Tarsiut P-45 and Amauligak I-65 projects simultaneous use of two and four dredges respectively was required and formal scheduling was necessary. Second, it was observed that attempting to fill in a low spot by bottom discharge increased rather than decreased the hole; presumably because of density-flow induced sour currents. The solution to both situations is a formal dump-zone scheduling for bottom valve discharge in a similar exactly similar manner to that used by ERCL. The scheduling sequence used at Amauligak F-24 is shown on Fig. 16.

Positioning and Surveying

The sand placement methods described above all require a precise (standard deviation better than 1.5 m) navigational/surveying system. The navigational network used has been predominantly a Syledis one, which is a shore based UHF radio ranging system. Positional accuracies for this equipment are +-4 m to a distance of 100 km offshore [44], using a minimum of 3, carefully located shore stations. On each vessel a syledis receiver is mounted whose output is transferred to an engineering workstation for enhancement and display to construction operatives (such as the vessels helmsman).

The display technology used was coloured graphic 'cartoons' which display both the current vessel or spigot position (in one colour) and that desired at the particular stage of construction operations (in a different colour). The barge master or helmsman then adjusts the vessel position to maintain an image match between what is desired and what is achieved.

In addition to the real time display of the position information, measurement of progress and verification of compliance with design geometry is necessary. Bathymetric surveys performed by independent survey vessels were to check that material was deposited in the correct location. Dual high frequency sounders (33 and 210 KH_Z), adequately calibrated with velocity profile measurements, and bar checks combined with real time offshore tide gauge measurements enabled accurate and consistent surveys to be performed. Radio modems enable the corrected and reduced data to be transmitted directly from the survey vessel to the post-processing system on the dredge. This data was then used to produce the survey drawings which provide the basis of subsequent construction directives.

CONSTRUCTION METHOD STATEMENT

1 Excavate subcut to remove clay.
2 Infill subcut with sand by bottom discharge method.
3 Construct perimeter sand bund, with dredge drag arm positioned as shown in detail.
4 Infill berm with sand by bottom discharge method.
5 Repeat sequence ③ & ④ for subsequent lifts.
6 Final bund to consist of gravel. Bund crest width to be 16.0 m.
7 Place final infill lift by combination of bottom discharge and pump out to −15.5 m prior to levelling.
8 Berm to be levelled to −16.0 m utilizing modified drag head to remove excess sand.

DETAIL OF BUND CONSTRUCTION METHOD

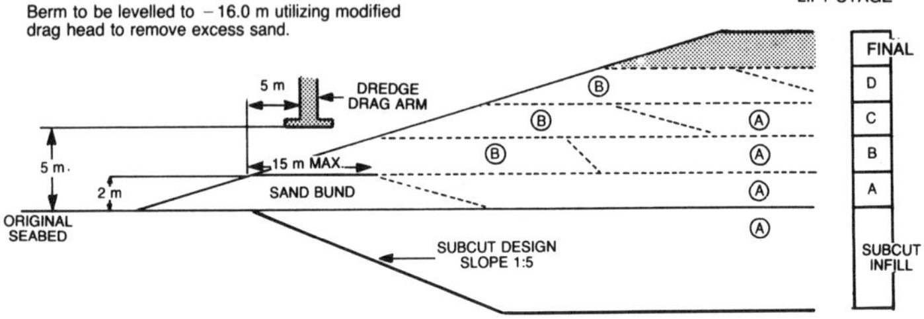

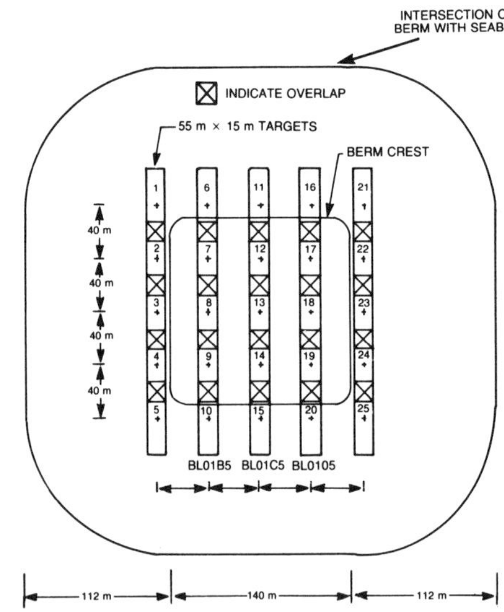

BERM LAYER #1

DETAIL OF BOTTOM DISCHARGE INFILL METHOD

Note:
1. Numbers indicate sequence of infill discharge.
2. Plan shown is typical of infill for other layers.
3. 6000 m^3 per discharge location.
4. A Sand type A. Fig. 12
   B Sand type B, Fig. 12

Figure 16

GCRL BERM CONSTRUCTION TECHNIQUE

Quality Control Methods

Early experience with in-situ densities achieved by purely hydraulic means led to the conclusion that better densities were achieved with coarser sand. In addition, it became apparent from both experience as well as theoretical studies [11] that steeper side slopes could be achieved with coarser sand. Considerable motivation therefore developed to construct with as coarse a sand as possible. Two sources of information on material gradation exist during construction:

- by drilling and sampling the intended borrow pit;
- by monitoring sand gradation on board the dredge.

It is usual to carry out both activities, regardless of whether suction or trailer hopper dredges are involved. Detailed borrow investigations have been made in previous years (eg: that reported in [45]). Recently, most attention has been given, to assessment of what grading of sand has been dredged, which is carried out in real-time on board the vessel. If out-of-specification material is encountered then that sand is wasted rather than placed in the island/berm.

There are a number of sampling procedures and protocols that have been used on the various dredges to date. Samples for gradation analysis have been obtained from a small spigot pipe off the main slurry intake or from sounding tubes in the hopper itself. The first sampling method gives reasonable estimates (when subsequently compared to borehole samples in the placed fill) of median grain size but markedly underestimates the silt content.

Concern over whether sampling methods were indeed representative caused GCRL to mount a drillrig to core the sand in the hopper of the dredge. Gradation analysis of samples was also advanced by using particle counting technology [46]. Together, these methods allow confidence in the material in the hopper; subsequent placement of the fill may result in minor reduction in silt content, which is desirable.

Results of these sand quality monitoring programs show a distribution in both median grain size and silt content; typical examples of these distributions are shown on Fig. 17.

Costs

Cost is an important issue that has received little attention in the published record of offshore island construction.

Some recent experience with the cost of constructing berms for CRI projects is summarized on Table 3, subdivided into accounting items. This table excludes any amortization cost of the caisson system itself and also excludes all costs of operating the drill rig. The cost picture is best understood by a simple picture, Fig. 18.

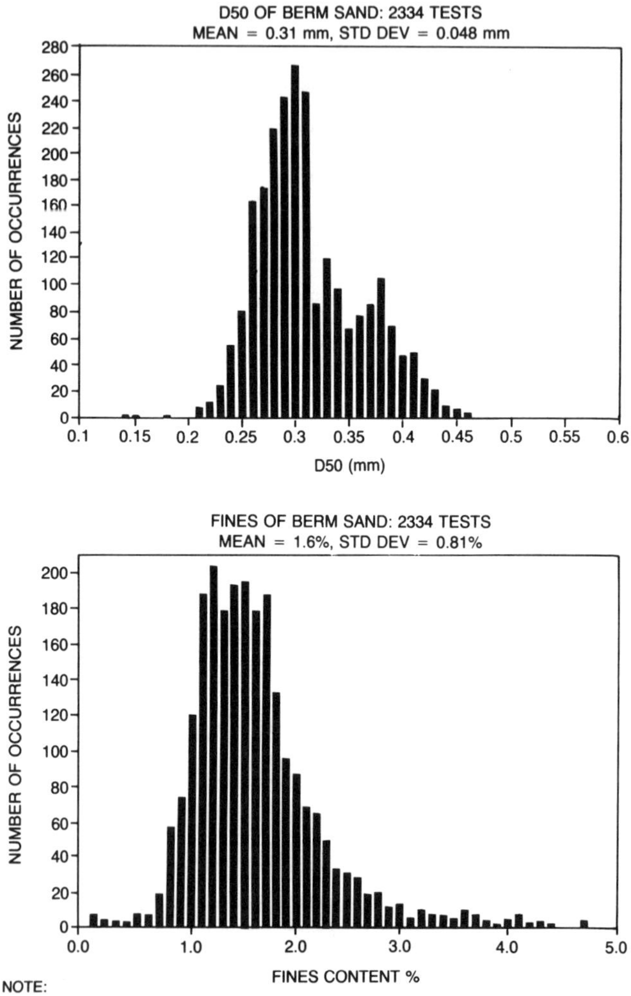

NOTE:

1 21 fines contents were in excess of 5% and are not shown on histogram (maximum was 13.5%).

2 2331 tests on Erksak, 3 tests on Amauligak I-65 sand.

HISTOGRAMS OF SAND QUALITY IN SUBCUT AND BERM

Figure 17

EXAMPLE OF PARTICLE SIZE DISTRIBUTION FROM QUALITY CONTROL PROGRAMS (AMAULIGAK F-24 BERM)

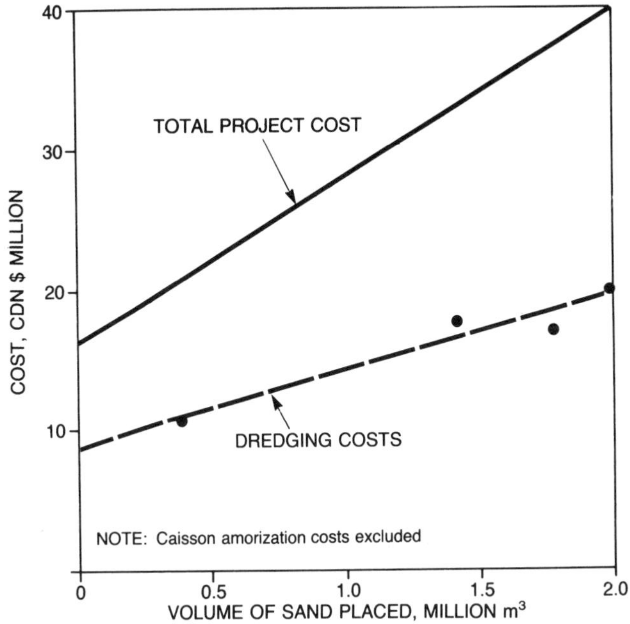

Figure 18
SUMMARY OF RECENT BERM COSTS

TABLE 3

Project / Activity	Tarsiut P-45	Amauligak I-65	Amauligak F-24
1. Berm Design	88	72	195
2. Site Investigation	427	821	805
3. Dredge Charter	10,617	17,631	19,756
4. Quality Control & Surveying	2,608	2,514	1,318
5. Densification & Dewatering	N/A	N/A	2,881
6. Support Costs	7,585	10,029	13,643
7. Post-Construction Verification	103	118	177
8. Miscellaneous Costs	466	795	1,091
TOTAL:	21,892	31,979	39,866

NOTE: 1. Costs include caisson core infill but exclude caisson amortization charges.
2. All costs are CDN$K.

It is clear from Fig. 18 that costs are not proportional to sand volume as assumed during the genesis of caisson retained islands [5]. Specifically, the cost experience shows that the project cost of earthworks in the Arctic offshore is approximated by:

$$\$ = 16 \text{ Million} + 12/m^3$$

That is, for three recent berms, the fixed cost have been in excess of 40% of total project cost without any capital amortization of the caisson system.

An identical pattern emerges for the dredging costs alone. The direct expenditures on dredge charter/operation are also plotted on Fig. 18 for four berm projects (3 GCRL, 1 ERCL) and it may be seen that the dredge costs approximate:

$$\$ = 9 \text{ million} + 5.5/m^3$$

Caution should be used if any extrapolation is made from Fig. 18 as it contains several hidden factors, the most important of which are:

- One site involved a long (>100 km) haul of sand.
- Another site involved substantial foundation treatment and costs were further increased by bad ice conditions in the open water season which forced the chartering of additional dredges.
- One project included densification and dewatering work of the hydraulically placed sand but offset this cost by particularly efficient use of dredge equipment.
- Dredge charter rates softened substantially (by nearly half) in the 1985-7 period compared to 1980-3. This reflected a worldwide surplus of dredging capacity.

In short, while Fig. 18 may provide a guide, there is no substitute for cost estimation from first principles using the concept of 'method related charges' [47] and allowing for the potential effect of external factors.

SANDFILL DENSITY ACHIEVED BY HYDRAULIC PLACEMENT

Need for Experience

The missing knowledge in the design and construction of hydraulic fill structures was how may in-situ density be controlled during hydraulic placement. No theory exists on the subject. The only available approach is to formalize existing experience and introduce this into design.

It should be appreciated that the onshore approach of specifying a design density (i.e. a 'performance specification') is unrealistic in the offshore. The dredging contractor does not have the option of making another pass with the equivalent of a compaction roller. Either acceptable results are to be achieved by the dredge on its own or a densification project is to be specified which has the effect of substantially increasing the capital cost and duration of the works. If owners attempt to lay this cost risk on contractors, then excessive bid prices are encountered. It is beholden on owners to recognize this fact, to understand the limitations and opportunities of hydraulic fill placement, and to structure construction contracts accordingly. Owners should use 'method specifications' and concept of method-related charges appears to provide the appropriate contract basis.

Measurement of Sandfill Density

Experience of sandfill density implies measurement, which is often done in terms of penetration resistance. Although the SPT is the commonest test for sandfill assessment, it has seen little use in the Beaufort Sea where the CPT has been preferred for several compelling reasons:

- Long rod lengths are required for the SPT in the offshore and the effects of these lengths on the SPT are uncertain. No such difficulty exists with the CPT which is a simple and precise test even in the offshore.

- Whereas only a few SPT's can be accomplished per hour, the same time is sufficient for as much as 20 m of CPT sounding. Further each metre of CPT contains approximately 70 individual readings. The CPT thus allows a wealth of data to be gathered in the limited time window available during offshore construction.

- Hydraulic fills are quite variable with regard to penetration resistance or density. For this variance to be understood and characterized it is necessary to have a test method that has reproducible results in standard conditions. Experience with the CPT around the world show a repeatability of end resistance, q_c, better than +-2% with occasionally +-0.5% being reported. Well controlled SPT work is at least an order of magnitude worse than this.

- Empirical correlations exist between CPT and SPT [49]. While this may appear to introduce additional uncertainty, this is not the case as using the CPT and a correlation to SPT equivalents has a lesser uncertainty that the scatter commonly encountered in SPT measurements.

These reasons were judged sufficiently compelling by all three exploration companies in the Canadian Beaufort Sea for the universal adoption of CPT soundings as the basic method of sandfill characterization. All testing since 1981 (which includes 96% of the data base) has been carried out with right-cylindrical, 10 cm^2 cones with 60 deg. tips.

CPT soundings have varied in number on individual projects from a low of 3 soundings per island to a maximum of 106. Generally, there

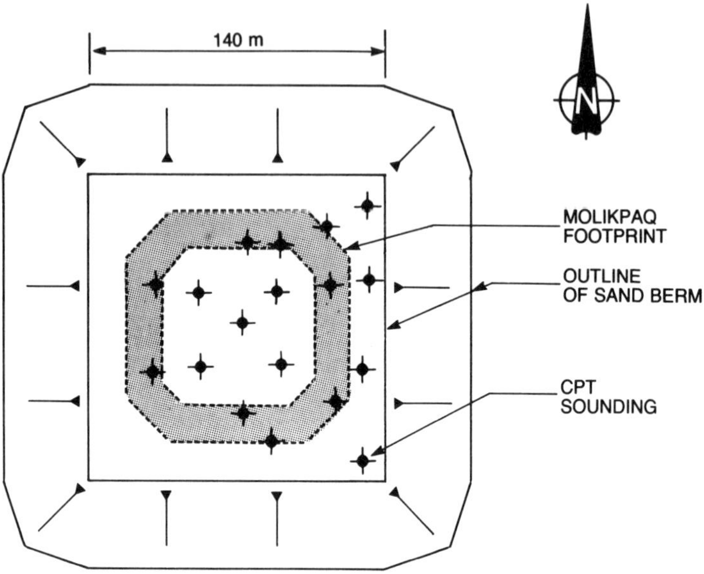

Figure 19

PLAN OF TYPICAL CPT PROGRAM FOR BERM DENSITY VERIFICATION

have been about 8 soundings at any particular stage of a project, a number which appears sufficient to provide a stable statistical estimate of sand state [50]. The layout of soundings is not necessarily regular, but rather is a regular pattern modified for convenience as required at the time. Figure 19 is typical and shows the arrangement of CPT soundings carried out to verify the density of the Tarsiut P-45 berm before placement and ballasting of the caisson onto it.

Individual CPT soundings in a 'uniform' sandfill show substantial variations in the various parameters measured, particularly end bearing resistance, q_c. A typical example of CPT data is shown on Fig. 20, which is taken from the Amauligak I-65 project. This figure shows a general increase in q_c with depth and display a local variation $\pm$30% of the mean value at any particular depth.

The variation in tip resistance during a sounding led to a questioning as to whether the low q_c values - which were associated with 'loose pockets' - had any significant lateral continuity. The issue was first examined in detail for the Kogyuk N-67 berm constructed in 1982 [42]. A total of 33 soundings were systematically arranged on a berm surface 140 x 140 m; it was found that no apparent lateral continuity of 'loose zones' existed at this scale. A further examination of the issue was carried out at Tarsiut P-45 [4] where a total of 32 CPT's were carried in an area 75 x 75 m. A proportion of these tests were at 9 m centres and Fig. 21 shows the data from this close-spaced portion of the testing program. It can be observed from Fig. 21 that the continuity of 'loose pockets' in hydraulic fills is apparently quite limited spatially.

With the understanding that 'loose pockets' were a random rather than discrete phenomenon, then it is reasonable to seek a statistical characterization of fills. The abundance of data from CPT soundings allowed the state probability density function (PDF) to be calculated directly from the CPT data without assumption about the nature of the PDF [50]. It turns out that it was fortunate that such an approach was adopted because a normal distribution - which experience with other materials would suggest as a first approximation to the PDF - is, in fact, a quite erroneous representation of sand property distribution.

The characterization approach used to date involves engineering workstation based processing of CPT data [50, 51]. Individual CPT soundings are read and each scan is classified into a two-dimensional domain: q_c value and depth. The q_c discretization is at intervals of 1 MPa (ie: from 0 to 50 MPa in 1 MPa steps); the depth discretization is from top of the sounding to the bottom in (usually) 1 m steps. A simple running total of each class then provides frequency histograms as a function of depth (or vertical effective stress). The results of this procedure are illustrated on Fig. 22 for the depth interval 10 to 11 m below berm surface at Amauligak I-65 based on 8 CPT soundings. It can be seen that - as previously stated - the distribution is not normal but is in fact well skewed. Skewness changes with placement technique and stress level (see, for example [42]).

A question then arises as to how the calculated frequency distributions may be summarized. Although the tendency in offshore soil mechanics is to use a 'conservatively assessed mean' [51] as the characteristic value for a particular soil property, it is more usual in general engineering practice to use the 90 or 95 per centile exceedence criteria. Ideally, one should explore the interplay of property distribution and overall behaviour with an advanced constitutive model properly implemented in a finite element code to define the characteristic value. This has not been done yet. It was proposed [42] as a compromise that the value exceeded by 80% of the berm volume at a given stress level be taken as characteristic; as this was the lower limit at which consistent trends existed between inferred density and log vertical effective stress. The location of the 80% exceedance value with regard to the PDF is shown on Fig. 22, the more traditional median (50 per centile) also being illustrated.

CPT Data on Underwater Sandfills

Cone resistance, q_c, is strongly affected by confining stress. Strictly, it is the mean effective stress that should be used to normalize cone resistance [53, 54]. However, mean stress involves use of horizontal geostatic stress which has often not been measured and where it has been measured the accuracy has been disputed.

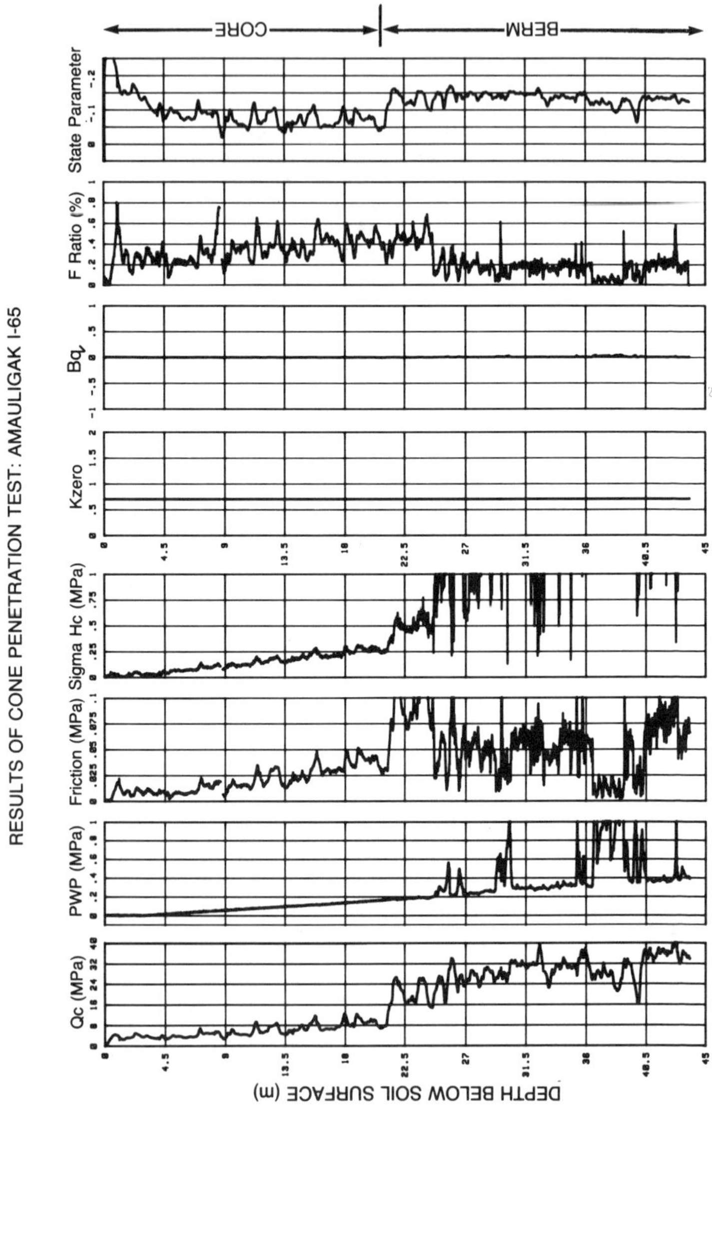

Figure 20
TYPICAL CPT DATA

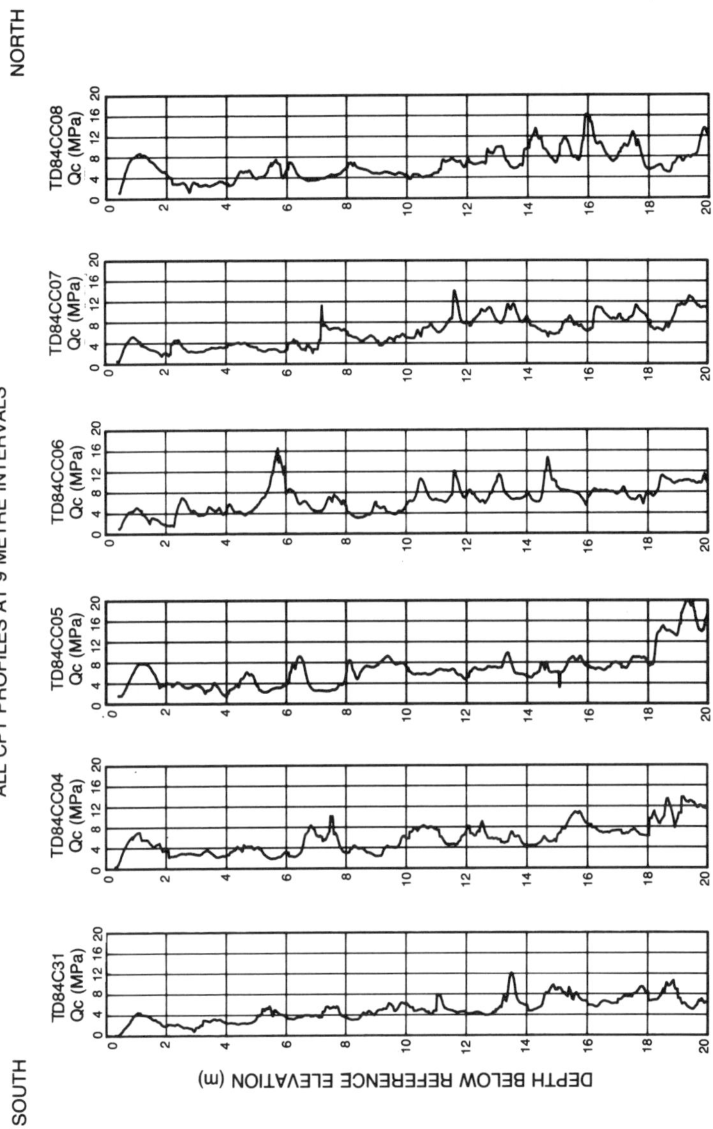

Figure 21

LATERAL CONTINUITY OF "LOOSE POCKETS" MOLIKPAQ CORE

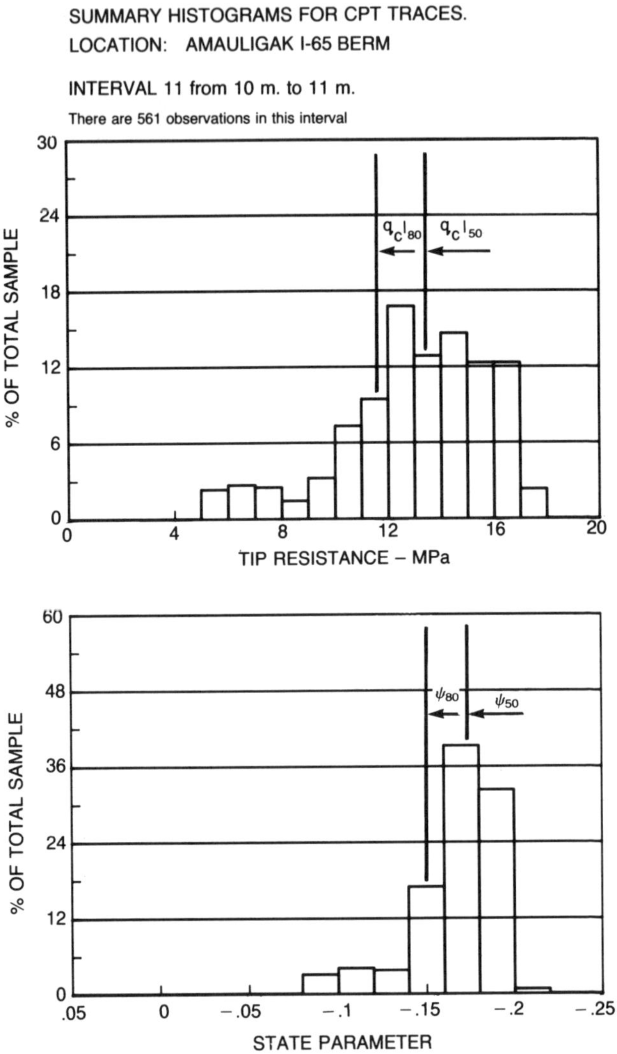

Figure 22

TYPICAL DISTRIBUTION FUNCTIONS OF q_c AND ψ

Accordingly, the q_c case history data will be presented against the vertical effective stress since that is a simple transform of piezometric conditions and the depth below sand surface subject to errors of only +-5%.

Nine of the islands constructed in the Canadian Beaufort Sea had sandfill density evaluated by CPT. In addition to these projects three trial fills have been constructed purely for evaluation of placement technique or sand quality on achieved density. The total amount of CPT data available is some 6000 m of sounding. Within this available data base, several different gradations of fill have been used as well as three distinct placement techniques. Each sand type or placement technique has been differentiated as a separate case record. The data base is summarized on Table 4. The measured CPT data that corresponds with Table 4 is presented on Fig. 23 for the 80% q_c trends.

Interpretation of CPT Data

The CPT q_c data presented above is not a direct measure of any particular sand property; rather the bearing resistance is the product of the sands initial stress state and constitutive behaviour. Extraction of a characterizing parameter for sand requires interpretation of the CPT data.

Earlier in the Paper the state parameter was introduced as an appropriate characterizing measure for sand. Considerable work was carried out on how state might be measured in-situ following introduction of the state parameter for normalization of laboratory data [53, 54, 55]. It was found that the state parameter may be inferred from CPT data regardless of sand type with a precision approximately +-10% of the range of state encountered with most sands [54,55].

The determination of state from CPT data uses the mean effective stress, p' as the stress measure and this requires knowledge of horizontal geostatic stress (or ko). Conventionally, one might calculate p' using an estimated ko from the relation ko = (1- sin Ø'). But, in-situ data obtained with the self-bored pressuremeter (SBP), has shown ko is rarely close to the conventionally expected value [56].

There is a tendency to dismiss SBP data as prone to the effects of imperfect self-boring (ie: disturbance). While it is true that perfect self-boring is rarely achieved, it is also hard to ignore the fact that similar results were obtained in some 200 tests carried out over five years by various testing companies and site investigation contractors. The average value from many SBP tests is likely to be an approximation of 'ground truth'; this average value is ko = 0.7 in hydraulically placed, undensified sandfill, about double what might be conventionally expected. Use of a conventionally expected value (ko = 0.45) is not conservative and indeed may lead to quite dangerous errors when dealing with loose fills. The interpretation of CPT data in this Paper uses a constant k_o = 0.7.

TABLE 4

PROJECT	YEAR	SAND GRAD-ATION	FILL PLACEMENT METHOD	NO. OF CPT SOUN-DINGS	DIS-TANCE SOUNDED (m)	KEY # FOR FIG. 23,24	REMARKS
Alerk P-23	1981	210/6	Pipeline Spigotted	3	52	a	undercut Delft cone
Tarsiut N-44	1981	350/4	Berm: Mainly Bottom Dumped	18	190	b	
			Infil: Spig-otted	8	96	c	
Uviluk P-66	1982	350/2	Berm: Bottom Dumped	18	368	d	wave densi fication
Ukalerk #1 Trial Fill	1982	350/2	Bottom Dumped	5	28	e	
Ukalerk #2 Trial Fill	1982	350/2	Pump-Out	10	62	f	
Kogyuk N-67	1982	350/2	Bottom Dumped Pump-Out	33 33	165 132	g h	
	1983	350/2	Bottom Dumped Pump-Out	3 3	15 15	j k	prior to 1983 const
Kogyuk N067	1983	350/2	1982 BD fill 1982 PO fill 1983 BD fill	9 9 9	36 45 54	l m n	midway through 1983 const
	1983	350/2	1982 BD fill 1982 PO fill 1983 BD fill	6 6 6	24 30 60	o p q	Carried out end of 1983 const
Kogyuk N-67	1983	350/2	1982 BD fill 1982 PO fill 1983 BD fill	6 6 6	24 30 60	r s t	With SSDC in place.
	1984	350/2	1982 PO fill 1983 BD fill	1 6	3 50	u v	After SSDC removal
Nerlerk	1983	320/2 260/2	Bottom Dump Spigotted	26 26	Total 280m	nb ns	Ukalerk s. Nerlerk sand
Isserk Trial Berm	1984	210/1	Bottom Dump	8	57	w	Silty fine sand
Amerk P-09	1984	290/1	Berm: Bottom Dump	3	58	x	Caisson in place
			Core: S	3	36	y	
Tarsiut P-45	1984	320/1	Berm: BD	18	180	aa	
	1984		Berm: Bottom Dump	7	63	ab	Caisson in place
Tarsiut P-45	1984 1985	320/1 320/1	Core: Cent. S Core: After 11 Mon. aging	32 5	784 140	ac ad	

TABLE 4

PROJECT	YEAR	SAND GRAD-ATION	FILL PLACEMENT METHOD	NO. OF CPT SOUN-DINGS	DIS-TANCE SOUNDED (m)	KEY # FOR FIG. 23,24	REMARKS
Amauligak I-65	1985	320/1	Berm: Bottom Dumped	9	182	ae	
	1985	320/1	Berm, Caisson in place	13	220	af	
	1985	320/1	Core: S	21	450	ag	
	1986	320/1	Core: After April 12 1986 event	9	190	ah	
Amauligak F-24	1987	320/2	Berm: Bottom-Dump	11	240	fa	
		320/2	Berm: Caisson in place	17	(see next row)	fb	
		350/1	Core: S	17	440	fc	
		350/1	Core: During Densification	8	210	fd	Series 'A' blasts
				18	530	fe	Series 'B' blasts
		350/1	Core: After Densification	29	845	ff	
	1988	350/1	Core: After Densification	20	430	fg	after vibro probing
	1988	350/1	Core: Check Aging	3	75	fh	
TOTAL					6949m		

S = Spigot
BD = Bottom Dump

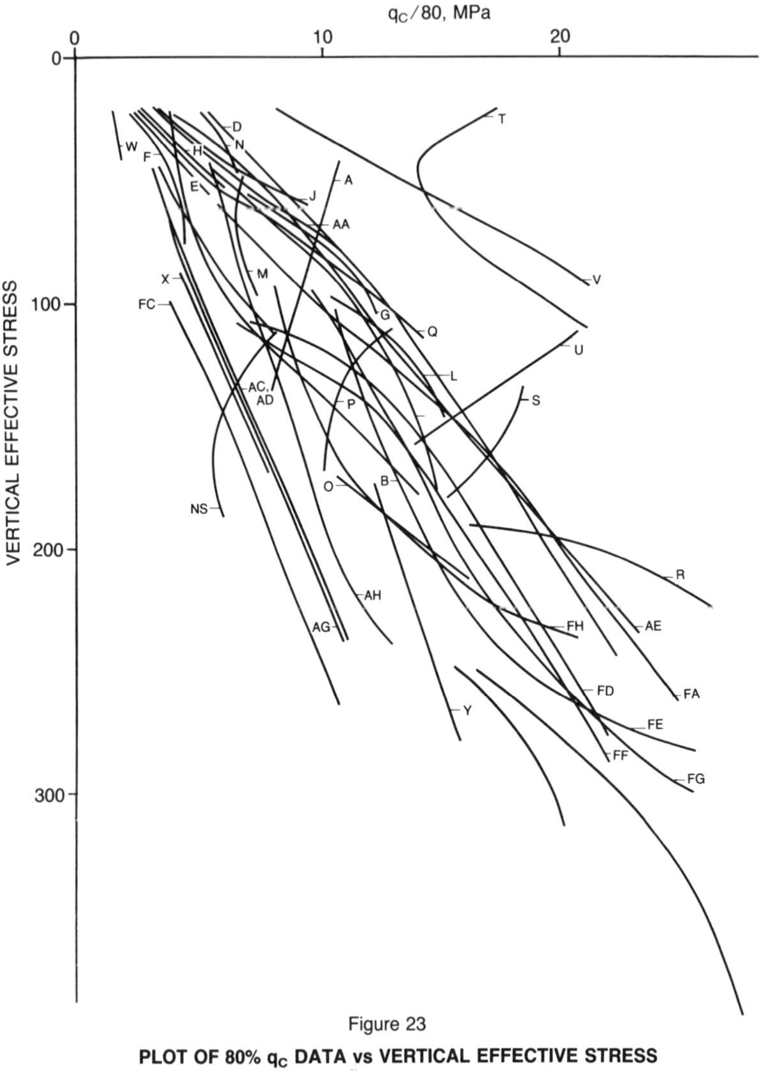

Figure 23

PLOT OF 80% q_c DATA vs VERTICAL EFFECTIVE STRESS

The method of determining state from CPT data [54] uses a scaling parameter which is the slope of the steady state line in e - log p' space, λ_{ss}. Although this parameter displays a wide range of numerical value with different sands [22, 31] considerable data on its value has been obtained for many gradations and several borrow sources for Beaufort Sea sands. Further, the general approach to CPT interpretation has been calibrated for one specific gradation of Erksak sand by chamber referrence testing [57].

The results of applying the procedure for determining from CPT data are shown on Fig. 24 which shows the same data as Fig. 23 but now transformed to state. It can be seen from Fig. 24 that a whole range of characteristic states have been obtained by purely hydraulic placement of sand. The densest fill has $\psi \approx -.18$ which corresponds to a substantially dilatant sand ($\phi' > 38°$ deg, Fig. 6). Equally, there are several cases where $\psi \approx 0.0$ which corresponds to a complete absence of dilatancy ($\phi' \approx 31°$ deg, Fig. 6). The picture is therefore a complicated one and the challenge is to identify those factors that produced substantially dilatant fills by purely hydraulic placement.

EFFECT OF CONSTRUCTION VARIABLES ON ACHIEVED SAND STATE

Factors Considered

Hydraulically placed sandfill makes the transition from a water supported condition without grain-grain contact to a grain-grain skeleton so it is natural to postulate that the process is related to consolidation. Consolidation parameters of sand vary with gradation, silt content in particular having a marked influence. However, the process of hydraulic transport also suggests that placement technique, which is a mixture of slurry density, energy, depth and erosion should be considered. To examine the influence of these various factors, the achieved state of underwater sandfills will be examined in the following steps:

i) Variation in state with water depth at constant stress level, placement technique and gradation.

ii) Variation in state with placement technique for constant water depth, constant sand type and constant stress level.

iii) Variation in state with stress level for constant placement method, constant sand type.

iv) Variation in state with sand type with constant placement technique, constant stress level, constant water depth.

Effect of Water Depth on Achieved State

Bottom dumping involves a free fall through the water column, which is a possible variable. However, several cases exist using nominally the same sand with different water depths. Specifically, the

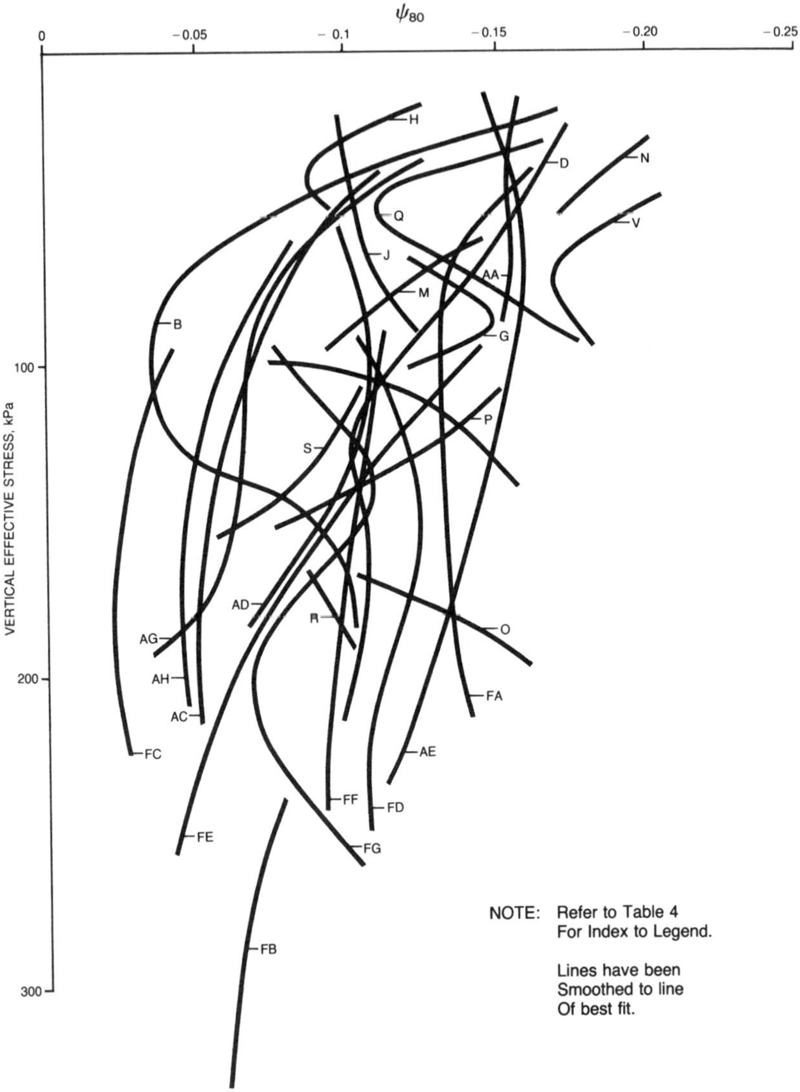

Figure 24

PLOT OF ψ_{80} vs VERTICAL EFFECTIVE STRESS

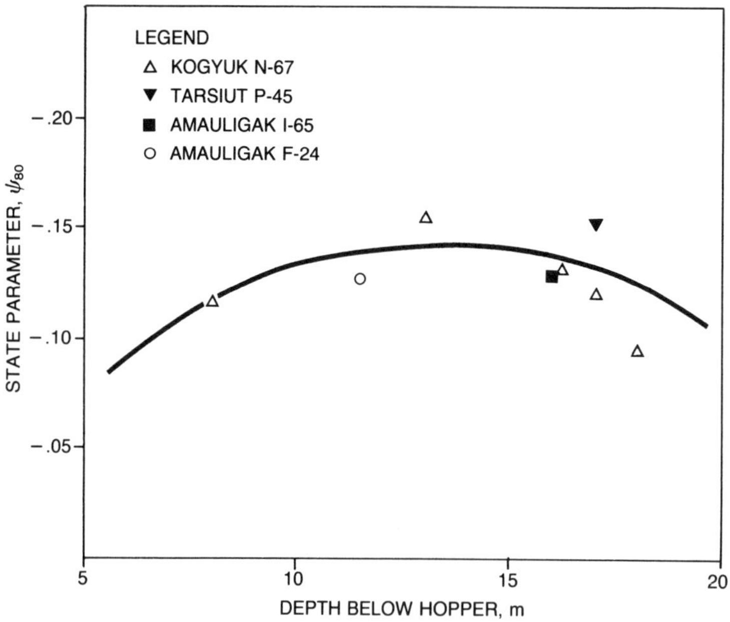

Figure 25

EFFECT OF DROP HEIGHT ON SAND STATE

1983 Kogyuk fill (10 m minimum depth) can be compared with 2 Molikpaq berms (20 m minimum depth) and the 1987 Molikpaq berm at Amauligak F-24 (15 m minimum depth). This is done on Fig. 25 in terms of the 80 per centile state. As can be seen, substantially the same state is achieved; water depth appears not to be a significant variable with the range 0 - 25 m, although there is the suggestion of possible shortfalls in greater water depths.

Effect of Placement Method on Achieved State

Three distinct placement methods have been used to date:

- Discharge directly from the base of a hopper using valves or gates; the sand descends to its final location as a dense plume. This technique is called bottom dumping.

- Discharge as a slurry from a pipe located only a few metres above the accretion surface at relatively low velocities. This may be achieved either from a floating dredge using its drag arm or by an anchored barge equipped with a tremie pipe; it is referred to as the 'pump-out' method.

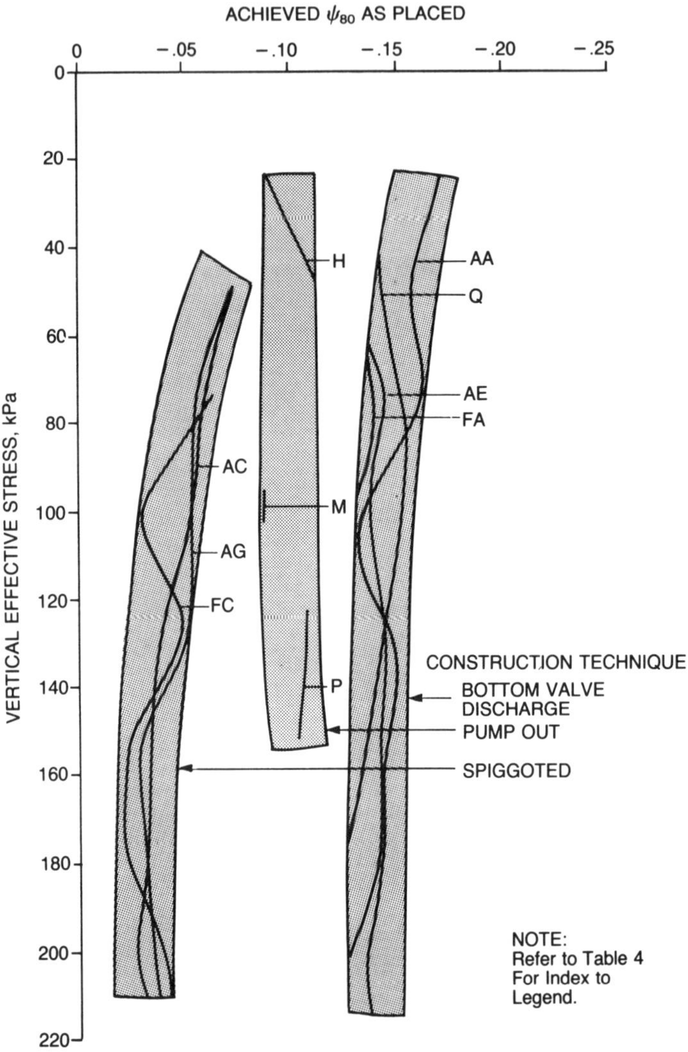

Figure 26

SUMMARY OF ACHIEVED STATES FOR ERKSAK SAND

- Discharge as a slurry at relatively high velocity (several m/s) some distance from the sand surface. Discharge may not even be underwater. This method is referred to as spigotting.

Substantial data exists on spigotted sands from four caisson infills all using to Erksak sand. These cases include Tarsiut P-45; Amauligak I-65 and Amauligak F-24 which specifically used the same gradation of sand and Amerk P-09 which used sightly finer sand from the same borrow source. The state achieved with these various case histories is remarkably consistent, as shown on Fig. 26. This figure also shows the corresponding data for 'bottom dumped' placement of the same sand. It may be observed that the difference in achieved state for the same sand is substantial; spigotting produces $\psi \simeq -.05$ at $\sigma_v' =$ 100 kpa where as bottom dumping produces $\psi \simeq -.14$ at the same stress level. Both procedures are repeatable over several case histories.

Although pump-out fill placement has seen substantial use in the three Molikpaq berms, little CPT data has been obtained because this method of fill placement is only used in the construction of retaining bunds which have no function in supporting the structure. The data base that allows comparison between pump-out and other placement procedures is restricted to the two 1982 trial fills and the upper layer of the 1982 Kogyuk berm, which was placed entirely by pump-out methods. This latter case was tested at several times as the 1982 berm was raised in 1983 to accommodate the SSDC, which provides information on pump-out fill at several stress levels. Both the trial fills and the 1982 Kogyuk berm, Erksak sand. The comparison is presented on Fig. 26 and shows that bottom dumped procedures produce a somewhat denser fill than pump-out but that pump-out is preferable to spigotting.

State as a Function of Stress Level

Soils consolidate with increases in confining stress and, if they are stressed in excess of any previous value, then consolidation will exhibit substantial irrecoverable (plastic) strains. The manner in which sands are deposited by hydraulic methods means that they are, by definition, normally consolidated and plastic strain may be expected.

All of the case histories for which CPT data exists have a significant range of stress. Unfortunately, the data was obtained at a single time and since construction in the case of bottom valve or spigot fills is carried out from near a constant elevation, the data contain the results of both a variable fall height as well as variable stress level.

Data on unarguable stress level effects does, nevertheless, exist. The Kogyuk N-67 case history had CPT testing carried out at three stages in construction, while the three Molikpaq berms were both tested on completion of construction and with the Molikpaq in place. The information on stress level effects is presented in Fig. 27, which shows that state becomes progressively more positive as stress level increases. This implies that normally consolidated sand exhibits a consolidation index, somewhat less than λ_{ss}. This conclusion corroberates the similar finding reported in [58].

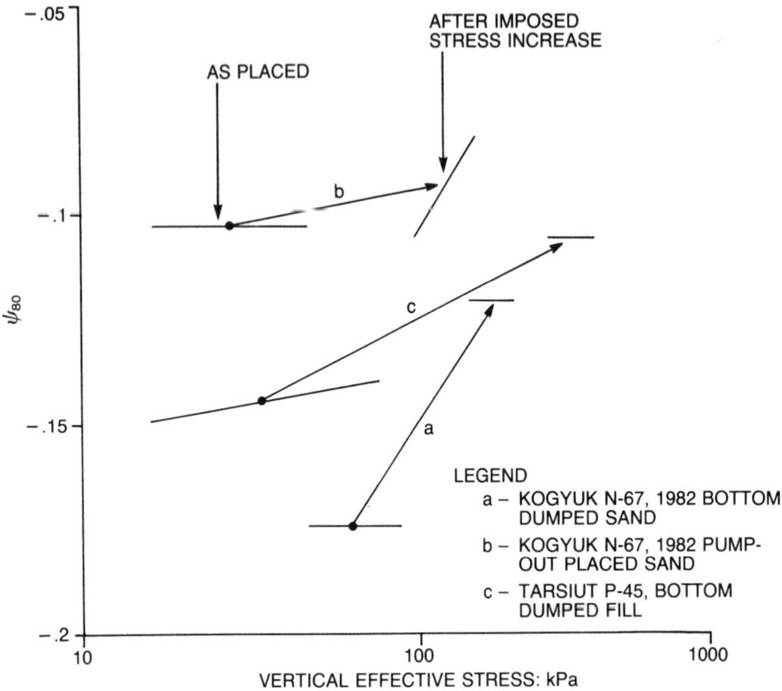

Figure 27

EFFECT OF IMPOSED STRESS INCREASE ON STATE OF HYDRAULICALLY PLACED SAND

A note of caution should be added regarding this conclusion. The analysis has been conducted with a constant k_o. There is insufficient precision in the existing data base to establish whether this is a reasonable assumption. If k_o becomes less with increasing stress level - as might be inferred from laboratory tests - then the influence of stress has been overstated.

State as a Function of Material Type

The final variable to be considered is the effect of material type. Material type is in itself not an appropriate property for comparison with achieved state; rather, material type influences permeability, compressibility and other mechanical attributes of the sand. As noted earlier, the process of changing from a slurry to sediment would be most naturally regarded as one related to consolidation; accordingly, the relationship of achieved state to a coefficient of consolidation will be explored.

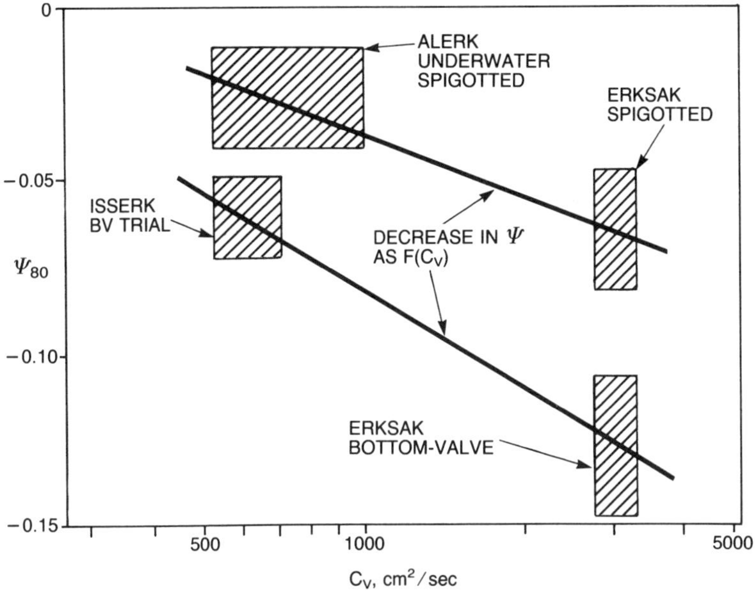

Figure 28

ACHIEVED Ψ AS A FUNCTION OF C_V

There are two comparisons we can make to gain insight on the influence of soil type. First, we can contrast bottom valve placement of Ukalerk and Isserk sands. Second we can contrast Erksak sand placed in CRI cores (spigoted) with the Alerk fill, which was spigotted from a pipeline. The results are shown of Fig. 28 and there appears to be a reasonable relation of state to coefficient of consolidation for a particular placement method. The change in state with coefficient of consolidation is similar for both placement methods.

AGING EFFECTS

There is evidence in the literature [59] to suggest that cone penetration resistance of sand increases significantly with time even when no settlement (densification) has occurred. This is a particularly interesting observation from both practical and theoretical points of view. It is interesting in the purely practical sense because doing nothing but waiting would appear to improve a fill. It is interesting from the theoretical point of view since it would imply the cone shows a substantial response to a variable other than state - which contradicts direct laboratory testing [54]. Careful tests were therefore carried out as part of the Tarsiut P-45 project to see whether any such aging effects existed with a clean Erksak sandfill over an 11 month period.

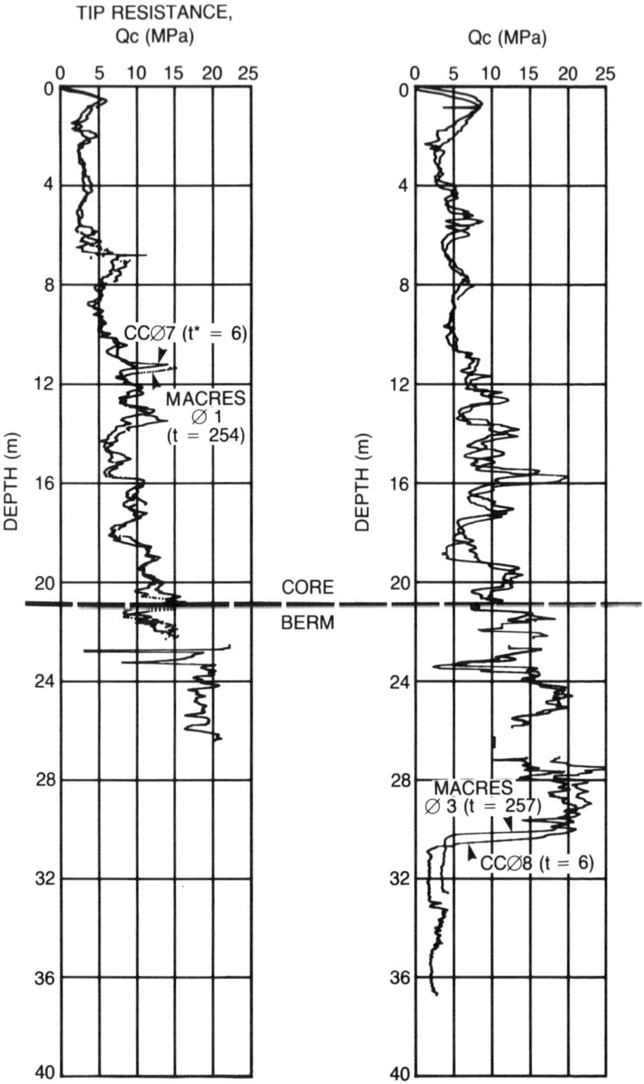

LEGEND:
– t * = time in days since completion of construction in Oct. 84.

Figure 29

EFFECT OF AGING ON SPIGOTTED ERKSAK SAND

The tests for aging were careful in the sense that CPT soundings were carried out as pairs with each member of a pair being less than 1.5 m horizontally from the other member. Five such sounding pairs were carried out. An example of the results obtained is shown on Figure 29 in terms of the raw q_c value; the data has not been subject to any interpretation. It can be seen that the q_c profile at 260 days differs from that at 6 days by less than 2%, which is essentially the repeatability between two different CPT transducers. Aging effects are not present on the timescale investigated for spigotted sandfills.

SUMMARY OF "UNTOWARD" INCIDENTS

Although 36 islands have been built, their construction has not been entirely a success story as eight islands have exhibited unexpected or untoward behaviour as summarized on Table 5. It should be remembered when reviewing these incidents that an "incident" is not necessarily a "failure". A "failure" means an occurrence which causes substantial loss of either money or life; only two such incidents might be regarded as such; Nerlerk B-67 and Minuk I-53. Nevertheless, "incidents" are a caution since a mechanism that only constitutes an incident in 5 m of water where remedial action may be trivial, can equally be a catastrophic failure in 50 m of water. Accordingly, much pertinent experience can be gained by reviewing the "untoward" incident record.

Excess Pore Pressures Induced by Construction

The first indication that hydraulic fill construction might be difficult using the sand and procedures available in the Beaufort Sea came from the Arnak L-30 project in 1976. This island was constructed using a typical Beaufort Sea sand which was a little siltier than those used more recently; the envelope of gradations for this island is shown on Fig. 30. The island construction was typical of the early years. The suction dredge "Beaver MacKenzie" was used to excavate local sand which was discharged via a floating pipeline at the island site. The above water portion of the island required the use of bulldozers and a backhoe to form a circular dyke which was then infilled with sand slurry. Trafficability of the fill was poor during dyke construction and the beach subsequently slumped downslope several tens of metres. This slumping was caused by transient excess pore pressures in the sand, as evinced by the sand boils observed on the beach [Fig. 30]. Since there were no outside disturbances it is reasonable to infer that these large excess pore pressures were entirely due to the sandfill and construction procedures used. The field estimated C_v based on the measured decay of excess pore pressure in a piezometer was 0.2 cm^2/sec [71]. It is difficult to explain the excess pore pressure even with this value of C_v using accreting layer theory [60]. Considerable caution should be used when assessing the transient stability of hydraulic sandfills.

TABLE 5
SUMMARY OF "UNTOWARD" INCIDENTS

PROJECT	YEAR	INCIDENT	CAUSE	ACTION
Arnak L-30	1976	Slumping of recently constructed beach	excess pore water pressure caused by consolidation of spigotted sand	Additional sand placed; island became stable as pwp dissipated
Issugnak 0-61	1979	Substantial erosion of the island fill when close to waterline	Storm waves on two separate occasions; wave induced lique-faction possible factor	Additional sandfill placed and island brought into service
Tarsiut N-44	1980	Caissons misplaced by 20 m (25% of island surface)	Surveying error	Toe berm constructed to avoid foundation failure and island brought into service
Alerk P-23	1981	Nearly half of island disappeared	Postulated as wave induced liquefaction [61] triggering slump ing of loose sand	Additional sandfill placed and island brought into service
Nerlerk B-67	1983	General slope failure during construction when at half height	see Text	Project abandoned with loss > $100 M
Adgo H-29	1984	Cracking and downward movement of sandfill in NE corner	Foundation failure in soft clay	Stabilizing toe berm placed and island brought into service
Minuk I-53	1985	Island erosion lead-ing to loss of rig and supplies	Storm substantially in excess of design conditions	Rebuilt and brought into service
Amauligak I-65	1986	Partial fluidization of sand	Extended cyclic loading of platform by ice crushing	Densification speci-fied for subsequent Molikpaq deployments

PHOTOGRAPH OF SAND BOILS – ARNAK ISLAND BEACH

(a) EVIDENCE

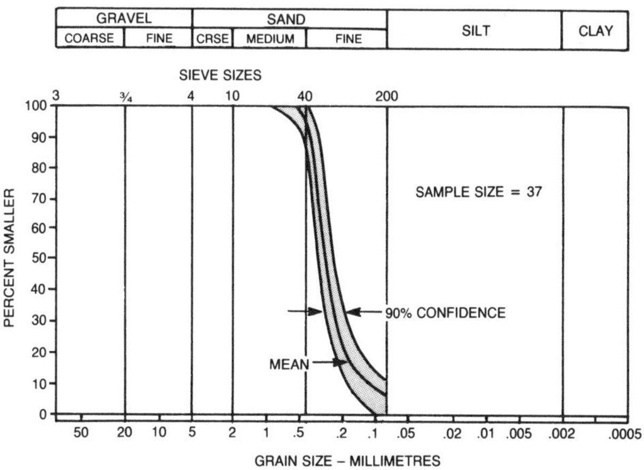

(b) GRADATION OF SAND INVOLVED

Figure 30

EVIDENCE OF LARGE EXCESS PWP INDUCED BY HYDRAULIC FILL CONSTRUCTION METHODS

Wave Induced Cyclic Mobility

A simple examination of Table 5 indicates that wave action was a common factor in several instances. Wave action can lead to the generation of excess pore pressure because of cyclic total stress reversals: Material type, expressed in terms of C_v, is the major factor in determining susceptibility to wave induced liquefaction [48]. Shown on Fig. 9 are the ranges of C_v for various borrow materials, and it is clear that Erksak sand is sufficiently clean and coarse (i.e. incompressible and with high permeability) to ensure that excess porewater pressures will not be developed under the "design" storm even if it is in a loose state.

Alerk and Issungnak islands were both constructed in the early 1980s by pipeline placement of local sand borrow which was relatively fine and dirty. As indicated on Fig. 9, the Issungnak sand had $C_v <$ 100 cm^2/sec and this material would have been expected to liquefy under a less severe storm than the design storm. During the construction of Issungnak, observations by field staff indicated that on at least two occasions the above water portion rapidly disappeared during storm action. Although the C_v for the Alerk material is greater than Issugnak sand, its performance in the design storm would be expected to be marginal; during 1981 half the island disappeared during a storm at the end of the construction stage. Analysis of failure survey data has led to the conclusion that Alerk experienced instability because of its near liquefiable state coupled with cyclic induced excess pore pressures [61].

Erosion

A late season storm in September 1985 resulted in major damage to the Minuk I-53 island. The storm eroded near half the island, toppled the drilling rig and swept part of the camp out to sea. However, the island had been evacuated and there was no loss of life [62]. Although the experience at Alerk and Issugnak might suggest that wave induced liquifaction was also the cause of the Minuk incident, review of available data suggests it was direct erosion by wave action.

Minuk I-53 was a sacrificial beach island. Such islands are designed by balancing the expected erosion in a design storm with the size of the beach, a very non-linear calculation that depends not only on significant wave height but also storm duration, sand size, modifications to the beach itself by the erosion process, etc. The complexity of design is further compounded by the limited history of environmental conditions in the Beaufort Sea (there are few records before the start of oil exploration in 1973). The design approach adopted for SBIs was to use physical model tests at 1:25 scale combined with numerical modelling to develop an adequate island for an estimated storm but also to recognize that the issue was one of economic risk for the operator provided drilling was not started prior to freeze-up and provided a means of evacuating personnel existed.

Post-failure analysis of the Minuk I-53 performance during the storm [63] indicated that the uncertainties in the extrapolation of model tests, the grain size of fill materials and the wave climate experienced were not critical to the incident. Excessive erosion was caused by a relatively rare storm with a duration approximating four times that used in the original design, although with a comparable wave heights.

The Minuk storm event is the first case where the design philosophy of accepting economic risk while preventing environmental damage/loss of life was tested. The investigation carried out after the event by the regulatory agency concluded that this design philosophy was still acceptable [63]. If anything, the lesson to be learned from Minuk is to be very cautious of probabilistic predictions of environmental conditions from short duration data bases.

Ice Induced Cyclic Mobility

Although ice loading is often considered as a quasi-static situation, much of the ice loading of the Molikpaq at significant load levels (ice Loads > 100 MN) has been associated with cyclic, dynamic loading [19]. These cyclic loads have induced vibration of structure at frequencies commonly in the order of 0.5 to 3 Hz and with durations as long as about 30 minutes. Although the ice induced vibrations were comparable to a moderate earthquake of extended duration, the view developed that the platform was behaving rather better than predicted in design.

Unlike the previous several years experience in the Beaufort Sea, ice conditions at Amauligak I-65 in winter 1985/86 included substantial amounts of multi-year ice whose thickness was commonly in the range 2 to 4 m. Ice loads on the structure were correspondingly greater. At 0800 hrs on Saturday April 12, 1986, a moderate size (2 km x 1 km) multi-year floe was continuously crushed against the side of the Molikpaq. This crushing caused peak loads on the structure marginally in excess of the design load (500MN) with large concurrent vibrations. The vibrations were sufficiently severe to induce fatigue of the sand core, the principal element by which the Molikpaq resists ice load, to the extent that the ability of the platform to withstand further ice loading was degraded.

The nature by which cyclic ice loads degraded platform stability can be observed on Fig. 31(a). This figure shows 10 cycles of ice loading on the structure and the corresponding response at a typical piezometer in the sand core. For any single load increase, a positive pore pressure is induced but the situation is stable in the sense that there is no 'run-away' generation of pore pressure. However, the induced pore pressure is not completely recovered during the unloading part of the cycle so that the effect of continuous load cycling is to ramp the average pore pressure upwards, as may be observed on Fig. 31(a). The overall pattern of behaviour at one piezometer is shown on Fig. 31(b). In this case, ramping continued in a linear manner at an average rate of accumulation of excess pwp of 0.8 kPa/cycle. The process terminated when the condition $p' = 0$ was reached

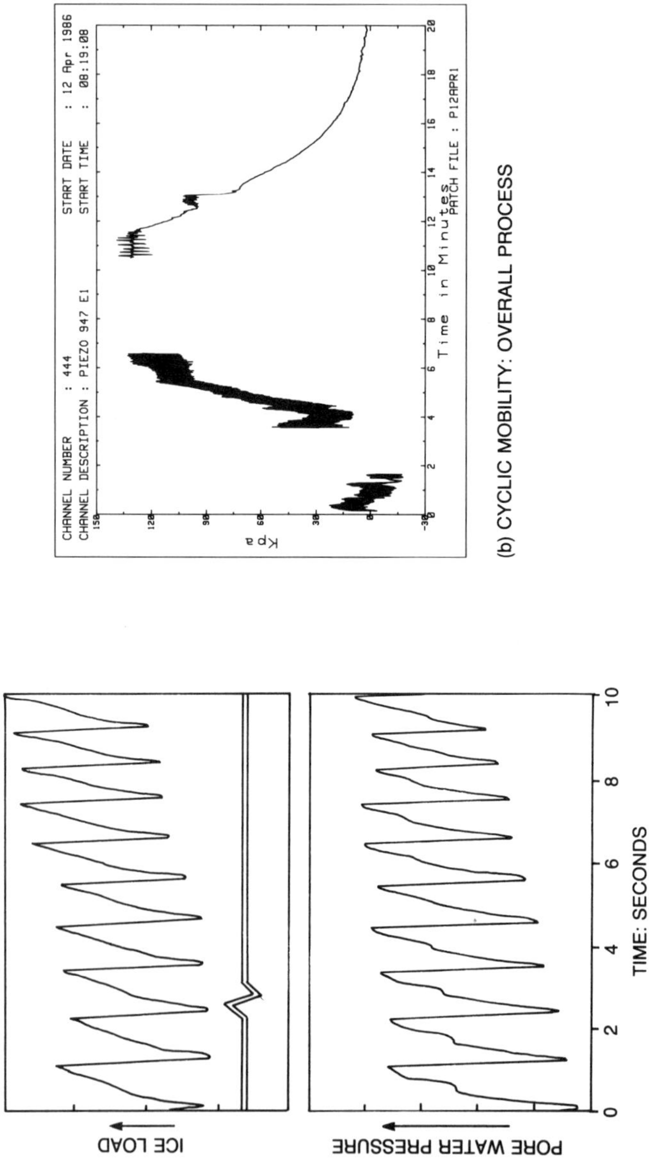

(b) CYCLIC MOBILITY: OVERALL PROCESS

(a) CYCLIC MOBILITY: DETAIL

Figure 31

DATA FROM AMAULIGAK I-65 SHOWING NATURE OF CYCLIC PORE PRESSURE BUILDUP

(i.e. liquefaction as defined by Seed [33]), as may be observed on Fig. 31(b). The increase in pore pressure that subsequently took place was entirely caused by an increase in hydrostatic pressure as the sand and the piezometer settled 1.5 m.

Fluidization of the sand core was a progressive phenomenon that initiated at mid-height of the loaded side and propogated rearwards and downwards, as documented by other piezometers. Fortunately, the ice action that caused fluidization ceased before pore pressures had risen sufficiently to cause overall platform instability.

The data obtained at Amauligak on the nature of excess pore pressure generation by cyclic loasing is presently being studied under a joint National Science Foundation/National Research Council project. The results of this study will be eventually published, but for the present it should be noted that the Amauligak data shows that:

i) Lightly dilatant sands ($\psi \approx -.05$) have substantial reserves of cyclic strength.

ii) Excess pore pressures continue to be generated as long as cyclic loading occurs.

iii) Redistribution of pore pressure occurs on the same timescale as cyclic loading; the assumptions inherent in the 'steady state' concept of minimum assured strength [31] are false.

Large Strain Strength of Sand

Nerlerk is the major failure in the Arctic offshore. Nerlerk was proposed as a 35 m high berm for the SSDC; unlike the previous successful berms it was to be built using the powerful suction dredge, Aquarius, and a local sand that was both finer and siltier than the precedent experience. It was also placed by a previously unused nozzle technique [65]. The Nerlerk sand was somewhat loose although examination of Figure 23 illustrates it was comparable to that at Alerk.

Construction at Nerlerk had to be abandoned because of large scale slope failures at a berm height well short of design with a direct loss exceeding $100 M. Specifically five slides occurred in the berm when it was at 25 M above seabed [63]. Fortunately, failure occurred before the SSDC was ballasted onto the berm.

A cross-section of the Nerlerk berm immediately prior to the first slide is shown on Fig. 32 together with a typical failure mechanism used in back analysis. Analysis is complicated by the difficulty in distinguishing between three potential failure modes:

° extrusion of the underlying soft clay which initiated sand slumping

° static liquefaction of the sand under self-weight

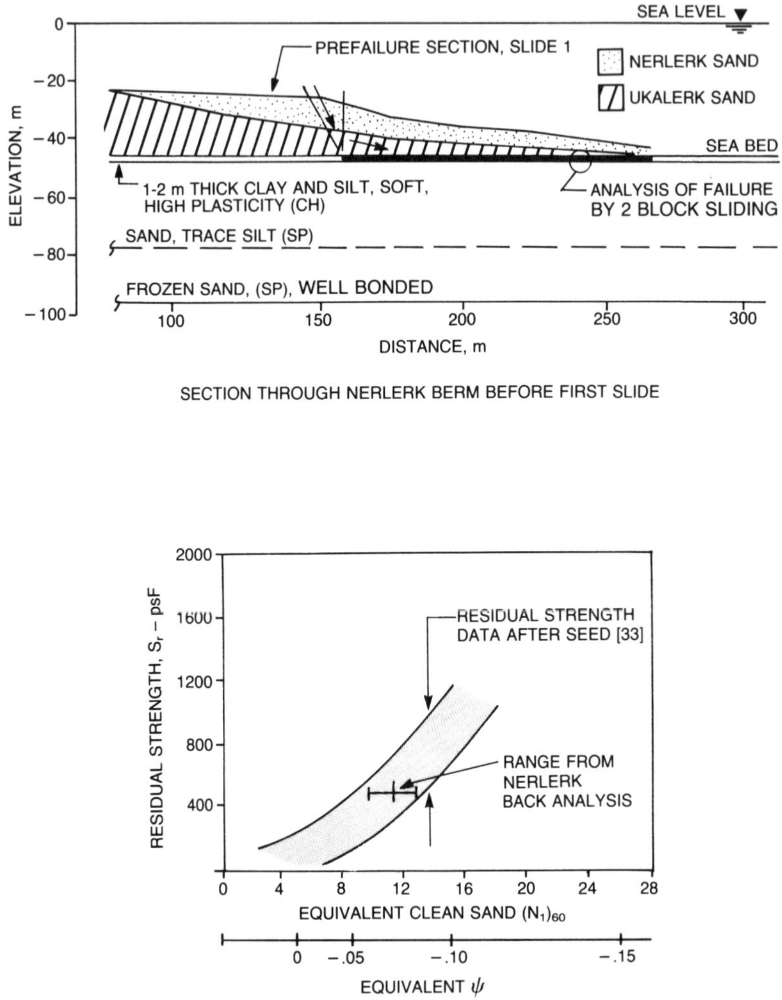

Figure 32

SUMMARY OF NERLERK BACK-ANALYSIS

- transient instability because of large excess pore water pressures induced by the construction method.

Considerable discussion has occurred in the literature on the Nerlerk failure [66, 67]. On one hand [66] it has been asserted that Nerlerk failed by static liquefaction of the sandfill in the sense defined by Castro & Poulos [30]. This assertion depends on the morphology of the failure as diagnostic of static liquefaction. The validity of morphology for this purpose has been disputed [67] and a recent review of the data base on fill failures [68] concludes that static liquefaction has actually never occurred to date for an underwater sand fill.

Caution should be used when considering the "results" of Nerlerk back analysis. As pointed out [67] there is no data on the undrained behaviour of Nerlerk sand at the lightly dilatant densities, there is no data on the drained behaviour of Nerlerk sand, there is no data on the clay strength in the soft foundation layer.

The clays and sands still exist at the Nerlerk site. There appears to be considerable merit in obtaining the relevant samples and conducting a thorough study of the failure. The Nerlerk case history is at least as important as the Amauligak ice event, as Nerlerk probably can define the limiting strength of sand with reasonable precision once all the soil data is known.

Until Nerlerk is thoroughly analysed, the fact that apparently lightly dilatant sand slumped in near-undrained loading suggests that prudent engineering should consider $\psi < -.1$ as the limiting condition for design of saturated sandfills. This is in accordance with a recent summary study of large scale sand strengths [33].

DENSIFICATION

Concern over the adequacy of undensified sand when exposed to cyclic ice loads existed from the early days of CRI type platforms. The approach adopted was to construct an undensified fill and then test it for adequacy, contingency plans being drawn up prior to construction so that densification could be undertaken if required. Examples of contingency plans are:

- 4 vibroprobes were mobilized to the Tarsiut N-44 site in 1981;

- an explosive densification method statement was engineered for the Molikpaq core at Tarsiut P-45;

- a high-powered vibroneedle was mobilized to Hay River and maintained on standby during construction of the Amauligak I-65 berm.

In all the above circumstances density verification testing showed the design target density had been achieved by purely hydraulic placement and no contingency plan was ever executed.

The view on what comprised an adequate density in the core of the Molikpaq changed substantially after the April/86 ice loading experience at Amauligak I-65. The manner in which pore pressures were generated during the April/86 ice load event strongly suggested that sliding of the platform would have occurred had cyclic ice loading continued. This conclusion about the adequacy of the fill under cyclic loading was at marked variance with the original reliance on the steady state concept supported by centrifuge models [4]. However, the original design had recognized uncertainty in behaviour of sand which led to the installation of fast response piezometers and a high speed data acquisition system. It was the data obtained by this system that forced the change in views, in itself an interesting example of the role of instrumentation in civil engineering works.

The Molikpaq was deployed at Amauligak F-24 in September, 1987 and densification of the core by blasting was adopted. Blasting was carried out in two 'passes', each pass consisting of holes drilled at 6 m c/c on a rectangular grid. The two passes were offset diagonally by 4.2 m . The plastic-cased holes were charged with 3 x 3.6 kg charges of 'Nitropel' explosive separated by gravel stemming. Holes were shot in groups of typically 14 using delay detonators to both minimize shock transmission to the steel structure and to prolong the duration of vibration. Extensive details of the densification methodology may be found in [69].

The costs for the explosive densification approximated $\$25/m^3$ and some 65,000 m^3 of sand were so treated. The work was substantially completed in only sixteen days on site, using two rigs operating 24 hours/day. Although these costs might seem quite significant, it should be noted that offshore Arctic construction usually shows cost factors between two to three times greater than expected costs in onshore, southern locations.

CPT soundings were carried out before, during and after blasting operations to provide data that blasting had produced the required densification. Some of this data is reproduced on Fig. 33 which shows the q_c and state profiles for the as-placed core and after the first pass of blasting (which produced most of the densification) was completed. As can be seen, the blasting has near-doubled the q_c values. The computed state profiles are also shown and it can be seen that state has typically changed by -0.07, a value which correlates well with the measured surface settlement of 0.6 m at the end of the first pass and a densification zone of about 16 m. Further details on settlements and achieved CPT resistance may be found in [69]. Perhaps the most important feature of the densification is that even after considerable effort the characteristic state of the undensified, hydraulically placed berm is more dilatant than the densified core as may be seen by comparing Fig. 33 with Figs. 26, 24 and 23.

Densification of the Molikpaq core was not limited by considerations of temporary instability. The fill densified was entirely contained by the caisson so that the sand fluidization caused by the densification process could not lead to short term instability.

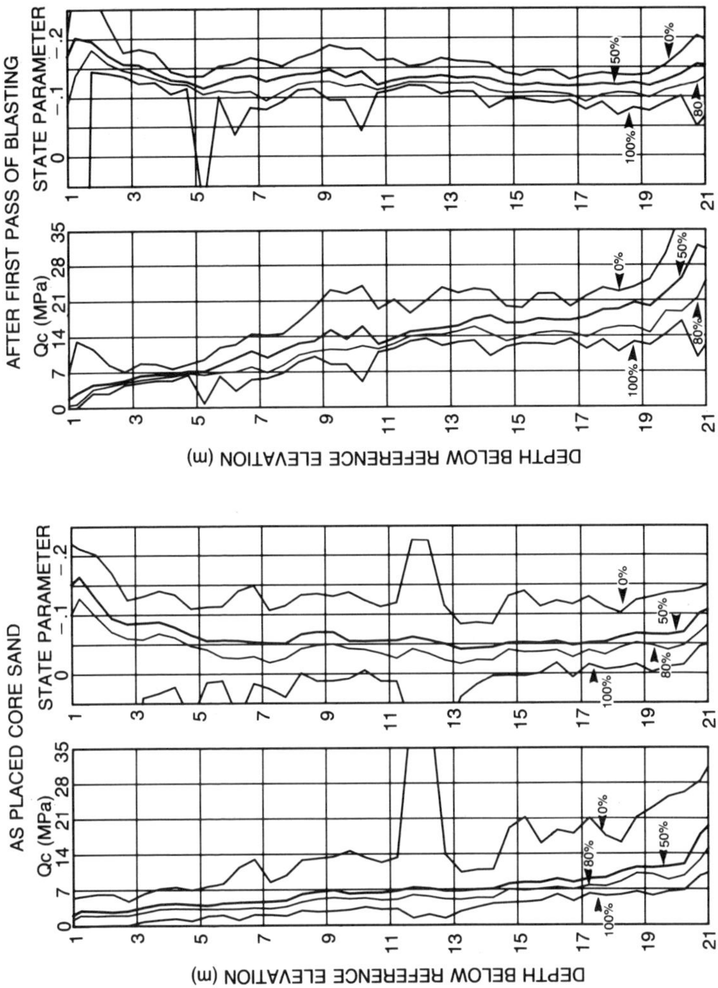

Figure 33

EFFECT OF EXPLOSIVE DENSIFICATION AT AMAULIGAK F-24

% BY VOLUME GREATER THAN VALUE SHOWN

Obviously, this would not be the case during densification of a berm or island where densification might well provoke a failure. Thus, although explosive densification was successfully applied to a clean hydraulically placed sand and gave the expected results, it should not be assumed to be an appropriate treatment for all hydraulically placed sands without proper consideration of temporary, transient conditions.

CONCLUSION

Experience gained in the construction of some 36 artificial islands in the Canadian Beaufort Sea has been reviewed and summarized. This experience shows that it is possible to repeatably construct quite dense fills purely by hydraulic placement of sand: specifically, this may be achieved by 'bottom dumping' of a medium sand with no more than five per cent silt from a trailer hopper dredge. Other placement techniques produce looser fills, also in an apparently systematic manner.

A central issue of hydraulic fills is that of adequate density, a particular concern when dealing with saturated hydraulic fills that may be subject to shock or cyclic loadings. The Amauligak I-65 and Nerlerk B-67 case histories were reviewed and force the conclusion that while dilatancy is a necessary criterion for stable behaviour of loose sands, it is not sufficient. The question then arises as to what is sufficient.

The state parameter approach was reviewed and its applicability to sandfill engineering summarized. This approach has particular relevance to steady state strength because the state parameter is defined with respect to the steady state locus in e - p' space. On the basis of the existing data and for vertical effective stress $\sigma'_v < 300$ KPa the condition that the steady state strength be at least as great as the drained strength will be met provided the characteristic state parameter is $\psi = -0.1$ or more negative (more dilatant). This corresponds to the condition that the total volumetric strain be dilatant at peak stress ratio, rather than just the dilatancy condition (which is the rate of volumetric strain change with shear strain change).

Both the existing data of full-scale steady state strengths as well as a simple critical state framework show that the strength may be expected to change exponentially with state or density. For typical quartz sands with only a few per cent silt content, the reduction in strength will be an order of magnitude for every 0.05 change to less dilatant states. This extreme sensitivity of strength to state, when combined with the state measurement precision of +-0.03, means that prudent engineering with saturated sandfills should involve characteristic states that are at least as dilatant as $\psi = -0.1$ regardless of any consideration of required steady state strength.

Selection of sand type to be used in construction is obviously an important decision. The Paper has illustrated the case for sand selection on the basis of the coefficient of consolidation, c_v. This presents a particular uniformity because, whether one assesses Beaufort Sea island construction in terms of construction induced pore pressure, or wave induced pore pressure, or achieved undensified state, the same value emerges as the criterion of a good construction sand: c_v > 2000 cm^2/sec. Fundamentally, this value should be project specific to the extent that the existence of excess pore pressure must also scale with fill size (drainage path), fill placement rate and slurry density. Nevertheless, these latter factors have a much smaller range of values in full scale construction than c_v for sand, so that use of a criterion that $c_v > 2000$ cm^2/sec is a reasonable starting point for other projects.

The uncertainty in extrapolating the Beaufort experience to other situations indicates the importance of developing a theoretical basis for the consolidation behaviour of sand. The experience in the Beaufort shows that a whole range of densities can be repeatably constructed, all of which comprise "normally consolidated" sands. While some of the important factors have been identified the fact remains that there is no theory for predicting the achieved density in terms of fundamental variables such as rate of accretion, drainage path, c_v, slurry density, etc. Until such a theory exists and has been calibrated against full scale data, a significant restriction will exist on the confident engineering of hydraulic sandfills.

The restriction on extrapolating experience was overcome in Beaufort island construction by a three step approach:

- build using best estimated materials/procedures;
- assess what state has been achieved on completion of construction by testing;
- have contingency plans in place for remedial activities in case they are required;

which in effect is an implementation of the much quoted "observational method".

Acknowledgement

The Authors gratefully acknowledge the support of their respective companies and colleagues in carrying out the work described in this paper.

REFERENCES

[1] Riley, J. G. (1975); The construction of artificial islands in the Beaufort Sea. Proc. 7th Annual Offshore Technology Conf., Houston, OTC 2167.

[2] de Jong, J. A. A., Stiger, C. and Steyn, B. (1975); Design and building of temporary artificial islands in the Beaufort Sea. Third Int. Conf. on Post and Ocean Eng. under Arctic Conditions, Fairbanks, University of Alaska.

[3] Boone, D. J. (1980); The construction of an artificial drilling island in intermediate water depths in the Beaufort Sea. Proc. 12th Annual Offshore Technology Conf., Houston. OTC 3873.

[4] Jefferies, M. G., Stewart, H. R., Thomson, R. A. A. and Rogers, B. T. (1985); Molikpaq deployment at Tarsiut P-45. Proc. ASCE Spec. Conf. Arctic '85, San Francisco.

[5] de Jong, J. J. A. and Bruce, J. C. (1978); Design and construction of a caisson retained island drilling platform for the Beaufort Sea. Proc. 10th Offshore Technology Conf., Houston. OTC 3294.

[6] Article in Canadian Consulting Engineer: 'Caisson-retained island for Esso Resources wins top honour'. October, 1984.

[7] 'Arctic Supplement', Offshore Engineer, Aug. 1981

[8] Robertson, F. P. (1985); Arctic island construction Proc. ASCE Spec. Conf. 'Arctic '85', San Francisco, p. 48-56.

[9] Fitzpatrick, J. and Stenning, D. G. (1983); Design and construction of Tarsiut Island in the Canadian Beaufort Sea. Proc. 15th Annual Offshore Technology Conf., Houston. OTC 4517.

[10] Janson, W. (1983); Artificial island construction in the Canadian Beaufort Sea. Proc. of World Dredging Conference, Singapore.

[11] Dikken, J. J. and Brakel, J. (1982); Dredging and construction techniques for steep slopes on artificial drilling islands in the Canadian Arctic. Proc. 14th Annual Offshore Technology Conf., Houston. OTC 4225.

[12] Bruce, J. C. and Harrington, G. (1982); Design aspects of a mobile Arctic caisson. Proc. 14th Annual Offshore Technology Conf., Houston. OTC 4333.

[13] Fitzpatrick, J. (1983); The single steel drilling caisson - a novel approach to bottom founded structures in Arctic waters. Proc. 58th Annual Tech. Conf., Soc. Petroleum Eng., San Francisco.

[14] Berzins, W. E. and Hewitt, K. J. (1984); The use of hydraulically placed sand in island construction. Proc. 16th Annual Offshore Technology Conf., Houston. OTC 4671.

[15] British Standard BS6235, British Standards Institution, London.

[16] Wright, B. D., and Weaver, J. S. (1983); Tarsiut Winter alert monitoring. Proc 15th offshore Tech. Conf., Houston. OTC 4518.

[17] Croasdale, K. R. (1988); Ice forces: Current Practices. Proc 7th OMAE, Houston, Vol. 4, p. 133-151 pbl ASME.

[18] Sanderson, T. J. O. (1986); a pressure-area curve for ice. Proc 8th Int. Symp. Ice, IAHR, Iowa City.

[19] Jefferies, M. G. and Wright, W. E. (1988); Dynamic response of Molikpaq to ice-structure interaction. Proc 7th Int. Symp. Offshore Mech. and Arctic Eng., Houston.

[20] Whitman, R. V. (1970); Hydraulic fills to support structural loads. Proc. ASCE SMI p 23-47.

[21] Rowe, P. W. (1962); The stress dilatancy relation for static equilibrium of an assembly of particles in contact. Proceedings of the Royal Society of London, A, 269: 500-527

[22] Been, K. and Jefferies, M. G. (1985); A state parameter for sands. Geotechnique 35, 2, p 99-112.

[23] Schofield, A. and Wroth, C. P. (1968); Critical state soil mechanics, pbl. McGraw-Hill.

[24] Been, K. and Jefferies, M. G. (1986); Discussion on 'A state parameter for sand'. Geotechnique, 36, 1, p 127-132.

[25] Jefferies, M. G. and Been, K. (1987); Use of critical state representations of sand in the method of stress characteristics. Can. Geot. J. 24, 3, p. 441-446.

[26] Arthur, J. R. F., and Menzies, B. K. 1972; Inherent anisotropy in a sand. Geotechnique, 22: p. 115-128.

[27] Oda, M. 1972a. Initial fabrics and their relations to mechanical properties of granular material. Soils and Foundations 12(1): 17-36.

[28] Oda, M., Konishi, J., and Nemmat-Nasser, S. 1980. Some experimentally based fundamenal results on the mechanical behaviour of granual materials Geotechnique, 30: 479-495.

[29] Tatsuoka, F. (1987); Discussion of 'The strength and dilatancy of sands', Geotechnique, 37, p. 219-226.

[30] Castro, G. and Poulos, S. J. (1977); Factor affecting liquefaction and cyclic mobility. J. Geotech. Eng. Div. ASCE, 103, 6, p 501-505.

[31] Poulos, S. J., Castro, G., and France, J. W. (1985); Liquefaction evaluation procedure. J. Geotech. Eng. Div. ASCE 111, 6, p 772-792.

[32] McCreath, D. R., Hodge, W.E., and Harrington, A. G., "Geotechnical Design Considerations for the Gulf Mobile Arctic Caisson", 2nd Canadian Conference on Marine Geotechnical Engineering, Halifax, 1982.

[33] Seed, H. B. (1987); Design problems in soil liquefaction. J. Geotech. Eng. Div. ASCE 113, 8, p 827-845.

[34] Mancini, C. V., Dowse, B. E. W. and Chevallier, J. M., (1983); Caisson retained island for Canadian Beaufort Sea - Geotechnical design and construction considerations. Proc. 15th Offshore Technology Conf., Houston. OTC 4581.

[35] Crooks, J. H. A., Becker, D .E., Jefferies, M. G., and McKenzie, K. 1984. Yield behaviour and consolidation. Part 1. Pore pressure response. Proceedings, ASCE Geotechnical Engineering Division Symposium on Sedimentation Consolidation Models Predictions and Validation, San Francisco, Edited by R.N. Yong and F.C. Townsend. pp. 356-381.

[36] Becker, D. E., Crooks, J. H. A., Jefferies, M. G., and McKenzie, K. 1984. Yield behaviour and consolidation. Part 2. Strength gain. Proceedings, ASCE Geotechnical Engineering Division Symposium on Sedimentation Consolidation Models, Predictions and Validation. San Francisco. Edited by R. N. Yong and F. C. Townsend. pp. 382-398.

[37] Becker, D. E., Jefferies, M. G., Shinde, S. B., and Crooks, J. H. A. (1985). Pore pressure in clays below caisson islands, Proc ASCE Spec. Conf. 'Arctic '85', San Francisco.

[38] Conlin. B. H., Jefferies, M. G. and Maddock, W. P. (1985); An assessment of the behaviour of foundaiton clay at Tarsiut N-44 caisson retained island. Proc 17th Offshore Tech. Conf., Houston. OTC 4884.

[39] Finn, W. D., Iai, S. and Ishihara, K. (1982); Performance of artificial offshore islands under wave and earthquake loading; field data analysis. Proc 14th Offshore Tech. Conf., Houston. OTC 4220.

[40] O'Connor, M. J. (1980); Development of a proposed model to account for the surficial geology of the Southern Beaufort Sea. Open file Report 954, Geological Survey of Canada.

[41] Fitzmaurice, S. J., Williams, S. E., Moor, R. and Moir, J. H. (1985); Construction of steep sloped underwater berms by bottom dumping using trailer suction hopper dredges. Proc. 17th Annual Offshore Technology Conf., Houston. OTC 4921.

[42] Stewart, H. R., Jefferies, M. G., and Goldby, H. M. (1983); Berm construction for the Gulf Canada mobile Arctic caisson. Proc. 15th Annual Offshore Technology Conf., Houston. OTC 4552.

[43] Stewart, H. R. and Brakel, J. (1986): 'Molikpaq Berm Construction At Amauligak I-65', World Dredging Conference, Brighton, U.K.

[44] Stewart, H. R., McPeak, K. and Jones, H. (1986); The challenge of precise positioning for offshore exploration and development in the Arctic. Proc. CPA/CHA Colloquium 4, Lake Louise, Alberta.

[45] MacLeod, N. R. and Butler, J. H. (1979); The evaluation of dredging materials for island construction in the Beaufort Sea. Proc. 11th Offshore Technology Conf., Houston.

[46] Goldby, H. M., Crooks, J. H. A., Harper, J. R. and Stuckert, B. (1986); A novel approach to fill material quality assessment: near real-time grading of dredged sand. Proc. 3rd Can. Conf. Marine Geotechnical Eng., St. Johns, Newfoundland.

[47] Barnes, M., (1977); 'Measurement in Contract Control'. pbl. Thomas Telford Ltd.

[48] Finn, W. D., Liam, Siddharthan, R. and Martin, G. (1983); Response of seafloor to ocean waves, Journal of the Geotech. Eng. Div. ASCE Vol. 109, GT-4, Proceedings paper 17880, April 1983, p. 556-572.

[49] Robertson, P. K., Campanella, R. G. and Wightman, A. (1982); SPT-CPT Correlations. ASCE Journal of Geotechnical Division.

[50] Jefferies, M. G., Rogers, B. T., Griffin, K. M. and Been, K. (1988); Characterization of sandfills with the core penetration test. Proc. Geotech conf. on penetration testing in the UK, pbl Thomas Telford Ltd., London.

[51] Hardy, M. D., Jonsson, L., Jefferies, M. G. and Griffin, K. M. (1987); The application of microcomputers to CPT testing in the Arctic Offshore. Proc. 1st Can. Symposium on Microcomputers in Geotech. Eng., Regina.

[52] Det Norske Veritas: Rules for design, construction and inspection of offshore structures.

[53] Been, K., Crooks, J. H. A., Becker, D. E. and Jefferies, M.G. (1986); The cone penetration in sands, part 1: state parameter inter- pretation. Geotechnique 36, 2, p 239-249.

[54] Been, K., Jefferies, M. G., Crooks, J. H. A. and Rothenburg, L.; The cone penetration test in sands, part 2: general inference of state. Geotechnique 37, 3, p 285-299.

[55] Jefferies, M. G., Jonsson, L. and Been, K. (1987); Experience with measurement of horizontal geostatic stress in sand during cone penetration test profiling. Geotechnique 37, 4, p 484-498.

[56] Graham, J. P. and Jefferies, M. G. (1986); Some examples of in-situ lateral stress determinations in hydraulic fills using the self-boring pressuremeter. Proc. 39th Can. Geot. Conf., Ottawa.

[57] Been, K., Lingnau, B. E., Crooks, J. H. A. and Leech, B. G. (1987); Cone penetration test calibration for Erksak (Beaufort Sea) sand. Can. Geot. J. 24, 4.

[58] Crooks, J. H. A., Shinde, S. B. and Been, K. (1985); In-situ state of underwater hydraulic sand fills, ASCE Arctic '85 Conference, San Francisco, March 1985, p 84-92.

[59] Mitchell, J. K. and Solymar, Z. V. (1984); Time-dependent strength gain in freshly deposited or densified sand. J. Geotech. Eng. Div. ASCE 110, 11, p 1559-1576.

[60] Gibson, R. E. (1958). The progress of consolidation in a clay layer increasing in thickness with time. Geotechnique, 8(4), pp. 171-182.

[61] Shinde, S. B., Crooks, J. H. A., James, D. A., Williams-Fitzpatrick, S. E. (1986); Geotechnical design for Beaufort Sea structures. Third Canadian Conference in Marine Geotechnical Engineering, St. John's, Newfoundland, June.

[62] Oilweek, Sept. 23, 1985

[63] Canadian Oil and Gas Lands Administration. Report on the Erosion Incident at Artificial Island Esso PCI Home et al Minuk I-53. May, 1986.

[64] Mitchell, D. (1984); Liequefaction slides in hydraulically placed sands. Proc. Int. Symposium on Landslides, Toronto, 1984.

[65] Berzins, W. E. and Hewitt, K. J. (1984); The use of hydraulically placed sand in island construction. Proc. 16th Offshore Tech. Conf. Houston. OTC 4671.

[66] Sladen, J. A., Hollander, R. D., Krahn, J. and Mitchell, D. E. (1983); Back-analysis of the Nerlerk berm liquefaction slides. Can. Geotech. J. 22, 4, p 579-588.

[67] Been, K., Conlin, B. H., Crooks, J. H. A., Fitzpatrick, S. W., Jefferies, M.G., Rogers, B.T. and Shinde, S. (1987); Back analysis of the Nerlerk berm liquefaction slides: Discussion, Can. Geot. J., 24, p 170-179.

[68] Been, K., Crooks, J. H. A., Conclin, B.H. and Horsfield, D. (1988); Liquefaction of hydraulically placed sandfills. Proc. ASCE Speciality Conf. on Hydraulic Fills, Fort Collins.

[69] Stewart, H. R. and Hodge, W. E. (1988); Molikpaq core densification with explosives at Amauligak F-24. Proc 20th Offshore Tech. Conf. Houston. OTC 5684.

[70] Kent, D. D., Graham, B. W. and Sangster, R. H. B. (1979); Geotechnical design of a caisson retained island for exploration drilling in the Beaufort Sea. Proc. 1st. Can. Conf. Marine Geotechnical Engineering, Calgary, p 429-437.

[71] Kent, D. D., and Hayley, D. W. (1977); Instrumentation and construction monitoring of Kugmallit and Arnak drilling islands. Report to Imperial Oil Limited, EBA Engineering Consultants Ltd.

REVIEW OF PRACTICE IN HYDRAULIC FILL CONSTRUCTION

By Serge Jih Chen Hsu[1], F.ASCE

ABSTRACT: This paper deals with the practical aspects and recent development in hydraulic fill embankments (other than tailings). Examples of early hydraulic fill dams from 1870 to 1940 in the U.S.A., their deficiencies, and several accidents that took place during construction are mentioned.A serious slide of the upstream slope of Fort Peck Dam on the Missouri River, nearing completion in 1938, and the rapid advent of heavy construction equipment produced a decline of hydraulic fill practice in the United States. Since 1940 there has been an extraordinary development of rolled fill dam construction in the U.S.A., Europe and Japan, while hydraulic filling has gradually become restricted to tailings and dredging of rivers and canals. This paper discusses experience learned from the early hydraulic filling, as well as improvements and recent developments in the Soviet Union of hydraulic fill techniques or hydromechanization, as it is commonly known there.Typical examples are briefly described. Examples of hydraulic fill dams built in North America, data on Soviet dams, bibliography, and conclusions are given.

INTRODUCTION

Generally speaking, hydraulic filling is associated with the process of transport, separation and deposition of solids using water or other liquids.A hydraulic fill dam is an earth dam in the construction of which the materials are transported and distributed to their final position by water. In some hydraulic fill dams part of the materials are transported to the dam by some means other than water, and then sluiced into the final position. Parts of a hydraulic fill can also be built as a rolled fill.

Starting in the 17th century, Holland developed rudimentary dredging equipment and successfully reclaimed large areas from the sea using the hydromechanization process.

On a large scale,dredging was used to excavate the Suez Canal in 1865; and for hydraulic mining to remove 120 million m^3 of overburden by monitors over the Steep Rock Iron Ore deposit in Canada, 1955-1960. In Arizona, 650 tons/hr of coal were moved hydraulically over a distance of 400 km through 450 mm diameter pipeline completed in 1967.

Between 1947 and 1973 over 100 structures in hydroelectric projects were built by hydromechanization process in the Soviet Union.

EARLY HYDRAULIC FILL DAMS

The hydraulic fill method of dam construction has been know for years

(1)Geotechnical Engineering Consultant.Rua Pedroso Alvarenga,260-apt.112 -CEP 04351,São Paulo-SP,Brazil.Phone:(011)280-8935.TELEX:1137472DGCGBR as an American technical process initiated during gold mining at the turn of the century in California. This process permitted economical excavation and transportation of sand and gravel using water.

In fact, until about 1930, it was the only economical way to build large earthfills. Thus over 100 hydraulic fill dams from 10 to 80 metres high involving more than 200 million m^3 were built in North America alone.

Before soil mechanics was introduced into engineering practice around 1925, each dam was designed and constructed on an individual basis, following rules of thumb.

The first principal rules and construction methods for building hydraulic fill dams were introduced by Schuyler, 1906. He described his experience and difficulties in building such early dams as the Lake Frances Dam and others in California; on the Snake Ravine in Idaho; Hawaii; New Mexico; as well as dams Nº 2,3,4 on the Necaxa River in Mexico; and Guarapiranga(Sto.Amaro) dam in Brazil. It is interesting to observe his remarks on two failures at the Snake Ravine Dam. He attributed them to lack of experience and supervision in building the dam on the top of mine tailings.

A typical early hydraulic fill dam would have a 3:1(H:V)upstream slope, a 2:1(H:V) downstream slope and a high clay content semiliquid core. It was often built without proper quality control and supervision.Impervious clay cores that were slow to consolidate, often combined with insufficiently strong fine-grained shells, resulted in several slope failures or slides during construction.

Calaveras Dam, (see Hazen, 1920, Table 1) near San Francisco was damaged while approaching completion in 1918. 600.000 m^3 of the upstream slope droped 30 metres and moved upstream over 100 metres. It was found that the effectives sizes of the core and filter materials were too fine - 0.002 mm and 0.3 to 0.4 mm respectively, while dry density of the core varied between 1.59 g/cm^3 to as low as 0.79 g/cm^3.

Hazen, 1920, pointed out that some of the accidents during construction could have been avoided. He said that if "... core material can be fully drained and consolidated, there is no reason why a dam built by hydraulic method... should not be safe..." he goes on to state that "... The number of failures with hydraulic fill dams growing out of this fundamental condition of fluid cores and inadequate toes has been such as to make it clear that this condition of instability has not been recognized in the design...".

A slide also occurred when the 80 m high Fort Peck Dam was nearing completion in September, 1938, (Creager et al., 1945).

A typical cross section is shown in Figure 1. This dam was built in Montana, U.S.A., between 1934 and 1941, by full hydraulic fill method, using large dredges and boosters from borrow pits up to 8 km away with average daily production of 95,000 m^3.

The total length of the dam (including the dike on the east abutment) is 6.4 km; the total volume of the dam is 94.8 x $10^6 m^3$, considered the largest dam of this type built at the time.

A core material grain size curve is shown in Figure 2. It contained fine sandy silt and had the angle of internal friction, $\emptyset = 31^{\circ}$ and

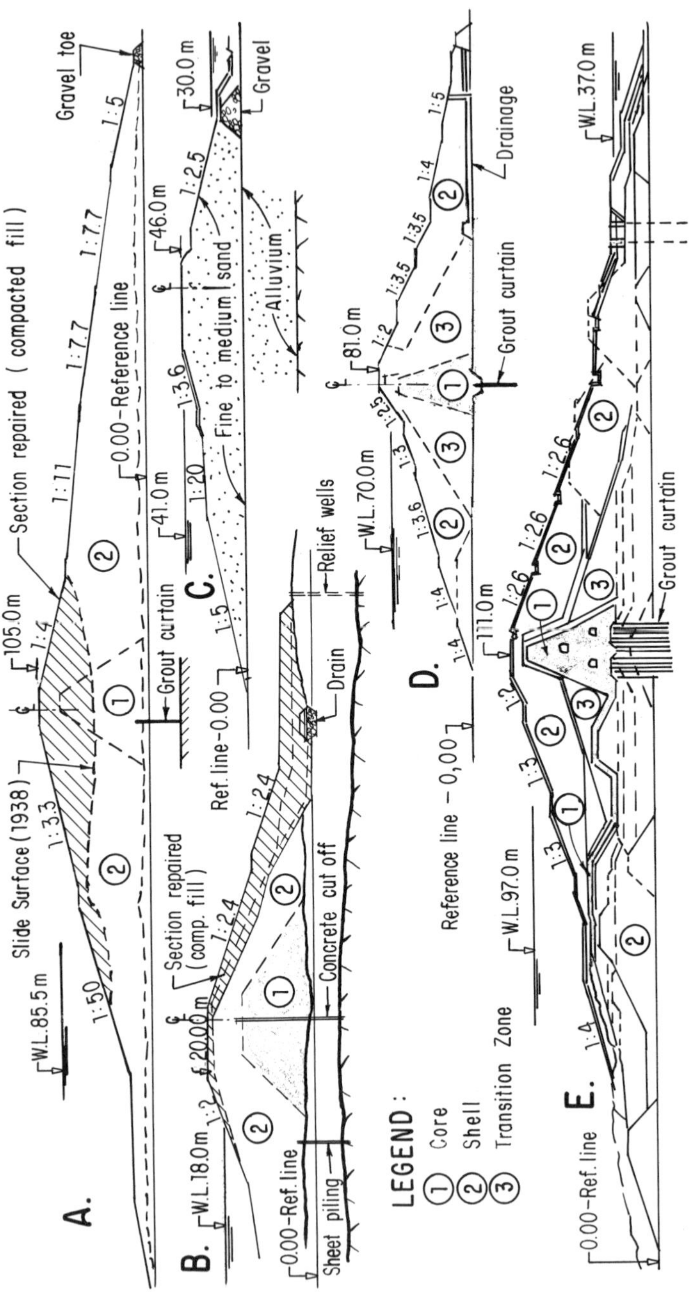

A. Fort Peck as reconstructed, Creager, et al, 1945; B. Guarapiranga (Santo Amaro), Guerra, 1986; C. XXII Party Reunion, D. Mingechaurskaya, E. Aswan High Dam, Hydroprojekt, 1973.

FIG. 1 – TYPICAL SECTIONS OF HYDRAULIC FILL DAMS

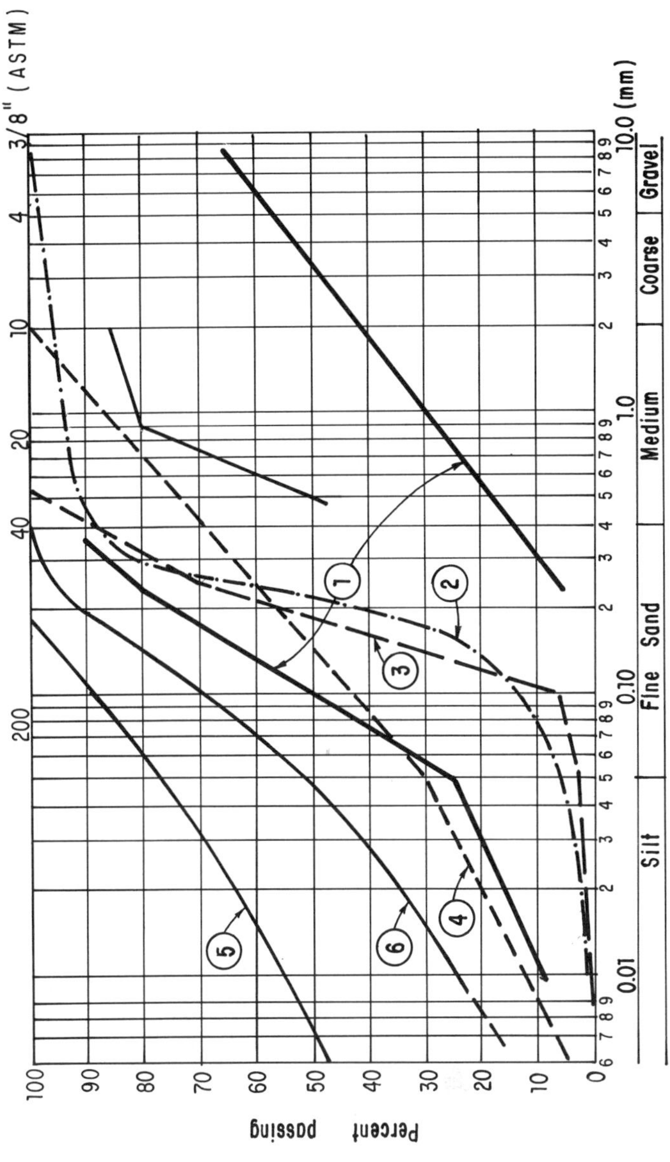

① Limits recommended by Soviet Union for "Homogeneous" sand dams (SNIP II-53-73, 1974) ② Porto Primavera, Brazil, (core), Hsu, 1980; ③ XXII Party Reunion, ④ Mingechaurskaya Dam, (cores), Hydroprojekt, 1973; ⑤ Guarapiranga (core), Hsu, 1980; ⑥ Fort Peck, (core), Creager et al, 1945.

FIG. 2 – GRAIN SIZE CURVES OF SOME HYDRAULIC FILL DAMS

no cohesion. The effective size of the shell material was rather fine, D_{10}= 0.13 mm; the uniformity coefficient Cu = D_{60}/D_{10} = 3.0 and ϕ' = 36°.

When the slide took place, the fill was a mere 6.7 m below the final crest elevation. Eight million m³ of the upstream portion moved outward and slid into the reservoir.

A postslide consulting board concluded that failure "... was due to the fact that the shearing resistance of the weathered shale and bentonite seams in the foundation was insufficient to withstand the shearing force... the slide progressed upstream... due, in some degree, to a partial liquefaction of the material in the slide..." (Creager et al.,1945).

The dam was subsequently reconstructed using very flat slopes,as shown in Figure 1 and the repaired core was completed by rolled fill.

Calaveras Dam contained clay particles of 0.001 to 0.002 mm size and had a very low dry unit weight as mentioned above. At Fort Peck Dam the core has over 20% of clay particles smaller than 0.01 mm (see the grain size curve in Figure 2). Hazen, 1920, recommended to discard all particles smaller than 0.01 mm. Slides occured at Calaveras and Fort Peck Dams during their construction.

It seems that the lessons of early slides during construction and Hazen's recommendations were not always taken into consideration.

Brazilian experience with hydraulic fills involves 12 low embankments, built between 1906 and 1945, by the former Brazilian Traction Light and Power Company, according to the prevailing American practice.

These structures are up to 35 metres high fills with silty clay central cores and fine grained shells, built without close supervision, using materials from borrow pits in residual gneiss soil sluiced into the dam. These dams and dikes, situated along the rims of the Billings and Lajes reservoirs,are located near the industrial cities of São Paulo and Rio de Janeiro.

During the construction of the Rio Grande Dam near São Paulo, Brazil, a mass involving 30,000 m³ of fill and 6,000 m³ of concrete core wall slid into the reservoir (1934). The accident was attributed to the pressure exerted by a liquid-like clay core on the concrete core wall, and to the high rate of fill placement. The damaged section was promptly repaired, Guerra, 1986.

Most of these dams were built on soft silty clay foundations with steel or wood sheet piling or concrete core walls extending into the fill, as shown on the typical section of Guarapiranga (Sto.Amaro)Dam in Figure 1.

The high void ratio and low density in core and shell materials alarmed Arthur Casagrande during his consultation visits in 1969-1971.Although the dams performed satisfactorily, with the growth of São Paulo and Rio de Janeiro, and because of the mentioned shortcomings, their safety was questioned. Upon consultant's urging, geotechnical investigations were undertaken and instrumentation was installed. Several of these hydraulic fills were reinforced in recent years, including Guarapiranga (Sto. Amaro) and Rio Grande (Billings) near São Paulo.

According to Moretti, 1986, a large test fill was executed in 1980 at the site of the Porto Primavera Hydroelectric Project on the Parana

River in Brazil, using a 16"-800 HP dredge. The 10 m high test fill was built using uniform fine to medium sand, (see grain size curve in Figure 2). The test was a part of an investigation carried out for an alternative design of a 22 m high, 9 km long section of the Porto Primavera Dam to be placed on 14 metres thick alluvial deposits.

The in situ test fill showed the following average results: dry density = 1.67 g/cm^3; relative density = 55.9%; Darcy's coefficient of permeability K = 1.5 m/s, considered as quite satisfactory.

HYDRAULIC FILL DAMS IN THE SOVIET UNION

Hydraulic fill dams, similar to those described above and using American techniques were introduced in the Soviet Union in 1935 and gradually evolved, especially after 1945, as follows.

According to Hydroprojekt-Coyne et Bellier Seminar, 1973, sponsored by the Soviet Hydromechanization Department, between 1947 and 1973 over 100 dams in 52 hydroelectric projects were built by hydraulic filling.600 km of structures as dams, dikes, levees, canal banks, jetties, cofferdams, and underwater fills with a total volume exceeding 800 million m^3 were constructed.

Since equipment for building rolled compacted fills was not so well developed in the Soviet Union, hydraulic filling was a good alternative to build large dams economically.

It seems that Soviet engineers have followed recommendations given by Hazen, 1920: they reduced the clay content and the width of cores,thereby the core would readily drain and consolidate,and provided flat slopes and ample external shells.

The Soviet experience is that the most important feature in hydraulic filling is the grain size distribution of the available borrow material at the site.

Accordingly, all soils are classified into the five main types shown in Figure 3:

Class I - is a fine to medium uniform sand ideally suited for homogeneous dams without cores;

Classes I and II - are also suitable for zoned embankments in which the core is built mostly of sand;

Classes III and IV soils are used for zoned embankments where core material consists of sandy clayey silt requiring gravel or rockfill shells;

Class V - is a coarse sand and gravel requiring cores of fine sandy silty clay.

The following three main types of hydraulic fill dams were developed in the Soviet practice:

1. HOMOGENEOUS(sand) dams have been the most widely used for the past 25 years.

These are built with Class I soils, and are most suitable for large rivers flowing on an alluvial foundation. Soviet experience in this type is such that the government regulated code, SNIP II-53-73, of 1974, requires no special design calculations for embankments up to 50 metres high. Also, wear on equipment and pipelines in transporting such mate-

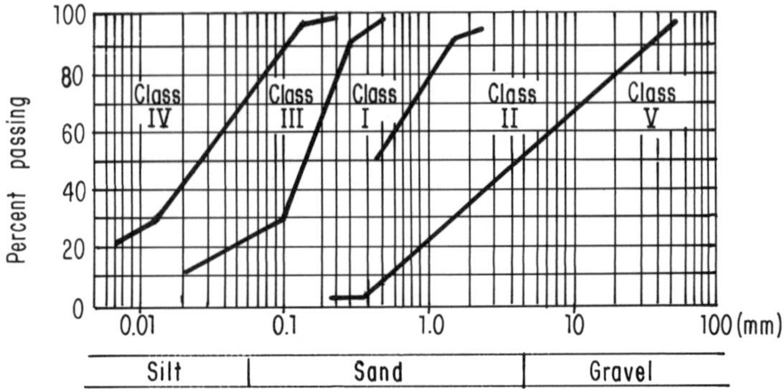

FIG. 3 – GRAIN SIZE DISTRIBUTION OF SOIL USED IN SOVIET UNION. SPECIFICATION SNIP II-53-73, 1974

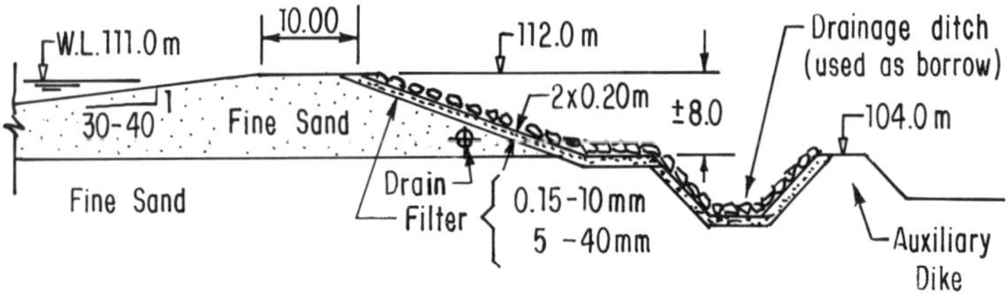

Seepage = 3.5 m^3/s (total)
Height = 8.00 m
Volume = 17×10^6 m^3 (fine sand)
Reservoir area = 12.5 Km^2

Dry density – γ_s = 1.62 g / cm^3
Length = 12 Km
Darcy's k = 5×10^{-3} cm/s to 10^{-2} cm /s
Reservoir volume = 52×10^6 m^3

FIG. 4 – CHERNOBYL DIKE, SOVIET UNION. TYPICAL SECTION

rial is minimal.

These types of dams, listed on Table 1, were built on Volga, Dniepre, Don and other rivers with large regulated flows, founded on thick alluvium deposits. Many of these fills have no cutoff provisions in the dam or in the foundations allowing a certain amount of seepage loss, usually small, such as XXII Party Reunion Dam on the Volga.See the typical cross section given in Figure 1, and the grain size distribution of the soil used in Figure 2.

In order to prevent the phreatic line from reaching the downstream slope, these embankments are provided with very wide crests,flat slopes, downstream filter layers, and toe drains or toe trenches. At the XXII Party Reunion Dam, a 100 m wide crest accommodates a highway and two railway tracks.

In homogeneous dams, particles less than 0.01 mm are discarded. The uniformity coefficient, $Cu = D_{60}/D_{10}$, is usually bellow 2.5, allowing placement without much segregation.

2. ZONED OR COMPOSITE type dams are constructed using Class III, IV and V materials with high uniformity coefficient to produce segregation.

Use of these types of materials produces a distinct central core,containing fines for water tightness, and pervious external shells which have to be strong enough to confine the soft core.

Usually not more than 5 to 10 percent of material less than 0.005 mm is allowed in the core, which permits it to consolidate readily.

Few dams of this type are built in the Soviet Union. A typical example is the Mingechaurskaya dam, 80 m high and 1.5 km long on the Kura River, Tashkent,see the typical cross section in Figure 1, and data in Table 1. This dam has fine to medium sandy silt in the core, flanked by wide intermediate zones of medium to fine sand, and confined by gravel shells with relatively flat external slopes. The dam is situated in a seismic region, Hydroprojekt, 1973.

Another example of the Zoned or Composite type dam is the 110 m high, 3.5 km long Aswan High Dam on the Nile River in Egypt. See the cross section in Figure 1. This dam, with elaborate zoning, was built on 210 m thick alluvium deposits. The confining rockfill shells and most of the core of coarse sand was placed under water. Extensive cement-clay grouting of the core (coarse sand) and the foundation alluvium deposits to a depth of 210 metres was carried out, Hydroprojekt, 1973.

The voids in rockfill shells were filled with fine dune sand.The foundation alluvium sand strata were densified under 37 m depth of water using special floating vibrators. Sandy core, transitions and rockfill shells above water level were placed as rolled fill and compacted with vibratory rollers and tractor treads.

3. FLAT SLOPED dams are usually built mostly of sand with a wide crest and 30:1(H:V) to 50:1(H:V) slopes. These are low dams, 8 to 15 m high, constructed in regions where other soils are not available.

Chernobyl dike is an example of this type of hydraulic fill. It was built to retain a reservoir for cooling a 2000 MW nuclear power plant near Kiev. See Figure 4. Completed in 1976 to form a 12.5 km long reservoir to retain 52 million m^3 of water, the dike is 12 km long, 8 m

TABLE 1. - DATA ON HYDRAULIC FILL DAMS IN SOVIET UNION

Name	Built (Compl.)	River	Location	Type	Hight (m)	Length (m)	Volume x 1000 (m³)	Foundation	Slopes US	(1 on..) DS	γ_s mean (g/cm³)	K (m/s)	Exit gradient (mean)	Placement (m³/day)	Cu= D_{60}/D_{10}
Tsimlianskaya	1952	Don		H	30	12.8km	40.600	fine sand clay	3.5-5-15	4-12	1.59		0.13	75,000	3.6
Mingechaurs-kaya	1955	Kura	Azerbeijan	Z	80	1,550	15,600	samdstone ardosia	3-3.6	core: 2-3.5-4	γ_s=1.55 ϕ'=27°	0.1 0.012			22
										(intermed. zone):	γ_s=1.7 ϕ'=33°	2.0			6-14
										(shells):	γ_s=1.95 ϕ'=35°	1.0			
Dubossarskaya	1955	Dnestr	Moldava	Z	26	130	95,6	sand gravel	3.5	2.5-3-3.5	core: γ_s=1.60- shell: γ_s=1.85-	0.09		2,400	3-0 16.4
Kairak-Kums-kaya	1956	Syr Daria	Tajik	H	32	1200	3100	fine sand clay; K= 0,3m/s	4	4	γ_s=1.51 ϕ'=34°	0.014	0.08	3,700	2.5
Kamskaya	1956	Kama	Perm	Z	33	1900	4975	argilite	-	(core): (shells):	γ_s=1.66 γ_s=1.68		0.14		
Gorkovskaya-3	1956	Volga	Gorky	H	28	18,600	17,000	sandy gravel	-		γ_s=1.64			6,800	2.8-4.6
-4	1956	Volga	Gorky	H	-	-	total: =56,100		-		γ_s=1.62				
-5	1956	Volga	Gorky	H	-	-			-		γ_s=1.66				
Kakhovskaya	1956	Dniepr	Ucraina	Z	30	3190	12,100	sand	4-6		γ_s=1.57	0.01	0.06	100,000	2.2
Novosibirsk	1957	Ob	Novo-Sibirsk	Z	30	4380	8241	sand over clay		(core):	γ_s=1.62 ϕ'=26°	0.002	0.12		2.6
										(shells):	γ_s=1.66 ϕ'=28°	0.007		8,000	2
Volga (Lenin)	1958	Volga	Volvograd	H	47	3249	23,400	clay sand	6-7	3-6.5-3,5	γ_s=1.56		0.06	300,000	2.2
Volga (XII Meeting)(1)	1959	Volga	Stalingr.	H	50	6500	21,700	fine,medium sand,limes-tone	5-20-5	5-9	γ_s=1.55 γ_s=1.35	0.01	0.08	310,000	2
Pavlovskaya	1961	Ufa	Bashkere	Z	45	232	720	sand/gravel limestone		(core): (shells):	γ_s=1.40 γ_s=1.80	0.012 0.046	0.24		3.2

Name	Built (Compl.)	River	Location	Type	Hight (m)	Lenght (m)	Volume x 1000 (m^3)	Foundation	Slopes Us	(1 on ..) DS	Ysmean (g /cm^3)	K (m/s)	Exit gradient (mean)	Placement (m^3/day)	Cu=
Golovnaya	1963	Vakhsh	Tajik	Z	33	1047	1977	alluvial gravel	3-2.25	4.5-5.5-2	core: Ys=1.50 Ø'=10º C =0.15 Shells: Ys=1.62 Ø'=29º C =0.15	0.00046 0.00058		2,600	7.2 10
Caratamar	1965	Tobol	Kazakh	Z	22	1400	2100	sand gravel	3.5-4	2.5-4	core: Ys=1.53 shells: Ys=1.65	0.001 0.00012 0.025		3,500	11 4.7
Bratsk(1)	1965	Angara	Irkutsk	H	37	3000	6500	diabase	3-4	2.5-3	Ys=1.56 Ø'=27º	0.012		3,000	2.6 3.0
Pliavinskaya	1965	Daugava	Latvia	C	45	440	1530H 635C	dolomite	3-3.5	2.5-3-1.5	core: Ys=1.53 shells: Ys=1.95	0.0058	0.26		
Kievskaya	1965	Dniepr	Ukraina	H	22	70km	75000	loose sandy clay over dolomite	4	4-30-3.75	Ys=1.63	0.005 0.012		30,000	
Chernobyl	1976	Pri-prietr	Kiev	H	8	12km	17000		30-40	4	Ys=1.62				

H - Homogeneous hydraulic fill
Z - Zoned
C - Composit (mixed) hydraulic and compacted fill
(1) - Slide occurred during construction
US - Upstream slope
DS - Downstream slope
Ys - Dry unit weight
K - Darcy's coefficient of permeability
Cu - Uniformity Coefficient ($Cu = D_{60}/D_{10}$)

After: Meletiev et al., 1973; Energomachexport, Moscow-(1979); Also, information obtained by Brazilian engineers during techinical visit to the Soviet Union in 1980.

TABLE 1.(CONT.)

high, and contains 17 million m^3 of hydraulic fill (sand). Using flat upstream slopes 30:1(H:V) to 40:1(H:V) and 4:1(H:V) downstream slopes, there is no cutoff provision through the dike or alluvial foundation.The upstream slope is not protected against waves. The downstream slope is covered by a double layer inverted filter of 0.15 - 10 mm and 5 - 40 mm sizes. The dry unit weight of the sand is 1.62 g/cm^3, and the total seepage loss is 3.5 m^3/s.

Information on geotechnical characteristics,such as grain size curves, dry density,permeability, seepage discharge, angle of internal friction, uniformity coefficient, rate of placement, etc. of several Soviet dams is given in Table 1.(See Melentiev, 1973; Hydroprojekt, 1973).

Soviet developments in hydromechanization have permitted substantial improvement in hydraulic filling by eliminating unconsolidated plastic cores, introducing homogeneous or flat-sloped sand embankments, and reducing the width of cores in zones dams, thereby increasing the stability of the core confined by external shell zones.

High permeability of homogeneous dams or sandy cores produced higher seepage flows and higher phreatic lines that had to be accounted for. Measures such as flat slopes, wide crests and effective downstream drainage had to be introduced.

There is limited available information on accidents suffered by hydraulic fills in the Soviet Union, especially the behavior of embankments built in seismic regions. It is known that a few slides took place during construction due to high rates of placement. One such event occurred at the XXII Party Reunion Dam on the Volga River, Kuznetsov, 1966.

Accidental blasting of a temporary upstream cofferdam caused a near failure of Svir dam in 1935. Within seconds, a large portion of the upstream slope, built of sandy material liquefied and flowed 150 m out, destroying concrete structures. Total breaching of the embankment was prevented due to the presence of a compacted clay core that remained intact, Maslov, 1959; Ivanov, 1973; Castro, 1969.

Soviet design and construction experience is given in the following publications:

1. Melentiev et al, 1973, provide basic theoretical considerations and practical experience, giving performance data and examples of completed dams. From the grain size distribution of borrow material and the particle angularity, the final grain size distribution of hydraulic fill can be estimated, as well as the permeability, phreatic line, density, etc. Laboratory model and in situ hydraulic test fills serve to estimate mechanical characteristics. An analogy method is frequently used to obtain preliminary parameters for design, using constructed embankments of similar materials for comparison.

2. Ufin, 1974, gives details on the use and installation of mechanical equipment, excavation, transport and filling technique, the soil-water mixture and consistency, and quality control.Considerable improvement in construction, was introduced by placing pipes, through which the soil mixture is pumped to the dam, directly on the fill, using a tractor driven crane. These pipes used to be placed over timber trestles in dams built prior to 1947.

The maximum size of solids that can be transported economically at present is 100 to 150 mm.

The unit price of hydraulic fill was estimated to be between 0.27 to 1.0 Ruble, 1980 (1 US$ = 0.70 Rubles), depending on the pumping distance and other working conditions, Meletiev et al., 1973.

LIQUEFACTION OF SANDS

Liquefaction or flow failure of loose, saturated sand deposits has been responsible for the failure of slopes,dikes, earth dams, and buildings founded on such deposits.

Liquefaction of sands is caused by a substantial reduction in its shear strength which, in turn, is caused by the development of high pore water pressures, induced by the static or cyclic stress application. During liquefaction the mass of sand actually flows spreading out until the shear stress acting within the sand become so small that it turns compatible with the reduced shear strenght of the sand.

Over the past 50 years this important but complex problem has been the subject of many studies and research works.

A considerable amount of research has been carried out by Casagrande, 1975, at Harvard and Seed, 1975, and his group at Berkeley Universities.

There has been no general agreement on the basic concepts of the liquefaction phenomenon, and certain differences still persist, Castro, 1969; Casagrande, 1975; Seed, et al., 1975; Peck, 1979; Seed et al., 1983;Discussions: Pilecki, 1985; Poulos et al., 1985; Discussions: Alarcon, Leonards, De Alba, Dennis, Pilecki, Poulos et al., 1988 (Feb.).

Research on the liquefaction problem was also carried out in the Soviet Union since 1935 by Maslov, 1959, and others.

Ivanov, 1978, concludes that for a complete liquefaction to take place, the following three conditions must be satisfied:

1.there must be a total destruction of the soil mass,(large deformation);
2.possibility of volume change (densification);
3.total or near total soil saturation(increase in pore water pressure).

It follows, that if one of the above conditions were not satisfied, liquefaction could not take place.

Casagrande, 1975, came to the following, apparently his final, conclusion concerning soil liquefaction: "... it is unlikely that a laboratory test can be devised which will produce in test specimens the type of uniform stress distribution that exists during cyclic loading in a typical element in situ, the author believes that... in situ response will require primarily comprehensive field investigations of many full-scale tests nature has performed in highly seismic regions...".

In spite of problems related to liquefaction, numerous hydraulic fill dams have been successfully built over the past 100 years.

A considerable technical advance in hydromechanization was made in the Soviet Union, especially in building homogeneous hydraulic fills using fine to medium uniform sand.

In seismic regions an additional compacted rockfill layer can be placed over the fill to provide stable zones, and to insure an adequate safety factor.

EVALUATION OF LIQUEFACTION PONTENTIAL USING FIELD PERFORMANCE DATA

Since the Alaska and Niigata earthquakes of 1964, studies on data for sites which are known to have liquefied or not liquefied and the standard penetration resistance, SPT tests,at these sites are being conducted in the U.S.A., Japan, China, and other countries.

Correlation, charts and the empirical methods for evaluating the liquefaction induced by actual earthquakes using SPT tests are given by Seed et al., 1983. These simplified practical procedures seem to be an important advance in engineering practice.

HYDRAULIC FILL DESIGN PROBLEMS

In evaluating the stability of hydraulic fill dams Ivanov, 1978, and Seed, 1987, arrived at conclusions that in order to design new construction or devise remedial measures it has to be done in such a way that high pore pressure cannot develop,or alternatively,the liquefiable soils have to be confined, so that no significant deformations can occur.

Seed and Ivanov further suggest that when large deformations can possibly be tolerated, it may be adequate and economical to ensure the stability of the embankment against major sliding after large movement or liquefaction has occurred.

Seed also suggests that empirical methods, correlated to standard penetration resistance, may be used to evaluate the residual stregth of soil after liquefaction has taken place where such values are required.

CONCLUSIONS

1. In spite of Hazen's early recommendations presented in 1920 to discard fines less than 0.01 mm in hydraulic fill cores, many dams were built in various parts of the world using liquid like cores with high clay content that remained unconsolidated and unstable over many years. Currently some of these dams still in use, with saturated soft cores, present stability and rehabilitation problems.

2. Such practice continued to about 1945.Accidents during constructions were observed at Calaveras Dam (1918), Fort Peck Dam (1938)in the U.S.A.; Rio Grande Dam(1934) in Brazil, and Svir Dam(1935) in the Soviet Union, described in the previous pages.

3. Starting about 1946 and over the past 40 years,a number of hydraulic fill dams were successfuly built in the Soviet Union without clay cores or the so called homogeneous sand dams.Data on these dams is given in the Table 1.

There seems to be no reason why similar dams cannot be built in other countries, especially in non-seismic regions, such as in Brazil, where very similar materials and foundation conditions exist.

4. In this connection perhaps the following suggestions by Casagrande on liquefaction can be used merely as a guideline(1975):

"... For typical alluvial sands containing not more than a few percent of silt sizes, the upper limit of relative density for which I consider actual liquefaction to be a possibility, lies in the range between about 40 and 50%, depending on confining pressure. Medium loose sands in the range between 40 and 60% may be slightly contractive or slightly dilative; and in situ they may respond to cyclic loading with strains of objectionable magnitude, but rarely with actual liquefaction.In strongly dilative, anisotropically consolidated sands in situ, with relative densities greater than about 70%, I consider it normaly impossible for cyclic pore pressures to approach or equal the confining pressure...; and I doubt that induced pore pressure will normally rise as high as 50% of the confining pressure".

5. Where a potential for soil liquefaction exists, it is suggested to design new structures or devise remedial measures in such a way that high pore water pressure cannot develop, or alternatively, to confine the liquefiable soils by means of stable zones.

6. Unfortunately, Soviet experience in building improved hydraulic fills is not readily available, and apparently has not been discussed among western engineers. Until this is done the improved tecnique cannot be readily accepted by the profession. Hopefully, this paper(by no means complete), may serve as a first preliminary basis for further discussions and comments.

APPENDIX - BIBLIOGRAPHY

- Casagrande, A., 1965, "The Role of the Calculated Risk in Earthwork and Foundation Engineering", Journal of the Soil Mechanics and Foundations Division, ASCE, Vol. 91, No, SM4, pp. 1-40.
- Casagrande, A., 1975, "Liquefaction and Cyclic Deformation of Sands, A Critical Review", Proceedings of the 5th Pan-American Conference on Soil Mechanics and Foundation Engineering, Vol. 5, Buenos Aires, Argentina, pp. 79-133.
- Castro, G., 1969, "Liquefaction of Sands",Harvard Soil Mechanics Series No. 81.
- Castro G., 1975, "Liquefaction and Cyclic Mobility of Saturated Sands", Journal of the Geotechnical Engineering Division, ASCE, Vol. 101, No. GT6, pp. 551-569.
- Castro, G., and Poulos, S.J., 1977, "Factors Affecting Liquefaction and Cyclic Mobility", Journal of the Geotechnical Engineering Division, ASCE, Vol. 103, No. GT6, pp. 501-516.
- Castro G., Poulos, S.J., Leaders F.D., 1985, "Re-Examination of Slide of Lower San Fernando Dam. J. Geotech. Engrg., ASCE, 111 (9).
- Christian, J.T., and Swiger, W.F., 1975, "Statistics of Liquefaction and SPT Results", Journal of the Geotechnical Engineering Division, ASCE, Vol. 101, No. GT11, Proc. Paper 11701, Nov., pp-1135-1150.
- Creager, W.P., Justin, J.D., Hinds J., 1945, "Engineering for Dams", Vol. 111, John Wiley & Sons, Inc., N.Y.
- Energomachexport, 1979, "The Dams of the USSR", Moscow. (in Russian).
- Ferreira, R.C., Peres, J.E.E., Monteiro, L.B., 1980, "Geotechnical Characteristics of Hydraulic Fill Modeling", XIII National Seminar, Brazilian Committee on Large Dams, Rio de Janeiro, Brazil, (in Portuguese).
- Ferreira, R.C., Peres, E.E., Mendes, L.C., "Model and Test Fill Studies of Hydraulic Fill", XVII National Seminar, Brazilian Committee on Large Dams, Brasilia, Brazil, 1987, (in Portuguese).
- Guerra, M.O., França, P.C.T., 1986, "Hydromechanization Brazilian Experience at Rio Grande and Guarapiranga Dams" Separata, ABMS/ELETROPAULO. (in Portuguese).
- Hazen, A., 1920,"Hydraulic Fill Dams", Transactions, ASCE, Vol. 83, pp. 1713-1745.
- Hsu, S.J.C., 1980, "Prospect of Building Hydraulic Fill Dams in Brazil", XIII National Seminar, Brazilian Committee on Large Dams, Rio de Janeiro, Brazil, (in Portuguese).
- Hydroprojekt, Moscou - Coyne et Bellier, Paris, 1973, "Construction Des Barrages en Terre par le Remblaiement Hydraulique", Moscow, Sept. 27 p. (in French).
- Ivanov, P.L., 1978, "Liquefaction and Compaction of Cohesionless Soils under DynamicLoading", Leningrad Polytechnical Institute (Kalinin), Leningrad, 49 p. (in Russian).
- Kuznetsov, A.Y., 1966, "Volga Hydro-Station in Honour of XXII PartyReunion", Chapter 2, Organization and Construction, Energia, Moscow, (in Russian), pp. 582.

- Lee, K.L., Seed, H.B., Idriss, I.M., and Makdisi, F.I., 1975, "Properties of Soil in the San Fernando, Hydraulic Fill Dams", Journal of the Geotechnical Engineering Division, ASCE, Vol. 101, No. GT8, pp.801-821.
- Leps, T.M., Strassburger, A.G., Meehan, R.L., 1978, "Seismic Stability of Hydraulic Fill Dams", Part 1 and 2, Water Power & Dam Construction, (Oct.), pp. 27-36 & 43-52.
- Liu, H., and Qiao, T., 1984, "Liquefaction potential of saturated sand deposits underlying foundation of structure", Proceedings, Eighth World Conference on Earthquake Engineering, Vol. III, San Francisco, Calif.
- Marcuson, W.F. III, Ballard, R.F., Jr., and Ledbetter, R.H., 1979, "Liquefaction failure of tailings dams resulting from the Near Izu Oshima Earthquake, 14 and 15 January", Proc. 6th Pan-American Conf. on Soil Mechanics and Foundations Engineerings, Lima, Peru.
- Maslov, N.N., 1959, "Stability of Saturated Sand", Gosezdat, Moscow, 327 p. (in Russian).
- Melentiev, V.A., 1960, "Sand and Gravel Soils for Hydraulic Fills", Gosezdat, Moscow, pp. 162, (in Russian).
- Melentiev, V.A., Kolpashnikof, N.P., Volnin, B.A., 1973, "Hydraulically Filled Hydrotechnical Structures" (Design), Energia, Moscow, pp. 245, (in Russian).
- Moretti, M.R., 1986, "Mechanical Parameters of Hydraulic Test Fill of Porto Primavera", VII Brazilian Congress on Soil Mech. & Found. Engrg., Porto Alegre, Brazil, (in Portuguese).
- Ogurtsov, A.I., 1974, "Hydromechanization of Earth Structures",Struyizdat, Moscow, pp. 365, (in Russian).
- Peck, Ralph B., 1979, "Liquefaction Potential: Science Versus Practice", Journal of the Geotechnical Engineering Division, ASCE, Vol. 105, No. GT3, Mar., p. 393.
- Poulos, S.J., Gonsalo Castro, France J.W., 1985,"Liquefaction Evaluation Procedure", Journal of the Geotechnical Engineering Division,ASCE, Vol. 111, No. 6, June, pp. 772-792.
- Santos, N.B., Meyer, P., 1980, "Prediction of Segregation (grain size), in Hydraulic Fills, XIII National Seminar, Brazilian Committee on Large Dams, Rio de Janeiro, Brazil, (in Portuguese).
- Schuyler, J.D., 1906, "Recent Practice in Hydraulic-Fill Dam", Transaction, ASCE, Paper No. 1043, pp. 195-277.
- SNIP II-53-73, 1974, "Building Code and Regulations", Part 11,Design Code, Ch. 53, Embankment Dam materials, Moscow (in Russian).
- Seed,H.B.,and Lee,K.L.,1966,K L.,"Liquefaction of Saturated Sands During Cyclic Loading", Journal of the Soil Mechanics and Foundation Division, ASCE, Vol. 92, No. SM6, pp. 105-134.
- Seed, H. Bolton, and Idriss, I.M., 1971, "Simplified Procedure for Evaluation Soil Liquefaction Potential", Journal of the Soil Mechanics and Foundations Division, ASCE, Vol. 97, No. SM9, Sept., p. 1249.
- Seed, H.B., Lee, K.L., Idriss, I.M., and Makdisi, F.I.,1975,"The Slides in the San Fernando Dams During the Earthquake of February 9,1971,Journal of the Geotechnical Engineering Division, ASCE, Vol. 101, No. SM7, pp. 651-688.

- Seed, H.B., Arango, I., Chan, C.K., 1975,"Evaluation of Soil Liquefaction Potential During Earthquakes", Report No. EERC 75-28, Univ.of California, Berkeley, Cal. (October).
- Seed, H.B., Mori, K., Chan, C.K., 1975, "Influence of Seismic Hystory on the Liquefaction Characteristics of Sands "Report No. EERC 75-25, Univ. of Berkeley, California.
- Seed, H. Bolton, Martin, Philippe P. and Lysmer, John, 1976, "Pore-Water Pressure Changes during Soil Liquefaction", Journal of the Geotechnical Engineering Division, ASCE, Vol. 102, No. GT4, Proc. Paper 12074, April, pp. 323-346.
- Seed, H.B., 1979, "Considerations in the earthquake-resistant design of earth and rockfill dams", Rankine Lecture presented in England and published in Geotechnique, XXIX(3), 215-263.
- Seed, H.B., Idriss, I.M., and Arango, I., 1983,"Evaluation of liquefaction potential using field performance data", J. Geotech.,Engrg.,ASCE, 109(3), 458-452.
- Seed, H.B., et al., 1985,"Influence of SPT procedures in soil liquefaction resistance evaluations", J. Geotech. Engrg., ASCE, 111(12), 1425-1445.
- Seed, H.B., 1987, "Design Problems in Soil Liquefaction", J. Geotech. Engrg., ASCE, 113(8), August 1987, pp. 827-845.
- Skempton, A.W., 1986, "Standard penetration test procedures and the effects of overburden pressure, relative density, particle size, aging and overconsolidation", Geotechnique, 36(3), 425-447.
- Terzaghi, K., 1956, "Varieties of Submarine Slope Failures", Proceedings 8th Texas Conference on Soil Mechanics and Foundation Engineering, Harvard Soil Mechanics Series No. 52, p.24.
- Vaid, Y.P., Chern, J.C., Tumi, H., 1985, "Confining Pressure,Grain Angularity, and Liquefaction", J. Soil Mech. and Found., ASCE, 110 (10), Oct., pp. 1229-1235.
- Valenzuela, L.A., Negro Jr., A., Santos, A.B., Santos, M.G., 1979,"Hydraulic Fill Method: A Feasible Alternative for Earth Dam Construction", VI Panamerican Congress on Soil Mech. & Found Engrg.,Lima,Peru.
- Volnin, B.A., 1965, "Hydromechanization Technology in Hydraulic Construction, Energia, Moscow, (in Russian).
- Whitman, R.V., 1985, "On liquefaction", Proceedings XIth Int'l. Conf. on Soil Mech. and Foundation Engrg., San Francisco, California.
- Wang, Wenshão, 1979, "Some Findings in Soil Liquefaction", Water Conservancy and Hydroelectric Power Scientific Research Institute",Beijing, China,(August).
- Xie, Junfei, 1979, "Emperical Criteria for Sand Liquefaction", Proceedings of the 2nd U.S.National Conference on Earthquake Engineering, Stanford University (August).
- Yufin, A.P., 1974, Hydromechanization, Stroizdat,Moscow,(in Russian), pp.223.

- Zhou, S.G., 1981, "Influence of Fines Evaluating Liquefaction of Sand by SPT, Proceedings of the International Conference on Recent Advances in Geotechnical Earthquake Engineering & Soil Dynamics,Vol. 2, pp.167-172.

- Discussions on - "Evaluation of Liquefaction Potential Using Field Performance Data", Seed et al., 1983.J.Geotech.Engrg. ASCE, 111(11), 1343-1346, 1985.

- Discussions on - "Liquefaction Evaluation Procedure", Poulos, et al., 1985. J.Geotech.Engrg. ASCE, 114(2) pp.232-259, 1988.

HISTORY AND INVESTIGATION OF JACKSON LAKE DAM

David G. Miedema, M. ASCE*; and Jeffrey A. Farrar, A.M. ASCE**

ABSTRACT

Jackson Lake Dam, located on the Snake River in Grand Teton National Park, was constructed in the early 1900's by the Bureau of Reclamation. Between 1913 and 1916, the dam was enlarged and the north embankment raised primarily by hydraulic fill methods. Investigations of the dam in the mid-1970's included standard penetration tests, shear wave velocity, and sampling and density tests using a 48-inch-diameter steel casing to support seven vertical shafts. This paper summarizes the history of construction, embankment investigations, hydraulic fill properties, and lessons learned.

INTRODUCTION

Jackson Lake, fed by the Snake River which drains the southern section of Yellowstone National Park, is the largest lake in scenic Grand Teton National Park, Wyoming (fig. 1). Jackson Lake, formed by glacial action, existed as a natural lake prior to construction of a dam in 1906 by the Reclamation Service, which later became the Bureau of Reclamation. This dam was modified in 1911 and again in 1916. The dam was part of the Minidoka Project in which the Snake River supplies irrigation water to farms in Idaho. The 1916 embankment was constructed primarily with hydraulic fill. Investigations of the embankment, starting in 1975, included SPT (standard penetration tests), shear wave velocity tests, and in situ density tests inside seven 48-inch-diameter, steel-lined, vertical shafts. Restrictions were placed on the reservoir water level in 1978 when analyses indicated that embankment and foundation soils could liquefy during a major earthquake. A complete rehabilitation of the dam is now underway (Ralston, 1987). This paper summarizes the history of the dam, embankment investigations, hydraulic fill properties, and lessons learned.

---

*Civil Engineer, U.S. Department of the Interior, Bureau of Reclamation, Denver, Colorado.
**Supervisory Civil Engineer, U.S. Department of the Interior, Bureau of Reclamation, Denver, Colorado.

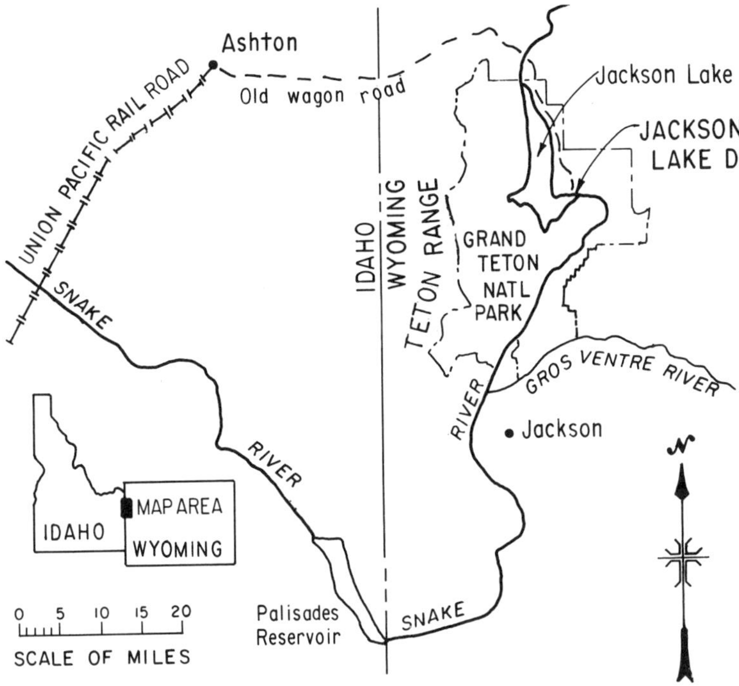

Figure 1. - Location map. - Freight for the project was transported 66 miles by wagon from the railroad terminal at Ashton, Idaho.

HISTORY

The location and extent of the 1906, 1911, and 1916 dams are shown on figure 2. Their construction is summarized below and in table 1.

Timber Crib Dam of 1906

A timber crib was floated to the natural outlet of Jackson Lake, filled with rock, and sunk in 13 feet of water. The crib had nine gate openings capable of discharging 2,000 ft^3/s. A double row of timber piles was driven along the upstream face of the crib. An earth and rockfill embankment, with slopes of 1.5:1, extended 1,500 feet to the north.

In 1909, the upstream and downstream slopes were flattened to 3:1 and 2:1, respectively (fig. 3). On July 6, 1910, the gated outlet structure failed.

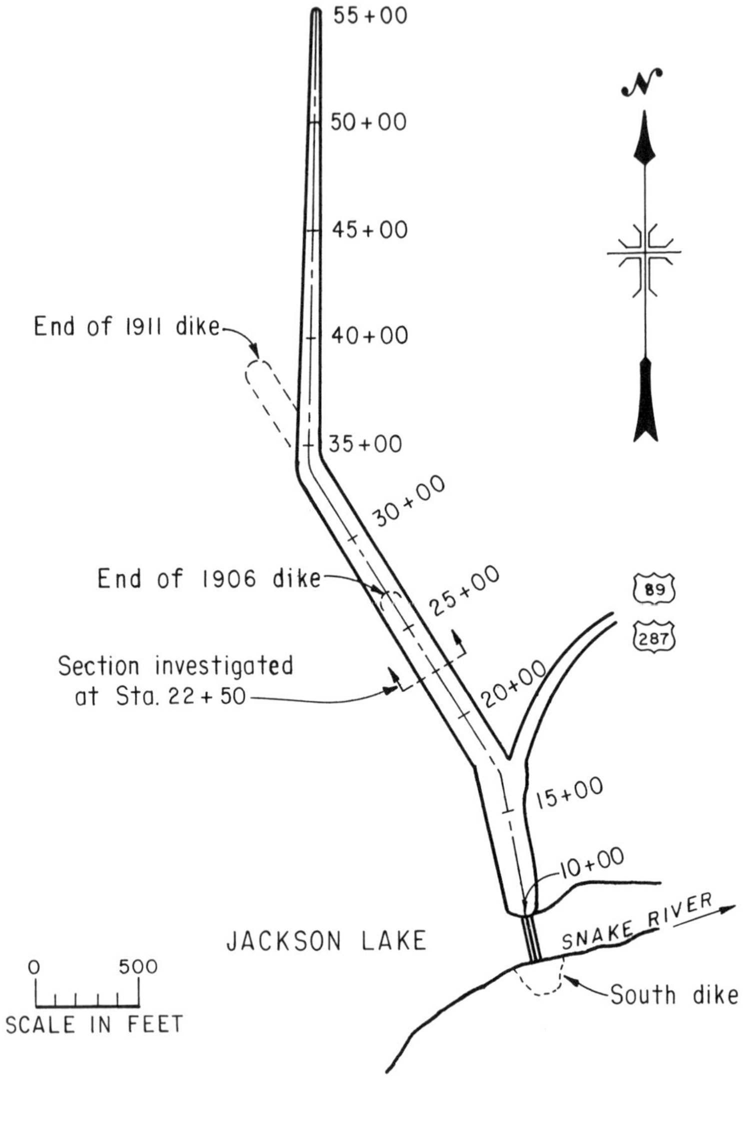

Figure 2. - Construction and exploration of Jackson Lake Dam

Table 1. - Embankment construction history
("Design Summary," 1987)

Year	Crest elev. (feet)	Upstream slope	Down-stream slope	Construction details
1906-08	6751	1.5:1	1.5:1	Earth and rockfill embankment.
1909	6751	3:1	2:1	Upstream slope protected by stones placed over a willow mattress.
1910-11	6760	3:1	2:1	Fill placed by dragline and horse-drawn equipment; "earth and gravel compacted by rollers in thin layers."
1913-16	6777	3:1	2:1	Primarily hydraulic fill, supplemented by dragline and dumpcar methods.
1917	6777	3:1	2:1	Clay blanket placed on upstream face to reduce seepage. Toe drain constructed in high seepage areas.
1918	6777	3:1	1-7/8:1	Gravel blanket placed on downstream toe and foundation.
1921-22	6777	3:1	1-7/8:1	Gravel placed on upstream face to reduce seepage.
1928-29	6777	3:1	1-7/8:1	Gravel placed downstream to "partially suppress the seepage and to prevent slippage of the more or less saturated material."

Concrete Dam of 1911

Using the undamaged section of the crib dam as a cofferdam, a 202-foot-long concrete spillway with 20 radial gates was constructed (fig. 4). A concrete corewall extended 300 feet north and was tied into sheetpiles driven along the entire length of the 1906 dike. The embankment was constructed over the 1906 dike and extended 2,600 feet to the north. With a maximum height of 20 feet, the earth and gravel fill was placed by horse-drawn equipment and "compacted by rollers in thin layers." A description of these rollers could not be found.

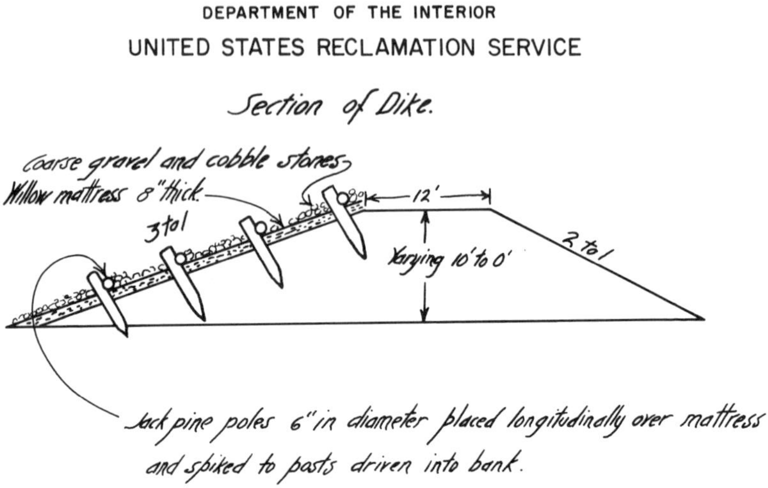

Figure 3. - Section of 1909 dike. - Upstream slope was protected by coarse gravel and cobble stones placed over a willow mattress.

Figure 4. - Jackson Lake Dam of 1911, prior to 1914-16 enlargement.

Enlargement of 1916

In order to provide more reservoir capacity for irrigation, the dam was enlarged between 1914 and 1916. A new 65-foot-high, 222-foot-long concrete spillway and outlet works was built 80 feet south of the old one because foundation investigations showed the 1911 spillway rested partly on soil and partly on rock. Consequently, 80 feet of the old concrete section was buried by rockfill. The north dike was extended about 2,000 feet and realigned beginning at station 35+00 to include drainage from Pilgrim Creek (fig. 2).

Hydraulic fill operations began in September 1914, after the foundation of the dike was stripped of a thick mat of grass, willows, and other organic matter using a 20-hp hoist-operated power scraper and heavy rake. Fill material was taken from a borrow area in the reservoir using a suction dredge (fig. 5). This material was pumped to the dike through a 12-inch-diameter steel discharge pipe supported on two-legged bents.

Figure 5. - Dredge used in the hydraulic fill construction of Jackson Lake Dam. - The project history records a cobble measuring 13 inches long, 9-1/2 inches wide, and weighing 33 pounds that passed through the 12-inch dredge pump on May 14, 1915 (photographed on October 1, 1914).

Hydraulic filling began at the north end of the dike where the final fill was to be only 4 feet high. The first method of distribution consisted of using a wooden

> "flume from which the fine material was trapped to the upstream slope and the coarse to the down stream slope. The water was turned towards the center and an effort made to retain as large a percentage of the finest grade there. The portion of the dike under construction then assumed a slope of about ten percent along the center line. It was found by this method a large proportion of the clay and fine sand was being carried away" (Banks, 1914).

Approximately 2,000 yd^3 of material were placed by this method in 1914.

The next year, dredging operations began on April 27. Because too great a percentage of fines were being lost, the "summit pool" method was begun on May 24. In the summit pool method, settling basins or pools, varying in length from 500 to 1,000 feet, were formed by longitudinal berms "built along both toes, the slopes being retained behind vertical [wooden] 'fences'" (Banks, 1914) and by temporary transverse berms. (The 1911 dike served as an upstream berm for summit pools in that area.) The coarser material was used to raise the berms, the height of which were kept approximately level. Water from the pool was discharged through a drop or chute. Because the pool was large and relatively quiet, in theory, the finer material had time to settle and form a puddled, central core. Following completion of one pool, the temporary transverse berm was sluiced into the adjacent pool to form a continuous core (fig. 6).

As the fill became higher, the pool width decreased and it became increasingly difficult to maintain a quiet pool; consequently, fewer fines settled out. A modification was to use teams of horses to construct a berm along the centerline of the dike and south of the section under construction, and the space between it and the old dike (which formed the upstream face of the new dike) was filled with fines that otherwise would have been wasted.

The original plan was to proceed from north to south, constructing the dam to fill height in successive stages. But in order to store as much water as possible for the 1916 irrigation season, it was decided to raise the fill uniformly along the length of the dike. By the end of the 1915 construction season, fill was at elevation 6760 (fig. 7).

To assure completion of the dike during 1916, three dragline cableways were erected, two just north of the concrete spillway (fig. 8) and one south of the river. Material (consisting of sand, clay, and loam) was dug from borrow pits in the reservoir, using 3/4- and 1-yd^3 buckets, and placed in the central core of the north dike. Gravel from the dredge and the riverbed was placed in the upstream and downstream areas of the dike. The river gravel was dug by the south cableway, dumped through a hopper into narrow gauge cars (fig. 9), conveyed by gasoline dinkies across the dam (fig. 10), deposited wet in the dike, and finally "compacted by teams and sluicing" (Banks, 1916).

Hydraulic fill operations ended on June 25, 1916, because the dredged material contained so little gravel that it was difficult to retain; and consequently a large amount of it was being wasted. On the north, cableway work continued until July 12. The south cableway closed down on August 31 with the placing of the last load of "dirt" in the dike. The volume of fill placed from 1913-16 was 301,667 yd^3 at a unit cost of about \$0.64/yd^3.

Figure 6. - Dredged material is being discharged into a setting basin after being pumped through a 12-inch-diameter steel pipe. Upstream row of piling is being driven on the left (photographed October 6, 1915).

Figure 7. - Dike is completed from far end to first curve with hydraulic fill at elevation 6760. Dredge is operating at far left. Forms for concrete corewall are just beyond spillway section (photographed November 1, 1915).

Figure 8. - North dragline cableways in operation. - River gravel from south cableway (not shown) is being dumped onto the fill. Dredge is still in operation (photographed on May 5, 1916).

Figure 9. - South slack-line cableway dumping material through a hopper into the narrow-gauge, side-dump cars.

Figure 10. - Henry 1 gasoline dinkey, powered by a Ford engine, moving fill across the dam.

Historical Perspective

A glimpse of construction life, as recorded in the project history (Banks, 1913-1916), gives one a fuller appreciation of early dam building in the west. The project was 66 miles from the railroad terminal at Ashton, Idaho, from which all freight was transported by wagon (fig. 1). Twenty-eight miles of the road were built by the Reclamation Service in 1910, crossing "through the boggy meadows and across the Teton Range to intercept the military road that runs from Fort Yellowstone past Moran and on to Fort Washakie." When work was first started in 1913, the roads were nearly impassible; but during 1914, 6,495,054 pounds of freight were received. Spruce and red fir were felled near the head of the lake and towed about 12 miles with a paddle-wheel boat. Good quality bituminous coal was mined 8 miles from camp and hauled there on sleds.

Buildings in the Government camp included 14 bunkhouses, 2 granaries, a root cellar, a blacksmith shop, and a mercantile store. Two shower baths and two "set tubs" were provided with hot and cold water in the heated pump house. The facilities were "well patronized." All 39 buildings in the camp were heated by wood stoves and lighted by kerosene lamp.

During January 1914, 1,540 cords of firewood were hauled to camp. A minimum temperature of -52 °F (-47 °C) was recorded in February, and 35 tons of ice were cut and stored. Dredge machinery was freighted in March, but six boilers (each weighing 14,700 lb) were left at the foot of the pass because the road was in such poor condition. By May 6, the saw mill was running; but on May 31, it caught fire and burned down. Equipment used that year included 12 boilers, 13 steam engines, 28 horses and mules, 26 wagons, and 4 tongue-, 24 wheel-, and 74 slip-scrapers.

The number of men employed during 1915 ranged from 28 in January to 224 in May. The wage scale ranged from $40 per month for "flunkeys" (kitchen help) to $5 per day for a sawyer. Except for kitchen help, a deduction of $0.25 for each meal was made; all were charged $1 per month for hospital service.

Remedial Measures

After filling of the reservoir in 1916, seepage appeared on the embankment (up to 4 feet above the toe) and from the natural ground 25 feet downstream of the toe. In 1917, a 3-foot-deep trench was dug at the downstream toe of the dam and the salvaged dredge pipe and gravel were placed in the trench to control seepage; in addition, a clay blanket was placed on the upstream face of the dam. However, seepage continued and gravel blankets were placed over areas of seepage in 1918, in 1921-22, and finally in 1928-29.

EMBANKMENT INVESTIGATIONS

As part of the Bureau of Reclamation's Safety of Dams Program, the embankment and foundation of Jackson Lake Dam were investigated, beginning in 1975. Based on a review of field and laboratory test results, it was determined that not only the embankment but also the foundation was susceptible to liquefaction (Seed and Lee, 1977). After 1977, efforts were concentrated on the foundation. Only the embankment investigations are reported here.

1975 and 1976 Investigations

The 1975 pilot drill program consisted of a pair of auger and rockbit drill holes each, at stations 16+50 and 21+00. In 1976, near station 16+50, two holes were drilled by rockbit using clear water for down-hole shear wave velocity testing. Due to the presence of coarse-sized gravel, the embankment was difficult to drill and sample; and little definitive information was obtained. However, a few borings encountered wood, which probably was railroad trestle timber buried in the fill; it is likely that the dike at this location was constructed primarily by dumped fill and dragline methods.

1977 Investigation

To investigate the hydraulic fill portion of the dike, investigations were made further out on the dike in the vicinity of station 22+50. (This location contained the 1906 and 1911 dikes, fig. 2.) Because of sampling difficulties experienced the previous year and the desirability of obtaining in situ density tests, large (48-inch) diameter, vertical shafts were excavated to gain access for sampling and testing. Seven shafts, numbered CM-110 through -116, were excavated between October 10 and November 2, 1977. Using a 15-ton crane and a 1/4-yd^3 clamshell bucket, excavation was rapid and preceded the steel casing by about 1 to 2 feet. At significant changes in material, the soil was logged. Sampling and density tests were performed at about 5-foot intervals. The shafts were excavated as deep as allowed by ground water, which was near foundation contact.

The in situ density tests were taken in the soil at the bottom of the hole using sand to determine the volume of the 8-inch-diameter test hole (Earth Manual, 1980). Precautions were taken to avoid deforming the hole; movement of heavy equipment in the test area was stopped, and the operator was supported by a piece of plywood. Working space was limited but adequate.

Sampling for each interval consisted of two to four sacks of material. Several attempts were made to obtain hand-trimmed, 8-inch-diameter sleeve samples (Earth Manual, 1980); because of the loose, cohesionless nature of the soil, no samples were obtained in the hydraulic fill. Two such samples of finer grained soil were obtained, probably from the 1911 or earlier dike.

In addition, nine 8-inch-diameter auger holes, DH-101 through -109, were drilled as pilot holes in preparation for the large diameter holes. SPT and gravity water tests were performed in these holes.* A downhole shear wave velocity test was performed in another hole.

1984 Investigation

During 1984, seven 24-inch bucket auger holes were drilled between stations 17+04 and 36+00 in the embankment. Samples were taken and gradation tests performed to determine usefulness of the material for the present dam rehabilitation.

HYDRAULIC FILL PROPERTIES

The 1977 investigation provided the most definitive information about the embankment. Figure 11, which summarizes much of this information, is a cross section at station 22+87 and includes soil classification, relative density, and SPT data.

An indication of the complexity of the hydraulic fill is found in the geologic log for hole CM-110 where the upper 20 feet is a loose, poorly graded gravel that is "easily excavated by hand." Scattered clayballs were found at a depth of 14 to 15 feet. However, from 20 to 22 feet, a varved clay unit was observed and described as "soft; thumb penetrates 1/4 inch with light pressure; moist, brown to grey (SC). Varved, lenses of (SP), (SC), and (CL) to 0.02 foot in thickness" (Lockhart, et al., 1986). The varved clay unit was probably formed at a time when summit pool operations were working successfully.

Data from gradation and relative density tests were found to be the most useful in understanding the properties of the embankment. Because of uncertainties about where hydraulic fill ended and

---

*The permeability tests are not considered to be reliable and, therefore, are not reported.

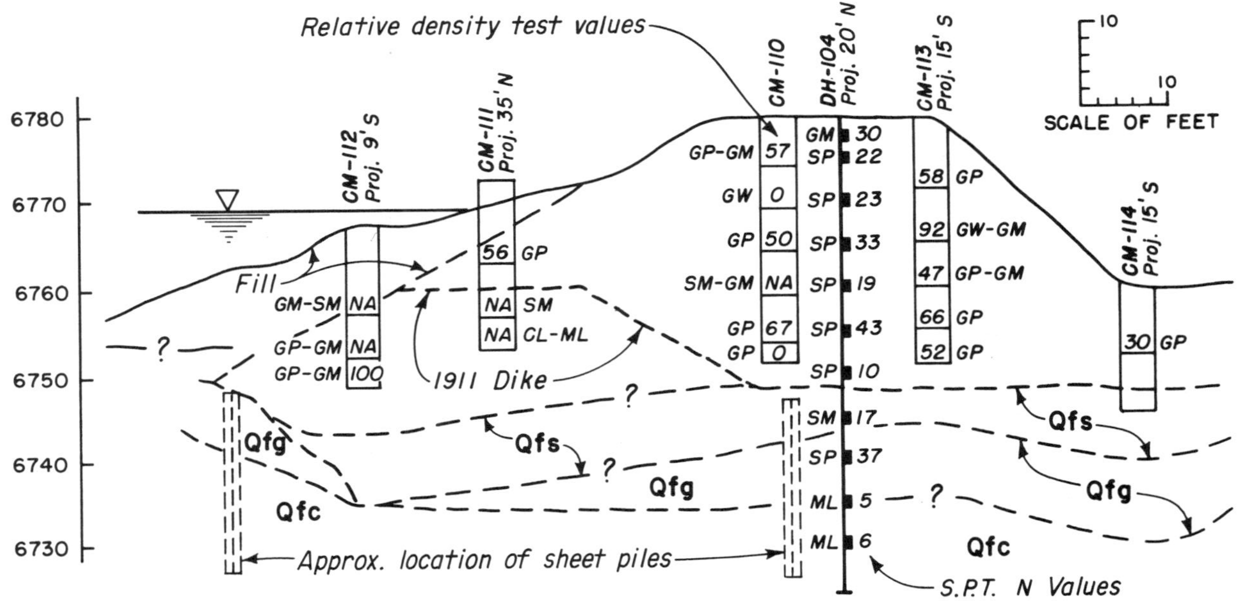

Figure 11. - Cross section at station 22+87 showing sample location, classification, relative density, and SPT N values. Fluviolacustrine sediments of the foundation are identified as predominantly gravel (Q_{fg}), sand (Q_{fs}), and silt and clay (Q_{fc}).

dragline/dumped fill began, only data obtained north of station 18+00 are included in the analysis. In all likelihood, the upper 5 feet of the crest is not hydraulic fill and was densified by vehicular traffic over the years; therefore, these data are excluded.

Gradation Tests

Gradations of 10 samples from holes CM-110 and -113 are plotted on figure 12. Except for the silty sand (SM-GM) from a depth of 20 feet (the varved clay unit described earlier), the predominantly poorly graded gravels (GP) are very similar containing only 2 to 6 percent fines and 58 to 76 percent gravel. The gradation of the silty sand foundation (Q_{fs}) was plotted to show its dissimilarity.

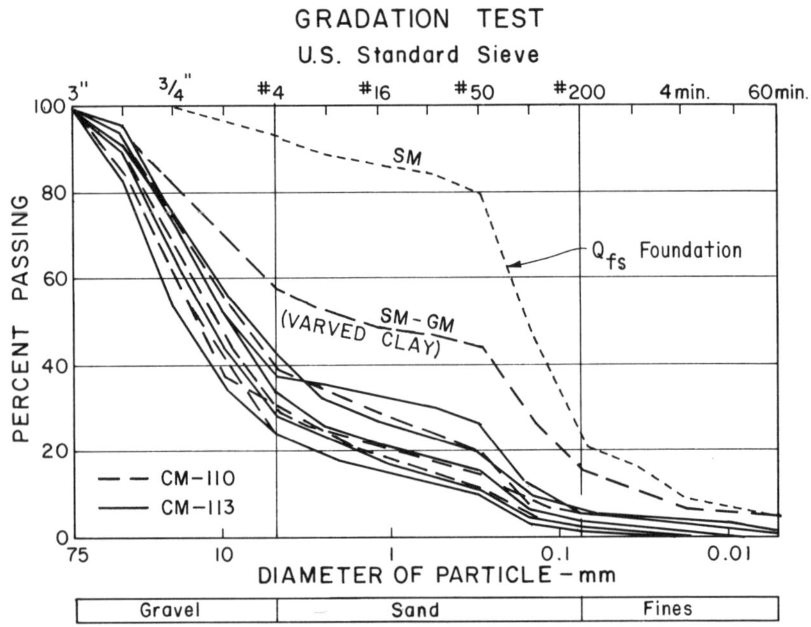

Figure 12. - Hydraulic fill from large diameter holes CM-110 and -113 is very similar; the exception is the SM-GM material taken from the varved clay unit.

Samples from the 1911 or earlier dike (fig. 13) help explain its role. With the silty sand (SM) and silty clay (CL-ML) containing 36 and 77 percent fines, respectively, the 1911 dike was the impervious core of the dam.

The average gradation of hydraulic fill material from the CM holes (fig. 14) is strikingly similar to the bucket-auger (AP) material (fig. 15).

Based on these gradation analyses, the hydraulic fill is a predominantly poorly graded gravel containing about 50 to 60 percent gravel, about 40 percent sand, and very few fines. There is little indication of a central core of fines in the hydraulic fill, and difficulties recorded during construction and postconstruction seepage problems substantiate this.

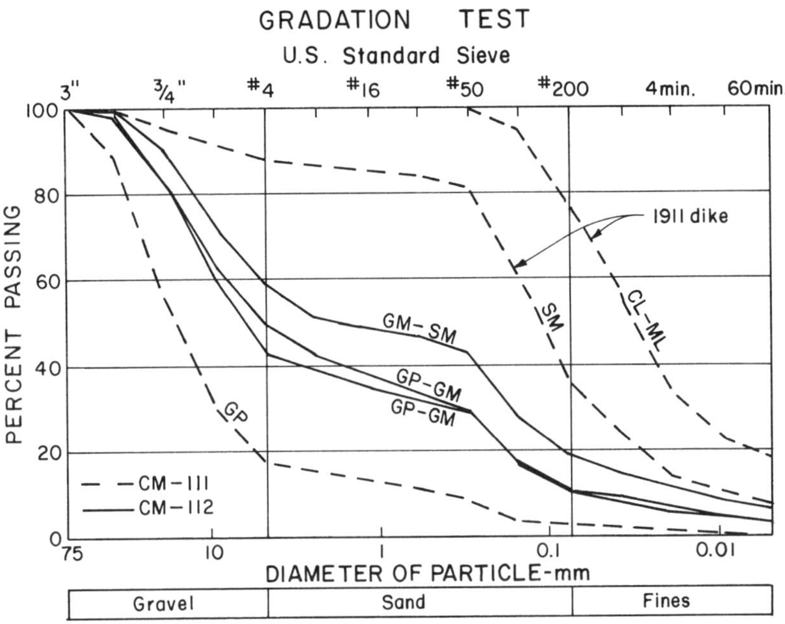

Figure 13. - Soil from large diameter holes CM-111 and -112. - The 1911 dike acted as an impervious core for the 1916 dike.

Relative Density Tests

Of 15 in situ density tests taken in hydraulic fill, 12 were in free-draining, cohesionless soil: 7 in poorly graded gravels (GP); 3 in poorly graded gravels with silt (GP-GM); 1 in well-graded gravel (GW); and 1 in well-graded gravel with silt (GW-GM). For the 12 soils, percentages of gravel, sand and fines were as follows:

	Range (%)	Average (%)
Gravel	50 - 82	67
Sand	15 - 44	29
Fines	2 - 6	4

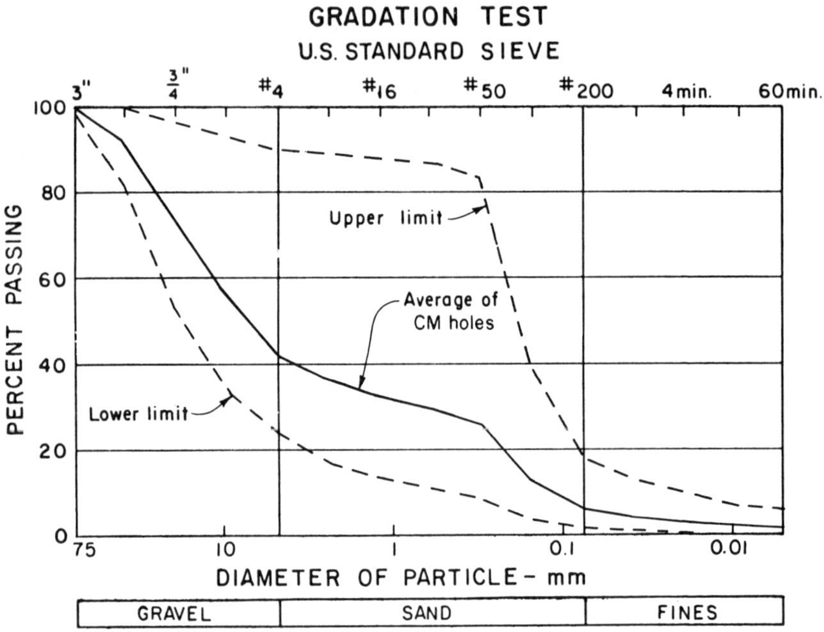

Figure 14. - Limits and average gradation of 14 samples of hydraulic fill from large diameter CM holes at stations 22+58, 22+87, and 26+40.

With the GP and GW soils having less than 5 percent fines and the GP-GM and GW-GM soils having less than 8 percent fines, the relative density test was applicable (Earth Manual, 1980). Maximum index unit weights determined by the wet method ranged from 137 to 148 lbf/ft^3 (fig. 16). Minimum index unit weights ranged from 109 to 120 lbf/ft^3. In situ dry unit weights ranged from 102 to 137 lbf/ft^3; they were generally lower in hole CM-115 at station 27+85 where only one operator worked and consistently obtained zero relative density values. Computed relative densities ranged from 0 to 92 percent, with the majority between 47 and 67 percent; the average was about 55 percent, which is typical of hydraulic fills (Lepps, et al., 1978; and Poulos and Hed, 1973).

Low values of relative density may have resulted from layering within the test interval. If a seam of sand, varved clay, or openwork gravel was present, the measured unit weight and hence the relative density would be lower than for an assumed homogeneous gravel deposit. High values of relative density may have resulted from hole squeezing. Sources of error include (a) size of sand cone (today a 20- instead of an 8-inch-diameter hole would be recommended for this material); and (b) human errors, perhaps caused by limited workspace in the shaft. The sensitivity of the test itself (two index tests must be performed to obtain relative density) may have affected the accuracy of the results.

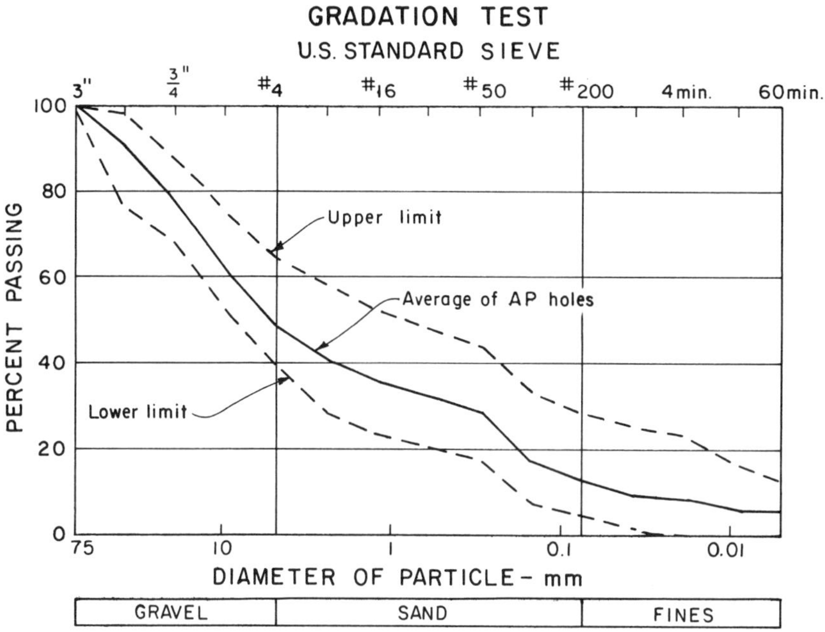

Figure 15. - Limits and average gradation of eight samples of hydraulic fill from the bucket auger holes drilled at stations 30+85, 33+53, 41+30, 44+69, and 46+80.

Standard Penetration Tests

SPT's were performed in 14 auger holes from 1975 to 1980. Out of 70 tests performed in the embankment, only 9 tests were in hydraulic fill sands (SP, SM) containing less than 25 percent gravel and less than 20 percent fines.

SPT values are related to relative density using the relationship developed by Gibbs and Holtz (1957), and assuming a moist unit weight of 130 lb/ft^3 for the embankment (fig. 17). Although there is considerable scatter in the data, it averages about 50 percent relative density, which is similar to the relative density tests.

Shear Wave Tests

Shear wave velocies from down-hole measurements in the hydraulic fill ranged from 600 to 700 ft/s. One way to compare shear wave velocity with other data is to use a shear modulus term $K_{2\ max}$. Empirical correlations for many granular soils show that shear modulus (G) increases with effective mean normal stress (σ_m') as follows:

$$G = 1000\ K_{2\ max}\ (\sigma_m')^{0.5}$$

The shear wave velocity (V_s) is related to shear modulus by $G = \gamma/g\ V_s^2$ where γ is the total unit weight of the soil and $g = 32.2\ ft/s^2$. Computed values of $K_{2\ max}$ range from about 42 to 57, which relates to relative densities of about 45 to 70 percent for sands (Seed and Idress, 1970).

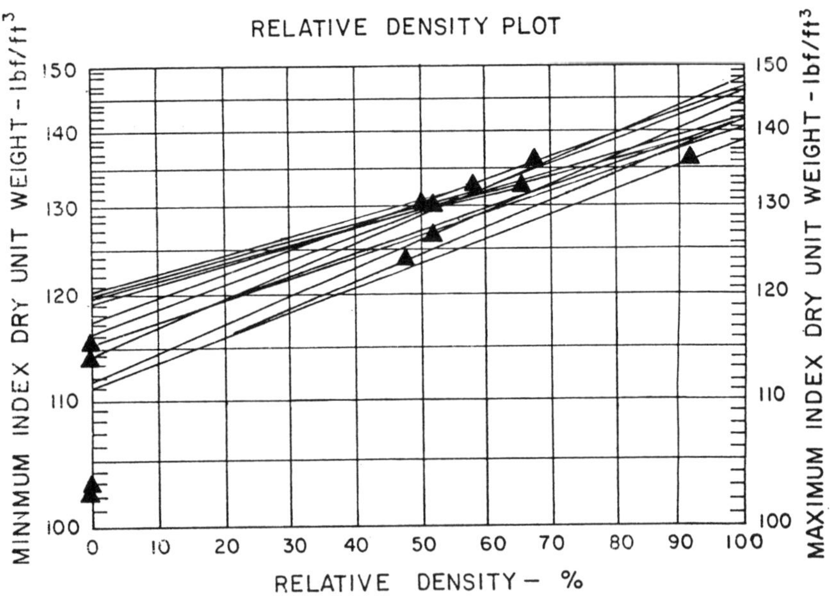

Figure 16. - In situ relative density test results. - The average relative density of 12 free-draining, cohesionless samples was about 55 percent.

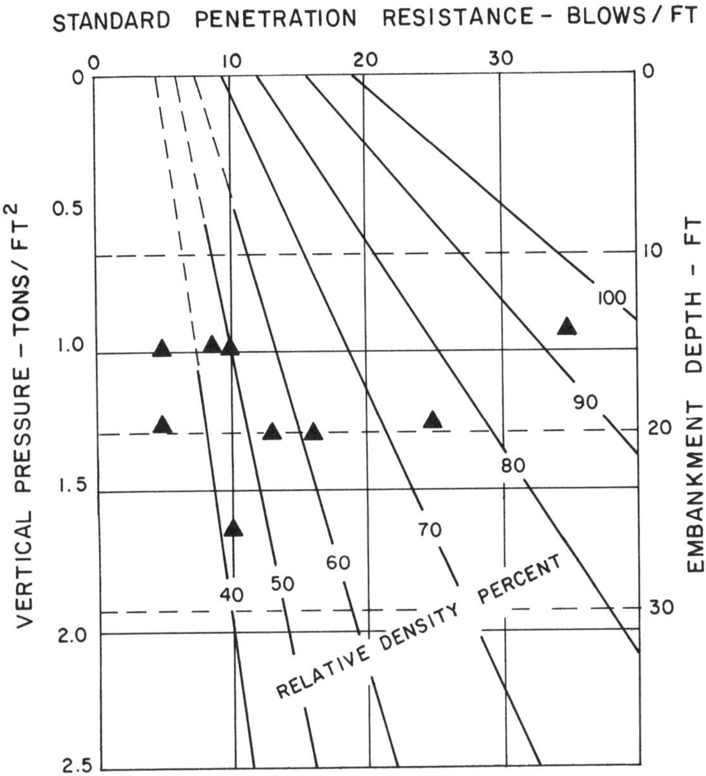

Figure 17. - SPT N values in hydraulic fill sands (SP, SM) containing less than 25 percent gravel and less than 20 percent fines average about 50 percent relative density.

SUMMARY

The hydraulic fill is a heterogeneous mixture of gravel, sand, silt, clay, and some cobbles. The results of the investigations do not indicate a central core of fine-grained material as is customary with hydraulic fill embankments and difficulties recorded during construction and postconstruction seepage problems confirm this. Instead, the fill is predominantly poorly graded gravel with an average of about 50 to 60 percent gravel, 40 percent sand, and 10 percent fines. Results based primarily on relative density tests, and to a lesser degree on correlations relating SPT and shear wave velocity tests to relative density, indicate the poorly graded gravels and sands average about 50 to 60 percent relative density.

LESSONS LEARNED

1. The construction history, especially the photographs, were valuable but incomplete as the limits of hydraulic fill operations were not recorded. During the 1977 investigations, more photographs and better records should have been taken inside the large-diameter exploratory shafts.

2. The difficulty, expense, and time required to investigate the dam were underestimated. The presence of gravels in the embankment made it very difficult to sample, but use of large-diameter shafts aided the investigation. Because of the gravels, a 20- instead of an 8-inch-diameter hole should have been used for the in situ density test.

3. Larger variability in the test results should have been expected because the dike construction was complex and because of inherent variables in the tests.

APPENDIX I. REFERENCES

Banks, F. A., "Annual Project History - Jackson Lake Enlargement Project - Wyoming, Minidoka Project," Bureau of Reclamation, Denver, Colorado, 1913-1916.

"Design Summary - Jackson Lake Dam Modification - Minidoka Project, Wyoming," Bureau of Reclamation, Denver, Colorado, 1987.

Earth Manual, 2d edition, U.S. Department of the Interior, Bureau of Reclamation, U.S. Government Printing Office, Washington, D.C., reprint 1980.

Gibbs, H. T., and W. G. Holtz, "Research on Determining the Density of Sand by Spoon Penetrator Test," Proceedings of Fourth International Conference on Soil Mechanics and Foundations Engineering, London, England, vol. 1, pp. 35-39, 1957.

Leps, T. M., A. G. Strassburger, and R. L. Meehan, "Seismic Stability of Hydraulic Fill Dams," Water Power & Dam Construction, part 1, October 1978.

Lockhart, A. C., R. A. Link, and J. B. Buehler, Geologic Report for Final Design - Jackson Lake Dam Modification - Minidoka Project, Wyoming, Bureau of Reclamation, Boise, Idaho, 1986.

Poulos, S. J., and A. Hed, "Density Measurements in a Hydraulic Fill," Evaluation of Relative Density and Its Role in Geotechnical Projects Involving Cohesionless Soils, ASTM STP 523, American Society for Testing and Materials, Philadelphia, Pennsylvania, pp. 402-424, 1973.

Ralston, M., "Dam Rebuilt to Save It from Quakes," Engineering News Record, pp. 46-47, September 17, 1987.

Seed, H. B., and K. L. Lee, "Seismic Stability Considerations - Jackson Lake Dam, Wyoming," Consultant's Report to the Bureau of Reclamation, December 1977.

Seed, H. B., and I. M. Idress, Soil Moduli and Damping Factors for Dynamic Response Analyses, Report No. EERC 70-10, University of California, Berkeley, California, 1970.

Appendix II. U.S. Customary - SI Conversion Factors

To convert	To	Multiply by
inches (in.)	millimeters (mm)	25.4
feet (ft)	meters (m)	0.305
yards (yd)	meters (m)	0.914
miles (miles)	kilometers (km)	1.61
cubic yards (cu yd)	cubic meters (m^3)	0.765
pounds (lb) mass	kilograms (kg)	0.453
tons (ton) mass	kilograms (kg)	907
pound force (lbf)	newtons (N)	4.45

PORE WATER PRESSURE CONDITIONS IN TAILING DAMS

Peter A. Stauffer, Associate Member, ASCE*, and
James R. Obermeyer, Member, ASCE**

ABSTRACT

Typically, effective stress slope stability analyses of tailing dams are performed assuming a location for the phreatic surface with hydrostatic pore water pressures calculated below this surface. Depending on numerous factors including tailing properties, deposition-raise rates and boundary conditions, these assumptions may be either conservative or unconservative. Pore water pressure conditions within several tailing dams at a copper mine in the southwestern United States are examined with respect to the factors listed above. For the structures considered, the assumption of hydrostatic pore water pressures was found to give conservative results with respect to calculated factors of safety.

Pore water pressure conditions within several operating dams at the site have been monitored for many years using an extensive piezometer system which includes open well and isolated tip piezometers. It was found that the pore water pressures below the phreatic surface are less than hydrostatic, a result of the favorable drainage conditions and the relatively slow loading rates experienced by these dams.

The methods used to measure pore water pressures and the observed conditions are presented. Characteristics of the tailing dams including size, historic raise rates and subsurface conditions along with tailing properties (index and engineering properties), are also presented to give an overall understanding of these structures, the materials of which they are comprised and the observed behavior of the structures in terms of observed pore water pressures.

---

*Assistant Project Engineer, Woodward-Clyde Consultants, 4582 South Ulster Street Parkway, Suite 1000, Denver, CO 80237

**Vice President, Woodward-Clyde Consultants, 4582 South Ulster Street Parkway, Suite 1000, Denver, CO 80237

INTRODUCTION

The complexity and variability inherent in tailing dams makes prediction as well as evaluation of the performance of these structures difficult. Many factors are found to influence the performance of tailing dams including: geologic and topographic setting; dam geometry; tailing properties; deposition methods and operation of the dam. Nelson, Shepherd and Charlie (1977) have investigated operation and construction factors and their affect on the stability of tailing dams. Modeling of seepage through tailing dams has been investigated by Van Zyl and Harr (1977) as well as many others. The purpose of this paper is to evaluate the performance of several tailing dams where data are available that reflect the behavior of these structures during normal operation. Given this understanding, important aspects of tailing dam behavior can be addressed when assessing the performance of existing or proposed dams.

Four upstream method tailing dams and one rockfill dam are currently used for tailing deposition at this mine. The behavior of the upstream method tailing dams is monitored by an extensive system of piezometers, movement markers and staff gauges installed in each of the dams. This paper presents a description of the tailing dams and examines their behavior as indicated by pore water pressure conditions within the dams.

PHYSICAL DESCRIPTION OF TAILING DAMS

Operation of the mine began in the late 1960's when construction of the original tailing dams was initiated. Since that time new tailing disposal facilities have been constructed as needed. The crest lengths of the tailing dams currently range up to 2 1/2 miles (4 km) and the dam heights range between about 130 feet (40 m) and 250 feet (76 m) with ultimate planned heights approaching 400 feet (122 m).

The tailing dams are located in the topographic expression of a fault-bounded structural trough. The dams are founded primarily on Bolson Fill that is derived from materials eroded and transported by water from the surrounding mountains. The Bolson Fill consists of relatively clean to clayey and silty sands with gravelly and cobbly lenses. In valley bottoms, the Bolson Fill is overlain by up to about 20 feet (6 m) of alluvial materials consisting primarily of relatively clean to slightly silty sands and gravels.

Tailing production at the mine is generally about 40,000 to 50,000 tons per day. Tailing deposition is normally restricted to one dam at a time, and deposition is rotated from dam to dam to provide alternating periods of deposition and rest on each dam. Recently, deposition periods have been on the order of two to six months, and rest periods on the order of six months. Raise rates for the tailing dams have varied and have typically been relatively high during the first few years of deposition on a dam. Raise rates for the smaller dams have recently been on the order of 10 feet per year (3 mpy), and for the larger dams on the order of 3 to 5 feet per year (1 to 2 mpy).

The tailing dams are raised using the upstream method of construction with tailing deposited by cycloning. Tailing is currently delivered to the dams by pumping the tailing slurry in a 30-inch (76 cm) diameter delivery pipe that extends across the dam on a bench constructed on the downstream slope. The tailing is deposited using 12-inch (30 cm) diameter cyclones spaced about every 52-feet (16 m) along the dam. Tailing is fed to the cyclones by a 6-inch (15 cm) diameter line that is extended as needed to raise the dam. The tailing delivery pipe is raised about every 45 to 50 feet (14 to 15 m) of dam raise. The cyclone underflow, coarser tailing, is deposited in cones that form the "sand beach" or downstream portion of the dam, and the cyclone overflow, finer tailing, is deposited using lines that extend upstream from the cyclones into the storage area. A schematic cross-section is shown on Fig. 1, and a photograph of deposition on Fig. 2.

The whole tailing slurry is typically about 45 percent solids by weight. The typical range in grain size distributions for whole tailing, underflow tailing and overflow tailing are shown on Fig. 3. The percent passing the No. 200 sieve for whole tailing ranges from about 50 to 70, for underflow tailing from about 20 to 45, and for overflow tailing from about 63 to 88 percent. Atterberg limits for tailing materials are also shown on Fig. 3. The whole tailing and cyclone overflow tailing classify as either a ML, CL-ML or CL, and the cyclone underflow tailing is generally non-plastic and classifies as an SM according to the Unified Soil Classification System.

DESCRIPTION OF MONITORING SYSTEM

Many elements comprise the monitoring system utilized on these tailing dams. The following is a description of the piezometer and staff gauge systems, the two elements that are pertinent to this study.

Piezometers are strategically located at study sections established on each of the tailing dams. Two types of piezometers are used: open well and isolated tip piezometers. The open wells are slotted essentially along the entire length of the piezometer so that they reflect the uppermost water level intercepted by that well. Isolated tip piezometers are designed to measure water pressures at the tip location; both porous and pneumatic isolated tip type piezometers are used. The piezometers have been installed in

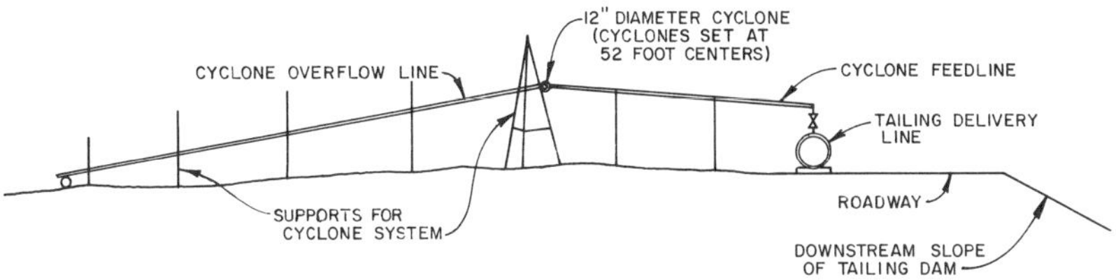

Fig. 1. Typical Cyclone Setup

Fig. 2. Photograph of Cyclone in Operation

groups, each group typically including one open well and one or more isolated tips. The piezometers are grouped in areas along the study sections between the toe and crest of the dam, generally at each bench. In addition, isolated pneumatic tip piezometers are located at varying distances and elevations upstream of the crest in the storage area.

A total of about 20 to 110 piezometers are used to monitor internal water conditions in each dam. The piezometers are typically read every one to two weeks. Computerized data management systems are used to facilitate timely review of these data.

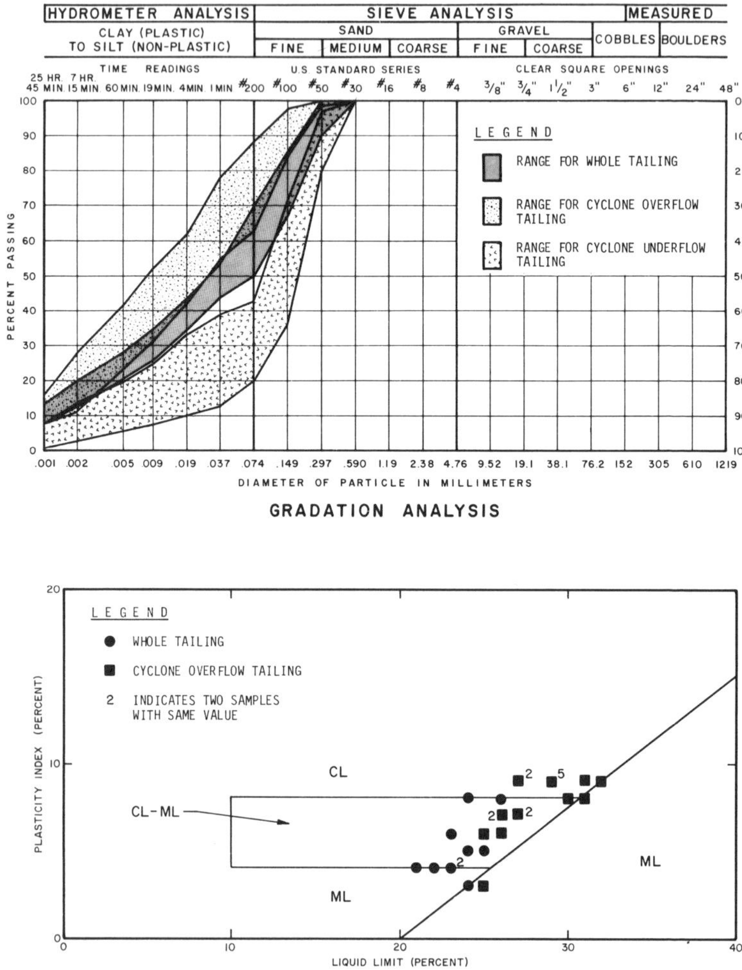

Fig. 3. Typical Grain Size Distributions and Atterburg Limits for Tailing Materials

Staff gauges in the storage areas are used to monitor the tailing level in each dam. The gauges are typically secured in the natural ground at several locations around the perimeter of the storage area. These gauges, along with records of tonnage of tailing deposited, allow for accurate monitoring of dam usage and raise rates.

EXAMINATION OF PIEZOMETER DATA

Piezometer data offers a means of assessing the performance of a tailing dam during operation of the dam. Interpretation and analysis of these data provides a reasonable basis for modification of operation of the dam to optimize deposition while maintaining an acceptable margin of safety. Monitoring of the dams is accomplished by comparing measured water levels to design assumptions and to threshold water levels that have been established for the piezometers.

Following is a presentation of the piezometer data from one study section located near the maximum section on one of the tailing dams. This section was selected because it is well instrumented and illustrates many of the important aspects related to pore water conditions in the dams. The height of this dam is about 200 feet (61 m) and the crest length about 9600 feet (2900 m). Fig. 4 shows the cross-section including piezometer types and locations as well as water level readings for the piezometers. Additional isolated pneumatic tip piezometers located further upstream in the storage area are not shown.

As shown on Fig. 4, the water levels for isolated tip piezometers are in all cases lower than the water level measured in the adjacent open well for each piezometer group. The line of saturation or phreatic surface at this study section, and in general in the dam, is located about 20 to 40 feet (6 to 12 m) below the slope surface. Fig. 5 presents the piezometer data from the piezometer group at the first raise bench in a different form. The isolated tip piezometer readings are plotted as pore water pressures, and the open well reading is used to estimate the location of the phreatic surface. The pore water pressure profile corresponding to hydrostatic conditions (assuming the phreatic surface is located at the level indicated by the open well piezometer), and the actual pore water pressure profile are shown on Fig. 5. It is clear that the actual pore water pressures are less than hydrostatic, and approach zero at the base of the dam.

Fig. 6 shows pore water pressure contours on the same cross-section of the dam. The pore water pressure contours are based on piezometer data from all of the piezometers at this section. Examination of the pore water pressure contours indicates that pore water pressure conditions throughout the dam cross-section are less

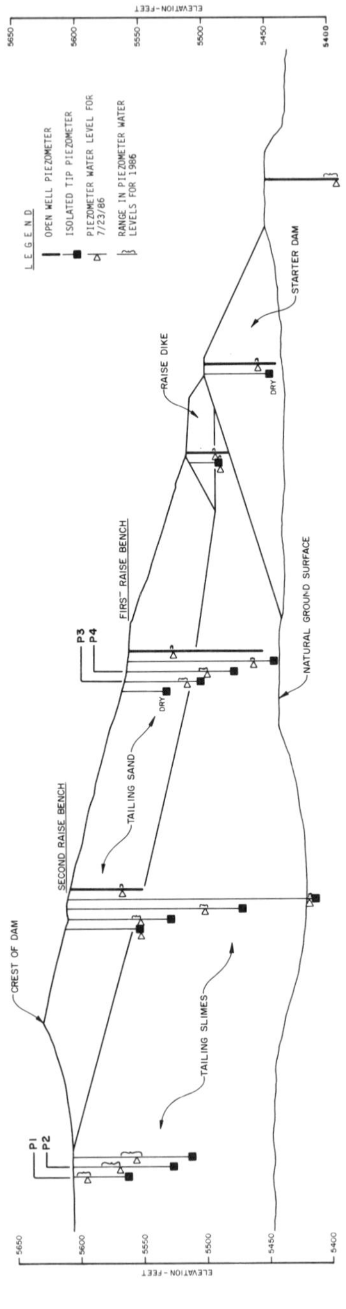

Fig. 4. Cross-Section of Study Section

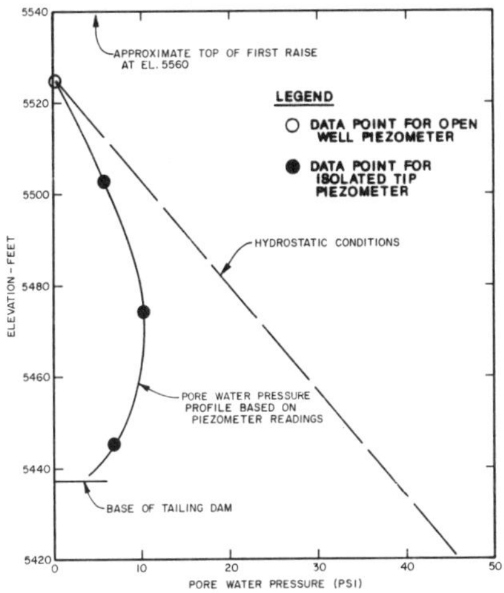

Fig. 5. Pore Water Pressure Profile at First Raise Bench

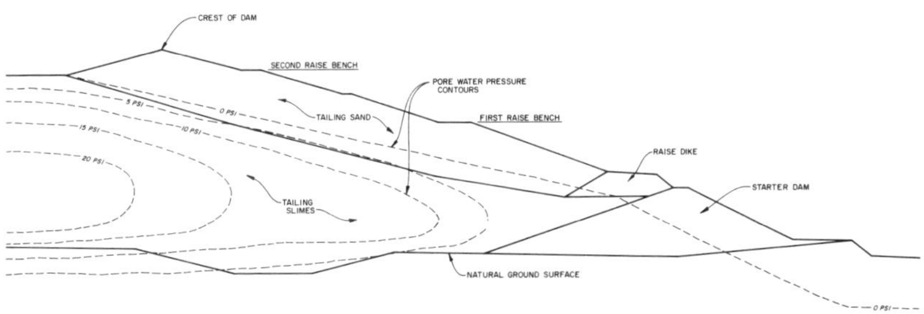

Fig. 6. Cross-Section Showing Pore Water Pressure Contours

than hydrostatic. These conditions result from the permeability of the underlying natural materials being greater than that of the tailing. Thus, the foundation materials act as a nearly free draining boundary. These drainage conditions are favorable from the standpoint of dam stability. The importance of this finding with respect to the stability of this dam is considered later in this paper.

Investigations that we have performed using piezometer cone penetration testing methods indicate that similar less than hydrostatic pore pressure conditions exist in other tailing dams. We found that the piezocone works quite well as an investigation tool for tailing dams; static pore water pressures can be measured by performing pore pressure dissipation tests that are relatively short in duration. Campanella, Robertson and Gillespie (1984) describe a piezometer cone investigation of tailing dams in British Columbia where equilibrium water pressures indicated a downward gradient of pore water pressures.

Data shown on Figs. 4, 5 and 6 correspond to one set of piezometer readings and thus represent the "static" pore water pressure conditions within the dam. The range in typical piezometer readings for each of the piezometers at this study section is shown on Fig. 4 using brackets. The ranges in water level readings are greater for piezometers upstream of and near the dam crest compared to those further down the slope. Increases in the piezometer readings generally occur in response to tailing deposition on the dam, while decreases in piezometer readings occur during rest periods.

Tailing deposition on the dam during the first six months of 1986 is illustrated graphically on Fig. 7. The only significant deposition during this time occurred over about a month long period in late January and early February. No significant deposition had occurred on the dam for about four months prior to this period. Examination of the staff gauge readings indicates that the average increase in the tailing slimes surface in the storage area as a result of this deposition was about 1.8 feet (0.5 m). Assuming an even distribution of tailing across the storage area and a total unit weight for deposited tailing of 90 pounds per cubic foot (14 kN/m^3), the total tonnage of tailing deposited during this period amounts to about 2.2 feet (0.7 m) thickness of tailing, a good agreement with staff gauge readings.

Examination of piezometer water level history plots indicates that the response of the piezometers to this period of deposition depends on location and type of piezometer. The response of two isolated pneumatic tip piezometers located beneath the first raise bench and two isolated pneumatic tip piezometers located beneath a point just upstream of the dam crest are shown on Fig. 8. It is clear that the piezometers located near the crest of the dam show an almost instantaneous and marked response to deposition on the dam. The total increase in water level is about 10 feet (3.0 m) with the highest water level occurring about one to three weeks after tailing deposition has ceased. Water levels for these two piezometers return to near the pre-deposition levels about two months after the peak reading. For the two piezometers located on

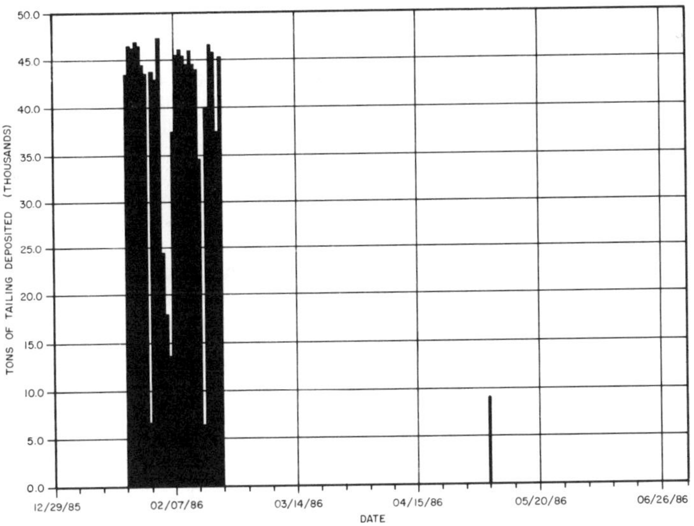

Fig. 7. Tailing Deposition on Dam

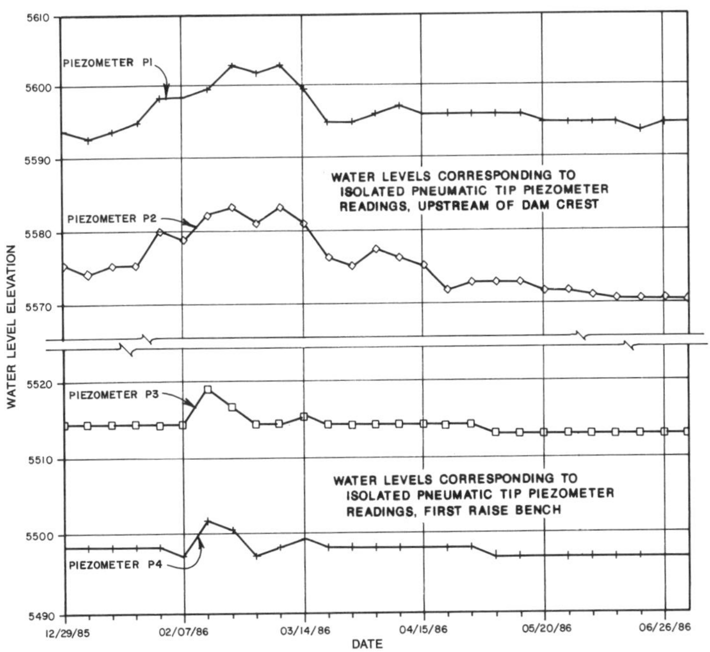

Fig. 8. Response of Piezometers to Tailing Deposition

the first raise bench, response begins about two weeks after deposition was initiated, and the increase in water level was about 5 feet (1.5 m) with the highest level measured occurring at about the end of the tailing deposition period. Water levels for these two piezometers return to near the pre-deposition levels about two weeks after the peak reading. The open well piezometers at this section generally showed one to two foot increases in water levels, with gradual returns to pre-deposition levels.

The difference between the behavior of piezometers located upstream of the dam crest and those located beneath the downstream slope of the dam is explained in part by the fact that previously deposited tailing material upstream of the crest of the dam is experiencing an increase in load from the recently deposited tailing. Initially this increased load generates excess pore water pressures as shown on Fig. 8 (excess with respect to the static pore water pressure). With time these excess pore water pressures dissipate and the increased load is carried by the tailing materials. Based on estimates of the amount of tailing deposited (about 2 feet (0.6 m) thickness) and assuming a total unit weight for the recently deposited tailing of 90 pounds per cubic foot (14 kN/m^3), this deposition corresponds to an increase in pore water pressure of about 3 feet (0.9 m) of water. This is about one third of the increase in water level readings observed for the isolated tip piezometers located upstream of the crest as shown on Fig. 8. This difference is probably caused by deposition in the vicinity of these piezometers being greater than elsewhere on the dam, or by deposition of cyclone underflow sand over these piezometers.

As shown on Fig. 8, piezometers located beneath the downstream slope of the dam generally showed a smaller increase in response to the deposition. This response was probably generated in part when the excess pore water pressures generated beneath the deposited tailing, as described above, began to dissipate. In addition, some of the water from the deposited tailing was probably flowing through the cyclone underflow sands beneath the slope of the dam. Examination of piezometer responses to longer deposition periods on this dam and on other dams indicates that the phreatic surface beneath the downstream slope of the dam continues to rise during prolonged periods of deposition.

The pore water pressure response at one of the other study sections on the dam during the same deposition period shown on Fig. 7 is illustrated by readings from three additional isolated pneumatic tip piezometers. These piezometers are located at varying depths just upstream of the crest of the dam. The pore water pressure profiles as indicated by the piezometer readings at four points in time are shown on Fig. 9. Hydrostatic pore water pressure conditions are also shown assuming that the line of saturation is at the tailing surface, generally a reasonable assumption at locations upstream of the crest. The pore water pressure profiles before deposition (1/7/86) and about five months after deposition (7/9/86) are about the same and appear to represent pore water pressure conditions during "resting" of the dam. The pore water pressure profile at the time when deposition has ceased (2/18/87) is approaching the hydrostatic line in the upper 40 feet (12 m) of the profile. The pore water pressure profile about one month after deposition has ceased (3/19/87) is somewhat higher and appears to be at about hydrostatic conditions in the upper 40 feet (12 m) of the profile.

While the fact that the highest pore water pressures are not observed until after tailing deposition has ceased is difficult to explain, this observation points out the importance of understanding how the dam responds to tailing deposition periods of

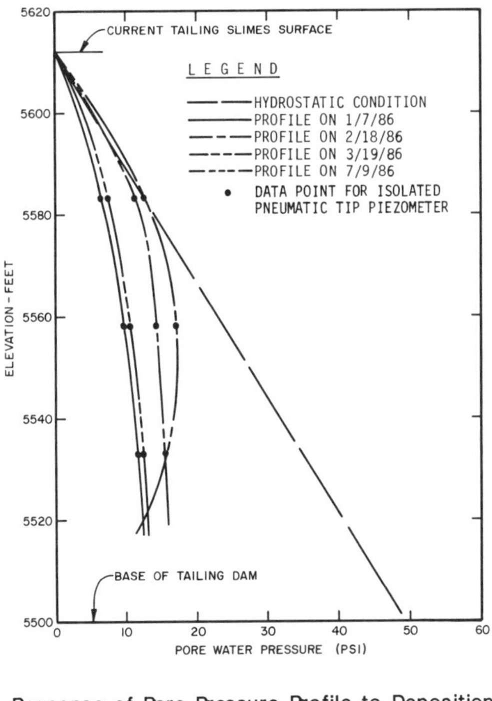

Fig. 9. Response of Pore Pressure Profile to Deposition on Dam

varying duration. This understanding can only be gained by instrumentation of the dams and analysis of piezometer data by personnel who are experienced in operation of these complex structures.

SLOPE STABILITY CONSIDERATIONS

Given the understanding of pore water pressure conditions described above, stability analyses were performed to assess the importance of these conditions with respect to slope stability. The slope stability analyses performed were limit equilibrium analyses using effective stress strength parameters with pore water pressures defined as described below.

Tailing materials were modeled using two material types, one to represent the tailing sands (predominantly cyclone underflow materials), and the other to represent the tailing slimes (predominantly cyclone overflow materials). The width of the tailing sand (sand beach) was based on drill hole and surface sampling information and was generally about 200 feet (61 m).

Two definitions of pore water pressures were used for these analyses. For the first, the phreatic surface was located based on water levels measured in the open well piezometers, and the pore water pressures were calculated in the usual manner as depth below the phreatic surface times the unit weight of water. Pore water pressures defined in this manner are referred to as hydrostatic. For the second definition of pore water pressures, the contours shown on Fig. 6 were used. As discussed above, pore water pressures defined by these contours are less than hydrostatic. The same phreatic surface was used as a contour line along which the pore water pressure is zero, and pore water pressure data were input to define the other contours. Pore water pressures were assumed to be zero above the phreatic surface for both cases analyzed.

Results for the slope stability calculations assuming hydrostatic conditions, presented on Fig. 10a, indicate a factor of safety of 1.8. The critical circular shear surface passes through the entire slope height with a maximum depth of about 75 feet (23 m). Results for the slope stability calculations using pore water pressure contours, presented on Fig. 10b, indicate a factor of safety of 2.2. The critical circular shear surface again passes through the entire slope height with a slightly deeper maximum depth of about 90 feet (27 m). The difference in the two calculated factors of safety is about 20 percent in this instance. Similar analyses that we have performed for other dams where pore water pressures are less than hydrostatic indicate that the difference in factors of safety can range up to 50 percent.

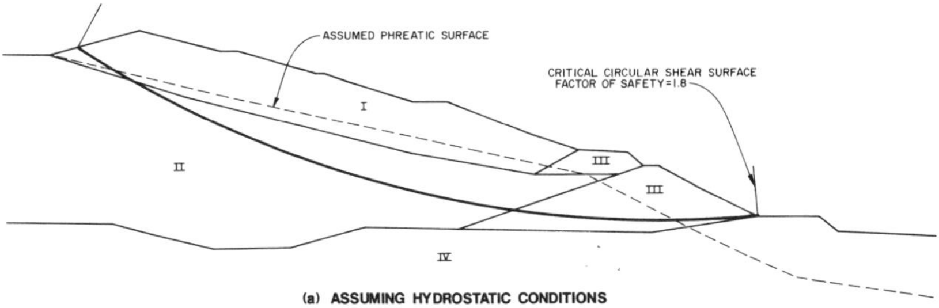

(a) ASSUMING HYDROSTATIC CONDITIONS

ASSUMED MATERIAL PROPERTIES			
MATERIAL	γ (PCF)	$\bar{\phi}$ (DEGREES)	$\bar{C}$ (PSF)
I. TAILING SAND	122	35	0
II. TAILING SLIMES	120	25	0
III. STARTER DAM FILL	130	37	0
IV. NATURAL	135	37	0

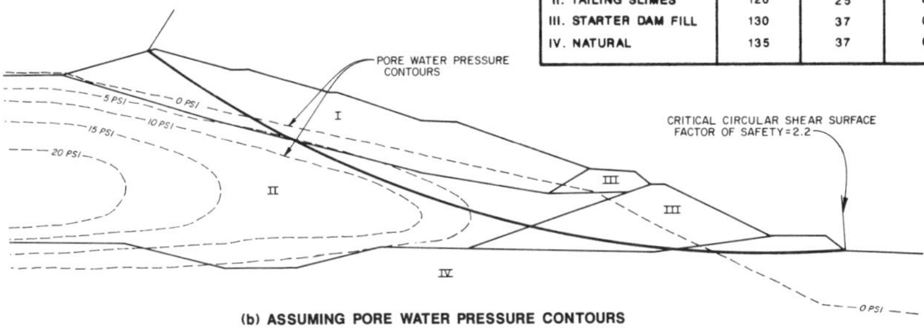

(b) ASSUMING PORE WATER PRESSURE CONTOURS

Fig. 10. Results of Slope Stability Analyses

Slope stability calculations described above indicate that the factor of safety calculated based on observed pore water conditions in this instance is about 20 percent higher that that calculated assuming hydrostatic conditions. It is important to point out, however, that the pore water pressure contours used for these calculations (shown on Fig. 6) represent conditions within this dam during "resting". The pore water pressures within the dam were observed to increase as a result of tailing deposition, and in the upper reaches of tailing material upstream of the crest, to reach hydrostatic conditions. Measured pore water pressure increases in the lower portion of the dam slope were less.

Results of slope stability analyses performed on these tailing dams under normal operating conditions indicate that the assumption of hydrostatic pore water pressure conditions is conservative. Considering the complexity of observed pore water pressure conditions within these dams, and the difficulty associated with measuring and modeling these conditions, the assumption of hydrostatic conditions for slope stability analyses in this case makes good sense. Analyses using pore water pressure contours, however, provide further insight into the stability of these dams and in some instances may be warranted.

CONCLUSIONS

The complexity of pore water pressure conditions within operating tailing dams is clearly illustrated by examination of the piezometer data for these dams where an extensive piezometer system and a regular reading schedule have been implemented. The phreatic surface is generally located about 20 to 40 feet (6 to 12 m) below the slope surface. Pore water pressure conditions within these dams are found to be less than hydrostatic as a result of drainage that is occurring at the base of the dams and prudent operations of the dams (i.e., limiting raise rates and providing rest periods). The pore water pressure conditions at a given instant are found to vary throughout the cross-section and to be dependent on the deposition history and characteristics of the tailing and foundation materials. Slope stability analyses performed for these dams indicate that with pore water pressures corresponding to actual measured values (less than hydrostatic) the factor of safety was about 20 percent higher than the factor of safety corresponding to hydrostatic conditions. These findings point out the importance of considering drainage conditions and dam operation when assessing the performance and stability of tailing dams.

The writers wish to emphasize that the properties of the tailing and foundation materials at every mining facility are unique and that it would be improper to assume that conclusions presented in this paper are applicable to tailing dams in general. Each dam at each mine must be evaluated independently considering all factors that influence its behavior. It is also important to note that poor drainage at the base of a tailing dam can result in pore pressures within the dam that are in excess of hydrostatic pressures. Furthermore, operation of the dams (i.e., raise rates and rest periods) have major effects on internal pore pressures and slope stability as indicated by this study.

ACKNOWLEDGMENTS

The authors would like to thank the many individuals within Woodward-Clyde Consultants who have contributed to this study.

APPENDIX. REFERENCES

Campanella, R.G., Robertson, P. K., and Gillespie, D. (1984). "Piezometer-Friction Cone Investigation of a Tailings Dam." Can. Geotechnical J. Vol. 21, pp. 551-562.

Nelson, John D., Shepherd, Thomas A., and Charlie, Wayne A. (1977). "Parameters Affecting Stability of Tailings Dams." Proceedings of the Conference on Geotechnical Practice for Disposal of Solid Waste Materials, ASCE, June 13-15, pp. 444-460.

Van Zyl, Dirk J. A., and Harr, Milton E. (1977). "Modeling of Seepage Through Mine Tailings Dams." Proceedings of the Conference on Geotechnical Practice for Disposal of Solid Waste Materials, ASCE, June 13-15, pp. 727-743.

HYDRAULIC FILL PERIMETER DIKE, GAILLARD ISLAND DREDGED MATERIAL DISPOSAL AREA, MOBILE, ALABAMA

Jack Fowler,[1] M. ASCE and S. J. Spigolon,[2] M. ASCE

ABSTRACT

Barge hauled and hydraulically placed dredged material was used to construct the 6 mile (9.7 km) long perimeter dike for a dredged material disposal island in Mobile Bay. The foundation soils were very soft, highly organic, plastic, bay-bottom clays. Dike materials were from two sources: a new land-cut channel and a new bay-cut channel. About 95 percent of material from the bay-cut and 52 percent of the material from the land-cut was very soft to stiff clay. Silt, sand, and small amounts of gravel and shell made up the remainder of the fill soils.

INTRODUCTION

Background

Very few earth dikes have been successfully constructed below water on very soft foundations using new work dredged material that consisted of less than 50 percent sand and over 50 percent silts and clays.

The Gaillard Island Dredged Material Disposal Area is in Mobile Bay about 2 miles (3 km) east of Theodore, Alabama and about 8 miles (13 km) south of Mobile. The Theodore Ship channel extends from the main Mobile Bay Ship Channel to the Theodore Industrial Park. During excavation of the channel, bay-cut dredged materials and land-cut excavated materials were dredge-pumped and barge hauled to the perimeter dike of Gaillard Island. The 6 miles (9.7 km) long perimeter dike surrounds an area of about 1300 acres (526 hectare) in a triangular shape.

Mobile Bay is on the Gulf coast of Alabama (Figure 1). It is about 30 miles (48 km) long north to south, about 8 miles (13 km) wide at the north end and about 20 miles (32 km) wide at the south end. The depth of the bay ranges from about 8 to 10 ft (2.4 to 3 m) and the tides range from 1 to 2.5 ft (0.3 to 0.8 m). The main Mobile Bay Ship Channel extends from the entrance bar to Mobile, a distance of over 33 miles (53 km).

---

[1] Research Civil Engineer, USAE Waterways Experiment Station, Vicksburg, Mississippi 39180

[2] Engineering Consultant, Coos Bay, Oregon 97420

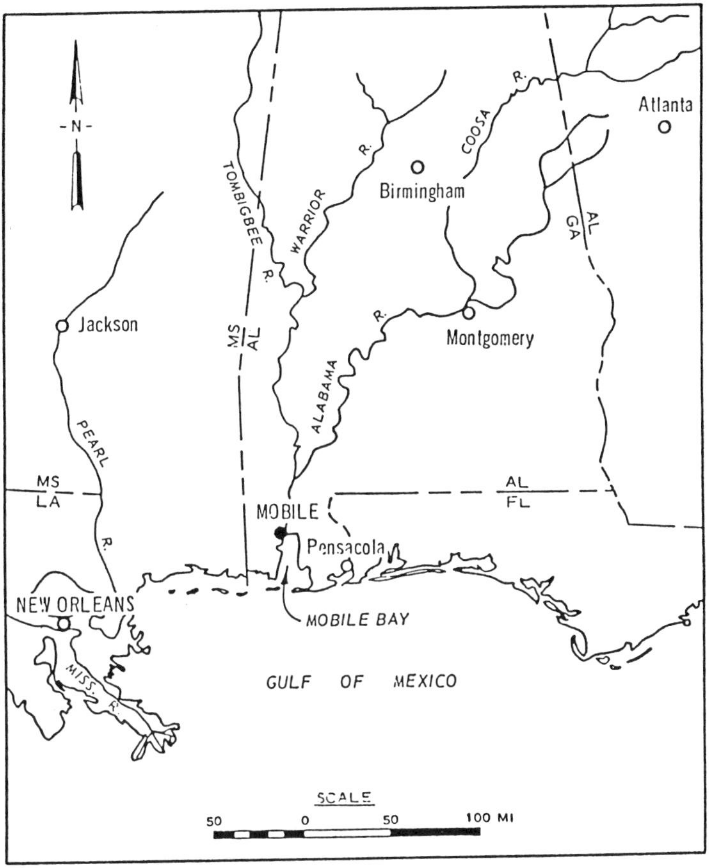

Figure 1. Vicinity Map

A 5.2 miles (8.3 km) long ship channel was dredged from the Mobile Ship Channel, at its mid-length, to the western shore of Mobile Bay, as shown in Figure 2. This part of the channel is 40 ft (12.2 m) deep and 400 ft (122 m) wide. The channel extends inland 1.9 miles (3.0 km) to join a 42 acre (17 hectare) ship turning basin at the Theodore Industrial Park. The land-cut part of the channel is 300 ft (91.4 m)

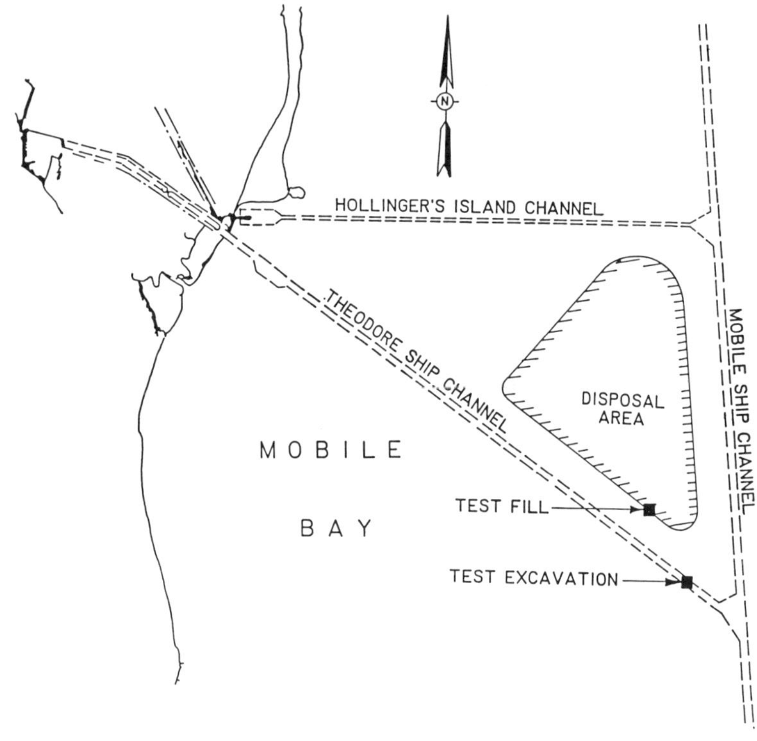

Figure 2. Ship Channel and Disposal Area

wide, and 40 ft (12 m) deep. Most of the material from the land-cut excavation and all of the bay-cut dredged material was used to construct the perimeter dike of the disposal area. The bulk of the dike fill material was very soft to stiff clay, with the remainder consisting of silt, sand, gravel, and shells.

Statement of Problem

Above water dike construction has been successful using hydraulic filling methods with new work dredged sand and clay as the fill material. These methods were used for flood control levees in Louisiana and Texas. Dike construction with clay balls, i.e., balls formed when large clay cuttings roll along the dredge pipe and develop into various sizes and shapes, has also been successfully used above water. The clay balls stack on very steep slopes and sand, silt, and clay fill the voids between the clay balls. These clay matrices are very strong immediately after they are formed and are capable of supporting lightweight equipment for reshaping.

The Gaillard Island dike was to be constructed in 8 to 10 ft (2.4 to 3.0 m) of water with hydraulically placed new work dredged and excavated material that consisted of various combinations and consistencies of sand, silt, clay, and clay balls. The foundation soils for the dike consisted of very soft, highly organic clay overlying firm to stiff clay. About 95 percent of the material from the bay-cut and about 52 percent of the material from the land-cut was clay. A large percentage of the clay materials formed into clay balls with consistency varying from very soft to stiff. Silts, sands, and some shell and gravel fill made up the remainder of the dredged material.

There was considerable concern whether an embankment could be successfully constructed without large foundation failures that might block the Mobile Bay Ship Channel as well as concern over the possible environmental effects caused by turbidity and mud flows during construction. The basic questions to be answered by the design of the project were the dike slopes, both initial and final, that can be constructed with different soil types and consistencies; the amount of bottom displacement and consolidation with varying heights of dike and foundation conditions; and the performance of various dredging equipment and dredged material placement techniques.

Scope

This paper contains a summary of the project. It reviews the data available during design, including a characterization of the soils in the channel excavation and the dike foundation. The various dredging techniques used to construct the embankments are presented. Embankment slopes, embankment erosion protection, and observations of mud flow of fine dredged materials are addressed.

DESIGN OF THE PERIMETER DIKE

Construction Experience by Others

Construction experience by personnel of the Coastal Engineering Research Laboratory, WES, indicated several embankments successfully constructed above and below water with new work dredged material, but no information was available concerning design and construction procedures for these dredged material containment dikes. Construction personnel indicated that if enough sand was available and if clay balls would form and stay together without eroding and abrading into small pieces, then a dike could be constructed in water 8 to 10 ft (2.4 to 3.0 m) deep with moderate current velocities.

Hydraulic Model Testing

A fixed-bed hydraulic model of Mobile Bay was constructed at the test facilities of the Hydraulic Laboratory, WES, Vicksburg, Mississippi. The hydraulic model was equipped with the necessary instrumentation to provide an accurate reproduction and measurement of tides, tidal currents, salination, freshwater inflows, density effects, and other important prototype phenomena. The purpose of the study was to determine the effect of the ship channel and dredged material containment area on flow patterns that would affect subsequent maintenance dredging and the oyster industry at the lower end of Mobile Bay. Analysis of the resulting data indicated that the most desirable configuration was a triangular shaped island, west of the Mobile Ship Channel and north of the proposed Theodore Ship Channel, as shown in Figure 2.

Environmental Considerations

Before construction of Gaillard Island, an environmental impact statement was prepared by the Mobile District that addressed the environmental concerns of construction of the project related to the physical alteration of Mobile Bay, loss of natural resources, and effect on water quality. The hydraulic model indicated that the proposed Gaillard Island configuration and location would not materially affect circulation and salinity patterns. Water quality would be affected only briefly from suspended solids and turbidity during the construction and stabilization period.

Geology

The area around Mobile Bay lies in the East Gulf Coastal Plain physiographic province. The region is characterized by low, smooth hills developed on Quaternary period terrace alluvial and beach deposits. These deposits overlie older Miocene and Pliocene beds which form the high ground that flanks Mobile Bay to the east and west. Lithologically, the Quaternary Holocene and Pleistocene deposits consist of light colored, very fine to coarse-grained, gravelly sand interbedded with very soft to firm beds of clay. The Holocene and Pleistocene deposits are unconsolidated and are typically 20 ft (6 m) thick except in Mobile Bay where the bay-bottom clays are up to 150 ft (46 m) thick. The Quaternary deposits formed the source of bay-cut and land-cut dredged materials for the dike.

Geotechnical Investigation

A subsurface soil investigation was conducted by the US Army Engineer District, Mobile, before construction. One hundred and seven test borings were made, located about 500 to 1000 ft (150 to 300 m) apart along the proposed alignment of the Theodore Ship Channel and the Gaillard Island dike. Details of the investigations are given elsewhere (Mobile District, 1977).

Samples were obtained with either Vibracore, split-barrel, or undisturbed tube samplers. In the bay, samples were obtained with a Vibracore sampler to a depth of 30 ft (9 m) below bay bottom and with a split-barrel sampler below that depth. Samples in the land-cut borings were taken in the same sequence used in the bay-cut borings or, when water was above the ground surface, the split-barrel sampler was used full depth of the boring. Undisturbed tube samples were taken at selected locations and depths.

Laboratory tests on the disturbed samples consisted of soil classification tests, i.e., water content, Atterberg limits, and grain size distribution. Tube samples were tested for Q, R, and R-bar triaxial tests, direct shear S-tests, and consolidation tests. Detailed results of all laboratory tests are given elsewhere (Mobile District, 1977). The boring log and laboratory test data were combined with the geologic studies to develop a generalized profile of the soils to be cut along the proposed Theodore Ship Channel and used to form the Gaillard Island dike.

Land-cut Soils. Soils within the land-cut ship channel area ranged from highly plastic clay (CH) to poorly graded sands (SP). There was no consistent, clearly defined stratigraphy along the land-cut alignment. A generalized soil profile to the maximum depth of cut, el -44 ft mlw (-13 m), in the land-cut area consists of about 10 ft (3 m) of silty sand (SM) or clayey sand (SC), underlain by 10 to 25 ft (3 to 7.6 m) of poorly graded silty sand (SP-SM), or a soft, high plasticity clay (CH), or interbedded layers of both. The remainder of the cut profile consists of firm to very stiff, high plasticity clay (CH), with the clay becoming stiffer with depth. However, a few borings showed a poorly graded sand (SP) instead of clay for the final 20 to 30 ft (6 to 9 m) to final cut grade. About 48 percent of the material from the land-cut ship channel area was sandy (SC, SM, SP-SM, SP), with the other 52 percent consisting mostly of high plasticity clay (CH), with some low plasticity clay (CL).

Bay-cut Soils. The soils along the bay-cut ship channel alignment were mostly high plasticity clays (CH) with occasional pockets of shells and fine sand. The clays were normally consolidated with a high degree of sensitivity. Typically, the first 4 to 6 ft (1 to 2 m) below bay bottom consisted of flocculated silts and clays having the consistency of "thick soup". This is underlain by 15 to 20 ft (4.5 to 6 m) of grey, soft, high plasticity clay (CH) with pockets of sand and oyster shells. The remaining material, extending down to final cut grade, is a medium to stiff, high plasticity clay (CH) that became firmer with depth. Sand and silt lenses and layers, up to 2 ft (0.6 m) thick were found randomly within the clay bed.

Gaillard Island Test Fill

In an attempt to confirm some of the proposed construction techniques during the design phase of Gaillard Island, the Mobile District decided to construct a test section along the proposed dike alignment. An area was chosen where foundation conditions and dredged material from the proposed bay-cut were considered to be worst.

Early attempts to construct a dike with a bucket dragline failed because of rapid subsidence. The final test fill was constructed with a hydraulic dredge; the fill material consisted of very soft bay bottom deposits and considerable oyster shells that were later mined by local oyster shell suppliers. The locations chosen for the test fill and for the cut sections were considered to be worse than anything expected to be met during construction of the dike. The test embankment was constructed to about +2 ft (+0.6 m) above mean low water (mlw). Even after a heavy windstorm and the consolidation and spreading of the soft dike and foundation, the dike settled to only about one foot (0.3 m) below water during the first month after construction and only 3 ft (0.9 m) below water during the first year. Two years after construction the dike was 5 ft (1.5 m) below water due to a combination of settlement and material being mined for oyster shells. It was concluded from this test that an embankment could be constructed under the worst conditions and with the worst fill materials and foundation conditions that existed in the proposed construction area. Even though the embankment test section settled or eroded below water surface, the embankment slopes before and after construction were about parallel, showing that a stable slope between 1V to 40H and 1V to 50H could be constructed.

Proposed Dike Design

A cross-section of the proposed dike is shown in Figure 3. The initial design concept was to place land-cut material hydraulically along the peripheral rim of the 1300-acre (530 hectare) dredged material disposal island to about el +2 ft (0.6 m) mlw as the first phase of construction. This material was then to be shaped to about el +4 ft (1.2 m) mlw with draglines placed on the fill. It was assumed that the embankment base width would vary from 1000 to 1200 ft (300 to 370 m) with side slopes varying from 1V:30H to 1V:50H. The dike constructed during the first phase of construction would serve as the exterior toe retaining dike for the poorer quality dredged material expected later from the soft bay bottom deposits of the bay-cut and from future maintenance dredging. It would also serve as the foundation for the second phase of construction, raising the dike to el +10 ft (3 m) mlw. A 500-ft (150 m) wide gap or weir was proposed to be left open in the dike during construction and closed at the end of construction.

Proposed Shore Line Protection

As designed, the shore line protection was to consist of riprap and vegetation protection. The rip rap was to be placed on the segments of the island closest to long reaches of Mobile Bay and ship

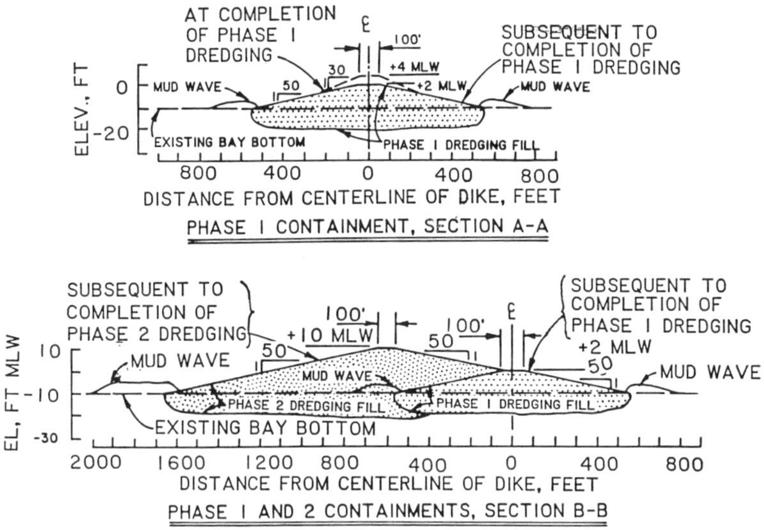

Figure 3. Proposed Dike Design

traffic in the Mobile Ship Channel. This included the eastern leg and a short distance around the northern and southern points of the dike. The two other legs of the dike, located to the west, were to be protected by planted vegetative cover only.

CONSTRUCTION METHODS & EXPERIENCES

General

Construction of the Theodore Ship Channel and the Gaillard Island dike began with clearing and grubbing of the land-cut areas on May 24, 1979, and was completed on August 28, 1981, 828 days later, with the deposition of new work dredged material on the dike section. About 33.5 million cubic yards (25.6 million cubic meters) of dredged material was either hauled by barge or hydraulically pumped over 7 miles (11 km) to construct the dike. The contractor for the project was Bean Dredging Corporation, New Orleans, Louisiana.

Dredging Equipment

Major items of equipment used on the project consisted of two cutter-head dredges, a dust-pan dredge, three booster pumps, and about 7 miles (11 km) of 27-inch (69 cm) inside diameter dredge pipe. In addition there was the typical assortment of crew boats, dozers, draglines, cranes, skidders, barges, spud barges, and spill discharge barges. The two cutter-head dredges were used for cutting all materials except the upper, fluffy silts and clays of the bay-cut where the dust-pan dredge was used effectively.

The Jim Bean dredge is a modern 9200-BHP diesel powered cutter-head dredge, 65 ft (20 m) wide and 262 ft (79.9 m) long. Estimated dredging capacity is about 1450 cu yd (1110 cu m) per hour at a discharge pressure of about 220 to 240 psi (1500 to 1650 kPa). Three on-board centrifugal pumps and one booster pump were used to pump dredged material from the ship turning basin to Gaillard Island, a distance of over 7 miles (11 km).

The Dave Blackburn dredge is a 3750-BHP diesel powered cutter-head dredge capable of pumping about 1200 cu yd (920 cu m) of dredged material per hour at a discharge pressure of 7 to 150 psi (50 to 1000 kPa). This dredge is 52 ft (16 m) wide and 180 ft (55 m) long. One on-board centrifugal pump and two booster pumps were used to pump dredged material from the ship turning basin to Gaillard Island, a distance of over 7 miles (11 km).

The Lenel Bean Dust-Pan dredge was first used experimentally at this site. Dust-pan dredges were originally designed to remove point bar deposits in the Mississippi River by using high pressure water jets to stir up or break up the sand layer before it is sucked into the dust pan. Because of potential environmental problems with the stirring up of the fluffy clay size particles found on the bay bottom, which were then expected to drift downstream and cover valuable oyster beds, the water jets were not used. Dust pan dredges work while moving. The Lenel Bean would advance about 1000 ft (300 m) on each pass, scooping up a path of soft bay bottom material 1 to 2 ft (0.3 to 0.6 m) and 30 ft (9 m) wide and continuously pumping this material about 1500 ft (460 m) into the Gaillard Island containment area. The Lenel Bean is a modern 3600-BHP diesel dredge with one centrifugal pump, and is 40 ft (12 m) wide and 252 ft (76.8 m) long.

Two specially designed discharge barges were used to assist in placement of the dredged material fill. The discharge barges were 220 ft (67 m) long by 60 ft (18 m) wide and were capable of discharging over a distance of 280 ft (85 m) through a 27 inch (69 cm) ID pipe. Mounted on each barge was a 140 ft (43 m) long boom that was winch operated to move the discharge pipe up and down. Positioning and locating of the barges was by anchored cables and by vertical spud poles on the bay bottom.

Construction Operations

As described above, the initial plan was to construct Phase I, the outside perimeter dike, hydraulically, then raise it slightly using draglines, and then, as Phase II, hydraulically construct a higher dike inside the first one using the more clayey, bay-cut materials. However, as dike construction progressed and different sections of the Phase I dike either settled or were overtopped from wave wash, the initial plan was altered. The use of draglines to raise the Phase I dike was abandoned and the dredged material for the second pass, Phase II, was placed on top of and to the inside of the Phase I dike. The weight of the materials caused lateral displacement of the soft bottom clays, causing small mud waves. However, no major foundation failures were experienced.

Dredging started at the western shore of Mobile Bay, at the mouth of Deere Creek, and proceeded westerly using both of the cutter head dredges in the land-cut area. Later the cutter head dredges were used in the deeper bay-cut. The dust pan dredge was used to cut the upper, fluffy soft clays of the bay bottom. The contractor also elected to use land-based and floating draglines on some of the land-cut, loading the material onto barges, and ferrying it to Gaillard Island for off loading.

Cutter Head Dredging

The two cutter head dredges were stationed so that the stronger Jim Bean dredge would pump the longer distances. The maximum pipe length recorded was 34,565 ft (10,500 m). During excavation, a dredge would swing the cutter head from side to side in a sweeping motion controlled by winches connected to large side anchors embedded in the channel bottom. The sandy materials tend to flow toward the cutter head, but in the stiff clays vertical faces as high as 10 to 20 ft (3 to 6 m) remain standing. These slopes would eventually slough off. The toe of the channel slope was overcut during each pass so, after slumping, the required slope of the channel sides could be formed.

Dredged material placement from the a cutter head dredge was controlled by personnel on the discharge barge. Material was pumped through the 240 ft (73 m) pipe up to 280 ft (85 m) away from the barge. The material was placed into three large mounds, one mound on the dike center line and the other two mounds at each side of center line. The dredge pipe was maintained about 3 to 8 ft (1 to 2.4 m) above water. When the center line of the dike was about 2 to 3 ft (0.6 to 0.9 m) above mlw, the spill barge operator would winch the barge over about 100 ft (30 m) to either side of the center line and raise that portion of the dike. An energy dissipater was used on the end of the dredge pipe so the dredged material would not wash out a large hole in the dike fill; however, it was not very successful. At the end of the dredge pipe the sandy materials often left large holes in the dike surface, but the clay balls stacked up in large mounds with slopes of about 1V:1H. Where clay balls formed the dike surface, they would be sticky for a few days, after which personnel and light equipment could operate on the surface.

During the first phase of construction, the discharge barge was able to float along the dike center line, but during subsequent fill placement the discharge barge had to be positioned on the outside perimeter. It proved difficult to get the barge close to the first phase dike because of the shallow water depths and the extremely flat slopes. Towboats were used to push the discharge barge ashore during high tide and then to reposition the barge to a new dump location during another high tide.

Survey control points were established by contractor personnel using 30 ft (9 m) long, 4 in. (10 cm) diameter bamboo poles pushed into the soft bay bottom mud. The bamboo poles had to be water filled, by having holes drilled into the cellular compartments, to keep them from floating out of the mud. Three survey poles were located every 500 ft (150 m) along the proposed dike alignment, with one pole on the center line and the other two located about 250 ft (75 m) to either side on a transverse line. Once the embankment was constructed to grade at a given location, the contractor was required to replace the three survey control points on the dike and provide the Corps of Engineers a profile of embankment construction above and below the water surface.

Dust-Pan Dredge Operations

The Dust-Pan dredge was used like an underwater elevator scraper loader that moved along the bay bottom scooping up about a foot (0.3 m) of soft material which was then sucked into the dust pan. Pumping distances varied from a minimum of about 1625 ft (495 m) to a maximum of about 3265 ft (995 m). Since the dredged material was very soft, it was not suitable for dike construction. However, the dust pan was capable of dredging soft clay balls. Small row dikes had to be constructed along the crest of the existing dike to prevent the soft soils from flowing back into Mobile Bay. The dredged clay was too soft to stack up; rather, it flowed out into very long and flat slopes of about 1V:500H.

The duct pan deposits filled about half of the surface area of the lower southeast corner of the Gaillard Island containment area. This operation was considered very successful; it not only saved the contractor money but expedited the work and was a more environmentally acceptable method of construction than the cutter head. Cutter head dredging would have stirred up more of the bay bottom materials, creating much greater turbidity, and would have been considerably inefficient because of the shorter pumping distances.

Barge Haul Operations

Land-based draglines were used at various times to excavate some of the land-cut area during barge haul operations. One, or sometimes two, draglines worked on the bank and one floating dragline on a self-propelled spud barge was used. As deck barges were loaded, tow boats were used to ferry them to the dike for off loading. A dragline, loaded on a spud barge, equipped with a dozer blade instead of a bucket, was used to unload the barges. The dozer blade was positioned on the haul barge and then pulled to scrape the material off the barge. Special 4-ft long standoffs were welded to the spud barge so the blade-moved material on the haul barge would fall between the two barges. Tow boats tied to the haul barges continuously re-positioned the two combined units to the desired material deposition locations.

It was very difficult to get an even distribution of dredged material over the area to be covered. Trying to keep the spud barge on the dike alignment at night or during bad weather left a lot of gaps in the fill. The inability to identify the depth of the previously placed fill and the loss of survey markers disturbed by the tow boats and barges also contributed to uneven fill distribution by barge haul operations.

Turbidity Plume Studies

Open water disposal of hydraulically placed clayey and silty dredged material in moving water has always been a problem. Aerial photographs taken during construction showed a continuous turbidity plume, a stream of murky and discolored water, around and south of Gaillard Island. Local oyster fishermen were concerned that the fine soil particles would drift into areas that contained highly productive oyster bed and destroy them. The Corps of Engineers' Mobile Area Office conducted a study of the potential problem.

Depth measurements using both lead line and depth sounding sonar were made along a number of profiles. The lead line measured the actual bay bottom. The sonar showed the top of the turbidity plume. Samples of the layer of suspended fine silt and clay had wet densities of 62.9 to 80.5 lb/cu ft (1.01 to 1.29 g/cc) with an average density of 70 lb/cu ft (1.12 g/cc). The suspended material plume lay along the bottom, had a smooth upper surface, and was about 4 ft (1.2 m) thick near the island and about 1 ft (0.3 m) thick at 2500 ft (760 m) south of the dike.

It was concluded that the suspended fine grained soil particles were only temporarily dispersed in the water south of the Gaillard Island, and that this material would eventually flocculate out in the salt water of Mobile Bay. Therefore no down-bay problems from turbidity should exist after dredging was completed. Mobile Bay is very shallow and exhibits this same type of behavior in areas that have never been dredged, especially after a small storm or a large flow of fresh water into the bay. Then, depending on the relative amounts of fresh and salt water present, parts of Mobile Bay may stay murky over a period of months.

POST CONSTRUCTION OBSERVATIONS

Immediately after completion of the dike the contractor surveyed cross-section profiles at a number of locations along the dike. Later, a series of borings was made along the dike center line to locate the original bay bottom and establish the amount of dike vertical displacement. Slope protection was established at the same time. The eastern slope and the north and south corners of the dike received a 4 ft (1.2 m), 1V:3H sloped riprap embankment on a geotextile base. Vegetation plots, using marsh grass and other procedures, were started successfully on the northwest and southwest dike slopes.

Table 1 is a listing of measured slopes and observed soil types. The dike cross-sections showed the fill had a rough, hummocky surface. This was attributed to the large quantity of clay balls present and to the erosion produced during dredged fill placement. They were determined from the long flat surfaces normally found below the waterline. Slopes above water were usually steeper.

Table 2 contains a listing of the results of post construction borings. The top of the dike, the original bay bottom, and the bottom elevation determined in the borings are shown, along with the dredged material types. Apparently during construction the bay bottom materials not only moved down and out, but also moved the adjacent dike upwards in some cases.

Undisturbed tube samples taken during the design phase borings were consolidation tested. Using a consolidation coefficient of 0.015 ft^2/day, and assuming one drainage surface, 100% consolidation is expected in over 800 years. Total settlement, as a function of embankment height and clay layer thickness, is expected to be as little as 2 ft (0.6 m) to over 9 ft (2.7 m). Twelve subsidence observation stations were established on the surface of the dike. Fowler (1983) has reported measured consolidation settlement ranging from 2 to 20 inches (5 to 50 cm), averaging 13 inches (33 cm), and continuing about as predicted by the theoretical consolidation curve.

A discussion of the vegetation used to stabilize the northwest and southwest slopes is beyond the scope of this paper. The interested reader is referred to reports by Allen et al (1984) and Landin (1984).

CONCLUSIONS

It is concluded that the dike design and construction techniques used to build the Gaillard Island dike were satisfactory. Although some small displacement failures and mud waves were encountered in construction, no catastrophic failures occurred.

The firm to stiff clays in the channel cut formed clay balls as they rolled down the dredge pipe and stacked up on slopes of about 1V:1H. Sand, silts, clays, and shells filled the interstices. Clay ball sizes ranged from two to four inches (5 to 10 cm) and larger.

It is concluded that the barge haul operation used for moving sand and stiff clay from the land-cut was successful. It is also concluded that the dust pan dredge was very effective in excavating the soft clay deposits in the bay-cut in Mobile Bay.

Table 1

Dike Slopes and Material Types

Station	Inside Offset ft	Inside Slope V:H	Outside Offset ft	Outside Slope V:H	Type of Material**
0+06	---	1:35	5	1:35	Clay Balls
8+37	340	1:80	---		Clay Balls
9+91	---	---	205	1:35	Sand
10+02	---	1:80	15	1:35	Sand
20+00	0	1:55	0	1:30	Sand
30+00	---	1:55	15	1:40	Clay Balls
40+00	0	1:55	0	1:70	Sand
50+06	90	1:55	---	---	Sand & Shell
50+06	---	---	290	1:60	Sand
50+08	---	---	90	1:55	Sand
60+00	0	1:35	0	1:85	Clay Balls
70+00	0	1:20	0	1:70	Clay Balls
80+00	0	1:20	0	1:40	Clay Balls
101+50	---	---	40	1:45	Clay Balls
113+85	---	---	225	1:55	Clay Balls
120+00	---	---	0	1:65	Clay Balls
130+00	---	---	0	1:55	Clay Balls
140+00	0	1:25	0	1:30	Clay Balls
150+00	---	1:15	40	1:40	Clay Balls
160+00	0	1:170	0	1:90	Sand
169+95	90	1:250	---	---	Sand
169+97	---	1:250	5	1:40	Sand
170+12	---	---	120	1:40	Sand
180+00	0	1:85	0	1:45	Sand
190+00	0	1:75	0	1:55	Sand
200+00	0	1:80	0	1:45	Sand
210+00	0	1:45	0	1:50	Sand
220+00	0	1:25	0	1:35	Clay Balls
240+00	0	1:25	0	1:30	Clay Balls
250+00	0	1:15	0	1:40	Clay Balls
260+00	0	1:25	0	1:25	Clay Balls
270+00	0	1:70	0	1:75	Sand
282+15	210	1:35	---	---	Clay Balls
282+25	---	---	80	1:30	Clay Balls
282+65	---	---	230	1:70	Clay Balls
290+00	0	1:30	0	1:70	Sand
300+00	0	1:30	0	1:60	Clay

** Based on visual observation at surface.

Table 2

Fill Depths and Material Types

	Elevation, Ft., Mlw			Ft.		
Station	Dike Center Line	Original Bay Bottom	Final* Bay Bottom	Fill Thick.	Total Fill Settle.	Type of Material**
0+06	4.9	- 9.0	-14.5	13.9	5.5	Clay Balls
8+37	3.2	- 9.0	-10.0	13.2	3.2	Clay Balls
9+91	2.7	-16.8	-15.4	18.1	-1.4	Sand
10+02	8.0	- 9.6	-21.2	29.2	11.6	Sand
20+00	8.2	-10.0	---	---	---	Sand
30+00	---	-11.8	-10.1	---	-1.7	Clay Balls
40+00	6.4	-12.0	-17.0	23.4	5.0	Sand
50+06	6.4	-11.0	- 8.6	15.0	-1.4	Sand&Shell
50+06	1.5	-12.0	- 8.2	9.7	-3.8	Sand
50+08	5.0	-11.8	- 9.2	14.2	-2.6	Sand
60+00	11.2	-11.0	-19.4	30.6	8.4	Clay Balls
70+00	8.0	- 9.0	- 9.7	17.7	0.7	Clay Balls
80+00	8.7	- 9.0	-21.0	29.7	12.0	Clay Balls
101+50	7.9	- 9.0	- 8.4	16.3	-0.6	Clay Balls
113+85	2.2	-10.0	- 5.0	7.2	-5.0	Clay Balls
120+00	6.4	-11.0	-13.0	19.4	2.0	Clay Balls
130+00	5.1	- 9.0	-12.8	17.9	3.88	Clay Balls
140+00	9.0	- 9.0	- 8.8	17.8	-0.2	Clay Balls
150+00	8.0	- 8.8	- 5.8	13.8	-3.0	Clay Balls
160+00	4.6	- 9.6	- 7.0	11.6	-2.6	Sand
169+95	3.5	- 9.0	- 5.4	8.9	-3.6	Sand
169+97	3.5	- 9.0	-10.0	13.5	1.0	Sand
170+12	2.5	- 9.0	-11.0	13.5	2.0	Sand
180+00	7.7	- 9.0	- 7.0	14.7	-2.0	Sand
190+00	4.1	- 9.3	-11.1	15.2	1.8	Sand
200+00	4.2	- 9.0	-11.0	15.2	2.0	Sand
210+00	5.4	- 9.0	- 7.9	13.3	-1.1	Sand
220+00	6.0	- 8.0	- 9.0	15.0	1.0	Clay Balls
240+00	12.5	-10.0	-10.0	22.5	0	Clay Balls
250+00	11.3	-10.0	---	---	---	Clay Balls
260+00	12.9	-10.0	---	---	---	Clay Balls
270+00	5.0	-10.0	-13.0	18.0	3.0	Sand
282+15	2.5	-10.1	-10.8	13.3	0.7	Clay Balls
282+25	2.2	-10.0	-13.0	15.2	3.0	Clay Balls
282+65	0.7	-10.0	-15.8	16.5	5.8	Clay Balls
290+00	8.2	-10.0	-18.8	27.0	8.8	Sand
300+00	5.8	-10.4	---	---	---	Clay

* Determined by borings made at end of construction.

** Based on visual observation at surface.

APPENDIX - REFERENCES

Allen, H.H., Webb, J.W., and Shirley, S.O. (1984). "Wetlands Development in Moderate Wave-Energy Climate",Dredging and Dredged Material Disposal, Vol. 2, Proceedings of the Conference Dredging '84, American Society of Civil Engineers, New York, NY, November, p. 943.

Champion, R.W. (1984). "Design and Construction, Theodore Ship Channel", Dredging and Dredged Material Disposal, Vol. 2, Proceedings of the Conference Dredging '84, American Society of Civil Engineers, New York, NY, November, p. 766.

Fowler, Jack (1986). "Verification of Design and Construction Techniques for Gaillard Island Dredged Material Disposal Area, Mobile, Alabama", Miscellaneous Paper GL-86-77, USAE Waterways Experiment Station, Vicksburg, Mississippi, August.

Fowler, J., Blakney, H. N., and Hayden, M. L. (1986). "Verification of Design and Construction Techniques for Gaillard Island Dredged Material Disposal Area, Mobile, Alabama", Proceedings, Conference of the World Organization of Dredging Associations, Brighton, England, March 4.

Fowler, Jack and Hayden, M.L. (1986). "Constructed Dike Slopes at Gaillard Island Dredged Material Disposal Area, Mobile, Alabama", World Dredging and Marine Construction, Vol. 22, No. 6, June.

Landin, M.C. (1984). "Habitat Development Using Dredged Material" Dredging and Dredged Material Disposal, Vol. 2, Proceedings of the Conference Dredging '84, American Society of Civil Engineers, New York, NY, November, p 907.

Lawing, R.J., Boland, R.A., and Bobb, W.H. (1975). "Mobile Barge Model Study Report 1, Effects of Proposed Theodore Ship Channel and Disposal Areas and Tides, Currents, Salinities, and Dye Dispersion", Technical Report H-75-13, USAE Waterways Experiment Station, Vicksburg, Mississippi.

Mobile District (1977). "Theodore Ship Channel and Barge Channel Extension, Mobile Harbor, Alabama", General Design Memorandum No. DM-1, Phase II, US Army Engineer District, Mobile, Alabama.

LATERAL MOVEMENTS AND MATERIAL PROPERTIES OF A POTASH TAILINGS PILE IN SASKATCHEWAN CANADA

D.E. Pufahl*
and
D.G. Fredlund**

Abstract

Saskatchewan Potash mines store enormous quantities of tailings and saturated brine, within earthen dykes, on the surface of the ground adjacent to the mine sites. The tailings consist of approximately 90% NaCl, 2% to 3% KCl, and 8% insolubles and other trace minerals. Grain size analyses indicate that the tailings have a gradation corresponding to a uniform medium to coarse sand. Direct shear and triaxial compression tests yield angles of shearing resistance of 45° or greater. Deformation moduli determined from laboratory triaxial tests indicate values of 115 to 700 MPa depending on the initial density and confining pressure. Poisson's ratio has an average value of 0.34.

General limit equilibrium analyses that have been used to assess the short term stability of the piles indicate that the properties of the foundation soils and the pore fluid pressures existing in the foundations are more likely to control the stability of the piles rather than the properties of the tailings.

Long term lateral movements, temperatures and pore pressures were studied by instrumenting a tailings pile with slope indicator access tubes, thermistor sensors and piezometers. Plane strain finite element analyses using assumed material properties established a best-fit simulation of the deformations that had been measured by using the slope inclinometers. Deformation properties obtained in this manner are one to two orders of magnitude lower than those obtained from laboratory tests.

The results indicate that most of the lateral movement occurs in the upper portion of the pile, within the freshly placed tailings. The movements in the pile appear to correlate with temperature, apparently increasing with increasing summer temperatures. However the movements are not large and presently are not reason for concern.

The present tailings disposal methods tend to trap deposits of unconsolidated slimes within the pile. Since the slimes are of low permeability and high compressibility their presence may represent some threat to the stability of the piles.

*Professor of Civil Engineering, University of Saskatchewan, Saskatoon, Saskatchewan, Canada, S7N OWO

**Professor of Civil Engineering, University of Saskatchewan, Saskatoon, Saskatchewan, Canada, S7N OWO

Introduction

Saskatchewan potash mines have generated approximately 250 M tonnes of tailings since they began operating in 1962, all of which are stored in piles on the surface of the ground adjacent to the mine sites. In addition, the nine producing mines in the province continue to generate approximately 30 M tonnes of tailings per year. With ore reserves that have been conservatively estimated to last at least 100 years, the projected quantity of tailings is indeed enormous.

The mining companies would like to construct these piles as high as is practicable in order to minimize the storage area. The need for prudent engineering becomes more important as the height of the piles exceed 50 m. A massive failure of one of these piles could rupture the dykes surrounding the brine ponds which in turn would release large quantities of saturated brine to the environment.

A study was undertaken to ascertain the stability of the tailings piles both in the short and long term periods (Chiu and Fredlund, 1986). The short term stability was addressed by applying general limit equilibrium methods of slope stability analysis to establish the factor of safety for the future geometry of the piles as predicted or anticipated by the mine operators. A variety of potential failure surfaces and piezometric conditions were investigated. Based on preliminary results, the stability of the piles in terms of a sudden or catastrophic failure depends largely on the characteristics of the foundation soils and the pore fluid pressures in the foundations rather than on the properties and the behavior of the tailings themselves. As will be seen subsequently, the tailings are a strong and competent material, capable of sustaining high shear stresses at least in short term loading conditions.

Previous investigations of the behavior of potash tailings indicated that creep under sustained stress could constitute a substantial amount of the total deformation of samples loaded in the laboratory (Pufahl, 1982). Furthermore, potash ore can undergo large deformations under constant stress (King and Acar, 1970). This background suggested that continuous movements within the tailings pile caused by sustained shearing stresses could be an important consideration in assessing the long term stability.

As a result, a tailings pile some 22 m in height was instrumented with slope inclinometer access tubes, thermistor cables to monitor temperatures, and piezometers to measure pore fluid pressures. This instrumentation which was installed near the edge of the pile extended throughout the entire depth of the pile and down into the foundation soils.

This paper describes the lateral movements in the pile as determined from the results of the field instrumentation. Some physical and chemical properties of the tailings are provided as well as some of the mechanical properties. A plane strain finite element program employing linear elastic theory was used to establish a best-fit simulation of the results obtained from the slope inclinometer data. The deformation properties corresponding to the best-fit simulation are compared to those obtained from triaxial tests that were performed on laboratory samples.

Site Description

The tailings management site that was selected for the study is the Potash Corporation of Saskatchewan Mining Ltd., (PCS), Lanigan Division, Potash Mine. This mine, which is located approximately 100 km east of Saskatoon, Saskatchewan on No. 16 highway, has been in operation since 1967. In excess of 20 M tonnes of salt tails are presently being stored on the surface of the ground adjacent to the refinery and other surface facilities. This waste disposal site is located on a 270 hectare area of flat to gently undulating fluvial lacustrine terrain consisting of sand, silt and clay which is underlain by thick deposits of clay till (Goodall et al. 1983). Figure 1 shows an aerial view of the site.

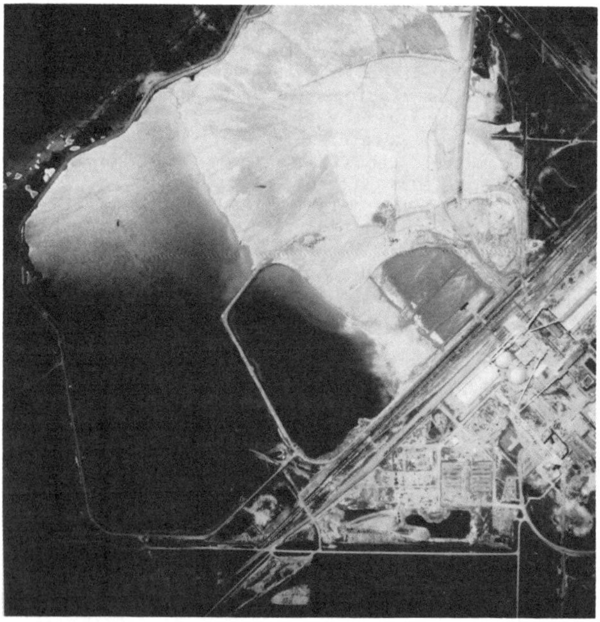

Figure 1 Aerial view of waste disposal area (Courtesy PCS Mining)

After refining; the halite and the slimes are delivered to the disposal area through a pipeline in the form of a brine slurry. The coarse fraction of the slurry is deposited near the exit of the pipe while the finer fraction settles out as the slurry proceeds down the slope of the tailings pile to a settling brine pond. The discharge pipes are moved around the disposal area and low tailings dykes are occasionally built with a crawler tractor and dozer so as to confine the slurry to small areas or cells. This procedure is a combination of the upstream and center-line method of construction. The tailings are pushed over the side of the pile and typically form a natural angle of repose of approximately 35° to the horizontal (Fig. 2).

Figure 2 Potash tailings; Angles of repose

The natural slope formed on the surface of the tailings pile is approximately 4% to 5% in the upper portion of the pile and as little as 0.5% as the slurry, consisting primarily of slimes, proceeds down the slope to the settling brine pond (Tallin and Pufahl, 1983). Even though the total percentage of slimes within the tailings is on average less than 10% to 12%, these disposal methods may cause layers of slimes of substantial thickness to be deposited within or adjacent to the piles which may then be covered with additional tailings. The extent to which these layers of slimes influence the stability of the pile has not been determined.

During the period of monitoring, the tailings discharge pipe was generally located in the northeast quadrant of the pile where tailings had previously been deposited to a height of 22 m or more above the surface of the ground.

Material Properties

The potash tailings typically consist of 85% to 90% NaCl, 2% to 3% KCl, 1% to 2% $MgCl_2$ and 5% to 12% water insolubles. The tailings have a texture and gradation that is equivalent to a uniform medium to coarse sand (Fig. 3). The relative density (specific gravity) of the individual particles ranges from 2.100 to 2.186. The in situ density at the surface is approximately 1.50 Mg/m^3, 1.65 Mg/m^3 at 1 m of depth and increases to about 1.95 Mg/m^3 near the base of the pile (Pufahl, 1983).

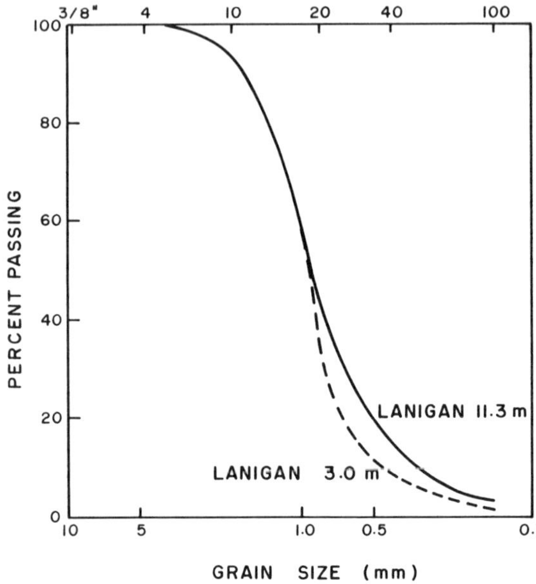

Figure 3 Gradation of potash tailings

In situ infiltration studies have been performed on this tailings pile using brine as the infiltration fluid. The results indicate that the saturated permeability varies from about 1.5 x 10^{-5} m/s to 4.0 x 10^{-5} m/s (Wong and Barbour, 1985). The low values of permeability were obtained from shallow pits while the larger values were established from large diameter bore holes (0.6 m in diameter x 1.5 m deep). The difference may be caused by the predominantly lateral flow from the bore holes compared with the predominantly vertical flow from the shallow trenches, indicating a substantial degree of anisotropy with respect to permeability within the tailings pile.

The strength characteristics under short term loading were evaluated by using both direct shear and triaxial compression tests (Figs. 4 and 5). The material exhibits a large angle of shearing resistance (i.e., greater than 45°). This infers a shearing resistance greater than the normal stress applied to the sample. Such behavior is characteristic of extremely dense, angular, granular soils, where the component of strength caused by interlocking and crushing is important. A substantial difference between the angle of repose from field

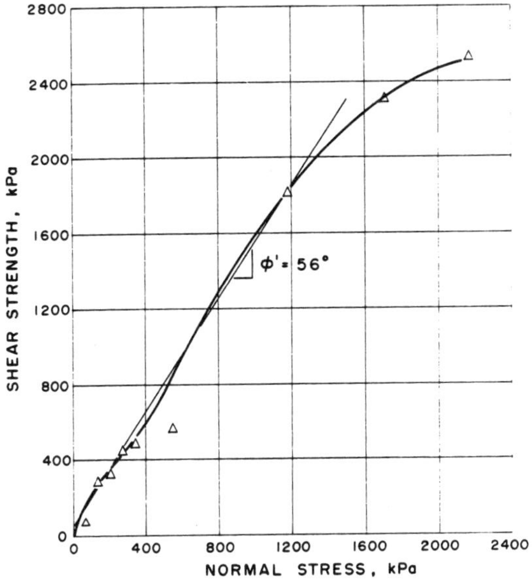

Figure 4 Direct shear test results

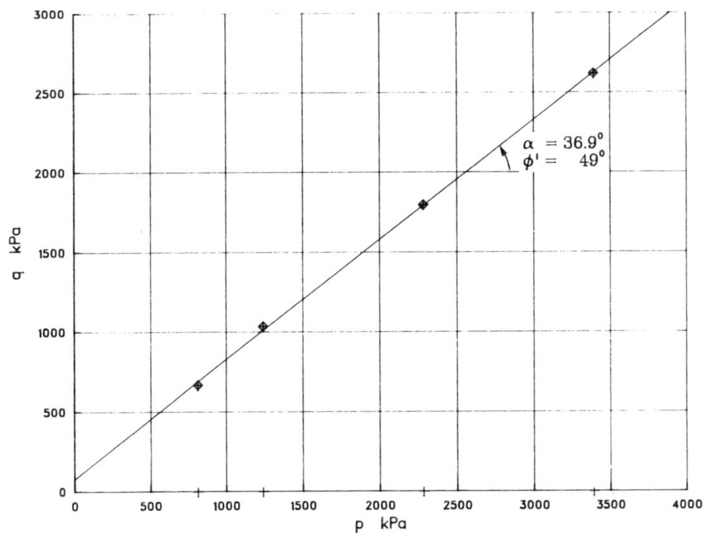

Figure 5 Triaxial test results

observations and the angle of shearing resistance from the laboratory tests is apparent. The source of this difference must be attributed to the strongly dilatant behavior of the tailings when they were tested at the initial density and confining pressures selected for the laboratory program.

Drained triaxial tests were performed in order to evaluate the deformation modulus and Poisson's ratio. Figure 6 shows a typical stress-strain relationship for this material. The stiffness of the tailings, as with any granular material, depends upon the initial density and the confining pressure. In addition, significant cementation occurs between the salt particles which further contributes to the rigidity of the material, especially at low strains.

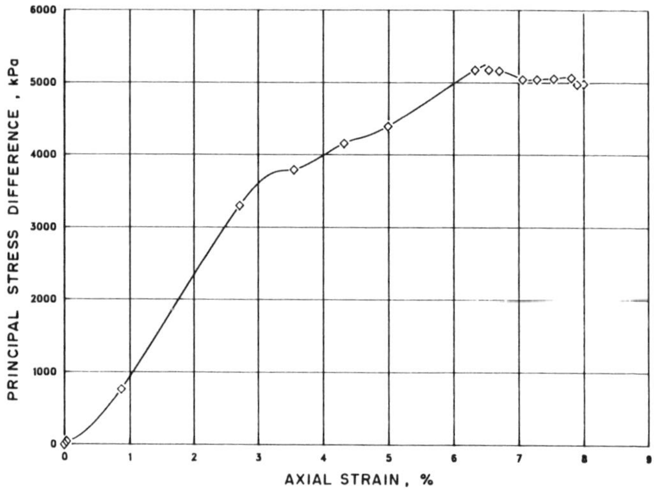

Figure 6 Stress vs strain: Potash tailings

In determining the stress-deformation characteristics of the tailings, it is important to be aware that the tailings undergo substantial creep under sustained loads. Therefore, the rate of loading has important implications in establishing the deformation modulus. The results presented in Figure 6 were obtained from a sample that was brought to failure over a period of six to eight hours representing a strain rate of approximately 1.3 x 10^{-2}%/min. Values of Poisson's ratio were obtained by measuring the volume of brine expelled from the sample in the initial stages of loading. By assuming uniform lateral deformation, the Poisson's ratio was computed. The density and confining pressure also determine the volume change behavior of granular materials. It is only at small strains where volume change is negative (i.e., decreases) that the results are meaningful in terms of elastic theory. The initial tangent modulus and Poisson's ratio are tabulated in Table 1. It is likely that the low values of Poisson's ratio are more realistic for short term loading conditions where elastic conditions are likely to apply.

Table 1: Tailings Properties from Laboratory Triaxial Tests

Sample Number	Density Mg/m^3	Deformation Modulus MPa	Poisson's Ratio
R-3	1.50	115	0.46
R-13	1.60	635	0.33
R-14	1.54	575	0.25
R-15	1.57	705	0.26
R-16	1.55	588	0.39
Average	1.55	524	0.34

Instrumentation

Figure 7 shows a plan view of the field instrumentation. Instrumentation consisting of piezometers, slope inclinometer access tubes, and thermistor cables was installed in the fall of 1982. The slope inclinometer access tubes and a number of piezometers were placed approximately 2 m from the edge of the pile. The thermistor cable and another set of piezometers were installed toward the center of the pile, approximately 165 m in from the crest. The position of the thermistors relative to the slope indicator access tubes prevents a complete understanding of the temperatures at the slope inclinometer access tubes.

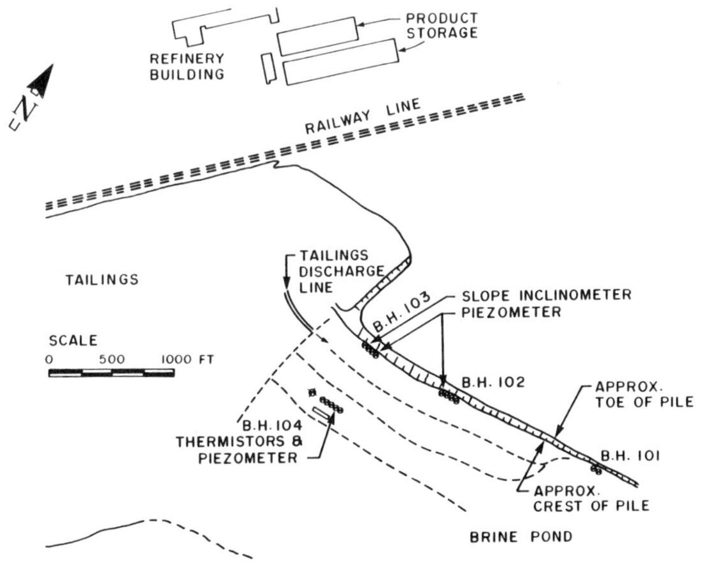

Figure 7 Plan view of instrumentation location

An initial set of slope indicator readings was taken on November 16, 1982. Regular monitoring began in May 1983 when readings were taken monthly to December 1983. Readings were discontinued at that time because excessive deflections of the access tube would not allow passage at the inclinometer sensor. Additional inclinometer access tubes were installed at the site in November of 1984 and monitoring was undertaken until January 1985.

Presentation of Data

Figure 8 shows the profile of deflections perpendicular to the strike of the slope that occurred over the period of a year. The contour of maximum deflection occurs at a depth of approximately 3 m to 8 m decreasing to the lesser value with increasing time. Figure 9 indicates the temperature profiles that were established from the thermistor readings (Chiu and Fredlund, 1986).

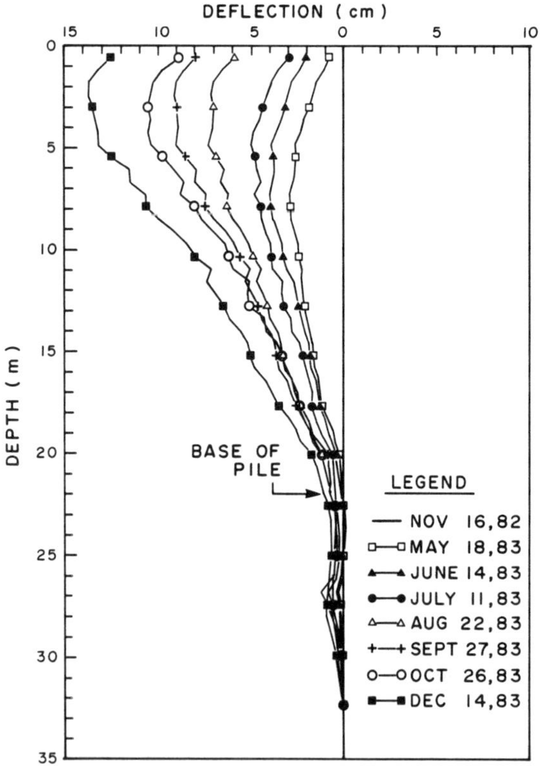

Figure 8 Deflections from slope inclinometer results

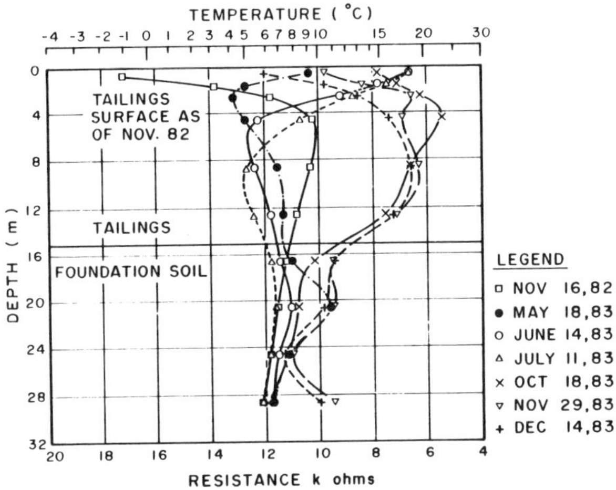

Figure 9 Temperature profile of tailings pile

Figure 10 shows the deflections and temperature at the 5 m depth with time.

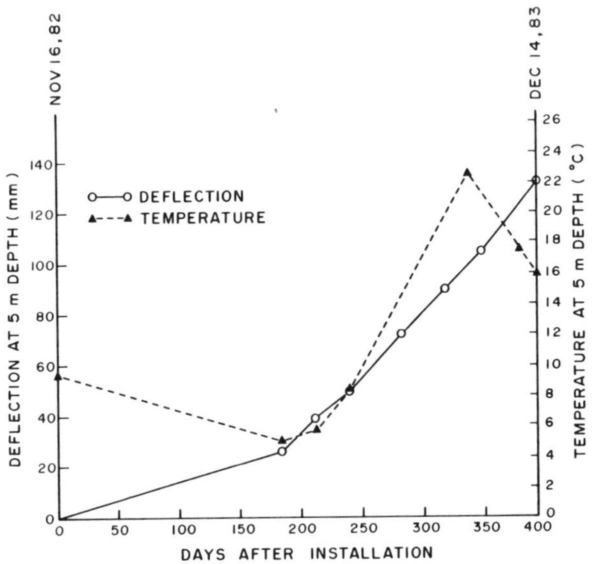

Figure 10 Deflection and temperature vs time at 5 m below surface

The rate of deflection over the winter months (1982-1983) is extremely small; however, the deflections begin to increase in a more or less linear fashion from April through to December 1983. Assuming the load has remained essentially constant with time, the steady increase in deflection with time appears to bear a relationship with the increase in temperature with time.

Analysis and Deformation Properties

A numerical analysis consisting of a two-dimensional, linear elastic, plane strain finite element program with a constant strain triangular element mesh was used in an attempt to model the lateral deformations that had been measured by the slope inclinometer. It is recognized that potash tailings will creep under sustained stress (Fig. 11) and, therefore, the input parameters to the finite element program are pseudo-elastic parameters because they implicitly contain a function of time. The stress analysis requires only the density of the tailings, a deformation modulus and Poisson's ratio as input. The density is input only for the layers of tailings which are applying a load that is producing the deformations being measured. The density is known from field and/or laboratory measurements, but the deformation modulus and Poisson's ratio can be varied so as to obtain the best-fit between the analysis and the observed behavior in the field.

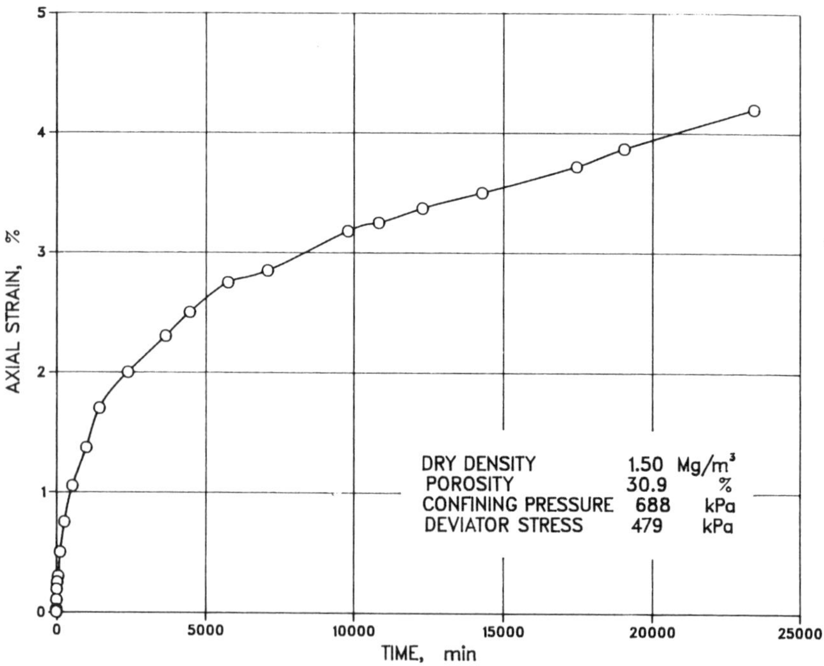

Figure 11 Creep under constant deviator stress

Figure 12 shows the schematic cross-section of the tailings pile. The pile was divided into 4 identifiable layers and different properties were assigned to each layer. Layer 3 and layer 2 represent the original tailings pile prior to installation of the instrumentation. Layer 3 is assigned a density and modulus larger than layer 2 because it was deposited a number of years ago and field measurements indicate that it is denser than the upper part of the pile. However, the analysis assumes that deformation due to the self weight of layers 2, 3 and 4 is complete and that additional deformation is caused by the surcharge load applied by layer 1. Accordingly, the density of layers 2, 3, and 4 is assumed to be zero and only the modulus and Poisson's ratio are of relevance.

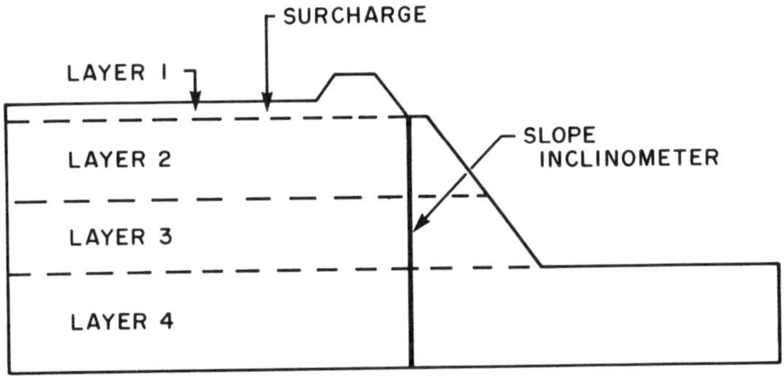

Figure 12 Schematic cross section of tailings pile

By varying the material properties within each zone, a best fit relationship was established. The final selection of material properties is tabulated in Table 2. The comparison between the best-fit simulation and measured values of deflection is presented in Figure 13. Other combinations of material properties were used in the analysis (Chiu and Fredlund, 1986); however, only the best-fit results are reported here.

Table 2: Properties of Tailings Layers

Material	Density	Deformation Modulus	Poisson's Ratio
Foundation Soil	0	30 MPa	0.3
Potash Tailings			
Layer 3	0	6 MPa	0.4
Layer 2	0	3 MPa	0.4
Layer 1 Surcharge	1.60 Mg/m^3		

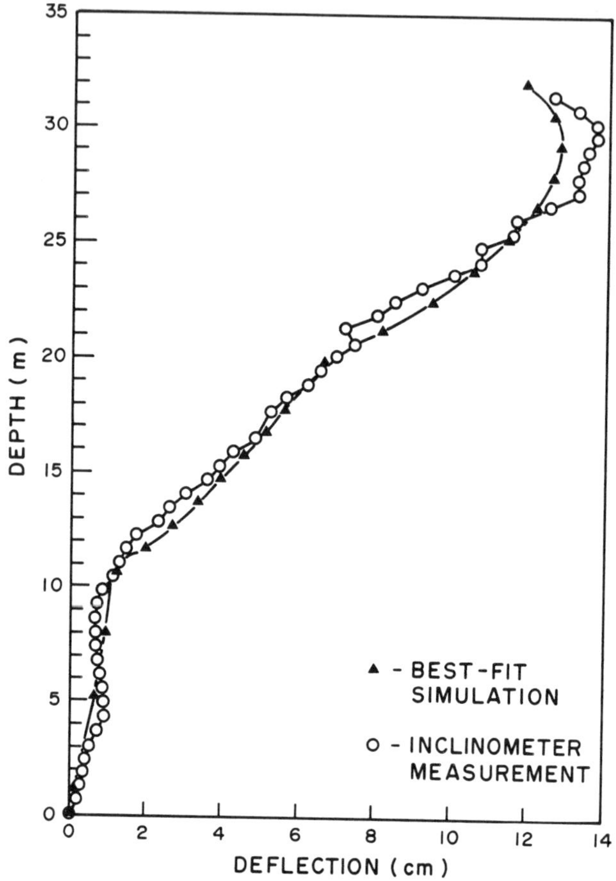

Figure 13 Comparison of best-fit simulation and measured deflections

Discussion and Conclusions

A large disparity, as much as two orders of magnitude, exists between the deformation modulus obtained from laboratory triaxial tests, tabulated in Table 1, and those used for the best-fit simulation of the measured deflections. The laboratory tests were performed using confining pressures in the order of 900 kPa. The magnitude of this pressure is well in excess of that acting in the tailings pile around the inclinometer access tubes and represents a value associated with a depth of 85 to 100 m of tailings. These triaxial tests were performed to obtain data for a somewhat different application, and that is the reason that such high confining pressures were used. As stated previously the stiffness of granular materials is dependent on the level of confining pressure. Therefore, the modulus obtained from the laboratory tests would be expected to be in excess of that obtained in the field. Perhaps more importantly the rate of loading of the laboratory samples was extremely large compared to the field situation. This difference in rate of loading is thought to be the major cause of difference in the two values. However, the comparison is useful because the data emphasize the extremely large range of values that may be anticipated under different conditions of loadings.

Poisson's ratio measured in the laboratory varied from 0.25 to 0.46 with an average value of 0.34. This average value is somewhat lower than the value of 0.4 that was selected for the analytical study. As indicated earlier, Poisson's ratio is also somewhat influenced, for a given density, by the level of confining pressure. The smaller values of Poisson's ratio tend to be associated with the higher values of confining pressure because the tendency for dilation in the sample is suppressed.

The extent of the field data does not permit a complete appraisal of the deformation characteristics and the movements in the tailings pile that was investigated in this study. However, a number of observations can be made.

i) The movements are relatively small. The maximum deformation measured over the period of a year was approximately 130 mm.

ii) The movement is apparently caused by creep under constant deviator stress.

iii) The depth below the surface of the pile at which maximum movements are occurring is quite shallow, approximately the upper 1/3 point of the pile. This infers that the maximum safe height of the pile should not be governed by this type of movement.

iv) The rate of movement appears to be related to the temperature, with increases in rate occurring with increases in temperature.

v) The rate of movement did not decrease with time over the period of measurement as might be expected. Further readings could not be taken because of the slope inclinometer access tubing collapsed.

vi) Precise records of loading associated with the tailings disposal activity in the vicinity of the instrumentation were not established over the period of measurement. As a result, the observations contained in items ii), iv) and v) may have been affected by the addition of tailings and/or of brine to the pile near the instrumentation.

vii) Additional studies could provide information on the correlation of temperature changes with deformation rate. Precise measurements of deposition rate and other details of tailings disposal activity would be required over the monitoring period.

The amount of creep or lateral deformation in the tailings pile measured in this study does not appear to be alarming. Most of the movement occurs in the freshly deposited tailings and it seems to decrease as the tailings age and become denser.
Perhaps of greater concern in determining the stability of the tailings piles is the effect of deposits of slimes that can collect within or adjacent to the piles. The slimes, consisting mainly of silt size particles with traces of chemicals used in the refining process have low permeabilities and relatively high compressibilities. This combination of material properties indicates that consolidation and the concommittant gain in strength will occur rather slowly when they are loaded with additional tailings (Pufahl, 1987). Until more information is acquired regarding the properties of the slimes and how they affect the stability of the piles, definitive statements regarding the stability of the piles are not possible.

Acknowledgements

The authors express their appreciation to the Potash Corporation of Saskatchewan for the financial and field support in obtaining this data and for their permission to allow publication.

References

Chiu J. and Fredlund D.G. 1986. The analysis for the stability of potash tailings piles. University of Saskatchewan research report for Potash Corporation of Saskatchewan, Saskatoon, Saskatchewan.
Goodall D.G., Fredlund D.G., Lamb K.N. 1983. Methods and implications of a state of the art geotechnical field investigation of potash tailings near Lanigan, Saskatchewan. Proceedings, First International Potash Technology Conference, Saskatoon, Saskatchewan. pp. 787-792.
King, M.S. and Acar, K.Z. 1970. Creep properties of Saskatchewan potash as a function of changes in temperature and stress. Third Symposium on Salt, Northern Ohio Geological Society, Inc., Vol. 2, pp. 226-235.
Pufahl, D.E. 1983. Properties of potash tailings for tailings pile stability analysis. Proceedings First International Potash Technology Conference, Saskatoon, Saskatchewan, pp. 787-792.
Tallin, J.E. and Pufahl, D.E. 1983. Waste management schemes of potash mines in Saskatchewan. Proceedings, First International Potash Technology Conference, Saskatoon, Saskatchewan, pp. 755-760.
Wong, D. and Barbour, S.L. 1985. Studies of the infiltration and migration of brine in salt tailings. Proceedings, Canadian Society for Civil Engineering, Annual Conference, Saskatoon, Saskatchewan, pp. 357-374.

Overview of Hydraulic Construction Techniques
Utilized at the Syncrude Canada Limited
Tailings Pond

L.T. Yano and A.E. Fair*

The Syncrude Canada Ltd. open pit mine located near Fort McMurray in northern Alberta produces approximately 300,000 tonnes per day of oil sand feed. After removal of the bitumen, approximately 240,000 tonnes per day of tailings solids are generated. The sand is slurried with process water and pumped to the tailings pond, where it is used to hydraulically construct the containment dykes and supporting beaches.

The dyke structure is constructed using a modified centerline technique. The dyke is built up by hydraulically sluicing the tailings slurry into various cells formed by dozing three meter high sand dykes. The sand settles out and is then spread and compacted using wide track dozers. Any sand not captured, as well as the fines and water, drain through a spillbox positioned at the end of the cell. This diluted slurry mixture then flows over the beach and into the pond. The tailings pond water is continually recycled back to the plant to provide water for a number of process streams. Other than the imported fill required to construct starter dykes, the entire tailings dyke has been constructed with tailings sand. The design parameters of the dyke structure are mainly dictated by the foundation geology and the in-place density of the tailings sand which forms the compacted shell. The tailings sand shear strength properties have little impact on the design parameters largely due to tr..ir uniform nature.

Approximately 550 Mm^3 of sand, 370 Mm^3 of thick sludge and 50 Mm^3 of freewater will require storage within the Syncrude Tailings Pond. To accommodate these volumes, approximately 18 kilometers of dyke ranging from 32 to 90 meters in final height are being constructed. At completion the tailings disposal area will have a surface area of 22 square kilometers. This paper documents both the construction techniques and history of the Syncrude Tailings Pond.

Introduction

The Syncrude Canada Limited project, located 40 km north of Fort McMurray in northeastern Alberta (see Figure 1) operates one of the largest open pit mine and tailings disposal operations in the world. Operating at design capacity, the mine produces 100 million tonnes of oil sand per year. The oil sand is conveyed to an extraction facility which utilizes a hot water process to separate the bitumen and sand. Approximately 240,000 tonnes per day of tailings solids are produced, which are hydraulically transported to the tailings disposal area.

*Syncrude Canada Limited, P.O. Bag 4009, Fort McMurray, Alberta T9H 3L1

Tailings disposal operations began in May 1978 with the start-up of the Extraction Plant. Currently tailings are accommodated by an external tailings disposal area located north of the plant site. The general layout of the tailings pond and perimeter dyke is illustrated in Figure 2. For planning purposes, the dyke perimeter has been divided into 700 m long segments which are referred to as cells. The cell locations with respect to the dyke perimeter and numbering sequence are also shown in Figure 2. A typical design section for the tailings dyke is illustrated in Figure 3 which shows the compacted shell, the beached sand and fluid storage. The existing tailings pond provides initial storage to allow sufficient mine development to meet long term storage requirements. Longer term plans show 'dry' tailings sand storage in another area adjacent to the mine and mature sludge storage within the mined out pit. The tailings pond will continue to operate as a 'settling pond' to facilitate initial consolidation of the oil sand sludge and to continue providing recycle water requirements.

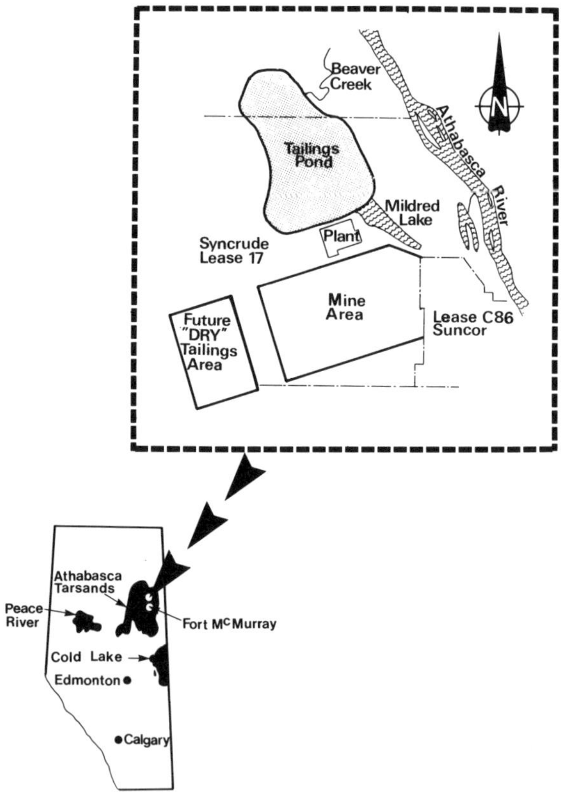

Figure 1. Location Map of Syncrude Canada Ltd.

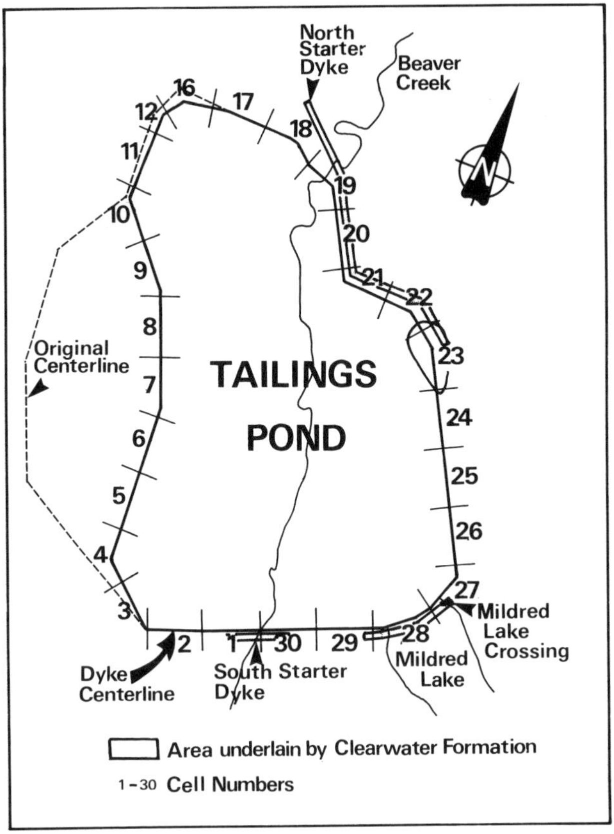

Figure 2. Layout of Syncrude Tailings Pond

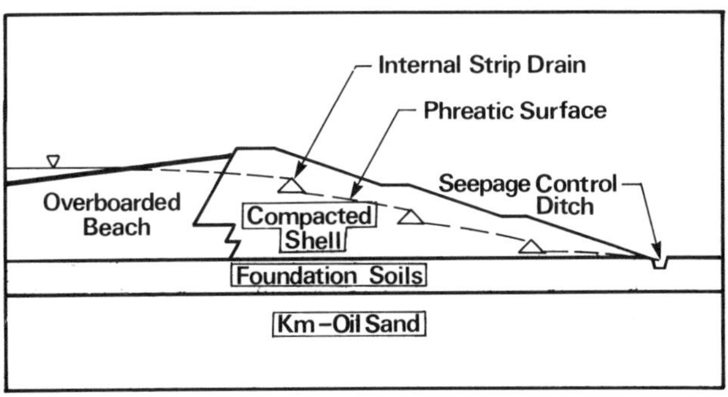

Figure 3. Typical Design Cross-Section of Tailing Dyke

Hydraulic Transport System

The Extraction Plant uses a hot water bitumen extraction technique to process the oil sand. The ore is fed into large tumblers where steam, hot water and caustic are added to condition the oil sand in preparation for the separation process. The oil sand slurry then goes into the primary separation vessels where density differences result in the separation of the bitumen and solids. The solids move to the bottom of the vessel while the bitumen is removed from the surface as a froth. The bitumen froth goes through an additional cleansing stage to improve the product quality and prepare it for upgrading to synthetic crude oil. From the bottom of the vessel the relatively bitumen free tailings are routed to a distributor box where they can be pumped out of the Extraction Plant through any one of five independent pumping and piping systems.

The tailings slurry has an exit temperature of 85°C and consists of approximately 0.3% bitumen, 39.7% water and 60.0% solids by weight. The total slurry flow under normal plant operating conditions is 5600 kg/s. This flow is distributed to the five tailings lines for transport to the tailings area. Each tailings line has a flow of 1400 kg/s with a velocity of 3.2 m/s. Under these conditions the slurry moves as a bedded flow with the coarser solids moving along the lower portion of the pipeline. The transport water, fines and bitumen travel along the upper portion. This bedded flow condition results in uneven pipeline wear patterns with the highest wear points occurring in the 10 o'clock and 2 o'clock positions. The combination of both the presence of abrasive solids and increased flow velocities above the moving bed results in a more rapid wear rate at these locations. In order to maximize its life, the pipe is rotated 180 degrees when the thickness reaches 8 mm. The pipe is then rejected when any portion has reached a thickness of 5 mm. Regular inspections of the pipe spools are carried out to monitor wear and to schedule pipe rotations. Pipe life is highly variable depending on location (i.e. after bends, on ramps, after pumps etc.), however on flat run sections up to 5 years service has been observed.

The tailings distribution system consists of a network of 600 mm carbon steel piping in combination with 20 x 20 inch centrifugal pumps driven by 800 HP motors. Coverage of the disposal site by the five tailings lines is shown in Table 1. (Note: Cells 13-15 were eliminated by the dyke centerline re-alignment in 1981)

TABLE 1. TAILINGS LINE DISTRIBUTION SYSTEM

LINE #	CELLS COVERED	NO. OF PUMPS	MAXIMUM PIPE LENGTH
1	16 - 21	10	12000 m
2	22 - 27	7	8700 m
3	28 - 30, 1,2	5	5500 m
4	3 - 7	5	5700 m
5	8 - 12	9	10500 m
TOTAL		36	42400 m

Only four tailings lines are required to meet production demands therefore one, and in many instances, two lines are idle. This operational flexibility permits routine pump and pipeline maintenance or emergency shutdowns without limiting production in the Extraction Plant. Scheduling of line operation also provides the flexibility to utilize the tailings lines for sand placement in priority areas.

Dyke Construction Techniques

The compacted shell and the supporting beach are placed using hydraulic construction techniques. The technique has also been successfully used at the adjacent Suncor oil sand mine, formerly Great Canadian Oil Sands, since 1967. During the summer hydraulic cell construction period from mid-April to mid-November the emphasis on sand placement is directed to the compacted shell. The cell construction provides a higher density fill in the compacted shell and provides a discharge platform for winter overboarding (see Figure 3). This technique involves preparing a 500 to 700 m long area to the width of the compacted shell by dozing up 3 m high sand dykes. The total cell width is a function of the downstream slope angle and dyke elevation. For wide cells, up to five smaller sub-cells approximately 50m wide have been used to faciliate the distribution and spreading of the sand. The tailings is sluiced into the cell where the coarse sand settles out and is then spread and compacted by wide track dozers operating within the active cell. The excess transport water and the solids which do not settle out, flow to the end of the cell and out onto the beach through a moveable spillbox structure. The spillbox is simply a rectangular weir with drainage pipes connected at its base, and removable retaining boards in the front face. As the sand builds up to a 3 m lift height at the discharge end of the cell, the pipelines are advanced and the weir height within the spillbox is increased to maximize the solids capture. Once the lift has advanced to within 50m of the spillbox, the end dyke is breached and pouring continues into another previously prepared construction cell. After the lift has advanced beyond the spillbox it is removed for future re-installation. Figure 4 schematically illustrates the cell construction operation. During pipeline advances in the cell or when dozers are not available for spreading and compacting, an overboard leg discharges the tailings slurry directly to the beach off the upstream crest of the compacted shell. Because of the high slurry temperature, cell construction is not possible during cold weather due to excess fogging which creates unsafe working conditions for the dozers in the cell. From mid-November to mid-April all tailings streams are overboarded to build up the supporting beach. Previous surveys of the beach have shown that, for the Syncrude tailings grain size characteristics, sand settles out on the beach to form a 2% above water slope. Once the tailings stream reaches the water's edge, which is typically 300 to 500m from the dyke crest, the beach continues to form at a steeper 4% below water slope.

The tailings sand characteristics are such that the majority of the solids settle out in either the compacted shell or the beach. A typical grain size curve for Syncrude tailings is shown in Figure 5. Because of the uniform nature of the tailings, the coarse solids (SG = 2.65) exhibit relatively good settling rates. The fines fraction of

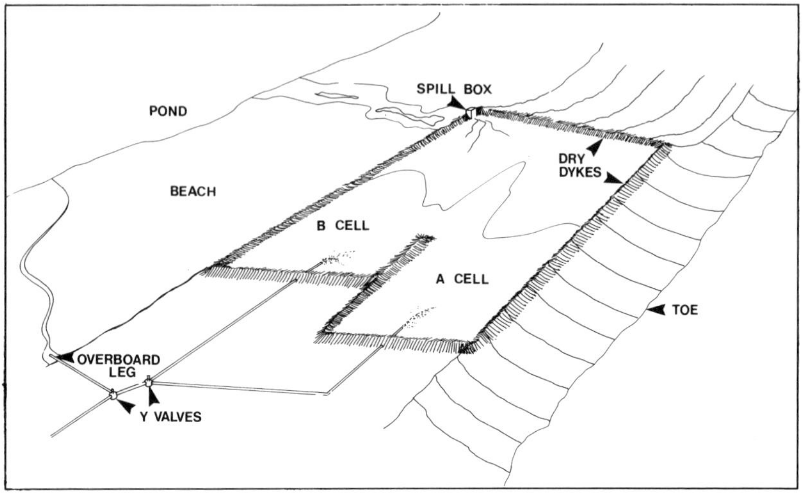

Figure 4. Hydraulic Cell Construction

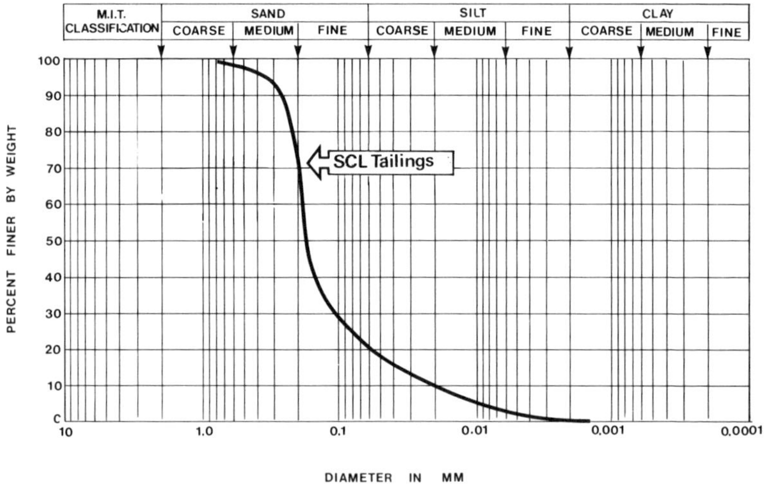

Figure 5. Syncrude Tailings Grain Size Distribution

the tailings, defined as the solids less than 22 micron, represent approximately 10% of the total solids. During the beaching process roughly one half of the fines are captured in the sand voids. The remaining fines run off to the pond interior along with the excess transport water. In the pond, the fines quickly settle out to form the uppermost sludge layer.

Dyke Construction Scheduling

Tailings dyke construction involves an extensive field program during the months from April to November. Compacted sand volumes of up to 14 Mm^3 are hydraulically placed in the dyke using the cell construction technique. The target dyke volume to be placed each year is derived from the long term plan for ultimate completion of the tailings dyke and beach. The volume placed in a given year is distributed between the five tailings lines, each of which covers a specific area. As the dyke construction season progresses the priorities for the lines are continually revised to ensure the sand is evenly distributed to the various cells around the dyke.

The amount of tailings sand which can be placed in the dyke is calculated on the basis of projected tonnage of oil sand mined. Approximately 75% of the oil sand tonnage is assumed to be available for placement in either the dyke or beach. Of the total sand available the volume which is placed in the dyke is dependent on:

1) the percentage of slurry which is poured to the cells as opposed to the slurry which is discharged to the beach (i.e. percent pour) and
2) the percentage of the sand which remains in the dyke during active cell construction as opposed to the amount which runs out onto the beach through the spillbox (i.e. percent cell capture).

The percent pour is dependent on weather, equipment availability and cell availability. The percent cell capture is mainly a function of the efficiency of the spillbox to retain the solids during cell construction. Historical averages for percent pour and percent cell capture are used to forecast the dyke construction volume available for a given production level and the time of year. Typical averages for the peak construction months are 80% pour and 60% cell capture. This results in an overall capture of sand to the compacted shell equal to 50% of the total sand tonnage produced from extraction.

Early Tailings Development

Based on the hydraulic construction technique described in the preceding section, the tailings pond was initially sized to sustain 25 years of mine production. Various assumptions were made in terms of sand and sludge production as well as a rough water balance in order to provide the necessary volumes to generate the pond/dyke size requirements. Initial estimates suggested a total pond capacity of approximately 800 Mm^3 would be required. The initial water balance assumed that 0.9 m^3 of raw water was required for every 1 m^3 of oil sand mined. Recycled water from the pond was used to partly offset the amount of raw water required for extraction processing. Pond water reclaim was planned to start-up three years after hydraulic construction began. The reclaim rate was projected to be in the order of 55% of the total tailings production. These initial assumptions led to a tailings disposal area which covered approximately 27 km^2 and which would be built to a final design elevation of 347m (i.e. maximum dyke height of approximately 85 m).

The tailings pond was located directly north of the plant site, largely on the basis of economics, extent of underlying oil sand reserves and foundation conditions. The actual dyke perimeter was chosen primarily on the basis of physical constraints (i.e. lease boundaries, Highway 63 and the plant site to the north, east and south, respectively) and the extent of underlying oil sand reserves to the west. The additional costs which would be incurred, due to greater pipeline lengths, with remote sites tended to limit the location options to nearby sites. The site chosen minimized the pipeline lengths, was deemed to have minimal underlying oil sand reserves and acceptable foundation conditions.

Although the regional topography of the chosen site slopes gently downward from west to east toward the Athabasca River, two more prominent geographical features existed within the confines of the proposed dyke perimeter. The first was Beaver Creek Channel which flowed from south to north through both the mine and tailings area. A dam was constructed across Beaver Creek at the southern limit of the mine site which eliminated the flow of Beaver Creek through the Syncrude leases. The second feature was Mildred Lake. The dyke perimeter crossed Mildred Lake at its approximate midpoint, some 700 m across. Three starter dykes were required in these topographical low areas to provide initial containment for tailings discharge. By late 1977 the North and South Starter Dykes across Beaver Creek Channel and the Mildred Lake Crossing had been completed (see Figure 2). All three dykes were constructed with mobile equiment utilizing local fill materials. The Mildred Lake Crossing Dyke required special attention to ensure end-dumped fill material met construction placement criteria. Coarse grained materials were used in order to achieve adequate fill densities. An additional downstream toe berm was added to ensure dyke stability in this area. In addition to the starter dykes, mobile equipment was used to provide initial containment along portions of the south and east perimeter dykes.

Disposal of tailings solids began in May 1978 with the start-up of the Extraction Plant. The hydraulic cell construction technique described earlier was used to construct the perimeter dykes. A form of upstream construction was used in the initial years to construct the compacted shell. However, during the early phases of dyke construction it was not always possible to construct to the ultimate centerline. Consequently portions of the east dyke contain areas where tailings sand was placed into ponded water at lower elevations. Within two to three years all these problem areas were brought under control and hydraulic cell construction techniques could be fully implemented to the design centerline. As the pond level rose the dyke perimeter began to progress westward, necessitating dyke construction along the remaining west side dyke perimeter. Closure of the dyke, in

Cell 12, occurred in 1983.

Tailings Pond Modifications

As the tailings pond developed, refinements to the design were required to optimize both the geometry and pond life. The ability of the overall structure to meet a variety of mining requirements necessitated three major revisions to the initial design. They included, in order of occurrance, a realignment of the dyke centerline, a change in the dyke construction method and an increase in the ultimate dyke crest elevation.

In 1981 a decision was made to realign the west tailings dyke from its original position to a more easterly location, as illustrated in Figure 2. Since the west side of the tailings pond area had a higher original ground topography, construction of the dyke in Cells 4 to 16 had not yet started. The reduction in total dyke length resulted in the elimination of Cells 13 to 15. The driving force behind the dyke realignment was the desire to minimize the sterilization of mineable oil sands on the west side of the pond while maintaining sufficient storage capacity to meet coarse tailings and sludge volume demands over the next 25 years. The tailing pond and dyke surface area was reduced from 27 km^2 to 22 km^2. Since the design dyke crest remained at elevation 347 m the total design storage volume for sand, sludge and water was reduced from 816 Mm^3 to 697 Mm^3. The dyke realignment was made viable by a reduction in the long term water storage requirements and a fine tuning of the tailings and volumetric parameters, based on field performance since 1978. Tailings water balances were revised in the following key areas:

1) By better water management in the plant there was a reduction in effluent water pumped to the tailings pond.
2) A number of plant water systems reduced their usage of raw water by using more tailings reclaim water.
3) Long term water storage within the tailings sand was increased by revising the original sand void ratio based on field measurements of the placed sand (i.e. dry density). Water entrapment in the dyke and beach is expected to be approximately 200 Mm^3.

As noted earlier, the original dyke and beach construction method utilized a modified upsteam approach where each year's compacted shell was partially placed on beached sand. The primary benefit of this method stems from the relatively low compacted dyke volumes required for the initial development of the tailings pond. In 1982 the construction method was changed to a centerline approach where the dyke was built independent of the beach. Figure 6 illustrates both construction methods in cross-section. Using centerline construction the entire dyke structure defined by a 2H:1V upstream slope and the design downstream slope is compacted. The change to this method required a large step-out of the compacted shell onto the beach to allow future dyke development to the ultimate 347 m crest elevation. The step-out was achieved by utilizing cell construction over the beached sand. A modification to the normal cell construction method was required because of problems encountered while operating dozers

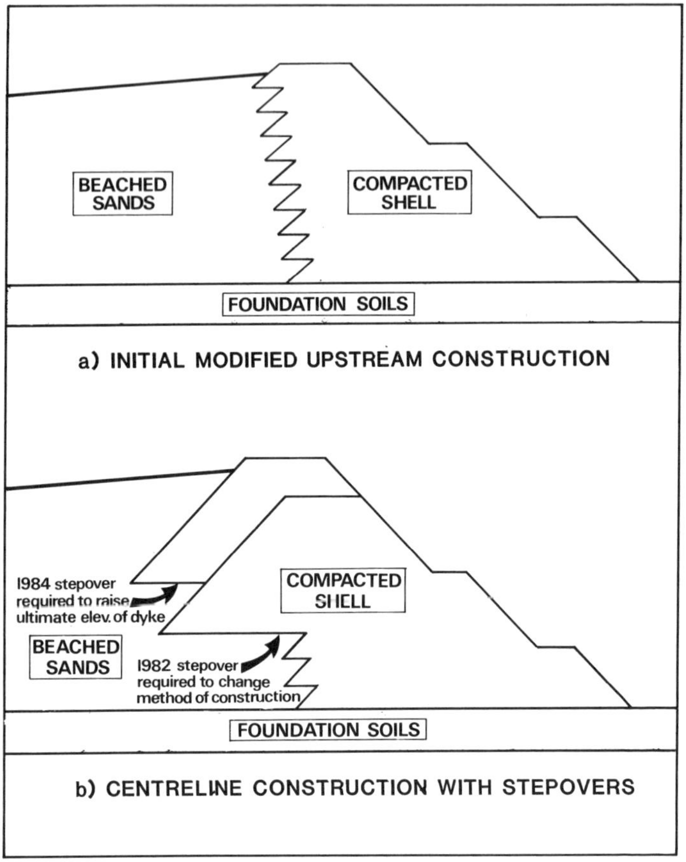

Figure 6. Schematic Cross-Sections of Syncrude Tailings Dyke Illustrating the Different Construction Techniques

over the looser, more saturated beach material. The cells were poured open-ended with short, 'pond side' dry dykes and no end dyke or spillbox. Although sand capture in the cells was much lower the step over construction was successfully achieved. A significant increase in the total dyke volume was required, therefore independent dyke and beach construction had advantages which justified the change. The additional dyke volume was obtained by placing less beach volume during the summer construction period. Operational experience indicated that the increased annual dyke volume demand could still be met based on the long term sand schedule for dyke and beach development to complete the tailings pond structure. The sand placement schedule was revised by increasing the amount of sand placed in the compacted shell in the early years and reducing the beaching volume requirements in the later years. The advantages of the centerline construction method included:

1) Increased dyke stability with a greater proportion of the total sand placed in the compacted shell.
2) Completion of the dyke to elevation 347 m well ahead of beach construction to elevation 346 m. This allows optimization of the total sludge and water storage volume by providing sufficient lead time to maximize the final above water beach slope angle. Completion of the beach to 346 m was planned to be placed by spigotted discharge (described later) which creates a steeper above water beach slope.

The original tailings schedule showed completion of the out-of-pit sand structure to elevation 347 m in 1987 followed by sand disposal in the mined out pit. In 1984 a decision was made to delay completion of the out-of-pit pond to 1990 by raising the ultimate dyke crest elevation from 347 m to 352 m. Economics was the key factor in the decision. During 1984 Syncrude was considering a major increase in plant capacity which was scheduled to come on stream in 1987. To avoid two large capital projects occurring over the same time frame, the tailings plan was modified. Given a three year delay, a detailed study was carried out to determine the optimum dyke crest elevation. Elevation 352 m was chosen as the best alternative which balanced the economics of capital cost delays with a geometry that maximized sludge storage volume and hence, tailings pond life. Since the increase in dyke elevation affected overall dyke stability, a redesign of the downstream slopes and future internal drainage filters was completed. The previous 3H:1V slopes were flattened to 4H:1V and the 7H:1V slopes were changed to 9H:1V. Existing 4H:1V slopes were not changed. To implement the redesign, an additional step-out onto the beach was required to accommodate the increase in dyke elevation and the flatter downstream slope (see Figure 6b). On the west side of the pond the step-out was accomplished by constructing cells over the existing beaches to widen the dyke. Along the east side the step-out required was narrower because of the lower impact of the steeper downstream slopes.

To quickly build the beach elevation up to the existing upstream dyke crest, spigotting of the tailings slurry through a series of 150 mm pipes was employed. This procedure involved connecting eight 150 mm pipes to the bottom of the 600 mm main discharge pipe at a spacing of 25 m. By drawing off the higher density slurry from the bottom of the main pipe and discharging it at a lower flow rate, a steeper beach slope was achieved, which required less beach volume. Once the beach was raised, the next dyke construction lift was widened to include the required step-out. The current tailings pond design has the ultimate dyke crest at elevation 352 meters with west dyke slopes at 9H:1V while east dyke slopes are generally at 4H:1V.

Geotechnical Considerations

Early in the initial planning stages of the Syncrude Tailings Pond it was recognized that the 'observational approach', originally proposed by Terzaghi, was the only practical method to ensure an adequate design. Syncrude has implemented an extensive monitoring and testing program to support the successful and ongoing use of the observational approach (Fair and Handford 1986, Handford et al 1982). Geotechnical design input was required in three fundamental areas: overall dyke stability, seepage control and liquefaction potential. The following paragraphs briefly discuss the geotechnical considerations in each of these three areas as they relate to dyke construction.

Adverse experience at other sites had alerted the designers to potential foundation instability problems associated with the clay shale foundation materials (Clearwater Formation) which underlie much of the tailings pond area. Figure 2 illustrates the foundation conditions underlying the dyke perimeter (note that the majority of the west side is underlain by clay shales). Beginning in the early years and continuing up to the present, considerable work has been devoted to developing a better understanding of the properties of the clay shale units. Two areas, referred to as Cell 18 and Cell 23, have experienced considerable foundation deformation as a result of dyke construction. The monitoring program referenced earlier has been very successful in identifying these potential problem areas. Various slope angle geometries and rates of construction have been used to control foundation stability problems. As noted earlier, current dyke slope angles range from 4H:1V to 9H:1V around the dyke perimeter. Rates of construction have ranged from 6 m in one year to as low as 3 m in two years in problem areas. Considerable experience has been gained over the past ten years in terms of implementing design changes to respond to any foundation deformation.

Seepage control, both within the dyke and through the underlying foundation sands, is also a major design consideration. Internal strip drains are used to control the phreatic surface within the dyke as well as any potential internal erosion or piping. Offtake pipes spaced at 100 m intervals drain the water from the trunkline, out to the downstream slope and then down to the seepage control ditch. Figure 7 shows the two different strip drain designs used in the Tailings Dyke. The hydraulic cell construction technique has two features associated with it which merit special consideration when designing the seepage control system. Firstly, the horizontal layering of the hydraulically placed sand, within the compacted shell, and the unrecovered bitumen associated with it, result in a much higher horizontal than vertical permeability. Extensive field and laboratory testing as well as computer modelling has been done to establish these values. The results indicate the ratio of horizontal to vertical permeabilities ranges from 9:1 to 16:1. The actual valves are: horizontal permeability of 10^{-3} cm/sec and vertical permeability of 10^{-4} cm/sec. As a result the seepage control system must be designed to accommodate this by placing filters in the appropriate location. Secondly, during cell construction, there is a substantial surge of seepage water which must be handled by the seepage control system. As a result the system must be over-sized to accommodate short-term surges of construction water to ensure the additional water does not adversely affect dyke performance. The seepage control system is currently performing well. When all dyke construction is complete and at ultimate pond elevation the system

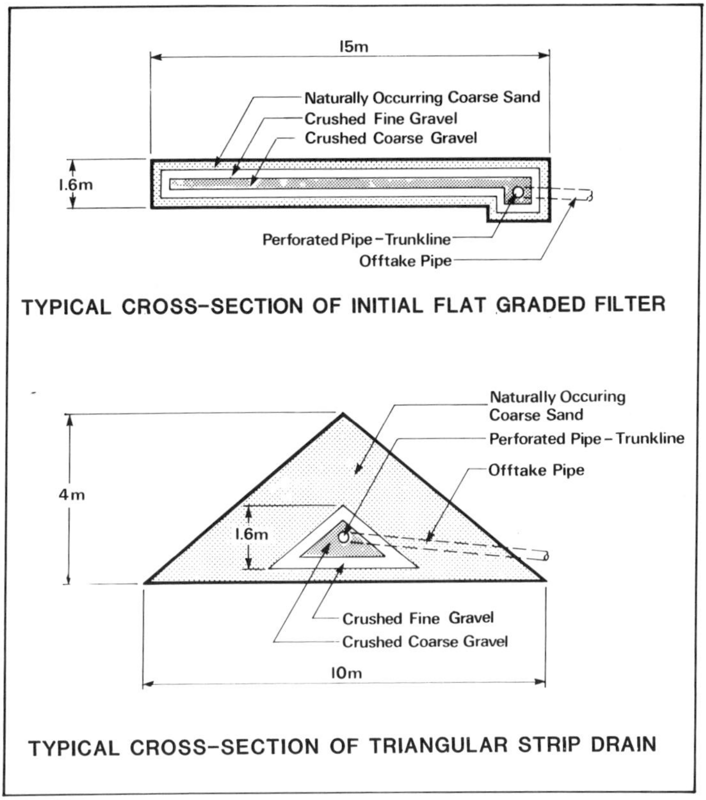

Figure 7. Typical Cross-Sections of Internal Strip Drains

will be required to handle over 11 Mm^3 of seepage water per year.

As discussed earlier, a form of upstream construction was used in the initial years to construct the compacted shell. Experience at Suncor suggested a minimum relative density of 70% could be achieved within the compacted shell, reducing to 30 to 40% within the upstream beach areas (Mittal and Hardy, 1977). The step-overs, associated with the upstream construction method, caused some initial concern regarding the potential for liquefaction of the underlying beached sands. The in-situ densities of the compacted shell and beach areas are assumed to be 100 and 97 pcf respectively, based on extensive field testing. The prevalent thinking was that the potential for large scale liquefaction failures was minimal because the dyke was located in a low seismic risk area. The rapid rate of beach construction required in the early years was however judged to increase the likelihood of localized upstream beach liquefaction failures. More recently, the potential for liquefaction of various portions of the dyke has again come under review. Since construction began, various design changes have been implemented which required significant step-outs onto the uncompacted beach (see Figure 6b). Although the seismic risk classification for the Syncrude site remains low, recent liquefaction concerns stem from the potential for additional triggering mechanisms, largely related to possible strains induced by foundation deformations and/or additional loading due to dyke construction. Consequently significant portions of the compacted shell in areas where step-outs occurred have since been densified by using explosives (Handford, 1988). To a lesser extent, explosive densification has also been used to density other areas within the compacted shell. Specifically where field testing has shown that inadequate densities were achieved during initial cell construction into areas of ponded water. Work continues in this area to ensure the ongoing integrity of the dyke.

Tailings Pond Life

As pointed out earlier, the tailings pond provides a site for sand storage during active dyke and beach development. When the dyke is completed to elevation 352 m in 1990 and the beach to elevation 351 m in 1991, the tailings slurry will be transferred to another site. The new site will essentially be used for sand storage with the fines and excess transport water being pumped back to the tailings pond. The tailings pond will then serve as a storage basin for sludge and a holding area for the clarification of process water to be recycled back to the plant. Since there is only a finite volume available for fluid storage when the beach is completed, the life of the pond will depend on subsequent sludge and water accumulation rates. In order to limit the growth of freewater, Syncrude has adopted a water management program which maximizes the life of the pond as a sludge storage basin. The tailings pond has a profile as illustrated in Figure 8 which includes a freewater zone, a consolidating sludge zone and a mature sludge zone.

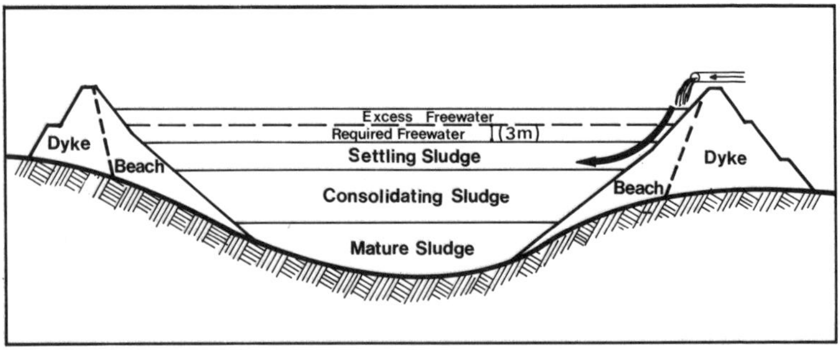

Figure 8. Typical Schematic Tailings Pond Cross Section

The sludge volume is affected by the fines input rate and the sludge density. In turn the fines input rate is directly related to the fines content of the oil sand feed and the amount of fines captured during the beaching process. Fines originate from silty clay bands within the ore body, therefore a high fines content is generally associated with lower grade ore. In order to project future sludge growth volumes the fines content of the oil sand mined must be known in advance. During beaching, a portion of the total fines in the tailings will be trapped within the coarse solid voids. A fines balance around the existing tailings pond indicates that approximately one-half of the fine solids in the tailings are captured in the placed sand. The other half reports to the pond to form sludge. Both the consolidation rate and the ultimate consolidation density determine the long term sludge storage requirement. The immature sludge has a density in the order of 25% solids by weight, whereas mature sludge achieves a density of approximately 35% after 10 years and is projected to ultimately consolidate to 40%. Current projections for the tailings pond indicate that 370 Mm^3 of sludge can be stored when the beaches are completed. Using observed and projected sludge consolidation rates the tailings pond could provide sludge storage until the year 2010.

A simplified water balance for the Syncrude site can be determined by the total raw water imported to the plant less the water removed through plant systems evaporation, storage in the sand voids and storage in the sludge. This balance assumes that any water introduced to the site, if not removed by one of the above three processes, will eventually require storage in the tailings pond as part of the freewater layer. To maximize pond life the freewater layer must be minimized to a level which allows continuing clarification and recycle to the plant. Effective management of the raw water import rate is the most effective method of optimizing the freewater volume since evaporation and storage in the sand and sludge are less controllable. The optimum volume of freewater when the pond is full to its design capacity is 50 Mm^3.

The effective life of the tailings pond with respect to its sludge and water storage capacity has a significant impact on future lease development plans. Long term storage of overburden, sand and sludge are dictated by the timing of tailings pond infilling.

Future Tailings Disposal

The primary objectives of future tailings disposal are largely dictated by the need to create safe, long term containment for sludge and to provide it economically. As noted earlier, sludge storage within the existing pond can only be accommodated until approximately 2010. In order to sustain the pond as a process water clarifier and a sludge consolidation basin, mature sludge from the base of the pond will be removed as required to maintain a constant pond elevation. The volume of new sludge entering the pond will be balanced by mature sludge removal. The sludge will be pumped from the Tailings Pond to the mined-out pit once adequate sand and overburden containment dykes have been established. Eventually, an in-pit sludge storage cell large enough to accommodate the entire Tailings Pond will be created. At that time the Tailings Pond will be emptied to allow mining of oil sand reserves below the site. Removal of the sand dykes and beaches will coincide with the water and sludge drawdown thereby minimizing the interior stability concerns.

Current plans call for the transfer of mature sludge to the mined out pit area beginning in 1997. Hydraulically constructed tailings dykes will be used for containment to ensure the sludge storage operation does not interfere with highwall mining. The philosophy for long term sludge storage requires that the sludge be placed below ground level as much as possible to minimize both risks and ongoing maintenance costs associated with above ground storage.

Additional sand storage beyond 1991, will be accommodated by creating a 'dry' sand storage area within the southwest corner of the Syncrude lease (see Figure 1). The southwest sand storage area will ultimately provide 1500 Mm^3 of sand storage capacity. The sand storage area will be constructed utilizing hydraulic construction techniques. Only minimal hydraulic cell construction, initiated from mobile equipment placed overburden starter dykes, will be used to establish the lower slope portion of the sand storage area. The majority of the sand will be placed using an overboarding mode where the sand is simply discharged through an open 600 mm line onto the beach area. Minimal stability concerns are anticipated as development of an internal pond will be minimized by returning the fluid portion (i.e. water and sludge) back to the main tailings pond.

Tailings disposal management is a key component of the Syncrude oil sand mining operation. The successful application of hydraulic construction techniques, with a cost of $.15 per tonne of oil sand mined, is essential to ensure a continued cost effective mining operation. Syncrude has been very successful in using the hydraulic construction techniques described in this paper in the construction of the world's largest hydraulically constructed dam. This success is very dependent upon careful planning both in the short and long term. Geotechnical problems associated with hydraulic construction have been dealt with by implementing various construction and/or design constraints. Future tailings disposal at Syncrude continues to rely heavily on hydraulic construction techniques.

References

Fair, A.E., and Handford, G.T. (1986). Overview of the Tailings Dyke Instrumentation Program at Syncrude Canada Ltd. International Symposium on Geotechnical Stability in Surface Mining, Calgary, Alberta, pp. 245-253 (Proceedings).

Handford, G.T., Buckles, J., Ellis, H., and Khan, F.Z., (1982). The Use of the Observational Approach in the Design and Construction of the Syncrude Tailings Dyke. Seminar on Geotechnical Aspects of Mine Design and Tailings Containment, Calgary, Alberta (Proceedings).

Handford, G.T. (1988). Densification of an Existing Dam with Explosives. Hydraulic Fill Structures '88, Colorado State University.

Mittal, H.K., and Hardy, R.M. (1977). Geotechnical Aspects of a Tar Sand Tailings Dyke. Geotechnical Practice for Disposal of Solid Waste Materials, ASCE, Ann Arbor, Michigan (Proceedings).

CYCLONE DEPOSITED TAILINGS DAM FOR 300 MILLION TONS

K.A. Lyell M.S.A.I.C.E.*, H.K. Prakke M.S.A.I.C.E.**

ABSTRACT

A gold tailings dam for 1 000 000 tons per month and a life of 25 years was recently commissioned. Wall building, using 30 ton per hour cyclones, comprises both upstream and downstream methods to ensure a geometry to allow early completion of the outer wall profile for grassing for environmental reasons. The filter drains are constructed using graded mine waste materials covered by the underflow material in the first upstream cycle. The decant system involves a floating pumpstation, and trials using geomembranes and cycloned tailings are currently underway to develop a cost effective means of road access across the fines beach to the pool.

1. INTRODUCTION

A new gold retreatment plant situated east of Johannesburg on the Witwatersrand of South Africa was commissioned in April 1987. The process involves the re-mining of tailings dams deposited decades ago and re-treating the tailings using a carbon in leach process to extract gold at a recovery grade of 0.25 grams per ton for a production of 1 000 000 tons per month.

The re-mining process is effected by sluicing the old deposits using high pressure water jets and then pumping the slurry over distances ranging between 5 and 10km to the new treatment plant. The re-treated tailings are then pumped to a new tailings dam site situated 3km east of the plant.

The disposal site covers an area of 430 ha which includes areas for roads, pipelines, and pollution control paddocks, canals and reservoirs. The nett area for disposal is 380 ha and caters for the total production over 25 years of 300 million tons. The ultimate wall height will vary between 46m and 80m above natural ground level.

The area lies within a summer rainfall region with a mean annual rainfall of 720mm and an annual average evaporation of 1480mm. The site is in a maize farming area and is constrained by a river and bird sanctuary, powerlines, and proclaimed future freeway and railway routes as shown in the locality sketch figure 1.1. It was therefore necessary to consider environmental aspects throughout the design to create features to minimise pollution where possible.

*Divisional Civil Engineer, Johannesburg. Republic of South Africa

**Senior Civil Engineer, Johannesburg. Republic of South Africa

As far as soil conditions are concerned, transported soils, 2 to 3m thick and of colluvial origin, overlie the residual silty clays and clayey silts or dense clayey sands derived from mudrock and sandstones. A dolomite contact to the north east forms a natural limitation to the positioning of return water dams.

The plant was commissioned in April 1987 and by February 1988 a total of 11 million tons had been deposited. This has resulted in the tailings rising to approximately 14,5m above the lowest natural ground level and covering an area of 140ha.

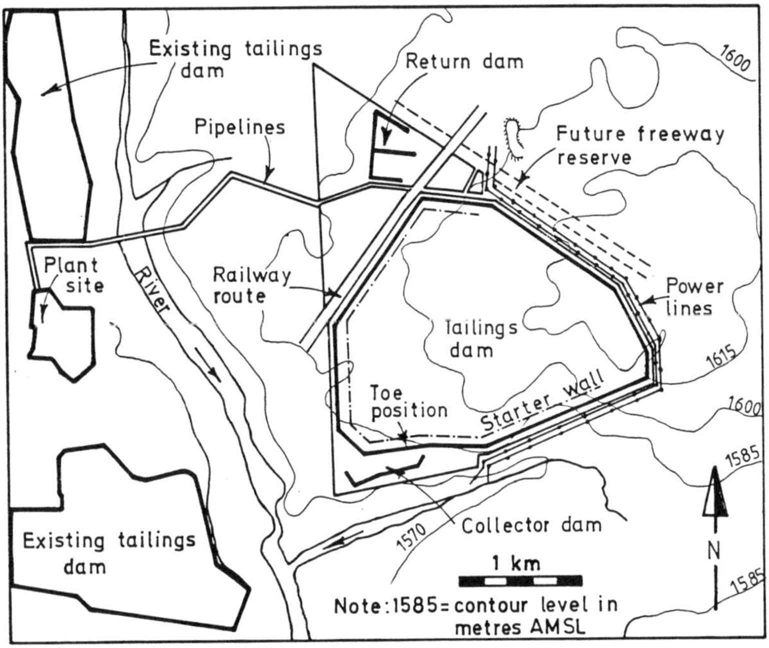

Figure 1.1 - Locality plan

2. WALL BUILDING

2.1 Method of Deposition

Traditionally most gold tailings dams in South Africa are constructed using the paddock ring dyke system as described by Blight and Steffen (1979)[2] with an acceptable rate of rise in the region of 2 metres per year. The high deposition rate and the sloping site required in this instance a higher rate of rise and therefore necessitated the construction of the outer ring dyke using cyclones which from experience at previous projects are capable of sustaining a rate of rise in the region of 6 metres per year.

The cyclones have a capacity of 30 tons per hour and are fed from a perimeter ring feed. These small cyclones can easily be moved manually and provide a great deal of flexibility in wall building. The coarse underflow is shaped into berms which support the overflow tailings. Typically these cyclones produce the average gradings and properties as shown in table 1 and figure 2.1.

Table 1
Tailings Properties

	Feed	Coarse	Fines
SG solids :	2.7	2.7	2.7
Slurry density:	1.42	1.83	1.4
Cyclone split:		25%	75%
In situ dry density (t/m^3):		1.5	1.3
Moisture content:		15%	30%
Permeability (m/year):		32	0.3

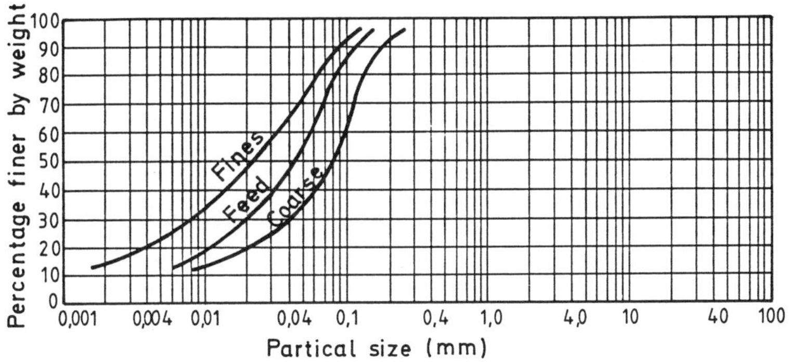

Figure 2.1 Average cyclone feed and product

The slurry is transported to the site via two 450mm diameter pipelines and each pipeline serves a bank of 46 cyclones spaced 24m apart. At any time approximately 23 cyclones are operative on each pipeline, which allows a degree of flexibility in wall building within the cyclone bank. The natural ground slopes are such that the height of the coarse wall will vary by as much as 37m. This means that some areas will require the deposition of more coarse material than others, and this entails more cycloning at these locations. A typical cyclone in operation is shown in figure 2.2.

The rapid rate of rise in the low lying areas during the initial deposition necessitated the construction of a starter wall to contain the fine overflow slurry. The starter wall, which in places is as much as 18m high, was constructed using selected residual soils. This height was selected to limit the rate of rise of the contained fine slurry and to provide adequate freeboard in the early stages.

2.2 Wall Cross-section

The limitation of dust pollution was an important requirement for the design of the wall. Therefore it was necessary to provide a final sideslope which could be grassed as early as possible during the life of the dam. An overall sideslope of 3:1 (horizontal to vertical) was selected with horizontal berms at regular intervals to limit erosion as well as provide access and support for the delivery ringmain. The sideslope between berms is 2.7:1 which corresponds to the average natural angle of repose for the newly deposited underflow tailings.

The cyclone material split together with the selected cycloning sequence (i.e. upstream or downstream sequence) and the starter wall position influence the shape of the wall, the freeboard, and the interface between the coarse and fine sections of the dam. This interface in turn directly determines the phreatic surface which has a direct influence on the overall slope stability.

Figure 2.2 - Typical Cyclone Operation

In order to meet the requirements of stability, freeboard and early grassing, volumes were calculated for various cyclone sequence and starter wall positions. In all these trials a conservative cyclone coarse to fine split of 20%:80% was used together with material dry densities of 1500 kg/m^3 for coarse and 1300 kg/m^3 for fines. The resulting cross section that met these criteria is shown in figure 2.3.

From figure 2.3 it can be seen that a pattern of 2 upstream and 1 downstream sequence will be adopted until the outer face is complete after some 8 years and thereafter the wall building will continue on an upstream basis until completion. A plot of the coarse and fine heights is shown in figure 2.3.

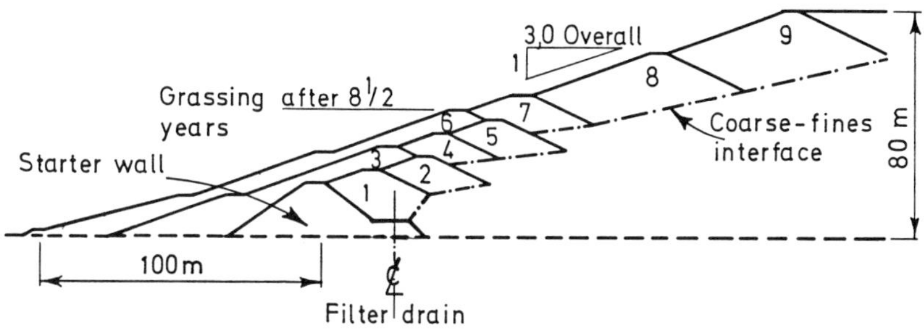

Figure 2.3 - Typical wall cross section

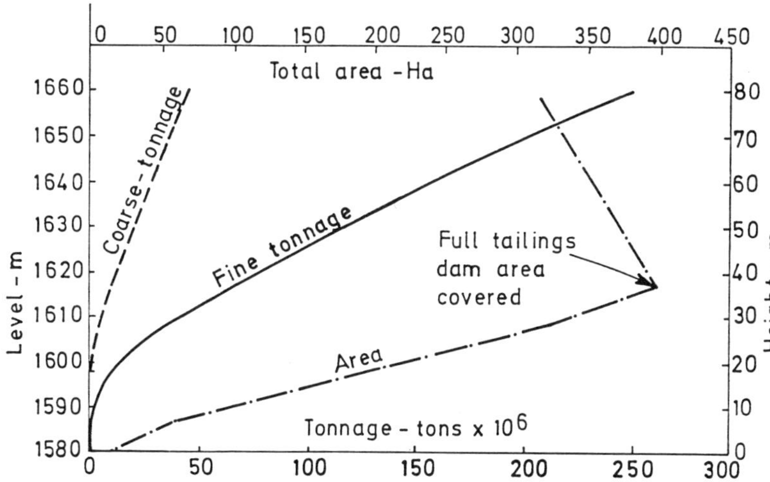

Figure 2.4 - Wall construction height versus tonnage

The final toe position was selected to allow sufficient space to accommodate external pollution control features such as silt catchment paddocks, seepage cutoff trenches, roads and pipelines.

The catchment paddocks together with the solution trench in figure 2.5 are adequate for the life of the dam, but will require alteration to suit the conditions for dam closure as described by the Chamber of Mines (1983)[3]. After closure, the return water and collector dams and seepage cutoff trenches will be abandoned and runoff from the sideslopes will be collected in larger evaporation paddocks as shown in figure 2.6.

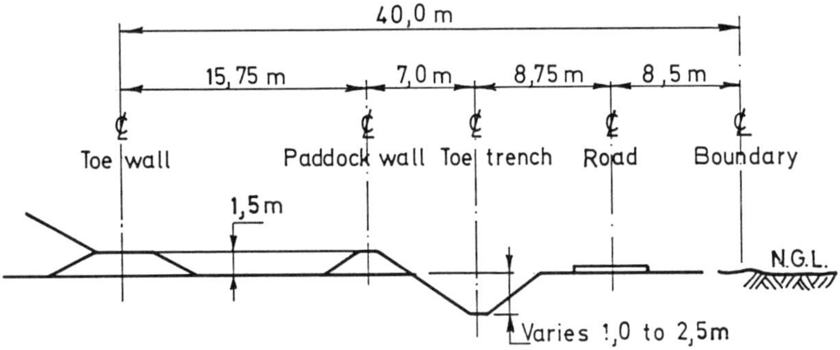

Figure 2.5 - Toe configuration before closure

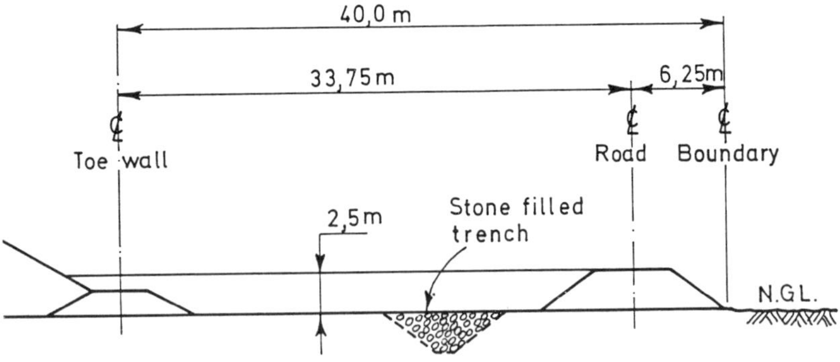

Figure 2.6 - Toe configuration after closure

2.3 Slope Stability

The effective stress stability analysis consisted of analysing critical sections during the life of the dam, using computer programmes based on the method of slices for circular failure surfaces after Bishop (1955)[1] and the method for wedge failure modes described by Morgenstern and Price (1965)[4]. The phreatic surface selected was that at the interface between the coarse and fines. This was considered to be conservative because there is at least an order of magnitude difference in permeability between the fines and the coarse material and a build-up in phreatic surface above this interface is expected only in the early stages. The typical shear strength values used in the analysis are given in table 3. The coarse and fine tailings friction angle have long been well established. The shear strength value of the foundation is based on the lower range of values of triaxial test results conducted on undisturbed samples of the insitu soils. The strength of the starter wall fill was based on shear box tests on recompacted selected insitu material. The maximum factor of safety calculated is approximately 1.8.

Table 3 Shear Strength properties

Material	Effective Cohesion (kPa)	Effective Friction Angle
Coarse tailings	0	33°
Fine tailings	0	29°
Starter wall fill	50	35°
Foundation	20	30°

The site is not within an area of naturally occurring seismic activity, but some shocks due to underground mining can be expected. Accordingly the stability was checked by means of calculating the critical seismic acceleration which would induce failure. This method after N.M. Newmark (1965)[5] revealed that an acceleration exceeding 0.25g would be critical. Considering that the recommended seismic acceleration given by the South African Chamber of Mines (1983)[3] is 0.1g it was concluded that the wall is unlikely to be affected by seismic events.

3. FILTER DRAINS

3.1 Primary Function

The main roles of the filter drains are to

- prevent slumping of the coarse underflow cones by removal of excess water in the cyclone split and
- maintain long term overall slope stability by keeping the water table depressed.

Arising from experience gained from operations at a sister plant commissioned in 1978 (and now treating 2 million tons per month) it has become evident that by careful positioning of the drains, the long term requirements will easily be met if the drains are sized for the first requirement of removal of excess water in the cyclone split. It was noticed that flow from the drains took place strongly in areas where cycloning was taking place. In due course flow stopped almost entirely when cycloning had ceased for any length of time. The heaviests flows to cater for occur right at start up when the rate of rise is highest. This concept is illustrated in figure 3.1.

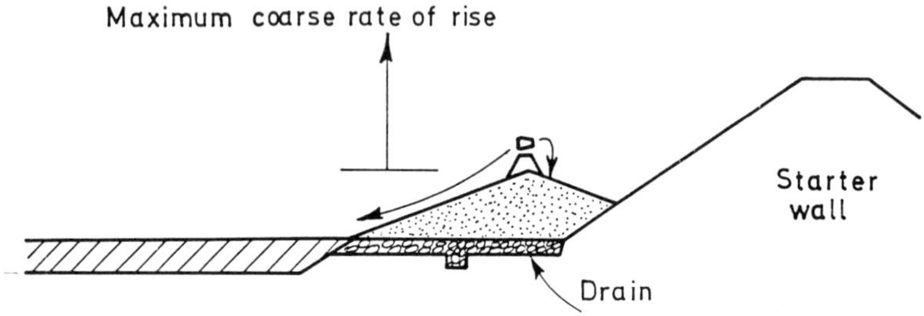

Figure 3.1 - Drainage of underflow cones at start of deposition

The main drain was sized by considering the excess water to be drained under operating conditions. Initially cyclones spaced at 24m centres discharge directly onto the main drain. However, due to the operating requirements of moving cyclones (which require short drying spells to allow access to the cones), and of having to operate on the entire available length of wall at startup, cyclones only discharge at any particular point, on average, for about 30% of the time. The following design criteria were used to size the drain.

permeability of underflow	k	= 32m/year
initial water content of underflow	w_i	= 0.38
final water content of underflow	w_f	= 0.15
solids underflow rate 25% of total of 30 ton/hr, i.e. 7,5 tons/hr		

Arising from these criteria, a maximum flow rate of 72 litres per hour per longitudinal metre (ℓ/hour/m) of drain was calculated. The constriction of flow within the slimes itself dictates the drain width. The width of 7,4 metres finally chosen assumed 2/3 of the flow through the top of the drain and 1/3 of the flow through the starter wall and into the underside of the drain, allowing for cycloning for 50% of the time compared to the average of 30%.

By contrast, the total long term outflow, as depicted in figure 3.2 is estimated at only 0.7 ℓ/hour/m i.e. only about 1% of the flow generated by the cyclones during the initial stages. This bears out the practical observations made earlier that the capacity of the filter drains is far larger than the long term flows. Furthermore, the entire coarse wedge, being at least an order of magnitude coarser than the contained fines, will in fact act as an adequate buttress type drain.

In addition to the main blanket drain, toe drains are required as shown in figure 3.2 to limit local slumping at the downstream face of the wall so as to contain local spread of the toe.

3.2 Materials Used

The cross section of the main blanket drain is shown in figure 3.3.

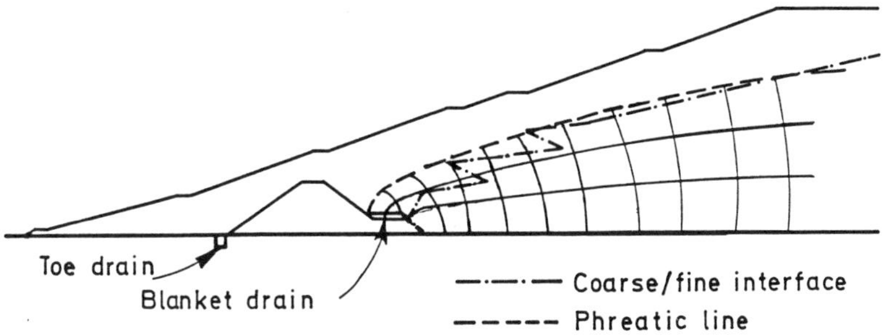

Figure 3.2 - Drainage of fines during later years of dam's life

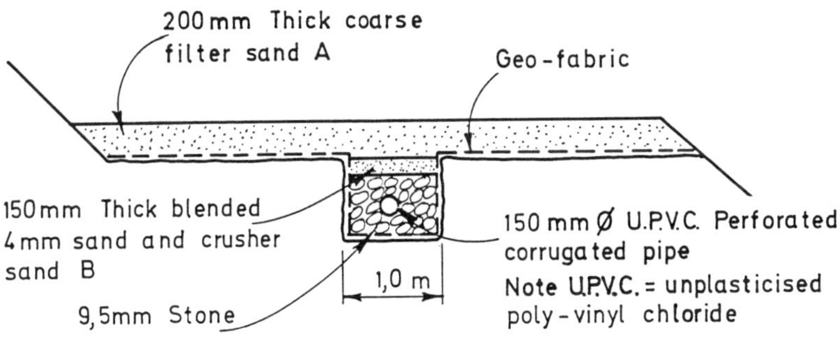

Figure 3.3 - Typical blanket drain cross section

At the sister plant previously mentioned, geofabrics used in drains had been observed to clog up with insoluble salts, probably due to oxidation. It was decided for this new dam to use geofabrics as a construction aid only and to use graded materials for the top of the blanket. Waste rock and sand were available locally from previous mining operations, and with due crushing, selection and blending these quartzitic materials provided a very convenient source. Filter sand A was available from a very large dump; sand B was blended from crusher run and 4mm stone; the 9,5mm stone was available from a commercial source which sells crushed mine waste for concrete and road building aggregate.

3.3 Prevention of Oxidation within the Drain

As a further precaution against clogging of the filter by oxidation, anaerobic conditions are imposed on the drainage system by the use of 'p' traps at lateral filter outlets. The layout is shown in figure 3.4, where the traps are housed within 3 sided open manholes at the side of the toe drain, easily available for inspection.

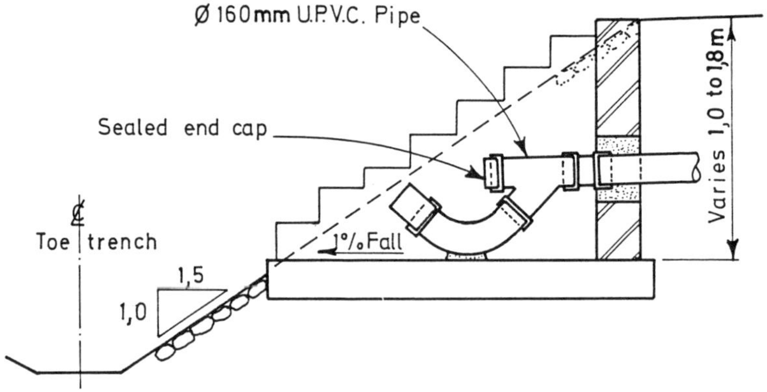

Figure 3.4 - Typical 'P' trap drain outlet

To date drains on this project have performed well with measured flow rates in the range 12 to 36 ℓ/hour/m and it has been found that only minimal cleaning out of the 'p' traps was necessary just after the start of cycloning.

5. DECANT SYSTEM

5.1 General Features

Except for a relatively short initial period, the entire outer perimeter of the wall will be cycloned, with beaches sloping towards a central pool. Due to height considerations and relatively deep soils, a fixed penstock was discounted in favour of a decant barge containing electrical transformers, switchgear and 8 No 75 kW vertical spindle pumps as shown in figure 5.1. The pipes from the barge to the dam perimeter are twin 500mm high density polyethylene (HDPE) pipes floating in the pool and twin 450mm diameter steel pipes which require support, across the beach.

5.2 Access to the Barge

As the pool is entirely surrounded by the beach, special access difficulties were encountered. The pumps and motors need to be removed from the barge and transported to the workshop on a regular basis and planning provides for annual routine servicing. As the pumps weigh approximately 2000 kg and the motors 500 kg provision for access from the outside of the dam to the barge posed unusual problems, particularly so across the fines beach. In addition it is necessary to support the pipes across the beach in such a manner that the supports could be continually raised as the beach rises. Access is required to the barge at all times by either vehicle or boat, irrespective of where the pool moves during its life.

The following options, or a combination of some of them, were initially considered and accepted or rejected for the reasons stated.

Coarse tailings berm This would consist of a berm constructed across the beach using cyclones to create cones of relatively coarse dry solids which would be levelled to provide a roadway along the top. Because such a berm is sensitive to wave action this was acceptable for use along that area of the beach only which would never be inundated by the pool, even under storm conditions. The use of this method was also limited due to the local concentration of fines overflow along the berm which would tend to push the pool further away from the end of the berm.

Figure 5.1 Decant Barge

Floating structure This would consist of bridge-like structures supported on floats 45m apart and could traverse the wet beach and the pool. Transport of heavy machinery items would be along a lightweight rail system. Apart from an adverse cost point of view, this system was discarded due to problems foreseen in articulating both the structure and the pipes to cater for movement and rotations at the joints caused by wave action. Another problem was the perceived inflexibility in accommodating horizontal movements of the pool when the barge may have to be shifted significantly.

Jacked structure This would use the same principle as the floating structure but the ends of the spans would be supported on "islands" and would be continuously lifted. Cost, inflexibility and the necessity for specialist skills led to the abandonment of this alternative.

Reinforced berm This would consist of a vertical sided berm made of coarse cycloned tailings or preferably of whole tailings feed, contained within geocell or geofabric wrapped layers deriving stability from the reinforced earth principle. The advantages over the previously mentioned coarse tailings berm lie in the vastly superior resistance to wave action and relatively far less fines generated in the vicinity of the berm which could adversely affect the pool position. The cost is however higher than a coarse tailings berm, and this method is being adopted only over those areas of the beach subject to inundation. The berm will be terminated about 100 metres into the pool, with final access to the decant by boat and tender barge.

At present the geocell and geofabric alternatives are both being subjected to full scale field trials to develop and evaluate the most practical and economical means of dealing with pipe lift and other operational difficulties, such as achieving early access over recently raised areas of the berm. The layout of the proposed access is depicted in concept in figure 5.2.

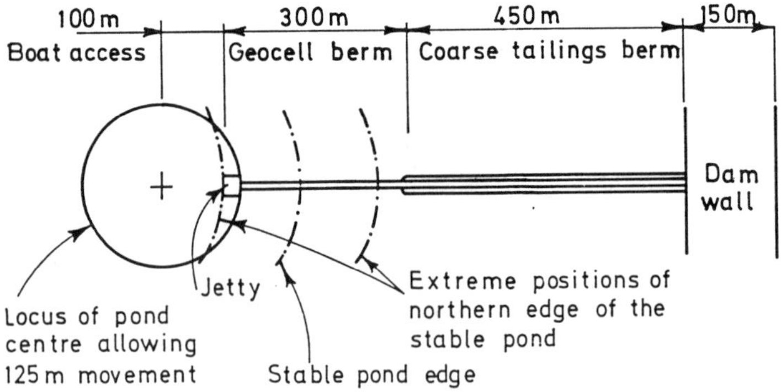

Figure 5.2 - Access to the floating decant

6. COMMISSIONING

Commissioning of the treatment plant commenced in March 1987 with full production being achieved for the first time in July 1987. By February 1988 a total of 11 million tons has been deposited to a maximum depth of 14,5m. The initial priority of covering the blanket drain with coarse underflow was achieved and the first upstream cyclone sequence has progressed with cycloning taking place above the starter wall level.

Initially two separate supernatant pools were established but these merged at the end of September 1987. Decant volumes measured represent in excess of 60% of the feed water volumes which is greater than the predicted return of 55%. This occurred due to the lower slurry feed relative density of 1.38 encountered in the early months. The lower feed density has also resulted in marginally higher rates of rise but this is expected to normalise soon. Surface water samples from the water courses and ground water analysis from monitoring boreholes around the site perimeter indicate no change in the water quality status to date.

7. CONCLUSION

In this paper some of the practical features of a recently commissioned one million ton per month cycloned tailings dam are described. Environmental considerations have influenced the design insofar as the wall design and wall building technique are concerned. Where possible derivatives of mining waste rock have been utilised in the graded filter drains. The decant system entails a floating pumpstation and trials utilising geomembranes and cycloned tailings are in progress to develop a cost effective access berm system.

REFERENCES

1. Bishop, A.W. The use of the slip circle in the stability analysis of slopes. Geotechnique, vol. 57, no. 1, 1955.

2. Blight, G.E. and Steffen O.K.H. Geotechnics of Gold Mining Waste Disposal. Current Geotechnical Practice in Mine Waste Disposal : ASCE, 1979.

3. Chamber of Mines of South Africa, Handbook of Guidelines for Environmental Protection; The Design, Operation and Closure of Metalliferous and Coal Residue Deposits: Volume 1/1979 (revised 1983).

4. Morgenstern, N. and Price, V.E. The analysis of the stability of general slip surfaces. Geotechnique, vol. 15, no. 1, 1965.

5. Newmark, N.M. Effects of earthquakes on dams and embankments. I.C.E. Fifth Rankine lecture, Geotechnique 15,iv. 1965.

SOME LESS FAMILIAR ASPECTS OF HYDRAULIC FILL STRUCTURES

by

Geoffrey E Blight[a], MASCE

ABSTRACT

There are many aspects of hydraulics fill structures that may be of great importance in particular cases and circumstances, even though they may not be considered of general importance and may be less familiar to the geotechnical engineer. This paper selects ten such aspects and gives a brief discussion of each. The object is to widen the readers' perspective and knowledge of hydraulic fill structures beyond what is usually covered in papers and conferences dealing with this subject.

1. INTRODUCTION

Hydraulic fill structures form an important class of geotechnical construction in the fields of civil and mining engineering, in the mineral extraction industry, the electricity generating industry and to a lesser extent, in the chemical processing industry. Within these fields, hydraulic fill structures of various types are used for the following general functions:

.1 To provide access and support, e.g. reclamation fills, artificial islands for off-shore drilling, backfill in underground mines;

.2 To provide storage for low grade mineral ores and for water.

It will be noted that no direct reference has been made to "waste disposal". This is because, with rapidly advancing techniques of mineral extraction and energy utilization, most deposits of what was formerly considered to be mineral waste must now be regarded as low grade ore or a mineral resource awaiting a favourable economic and technological opportunity for its re-exploitation.

---

[a] Professor of Construction Materials, Department of Civil Engineering, University of the Witwatersrand, Johannesburg, South Africa.

Many papers have been written describing the macro-aspects of the design and construction of hydraulic fill structures of various types. For example, the design, construction, maintenance and reclamation of what are termed mine waste deposits have been codified and extensively described by authorities such as the British National Coal Board(1970), the Canada Centre for Mineral and Energy Technology(1977), the Australian Department of Home Affairs and Environment(1982) and the Chamber of Mines of South Africa(1979). However, there are aspects of hydraulic filling that have received less attention in the literature, but which may often be of equal importance to the macro-aspects of a particular operation.

This paper will select ten of these less well known and researched aspects and examine them in limited detail with the purpose to inform and broaden the perspective of the engineer with an interest in hydraulic fill construction.

The aspects that have been selected for examination are the following:

.1 the flow properties of mineral slurries;
.2 the flow of slurry escaping from a breached containment;
.3 the beaching behaviour of slurries;
.4 the effects of desiccation on the construction of hydraulic fills;
.5 tailoring the composition of a fill to improve its qualities;
.6 stiff underground fills;
.7 horizontally reinforced hydraulic fill mine supports;
.8 effects of underground mine filling on surface storage of residue;
.9 the effect of desiccation on hydraulic fill structures;
.10 surface erosion of hydraulic fill mineral deposits.

The material for these discussions has mainly been taken from the writer's own published experience, [e.g. Blight(1979), Blight(1980), Blight(1984), Blight(1987), Blight and Bentel(1983), Blight and Clarke(1983), Blight, McPhail and Fourie(1986), Blight, Robinson and Diering(1981), Blight, Thomson and Vorster(1985), Hahn, Blight and Dison(1982)]. It is freely admitted that the selection of topics and the material itself has been strongly biased by the writer's background and the practice in the area in which he works.

2. ASPECTS OF HYDRAULIC FILLING

.1 FLOW PROPERTIES OF MINERAL SLURRIES

The construction of hydraulic fill structures involves the transportation and deposition of mineral material in the form of water-borne slurries. Hence the flow of slurry through pipelines, launders or other channels, as well as the distribution of the slurry to build the structure are subjects of great importance. All of these operations depend on the rheological behaviour of mineral slurries. The water content at which slurries are transported and placed varies widely, depending on the material being moved and the application. As examples, the water content of mine tailings during transportation from the mill to the storage dam is typically in the range from 150 to 100 per cent by mass of dry solids[b]. For the hydraulic filling of underground mine excavations, the water content is typically 50 per cent or less.

The rheological behaviour of mineral slurries is complex, but can be approximated by that of a Bingham plastic for which the Newtonian viscosity η is related to the shear strength τ by

$$\tau = \tau_0 + d\gamma/dt \cdot \eta \qquad [1]$$

Where τ_0 is a yield shear strength that applies when the rate of shear strain $d\gamma/dt$ is zero.

Slurry viscosities are commonly measured by observing the variation of torque with water content and rate of shear strain using a co-axial cylinder viscometer. Any single torque measurement can be interpreted in terms of an apparent viscosity η_a where

$$\eta_a = \eta + \frac{\tau_0}{d\gamma/dt} \qquad [2]$$

The apparent viscosity at a specific water content and rate of shear strain is usually sufficient for the purpose. If the true viscosity is required, the value of τ_0 has first to be established.

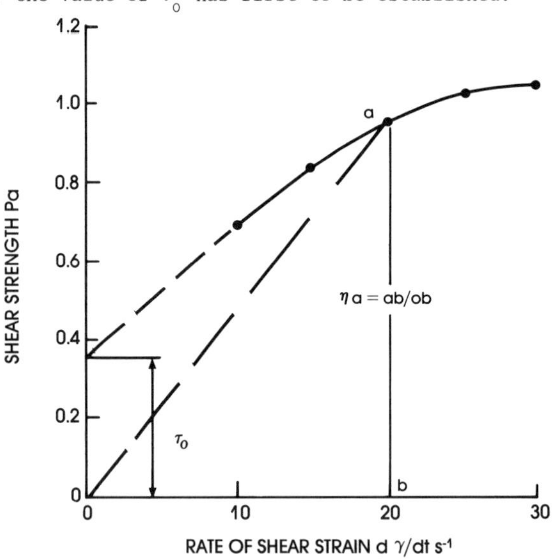

Figure 1a: Relationship between shear strength and rate of shear strain for diamond tailings at a water content of 125 per cent

---

[b] It should be noted that in the mining and mineral industries, it is more common to express water contents as a percentage of the wet mass.

Figure 1a illustrates the concepts of η_a and τ_0 as applied to a slurry of diamond tailings. For a true Bingham plastic the relationship between τ and $d\gamma/dt$ would be linear and the slope of this line would represent the true viscosity η. Figure 1b shows the relationship between τ and $d\gamma/dt$ calculated for Figure 1a. Calculated values of η are also shown. For a true Bingham plastic, η would, of course, be independent of $d\gamma/dt$. Figure 1 relates to a single water content whereas one is usually interested in the effect of varying water content on flow properties. This effect is shown in Figures 2a and 2b for a platinum tailings. Further data on the rheological properties of mineral slurries have been published by Whitney et al(1977), Jeyapalan(1980) and Lucia et al(1981).

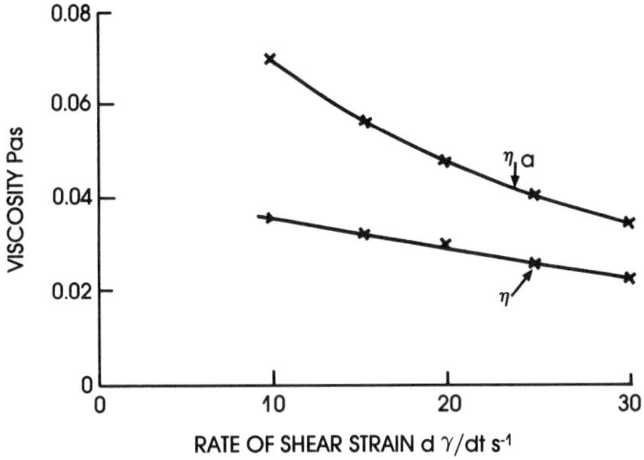

Figure 1b: Relationship between viscosity and rate of shear strain for diamond tailings at a water content of 125 per cent
η_a = apparent viscosity η = 'true' viscosity

Once the rheological properties of a slurry have been established, they can be used to estimate friction losses in pipelines, flow rates in open channels, etc. via established flow theory [e.g. Graf(1971) and Verkerk(1987)]. It is interesting to note that none of the accepted flow equations actually use viscosity as a basic parameter. Viscosity is always converted back to an estimated shear resistance. From the geotechnical engineers' viewpoint, therefore, it would be preferable to use rheological data in terms of shear strength, as given in Figures 1a and 2a rather than to convert it to terms of viscosity, a far more difficult concept to grasp and use.

Note also that the range of shear strengths for slurries is smaller than the range of strengths of solid soils by a factor of a thousand. The increase of strength from Pascals to kiloPascals takes place after deposition and consolidation of the slurry in the hydraulic fill structure.

Perhaps the most important point to note is that the shear strength of slurries and the soil-like materials that they form after draining are affected in a complex way by both water content and rate of shearing.

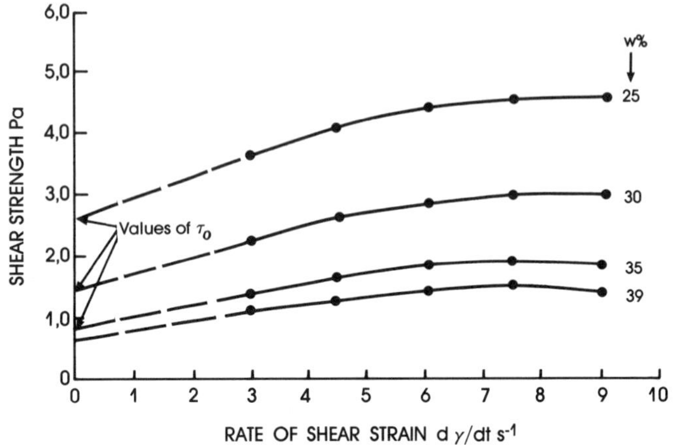

Figure 2a: Measurements of shear strength on platinum tailings, showing the influence of varying water content
w = water content of the slurry

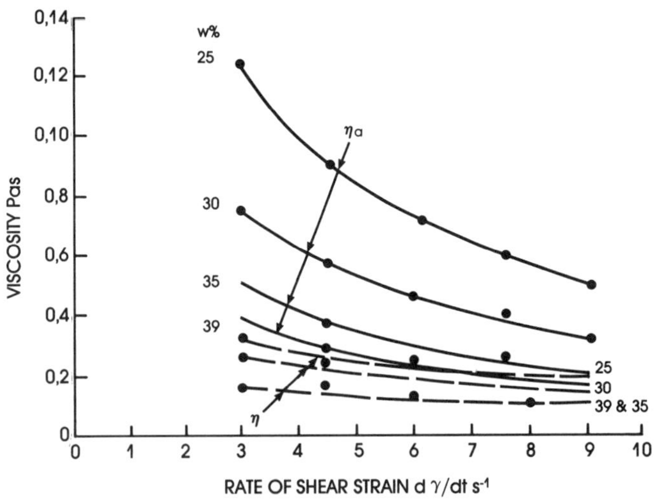

Figure 2b: Measurements of viscosity on platinum tailings, showing the influence of varying water content
w = water content of the slurry ηa = apparent viscosity η = 'true' viscosity

The effect of water content is well known, but the effect of rate of shearing is far less recognized.

.2 THE FLOW OF SLURRIES ESCAPING FROM BREACHED CONTAINMENTS

In the past, and even very recently, hydraulic fill structures have suffered from their share of failures. The usual consequence of such an accident is that the outer containment has breached and the liquid slurry contents have escaped to wreak terrible damage. Examples of such failures in recent years have been the breaching of the Mochikoshi tailings dam in Japan in 1978 [Marcuson et al(1979)], the Stava tailings dam in Italy in 1985 [Berti et al(1988)] and the Bafokeng tailings dam in South Africa in 1974 [Jennings(1979)]. All of these events were accompanied by tragic loss of life and damage to property.

To give an idea of the potential mobility of a mineral slurry accidently released in this way, the 3 million m^3 of slurry that escaped from the Bafokeng tailings dam entered a small water course and flowed down it for 42km. The flow was only stopped because it entered a water storage reservoir at that point. An estimated 2 million m^3 of slurry reached and almost filled the reservoir, slopping the water it contained over the dam wall.

Relatively little research seems to have been done on this problem. Jeyapalan et al(1983), Lucia et al(1981) and the writer and colleagues are among the few that have investigated the matter, although work along similar lines has been done on natural mudflows [Hutchinson and Bhandari(1971)] and submarine slurry flows [Morgenstern(1967)].

Blight et al(1981) concentrated on finding a method of predicting flow depths and flow velocities downstream of a hypothetical breach. The object was to be able to assume that a ringdyke tailings dam would breach in various positions and then to analyse the effect of the ensuing flood of slurry on surface installations, shafts, nearby housing, etc. If, as a result of the analysis, it became clear that a particular structure or group of structures was at risk, a protective bund could be constructed to deflect the possible flood, or the installation could be re-located.

The proposed analysis proceeds step-wise from the breach by applying the general equation of motion to the element of moving slurry depicted by Figure 3:

Downstream forces on element - upstream forces
= (mass of slurry in element) x acceleration. [3]

This, together with the equation of continuity, namely that the mass of slurry passing any cross-section is the same at any instant and an assumed surface profile for the flow, enables the flow depth and velocity to be estimated progressively along its course.

The analysis was calibrated by applying it to the failure of the Bafokeng tailings dam and comparing calculated flow depths and velocities with slurry depths observed on various structures affected by the flow and with eye-witness accounts. Figure 4 is a plan showing the complex form taken by the flow as it issued from the breach in the dam. The figure shows the sections analysed and some of the structures affected by the flood.

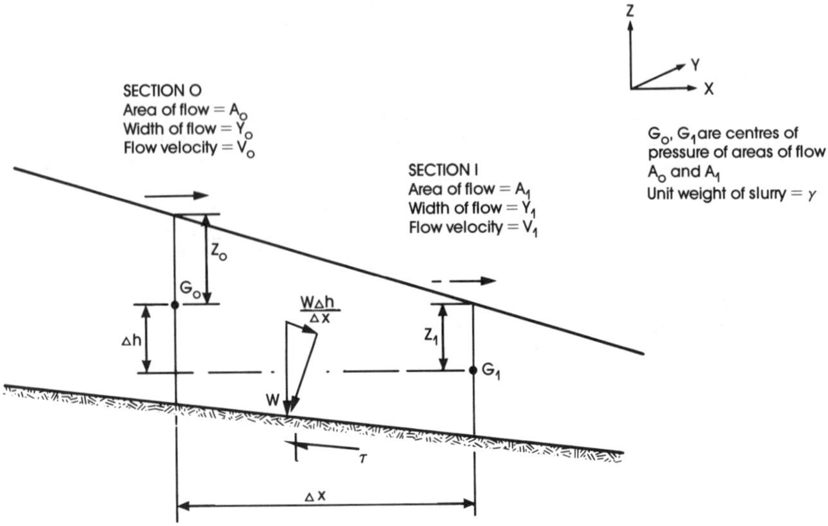

Figure 3: Free body diagram of the flow of slurry at steady state

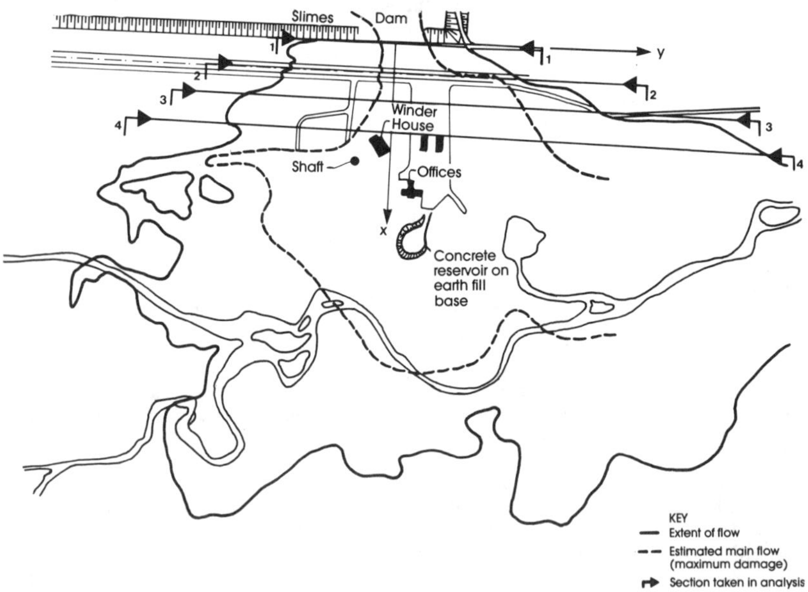

Figure 4: Failure of Bafokeng slimes dam

Agreement between analysis and visual evidence was reasonably good. For example the predicted flow depth at one of the affected structures, the winder house, was 1.1m and the flow velocity 9.5m/s. The evidence was that the flow depth must have been between 1m and 1.5m and the flow velocity about 10m/s.

In the light of .1 above, it is interesting to note that the shear stress on the base of the element shown in Figure 3 played a negligible part in deciding the flow dimensions. Inertial forces proved far more important that the flow resistance.

.3 THE BEACHING BEHAVIOUR OF SLURRIES

When the hydraulic fill material reaches the site of deposition it is discharged in one of a number of possible ways. Perhaps the fullest description of current discharge methods is given by the International Commission on Large Dams(1982). Whatever the method used, the slurry forms a beach, cone or fan and gravitates towards a pool where the excess decant water is drawn off.

The profile assumed by the beach is characteristic for a particular product and slurry water content, as has been described by Melent'ev et al(1973), Abadjiev(1985), Smith, Abt and Nelson(1987), Blight(1987) and Fourie(1988). Figures 5a and 5b for example, show how beach profiles measured on six separate platinum tailings dams (Figure 5a) all reduce to a single master profile (Figure 5b) if plotted as non-dimensional height versus non-dimensional distance down the beach. An accurate prediction of the profile of an hydraulic fill beach is important if the solids storage capacity of the dam as well as the water storage capacity of its top surface is to be estimated with any accuracy. However, other properties of the beached material are equally important. Because the beach forms by a process of gravitational sorting of particles, the properties of the deposited fill vary progressively from the point of deposition to the edge of the pool. In general, the material will become finer and less pervious towards the pool. Figure 6 shows how the particle size has been observed to decrease down a beach of diamond tailings. Because the permeability increases progressively from the pool towards the outside of the deposit, the phreatic surface becomes depressed below the position it would occupy in a homogeneous deposit of similar material. This has a positive effect on the shear stability of the outer slopes. The depressed phreatic surface calculated for the particle size distribution curve shown in Figure 6 is illustrated by Figure 7.

The scatter of observations in Figure 6 shows that the particle sorting process is somewhat variable. This arises because flow down a hydraulic beach becomes unsteady. The unsteadiness results in the formation of a changing pattern of rills which, as they vary, often overflow depositing material that is finer than it should be at that distance from the point of deposition. In extreme cases the result is a sub-horizontally layered deposit with negligible vertical permeability in comparison with its horizontal permeability. In a deposit like this a true phreatic surface does not form, but seepage emerges from the face of the deposit as a series of perched watertables. The effect on shear stability of the force is extremely adverse. Walls in this condition have failed at heights as low as 3m.

Recent work on very fine tailings (cyclone overflow gold mine tailings) has shown that material deposited on a beach settles appreciably after it has come to rest. This has the effect of

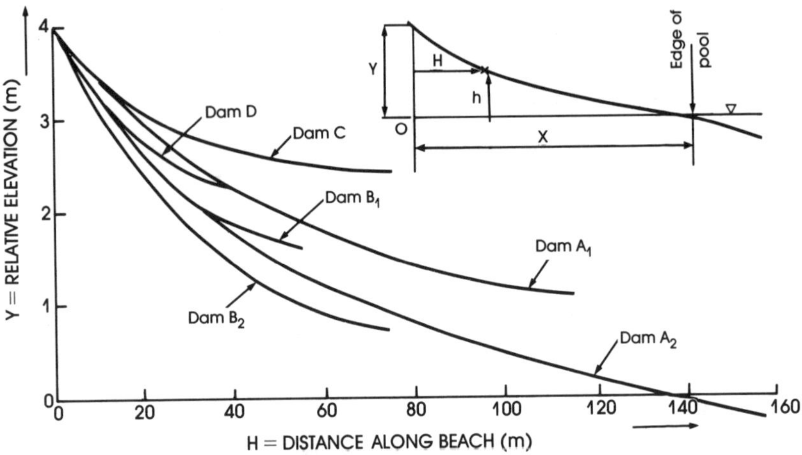

Figure 5a: Measured beach profiles on six platinum-tailings dams

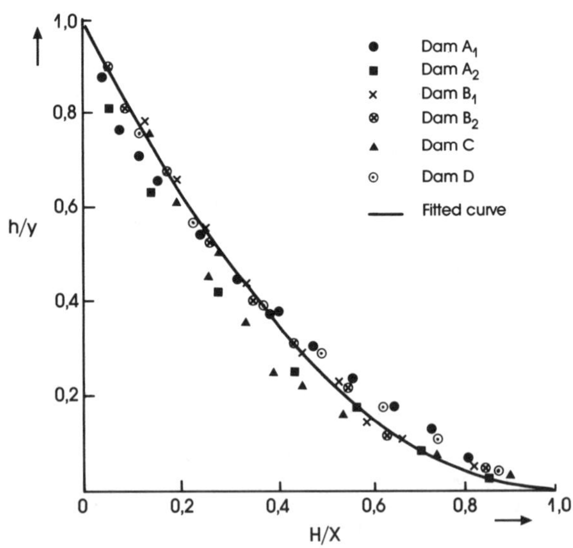

Figure 5b: Dimensionless beach profiles for platinum tailings

flattening the beach profile. Hence it is not just particle sorting that determines the profile, but settlement as well. This conclusion is supported by Wates et al(1987).

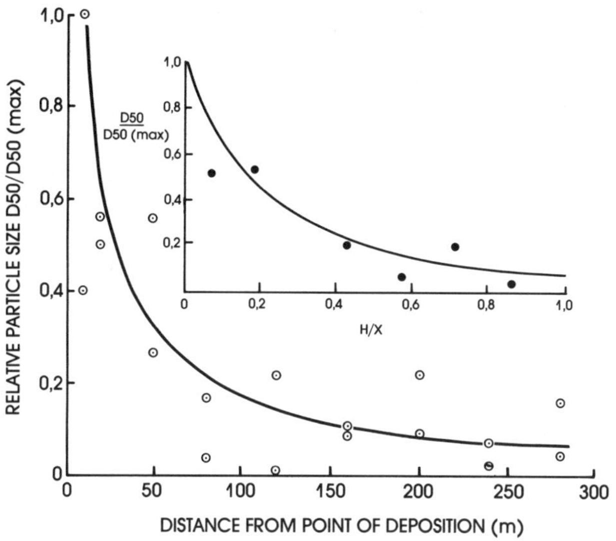

Figure 6: Particle size sorting that occurs on a hydraulic beach

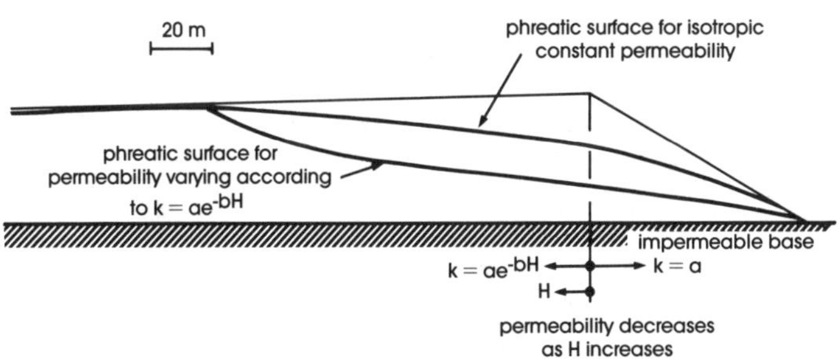

Figure 7: Effect of varying permeability on position of phreatic surface in a tailings dam

4 THE EFFECT OF DESICCATION DURING CONSTRUCTION

In many climates of the world, evaporation from a free water surface considerably exceeds precipitation in the form of rainfall. The water deficit may apply either seasonally or throughout the year. This favourable climatic condition is often deliberately exploited by allowing each layer of hydraulically deposited material to dry out before placing the next layer. This is known as the "semi-dry" process of deposition [Blight and Steffen(1979)]. The practice of sundrying has the effect of considerably reducing the void ratio of the deposited material and correspondingly increasing the solids storage capacity of the impoundment. Figure 8 shows the effect of desiccation on reducing the void ratio of a gold slimes dam.

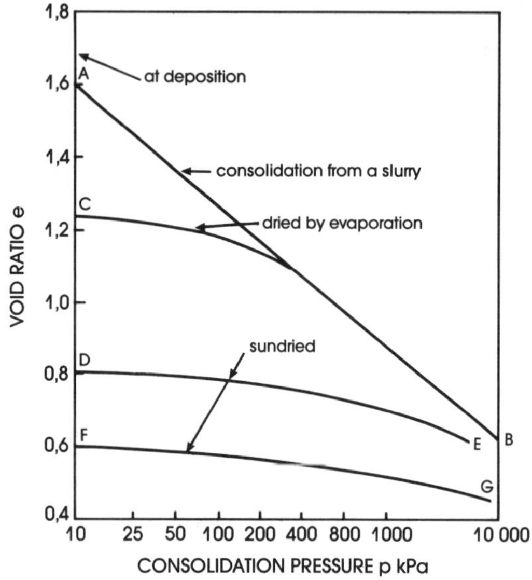

Figure 8: Effect of sundrying on void ratio and consolidation characteristics of slimes

It may be unwise to design and rely on the volume reduction caused by dessication if there is any possibility that as a result of operating errors, the sundrying will not take place. In one case known to the writer, the volumetric design of a tailings deposit (a valley dam behind a pre-built earth embankment) depended for its five-year life on volume reduction by sundrying. When the mill was commissioned, loss of control resulted in the tailings impoundment almost filling with tailings at a very high water content. Decanting the water once the solids had settled did not help, as only the surface of the saturated tailings could be affected by subsequent sundrying. The effect of this failure to operate the deposit as designed was the loss of half its useful life.

The practice of allowing successive layers to shrink to their maximum may also have negative effects, as vertical desiccation cracks may open up in each layer. These cracks may form a network of incipient piping channels within the deposit and later result in piping failures that could put the deposit out of commission temporarily, or in severe cases result in a failure of the complete impoundment. The Bafokeng failure is believed to have resulted from a combination of desiccation cracking and horizontal layering by fine material followed by a piping failure.

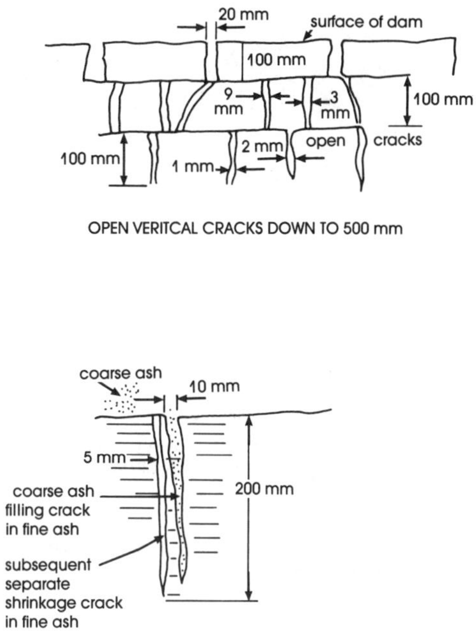

Figure 9a: Systems of shrinkage cracks near the surface of hydraulically deposited power-station fly ash

Figure 9a illustrates the sort of cracking that occurs in a deposit of pulverized fuel ash that has suffered repeated piping failures over the course of its life. The most dangerous situation arises when the pool on the deposit becomes enlarged in size either because of rain, lack of control by the operating staff or the deliberate storage of water on the dam. Once water enters the crack network, the pool empties into the cracks and erodes long deep subterranean channels that every so often emerge from the outer slopes of the deposit and result in a localized failure. Figure 9b shows some of the piping channels that have been observed to form in the material of Figure 9a.

The only remedy for this situation appears to be to avoid desiccation of the beaches by a more frequent deposition cycle, tight control over the size of the pool, or, as a more certain solution, by changing the method of deposition or the properties of the material being deposited.

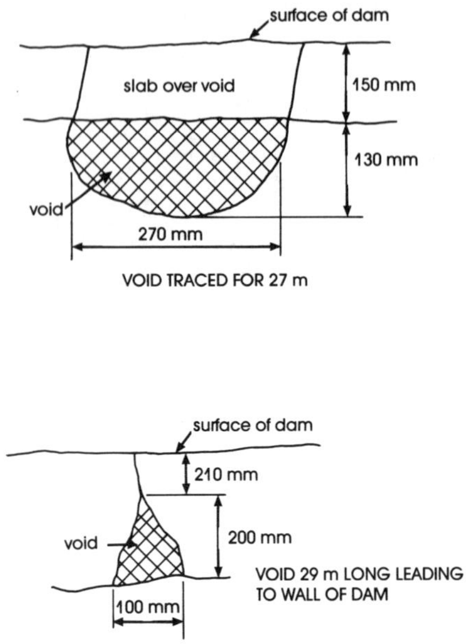

Figure 9b: Erosion pipes located beneath the surface of hydraulically deposited fly ash

.5 TAILORING THE PROPERTIES OF A FILL

It is a well-established geotechnical principle that the strength and compressibility of a soil-like material depend on its density. Strength increases and compressibility reduces as density is increased. It is perhaps not always remembered that this principle is the sole reason for the geotechnical engineer's pre-occupation with moisture and density control in earthworks construction. The density of a particulate material is dependent on its particle size distribution and it is often possible to tailor the size distribution by combining available materials or by removing certain size fractions so that the fill will settle into a maximum or improved density state.

Size distribution curves may be subdivided into two general categories:

.1 Straight line gradings that give approximate straight line plots on the conventional percentage passing versus log particle size chart, and

.2 gap gradings which consist of two sections of straight lines separated by a size gap in which there are no or few particles present.

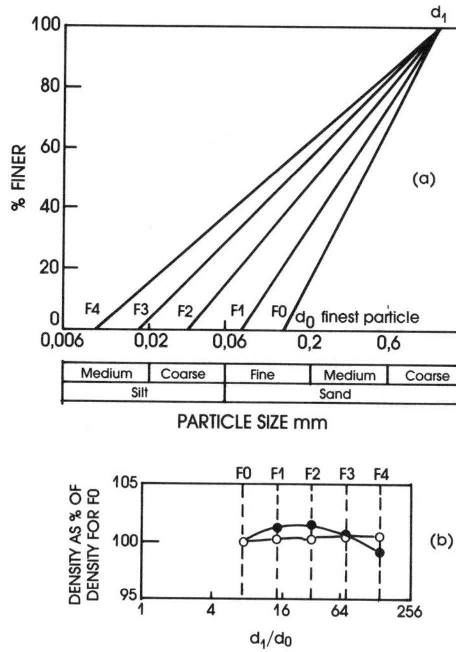

Figure 10: (a) Series of continuous gradings F0, F1, F2, F3, F4
(b) Settled densities of two materials made up in gradings F0 to F4

Figures 10a and 10b illustrate the effect on settled density of varying the slope of a straight line grading while keeping the maximum particle size constant. The various materials were all allowed to settle out from a slurry under identical conditions. It is obvious that the slope of the grading has a relatively minor effect on the density.

Figures 11a and 11b show the effect of varying the maximum particle size while keeping the slope of the grading constant. It appears that the maximum particle size has a far greater influence on density than does the slope of the grading.

Thirdly, Figure 11b also shows a number of gap gradings in which the gap between the uniform gradings F0 and F5 is bridged at various percentages finer, to give, for example grading F0-G0-G0-F5. Figure 12 shows the effect on settled density of varying the percentage of finer

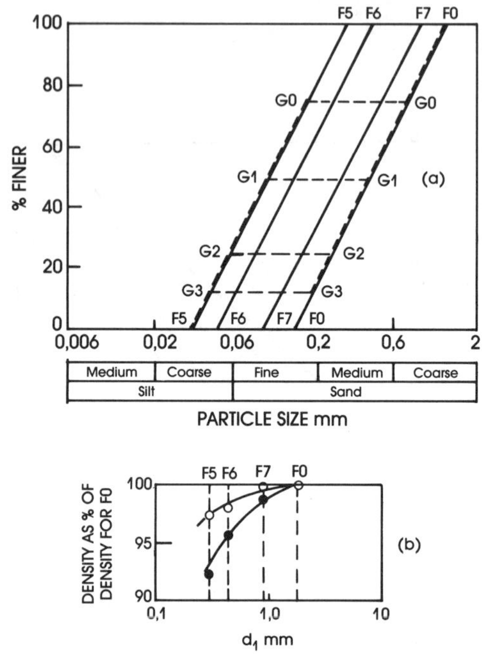

Figure 11: (a) Series of continuous gradings F0, F5, F6, F7, and gap gradings G0, G1, G2, G3
(b) Settled densities of two materials made up in gradings F0, F5, F6, F7

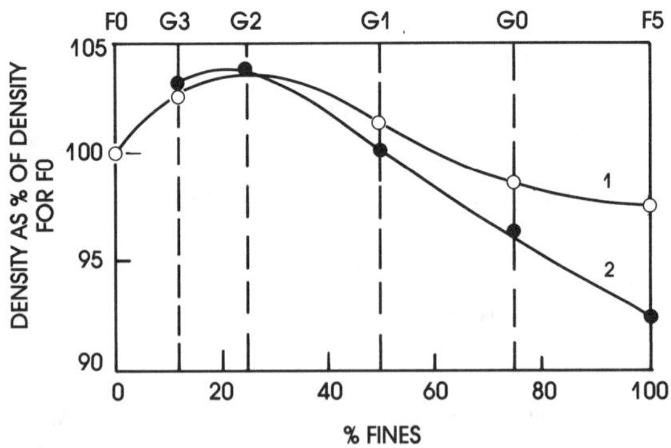

Figure 12: Settled densities of two materials made up in gradings F0, G0 to G3, and F5

material in this way. This manipulation produced the greatest variation of density. For example, the addition of 20 per cent of fines to a material with grading F0 can increase the density by 4 per cent, whereas the density of the fines alone would be 5 per cent less than the density obtainable with grading F0.

These illustrations of the effect of tailoring the properties of a fill do not provide a set of rules for obtaining maximum settled density in a hydraulic fill, as there are other variables such as particle shape, particle surface texture and particle density that also affect the settled density of a slurry. The examples do, however, illustrate the scope that exists for improving the properties of a fill by manipulating the grading.

As a ridiculous example, cement is often added to pumped mining fills with a resultant improvement of density and strength which is ascribed to cementing action. In reality, the water/cement ratio may be so large that the cement bonds have a negligible strength. The improvement of properties arises mostly from the modification of the material grading by the addition of the cement fines. An equal percentage of similar, but chemically inert fines would have almost be same effect.

Density is obviously not the only property of an hydraulic fill that it may be desirable to optimize. Permeability may be improved or even optimized by cycloning to remove much of the fine fraction of a fill. This process also has the effect of reducing in place shrinkage (and hence the piping problems referred to in .4 above) by reducing the water content at placement and removing most of the material that is the source of the shrinkage.

.6 STIFF UNDERGROUND FILLS

The concept of tailored fills is taken to an extreme in the design of stiff underground mining fills.

The stiff fill is designed primarily to provide lateral restraint to ore pillars, thus strengthening the pillars and enabling them to be made smaller. Smaller ore pillars mean a greater ore extraction ratio and hence a more profitable mining operation. The design concept is to create a rigid in situ skeleton of rock particles with the voids just filled with cemented tailings. In addition to the requirement of in situ rigidity, the fill has to be pumped into place and the fresh mix must therefore have a low enough viscosity to be pumped and placed. This is done by means of positive displacement concrete pumps [see Verkerk(1987)]. The as-placed density of such a fill is extremely high and can be more than 90 per cent of the density of the solid constituents. The maximum possible volumetric compression of such a stiff fill is thus less than 10 per cent.

Figure 13 compares the confined (i.e. K_o) compression characteristics of stiff fills with those of conventional "soft" cemented tailings fills. The cement content is similar in each case. It will be seen that the stiff fills become virtually incompressible beyond a limit of

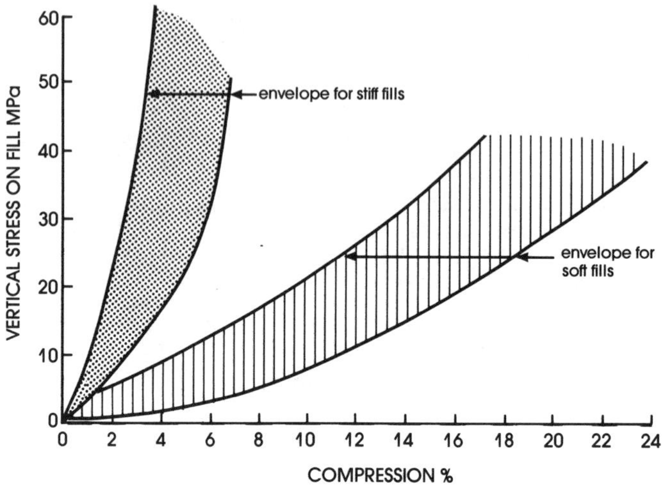

Figure 13: Confined compression tests on stiff and soft fills: stress-compression envelopes

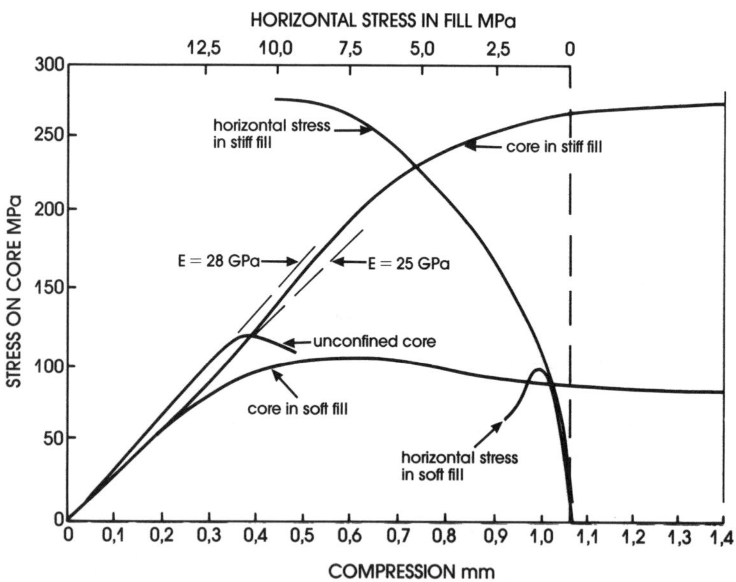

Figure 14: Comparison of stress-compression behaviour of unconfined core with fill-supported cores

about 7 per cent compression. Note in particular the stress range for the figure which is roughly 1000 times larger than that for conventional surface geotechnics.

The extent to which a stiff fill can provide lateral support and strengthening to a rock pillar is illustrated by the results of model tests shown in Figure 14. The ore pillars were represented by drill cores of quartzite conglomerate ore and the cores were embedded in either stiff or conventional soft fill contained in cylindrical steel moulds to provide lateral constraint. The composite core-fill system was loaded by means of a rigid piston that caused fill and model pillar to strain by the same amount. Figure 14 shows that the soft fill provided very little lateral support to the model pillar, as the strengths of the unconfined core and the core supported by the soft fill were almost identical. However, enough lateral support was generated to maintain a post-failure strength in the pillar of 85 per cent of the peak strength.

The presence of the stiff fill considerably augmented the strength of the model pillar. As the figure shows, a peak strength was not reached by the model pillar/stiff fill system even at a compression of 1.75 per cent which is a large strain for a system as rigid as this.

It therefore appears that surrounding a highly stressed pillar with a soft fill will not materially improve the strength of the pillar, but will ensure a high level of residual resistance. A stiff fill, in contrast, can be used considerably to increase the strength of a pillar.

.7 HORIZONTALLY REINFORCED HYDRAULIC FILL MINE SUPPORTS

Apart from providing regional roof or hanging wall support and lateral support to ore pillars, hydraulic fill methods have also been adapted to provide point supports in the form of yielding horizontally reinforced cemented tailings packs in the narrow tabular stopes of South African gold mines.

The concept of such a support is illustrated by Figure 15 [Dison and Blight(1986)]. The porous bag, with the steel mesh mats sewn into it at a pre-determined spacing, is transported underground and positioned in the stope it is intended to support. The sides of the bag are braced with steel mesh shutters and the bag is then pumped full of a carefully designed cemented tailings fill. Excess water drains from the bag and the fill hardens within 24 hours. The support now carries an increasing load as the working face advances away from it and the stope converges.

The pack carries the load in the following manner:

The vertical load on the pack tends to cause a horizontal strain that is resisted by tension in the horizontal reinforcing. If the reinforcing steel has a cross-section of area A_R, a vertical spacing v and a horizontal spacing h, then

$$\sigma_h vh = A_R \sigma_R \qquad [4]$$

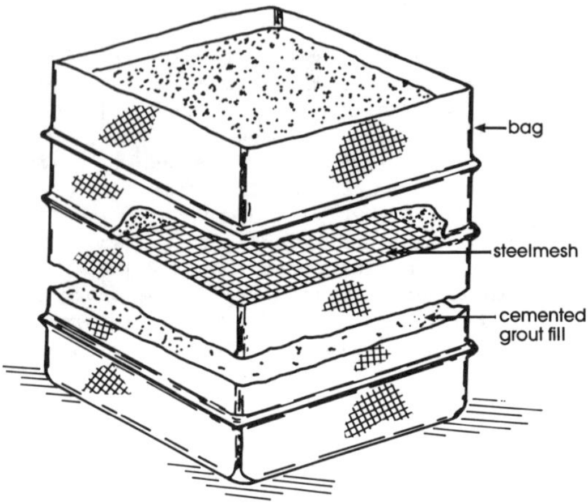

Figure 15: Diagrammatic view of pack assembly

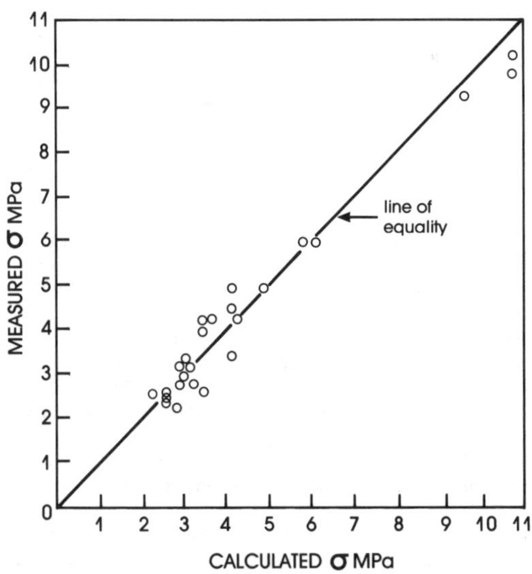

Figure 16: Relationship between calculated and measured strengths of walls built of horizontally reinforced granular material

where σ_h is the horizontal stress in the cemented tailings fill and σ_R is the reinforcing.

Ultimately, the fill fails in shear and the reinforcing yields. Equation[4] then becomes

$$\sigma_h = \frac{A_R \sigma_{yR}}{vh} \qquad [4a]$$

in which σ_{yR} is the yield strength of the reinforcing. σ_v, the vertical stress carried by the pack is then related to σ_h by

$$\sigma_v = \sigma_h \tan^2(45+\phi/2) + 2c\tan(45+\phi/2) \qquad [4b]$$

where 2c is the unconfined compression strength of the cemented fill and ϕ is the angle of shearing resistance. There is a considerable capillary stress component in the strength of the cemented fill and c and ϕ are total stress parameters that incorporate this component of strength.

The validity of equations[4a] and [4b] has been established by means of an exhaustive series of tests using a wide variety of fill materials and types and specimens of steel reinforcing. Figure 16 shows the excellent correlation between measured and calculated strengths for some of these tests.

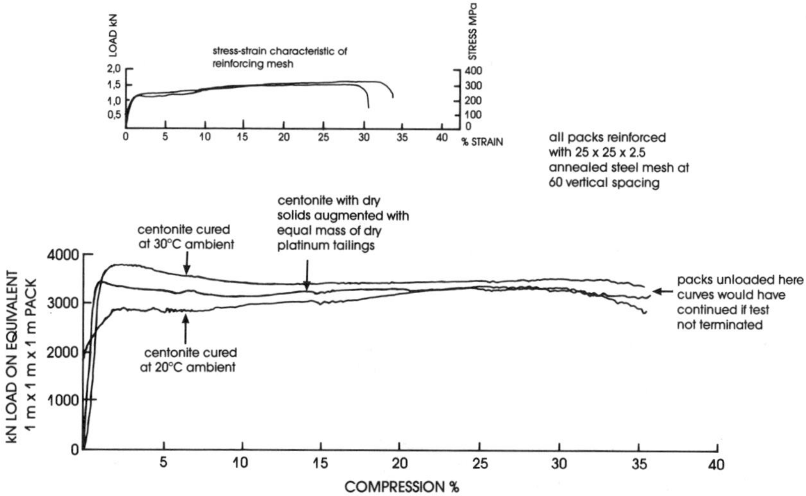

Figure 17: Compression characteristics of 300mm cubes of centonite reinforced with annealed steel mesh

The performance required from yielding supports is that they initially be as stiff as possible, reach a predetermined load at which the reinforcing yields, and then maintain that load over as great a range of compression as possible.

To obtain the long-sustained yield load, the mesh reinforcing is annealed after fabrication. Figure 17 shows typical stress-strain curves for annealed mesh reinforcing and also for three packs reinforced with annealed steel mesh. It will be noted that whereas normal mild steel fails at a strain of about 20 per cent, annealed mild steel fails at strains of up to 50 per cent. The load-compression characteristic illustrated in Figure 17 is nearly ideal. The filling material in this case was a new development known as Hydropak which consists largely of water, bentonite and a quick-setting cement. With this material, strengths of up to 3MPa are attainable 4 hours after casting. The tests illustrated in Figure 17 were made 24 hours after casting.

.8 EFFECTS OF HYDRAULIC FILLING OF UNDERGROUND EXCAVATIONS ON SURFACE STORAGE OF TAILINGS

Many mines engaged in deep level (3 to 4km below surface) gold mining in South Africa are experimenting on a large pilot scale with the use of pumped cycloned tailings fills to provide regional underground support. The cyclone underflow is placed underground and the finer overflow, together with most of the water from the tailings stream must be stored on surface.

The density of solid gold ore is about 2700kg/m^3 and the dry density of deposited cyclone underflow tailings is only 1400kg/m^3. Hence the volume of tailings required to fill a given rock excavation is about half of that actually available, and the remainder must continue to be stored on surface.

A closer analysis of the situation shows that, allowing for space underground that cannot be backfilled for reasons of maintaining permanent access, the residue requiring storage on surface will comprise at least 60 per cent by mass of run of mill tailings and at most 40 per cent of cyclone overflow material.

Hence if large scale, complete underground filling is adopted, the problems of surface storage of residues do not go away. Rather, they become exacerbated by the finer grade of material to be handled and the great proportion of water it will inevitably contain.

A recent investigation [Blight, McPhail and Fourie(1986)] has shown that the geotechnical properties of the surface tailings stream do not change dramatically, but do become less favourable. Shear strength parameters in terms of effective stresses remain almost unchanged. Coefficients of consolidation, however, reduce by a factor of 3, indicating that lower rates of rise for outer impoundments slopes may be necessary.

In the field, beach profiles will flatten, which means that pool areas will enlarge. This in turn may mean higher phreatic surfaces and possibly require flatter outer slopes to achieve the same factor of safety against shear failure.

In the worst analysis, the storing of tailings underground may actually require more, rather than less, surface area to be devoted to tailings storage.

.9 THE EFFECTS OF DESICCATION ON THE SAFETY OF HYDRAULIC FILL STRUCTURES

A previous section (.4) considered the benefits and some of the problems of desiccation of hydraulic fill tailings during deposition. This section will consider the benefits and effects of desiccation on the stability of hydraulic fill structures built in favourable climates - those where there is a seasonal or perennial water deficit.

Hydraulic fill gold tailings impoundments in South Africa are all operated by the semi-dry process whereby a long rotation of deposition is adopted that allows each layer to dry out before the next is deposited over it. Outer impoundment slopes are usually underdrained and the pool is kept to an absolute minimum size. The effect is considerably to depress the phreatic surface in the outer slopes with the result that these dams are remarkably stable.

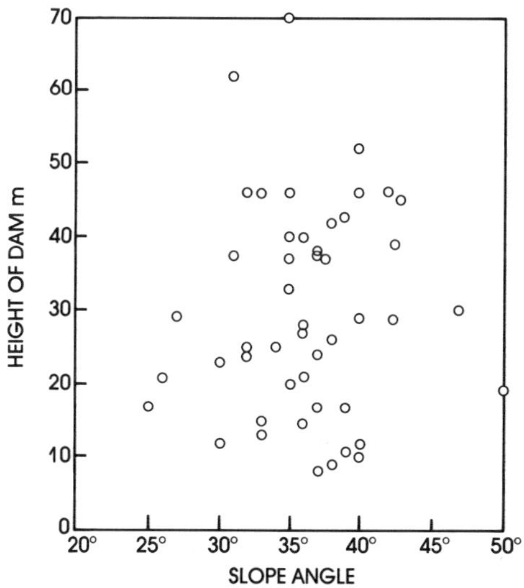

Figure 18: Heights and slope angles of 46 stable gold mine tailings dams

Figure 18 shows a plot of height versus slope angle for 46 stable gold tailings dams in South Africa. The average slope angle is 35° while heights range up to 70m and some slopes exceed 40° at heights in excess of 40m. What makes these slopes remarkable is that they are constructed of a cohesionless material having an effective stress angle of shearing resistance of between 27° and 35°. Their stability is due almost entirely to the permanent capillary tensions that act within the tailings.

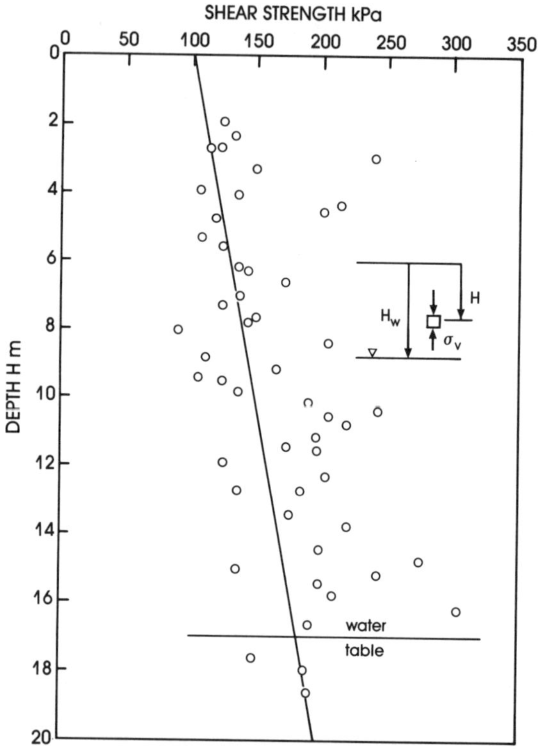

Figure 19: Profile of vane shear strength in a tailings deposit where the water table was 17 m deep

The effect of the capillary stresses on enhancing the strength of a tailings deposit is illustrated by Figure 19 which shows a vane strength profile measured in a hydraulic fill gold tailings deposit in which the phreatic surface was at a depth of 17m. The line of best fit to the measured vane strengths has the equation

$$\tau = [\gamma H + \gamma_w (H_w - H)]\tan\phi' \qquad [5]$$

where H and H_w are as defined in Figure 19 and ϕ' has a value of 29°. This value of ϕ' is well within the range of 27° to 35° mentioned earlier for the angle of shearing resistance of gold tailings. Hence the capillary stresses in the tailings are virtually fully effective in their contribution to effective stresses in the tailings.

.10 SURFACE EROSION OF HYDRAULIC FILL TAILINGS DEPOSITS

Pollution of air and water by particulate matter eroded off the surface of tailings deposits constitutes a serious threat to the environment, and as such must be minimized. The problem is especially severe with ringdyke structures that project above the general level of the countryside. Rates of erosion from the slopes of gold tailings dams of as much as 500T/ha/y (dry tons of tailings per hectare per year) have been published [Blight(1987)]. However, rates as high as 950T/ha/y have since been measured. Under South African conditions where most of the rainfall occurs in summer and winters are dry, erosion by wind appears to constitute about one third of the total loss of solids.

Erosion from the steep slopes of tailings dams appears to be affected by much the same factors as are incorporated in the Universal Soil Loss Equation [e.g. Blight(1987)], principally slope length and slope angle, as well as a measure of surface shear strength. Figure 20, for example, shows a correlation between surface shear strength and erosion loss for gold tailings dams.

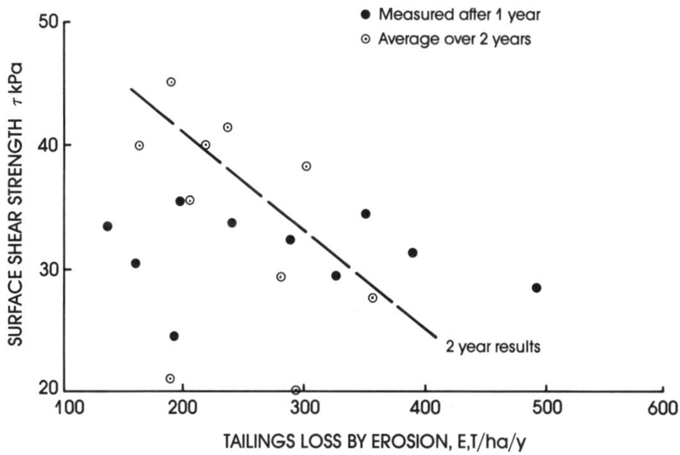

Figure 20: Correlation between surface shear strength and tailings loss by erosion

Experiments have been made to stabilize the top surfaces of tailings dams with cement and lime in order to reduce rates of water and wind erosion [Blight and Caldwell(1984)]. Initially, it was thought that this treatment was unsuccessful as the strength of the stabilized tailings deteriorated rapidly in the first three years as a result of attack by sulphate and salt crystallization [e.g. Blight(1976)]. However, recent measurements on 9 year old horizontal areas of cement and lime stabilized tailings have shown zero erosion loss over a full year, while an adjacent untreated control area lost 65T/ha/y over the same period.

Pebble mulches consisting of a single particle layer of 10mm or 13mm gravel spread over the surface have also shown some promise as a means of reducing rates of erosion on horizontal or gently sloping surfaces. Two pebble mulched surfaces sloping at 5° lost an average of 70T/ha/y over a full year period. An untreated control area between the two pebble mulched surfaces lost no less than 900T/ha/y over the same period.

The outer slopes of most tailings dams in South Africa are too steep to allow of access by mechanical plant and hence cannot be treated by, for example, cement stabilization. They are also too steep to retain a pebble mulch. The only alternatives at present are to leave the slopes untreated and catch the eroded material at the foot of the slope, or to cover them with vegetation - this may be the obvious course in humid climates, but is not easy in arid areas.

Measurements of rates of erosion from grassed slopes indicate that losses may be just as high as from untreated slopes. However, this preliminary conclusion needs confirmation.

3. CONCLUDING REMARKS

It is hoped that the material presented in this paper has proved of interest to the reader. As stated earlier, the object was to bring to attention some of the less familiar aspects of hydraulic fill structures. All of the aspects described above are of importance although not of general importance. If the readers' perspective of hydraulic fill structures has been widened, the object of the paper will have been achieved.

4. REFERENCES

Abadjiev, C.B., "Estimation of the physical characteristics of deposited tailings in the tailings dam of non-ferrous metallurgy", Proceedings, 11th International Conference on Soil Mechanics and Foundation Engineering, San Francisco, 1985, vol 3, pp 1231-1234.

Berti, G., Villa, F., Dovera, O., Genevois, R and Brauns, J., "The disaster of Stava/Northern Italy", ASCE Geotechnical Division Specialty Conference on Hydraulic Fill Structures, Fort Collins, Colorado, USA, 1988.

Blight, G.E., "Properties of pumped tailings fill", Journal of the South African Institute of Mining and Metallurgy, vol 79, no 10, 1979, pp 446-453.

Blight, G.E., "Partial saturation can assist the soil engineer", Proceedings, 6th South East Asian Conference on Soil Engineering, Taipeh, Taiwan, 1980, vol 1, pp 15-29.

Blight, G.E., "The concept of the master profile for tailings dam beaches", Prediction and Performance in Geotechnical Engineering, Balkema, 1987, pp 361-368.

Blight, G.E., "Erosion of the slopes of gold tailings dams", Geotechnical Practice for Waste Disposal '87, ASCE Geotechnical Special Publication No 13, 1987, pp 294-305.

Blight, G.E. and Bentel, G.M., "The behaviour of mine tailings during hydraulic deposition", Journal of the South African Institute of Mining and Metallurgy, vol 83, no 4, 1983, pp 73-86.

Blight, G.E. and Caldwell, J.A., "The abatement of pollution from abandoned gold-residue dams", Journal of the South African Institute of Mining and Metallurgy, vol 84, no 1, 1984, pp 1-9.

Blight, G.E. and Clarke, I.E., "Design and properties of stiff fill for lateral support of pillars", Proceedings, International Symposium on Mining with Backfill, Lulea, Sweden, 1983, pp 303-307.

Blight, G.E., McPhail, G.I. and Fourie, J., "Possible difficulties with building tailings dams of cyclone overflow product", Association of Mine Managers of South Africa, Symposium on Backfill, 1986, 14pp.

Blight, G.E., Robinson, M.J. and Diering, J.A.C., "The flow of tailings slurry from a breached dam", Journal of the South African Institute of Mining and Metallurgy, vol 81, no 1, 1981, pp 1-8.

Blight, G.E., and Steffen, O.K.H., "Geotechnics of gold mining waste disposal", Current geotechnical Practice in Mine Waste Disposal, ASCE, 1979, pp 1-52.

Blight, G.E., Thomson, R.R. and Vorster, K., "Profiles of hydraulic fill beaches and seepage through hydraulically sorted tailings", Journal of the South African Institute of Mining and Metallurgy, vol 85, no 5, 1985, pp 157-161.

Canada Centre for Mineral and Energy Technology, "Pit and Slope Manual, Chapter 9, Waste Embankments", Canmet Report 77-01, 1977.

Chamber of Mines of South Africa, "Handbook of guidelines for environmental protection", vol 1/1979, revised 1983, "The design, operation and closure of metalliferous and coal residue deposits", The Chamber, 1983.

Commonwealth of Australia, Department of Home Affairs and Environment, "Code of practice on the management of radioacative wastes from the mining and milling of radioactive ores", Australian Government Publishing Service, Canberra, 1982.

Dison, L. and Blight, G.E., "Reinforced cemented tailings fill", Proceedings, Association of Mine Managers of South Africa Symposium on Backfill, Johannesburg, South Africa, 1986, 13pp.

Fourie, A.B., "A laboratory study of the beaching and consolidation characteristics of two hydraulically placed soils", ASCE Geotechnical Division Speciality Conference on Hydraulic Fill Structures, Fort Collins, Colorado, USA, 1988.

Graf, W.H., "Hydraulics of sediment transport", McGraw-Hill, New York, 1971, pp 455-461.

International Commission on Large Dams, "Manual on tailings dams and dumps", ICOLD Bulletin 45, 1982.

Hahn, J.A., Blight, G.E. and Dison, L., "Support in shallow mines using horizontally reinforced systems", Journal of the South African Institute of Mining and Metallurgy, vol 82, no 10, 1982, pp 277-290.

Hutchinson, J.N. and Bhandari, R.I., "Undrained loading: A fundamental mechanism of mudflows and other mass movements", Geotechnique, vol 21, no 4, 1971, pp 353-358.

Jennings, J.E., "The failure of a slimes dam at Bafokeng", The Civil Engineer in South Africa, vol 21, no 6, 1979, pp 135-141.

Jeyapalan, J.K., Duncan, J.M. and Seed, H.B., "Analysis of flow failures of mines tailings dams", Journal of Geotechnical Engineering, ASCE, vol 109, no 2, 1983, pp 150-171.

Jeyapalan, J.K., "Analysis of flow failures of mine tailings impoundments", PhD Thesis, University of California, Berkley, 1980.

Lucia, P.C, Duncan, J.M. and Seed, H.B., "Summary of research on case histories of flow failures of mine tailings", Mine Waste Disposal Technology, U.S. Bureau of Mines Information Circular IC8857/1987, pp 46-53.

Marcuson, W.F., Ballard, R.F. and Ledbetter, R.H.G., "Liquefaction failure of tailings dams resulting from the new Izu Oshima earthquake, 14 and 15 February, 1978", Proceedings, 6th Pan American Conference on Soil Mechanics and Foundation Engineering, Lima, Peru, 1979.

Melent'ev, V.A., Kolpashnikov, N.P. and Volnin, B.A., "Hydraulic fill structures", Energy (Moscow) 1973. (English translation of original Russian).

Morgenstern, N.R., "Submarine slumping and initiation of turbidity currents", Marine Geotechnique, A.R. Richards, ed., University of Illinois Press, Urbana, Illinois, 1967, pp 189-220.

National Coal Board (UK), "Spoil heaps and lagoons", 1970.

Smith, G.M., Abt, S.R. and Nelson, J.D., "Profile prediction of hydraulically deposited tailings", Transactions, Society of Mining Engineers of AIME, vol 80, 1987, pp 2024-2027.

Verkerk, C.G., "The hydraulic transportation of gold mine waste", Proceedings, International Conference on Mining and Industrial Waste Management, Johannesburg, South Africa, 1987, pp 139-145.

Wates, J.A., Stevenson, C. and Purchase, A.R., "The effect of relative densities on beaching angles and segregation on gold and uranium tailings", Proceedings, International Conference on Mining and Industrial Waste Management, Johannesburg, South Africa, 1987, pp 89-93.

Whitney, E.D., Moudgil, B.M. and Onoda, G.Y., "Dewatering of phosphate clay wastes", Department of Material Science and Engineering, University of Florida, 1977.

REPORT FOR SESSION 1
EXCAVATION, TRANSPORT AND DEPOSITION

Don Banks*, M. ASCE

Active excavation projects, both surface and subsurface, either for mineral or ore extraction or for tunneling purposes, result in large volumes of materials that must be transported and deposited in a safe, economical and environmentally acceptable manner. The resulting deposit of material in many cases serves, not merely as a waste pile, but rather for some engineering purpose. Thus, a knowledge of or control of engineering properties is desirable.

While no papers are presented in the subject area of excavation, the nine papers grouped for this Session describe work in transportation and deposition. The order of papers reviewed for this report was purposely chosen to progress from slurry transport to pneumatic transport of mine tailings; then from underground to above ground deposition; and finally considerations of underwater deposition. This order also shows a transition in the topic from mining engineering to civil engineering interests. Two observations are made. First, there are obvious areas where geotechnical approaches (specifically soil mechanics) have and can contribute to the understanding of transportation and deposition processes and the resulting material properties. Secondly, much has been learned through the use of small-scaled laboratory physical models.

The thrust of the papers for this Session are briefly described in the remainder of this report.

In his paper, "Slurry Pipelines: Present and Future", Henderson points out while solid materials have been hydraulically conveyed for almost 100 years, little is known about the physical processes involved or their mathematical formulation. The designer of a slurry pipeline system is cautioned to carefully examine the data upon which the system is sized and operating parameters determined. Much useful data are available, as Henderson shows by a lengthy bibliography and discussion of the operation of the West Tailings Delivery Line at the Climax Molybdenum Company and findings at the Hydraulic Transport Research Facility. Finally, Henderson makes the point that slurry systems should be considered as a means of transporting low or high level radioactive wastes, tunnel muck and in open pit mining operations.

---

*Chief, Engineering Geology and Rock Mechanics Division,
Geotechnical Laboratory
US Army Engineer Waterways Experiment Station
Vicksburg, MS 39180-0631

A research effort that resulted in a system to transport dense mill tailings is described by Archibald in his paper "Pneumatic Placement of High Density Mill Tailings at 25 Tons/Hour". Whereas the overview given by Henderson looked at systems that transported materials in concentration ranges of from 10 to 60 percent by weight, using water and over distances measured in miles, Archibald's work lead to a system that uses air as the transporting medium at up to distance of about 300 ft., at velocities of up to 240 ft/sec; also of interest was the tailings concentration of 80 percent solid, 20 percent water. The successful design, according to Archibald, resulted from treating the air/water/solid mixture as a gas. The pneumatic stowing machine, called the Stopepak, permitted stopes to be filled on a near continuous basis, thus smoothing-out the somewhat cyclic mining activity.

Brown, Murphy and Bartell in their paper, "Use of Hydraulic Fill to Stabilize an Abandoned Underground Coal Mine", describe the planning, performance and effectiveness of a subsidence control project in Hanna, Wyoming. In the project, described by Archibald, the backfilling was done in an active mine with the requirement that the fill consolidate in a few days to support mining equipment. In the Hanna project, backfilling was required to stop the occurrence of surface sinkholes and subsidence troughs. The project required that the areal extent of the then flooded mine be determined, an assessment of the subsidence risk be made and a remote slurry backfilling system be designed and constructed. Importantly, to be effective in reducing or eliminating subsidence the backfill must fill the voids, as well as possess adequate load carrying capacity. In total almost 800,000 tons of backfill was placed in about 8 months. Originally a solid/water ratio of about 0.2 was used with slurry flow velocity varying from eight to twelve feet per second; later the solid/water ratio was increased to about 0.5. It was noted that only a few injection wells ever "refused" to take slurry, thereby indicating a full room. Rather, volumetric calculations were employed to determine that a room was full; in addition observation holes were drilled in selected rooms. Both laboratory and field tests were made to determine the load-carrying capacity of the backfill. The field penetration tests showed rapid increases in penetration pressures with depth of fill and the penetration pressures decreased as the distance from the injection well increased.

Faddick and Eby, in their paper, "Effects of Slurry Backfilling - Lessons Learned", also reviewed the Hanna project. The paper is divided into two parts; the first part deals with the backfilling process itself, the second part describes observations drawn from the Hanna experience. Faddick and Eby's discussion of the backfilling process expands on specific compontents described in the papers by Henderson and Archibald. The review of the Hanna project describes the problems (and solutions) in a slightly different perspective than offered in the paper by Brown, et al; however the two papers are not contradictory.

In his paper, "Tailings Management in Australia: A Geotechnical Engineer's Viewpoint", Cooper describes some of the history of mine waste management (primarily for gold mining operations) and suggests some future directions that the industry may take and the part that the civil/geotechnical engineer may play. He points out that the cost of labor has caused increased volumes of machine produced ores, more efficient grinding has caused finer particle sizes to be managed and increased emphasis on environment aspects has caused more attention being given to carefully plan and manage large tailings storage areas. In particular, Cooper points out that successful management requires an understanding of the mechanics of slurry flow, principals of drainage and consolidation and the development of the strength of tailings. This requirement has both short-term as well as long-term implications with this view. Cooper sees an increasing role for the geotechnical engineer and an increasing use of such devises as the Rowe consolidation cell as well as increasing monitoring system requirements.

Minns, in his paper, "A Review of Tailings Disposal Practices in North America and Australia", discusses a wide variety of tailing disposal methods currently being used. He notes that many factors, such as physical and chemical properties of the tailings, costs, availability of dam wall construction materials, type of mining operation, topography and environmental constraints, all influence the manner of disposal. Minns presents a rather comprehensive bibliography from which numerous examples are drawn to illustrate different methods that are being used. These methods are described as inpit disposal, conventional zoned earth embankments, upstream construction techniques, centerline construction techniques, downstream construction techniques, thickened tailings disposal methods, and dry disposal techniques. Minns concludes that it is important that mining companies be aware of the range of available disposal methods so that planning, design and construction will lead to safer and environmentally efficient disposal facilities.

Both Cooper and Minns speak of the need for engineering decision-making when designing tailing disposal works. Fourie, in his paper, "Beaching and Permeability Properties of Tailings", emphasizes that the sizing (capacity) of an impoundment area will depend upon the beach profile. Since the profile may depend upon a particular set of conditions that is outside the designer's experience, Fourie describes laboratory scale flume tests by which the influence of different variables can be assessed. The flume was similar to one previously described by Blight, Thomson and Vorster* (1985). The test results reported by Fourie indicated that normalized profiles were reproducible. As further test results are obtained important conclusions may be reached that have equally important design implications. For example, for a given type of tailings, Fourie found the normalized profile to be relatively insensitive to the solid/tailing water ratio. Fourie found important changes in the coefficient of permeability depending upon whether the measurement was made near the discharge point or near the edge of the

---

*Referenced in Fourie's paper.

tailing beach (i.e, an order of magnitude decrease at the edge). The horizontal and vertical coefficients of permeability were approximately the same for poorly-graded course bauxite tailings but were an order of magnitude different for well-graded fine-grained coal tailings (i.e., $k_h \simeq 10k_v$).

The previous portion of this report has dealt with the practice of material disposal. There has been the view that the material is waste with respect the purpose of the project, i.e., the production of ore in mining (or the achievement of a completed opening in tunneling). There is a parallel endeavor where material is the desired product of the project, i.e., sands dredged to build underwater fills for such use as artificial islands, support for mobile drilling platforms, breakwaters, harbor extensions, etc.

In their paper, "The Construction of Sand Bodies in the Sea or Rivers", de Groot, Heezen, Mastbergen and Stefess describe exhaustive investigations concerning slope angles and density of fine-sand fills in estuaries both in the field, during and following construction, and in the laboratory. Their paper describes the sedimentation process above the high water level, beneath the low water level, and in the intermediate or tidal zone. In general terms, coarse sands and low specific flow rates caused relative steep slopes to be formed above the high water levee. The relative density of the sand is relative high and the deposit is seldom susceptible to liquefaction. Below the low water level the slopes are only slightly steeper than the above water slopes because the sand is deposited in a looser state and is susceptible to flow slides. This looser state also makes the deposit to be more susceptible to liquefaction The flattest slopes were found in the tidal zone particularly when the water depth was small and the tidal difference large.

It was particularly interesting to note that de Groot, et al, made extensive use of laboratory models explaining observations in terms of theoretical (geotechnical or soil mechanics) considerations to explain results in the field. Importantly the laboratory tests led to the development of empirical formulae where field behavior could be quantified.

The observation by de Groof, et al, that sands deposited beneath the low water level are loose, resulting in low strengths and poor deformation properties have similarly been noted by others. Hodge, in his paper, "Underwater Sand Fills", described experimental efforts at laboratory scale to increase the density. The experimental efforts are based on the concept of pumping water out of the submerged fill, during deposition, creating an inward flow gradient. The model tests showed that steep side slopes could be produced and that the erodibility of the slope surface could be reduced while suction was being applied. Since this induced improvement was lost when pumping ceased, in practice, pumping would have to be continued until after slope protection materials were placed and possibly later reactivated as an added safety measure during such events as a major storm.

It would appear from the above review that Specialty Conferences such as this one afford excellent opportunities for effective technology transfer between workers in the mining and civil engineering fields. The papers in this Session are a step in that direction; discussions among Conference attendees will further the process.

The Essence of Soil Properties in Today's Dredging Technology

Authors: ir. F.A. Verhoeven - Head R & D Department, Royal Boskalis Westminster
ir. A.J. de Jong - Head Estimating & Research, Volker Stevin Dredging
ir. P. Lubking - Head Hydraulic and Marine Structures Department, Delft Geotechnics

ABSTRACT

Since the early sixties dredging technlogy progressed considerably. Dredging equipment was improved and scaled up, working methods were ameliorated. Besides the role of soil properties in the interaction between equipment and soil in all stages of the dredging process (loosening, mixing, transport and depositon) was studied.
The relevant parameters of sand, clay and rock material are described in relation to these dredging stages.
Next the determination of the soil parameters by direct (in situ or laboratory) tests or by indirect methods is treated.
Particularly the use of correlations between dredging parameters and relatively simple and fast tests is mentioned; some examples of well known correlations are given.

INTRODUCTION

In the last two decades there has been a considerable progress in dredging technology. We find this progress in all the different fields of dredging technology.
In the first place we clearly see these improvements when we look at the various types of dredging equipment. Secondly also the operational techniques improved considerably. As in every up to date industry instrumentation and computerisation in dredging progressed and became a major item; hardly any dredging can be imagined without automation. Dredging turned into a highly sophisticated and scientific operation.

To a large extent the basis of all these rapid changes has been the basic research done in the last two decades. Developments both in the field of equipment and working methods have taken place based on the insight into the physical processes that occur during the different dredging processes.

Dredging processes and the role of soil properties in dredging have been studied for more than 25 years by Delft Hydraulics and Delft Geotechnics under the sponsorship of a consortium of the major Dutch dredging companies, CSB (Dredging Research Association) in which participate Amsterdam Ballast Dredging, HAM, Royal Boskalis Westminster, Volker Stevin Dredging and Rijkswaterstaat (Public Works Department, Ministry of Transport and Public Works).

Often the soil information in tender documents is very limited and/or does not give the relevant parameters to determine the working method and production.
In this paper therefore special attention is paid to the relevant soil parameters for the different dredging processes.
Before coming to this subject firstly in chapter 2 the development in dredging equipment during the last two decades is discussed.

Note: This paper can be regarded as an extra background to the soil classification for dredging purposes of the PIANC. (PIANC, 1984)

DEVELOPMENTS IN DREDGING EQUIPMENT

When we study the developments in dredging equipment and place them on a time scale, we can read exactly in which field the major dredging operations have been concentrated in a certain period. For example, in the seventies, during the Middle East boom, dredging industry concentrated on cutter suction dredgers and a development hardly to follow took place in cutter suction dredgers, especially suited for dredging of rock.
As so much effort was put into the development of cutter dredgers, very little attention was paid to the hopper suction dredgers.

At the end of the Middle East boom, capital dredging reduced considerably and dredging activities moved to maintenance dredging and dredging in severe weather conditions, such as in areas of rough sea conditions and in the arctic areas (Beaufort Sea).

The impact of the concentration of dredging activities in certain areas is shown when we discuss the developments in the various types of equipment.

Trailing suction hopper dredgers

To discuss the improvements in the recently built trailing suction hopper dredgers we look at the dredging cycle of a hopper dredger.
We distinguish the following processes:

- loading
- sailing
- unloading

- The loading stage

  The loading stage is the most significant period of the dredging cycle, as this is the part where the basic activity of the dredging, removal of the soil takes place. In the development of hopper dredgers a lot of attention has therefore been given to the loading process.
  Especially in the last five years we see an increase in the maximum waterdepth in which the hopper dredger can dredge and several of today's hopper dredgers can dredge in water depths of up to 80 metres. This technique is of course not based on ideas on future depth requirements of harbours or entrance channels, but it is essential in a number of offshore oil projects.

The modern hopper dredgers have therefore the connection of the suction pipes on the frontside of the vessel to give space for the long suction tubes.
Sometimes we see a combination of long suction tubes with underwater pumps on the tubes in order to guarantee the required suction production capacity. To get a high suction capacity it is essential to feed the suction system with soil which has been loosened from the seabed. Therefore during the last five years a lot of effort has been put into the improvement of the dragheads. Different shapes, jets in the draghead, double row of knives, remotely adjustable water-inlets are examples of these improvements fundamentally based on a thorough knowledge of the principal process in the dredging companies concerned..

As maintenance dredging became more and more important, the need for an improvement of the loading production in mud and silt grew.
The existence of gas in the dredged slurry ruined the suction capacity of the dredge pumps and the solution has been found in the developing of the degassing systems, an old idea but made effective with today's techniques as automated control systems.
Another improvement in the loading technique has been found in the swell compensating and winch systems, resulting in the possibility to operate in quite rough sea conditions. It has also resulted in dredging with a remarkably improved accuracy which is a major contribution to the efficiency improvement.

- The sailing stage

  As sailing is at least ten times as old as dredging no major changes in sailing itself could be expected. However, the sailing part of the dredging cycle has been made more efficient. In the first place this has been by an increase of the hopper capacity, which makes the sailing part relatively less important in the dredging cycle. Furthermore an optimisation in the propulsion has to be mentioned. The improved technique of controllable pitch propellers makes it possible to use the extra power available on board when the dredge pumps have stopped. This results in a more optimum use of the diesel engines on board of the vessel and a higher sailing speed.

  As to use this potential higher sailing speed also in remote areas with extreme climates such as the Beaufort Sea, a number of today's hopper dredgers are ice-strengthened, enabling them to sail with higher speeds in areas with small ice-fields.

- Unloading

  Most of the improvements in the unloading stage are based on reducing the unloading time, especially when working in clay.

  In the new hopper dredgers the originally common-used conical valves have been replaced by large bottom doors, increasing the ratios bottom opening/hopper area considerably. An extreme progress in this field is the expensive but efficient solution of the split hopper dredger, which not only makes sure that the unloading process is quick but also makes it possible to unload in shallower waters.

  The most recent, but already very popular development in the unloading technique is the use of a self-emptying system, using suction pumps. This enables the dredging contractors to use trailing hopper dredgers for land reclamation projects, trench filling at sea etc.

FIG. 1 MODERN TRAILING SUCTION HOPPER DREDGER "GEOPOTES 14" (VOLKER STEVIN DREDGING)

Besides the technical improvements especially for parts of the dredging cycle, we certainly have to mention separately the efficiency improvement of all technical systems on board of the dredger by using maintenance friendly systems, automation of most of the processes on board and of the process control systems.
On today's large hopper dredgers all operations are executed by a crew of 15 to 20 high-skilled people (in 24 hours 7 days a week operation) instead of the approx. 30 people necessary on the older vessels.

The cutter suction dredger

As mentioned before the major developments of cutter suction dredgers took place during or where inspired by the boom in the Middle East. Hardly any dredging of rock or other hard material had to be done before. In the Middle East the dredging contractors were fully confronted with the problems of dredging caprock and limestone, which led to a number of developments:

- Dredger dimensions and installed power, and power of the cutterdrive increased very rapidly. In 1976 the largest cutter dredger had a total installed power of approx. 8000 kW of which 1500 kW on the cutterdrive. Ten years later the dredger "Leonardo da Vinci" had installed more than 20.000 kW of which more than 4000 kW on the cutterdrive.

- The new large cutter suction dredgers are self-propelled. This relatively cheap addition to the cutter dredger (the power was already available) makes these cutter dredgers more flexible, cheaper in mobilization costs and enables them to escape quickly in case of extreme sea conditions.

- Use of wear resistant material is also a development which started in the seventies. By using these materials in pumps and decklines and for pump impellers the time lost for maintenance in these areas was reduced tremendously and so did the costs of wear on pipes and pumps.

However, there are also other developments not based on experiences in the Middle East, but based on efforts in the laboratories and research departments.
In this context we have to mention the development of the cutterheads which changed in shape and are now equipped with completely different cutting tools than the original pick points. We can also mention the development of the bucket wheel as replacement for the cutterhead.

Both the development of the bucketwheel and the improvement of the cutterhead, do not only improve the cutting efficiency but are also meant to improve the mixing of water and soil, to increase the density of the mixture to be pumped. The developments of the cutterheads resulted in different cutterheads for different soil types (sand, clay, rock). The use of jets for loosening soils before dredging is another recent development, which improved considerably the dredging efficiency especially in sand.
(Note: some of the cutter dredgers today can handle 500.000 m^3 per week !).
Also the cutter dredgers are highly automated today. The knowledge of the mechanisms which influence the cutting, mixing and pumping process makes it possible to design and build highly computerised automation systems, which take over the control of winches and dredge pumps from the dredge master; this system performs his work more accurate, with less mistakes. The former dredge master, working on a basis of his feelings and experiences has now become a well trained process operator.

FIG. 2 MODERN 15.000 kW CUTTER DREDGER "ORANJE" (BOSKALIS WESTM.)

Grabs, dippers and backhoes

As grabs and dippers are not such complex dredging tools as trailing suction hopper dredgers and cutter suction dredgers, developments have been limited.
Grabs not having been popular for many years, now get some more attention in the last five years, mainly because the oil industry is working in greater water depths. Developments have been in the size of the grabs (up to 35 m^3 today) and in the way of closing. Today quite often hydraulic cylinders are used for closing the grab, instead of a closing wire, especially when high closing forces are required f.i. when dredging hard clay.

Placing the grabs in self-propelled pontoons equipped with self-tensioning winches for mooring under favourable environmental conditions today's grab is very useful for dredging alongside quaywalls, but even more for trenches in deep water.
The conventional dipper dredgers have hardly been built the last two decades. The hydraulic excavator has taken over its work such as the removal of relatively small amounts of hard material or fragmented rock. These hydraulic excavators are mostly placed on jack-up pontoons and in 90% of the cases built as backhoes instead of as a dipper dredger.
In the last five years those backhoes can be equipped with accurate position measuring systems and therefore can dredge (from the jack-up pontoon) at least as accurate as the bucket dredgers used to do. Besides, especially in shallow water they can work far more economically.

SOIL BEHAVIOUR DURING THE DIFFERENT PARTS OF DREDGING

In this chapter the behaviour of the soil during the different stages of dredging is discussed.
The following basic division into dredging processes is used:

- Loosening process
  - mechanical (blades, teeth)
  - hydraulic (suction, erosion by draghead, waterjets)
- Mixing process
- Transport
- Deposition
  - settling
  - differences in volume (bulking)
  - trafficability

An aspect that has not been covered by the above mentioned processes is wear. This will be discussed here as a separate item because soil plays often a dominant role in the wear process. A distinction can be made between wear of cutting elements and wear of pumps and pipes.

Based on the behaviour of soil with respect to the loosening process a subdivision is made into the following five soil types:

soil type 1: sand
soil type 2: mixed type of soil between sand and clay; sandy silty clay; clayey silty sand
soil type 3: clay
soil type 4: mixed type of soil between clay and rock
soil type 5: rock

The subdivision between the soil types 1, 2 and 3 is made on a basis of soil composition as characterized by the grain size distribution and the plasticity index.
The separation between the soil types 3, 4 and 5 is based on strength criteria, viz. the undrained shear strength and the compressive strength of the soil.
The boundaries between the five soil types have been chosen arbitrarily:

boundary 1-2	:	d_5 = 20 μm and/or d_{20} = 60 μm
" 2-3	:	P.I. = 8%
" 3-4	:	c_u = 100 kPa
" 4-5	:	σ_c = 2 MPa

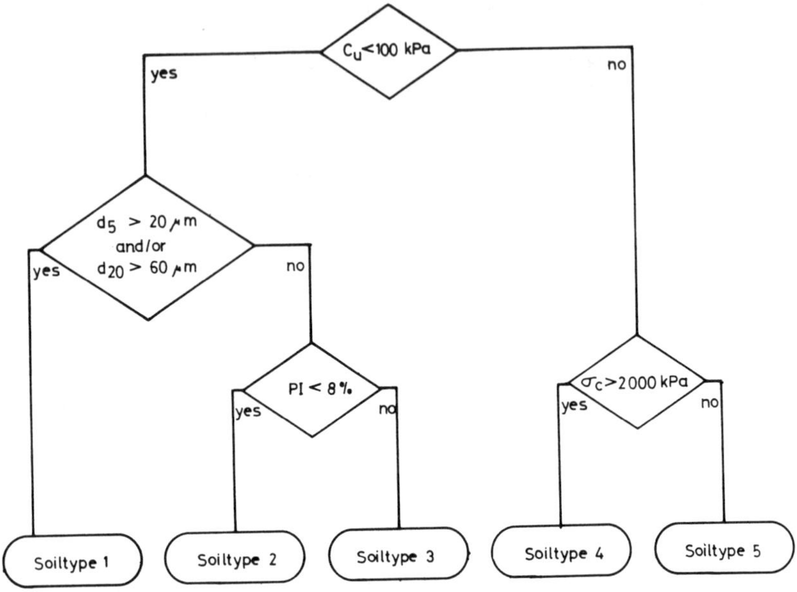

FIG. 3 SOIL CLASSIFICATION CRITERIA FOR LOOSENING PROCESS

Note: The soil type peat is not taken into consideration.

Loosening process

- Sand
  In mechanical cutting of sand a dominant role is played by the negative pore water pressures, caused by the dilatancy of the sheared sand. Densely packed sand will tend to increase in volume during shearing and causes negative pore pressures (van Os, 1987).

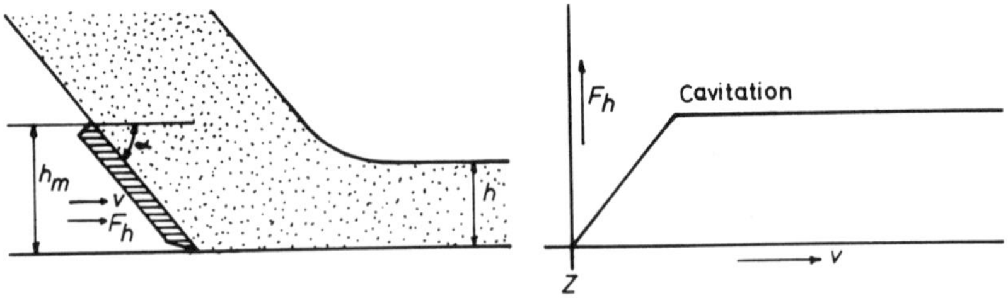

FIG. 4 CUTTING OF SAND

In well permeable sand the pore water pressures can dissipate to a large extent.

The shearing of the sand mass in the shear zone and the shearing of the sand over the cutting blade are two essential deformation processes in sand cutting. The forces in both shearing processes are dominated by the pressure decrease in the pore water, caused by the dilatancy of the sheared sand. Regarding these processes the following properties of the sand mass proved to play an important role (van Leussen, 1984):

1. angle of internal friction (φ)
2. angle of friction between sand and steel (δ)
3. permeability (k)
4. dilatancy

In practice it is mostly impossible to determine all the above mentioned parameters. Therefore often the relative density (D_R) is used, because this parameter has a reasonable relationship with the angle of internal friction (φ), the friction angle sand-steel (δ) and the dilatancy.

In the hydrualic loosening processes the permeability appears to be a very important parameter.

- Clay

  The failure behaviour of the clay during cutting is to a large extent determined by the stiffness. Generally the softer clay types show a plastic failure behaviour.

  From the soil point of view the mechanical cutting forces in softer clay are governed by adhesion of the clay in contact with the steel blade and the undrained shear strength in the failure plane.
  Assuming a unique relation between adhesion and cohesion the undrained shear strength can be considered as the only relevant parameter.
  The stiffer clays deviate from this in two ways, viz.:

  - failure starts often from a tensile crack
  - shearing occurs along preferring planes due to discontinuities.

  The specific energy to cut stiff clays will be relatively less than for the softer clay types.

- Rock

  Rock can be loosened by dredging in several ways:
  - cutting of the intact rock
  - shear failure along the discontinuities in the rock
  - combination of the above

  These modes of loosening are determined by:
  - strength of the intact rock
  - nature of fractures in the rock mass
  - type of dredger (peel thickness); as an example the difference in excavated lump size using a bucket dredger or a cutter dredger can be mentioned.

  The cutting tests in the laboratory indicate that the most important parameters in cutting of intact rock are:
  - compressive strength (σ_c)
  - porosity (n)

  In dredging practice the ratio between the compressive strength (σ_c) and the tensile strength (σ_t) is used in order to get information concerning the brittleness of the rock.
  The porosity determines the crushing degree of the rock at the tooth point.

  In relation to discontinuities of the rock mass, information concerning the following aspects is essential:
  - type (e.g. fracture, bedding planes etc.)
  - spacing
  - orientation (e.g. random)
  - openness

Mixing

Especially for cutter dredgers the mixing process plays an important role because a poor mixing process leads to a considerable spill.
The spill is mainly influenced by the used working method, as
- number of cutter revolutions per second (rpm)
- suction velocity
- geometry of the soil layer to be dredged

The process is mainly governed by the settling velocity of the soil particles in water. Very fine sand comes easier in suspension than coarse sand.

To judge the process the grain size distribution is sufficient in case of sand.
In clay and rock the phenomenon is somewhat different. Just after cutting both clay and rock consist mainly of lumps (and a very small amount of dissolved material), which limit the concentration in the pipeline due to the difficulty to suction them.
So, for clay and rock the mixing process is hardly sensitive to the soil properties.

Transport

Only hydraulic transport will be discussed because other ways of transport (as barges) are less important in relation to soil mechanical aspects.

Generally the transport mechanism in a pipeline is of major influence on the friction losses.
The size of the soil particles (or lumps) in co-ordination with the velocity of the mixture in the pipe determines the transport mechanism, which can consist of:
- complete suspension
- jumping of the soil particles over the bottom
- dragging of the particles over the bottom

- Sand
  The settling velocity is the essential parameter for the head losses. This parameter is influenced by each of the following characteristics:
  - grain size
  - angularity
  - grain density

  Assuming an average angularity (only extremely angular or round sand has a major influence) only a grain size distribution and a grain density (or mineral type) is necesary.

- Clay
  The friction losses of clay-water mixtures will depend on:
  - the rheological properties of the slurry
  - the density of the slurry
  - the percentage of clay lumps

  The clay slurrifies during transport in the pipeline. In the beginning of the pipeline (just after the dredge) hardly any clay has been dissolved. The degree of slurrification in the pipeline is a function of the clay type:

- the undrained shear strength
- plasticity index

The above mentioned shows, that at the beginning of the pipeline a higher percentage of clay lumps is present in the mixture than at the end; therefore the friction losses are generally the highest at the beginning.

In order to calculate the friction losses of the slurry the rheological properties have to be determined on slurries with densities between 1.2 - 1.4 t/m^3.
Densities lower than 1.2 t/m^3 cause no considerable increase in friction losses in relation to water.
The rheological properties of a slurry are measured by a viscotester.

- Rock
  For transport of rocky materials generally a fixed amount of lumps (± 50% vol.) is assumed.
  Only special rock types can lead to other assumptions.

Deposition

Regarding deposition, different aspects can be mentioned for which accurate information of the soil is necessary. The main items on a reclamation area are the bulking, the bearing capacity and the trafficability.

- Sand
  From various measurements it appeared that in general on hydraulic fills above the waterlevel the relative density* is 60-70%, which can be considered as approximately 98% of the normal maximum Proctor density.
  With this information the difference in volume between the pit and the fill can be calculated, when there is a good indication of the density in the pit.

  The trafficability of the fill is governed by the permeability.
  Hence the bearing capacity of wet sands can be insufficient if the amount of fines is increasing. A part of the hydraulic fill can have a lower bearing capacity due to local segregation of fines.

* The relative density is here defined as follows:

$$D_R = \frac{n_{max} - n_{situ}}{n_{max} - n_{min}} * 100\%$$

A special problem can be met in calcareous sands. For these materials the Proctor test appears to be unsuitable.
Due to the impact of the Proctor hammer crushing of the sands occurs. The increase of the percentage of fines causes a higher maximum Proctor density. As a result the densification requirements is aggravated.
In this way hydraulic fills consisting of calcareous sands are sometimes not approved although the bearing capacity appears to be sufficient.
Moreover the generally assumed correlation between compaction degree (% Proctor density) and bearing capacity certainly does not fit calcareous sands.
Generally calcareous fills have a relatively high bearing capacity due to:

- mostly higher φ-values than quartz sands
- a certain cementation which often occurs shortly after deposition

- Clay

  The behaviour of a clay fill is predominantly determined by the percentage of lumps in relation to the percentage of slurry.
  All main items are influenced by this ratio.

  The bulking of the slurry can be estimated from the in situ consistency of the excavated soil and the Atterberg limits. It is assumed that some days after the reclamation, at the start of the consolidation under its own weight, the water content of the slurry equals 2 to 3 times the liquid limit. The bulking from a clay with its natural water content to this slurry is simple to calculate.
  The bearing capacity and the trafficability of clay fills are generally negligible. To judge the ripening capability by evaporation the main parameters are the composition of the material (percentage of clay, peat, silt) and the type of clay minerals.

- Rock

  The main problem with the depositon of rock material is the amount of fines. Without preventive measures a large portion of the fines can be transported by the transport water outside the reclamation area causing a pollution problem.
  The bulking of the lumps can be estimated. The bearing capacity and the trafficability are good.

Wear

Wear is a process that occurs during cutting and transport of the soil-water mixture through pumps and pipes.
It is clear that considerable wear occurs in coarse sand/gravel and harder rock types.

In the table below the relevant soil parameters have been mentioned in relation to wear for cohesionless soils (sands) and strongly cohesive soils (rock).

	sand/gravel	rock
wear of cutting elements	- mineralogical comp. - density	- mineralogical comp. - compressive strength - grain size
hydraulic wear	- mineralogical comp. - grain size distr.	- mineralogical comp. - compressive strength

Fig. 5 SOIL PARAMETERS RELATED TO WEAR.

In the past the mineralogical composition was often a forgotten parameter but both from laboratory and field tests it appeared to be of utmost importance.

In clay the wear is generally very low.

At the end of this chapter some general notes are made which do not clearly belong to one of the discussed subjects.

1. Only the soil types sand, clay and rock have been discussed. For the mixed soil types (as mentioned in 3.1) still insufficient know-how is available and for the time being it is required to determine the soil parameters as mentioned for the "clean" soils (sand and clay) and rock.

2. In areas where possibly boulders can be found it is of utmost importance that the soil survey gives an indication concerning the frequency and size of boulders because these determine to a large extent the equipment and working method.

The matrix on the following page virtually summarizes the result of this chapter.

In the next chapter the determination of the different parameters from the matrix will be further discussed.

	SAND	CLAY	ROCK
Loosening	- permeability - angle of internal friction - friction sand-steel - dilatancy -(relative density)	- undr. shear strength - (adhesion)	- compr. strength - porosity - discontinuities
Mixing	←	grain/lump size distribution density of grain/lump	→
Transport	- grain size distr. - density - (angularity)	- rheol. prop. - undr. shear strength - plasticity index - wet density	
Deposition	- relative density - permeability	- undr. shear strength - plasticity index	
Wear	- mineral. comp. - grain size distr. - relative density - (angularity)		- mineral. comp. - compress. strength - grain size of min.

In fact the soil parameters between parentheses are relevant for the processes mentioned as well, but generally the other parameters can be regarded as sufficient.

FIG. 6 SUMMARY OF SOIL PARAMETERS RELATED TO DREDGING PROCESSES

DETERMINATION OF SOIL PARAMETERS

The soil properties and/or soil characteristics described in the matrix of chapter 3 can be distinguished in 3 categories:

1. The mechnical/physical properties of the soil mass, dealing with resistance during loosening, with strength and rigidity after deposition and with friction resistance causing wear; also the rheological properties during transport must be mentioned.

2. The conditional properties like density and moisture content, stress level/history are important as well in the stages of loosening and deposition; on the other hand the conditional properties play an essential role in mixing and transport of individual particles or lumps because of their resistance to desintegration and weight.

3. The parameters, which are necessary for the classification or identification of the soil material under consideration: parameters from the grain size distribution, the Atterberg limits, the density of the solid particles, the maximum/minimum density etc.

The determination of these properties and/or characteristics is realised by carrying out in situ and laboratory tests.
Dredging projects usually are extensive. Hence they ask for large survey operations, which are expensive and also time consuming.

Therefore an economic set-up and organization of the survey is necessary and inevitable; attention must be paid to the following aspects:

- Use of all available information of the site (geologic, historic, hydrologic, geotechnic); this enables the dredging contractor to make an optimum planning and set-up of the soil survey. In this stage the contribution of an engineering-geologist is more and more considered as very essential.

- In dredging optimum organization of the survey always means collecting a maximum amount of relevant soil data with a minimum effort and above all things in the shortest possible time.
  - This can be done be sensible selection of soil investigation methods and careful planning of the investigation locations.

- Next the gained data must be evaluated and a statistical or geostatistical analysis is made in order to calculate the risks of executing the dredging job.

- However in many cases the information on soil data is lacking: there is a shortage of soil data and/or the accuracy of the gathered data is insufficient to make a justified estimation of the configuration, condition and composition of the soil layers.
  In those cases the utilization of correlations is recommended. Correlations can be defined as the well known, approved and often mentioned (in literature) relations between the results of simple, cheap and fast soil tests (mostly classification/identification parameters) and the outcomes of more complicated tests (mechanical/physical parameters).

In the matrix of figure 7 an overview is given of the parameters which can be determined by direct test (in situ and in the laboratory; in A, B and C) and which can be derived from indirect methods (correlations; in a, b and c). Also a distinction is made between tests on sand, clay and rock material.

Direct tests in situ and laboratory

Although in situ and laboratory tests both are classified as direct tests the in situ tests are virtually more direct than the laboratory tests.
For laboratory tests samples are needed and the quality of the test results is greatly dependent on the sample quality (degree of disturbance) Furthermore the reconstruction of the in situ condition (density, stress level cementation) is often very difficult (Lubking, 1979).

A. Direct tests- mechanical/physical.
Only very few in situ tests are available for the determination of the mechanical/fysical properties.
In sand in situ permeability tests can be done. Apart from the well-known constant head and falling head tests a special permeability probe (the so-called dipole probe) was developed within the Dredging Research in Delft, committed by the Dredging Research Association.
The principle of this device is similar to the electric resistivity measurement of the density probe. (Rietsema, 1979)
In clay the in situ vane test is used for the determination of the undrained shear strength; this test is especially popular in Scandinavian countries.

PROPERTIES / DIRECT TESTS	MECHANICAL/ PHYSICAL A	IN SITU CONDITION B	CLASSIFICATION IDENTIFICATION C
in situ	perm. probe vane	dens. probe pressiometer/ dilatometer	field classif.
laboratory	perm.) triax.) test adhes.) rheol.)	dens./moist.c.	grain size max/min. density spec. density mineral comp. Atterberg limits
CORRELATIONS	a	b	c
	D_r) → φ δ CPT/SPT) dil. D_r/fines → k SPT/CPT) p. pen.) LI) → c_u PI/ σ_v) CPT/FR) Bras. tens.) → σ_c point load)	CPT/SPT → D_r	CPT/FR → comp.

FIG. 7 SUMMARY OF (IN)DIRECT TEST METHODS FOR DETERMINATION OF SOIL PARAMETERS

With laboratory testing the in situ conditions of the sample must be restored before the actual test can be performed. This fact makes the execution of relevant tests on representative samples mostly very complicated and therefore expensive and/or time consuming.

Particularly <u>sand</u> samples cause difficulties as they have to be rebuilt having the same density or even grain skeleton of the in situ condition.
So for laboratory tests for the determination of the permeability, the angle of internal friction, the sand-steel adhesion and the dilatancy not only the in situ stress conditions are required but also information about the configuration and packing of the grains. This information can be delivered roughly by the determination of the density or relative density. This means that in situ density measurements are required.
The presence of cohesion in <u>clay</u> and <u>rock</u> samples makes the execution of laboratory tests relatively easy. Triaxial tests can produce relevant data, provided that they are carried out at the right stress level.

B. Direct test - conditional.
The characterisation of the in situ conditions has to be established by in situ tests. In soft soils density measurements present information concerning density or water content.
The measuring principles are very different: the electric resistivity method and the radiometric transmission or back scatter probes are both used in practice. (Nieuwenhuis, 1982) (Nieuwenhuis, 1978).
The stress situation can be read from some types of pressiometer or dilatometer tests; in normally consolidated soils the stress level can also be derived from depth data.

C. Direct test - classification
Classification of the material in situ can be realised by certain standardized classification procedures, for instance the Unified Soil Classification System (U.S.C.S.)
They produce only very global information.

Further classification in the laboratory generally is carried out by tests on undisturbed, but mostly on disturbed or remoulded samples.
Usually a grain size distribution is produced from <u>sand</u> samples. The determination of the maximum and minimum density is needed in case of relative density characteristics of the in situ condition of the sand.

Sometimes the specific density of the grains and/or the mineralogic composition of the material is determined; the last tests are also performed on rock samples. The determination of Atterberg limits is a very common test on clay samples.
The above mentioned tests together with some other simple laboratory tests are commonly used in order to establish a more extensive classification of the soil.

Many countries but also many institutions have their own soil classification system, depending on branch and local circumstances concerning soil material, laboratory facilities and require ments. From the classification charts indications and/or recommendations can be read concerning the execution of works, for instance excavation, trafficability, densification, etc.
In this context it is useful to refer to the "PIANC classification (PIANC, 1984).

Indirect methods or correlations

As a matter of fact all geophysical (seismic geo-electric, electro-magnetic) measurements are indirect methods to produce relevant parameters with respect to the dredging process.
Up to now the derivation of mechanical/physical dredging parameters from geophysical measurements is too global to be useful in practice.

Geophysical measurements certainly produce information concerning global material classification, in situ density, mineralogical composition, determination of boundaries between different soil layers or detection of discontinuities (Lubking, 1987).
Another indirect method to produce dredging parameters is the so-called application of "correlations".
Although the employment of correlations holds many risks particularly in "strange" soils, this practice is wide spread and popular.
However, the application of correlations is mostly economic provided that the concerned relation is relevant and logical and taking into account a certain accuracy.

a. Correlations - mechanical/physical.
As stated above the mechanical/physical properties of the sand material have a close relationship with density or relative density.

If density measurements are not available relative density can also easily be derived from cone penetration test. In literature many correlations can be found between the cone resistance of the standardized cone penetration test (CPT) resp. the number of blows for 12 inch or 0,3 m penetration (N_{30}) of the Standard Penetration Test (SPT) on one side and the relative density on the other.
More recently it was recognized that the relationship is also dependent on the stress level; so the effective soil stress is given in the diagrams as well (Mitchell, 1975)

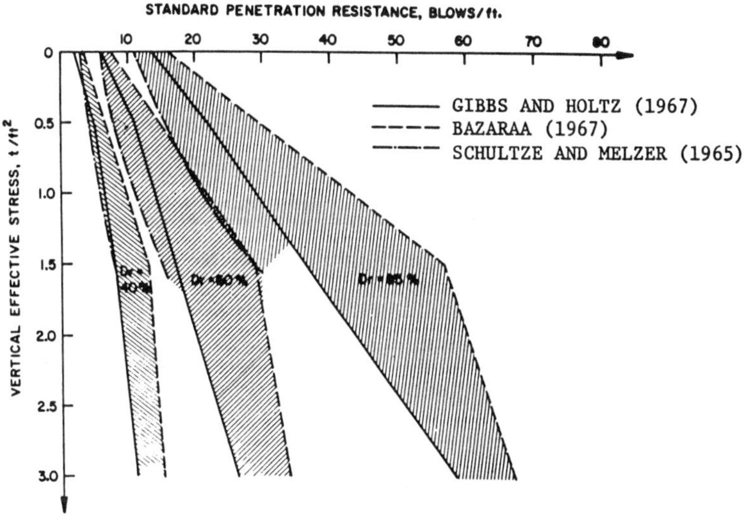

FIG. 8 CORRELATION BETWEEN SPT-RESISTANCE, EFFECTIVE STRESS AND RELATIVE DENSITY IN SAND (Mitchell, 1975)

Every relative density line represents the test result in sand of that specific density.
It must be noted that the accuracy of this correlation is not very great; there is also a substantial difference between the result of the respective researchers.
Besides, most SPT-results are not very reliable.

For CPT-results similar correlations are published. The recommendation of Baldi can be regarded as a fairly reliable directive for normally consolidated quartz sand. (Robertson, 1983)

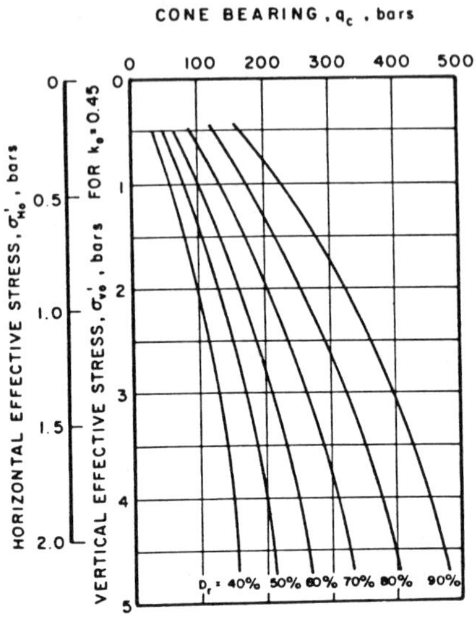

FIG. 9 CORRELATION BETWEEN CPT-RESISTANCE, EFFECTIVE STRESS AND RELATIVE DENSITY IN SAND (Robertson, 1983).

This graph was designed for the standardized electric 36 mm cone. It must be emphasized that in practice many differing cones are employed; adaptation of this graph for other cones can be very risky.

The correlations CPT-D_r and SPT-D_r are generally utilized in order to estimate the dredging parameters (angle of internal friction, adhesion, dilatancy).

For that purpose the individual dredging contractors employ the knowledge and experience gained in the joint research efforts.

Theoretically the correlation of the cone resistance and the angle of internal friction can be derived as well. Many researchers found a relation between the bearing capacity factor $N_q = \frac{q_c}{\sigma_v'}$ where q_c represents the cone resistance and σ_v' the vertical effective stress on the same level. (Janbu, 1979), (Lunne, 1983)

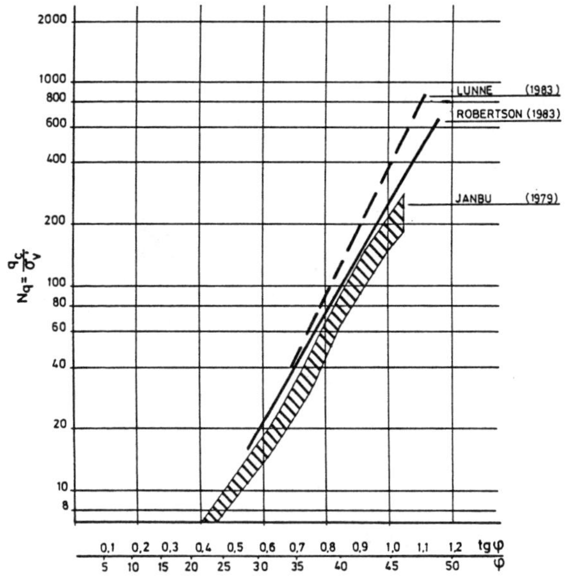

FIG. 10 CORRELATION BETWEEN CPT-RESISTANCE, EFFECTIVE STRESS AND ANGLE OF INTERNAL FRICTION IN SAND (Mitchell, 1975), (Janbu, 1979), (Lunne, 1983).

De Mello found a similar graph for SPT-data (De Mello, 1971)

The coëfficient of permeability k has a certain correlation with the relative density but the correlation with grain size characteristics is dominating, particularly with the percentage of fines. (% particles smaller than 16 um or d_{10}).(SCW, 1979).

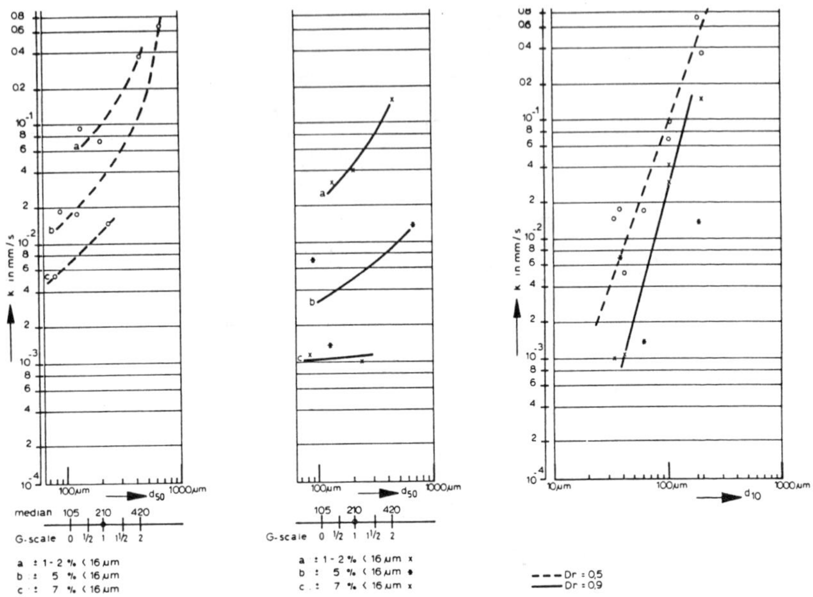

FIG. 11 CORRELATION BETWEEN GRAIN SIZE CHARACTERISTICS, RELATIVE DENSITY AND PERMEABILITY IN SAND (SCW, 1979).

In clay material the correlation between the undrained shear strength and penetration test results is quite obvious and theoretically proven.

The following formula is found in literature (de Ruyter, 1982):

$$q_c = N_c \cdot c_u + \sigma_v'$$

where q_c represents the cone resistance, c_u the undrained shear strength, σ_v' the vertical effective stress and N_c a bearing capacity factor dependent on clay type, in situ stress situation and test characteristics. In normally consolidated high plasticity clays N_c varies between 9 and 14 for standardized electrical cones; in overconsolidated clays N_c usually lies between 12 and 18. For standardized mechanical cones the N_c values are higher (approx. + 3). More silty clays produce much higher values for N_c.

A specific application of this correlation is the pocket penetration test on clay samples. This test immediately produces a c_u-value.

The correlation of SPT-values and unconfined compressive strength is often presented in dependence on the plasticity of the clay (NAVFAC, 1971).

Less accurate are the correlations of the undrained (remoulded) shear strength and the liquidity index LI, which is defined as

$$LI = \frac{w - PL}{PI}$$

where W represents the moisture content of the clay soil, PL the plastic limit and PI the plasticity index (Mitchell, 1976)

In literature also a very global relation can be found, which is basCd on the plasticity index and the stress level

$$\frac{c_u}{\sigma_{v'}} = 0.11 + 0.0037\ PI \qquad \text{(Skempton, 1957)}$$

The undrained shear strength can also be derived from the friction sleeve measurements.
Usually the undrained shear strength in soft clays is found to be approximately equal to the sleeve friction.
For stiffer overconsolidated clays the shear strength may increase to approximately 2 times the friction value.

In rock material the unconfined compressive strength can be derived from the Brasilian tensile test. According to the CSB (Dredging Research Association) this correlation produces good results (Cools, 1984).

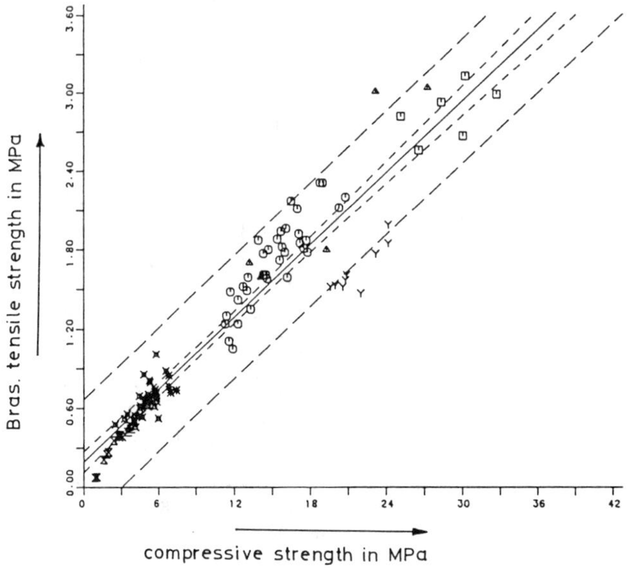

FIG. 12 CORRELATION BETWEEN COMPRESSIVE STRENGTH AND BRASILIAN TENSILE STRENGTH FOR DREDGED ROCK (Cools, 1984)

In literature another well-known correlation can found between the unconfined compressive strength:$_c$ and the point load strength $\sigma_{p.l.}$ (Bieniawski, 1974):

$$\sigma_c = 24\,\sigma_{p.l.}$$

This relatively rough correlation holds for rock materials with compressive strengths up to 300 MPa.

The tensile strength is usually derived from the Brasilian tensile test or the point load test using the respective formulae:

$$\sigma_t = \frac{F}{\pi r_o t} \qquad \text{and } \sigma_t = \frac{0.255F}{r^2}$$

where σ_t represents the tensile strength, F the applied force, r_o the radius of the disk, t the thickness of the disk and r the radius of the sample.

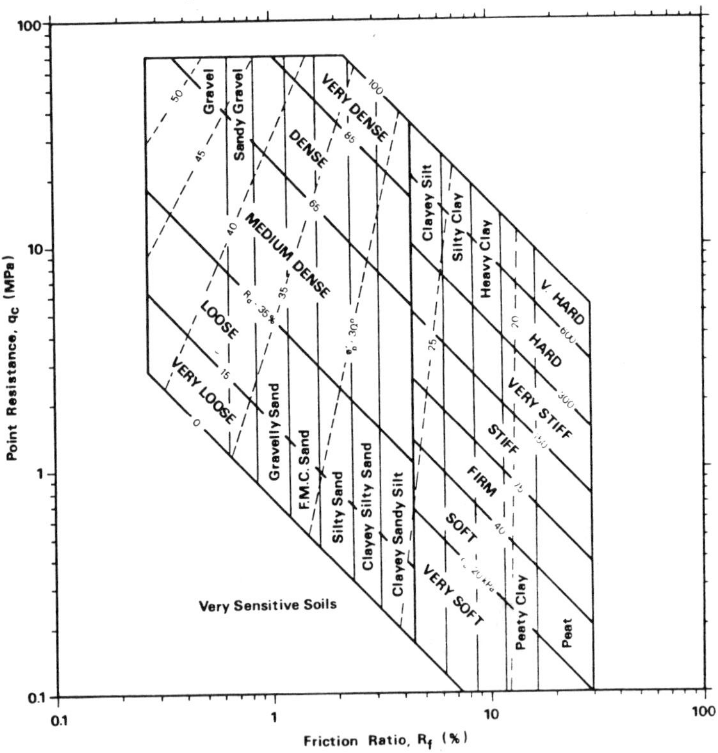

FIG. 13 CORRELATION BETWEEN CPT-RESISTANCE/FRICTION RATIO AND SOIL COMPOSITION (Searle, 1979).

b. Correlations - conditional.
The in situ condition of the sand material can be characterized by cone penetration tests, which produce a fair correlation with relative density (see a.).

c. Correlations - classification.
The friction measurement of the CPT offers a very useful correlation of cone resistance and soil composition.
The charts of Begemann and Schmertmann were combined (Searle, 1979) to a useful chart containing additional information about D_r and

In all charts the friction ratio is used as a parameter; it is defined as $FR = \frac{c_a}{q_c}$

where c_a represents the friction between sleeve and soil and q_c the cone resistance.

However, it must be emphasized that there is a difference in friction values between the mechanical "Begemann" cone and the standardized electrical cone.

CONCLUSIONS AND RECOMMENDATIONS

From the early sixties very important developments occured in dredging industry.
Dredging activities were scaled up immensely and research was stimulated in several disciplines.
The development of new dredging equipment on one side and the soil mechanic research into the dredging behaviour of the soil on the other resulted in the construction of very modern, fast working dredging equipment and also in the development of new soil mechanic theories for the respective stages of dredging (loosening, mixing, transport and deposition).
The soil mechanic research produced a good insight into the relevant soil parameters during these stages. Thus the interation of equipment and sand, clay or rock material could be described and understood. The outcomes of the studies are used in normal dredging practice as well as in the design of new equipment.

However, the soil investigation activities did not keep step with these developments. Usually the contractor does not have the disposal of qualitativily and quantitatively sufficient information. Particularly in the tender phase there is only rarely enough time and opportunity to carry out a complete survey with the aid of adequate soil investigation methods.
Furthermore but a few in situ methods are available producing reliable and accurate parameters for direct or indirect (via correlations) use in the tender phase resp. during actual execution of the dredging project.
In this context the further development of CPT equipment is very promising; the simultaneous measurement of cone resistance, sleeve frictions, density, stress situation, water pressure etc. in one penetration operation will undoubtedly improve the present unfavourable situation. More frequent application of the CPT-equipment will produce more experienced and better correlations as well.
Generally it can be concluded that soil investigations for the dredging operations should claim more time and attention in order to improve the present reprehensible situation.
Besides the development of adequate in situ methods, as well as the determination of more reliable correlations between in situ soil investigation results and dredging parameters deserves more attention.

REFERENCES

Bieniawski, Z.T. (1974). Geomechanics classification of rock masses and its application in tunneling. Proc. 3rd Congr. Int. Rock Mech., Denver.

Cools, P.M. (1984). Correlation, regression and reliability analysis of the material parameter data base (in Dutch), Nota M1325/M1393 CSB (Dredging Research Association).

Janbu, N. and Senneset, K (1979). Effective stress interpretation of in situ static penetration tests. Proc. ESOPT I, Stockholm.

Leussen, W van and Nieuwenhuis, J.D. (1984). Soil mechanics aspects of dredging Geotechnique 34, no. 3

Lubking, P. (1979). In situ soil investigation methods for soil mechanics and foundation engineering in delta regions. Geologie en Mijnbouw, Vol. 58 (4).

Lubking, P. e.a. (1987). Use of geophysical methods in earthworks. Report Techn. Com. 12 (Earthworks, drainage and subgrade) PIARC, Brussels.

Lunne, T. and Christoffersen, H.P. (1983). Interpretation of cone penetrometer data for offshore OTC, Houston.

Mello, V.F.B. de (1971). The standard penetration test. Proc. 4th Pan Am Conf. SMFE, Puerto Rico.

Mitchell, J.K. (1976). Fundamentals of soil behaviour J. Wilne and Sons, New York.

Mitchell, J.K. and Gardner W.S. (1975). In situ measurement of volume change characteristics Proc. Conf. on in situ measurement of soil properties, Raleigh, North Carolina.

NAVFAC (1971). Design manual (DM-7) Soil mechanics, Foundations and Earth Structures Naval Facilities Engineering Command, Alexandria.

Nieuwenhuis, J.D. and Ruygrok, P.A. (1978). Measurement of in situ density of thick fill layers by radiometric methods. IAEG Int. Congr. Madrid.

Nieuwenhuis, J.K. and Smits, F.P. (1982). The development of a nuclear density probe in a cone penetrometer. Proc. ESOPT II, Amsterdam.

Os, A.G. van and Leussen, W. van (1987). Basic research on cutting forces in saturated sand. Journal of Geotechn. Eng. Vol. 113 no. 12.

PIANC, (1982) Classification of soils and rocks to be dredged. Report of a working group of the Permanent Technical Committee II Supplement to Bulletin no. 47.

Rietsema, P.A. and Viergever, M.A. (1979). In situ measurement of permeability. Proc. Europ. Conf. on SMFE, Brighton.

Robertson, P.K. and Campanella, R.G. (1983). Interpretation of cone penetration tests, Part 1: sand, Can. Geotechn. Journal 20.

Ruyter, J. de (1982). The static cone penetration test, state - of - the - art report. Proc. ESOPT II, Amsterdam.

SCW (1979). Various properties of natural sands for Netherlands highway engineering. Record 4, Study Centre for Road Construction, Arnhem, The Netherlands.

Searle, I.W. (1979). The interpretation of Begemann friction jacket cone results to give soil types and design parameters. Proc. Europ. Conf. SMFE, Brighton.

Skempton, A.W. (1957). Discussion: The planning and design of the new Hong Kong airport. Proc. Inst. of Civ. Eng., London, 7.

SUBJECT INDEX

Page number refers to first page of paper.

AUTHOR INDEX

Page number refers to first page of paper.